AMERICAN COLLEGE OF LABORATORY ANIMAL MEDICINE SERIES

Steven H. Weisbroth, Ronald E. Flatt, and Alan L. Kraus, eds.:
The Biology of the Laboratory Rabbit, 1974

Joseph E. Wagner and Patrick J. Manning, eds.:
The Biology of the Guinea Pig, 1976

Edwin J. Andrews, Billy C. Ward, and Norman H. Altman, eds.:
Spontaneous Animal Models of Human Disease, Volume I, 1979;
Volume II, 1979

Henry J. Baker, J. Russell Lindsey, and Steven H. Weisbroth, eds.:
The Laboratory Rat, Volume I: Biology and Diseases, 1979;
Volume II: Research Applications, 1980

Henry L. Foster, J. David Small, and James G. Fox, eds.:
The Mouse in Biomedical Research, Volume I: History,
Genetics, and Wild Mice, 1981; Volume II: Diseases, 1982;
Volume III: Normative Biology, Immunology, and Husbandry, 1983;
Volume IV: Experimental Biology and Oncology, 1982

James G. Fox, Bennett J. Cohen, and Franklin M. Loew, eds.:
Laboratory Animal Medicine, 1984

Laboratory Animal Medicine

EDITED BY

James G. Fox
Division of Comparative Medicine
Massachusetts Institute of Technology
Cambridge, Massachusetts

Bennett J. Cohen
Unit for Laboratory Animal Medicine
University of Michigan Medical School
Ann Arbor, Michigan

Franklin M. Loew
Tufts University School of Veterinary Medicine
Boston, Massachusetts

ACADEMIC PRESS, INC.

(Harcourt Brace Jovanovich, Publishers)

*Orlando   San Diego   San Francisco   New York   London
Toronto   Montreal   Sydney   Tokyo   São Paulo*

ACADEMIC PRESS, INC.
Orlando, Florida 32887

United Kingdom Edition published by
ACADEMIC PRESS, INC. (LONDON) LTD.
24/28 Oval Road, London NW1 7DX

Library of Congress Cataloging in Publication Data
Main entry under title:

Laboratory animal medicine.

   (ACLAM)
   Includes index.
   1. Laboratory animals--Diseases. 2. Laboratory animals.
I. Fox, James G. II. Cohen, Bennett J. III. Loew,
Franklin M., Date   . IV. Series: ACLAM (Series)
SF996.5.L33 1984   636.089   83-21477
ISBN 0-12-263620-1 (alk. paper)

PRINTED IN THE UNITED STATES OF AMERICA

84 85 86 87   9 8 7 6 5 4 3 2 1

Contents

List of Contributors

Numbers in parentheses indicate the pages on which the authors' contributions begin.

Miriam R. Anver (427), Clement Associates, Inc., Arlington, Virginia 22209

W. Emmett Barkley (595), Office of Research Safety, National Cancer Institute, National Institutes of Health, Bethesda, Maryland 20205

Stephen W. Barthold (91), Section of Comparative Medicine, Yale University School of Medicine, New Haven, Connecticut 06510

S. L. Bernard (385), Agricultural Research Service, U.S. Department of Agriculture, and Department of Veterinary Microbiology and Pathology, Washington State University, Pullman, Washington 99164

W. Sheldon Bivin (563), Veterinary Anatomy, Laboratory Animal Facility, School of Veterinary Medicine, Louisiana State University, Baton Rouge, Louisiana 70803

Nathan Brewer (207), Department of Physiology, University of Chicago, Chicago, Illinois

Dale L. Brooks (273), Animal Resources Service, School of Veterinary Medicine, University of California, Davis, California 95616

Everett Bryant (399), Department of Pathobiology, University of Connecticut, Storrs, Connecticut 06268

Charles C. Capen (667), Department of Veterinary Pathobiology, College of Veterinary Medicine, The Ohio State University, Columbus, Ohio 43210

J. Derrell Clark (183), Department of Medical Microbiology, College of Veterinary Medicine, University of Georgia, Athens, Georgia 30602

Thomas B. Clarkson (697), Department of Comparative Medicine, Bowman Gray School of Medicine, Wake Forest University, Winston-Salem, North Carolina 27102

Donald H. Clifford (527), Division of Laboratory Animal Medicine, Medical College of Ohio, Toledo, Ohio 43699

Bennett J. Cohen (1), Unit for Laboratory Animal Medicine, University of Michigan Medical School, Ann Arbor, Michigan 48109

Ronald E. Flatt[1] (207), Veterinary Pathology, and Laboratory Animal Resources, College of Veterinary Medicine, Iowa State University, Ames, Iowa 50011

James G. Fox (31, 613), Division of Comparative Medicine, Massachusetts Institute of Technology, Cambridge, Massachusetts 02139

J. R. Gorham (385), Agricultural Research Service, U.S. Department of Agriculture, and Department of Veterinary Microbiology and Pathology, Washington State University, Pullman, Washington 99164

John E. Harkness (149), Laboratory Animal Resources, Centralized Biological Laboratory, The Pennsylvania State University, University Park, Pennsylvania 16802

Roy V. Henrickson (297, 301), School of Veterinary Medicine, California Primate Research Center, University of California, Davis, California 95616

J. R. Hessler (505), Animal Resource Division, University of Tennessee Center for the Health Sciences, Memphis, Tennessee 38163

Chao-Kuang Hsu (603), Central Animal Facility, Comparative Medicine Program, University of Maryland School of Medicine, Baltimore, Maryland 21201

Elliott R. Jacobson (449), Department of Special Clinical

[1]Deceased January 26, 1984.

Sciences, College of Veterinary Medicine, University of Florida, Gainesville, Florida 32610

Robert O. Jacoby (31), Section of Comparative Medicine, Yale University School of Medicine, New Haven, Connecticut 06510

Dennis F. Kohn (91), Department of Comparative Medicine, University of Texas Medical School, Houston, Texas 77030

Alan L. Kraus (207), Division of Laboratory Animal Medicine, University of Rochester School of Medicine and Dentistry, Rochester, New York 14642

Warren C. Ladiges (123), Animal Health Resources, Fred Hutchinson Cancer Research Center, Seattle, Washington 98104

Noel D. M. Lehner (297, 321), Department of Comparative Medicine, Bowman Gray School of Medicine, Wake Forest University, Winston-Salem, North Carolina 27103

Franklin M. Loew (1), Tufts University School of Veterinary Medicine, Boston, Massachusetts 02111

Yue-Shoung Lu (649), Department of Comparative Medicine, Animal Resources Center, University of Texas Southwest Medical School, Dallas, Texas 75235

Frances M. Lusso (697), Department of Comparative Medicine, Bowman Gray School of Medicine, Wake Forest University, Winston-Salem, North Carolina 27103

Patrick J. Manning (149), Division of Comparative Medicine and Research Animal Resources, Department of Laboratory Medicine and Pathology, Medical School, University of Minnesota, Minneapolis, Minnesota 55455

Charles W. McPherson (19), Animal Resources, School of Veterinary Medicine, North Carolina State University, Raleigh, North Carolina 27606

Paul C. Meunier (649), Department of Comparative Medicine, Animal Resources Center, University of Texas Southwest Medical School, Dallas, Texas 75235

George Migaki (667), Universities Associated for Research and Education in Pathology, Inc., and Registry of Comparative Pathology, Armed Forces Institute of Pathology, Washington, D.C. 20306

A. F. Moreland (505), Animal Resources Department, J. Hillis Miller Health Center, University of Florida, Gainesville, Florida 32610

Christian E. Newcomer (613), Division of Comparative Medicine, Massachusetts Institute of Technology, Cambridge, Massachusetts 02139

Steven M. Niemi (273), Division of Comparative Medicine, Massachusetts Institute of Technology, Cambridge, Massachusetts 02139

Steven P. Pakes (649), Department of Comparative Medicine, Animal Resources Center, University of Texas Southwest Medical School, Dallas, Texas 75235

Gregory K. Peter[2] (241), Unit for Laboratory Animal Medicine, University of Michigan Medical School, Ann Arbor, Michigan 48109

Cynthia L. Pond[3] (427), Unit for Laboratory Animal Medicine, University of Michigan Medical School, Ann Arbor, Michigan 48109

John H. Richardson[4] (595), Office of Biosafety, Centers for Disease Control, Atlanta, Georgia 30333

Conrad B. Richter (297, 353), Comparative Medicine Branch, National Institute of Environmental Health Sciences, National Institutes of Health, Research Triangle Park, North Carolina 27709

Daniel H. Ringler (241), Unit for Laboratory Animal Medicine, University of Michigan Medical School, Ann Arbor, Michigan 48109

Harry Rozmiarek (613), Department of Preventive Medicine, Laboratory Animal Center, The Ohio State University, Columbus, Ohio 43220

L. M. Ryland (385), Halecrest Veterinary Hospital, Seattle, Washington 98133

J. David Small (709), Diagnostic and Research Laboratory, Comparative Medicine Branch, National Institute of Environmental Health Sciences, National Institutes of Health, Research Triangle Park, North Carolina 27709

Gerald D. Smith (563), Laboratory Animal Resources, School of Veterinary Medicine, Louisiana State University, Baton Rouge, Louisiana 70803

Philip C. Tillman (273), Animal Resources Service, School of Veterinary Medicine, University of California, Davis, California 95616

G. L. Van Hoosier, Jr. (123), Division of Animal Medicine, School of Medicine, University of Washington, Seattle, Washington 98195

Joseph E. Wagner (149), Veterinary Pathology, Research Animal Diagnostic Laboratory, College of Veterinary Medicine, University of Missouri, Columbia, Missouri 65211

Alexander H. Walsh (477), Pfizer Central Research, Groton, Connecticut 06340

David S. Weaver (697), Department of Anthropology, Wake Forest University, Winston-Salem, North Carolina 27109

Steven H. Weisbroth (207), AnMed Laboratories, New Hyde Park, New York 11040

[2]Present address: Experimental Pathology and Toxicology, Warner Lambert—Parke Davis, Ann Arbor, Michigan 48105.

[3]Present address: Comparative Pathology Section, Veterinary Resources Branch, National Institutes of Health, Bethesda, Maryland 20205.

[4]Present address: Department of Community Medicine, School of Medicine, Emory University, Atlanta, Georgia 30322.

Preface

The American College of Laboratory Animal Medicine (ACLAM) was founded in 1957 to encourage education, training, and research in laboratory animal medicine and to recognize veterinary medical specialists in the field by certification and other means. Continuing education has been an important activity in ACLAM from its inception. This teaching text, "Laboratory Animal Medicine," reflects the College's continuing effort to foster education. It is, in part, a distillation for teaching purposes of a series of volumes on laboratory animals developed by ACLAM over the past decade: "The Biology of the Laboratory Rabbit" published in 1974, "The Biology of the Guinea Pig" in 1976, and a two-volume work "Biology of the Laboratory Rat" in 1979 and 1980. Also, in 1979 the college published a two-volume text on "Spontaneous Animal Models of Human Disease." In 1981–1983, four volumes of "The Mouse in Biomedical Research" were published.

Most major advances in biology and medicine in one way or another have depended on the study of animals. During the past generation, the health, genetic integrity, and environmental surroundings of the animals have been recognized as important factors to be taken into account in planning animal studies. The ultimate responsibility for insuring the validity of scientific results, together with humane and scientifically appropriate animal care, resides with two categories of scientists: veterinarians responsible for the acquisition, care, nutrition, anesthesia, and other aspects of humane animal use and scientific investigators who use animals as subjects of study. This book therefore is intended for students of veterinary medicine and others in the fields of biology and medicine who utilize animals in biomedical research. The editors and contributors hope it will prove useful in introducing students to important concepts related to animals in research.

The contents of this book are presented in twenty-six chapters that provide information on the diseases and biology of the major species of laboratory animals used in biomedical research. The history of laboratory animal medicine, legislation affecting laboratory animals, experimental methods and techniques, design and management of animal facilities, zoonoses, biohazards, animal models, and genetic monitoring are also covered. The editors acknowledge the contributors' outstanding efforts to follow the guidelines on content and accept sole responsibility for any significant omissions.

As with all volumes of the ACLAM series texts, the contributors and editors of this book have donated publication royalties to the American College of Laboratory Animal Medicine to foster continuing education in laboratory animal science. It could not have been completed without the support and resources of the editors' parent institutions. A special thanks also is extended to the reviewers of each chapter whose excellent and thoughtful suggestions helped the authors and editors present the material in a meaningful and concise manner. We acknowledge and thank Rosanne Brown and Kathi Edelson for their secretarial assistance. The assistance of the staff of Academic Press also is greatly appreciated and acknowledged.

James G. Fox
Bennett J. Cohen
Franklin M. Loew

List of Reviewers for Chapters in This Volume

Adams, Robert J.	The Johns Hopkins School of Medicine
Andrews, Edwin J.	Extracorporeal, Inc.
Baker, Henry J.	University of Alabama
Balk, Melvin W.	Charles River Breeding Laboratories, Inc.
Baum, Michael	Massachusetts Institute of Technology
Brownstein, David	Yale University School of Medicine
Burek, Joe D.	Merck Institute for Therapeutic Research
Calnek, Bruce W.	Cornell University
Cohen, Bennett J.	University of Michigan Medical School
Cotter, Susan	Tufts University School of Veterinary Medicine, Boston
Crowell, James S., Jr.	National Institutes of Health
Eyestone, Willard	University of Missouri, Columbia
Gay, Clive C.	Washington State University
Gay, William I.	National Institutes of Health
Harkness, John E.	The Pennsylvania State University
Hickey, Thomas E.	Bristol-Meyers Company
Hughes, Howard C., Jr.	The Pennsylvania State University
Jakowski, Richard M.	Tufts University School of Veterinary Medicine
Jones, Thomas C.	New England Regional Primate Center
Kastello, Michael D.	Battelle Columbus Laboratories
Kraft, Lisbeth M.	NASA/AMES Research Center
Lang, C. Max	The Pennsylvania State University
La Regina, Marie C.	St. Louis University Medical School
Lawson, Robin	Louisiana State University
Lindsey, Russell J.	University of Alabama Medical Center
Manning, Patrick J.	University of Minnesota
Montali, Richard J.	National Zoological Park
Moses, John M.	Massachusetts Institute of Technology
Mulder, John B.	University of Kansas

Newcomer, Christian   Massachusetts Institute of Technology
Patterson, D. Reid   Shell Development Company
Potkay, Stephen   National Institutes of Health
Quimby, Fred W.   Cornell University
Ribelin, W.   Monsanto Environmental Health
    Laboratories
Richter, Conrad B.   NIEHS
Ringler, Daniel H.   University of Michigan Medical School
Rush, Howard G.   University of Michigan Medical School
Russell, Robert J.   NCI–FCRF
Shadduck, John A.   University of Illinois at Urbana–
    Champaign
Simmonds, Richard C.   LAM–USUHS
Stark, Dennis M.   The Rockefeller University
Streilein, Wayne J.   University of Texas Health Science
    Center at Dallas
Wolke, Richard E.   University of Rhode Island
Womack, James E.   Texas A & M University
Yager, Robert H.   Columbia, South Carolina

Laboratory Animal Medicine

Chapter 1

Laboratory Animal Medicine: Historical Perspectives

Bennett J. Cohen and Franklin M. Loew

I. INTRODUCTION

Five key terms identify the fields or activities that relate to the care and use of animals in research, education, and testing. *Animal experimentation* refers to the scientific study of animals, usually in a laboratory, for the purpose of gaining new biological knowledge or solving specific medical, veterinary medical, dental, or biological problems. Most commonly, such experimentation is carried out by or under the direction of persons holding research or professional degrees (e.g., Ph.D., D.D.S., D.V.M., M.D.). *Laboratory animal care* is the ap-

plication of veterinary medicine and animal science to the acquisition of laboratory animals and to their management, nutrition, breeding, and diseases. The term also relates to the care that is provided to animals as an aid in managing injury and pain. Laboratory animal care usually is provided in scientific institutions under veterinary supervision or guidance. *Laboratory animal medicine* is recognized by the American Veterinary Medical Association as the specialty field within veterinary medicine that is concerned with the diagnosis, treatment, and prevention of diseases in animals used as subjects in biomedical activities. Laboratory animal medicine also encom-

passes the methods of minimizing pain or discomfort in research animals and the identification of complicating factors in
animal research. *Comparative medicine* is "the study of the
nature, cause and cure of abnormal structure and function in
people, animals and plants for the eventual application to and
benefit of all living things" (Bustad *et al.*, 1976). *Laboratory
animal science* is the body of scientific and technical information, knowledge, and skills that bears on both laboratory animal care and laboratory animal medicine and that is roughly
analogous to "animal science" in the agricultural sector.

Laboratory animal medicine has grown rapidly because of its
inherent scientific importance and because good science and
the public interest require the best possible care for laboratory
animals. In this chapter, we trace briefly the historical evolution of laboratory animal medicine and consider its relationship
to other areas of biology and medicine.

II.   ORIGINS OF ANIMAL
EXPERIMENTATION

The earliest references to animal experimentation are to be
found in the writings of Greek philosopher–physicians of the
fourth and third centuries BC. Aristotle (384–322 BC), characterized as the founder of biology, was the first to have made
dissections which revealed internal differences among animals
(Wood, 1931). Erasistratus (304–258 BC) probably was the
first to perform experiments on living animals, as we understand them today. He established in pigs that the trachea was
an air tube and the lungs were pneumatic organs (Fisher,
1881). Later, Galen (130–200 AD) performed anatomical dissections of pigs, monkeys, and many other species (Cohen and
Drabkin, 1948; Cohen, 1959a). Galen justified experimentation as a long arduous path to the truth, believing that uncontrolled assertion that was not based on experimentation could
not lead to scientific progress. Dogma replaced experimentation in the dark centuries following Galen's lifetime. Whereas
anatomical dissection of dead animals and people had been
among the earliest types of experimentation, in medieval times
this was prohibited by ecclesiastical authorities who wanted to
prevent acquisition of knowledge about the natural world that
could be considered blasphemous. Not until the 1500s was
there a reawakening of interest in science. Andreas Vesalius
(1514–1564), the founder of modern anatomy, used dogs and
pigs in public anatomical demonstrations (Saunders and
O'Malley, 1950) (Fig. 1). This "vivisection" led to great
leaps in understanding of anatomy's correspondence with
physiology. In 1628, Sir William Harvey published his great
work on the movement of the heart and blood in animals (Singer, 1957). By the early 1700s, Stephen Hales, an English cler-

Fig. 1.   Illustration by Andreas Vesalius. Pig tied to dissection board "for
the administration of vivisections." (From Saunders and O'Malley, 1950.)

gyman, reported the first measurement of blood pressure,
using as his subject a horse "fourteen hands high, and about
fourteen years of age, (and with) a fistula on her withers. . . ."
(Hoff *et al.*, 1965) (Fig. 2). During the 1800s France became a
primary center of experimental biology and medicine. Scientists, such as François Magendie (1783–1855) and Claude Bernard (1813–1878) (Fig. 3) in experimental physiology and
Louis Pasteur (1827–1895) in microbiology, contributed enormously to the validation of the scientific method which included the use of animals. Bernard (1865) commented:

> . . . it is proper to choose certain animals which offer favorable anatomi-
> cal arrangements or special susceptibility to certain influences. For each
> kind of investigation we shall be careful to point out the proper choice of
> animals. This is so important that the solution of a physiological or patho-
> logical problem often depends solely on the appropriate choice of the
> animal for the experiment so as to make the result clear and searching.

Pasteur studied infectious diseases in a variety of animals, such
as silkworms ("pebrine"), dogs (rabies), and sheep (anthrax).
"Pebrine" (pepper) was an economically important disease of
silkworms in France when silk was a major fabric; Pasteur and
others demonstrated the parasite which caused the disease
Duclaux, 1920). As pathogenic organisms were identified that
could be related to specific human diseases, their animal disease counterparts also were studied. Pasteur and others perceived that the study of animal diseases benefited animals and
enhanced understanding of human diseases and pathology. The
extraordinary power of the experimental approach, including
experiments on animals, led to what has been called the Gold-
en Age of scientific medicine. Despite advances in physiological and bacteriological understanding, however, criticisms of
the use of animals in science began, particularly in England
(Loew, 1982). The first Society for the Prevention of Cruelty
to Animals (SPCA) was established in England, followed in
the 1860s by an American SPCA in New York, a Philadelphia
SPCA, and a Massachusetts SPCA. Objections to the use of
animals in science were part of the concerns of these societies,
particularly since Darwin's findings on evolution made "differences" between animals and men less sure in many persons'
minds (Loew, 1982).

Most American and Canadian scientists, physicians, and vet-

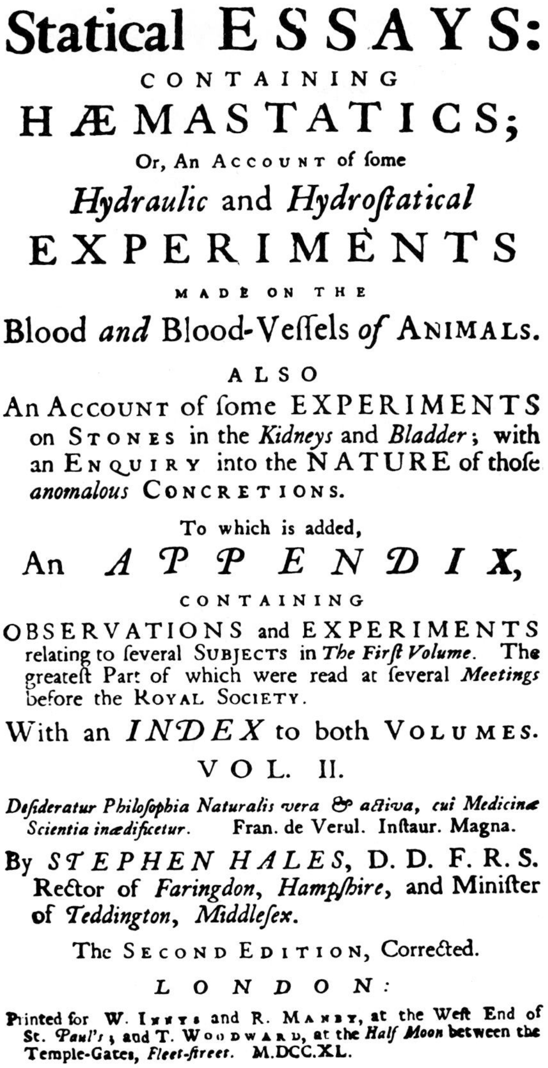

Statical ESSAYS:

CONTAINING

HÆMASTATICS;

Or, An Account of some

Hydraulic and *Hydroſtatical*

EXPERIMENTS

MADE ON THE

Blood *and* Blood-Veſſels *of* Animals.

ALSO

An Account of some EXPERIMENTS
on Stones in the *Kidneys* and *Bladder*; with
an Enquiry into the Nature of thoſe
anomalous Concretions.

To which is added,

An *APPENDIX*,

CONTAINING

OBSERVATIONS and EXPERIMENTS
relating to ſeveral Subjects in *The Firſt Volume*. The
greateſt Part of which were read at ſeveral *Meetings*
before the Royal Society.

With an *INDEX* to both Volumes.

VOL. II.

Deſideratur Philoſophia Naturalis vera & actiua, cui Medicinæ
Scientia inædificetur. Fran. de Verul. Inſtaur. Magna.

By *STEPHEN HALES*, D. D. F. R. S.
Rector of *Faringdon, Hampſhire*, and Miniſter
of *Teddington, Middleſex*.

The Second Edition, Corrected.

LONDON:

Printed for W. Innys and R. Manby, at the Weſt End of
St. *Paul's*; and T. Woodward, at the *Half Moon* between the
Temple-Gates, *Fleet-ſtreet*. M.DCC.XL.

Fig. 2. Title page from "Statical Essays" (Hales, 1740).

erinarians soon applied emerging scientific concepts in their research. D. E. Salmon, recipient of the first D.V.M. degree awarded in the United States (by Cornell University in 1879), studied bacterial diseases, and the genus *Salmonella*, a ubiquitous human and animal pathogen, was named for him. Cooper Curtice (Fig. 4), Theobald Smith, and others first demonstrated the role of arthropod victors in disease transmission in their studies of bovine Texas fever (Schwabe, 1978). The first paper published at the then-fledging Johns Hopkins Hospital and School of Medicine was by the physician William H. Welch, for whom *Clostridium welchii* was named and was entitled "Preliminary Report of Investigations concerning the Causation of Hog Cholera" (Welch, 1889). Thus, it became evident that the study of the naturally occurring diseases of

animals could illuminate principles applicable to both animals and mankind and lead to improved understanding of biology in general.

John Call Dalton, M.D. (1825–1889), an American physiologist, spent a year in Bernard's laboratory in Paris, about 1850. He was highly impressed with Bernard's instructional methods, which included demonstrations in living animals of important physiological principles. Subsequently, Dr. Dalton included such demonstrations in his teaching at the College of Physicians and Surgeons in New York City (Mitchell, 1895), the forerunner of the "animal labs" in which generations of students in biology and medicine have been trained. When Alexis Carrel received the Nobel Prize in 1912, the citation stated in part (Malinin, 1979): ". . . you have . . . proved once again that the development of an applied science of surgery follows the lessons learned from animal experimentation." Thus, starting in ancient times and continuing to the present day, animal experimentation has been one of the fundamental approaches of the scientific method in biological and medical research and education.

Fig. 3. Claude Bernard, often referred to as the founder of experimental medicine, developed and described highly sophisticated methods of animal research in his laboratory in Paris. (Photograph from Garrison, 1929.)

Fig. 4. Dr. Cooper Curtice examining ticks on a cow dead of Texas fever. Curtice contributed importantly to the demonstration that arthropods can act as carriers of mammalian diseases. (Courtesy of *The Nation's Business.*)

III. EARLY VETERINARIANS IN LABORATORY ANIMAL SCIENCE AND MEDICINE

On September 15, 1915 Dr. Simon D. Brimhall (1863–1941, V.M.D., University of Pennsylvania, 1889) (Fig. 5) joined the staff of the Mayo Clinic in Rochester, Minnesota, the first veterinarian to fill a position in laboratory animal medicine at an American medical research institution (Cohen, 1959b; Physicians of the Mayo Clinic and the Mayo Foundation, 1937). There was no such recognized field at the time, of course; but Dr. Brimhall's activities—management of the animal facilities, development of animal breeding colonies, investigation of laboratory animal diseases (Brimhall and Mann, 1917; Brimhall and Hardenbergh, 1922), and participation in collaborative and independent research (Brimhall *et al.,* 1919–1920)—were the prototype of the present role of "laboratory animal veterinarians" in scientific institutions throughout the world.

The decision to employ a veterinarian at the Mayo Clinic in 1915 appears to have resulted from a unique juxtaposition of institutional needs and personalities. Although the Mayo Clinic was already world renowned, organized research was only in a rudimentary stage of development. About 1910, there was an unsuccessful effort to convert an old barn, belonging to the Chief of Surgical Pathology, Dr. Louis B. Wilson, for animal experimentation (Braasch, 1969). Then, in 1914, with Dr. William J. Mayo's active encouragement, a Division of Experimental Surgery and Pathology was created, the first real research laboratory at the Clinic. Dr. Frank C. Mann, a young medical scientist from Indiana, was invited to head the Division, with the primary assignment of developing a first class animal research laboratory. Dr. Brimhall's employment followed within a year and was accompanied by the planning and ultimate construction of new animal facilities (Figs. 6 and 7). Christopher Graham, M.D., then head of the Division of Medicine, greatly influenced the decision to employ Dr. Brimhall. Perhaps the fact that Dr. Graham also was a veterinarian (V.M.D., University of Pennsylvania, 1892) provided insights into the contributions that veterinary medicine could make to experimental surgery and pathology. Certainly, the concept of mutual support among the professions was not at that time widely held; there was, in fact, relatively little interprofessional communication between medicine and veterinary medicine then.

Dr. Brimhall retired in 1922 and was succeeded by Dr. John G. Hardenbergh (1892–1963; V.M.D., University of Pennsyl-

Fig. 5. Simon D. Brimhall, V.M.D., the first veterinarian in laboratory animal medicine at an American medical research institution, worked at the Mayo Clinic from 1915 to 1922. (Courtesy of University of Minnesota Press and Dr. Paul E. Zollman.)

vania, 1916). During his 5-year tenure at the Mayo Clinic, Dr. Hardenbergh was an active clinical investigator (Hardenbergh, 1926–1927) as well as animal facility manager. In a stout defense of animal experimentation, he also demonstrated the communication skills in the public arena that were to serve him well later in his career (1941–1958) as Executive Secretary of the American Veterinary Medical Association (Hardenbergh, 1923).

Dr. Carl F. Schlotthauer (1893–1959, D.V.M., St. Joseph Veterinary College, 1923), who had joined the Mayo Clinic staff in 1924 as assistant in veterinary medicine, succeeded Dr. Hardenbergh in 1927. By this time, the Mayo Foundation was functioning as the graduate medical education and research arm of the Mayo Clinic, and had become formally affiliated with the University of Minnesota. Dr. Schlotthauer ultimately (1952) became head of the Section of Veterinary Medicine at the Mayo Foundation and Professor of Veterinary Medicine at the University of Minnesota Graduate School (1945). Thus,

Dr. Schlotthauer was the first veterinarian to attain a full professorship for laboratory animal medicine–related academic activities. Dr. Schlotthauer vigorously opposed antivivisectionist attacks on animal research. He was a leader in the statewide campaign that led to adoption of the Minnesota "pound law" in 1950, i.e., a law authorizing the requisitioning for research and education by approved scientific institutions of impounded, but unclaimed dogs and cats. Dr. Schlotthauer believed that open and honest communication between medical scientists and humane society workers could lead to better public understanding and support of animal research. Consequently, he was also active in humane society activities, serving, for many years, on the Board of Directors of the Minnesota Society for the Prevention of Cruelty to Animals. Dr. Schlotthauer also was an important figure in the early years of the American Association for Laboratory Animal Science (AALAS). He was a founding member of the Board of Directors and presented a paper on animal procurement at the first meeting in 1950 (Schlotthauer, 1950).

While other veterinarians also held appointments at the Mayo Foundation between 1915 and 1950, Dr. Brimhall, Dr. Hardenbergh, and Dr. Schlotthauer were the ones most closely associated with activities that today are identified with laboratory animal medicine. It is noteworthy that the Mayo Clinic/Foundation has maintained a program in animal medicine continuously for more than 65 years, having initiated it long before most medical research institutions were prepared even to consider the possible value of adding veterinarians to their professional staff (P. E. Zollman, personal communication, 1982).

Karl F. Meyer [1884–1974; D.V.M., University of Zurich, 1924, M.D. (honorary), College of Medical Evangelists, 1936] was an internationally known epidemiologist, bacteriologist and pathologist. Dr. Meyer was intensely interested in matters related to laboratory animals for most of his professional life. He was the author of an early review of laboratory animal diseases (Meyer, 1928), one of the first publications of its kind in the United States. Dr. Meyer was a unique personality—vigorous, dynamic, active—a world traveler on missions related to international health; a scientist who engendered in his students respect, admiration, love, and fear in varying proportions. Together with his long-time associate Bernice Eddy (Ph.D.), a bacteriologist, Dr. Meyer developed a model animal facility at the George Williams Hooper Foundation at the University of California, San Francisco during a 30-year tenure as director (1924–1954). Dr. Meyer often was away from the laboratory, and it fell to Dr. Eddy to supervise the animal facility, which she did with great skill and dedication. Dr. Meyer foresaw the need for and was an early advocate of the participation of veterinarians in the operation of institutional laboratory animal colonies (Meyer, 1958). He figured importantly in the planning that led the University of Cal-

Fig. 6. Dog breeding facility, Institute of Experimental Medicine, Mayo Clinic, constructed mid-1920s. (Courtesy of Dr. Paul E. Zollman.)

Fig. 7. Interior of guinea pig breeding house, Institute of Experimental Medicine, Mayo Clinic, constructed early 1920s. (Courtesy of Dr. Paul E. Zollman.)

ifornia to create the position of "statewide veterinarian" in 1953, which subsequently was superseded by the appointment of veterinarians at each of the University's major campuses. Among his many honors, Dr. Meyer received the Charles A. Griffin Award of AALAS in 1959.

Charles A. Griffin (1889–1955; D.V.M., Cornell University, 1913) was a bacteriologist at the New York State Board of Health, Division of Laboratories, Albany, New York from 1919 to 1954. Dr. Griffin pioneered the concept of the development of "disease-free" animal colonies long before gnotobiotic technology had evolved (Brewer, 1980). In the 1940s, Dr. Griffin utilized progeny testing to establish a rabbit colony free of pasteurellosis. Additionally, he showed that *Salmonella* spp. could be transmitted in meat meal (Griffin, 1952). This led feed manufacturers to improve the processing of laboratory animal diets so as to eliminate *Salmonella* contamination. The Charles A. Griffin Award of AALAS was established and named in Dr. Griffin's honor. He received the award posthumously in 1955, the first recipient of this prestigious award. The Griffin Laboratory at the New York State Board of Health central facility in Albany, New York also is named in his honor.

Nathan R. Brewer (D.V.M., Michigan State University, 1937; Ph.D. University of Chicago, 1936) headed the laboratory animal facilities at the University of Chicago from 1945 until his retirement in 1969 (Fig. 8). Dr. Brewer's interest in laboratory animals originated in the mid-1920s when he started veterinary school and continued during his graduate student years in the Department of Physiology at the University of Chicago. About 1935, Professors Anton J. Carlson (Ingle, 1979) and A. B. Luckhardt first approached Dr. Brewer about managing the University of Chicago animal facilities. They saw merit in the concept of a veterinarian, well-grounded in the scientific method, as animal facility manager. They believed this would contribute to public confidence in the care and treatment of animals in research, and would help to turn aside antivivisection activists. However, many investigators at the University feared that a veterinarian would dictate the conditions of care and use of animals, and they opposed the creation of this position. It was not until 1945 that this opposition was overcome, and Dr. Brewer became supervisor of the Central Animal Quarters. Laboratory animal medicine began its modern evolution in the following years. Dr. Brewer's role was seminal—as a founder of the American Association for Laboratory Animal Science, as first President of AALAS (1950–1955) and as a "father-figure" for the then youthful group of veterinarians that had been employed by other medical schools and medical research institutions in the Chicago area between 1945 and 1949. Dr. Brewer received the AALAS Griffin Award in 1962. Although retired since 1969, he remains vigorously active as his most recent publications in comparative physiology attest (Brewer, 1982; Kaplan *et al.*, 1983).

Fig. 8. Dr. Nathan R. Brewer, Director of The Central Animal Quarters at the University of Chicago (1945–1969) and first President of the American Association for Laboratory Animal Science. Photograph taken in the late 1940s. (Courtesy of Dr. N. R. Brewer.)

Other personalities that played important roles in the early history of laboratory animal science and medicine have been characterized and their roles assessed by Brewer (1980).

IV. THE ORGANIZATIONS OF LABORATORY ANIMAL SCIENCE

A. Background

Organizations are important in scientific life as a means of implementing the content and activities of the fields they represent. Present-day students of laboratory animal science are confronted with a confusing array of organizational acronyms: NSMR, ABR, AALAS, CALAS, AAALAC, ILAR, ACLAM, ASLAP, ICLAS, and so on. It is instructive to examine why organizations such as these came into being and to evaluate their impact on laboratory animal science.

Consider the milieu for research in biology and medicine in the United States about 1945. A new national policy was just being initiated to provide increased federal support of science. The use of laboratory animals began to expand rapidly as the funding of medical and biological research increased, and a host of problems as well as challenges accompanied this development. The base of knowledge about the care and diseases of laboratory animals was small. Published information was scattered and sparse. Few veterinarians were devoting themselves to "laboratory animal care," which was not yet recognized as a special field. In many institutions, animal facilities and administrative arrangements for operating them were poor. Institutions were ill-prepared to accommodate increasingly large animal colonies. Simultaneously, medical scientists were under increasingly vigorous attack from antivivisectionists whose objective was to stop or limit animal research. It became essential for scientists both to confront their persistent critics and to face up to the problems they knew existed.

The Chicago area was a hotbed of antivivisection activity in 1945 (Fig. 9). The National Antivivisection Society, based in Chicago, was distributing its literature widely and working for legislative abolition of animal research in Illinois and elsewhere. Orphans of the Storm, a humane society with a strong antivivisection outlook, was headed by it founder, Irene Castle McLoughlin, a famous dancer of the World War I era. Mrs. McLoughlin had been appointed to the Animal Advisory Committee of the Arvey Ordinance. The Ordinance permitted the medical schools in Chicago to obtain unclaimed dogs and cats from the public pound. On one occasion, during an inspection of the animal facilities at Northwestern University Medical School, Mrs. McLoughlin deliberately removed a dog from its cage because she felt the animal was not receiving adequate treatment. She planned to take the dog to her shelter in Winnetka. Dr. Andrew C. Ivy, then Professor and Chairman of the Department of Physiology and Dr. J. Roscoe Miller, then Dean of the Medical School, were notified hastily. They intercepted Mrs. McLoughlin and the dog at the entrance of the Medical School. At this point, Mrs. McLoughlin made a citizen's arrest of Dr. Ivy and Dean Miller and the protagonists proceeded to the Chicago Avenue police station. The dog was returned to the Medical School and the arrests subsequently were nullified. However, the incident was given wide publicity in the media, especially in the *Chicago Herald-Examiner,* reflecting the antivivisection sentiments of publisher William Randolph Hearst and Mr. Hearst's close friends, Mrs. McLoughlin and actress Marian Davies. This incident illustrates the "flavor" of the relationships between animal research sci-

Fig. 9. Cover page of antivisection brochure, late 1940s.

entists and their critics in the mid- and late 1940s. Without realizing it, Mrs. McLoughlin had alerted the scientific community to the significant and determined opposition it faced. An organized response was a clear necessity.

B. The National Society for Medical Research

The National Society for Medical Research (NSMR) was created in 1946 by the Association of American Medical Colleges (AAMC) and about 100 supporting groups (Grafton, 1980). AAMC had become concerned that progress in medical science could be jeopardized if antivivisectionists were successful in their numerous campaigns to prohibit or restrict animal experimentation. It was deemed essential to establish a separate organization to counter these antiscience activities and, especially, to promote better public understanding of the needs and accomplishments of animal experimentation. Public support of animal research depended upon such understanding. NSMR headquarters were established in Chicago. Dr. Anton J. Carlson was elected the organization's first president (Fig. 10). From its inception, NSMR contributed importantly to campaigns conducted at the state, city, and county levels to win public support for the use of public pounds as a source of unclaimed dogs and cats for research (Fig. 11). Antivivisection efforts to restrict or prohibit animal experimentation were fought successfully in several states. NSMR also developed much educational material about animal research and distributed it throughout the country. In the late 1940s, NSMR provided legal counsel to several Chicago area research scientists who had been attacked by the Hearst newspapers. Dr. Nathan Brewer was among this group. Libel suits were filed which dragged on for several years, until shortly before William Randolph Hearst's death in 1951. The suits were concluded in favor of the scientists, but without significant monetary settlement. The Hearst publications agreed to stop publishing statements tending to damage the reputations of scientists involved in animal research. The suits and Mr. Hearst's death brought to an end the extremist approach of the Hearst publications to the vivisection–antivivisection issue.

In 1952, a cause celebre developed within the American Physiological Society (APS) that also involved other constituent societies of the Federation of American Societies for Experimental Biology (FASEB) and the NSMR. Robert Gesell, M.D., Professor and Chairman, from 1923 to 1954, of the Department of Physiology at the University of Michigan (Fig. 12), made the following statement at the APS business meeting on April 15:

> The National Society for Medical Research would have us believe that there is an important issue in vivisection versus antivivisection.
>
> To a physiologist there can be no issue on vivisection per se.
>
> The real and urgent issue is humanity versus inhumanity in the use of experimental animals.
>
> But the NSMR attaches a stigma of antivivisection to any semblance of humanity.
>
> Antivivisection is their indispensable bogie which must be kept before the public at any cost.
>
> It is their only avenue towards unlimited procurement of animals for unlimited and uncontrollable experimentation.
>
> The NSMR has had but one idea since its organization, namely, to provide an inexhaustible number of animals to an ever growing crowd of career scientists with but little biological background and scant interest in the future of man.
>
> Consider what we are doing in the name of science, and the issue will be clear.
>
> We are drowning and suffocating unanaesthetized animals—in the name of science.
>
> We are determining the amount of abuse that life will endure in unanaesthetized animals—in the name of science.
>
> We are producing frustration ulcers in experimental animals under shocking conditions—in the name of science.
>
> We are observing animals for weeks, months, or even years under infamous conditions—in the name of science.
>
> Yet it is the National Society for Medical Research and its New York satellite that are providing the means to these ends.
>
> And how is it being accomplished?
>
> By undermining one of the finest organizations of our country, THE AMERICAN HUMANE SOCIETY.
>
> With the aid of the halo supplied by the faith of the American people in medical science, the NSMR converts sanctuaries of mercy into animal pounds at the beck and call of experimental laboratories regardless of how the animals are to be used.
>
> What a travesty of humanity!
>
> This may well prove to be the blackest spot in the history of medical science.

Fig. 10. Dr. Anton J. Carlson, Professor of Physiology at the University of Chicago and first President of the National Society for Medical Research. (Courtesy of the University of Chicago Archives and Dr. N. R. Brewer.)

Dr. Gesell had supported the formation of NSMR, but subsequently took issue with Dr. Carlson on NSMR involvement in pound legislation. He also was dissatisfied with what he perceived to be a lack of interest by NSMR in promulgating more detailed humane criteria for the care and use of animals than existed at that time. Dr. Gesell knew of the formation of the American Association for Laboratory Animal Science

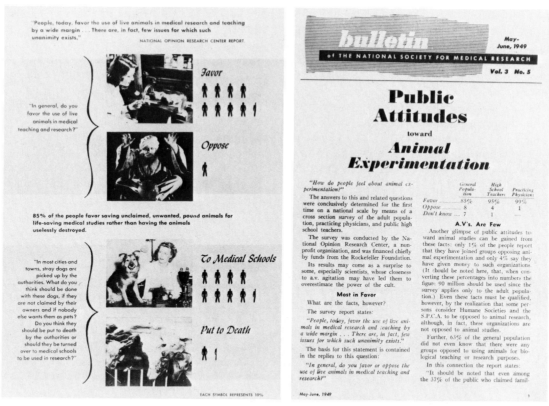

Fig. 11.   Inside cover and p. 1 of *NSMR Bulletin*, May–June, 1949, reporting a national opinion poll on favorable public attitudes toward animal research.

(AALAS) in 1950 and of the assistance that NSMR provided to AALAS in its formative years. This did not soften his view that NSMR was not constructively dealing with the issue of humane use of animals in research.

Dr. Gesell asked that his statement be made part of the minutes. Vigorous discussion followed. Dr. Ralph Gerard, APS President, explained that it was not APS policy to include all statements by APS members in the minutes. Finally, a motion by Dr. Maurice Visscher, then Professor and Chairman of the Department of Physiology at the University of Minnesota (later, President of NSMR), was adopted, to be included in the minutes, ''that Dr. Gesell had made a statement concerning animal experimentation which criticized physiologists and the NSMR and that the statement had been challenged'' (APS minutes, April 15, 1952).

At a second business meeting, on April 17, 1952, APS adopted the following formal response:

The American Physiological Society reaffirms its sincere belief that the moral justification for humane animal experimentation, for the purpose of furthering biological and medical knowledge, in the interest of both human and animal welfare, is completely established.

The American Physiological Society rejects the sweeping allegations made by Dr. Gesell in a recent business meeting.

The American Physiological Society rejects unequivocally the inference

that its members are insensitive to the moral responsibilities which they have in protecting the welfare of man and animals.

The American Physiological Society expresses the hope that in the future all of its members will act in unison in promoting conditions facilitating humane animal experimentation.

Despite efforts by Dr. Gesell to prevent and suppress use of his statement by antivivisection groups, it was distributed widely by these groups in their campaigns for legislative restriction of animal research. After all, it reflected the views of a respected American physiologist. The APS response was not similarly distributed by these groups. Dr. Carlson prepared a lengthy and thoughtful rebuttal of the Gesell statement for members of FASEB; but it too had only a limited distribution (Carlson, A. J., Letter to FASEB members, September 17, 1952). After Dr. Gesell's death in 1954, his daughter, Christine Stevens, a founder and President since 1950 of the Animal Welfare Institute, continued to espouse her father's views and her own strong opinion that too many scientists are insufficiently concerned about humane treatment of animals in research. These views have included critical commentary about NSMR and AALAS (Stevens, 1963; 1976; 1977).

The Gesell–APS–NSMR controversy highlighted issues that, to this day, underlie the difficult relations between the

Fig. 12. Dr. Robert Gesell, Professor and Chairman, 1923–1954, Department of Physiology, University of Michigan. Dr. Gesell's statement at the APS business meeting in 1952 became a cause celebre. Photograph taken in the late 1930s. (Courtesy of the Bentley Historical Library, University of Michigan.)

scientific community and the animal welfare movement. Perhaps a positive result has been that it also contributed to the climate of opinion that led additional numbers of medical research institutions to employ veterinarians to care for research animals. Ultimately, the controversy raised questions that influenced and should continue to influence all those having a constructive concern both for science and animal welfare. What, if any, are the appropriate limits on scientific freedom in animal research? Who is best qualified to make judgments about the propriety of animal studies? Can "humaneness" be legislated? Is there not a moral imperative to conduct animal studies in the interest of human and animal welfare? How best can refinement of animal studies, reduction in the numbers of animals used, and replacement of animals, where appropriate, best be incorporated into the design of experiments (Russell and Burch, 1959).

The National Society for Medical Research has remained active to the present day in educating Congress and the public about the needs and accomplishments of animal research. Other organizations also have evolved recently to work in related areas. The Association for Biomedical Research (ABR) was organized in 1979, with Dr. Edward C. Melby as its first President, to foster a framework for animal use that is acceptable both to the scientific community and the public. Scientists Center for Animal Welfare (SCAW) was formed about the same time to contribute scientific perspectives to laboratory animal welfare.

C. The American Association for Laboratory Animal Science

By 1949, veterinarians were managing the laboratory animal facilities at five Chicago area institutions: The University of Chicago (Nathan R. Brewer), The University of Illinois (Elihu Bond), Northwestern University (Bennett J. Cohen), the Argonne National Laboratory (Robert J. Flynn) and the Hektoen Institute for Medical Research of Cook County Hospital (Robert J. Schroeder). The veterinarians sought one another out to exchange information and experience on the day to day problems they were encountering. The group met at least monthly, starting during the summer of 1949. Among the subjects reviewed at these meetings were husbandry and diseases of laboratory animals, the need to develop basic standards of animal care, and the need to counter the strident antivivisection attacks on medical science in the Chicago area. The Chicago veterinarians knew that few other veterinarians elsewhere in the country were engaged in the activity they had begun to identify as "laboratory animal care." For example, in reviewing the proceedings of a symposium on Animal Colony Maintenance, held under the sponsorship of the New York Academy of Sciences in 1944 (Farris *et al.*, 1945), they noted that not a single veterinarian had presented a paper. Their perception was that the problems of laboratory animal care merited organized attention and they wondered whether others felt the same way.

Special meetings were arranged when colleagues from other institutions visited Chicago. Among these were C. F. Schlotthauer, D.V.M., Mayo Clinic; Charles A. Slanetz, Ph.D., Director of the Central Animal Facility at the College of Physicians and Surgeons, Columbia University; Mr. Harry Herrlein, Rockland Farms, New City, New York (then a major commercial rodent and rabbit breeding facility); Mr. C. N. W. Cumming, Carworth Farms, New City, New York (also a major rodent breeding facility at the time) and W. T. S. Thorp, D.V.M., then Chief of the Laboratory Aids Branch, National Institutes of Health. These meetings were exciting, interesting, and rewarding to the participants. They demonstrated that interest in laboratory animal problems extended well beyond the Chicago area and included individuals having a broad range of scientific, professional, and technical backgrounds.

In a letter signed by the five Chicago veterinarians and sent in May, 1950 to individuals in the United States and Canada thought to have an interest in the care of laboratory animals, the development of a national organization was proposed "to be open to all individuals interested in animal care work on an institutional scale" (Flynn, 1980). The response was overwhelmingly favorable, and the first meeting was convened in Chicago on November 28, 1950, with an attendance of 75. The founding members named the organization the Animal Care Panel (ACP), reflecting their broad concern with the *care* of laboratory animals (Fig. 13). "Panel" was used in the name to emphasize the organization's purpose as a forum for the exchange of information on all aspects of animal care. Dr. Brewer was elected the first President, a post he held until 1955. During the early meetings of the ACP, its programs were dominated by papers on animal colony management, design of facilities and equipment, and descriptions of common

Fig. 13. Cover of early descriptive brochure about the Animal Care Panel, now the American Association for Laboratory Animal Science.

diseases. This reflected the relatively underdeveloped "state of the art" with respect to the technology of animal care. ACP meetings became more sophisticated with each passing year. By the sixth meeting in 1955, original research was being presented (Flynn, 1980). Shortly thereafter, the annual *Proceedings of the Animal Care Panel* was transformed into the scientific journal, *Laboratory Animal Care,* and subsequently renamed *Laboratory Animal Science.* The ACP grew rapidly in its institutional and individual membership and was characterized by the unique diversity of scientific, professional, and technical backgrounds of its members. By 1960, the ACP was able to employ a full time Executive Secretary, Mr. Joseph J. Garvey. Mr. Garvey had served earlier as Assistant Executive Secretary of NSMR and, in this position, had assisted with ACP administration, reflecting the support and encouragement the ACP received from NSMR in its formative years.

From its inception, the ACP also worked to enhance the stature and training of laboratory animal technicians. This is exemplified in the career and contributions of George Collins (1917–1974) (Brewer, 1980). Mr. Collins served successively as Supervisor of the animal facilities at the Argonne National Laboratory, Rockefeller University, and the AMA Education and Research Foundation. He was a founding member of AALAS and, in 1963, received the AALAS Animal Technician Award. In 1967, he edited the first edition of the AALAS Manual for Animal Technicians, a landmark in its time (Collins, 1967).

Development of standards was another early activity of the ACP. Indeed, the first edition of the "Guide for Laboratory Animal Facilities and Care" (Cohen, 1963), now known as the "Guide for the Care and Use of Laboratory Animals" (Moreland, 1978) was prepared under ACP auspices. The Guide has become the basic standard guiding use and care of animals in American research institutions.

In 1967, the name of the Animal Care Panel was changed to the American Association for Laboratory Animal Science. Today, AALAS has nearly 2500 individual and institutional members and more than 30 local branches; its annual meeting and scientific journal, *Laboratory Animal Science,* are the principal means of scientific exchange in the field.

D. The Institute of Laboratory Animal Resources

Many problems of supply, standardization, and procurement of animal resources accompanied the rapid growth of medical and biological research after World War II. Concerns surfaced about these problems within the National Academy of Sciences (NAS). These concerns developed independent of those that led to the formation of AALAS. The NAS is a private organization with a federal charter. Since 1863, it has been a principal advisor to the federal government on matters related to science and science policy (Seitz, 1967). Election to membership in NAS is among the highest honors a scientist can receive. It is a prestigious organization and therefore NAS advisory groups, all of which serve without compensation, have a standing and authority they might not otherwise have.

In the early 1950s, organized efforts to improve and standardize animal supply and quality had barely been initiated. Scientific standards for laboratory animal production, genetics, breeding, husbandry, and transportation did not exist. There were no good mechanisms to facilitate information exchange about laboratory animals internationally. Education and training in laboratory animal science were in an undeveloped state, and no guidelines for such training existed. Problems such as these led Dr. Paul Weiss, then Chairman of the Division of Biology and Agriculture of the National Research Council (the NAS advisory arm), to appoint a Committee on Animal Resources in 1952. Dr. Weiss appointed Dr. Clarence Cook Little, the eminent geneticist and founder of the Jackson Laboratory (Bar Harbor, Maine), to be chairman. The Committee on Animal Resources recommended establishment of an Institute of Animal Resources (IAR). IAR commenced fulltime operation in July, 1953 (Hill, 1980). In 1956, it was renamed Institute of Laboratory Animal Resources (ILAR).

Through the years, the ILAR office has been headed by an Executive Secretary, with oversight from an Advisory Council and Executive Committee that is appointed in accordance with NAS–NRC procedures. Dr. Orson Eaton, a geneticist from the Bureau of Animal Industry, was the first Executive Secretary. He was succeeded by the vigorous and energetic Berton F. Hill, who also had a background in genetics. During Mr. Hill's tenure (1955–1965), ILAR became established as the major standards-development organization within laboratory animal science. In 1965, Mr. Hill was succeeded by Dr. Robert H. Yager, former director of the animal facilities at the Walter Reed Army Institute of Research. Dr. Yager was one of the "founding fathers" of ILAR, having served on the Committee on Animal Resources in 1952. Among many important ILAR accomplishments during Dr. Yager's tenure were the development of the first guidelines for education and training in laboratory animal medicine (Clarkson, 1967); the publication of an important national survey of animal facilities in the United States (Trum, 1970), following up on the first such survey, during Mr. Hill's tenure (Thorp, 1964); and enlargement of United States participation in international laboratory animal activities through support of the International Council on Laboratory Animal Science (ICLAS), known then as the International Committee on Laboratory Animals (ICLA).

During the formative years of ILAR and AALAS, there were obvious areas of overlap. Both organizations had been involved in standards development; both were holding scientific meetings, and in many other areas their interests coincided. In 1962, the Executive Committees agreed on a division of re-

sponsibility which solidified ILAR's role in standards development (Garvey and Hill, 1963). This proved to be an important agreement because it enabled ILAR and AALAS to concentrate on the things each could do best. It also was important because of the standing ILAR standards subsequently achieved under the umbrella of the National Academy of Sciences.

E. The American College of Laboratory Animal Medicine

Formal recognition of veterinary medical specialty fields by the American Veterinary Medical Association (AVMA) began in 1951 with the establishment of the American Board of Veterinary Public Health and the American College of Veterinary Pathologists (Grafton, 1974). In 1957, laboratory animal medicine was accorded the same recognition, when the American Board of Laboratory Animal Medicine (ABLAM) was incorporated under the laws of the state of Illinois, with 18 "Charter Fellows." In August, 1961, the name was changed to American College of Laboratory Animal Medicine (ACLAM) and the designation "Fellow" was discontinued in favor of "Diplomate," a term used by other specialties.

The American College of Laboratory Animal Medicine (ACLAM) (1981) was established to encourage education, training, and research in laboratory animal medicine, to establish standards of training and experience for qualification of specialists, and to certify specialists by examination. These objectives, which today are well understood and accepted, were but a vague concept in the early 1950s. On June 23, 1952, thirty-four veterinarians assembled in a meeting room at the Ambassador Hotel in Atlanta City, during the AVMA meeting, to consider the role of veterinarians in laboratory animal care. There was a lively discussion about this rapidly developing field, with special emphasis on defining activities that veterinarians were uniquely qualified to pursue (News Reports, 1952). Those in attendance noted that an increasing number of veterinarians were being employed by medical schools and that further growth seemed likely. They felt that more specific definition of this newly developing field was needed. The group organized as a Committee on the Medical Care of Laboratory Animals, with Dr. Nathan R. Brewer as Chairman, Dr. Mark Morris as Vice Chairman, and Dr. W. T. S. Thorp as Secretary. The decision was made to organize programs of special interest to laboratory animal veterinarians at future AVMA meetings. During the ensuing four years, the term "laboratory animal medicine" came into use to differentiate the activities of veterinarians from other professional or technical people working in the broad area of laboratory animal science. Additionally, within this period, a number of laboratory animal veterinarians were able to establish academic units in their institutions (Clarkson, 1961a), some of which were identified as

Sections, Divisions, or Departments of Laboratory Animal Medicine or Comparative Medicine. With this development, laboratory animal medicine began to establish its separate identity. The Committee felt it appropriate to seek recognition of specialization in the field by the AVMA. Late in 1956, the Committee disbanded in favor of the American Board of Laboratory Animal Medicine and the new specialty was born.

V. EDUCATION AND TRAINING IN LABORATORY ANIMAL MEDICINE

Veterinarians entering "laboratory animal care" in the 1940s and early 1950s had to be largely self-trained. They relied on their basic education in veterinary medicine and on what they could learn from one another at AALAS and AVMA meetings (Clarkson, 1980). There were no post-D.V.M. training programs. The establishment of ACLAM in 1957, with its strong commitment to fostering education and training, stimulated more specific discussion of training needs in this new field. At this same time, NIH-supported training programs were being initiated in basic medical science and clinical fields in leading scientific institutions throughout the country. As mentioned in Section IV,E, by the late 1950s "laboratory animal medicine" was being conducted as an academic program in a few medical schools (Clarkson, 1961a). In such settings it became possible to consider establishing postdoctoral training. The animal resources program in the Division of Research Resources (DRR) at NIH had not yet matured and the Animal Resources Branch (ARB) did not have training authority. However, with great insight about the underlying significance of laboratory animal medicine, the Physiology Training Committee, within the Division of General Medical Sciences [later to become the National Institute of General Medical Sciences (NIGMS)], decided to accept applications to establish a few research training programs in this new field. During the time this matter was under consideration, Dr. Howard Jenerick and Dr. J. H. U. Brown, both physiologists, served as Secretaries of the Committee. The Committee Chairmen were Dr. T. C. Ruch, Professor and Chairman of the Department of Physiology at the University of Washington, and Dr. Wallace O. Fenn, Professor and Chairman of the Department of Physiology at the University of Rochester. The Committee's decision to sponsor such training was of paramount importance, since for the first time, training in laboratory animal medicine was placed on a par with other areas of research training in the medical and biological sciences. In the ensuing years, the specialists that now comprise the academic core of present-day laboratory animal medicine were trained in such programs.

The first training program was established in January, 1960, at the Bowman Gray Medical School, directed by then Assistant Professor of Laboratory Animal Medicine, Thomas B. Clarkson. In July, 1960, a second program was started at UCLA Medical School, directed by Bennett J. Cohen, then Assistant Professor of Physiology and Director of the Vivarium. The program moved with Dr. Cohen to the University of Michigan in 1962. Later, programs were established at other Medical Schools and Universities: Tulane University (1963, Dr. K. F. Burns), Stanford University (1965, Dr. O. A. Soave), University of Florida (1965, Dr. A. F. Moreland), Johns Hopkins University (1968, Dr. E. C. Melby) and University of Missouri (1968, Dr. C. C. Middleton). Edgewood Arsenal, Maryland and Brooks Air Force Base, Texas became the sites of training programs for military veterinarians. With strong encouragement from Dr. Jules S. Cass, Chief Veterinary Medical Officer at the Veterans Administration, a program was established in the mid-1960's at the Hines Veterans Administration Medical Center in the Chicago area. This program was guided initially by Dr. Robert F. Locke.

The "core of knowledge" comprising laboratory animal medicine was not well defined at the time these early programs were started. The curricula of the training programs simply reflected the outlook of the directors and the settings in which they were conducted. Some were formal graduate programs leading to a Master of Science degree. Others stressed residency training analogous to that of residency programs in the medical specialties. Thus, there were research-oriented programs and others that focused more on the clinical or managerial aspects of laboratory animal medicine (Clarkson, 1961b). By 1964, the need for better definition of the field had become apparent. An ILAR-sponsored workshop held in that year pointed to the need for educational guidelines to be used by all training programs (Clarkson, 1965). The first such guidelines subsequently were published (Clarkson, 1967). The most recent edition appeared in 1978 (Cohen, 1978, 1979).

In the mid-1960s, ARB received training authority, and the training grants in laboratory animal medicine were transferred there. Other aspects of the NIH extramural animal resources program also grew significantly, as for example, the laboratory animal science program. Overall, the impact of these programs on laboratory animal medicine has been enormously beneficial. Some of the early training programs have been terminated, but most have continued, and several new programs have been started, as for example, at the Hershey Medical Center, Pennsylvania State University, Dr. C. M. Lang; at the University of Alabama, Birmingham, Dr. H. J. Baker and Dr. J. R. Lindsey; and at the University of Washington, Dr. G. L. Van Hoosier, Jr. Furthermore, a number of institutions have initiated residency programs independent of NIH sponsorship. Post-D.V.M. training is recognized today as necessary for a career in academic laboratory animal medicine.

The American Society of Laboratory Animal Practitioners (ASLAP) was founded in 1967 to promote dissemination of knowledge about laboratory animal medicine, foster research, and serve as a spokesman for veterinarians in laboratory animal practice. ASLAP, together with ACLAM, has played an important role in encouraging continuing education programs in the field. Continuing education has become an important adjunct to the formal training programs in laboratory animal medicine, and such activities now are a regular component of the scientific program at AVMA and AALAS meetings.

VI. IMPACT OF LAWS, REGULATIONS, AND GUIDELINES ON LABORATORY ANIMAL MEDICINE

Prior to 1966, no federal law existed in the United States specifically regulating the acquisition or care of research animals. Pressure for federal legislation mounted steadily in the late 1950s and early 1960s. Animal welfare organizations, such as the Humane Society of the United States, the Society for Animal Protective Legislation, and the Animal Welfare Institute, argued for legislation to curb alleged "pet stealing" and abuse of animals in laboratories. They used the media effectively to generate public interest in their causes (e.g., cover headline of *Life,* February 4, 1966: "Concentration Camps for Lost and Stolen Pets: Your Dog Is in Cruel Danger"). Organizations of the scientific community, such as AALAS, NSMR, and FASEB, argued against the proposed legislation. Their position was that the best way to foster the humane use and care of animals was to provide better support of research and training, provide funds to upgrade animal facilities, and strengthen self-regulation through mechanisms such as the newly organized (1965) American Association for Accreditation of Laboratory Animal Care (AAALAC) and institutional committees to assess the adequacy of animal care and use programs (Galton, 1967).

After holding hearings on a series of bills in the House of Representatives and the Senate dealing with the regulation of animal research, Congress passed the Laboratory Animal Welfare Act in 1966. The principal purposes were to regulate commercial traffic in dogs, cats, monkeys, rabbits, guinea pigs, and hamsters and to establish standards for their housing, transportation, and "adequate veterinary care." The Act, administered by the Department of Agriculture, established a legal requirement for scientific institutions to provide appropriate care for research animals by or under the direction of a veterinarian. Since its initial passage, the Act has been broadened and its name changed to Animal Welfare Act (see Chapter 2). Today, more than 1200 research facilities are registered.

Thus, the Act has contributed to the betterment of animal care through its requirement for participation of veterinarians in institutional animal medicine programs.

The National Institutes of Health (NIH) has long recognized that good research requires animals that are healthy and well cared for. In 1963, NIH published the first edition of the "Guide for Laboratory Animal Facilities and Care" (Cohen, 1963), as developed by the Standards Committee of the Animal Care Panel. Revised several times since 1963 by ILAR's Committee on Revision of the Guide, it is now the "Guide for the Care and Use of Laboratory Animals" (Moreland, 1978). Since 1963, NIH and other granting agencies have required scientific institutions to provide assurance of compliance with the standards in the Guide as a condition for receiving funds for research. The Guide also is used as the basis for accreditation by AAALAC. One of the basic requirements established in the Guide is for the provision of adequate veterinary medical care, a concept also expressed in the regulations of the Department of Agriculture.

In 1978, the Food and Drug Administration (FDA) promulgated regulations for the conduct of animal experiments relating to new or existing pharmaceutical medicinal substances, food additives, or other chemicals. These regulations, known as the Good Laboratory Practice (GLP) regulations, also specify the need for adequate diagnosis, treatment, and control of diseases in animals used in such studies. Thus, the standards of AAALAC, NIH, USDA, and FDA all include specific references to veterinary medical participation in the care of laboratory animals. In fact, these standards provide the basis for implementation of the legal requirement that research animals receive "adequate veterinary care."

VII. REGULATION OF ANIMAL RESEARCH IN THE UNITED KINGDOM AND CANADA

The use of animals in the United Kingdom is governed by the "Cruelty to Animals Act of 1876" (French, 1975) (see Chapter 2). Major debates about "vivisection" occurred in Parliament in the late 1860s and early 1870s. Finally, the Act was passed with the active support of leading scientists of that time. Antivivisectionists had been working for a law that would have prohibited animal research or regulated it more strictly than called for in the Act. The Act requires the licensing of scientists using animals. There also is provision for certificates to be issued to scientists depending upon the species used and the nature of the experimentation. The legal regulatory relationship is between the government and the scientist, not the government and the institution, as it is in the United States. Veterinarians, as such, do not have legal standing in the implementation of the Act, although within recent years veterinarians have been added to the Home Office inspectorate. In the private sector, a British Laboratory Animals Veterinary Association was organized in the early 1970s, and is affiliated with the British Veterinary Association. The specialty of laboratory animal medicine does not have standing comparable to that accorded ACLAM by the AVMA in the United States. Nevertheless, laboratory animal medicine clearly is an emerging field in Great Britain, and excellent animal medicine programs are evolving under the leadership of British laboratory animal veterinarians.

The use of animals in Canada is not specifically regulated by federal law. However, in 1968, the Canadian Council on Animal Care (CCAC) was established by the major agencies that fund animal research (see Chapter 2). Dr. Harry Rowsell has played an instrumental role in the founding of CCAC and in its operation over the years. The CCAC assesses animal care in Canadian research laboratories based on the standards in the "Guide to the Care and Use of Experimental Animals" (CCAC, 1980). This program is analogous in many respects to that of AAALAC and has been a major factor in the elevation of animal care standards and the employment of veterinarians in Canadian research laboratories. In addition, some of the provinces have laws which apply to the requisition or use of animals in research in those provinces (e.g., Ontario and Alberta).

VIII. COMMERCIAL RODENT BREEDING

The development of gnotobiology in the 1950s represented a major conceptual and technological advance in the commercial breeding of healthier rodents for research (Foster, 1958). This had been preceded by laboratory research (Trexler and Reyniers, 1957; Reyniers, 1957) and years of attempts to breed animals which would lead to the most unambiguous results in research as possible. The rearing of breeding stock in plastic isolators is now a commercial necessity in many cases (see Fig. 6 of Chapter 17).

IX. CONCLUSION

A complete history of the individuals and organizations that have influenced the development of laboratory animal science and medicine would require a separate volume. Many other relevant topics could not be dealt with here because of the page limits that had to be set. Some of these are the history of labo-

ratory animal science internationally; the history of the major commercial and institutional animal colonies and of the important genetic stocks and strains of laboratory animals; the evolution of animal technology, including the field of gnotobiology; the contributions of animal technicians to laboratory animal science; the origins of the NIH extramural and intramural laboratory animal science programs; and reviews, in historical perspective, of the major diseases of laboratory animals. Some of these topics are dealt with elsewhere (Foster, 1980; Lindsey, 1979; McPherson, 1980). Others must await documentation of the historical record.

Laboratory animal science and medicine are fields of expanding horizons that provide challenging opportunities for satisfying professional careers. The following chapters clearly document the progress that has occurred while pointing to many challenges that lie ahead. It remains for each generation of laboratory animal scientists to build on the base of knowledge established by its predecessors and so determine its own future.

REFERENCES

American College of Laboratory Animal Medicine. (ACLAM) (1981). "Directory of Diplomates," p. 1.

Bernard, C. (1865). "An Introduction to the Study of Experimental Medicine" (transl. by H. C. Greene, p. 123. Reprinted by Dover, New York, 1957).

Braasch, W. F. (1969). "Early Days of the Mayo Clinic," p. 141. Thomas, Springfield, Illinois.

Brewer, N. R. (1980). Personalities in the early history of laboratory animal science and medicine. *Lab. Anim. Sci.* **30**, No. 4, Part II, 741–758.

Brewer, N. R. (1982). Nutrition of the cat. *J. Am. Vet. Med. Assoc.* **180**, 1179–1182.

Brimhall, S. D., and Hardenbergh, J. G. (1922). A study of so-called kennel lameness: Preliminary report. *J. Am. Vet. Med. Assoc.* **61**, 145–154.

Brimhall, S. D., and Mann, F. C. (1917). Pathologic conditions noted in laboratory animals. *J. Am. Vet. Med. Assoc.* **52**, 195–204.

Brimhall, S. D., Mann, F. C., and Foster, J. P. (1919–1920). The relation of the common bile duct to the pancreatic duct in common domestic and laboratory animals. *J. Lab. Clin. Med.* **5**, 203–206.

Bustad, L. K., Gorham, J. R., Hagreberg, G. A., and Padgett, G. A. (1976). Comparative medicine: Progress and prospects. *J. Am. Vet. Med. Assoc.* **169**, 90–105.

Canadian Council on Animal Care (CCAC) (1980). "Guide to the Care and Use of Experimental Animals," Vol. 1, p. 112. CCAC, Ottawa, Ontario.

Clarkson, T. B. (1961a). Laboratory animal medicine and the medical schools. *J. Med. Educ.* **36**, 1329–1330.

Clarkson, T. B. (1961b). Graduate and professional training in laboratory animal medicine. *Fed. Proc., Fed. Am. Soc. Exp. Biol.* **20**, 915–916.

Clarkson, T. B., Chairman (1965). Committee on Professional Education, Institute of Laboratory Animal Resources. Laboratory Animals. IV. Graduate education in laboratory animal medicine. Proceedings of a workshop. *NAS–NRC, Pub.* **1284**, 33.

Clarkson, T. B., Chairman (1967). Committee on Professional Education. Institute of Laboratory Animal Resources. A guide to postdoctoral training in laboratory animal medicine. *NAS–NRC, Publ.* **1483**, 9.

Clarkson, T. B. (1980). Evolution and history of training and academic programs in laboratory animal medicine. *Lab. Anim. Sci.* **30,** No. 4, Part II, 790–792.

Cohen, B. J. (1959a). The early history of animal experimentation and animal care. I. Antiquity. *Lab. Anim. Sci.* **9,** 39–45.

Cohen, B. J. (1959b). The evolution of laboratory animal medicine in the United States. *J. Am. Vet. Med. Assoc.* **135,** 161–164.

Cohen, B. J., Chairman (1963). "Guide for Laboratory Animal Facilities and Care," Anim. Facilities Stand. Comm., Anim. Care Panel, Public Health Serv. Publ. No. 1024, p. 33. U.S. Dept of Health, Education and Welfare, Washington, D.C.

Cohen, B. J., Chairman (1978). "Laboratory Animal Medicine: Guidelines for Education and Training." Committee on Education, Institute of Laboratory Animal Resources, National Academy of Sciences, Washington D.C.

Cohen, B. J. (1979). *ILAR News* **22,** No. 2, 26.

Cohen, M. R., and Drabkin, I. E. (1948). "Sourcebook in Greek Science," p. 479. McGraw-Hill, New York.

Collins, G. R., Chairman (1967). "Manual for Laboratory Animal Technicians." Animal Technician Training Committee, American Association for Laboratory Animal Science. Publ. 67-3, Joliet, Illinois.

Duclaux, E. (1920). "Pasteur—The History of a Mind" (transl. by E. F. Smith and F. Hedges), p. 363. Saunders, Philadelphia, Pennsylvania.

Farris, E. J., Carnochan, F. G., Cumming, C. N. W., Farber, S., Hartman, C. G., Hutt, F. B., Loosli, J. K., Mills, C. A., and Ratcliffe, H. L. (1945). Animal colony maintenance. *Ann. N.Y. Acad. Sci.* **46,** 1–126.

Fisher, G. J. (1881). Historical and bibliographical notes. XII. Herophilus and Erasistratus. The Medical School of Alexandria. BC 320–250. *Ann. Anat. Surg.* **4,** 28, 67.

Flynn, R. J. (1980). The founding and early history of the American Association for Laboratory Animal Science. *Lab. Anim. Sci.* **30,** No. 4, Part II, 765–779.

Foster, H. L. (1958). Large scale production of rats free of commonly occurring pathogens and parasites. *Proc. Anim. Care Panel* **8,** 92–99.

Foster, H. L. (1980). The history of commercial production of laboratory rodents. *Lab. Anim. Sci.* **30,** No. 4, Part II, 793–798.

French, R. D. (1975). "Antivivisection and Medical Science in Victorian Society," p. 425. Princeton Univ. Press, Princeton, New Jersey.

Galton, L. (1967). Pain is cruel, but disease is cruel too. *N.Y. Times* Sect. 6 (Magazine), February 26, p. 30.

Garrison, F. H. (1929). "An Introduction to the History of Medicine," 4th ed. Saunders Co., Philadelphia, Pennsylvania.

Garvey, J. J., and Hill, B. F. (1963). Cooperation for progress. *Lab. Anim. Care* **13,** 179–180.

Grafton, T. S. (1974). The veterinary profession: A review of its progress in the United States and some indications for the future. *Vet. Rec.* **94,** 441–443.

Grafton, T. S. (1980). The founding and early history of the National Society for Medical Research. *Lab. Anim. Sci.* **30,** No. 4, Part II, 759–764.

Griffin, C. A. (1952). A study of prepared feeds in relation to *Salmonella* infection in laboratory animals. *J. Am. Vet. Med. Assoc.* **121,** 197–200.

Hales, S. (1740). "Statical Essays." Innys & Manby, London.

Hardenbergh, J. G. (1923). The value of animal experimentation to veterinary medicine. *J. Am. Vet. Med. Assoc.* **62,** 731–735.

Hardenbergh, J. G. (1926–1927). Epidemic lymphadenitis with formation of abscess in guinea pigs due to infection with hemolytic streptococcus. *J. Lab. Clin. Med.* **12,** 119–129.

Hill, B. F. (1980). The founding and early history of the Institute of Laboratory Animal Resources. *Lab. Anim. Sci.* **30,** No. 4, Part II, 780–785.

Hoff, H. E., Geddes, L. A., and McCrady, J. D. (1965). The contributions of the horse to knowledge of the heart and circulation. *Conn. Med.* **29,** 795–800.

Ingle, D. J. (1979). Anton J. Carlson: A biographical sketch. *Perspect. Biol. Med.* **29,** Part 2, 114–136.

Kaplan, H. M., Brewer, N. R., and Blair, W. H. (1983). Physiology. *In* "The Mouse in Biomedical Research" (H. L. Foster, J. D. Small, and J. G. Fox, eds.), Vol. III, pp. 248–292. Academic Press, New York.

Lindsey, J. R. (1979). Origin of the laboratory rat. *In* "The Laboratory Rat" (H. G. Baker, J. R. Lindsey, and S. H. Weisbroth, eds.), Chapter 1, pp. 2–36, Academic Press, New York.

Loew, F. M. (1982). Animal experimentation *Bull. Hist. Med.* **56,** 123–126.

McPherson, C. W. (1980). The origins of laboratory animal science at the National Institutes of Health. *Lab. Anim. Sci.* **30,** No. 4, Part II, 786–789.

Malinin, T. I. (1979). "Surgery and Life," p. 242. Harcourt Brace Jonanovich, New York.

Meyer, K. F. (1928). Communicable diseases of laboratory animals. *In* "The Newer Knowledge of Bacteriology and Immunity" (E. O. Jordan and I. S. Falk, eds.), pp. 607–638. Univ. of Chicago Press, Chicago, Illinois.

Meyer, K. F. (1958). Introductory address. *Lab. Anim. Sci.* **8,** 1–5.

Mitchell, S. W. (1895). Memoir of John Call Dalton. 1825–1889. Biographical memoirs. *Natl. Acad. Sci.* **3,** 177.

Moreland, A. F., Chairman (1978). "Guide for the Care and Use of Laboratory Animals," Committee on the Care and Use of Laboratory Animals of the Institute of Laboratory Animal Resources, DHEW Publ. No. (NIH) 78–23, p. 70. U.S. Dept. of Health and Human Services, Public Health Service, National Institutes of Health, Bethesda, Maryland.

News Reports (1952). *J. Am. Vet. Med. Assoc.* **121,** 257.

Physicians of the Mayo Clinic and the Mayo Foundation (1937). Univ. of Minnesota Press, Minneapolis (p. 184).

Reyniers, J. A. (1957). The control of contamination in colonies of laboratory animals by use of germfree techniques. *Proc. Anim. Care Panel* **7,** 9–29.

Russell, W. M. S., and Burch, R. L. (1959). "The Principles of Humane Experimental Technique," p. 238. Methuen, London.

Saunders, J. B., de, C. M., and O'Malley, C. D. (1950). "The Illustrations from the Works of Andreas Vesalius of Brussels," p. 128. World Publ. Co., New York.

Schlotthauer, C. F. (1950). Procurement of animals. *Lab. Anim. Sci.* **1,** 20–25.

Schwabe, C. W. (1978). "Cattle, Priests and Progress in Medicine," p. 277. Univ. of Minnesota Press, Minneapolis.

Seitz, F. (1967). The National Academy of Sciences. *J. Wash. Acad. Sci.* **57,** 38–41.

Singer, C. (1957). "A Short History of Anatomy and Physiology form the Greeks to Harvey," p. 209. Dover, New York.

Stevens, C. (1963). Letter to the Editor. *Perspect. Biol. Med.* **7,** No. 1 (Autumn), 129–131.

Stevens, C. (1976). Humane considerations for animal models. *In* "Animal Models of Thrombosis and Hemorrhagic Diseases." DHEW Publ. No. (NIH) 76–982, pp. 151–158. U.S. Dept. of Health and Human Services, National Institutes of Health, Bethesda, Maryland.

Stevens, C. (1977). Humane perspectives. *In* "The Future of Animals, Cells, Models, and Systems in Research, Development, Education, and Testing," pp. 16–24. National Academy of Sciences, Washington, D.C.

Thorp, W. T. S., Chairman (1964). "ILAR Committee on the Animal Facilities Survey Animal Facilities in Medical Research," Final Report and Tabular Appendix, p. 157.

Trexler, P. C., and Reynolds, L. I. (1957). Flexible film apparatus for the rearing and use of germfree animals. *Appl. Microbiol.* **5,** 406–412.

Trum, B. F., Chairman (1970). ILAR Committee on Laboratory Animal Facilities and Resources Survey. Laboratory animal facilities and resources supporting biomedical research. *Lab. Anim. Care* **20,** 795–869.

Welch, W. H. (1889). Preliminary report of investigations concerning the causation of hog cholera. *Johns Hopkins Hosp. Bull.* **1,** (1), 9–10.

Wood, C. A., ed. (1931). "An Introduction to the Literature of Vertebrate Zoology," Chapter IV. Oxford Univ. Press, London and New York.

Chapter 2

Laws, Regulations, and Policies Affecting the Use of Laboratory Animals

Charles W. McPherson

I. INTRODUCTION

The purpose of this chapter is to acquaint those who care for or use animals in biomedical activities with the principal laws, regulations, and policies that govern these activities. The content of these rules is subject to change, and in some instances the information presented here might not be sufficiently detailed to meet everyone's specific needs. Therefore, for each major category of rules, we have tried to indicate sources of updated and additional information.

Rules directly relating to laboratory animals generally fall into two categories. The first category deals with animal welfare and methods of acquisition of animals for use in laboratories. These laws, regulations, and policies are designed to assure that animals obtained for or used in laboratories are not mistreated and receive appropriate care. The second category of rules deals with the importation and shipment of animals. These rules are designed to protect human beings, other animals, or the environment from harm or diseases from imported animals or to prevent exploitation of endangered species.

Other rules that govern biomedical activities in general also impact on the use of animals. Some of these rules are designed to protect laboratory workers and the general public from hazardous substances used in experiments, while others are designed to assure good laboratory practices especially in the area of drug testing.

In summary, the laws and guidelines discussed in this chapter simply say that a researcher should provide the best type of animal care, keep clear and complete records, and refrain from practices which exploit animals or the environment. With these guiding principles in mind compliance with the laws and guidelines should be easier.

II. ANIMAL WELFARE

A. Background

The welfare of animals used in biomedical activities is a matter of concern not only to those who care for and use animals, but also to significant numbers of the general public. The laws, regulations, and policies that are in effect in various nations reflect this concern.

Some opponents of animal research or antivisectionists (see Chapter 1) believe that man has no right to use animals in painful or even painless experimentation, no matter what the potential or actual benefit to mankind or animals. No society has accepted this extreme position and none has totally banned the use of animals in laboratory activities. In recent years there has been a growing advocacy for the use of alternatives to replace or reduce numbers of animals used. Some of the alternatives advocated include computer simulation, tissue and organ culture, statistical analysis, and *in vitro* assays. These techniques are in widespread use, and most scientists agree that their application has reduced animal use to some extent. However, virtually no knowledgeable biological scientist foresees the complete replacement of animal tests and experiments. Furthermore, scientists are concerned that legislative or regulatory requirements to use alternatives would be extremely detrimental to research progress. Thus far no nation has enacted laws that have as their principal focus the requirement to use alternatives.

Many individuals take the position that it is fully justified to use animals in laboratories under certain circumstances. The debate then focuses on the definition of those circumstances, such as the degree of requirement to use an animal *vis à vis in vitro* methods, restriction of painful procedures, alleviation of pain, and standards of animal care. There is nearly unanimous support for the thesis that animals used in biomedical activities should be well cared for and not subject to unnecessary pain or distress. The laws, regulations, and policies that are in effect in the United States and some other countries are for the most part a codification of this principle. It should be kept in mind that animal welfare is a matter under active political consideration, and the laws and policies outlined below are subject to change.

B. National Laws and Policies

1. United States

a. The Animal Welfare Act. In the United States federal laws for the humane treatment of animals have been on the books since 1873, when a law was passed governing the treatment of farm animals during shipment for export. This law was called the "28 hour law" after the maximum length of time animals could be transported before receiving food, water and rest. The first law protecting nonfarm animals, however, was not passed until 1966. It was commonly referred to as the Laboratory Animal Welfare Act or P.L. 89-544. It was administered by the United States Department of Agriculture (USDA). The 1966 law was primarily a dog and cat dealer law. It required that individuals or corporations that bought or sold dogs or cats for laboratory activities be licensed and adhere to certain minimum standards for the care of the animals. It also required that organizations which used dogs or cats in biomedical activities register with the USDA and adhere to the standards promulgated for the care of animals. A feature of the law was that it required only dealers in dogs and cats to be licensed and only biomedical users of dogs and cats to register, but licensed or registered persons and organizations were required to care for hamsters, guinea pigs, rabbits, and nonhuman primates according to standards which were promulgated. Within biomedical organizations the law applied only to animals being held prior to, or after, actual research and testing and *not* during the time the animals were being used in these activities.

The 1966 Act was amended in 1970 (P.L. 91-579) and given the official title of "The Animal Welfare Act." The 1970 amendments markedly broadened the coverage of the law. Animals covered included not only dogs, cats, hamsters, guinea pigs, rabbits, and nonhuman primates but also, with certain exceptions, any other warm-blooded animal designated by the Secretary of Agriculture. The term dealer was broadened to cover not only persons who bought and sold dogs for biomedical activities, but any person who bought or sold any dog or other animal designated by the Secretary of Agriculture for research, teaching, exhibition, or use as a pet at the wholesale level. Retail pet stores were specifically exempt from coverage by the Act. Exhibitors of animals such as zoos were covered under the Act and were required to be licensed.

Organizations (except elementary and secondary schools) that used any of the specified animals in research, tests, or experiments were required to register as a research facility. The Secretary of Agriculture could exempt from registration research facilities that used no dogs or cats and only small numbers of other animals. A significant change with respect to research organizations was that the standards for animal care were to apply to animals throughout their stay in the facility. The Act did not allow the Secretary of Agriculture to promulgate rules, regulations, or orders with regard to the design or performance of actual research or experimentation. It did require, however, that every research facility show at least annually that professionally acceptable standards governing the care, treatment, and use of animals including appropriate use of anesthetic, analgesic, and tranquilizing drugs were being followed.

The Animal Welfare Act was amended again in 1976 by P.L. 94-279. The 1976 amendments brought common carriers such as airlines under coverage of the Act. Subsequently, standards were developed for the containers and conditions of shipment of animals covered by the Act. The 1976 amendments also contained provisions to outlaw interstate promotion or shipment of animals for animal fighting ventures.

The principal features of the Animal Welfare Act and the relevant regulations as they are currently constituted are as follows:

1. Animals that are covered include any live or dead dog, cat, nonhuman primate, guinea pig, hamster, rabbit, aquatic mammal, or any other warm-blooded animal that is intended to be used in research, testing, experimentation, exhibition, or as a pet except birds, rats, mice, horses and farm animals intended for use as food or fiber or used in studies to improve production of food and fiber.

2. Animal dealers, operators of auction sales where dogs or cats are sold affecting commerce, and exhibitors of animals are required to have a license issued by the USDA. Exempt from the licensing requirements are retail pet stores; any person who does not sell any wild animal, dog, or cat and who derives no more than $500 gross income from the sale of other animals during any calendar year; any person who breeds and raises dogs or cats on his own premises and who derives less than $500 per calendar year from the sale of such animals to dealers and research facilities; animal pounds operated by state and local governmental units; and state and county fairs, livestock shows, and dog and cat shows.

3. Research facilities, intermediate handlers, and common carriers of animals must register with the USDA. A research facility is defined as any school (except an elementary or secondary school), institution, organization, or person that uses animals covered by the Act in research, tests, or experiments and purchases or transports animals in commerce or receives funds from the United States Government. The USDA may exempt from registration research facilities that do not use dogs or cats or substantial numbers of other animals.

4. Animal dealers or exhibitors must be licensed to sell or transport animals in commerce to any dealer, exhibitor, or research facility.

5. Dealers and exhibitors who initially acquire a dog or cat must hold these animals for 5 business days before they are sold or disposed of. Dogs or cats that have completed the 5-day holding period may be disposed of by subsequent dealers or exhibitors after a minimum holding period of 1 day. Operators of auction sales do not have to comply with this provision.

6. Research facilities and United States Government agencies must purchase all dogs and cats from licensed sources unless the source is exempted from obtaining a license.

7. Research facilities and United States Government agencies must submit an annual report each year on or before December 1 covering the previous year of October 1 through September 30. This report must show that professionally acceptable standards governing the care, treatment, and use of animals, including appropriate use of anesthetic, analgesic, and tranquilizing drugs during actual research, testing, or experimentation were followed. The attending veterinarian or an institutional committee of at least three members, one of whom must be a veterinarian, must certify that the type and amount of anesthetic, analgesic, and tranquilizing drugs used was appropriate. The annual report must also include the common names and numbers of animals used in activities involving no pain or distress; the numbers of animals used in activities that otherwise would involve pain or distress and for which appropriate anesthetic, analgesic, or tranquilizing drugs were used; and the numbers of animals used in activities involving pain or distress and for which the use of anesthetic, analgesic, or tranquilizing drugs would adversely affect the procedures or results. The latter must be explained. The USDA provides forms for making the annual report.

8. Dealers and exhibitors must individually identify weaned dogs and cats by means of an official tag,* permanent tattoo, or in the case of animals under 16 weeks of age with a plastic identification collar. Animals, except dogs and cats, may be identified on a group basis with a record or label of the number of animals, species, and description of distinctive physical features.

9. Research facilities must identify dogs and cats individually by use of a collar, tattoo, or official tag.

10. All official tags that are removed from animals for any reason or from animals that die must be held until called for by the USDA or for a period of 1 year.

11. Dealers and exhibitors are required to maintain individual records on dogs and cats, including source, description, and disposition of the animal. Similar records must be kept on the source and disposition of other animals. The USDA provides forms for these records. A copy of these records must accompany each shipment of animals.

*A list of commercial manufacturers who produce official tags is available from the USDA Veterinarian in Charge in each state or states.

12. Research facilities must maintain individual records on dogs and cats including the source, description, and disposition of the animals. The USDA will provide forms for these records. Records are to be retained for 1 year. Research facilities do not need to maintain individual records on animals other than dogs or cats.

13. Dogs, cats, and nonhuman primates going through intermediate handlers or carriers for transportation must be accompanied by a health certificate.

14. No C.O.D. shipments of animals covered are allowed unless the consignor guarantees payment.

15. All licensees, registrants, and United States Government agencies must care for animals according to standards issued by the USDA. Specific standards have been promulgated for dogs, cats, guinea pigs, hamsters, rabbits, nonhuman primates, and marine mammals. For other animals covered by the Act general standards have been issued. The standards set forth minimum requirements with respect to facilities, shelter from extremes of weather, primary enclosures (cages), feeding, watering, sanitation, employees, grouping of animals, veterinary care, and transportation of animals. The scope of this chapter does not allow for complete detailing of the standards, but several points will be emphasized.

For all animals the standards require that the primary enclosure (cage, pen, run) provide sufficient space to allow each animal to make normal postural adjustments with freedom of movement. For a number of animals additional specific minimum enclosure sizes have been set forth. They are summarized in Table I.

The standards require that veterinary care and programs of disease control and prevention and euthanasia be established and maintained under the supervision and assistance of a doctor of veterinary medicine. Animals must be observed on a regular basis and veterinary care obtained for diseased or injured animals. In the case of a research facility, the program of veterinary care shall include appropriate use of anesthetic, analgesic, and tranquilizing drugs.

16. The transportation standards include rules for consignment to carriers and intermediate handlers; standards for construction of and space provided by primary enclosures; and standards for conveyance vehicles, food and water, care in transit, terminal facilities, and handling. Allowable temperature ranges are specified for terminal facilities where animals are held for an intermediate stop or before or after shipment. The temperature measurements are to be taken within 3 ft of the primary enclosure. For all animals covered by the Act, these temperature measurements may not be above 85°F or below 45°F unless the animals are acclimated to lower temperatures (except guinea pigs). A USDA accredited veterinarian must provide a certificate signed not more than 10 days prior to shipment stating that the animals are acclimated to temperatures below 45°F. Because some common carriers are not confident of their ability to meet these standards, especially the terminal temperature requirements, animal delivery schedules are occasionally disrupted especially during periods of extreme temperatures.

Detailed implementing rules and regulations for the Animal Welfare Act are published in the *Code of Federal Regulations* (CFR), Title 9, Animals and Animal Products, Subchapter A, Animal Welfare, Parts 1, 2, and 3. All proposed and final

Table I

Animal Welfare Act Primary Enclosure Space Requirements

Animal	Floor space	Height	Maximum number per enclosure
Dogs	Mathematical square of the sum of the length of the dog from the tip of the nose to base of the tail plus 6 in.	To stand easily	12
Cats			
Adults	2½ ft²	To stand easily	12
Guinea pigs			
<350 gm	60 in.²	6½ in.	—
>350 gm	90 in.²	6½ in.	—
Breeders	180 in.²	6½ in.	—
Syrian hamsters			
< 5 weeks old	10 in.²	5½ in.	20
5–10 weeks old	12.5 in.²	5½ in.	16
>10 weeks old	15 in.²	5½ in.	13
Nursing female with litter	121 in.²	5½ in.	—
Chinese hamsters			
<5 weeks old	5 in.²	5 in.	20
5–10 weeks old	7.5 in.²	5 in.	16
>10 weeks old	9 in.²	5 in	—
Rabbits			
3–5 lb			
In groups	144 in.²	—	—
Individuals	180 in.²	—	—
Nursing females	576 in.²	—	—
6–8 lb			
In groups	288 in.²	—	—
Individuals	360 in.²	—	—
Nursing females	720 in.²	—	—
9–11 lb			
In groups	432 in.²	—	—
Individuals	540 in.²	—	—
Nursing females	864 in.²	—	—
12 or more lb			
In groups	432 in.²	—	—
Individuals	720 in.²	—	—
Nursing females	1080 in.²	—	—
Nonhuman primates	Three times the floor space occupied by the animal when it is standing on four feet	—	—

amendments to the rules and regulations are published in the Federal Register under the heading, Department of Agriculture, Animal and Plant Health Inspection Service. Copies of the rules and regulations can be obtained from the USDA Veterinarian in Charge in each state or states or from the Senior Staff Veterinarian, Animal Welfare Staff, Veterinary Services, Animal and Plant Health Inspection Service, Federal Building, USDA, 6505 Belcrest Road, Hyattsville, Maryland 20782.

b. National Institutes of Health and Public Health Service Policy. The United States Public Health Service has an animal welfare policy relating to the use of animals in research and other biomedical activities supported by grants, contracts, and awards from the Service (U.S. Public Health Service, 1979). Included are all relevant National Institutes of Health (NIH) awards and awards issued by any other component of the Public Health Service (PHS) such as the Alcohol, Drug Abuse and Mental Health Administration and the Food and Drug Administration. The policy requires that all awardee institutions submit a written assurance to the Office for Protection from Research Risks, NIH, that they are committed to follow a set of Principles for the Use of Animals and the "Guide for The Care and Use of Laboratory Animals" (Moreland, 1978). The Principles are contained in the policy statement and are reproduced in Table II. The Guide sets forth comprehensive guidelines for the care of laboratory animals. Furthermore, the institution must have a committee to maintain oversight of its animal facilities and procedures. The committee should be composed of at least five members who are knowledgeable regarding the care and use of animals in research, including at least one veterinarian. Institutions are required to establish a mechanism to review their animal facilities and procedures for conformance with provisions of the Guide. Accreditation by The American Association for Accreditation of Laboratory Animal Care (AAALAC)* is deemed the best means of demonstrating conformance with the Guide. An alternative to AAALAC accreditation is annual review of animal facilities and procedures by the institution's committee. The committee determines whether improvements are needed in the institution's animal facilities or procedures. If improvements are needed, they are to be reported in the assurance statement, and an annual report of progress must be submitted to The Office for Protection from Research Risks, NIH.

Grant applications and contract proposals to NIH and the Public Health Service should indicate whether animals are involved in the proposed activity and should state the rationale for using animals. Information should be provided to confirm that the species and numbers of animals are appropriate; that

*The American Association for Accreditation of Laboratory Animal Care, 208A North Cedar Rd., New Lenox, Illinois, 60451 is a private, nonprofit body that accredits laboratory animal facilities and care through peer review based on "The Guide for the Care and Use of Laboratory Animals."

Table II

National Institutes of Health and Public Health Service Principles for Use of Animals

The personnel

1. Experiments involving live, vertebrate animals and the procurement of tissues from living animals for research must be performed by, or under the immediate supervision of, a qualified biological, behavioral, or medical scientist

2. Housing, care, and feeding of all experimental animals must be supervised by a properly qualified veterinarian or other scientist competent in such matters

The research

3. The research should be such as to yield fruitful results for the good of society and not random or unnecessary in nature

4. The experiment should be based on knowledge of the disease or problem under study and so designed that the anticipated results will justify its performance

5. Statistical analysis, mathematical models, or *in vitro* biological systems should be used when appropriate to complement animal experiments and to reduce numbers of animals used

6. The experiment should be conducted so as to avoid all unnecessary suffering and injury to the animals

7. The scientist in charge of the experiment must be prepared to terminate it whenever he/she believes that its continuation may result in unnecessary injury or suffering to the animals

8. If the experiment or procedure is likely to cause greater discomfort than that attending anesthetization, the animals must first be rendered incapable of perceiving pain and be maintained in that condition until the experiment or procedure is ended. The only exception to this guideline should be in those cases where the anesthetization would defeat the purpose of the experiment and data cannot be obtained by any other humane procedure. Such procedures must be carefully supervised by the principal investigator or another qualified senior scientist

9. Postexperimental care of animals must be such as to minimize discomfort and the consequences of any disability resulting from the experiment, in accordance with acceptable practices in veterinary medicine

10. If it is necessary to kill an experimental animal, this must be accomplished in a humane manner, i.e., in such a way as to ensure immediate death in accordance with procedures approved by an institutional committee. *No animal shall be discarded until death is certain*

The facilities

11. Standards for the construction and use of housing, service, and surgical facilities should meet those described in the publication, "Guide for the Care and Use of Laboratory Animals," DHEW No. (NIH) 78-23, or as otherwise required by the United States Department of Agriculture regulations establishd under the terms of the Laboratory Animal Welfare Act (P.L. 89-544) as amended 1970 and 1976 (P.L. 92-579 and P.L. 94-279)

Transportation

12. Transportation of animals must be in accord with applicable standards and regulations, especially those intended to reduce discomfort, stress to the animals, or spread of disease. All animals being received for use as experimental subjects and having arrived at the terminal of a common carrier, must be promptly picked up and delivered, uncrated, and placed in acceptable permanent facilities

unnecessary discomfort and injury to animals will be avoided; and that analgesic, anesthetic, and tranquilizing drugs will be used where indicated to minimize pain or stress to the animals. Peer review groups as well as PHS grant and contracts staff are required to consider animal welfare policies and principles in their review of all pertinent applications. The staff and reviewers are directed to pay special attention to the Principles for the Use of Animals.

The Public Health Service policy on the care and use of animals is administered by The Office for Protection from Research Risks, Office of the Director, NIH, Bethesda, Maryland 20205. Copies of the full policy statement, the Principles, and sample assurance forms can be obtained from them.

2. Canada

In Canada there is no specific national legislation pertaining to the use of animals in research, testing, or teaching; however, the Canadians have established an effective voluntary program for the surveillance of the care and use of experimental animals (Rowsell, 1980). Following a study into the state-of-the-art of experimental animal care, and with the endorsement of universities and government departments utilizing animals, the Canadian Council on Animal Care (CCAC) was founded in 1968. Functioning under the administrative services of the Association of Universities and Colleges of Canada (AUCC), this autonomous organization is also funded by the Medical Research Council (MRC) and the Natural Sciences and Engineering Research Council (NSERC) (CCAC, 1980b). The Council consists of representatives from 15 agencies. The AUCC appoints four members including the chairman. The Canadian Federation of Humane Societies appoints two members. The remaining agencies, which are all governmental or professional scientific organizations, have one representative each. The Council makes recommendations for improvements in the procurement and production of experimental animals, the facilities and care of experimental animals, and the control over experiments involving animals.

A fundamental concept of the CCAC program is that of control from within institutions exercised by animal care committees with specific authority, terms of reference, and responsibilities. Essentially all Canadian institutions have established such animal care committees and have adopted CCAC guidelines. These include the designation of authority in the chairman and/or director of animal care to stop an experiment or to order an animal removed from an experiment and euthanized if in his/her opinion there is undue or unwarranted suffering. Protocols for projects involving experimental animals are submitted to the institutional animal care committee for approval before a project is started.

The facilities, animal care practices, and effectiveness of the institutional animal care committee are subject to periodic assessment by CCAC panels. These panels are drawn from a pool of scientists with experience and special knowledge in various aspects of animal care, and from representatives nominated by the Canadian Federation of Humane Societies (CFHS). Panelists (scientific members) are selected, as much as possible, with reference to the predominant thrust of the research at the institution to be assessed. Most panels are comprised of three scientists, a CFHS representative, and the CCAC Director of Assessments, as an *ex officio* member.

Assessments are based on standards set forth in the "Guide to the Care and Use of Experimental Animals" (CCAC, 1980a). Included in the Guide is CCAC's monograph "Ethics of Animal Experimentation," which has received the wholehearted support of the major granting councils. Continued noncompliance with the Guide can result in withdrawal of all research funding to an institution. Panels undertake in-depth site visits to all institutions using laboratory animals generally every 3 years; however, the visits may be scheduled more frequently if the panel and the Council feel that conditions at the institution so warrant, or at the request of the institution itself. An average of 35 site visits are conducted annually, throughout Canada.

Prior to an assessment visit, the Council seeks and receives completed questionnaires pertaining to administrative organization, animal care personnel, and space allocation and locations for animal housing and care. Of particular value to the panels is a CCAC Animal Utilization Summary, on which are noted the number and species of animals being used, and an indication of the research or teaching project in which they are involved. Supplied with this information, panelists may give notice of specific projects members may wish to examine in depth, including direct discussion with the principal investigator (Flowers, 1982).

During the site visit the panel inspects facilities and meets with the animal care committee and senior administrative officers to review all aspects of animal care and use. A detailed confidential report is prepared by the panel for the guidance of the institution and for the information of the member agencies of the CCAC. A letter of implementation is requested from the institution within 6 months of receipt of the assessment report.

The CCAC assessment program has enjoyed clearly expressed support of its member agencies responsible for funding research grants and contracts. Where a situation arising out of an assessment visit so warrants, the CCAC requests the agency (or agencies) whose research or contract interests are most involved to participate in any subsequent special visit. Through peer review and optimal communication, these methods have proved both effective and acceptable to all concerned (Rowsell, 1980). Additional information on the Canadian program can be obtained from the Canadian Council on Animal

Care, 1105-151 Slater Street, Ottawa, Ontario KIP 5H3. See Section II,C,2 for information on Canadian provincial legislation.

3. Great Britain

The British Cruelty to Animals Act of 1876 is the oldest of the national laws regulating the use of animals in scientific activities. Although this Act is currently under possible revision, amendment, or replacement, its thrust will almost certainly remain the same as it has been since 1876. It requires that persons in charge of experiments or tests involving the use of animals be licensed. The basic license allows a person to conduct only experiments where the animal is under anesthesia throughout the experimental period. Various certificates in addition to the license are necessary for the scientist to conduct other procedures. A certificate is required for a scientist to perform procedures without anesthesia, and even these are restricted. Another certificate is required for procedures conducted under anesthesia where the animal is not killed before it recovers. Other certificates are required to use an animal in illustration of a lecture or to use a dog or cat or a horse, ass, or mule. The license and certificates must be applied for by an individual investigator, but the application must be signed by the president of a learned society and a university professor in a branch of medical science. Only persons who are judged to be appropriately qualified are granted licenses or certificates. The owner of a license must carry out the experiments in a recognized institution. Institutions are inspected on a regular basis. The individual scientist is required to make an annual report of the animals used in different types of tests and experiments.

Loew (1981) studied the British system of laboratory animal care and use regulation and contrasted it with the American and Canadian systems. He concluded that British scientists were more aware of their law and regulations than are Canadian or American scientists. On the other hand, institutional concern for animal care as reflected in the state of animal facilities and animal care technology seems to be greater in the United States than in the United Kingdom. In the United States, followed by Canada, veterinarians are far more numerous, experienced, and trained in the specialty of laboratory animal medicine than in Britain. Loew (1981) attributes this as a logical outcome of the American law, which places institutions employing scientists, rather than the scientists themselves as in the United Kingdom, in the legally responsible position. The Canadian system and situation tends to occupy a middle ground between that of the United States and that of the United Kingdom.

The British law is administered by The Secretary of State,

The Home Office, Cruelty to Animals Inspectorate, 50 Queen Anne's Gate, London SW1 H9AT.

4. Other European Countries

Most other European countries including Belgium, Denmark, Finland, Federal Republic of Germany, France, Italy, Luxembourg, Netherlands, Norway, Sweden, and Switzerland have national laws that regulate the use of animals for tests and experiments. Generally these laws contain provisions to minimize or prohibit painful experiments, require licenses or permits for experiments, and incorporate a reporting system. Typically they are administered by the agricultural ministry or the health ministry.

5. Japan

The Japanese law concerning protection and control of animals was enacted in 1974. It contains provisions that when animals are used in scientific activities the procedures must be carried out by painless methods insofar as this does not frustrate the objective of the project. The law also establishes a Council for the Protection of Animals in the Prime Minister's Office.

6. Council of Europe

The Council of Europe is an organization of 21 European countries. In 1971, the Council appointed an Ad Hoc Committee of Experts for the Protection of Animals. This committee has been drafting a convention for the protection of all vertebrate animals used in any experimental or scientific procedure where that procedure may cause pain, suffering, distress, or lasting harm. If adopted, signatory parties to the convention would be required to regulate the care and treatment of laboratory animals prior to, during, and following experimentation; regulate the procurement of animals for use in experiments; register organizations using animals in experiments; establish inspection, monitoring, and record-keeping procedures; and establish mechanisms to determine if particular procedures are permissible.

C. State and Provincial Laws and Policies

1. State

All 50 states and the District of Columbia in the United States have animal anti-cruelty laws. Most of these laws (a) protect animals from cruel treatment, (b) require that animals have access to wholesome food and water, and (c) require that animals have shelter from extreme weather. According to Erbe (1966) the anti-cruelty laws of 15 states (Alaska, California, Florida, Hawaii, Idaho, Missouri, Nebraska, Nevada, New Jersey, New York, Pennsylvania, South Dakota, Texas, Washington, and Wisconsin) have special provisions for the use of animals in *bona fide* medical research. Generally, these provisions provide that no part of the anti-cruelty act should be construed to interfere with properly conducted scientific experiments or investigations performed under the authority of a regularly recognized medical college, university, or scientific institution.

In Maryland there was an attempt to include research animals under its animal anti-cruelty laws. In a recent case (Maryland versus Taub) a scientist was convicted of failure to provide adequate veterinary care to research monkeys (Holden, 1981). This conviction was overturned by the Maryland Court of Appeals. The Court found that because the research facility was subject to the Federal Animal Welfare Act, it was not subject to the state's animal cruelty statute.

A number of states have laws regulating the release of impounded animals for medical research. Ten states (Connecticut, Florida, Hawaii, Maine, Massachusetts, Montana, New Jersey, New York, Pennsylvania, and Rhode Island) have laws that prohibit the release of impounded animals to scientific institutions. There are five states (Iowa, Minnesota, Oklahoma, South Dakota, and Utah) that provide for the release of impounded animals under certain conditions. Generally these conditions are inspection or licensure by an appropriate state agency or by the USDA under the Animal Welfare Act. Virginia has a law that prohibits out-of-state animal dealers from purchasing impounded animals in Virginia. North Carolina prohibits sales to dealers but not to in-state scientific organizations.

Other states either have no laws governing the release of impounded animals or have laws that provide for local option. Some states have adopted a Model State Law developed by the United States Animal Health Association. This law provides standards for the care of animals in shelters, pounds, pet shops, and regulates dealers and dog wardens.

2. Canadian Provincial Legislation

Several Canadian provinces have enacted laws governing the acquisition and use of laboratory animals. Saskatchewan has a law that gives the University of Saskatchewan the right to claim unwanted dogs in municipal pounds (Urban Municipalities Act, 1970).

Alberta also has legislation that gives universities the right to claim unwanted animals from municipal pounds (Universities Act of 1966, 1972). The Alberta Act has provisions that are designed to facilitate reclaiming of animals by owners. There is a mandatory holding period of 10 days. Individuals are allowed to visit and reclaim their animals from the quarantine and isolation facility of the university. In addition, the regulations set forth conditions for the transportation, care, and use of research animals; conditions under which surgical or other painful experiments may be performed including the use of anesthesia; and qualifications for persons allowed to use animals for scientific purposes.

The Province of Ontario also has comprehensive animal welfare legislation (Animals for Research Act, 1977). The Ontario law requires the registration of research facilities. It allows registered facilities to claim animals from public pounds. Laboratory animal supply facilities where animals are bred and reared are required to meet certain standards and be licensed. The Act requires that research facilities administer anesthetics and analgesics to prevent unnecessary pain. It mandates the establishment of local institutional animal care committees to monitor the care and use of animals.

III. COLLECTION, IMPORTATION, EXPORTATION, AND SHIPMENT OF ANIMALS

Laws regulating the importation and shipment of animals generally are designed to protect human beings and other animals from diseases carried by animals or are designed to protect the environment including endangered species from harmful effects. For many species of animals, including the common laboratory rodents, the only United States importation requirements are that they be legally exported from the country of origin, that an export permit be issued if required, that the shipment be properly marked and declared, that they appear healthy, and that the animals enter through one of nine designated ports of entry (New York, Miami, Chicago, New Orleans, Dallas/Fort Worth, Seattle, San Francisco, Los Angeles, or Honolulu). Additional ports of entry are allowed for animals coming from Canada and Mexico. For animals imported through other ports and for importation of certain species of animals and related vectors and agents, special conditions must be met and/or permits must be obtained. These requirements are discussed in Sections III,A–C and are summarized in Table III.

A. Public Health

Special requirements must be met if animals known to be potential carriers of diseases contagious to human beings are to be imported. All dogs, cats, turtles, rodents, psittacine birds, and nonhuman primates brought into the United States from any foreign country are inspected at the port of entry in accor-

Table III

Sources of Information on Permits and Other Requirements for Importation and Shipment of Certain Animals, Vectors, and Agents

Activity	Responsible agency
Import of animals known to be potential carriers of diseases contagious to humans (dogs, primates, turtles and others)	Director, Quarantine Division Centers for Disease Control Atlanta, Georgia 30333
Import or transfer of etiologic agents or vectors of human disease (insects, bats, wild rodents, and others)	Biohazards Control Officer Office of Biosafety Centers for Disease Control Atlanta, Georgia 30333
Import of ruminants, swine, poultry and other birds, and certain other livestock. Import of organisms, vectors, tissue cultures, cell lines, blood, serum, and other animal parts or products pathogenic to livestock or that could serve as a vector of such pathogens	Senior Staff Veterinarian Import/Export Animals and Products Staff VS/APHIS U.S. Department of Agriculture Federal Building 6505 Belcrest Road Hyattsville, Maryland 20782
Import/export of wildlife through a nondesignated port of entry	Director U.S. Fish and Wildlife Service Department of the Interior Washington, D.C. 20240
Import, sell, donate, trade, loan, or transfer any live injurious wildlife (fruit bats, mongooses, European rabbits, multimammate rats, and others) or any of its eggs or progeny	Director U.S. Fish and Wildlife Service Federal Wildlife Permit Office (WPO) Department of the Interior Washington, D.C. 20240
Import/export, take, or ship in interstate or foreign commerce endangered or threatened species and parts thereof (great apes, lemurs and others)	Director U.S. Fish and Wildlife Service Federal Wildlife Permit Office (WPO) Department of the Interior Washington, D.C. 20240
Import/export/re-export of species covered by the Convention on International Trade in Endangered Species	Director U.S. Fish and Wildlife Service Federal Wildlife Permit Office (WPO) Department of the Interior Washington, D.C. 20240
Import, take, possess, purchase, sell or transport certain marine mammals (walrus, sea and marine otters, polar bears, manatees, dugongs)	Director U.S. Fish and Wildlife Service Federal Wildlife Permit Office (WPO) Department of the Interior Washington, D.C. 20240
Import, take, possess, purchase, sell or transport other marine mammals (cetaceans and pinnipeds except walrus)	Permit and Documentation Division Office of Marine Mammals and Endangered Species NMFS/NOAA Department of Commerce Washington, D.C. 20235

dance with United States Public Health Service regulations. Only animals with no evidence of communicable disease are admitted. In addition, dogs must have a certificate of valid rabies vaccination unless they are coming from a rabies-free country. Imported nonhuman primates must undergo a 31-day surveillance period in approved facilities in the United States and then they and their progeny are limited to use for scientific, educational, or exhibitive purposes. Small turtles are known to be carriers of salmonellosis. No more than six turtles less than 4 in. in carapace length or six turtle eggs or combination of turtles and eggs, may be imported in a shipment without a permit. Turtles greater than 4 in. in carapace length are not restricted. Psittacine birds are quarantined for 30 days by the United States Department of Agriculture. During this time they are given chlortetracycline-treated feed. Upon release from quarantine, the United State Public Health Service requires that they be given chlortetracycline for an additional 15 days in order to reduce the carrier rate for psittacosis. Additional information and applications for permits may be obtained from the Director, Quarantine Division, Centers for Disease Control, Atlanta, Georgia 30333.

Etiological agents and animal vectors of human disease, such as insects, other arthropods, snails, bats, wild rodents, are admissible only by special permit. Application for these permits should be submitted to the Biohazards Control Officer, Office of Biosafety, Centers for Disease Control, Atlanta, Georgia 30333.

B. Animal Health

The USDA controls importation of animals that are capable of transmitting diseases of domestic livestock and poultry. These animals include all ruminants and swine, members of the equine family, poultry and other birds, hatching eggs, and certain wild animals. Special permits must be obtained prior to importation of these animals and parts thereof. In addition, animals are subject to quarantine at a United States Customs port of entry designated by the USDA. The USDA is especially concerned about exclusion of diseases exotic to the United States, such as foot-and-mouth disease, rinderpest, African horsesickness, and exotic viscerotrophic velogenic Newcastle disease. For additional information and applications for permits, contact the Senior Staff Veterinarian, Import–Export Animals and Products Staff, Veterinary Services, Animal and Plant Health Inspection Service, USDA, Federal Building, 6505 Belcrest Rd., Hyattsville, Maryland 20782.

All states have laws that govern shipment of animals into them. The USDA and the states also have regulations pertaining to the intrastate sale of animals. In general, these laws do not specifically regulate sale and shipment of common laboratory animals, except dogs and in some states cats which might require vaccination against rabies. Domestic livestock generally requires a USDA accredited veterinarians's certificate of health that requires tests to ensure freedom from tuberculosis, brucelosis, and certain other specific diseases. Common laboratory rodents do not require health certificates, but almost all states prohibit the importation of any bird or animal infected with or recently exposed to any transmissible disease. Several state game departments have regulations prohibiting or controlling importation of wild animals.

C. Endangered Species and Wildlife Protection

The Lacey Act of May 25, 1900, governs the import, export, and interstate commerce of foreign wildlife. The importation of live specimens of harmful species, such as fruit bats (genus *Pteropus*), mongooses, or meerkats (genera *Atilax, Cyinictis, Helogale, Herpestes, Ichneumia, Mungos,* and *Suricata*), European rabbits (genus *Oryctolagus*), Indian wild dogs (genus *Cuon*), multimammate rats or mice (genus *Mastomys*), pink starlings (*Sturnus roseus*), dioch (*Quelea quelea*), Java rice sparrows (*Padda oryzivora*), red-whiskered bulbuls (*Pycnonotus jocosus*), live or dead fish or eggs of salmonids (family Salmonidae), and live fish or viable eggs of catfish (family Clariidae) are restricted. All of these have been deemed injurious or potentially injurious to public health and welfare, to agricultural interests, or to the welfare and survival of wildlife and wildlife resources of the United States. When there has been a proper showing or responsibility and continued protection of the public interest and health, these species may be imported for zoological, educational, medical, or research purposes by permit. Applications for permits and additional information may be obtained from the Director, United States Fish and Wildlife Service, Federal Wildlife Permit Office (WPO), Department of the Interior, Washington, D.C. 20240.

Nonhuman primates are probably the most widely used group of imported laboratory animals. In the past few years, regulatory factors have had a marked influence on importation of primates. For the most part, these regulations have come from the governments of exporters. Because of declining populations of many species of primates, a number of countries have either banned or severely restricted their export. Former suppliers that now do not allow commercial export of primates include India, Bangladesh, Thailand, Columbia, Peru, Brazil, and Guyana.

The United States has several laws designed to protect endangered species. The United States Endangered Species Act requires that a permit be obtained to engage in the following activities with listed endangered or threatened species and their parts or products: (1) import and export, (2) take (e.g., capture from the wild, kill), and (3) interstate or foreign commerce. The Act also prohibits (1) possession and other acts with *unlawfully* taken wildlife and (2) sale or offer for sale in interstate or foreign commerce without a warning to the effect that no sale may be consummated until a permit has been obtained. The United States is also party to the Convention on International Trade in Endangered Species (CITES). This international agreement provides procedures to regulate the import, export, or reexport of imperiled species covered by the treaty. Animals listed in the Convention's Appendix I are deemed threatened with extinction and require both export and import permits. Animals listed in Appendix II are deemed less threatened, and only export permits are required by the Convention. All species of nonhuman primates that are not listed in Appendix I are listed in Appendix II. Several primate species used in research are currently classified as endangered under the Endangered Species Act and/or are listed in Appendix I of CITES. These species include all gibbons, all great apes (gorilla, orangutan, chimpanzee), all species of lemurs, and cotton-topped marmosets. Some 30 other primate species and many species in other orders are listed under the United States Endangered Species Act and/or CITES. Current listings of protected species and applications for permits to import, export, or reexport these animals can be obtained from the Director, United States Fish and Wildlife Service, Federal Wildlife Permit Office (WOP), Department of Interior, Washington, D.C. 20240.

The Marine Mammal Protection Act requires that all marine mammals and parts thereof be imported, taken, possessed, purchased, sold, and transported under permit. The United States Fish and Wildlife Service (address in preceding paragraph) handles permits for walrus, sea and marine otters, polar bears, manatees, and dugongs. Applications for permits for all other marine mammals (cetaceans and pinnipeds except walrus) should be directed to Permit and Documentation Division, Office of Marine Mammals and Endangered Species, National Marine Fisheries Service, National Oceanic and Atmospheric Administration, Department of Commerce, Washington, D.C. 20235.

IV. HAZARDOUS SUSTANCES

A. Drugs with Potential for Abuse

Some drugs that are useful in laboratory animal medicine or are used in research have the potential to be abused. These drugs are termed "controlled substances" by the United States Drug Enforcement Administration. Every person who manufacturers, imports, distributes, dispenses, prescribes, or administers any controlled substance must register annually with the Drug Enforcement Administration. The registration is specific to the type of activity to be undertaken, i.e., manufacturer, research, distribution, etc. The registration is also specific to certain schedules or classes of drugs. There are five schedules of controlled substances. Substances on Schedule I have the greatest potential for abuse or there is no accepted medical use for them. Substances on Schedule V have the lowest potential for abuse. Anesthetics such as fentanyl and pentobarbitol are on Schedule II.

Depending on the activity and schedule for the drugs involved, the registrant is required to pay certain fees, maintain inventory and other records, and store the substances in a secure manner. For application forms for registration and more information contact the Drug Enforcement Administration, United States Department of Justice, P.O. Box 28083, Central Station, Washington, D.C. 20005.

B. Biohazards

The National Institutes of Health, the Centers for Disease Control, and the United States Department of Health and Human Services have developed guidelines for the use of biohazardous substances. See Chapter 20 for more information on biohazards and these guidelines.

V. GOOD LABORATORY PRACTICES

The Good Laboratory Practice Regulations (GLPRs) of the Food and Drug Administration (FDA) also affect the use of research animals. These regulations were published in the December 22, 1978 issue of the Federal Register (43 FR 59986–60025) under the title of Nonclinical Laboratory Studies—Good Laboratory Practice Regulations. All nonclinical studies requesting research or marketing permits from the FDA must conform to the GLPRs. The regulations have no force for studies conducted for purposes other than obtaining FDA research or marketing permits.

The GLPRs have a number of provisions that concern laboratory animals. For the most part these requirements are similar in substance to portions of the "Guide for the Care and Use of Laboratory Animals" (Moreland, 1978). Among these provisions are that animal facilities must have a sufficient number of rooms to assure that species, projects, and specially housed animals are properly separated and a quarantine area is provided. Similarly, isolation is to be provided for studies involving biohazards and diseased animals. Separate areas are to be provided for diagnosis, treatment, and food and bedding storage. Refrigeration must be provided for perishable items. The facility should be designed to minimize disturbances of the animal population.

The health of laboratory personnel and the clothing they wear is expected to preclude the introduction of unwanted variables to a study, such as the transmission of diseases to laboratory animals. Cleaning and sanitization of equipment is called for, as well as appropriate disposal of wastes. Although diseased animals may be treated if study results are not compromised, animals are expected to be free of complicating diseases prior to use.

Standard operating procedures are to be developed and followed for animal care and facility maintenance procedures, postmortem examinations, collection of specimens, handling of moribund or dead animals, and animal identification. Periodic food and water analysis is required to assure that contaminants do not interfere with a study and that their levels are less than specified in the study protocols. Pest control or bedding materials that might interfere with a study are not to be used.

Much emphasis is placed on keeping thorough, well-documented records. Included in the records to be kept are those of pest control materials used and diagnostic procedures and treatments administered to study animals.

Other provisions of the GLPRs will indirectly influence how animal care and use services are provided. In addition to the basic responsibility that a study director has for properly conducting a study, a quality assurance unit is required for each testing facility to monitor conformance with protocols and standard operating procedures. This unit is to be independent from the study director, and its "watch dog" function will include periodic inspections and reports to the management of the testing facility that become part of the study record. Management bears the responsibility for assuring that corrective actions are taken as well as providing adequate physical and personnel resources, including quality testing and training functions. For further information on the GLPRs, contact the Office of Regulatory Affairs (HFC-30), Food and Drug Administration, Department of Health and Human Services, 5600 Fishers Lane, Rockville, Maryland 20851.

REFERENCES

Animals for Research Act (1977). Statutes of Ontario, 1970 Chapter 22; as amended by 1971 Chapter 50, S.6; 1972 Chapter 1, S.1. Province of Ontario.

Canadian Council on Animal Care (CCAC) (1980a). "Guide to the Care and Use of Experimental Animals," Vol. 1, p. 112. CCAC, Ottawa, Ontario.

Canadian Council on Animal Care (CCAC) (1980b). "Surveillance over the Care and Use of Experimental Animals in Canada," p. 4. CCAC, Ottawa, Ontario.

Erbe, N. A. (1966). "The Law and Science. Cooperation or Conflict?" National Society for Medical Research, Washington, D.C. (unpublished thesis).

Flowers, F. H. (1982). Research animal care in Canada: Its control and regulation. Presented at the New York Academy of Sciences workshop on "The Role of Animals in Biomedical Research."

Holden, C. (1981). Judge finds six monkeys had inadequate veterinary care; NIH yet to decide on reinstatement of suspended grant. *Science* **214,** 1218–1219.

Loew, F. M. (1981). "British, Canadian and U.S. Patterns in Regulating the Use of Animals in Research: A Report to the Animal Welfare Foundation of Canada." Animal Welfare Foundation of Canada, Thornhill, Ontario.

Moreland, A. F., Chairman (1978). "Guide for the Care and Use of Laboratory Animals," Committee on the Care and Use of Laboratory Animals of the Institute of Laboratory Animal Resources, DHEW Publ. No. (NIH) 78-23, U.S. Dept. of Health and Human Services, Public Health Service, National Institutes of Health, Bethesda, Maryland.

Rowsell, H. C. (1980). The voluntary control program of the Canadian Council on Animal Care. *J. Med. Primatol.* **9,** 5–8.

Universities Act of 1966 (1972). "Dog Control and Procurement Regulations for the Treatment of Animals," Sect. 50, Alberta Regul. 33-72. Province of Alberta.

Urban Municipalities Act (1970). "Saskatchewan Dog Legislation, Sect. 177, Paragraph 32. Province of Saskatchewan.

U.S. Public Health Service (1979). "Grants Administration Manual," Chapter 1–43. U.S. Department of Health and Human Services, Washington, D.C.

Chapter 3

Biology and Diseases of Mice

Robert O. Jacoby and James G. Fox

I. INTRODUCTION

A. Origin and History

The mouse is assigned to the genus *Mus*, subfamily Murinae, family Muridae, order Rodentia. Anatomical features of the molar teeth and cranial bones help to differentiate it from other murids. The house mouse of North America and Europe, *Mus musculus,* is the species commonly used for biomedical research. It was employed in comparative anatomical studies as early as the seventeenth century, but the acceleration of biological research in the nineteenth century, a renewed interest in Mendelian genetics, and the requirement for a small, economic mammal that was easily housed and bred were instrumental to the development of the ''modern'' laboratory mouse. These studies have grown exponentially during the current century and have made the laboratory mouse, in genetic terms, the most thoroughly characterized mammal on earth (Morse, 1979).

B. Genetics

Genetic mapping in mice began in the early 1900s. Extensive linkage maps and an impressive array of inbred strains are now available to expedite sophisticated genetic research. Mice have 40 chromosomes that are differentiated by the size and pattern of transverse bands. The chromosomes are designated in order of decreasing size by arabic numbers. Chromosome rearrangements serve as markers for assigning known linkage groups to specific chromosomes and for determining locus order with respect to the centromere.

One of the most thoroughly studied genetic systems of the mouse is the *histocompatibility complex*. Histocompatibility

(*H*) loci control expression of cell surface molecules that modulate major immunological phenomena, such as the recognition of foreign tissue. For example, the time, onset, and speed of skin graft rejection are controlled by two groups of *H* loci. The major group is called *H-2* and is located on chromosome 17. These genes cause rapid rejection (10–20 days) of grafts that display foreign *H-2* antigens. Minor *H* loci groups are scattered throughout the genome and are responsible for delayed graft rejection. Genes associated with the *H-2* complex also control other immunological functions, such as cell–cell interactions in primary immune responses and the level of response to a given antigen. Immune-mediated responses to infectious agents such as viruses and complement activity are influenced directly or indirectly by the *H-2* complex (Klein, 1975).

Inbred mice are also valuable for research in other fields such as immunology, oncology, microbiology, biochemistry, pharmacology, physiology, anatomy, and radiobiology. For example, inbred histocompatible strains are used extensively as donors of plasma cell tumors to immortalize cell lines (hybridomas) that secrete highly uniform, monospecific immunoglobulin *in vitro,* theoretically in unlimited quantities. This technology has made the full range of functional mouse antibody molecules available for study. Transplantable murine plasmacytomas also have provided specific components of immunoglobulin biosynthesis, including the immunoglobulin genes, mRNA, and intermediates in the assembly of immunoglobulin genes (Potter, 1983).

C. Nomenclature and Breeding Systems (Lyon, 1981; Green, 1981)

Laboratory mice are identified by strain or by breeding system. The geneology of major inbred mouse strains is presented in Figs. 1–3. Seven breeding system designations have been developed (Table I) and each requires technical skill and a firm understanding of mammalian genetics. *Inbred* strains were first developed in 1909 by C. C. Little and offer a high degree of genetic uniformity. They are, for practical purposes, genetically identical to other mice of the same strain and sex. They are produced by brother–sister matings for at least 20 generations. By contrast, *random bred* mice are genetically heterogeneous, and are often produced by breeding systems that minimize inbreeding. Individual random bred mice may differ in coat color, histocompatibility loci, enzyme polymorphisms, and other characteristics. Random pairing is best planned with the aid of tables of randomized numbers or a randomizing device. However, true random breeding may be difficult to achieve in a small research colony by random pairing. In a population of 25 breeding pairs, for example, heterozygosity will decrease at 1% per generation with standard randomiza-

tion techniques. A random breeding program that is easy to manage is the circular pair mating system, where each pair is mated only once. Conceptually, cages are visualized in a circle and each cage contains one breeding pair in the *n*th generation. Another "circular" set of cages serves as the nucleus for the *n* + 1 generation. Each mated pair from the *n*th generation contributes one female and one male for the *n* + 1 generation. Random breeding is accomplished by assigning the male and female from the *n*th generation to different cages rather than to the same cage. Other useful random mating systems are the cousin system and another circular system (see Table I).

Recombinant inbred strains are developed by single-pair random matings of mice from an F_2 generation by crossing two inbred strains. Lines selected for desired characteristics are perpetuated by brother–sister matings for 20 generations to obtain homozygosity. Recombinant inbred strains may take as long as 7 years to produce. However, they are essential for detecting new genetic linkages and inherited traits.

There are currently more than 500 separate outbred stocks and inbred strains, some with various sublines. Mutant stocks have also been established and are used as models of human diseases. Therefore, strain or stock designations must be as complete to avoid semantic and genetic confusion. As an example of subline variation, CBA/J carries the gene for retinal degeneration, while CBA/Ca does not.

Specific nomenclatures have been developed for both inbred and noninbred strains. Strains are designated by a series of letters and/or numbers, a shorthand description of the strain's origin and history. For example, the inbred strain C57BL/6J originated from female 57 at the Cold Spring Harbor Laboratory (C) and was the black (BL) line from this female. The 6 indicates that it is subline number 6, and the J that it was bred at the Jackson Laboratory. A noninbred stock designated A5 (S) indicates that it was produced by High Quality farms, stock A5 of Swiss (S) stock origin. Specific designations also apply to substrains and to coisogenic, congenic, and segregating inbred strains. The Committee on Standardized Genetic Nomenclature for Mice publishes a periodic listing of strains and stocks and updated rules for nomenclature. Inbred strains are also listed in the biennial *Inbred Strains of Mice* (Staats, 1977) and every 4 years in *Cancer Research* (Staats, 1976). *The Mouse News Letter* (issued twice yearly by MRC Laboratories in England and The Jackson Laboratory in the United States) lists genetic symbols currently in use and all changes in nomenclature.

D. Housing and Husbandry

Housing and husbandry for mice are often guided by microbiological requirements. A colony can be maintained in a "conventional" environment or behind a barrier where the

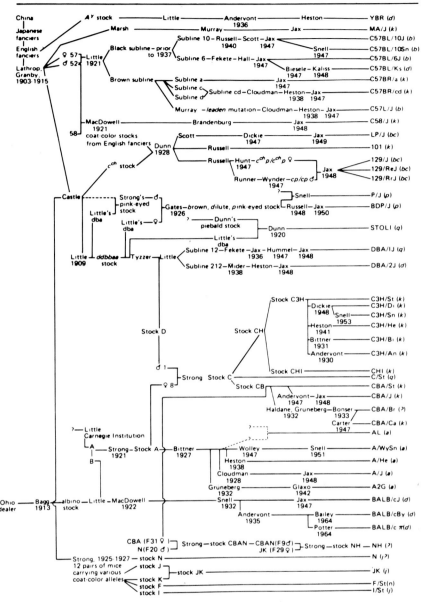

Fig. 1. Genealogy of current inbred strains of mice. The *H-2* haplotypes are in parentheses. (From Morse, 1981.)

mice are protected from specific microorganisms. Examples of barrier housing include positive pressure isolators and mass air-flow rooms that provide sterile air through high-efficiency particle-arresting (HEPA) filters.

Mouse cages vary in design, size, and composition. The popular shoebox cage used for housing and breeding mice is usually made of polycarbonate, polypropylene, or polystyrene plastic (in order of decreasing cost and durability). Stainless steel is considered by some to be the most satisfactory caging material since it is smoother, more durable, and easier to clean than plastic. Mice are sometimes housed in suspended cages with open-mesh bottoms that allow excrement to fall through to a collecting pan. Suspended caging is rarely used for breeding because neonatal thermoregulation is difficult to maintain without nesting material. Cage mesh should be stainless steel rather than galvanized steel, since the former is easier to clean and less susceptible to rust. Cages should keep animals dry and clean, maintain a comfortable ambient temperature, allow freedom of movement and normal postural adjustments, avoid unnecessary physical restraints, provide convenient access to feed and water, and prevent overcrowding.

Solid bottom cages should contain sanitary bedding, such as

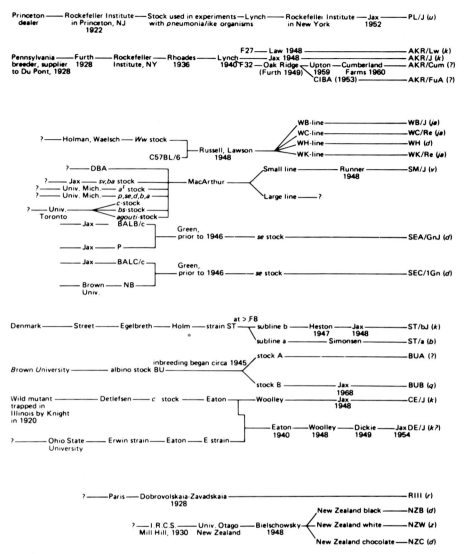

Fig. 2.   Genealogy of current inbred mouse strains derived from widely varying ancestral stocks. The haplotypes are in parentheses. (From Morse, 1981.)

wood chips or ground corn cob. Criteria for selecting bedding vary with experimental and husbandry needs (Table II). It is preferable to autoclave bedding prior to use, but if this is not convenient, it should be used only after its origin and microbial content have been evaluated (Table III).

Nutrient requirements for the mouse are influenced by genetic background, disease status, pregnancy, and environment. The best current estimate of nutritional requirements is shown in Table IV. Nutritional requirements for laboratory mice are also published periodically by the National Research Council. Feed intake and weight gain data are used to estimate the nutritional needs of a particular stock or strain. Mice consume about 3–5 gm of feed per day after weaning and maintain this

intake throughout life. Outbred mice tend to gain weight faster than inbred mice and are heavier at maturity (Figs. 4 and 5).

Diet is often neglected as a variable in animal-related research. Diet can influence responses to drugs, chemicals, or other factors and lead to biased research results. Therefore, diet must provide a balance of essential nutrients, and contaminants must be kept to a minimum (see also Chapter 23 by Pakes *et al.*). Natural product commercial diets for mice are satisfactory for breeding and maintenance. Fresh produce, grains, fishmeal, or other supplements may expose colonies to pathogenic bacteria or harmful chemicals and should be avoided.

Mice should have continuous access to potable water even if

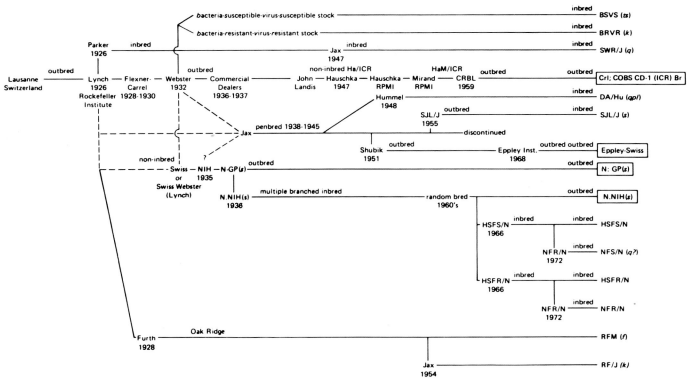

Fig. 3. Genealogy of current inbred and outbred strains of mice of Swiss ancestry. The haplotypes are in parentheses. (From Morse, 1981.)

Table I

Kinds of Mice Used in Research[a]

Definition of breeding system	Perpetuation of breeding system	Reference
1. *Random bred stock:* Random mating within a large, heterogeneous population	Continue random mating	Poiley (1960) Kimura and Chow (1963)
2. *Inbred strain:* Brother–sister mating for 20 or more generations	Continue brother–sister mating	Green (1981)
3. *F₁ hybrids:* Crosses between inbred strains	Cannot be perpetuated	Green (1981)
4. *Recombinant inbred strains:* Brother–sister mating for 20 or more generations from random and selected pairs in F_2 of a cross between two inbred strains	Continue brother–sister mating	Bailey (1971)
5. *Coisogenic inbred strains:* Occurrence of a mutation within an inbred strain	Perpetuate the mutation by (1) brother–sister mating within strain or origin, (2) backcross system or cross-intercross system with strain of origin as parent strain, (3) brother–sister mating with heterozygosity forced by backcrosses or intercrosses or (4) brother–sister mating with homozygosity forced by incrosses	Flaherty (1981)
6. *Congenic inbred strains:* (A) Backcross system for 10 or more generations or (B) cross-intercross system for 10 or more cycles with an inbred parent strain	Perpetuate the transferred mutation by (2), (3), or (4) as above	Flaherty (1981)
7. *Segregating inbred strains:* Brother–sister mating systems for 20 or more generations with heterozygosity forced by (1) backcrossing, (2) intercrossing, (3) crossing and intercrossing, or (4) backcrossing and intercrossing	Continue brother–sister mating with heterozygosity forced by one of the four methods, as at left, or with homozygosity forced by intercrossing	Green (1981)

[a]Modified from Green (1981).

Table II

Desirable Criteria for Rodent Contact Bedding[a]

Moisture absorbent	No microbial and chemical
Dust-free	contamination
Does not promote microbial growth	Nonpalatable
Nonstaining	Nonallergenic
Atraumatic	Nontoxic
Ammonia binding	Nonenzyme inducing
Sterilizable	Nestable
Deleterious products not formed as	Readily available
a result of sterilization	Inexpensive
Easily stored	Chemically stable during use
Uniform from batch to batch	Animal behavior is not adversely
	affected

[a]Modified from Kraft (1980).

Table III

Tests of Bedding Quality[a]

Chemical properties
  Pesticides and polychlorinated compounds
  Mycotoxins
  Nitrosamines
  Detergent residues
  Ether-extractable substances
  Heavy metals
Physical properties
  Particle uniformity
  Absorptivity
  Ammonia evolution
  Visible trauma and irritant potential
Microbiological properties
  Standard plate count
  Yeasts and molds
  Coliforms and *Salmonella*
  Pseudomonas

[a]Modified from Kraft (1980).

a high moisture diet is fed. Water is needed for lubrication of dry food and for hydration. Adult mice drink 6 to 7 ml of water per day. Decreased water intake will decrease food consumption. Water imbalance may occur during disease, since sick mice commonly drink very little water. Because of this, it may be unsuitable to administer medicine by this route.

II. BIOLOGY

A. Physiology and Anatomy (Cook, 1983; Kaplan *et al.*, 1983)

1. Temperature and Water Regulation

Mice have a relatively large surface area per gram of body weight. This results in dramatic physiologic changes in response to fluctuations in the ambient temperature (T_A). The mouse responds to cold exposure, for example, by nonshivering thermogenesis. A resting mouse acclimated to cold can generate heat equivalent to about triple the basal metabolic rate, a change that is greater than for any other animal. A mouse must generate about 46 kcal/m^2/24 hr to maintain body temperature for each 1°C drop in T_A below the thermoneutral zone. Mice cannot tolerate nocturnal cooling as well as larger animals that have a greater heat sink. Therefore, conserving energy in animal quarters at night by lowering thermostats is not prudent management.

Because of its great ratio of evaporative surface to body mass, the mouse has a greater sensitivity than most mammals to water loss. Its biological half-time for turnover of water (1.1 days) is more rapid than for larger mammals. Water conservation is embellished by cooling of expired air in the nasal passages and by the highly efficient concentration of urine.

The conservation of water can preempt thermal stability. If

the mouse had to depend on the evaporation of body water to prevent elevations of body temperature, it would go into shock from dehydration. The mouse has no sweat glands, it cannot pant, and its ability to salivate is severely limited. Mice can partially compensate for changes in T_A by varying body temperature (T_B). For example, the T_B of a mouse can increase from 33.8° to 37.2°C when the T_A increases from 20° to 35°C. It adapts to moderate but persistent increases in environmental temperature by a persistent increase in body temperature, a persistent decrease in metabolic rate, and increased vascularization of the ears to increase heat loss. Its primary means of cooling in the wild is behavioral—retreat into a burrow. In the confinement of a cage, truck, or plane, mice do not survive well in heat and begin to die at an ambient temperature of 37°C. Thus, the mouse is not a true endotherm. In fact, the neonatal mouse is ectothermic and does not have well-developed temperature control before 20 days of age.

The thermoneutral zone for mice varies with strain and with conditioning, but is about 29.6°–30.5°C, narrower than that of any other mammal thus far measured. Thermoneutrality should not be equated with comfort or physiological economy. There are repeated studies to show that mice in a T_A range of 21°–25°C grow faster, have larger litters, and have more viable pups than those maintained in the thermoneutral zone.

2. Respiratory System

The respiratory tract has three main portions: the anterior respiratory tract consists of nostrils, nasal cavities, and nasopharynx; the intermediate section consists of the larynx, trachea, and bronchi, all of which have cartilagenous support; and the posterior portion of the respiratory tract consists of the

Table IV

Nutrient Requirements of Mice[a]

Nutrient	Concentration in a diet (%)
Protein (as crude protein)	20–25
Fat[b]	5–12
Fiber	2.5
Carbohydrate	45–60

Estimated Dietary Amino Acid Requirement

Amino acid	Natural ingredient open formula diet (%)[c]	Purified diet (%)[d]
Arginine	0.3	—
Histidine	0.2	—
Tyrosine	—	0.12
Isoleucine	0.4	0.2
Leucine	0.7	0.25
Lysine	0.4	0.15
Methionine	0.5	0.3
Phenylalanine	0.4	0.25
Theronine	0.4	0.22
Tryptophan	0.1	0.05
Valine	0.5	0.3

Mineral and Vitamin Concentrations of Adequate Mouse Diets

Mineral	Natural ingredient open formula diet[e]	Purified diet[f]	Purified diet[g]	Chemically defined diet[h]
Calcium (%)	1.23	0.52	0.81	0.57
Chloride (%)	—	0.16	—	1.03
Magnesium (%)	0.18	0.05	0.073	0.142
Phosphorus (%)	0.99	0.4	0.42	0.57
Potassium (%)	0.85	0.36	0.89	0.40
Sodium (%)	0.36	0.1	0.39	0.38
Sulfur (%)	—	—	—	0.0023
Chromium (mg/kg)	—	2.0	1.9	4.0
Cobalt (mg/kg)	0.7	—	—	0.2

Table IV (*Continued*)

Mineral	Natural ingredient open formula diet[e]	Purified diet[f]	Purified diet[g]	Chemically defined diet[h]
Copper (mg/kg)	16.1	6.0	4.5	12.9
Fluoride (mg/kg)	—	—	—	2.3
Iodine (mg/kg)	1.9	0.2	36.0	3.8
Iron (mg/kg)	255.50	35.0	299.0	47.6
Manganese (mg/kg)	104.0	54.0	50.0	95.2
Molybdenum (mg/kg)	—	—	—	1.55
Selenium (mg/kg)	—	0.1	—	0.076
Vanadium (mg/kg)	—	—	—	0.25
Zinc (mg/kg)	50.3	30.0	31.0	38.0

Vitamin	Natural ingredient open formula diet[e]	Purified diet[f]	Purified diet[g]	Chemically defined diet[h]
A (IU/kg)	15,000	4,000	1,100	1,730
D (IU/kg)	5,000	1,000	1,100	171
E (IU/kg)	37	50	32	1,514
K₁ equiv. (mgkg)	3	0.05	18	10.7
Biotin (mg/kg)	0.2	0.2	0.2	1
Choline (mg/kg)	2,009	1,000	750	2,375
Folacin (mg/kg)	4	2	0.45	1.43
Inositol (mg/kg)	—	—	—	248
Niacin (mg/kg)	82	30	22.5	35.6
Calcium pantothenate (mg/kg)	21	16	37.5	47.5
Riboflavin (mg/kg)	8	6	7.5	7.1
Thiamine (mg/kg)	17	6	22.5	4.8
Vitamin B_6 (mg/kg)	10	7	22.5	6.0
Vitamin B_{12} (mg/kg)	0.03	0.01	0.023	0.58

[a]Modified from Knapka (1983).
[b]Linoleic acid: 0.6% is adequate.
[c]John and Bell (1976).
[d]Theuer (1971).
[e]Knapka *et al.* (1974).
[f]AIN 76 (1977).
[g]Hurley and Bell (1974).
[h]Pleasants *et al.* (1973).

lungs. The left lung is a single lobe. The right lung is divided into four lobes: superior, middle, inferior, and postcaval.

A mouse at rest uses about 3.5 ml O_2/gm/hr, which is about 22 times more O_2/gm/hr than is used by an elephant. To accommodate for this high metabolic rate, the mouse has a high alveolar P_{O_2}; a rapid respiratory rate; a short air passage; a moderately high erythrocyte (RBC) concentration; high RBC hemoglobin and carbonic anhydrase concentrations; a high blood O_2 capacity; a slight shift in the O_2-dissociation curve enabling O_2 to be unloaded in the tissue capillaries at a high P_{O_2}; a more pronounced Bohr effect, i.e., the hemoglobin af-

finity for O_2 with changes in pH is more pronounced; a high capillary density; and a high blood sugar concentration.

3. Urinary System

The kidneys, ureters, urinary bladder, and urethra form the urinary system (Fig. 6). The paired kidneys lie against the dorsal body wall of the abdomen on either side of the midline. The right kidney is normally located anterior to the left kidney. Kidneys from males of many inbred strains are consistently heavier than kidneys from females. The glomeruli of mice are

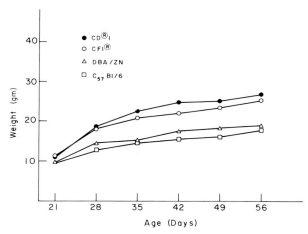

Fig. 4. Growth comparison: outbred (CD1 and CF1) and inbred mice (♀). (Courtesy of Charles River Breeding Laboratories.)

small, about 74 m in diameter, or about half the size of glomeruli in rats. There are, however, 4.8 times as many glomeruli in the mouse and the filtering surface per gram of tissue is twice that of the rat.

Mice excrete only a drop or two or urine at a time, and it is highly concentrated (Table V). The high concentration is made possible by long loops of Henle and by the organization of giant vascular bundles (vasa recta) associated with the loops of Henle in the medulla. The mouse can concentrate urine to 4300 mOsm/liter compared to a maximum permissible concentration of 1160 mOsm/liter for a human.

Mice normally excrete large amounts of protein in the urine. Taurine is always present in mouse urine, whereas tryptophan is always absent. Creatinine is also excreted in mouse urine, a trait in which mice differ from other mammals. The creatinine/creatine ratio for fasting mice is about 1:1.4. Mice excrete much more allantoin than uric acid.

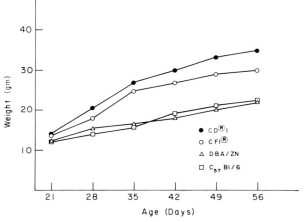

Fig. 5. Growth comparison: outbred (CD1 and CF1) and inbred mice (♂). (Courtesy of Charles River Breeding Laboratories.)

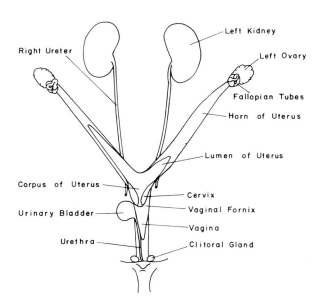

Fig. 6. Female reproductive tract. (From Cook, 1983.)

4. Gastrointestinal System

The submaxillary salivary gland, a mixed gland in most animals, secretes only one type of saliva (seromucoid) in the mouse. The tubular portion of the gastrointestinal tract consists of esophagus, stomach, small intestine, cecum, and colon.

The esophagus of the mouse is lined by a thick cornified squamous epithelium making gavage a relatively simple procedure. The proximal portion of the stomach is also keratinized, whereas the distal part of the stomach is glandular. Gastric secretion continues whether or not food is present.

The gastrointestinal flora consists of over 100 species of bacteria that begin to colonize the alimentary canal selectively shortly after birth. They form a complex ecosystem that provides beneficial effects, such as an increase in resistance to certain intestinal pathogens, production of essential vitamins, and homeostasis of important physiologic functions (Fig. 7).

5. Lymphatic System

The lymphatic system consists of lymph vessels, thymus, lymph nodes, spleen, solitary peripheral nodes (Fig. 8), and intestinal Peyer's patches. The typical lymph node is bean-shaped and consists of a cortex and a medulla. The cortex is divided into B lymphocyte domains, called primary follicles, and T lymphocyte domains, known as the diffuse cortex. The mouse does not have palatine or pharyngeal tonsils. The spleen lies adjacent to the greater curvature of the stomach. Different strains of mice have varying degrees of accessory splenic tissue. Age, strain, sex and health status can affect the size, shape, and appearance of the spleen. Male spleens, for exam-

Table V

Some Characteristics and Constituents of Mouse Urine[a]

Output	0.5–1 ml/24 hr
pH	7.3–8.5
Titerable acidity	4.68–5.67 mg/24 hr
Mean specific gravity	1.058
Osmolality	1.06–2.63 osmol/kg
Total solids	12.1–16.1 gm/100 gm
Chloride	5.75–5.79 mg/24 hr
Total sulfur	0.27%
Inorganic sulfate	0.15%
Inorganic phosphorus	0.43%
Glucose	1.98–3.09 mg/24 hr
Protein	6.8–25.8 mg/24 hr
Total nitrogen	40.2–40.8 mg/24 hr
Ammonia nitrogen	4.68–5.48 mg/24 hr
Urea nitrogen	24.3–29.8 mg/24 hr
Uric acid	0.04%
Creatine	0.86–1.02 mg/24 hr
Creatinine	0.57–0.67 mg/24 hr
Taurine	11.89(9.39–14.64)mg/kg/day
Allantoin	95(75–117) μmoles/kg/day
Homovanillic acid	40 μg/kg/day
Leucine	0.86 mg/day (on diet of 10% casein)
Valine	0.91 mg/day
Histidine	0.27 mg/day
Alanine	0.53 mg/day
Deoxycytidine	125–625 μg/kg/day
4-Amino-5-imidazole carboxamide	260 μg/kg/day

[a]Modified from Kaplan *et al.* (1983).

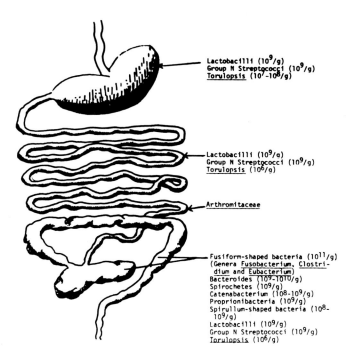

Fig. 7. Location of bacteria comprising the autochthonous microflora in the gastrointestinal tract. (From Schaedler and Orcutt, 1983.)

ple, may be 50% larger than those of females. Most lymphocytes enter and leave the spleen in the bloodstream. Cellular and humoral components of immunity are distributed to the bloodstream and tissues by efferent lymphatic vessels and lymphatic ducts, which empty into the venous system.

The thymus is a bilobed lymphoid organ lying in the anterior mediastinum. It reaches maximum size around the time of sexual maturity and involutes between 35 and 80 days of age. The thymus plays a major role in the maturation and differentiation of T lymphocytes. This function is not complete in newborn mice. Thymectomy is routinely performed in immunological research for experimental manipulation of the immune system. Thymectomy of newborn mice causes a decrease in circulating lymphocytes and marked impairment of certain immune responses, particularly cellular immune responses. Thymectomy in adult mice produces no immediate effect, but several months later mice may develop a progressive decline of circulating lymphocytes and impaired cellular immune responses. The mutant athymic nude mouse offers a powerful experimental tool in the study of the thymus in immune regulation. Some transmission of humoral immunity occurs *in utero,* but the majority of antibody is transferred after birth through colostrum.

Decay of passive immunity progresses with age; for example, the waning of protective maternal antibody in mice between 1 and 2 months of age makes these mice susceptible to enzootic Sendai infection in breeding colonies.

6. Cardiovascular System

The heart consists of four chambers, the thin-walled atria and the thick-walled ventricles (Fig. 9).

Mice conditioned to a recording apparatus have mean systolic blood pressures ranging from 84 to 105 mm Hg. An increase in body temperature does not lead to an increase in blood pressure. Heart rate, cardiac output, and the width of cardiac myofibers are related to the size of the animal. Heart rates from 310 to 840/min have been recorded for mice, and there are wide variations in rates and blood pressure among strains. Normal hematologic, blood chemistries, and other physiologic values are listed in Tables VI and VII.

7. Skeletal System

The skeleton is composed of two parts: the axial skeleton, which consists of the skull, vertebrae, ribs, and sternum, and the appendicular skeleton, which consists of the pectoral and pelvic girdles and the paired limbs. The normal vertebral formula for the mouse is C7T13L6S4C28, with some variations between strains, especially in the thoracic and lumbar regions.

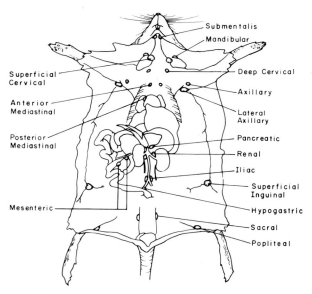

Fig. 8. Lymph nodes. (Modified from Cook, 1983.)

Normal mouse dentition consists of an incisor and three molars in each quadrant. These develop and erupt in sequence from front to rear. The third molar is the smallest tooth in both jaws; the upper lower third molar may be missing in wild mice and in some inbred strains. The incisors grow continuously and are worn down during mastication.

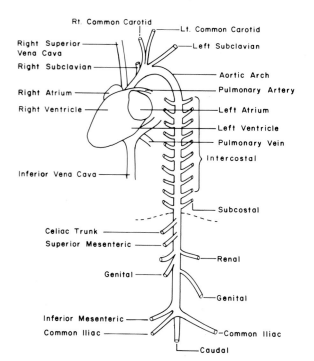

Fig. 9. Heart and major vessels. (Modified from Cook, 1983.)

Table VI

Normative Data

Adult weight	
Male	20–40 gm
Female	18–35 gm
Life span	
Usual	1–3 yr
Maximum reported	4 yr
Surface area	0.03–0.06 cm^2
Chromosome number (diploid)	40
Water consumption	6.7 ml/8 wk age
Food consumption	5.0 gms/8 wk age
Body temperature	98.8°–99.3°F
Puberty	
Male	28–49 days
Female	28–49 days
Breeding season	None
Gestation	19–21 days
Litter size	4–12 pups
Birth weight	1.0–1.5 gm
Eyes open	12–13 days
Weaning	21 days
Heart rate	310–840 beats/min
Blood pressure	
Systolic	133–160 mm Hg
Diastolic	102–110 mm Hg
Blood volume	
Plasma	3.15 ml/100 gm
Whole blood	5.85 ml/100 gm
Respiration frequency	163/min
Tidal volume	0.18 (0.09–0.38)ml
Minute volume	24 (11–36)ml/min
Stroke volume	1.3–2.0 ml/beat
Plasma	
pH	7.2–7.4
CO_2	21.9 moles mM
CO_2 pressure	40 + 5.4 mm Hg
Leukocyte count	
Total	8.4(5.1–11.6)×10^3/μl
Neutrophils	17.9 (6.7–37.2)%
Lymphocytes	69 (63–75)%
Monocytes	1.2 (0.7–2.6)%
Eosinophils	2.1 (0.9–3.8)%
Basophils	0.5 (0–1.5)%
Platelets	600(100–1000)×10^3/μl
Packed cell volume	44 (42–44)%
Red blood cells	8.7–10.5×10^8/mm^3
Hemoglobin	13.4 (12.2–16.2) gm/dl
Maximum volume of single bleeding	5 ml/kg
Clotting time	2–10 min
PTT	55–110 sec
Prothrombin time	7–19 sec

nant dams, but the prevalence of prepartum transmission in nature has not been determined.

The distribution of lesions reflects the pathway of infection. Macroscopically, the ileum and colon may be normal or red and dilated with watery, fetid contents. The liver often contains one or more gray-white foci. Histologically, lesions are characterized by necrosis. In the intestine, necrosis of mucosal epithelium may be accompanied by acute inflammation and hemorrhage. In the liver, foci of coagulation necrosis are generally distributed along branches of the portal vein; a finding compatible with embolic infection from the intestine. Peracute lesions are largely free of inflammation, but neutrophils and lymphocytes may infiltrate less fulminant lesions. Myocardial necrosis occurs in some species (e.g., rabbit and rat), but is not commonly seen in the mouse.

Bundles of long, slender rods can be found in the cytoplasm of viable hepatocytes bordering necrotic foci (Fig. 13). Organisms are found more easily during early stages of infection. Organisms in tissue sections do not stain well with hematoxylin–eosin stain. Special preparations such as silver stains (Warthin–Starry), Giemsa stains, or periodic acid–Schiff stains are usually required. Bacilli can also be found in the epithelial cells of the intestine especially in association with focal necrotizing enterocolitis.

Fig. 13. *Bacillus piliformis* organisms in hepatocytes at the border of a necrotic lesion in the liver of a mouse with Tyzzer's disease. Warthin–Starry stain. (From Ganaway, 1982.)

Tyzzer's disease is diagnosed most frequently by the demonstration of characteristic organisms in tissue sections of liver and intestine. Supplemental procedures include inoculation of cortisonized mice or embryonated eggs with suspect material followed by histological or immunocytochemical demonstration of organisms in tissues. Asymptomatic infection can be detected serologically with an immunofluorescence technique. The detection of organisms at the periphery of necrotic foci is key to differentiating Tyzzer's disease from other infections that can produce similar signs and lesions, especially mouse pox, corona viral hepatitis, reoviral hepatitis, and salmonellosis.

The control and prevention of Tyzzer's disease have not been adequately studied. Prevalence rates, reservoirs of infection, carrier states, and the mechanism of spread remain speculative. A build up of spores in the environment or intercurrent immunosuppression might lead to fatal infection. It has been suggested that administration of tetracycline can avert epizootic infection, but it is unlikely that a spore-forming bacterium can be eliminated by the use of antibiotics. Therefore, total replacement of affected or exposed stock must be considered. Vaccination is not yet available.

b. Transmissible Murine Colonic Hyperplasia

Citrobacter freundii can cause a natural epizootic disease of mice characterized by colonic mucosal hyperplasia and colitis. Typical *Citrobacter* organisms are motile, gram-negative aerobes that ferment lactose and utilize citrate as a sole source of carbon. The substrains associated with colonic hyperplasis are nonmotile and either fail to utilize citrate or do so marginally. Only the strain 4280, (API profile number)* has been associated with natural or experimental disease.

In natural outbreaks, retarded growth, ruffled fur, soft feces, rectal prolapse, and moderate mortality in late suckling and early weaning mice have been observed. Diarrhea has also been found with some regularity.

The prevalence of *C. freundii* infection in mouse colonies is unknown. Factors such as genotype, age, virulence of bacterial strains, and diet influence the course and severity of disease (Barthold *et al.*, 1977). For example, DBA, NIH Swiss, and C57BL mice are relatively resistant to mortality; whereas C3H/HeJ mice are relatively susceptible both as sucklings and as adults. Dietary factors also modulate susceptibility to infection, but they have not been identified.

The pathognomonic gross finding is severe thickening of the descending colon, which is rigid and either empty or contains semiformed feces (Fig. 14). The lesion lasts for 2 to 3 weeks in surviving animals. Histologically, mitotic activity in the

*API20E Strips, Analytab Products, 200 Express Street, Plainview, New York 11803.

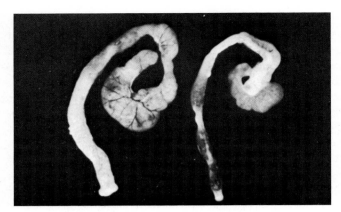

Fig. 14. Colons of a normal mouse (right) and a mouse with transmissible murine colonic hyperplasia (left). The descending colon is thickened and opaque due to mucosal hyperplasia. (From Barthold *et al.,* 1978 with permission.)

mucosa of the descending colon is accelerated and may be associated with inflammation. At the height of disease, the hyperplastic mucosa has crypt epithelium that resembles undifferentiated cells normally found only at crypt bases. In young mice, spotty mucosal necrosis and inflammation are more common. Inflammation may subside after several weeks, and there is a decrease in mitotic activity. Repair is rapid and complete in adults but slow in sucklings.

The diagnosis of colonic hyperplasia depends on clinical signs, characteristic histological changes, and isolation of *C. freundii* 4280 from the gastrointestinal tract or feces (Barthold *et al.,* 1978). *Citrobacter freundii* is relatively easy to culture during early phases of infection, whereas during later stages the intestine may be free of aerobic bacteria.

Some success in curtailing epizootics has resulted from adding sodium sulfamethazine (0.1%), tetracyclines (450 mg/liter), or neomycin sulfate (2 mg/ml) to the drinking water. Because *C. freundii* may contaminate food, bedding, or water, proper disinfection of such materials is prudent before they are used for susceptible animals. Surveillance for *C. freundii* 4280 can be incorporated into a quality assurance program.

The potential effects on research of colonic hyperplasia as a clinically severe disease are obvious. Colonic hyperplasia also has been shown to increase the sensitivity of colonic mucosa to chemical carcinogens and to decrease the latent period between administration of carcinogen and the appearance of focal atypical cell growth (Barthold and Beck, 1980).

c. Pseudomoniasis

Pseudomonas aeruginosa is a gram-negative, motile, non-spore-forming rod that inhabits moist, warm environments such as water, sewage, and skin.

It is an opportunist and may be shed in the feces or found in

the oropharynx, gastrointestinal tract, and nasopharynx. Infections are almost always silent, but immunologically compromised animals are prone to septicemia (Brownstein, 1978). *Pseudomonas* can, for example, cause severe or lethal infections in nude athymic mice. Sick mice may have equilibrium disturbances, conjunctivitis, serosanguinous nasal discharge, edema of the head, weight loss, and skin infections. Immunosuppressed mice may also develop gastrointestinal ulcers. Generalized infection is associated with severe leukopenia. Neurologic signs are rare, but there are reports of natural central nervous system infection. Chronic proliferative inflammation in the cochlea and vestibular apparatus with dissolution of surrounding bone may be responsible for clinical signs of torticollis.

Organisms enter at the squamocolumnar junction of the upper respiratory tract and, in some cases, the periodontal gingiva. Bacteremia is followed by necrosis of liver and spleen, and the liver may undergo fatty change. The tympanic bullae may contain green suppurative exudate. The bowel may be distended with fluid, and gastrointestinal ulcers can occur.

Infection is diagnosed on the basis of history (e.g., immunosuppression), clinical signs, lesions, and isolation of *P. aeruginosa* from affected mice. Carrier mice can be detected either by nasal culture of individual animals or by placing bottles of sterile, nonacidified, nonchlorinated water on cages for 24 to 48 hr and then culturing the sipper tubes for *Pseudomonas.*

Infection can be prevented by acidification or hyperchlorination of the drinking water. These procedures will not, however, eliminate established infections. Entry of infected animals can be prevented by surveillance of commercially procured colonies. Maintenance of *Pseudomonas*-free animals usually requires barrier quality housing and husbandry.

Pseudomonas infection will complicate experimental studies by causing rapidly fatal septicemia in immunosuppressed mice. Virus infections that alter host defense mechanisms, such as cytomegalovirus, may enhance susceptibility to the organism.

d. Pasteurellosis

Pasteurella pneumotropica is a common gram-negative bacterium whose pathogenicity for otherwise healthy mice is probably minor. Many early observations concerning its pathogenicity are questionable because they were made on colonies of varying microbiological and virological status. There is sentiment for the view that *P. pneumotropica* is an opportunistic pathogen rather than a primary invader. Studies of experimental *P. pneumotropica* suggest that the organism may complicate pneumonias due to *Mycoplasma pulmonis* or Sendai virus. *Pasteurella pneumotropica* can also cause local suppurative lesions in organs other than the respiratory tract. Conjunctivitis, panophthalmitis, dacryoadenitis, and infections of the bulbourethral glands have been attributed to this organism. It has been cultured from the uterus, but its role in metritis is unclear. There is evidence that it may cause infertility or abortion. *Pasteurella urea* has also been recovered from mice with metritis. *Pasteurella pneumotropica* also has been isolated from suppurative lesions of the skin and subcutaneous tissue of the shoulders and trunk. Cutaneous lesions can occur without systemic disease. *Pasteurella pneumotropica* can be isolated from preputial and orbital abscesses especially in nude mice.

Diagnosis is made by isolation of the organism. Serological techniques to identify carrier animals are not available. Antibiotic therapy with penicillin may be beneficial for individual animals, but it will not eliminate infection from an entire colony.

e. Salmonellosis

Salmonellosis has been studied thoroughly in mice as a natural and as an experimentally induced infection (see review by Ganaway, 1982). The mouse is also used to test the potency of vaccines used for protection of humans against typhoid fever. Modern production and husbandry methods have reduced the importance of *Salmonella* as a naturally occurring pathogen of laboratory mice. However, the organisms are ubiquitous in nature and cross-infection from other species or from feral mice remains possible.

There are approximately 1600 recognized serotypes of *Salmonella*. The *Salmonella* most commonly isolated from mice is *S. enteriditis* serovar *typhimurium,* a gram-negative, slow lactose-fermenting rod.

Salmonellae are primarily intestinal microorganisms and can contaminate food and water supplies. Infection occurs primarily by ingestion. In a colony where vermin, birds, and feral animals are excluded, human carriers may be a source of infection. The induction and course of infection are influenced by the virulence of the organism, route of infection, dose of organism, age, sex, genetic factors, nutrition, and intercurrent disease. Stresses that suppress immunity such as X irradiation, corticosteroid administration, and exposure to heavy metals, and environmental factors such as temperature can alter expression of disease. Resistance to natural infection is increased by the presence of normal gastrointestinal microflora. For example, more than 10^6 virulent *S. enteriditis* organisms are needed to establish lethal infection in conventional mice, whereas only a few organisms are required for lethal infection of germfree mice. Nutritional iron deficiency has an attenuating effect on *Salmonella* infection in mice, whereas iron overload appears to promote bacterial growth and enhance virulence. Resistance to infection also can be an inherited trait among inbred strains. Weanling mice are more susceptible to infection than older mice.

Frank salmonellosis is rare in mice. If acute disease occurs, it is especially severe in young mice and is characterized by anorexia, weight loss, lethargy, dull coat, humped posture, and occasionally, conjunctivitis. Gastroenteritis is a common sign, but feces may remain formed. Subacute infection can produce distended abdomens from hepatomegaly and splenomegaly. Chronic disease is expressed by anorexia and weight loss. If salmonellosis is enzootic in a production colony, there are alternating periods of quiescence and high mortality, the latter being associated with diarrhea, anorexia, weight loss, roughened haircoat, and reduced production.

The virulence of *S. enteriditis* serovar *typhimurium* depends on its ability to penetrate intestinal walls, enter lymphatic tissue, multiply and disseminate. Organisms reach Peyer's patches within 12 hr after inoculation and spread quickly to the mesenteric lymph nodes. Bacteremia results in spread to other lymph nodes, spleen, and liver within several days. In chronic infections, organisms persist in the spleen and lymph nodes as well as the liver and gallbladder and from the latter are discharged into the intestinal contents. Bacteria reaching the intestine can reinvade the mucosa and can be shed intermittently in the feces for months. Chronic arthritis associated with *S. enteriditis* infection has been reported.

In animals dying acutely there may be no gross lesions, but visceral hyperemia, pale livers, and catarrhal enteritis are common. If mice survive for up to several weeks, the intestine may be distended and reddened, while the liver and spleen are enlarged and contain yellow-gray foci of necrosis. Affected lymph nodes are also enlarged, red, and focally necrotic. Focal inflammation can develop in many organs including the myocardium.

Microscopically, lesions reflect bacterial invasion, and the extent of lesions is proportional to the course of disease and the number of bacteria in the tissues. Necrotic foci are found in the intestine, mesenteric lymph nodes, liver, and spleen. Neutrophilic leukocytes and histiocytes accumulate in lymphoid tissues. Thrombosis from septic venous embolism may occur especially in the liver. Granulomatous lesions are particularly characteristic of chronic salmonellosis.

Diagnosis is based on isolation of salmonellae together with documentation of compatible clinical signs and lesions. In mice with systemic disease, bacteria may persist in the liver and spleen for weeks. During acute stages, bacteria can also be isolated from the blood. Asymptomatically infected animals can be detected by fecal culture using selective enrichment media. Confirmation of infection can be made serologically. Serotyping reagents aid speciation and can be obtained commercially. Alternatively, samples can be sent to a reference laboratory for confirmation. Antibody to *Salmonella* can be detected in the serum of infected mice by an agglutination test, but this method is not entirely reliable because serological cross-reactivity is common even among bacteria of different genera.

Salmonellosis can be prevented by proper husbandry and sanitation. Contact between mice and potential carriers, such as nonhuman primates, dogs, and cats, should be prevented. Diets should be cultured periodically to check for inadvertent contamination. Humoral and cellular immunity appear to be important in host defenses. Nevertheless, live, virulent, or attenuated vaccines cannot prevent development of the carrier state and are not recommended. About 5% of vaccinated mice reportedly remain asymptomatic carriers. Antibiotic therapy can also enhance the prevalence of carriers. Therefore, contaminated colonies should be replaced to eliminate infection and its zoonotic potential.

f. Streptobacillosis

Streptobacillosis has historical importance as a disease of mice, but modern husbandry and production methods have reduced its impact dramatically. The causative agent, *Streptobacillus moniliformis*, is a gram-negative, pleomorphic bacillus that can exist as an L phase variant *in vivo* and *in vitro*. The L phase is considered nonpathogenic, but it can revert *in vivo* to the virulent bacillus form.

Streptobacillosis generally has a 1- to 3-day acute phase with high mortality, followed by a subacute phase and then a chronic phase that may persist for months. Signs of acute disease include a dull, damp haircoat and conjunctivitis. Variable signs include anemia, diarrhea, hemoglobinuria, cyanosis, and emaciation. In chronic infections, cutaneous ulceration, arthritis, and gangrenous amputation may occur. The arthritis can leave joints deformed and ankylosed. Hindlimb paralysis with urinary bladder distention, incontinence, kyphosis, and priapism may occur if vertebral lesions impinge on motor nerves. Breeding mice may have stillbirths or abortions.

Streptobacillus moniliformis is a primary pathogen for humans (rat bite fever, Haverhill fever) wherein the rat usually serves as a reservoir for dissemination of infection. Clinically healthy, but latently infected rats housed with mice have been implicated as a source of outbreaks. Transmission may occur from aerosol exposure, bite wounds, or contaminated equipment, feed, or bedding (see Chapter 22).

During acute disease in mice, necrotic lesions develop in parenchymal organs of the thoracic and peritoneal cavities. Histological lesions include septic thrombosis of small vessels, acute inflammation, fibrin deposition, and abscesses. Chronically infected mice develop purulent polyarthritis because of the organism's affinity for bones and joints.

Diagnosis depends on isolation of the organism. It has been recovered from joint fluid as long as 26 months after infection. Isolation from chronic lesions requires serum-enriched medium.

Clinical signs must be differentiated from mouse pox, Tyzzer's disease, corynebacteriosis, salmonellosis, mycoplasmosis, and lesions due to trauma.

Control is based on prevention of exposure to wild rodents or to carrier animals such as latently infected rats. Bacterins, antibiotic therapy, and other treatments have proved only partially effective.

g. Corynebacteriosis

Corynebacterium kutscheri, a short gram-positive rod, is a cause of pseudotuberculosis in laboratory animals. Latent infections may be common in conventionally housed mice. Active disease is precipitated by immunosuppression or environmental stresses and is expressed as an acute illness with high mortality or a chronic syndrome with low mortality. Clinical signs include inappetance, emaciation, rough haircoat, hunched posture, hyperpnea, nasal and ocular discharge, cutaneous ulceration, and arthritis.

Lesions develop in various internal organs, such as kidney, liver, lung, and brain, from hematogenous spread. They are characterized by coagulative or caseous necrosis bordered by intense neutrophilic infiltration. Colonies of short gram-positive rods can usually be demonstrated in caseous lesions. Mucopurulent arthritis of carpal, metacarpal, tarsal, and metatarsal joints are related to bacterial colonization of synovium accompanied by necrosis, cartilage erosion, ulceration, and eventually ankylosing panarthritis.

Corynebacterium kutscheri is not a primary skin pathogen, but skin ulcers or fistulae follow bacterial embolization and infarction of dermal vessels. Subcutaneous abscesses have also been reported.

Diagnosis depends on isolation and identification of *C. kutscheri*. Agglutination serology is available, and immunofluorescent and immunodiffusion tests have also been reported. However, the reliability of serological diagnosis has not been widely confirmed. The caseous nature of *C. kutscheri*-induced lesions help to separate them from necrotic changes or abscesses caused by other infectious agents of mice. Because the lung is frequently involved, corynebacteriosis must be differentiated from streptococcosis and mycoplasmosis, neither of which is associated with caseous necrosis. Recent evidence suggests that mice can sustain natural infections with *Mycobacterium avium*. Therefore, histochemical techniques for acid-fast bacilli and appropriate culture methods for mycobacteria should be considered when nodular inflammatory lesions of the lung are detected.

Because clinically apparent corynebacteriosis tends to occur sporadically, treatment and control are difficult. Antibiotic prophylaxis or therapy is probably ineffective. Culling of clinically ill animals may be useful for conventional colonies, but is of little use where animals will be severely stressed or immunosuppressed during experimentation. Replacing or rederiving infected colonies in a specific-pathogen-free environment can be effective in eliminating infection and in preventing reinfection.

h. Staphylococcosis

Staphylococci are found in the nasal passages of healthy mice and can occasionally invade adjacent bone and produce necrosis. The significance of staphylococcosis for mice rests predominantly, however, with its role in purulent dermatitis. Staphylococci produce lipolytic enzymes to counter the bactericidal actions of skin lipids. Species common to the flora of the skin and mucous membranes of mice are *Staphylococcus aureus*, which is pathogenic, and *Staphylococcus epidermidis*, which is generally nonpathogenic.

Staphylococcosis is probably an opportunistic infection. Some evidence suggests that staphylococci can produce primary cutaneous infections, but cutaneous lesions usually occur after contamination of skin wounds. Dermatitis develops primarily on the face and neck and can progress to abscessation, cellulitis, and ulcerative dermatitis. Because lesions are often pruritic, scratching causes additional trauma and autoinoculation. Staphylococcal infection in the genital mucosa of males may produce preputial gland abscesses. These occur as firm, raised nodules in the inguinal region or at the base of the penis and may rupture to spread infection to surrounding tissues. Histologically, superficial staphylococcal infections occur as fibrinonecrotic lesions with neutrophils, but lymphocytes, macrophages and fibroblasts are common in chronic lesions. Deep infections appear as coalescing granulomas with necrotic centers containing bacterial colonies.

The prevalence of staphylococcal dermatitis appears to be influenced by host genotype, the overall health of the animal and the amount of environmental contamination with *Staphylococcus*. C57BL/6, C3H, DBA, and BALB/c mice seem to be most susceptible. Age may also influence susceptibility, young mice being more susceptible than adults. Immunodeficient mice (e.g., athymic mice) contaminated with staphylococci often develop multiple abscesses. Once virulent staphylococci contaminate the environment, colonization of the gastrointestinal tract can occur and produce a carrier state. Human phage types of staphylococci can infect mice, but the zoonotic importance of this connection is not thoroughly understood.

Diagnosis is made by isolation of gram-positive, coagulase-positive cocci that produce β-hemolysis on blood agar. Phage typing can help to determine the source of infection. Gram stains of skin lesions should reveal gram-positive cocci, but diagnosis should not rest on histological findings alone.

Removal of affected animals, sterilization of food and bedding, and frequent changing of bedding may limit or reduce transmission. Affected animals may be helped by trimming or amputation of the hind toes to reduce self-inflicted trauma. This procedure has proved moderately effective in rats with *Staphylococcus*-associated ulcerative dermatitis.

i. Streptococcosis

Most streptococcal infections in laboratory mice are caused by organisms in Lancefield group C, but epizootics caused by group A and group D streptococci have occurred, and group G organisms have been isolated occasionally. Mice can carry streptococci asymptomatically in their upper respiratory tracts. Lethal epizootics can occur, but factors leading to clinical disease are unknown, although some infections may be secondary to wound contamination. Signs include depression, conjunctivitis, rough haircoat, hyperpnea, and emaciation. Enlarged spleens with multifocal abscesses from hematogenous dissemination have been reported. A common streptococcal syndrome in mice is cervical lymphadenitis with fistulous drainage to the neck complicated by ulcerative dermatitis. Diagnosis depends on isolation of organisms from infected tissues combined with microscopic confirmation.

j. Colibacillosis

Escherichia coli is not a major pathogen for mice. Serotypes pathogenic for some species are not necessarily pathogenic for others, but the enterotoxin plasmid can be transmitted between serotypes. This may account for occasional outbreaks of colibacillosis especially in conventionally housed mice.

k. Klebsiellosis

Klebsiella pneumoniae is a ubiquitous gram-negative opportunistic pathogen associated with urinary and respiratory tract infections in humans and in animals. Although capsule types 1 and 2 are highly virulent for mice, *K. pneumoniae* does not appear to be a significant cause of naturally occurring disease.

l. Clostridial Infection

Clostridia are large, rod-shaped gram-positive anaerobic bacteria. Naturally occurring clostridial infection in mice is rare. Epizootics of *Clostridium perfringens* type D infection has been reported in a barrier colony where heavy mortality occurred in 2- to 3-week-old suckling mice. Clinical signs included scruffy haircoats, paralysis of the hindquarters, and diarrhea or fecal impaction. Attempts to produce the disease experimentally with *Clostridia* isolated from naturally infected animals were unsuccessful.

m. Murine Leprosy

Mycobacterium lepraemurium is a gram-positive acid-fast, obligate intracellular bacterium. It has been isolated from healthy laboratory mice and can persist as a latent infection (see review by Harkness and Ferguson, 1982). On rare occasions, *M. lepraemurium* can cause a chronic granulomatous disease in mice exposed to infected wild rats, and there are reports of spontaneous clinical leprosy in mice. Murine leprosy can also be induced by inoculation with *M. lepraemurium*. C57BL/6 and BALB/c mice appear to be more susceptible than DBA/2 mice. Clinical signs include alopecia, thickening of skin, subcutaneous swellings, and ulceration of the skin. Disease can lead to death or clinical recovery. Gross lesions are characterized by nodules in subcutaneous tissues and in reticuloendothelial tissues and organs (lung, spleen, bone marrow, thymus, and lymph nodes). Lesions can also occur in lung, skeletal muscle, myocardium, kidneys, nerves, and adrenal glands. The histological hallmark is perivascular granulomatosis with accumulation of large, foamy epithelioid macrophages (leprae cells) packed with acid-fast bacilli. Research interest in murine leprosy stems from its similarities to human leprosy.

n. Proteus Infection

Proteus mirabilis is a gram-negative opportunistic pathogen that can remain latent in the respiratory and intestinal tracts of mice. Clinical disease can occur following stress or immunosuppression. For example, *Proteus* has been associated with ulcerative lesions in the gastrointestinal tract of mice that have been chemically immunosuppressed. Infected animals lose weight, develop diarrhea, and die within several weeks. If septicemia develops, suppurative or necrotic lesions may be found in many organs, but the kidney and especially the renal cortex is commonly affected. *Proteus* nephritis is characterized by abscessation and scarring. Ascending lesions, following urinary stasis, for example, tend to involve the renal pelvis with only secondary spread to the cortex. *Proteus mirabilis* and *Pseudomonas aeruginosa* have been isolated concomitantly from cases of suppurative nephritis or pyelonephritis.

o. Leptospirosis

Many leptospire species can infect laboratory mice, but infection is generally asymptomatic. Nevertheless, latent murine infections associated with active shedding present a zoonotic hazard for humans; therefore infected mice should be discarded (see review by Fox and Brayton, 1982). Zoonotic outbreaks from contact with mice have been associated primarily with *Leptospira ballum*, an organism that also infects rats. Because clinical signs and lesions of leptospirosis rarely develop, serological testing or isolation of organisms is the preferred means for diagnosis. Leptospires can occasionally be detected by dark field examination of body fluids or by staining histological sections by silver impregnation methods. Serological diagnosis can be made by a variety of tests, including indirect hemagglutination, macroscopic agglutination, complement fixation, and indirect immunofluorescence.

Seropositive dams do not transmit organisms to their offspring, but seronegative carriers can spread infection. The administration of chlorotetracycline hydrochloride (1000 gm per ton of feed) for 10 days is said to be an effective treatment. Mice are moved to autoclaved cages after 7 days and given sterile water. Nevertheless, elimination of infection usually requires rederivation or total replacement of infected colonies combined with measures to prevent entry of infected wild or feral rodents.

2. Mycoplasmal Diseases

a. Murine Respiratory Mycoplasmosis (MRM)

Murine respiratory mycoplasmosis is a syndrome characterized by suppurative rhinitis, otitis media, and chronic pneumonia (see review by Lindsey *et al.*, 1982). Viruses and bacteria have been implicated as causative agents, but the clinical and morphological characteristics of the disease can be duplicated by experimental infection with *Mycoplasma pulmonis*.

Mycoplasma pulmonis is a pleomorphic bacterium of the sterol-requiring family Mycoplasmataceae. It lacks a cell wall and has a single outer limiting membrane. Colonies growing on agar have a fine granular appearance and may resemble fried eggs when viewed at low magnification. *Mycoplasma pulmonis* can ferment glucose but not arginine. Most isolates have up to three common antigens.

Asymptomatic infection may occur, but mice commonly display chattering, inactivity, weight loss, rough haircoat, and dypsnea. Chattering and dypsnea are due to accumulations of purulent exudate in nasal passages together with inflammatory thickening of nasal mucosa. In the mouse, experimental MRM is highly dose dependent. Doses of 10^4 colony-forming units (CFU) or less cause mild transient disease involving the upper respiratory tract and middle ears, whereas higher doses often cause death from acute pneumonia that accompanies upper respiratory lesions. Survivors develop chronic bronchopneumonia with bronchiectasis and pulmonary abscesses and can spread disease to other mice. Intravenous inoculation of *M. pulmonis* can cause arthritis in mice, but arthritis is uncommon in natural infection. Rarely, abscesses in the brain and spinal cord may cause flaccid paralysis. Naturally occurring genital disease due to *M. pulmonis* has not been reported in mice, but experimental parenteral inoculation has caused oophoritis, salpingitis, and metritis. *Mycoplasma pulmonis* reportedly can also cause infertility or fetal deaths.

There is some evidence that mouse strains may differ in susceptibility to MRM, but this observation has not been verified experimentally. Systematic investigations of natural outbreaks are lacking. Transmission can occur between cage contacts and between adjacent cages. The offspring of infected dams probably acquire infection by aerosol transmission early in life. *In utero* infection of fetuses has been demonstrated in rats, but not in mice. *Mycoplasma pulmonis* is a serious pathogen of rats and has been isolated from hamsters, guinea pigs, and rabbits. Among these species only rats are significant reservoirs of infection for mice.

The primary lesion early in experimental or natural disease is suppurative rhinitis (Fig. 15), which can progress to prominent squamous metaplasia. Transient hyperplasia of submucosal glands may occur, and lymphoid infiltration of the submucosa can persist for weeks. Syncytia can sometimes be found in nasal passages in association with purulent exudate. Affected mice usually have suppurative otitis media and chronic laryngotracheitis with mucosal hyperplasia and lymphoid cell infiltrates.

The lung lesion is chronic bronchopneumonia that spreads from the hilus. Neutrophils accumulate in bronchial lumens, and large peribronchial clusters of lymphoid cells (primarily plasma cells) develop. Bronchial exudate can cause atelectasis, bronchiectasis, bronchiolectasis, and abscesses (Fig. 16). Advanced lesions may be expressed macroscopically as a cob-

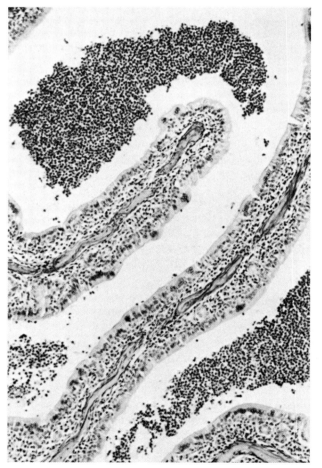

Fig. 15. Suppurative rhinitis in a mouse with *Mycoplasma pulmonis* infection. Note syncytia in surface epithelium. (From Lindsey *et al.*, 1982.)

blestone appearance of the lung. Cuboidal epithelium often lines alveoli immediately surrounding affected airways, but pleuritis is rare.

The pathogenesis of cell injury in MRM is not well understood, although it is known that *M. pulmonis* attaches or adheres to host cell membranes as an initial event. This attachment to respiratory epithelium occurs anywhere from the anterior nasal passages to the alveoli and may be mediated by surface glycoproteins. The organism may injure host cells through competition for metabolites such as carbohydrates and nucleic acids or by release of toxic substances such as peroxides. Ciliostasis, reduction in the number of cilia, and ultrastructural changes leading to cell death have also been described in MRM.

Prior infection with Sendai virus enhances the growth of *M. pulmonis* and the severity of lung lesions. *Pasteurella pneumotropica* is thought by some to exacerbate the severity of respiratory disease from *M. pulmonis,* but this has not been confirmed experimentally.

Mice mount an effective immune response to *M. pulmonis* as measured by their recovery from mild infection and their resistance to infection after active or passive immunization. Antibodies of various classes are produced locally and systemically, but their role in the disease is unclear. There is some evidence that antibody may facilitate phagocytosis of *M. pulmonis.* Classic cellular immunity, however, does not appear to play a major role in *M. pulmonis* infection in mice, since immunity cannot be transferred with immune cells. In addition, neither athymic nor neonatally thymectomized mice are more susceptible to *M. pulmonis* pneumonia than normal mice. By contrast, passive immunity in the rat can be transferred with immune spleen cells but not with serum. Mice given more than 10^4 CFU can, however, develop severe chronic disease even in the presence of well-developed immunity. This has led to the suggestion that *M. pulmonis* infection has an immunopathologic component.

Clinical signs may not be pathognomonic during acute outbreaks because they may be mimicked by other respiratory infections, including Sendai virus pneumonia. Lesions are fairly characteristic, especially in advanced disease. The upper respiratory tract should be cultured since it is a common site for natural infection. Buffered saline or *Mycoplasma* broth can be used to lavage the trachea, larynx, pharynx, and nasal passages. Genital tract infection may occur in the absence of respiratory infection, therefore, culturing at this site also is prudent.

Mycoplasmas may be difficult to grow on artificial medium but can generally be cultured in standard Hayflick's broth or *Mycoplasma* agar at 37°C. Cultural isolation can give good results if it is carefully done. Because the quality of the culture medium may vary from batch to batch, it is essential that it be pretested for its ability to support growth of stock strains. Culture vessels, water, and medium components should be of

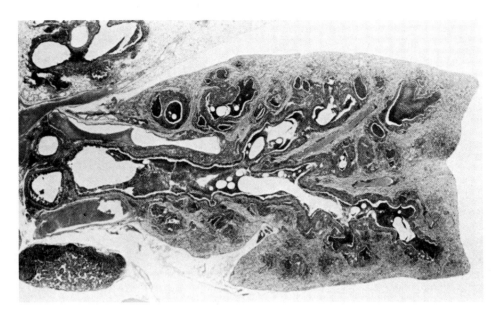

Fig. 16.   Mouse lung with advanced lesions of *Mycoplasma pulmonis* infection. Note massive accumulation of purulent exudate in the airways, bronchiectasis, bronchiolectasis, atelectasis, and peribronchial lymphoid cell infiltrates. (From Lindsey *et al.*, 1982.)

tissue culture quality. Growth can usually be detected in 1 week in agar or broth, but cultures should be held for 21 days before they are discarded as negative. Organisms can be identified by colony morphology, but *M. pulmonis* has little tendency to produce typical fried egg colonies under less than optimal growth conditions. Speciation can be accomplished by immunofluorescence or immunoperoxidase staining or by growth inhibition. A rapid presumptive test is the hemadsorption test, but some strains of *M. pulmonis* do not hemadsorb.

Immunofluorescence and immunoperoxidase techniques can also be used to identify mycoplasmal antigen in tissue sections or in cytological preparations of tracheobronchial or genital tract lavages. Conventional serological tests, such as complement fixation (CF), hemagglutination inhibition (HAI), and growth inhibition, have limited value for serological detection of mycoplasmosis because serum titers are usually low. More recently, a sensitive enzyme-linked immunosorbent assay (ELISA), a radioimmunosorbent assay, and a solid phase radioimmunoassay show promise of increasing the sensitivity and speed of serological detection (Cassell *et al.*, 1981).

Effective control of MRM depends on prevention of infection through maintenance of pathogen free colonies. This implies careful surveillance for infection by serology, microbiology, and histopathology and adherence to rigid principles of barrier management. Cesarean derivation can eliminate infection, but fetal membranes and offspring must be tested to rule out transmission *in utero*. Treatment with tetracyclines suppresses clinical disease but does not eliminate infection. Effective vaccines are not available.

Mycoplasma pulmonis can interfere with research by causing clinical disease or death. Experiments involving the respiratory tract, such as inhalation toxicology, can be compromised by chronic progressive infection. Experimentally induced immunosuppression can enhance the pathogenicity of *M. pulmonis*. Conversely, altered immunological responsiveness of mice infected with *M. pulmonis* is also a possibility because the organism is a mitogen for lymphocytes. In rats, *M. pulmonis* has been associated with an increased incidence of carcinogen-induced cancers.

b.   Neurotoxic Mycoplasmosis (Rolling Disease)

Mycoplasma neurolyticum is the etiological agent of rolling disease. Its natural prevalence in mice is rare. Clinical signs appear, however, within 1 hr after intravenous inoculation of *M. neurolyticum* exotoxin (Thomas *et al.*, 1966). They include spasmodic hyperextension of the head and raising of one foreleg followed by intermittent rolling on the long axis of the body. The rolling becomes more constant, but mice occasionally leap or move rapidly. After 1 or 2 hr of rolling, animals become comatose and usually die within 4 hr. Mice of all ages appear to be susceptible.

All published descriptions of rolling disease are associated with experimental inoculation of organisms or exotoxin. Large numbers of organisms are needed to produce disease, and there is no indication that, under natural conditions, organisms replicate in the brain to concentrations required for the induction of these signs. Because animals are frequently inoculated with biological materials by parenteral routes, contamination with *M. neurolyticum* may induce rolling disease inadvertently.

Mycoplasma neurolyticum exotoxin enters the brain from the vascular system and fixes to receptors on glial cells. Lesions are not striking in animals that die peracutely. If mice survive for 8 hr or more, astrocytes can undergo spongiform degeneration from intracellular accumulation of fluid. The disruption of fluid transport with concomitant compression of neurons by swollen astrocytes may be responsible for the neurological signs. *Mycoplasma pulmonis* has been recovered from the brain of mice, but it does not seem to cause overt neurological disease.

Diagnosis can be made from the appearance of typical clinical signs, astrocytic swelling, and isolation of the causative organism. Clinical signs must be differentiated from rolling associated with *Pseudomonas*-caused otitis.

There is a recent report of naturally occurring infection with *M. arthritidis* in mice. The significance of this finding for epizootiological studies of MRM in mouse colonies is unclear.

3. Rickettsial and Chlamydial Diseases

Rickettsia are small, rod-shaped, coccoid, and occasionally pleomorphic organisms with typical bacterial cell walls but lacking flagellae. They are gram-negative and multiply by binary fission in host cells. Most rickettsia are transmitted by insects. The organisms commonly infect reticuloendothelial cells, produce vasculitis, and destroy erythrocytes to cause hemolytic anemia. Few rickettsia cause natural disease in mice, although mice are susceptible to experimental inoculation with human rickettsia, including the organisms of scrub typhus and rickettsial pox.

Two rickettsia, *Eperythrozoon coccoides* and *Hemobartonella muris,* have historical importance for mice. The former is primarily an organism of mice and the latter, although infectious for mice, is more commonly associated with rats (see review by Baker *et al.,* 1971; Hildebrandt, 1982). Both are transmitted by insects and destroy erythrocytes. *Eperythrozoon* occurs as a ring-shaped or coccoid organism and occasionally as a rod either on erythrocytes or free in plasma (Fig. 17). It is enclosed by a single limiting membrane, but has no cell wall, and no nucleus or other membrane-bound organelles. The natural vector for transmission is the mouse louse, *Polyplax serrata.* Infected mice remain clinically normal or may develop anemia and splenomegaly, become febrile, and occasionally die. Hepatocellular degeneration and multifocal necrosis have

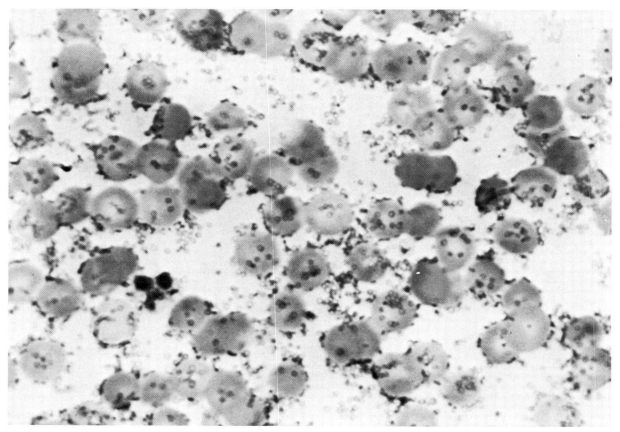

Fig. 17. Mouse blood infected with *Eperythrozoon coccoides.* Ring-shaped organisms are attached to erythrocytes and free in plasma. (From Baker *et al.,* 1971, with permission.)

been recorded in acute infections. *Hemobartonella muris* is not a major problem in mouse colonies. In nature, it is spread by the rat louse *Polyplax spinulosa*.

Rickettsial infections of rodents are long lived and are expressed clinically in one of two ways: acute febrile anemia and latent or asymptomatic infection. The latter can be reactivated by splenectomy, a procedure that ablates a major source of macrophages. The prevalence of carriers is not known, but the carrier state may be life long even when infection begins at an early age. Splenectomy is mimicked by irradiation, immunosuppressive therapy, anti-lymphocyte serum, and certain intercurrent diseases. Conversely, *E. coccoides* may convert asymptomatic mouse hepatitis virus infection into a fatal disease. Other mouse viruses thought to be potentiated by *E. coccoides* are lymphocytic choriomeningitis virus and lactic dehydrogenase-elevating virus.

Inoculation of test material into splenectomized mice is the most sensitive means of detection. Latent carriers can be uncovered by this procedure, and rickettsemia usually begins in 48 hr. Because rickettsemia may be brief, blood smears should be prepared every 6 hr beginning at 48 hr postsplenectomy to assure that transient blood-borne infection is not missed. Smears are stained with a Romanovsky procedure to visualize parasites, but indirect immunofluorescence can also be used.

Treatment of *E. coccoides* is not practical. Control is based on elimination of lice and on rederivation of infected stock. If replacement animals are readily available, euthanasia is a more prudent course. Biological materials destined for animal inoculation should be checked for rickettsial contamination by inoculation of splenectomized mice.

Chlamydia are intracellular parasites that cause disease in many species of animals and in humans. They multiply intracellularly and form membrane-bound cytoplasmic inclusions.

The human organism, *Chlamydia trachomatis*, causes trachoma, conjunctivitis, lymphogranuloma venereum, and urogenital tract infections. A strain of this organism, the so-called "Nigg agent," is thought to be responsible for an historically noteworthy pneumonitis in mice. It was characterized clinically by ruffled fur and hunched posture with labored respiration and death in 24 hr. Mice dying more slowly showed progressive emaciation and cyanosis of the ears and tail. Gross findings included focal elevated grayish lesions in the lung. Their size progressed until total pulmonary consolidation occurred. Microscopically, interstitial pneumonia developed. The disease resembled respiratory mycoplasmosis, and the Nigg agent was thought to be a causative factor for this condition.

Chlamydia psittaci infects many avian and mammalian species. Latent infections have been recorded in mouse colonies. The agent can produce ascites, splenomegaly, and serofibrinous exudate on the liver and spleen. Nevertheless, it does not cause serious natural disease in mice. The fact that latent infec-

tions have been demonstrated indicates that surveillance may be advisable.

4. Viral Diseases

a. Mousepox (Infectious Ectromelia)

Mousepox is a devastating disease of mice caused by ectromelia virus, an orthopoxvirus that is closely related antigenically and physicochemically to vaccinia virus (see review by Fenner, 1982). Mousepox was first reported by Marchal in England, and the causative virus was detected soon after by Barnard and Elford. Field strains of ectromelia virus have been isolated in many countries, but two strains, Hampstead (low virulence) and Moscow (high virulence), have been used extensively for laboratory study. Ectromelia virus grows well on the chorioallantoic membrane of embryonated chicken eggs and can also infect HeLa cells, human amnion cells, mouse fibroblasts (L cells), and chick embryo fibroblasts. The B-SC-1 cell line is particularly sensitive to ectromelia virus. Ectromelia virus produces an envelope hemagglutinin whose detection forms the basis for hemagglutination inhibition (HAI), an historically important serological test for mousepox.

Mousepox usually takes one of three clinical courses: acute infection with high mortality, chronic infection with variable mortality, or asymptomatic infection. The expression of clinical signs reflects an interplay between virus-related factors, such as virulence, and dose- and host-related factors, such as age, genotype, immunological competence, and portal of entry. Acute lethal infection occurs in genetically susceptible mice and may produce clinical signs, such as ruffled fur or prostration, for only a few hours before death. The rapidly fatal form of mousepox is associated with extensive necrosis of lymphoid tissue and liver and with intestinal hemorrhage.

Mice that survive acute infection often develop a skin rash whose severity depends on the extent of secondary viremia after infection of parenchymal organs. The pox rash can develop anywhere on the body and may be solitary or generalized (Fig. 18). In some mice, conjunctivitis also occurs. The rash often recedes within several weeks, but hairless scars can remain. Severe viral infection of the feet and tail can lead to amputation; hence the name infectious ectromelia.

Natural exposure is thought to occur through small abrasions of skin, but experimentally, oral exposure can cause chronic inapparent infection of Peyer's patches, prolonged excretion of virus in feces, and occasional chronic tail lesions. Mice with chronic intestinal infection appear not to readily transmit infection by contact, but carrier mice can be a source of contaminated tissue suspensions. Arthropod transmission is important for some poxviruses, but this appears not to apply to ectromelia virus, although the blood-sucking rat mite *Ornithonyssus bacoti* may be a passive vector. Intraperitoneal inoculation

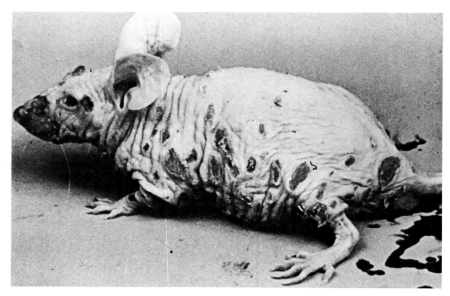

Fig. 18. A hairless mouse with mousepox. (From Fenner, 1982, and with permission of the Zentral Institut für Veruchstieire, Hannover, Federal Republic of Germany.)

usually results in acute visceral lesions and death before primary skin involvement occurs. Intranasal infection can produce necrotizing rhinitis and pneumonia when large doses of virus are inoculated. Intrauterine infection and fetal deaths have been reported. Therefore contamination of cesarean-derived progeny or embryo-derived cell cultures has to be considered.

The laboratory mouse is the primary host for ectromelia virus, although infection of wild mice has been reported. Natural infections in other laboratory animals have not been confirmed, but poxlike viruses have been found in rats. Genotype can modulate the course and severity of infection. DBA/1, DBA/2, BALB/c, A, and C3H mice are among the most susceptible inbred strains, whereas C57BL/6 and AKR are resistant to lethal infection (see review by Briody, 1959). C57BL/6 mice may be resistant by virtue of their ability to mount rapid immune responses to ectromelia viral infection. Highly susceptible mice die from visceral infections before a skin rash develops; therefore they are a relatively small hazard for dissemination of virus. Resistant mouse strains are dangerous because they can develop enzootic asymptomatic infections. Addition of susceptible strains to an enzootically infected colony can provoke explosive lethal outbreaks. Intermediately resistant (or susceptible) mice frequently survive long enough to develop skin lesions that, aside from being extensive and unsightly, also can shed virus and serve as a major reservoir for spread of infection. Very young and aged mice seem to be more susceptible to lethal infection than are young adult mice.

Mousepox has been a common disease of mouse colonies in Europe, Japan, and China since 1930. It has been imported, inadvertently, to North America several times, and outbreaks have occurred periodically in the United States. In Europe, enzootic infections prevail in breeding colonies of resistant mice where clinical signs may be mild or inapparent. Maternal immunity may perpetuate infection by protecting young mice from death, but not from infection. Such mice can infect other mice by contact exposure.

Ectromelia virus multiplies in the cytoplasm and produces two types of inclusion bodies. The A type (Marchal body) is acidophilic and is found primarily in epithelial cells of skin or mucous membranes (Fig. 19). The B type inclusion is basophilic and can be found in all ectromelia-infected cells if they are stained intensely with hematoxylin.

The pathogenesis of infection following skin invasion begins with viral multiplication in the draining lymph node and a primary viremia (Fig. 20). Splenic and hepatic involvement begin within 3 to 4 days, whereupon larger quantities of virus are disseminated in blood to the skin. This sequence takes approximately 1 week and, unless mice die of acute hepatosplenic infection, ends with the development of a primary skin lesion at the original site of viral invasion. The primary lesion is due ostensibly to the development of antiviral cellular immunity.

Focal or confluent hepatocellular necrosis occurs in susceptible mice during acute stages of mousepox. White spots indicative of necrosis are seen grossly throughout the liver. In nonfatal cases, regeneration begins at the margins of necrotic areas, but inflammation is variable. Splenic necrosis in acute disease commonly precedes hepatic necrosis but is at least equally se-

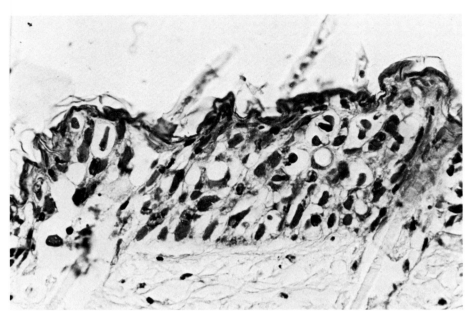

Fig. 19.   Skin with intracytoplasmic type A inclusions of mousepox.

vere. Necrosis and scarring of red and white pulp can produce a gross "mosaic" pattern of white and red-brown. Necrosis of the thymus, lymph nodes, Peyer's patches, intestinal mucosa, and genital tract also have been observed during acute infection, whereas resistant or convalescent mice can develop lymphoid hyperplasia.

The primary skin lesion, which occurs 6–10 days after exposure, is a localized swelling that enlarges from inflammatory edema. Necrosis of dermal epithelium provokes a surface scab and heals as a deep, hairless scar. Secondary skin lesions (rash) develop 2 to 3 days later, are often multiple and widespread, and can be associated with conjunctivitis, blepharitis, and, in severe cases, with buccal and lingual ulcers. Secondary skin lesions also ulcerate and scab before scarring.

Mousepox can be diagnosed from clinical signs, lesions, serological tests, and demonstration of virus or viral antigen in tissues. Characteristic intracytoplasmic eosinophilic inclusions aid histological confirmation, and typical poxvirus particles can be found in tissues by electron microscopy. Fenner has suggested that demonstrating poxvirions in homogenized tissues or scabs from suspected cages could be a rapid method to diagnose mousepox. Virus can be isolated from infected tissues by inoculation of cell cultures (B-SC-1) or embryonated eggs.

Several serological tests are available to detect mousepox. The standard test has been hemagglutination inhibition (HAI) using vaccinia antigen as a source of hemagglutinin. The HAI test has been valuable, but it has several potential pitfalls. Hemagglutinin-deficient mutants of vaccinia virus have been found (see below), and the possibility that similar ectromelia mutants can occur in nature has not been resolved. The HAI test also can give occasional false positive and false negative results, so it must be performed and interpreted by experienced personnel. Immunofluorescence, gel diffusion, or a recently developed enzyme-linked immunosorbent assay (ELISA) may eventually replace HAI.

Mousepox must be differentiated from other infectious diseases associated with high morbidity and high mortality. These include mouse hepatitis, Tyzzer's disease, and reovirus 3 infection. Each can be expressed by acute necrosis in parenchymal organs, but they can be differentiated by morphological, serological, and virological criteria. The skin lesions of chronic mousepox must be differentiated from other skin diseases caused by opportunistic or pathogenic bacteria, acariasias, and bite wounds.

Serological differentiation of mousepox from vaccinia infection in vaccinated mice is based on the lack of hemagglutinin in the vaccine strain of virus. Thus, serum from vaccinated mice may react in complement fixation (CF), immunofluorescent antibody (IFA), or ELISA tests, but should not react in HAI tests (Briody, 1959).

Because mousepox is a dangerous disease in laboratory mice, depopulation coupled with vaccination is the primary means for control. Infection is sometimes disregarded in enzootically infected colonies of resistant mice. Elimination is still desirable under these conditions because of potential spread to susceptible mice.

Infected colonies should be quarantined immediately, but confirmation of infection should be obtained before exposed mice are destroyed. Tissues, supplies, instruments, or other

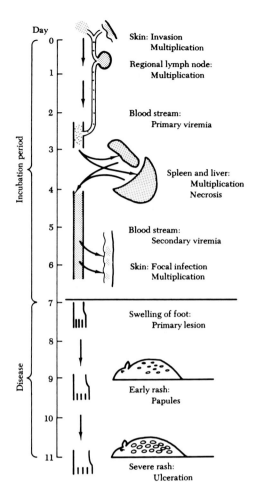

Day

Skin: Invasion
Multiplication

Regional lymph node:
Multiplication

Blood stream:
Primary viremia

Spleen and liver:
Multiplication
Necrosis

Blood stream:
Secondary viremia

Skin: Focal infection
Multiplication

Swelling of foot:
Primary lesion

Early rash:
Papules

Severe rash:
Ulceration

Fig. 20. Diagram illustrating the pathogenesis of mousepox. (From Fenner, 1948.)

items that have had potential contact with infected mice should be disinfected by heat or chemicals (formalin or sodium hypochlorite). Materials should be autoclaved or, preferably, incinerated. Disinfected rooms should be challenged with susceptible sentinel animals that are observed for clinical signs and tested for seroconversion after several weeks. Depopulation and disinfection must be done vigorously. Culling is not a currently accepted alternative. Valuable mice, such as irreplaceable breeding stock, can be retained in isolators for rederivation in lieu of prompt euthanasia.

Vaccination on a massive scale can effectively control or prevent clinical mousepox. A hemagglutinin-deficient strain of vaccinia virus (IHD-T) is used to scarify skin on the dorsum of the tail where ''takes'' can be easily read. Vaccination reactions (Fig. 21) should occur in all previously uninfected mice by 6–10 days, but not in infected mice. The latter should be eliminated. Although vaccination can prevent clinical disease, it may not prevent infection, however transient. Therefore,

vaccination should not lead to complacency in adherence to other preventive measures, such as strict controls on the entry of mice or mouse products combined with periodic serological monitoring.

The primary threat from ectromelia virus pertains to its lethality in susceptible mice. Infection from contaminated tissues, cell lines, or asymptomatically infected mice can produce explosive epizootics. Because the long-term epizootiological consequences of infection are still unresolved, especially in inbred mice, loss of time, animals, and financial resources are significant when current control measures are employed.

b. Herpesvirus Infection (see review by Osborn, 1982)

Two mouse herpesviruses have been identified: mouse cytomegalovirus (MCMV) and mouse thymic virus (MTV).

i. Mouse cytomegalovirus (MCMV). Several isolates of MCMV have been studied. Their biological properties differ, but they are biophysically and antigenically similar. Mouse cytomegalovirus is infectious only for mice, but it can replicate in cell cultures from several species, including mouse, hamster, rabbit, sheep, and nonhuman primate.

Mouse cytomegalovirus produces a spontaneous, persistent, subclinical infection in wild mice, but infection of laboratory mice is thought to be uncommon. The susceptibility of mice to experimental infection varies with age, dose, route, virus strain, and host genotype. The pathogenicity of MCMV for mice decreases with age, neonates being highly susceptible to lethal infection, and 4-week-old mice being highly resistant. Mouse cytomegalovirus can replicate in many tissues, and viremia commonly occurs. Mouse cytomegalovirus often establishes persistent infection in the salivary glands and in the pancreas. The persistence of salivary gland infection appears to be dose dependent. There is experimental evidence that MCMV can produce latent infection of B cells, probably T cells, and possibly prostate and testicle (including spermatogonia). Renal infection is less severe and does not persist in mice as long as in CMV infection. Latent infection can be reactivated by lymphoproliferative stimuli and by immunosuppression.

Lesions are not remarkable during natural infection and may be limited to an occasional enlarged cell (megalocytosis) containing an intranuclear inclusion together with some nonsuppurative interstitial inflammation, especially in the cervical salivary glands. Experimental infection can induce necrosis and inflammation in parenchymal organs such as the liver, and intranasal infection, which probably mimics natural exposure, can produce a severe and even lethal interstitial pneumonia.

Because MCMV can be latent, its epizootological status in a colony can be difficult to determine accurately. Close contact

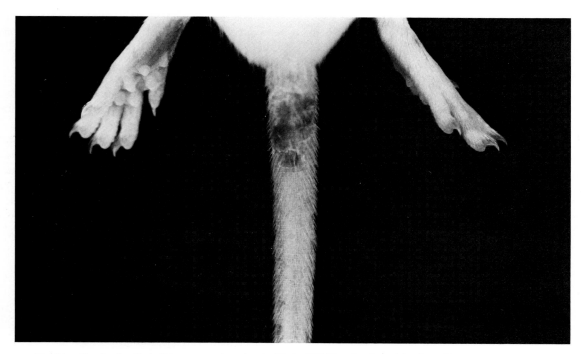

Fig. 21. Vaccination "take" in a mouse inoculated with the IHD-T strain of vaccinia virus. (From Jacoby *et al.*, 1983.)

is required for spread of infection in laboratory mice. Infected saliva contributes to horizontal transmission.

Mouse cytomegalovirus antigens appear to be weak stimuli for humoral antibody production. Complement fixing (CF) and neutralizing (NT) antibody titers are low during acute infection and difficult to find during chronic infection. Cellular immunity, by contrast, appears to be important in protection against severe or lethal MCMV infection. Mouse cytomegalovirus can be grown in mouse embryo fibroblasts or 3T3 cells, but cocultivation may be required to rescue latent virus. Detection of enlarged cells with intranuclear inclusions especially in salivary glands, can aid in diagnosis. A recent evaluation of MCMV antibody detection by serological techniques demonstrated that the ELISA technique was more sensitive for detecting persistent infection with MCMV, and nuclear anticomplement immunofluorescence was more useful for detecting acute MCMV infection (Anderson *et al.*, 1983).

Mouse cytomegalovirus infection must be differentiated from infection with mouse thymic virus. The latter virus can produce necrosis and atrophy of thymic and peripheral lymphoid tissue. Lytic lesions of lymphoid tissues are not a hallmark of MCMV. The viruses can also be distinguished from each other serologically.

Control measures for MCMV have not been established because it has not been considered an important infection of laboratory mice. Cage to cage transmission has not been demonstrated, but horizontal infection from contaminated saliva must be accounted for.

Mouse cytomegalovirus can suppress responses of mice to other antigens and may be autoimmunosuppressive and thus contribute to the virus' "antigenic weakness." Mouse cytomegalovirus-induced immunosuppression also can exacerbate the pathogenicity of opportunistic organisms such as *Pseudomonas aeruginosa.*

ii. Mouse thymic virus (MTV). Mouse thymic virus was discovered inadvertently during studies of mammary tumor virus and was so named for the thymic necrosis it causes in suckling mice. No suitable *in vitro* method for cultivation has been developed, therefore viral propagation depends on mouse inoculation.

Mouse thymic virus infection is usually asymptomatic, but it causes severe, diffuse necrosis of the thymus during the first week of life and persistent infection of salivary glands. Infected thymocytes display MTV-positive intranuclear herpetic inclusions. Necrosis is associated with granulomatous inflammation and formation of syncytia. Necrosis and inflammation can also occur in lymph nodes. Reconstitution of lymphoid organs may take up to 8 weeks.

The prevalence of MTV in laboratory mice is not accurately known. The mode of transmission also is obscure, but the virus' sialotropism and its presence in the oral cavity of infected adults suggests that horizontal spread occurs from dams to unprotected neonates.

Infection is usually diagnosed from acute morphological

changes together with intranuclear inclusions, but the severity of thymic and lymph node necrosis can be mouse strain dependent and can also occur in severe mouse hepatitis virus or epizootic diarrhea of infant mice infection. Serological confirmation of infection can be obtained from adult mice by CF or NT tests. Neonatal mice inoculated with tissues (thymus, lymph node, salivary gland) from infected mice will develop thymic necrosis in 2 to 7 days. Mice infected as neonates do not develop detectable antibody responses, but viral antigen can be demonstrated in the thymus by immunocytochemical staining.

Because MTV induces persistent salivary infection, control measures suitable to acute infections, such as quarantine coupled with temporary cessation of breeding, may not be effective. Rederiving or restocking affected colonies is a more appropriate alternative, if infection cannot be tolerated as a research variable.

Mouse thymic virus transiently suppresses cellular and humoral immune responses because of its destructive effects on neonatal T lymphocytes. Diminished responses to common antigenic stimuli, such as sheep erythrocytes and prolonged survival of allogenic skin grafts, have been demonstrated in experimentally infected mice.

c. Lactate Dehydrogenase-Elevating Virus (LDHV) Infection

Lactate dehydrogenase-elevating virus is a togavirus specific to mice and causes elevated levels of several serum enzymes, most notably lactate dehydrogenase (LDH), in infected mice (see review by Brinton, 1982). The primary mode of transmission is mechanical transfer of tissues or serum from infected mice. Infection is asymptomatic, and the only significant lesion thus far associated with infection is polioencephalitis in old C58 mice that were immunosuppressed during early stages of infection, although there is evidence that T cell–dependent areas of thymus and peripheral lymphoid tissue may undergo mild necrosis early in infection.

Natural transmission between cagemates or between mother and young is rare even though infected mice may excrete virus in feces, urine, milk and probably saliva. Infection by bite wounds or by ingestion of viral contaminated tissues cannot be ruled out.

Lactate dehydrogenase-elevating virus produces a lifelong viremia, and plasma LDH levels are permanently elevated, a response that is used to detect and titrate LDHV infectivity. Of the five isoenzymes of LDH in mouse plasma, only LDH V is elevated. SJL/J mice in particular show spectacular increases in LDH levels (15–20 times normal), a response controlled by a recessive somatic gene. Virus persists despite a modest humoral antibody response, the latter being difficult to detect due to formation of virus–antibody immune complexes.

Although neither clearance mechanisms nor their selectivity are fully understood, it is thought that viral interference with clearance functions of the reticuloendothelial network is a significant factor in elevated blood levels of enzyme. Lactate dehydrogenase-elevating virus also interferes with the clearance of other serum enzymes and results in their elevation in serum. The mechanisms associated with this selective increase also are unclear. Immune complex glomerular disease is not a significant complication of LDHV infection, despite the virus' propensity to form immune complexes.

Lactate dehydrogenase-elevating virus is detected by measuring LDH levels in mouse plasma before and 4 days after inoculation of specific pathogen free (SPF) mice with suspect material. It is important not to use hemolyzed samples because they will produce falsely elevated readings. Plasma enzyme levels are measured in conventional units/ml, 1 conventional unit being equivalent to 0.5 IU. Normal plasma levels are 400–800 IU, whereas in LDH virus infection, levels as high as 7000 IU can be detected.

Although LDHV can infect tumor cells, it does not replicate in them in vitro. Since LDHV is mouse specific, tumors can be freed of virus by passaging them several times in nonpermissive rodents (e.g., rat) before repassaging them in mice. All tumors or cell lines destined for mouse inoculation should be monitored for LDHV contamination by plasma enzyme bioassay.

Lactate dehydrogenase-elevating virus has numerous potential effects on immunological responsiveness. It reduces autoantibody production, causes transient thymic necrosis and lymphopenia, suppresses cell-mediated immune responses, and enhances or suppresses tumor growth.

Lactate dehydrogenase-elevating virus has been found in more than 50 transplantable mouse tumors. Contamination probably occurred during passage of tumors through infected mice.

d. Lymphocytic Choriomeningitis (LCM)

Lymphocytic choriomeningitis is caused by an arenavirus. It was initially isolated from three sources: monkeys inoculated with infected human tissues, albino mice, and humans, one of whom had worked with laboratory mice. It replicates by budding, and the viral envelope is formed by the cell outer membrane. The virus contains single-stranded RNA and is sensitive to lipid solvents. Replication of infectious LCMV both in vitro and in vivo is associated with production of interfering virus.

Lymphocytic choriomeningitis virus strains vary in their rate of replication, tissue tropism, pathogenicity, immunogenicity, and replication, but they are closely related antigenically. These properties can, however, be modulated by passage in vivo or in vitro. Different biological properties have even been detected in LCMV recovered from different organs of the same mouse. LCMV can infect insect cells as well as mammalian cells and can persistently infect naturally exposed mice and cultured cells.

Clinical signs of LCMV infection vary with age and strain of mouse, route of inoculation, and strain of virus. Four basic patterns of clinical infection are recognized (Fig. 22): (1) the *cerebral form*—induced in adult mice by intracerebral inoculation; (2) the *visceral form*—induced in adult mice by peripheral routes of inoculation; (3) *late onset disease*—found in previously asymptomatic carrier mice; (4) *runting and death*—

in neonatally infected mice (see review by Lehmann-Grube, 1982).

The *cerebral form* of LCM is characterized by sudden death beginning 5 to 6 days postinoculation or by subacute illness associated with one or more of the following signs: ruffled fur, hunched posture, motionlessness, and neurological deficits. Mice suspended by the tail will have coarse tremors of the head and extremities culminating in clonic convulsions and tonic extension of the hindlegs. Spontaneous convulsions also can occur. Animals usually die or recover in several days.

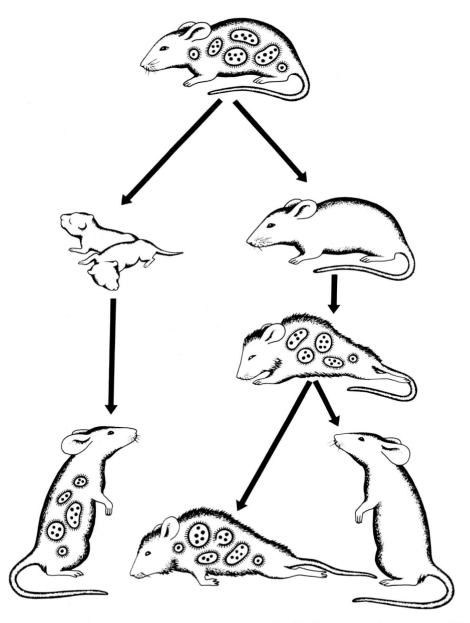

Fig. 22.   Basic phenomena associated with infection of mice with LCM virus. (From Lehmann-Grube, 1982.)

Signs of *visceral* LCM can vary from asymptomatic infection to ruffled fur, conjunctivitis, ascites, somnolescence, and death. If mice survive, recovery may take several weeks.

Late onset disease occurs in persistently infected mice when they are 9–12 months old. Clinical signs are nonspecific and include ruffled fur, hunched posture, weight loss, proteinuria, and ascites. Late onset disease is usually the result of prenatal or neonatal infection.

Runting and death from LCM virus infection may occur in neonatally infected suckling mice and can lead to transient illness or to death. Clinical signs are nonspecific, recovery is slow, and survivors may remain smaller than unaffected mice. The cause of early deaths is not known, but the syndrome does not occur among offspring of congenital carriers.

Mice, hamsters, and humans are natural hosts for LCMV. Wild mice are a natural reservoir of infection and carrier laboratory mice can infect animal colonies.

Lymphocytic choriomeningitis virus is distributed widely in North and South American and in Europe, but some areas of the world, such as Australia, appear to be free of infection. Although other common laboratory species, such as the guinea pig, are occasionally infected with LCMV, only the mouse and hamster are known to transmit virus. Carrier mice can have persistently high concentrations of virus in many organs, including kidney and salivary gland. Mice can shed virus in saliva, nasal secretions, and urine and can infect contact-exposed mice. Horizontal spread is facilitated by close contact, but rapid horizontal transmission is not characteristic. Self-limiting, immunizing infections occur in adults whereby further spread of virus is halted. If virus infects neonates, persistent infection is more commonly established. Persistently infected neonates usually reach breeding age and can perpetuate infection in a breeding colony. Thus introduction of a single LCMV carrier mouse to a breeding colony can eventually result in a high prevalence of persistently infected mice. Transmission *in utero,* is also a primary mode of spread. Horizontal transmission occurs commonly among hamsters. Mice transmit LCMV to hamsters, which in turn transmit virus to other hamsters, mice, and humans. Although hamsters can remain viremic and viruric for relatively long periods (months), most appear to eventually clear virus and recover (see Chapter 5).

Central nervous system lesions follow intracerebral inoculation of virus but are not usually seen during natural infection. Nonsuppurative leptomeningitis, choroiditis, and focal perivascular lymphocytic infiltrates are characteristic, but diffuse encephalitis does not occur. The mononuclear cell response during which host tissues are damaged reflects cellular immunity to the virus. The character of visceral lesions depends greatly on virus strain and mouse strain. The severity of cytolytic lesions seems to parallel the intensity of cellular immunity. In severe infection, inflammatory lesions may be widely distributed in many organs. Liver lesions can include hepatocytic necrosis accompanied by nodular infiltrates of lymphoid cells and Kupffer cells, activated sinusoidal endothelium, an occasional granulocyte or megakaryocyte, and fatty metamorphosis. Cytolysis, cell proliferation, and fibrinoid necrosis can develop in lymphoid organs. Necrosis of cortical thymocytes can lead to thymic involution. The ratio of cytolytic to proliferative responses in lymphoid organs is also mouse strain dependent.

Lesions of late onset disease are strain dependent and characterized by formation of immune complexes. Renal glomeruli (Fig. 23) and the choroid plexus are most severely affected, but complexes may also be trapped in synovial membranes, blood vessel walls, and skin. Lymphoid nodules can form in various organs.

Lesions associated with early deaths in neonatally infected mice have not been thoroughly described but include hepatic necrosis.

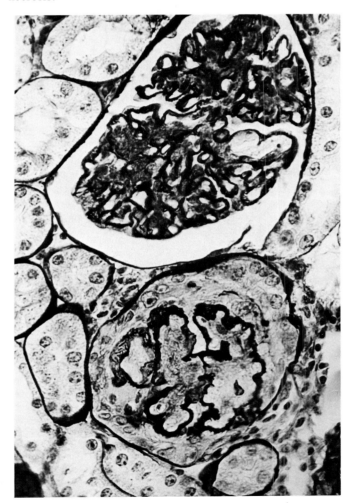

Fig. 23. Thickening of glomerular basement membrane in late onset LCM disease. From Lehmann-Grube (1982). (Photograph supplied to Dr. Lehmann-Grube by Dr. J. Löhler, Hamburg, Federal Republic of Germany.)

The lesions of acute and persistent LCM infection reflect separate immunopathological processes. In adult mice with acute LCM, virus multiples in B cells and macrophages whereas T cells are resistant. An internal viral antigen induces CF antibody, whereas surface antigens induce neutralizing antibody and probably cellular immunity. Thus elimination of virus and virus-associated immunological injury are both T cell mediated. This apparent paradox has been explained by the view that prompt cellular immunity limits viral replication and leads to host survival, whereas slower cellular immune responses permit viral spread and increases the number of virus-infected target cells subject to attack once immunity is fully developed. Antibody can be detected by 1 week after infection but does not play a significant role in acute disease.

Lesions of LCM appear to develop from direct T cell-mediated damage to virus-infected cells and may involve humoral factors released from immune effector T cells. Lymphocytic choriomeningitis versus can also suppress humoral and cellular immunity in acutely infected mice, but the mechanism is not fully understood.

Persistent infection commonly evolves from exposure early in pregnancy, and virus has been demonstrated in the ovaries of carrier mice. Prenatal or neonatal infection induces immunological tolerance to LCMV, which can then replicate to high titer in many tissues. Nevertheless, persistently infected mice develop humoral antibody to LCMV. Antibody can complex with persistent virus to elicit complement-dependent inflammation in small vessels. Immune complex glomerulonephritis, as described above, is a good example of this process.

Lymphocytic choriomeningitis virus infection can be diagnosed serologically. Complement fixation and NT tests have been used with varying success, but a specific and sensitive IFA test is currently the preferred serodiagnostic method. Since carrier mice are persistently viremic, small blood samples (0.3 ml) can be collected from live suspects and used to inoculate cultured cells or for intracerebral inoculation of SPF mice. Intracerebral inoculation of LCMV-positive tissues will kill adult mice, but not neonatal mice. Adults develop neurological signs within 10 days after inoculation or the donor may be assumed to be virus-free. Virus can be detected in brains of suckling and adult mice by immunofluorescence. Virus also can be grown and quantified in several continuous cell lines, including mouse neuroblastoma (N-18) cells, BHK-21 cells, and L cells.

Although acute LCMV is usually asymptomatic, neurological signs must be differentiated from those due to mouse hepatitis virus, mouse encephalomyelitis virus, *Pseudomonas aeruginosa* infection, or parasitism. Trauma, neoplasia, and toxicities also must be ruled out in neurological disease with low prevalence. Late onset disease is associated with characteristic renal lesions, including deposition of viral antigen in tissues. Early onset disease must be differentiated from other causes of early mortality, such as mouse hepatitis, mousepox, reovirus 3 infection, Tyzzer's disease, or husbandry-related insults.

Lymphocytic choriomeningitis virus is not prevalent in laboratory mice produced and maintained in modern quarters. Introduction of infection usually occurs through inoculation of virus-infected biologicals, such as transplantable tumors, or through incursions by feral carriers. Adequate safeguards for procurement of animals and animal products are essential to prevent entry of infection. Surveillance of animal colonies and animal products by serological or virological testing supplements this approach. Because mouse to mouse spread is slow, selective testing and culling for seropositive or carrier mice is possible. If mice are easily replaced, however, depopulation is a safer and more reliable option. Valuable stock can be re-derived, but progeny must be tested to preclude *in utero* transmission. Because infected hamsters can excrete large quantities of virus, exposed hamsters should be destroyed and hamsters should not be housed with mice. Lymphocytic choriomeningitis virus is the primary viral infection of laboratory mice from which humans can contract severe illness. Therefore, it is important to detect and eliminate carrier animals and other potentially contaminated sources, such as cell cultures, transplantable neoplasms, and vaccines, before human exposure occurs (see Chapter 22). Periodic serological testing of high-risk human populations is recommended.

Lymphocytic choriomeningitis virus has been used extensively to study the immunology and immunopathology of viral infection. Lymphocytic choriomeningitis virus may stimulate or suppress immunological responses *in vivo* and *in vitro,* and it can replicate in cells used as targets or indicators for immunological studies. Introduction of immune cells to a carrier animal may elicit an immunopathological response. Immune complex disease can complicate long-term experiments and morphological interpretations. Illness and death in mice and zoonotic risks to humans are obvious hazards of LCMV infection.

e. Sendai Viral Pneumonia

Sendai virus is a paramyxovirus that was isolated in Japan in the early 1950's and is closely related antigenically to parainfluenza 1 virus of humans. There is no firm evidence, however, that cross-infection occurs between humans and mice. All isolates appear to be antigenically homologous but differ from each other in their virulence for cultured cells. Viral particles are pleomorphic, contain single-stranded RNA, and have a lipid solvent-sensitive envelope that contains glycoproteins with hemagglutinating, neuraminidase, and cell fusion properties. Sendai virus grows well on embryonated hen's eggs and in several mammalian cell lines [e.g., monkey kidney, baby hamster kidney (BHK-21) and mouse fibroblast (L)]. Virus replicates in the cytoplasm and by budding through cell outer membranes.

Natural infections occur in mice (see review by Parker and Richter, 1982), rats, hamsters, and guinea pigs, but the latter three species rarely show clinical signs. Ferrets and nonhuman primates can be infected intranasally, but only the former species can develop severe pneumonia. Acute epizootics are common in previously uninfected mouse colonies and are characterized by respiratory distress, neonatal mortality, retarded growth, and prolonged gestation. Weaning rates can decrease dramatically until infection subsides.

Susceptible adult mice typically sit in a hunched position and have an erect haircoat. Rapid weight loss and dyspnea occur, and there may be crusting of the eyes and chattering. Severely affected adults may die.

Infection is commonly more lethal in suckling mice and, to some extent, is more severe in aged mice than in young adults. Sex differences in susceptibility have not been found. Sendai virus is highly infectious, and morbidity in infected colonies is commonly 100%. Mortality in natural epizootics can vary from 0 to 100% partly because strains of mice vary greatly in their susceptibility to lethal Sendai virus infection (Table X). Resistant mice usually have asymptomatic infection, especially if they are otherwise in good health. Athymic mice and immunosuppressed mice are at high risk. However, they develop illness later than their immunocompetent counterparts, and infection can persist for 10 or more weeks.

Sendai virus is transmitted by aerosol or by contact. Airborne infection is promoted by high relative humidity and by low air turnover.

Infection is perpetuated by the introduction of susceptible animals. It may be inapparent or it may be expressed as acute clinical disease or inapparent enzootic infection. Enzootic infection is commonly detected in postweaned mice (5–7 weeks old) and is associated with seroconversion and persistence of circulating antibody for at least a year (Fig. 24). There is no evidence for chronic or persistent infection in imunocompetent mice.

Gross lesions are characterized by partial to complete consolidation of the lungs. Individual lobes are meaty and plum-colored, and the cut surface may exude a frothy serosanguinous fluid. Demarcation between normal and pneumonic zones is usually distinct. Pleural adhesions or lung abscesses caused by secondary bacterial infection are seen occasionally, and fluid may accumulate in the pleural and pericardial cavities.

Histologically, typical changes begin with inflammatory edema of bronchial lamina propria, which may extend to alveolar ducts, alveoli, and perivascular spaces. Necrosis and exfoliation of bronchial epithelium ensues, frequently in a segmental pattern. Alveolar epithelium also may desquamate, especially in severe disease, and necrotic cell debris and inflammatory cells can accumulate in airways and alveolar spaces.

Table X

Susceptibility of Inbred and Outbred Strains of Mice to Sendai Virus Infection[a]

Mouse strain	No. of replicate titrations	$LD_{50} \pm SE$[b] (log 10)
129/ReJ	1	0.5
129/J	4	0.6 ± 0.4
Nude (Swiss)	1	0.7
DBA/1J	3	1.3 ± 0.4
C3H/Bi	1	1.4
DBA/2J	3	1.6 ± 0.3
DBA/2	1	2.0
A/HeJ	3	2.5 ± 0.1
A/J	2	2.5 ± 1.0
SWR/J	1	2.7
Swiss[c]	1	2.7
C57L/J	2	2.7 ± 0.5
C57BL/10Sn	2	2.8 ± 0.6
C3HeB/FeJ	2	2.8 ± 0.1
BALB/cJ	1	3.0
C57BL/6	1	3.0
Swiss[d]	1	3.1
C58/J	1	3.2
AKR/J	3	3.4 ± 0.2
Swiss[e]	1	3.4
Swiss[f]	4	4.4 ± 0.0
C57BL/6J	1	4.4
RF/J	2	5.0 ± 0.5
SJL/J	3	5.0 ± 0.4

[a]From Parker and Richter (1982).
[b]$TCID_{50}/LD_{50}$. SE, standard error.
[c]National Institutes of Health.
[d]Life Sciences.
[e]National Laboratory Animal Co.
[f]Microbiological Associates.

Alveolar septae are usually infiltrated by leukocytes to produce interstitial pneumonia. Lymphoid cells also invade epibronchial and perivascular spaces. Regeneration and repair begin shortly after the lytic phase and are characterized by hyperplasia and squamous metaplasia of bronchial epithelium, which may extend into alveolar septae (Fig. 25). Proliferation of cuboidal epithelium may give terminal bronchioles an adenomatoid appearance.

Repair of damaged lungs is relatively complete in surviving mice, but lymphocytic infiltrates, foci of atypical epithelium, and mild scarring can persist. Multinucleated syncytia are occasionally seen in affected sucklings, and inclusion bodies have been reported in infected athymic mice. In the latter, mortality is often extremely high, and acute phase lesions are prolonged.

After intranasal exposure, virus can be found transiently in extrahepatic tissues, but replication is nominally restricted to the respiratory tract. Parenteral inoculation of Sendai virus can

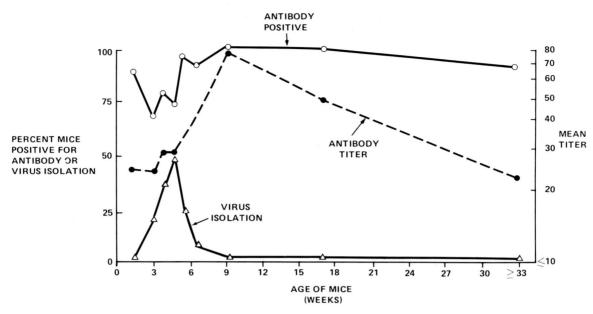

Fig. 24. Sendai virus infection pattern in an enzootically infected mouse breeder colony. Mean titer is the reciprocal geometric mean HAI antibody titer of positive mice. (From Parker and Reynolds, 1968.)

produce infection of parenchymal tissues, but this effect is considered an experimental artifact and is relevent only because such tissues may be inadvertently contaminated with Sendai virus during experimental procedures.

Sendai virus-infected mice are less able to clear bacteria from the lung. This may be due to defective intracellular killing of phagocytized bacteria by Sendai virus-infected pulmonary macrophages. It also helps to explain why secondary bacterial pneumonias can supercede primary Sendai virus infection. Sendai virus infections are self-limiting because of a vigorous immune response, but interferon may also play an early role in recovery. Suckling mice from immune dams are passively immune until

maternal antibody has decayed, usually shortly after weaning. The lymphocytic response to Sendai viral infection suggests that cellular immunity contributes to recovery, but the mechanism is not clear. Local immunoglobulin synthesis by infiltrating cells also occurs. The extent of inflammatory cell infiltration corresponds to the level of genetic resistance expressed by the infected host, susceptible hosts mounting a more florid response than resistant hosts. It has been suggested that this apparent paradox may indicate an immunopathological mechanism in the development of Sendai viral pneumonia.

Because only one serotype is known, serodiagnosis is an effective means to detect exposure to infection. Although several

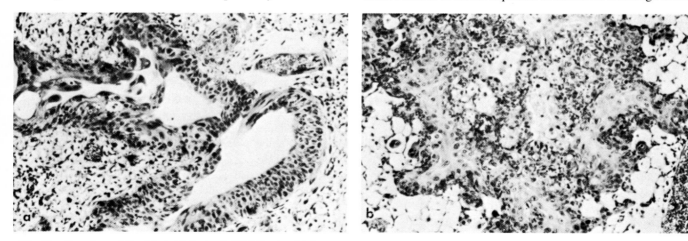

Fig. 25. Experimental Sendai viral infection in DBA/2 mice. (a) Severe squamous metaplasia in bronchioles. (b) Extension of squamous metaplasia from a terminal bronchiole into adjacent alveoli. (From Richter, 1973.)

serological tests are available (HAI or CF procedures have been used traditionally), the ELISA or an IFA procedure offer superior sensitivity and specificity. Antibody can be detected in sera by 7 days postinfection. It should be remembered, however, that seroconversion indicates exposure to virus but does not confirm the stage of infection. Therefore, criteria such as histopathology and virus isolation should be used, or sentinel animals should be added to seropositive colonies to determine if infection is active. Virus can be isolated from the respiratory tract for up to 2 weeks, with peak titers occurring at about 9 days postinfection (Fig. 26). Nasopharyngeal washings or lung tissue homogenates are most reliable and should be inoculated into embryonated hens' eggs or BHK-21 cell monolayer cultures. Alternatively, repeated serological sampling over several weeks can help establish the incidence and prevalence of infection. An increase in both parameters indicates active infection.

Sendai viral pneumonia must be differentiated from other pneumonias of mice. Pneumonia virus of mice (Section III,A,3,f) is generally milder and asymptomatic. Bacterial pneumonias of mice are sporadic and can be differentiated morphologically and culturally. The same is true of murine respiratory mycoplasmosis. Because Sendai viral pneumonia may predispose the lung to opportunistic bacterial infections, the presence of bacteria should not deter evaluation for a primary viral insult.

Control and eradication measures must eliminate exposure of susceptible animals so that infection can "burn itself out." This is most easily accomplished by a quarantine period of 4–6 weeks wherein no new animals are introduced either as adults or through breeding. Physical facilities that permit specific pathogen-free husbandry are preferred for prevention and for control of room to room spread. If Sendai virus-free mice are held in conventional quarters, laminar airflow cabinets or isolators can be used effectively when they are combined with proper sanitation and traffic patterns. The recent introduction of filter frame cages is a promising alternative for protecting animals housed in conventional environments.

Vaccination with formalin-killed Sendai virus can provide short-term protection of valuable mice (e.g., breeding stock). It is likely that the killed vaccine will be supplanted with a temperature-sensitive mutant strain of virus. The advantages of vaccination must be weighed carefully in light of experimental objectives and the ability to interpret serological results of surveillance procedures.

Sendai virus may have broad and prolonged effects on immune responsiveness, particularly in the direction of immunosuppression. For example, it seems to interfere with the ability of mouse lymphocytes to respond to mitogenic stimuli. Sendai virus can also inhibit growth of transplantable tumors. This effect has been attributed to virus-induced modification of tumor cell surface membranes. Pulmonary changes during Sendai viral pneumonia can compromise interpretation of experimentally induced lesions and may lead to opportunistic infection. The use of genetically susceptible mice and nonimmune young stock for research must allow for potential hazards of Sendai virus epizootics.

f. Pneumonia Virus of Mice (PVM) Infection

Pneumonia virus of mice is a heat labile pneumovirus of the family Paramyxoviridae. It is a single-stranded RNA virus that replicates in the cytoplasm and buds from cell outer membranes. All isolates appear to have similar physicochemical,

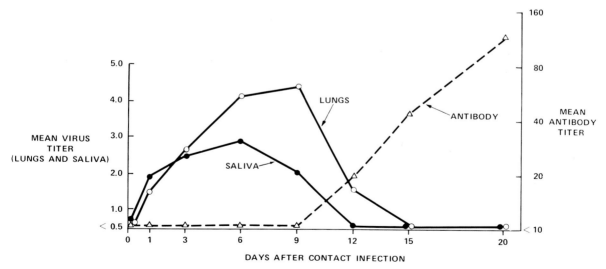

Fig. 26. Sendai virus titers in the lungs and saliva of infected mice. Mean antibody titer is the reciprocal geometric mean HAI antibody titer of positive mice. Virus titers of positive specimens are expressed as negative log $TCID_{50}$ per 0.05 ml. (From Parker and Reynolds, 1968.)

biologic, and antigenic properties. The virus agglutinates erythrocytes of several rodent species including mice. Pneumonia virus of mice replicates exclusively in the respiratory tract and reaches peak titers in the lung 6–8 days after infection. It replicates well in vitro in BHK-21 cells.

Pneumonia virus of mice causes natural infections of mice (see review by Parker and Richter, 1982), rats, hamsters, and probably other rodents. Serological evidence suggests that it also may be infectious for rabbits and nonhuman primates. Similarly, impressions that inbred strains of mice differ in susceptibility to infection have not been studied systematically. Natural PVM infection in mice is asymptomatic. Experimentally infected mice may lose weight, develop listlessness, and have clinical signs of pneumonia.

Serological data indicate that PVM is highly prevalent in mice and that it has a world-wide distribution. However, it appears to spread less rapidly than Sendai virus in a given colony. Infection is thought to be acute and self-limiting. Because the virus is rapidly inactivated, intimate contact between mice is probably required for effective transmission.

Pulmonary consolidation can occur in experimentally infected mice, but gross lesions are not easily detected during natural infection. A recent report describes lesions associated with a natural outbreak of PVM in mice. Histological lesions consisted of desquamation of bronchial epithelium with formation of bronchial plugs, hyperemia and edema of alveolar walls, and alveolar septal infiltration by neutrophils and macrophages. Immunofluorescence revealed viral antigen in bronchial epithelium, alveolar macrophages, and possibly alveolar epithelium.

The diagnosis of PVM is based primarily on serological detection that can be supplemented by histopathology and by virus isolation. The HAI test using mouse erythrocytes has proved highly reliable. Titers of 1:20 or greater from unpooled samples are considered significant. An ELISA has recently been developed and is purported to be up to eight times more sensitive than HAI for detection of seroconversion. This may be important for the detection of PVM since HAI titers are frequently low. An IFA test also is available. Virus isolation in BHK-21 cells is detected by cytopathic effect, hemagglutinin production, or hemadsorption. Viral antigen can also be detected by immunofluorescence.

The differential diagnosis of PVM is problematical, since it is clinically asymptomatic and lesions are not well characterized. It must be differentiated from other viral pneumonias (e.g., Sendai viral pneumonia). Because PVM is antigenically distinct from other murine viruses, serology is the most useful method to separate it from other respiratory infections.

Pneumonia virus of mice is an acute, self-limiting infection, therefore control and prevention follow guidelines used for Sendai virus. Vaccination is not available.

g. K Virus Infection

K virus is a papovavirus that is antigenically distinct from polyomavirus. It contains double-stranded DNA that is stable for more than 2 weeks at room temperature and that is relatively resistant to heat and chemical inactivation. Reliable cell culture systems for in vitro cultivation have not been developed, but the virus replicates in mice.

Natural disease is asymptomatic. Inoculation of virus or virus-contaminated materials into neonatal mice can, however, provoke dyspnea and lead to death in less than 24 hr (see review by Parker and Richter, 1982). Resistance to lethal experimental infection develops between 8 and 18 days postpartum.

K virus is thought to be distributed worldwide. The prevalence of infection in individual colonies can vary, but serological surveys suggest it is usually low. There is some sentiment, however, that serological testing may not accurately determine the prevalence of latent enzootic infection.

The natural history of K virus infection is not well understood. Immune mothers confer passive immunity on their litters, therefore infant mortality is rare. After weaning, mice sustain asymptomatic infection that may be prolonged and associated with excretion of virus in the urine, feces, and perhaps saliva. Virus can be detected in a variety of tissues and fluids, including the mammary glands, saliva, lung, liver, spleen, intestinal contents, blood, and urine. Oral–fecal transmission may be important, because virus can replicate in intestinal epithelium.

Gross lesions are limited to the lungs and can include hemorrhage, congestion, edema, atelectasis, consolidation, and pleural effusion. Histological changes occur in endothelial cells of small pulmonary and hepatic blood vessels. These cells have swollen nuclei with prominent intranuclear inclusion bodies that are amphophilic or basophilic when stained with hematoxylin and eosin (Fig. 27). Inclusion bodies may also be found in renal glomeruli and in jejunal villous endothelium. Infected endothelial cells swell and desquamate and can cause vascular occlusion. Pulmonary vascular lesions may be accompanied by alveolar septae thickened by mononuclear infiltrates. Alveoli in experimentally infected mice may fill with proteinaceous material. Peribronchiolar and perivascular lymphoid aggregates are not typical of K virus. Livers of experimentally infected mice can develop a ''Swiss cheese appearance'' due to formation of membrane-lined spaces. Hepatocellular necrosis and inflammation may occur, but they must be carefully differentiated from intercurrent infection with other agents such as mouse hepatitis virus and ectromelia virus. The presence of intranuclear inclusion bodies in K virus-infected endothelial cells is a helpful marker.

Diagnosis is made from clinical signs, virus isolation, histopathology, and serology. Complement fixation and HAI tests

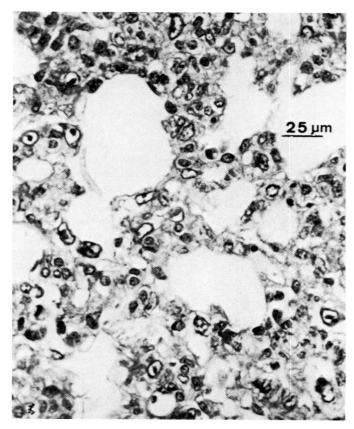

Fig. 27. Mouse lung infected with K virus. Alveolar walls are thickened with a sparse inflammatory infiltrate. Intranuclear inclusions are present in many cells within alveolar walls. From Parker and Richter (1982). (Photograph supplied by Dr. J. E. Greenlee and the American Society for Microbiology.)

can be used, but the former appears to be more sensitive. Antibody titers of 1 : 10 or greater are considered significant for either test. Recently developed immunofluorescence tests are also useful.

K virus infection is commonly latent and chronic with low prevalence and low antibody titers, therefore selection of mice for serological sampling must be done carefully. It may be necessary to test large numbers of animals at frequent intervals to conclude that a population is free of infection.

Virus isolation is relatively easy. Suckling mice are inoculated (usually intracerebrally) followed by demonstration of sheep erythrocyte agglutinins in suspensions of visceral tissue and by specific inhibition of hemagglutination with anti-K virus reference serum. A recently developed immunofluorescence technique also is useful for demonstrating viral antigen.

Because infection is persistent and asymptomatic, rederivation is preferred if K virus must be eliminated. Maintenance of animals in strict barrier conditions can prevent reinfection.

Transplantable neoplasms can become contaminated with K virus during passage in infected mice. K virus also may exacerbate mouse hepatitis virus infection.

h.   Epizootic Diarrhea of Infant Mice (EDIM)

The causative agent, EDIM virus, is a rotavirus of the family Reoviridae. The term rotavirus is derived from the wheel-like ultrastructural appearance of the virus. Currently, only a single antigenic strain is recognized, but experience with rotaviruses of other species suggests that antigenically distinct variants exist. The EDIM virus shares an inner capsid antigen with other rotaviruses but can be differentiated from other rotaviruses by serum neutralization tests. The virus contains RNA and replicates in epithelial cells of the small intestine by budding into cisternae of endoplasmic reticulum. Intestinal infection may be related to a receptor on the villous brush border of the intestine. The EDIM virus is difficult to grow *in vitro* despite attempts using cell lines from several species embryonated hen eggs or mouse intestinal organ cultures. The virus is heat labile.

Clinical signs are prevalent only in suckling mice (see review by Kraft, 1982). Mild disease is characterized by fecal soiling of the perineum and, in severe cases, of the entire pelage. Some mice may develop rectal impaction and die, but high morbidity and low mortality are more characteristic. Affected mice continue to nurse, but weight loss does occur and there may be delays in reaching adult weight.

Epizootic diarrhea of infant mice virus appears to be infectious only for mice. All ages and both sexes can be infected, but genetic resistance and susceptibility have not been determined. The virus is highly infectious and is transmitted by oral contact. Fomites, arthropods or humans, cannot be excluded as passive vectors. Transplacental transmission has not been demonstrated. Asymptomatically infected adult mice can shed virus in feces for at least 17 days. After oral inoculation, virus is essentially restricted to the gastrointestinal tract, particularly the small and large intestine, although small amounts of virus may be present in liver, spleen, kidney, and blood. Nursing dams can contract infection from their litters.

The virus is probably distributed worldwide, but its prevalence in mouse colonies has not been systematically investigated primarily due to the absence of reliable serological tests. Some seasonal periodicity in attack rates has been observed; the incidence being higher during winter.

Gross lesions are primarily limited to the gastrointestinal tract, but thymic atrophy is common. Fecal soiling of hair is caused by yellow to gray-green gaseous liquid or mucoid feces that also distends the colon. The stomach contains curdled milk except in terminal cases with anal impaction. Subtle histological changes occur in the small intestine. Increased vacuolation of villar epithelial cells with cytoplasmic swelling give villi a clubbed appearance and must be differentiated from normal absorption vacuoles in nursing mice. Fuchsinophilic intracytoplasmic inclusions have been reported in villar epithelium during early stages of disease, but are, at best, a variable finding. Necrosis and inflammation are not characteristic of EDIM.

Epizootic diarrhea of infant mice is diagnosed from clinical signs, histology, and demonstration of virus. Until recently, reliable serological tests have not been available, but a new immunofluorescence test and an ELISA may fill this deficiency (Sheridan *et al.*, 1983). Virus can be detected by inoculation of suckling mice with clarified intestinal suspensions or by placing sentinel litters in suspect colonies. Intestinal filtrates or smears can also be examined ultrastructurally by conventional or immune electron microscopy.

Epizootic diarrhea of infant mice must be differentiated from other diarrheal diseases of suckling mice such as intestinal coronavirus infection (mouse hepatitis), reovirus 3 infection, Tyzzer's disease, and salmonellosis. The possibility of dual infections must also be considered. Thymic atrophy in EDIM, although nonspecific, must be differentiated from that due to MHV or MTV infection.

The spread of EDIM is effectively controlled by the use of filter bonnets and good sanitation. Since disease appears to be acute and self-limiting, cessation of breeding for 4–6 weeks to allow immunity to build in adults while preventing access to susceptible neonates is also effective. Alternatively, litters with diarrhea can be culled in combination with the use of bonneted cages. Litters of immune dams are more resistant to infection. Prevention of EDIM infection depends on maintenance of sanitary barrier housing.

i. Reovirus 3 Infection

Reovirus 3 is an orthoreovirus of the family Reoviridae. A number of wild-type and laboratory strains have been characterized, and related viruses have been recovered from mammals, marsupials, birds, insects, and reptiles. The virion contains double-stranded RNA and is relatively heat stable, but temperature-sensitive mutants have also been developed. Reovirus 3 can be distinguished antigenically from reoviruses 1 and 2. Reoviruses replicate well in L cells and other continuous cell lines as well as in primary monolayer cultures from several mammals.

Clinical expression of reovirus infection is age dependent. Acute disease affects sucklings and weanlings, chronic disease is encountered in mice older than 28 days, and adults commonly have asymptomatic infection. Clinical signs of acute disease include emaciation, abdominal distension, and oily, matted hair due to steatorrhea from decreased lipase and amylase activity. Icterus may develop in the feet, tail, and nose. Incoordination tremors and paralysis occur just before death. Convalescent mice are often partially alopecic and are typically runted. Alopecia, runting, and icterus may persist for several weeks, even though infectious virus can no longer be recovered (see reviews by Kraft, 1982; Stanley, 1978).

Reovirus 3 can be transmitted orally or parenterally. There is no evidence that vertical transmission is important or that genetic resistance or sex influence expression of disease. The true prevalence of reovirus 3 infection in laboratory mice is unresolved, but it is known to exceed 80% in some colonies.

Infection can be pantropic in mice. After parenteral inoculation of sucklings, virus can be recovered from the liver, brain, heart, pancreas, spleen, lymph nodes, and blood vessels. Following ingestion, reoviruses gain entry by infecting intestinal epithelial cells (M cells) that cover Peyer's patches. Virus can be carried to the liver in leukocytes where it is taken up by Kupffer cells prior to infecting hepatocytes.

In acute disease, livers may be large and dark with yellow foci of necrosis. The intestine may be red and distended, and, in infants, intestinal contents may be bright yellow. Myocardial necrosis can evoke pale epicardial foci, and pulmonary hemorrhages have been reported. Myocardial edema and necrosis are especially prominent in papillary muscles of the left ventricle. The brain may be swollen and congested.

Central nervous system lesions are most prevalent in the brainstem and cerebral hemispheres. Neuronal degeneration and necrosis are followed quickly by meningoencephalitis and satellitosis. Severe encephalitis may evoke focal hemorrhage. In the chronic phase, wasting, alopecia, icterus, and hepatosplenomegaly may persist.

Gastrointestinal lesions are often mild and may be limited to lymphatic dilatation. On the other hand, orally infected suckling mice can develop multifocal hepatocytic necrosis, which may include the accumulation of dense eosinophilic structures resembling Councilman bodies. Hepatocytomegaly, Kupffer cell hyperplasia, and intrasinusoidal infiltrates of mononuclear cells and neutrophilic leukocytes can develop. In experimentally inoculated mice, necrotic foci can persist in the liver for at least 4 weeks. Chronic active hepatitis may develop after acute infection and result in biliary obstruction.

Acinar cells of the pancreas and salivary glands can undergo degeneration and necrosis. Because pancreatic duct epithelium is susceptible to infection, parenchymal lesions in pancreas may be caused by obstruction rather than by viral invasion of parenchyma. Pulmonary hemorrhage and degeneration of skeletal muscles also has been observed.

Both humoral and cellular immunity seem to participate in host defenses. How this influences chronic infection is not clear.

The diagnosis of infection is aided by detection of the oily hair effect, a phenomenom that is said to be typical for reovirus 3, especially if it is accompanied by jaundice and wasting.

Histological lesions characterized by necrosis and inflammation of liver, pancreas, salivary gland, heart, or brain are also helpful but must be differentiated from those caused by mouse coronaviruses, ectromelia virus, or *Bacillus piliformis*. Serological detection is used routinely because infection is often asymptomatic, especially in adults. Unfortunately, the commonly used HAI test is subject to false negative and false positive results. Neutralization tests are more accurate but are expensive and time consuming. Recently, ELISA and IFA tests have been developed and seem to offer improved specificity and sensitivity. Reovirus 3 can be recovered by inoculation of mouse L cells and several other continuous cell lines.

Since the course of typical infections is not well documented, especially for inbred mice, the potential for a carrier state is unresolved. Therefore, it may be necessary to rederive or replace infected stock. Prevention depends on adequate barrier husbandry coupled with regular serological monitoring.

Reovirus 3 infection can interfere with research in at least several ways. Infections in breeding colonies can result in high mortality among sucklings from nonimmune dams. Virus has been recovered from transmissible neoplasms and is suspected of being oncolytic. The potential exists for interference with hepatic, pancreatic, cardiovascular, or neurological research.

j. Mouse Coronavirus Infection (Mouse Hepatitis)

Mouse coronaviruses are pleomorphic, enveloped RNA viruses with radially arranged peplomers characteristic of coronaviruses. They were first recognized through studies of the neurotropic and hepatotropic strain, JHM. A number of strains have since been identified and differ from one another in virulence, tissue tropism, and antigenicity. They have been grouped for convenience under the name mouse hepatitis virus (MHV), even though hepatitis does not always occur during natural infection. Five prototype strains are commonly referred to: JHM (MHV-4), MHV-1, MHV-3, MHV-S, and MHV-A59. Mouse hepatitis virus strains share internal antigens that can be detected by CF tests, but the viruses can be distinguished from each other by neutralization tests for strain-specific envelop antigens. Mouse hepatitis virus shares complement-fixing antigens with rat coronaviruses, a finding that has been exploited to develop heterologous antigens for serological tests. Several strains of MHV also are related to human coronaviruses C38, OC43, and HCV-229E. It has been suggested that antibody to MHV in human sera can result from infection with antigenically related coronaviruses, but this has not been confirmed experimentally.

Murine coronaviruses develop exclusively in cytoplasm and bud into cytoplasmic cisternae. Several tissue culture systems are available for *in vitro* propagation of virus. A line of mouse liver cells, NCTC 1469, is particularly useful for growing most strains. Cytopathic effects may differ from strain to strain, but syncytium formation is a commonly observed change. Mouse hepatitis virus can also be grown in mouse macrophages, cells that have been used for genetic studies of resistance and susceptibility to MHV-2.

Clinical signs of MHV infection depend on a number of factors, including age and strain of mouse, virus strain and tropism, and the presence of enhancing or inhibitory factors in mice or their environment (see review by Kraft, 1982). Acute MHV is most prevalent in young mice. Suckling mice can develop diarrhea, inappetance, dehydration, weight loss, lassitude, and ruffled hair (Fig. 28). These signs are seen in various combinations and often terminate in death. A virus previously referred to as lethal intestinal virus of infant mice (LIVIM) is most likely an enterotropic variant of MHV and produces signs identical with many of those just listed. Neurotropic variants such as JHM may induce flaccid paralysis of the hindlimbs. Conjunctivitis, convulsions, and circling may be seen occasionally. Mouse hepatitis virus in athymic (*nu/nu*) mice can be particularly devastating and usually causes a wasting syndrome that may be accompanied by generalized progressive paralysis. Mouse hepatitis virus infection, especially in adults, also can be asymptomatic and is often detected by seroconversion or by its influence on research.

Mouse hepatitis is, for all practical purposes, an infection of mice. However, there is a recent report that suckling rats developed necrotizing rhinitis after intranasal inoculation with MHV-S. Resistance to severe or lethal infection decreases as mice mature, but sex differences in susceptibility or seasonal periodicity have not been found. Mouse hepatitis virus has a world-wide distribution, and infection rates for individual

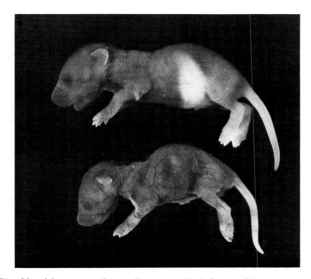

Fig. 28. Mouse pups from a litter naturally infected with an enterotropic strain of MHV. Upper mouse appears normal and stomach is filled with milk, whereas its littermate (lower pup) is runted, dehydrated, and has an empty stomach. (From Barthold *et al.*, 1982.)

colonies can vary from 20 to 100%. Natural transmission can occur through respiratory or oral routes. Feces, nasopharyngeal exudates, and perhaps urine can serve as sources of infection. Experimental data suggest that vertical transmission is possible, but natural vertical transmission has not been confirmed.

The course of natural MHV varies with virus strain and host strain. Prolonged infection in experimentally inoculated mice suggests that viral persistence in asymptomatic carriers is a potential means for spread of virus in a colony. The continuous introduction of susceptible offspring helps to propagate infection. Transmission of MHV by injection can also be an important epizootiological factor in the laboratory.

Strain differences in resistance and susceptibility can be inherited as an autosomal dominant trait. C3H/RV mice, for example, are resistant to MHV-2, but are "semi-susceptible" to MHV-3. During infection with the latter strain of virus, mice that survive acute disease develop chronic illness with wasting and occasionally become paralyzed. DBA/2 mice are highly susceptible to MHV-3 and die acutely even as adults, whereas A/J mice develop resistance to lethal infection shortly after weaning. Some work suggests that susceptibility to MHV is influenced by at least two major genes, one which modulates acute disease and a second, linked to the *H-2* complex, that modulates chronic disease. Mildly pathogenic strains may not cause acute disease in athymic mice, but rather a progressive wasting syndrome.

The distribution and severity of lesions in MHV depend on multiple factors, among which are viral tropism and host age. In susceptible weanlings and adults, yellow-gray foci of hepatic necrosis are seen with varying frequency. Icterus, san-

guinous peritoneal exudates, or intestinal hemorrhage may accompany hepatic lesions. In suckling mice, focal spotting of the liver may occur, but intestinal lesions are more common. The stomach is often empty, and the intestine is filled with watery to mucoid yellowish, sometimes gaseous contents. Hemorrhage or rupture of the intestine can occur.

Histologically, hepatic necrosis can be focal or confluent and may be infiltrated by inflammatory cells (Fig. 29). Syncytia commonly form at the margin of necrotic areas and, in mild infections, may develop in the absence of frank necrosis. Intestinal lesions can be found at all levels and range from syncytium formation to necrosis and inflammation with severe blunting of surviving villi. Syncytia can often be found in asymptomatic adults on careful examination of intestinal mucosa (Fig. 30). Necrosis and syncytia have also been detected in spleen, stomach, lymph nodes, and pancreas. In athymic mice, syncytia occur in many tissues, and hepatic necrosis can be extensive.

Neurotropic variants, such as JHM, produce central nervous system lesions after invasion of the nasal passages. Necrosis predominates in the hippocampus and olfactory lobes, whereas demyelination, secondary to viral invasion of oligodendroglia, occurs in brainstem and in periependymal areas.

Hepatotropic strains of MHV cause damage to liver after parenteral inoculation by extension from littoral cells to hepatocytes. Neurotropic strains reach the nervous system after parenteral inoculation but can also penetrate the cribiform plate to the olfactory bulbs after initial replication in nasal mucosa. Recent work also indicates that the respiratory tract is a major portal of entry for MHV and primary lesions (syncytia) in pulmonary vascular endothelium. Enterotropic strains can pro-

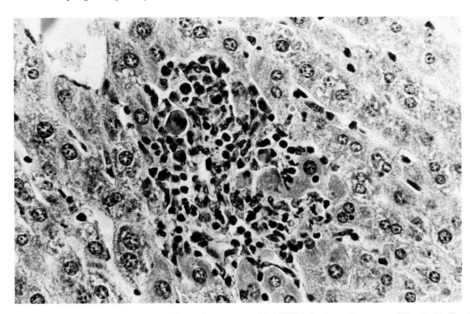

Fig. 29. Necrosis and inflammation in the liver of a mouse with MHV infection. (Courtesy of Dr. S. W. Barthold.)

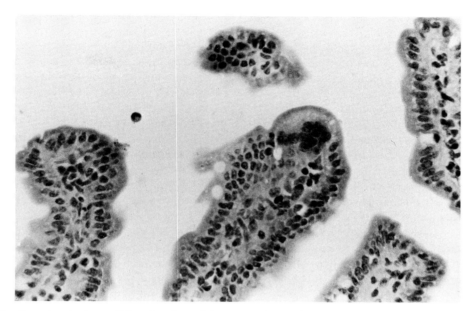

Fig. 30. Syncytium in the small intestine of an adult mouse with MHV infection. (Courtesy of Dr. S. W. Barthold.)

duce syncytia and necrosis in the intestine after oral exposure and can contaminate the hepatic portal system to produce lesions in the liver and elsewhere.

Humoral and cellular immunity appear to participate in host defenses to infection; a primary role being played by macrophages. Age-related resistance to MHV correlates with maturation of lymphoreticular subpopulations. The role of interferon in MHV infection is unclear, but it has been suggested that susceptibility of suckling mice is caused by low serum levels of interferon coupled with macrophage sensitivity to viral replication. Mouse hepatitis virus-induced interferon can also modify responses of mice to nonreplicating antigens such as sheep erythrocytes.

Clinical and epizootiological signs of MHV infection are not pathognomonic for diagnosis. Mouse hepatitis virus must be differentiated from other infectious diseases that cause diarrheal illness, runting, or death in suckling mice. These include EDIM, mousepox, reovirus 3, Tyzzer's disease, and salmonellosis. Neurological signs must be differentiated from mouse encephalomyelitis, *M. neurolyticum* toxicosis, or noninfectious CNS lesions, such as neoplasms or musculoskeletal degeneration.

Sentinel athymic mice can develop clinical disease if they are placed among asymptomatically infected mice. Immunosuppression with adrenal corticosteroids or alkylating agents can also unmask asymptomatic MHV. Asymptomatic MHV infection is, however, commonly detected and differentiated from other murine viral infections by serological tests. The CF test has been used traditionally, but it is relatively insensitive. Newer serological assays, such as ELISA and immunofluorescence, are significantly more sensitive and are easy to perform.

Serum neutralization can be used to detect envelope antigens specific for individual virus strains.

Morphologically, syncytium formation with or without necrosis is highly characteristic of MHV infection. Viral antigen can often be demonstrated in affected tissues by immunocytochemistry; a technique amenable to paraffin-embedded specimens (Brownstein and Barthold, 1982).

Virus can be isolated in mouse liver cell line NCTC-1469. Syncytium formation and necrosis occur within several days, and the careful observer can differentiate among prototype strains by the character of their cytopathic effects.

Control and prevention of mouse hepatitis can be difficult because of the spectrum of virus-related, host-related, and environment-related factors that influence the expression and course of infection. The increased sensitivity of newer serological tests presents an additional problem in that MHV is being detected in colonies previously thought to be free of infection. Another major but ill-defined variable is the potential for persistent infection. Some virus–host combinations are expressed as acute self-limiting infections, whereas other combinations appear to elicit chronic infection. Recent *in vitro* studies indicate that cells can be persistently infected with MHV in the presence of anti-MHV antibody. Therefore, control measures for individual outbreaks should account for the possibility of persistent infection.

Methods for control of MHV should be tailored to research objectives. Valuable breeding colonies may benefit from cesarean rederivation if vertical transmission of infection can subsequently be ruled out. Depopulation, disinfection, and restocking of infected rooms is suitable where replacement animals can be obtained commercially. Short-lived infection in a breeding colony would be amenable to a brief cessation of breeding to permit immunity to develop. However, this procedure would be relatively ineffective for persistent infection. Control of feral mouse populations, proper husbandry and sanitation, and strict monitoring of biological materials that may harbor virus (e.g., transplantable neoplasms) will help prevent entry of virus. The prevention of MHV follows recommendations common to many murine virus infections; procure animals from virus-free colonies and maintain them under strict barrier conditions that are monitored by a well-designed quality assurance program.

Because MHV is ubiquitous and often clinically silent, it is a major potential influence on research using mice. It may immunosuppress or stimulate immune responses, contaminate transplantable neoplasms, and be reactivated by treatment of asymptomatically infected animals with several classes of drugs including immunosuppressive agents and by intercurrent infections (K virus, *Eperythrozoon coccoides*). It also can alter liver enzyme levels.

k. Minute Virus of Mice (MVM)

Minute virus of mice is a parvovirus that was originally identified as a contaminant of a stock of mouse adenovirus. It is highly contagious and highly prevalent in wild and laboratory mice. Because it is small and contains single-stranded DNA, it has also been studied extensively at the molecular level as a model for viral genetics and for viral pathogenetic mechanisms (see review by Ward and Tattersall, 1982). It agglutinates mammalian erythrocytes, and each viral serotype displays specificity for the species of red cells it will agglutinate. The virus replicates in monolayer cultures of rat or mouse embryo cells, and resulting cytopathic effects include development of large intranuclear inclusion bodies. Minute virus of mice is antigenically distinct from rat parvoviruses. However, antisera to rat virus (Kilham) and to H-1 virus, can detect MVM-infected rat cells *in vitro*. This suggests that a common murine parvoviral antigen is expressed during infection of rat cells.

Natural MVM infection is essentially asymptomatic. Young mice in enzootically infected colonies are protected by maternal antibody, sustain active infection as weanlings, and develop active immunity. Virus is excreted in feces and urine and is very stable to drying. Transmission occurs by oronasal exposure. Continuous contact exposure produces seroconversion within 3 weeks, but limited contact exposure results in delayed seroconversion. Seroconversion is often accompanied by prolonged elevation of circulating antibody titers, an effect indicative of persistent and probably lifelong infection. Infectious virus can, in fact, be isolated from mice with high HAI antibody titers.

Minute virus of mice is only moderately pathogenic for mice, but it is teratogenic in neonatal hamsters and produces mongoloid deformities. Retarded growth and granuloprival cerebellar hypoplasia can develop after intracerebral inoculation of neonatal mice. Contact exposed neonates occasionally develop cerebellar lesions. Passively immune suckling mice can pass through the age of maximum susceptibility without ill effects. Infection of adult mice causes viremia, but lesions are not found. Virus also can replicate in fetal tissues without inducing lesions.

Because clinical signs are unreliable and because virus is essentially nonpathogenic for all but nonimmune neonatal mice, serology is the primary method used to detect infection. The HAI test is widely used but the serum of some inbred mice may contain nonspecific inhibitors that can elicit false positive reactions. Therefore nonspecific inhibition should be excluded during testing procedures. An IFA test can be used to confirm seroconversion since it appears not to suffer from nonspecific inhibition.

Minute virus of mice can be isolated from spleen, kidney, intestine, and other tissues by inoculation of C-6 cells, a rat glial cell line. Virus can also be detected by the mouse antibody production test (see Chapter 26).

Because MVM induces persistent infection, rederivation is a preferred means to eliminate the virus. Control and prevention also depends on strict barrier husbandry. Routine serological surveys should be made of populations at risk.

Minute virus of mice contamination of transplantable neoplasms is quite common, therefore infection can be introduced to a colony through inoculation of contaminated cell lines. Failures to establish long-term cell cultures from infected mice or a low incidence of tumor takes should alert researchers to the possibility of MVM contamination. Because lymphotropic, immunosuppressant strains occur in nature, it is important for experimental oncologists and immunologists to have their animals and cell cultures monitored regularly for MVM.

l. Mouse Encephalomyelitis

The causative agent, mouse encephalomyelitis virus (MEV), is a small RNA-containing enterovirus of the family Picornaviridae. It was discovered by Max Theiler during his studies of yellow fever. Several strains are recognized including TO (Theiler's original) and GD strains I–VII, which are named after George Martin (George's disease), a laboratory technician who worked in Dr. Theiler's laboratory. Mouse encephalomyelitis virus is thought to be moderately to highly prevalent in conventional mouse colonies.

The biological behavior of MEV allows its separation into two subgroups: highly virulent isolates that cause acute infection and less virulent isolates that produce persistent infection of the CNS (see review by Downs, 1982). The virus is rapidly destroyed by temperatures over 50°C and by hydrogen peroxide at 37°C, 50% acetone, or alcohol but MEV is not inactivated by ether. The virus has a hemagglutinin, that is useful for serological testing. It can be cultivated *in vitro* in several continuous cell lines, but BHK-21 cells are routinely used for isolation and propagation.

Clinical responses to MEV infection depend on the virus strain, the route of exposure, and the strain of mouse. The most common response is asymptomatic infection. The characteristic but rarely observed sign of natural infection is flaccid paralysis of the rear legs, while the tail remains mobile (Fig. 31). Paralysis may be preceded by weakness in the forelimbs or hindlimbs. Some mice may recover from paralysis, but death frequently ensues. The course of disease may be exacerbated by a failure to obtain food or water and by urinary incontinence. Mice that recover from the paralytic syndrome are disposed to a chronic demyelinating phase, which is expressed clinically as a mild gait disturbance. The SJL/J strain is more prone to develop severe chronic demyelinating disease than other inbred strains tested thus far. During the poliomyelitic syndrome, virus can be present in the brain and spinal cord for at least 1 year. Virus also replicates in intestinal mucosa and is excreted in feces. However, studies on persistent excretion from the gastrointestinal tract have not been done. Young mice acquire intestinal infection shortly after weaning, but virus can be recovered only irregularly in mice over 6 months of age. Immunity to one strain of MEV engenders cross-protection to the other strains. There are no reports of strain differences in mice with respect to susceptibility to infection under natural conditions or following inoculation.

Mouse encephalomyelitis viruses occur naturally only in colonies of laboratory mice with the exception of the MGH strain, which has been isolated from laboratory rats. There are no reports of MEV in wild mice. Infant cotton rats, hamsters, and laboratory rats are susceptible to intracerebral inoculation.

The poliomyelitis-like disease described above is characterized morphologically by acute necrosis of ganglion cells neurophagia and perivascular inflammation, particularly in the ventral horn of the spinal cord gray matter. In demyelinating disease, mononuclear cell inflammation develops in the leptomeninges and white matter of the spinal cord. Patchy demyelination is seen in areas of inflammation. The white matter lesions are similar to those seen in experimental allergic encephalomyelitis and can be halted by chemical immunosuppression.

Clinical diagnosis is made by recognition of paralyzed mice, an event that occurs rarely (about 1 in 10,000 infected animals). Neurotropic variants of MHV may, on occasion, elicit similar signs. Injury to the spinal cord can also produce posterior paralysis. Infection is usually detected serologically using the HAI test, a test prone to false positive and false negative reactions, or by the serum neutralization test, which is more precise, but which is also expensive and time consuming. A recently developed ELISA shows promise of increased spec-

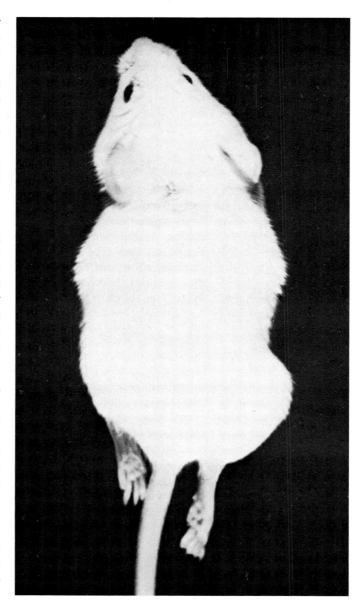

Fig. 31. Mouse with posterior paralysis due to naturally occurring MEV infection From Downs (1982). (Photograph supplied by Dr. R. O. Jacoby.)

ificity and sensitivity. Histological lesions in the CNS and especially the spinal cord are characteristic when present. Virus can be isolated by mouse inoculation with fecal material or by inoculation of BHK-21 cells with intestine or CNS tissue. Brains of paralyzed mice can be tested for MEV hemagglutinin using type O human red blood cells.

Disease-free stocks were originally developed by foster nursing infant mice. This technique or cesarean derivation techniques can be used successfully to eliminate infection. In either case, foster mothers should be surveyed in advance to ensure their MEV-free status. Mouse encephalomyelitis virus appears to spread slowly, and, under some conditions, selective culling can eliminate infection. Because the virus is hardy in the environment and resists chemical deactivation, the safest course is to destroy animals, wash and disinfect the room and equipment thoroughly, and reintroduce virus-free stock.

The principal hazard of MEV for research relates to its potential effects on the CNS.

m. Mouse Adenovirus Infection

Mouse adenovirus is a heat-labile DNA virus that produces intranuclear inclusions in infected cell cultures and animals. Many adenoviruses express hemagglutinins, but murine adenoviruses do not. Some strains of human adenovirus are oncogenic for hamsters, but mouse adenovirus isolates have not shown oncongenic potential.

Two distinct strains of mouse adenovirus have been identified. The FL strain produces a rapidly fatal infection in suckling mice characterized by necrosis of brown fat, myocardium, adrenal cortex, salivary gland, and kidney. Infected tissues also develop type A intranuclear inclusion bodies. Adult or weanling mice are resistant and merely seroconvert. A second strain of virus, designated K-87, is essentially nonpathogenic and tends to localize in the intestine, where it also produces intranuclear inclusions in intestinal epithelium. Both strains will replicate in mouse kidney tissue culture, but are antigenically distinct. Cross-neutralization tests have revealed, however, that antiserum to K-87 neutralizes both virus strains but antiserum to FL only weakly neutralizes K-87. Partial cross-reactivity has also been detected by complement fixation.

Experimentally inoculated mice may remain persistently infected and can excrete virus in the urine for at least 2 years. Spontaneous adenovirus infection has been recorded for athymic mice wherein amphophilic inclusion bodies were found in mucosal epithelium of the small intestine.

The overall prevalence of mouse adenovirus infection appears to be significantly lower than that of other common murine viral agents. Caesarean derivation of infected mice seems to abrogate enzootic infection.

n. Polyoma Virus

Polyoma virus is a small DNA virus in the papovavirus B group. It is highly antigenic in adult mice, whereas neonatally infected mice develop tumors. It is resistant to heat (60°C for 30 min) and chemical disinfectants.

Polyoma-induced tumors are primarily a laboratory phenomenon and seldom occur under conditions of natural infection. Inoculation of neonatal mice with contaminated biologicals or cell cultures is a potential source of spread. Tumors appear 2–12 months after inoculation, and, in most strains of mice, the salivary glands are prevalent sites for tumor development. However, tumors can occur at other sites, especially in skin adnexae, the upper gastrointestinal tract, and the kidneys. The location of tumors varies with virus strain and, to some extent, with the route of inoculation. Intranasal exposure is believed to be the primary natural route of infection, whereas ingestion is not an efficient method of transmission.

Polyoma virus exists in nature most commonly as a silent infection. Infected animals can be detected serologically with the HAI or CF test. Infection can be confirmed by virus isolation. Young mice are important in virus dissemination because virus can multiply to high titer before antibodies develop. Virus also can persist in food and soiled bedding.

Control depends on elimination of infected mice and material together with prevention of airborne spread. Polyoma infection can affect experiments by stimulating or repressing infection with other microorganisms or by eliciting tumors.

5. Parasitic Diseases

a. Invasive Protozoal Infections (see review by Hsu, 1982)

Toxoplasma gondii is a ubiquitous gram-negative coccidian parasite for which the mouse serves as a principal intermediate host, the definitive host being domestic and wild felids. Tachyzoites proliferate in mouse cells and form pseudocysts that lack a well-defined membrane. In contrast to the tachyzoite, the bradyzoite multiplies slowly within a true parasitic cyst. Therefore, it is less susceptible to proteolytic enzymes. These cysts occur in tissues and grow by endodyogeny and represent a resting stage of Toxoplasma. Cysts can contain numerous crescent-shaped bradyzoites and persist in muscular and neural tissues.

The pathogenicity of Toxoplasma infection depends on the number and virulence of infecting organisms and on the route of infection. Natural infections of laboratory mice are usually acquired by ingestion of food or water contaminated with sporulated oocytes from cats, but congenital transmission can also occur. Pregnancy, lactation, immunosuppression, or concurrent infection with other agents exacerbate susceptibility.

Toxoplasmosis can cause necrosis and granulomatous inflammation in the intestine, mesenteric lymph nodes, eyes, heart, adrenals, spleen, brain, lung, liver, placenta, and muscles. Animals commonly recover from natural infection and develop humoral and cellular immunity with cysts being formed in various organs.

The diagnosis of T. gondii infection is based on immunochemical methods (CF, hemagglutination, Sabin–Feldman dye tests, IFA, ELISA, toxoplasmin skin tests, or lymphocyte transformation). Histological examination and experimental inoculation of immunosuppressed animals are also useful.

The prevalence of *T. gondii* infection in laboratory mice is low. Control and prevention depend largely on precluding access of mice to cat feces or to materials contaminated with cat feces. Oocytes are very resistant to adverse temperatures, drying, and chemical disinfectants, therefore thorough cleaning of infected environments is required.

Research complications associated with toxoplasmosis in mice include prolonged immunosuppression, macrophage activation, and enhanced protection of mice against unrelated pathogens. Infection can also alter the development of leukemia, mammary tumors, or carcinogen-induced liver tumors.

Sarcocystis muris has a life cycle similar to that of *Toxoplasma gondii* and is considered extremely rare or nonexistent in mice produced and housed in modern colonies.

Klossiella muris is a renal coccidian of wild and laboratory mice. Mice are infected by ingestion of sporulated sporocysts. Sporozoites released from the sporocysts enter the bloodstream and are disseminated through the body. Sporozoites enter endothelial cells lining renal arterioles and glomerular capillaries where schizogony occurs. Mature schizonts rupture into Bowman's capsule to release merozoites into the lumen of renal tubules. Merozoites can enter epithelial cells lining convoluted tubules where the sexual phase of the life cycle is completed. Sporocysts form in renal tubular epithelium and eventually rupture host cells and are excreted in the urine, but oocysts are not formed. *Klosiella muris* infection is usually nonpathogenic and asymptomatic. Gray spots may occur in heavily affected kidneys and reflect necrosis, granulomatous inflammation, and focal hyperplasia. Destruction of tubular epithelium may impair renal physiology.

Diagnosis is based on detection of organisms in tissues. Prevention and control rest on proper sanitation and management techniques. No known treatment is effective.

Encephalitizoon cuniculi is a gram-positive microsporidian that is frequently found in tissues of mammals, including rabbits, mice, rats, guinea pigs, dogs, nonhuman primates, and humans. The life cycle is direct, and animals are infected by ingesting spores or by cannibalism. Organisms proliferate in peritoneal macrophages by asexual binary fission. They have a capsule that accepts Giemsa and Goodpasture's stain but is poorly stained by hematoxylin. Spore cells are disseminated in the blood to the brain and other sites. Infection can last more than 1 year, and spores shed in the urine serve as a source of infection. Vertical transmission has not been confirmed in mice. *Encephalitizoon cuniculi* is an obligate intracellular parasite but usually elicits no clinical signs of disease. Fulminating infection can, however, cause lymphocytic meningoencephalitis and focal granulomatous hepatitis. In contrast to encephalitozoonosis of rabbits, affected mice do not develop interstitial nephritis.

Encephalitizoon cuniculi is diagnosed by cytological examination of ascitic fluid smears, by histopathological examination of brain tissues stained with Goodpasture's stain, and by various skin tests and serological methods including immunofluorescence, immunoperoxidase, immuno-India ink, complement fixation, and microagglutination. No effective treatment has been reported. Prevention and control rest on rigid testing and elimination of infected colonies and cell lines.

Pneumocystic carinii is a ubiquitous opportunistic organism that is present as a latent infection in many species, including rats and mice. It is normally not pathogenic but can be activated by intercurrent immunosuppression and fills the lung with trophozoites, precysts, cysts, and intracystic bodies. Naturally occurring *P. carinii* infection has, for example, been detected in mice treated with corticosteroids or a low protein diet and spontaneous pneumocystosis has been reported in athymic mice.

Infection is diagnosed by histological examination of lung sections stained by hematoxylin and eosin and by methanamine silver. Activation of lesions usually requires pretreatment with corticosteroid. Serological tests, such as CF and immunofluorescence, also have been used for diagnosis.

Sulfadiazine/pyrimethamine combinations have been used to treat *P. carinii* infection. Because this regimen may arrest granulocyte maturation in the bone marrow through interference with folic and folinic acid biosynthesis, a folinic acid supplement should be added to the treatment regimen. Pneumocystosis control is based on rederivation of colonies and the selection of uninfected animals as breeders.

b. Noninvasive Protozoal Infections (see review by Hsu, 1982)

Giardia muris, a pear-shaped bilaterally symmetrical organism, resides in the upper small intestine of mice, rats, and hamsters. *Giardia muris* replicates in the lumen and adheres to microvilli of columnar crypt cells but does not cause significant enteritis. Nevertheless, heavily infected mice can have weight loss, rough haircoat, sluggish movement, and distended abdomens. Diarrhea is not common, but the small intestines may contain yellow or white watery fluid. C3H/He mice are particularly susceptible to giardiasis, whereas BALB/c mice are more resistant. Athymic mice may die during heavy infestation.

Diagnosis is based on detection of trophozoites in the small intestine or ellipsoidal cysts with four nuclei in the feces or on histological identification of the parasite in tissue sections. Organisms can be recognized in wet preparations by their characteristic rolling and tumbling movements. Murine giardiasis can be treated by the addition of 0.1% dimetridazole to drinking water for 14 days. Chlorocrine, quinacrine, and amodiaquin also are effective. Prevention and control depend on proper sanitation and management including adequate disinfection of contaminated rooms.

Spironucleus muris is a common flagellated protozoan of the small intestine of mice, rats, and hamsters, where it usually inhabits the crypts of Lieberkuhn (Fig. 32). *Spironucleus muris* is elongated, pear-shaped, and bilaterally symmetrical. It is more pathogenic for young, stressed, or immunocompromised animals than for adult animals, and the former group may develop chronic infection. Infected mice can have a poor haircoat, sluggish behavior, and weight loss. Dehydration, hunched posture, abdominal distension, and diarrhea can occur in young mice, and there may be some deaths. Infection has been associated with duodenitis, but it is not clear whether this lesion is a primary pathogenic effect of *S. muris* or if it represents opportunism secondary to a primary bacterial or viral enteritis.

The diagnosis of spironucleosis is based on identification of trophozoites in the intestinal tract. They can be distinguished form *G. muris* and *Tritrichomonas muris* by their small size, horizontal or zigzag movements, and the absence of a sucking disk or undulating membrane. Treatment consists of adding 0.1% dimetridazole in drinking water for 14 days as described for *Giardia*. Prevention and control require good husbandry and sanitation.

Infected mice may have enlarged lymph nodes with activated macrophages that kill tumor cells nonspecifically. Some workers believe that *S. muris*–infected mice are unsuitable for immunological studies because infection is associated with diminished responses to soluble and particulate antigens. Infected mice also have increased sensitivity to irradiation and increased mortality from cadmium exposure. These effects should, however, be interpreted cautiously to rule out intercurrent viral infection.

Tritrichomonas muris is nonpathogenic and occurs in the cecum, colon, and small intestine of mice, rats, and hamsters.

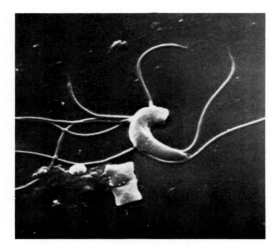

Fig. 32. Spironucleus muris. From Hsu (1982). (Photograph courtesy of Dr. J. E. Wagner.)

Cysts are not formed and transmission is by ingestion of trophozoites passed in the feces.

Eimeria falciformis is a pathogenic coccidian that occurs in epithelial cells of the large intestines of mice. It appears to be common in European mice but is seldom observed in the United States. Heavy infection may cause diarrhea and catarrhal enteritis.

Cryptosporidium muris is a sporozoan that adheres to the surface of the epithelium of the mouse stomach. It is uncommon in laboratory mice and is only slightly pathogenic. *Cryptosporidium parvum* inhabits the small intestine, but it also is nonpathogenic and uncommon in laboratory mice.

Entamoeba muris is found in the cecum and colon of mice, rats, and hamsters throughout the world. Organisms live in the lumen where they feed on particles of food and bacteria. They are considered nonpathogenic.

c. Helminth Infections (see review by Wescott, 1982)

Syphacia obvelata, the common mouse pinworm, is a ubiquitous parasite of wild and laboratory mice. It often occurs in combination with *Aspicularis tetraptera*. The rat, gerbil, and hamster are occasionally infected.

Female worms range from 3.4 to 5.8 mm in length, and male worms are smaller (1.1–1.5 mm). Eggs are flattened on one side and have pointed ends (Fig. 33). The nucleus fills the shell and is frequently at a larval stage when eggs are laid. The life cycle is direct and is completed in 11–15 days. Females deposit their eggs on the skin and hairs of the perianal region. Ingested eggs liberate larvae in the small intestine, and they migrate to the cecum within 24 hr. Worms remain in the cecum for 10–11 days where they mature and mate. The females then migrate to the large intestine to deposit their eggs as they leave the host. There is unconfirmed speculation that larvae may re-enter the rectum.

Infection usually begins in young mice and can recur, but adult mice tend to be more resistant. Because the life cycle of *S. obvelata* is much shorter than that of *Aspicularis tetraptera*, the number of animals that are apt to be infected with *S. obvelata* is correspondingly greater.

Infection is usually asymptomatic, and gross lesions are not prevalent aside from the presence of adults in the lumen of the intestine. Heavily infected mice can occasionally sustain various intestinal lesions including rectal prolapse, intussusception, enteritis, and fecal impaction.

Infestation is diagnosed by demonstrating eggs in the perianal area or adult worms in the cecum or large intestine. Since most eggs are deposited outside the gastrointestinal tract, fecal examination is not reliable. Eggs are usually detected by pressing cellophane tape to the perineal area and then to a glass slide that is examined by microscopy. *Aspicularis tetraptera* ova are not ordinarily found in tape preparations and are easily differ-

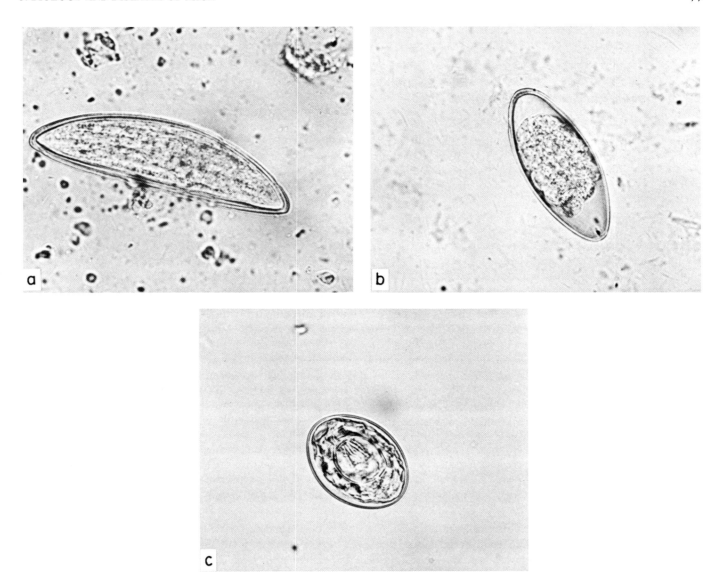

Fig. 33. Eggs of (a) *Syphacia obvelata*, (b) *Aspiculuris tetraptera*, and (c) *Hymenolepis nana*. (From Wescott, 1982.)

entiated from eggs of *S. obvelata* (see below). Adult worms can be found in cecal or colonic contents diluted in a petri dish of warm tap water. They are readily observed with the naked eye or with a dissecting microscope. Four- to 5-week-old mice should be examined because the prevalence is higher in this age group than in older mice.

Pinworm infestation can be treated effectively by adding piperazine citrate at 200–400 mg/kg to drinking water for 1 week followed by no treatment for a week and retreatment for a third week. High efficacy also has been reported for pyrivinium pamoate, stilbazium iodide, phosphate anthelminthics, mebendazole, trichlorphon, newer piperazines, thiabendazole, or dichlorvos.

Prevention of reinfection requires strict isolation because *Syphacia* eggs become infective as soon as 6 hr after they are laid and survive for weeks even in dry conditions. Strict sanitation, sterilization of feed and bedding, and periodic antihelminthic treatment are required to control infection. The use of filter bonnets can also reduce the spread of infective eggs.

Aspicularis tetraptera is the other major oxyurid of the mouse and may coinfect mice carrying *S. obvelata*. Females are 2.6–4.7 mm long and males are slightly smaller. The life cycle is direct and takes approximately 23 to 25 days. Mature females inhabit the large intestine where they survive from 45 to 50 days and lay their eggs. The eggs are laid at night and leave the host in a mucous layer of fecal pellets. They require 6–7 days at 24°C to become infective and can survive for weeks outside the host. Ingested eggs hatch and larvae reach the middle colon where they enter crypts and remain for 4 to 5 days. They move to the proximal colon about 3 weeks after infection of the host. Since the life cycle is 10–12 days longer than *S. obvelata*, infections appear in somewhat older mice; heaviest infestation is expected at 5–6 weeks and can produce signs similar to those discussed for *S. obvelata*. Light to moderate loads do not produce clinical disease.

Aspicularis tetraptera eggs can be detected in the feces and adult worms are found in the large intestine. Eggs are not deposited in the perianal area, therefore cellophane tape techniques are not useful.

Measures for treatment and control are similar to those described for *S. obvelata*. Since *A. tetraptera* takes longer to mature and since eggs are deposited in feces rather than on the host, adult parasites are more amenable to treatment by frequent cage rotations. Immune expulsion of parasites and resistance to reinfection is a hallmark of *A. tetraptera* infestation.

Hymenolepis nana, the dwarf tapeworm, can be a common cestode of mouse, rat and man. Infection rates, as with most parasitic infections, depend on the quality of husbandry. Adults are extremely small (25–40 mm) and have eggs with prominent polar filaments and rostellar hooks. The life cycle may be direct or indirect. The indirect cycle utilizes arthropods as intermediate hosts. Liberated oncospheres penetrate intestinal villi and develop into a cercocystis stage before reemerging into the intestinal lumen 10 to 12 days later. The scolex attaches to the intestinal mucosa where the worm grows to adult size in 2 weeks. The cycle from ingestion to patency takes 20 to 30 days.

Young adult mice are most frequently infected. Signs and lesions include weight loss and focal enteritis, but clinical disease is rare unless infestation is severe.

Hymenolepis nana can be diagnosed by demonstrating eggs in fecal flotation preparations (Fig. 33) or by finding adult worms in the small intestine. Adult worms can be found by opening the intestine in petri dishes containing warm tap water. *Hymenolepis nana* can be differentiated from another species of rodent tapeworm, *H. diminuta*, by the fact that *H. nana* has rostellar hooks and eggs with polar filaments, whereas *H. diminuta* has an unarmed rostellum.

Treatment is usually successful because eggs do not survive well outside the host and because the prevalence of infection is low in caged mice kept in sanitary facilities. Drugs recommended for treatment include quinacrine hypochloride, bunaminidine, niclosamide, and thiabendazole. Newer benzimidazoles have excellent activity against cestodes and nematodes, but they have not been tested in mice. Therefore, the present treatment of choice is bunaminidine at 1000 ppm administered for 18 days in the diet. Because *H. nana* can infect humans, proper precautions should be taken to avoid oral contamination during handling of rodents.

Several helminths infect mice occasionally, but they are not important agents in well-managed colonies. *Syphacia muris* is the common rat pinworm. It can be differentiated from *S. obvelata* because *S. obvelata* eggs are larger. Treatment is the same as for pinworms of mice. *Hymenolepis microstoma* is found in the bile ducts of rodents and could be confused with *H. nana* in the mouse. However, the location of the adult as well as the large size of the eggs compared to *H. nana* make differential diagnosis relatively simple. The mouse and the rat are intermediate hosts of the cestode *Taenia taeniaformis;* the definitive host is the cat. This parasite should not be found in laboratory mice housed separately from cats. A summary of more exotic parasites of the mouse is found in the ACLAM text "The Mouse in Biomedical Research" (Wescott, 1982).

d. Ectoparasitism (see review by Weisbroth, 1982)

The important ectoparasites of mice spend their lives on the host. Populations are limited by factors such as self-grooming, mutual grooming, the presence of hair (which is required for acariasis), and immunological responses, which tend to produce hypersensitivity to mites. Inherited resistance and susceptibility also affect clinical expression of acariasis. Mite populations, for example, vary widely among different stocks and strains of mice housed under similar conditions.

Lice and mites generally favor the dorsal anterior regions of the body, particularly the top of the head, neck, and withers (areas least amenable to grooming), but in severe parasitism, all areas of skin can be infested. Skin lesions of acariasis and pediculosis include pruritis, scruffiness, and patchy hair loss, and in severe cases, ulceration and pyoderma initiated or compounded by self-inflicted trauma. Histologically, hyperkeratosis, acanthosis, and chronic dermatitis may occur. Long-standing infestation provokes chronic inflammation, fibrosis and proliferation of granulation tissue. Ulcerative dermatitis associated with acariasis may have an allergic pathogenesis. Lesions resemble allergic acariasis in other species and are associated with mast cell accumulations. Some workers believe chronic mite infestation contributes to secondary amyloidosis.

Polyplax serrata (mouse louse) has five primary stages in its life cycle, which include the egg, three nymphal stages, and the adult (Fig. 34). Eggs attach near the base of hair shafts. They hatch in 5–6 days, and nymphs develop into adults in 7 days giving an average life cycle of about 13 days. Transmission is by direct contact. They are blood-sucking lice, thus clinical signs of heavy infestation are related to dermatitis, anemia, and debilitation and can occasionally be fatal. Lice should not be found, however, in well-managed colonies. Lesions associated with infestation include acanthosis and hyperkeratosis and allergic dermatitis. The role of immunological reactions in the pathogenesis and control of infection has, however, not been well-defined. These parasites can also transmit *Eperythrozoon coccoides* (see page 53).

Myobia musculi (mouse mite) has a world-wide distribution

Fig. 34. *Polyplax serrata,* female. From Weisbroth (1982). (Photograph courtesy of Robert J. Flynn and *Laboratory Animal Science.*)

in laboratory mice. It can be differentiated from *Radfordia,* another common mouse mite, in that *Myobia* has an empodial claw on the second pair of legs whereas *Radfordia* has two terminal claws on the terminal tarsal structure of leg 2 (Fig. 35). The life cycle can be completed in 23 days and includes an egg stage, first and second larval stage, protonymph, deutonymph, and adult. Eggs attach at the base of hair shafts and hatch in 7–8 days. Larval forms last about 10 days followed by nymphal forms on day 11. Adults appear by day 15 and lay eggs within 24 hr.

Myobia probably feed on skin secretions and are transmitted primarily by contact. With new infestations, mite populations increase followed by a decrease to equilibrium in 8–10 weeks. The equilibrium population can be carried for long periods, even years. Fluctuations in equilibrium populations may represent waves of egg hatchings. Because mites are thermotactic, they crawl to the end of hair shafts on dead hosts where they may live for up to 4 days. Treatment should account for the fact that acaracides are ineffective for eggs. Therefore, a second application should be made after the eighth day when residual eggs have hatched but sometime before the sixteenth day when new adults may have laid their eggs. Ideally, the

second treatment should be given between the tenth and twelfth day (see page 80).

Radfordia affinis (mouse mite) is a ubiquitous mite that is similar in appearance to *M. musculi.* The life cycle and pathogenesis of *Radfordia* infection have not been described.

Myocoptes musculinus (mouse mite) (Fig. 36) is the most common ectoparasite of the laboratory mouse but frequently occurs in conjunction with *Myobia musculi.* Its distribution is worldwide. The life cycle includes egg, larva, protonymph, tridonymph, and adult stages. Eggs hatch in 5 days and are usually attached to the middle third of the hair shaft. The life cycle has been reported as 14 days by one author and 8 days by another. Transmission requires direct contact, since mice separated by wire screens do not contract infections from infected hosts. Bedding does not seem to serve as a vector. Neonates may be infected within 4–5 days of birth, and parasites may live for 8–9 days on dead hosts.

Myocoptes appears to inhabit larger areas of the body than *Myobia.* It has some predelection for skin of the inguinal region, abdominal skin, and back, but it will also infest the head and neck. In heavy infestations, it tends to crowd out *Myobia.* It is a surface dweller that feeds on superficial epidermis. Infestation can cause patchy thinning of the hair, alopecia, or erythema. Lesions can be pruritic, but ulceration has not been reported. Chronic infestations induce epidermal hyperplasia and nonsuppurative dermatitis. Detailed pathogenetic studies have not been made.

Psorergates simplex (mouse mite) inhabits hair follicles. The life cycle is unknown, but developmental stages from egg to adult may be found in a single dermal nodule. Transmission is by direct contact. Infestation was common in laboratory mice 30 years ago but is now considered rare. It has not been reported as a naturally occurring infection in well-managed colonies for at least a decade.

Psorergates simplex invades the hair follicle and a cystlike nodule develops. Dermal cysts appear as small white nodules in the subcutis (Fig. 37). Histologically, they are invaginated sacs of squamous epithelium, excretory products, and keratinaceous debris. There is usually no inflammatory reaction, but healing may be accompanied by granulomatous inflammation. Diagnosis is made by examining the subcutis surface of the pelt grossly or by histological examination. Pouch contents also can be expressed by pressure with a scalpel blade or scraped and mounted for microscopic examination.

Trichoecius romboutsi has been reported in mice in the United States and Europe but infestation is believed to be rare. Its life cycle and pathogenesis are unknown.

Other acarid ectoparasites and others seen infrequently in laboratory mice (e.g., fleas and bedbugs) are shown in Table XI.

Gross observation of live, unrestrained mice may not detect ectoparasitism accurately. Therefore supplementary diagnostic tests are useful and the following techniques are suggested:

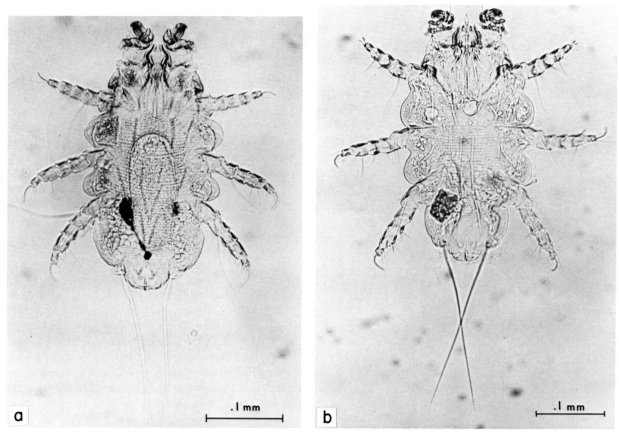

Fig. 35.   (a) *Myobia musculi*, female. (b) *Radfordia affinis*, female. From Weisbroth (1982). (Photographs courtesy of Dr. R. J. Flynn and *Laboratory Animal Science*.)

1. Direct observation of the hair and skin of dead or anesthetized mice. Hairs are parted with pins or sticks and examined with a dissecting microscope.

2. Cellophane tape is pressed against areas of the pelt of freshly euthanatized mice and examined microscopically. Alternatively, recently euthanatized mice can be placed on a black paper and double-sided cellophane tape can be used to line the perimeter to contain the parasites from leaving. As the carcass cools, parasites will vacate the pelage and crawl onto the paper. Sealed petri dishes can also be used.

3. Skin scrapings made with a scapel blade can be macerated in 10% KOH glycerine or immersion oil and examined microscopically. This method has the disadvantage of missing quickly moving species and low-level populations of slower moving forms. *Demodex, Sarcoptes,* and *Notoedres,* species for which scrapings are most useful, are not reported for the laboratory mouse (*Mus musculus*).

Eradication of ectoparasitism is costly and rarely completely effective. In fact, the only effective way to eliminate infestation is by gnotobiotic rederivation. Therefore, once infestation has been diagnosed, periodic control is a more pragmatic goal.

Treatment and control programs should be carried out on a colony-wide basis. A summary of published control programs for ectoparasites of laboratory mice is given in Table XII. Each has advantages and disadvantages. Dusts either as desiccating sorptives or insecticides have generally given poor results because of toxicity, high labor costs, and variable efficacy. Dips such as malathion also result in variable efficacy and in toxicity. Aramite has been widely used as a dip in the United States and with generally good results but its use is labor intensive.

Dichlorvos, an organophosphate, has received widespread use against mites. It is effective and labor saving. Cage treatments can include taping of resin strips inside filter bonets, or addition of 2 gm of tablets or 2 ml of liquid concentrate per cage. Dichlorvos has not been used specifically against *Polyplax* but is effective against avian lice, which suggests it may also be useful for rodent lice. It causes transient lowering of serum acetylcholinesterase, but it is rapidly cleared and metabolized. It is neither teratogenic nor carcinogenic, but it may delay breeding transiently during extensive exposures.

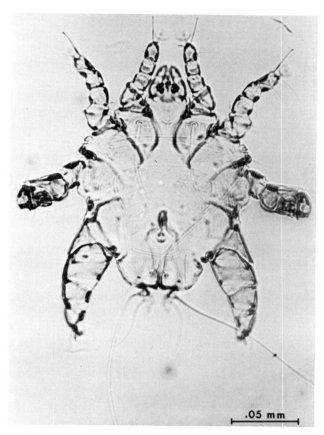

Fig. 36. *Myocoptes musculinis,* male. From Weisbroth (1982). (Photograph courtesy of Dr. R. J. Flynn and *Laboratory Animal Science.)*

6. Fungal Diseases

Trichophyton mentagrophytes is the major cause of ringworm in mice. It may cause clinical disease, but it is more commonly present in the skin of mice as an asymptomatic infection. Clinical signs include sparse haircoats or well demarcated crusty lesions with a chalky surface. *Microsporum* infections of mice are not well documented. The diagnosis of ringworm depends on effective specimen collection. Hairs should be selected from the periphery of the lesion. Hairless skin should be scraped deeply to obtain diagnostic specimens. *Trichophyton mentagrophytes* rarely fluoresces under ultraviolet light, and hyphi must be differentiated from bedding fibers, food particles and epidermal debris. *Trichophyton* can be cultured on Saboraud's agar. Plates are incubated at room temperature (22°–30°C) and growth is observed at 5–10 days. The zoonotic potential of ringworm is discussed in Chapter 22 by Fox *et al.*

Ringworm is not easily eradicated from laboratory mice. It may be necessary to destroy affected mice and sterilize all cages and equipment before reuse. Concurrent infection with ectoparasites also must be considered. The use of griseofulvin

in mice must be approached with caution because of metabolic differences between these animals and dogs and cats where this drug is often used. It should not be used in breeding females since it is potentially embryotoxic and teratogenic. Dip solutions should be kept at 37°C, and animals should be allowed to dry in a warm environment.

Pathogenic fungi such as *Candida albicans* are not an important cause of gastrointestinal lesions in the mouse, but they can be opportunistic pathogens in immunodeficient mice. Fungal infections of other organ systems are extremely rare.

B. Neoplastic Diseases

a. Lymphoid and Hematopoietic (see review by Furmanski and Rich, 1982)

Neoplasms of *lymphoid* and *hematopoietic* tissues are estimated to have a spontaneous prevalence of 1–2%. There are, however, some strains of mice that have been specifically inbred and selected for susceptibility to spontaneous tumors. Leukemogenesis in mice may involve viruses and chemical or physical agents. Viruses associated with lymphopoietic and hematopoietic neoplasia belong to the family Retroviridae

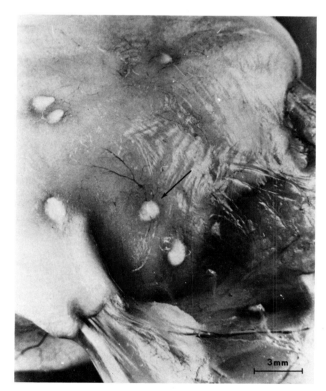

Fig. 37. *Psoregates simplex*-infested skin inverted to show pouches (arrow). From Weisbroth (1982). (Photograph courtesy of Dr. R. J. Flynn and *Laboratory Animal Science.)*

Table XI

Ectoparasites of Laboratory Mice[a]

Order	Genus	Species	Common name
Hemiptera	*Cimex*	*lectularis*	Bedbug
Siphonaptera	*Xenopsylla*	*cheopis*	Oriental rat flea
	Nosopsyllus	*fasciatus*	Northern rat flea
	Nosopsyllus	*londsniensis*	
	Leptosylla	*segnis*	Mouse flea
Anoplura	*Polyplax*	*serrata*	Mouse louse
	Hoplopleura	*acanthopus*	
	Hoplopleura	*captiosa*	
Acarina	*Ornithonyssus*	*bacoti*	Tropical rat mite
	Ornithonyssus	*sylviarum*	Northern fowl mite
	Liponyssoides	*sanguineous*	House mouse mite
	Haemogamasus	*pontiger*	
	Eulaelaps	*stabularis*	
	Laelaps	*echidnus*	Spiny rat mite
	Haemolaelaps	*glasgowi*	
	Haemolaelaps	*casalis*	
	Myobia	*musculi*	Fur mite
	Radfordia	*affinis*	Fur mite
	Psorergates	*simplex*	Hair follicle mite
	Notoedres	*musculi*	
	Demodex	*musculi*	
	Myocoptes	*musculinus*	
	Trichoecius	*romboutsi*	

[a]Modified from Weisbroth (1982).

(type C oncornaviruses) and contain RNA-dependent DNA polymerase (reverse transcriptase). These viruses are generally noncytopathogenic for infected cells, and mice appear to harbor them as normal components of their genetic apparatus. Although they may be involved in spontaneous leukemia, they are not consistently expressed in this disease. Recombinant viruses have recently been discovered that can infect mouse cells and heterologous cells and are associated with spontaneous leukemia development in high leukemia strains such as AKR mice. Their phenotypic expression is controlled by mouse genotype. Endogenous retroviruses are transmitted vertically through the germ line. Horizontal transmission is inefficient but can occur by interuterine infection or through saliva, sputum, urine, feces, or milk. The leukemia induced by a given endogenous virus is usually of a single histopathological type.

Chemical carcinogens, such as polycyclic hydrocarbons, nitrosoureas, and nitrosamines, can also induce hematological malignancies in mice. Physical agents such as X irradiation are also effective inducers of cancer.

The most common hematopoietic malignancy in the mouse is *lymphocytic leukemia* that originates in the thymus. Disease begins with unilateral atrophy and then enlargement of one lobe of thymus as tumor cells proliferate. Cells can spread to the other lobe and then to other hematopoietic organs, such as the spleen, bone marrow, liver, and peripheral lymph nodes. Clinical signs include dyspnea and protrusion of the eyeballs. The latter sign is due to compression of venous blood returning from the head. Tumor cells spill into the circulation late in disease. Most of these tumors originate from T lymphocytes, but there are leukemias of B lymphocyte or null cell lineage. In the last two syndromes, the lymph nodes and spleen are often involved, but the thymus is generally normal, and each form can be induced in athymic mice.

Myeloid leukemia is not common in mice, but it can be virus induced. Myeloblasts accumulate in the liver and in the circulation and infiltrate bone marrow and other organs. Leukemic cells in various stages of differentiation can be found in peripheral blood. In older animals, affected organs may appear green due to myeloperoxidase activity giving rise to the term chloroleukemia. The green hue fades on contact with air.

Erythroleukemia is rare in mice. The major lesion is massive splenomegaly, which is accompanied by anemia and polycythemia. Hepatomegaly can follow, but there is little change in the thymus or lymph nodes.

Reticulum cell sarcomas are common in older mice, especially in inbred strains such as C57BL and SJL. Primary tumor cell types have been divided into several categories based on morphological features. Type A sarcomas typically involve the spleen and liver and are often accompanied by ascites. Type B tumors produce a Hodgkin's-like syndrome characterized by mesenteric lymph node enlargement often involving Peyer's patches and spleen. A spontaneous reticulum cell sarcoma that occurs in more than 90% of inbred SJL/J mice begins in the mesenteric lymph nodes. This tumor can also affect other organs, such as the spleen, thymus, and liver.

Natural *plasma cell tumors* are infrequent in the mouse. They can, however, be induced by intraperitoneal inoculation of granulomatogenic agents such as plastic filters, plastic shavings, or a variety of oils.

Mast cell tumors are also very rare in mice. They are found almost exclusively in old mice and grow slowly. They should not be confused with mast cell hyperplasia observed in the skin following painting with carcinogens or X irradiation.

b. Mammary (see review by Medina, 1982)

Mammary tumors can be induced or modulated by a variety of factors, including viruses, chemical carcinogens, radiation, hormones, genetic background, diet, and immune status. Certain inbred strains of mice, such as C3H, A, and DBA/2, have a high natural prevalence of mammary tumors. Other strains, such as BALB/c, C57BL, and AKR, have a low prevalence.

Among the most important factors contributing to the development of mammary tumors are mammary tumor viruses. Several major variants are known. The primary tumor virus MMTV/S (Bittner virus) is highly oncogenic and is transmit-

Table XII

Reported Treatment for Ectoparasitism of Laboratory Mice[a]

Chemical or generic name	Trade or common name	Method of administration	Parasite treated and efficacy
Di-(*p*-chlorophenyl)methyl carbinol and tetraethylthiuram monosulfide	DMC or Dimite (2 gm/liter) and tetmosol (67 mg/liter of 25% solution)	Two dippings 1 hr apart by immersion and swimming, 30 sec; repeated 1–3 weeks later	*Myobia, Myocoptes, Psorergates, Polyplax;* eradicated
Di-(*p*-chlorophenyl)methyl carbinol	DMC (Dimite), 2 gm/liter	Single dipping	*Myocoptes;* eradicated
Tetraethylthiuram monosulfide	Tetmosol (67 gm/liter) of 25% solution	Single dip with swimming	*Myobia;* "usually sufficient"
	Tetmosol, 2.5%	Two dippings with immersion at 14-day intervals	*Myobia;* not effective
2-(*p-tert*-butylphenoxy)isopropyl-2-chloroethyl sulfite plus Nacconal	Aramite-15W, 2% plus wetting agent, 0.1%	Single dip by immersion and swimming, 17 sec	*Myobia, Myocoptes, Psorergates, Polyplax, Trichoecius, Radfordia;* eradicated
Benzene hexachloride	BHC, 0.625%	Two dippings with immersion at 14-day intervals	*Myobia;* not effective
Benzene hexachloride	BHC	Dusting	*Myocoptes,* "cures condition"; *Myobia;* ineffective
O-1-Dimethyl-5-(dicarbethoxy-ethyl)dithiophosphate	Malathion, 2.5%	Two dippings with immersion at 14-day intervals	*Myobia;* not effective
	Malathion, 1% and 4%	Five dippings at 3-day intervals	*Myobia;* not effective
	Malathion, 2%	Single dipping	*Myocoptes;* eradicated
	Malathion, 2%	Single dip by immersion and swimming, 17 sec	*Myobia, Myocoptes;* effective
	Malathion, 0.125% (Malastan E-C50, emulsifiable)	Single dip by immersion of rats	*Polyplax;* eradicated
Pyrethrins 5% with 2.5% bucarpolate	Pyractone, 1%	Five dippings at 3-day intervals	*Myobia;* eradicated
Pyrethrins 5% with 25% bucarpolate	Super Pyractone M.429, diluted to give 0.06% pyrethrin, 0.3% bucarpolate	Two dippings with immersion at 14-day intervals	*Myobia;* effective
SG-67	Dri-Die (0.5 gm per mouse)	Applied individually as a dust	*Myobia;* not effective
Silica sorptive dust	Dri-Die 67	Applied individually	*Polyplax*
Silica dust with pyrethrin	Dri-One	Applied individually	*Polyplax*
Formel 5-brommethyl-1,2,3,4,7,7-hexachlorobicyclo-(2,2,1)hepten-(2)	Alugan, 0.6% solution	Two dippings at 11-day intervals	*Myobia;* eradicated
Sulfur	325 mesh dusting sulfur	One pinch dusted onto animal	*Myobia, Myocoptes,* 95–99% effective
	Chlordane, 2 gm 5% dust	Sprinkled over contact bedding	*Polyplax;* eradicated
	Lindane, 10 gm 1% dust	Sprinkled over contact bedding; several applications	*Myobia, Myocoptes;* 90–98% effective
	Kelthane, 5% wettable powder	Used as dust, 0.5 gm per mouse	*Myobia;* effective
Gamma HCH	Jacutin spray (20% in oil)	Individually with small brush; twice at 12-day intervals	*Myobia;* effective
Dibutyl phthalate		Applied topically to affected areas	*Psorergates;* effective
Dichlorvos (DDVP)	Vapona (18% in resin strip)	1 × 2.5-in. resin strip on cage top, under filter cap	*Myocoptes;* effective
		2 × 2.5-in. resin strip on cage top, under filter	*Myobia;* effective
	Atgard (anthelmintic) pellets	2 gm sprinkled in bedding; on day 8 cage changed, 2 gm added (14-day exposure)	*Myobia;* effective
	Vaporal (4.65% DDVP, 5% ronnel) liquid dip	2 ml per cage on contact litter	*Myobia;* effective

[a]From Weisbroth (1982).

ted through the milk of nursing females. A summary of other virus strains is given in Table XIII. Infected mice typically develop a precursor lesion, the hyperplastic alveolar nodule, which can be serially transplanted.

Spontaneous mammary tumors metastasize with high frequency, but this property is somewhat mouse strain dependent. Metastases go primarily to the lung. Some mammary tumors are hormone dependent, some are ovary dependent, and others are pregnancy dependent. Ovary-dependent tumors contain estrogen and progesterone receptors, whereas pregnancy-dependent tumors have prolactin receptors. Ovariectomy will dramatically reduce the incidence of mammary tumors in C3H mice. If surgery is done in adult mice 2–5 months of age, mammary tumors will develop, but at a later age than normal.

Mammary tumors have been categorized morphologically into three major groups; carcinomas, carcinomas with squamous cell differentiation, and carcinosarcomas. The carcinomas are divided into adenocarinoma types A, B, C, Y, L, and P. Most tumors are type A or B. Type A consists of adenomas, tubular carcinomas, and alveolar carcinomas. Type B tumors have a variable pattern with both well-differentiated and poorly differentiated regions. They may consist of regular cords or sheets of cells or papillomatous areas. Type C tumors are rare and are characterized by multiple cysts lined by low cuboidal to squamous epithelial cells, and they have abundant stroma. Type Y tumors, which are also rare, are characterized by tubular branching of cuboidal epithelium and abundant stroma. Adenocarcinomas with a lace-like morphology (types L and P) are hormone dependent and have a branching tubular structure.

The control or prevention of mammary neoplasms depends on the fact that some strains of mammary tumor virus are transmitted horizontally whereas others are transmitted vertically. Therefore, although one can rid mice of horizontally transmit-

ted virus such as MMTV-S by cesarean rederivation or by foster nursing, endogenous strains of tumor virus may remain. Fortunately, these latter tumor viruses have generally low oncogenicity compared to the Bittner virus.

c. Liver (see review by Newberne and McConnell, 1982)

Liver tumors can develop in mice exposed to environmental chemicals, many of which are carcinogenic or potentially carcinogenic. Almost all strains of mice have a significant prevalence of hepatic tumors, some of which appear to result from dietary contamination or deficiency. The prevalence of spontaneous liver tumors in B6C3F$_1$ hybrids can, for example, be increased from between 5 and 58% up to 100% by feeding choline-deficient diets.

Spontaneous liver tumors in mice are usually derived from hepatocytes, whereas cholangiocellular tumors are rare. The morphological characteristics of each of these stages have been described in detail. Hepatocellular carcinoma is an end-stage lesion and generally displays local invasiveness or metastasis. It can have a variety of histopathological patterns ranging from medullary to trabecular. Although proliferative lesions of the mouse liver have been described morphologically, their biological behavior and their relationship to one another are not clearly understood. One prevalent view is that tumorigenesis proceeds from foci of altered cells to hyperplastic nodules to adenomas to hepatocellular carcinomas.

d. Respiratory (see review by Theiss and Shimkin, 1982)

Primary respiratory tumors of mice occur in relatively high frequency and are largely of alveologenic origin. It has been estimated that more than 95% of these tumors are pulmonary

Table XIII

Characteristics of the Major Types of Mammary Tumors Induced by MMTV or Chemical Carcinogens[a]

Oncogen	Mouse strain	Transmission	Tumor type	Tumor incidence	Latency period	Hormone responsiveness[b]
MMTV-S	C3H, A, DBA, BALB/cfC3H	Milk	Adenocarcinoma types A and B	≥70% (breeders)	<1 year	Ovarian, adrenal, pituitary hormone independent
MMTV-L	C3Hf	Germ cells	Adenocarcinoma types A and B	<40% (breeders)	>1 year	Ovarian, adrenal, hormone independent
MMTV-P	GR	Germ cells, milk	Carcinoma type P	>70% (breeders)	<1 year	Pregnancy dependent
MMTV-P	DD, RIII, DDD	Milk	Carcinoma type P pale cell carcinoma	>70% (breeders)	<1 year	Ovary dependent
Chemical carcinogens	BALB/c, IF, LAF$_1$, C3Hf$_1$, (C57BL × DBA/2f)F$_1$	—	Adenocarcinoma, type B, adenoacanthroma, pale cell carcinoma	30–70% (virgins)	<1 year	Ovary dependent (C57BL × DBA/2f/)F$_1$, ovary independent (BALB/c)

[a]From Medina (1982).

[b]Hormone responsiveness refers to the general hormonal requirements for growth of palpable mammary tumors.

adenomas that arise either from type 2 pneumocytes or from Clara cells lining terminal bronchioles. Pulmonary adenomas usually appear as distinct whitish nodules that are easily detected by examination of the lung surface. (Clara cell tumors commonly protrude into bronchioles.) Malignant alveologenic tumors are infrequent and consist of adenocarcinomas and squamous cell carcinomas. They invade pulmonary parenchyma and are prone to metastasize.

The prevalence of spontaneous respiratory tumors is mouse strain dependent. For example, by 2 years of age, lung tumors have been found in more than 70% of A strain mice but in less than 10% of C57BL mice. The number of tumors per lung is also higher in susceptible mice.

Lung tumors can be induced by a variety of chemical carcinogens, but, in contrast to mammary tumors, the incidence of respiratory tumors does not seem to be influenced by steroid hormones.

Neoplasms of other organ systems are less prevalent. They are reviewed in Volume IV of "The Mouse in Biomedical Research" (Foster et al., 1982b).

C. Noninfectious Diseases

The prevalence of noninfectious diseases is influenced by genetic, nutritional, environmental, age-related, and sex-related factors. Many of these diseases appear with increasing frequency as animals age.

a. Liver

Age-associated lesions are common in the livers of mice. Cellular and nuclear pleomorphism, including binucleated and multinucleated cells, are detectable by 6 months. Mild focal necrosis occurs with or without inflammation, but an association of mild focal hepatitis with a specific infectious disease is often hard to confirm. Other geriatric hepatic lesions include biliary hyperplasia with varying degrees of portal hepatitis, hepatocellular vacuolization, amyloid deposition especially in periportal areas, strangulated or herniated lobes, hemosiderosis, lipofuscinosis, and fibrosis. Extramedullary hematopoiesis occurs in young mice and in response to anemia.

b. Gastrointestinal Tract

Abscesses or necrosis may occur in the lips or oral cavity because of fighting injury or mechanical trauma from cages. Overgrown incisor teeth can cause malocclusion, and caries can develop in molar teeth. Gastric lesions include crypt dilatation, submucosal fibrosis, adenomatous gastric hyperplasia, mineralization, and erosion or ulceration. Gastric ulcers may be stress related, especially in mice with prolonged illness.

Diffuse or segmental intestinal amyloidosis is common in some strains of mice, and the lamina propria may be greatly distended by accumulations of amyloid. Germfree mice may have lower muscle tone in the intestinal tract. Cecal volvulous is a common finding in germfree mice and is caused by rotation of the large cecum.

c. Pancreas

Exocrine pancreatic insufficiency has been reported in CBA/J mice. Acinar cell atrophy is common but is strain and sex dependent.

d. Cardiovascular System

Atrial thrombosis appears to be strain related with a high prevalence in RFM mice. Dystrophic mineralization seen as pale focal or linear discolorations on the epicardium has been described in many strains of mice. A high incidence has been reported for DBA/2, BALB/c, and C3H strains, whereas C57BL/6 mice are essentially free of this lesion. The etiology is unknown. Peripheral vascular lesions are uncommon aside from mice with lupus-like disease where vasculitis and myocardial infarction can occur. Periarteritis appears in aged mice.

e. Genital System

Parvovarian cysts are observed frequently and may be related to the fact that mouse ovaries are enclosed in membranous pouches. Amyloidosis is also common in the ovaries of old mice. Cystic endometrial hyperplasia may develop unilaterally or bilaterally and may be segmental. In some strains, the prevalence in mice older than 18 months is 100%. Endometrial hyperplasia is often associated with ovarian atrophy. Testicular atrophy, sperm granulomas, and tubular mineralization occur with varying incidence. Inflammation of accessory sex glands may occur, especially in males.

f. Urinary System

Glomerulonephritis is one of the most common renal diseases of mice. It is commonly associated with persistent viral infections or immune disorders rather than with bacterial infections. Its prevalence in some strains approaches 100%. The NZB mouse and the F_1 hybrid of NZB × NZW crosses, for example, develop glomerulonephritis as an autoimmune disease resembling human lupus erythematosus, whereas glomerular disease is relatively mild in NZB mice (NZB mice have a high incidence of autoimmune hemolytic anemia) (see review by Talal, 1983).

Renal changes occur as early as 4 months of age but clinical signs and severe disease are not present until 6–9 months. The disease is associated with wasting and proteinurea, and lesions progress until death intervenes. Histologically, glomeruli have proteinaceous deposits in the capillaries and mesangium. Later, tubular atrophy and proteinaceous casts occur throughout the kidney. Immunofluorescence studies show deposits of immunoglobulin and the third component of complement, which lodge as immune complexes with nuclear antigens and antigens of murine leukemia virus in glomerular capillary loops. Mice infected with LCMV or with retroviruses can also develop glomerulonephritis of immune complex origin.

Ascending pyelitis occurs in mice secondary to urinary tract infection. Amyloidosis in older mice is associated with papillary necrosis, but amyloid deposition in glomeruli and in the renal interstitium is also common. Urethral obstruction in secondary hydronephrosis is seen especially in CDF_1 mice. Chronic progressive nephropathy is seen in varying incidence and appears to be an age-related lesion in mice as it is in the rat.

g. Endocrine System

Accessory adrenal cortical nodules are found in periadrenal and perirenal fat, especially in females. They have little functional significance other than their potential effect on failures of surgical adrenalectomy. Lipofuscinosis and subcapsular spindle cell hyperplasia and cystic dilatation of cortical sinusoids are found in the adrenal cortices of old mice.

Some inbred strains have deficiencies of thyrotropic hormone resulting in thyroid atrophy. Thyroid cysts lined by stratified squamous epithelium and generally of ultimobranchial origin may be seen in old mice. Amyloid can be deposited in the thyroid and parathyroid glands as well as in the adrenal glands. Spontaneous diabetes mellitis-like disease has been reported in genetic variants of several strains (see Section III,C,o).

h. Integumentary System

Skin lesions can be caused by fighting, tail biting, and whisker chewing. Fighting is not limited extensively to males, but they tend to be more aggressive. Bite wounds are usually located on the head, neck, shoulders, peritoneal area, and tail. Often one animal per pen is free of lesions and is considered the aggressor or dominant animal. Removal of the unaffected male usually ends the fighting, and the wounded animals recover. However, a previously submissive male may become dominant when an aggressive male is removed and fighting may resume. Fighting has some strain predilection, and is especially notorious among BALB/c males, but tail lesions resembling bite wounds have been reported in other strains.

Hair nibbling or whisker chewing (barbering) is also a manifestation of social dominance. Dominant animals retain whiskers, whereas cagemates have "shaved faces" (Fig. 12). Chronic hair chewing can produce histological abnormalities such as poorly formed or pigmented club hairs. Once chewing has ceased, many mice regrow previous hair loss in several weeks. Both sexes may engage in this activity, and sometimes females may be dominant. Regional alopecia, especially around the muzzle, may result from abrasion against cage surfaces. Improperly diluted disinfectants may also cause regional hair loss. Metal tags used for animal identification may cause pruritis and self-induced trauma. Clipping prior to application of experimental compounds to the skin may cause pruritic responses and can augment lesions that interfere with test results. Dermatophytosis or ectoparasitism must be considered in the differential diagnoses for muzzle or body alopecia.

High levels of estrogen in pregnancy may influence postpartum shedding. Various endocrine effects on hair growth have also been described. Spotty alopecia of unknown etiology may occur in some strains of mice such as C57BL/6. Abdominal and thoracic alopecia have been reported in $B6C3F_1$ mice.

i. Musculoskeletal System

Dystrophic mineralization of soft tissues occurs frequently in old mice, a high prevalence being reported in DBA/2 mice. It is found in the myocardium of the left ventricle, in the intraventricular systems, and in skeletal muscle, kidneys, arteries, and lung. It is associated with fibrosis and often with mononuclear inflammatory infiltrates. Diets high in calcium increase the prevalence of this lesion. Age-associated osteoporosis or senile osteodystrophy can occur in some mice. It is not associated with severe renal disease or parathyroid hyperplasia. Nearly all strains of mice develop some form of osteoarthrosis. It is generally noninflammatory, affects articulating surfaces, and results in secondary bone degeneration.

j. Respiratory System

Hyperplasia of alveolar or bronchial epithelium occurs in old mice and must be differentiated from pulmonary tumors.

k. Nervous System

Symmetrical mineral deposits commonly occur in the thalamus of aged mice. They may also be found in the midbrain, cerebellum, and cerebrum and are particularly common in A/J mice. Lipofuscin accumulates in the neurons of old mice. Deposits of melanin pigment occur in heavily pigmented strains, especially in the frontal lobe. Age-associated peripheral neuropathy with demyelination can be found in the nerves of the hindlimbs in B6 mice. A number of neurologically mutant mice have been described. They commonly have correlative anatomical malformations or inborn errors of metabolism.

l. Hematopoietic and Lymphoreticular System

Aggregates or nodules of mononuclear cells are found in many tissues of aged mice, including the salivary gland, thymus, ovary, uterus, mesentery and mediastinum, urinary bladder, and gastrointestinal tract. These nodules should not be mistaken for lymphosarcomas. The spleen is subject to amyloidosis and hemosiderin deposition. Lipofuscin deposition is common especially in older mice. The thymus undergoes age-associated atrophy.

m. Eyes

Cataracts can occur in old mice and have a higher prevalence in certain mutant strains. Inherited retinal degeneration is observed in strains such as C3H and light-associated retinal degeneration can occur, especially in nonpigmented mice.

n. Nutritional Diseases (see review by Knapka, 1983)

Dietary restriction increases the life span of mice and may retard development of renal disease and of tumors. Severe dietary protein deprivation, however, inhibits humoral and cellular immunity.

Chronic essential *fatty acid deficiency* may cause hair loss, dermatitis with scaling and crusting of the skin, and occasional diarrhea. Infertility has also been associated with this syndrome. Mice have an absolute requirement for a dietary source of linoleic and/or arachadonic acid.

Mineral deficiencies have only been described for several elements, and the consequences of the deficiencies are similar to those observed for other species. For example, iodine-deficient diets produce thyroid goiters; magnesium-deficient diets may cause fatal convulsions; maganese deficiency may cause congenital ataxia from abnormal development of the inner ear, and zinc deficiency may cause hair loss on the shoulders and neck, emaciation and decreased liver and kidney catalase activity, and immunosuppression.

Vitamin deficiencies in mice have not been thoroughly described. Unfortunately, much of the information that does exist reflects work done 30–50 years ago, thus the reliability and specificity of some of these syndromes is questionable. *Vitamin A* deficiency may produce tremors, diarrhea, rough haircoat, keratitis, poor growth, abscesses, hemorrhages, and sterility or abortion. *Vitamin E* deficiency can cause convulsions and heart failure as well as muscular dystrophy and hyaline degeneration of muscles. Deficiency of *B complex vitamins* produces nonspecific signs such as alopecia, decreased feed consumption, poor growth, poor reproduction, and lactation,

as well as a variety of neurological abnormalities. *Choline* deficiency produces fatty livers and nodular hepatic hyperplasia as well as myocardial lesions, decreased conception, and decreased viability of litters. *Folic acid*-deficient diets cause marked decreases in red and white cell blood counts and the disappearance of megakaryocytes and nucleated cells from the spleen. *Pantothenic acid* deficiency is characterized by nonspecific signs, such as weight loss, alopecia, achromotrichia, and posterior paralysis as well as other neurological abnormalities. *Thiamin* deficiency is associated with neurological signs, such as violent convulsions, cartwheel movements, and decreased food consumption. Dietary requirements for *ascorbic acid* have not been shown in mice, and mouse diets are generally not fortified with ascorbic acid.

o. Miscellaneous Genetic Abnormalities

Many spontaneous *genetic abnormalities* of mice have been exploited as animal models. Perhaps the most widely known of these is the athymic nude mouse that lacks a significant haircoat and more importantly, fails to develop a thymus and thus has a severe deficit of T cell–mediated immune function (see review by Rygaard and Polvsen, 1982). Because of this deficit, it is an excellent model to study the ontogeny and mechanisms of immune responsiveness. Athymic mice also serve as *in vivo* carriers for allogeneic and xenogeneic cells, including human tumor cells.

Mutations in ophthalmologic development, anatomy, and physiology have been described for mice (see review by Robison *et al.*, 1982). Albino mice are particularly sensitive to intense light and are susceptible to a light-accelerated retinal degeneration. Mice also display inherited convulsive disorders, including audiogenic seizures that have been advanced as animal models for human epilepsy.

Mouse mutations for obesity have been described, and diabetes is inherited as an autosomal recessive mutation in a substrain of C57BL mice (see review by Coleman, 1982). A recessive gene for obesity is also maintained in the C57BL strain. Some strains of mice are deficient in the C5 fraction of complement (A/Jax and B10.D2/0).

p. Environmental Variables

Environmental variables can affect responses of mice in experimental situations. Changes in respiratory epithelial physiology and function from elevated levels of ammonia, effects of temperature and humidity on metabolism, effects of light on eye lesions and retinal function, and effects of noise on neurophysiology are examples of complications that can vary with the form of insult and the strain of mouse employed. Chloroform has extreme nephrotoxicity for male mice but not for female mice of certain strains (see Chapter 23).

REFERENCES

American Institute of Nutrition (AIN) (1977). *Ad Hoc* Committee on Standards for Nutritional Studies. Report of the Committee. *J. Nutr.* **107,** 1340.

Anderson, C. A., Murphy, J. C., and Fox, J. G. (1983). Evaluation of murine cytomegalovirus antibody detection by serological techniques. *J. Clin. Microbiol.* **18,** 753–758.

Austin, C. R., and Short, R. V., eds. (1982). "Reproduction in Mammals," 2nd ed., Vols. 1–5. Cambridge Univ. Press, London and New York.

Bailey, D. W. (1971). Recombinant inbred strains. *Transplantation* **11,** 325–327.

Baker, H. J., Cassell, G. H., and Lindsey, J. R. (1971). Research complications due to *Hemobartonella* and *Eperythrozooan* infections in experimental animals. *Am. J. Pathol.* **64,** 625–656.

Barthold, S. W., and Beck, D. (1980). Modification of early dimethylhydrazine carcinogenesis by colonic mucosal hyperplasia. *Cancer Res.* **40,** 4451–4455.

Barthold, S. W., Osbaldiston, G. W., and Jonas, A. M. (1977). Dietary, bacterial and host genetic interactions in the pathogenesis of transmissible murine colonic hyperplasia. *Lab. Anim. Sci.* **27,** 938–945.

Barthold, S. W., Coleman, G. L., Jacoby, R. O., Livstone, E. M., and Jonas, A. M. (1978). Transmissible murine colonic hyperplasia. *Vet. Pathol.* **15,** 223–236.

Barthold, S. W., Smith, A. L., Lord, P. S., Bhatt, P. N., Jacoby, R. O., and Main, A. J. (1982). Epizootic coronaviral typhlitis in suckling mice. *Lab. Anim. Sci.* **32,** 376–383.

Brinton, M. A. (1982). Lactate dehydrogenase-elevating virus. *In* "The Mouse in Biomedical Research" (H. L. Foster, J. D. Small, and J. G. Fox, eds.), Vol. 2, p. 194. Academic Press, New York.

Briody, B. A. (1959). Response of mice to ectromelia and vaccinia viruses. *Bacteriol. Rev.* **23,** 61–95.

Bronson, F. H., Dagg, C. P., and Snell, G. D. (1966). Reproduction. *In* "Biology of the Laboratory Mouse" (E. L. Green, ed.), 2nd ed., Chapter II, pp. 187–204. McGraw-Hill, New York.

Brownstein, D. G. (1978). Pathogenesis of bacteremia due to *Pseudomonas aeruginosa* in cyclophosphamide-treated mice and potentiation of virulence of endogenous streptococci. *J. Infect. Dis.* **137,** 795–801.

Brownstein, D. G., and Barthold, S. W. (1982). Mouse hepatitis virus immunofluorescence in formalin- or Bouin's-fixed tissues using trypsin digestion. *Lab. Anim. Sci.* **32,** 37–39.

Cassell, G. H., Lindsey, J. R., Davis, J. K., Davidson, M. K., Brown, M. B., and Mayo, J. G. (1981). Detection of natural *Mycoplasma pulmonis* infection in rats and mice by an enzyme linked immunosorbent assay (ELISA). *Lab. Anim. Sci.* **31,** 676–682.

Coleman, D. L. (1982). Diabetes-obesity syndromes. *In* "The Mouse in Biomedical Research," (H. L. Foster, J. D. Small, and J. G. Fox, eds.), Vol. 4, p. 125. Academic Press, New York.

Cook, M. J. (1983). Anatomy. *In* "The Mouse in Biomedical Research" (H. L. Foster, D. G. Small, and J. G. Fox, eds.), Vol. 3, p. 102. Academic Press, New York.

Daniel, J. C., Jr., ed. (1978). "Methods in Mammalian Reproduction." Academic Press, New York.

Downs, W. G. (1982). Mouse encephalomyelitis virus. *In* "The Mouse in Biomedical Research" (H. L. Foster, J. D. Small, and J. G. Fox, eds.), Vol. 2, p. 341. Academic Press, New York.

Everett, R. M., and Harrison, S. D., Jr. (1983). Clinical biochemistry. *In* "The Mouse in Biomedical Research" (H. L. Foster, J. D. Small, and J. G. Fox, eds.), Vol. 3, p. 313. Academic Press, New York.

Fenner, F. (1948). The pathogenesis of the acute exanthems. An interpretation based on experimental investigations with mouse-pox (infectious ectromelia of mice). *Lancet* **2,** 915–920.

Fenner, F. (1982). Mousepox. *In* "The Mouse in Biomedical Research" (H. L. Foster, J. D. Small, and J. G. Fox, eds.), Vol. 2, p. 209. Academic Press, New York.

Flaherty, L. (1981). Congenic strains. *In* "The Mouse in Biomedical Research" (H. L. Foster, J. D. Small, and J. G. Fox, eds.), Vol. 1, p. 215. Academic Press, New York.

Foster, H. L., Small, J. D., and Fox, J. G. eds., (1982a). "The Mouse in Biomedical Research," Vol. 2. Academic Press, New York.

Foster, H. L., Small, J. D., and Fox, J. G., eds. (1982b). "The Mouse in Biomedical Research," Vol. 4. Academic Press, New York.

Fox, J. G., and Brayton, J. B. (1982). Zoonoses and other human health hazards. *In* "The Mouse in Biomedical Research" (H. L. Foster, J. D. Small, and J. G. Fox, eds.), Vol. 2, p. 407. Academic Press, New York.

Furmanski, P., and Rich, M. A. (1982). Neoplasms of the hematopoietic system. *In* "The Mouse in Biomedical Research" (H. L. Foster, J. D. Small, and J. G. Fox, eds.), Vol. 4, p. 352. Academic Press, New York.

Ganaway, J. R. (1982). Bacterial and mycotic diseases of the digestive system. *In* "The Mouse in Biomedical Research" (H. L. Foster, J. D. Small, and J. G. Fox, eds.), Vol. 2, p. 7. Academic Press, New York.

Ganaway, J. R., Allen, A. M., and Moore, T. D. (1971). Tyzzer's disease. *Am. J. Pathol.* **64,** 717–732.

Green, E. L. (1981). Breeding systems. *In* "The Mouse in Biomedical Research" (H. L. Foster, J. D. Small, and J. G. Fox, eds.), Vol. 1, p. 89. Academic Press, New York.

Harkness, J. E., and Ferguson, F. G. (1982). Bacterial, mycoplasmal and mycotic diseases of the lymphoreticular, musculoskeletal, cardiovascular and endocrine systems. *In* "The Mouse in Biomedical Research" (H. L. Foster, J. D. Small, and J. G. Fox, eds.), Vol. 2, p. 88. Academic Press, New York.

Hildebrandt, P. K. (1982). Rickettsial and chlamydial diseases. *In* "The Mouse in Biomedical Research" (H. L. Foster, J. D. Small, and J. G. Fox, eds.), Vol. 2, p. 99. Academic Press, New York.

Hsu, C.-K. (1982). Protozoa. *In* "The Mouse in Biomedical Research" (H. L. Foster, J. D. Small, and J. G. Fox, eds.), Vol. 2, p. 359. Academic Press, New York.

Hurley, L. S., and Bell, L. T. (1974). Genetic influence on response to dietary manganese deficiency. *J. Nutr.* **104,** 133.

Jacoby, R. O., Bhatt, P. N., and Johnson, E. A. (1983). The pathogenesis of vaccinia (IHD-T) virus infection in BALB/cAnN mice. *Lab. Anim. Sci.* **33,** 435–441.

John, A. M., and Bell, J. M. (1976). Amino acid requirements of the growing mouse. *J. Nutr.* **106,** 1361.

Kaplan, H. M., Brewer, N. R., and Blair, W. H. (1983). Physiology. *In* "The Mouse in Biomedical Research" (H. L. Foster, J. D. Small, and J. G. Fox, eds.), Vol. 3, p. 248. Academic Press, New York.

Kimura, M., and Crow, J. F. (1963). On maximum avoidance of inbreeding. *Genet. Res.* **4,** 399–415.

Klein, J. (1975). "Biology of the Mouse Histocompatibility – 2 – Complex." Springer-Verlag, Berlin and New York.

Knapka, J. J. (1983). Nutrition. *In* "The Mouse in Biomedical Research" (H. L. Foster, J. D. Small, and J. G. Fox, eds.), Vol. 3, p. 52. Academic Press, New York.

Knapka, J. J., Smith, K. P., and Judge, F. J. (1974). Effect of open and closed formula rations on the performance of three strains of laboratory mice. *Lab. Anim. Sci.* **24,** 480.

Kraft, L. M. (1980). The manufacture, shipping, receiving and quality control of rodent bedding materials. *Lab. Anim. Sci.* **30,** No. 2, Pt. II, 366–377.

Kraft, L. M. (1982). Viral diseases of the digestive system. *In* "The Mouse in Biomedical Research" (H. L. Foster, J. D. Small, and J. G. Fox, eds.), Vol. 2, p. 159. Academic Press, New York.

Lehmann-Grube, F. (1982). Lymphocytic choriomeningitis virus. *In* "The Mouse in Biomedical Research" (H. L. Foster, J. D. Small, and J. G. Fox, eds.), Vol. 2, p. 231. Academic Press, New York.

Lindsey, J. R., Cassell, G. H., and Davidson, M. K. (1982). Mycoplasma and other bacterial diseases of the respiratory system. *In* "The Mouse in Biomedical Research" (H. L. Foster, J. D. Small, and J. G. Fox, eds.), Vol. 2, p. 21. Academic Press, New York.

Lyon, M. F. (1981). Nomenclature. *In* "The Mouse in Biomedical Research" (H. L. Foster, J. D. Small, and J. G. Fox, eds.), Vol. 1, p. 28. Academic Press, New York.

Medina, D. (1982). Mammary tumors. *In* "The Mouse in Biomedical Research" (H. L. Foster, J. D. Small, and J. G. Fox, eds.), Vol. 4, p. 373. Academic Press, New York.

Morse, H. C., III, ed. (1979). "Origins of Inbred Mice." Academic Press, New York.

Morse, H. C., III (1981). The laboratory mouse—A historical perspective. *In* "The Mouse in Biomedical Research" (H. L. Foster, J. D. Small, and J. G. Fox, eds.), Vol. 1, p. 1. Academic Press, New York.

Newberne, P. M., and McConnell, R. G. (1982). Neoplasms of the digestive system. *In* "The Mouse in Biomedical Research" (H. L. Foster, J. D. Small, and J. G. Fox, eds.), Vol. 4, p. 490. Academic Press, New York.

Osborn, J. E. (1982). Cytomegalovirus and other herpesviruses. *In* "The Mouse in Biomedical Research" (H. L. Foster, J. D. Small, and J. G. Fox, eds.), Vol. 2, p. 267. Academic Press, New York.

Parker, J. C., and Reynolds, R. K. (1968). Natural history of Sendai infection in mice. *Am. J. Epidemiol.* **88,** 112–125.

Parker, J. C., and Richter, C. B. (1982). Viral diseases of the repiratory system. *In* "The Mouse in Biomedical Research" (H. L. Foster, J. D. Small, and J. G. Fox, eds.), Vol. 2, p. 110. Academic Press, New York.

Pleasants, J. R., Wostmann, B. S., and Reddy, B. S. (1973). Improved lactation in germfree mice following changes in the amino acid and fat components of a chemically defined diet. *In* "Germfree Research" (J. B. Heneghan, ed.), p. 245. Academic Press, New York.

Poiley, S. M. (1960). A systematic method of breeder rotation for non-inbred laboratory animal colonies. *Proc. Anim. Care Panel* **10,** 159–166.

Potter, M. (1983). Immunoglobulins and immunoglobulin genes. *In* "The Mouse in Biomedical Research" (H. L. Foster, J. D. Small, and J. G. Fox, eds.), Vol. 3, p. 347. Academic Press, New York.

Richter, C. B. (1973). Experimental pathology of Sendai virus infection in mice. *J. Am. Vet. Med. Assoc.* **163,** 1204.

Robison, W. G., Jr., Kuwabara, T., and Zwaan, T. (1982). Eye research. *In* "The Mouse in Biomedical Research" (H. L. Foster, J. D. Small, and J. G. Fox, eds.), Vol. 4, p. 69. Academic Press, New York.

Rygaard, J., and Polvsen, C. O. (1982). Athymic (nude) mice. *In* "The Mouse in Biomedical Research" (H. L. Foster, J. D. Small, and J. G. Fox, eds.), Vol. 4, p. 51. Academic Press, New York.

Schaedler, R. W., and Orcutt, R. P. (1983). Gastrointestinal microflora. *In*

"The Mouse in Biomedical Research" (H. L. Foster, J. D. Small, and J. G. Fox, eds.), Vol. 3, p. 327. Academic Press, New York.

Sheridan, J. F., Eydelloth, R. S., Vonderfecht, S. L., and Aurelian, L. (1983). Virus-specific immunity in neonatal and adult mouse rotavirus infection. *Infect. Immun.* **39,** 917–927.

Shorey, H. H. (1976). "Animal Communication by Pheromones." Academic Press, New York.

Sidman, R. L., Angevine, J. B., and Taber-Pierce, E. (1971). "Atlas of the Mouse Brain and Spinal Cord." Harvard Univ. Press, Cambridge, Massachusetts.

Staats, J. (1980). Standardized nomenclature for inbred strains of mice: 7th listing. *Cancer Res.* **40,** 2083–2128.

Staats, J. (1983). "Inbred Strains of Mice 13" *Companion Issue to Mouse Newsletter,* no. 69, p. 94. Jackson Lab., Bar Harbor, Maine.

Stanley, N. F. (1978). Diagnosis of reovirus infection: Comparative aspects. *In* "Comparative Diagnosis of Viral Diseases" (E. Kurstak and C. Kurstak, eds.), Vol. 1, Part A, pp. 385–421. Academic Press, New York.

Talal, N. (1983). Immune response disorders. *In* "The Mouse in Biomedical Research" (H. L. Foster, J. D. Small, and J. G. Fox, eds.), Vol. 3, p. 391. Academic Press, New York.

Theiss, J. C., and Shimkin, M. B. (1982). Neoplasms of the respiratory system. *In* "The Mouse in Biomedical Research" (H. L. Foster, J. D. Small, and J. G. Fox, eds.), Vol. 4, p. 477. Academic Press, New York.

Theuer, R. C. (1971). Effect of essential amino acid restriction on the growth of female C57BL mice and their implanted BW 10232 adenocarcinomas. *J. Nutr.* **101,** 223.

Thomas, L., Aleu, F., Bitensky, M. W., Davidson, M., and Gesner, B. (1966). Studies of PPLO infection. II. The neurotoxin of *Mycoplasma neurolyticum. J. Exp. Med.* **124,** 1067–1082.

Ward, D. C., and Tattersall, P. J. (1982). Minute virus of mice. *In* "The Mouse in Biomedical Research" (H. L. Foster, J. D. Small, and J. G. Fox, eds.), Vol. 2, p. 313. Academic Press, New York.

Weisbroth, S. H. (1982). Arthropods. *In* "The Mouse in Biomedical Research" (H. L. Foster, J. D. Small, and J. G. Fox, eds.), Vol. 2, p. 385. Academic Press, New York.

Wescott, R. B. (1982). Helminths. *In* "The Mouse in Biomedical Research" (H. L. Foster, J. D. Small, and J. G. Fox, eds.), Vol. 2, p. 373. Academic Press, New York.

Whittingham, D. G., and Wood, M. J. (1983). Reproductive physiology. *In* "The Mouse in Biomedical Research" (H. L. Foster, J. D. Small, and J. G. Fox, eds.), Vol. 3, pp. 137–164. Academic Press, New York.

Wilson, E. O. (1970). Chemical communication within animal species. *In* "Chemical Ecology" (E. Sondheimer and J. B. Simeone, eds.), pp. 133–155. Academic Press, New York.

Chapter 4

Biology and Diseases of Rats

Dennis F. Kohn and Stephen W. Barthold

I. INTRODUCTION

A. Origin and History

The diversity of research for which the laboratory rat is used is probably greater than that associated with any other animal. The laboratory rat is a descendent of the wild rat, *Rattus norvegicus,* which originated in Asia and reached Europe in the early 1700s. Wild and albino mutants were first used for experimental purposes in Europe in the mid-1800s and in the United States shortly before 1900. The Wistar Institute in Philadelphia was prominent in the development of the rat as a labo-ratory animal, for here originated many of the rat strains now used worldwide. Henry Donaldson and his colleagues at the Wistar Institute used these early rat strains for a variety of studies dealing with neuroanatomy, nutrition, endocrinology, genetics, and behavior. The history and evolution of the many rat strains used today have been recently summarized (Lindsey, 1979).

The most commonly used outbred rat stocks in North America are the Wistar, Sprague-Dawley, Long-Evans, and Holtzman. All are albino except the Long-Evans stock, which is usually marked with a black or gray hair coat over the shoulders and is sometimes referred to as a "hooded rat." There are numerous inbred and mutant rat strains, although the number is less than that in the mouse. Table I lists the more commonly used strains.

B. Sources and Nomenclature

There are a rather large number of commercial vendors of laboratory rats in the United States. Most of the stocks and strains mentioned above can be obtained from more than one source. Although the origin of an outbred stock, such as the Sprague-Dawley, may have been the same for a number of vendors, in many cases it has been 20 to 30 years since such a stock has been removed from its original breeding colony. Accordingly, the genotype of outbred stocks and inbred strains may vary among sources and be reflected by differences in data when multiple sources of rats are used. A standardized scheme of identifying stocks and strains of rats has been developed and is now used by nearly all commercial vendors. Moreover, it is important that authors correctly identify stocks and strains that are used in their research since the success in repeating the work in another laboratory may be dependent upon the genotype (source of the rat). Table II summarizes the standardized nomenclature for outbred stocks as developed by the

Table I

Commonly Used Strains

Inbred strains	Usefulness as models[a]
ACI	Congenital genitourinary anomalies, prostatic adenocarcinomas
BN (Brown-Norway)	Inducible, transplantable myeloid leukemia, hydronephrosis
BUF (Buffalo)	Spontaneous autoimmune thyroiditis, host for transplantable Morris hepatomas
F344 (Fischer 344)	Long-term xenobiotic toxicity, gerontology, esophageal and urinary bladder carcinoma
LEW (Lewis)	Multiple sclerosis, various experimentally induced autoimmune disease
SHR (spontaneous hypertensive rat)	Hypertension, myocardial infection
WF (Wistar-Furth)	Mononuclear cell leukemia

Mutant strains	Characteristics
Brattleboro	Diabetes insipidus (autosomal recessive)
Gunn	Jaundice, kernicterus (autosomal recessive)
Nude	T cell deficient (autosomal recessive)
Obese SHR	Type 4 hyperlipoproteinemia (autosomal recessive)

[a]National Institutes of Health (1981).

Table II

Nomenclature for Outbred Stocks

1. Letters preceding the colon designate the supplier/breeder code consisting of a capital and two or three lowercase letters
2. Capital letters following the colon are used by a breeder to identify his stock
3. Letters in parentheses denote origin of stock
4. Subscript symbols indicate rearing by means other than natural mother (f, fostered; fh, fostered by hand)

International Committee on Laboratory Animals (ICLA). Table III contains the scheme for designating inbred strains of rats (National Institutes of Health, 1981). "Animals for Research" (National Academy of Sciences, 1979), a directory of sources for laboratory animals sold in the United States and Canada, lists all rodents according to standard nomenclature, and is a valuable aid in purchasing laboratory animals.

C. Housing

Commercial production of rats has markedly changed since the 1960s due to the development of hysterectomy-derived and barrier-maintained breeding colonies. Prior to the application of this technology to production colonies, infectious diseases were ubiquitous in rats from most sources. Today, vendors can be selected who offer pathogen-defined animals for most stocks and strains. Concomitant with changes in commercial sources of rats are the major advances made in the design and construction of institutional animal resources and husbandry practices within them. Optimum housing of rats today includes provisions for quarantining and isolation of animals according to vendor subpopulations that have a similar microbial flora.

There are various levels of sophistication to provide barriers to the spread of infections in rat colonies. Since many rat pathogens are spread by aerosol, ventilation control is very important. Nonrecirculating room air or high-efficiency particulate air (HEPA)-filtered air has become a design standard in modern animal facilities. As discussed in Chapter 17, clean/contaminated corridor-designed facilities aid in containment against the spread of pathogens by aerosol, personnel,

Table III

Nomenclature for Inbred Rats

1. The strain designation is given in capital letters followed by a slash
2. The substrain designation follows the slash and is given as numbers or as individual or company codes. Numbers are used to denote substrains that were derived from a common strain but separated before the completion of inbreeding
3. Subscript symbols indicate rearing by means other than natural mother

and contaminated equipment. A more complete barrier system may include an entry area in which incoming supplies and equipment are sterilized and in which personnel shower and don sterile clothing and filter masks before entering animal rooms. More recently, laminar-flow (mass air displacement) rooms and mobile units have become popular because they can be incorporated in existing buildings that lack design characteristics mentioned above.

Environmental control in rat rooms is important to the comfort and health of the animals, as well as to the consistency of data derived from the rats. Room temperatures between 72 and 76°F are desirable, and the relative humidity should range between 40 and 60%. Daily fluctuations in temperature and humidity act as significant stressors. These fluctuations may be associated with the environmental control system of a building or may be induced by procedures such as cleaning floors with a water hose or high pressure sprayer. Twenty-four-hour temperature/humidity recorders are useful in detecting changes in environmental conditions. Light intensity should be evenly distributed to all animals within a room. Seventy-five to 125 fc have often been suggested as an optimal range for light intensity. However, recent evidence indicates that this intensity can induce retinal degeneration in albino rats (Anver and Cohen, 1979). Light-timing devices are a convenient means to provide desired day/night cycles such as 12–12 or 14–10 hr.

Caution should be exercised in the use of insecticides and air-deodorizing chemicals, since some have been shown to induce hepatic microsomal enzymes in rats. Accordingly, their use in animal rooms is not usually recommended (Baker et al., 1979a).

Rats can be housed in either wire- or solid-bottom cages. Wire-bottom cages are more frequently used since they are less labor-intensive. Frequency of cage and litter pan changing is a function of animal density. Solid-bottom cages should be sanitized two to three times per week, while wire-bottom cages should be sanitized on a weekly or biweekly schedule with litter pans changed two or three times per week. Feed should be contained in hoppers. Either automatic systems or bottles are satisfactory for providing water to rats. Some caution is necessary when using automatic systems, since weanling and newly arrived rats may not drink initially from such devices. To avoid undesirable microbial contamination, water bottles should be sanitized before they are refilled and automatic systems should be drained and flushed when racks are sanitized. Acidification of water to a pH of 2.5 to 2.8 or chlorination at 8 to 12 ppm will control *Pseudomonas aeruginosa* contamination of water (Weisbroth, 1979). However, this treatment is not necessary for immunocompetent animals. Wood shavings or chips are the most commonly used contact bedding materials. Hardwoods are preferred to softwoods, since the latter are capable of inducing hepatic microsomal enzymes (Baker *et al.*, 1979a).

II. BIOLOGY

A. Anatomy

This section summarizes some of the anatomical characteristics of the rat with emphasis on characteristics that are unique. The reader is advised to refer elsewhere in the literature for comprehensive descriptions (Bivin *et al.*, 1979; Caster *et al.*, 1956; Hebel and Stromberg, 1972; Smith and Calhoun, 1972; Zeman and Innes, 1963).

1. Digestive System

The rat dental formula is 2(I 1/1, C 0/0, PM 0/0, M 3/3) = 16. The incisors are well developed and grow continuously. The rat lacks tonsils and water taste receptors.

The major pairs of salivary glands are the parotid, submandibular (submaxillary), and sublingual. The parotid gland is a serous gland consisting of three to four lobes and is located ventrolaterally from the caudal border of the mandible to the clavicle. The submandibular glands are mixed glands located ventrally between the caudal border of the mandibles and the thoracic inlet. The sublingual glands are mucous glands and are much smaller than the parotid and submandibular glands. They are located at the rostral pole of the submandibular glands to which they are closely associated. Brown fat deposits are present in the ventral cervical region. These multilocular deposits are well demarcated and can be confused with salivary glands or lymph nodes.

The stomach of the rat is divided into two parts; the forestomach (nonglandular) and the corpus (glandular). The two portions are separated by a limiting ridge. The esophagus enters at the lesser curvature of the stomach through a fold of the limiting ridge. This fold is responsible for the inability of the rat to vomit. The forestomach, which is thinner than the corpus, is linked with an epithelium similar to that of the esophagus and extends from the cardia to a narrow band of cardiac glands at the junction of the glandular portion.

The small intestine is composed of the duodenum (10 cm), jejunum (100 cm), and ileum (3 cm). The cecum is a thin-walled, comma-shaped pouch that has a prominent lymphoid mass in its apical portion. The colon is composed of the ascending colon, with prominent oblique mucosal ridges, transverse and descending colons, with longitudinal mucosal folds; followed by a short rectum that is confined to the pelvic canal.

The liver has four major lobes (median, right lateral, left, and caudate) and is capable of regeneration subsequent to partial hepatectomy. The rat has no gallbladder. The bile ducts from each lobe form the common bile duct that enters the duodenum 25 mm from the pyloric sphincter.

The pancreas is a lobulated, diffuse organ that extends from the duodenal loop to the gastrosplenic omentum. It can be differentiated from adjacent adipose tissue by its darker color and firmer consistency. Up to 40 excretory ducts fuse into 2–8 large ducts, which empty into the common bile duct.

2. Respiratory System

The nasal cavity is not markedly different from that of other mammals. The rat has a maxillary recess (sinus) located between the maxillary bone and the lateral lamina of the ethmoid bone. The recess contains the lateral nasal gland (Steno's gland) that secretes a watery product that is discharged at the rostral end of the nasal turbinate. It has been postulated that the nonviscous secretion contributes to the humidification of inspired air and acts to regulate the viscosity of the mucous layer overlying the nasal epithelium.

The left lung has one large lobe, and the right lung is divided into four lobes (cranial, middle, accessory, and caudal). The pulmonary vein in the rat has cardiac striated muscle fibers within its wall that are contiguous with those in the heart. The rat does not have an adrenergic nerve supply to the bronchial musculature, and bronchoconstriction is controlled by vagal tone. Unlike the guinea pig, the rat lung has a low concentration of histamine (Bivin *et al.*, 1979).

3. Cardiovascular System

The heart and peripheral circulation in the rat differ little from that of other mammals. The blood supply to the heart is derived from both coronary and extracoronary arteries. The latter arise from the internal and subclavian arteries.

4. Genitourinary Systems

The right kidney, which is more craniad than the left, has its cranial pole at the L_1 vertebra and its caudal pole at the level of L_3. The rat kidney is unipapillate as are kidneys of other rodents, lagomorphs, and insectivores. Having only one papillus and calyx makes the rat useful for studies in which cannulization of the kidney is done. The presence of superficial nephrons in the renal cortex has made the rat widely used as a model for studying nephron transport in an *in vivo* micropuncture system.

The male reproductive system has a number of highly developed accessory sex glands. These include large seminal vesicles, a bulbourethral gland, and a prostate gland composed of the coagulation gland (dorsocranial lobe) and ventral and dorsolateral lobes. The inguinal canal remains open throughout the life of a rat and testes descend initially by 40 days of age.

The female rat has a bicornate uterus that is classified as the duplex-type because the lumina of the uterine horns are completely separate with paired ossa uteri and cervices. The female urethra does not communicate with the vagina or vulva, but rather exits at the base of the clitoris.

5. Central Nervous System

The brain of the rat has very large olfactory bulbs and a nonconvoluted cerebrum. The hypophysis is behind the optic chiasma and is attached to the base of the brain by a thin hollow stalk, the infundibulum. The blood supply to the brain is from the internal carotid and vertebral arteries. Blood leaves the brain via a system of sinuses that are enclosed in the dura mater. The ventricular system is similar to that of other animals, but the rat lacks a foramen of Magendie.

B. Normal Physiological Values

It must be recognized that many of the normal values determined for a specific group of rats may be accurate for only that rat stock/strain, source, and conditions under which they are held. Selected physiological, hematological, and clinical biochemical parameters are listed in Tables IV–VII. More complete information on biological values is available (Mitruka and Rawnsley, 1977; Ringler and Dabich, 1979).

C. Nutrition

Nutritionally adequate diets are readily available from commercial sources. These standard rations are quite satisfactory for most applications. However, for some types of experimentation there are factors, other than nutritional adequacy, which must be considered. The nutrient composition of diets and the contamination of feed components by mycotoxins, antibiotics, synthetic estrogens, heavy metals, and insecticides may have a profound impact on many studies. For instance, caloric intake and the percent of fat and protein in the diet of rats influence the incidence of neoplasia (Altman and Goodman, 1979). Similarly, various contaminants have an adverse effect on data from toxicologic, gerontological, and reproductive studies. Standard commercial diets are formulated from natural ingredients and will vary in nutrient composition on a batch-to-batch basis due to differences in type and quality of ingredients used. Commercial makers of rodent feeds take precautions to preclude the presence of contaminants in feeds, but only a few products have a defined profile of maximal levels of heavy metals, aflatoxins, chlorinated hydrocarbons, and organophosphates.

For some investigative purposes, feeds formulated with refined ingredients (purified diets) or with chemically defined compounds are useful when control of nutrient concentrations is essential (National Research Council, 1978). These diets are, however, too expensive for general use.

Table IV

Selected Normative Data[a]

Adult weight	
Male	300–400 gm
Female	250–300 gm
Life span	2.5–3 years
Body temperature	37.5°C
Basal metabolism rate (400 gm rat)	35 kcal/24 hr
Chromosome number (diploid)	42
Puberty	50 ± 10 days
Gestation	21–23 days
Litter size	8–14
Birth weight	5–6 gm
Eyes open	10–12 days
Weaning	21 days
Food consumption/24 hr	5 gm/100 gm body weight
Water consumption/24 hr	8–11 ml/100 gm body weight
Cardiovascular	
Arterial blood pressure	
Mean systolic	116 mm Hg
Mean diastolic	90 mm Hg
Heart rate	300–500 beats/min
Cardiac output	50 ml/min
Blood volume	6 ml/100 gm body weight
Respiratory	
Respirations/min	85
Tidal volume	1.5 ml
Alveolar surface area (400 gm rat)	7.5 m^2
Renal	
Urine volume/24 hr	5.5 ml/100 gm body weight
Na$^+$ excretion/24 hr	1.63 mEq/100 gm body weight
K$^+$ excretion/24 hr	0.83 mEq/100 gm body weight
Urine osmolarity	1659 mOsm/kg of H$_2$O
Urine pH	7.3–8.5
Urine specific gravity	1.04–1.07

[a]Data from Baker et al. (1979b) and Bivin et al. (1979).

Rats are commonly fed *ad libitum,* and food intake will vary according to requirements for growth, gestation, and lactation. The nutritive requirements for the rat are listed in Table VIII.

The duration of storage and the temperature at which feeds are stored prior to use effect the nutritive quality of diets. Commercial diets are formulated to have a shelf life of up to 6 months. However, storage in a hot or damp environment will reduce this shelf-life. To help assure that only fresh diets are used, products should be used which have milling dates identified on their containers (see Chapter 17).

D. Biology of Reproduction

1. Reproductive Physiology

Sexual maturity occurs between 6 and 8 weeks for both sexes, although the onset of first estrus in females occurs at about 5 weeks. The vagina opens between 34 and 109 days, and the testes descend between 15 and 51 days, although they remain fully retractable in adults. Rats ovulate spontaneously, but ovulation can also be induced by forced coitus during non-estrous intervals. Vaginal stimulation during mating is important in rat reproductive physiology. The more often a male inserts his penis into the vagina prior to ejaculation, the greater the probability of a resulting pregnancy. However, natural or artificial stimulation of the vagina within 15 min of a first mating will abrogate pregnancy from the first mating by inhibition of sperm transport. A 12-hr estrous period recurs every 4 to 5 days and after parturition, without seasonal variation. Estrus can be suppressed when females are housed in groups and synchronized in the presence of a male or its excreta (Whitten effect), but this effect is not as pronounced as in the mouse. Female fertility wanes at 600 to 650 days, but estrous cycles may continue through 32 months. Male fertility is lost between 16 and 20 months. Fertility of both sexes is generally regarded as maximal between 100 and 300 days of age (Adler and Zoluth, 1970; Baker, 1979; Farris, 1963; Lane-Petter, 1972; Leathem, 1979).

Males will mount estrous females numerous times with one or two rapid ejaculations in the course of 15 to 20 minutes. Ejaculated semen coagulates, forming a copulatory plug that remains in the distal vagina for a few hours, after which time it dissolves or is extruded. Copulation is usually nocturnal. Duration of gestation varies with strain, age, litter size, and other variables, and ranges from 19 to 23 days, with an average of 21 or 22 days. Primiparous females tend to have a slightly longer gestation than multiparous females (Farris, 1963).

2. Detection of Estrus and Pregnancy

Estrus can be detected in a number of ways. Females in estrus are hyperactive and brace themselves when touched. Their ears quiver when they are stroked on the head or back, and stimulation of the pelvic region induces lordosis (Farris, 1963). The vulva becomes swollen, and the vagina becomes dry in contrast to the moist pink wall during metestrus or diestrus. As proestrus occurs (approximately 12 hr), smears of vaginal cells contain nucleated epithelium, leukocytes, and occasional cornified cells. Estrus (approximately 12 hr) begins with about 75% nucleated and 25% cornified cells, with cornified cells predominating as estrus continues. Metestrus follows (approximately 21 hr) with large numbers of leukocytes and cornified cells, which form abundant caseous vaginal detritus. Metestrus is characterized by the presence of large flat nucleated (pavement) cells. Diestrus persists for approximately 57 hr (Baker, 1979; Farris, 1963).

Breeding dates can be established by examination of vaginal swabs for spermatozoa or examining the distal vagina or cage pan for copulatory plugs. Timed pregnancies are best achieved by placing the female in the male's cage in the afternoon and examining her for a plug or spermatozoa the following morning.

Table V

Hematological Parameters in the Rat[a,b]

Age (weeks)	Stock/strain	Erythrocyte parameters											
		Erythrocyte (×10^6/μl)		PCV[c] (%)		Hemoglobin (gm/dl)		MCV[d] (fl)		MCHC[e] (%)		Reticulocytes (%)	
		M	F	M	F	M	F	M	F	M	F	M	F
6	Hsd:SD(SD)BR	6.31	5.93	42	39	13.9	13.6	ND		ND		ND	
8	Crl:CD(SD)BR	7.69	7.25	41.7	40.7	15.6	14.9	54.5	55.6	40.1	40.8	ND	
16	Crl:CD(SD)BR	8.27	7.62	40.5	39.4	16.0	15.7	49.8	51.9	39.5	40.7	ND	
26	Hsd:SC(SD)BR	5.63	6.36	40	44	13.7	15.6	ND		ND		ND	
26	CDF(F344)/CrlBR	9.17	9.03	48.1	49.3	17.5	16.9	52.6	54.6	36.5	34.4	1.66	0.81
40	Crl:CD(SD)BR	8.45	7.49	41.1	38.5	16.1	15.4	48.4	51.2	38.9	39.9	ND	
52	CDF(F344)/CrlBR	9.13	8.41	47.2	46.8	16.2	15.7	51.8	55.9	34.3	33.5	0.46	1.69
78	CDF(F344)/CrlBR	9.60	8.32	54.5	46.7	18.9	16.3	56.8	56.3	34.7	35.1	1.99	1.80
104	CDF(F344)/CrlBR	9.26	8.20	57.8	45.8	19.5	15.1	62.4	55.9	33.7	32.9	3.08	1.25

Age (weeks)	Stock	Leukocyte parameters											
		WBC (×10^3 μl)		Neutrophil[d]		Lymphocyte[f]		Monocyte[f]		Eosinophil[f]		Basophil[d]	
		M	F	M	F	M	F	M	F	M	F	M	F
6	Hsd:SD(SD)BR	9.8	6.3	16	18	80	75	3	4	1	3	0	0
8	Crl:CD(SD)BR	17.2	10.0	15.7	19.3	81.2	77.7	2.2	1.9	0.7	1.0	0.2	0.3
16	Crl:CD(SD)BR	14.9	10.3	13.9	14.1	82.8	83.1	2.4	2.0	0.8	0.8	0	0
	CDF(F344)/CrlBR	10.2	13.8	23.8	31.7	73.4	66.0	2.3	1.8	0.9	0.5	0	0

[a]T. E. Hamm, unpublished data.
[b]Ringler and Dabich (1979).
[c]PCV, packed cell volume.
[d]MCV, mean corpuscular volume.
[e]MCHC, mean corpuscular hemoglobin content.
[f]Number per 100 cells counted.

Table VI

Clinical Chemistry Parameters in the Rat[a,b]

Age (weeks)	Stock/strain	Fasting blood glucose (mg/dl)		BUN[c] (mg/dl)		SGPT[d] (IU/liter)		ALK PHOS[e] (IU/liter)	
		M	F	M	F	M	F	M	F
6	Hsd:SD(SD)BR	163	169	8	8	89	92	ND	
26	Hsd:SD(SD)BR	219	217	7	6	75	61	ND	
26	CDF(F344)/CrlBR	ND		17.9	16.1	28.6	19.8	63.6	40.0
30	Crl:CD(SD)BR		134	15.4		ND		22	16
52	CDF(F344)/CrlBR	ND		22.9	20.0	39.5	26.5	95.9	58.5
78	CDF(F344)/CrlBR	ND		16.6	18.0	24.0	23.5	53.4	59.5
104	CDF(F344)/CrlBR	ND		23.9	19.3	21.2	21.2	66.7	53.9

[a]T. E. Hamm, unpublished data.
[b]Ringler and Dabich (1979).
[c]BUN, blood urea nitrogen.
[d]SGPT, serum glutamic pyruvic transaminase.
[e]ALK PHOS, alkaline phosphatase.

Table VII

Growth Rates in Rats[a]

Weight (gm)	Crl:CD(SD)BR (days old)		Hla:(SD)BR (days old)		Crl:(LE)BR (days old)		Crl:(WI)BR (days old)		Hla:(WI)BR (days old)		CDF(F344)/ CrlBR (days old)		NIH:(F344)/ HlaBR (days old)	
	M	F	M	F	M	F	M	F	M	F	M	F	M	F
51–75	22–26	22–30	21–26	21–27	25–29	25–30	22–30	21–25	21–25	21–26	29–35	28–32	29–35	32–39
76–100	27–30	31–35	26–30	27–32	30–33	31–34	31–35	26–30	25–28	26–30	36–42	33–39	35–42	39–46
101–125	31–35	36–40	30–33	32–35	34–37	35–40	36–39	31–35	28–32	30–33	43–49	40–46	42–48	46–58
126–150	36–42	41–47	33–37	35–38	38–41	41–48	40–43	36–42	32–35	33–36	50–56	47–58	48–56	58–70
151–175	43–46	48–54	37–40	38–43	42–45	49–55	44–48	43–48	35–37	36–41	57–63	59–70	56–63	
176–200	47–50	55–65	40–43	43–51	46–49	56–70	49–51	49–57	37–40	41–48	64–69		63–69	
201–225	51–55	66–75	43–47	51–60	50–53	71–90	52–56	58–70	40–43	48–53			69–75	
226–250	56–60	76–84	47–50	60–70	54–58	91–107	57–61	71–85	43–46	53–62				
251–275	61–65		50–52		59–66		62–68	86–98	46–48					
276–300	66–70		52–55		67–76		69–77		48–51					
301–325	71–74		55–60		77–88		78–85		51–56					
326–350	75–80		60–64		89–101		86–94		56–59					
351–375	81–87		64–70		102–117		95–103		59–63					

[a]Adapted from vendor data.

Abdominal enlargement becomes evident at about 2 weeks. Pseudopregnancy is rare (Lane-Petter, 1972).

3. Husbandry Needs

Rats reproduce successfully under a variety of conditions, but husbandry practices can significantly influence fecundity. Rats can be bred as monogamous pairs, taking advantage of postpartem estrus for maximal breeding efficiency. Polygamous breeding is more economical, since only one male can be kept with 6 to 9 females. Females are often removed to a separate cage prior to whelping, since they may not tolerate other females in the cage while nursing. They will tolerate their mates, however. Females with litters do best on clean dust-free wood shavings in solid-bottom cages. Due to heat regulation, pups neither thrive in overly spacious cages with wide flutuations in ambient temperature, nor in overly crowded cages where they cannot dissipate heat. The recommended cage floor area for a female and her litter is 150 in.[2]. Ambient room temperature and humidity should be within the acceptable range with minimal fluctuation. High ambient temperature can cause male infertility (Baker, 1979; Baker et al., 1979a; Lane-Petter, 1972).

The rat estrous cycle is particularly sensitive to variations in light. Daily lighting at an average of 100 fc with a spectrum approximating natural light for 12 to 16 hr is best for breeding. Constant light for as few as 3 days may induce persistent estrus, hyperestrogenism, polycystic ovaries and endometrial hypertrophy or metaplasia (Baker et al., 1979a; Gralla, 1981).

Nutrition may also affect reproductive performance. Requirements for certain components are increased during pregnancy and growth, but overfeeding is deleterious. Caloric restriction may actually improve fertility and possibly reproductive life of the female (Leathem, 1979). Excess dietary protein can adversely affect female sexual development. Vitamin deficiencies can cause infertility, particularly those vitamins (A, E, riboflavin, and thiamin) that are most labile to autoclaving or deterioration (Baker, 1979).

4. Parturition

It is not necessary to add nesting material to bedding for successful breeding, but rats will utilize it if offered. Shredded paper or cotton nesting material will be readily accepted and used by prepartem and nursing dams. Parturition is heralded by pronounced postural stretching and rear leg extension. A vaginal discharge may be noted $1\frac{1}{2}$–4 hr prepartum. Parturition is usually complete in 1 or 2 hr, but can range from a few minutes to several hours depending on litter size. Dystocia is exceedingly rare. Litters average between 6 and 12 pups, with highest fecundity through the sixth litter. Inbred rats tend to produce smaller litters. Although infrequent, cannibalism is most apt to occur with nervous or primiparous females subjected to stress (Farris, 1963; Lane-Petter, 1972; Leathem, 1979).

5. Early Neonatal Development

The neonate weighs about $5\frac{1}{2}$ gm, depending on litter size, sex, strain, and physical condition of the dam. Pups are born hairless, blind, with closed ears, undeveloped limbs, and short

Table VIII

Nutrient Requirements of Rats[a]

Nutrient	Concentration in a diet[b]	
	Growth, gestation, or lactation	Maintenance
Protein (as ideal protein)	12.00%	4.20%
Fat[c]	5.00%	5.00%
Digestible energy	3800.00 kcal/gm	3800.00 kcal/gm
L-Amino acids		
Arginine	0.60%	—
Asparagine	0.40%	—
Glutamic acid	4.00%	—
Histidine	0.30%	0.08%
Isoleucine	0.50%	0.31%
Leucine	0.75%	0.18%
Lysine	0.70%	0.11%
Methionine	0.60[d]	0.23%
Phenylalanine-tyrosine	0.80[e]	0.18%
Proline	0.40%	—
Threonine	0.50%	0.18%
Tryptophan	0.15%	0.05%
Valine	0.60%	0.23%
Nonessential[f]	0.59%	0.48%
Minerals		
Calcium	0.50%	
Chloride	0.05%	
Magnesium	0.04%	
Phosphorus	0.40%	
Potassium	0.36%	
Sodium	0.05%	
Sulfur	0.03%	
Chromium	0.30 mg/kg	

Table VIII (*Continued*)

Nutrient	Concentration in a diet[b]	
	Growth, gestation, or lactation	Maintenance
Copper	5.00 mg/kg	
Fluoride	1.00 mg/kg	
Iodine	0.15 mg/kg	
Iron	35.00 mg/kg	
Manganese	50.00 mg/kg	
Selenium	0.10 mg/kg	
Zinc	12.00 mg/kg	
Vitamins		
A[g]	4000.00 IU/kg	
D[g]	1000.00 IU/kg	
E[g]	30.00 IU/kg	
K_1	50.00 gm/kg	
Choline	1000.00 mg/kg	
Folic acid	1.00 mg/kg	
Niacin	20.00 mg/kg	
Pantothenate (calcium)	8.00 mg/kg	
Riboflavin	3.00 mg/kg	
Thiamin	4.00 mg/kg	
Vitamin B_6	6.00 mg/kg	
Vitamin B_{12}	50.00 μg/kg	

[a] From National Research Council (1978).

[b] Adequate to support growth, gestation, and lactation; based on 90% dry matter.

[c] Linoleic acid, 0.6%, is required.

[d] One-third to one-half can be supplied by L-cystine.

[e] One-third to one-half can be supplied by L-tyrosine.

[f] Mixture of glycine, L-alanine, and L-serine.

[g] Vitamin A, 1 IU = 0.300 μg retinol, 0.344 μg retinyl acetate, 0.550 μg retinyl palmitate. Vitamin D, 1 IU = 0.025 μg ergocalciferol. Vitamin E, 1 IU = 1 mg DL-α-tocopheryl acetate.

tail. The ears open between $2\frac{1}{2}$ and $3\frac{1}{2}$ days; incisors erupt between 8 and 10 days; and eyes open between 12 and 16 days. They are fully haired between 7 and 10 days (Baker, 1979; Farris, 1963; Lane-Petter, 1972). Maternal antibody is transferred *in utero,* via the yolk sac and by intestinal absorption of colostrum by the neonate for up to 18 days after birth (Cheville, 1976). Optimal weaning age is 20–21 days, although pups can be weaned as early as 17 days.

6. Sexing

Differentiation of sex in adult rats is relatively easy after the testes descend. The adult testes can be readily retracted through large inguinal canals. Male neonates have a larger genital papillus and the anogenital space is greater in males than females.

7. Artificial Insemination

Artificial insemination can be achieved in rats, but the major obstacle is the coagulative properties of their semen. Sperm can be obtained by maceration of the epididymis and vasa or by electroejaculation, although the latter method is unreliable and the semen often rapidly coagulates. Coagulation can be eliminated by prior surgical extirpation of the seminal vesicles and coagulating glands without significant effect on fertility. Semen can be diluted with a number of media but frozen storage of rodent semen has met with little success. Insemination can be achieved surgically by direct injection of seminal fluid into the uterus and by nonsurgical means. Successful conception seems to require not only insemination during estrus but also induction of pseudopregnancy by mating with a vasectomized male or mechanical stimulation within a few hours (before or after) insemination. Egg harvest for transfer can be accomplished by excision of the preovulatory ovaries and teasing from gravid follicles or recovery from the oviduct or uterus by flushing with transfer medium. Superovulation by injection of gonadotropisms may enhance yield, but is usually not necessary. Eggs are generally injected directly into the uterus but the recipient uterus must be at the same stage of the uterine cycle (Bennet and Vickery, 1970).

8. Synchronization

Synchronization of estrus can be achieved by vaginal insertion of polyurethane sponges containing 0.75 mg medroxyprogesterone for 7 days. Females are then put in a cage previously occupied by male rats, sponges are removed, and the rats are injected with 3 IU of pregnant mare's serum. Within 34 hr, 93% will be in estrus. This can also be attained by administering 40 mg medroxyprogesterone in 200 ml ethanol/liter drinking water, prepared fresh daily for 6 days, then intramuscular injection of 1 IU of pregnant mare's serum (Bennet and Vickery, 1970).

E. Behavior

The rat has been utilized extensively in a variety of research fields, including behavioral science. Rats are docile, adapt to new surroundings, tend to explore, and are easily trained to a variety of sensory cues by positive or negative reinforcement. Rats sleep during daylight hours and activity, including feeding, is greater during the night and early morning. Laboratory rats are easily handled, but strain differences exist. Sprague-Dawley and LEW rats tend to be less fractious than Long Evans or F344 rats. Docility is improved with routine and proper handling. Rats become nervous and refractory to handling when they hear others squeal. Nutritional deficiency, particularly hypovitaminosis A, and mishandling can make rats vicious. Rats seek entry into small openings, a trait that is utilized for coaxing them into restrait apparatus. Like other rodents, rats are coprophagic, which must be taken into consideration when administering drugs, measuring fecal output, or performing nutritional studies.

Unlike mice, rats are less apt to fight, and males can be housed together. In addition, rats are not gregarious like mice, and seem to tolerate single caging well. Experimental studies indicate significant changes in plasma corticosteroid levels, depending on cage cohort size. Levels tend to be least in rats housed singly, to increase in groups up to 5, to decrease in larger groups up to 10–12, then rise again in groups up to 30 (Lane-Petter, 1972).

III. DISEASES

A. Infectious Diseases

Infectious agents constitute a significant environmental variable that impacts on research data derived from laboratory rats. As is the case with other species, infectious agents induce a wide range of diseases in the rat that vary from inapparent to overt clinical disease. Most investigations use large numbers of rats in which a specific group or colony consists of several to hundreds of rats. Accordingly, emphasis on disease is one of prevention and placed at the colony level rather than on a single or a few animals. Curative use of antibiotics, which is important in the treatment of bacterial diseases of nonrodent species, is rarely useful in the laboratory rat. Administration of drugs to obtain therapeutic blood levels is difficult to achieve in a colony; also some animals may improve clinically but remain colonized by the pathogen and serve as carriers, reinfecting other animals.

Rats seldom show clinical signs of disease upon arrival to the laboratory from commercial sources. However, these rats may harbor pathogens that are of low to moderate virulence and that are capable of severely compromising the health of animals when the rats are exposed to various types of experimental stress. Moreover, some of these pathogens may never cause clinical disease, yet induce microscopic lesions or biochemical aberrations that can have profound effects on research data. For these reasons, investigators and clinicians should be aware of the pathogen status of the animals used in studies, both initially and throughout the course of the studies. This section on infectious diseases contains those agents that are of principal importance to the investigative use of the rat.

1. Bacterial Diseases

a. Streptococcosis. The causative organism, *Streptococcus pneumoniae,* is a gram-positive coccus that is rather ubiquitous among humans and animals. *Streptococcus pneumoniae* is frequently recovered from respiratory tract lesions in guinea pigs, nonhuman primates, and some domestic animals. In humans, it is often present in the nasopharynx in the absence of clinical symptoms of infection. Upper respiratory tract infection of conventionally raised rats has been reported to be common. However, it is seldom present in barrier-maintained, commercial rat sources. As in pneumococcal disease in humans, a number of serological types have been associated with respiratory disease in rats.

Streptococcus pneumoniae infection in rats often remains localized in the nasopharynx without the development of overt disease. A shift in the host–parasite balance due to stress or concurrent infection with another pathogen may result in bronchopneumonia and bacteremia. The most common signs of respiratory disease are serous to mucopurulent nasal discharge and "red tears" due to porphyrin pigments secreted from the Harderian glands, dyspnea, rales, and depressed activity. Animals will often die within a few days after the onset of pneumonic signs. The severity and prevalence of clinical disease within an infected colony are associated with environmental conditions that induce stress (e.g., experimental manipulation, overcrowding, fluctuations in ambient temperature and humidity, and copathogens). Although all age groups are susceptible to infection and clinical disease, young animals are more apt to be clinically affected. Transmission between rats is by aerosol droplet. Although both humans and rats can carry the same serotypes of *S. pneumoniae,* the authors are unaware of evidence indicating zoonotic or human-to-animal transmission.

The most characteristic gross lesions are pulmonary consolidation and fibrinopurulent pleuritis and pericarditis (Fig. 1). An extensive fibrinopurulent peritonitis, orchitis, or meningitis may occur as well. If a bacteremia occurs early, the disease may be acute with few gross lesions. *Streptococcus pneumoniae* induces an outpouring of exudate rich in fibrin, neutrophilic leukocytes, and erythrocytes into the alveoli. Bronchioles are filled with neutrophilic leukocytes. Embolic lesions may occur in multiple tissues which include the spleen, liver, kidneys, joints, and brain.

Streptococcosis is diagnosed by clinical signs, characteristic

lesions, and isolation of *S. pneumoniae* from lesions. The pericarditis, pleuritis, and pleural effusion noted above differentiate pneumococcal disease from pneumonia due to *Mycoplasma,* although the two pathogens often are superimposed. This organism produces an α-hemolysis on blood agar plates similar to that of the *Streptococcus viridans* group. *Streptococcus pneumoniae* isolates are most commonly differentiated from nonpathogenic *S. viridans* by the sensitivity of the former organism to Optochin (hydrocuprein hydrochloride). Optochin-impregnated discs are placed on a blood agar plate which has been inoculated with a pure culture of the clinical isolate. If the isolate is *S. pneumoniae,* a distinct zone of growth inhibition will be present around the disc. Although typing of *S. pneumoniae* isolates is seldom done today, one can type an isolate by reacting known specific *S. pneumoniae* antisera with *S. pneumoniae* isolates. This serological test is the Neufeld–Quellung reaction and is based on the capsular swelling that is induced by specific antiserum.

There is no effective means to control *S. pneumoniae* infection once it is enzootic in a colony. Benzathine penicillin (30,000 units/200 gm body weight) may be helpful in reducing the severity of the disease and as an aid in limiting infections to a subclinical mode in some animals. However, antibiotics will not eliminate the organism from rat colonies. Hysterectomy rederivation of breeding stock from infected colonies is an effective method of initiating new stock free from pneumococcal infection (Weisbroth, 1979).

b. Pseudotuberculosis (Corynebacteriosis). The causative agent of pseudotuberculosis is the gram-positive bacillus, *Corynebacterium kutscheri.* On occasion, other *Corynebacterium* species can cause similar syndromes in rats. Typically, the organism causes inapparent infections in rats, with exacerbation of respiratory disease under conditions of stress. When clinically ill, the most commonly seen signs include serous oculonasal discharge, dyspnea, anorexia, and loss of weight or retarded growth. Animals with severe pulmonary signs usually succumb within several weeks, while rats with less severe signs often survive much longer. Most rats will have inapparent infections in which *C. kutscheri* cannot be isolated from internal organs. Little is known concerning how *C. kutscheri* is carried or transmitted within a colony. It has been suggested that the organism is transmitted via aerosol droplet or direct contact. Once rats are infected, a hematogenous spread may be involved, since lung lesions are initially interstitial and not bronchial.

Gross lesions are characterized by a variable number of grayish-yellow foci surrounded by red zones, particularly in the lung (Fig. 2). In longer-standing cases, individual foci coalesce into raised lesions 1 cm or larger in diameter. Occasionally, fibrous adhesions occur between the lungs and thoracic walls. Similar lesions may be seen in other organs, including

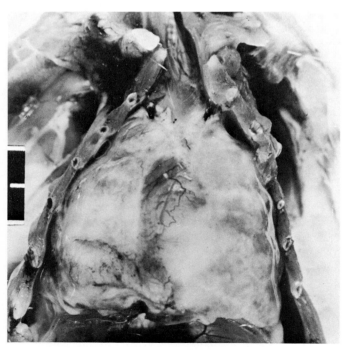

Fig. 1. *Streptococcus pneumoniae* infection. Thoracic viscera are covered with fibrinopurulent exudate.

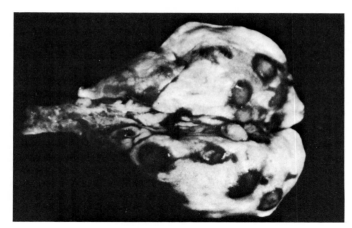

Fig. 2. Corynebacterium kutscheri infection. Lungs contain multiple pale nodular foci surrounded by zones of hyperemia.

the liver, brain, and kidneys. The hepatic lesions resemble tubercles and have caseous centers and fibrous capsules. Preputial adenitis, arthritis, and otitis media may also be caused by *C. kutscheri.*

The lesions in various target organs appear to be due to septic emboli. Pulmonary lesions initially consist of a polymorphonuclear cell and macrophage infiltrate of the bronchioles and interstitial tissue with a round cell infiltrate occurring later. Bronchi become impacted with polymorphonuclear cells and necrotic leukocytes. Giemsa or Gram staining of infected tissues will reveal the rod-shaped *C. kutscheri* organisms.

Diagnosis of *C. kutscheri* infection is made on clinical signs, gross and microscopic lesions, and isolation of the bacterium from infected tissues. Although the respiratory signs are similar to those present with mycoplasmosis, the rapidity with which *C. kutscheri* clinically affected rats succumb helps differentiate it from *Mycoplasma pulmonis*-induced disease. Unlike streptococcosis, fibrinopurulent pericarditis, peritonitis, and pleural effusion are not seen. Whereas peribronchial lymphoid hyperplasia is a dominant lesion in mycoplasmosis, it is unremarkable in *C. kutscheri* infections. *Corynebacterium kutscheri* is easily recovered from lesions and upper respiratory tract exudates by culturing on blood agar plates incubated aerobically at 37°C.

Epizootics of pseudotuberculosis may occur in conventionally raised breeding colonies, but rarely occur in barrier-raised colonies. Epizootics often can be retrospectively associated with an environmental stress (e.g., fluctuation in ambient temperature or ventilation). Culling of ill animals will not eliminate *C. kutscheri* from animals remaining in a colony. Isolation of the organism from animals with subclinical infections is not usually successful. For this reason, cortisone administration has been advocated as a means for surveillance of infection in colonies prior to necropsy and culturing for *C. kutscheri.* In the past, most serological methods have been un-

satisfactory in detecting antibody in animals with inapparent infections (Weisbroth, 1979). Recently, however, enzyme-linked immunoabsorbant assay (ELISA) has been shown to be capable of detecting antibody in animals without clinical signs of infection (Ackerman *et al.,* 1984). Hysterectomy derivation is an effective means to establish a *C. kutscheri*-free colony. Antibiotic therapy will not eliminate *C. kutscheri* from a colony, but a 7-day regimen of penicillin has been reported to be effective in curtailing an epidemic of *C. kutscheri*-induced pneumonia (Fox *et al.,* 1979).

Since *C. kutscheri* infection is, in most cases, inapparent and manifests itself whenever the host is sufficiently stressed, it can be a significant problem in experimentally stressed rats.

c. Tyzzer's Disease. Tyzzer's disease is caused by the gram-negative, spore-forming rod, *Bacillus piliformis.* This organism, which is not a true *Bacillus,* is an intracellular pathogen that has not been cultivated on artificial media, and is, as yet, taxonomically undefined. In the laboratory, *B. piliformis* is propagated in the yolk sac of embryonated chick eggs.

This disease occurs in other rodent species and appears to be widely distributed in many nonrodent species, but there appears to be a degree of species specificity among *B. piliformis* strains. It occurs occasionally in conventionally raised rat colonies. The vegetative form of *B. piliformis* is unstable in the environment. However, spores of the organism are relatively stable and are believed to be the source of transmission among animals.

Clinical signs associated with Tyzzer's disease are not particularly distinctive and, accordingly, only suggestive in making a diagnosis. Typically, affected rats are apt to be adolescents with signs such as lethargy, weight loss, and distended abdomens. Diarrhea is not a common sign in rats with *B. piliformis* infection. Animals displaying clinical signs generally die within several weeks. Clinically inapparent infections occur and are most probably responsible for transmission of the organism within a colony. Clinically evident Tyzzer's disease is usually associated with experimentation that compromises the immunocompetence of rats.

The most remarkable gross lesions involve the liver, ileum, and myocardium. Hepatic lesions consist of numerous small, pale foci on the surface and within the parenchyma. The intestinal lesion has been termed ''megaloileitis'' due to a segmental dilatation and inflammation of the ileum (Fig. 3) (Jonas *et al.,* 1970). Ileal distension is not always present. In some rats, circumscribed gray foci also occur in the myocardium.

The pathogenesis of the disease is believed to involve a primary intestinal infection with spread to the liver via the portal circulation. *Bacillus piliformis* invades enterocytes, resulting in villus shortening, inflammation, necrosis, and hemorrhage. Intracellular organisms are demonstrable in epithelium of

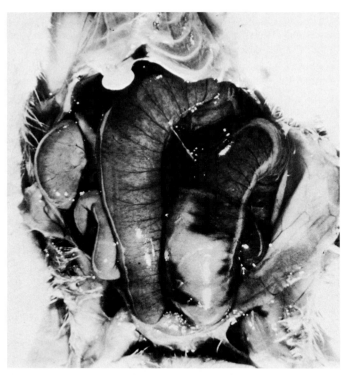

Fig. 3. *Bacillus piliformis* infection (Tyzzer's disease). The distal small intestine is dilated, congested, and hemorrhagic; also referred to as megaloileitis. (From A. M. Jonas, D. H. Percy, and J. Craft. Tyzzer's disease in the laboratory rat. *Archives of Pathology* **90**, 516–528. Copyright 1970, American Medical Association.)

crypts and villi. The necrotic foci in the liver are most often present near vessels. Surrounding these foci are varying numbers of leukocytes, macrophages, and fibroblasts. Intracytoplasmic bacteria may be seen in hepatocytes at the periphery of the lesions, but may be present in very small numbers and thus be hard to find. Organisms are also found in myocardium around foci of necrosis (Weisbroth, 1979).

A presumptive diagnosis can be made by the gross lesions, but a definitive diagnosis is dependent upon observation of the organism within hepatocytes, intestinal epithelium, or myocardium. Impression smears of liver taken at necropsy and stained with Gram, Giemsa, or methylene blue stains may be useful for a rapid diagnosis. However, formalin-fixed specimens stained by Giemsa or Warthin–Starry methods are usually performed to confirm a diagnosis. The ileal distension seen in rat Tyzzer's disease must be differentrated from other causes of adynamic ileus, particularly chloral hydrate-induced lesions.

Prevention of Tyzzer's disease in a colony is dependent upon a barrier that excludes entry of the agent by contaminated cages, equipment, and infected animals. Routine cage sanitation probably is ineffective in killing the spores of *B. piliformis*, but exposure of spores to 80°C for 30 min has been shown to inactivate them. Sodium hypochlorite (0.3%) is an effective disinfectant (Ganaway, 1980). Although antibiotics have been shown to be effective under experimental conditions in mice, there is no evidence to indicate that antibiotic therapy can be of value under natural conditions within a colony of rats (Weisbroth, 1979).

d. Pasteurellosis. *Pasteurella pneumotropica* frequently infects conventionally raised rats and has been recovered occasionally in rats from barrier- and axenic-maintained colonies. It is a pathogen of very low virulence, and most infections remain clinically inapparent. Only a relatively few reports document *P. pneumotropica* as a primary pathogen in cases of penumonia, otitis media, and conjunctivitis. As a copathogen with either *Mycoplasma pulmonis* or Sendai virus, it has a contributory role in the resultant respiratory lesions and otitis.

Its localization is not limited to the respiratory tract, since it is frequently isolated from the oral cavity, intestinal tract, and uterus. It also has been associated with mastitis and furunculosis in rats. It has been suggested that *P. pneumotropica* is essentially an enterotropic rather than a pneumotropic organism. The intestinal tract is probably the primary site for localization of the organism in subclinical infections.

Horizontal transmission is by the oral–fecal route and direct contact. Since *P. pneumotropica* is frequently carried in the uterus, vertical transmission can occur, and, accordingly, this can compromise the microbial status of axenic and gnotobiotic colonies.

Distinctive clinical signs and lesions do not occur with *P. pneumotropica*-induced disease. Accordingly, a diagnosis must be based upon its isolation as the sole pathogen or, as in many cases, as a copathogen within lesions. Blood agar medium is satisfactory for primary isolation from nonenteric sites. However, for recovery from the intestinal tract, enrichment in a medium such as GN broth is recommended before isolation is attempted on blood agar plates (Weisbroth, 1979).

Hysterectomy derivation and barrier maintenance are the only means to control infection. However, particular attention must be made to ensure that hysterectomy-derived young came from dams that had culturally negative uteri. Antibiotic therapy is not effective in eliminating the organism from a colony.

e. Salmonellosis. *Salmonella* species that infect rats include *Salmonella enteritidis*, *S. typhimurium*, *S. dublin*, and *S. meleagridis*. Salmonellosis, which was once a major cause of disease in laboratory rat and mouse colonies, is rarely reported in either species today. However, it still exists in wild populations of rodents and, therefore, remains a potential threat to laboratory rodents.

Infection in an immunologically naive colony typically results in an epizootic of clinically affected rats and a varying proportion of animals with inapparent infection. These latter animals act as subclinical carriers to render the infection as enzootic in a colony. Acute outbreaks will occur intermittently whenever immunological and other host defense mechanisms are altered. Signs associated with salmonellosis in the rat are anorexia, depressed activity, starry hair coats, and soft to formless feces. Affected animals die in 1 to 2 weeks.

Lesions that occur in salmonellosis differ depending on the stage of the disease. Salmonellae penetrate the intestinal mucosa at the level of the ileum and cecum. The earliest lesions occur in this locale and consist of a mild dilatation, thickened intestinal walls, and a granular mucosal surface. Involvement of the reticuloendothelial system is reflected by enlarged Peyer's patches, mesenteric lymph nodes, and spleen. In some infected animals, a bacteremic state occurs that results in the demise of the host before the development of further lesions. However, in animals not succumbing to septicemia, ulceration of the ileal, colonic, and cecal mucosa occurs. Histologically, the villus epithelium of the ileum is markedly degenerated, and the lamina propria is infiltrated with neutrophils and macrophages. Concomitant with intestinal lesions is the development of focal necrosis and granulomas in the spleen and liver due to hematogenous spread of the organism (Buchbinder *et al.*, 1935; Maenza *et al.*, 1970).

In rats who are intermittent or chronic shedders of salmonella, the most remarkable lesions are lymphadenitis of the mesenteric lymph nodes and ulceration of the cecal mucosa. Rats from which salmonella is chronically shed have more advanced lesions than do intermittent shedders of the organism.

A diagnosis of salmonellosis relies upon identification of an isolate as a *Salmonella* sp. Recovery of salmonella from the intestines, spleen, and liver is readily accomplished in rats clinically affected during an epizootic. However, this is not true for asymptomatic carriers, since some will shed the organism intermittently in the feces, and recovery from tissues is difficult. Recovery in carrier animals is best accomplished by initial incubation of fecal pellets in an enrichment broth, such as selenite F plus cystine broth, followed by streaking onto brilliant green agar (Weisbroth, 1979). From this medium, possible salmonella colonies are inoculated into triple-sugar-iron slants. Final identification is then made by biochemical tests and serotyping.

Prevention of this disease is based upon the exclusion of wild rodents from laboratory animal facilities and the use of only feed and bedding that has been properly processed and packaged to ensure against salmonella contamination.

f. Pseudomoniasis. *Pseudomonas aeruginosa*, a ubiquitous gram-negative bacterium found in soil and water, colonizes plants, insects, animals, and humans. It often colonizes the oropharynx and can be isolated from the intestinal tract of rodents. Infection with this organism in immunocompetent rats is nearly always inapparent. However, when rats are immunosuppressed, *P. aeruginosa* invades the upper respiratory mucosa and cervical lymph nodes, becomes bacteremic and induces an acute, lethal disease. In some cases, rats develop facial edema, conjunctivitis, and nasal discharge. In genet-ically thymic-deficient rats (nude), retro-orbital abscesses may occur prior to bacteremia.

Transmission in laboratory rodents occurs primarily by direct contact and contaminated water bottles and automatic watering systems. Phenolics are usually effective disinfectants, but quaternary ammonium compounds may actually support its growth.

Diagnosis of pseudomoniasis is based upon a history of immunosuppression associated with an epizootic of acute disease and isolation of *P. aeruginosa* from the blood and organs of affected rats. Facial edema in affected rats must be differentiated from viral sialodacryoadenitis.

Pseudomonas aeruginosa grows well on blood agar and most other standard laboratory media. Most strains are β-hemolytic and produce a bluish-green pigment, pyocyanin, as well as fluorescein. The use of specialized media (Pseudomonas P agar) enhances pigment production. The organism derives energy from carbohydrates via oxidation rather than fermentative metabolism. Identification of isolates as *P. aeruginosa* is easily made by the above characteristics and appropriate biochemical reactions (Weisbroth, 1979).

In most research applications, *P. aeruginosa*-free rats are not necessary for the conduct of the work. It is a major problem, however, in rats used for burn research and in studies in which drugs or radiation induce immunosuppression. Infection can be relatively well controlled in a colony by hyperchlorinating drinking water at 12 ppm or by acidification of water to a pH of 2.5–2.8. In a closed colony, it is also advisable to remove rats that remain culturally positive after water treatment has been instituted. In studies requiring pseudomonas-free rats, isolators are useful in which a gnotobiotic environment can be achieved. Alternatively, laminar flow units may suffice if supplies and equipment are sterilized and personnel wear sterile garments.

g. Streptobacillosis. *Streptobacillus moniliformis* is a commensal bacterium often present in the nasopharynx of conventionally raised rats. Although it may be involved occasionally as a secondary invader within inflammatory lesions of the rat, the chief importance of *S. moniliformis* is that it is the principal agent causing rat-bite fever in humans (Anderson *et al.*, 1983). The other bacterium associated with this clinical syndrome is *Spirillum minus*. Clinical signs in humans usually occur within 10 days of a rat bite and consist of headache, weakness, fever, a generalized rash, and arthritis. Often clinical signs subside in several days but then recur at irregular intervals for weeks or months (see Chapter 22).

2. Mycoplasmal Diseases

a. Murine Respiratory Mycoplasmosis. Murine respiratory mycoplasmosis (MRM) is the term now accepted for a disease which, for many years, had an undefined etiology and a number of synonyms [i.e., infectious catarrh, enzootic bronchiectasis, chronic respiratory disease (CRD), and chronic murine pneumonia]. Since 1969, the causal relationship of *Mycoplasma pulmonis* with this disease has become well established (Kohn and Kirk, 1969; Lindsey *et al.*, 1971; Whittlestone *et al.*, 1972). Of all the pathogens occurring in laboratory rats, *M. pulmonis* has had the greatest negative impact on studies. This has been primarily due to the chronicity of the disease, which often manifests itself only after months of infection. Long-term studies in areas of toxicology, carcinogenesis, nutrition, and gerontology, in particular, have been affected. Prior to the use of gnotobiotic techniques and barrier maintenance in rat production colonies, *M. pulmonis* was enzootic in nearly all commercial and institutional colonies. Today, vendors can be selected who offer mycoplasma-free rats. *Mycoplasma pulmonis* is highly contagious and induces a disease that frequently results in debilitation or demise of the host after a long period of time.

The clinical signs associated with MRM range from negligible upper respiratory tract signs to systemic signs associated with pneumonia. The earliest and most common signs include snuffling and serous or mucopurulent oculonasal discharge. Extension of *M. pulmonis* infection from the nasopharynx via the eustachian tubes to the middle ears is common. However, torticollis and circling due to involvement of the inner ear are infrequently observed, even though one or both middle ear bullae may be impacted with exudate. The onset of upper respiratory signs is variable, but often occurs within several weeks postinfection. Signs of penumonia include dyspnea, rales, and systemic effects such as weight loss, starry hair coat, and hunched posture. Characteristically, signs of pneumonia occur 3–6 months postinfection, but this is quite variable and is a function of environmental influences, such as intracage ammonia levels and the immune competence of the host. In a small percentage of cases, the disease will be nearly subclinical even in the presence of extensive pulmonary lesions.

Mycoplasma pulmonis is transmitted both horizontally and vertically from dams to their litters. In most instances, transmission from the female occurs postpartum by direct contact, but if the genital tract of the dam is infected, antenatal infection can occur. Horizontal transmission between postweanling rats of any age readily occurs, and there appears to be no significant age-related resistance to either infection or disease. Although little is known about differences in resistance among rat stocks and strains, the LEW rat has been shown to be more susceptible to MRM than the F344 rat. There is little evidence available to indicate that transmission occurs through fomites such as caging equipment and garments worn by personnel. Since aerosol droplet and direct contact appear to be the primary modes by which *M. pulmonis* infections are spread, the rapidity with which the organism is transmitted is dependent upon environmental factors, such as ventilation rates, degree of recirculation of air, and animal density within rooms.

The basis for the pathogenicity of *M. pulmonis* is not well understood. *Mycoplasma pulmonis* adsorbs to the cell membrane of the ciliated, columnar or cuboidal epithelia in the respiratory tract (Fig. 4). It has been suggested that adsorption is a means by which mycoplasmas damage host cells by uptake of essential cellular metabolites; release of cytotoxic products, such as H_2O_2; or cross reaction of antibody with cell membrane components that are antigenically similar to or altered by mycoplasmas. Infection severely distorts or ablates ciliary structures (Fig. 4), interfering with mucociliary clearance mechanisms.

The gross lesions in the upper respiratory tract include mucopurulent exudate in the nasal cavity, sinuses, and middle ear bullae. Later, the exudate becomes caseous within the bullae. Lesions in the lower respiratory tract reflect those of a bronchopneumonia. The earliest lesion is a mucopurulent exudate within the trachea, bronchi, and bronchioles. This precedes grossly evident lesions of the lung parenchyma that initially consist of atelectasis due to bronchial occlusion. Later, bronchiectatic lesions appear as numerous cream-colored nodular abscesses on the surface of the lung. These lesions may be restricted to only a portion of a lobe or may involve nearly all of the parenchyma (Fig. 5).

Microscopically, the inflammatory response is characterized by a lymphocyte and plasma cell infiltrate in the submucosa and neutrophilic leukocyte response within the lumina of the

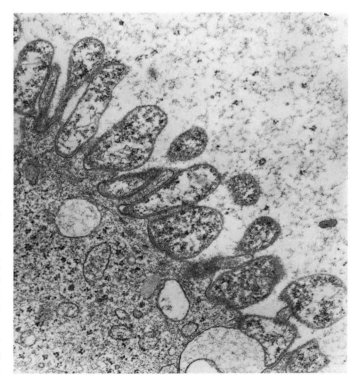

Fig. 4. Electron micrograph of *Mycoplasma pulmonis* attached to tracheal epithelium.

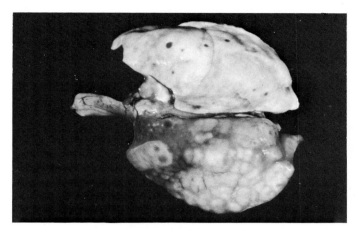

Fig. 5. Mycoplasma pulmonis infection (murine respiratory mycoplasmosis). Lungs are unevenly consolidated, and the pleural surface is elevated due to bronchiolectasis and abscess formation.

nasal cavity, eustachian tubes, middle ears, and tracheobronchial tree. A consistent and prominent lesion in the lung is the peribronchial lymphoid hyperplasia that often becomes quite massive. Within the lumina of the bronchi and bronchioles, mucin and neutrophil exudation increases during the course of the disease to the point of bronchiectasis. Concomitant with the impaction of bronchi is a change in the epithelia from a ciliated, columnar type to a squamoid type. This change in epithelial architecture is likely associated with cytotoxic enzymes from autolyzed neutrophils, although a direct cytotoxic effect from mycoplasmas could be involved.

A tentative diagnosis of MRM can usually be made by observance of the clinical signs and gross lesions described above. Clinical signs alone are not particularly helpful, since nasal exudates are present in bacterial infections such as *S. pneumoniae*. In addition, the reddish porphyrin deposition seen in the nares and periorbitally in sialodacryoadenitis virus infection and water deprivation may be confused with exudation. The gross lesions of otitis media and bronchiectasis are rather distinct. However, *C. kutscheri* lung lesions may grossly mimic those of MRM. Histopathology and serological evidence will differentiate MRM from Sendai virus infection, although the two infections are often superimposed. Recently a filamentous bacterium has been associated with bronchiectasis in wild and laboratory rats (MacKenzie *et al.*, 1981). However, the causal relationship of this organism with lesions is undefined since the rats were also infected with *M. pulmonis*. This filamentous bacterium has not been successfully grown on artificial media, and its presence is best verified by either histology, using the Warthin-Starry Stain, or electron microscopy (Fig. 6). Although a definitive diagnosis of MRM is made by isolation of *M. pulmonis* from involved tissues, it is evident that the existence of other agents must be evaluated to determine if copathogens are contributory to lesions.

Prevention of MRM in either breeding or experimental colo-

nies is dependent upon barrier systems that preclude the entry of *M. pulmonis* into the facility. Hysterectomy derivation is the only means of establishing an *M. pulmonis*-free breeding colony from a previously infected stock. Due to the frequent localization of this microorganism in the uterus, it is necessary to ensure that neonates taken by hysterectomy have not been infected *in utero*. Rats used in research animal facilities are obtained from various commercial and institutional sources. Accordingly, it is essential that the mycoplasma status of these sources is known and that the rats are housed by vendor or in groups with a similar microbial status.

For assessment of whether a group of rats is *M. pulmonis*-free, the best sites for isolation in animals without gross lesions are the nasal cavity, middle ear, trachea, and uterus–oviduct. *Mycoplasma pulmonis* is not particularly fastidious and grows well in several types of mycoplasma media (Cassell *et al.*, 1979; Lentsch *et al.*, 1979). Most formulations have a pH indicator that is useful since *M. pulmonis* ferments glucose. In broth media, moderate to heavy growth is reflected by pH and color of the broth. In broth cultures in which the titer is low, a perceptible pH change may not occur. Tissue and washing samples should be placed in broth rather than agar media, since recovery of the organism is more likely in those samples containing few mycoplasmas. Samples from broth cultures are transferred to agar media when a pH change is readily evident or at 7–10 days if no pH change occurs. Mycoplasma colonies are evident in 3–4 days by observation with 40× stereoscopic microscopy.

Although culturing and histopathology have been the usual means to survey rat colonies, ELISA testing has recently been shown to be a very sensitive serological assay and one that can be performed quickly in most clinical laboratories (Cassell *et al.*, 1981a). *In vitro* sensitivity tests show *M. pulmonis* to be susceptible to tetracycline and tylosin. Tetracyline, given at 5 mg/ml drinking water, may be useful in some situations (Lindsey *et al.*, 1971). However, treatment with antibiotics seldom influences the disease course of MRM in a colony situation.

b. Murine Genital Mycoplasmosis. *Mycoplasma pulmonis* recently has become recognized as an important pathogen in the female genital tract of rats, and thus is being treated here as a distinct disease rather than as a sequella to MRM. Infection of the genital tract is usually inapparent. However, reduced fertility and fetal deaths can occur. Infection of the oviduct and uterus occurs frequently in rats who have respiratory mycoplasmosis. It is unknown whether localization in the genital tract occurs due to a hematogenous spread or to an ascending infection of the genital tract. It has been shown that subsequent to intravenous inoculation, *M. pulmonis* almost invariably localizes in the female oviduct–uterus.

Gross lesions, when present, consist of a purulent oophoritis, salpingitis (Fig. 7), and pyometra. The LEW strain is particu-

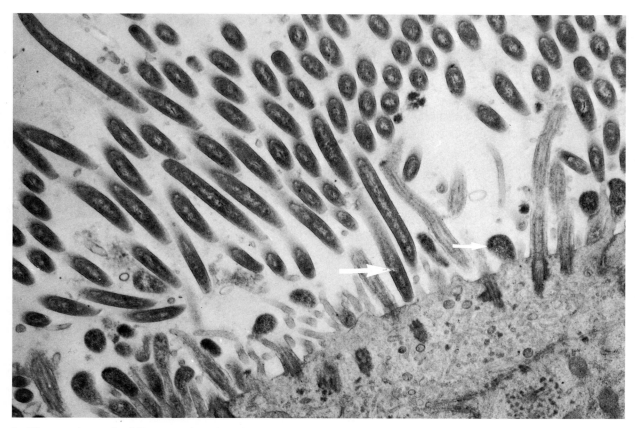

Fig. 6. Electron micrograph of filamentous bacterium (large arrow) and *M. pulmonis* (small arrow) attached to epithelium of respiratory mucosa. The morphology of size of the filamentous bacterium are similar to that of the cilia. (Courtesy of Dr. W. F. MacKenzie.)

larly prone to develop gross lesions. *Mycoplasma pulmonis* adsorbs to the epithelial cells in the genital tract in a manner similar to that seen in the respiratory tract. Salpingitis occurs most frequently and is characterized by exudation of neutrophils into the lumen, hyperplasia of oviductal epithelium, and a lymphoid response in the submucosa. The lesions in the ovarian bursa include edema and inflammation. Uterine lesions can vary from a mild inflammatory change to pyometra (Cassell *et al.,* 1981b).

Genital mycoplasmosis in the male rat has not been well documented. However, it is known that experimental inoculation can include an inflammatory response in the ductus efferens and epididymis. Moreover, it is known that *M. pulmonis* is capable of adherence to spermatozoa in an *in vitro* system.

Since *Pasteurella pneumotropica* can also induce similar lesions in the female rat, a diagnosis of mycoplasmosis is dependent upon isolation of *M. pulmonis* from the lesions. Methods for culturing and identification are similar to those used for respiratory mycoplasmosis.

Because the rat is widely used in various types of reproductive biology research, *M. pulmonis* colonization, even without

gross lesions, would probably impact on the validity of data. The grossly evident caseous lesions in the ovary and oviduct can be mistaken for neoplasia if microscopy is not done.

c. Mycoplasmal Arthritis. The etiological agent of this disease is *Mycoplasma arthritidis.* This mycoplasma species colonizes the pharynx, middle ears, and lungs of rats, although few studies have been done to document the relative frequency of this mycoplasma in rat sources. Within the respiratory tract, *M. arthritidis* colonization is thought to induce negligible lesions, and it has been shown to coexist with *M. pulmonis.*

Although it is often considered to be the principal agent involved in arthritis in rats, the disease has been rarely reported. Nearly all reports of its involvement in clinically apparent arthritis have been made prior to 1960. It has been suggested that poor cage sanitation and abrasions of the extremities are involved in entry of the organism to the joints by hematogenous spread or extension from surrounding tissues (Ward and Cole, 1970). Since the organism appears to be of low virulence, the immunocompetence of the host may be a major factor in the outcome of infection.

Arthritic animals limp and move with difficulty due to pain associated with the polyarthritis. Any of the joints in the limbs and vertebrae can be affected, but the tibiotarsal and radiocarpal joints are most often involved. Affected joints are hyperemic and swollen. Incised joints reveal a purulent exudate in both articular and periarticular tissues. Microscopically, there is exudation of neutrophils into the synovial spaces, and a lymphocyte and plasma cell infiltration in the synovial membranes. Destruction of the articular cartilage occurs subsequent to the inflammatory response.

Since polyarthritis can occur subsequent to septicemias associated with other bacteria, particularly *C. kutscheri*, a diagnosis of *M. arthritidis*-induced arthritis is contingent upon the demonstration of *M. arthritidis* by isolation or immunofluorescence techniques. This mycoplasma species grows well in media used to isolate *M. pulmonis* if arginine is added to the formulation (Cassell *et al.*, 1979).

Tetracyclines have been used to prevent the onset of arthritis when the organism has been inoculated intravenously, but there are no reports of its efficacy in spontaneous cases. *Mycoplasma arthritidis*, like *M. pulmonis*, may contaminate transmissible tumors and caution should be exercised to ensure transplanted tissues are not contaminated.

3. Rickettsial Diseases

Hemobartonellosis. The causative agent of this rickettsial disease is *Hemobartonella muris*. This organism is an extracellular parasite of erythrocytes and induces inapparent infections that may persist for long periods. The ability of the host to restrict the infection to a subclinical mode rests with the integrity of the reticuloendothelial system. Evidence of infection is usually limited to splenomegaly and laboratory findings of mild parasitemia and reticulocytosis.

Transmission of *H. muris* involves the blood-sucking louse, *Polyplax spinulosa*. Transmission can occur during a blood meal or when rats crush infected lice and are inoculated via pruritis-induced abrasions. The organism can also be transmitted inadvertently with transplantable tumors and other biological products.

Diagnosis of hemobartonellosis is dependent upon identification of the organism in the peripheral blood of infected animals. The usual method of detection is by splenectomizing rats suspected of harboring the organism. In these rats, severe parasitemia and hemolytic anemia occur within 2 weeks after surgery. *Hemobartonella muris* can be visualized on the surface of erythrocytes in Romanowsky-stained blood smears as coc-

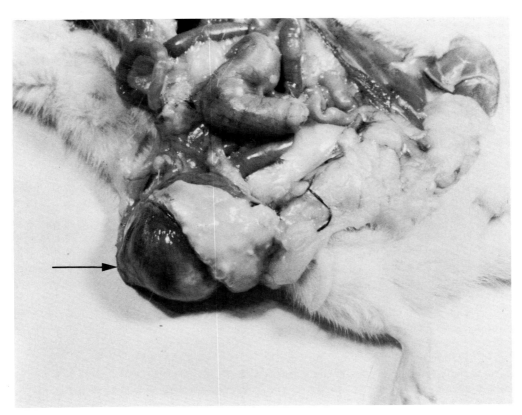

Fig. 7.   Genital *Mycoplasma pulmonis* infection. Uterine wall (arrow) has been excised to show exudation within lumen of uterus.

coid bodies arranged singly, in clusters, or chains (Cassell *et al.*, 1979).

The rarity of reported cases would indicate *H. muris* is no longer a significant problem in barrier-maintained colonies. However, conventionally maintained colonies may be exposed to infected wild rats and *P. spinulosa* and, accordingly, the disease still is of importance in the laboratory rat. The disease has had a negative impact on investigations of various types, but principally with those in which the host's immune competence has been impaired.

4. Viral Diseases

a. Parvoviral Syndromes. Parvoviruses that can infect rats include rat virus (RV), Toolan H-1 (H-1) virus, and minute virus of mice (MVM). Parvoviruses are small nonenveloped viruses that resist extremes in temperature, pH, and drying. Rat virus, or Kilham rat virus (KRV), has several antigenically related strains (RV, H-3, X-14, L5, HB, SpRV, HER, HHP, Kirk), all of which have been isolated as inadvertent contaminants of rat tissue or rat-passaged biological material. Toolan H-1 related serotypes (H-1 and H-T) are antigenically distinct from RV serotypes. Both RV and H-1 are experimentally pathogenic, producing similar lesions, but only RV has been associated with natural disease. Neonatal rats can be experimentally infected with MVM, but the virus does not seem to cause natural infection. Minute virus of mice antibody reactivity can be present in rat serum, but this is probably nonspecific, since it can be found in germfree rat serum and is reduced or eliminated by receptor destroying enzyme.

Rat virus infection is usually subclinical or latent, but a number of clinical syndromes have been associated with it. Infection of pregnant females can cause fetal resorption and birth of small litters. Pups are runted, atactic, or jaundiced. Neonates develop similar signs following postpartem exposure. Rats introduced to an infected colony can develop ruffled fur, dehydration, and sudden high mortality. A similar syndrome occurs in latently infected adults subjected to immunosuppressive regimens.

The rat is the only natural host for RV and H-1, although experimental infection can be established in a number of other species. Seroconversion to both RV and H-1 virus is common, with a high prevalence of infection within an enzootically infected colony. Horizontal transmission is achieved by the oral and probably respiratory routes, with virus excretion primarily in the feces. Some strains of RV can be excreted in the milk or *in utero*. Clinical signs are manifest transiently upon introduction of RV into a previously uninfected population, but, thereafter, the virus spreads rapidly to produce subclinical or inapparent enzootic infection. Rat virus can persist as a true latent infection in the presence of high circulating antibody, but disease can be activated by immunosuppression. It must, there-

fore, be assumed that seropositive rats are persistently infected and can serve as a source of infection to other rats.

Pups infected *in utero* or as neonates develop intranuclear inclusions and necrosis in the outer germinal cell layer of the cerebellum. The recovered animal has severe depletion of the internal granular layer and disorganized Purkinje cells. Intranuclear inclusions are also in hepatocytes, Kupffer cells, endothelial cells, and biliary epithelial cells, resulting in necrotizing hepatitis and the sequellae thereof (bile retention, jaundice, peleosis, bile ductal hyperplasia, parenchymal collapse, nodular hyperplasia). In adults, infection is usually inapparent, but when acute disease is precipitated, RV injures vascular walls and hematopoietic elements, causing coagulative disorders, thrombosis, hemorrhage, and infarction within the central nervous system (hemorrhagic encephalomyelopathy). Hemorrhagic and necrotic lesions have also been noted in the peritoneum, testis, and epididymis. Rat virus has broad tissue tropism and lesions or clinical signs may potentially be varied, depending on virus and host factors (Coleman *et al.*, 1982; Jacoby *et al.*, 1979).

Infertility and unthrifty pups caused by RV must be differentiated from environmental and husbandry factors or infectious agents such as *Mycoplasma* or Sendai virus. Adult disease must be differentiated from toxicity, nutritional deficiency, and trauma. Diagnosis is made by the typical lesions, if present, virus isolation, and serology. Seroconversion to each virus (RV or H-1) can be detected by serum neutralization, hemagglutination inhibition, complement fixation, and immunofluorescence. Hemagglutination inhibition is currently the most commonly used means of antibody determination (Jacoby *et al.*, 1979).

Since RV infection is usually silent and persistent and can be transmitted either vertically or horizontally, effective control is best achieved by destroying the entire population, decontaminating, and repopulating with clean stock. Virus-free rats can be obtained from selected commercial vendors or by caesarean rederivation. Rederived progeny must be tested for vertically transmitted strains of virus. Colonies can be kept virus-free by limiting entry to seronegative, virus-free rats (as well as transplantable rat neoplasms or tissues), periodic serological testing, and adequate physical containment.

Although parvovirus infection of rats is usually inapparent, there can be adverse effects on the research usefulness of infected rats. Immunosuppression may exacerbate illness and mortality in latent carriers. The viruses often contaminate transplantable tumors and cell lines, can modify immune responsiveness or cause teratological effects. A decision to work with infected animals should be made carefully.

b. Other DNA Virus Infections. Rats are susceptible to rat cytomegalovirus, which has a predilection for the salivary and lacrimal glands. Infection is widespread among wild, but not laboratory rats (Jacoby *et al.*, 1979). Rats also seroconvert to

of *Syphacia*, cellophane tape impressions of the anus for ova. Speciation is easily achieved by the differential morphology of ova. *Aspicularis* ova are symmetrically ellipsoidal, while *Syphacia* ova are flattened on one side. Since eggs are extremely resistant to environmental factors and disinfectants, and autoinfection can occur with *S. muris,* infection is difficult to eradicate. Anthelminthic treatment ameliorates infection (Table IX), but all rats must be treated simultaneously and the room must be thoroughly cleaned. Despite these actions, reinfection frequently occurs. Caesarean rederivation and subsequent strict sanitation are the most effective means of control. Pinworms are generally considered nonpathogenic, but heavy parasite loads can be deleterious to the host. Concern has been raised regarding the effects of these agents on research but no effects have been documented in the rat.

Trichosomoides crassicauda is common in some rat colonies but is not generally encountered in commercially raised animals. It inhabits the transitional epithelium and lumen of the urinary tract. The small male resides within the reproductive tract of the female, whose anterior end embeds within the mucosa. Embryonated ova with bipolar opercula are shed in the urine. Following ingestion, eggs hatch in the stomach, larvae penetrate the gastic wall, then migrate via the lungs, kidney, and ureter to the urinary bladder. Migrating larvae can incite eosinophilia and formation of granulomata, particularly in the lung. Drug treatment (Table IX) is effective, as is cesarean rederivation. Parasites do not incite an inflammatory response in the bladder, but can serve as a nidus for calculi or enhance natural and experimental bladder carcinogenesis.

Other nematodes are common in wild but rare in laboratory rats. *Trichinella spiralis* adults occur in the intestine, and larvae migrate extensively, encysting in muscle. Transmission is effected by ingestion of contaminated meat. *Capillaria hepatica* eggs are ingested, hatch, and migrate through the cecum to the liver through the portal veins. Adults mature and lay characteristic bipolar operculate ova in the liver, where they lie dormant until ingested by a carnivore. *Gongylonema neoplastium* burrows in the squamous epithelium of the tongue and anterior stomach, where it incites a proliferative response or ulceration. The life cycle involves an intermediate insect host. *Angiostrongylus cantonesis*, which requires the ingestion of a mollusk intermediate host, infects the brain and lung. The intestinal nematodes *Heterakis spumosa, Nippostrongylus brasiliensis, Strongyloides ratti,* and *Trichuris muris* lack intermediate hosts, but rarely occur naturally in laboratory rats. None is known to be very pathogenic. Further details on nematode diagnosis and ova morphology are available (Flynn, 1973; Hsu, 1979; Oldstone, 1967).

c. Cestodiasis.

Rats can serve as the intermediate host for *Taenia taeniaformis,* the cat tapeworm. Following ingestion of ova, oncospheres migrate to the liver and form strobilocerci

(*Cysticercus fasiolaris*). Host connective tissue capsules can give rise to sarcomas. This parasite has been used as a model of parasite-induced oncogenesis in the rat. *Cysticercus fasciolaris* is on occasion encountered in laboratory rodents with contaminated food or bedding.

Hymenolepis nana and *diminuta* infect the small intestine of several species, including the rat. *Hymenolepis nana,* the dwarf tapeworm, has a 1-mm wide segmented body ranging from 7 to 100 mm in length. The life cycle may be either direct or indirect. Following ingestion of ova, oncospheres penetrate the intestinal mucosa, form cysticercoid larva, then emerge into the lumen and mature into adults that shed eggs in the feces. Nonimmune hosts can be autoinfected; eggs are produced, hatched, and complete their life cycle within the intestine of a single host. The indirect cycle involves intermediate hosts, such as grain beetles or fleas, in which cysticercoid larvae form and await ingestion by the definitive host. *Hymenolepis diminuta* is 3–4 mm wide and 20–60 mm long. The eggs are passed in the feces but must be ingested by an intermediate host (flour beetle, flea, moth), which in turn must be ingested by the definitive host to complete the life cycle. Enteritis can occur in heavy infestations, which are most likely to be encountered with *H. nana.* Both can infect primates, including humans. *Hymenolepis nana,* which does not need an intermediate host and can cause autoinfection, is a particular public health hazard. Anthelmintic treatment is effective (Table IX). Diagnosis of *Hymenolepis* is made by finding ova containing hexacanth embryos in the feces or adults in the lumen of the bowel. Differential features are detailed elsewhere (Hsu, 1979).

d. Insect Infestation.

Two anopluran (sucking) lice infest wild rats, *Polyplax spinulosa* (the spined rat louse) and *Hoplopleura pacifica* (the tropical rat louse). The latter has not been reported among laboratory rats. *Polyplax spinulosa* is now rare in most laboratory rat populations. It completes its entire life cycle within the fur of the host and is transmitted by direct contact. Infested rats are unthrifty, irritable, pruritic, restless, and anemic. *Polyplax spinulosa* can transmit a number of infectious agents, which include *Hemobartonella muris* and *T. lewisi.* Diagnosis is made by identifying organisms in the fur.

Rat fleas (*Xenopsylla* sp., *Leptopsylla* sp., and *Nosopsyllus* sp.) are rare in laboratory rats, since their life cycles are usually interrupted by proper sanitation. Fleas can transmit other agents and can serve as the intermediate host of *H. nana* and *diminuta* (Flynn, 1973; Hsu, 1979; Oldstone, 1967). See Section III,A,5,e for methods of control.

e. Arachnid Infestation.

Mesostigmate mites can be encountered on occasion in laboratory rats. They are rapid bloodsuckers and have a nonselective host range. They are on the host only during feeding, spending much of their life cycle secreted in the environment. Diagnosis must be attained by finding engorged mites on bedding, cages, and crevices. Other hosts, including humans, fall prey to their painful and irritating bites. The most important mesostigmate is *Ornithonyssus bacoti* (tropical rat mite). Heavily infested colonies contain debilitated, anemic rats with reduced reproductive efficiency and occasional deaths. *Liponyssoides sanguineus* has not been reported in laboratory rats, but may be unrecognized because of its similarity to *Ornithonyssus* sp. *Laelaps echidninus* (spiny rat mite) does not adapt well to the laboratory rat environment unless husbandry is poor. It is a vector of *Hepatozoon*.

Prostigmate mites have a more selective host range and spend their life cycle in the fur (or follicles) of the host. *Radfordia ensifera,* the rat fur mite, can be common in some rat colonies. This mite produces few ill effects, but heavy infestations can induce self-inflicted trauma (Flynn, 1973; Hsu, 1979; Oldstone, 1967). *Demodex* sp. is also encountered, but its prevalence is unknown, since it lives deep within hair follicles and sebaceous glands where it produces minimal lesions (Walberg *et al.*, 1981).

Astigmate mites include the mange mites, which have a selective host range and complete their life cycle on the host. *Notoedres muris,* the ear mange mite, infests the skin of the ear and to a lesser extent, the nonglabrous regions of the body. This mite burrows in the cornified epithelium, and elicit a pruritic dermatitis (Flynn, 1973; Hsu, 1979). It is seldom seen in contemporary rat colonies.

Control of ectoparasites is gained by preventing the introduction of infected animals (including wild rodents), sanitation, treatment of rats, and, in the case of mesostigmates and fleas, treatment of the environment (Table IX). Treatment seldom completely eliminates infestation and can have an adverse effect on the usefulness of the treated rats for research. Repopulation or caesarean rederivation and subsequent prevention by proper management is the best approach.

6. Fungal Diseases

Deep mycoses are generally not considered to be significant natural diseases in laboratory rats. Pulmonary aspergillosis is occasionally observed in rats and mice, particularly when immunocompromised. *Aspergillus* sp. can be readily isolated or observed in affected lung tissue with selected histochemical techniques. Pulmonary lesions consist of miliary granulomata containing Langhans giant cells in areas with characteristic uniformly branched, septate hyphae. Phycomycotic encephalitis has been reported as a natural disease in 2- to 4-week-old rats. Phycomycotic fungi have thick, irregularly branched, nonseptate hyphae in tissue.

Dermatomycosis is probably more likely to be encountered than deep mycosis, but is also rare in laboratory rats. Ring-worm in rats is restricted largely to infection by *Trichophyton mentagrophytes*. Infected rats can be lesionless carriers or display patchy alopecia and erythema with scale, papule, or pustule formation. Fungal elements are visible within the stratum corneum and hair follicles. Ecothrix spore formation in hair shafts can be present. The disease can be diagnosed by examination of skin scrapings treated with 10% potassium hydroxide, histological sections of skin prepared with fungal stains, or isolation (Weisbroth, 1979).

B. Noninfectious Diseases

1. Metabolic/Nutritional Diseases

a. Genetic Anomalies. Genetic anomalies in coat color and character, organ development, immune responsiveness, obesity, hypertension, metabolism, and others have been identified and in many cases developed as specific characteristics of various stocks and strains. The reader is referred to several more reviews for further information (Altman and Katz, 1979; Hansen *et al.*, 1981; Robinson, 1965, 1979). Some strains naturally develop disease syndromes, such as Brattleboro rats with diabetes insipidus, while other strains have less obvious characteristics but react uniquely to different research variables. The researcher must be aware of these variations and judiciously select the appropriate stock or strain to accomplish the goals of the experiment.

Each stock, strain, substrain, and even group of rats that has drifted from its parental stock can develop its own unique diseases, manifest its own incidence and rate of development of various spontaneous lesions, and react in different ways to infectious and research variables. Expression of genetic traits is further influenced by extraneous factors such as husbandry and diet.

b. Nutritional Deficiencies. More is known about the nutritional requirements and deficiencies of rats than perhaps any other species. Natural deficiencies of single dietary components are rare and seldom recognized. This is because rats effectively store fat-soluble vitamins (and B_{12}), manufacture vitamin C, and fulfill much of their requirement for B vitamins by coprophagy. Furthermore, some deficiencies are influenced or compensated for by other dietary elements. The most common signs of nutritional deficiency are vague, such as poor hair coat, reduced weight gain or growth, diminished fertility, and susceptibility to infectious disease. Commercially available rodent diets effectively supply balanced diets to rodents, but diets should not be utilized beyond their expiration date and ideally should be kept in cold storage. Several components deteriorate with prolonged storage, heating, or sterilization, including lysine, vitamins A and E, and, to a lesser extent, riboflavin and thiamin. Nutritional requirements vary with the genetic strain, stage of life, and microbiological association. Axenic or specific pathogen-free (SPF) animals can lack gut microflora necessary for supplying certain vitamins, particularly vitamin K. They must receive supplemented diets both because they need more and because autoclaving or pasteurization reduce the levels of certain essential nutrients. Antibiotic treatment, which affects intestinal microflora, can reduce the bacterial source of vitamins by coprophagy. Nutritional adequacy goes far beyond the needs of the rat, since the composition of nutritionally adequate diets is known to profoundly influence the biological responsiveness to research manipulation, infectious disease, and expression of spontaneous lesions (National Research Council, 1978; Rogers, 1979).

The signs of nutritional deficiency are often vague and must be differentiated from other syndromes. Hemorrhage due to hypovitaminosis K can be confused with RV disease (which in itself can be precipitated by nutritional deficiency). Infertility or poor production can be caused by RV, SDAV, Sendai virus, *M. pulmonis,* or nutritional deficiency. Squamous metaplasia in the salivary and lacrimal glands of rats recovering from SDAV or in the respiratory tract of rats infected with Sendai virus or *M. pulmonis* must be differentiated from hypovitaminosis A, which in turn predisposes to respiratory infections. Environmental factors, such as excess heat or low humidity, contribute to infertility or skin disease which mimic hypovitaminosis E or other deficiencies. Ordinarily mild infectious diseases can become severe, such as pseudotuberculosis, due to nutritional deficiency. The diagnosis of many rat diseases therefore requires review of nutritional practices.

2. Management-Related Diseases (Nonnutritional)

Practices other than nutrition can also influence or cause disease in rats. Sanitation is of obvious import. Cage design, population density, and temperature influences have been mentioned in Sections II,D and E. Outbreaks of opportunistic infectious disease, such as pseudotuberculosis, can be precipitated by sudden fluctuations in temperature and humidity. *Pseudomonas aeruginosa* can be introduced and spread within a colony through inappropriate sanitation, particularly of water bottles. Refilling water bottles at a faucet in the animal room is not advised for this reason. Bottles and sipper tubes should be sanitized and water should be acidified or chlorinated. Excessive ammonia from dirty bedding, overcrowded cages, or poor ventilation predisposes to respiratory infections, particularly mycoplasmosis.

A number of management-related syndromes can occur that are not related to sanitation or infectious disease. Rats adapt well to automatic watering systems, but animals unfamiliar with these devices can become dehydrated. Similarly, malfunction of these systems and blockage of water bottle sippe

tubes with metal filings or other detritus can have a similar effect. Rats maintained for long periods on wire-bottom cages develop foot sores, which can result in severe lesions, discomfort, hemorrhage, and anemia. Aging rats must be examined periodically for overgrowth of their incisors, a very common problem in rats held long term. Teeth must be carefully clipped, avoiding splitting or shattering. Dusty bedding and food can result in foreign body pneumonia. Plant, mineral, and even bone fragments can be found in sections of lung obtained from rats under these circumstances. Softwood shavings used as bedding can cause intestinal impaction, particularly in young rats. High ambient light, or even light levels within the recommended range, can cause retinal degeneration and cataracts in albino rats. Rats maintained near ceiling light fixtures develop retinal degeneration, while those near the floor are spared. Low relative humidity (<40%) in concert with high temperature and other factors can cause one or more annular constrictions of the tail skin (ringtail) or toes. The segments distal to the constrictions swell or undergo dry gangrene (Fig. 10). This is most apt to occur in suckling or preweanling rats and in rats kept in wire-bottom cages. Increasing humidity, reducing air turnover rate, placing rats in solid-bottom cages, and providing nesting material for nursing dams and their litters are all ameliorative for this condition.

Axenic rats normally have enlarged ceca, which become approximately five times normal size. There is net efflux of fluid into the lumen in response to increased osmotic pressure of luminal content, reduced availability of ions needed for mucosal solute-coupled water resorption and vasoactive compounds which exert their effect on smooth muscle tone. Cecal enlargement can interfere with reproduction. Dietary manipulation can reduce the enlargement substantially (Foster, 1980). The effects on smooth muscle tone become pronounced in aging axenic rats, which can develop progressive cecal enlargement due to atony. Volvulus and torsion can result.

The reader must be cognizant of the much larger list of en-

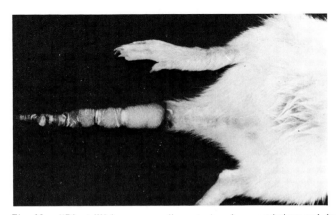

Fig. 10. "Ringtail" in a preweanling rat. Annular constrictions and dry gangrene of the tail.

vironmental variables that significantly alter physiological responsiveness of test animals (Baker *et al.*, 1979a; Gralla, 1981).

3. Traumatic and Iatrogenic Diseases

Unlike the mouse, traumatic skin disease in the rat is more likely to be due to self-inflicted trauma than fight wounds. Rats infested with ectoparasites or dermatophytes can develop excoriative dermatitis due to self-inflicted trauma. Roughly symmetrical ulcerative dermatitis over the dorsal and lateral neck and shoulders has been reported (Fig. 11). The etiology remains to be determined, but coagulase-positive *Staphylococcus aureus* is consistently isolated from the skin lesions, and colonies of gram-positive cocci are microscopically visible on the surface of the lesion. Attempts to induce the disease in other rats with this agent have yielded inconsistent results (Fox *et al.*, 1977).

The rat is often considered "resistant" to infection and accordingly submitted to a variety of experimental procedures without regard to aseptic techniques. Peritonitis and wound infections are often found under such conditions. Inappropriate handling of large rats by the tips of their tails will cause the skin to tear and slip off when they struggle.

Adynamic ileus occurs in rats given intraperitoneal injections of anesthetic preparations containing chloral hydrate or related compounds (Fig. 12). For variable periods up to 5 weeks after treatment, rats become lethargic, anorectic, and constipated with distended abdomens that may lead to death. Gross necropsy findings reveal segmental atony and distension of the bowel, usually jejunum, ileum, and cecum. Focal serosal inflammatory changes and fibrosis are seen microscopically (Fleischman *et al.*, 1977). This syndrome mimics megaloileitis (Tyzzer's disease), but can be differentiated on the basis of history, lack of characteristic bacteria in tissues, and lack of

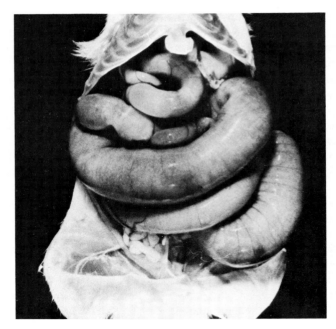

Fig. 12. Adynamic ileus in a rat which received a previous intraperitoneal inoculation of chloral hydrate.

liver and heart lesions. In addition, it should not be confused with the bowel dilatation observed in axenic rats.

C. Neoplastic Diseases

Space exists in this chapter to only summarize the more frequently reported tumors in the rat, and the topic will not include experimentally induced tumors. The reader is encouraged to refer to the articles and monographs referenced in the text for more complete information.

There have been numerous studies done to assess the incidence of neoplasia in various stocks and strains of rats (Burek, 1978; Coleman *et al.*, 1977; International Agency for Research on Cancer, 1973, 1976; MacKenzie and Garner, 1973; Ringler and Dabich, 1979; Sher, 1982). Some of these have led to the development of animal models and provided baseline data for experimental carcinogenesis work for which the rat is regularly used. There is a wide range of values reported in regard to the prevalence and type of tumors found in a particular rat stock or strain. These differences are a function of genetic and environmental influences and differences in sample selection. Nutrition is well known as a factor that influences the development of neoplastic disease. For instance, it has been shown that a high fat diet enhances tumorigenesis, that the protein–calorie ratio influences the occurrence of some tumors, and that restricted food intake of postweanlings for 7 or more weeks decreases tumor risk (Ross and Bras, 1971).

Most of the surveys on tumor occurrence in rats either state

Fig. 11. Ulcerative dermatitis in a rat, depicting the typical location for lesions in this syndrome.

nothing in regard to the presence of enzootic infectious disease or acknowledge that infectious diseases were present in the population. In many instances, infectious diseases can significantly influence data on tumor risk because of their effect on longevity, preneoplastic changes, and masking of small tumors.

With the exception of mammary fibroadenomas in many stocks and testicular tumors in F344 rats, the prevalence of tumors is quite low in rats under 18 months of age. Accordingly, the age at which rats are surveyed is very important in defining the incidence of neoplasia in a particular stock or strain.

The F344 rat is used in the bioassay program of the National Cancer Institute and is widely used in toxicology studies. It has been selected for these types of studies because of its small size, longevity, and low incidence of most tumors. However, this strain has a moderate to high incidence for some tumors, such as those of the testis, mammary gland, uterus, and hematopoietic and endocrine systems (Sher, 1982). A Sprague-Dawley-derived stock that is commonly used in carcinogenesis studies is the CRL:CD(SD)BR rat. This stock has a high incidence of mammary tumors and pituitary adenomas.

1. Skin and Mammary Gland

Papillomas and squamous cell carcinomas occur infrequently, but have been reported in many stocks of rats. Squamous cell carcinomas are usually located on the face and head, and tend to be highly invasive but not metastatic. Most of these arise as sebaceosquamous tumors from the glands of Zymbal in the ear. Similar tumors occur on the prepuce and clitoris. In females from most stocks and strains, the mammary gland is the most frequent site of neoplasia, with incidences of 30–60% being common. These tumors occur in males but much less frequently. Benign fibroadenomas are the most common type, but adenocarcinomas may also occur. Fibroadenomas have ductal epithelium and periductular connective tissue components and tend to be less vascular than carcinomas. Adenocarcinomas can metastasize to regional lymph nodes and the lung. Tumors of the mammary gland, which may arise at any site from the neck to the inguinal region, often attain a very large size (Fig. 13).

2. Digestive System

The prevalence of tumors in the alimentary tract is extremely low, with most cases involving the colon. Although hepatic tumors are readily induced by chemical carcinogens, very few spontaneously occurring malignant neoplasms have been reported. More commonly occurring are neoplastic nodular lesions. The term "neoplastic nodule" is used to describe circumscribed nodules of proliferating hepatocytes that compress the parenchyma (Altman and Goodman, 1979). There is some

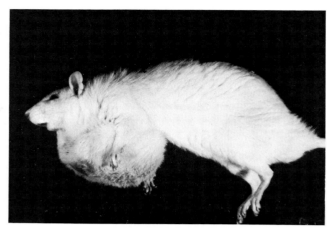

Fig. 13. Mammary fibroadenoma in an aged rat.

controversy as to whether these nodules are in all cases neoplastic. However, there is evidence that some develop into carcinomas.

3. Hematopoietic System

Leukemia is rare in many stocks and strains of rats. However, mononuclear cell leukemia is very frequently seen in F344 and WF rats. In one survey of male F344 rats, an incidence of 16% proved leukemia second in occurrence only to the testicular interstitial cell tumors. The leukocyte count in 9 leukemic rats ranged from 68,400 to 323,000 cells/μl with an average of 143,190 (Coleman *et al.*, 1977). Other studies indicate a similar incidence for the WF rat. In the BN/Bi rat, myelomonocytic leukemia has been reported to be the most common lymphoreticular tumor, with female and male incidences of 5 and 11%, respectively (Burek, 1978). Both types of leukemia occur in rats over 18 months of age.

4. Endocrine System

Neoplasms of the pituitary occur frequently with a higher proportion in females. Surveys have shown an incidence of 15–30% in F344 rats, 18% in OM (Osborne-Mendel) rats, and 3–13% in Sprague-Dawley rats. The most common pituitary tumor is the chromophobe adenoma. These are benign tumors that frequently become quite large, and, due to compression of the brain by the tumor, hydrocephalus and neurological signs may occur. They appear dark due to hemorrhagic areas arising from a vascular sinusoid component of the tumor. Clinical manifestations may include neurological signs or cachexia, but these are often absent.

Pancreatic islet cell tumors are common in some stocks and strains. Surveys have reported a 6% incidence in F344 rats, 4.4% in Holtzman rats, and a 1–2.9% in Sprague-Dawley stocks.

Tumors of the thyroid are usually parafollicular cell adenomas. These benign tumors occur in many stocks with an incidence varying between 1 and 7%. Follicular cell carcinomas, which occur infrequently, can metastasize to the lung.

Adenomas of the adrenal cortex are very common in several strains. In at least three strains, BUF/N, M520/N, and OM/N, over 40% of the females and 20% of the males have cortical tumors. Pheochromocytomas are the most common medullary tumors. They are particularly common in the BUF/N, F344/N, M520/M, and WN/N strains, and, unlike cortical tumors, they appear more often in males.

5. Central Nervous System

Tumors of the brain have been found to occur rarely in those studies that included examination of the CNS. Less is known about the prevalence of tumors in the CNS than in other systems because many studies have not included brain and spinal cord examination. The astrocytoma is the most commonly reported tumor of the CNS. Less often reported tumors include oligodendroglioma, meningioma, ependymoma, and granular cell tumors.

6. Respiratory System

Primary tumors of the lung are rare in the rat. The most frequently reported types are bronchogenic carcinomas, carcinomas, sarcomas, and adenomas.

7. Urinary System

Tumors of the bladder are rare in most stocks and strains of rats. However, the BN/Bi strain has an unusually high incidence of carcinomas of the bladder and ureter. Males of this strain have been reported to have a 35% incidence of bladder carcinoma, while 22% of females have carcinoma of the ureter (Burek, 1978). Nephroblastomas, renal tubular adenomas, and adenocarcinomas have been reported in the rat kidney.

8. Reproductive System

Tumors of the prostate are common in the AXC rat. In one report, adenocarcinoma of the ventral prostate was histologically detected in 70% of 30- to 46-month-old AXC rats. Prostatic tumors are reportedly uncommon in most stocks and strains of rats. Reportedly, however, some of these surveys did not include adequate sectioning to locate small adenomas and carcinomas.

Interstitial cell tumors are the most common testicular tumor type. Nearly all aged F344 rats and the majority of ACI/N rats

develop these tumors, but the tumor is rare in most other strains.

Tumors of the vagina and cervix are rare, except in aged BN/Bi rats. Tumors of the uterus and ovary are common in several strains of rats. One OM strain was reported to have a 33% incidence of granulosa cell tumors, and a high incidence of endometrial tumors has been reported in the F344 and M520 strains.

D. Miscellaneous Diseases and Lesions

1. Congenital/Hereditary Lesions

Congenital disorders in the rat are infrequently reported. This is probably a reflection of two factors; the first being that evidence indicates the incidence is extremely low in outbred stocks, and the second being that commercial/institutional suppliers select against breeders who have produced abnormal young. The genitourinary system and the eye have been most associated with congenital disorders.

One substrain of the AXC rat has a greater than 25% incidence of unilateral renal agenesis and hydronephrosis (Fujikura, 1970). A high incidence of unilateral and bilateral hydronephrosis has been found in BN/Bi rats (Cohen *et al.*, 1970). Pseudohermaphroditism has been reported in rats that are phenotypically female but who have an XY karyotype. Testicles are present either in the abdomen or inguinal canal. This condition is due to a sex-linked recessive gene that is expressed by an insensitivity of tissues to androgens (Bardin *et al.*, 1970).

A number of ocular disorders have been reported in the rat. Retinal dystrophy is inherited as a single autosomal recessive gene (*rdy*). Homozygotes have a progressive loss of the retinal photoreceptor cells and an overproduction of rhodopsin early in the disease process (LaVail *et al.*, 1972). A retinal degeneration reported in the Wag/Rij rat appears to be an expression of an autosomal dominant gene. End-stage lesions include disappearance of photoreceptor cells, migration of the pigment epithelium, and disorganization of remaining retinal layers. This retinal disease appears to mimic the pathogenesis of retinitis pigmentosa in humans (Lai *et al.*, 1980). Other reported developmental disorders of the eye include colobomas and cataracts.

Reports of congenital heart disease are rare. Among the few reports is one involving an inbred strain of Long-Evans rats (Fox, 1969). In this strain, approximately 25% of the neonates have interventricular septal defects. This malformation is believed to be of polygenic origin.

Unlike the mouse, there are few reported mutants that have disorders of the central nervous system. Recently, a partially inbred strain has been developed in which one subline has nearly a 50% incidence of hydrocephalus (Kohn *et al.*, 1981).

The mode of inheritance appears to be polygenic. The hydrocephalus is classified as communicating and the pathogenesis may be due to poorly developed veins in the periosteal and dural layers, and to underdeveloped pia-arachnoid cells.

2. Age-Related Diseases

Neoplastic and nonneoplastic lesions vary in type, onset, and prevalence due to the hereditary and environmental factors associated with a specific colony of laboratory rat. The rat is widely used in gerontological research. Those who use the rat must be clearly aware of the factors that can influence lesions and death in aged rats (Anver and Cohen, 1979; Burek, 1978). Neoplastic disease, a cause of death in many aged rats, has been discussed previously, and this section will summarize the more commonly seen nonneoplastic lesions in aging rats.

a. Chronic Progressive Nephropathy (CPN). Chronic progressive nephropathy is among the most commonly encountered lesions in the rat, particularly in the aged animal where it is a major life-limiting disease. Because of its complex nature, CPN has a large number of synonyms, often with inaccurate descriptive titles. Sequential development of lesions and factors that modify the course of CPN have been thoroughly studied, but the actual pathogenesis remains undetermined. Chronic progressive nephropathy occurs in most rats, but its rate of development is significantly influenced by numerous factors, including sex, genetics, age, diet, association with microflora, hormone treatment (particularly testosterone), and others. Male rats have an earlier onset and more rapid progression of CPN than females, and albino strains and stocks seem to be more predisposed. Diet is an extremely important predisposing factor, particularly the quality and quantity of dietary protein. Significant differences in severity of CPN can be observed among rats of the same sex, strain, age, and source when fed different diets for their lifetime. Axenic rats do not seem to develop significant CPN.

Rats with CPN develop progressively severe proteinuria, first with the selective excretion of only small molecules such as albumin. In late CPN there is nonselective excretion, and the electrophoretic pattern of urinary protein resembles that of serum. Proteinuria related to CPN should not be confused with the normal urinary excretion of α-globulin in male rats, probably of tubular origin. The histopathology of CPN is characterized as progressive basement membrane thickening of the glomerulus, Bowman's capsule, and proximal tubule. Basement membranes split and wrinkle as tubules become atrophic and collapse. Glomeruli enlarge, with mesangial deposition of protein and lipid, adhesion to Bowman's capsule, and segmental sclerosis. Tubular epithelium may contain granules of resorbed protein, which later is mixed with lipofuscin and hemosiderin. As glomerular protein leakage becomes more severe,

tubules dilate and accumulate hyaline eosinophilic castes of protein within their lumina, resembling colloid. Tubular epithelium may undergo atrophy with collapse or dilatation. Interstitial fibrosis and leukocytic infiltration is frequently observed. Grossly, the kidneys become enlarged, pale with irregular surfaces and pigmented (Fig. 14). Cystic tubules can be observed. Hypoproteinemia, parathyroid hyperplasia with serum chemical changes indicative of hyperparathyroidism, and nephrotic syndrome are observed in rats with severe CPN (Barthold, 1979).

b. Nephrocalcinosis. This disease, which occurs more frequently in females, has been reported in numerous strains of rats. In most cases, affected rats have been fed purified diets (e.g., AIN-76). Renal lesions usually involve the corticomedullary junction and consist of mineral deposition in the lumina of the proximal tubules (Nguyen and Woodward, 1980).

c. Polyarteritis nodosa. This disease is of unknown etiology and affects muscular arteries, most commonly the mesenteric, pancreatic, and spermatic. The lesions may be either acute or chronic. In the acute stage, there is intimal and medial fibrinoid necrosis, focal thrombosis, disruption of the elastic

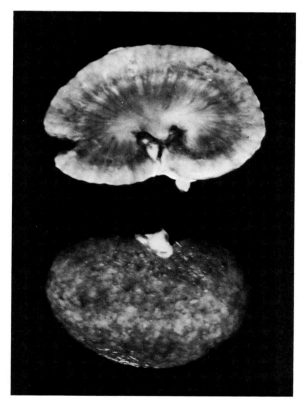

Fig. 14. Chronic progressive nephropathy in an aged rat. The kidneys are pale with irregular, pitted surfaces.

lamellae, and an infiltration of polymorphonuclear leukocytes and mononuclear cells. In the chronic stage, the arteries are nodular, tortuous, and thick-walled. Affected vessels frequently have aneurysms and thrombi. Both acute and chronic phases may be observed in the same artery in an animal.

d. Myocardial degeneration. Degeneration is a commonly observed lesion that occurs in most stocks and strains, with an onset at 12–18 months of age. It occurs more frequently in males. The lesions are usually microscopic but, in advanced stages, can appear grossly as grayish foci. The papillary muscles and their attachment sites in the wall of the left ventricle are the most frequent sites of the lesion (Anver and Cohen, 1979).

e. Radiculoneuropathy. Spinal nerve root degeneration has been reported in a number of rat stocks and strains. Lesion distribution may vary among strains; however, the cauda equina and ventral spinal nerve roots are commonly involved. Posterior paresis and paralysis have been associated with these lesions, but some suggest that these signs are associated with a separate degenerative process of the skeletal muscle (Anver and Cohen, 1979).

ACKNOWLEDGMENTS

Original Figs. 1, 2, 4, 5, and 10–14 are from the Section of Comparative Medicine, Yale School of Medicine, under support provided by Grant RR00393 from the Division of Research Resources, National Institutes of Health, Bethesda, Maryland. The authors gratefully acknowledge the contribution of Janice A. Halpin in the preparation of this manuscript.

REFERENCES

Ackerman, J. I., Fox, J. G., and Murphy, J. C. (1984). An enzyme linked immunosorbent assay for detection of antibodies to *Corynebacterium kutscheri* in experimentally infected rats. *Lab. Anim. Sci.* **34**, 38–43.

Adler, N. T., and Zoluth, S. R. (1970). Copulatory behavior can inhibit pregnancy in female rats. *Science* **168**, 1480–1482.

Altman, H. N., and Goodman, D. G. (1979). Neoplastic diseases. *In* "The Laboratory Rat" (H. J. Baker, J. R. Lindsey, and S. H. Weisbroth, eds.), Vol. 1, pp. 334–376. Academic Press, New York.

Altman, P. L., and Katz, D. D. (1979). "Inbred and Genetically Defined Strains of Laboratory Animals. Part 1. Mouse and Rat." Fed Am. Soc. Exp. Biol., Bethesda, Maryland.

Anderson, L. C., Leary, S. L., and Manning, P. J. (1983). Rat-bite fever in animal research laboratory personnel. *Lab. Anim. Sci.* **3**, 292–294.

Anonymous (1982). Muroid virus nephropathies. *Lancet* **II**, 1375.

Anver, M. R., and Cohen, B. J. (1979). Lesions associated with aging. *In* "The Laboratory Rat" (H. J. Baker, J. R. Lindsey, and S. H. Weisbroth, eds.), Vol. 1, pp. 377–399. Academic Press, New York.

Baker, D. E. J. (1979). Reproduction and breeding. *In* "The Laboratory Rat" (H. J. Baker, J. R. Lindsey, and S. H. Weisbroth, eds.), Vol. 1, pp. 154–168. Academic Press, New York.

Baker, H. J., Lindsey, R., and Weisbroth, S. H. (1979a). Housing to control research variables. *In* "The Laboratory Rat" (H. J. Baker, J. R. Lindsey, and S. H. Weisbroth, eds.), Vol. 1, pp. 169–192. Academic Press, New York.

Baker, H. J., Lindsey, J. R., and Weisbroth, S. H. (1979b). Selected normative data. Appendix I. *In* "The Laboratory Rat" (H. J. Baker, J. R. Lindsey, and S. H. Weisbroth, eds.), Vol. 1, pp. 412–413. Academic Press, New York.

Bardin, C. W., Bullock, L., Schneider, G., Allison, J. E., and Stanley, A. J. (1970). Pseudohermaphroditic rat: End organ insensitivity to testosterone. *Science* **167**, 1136–1137.

Barthold, S. W. (1979). Chronic progressive nephropathy in aging rats. *Toxicol. Pathol.* **7**, 1–6.

Barthold, S. W. (1984). Research complications and state of knowledge of rodent coronaviruses. *In* "Complications of Viral and Mycoplasmal Infections in Rodents to Toxicology Research and Testing" (T. E. Hamm, D. A. Hart, and A. McCarthy, eds.). Hemisphere Publ. Corp., Washington, D.C. (in press).

Bennet, J. P., and Vickery, B. H. (1970). Rats and mice. *In* "Reproduction and Breeding Techniques for Laboratory Animals" (E. S. E. Hafez, ed.), pp. 299–315. Lea & Febiger, Philadelphia, Pennsylvania.

Bivin, W. S., Crawford, M. P., and Brewer, N. R. (1979). Morphophysiology. *In* "The Laboratory Rat" (H. J. Baker, J. R. Lindsey, and S. H. Weisbroth, eds.), Vol. 1, pp. 74–94. Academic Press, New York.

Brownstein, D. G., Smith, A. L., and Johnson, E. A. (1981). Sendai virus infection in genetically resistant and susceptible mice. *Am. J. Pathol.* **105**, 156–163.

Buchbinder, L., Hall, L., Wilens, S. L., and Slanetz, C. A. (1935). Observations on enzootic paratyphoid infection in a rat colony. *Am. J. Hyg.* **22**, 199–213.

Burek, J. D. (1978). Age-associated pathology. *In* "Pathology of Aging Rats," pp. 29–168. CRC Press, West Palm Beach, Florida.

Carthew, P., and Sparrow, S. (1980). A comparison in germ-free mice of the pathogenesis of Sendai virus and pneumonia virus infection. *J. Pathol.* **130**, 153–158.

Cassell, G. H., Lindsey, J. R., Baker, H. J., and Davis, J. K. (1979). Mycoplasmal and rickettsial diseases. *In* "The Laboratory Rat" (H. J. Baker, J. R. Lindsey, and S. H. Weisbroth, eds.), Vol. 1, pp. 243–269. Academic Press, New York.

Cassell, G. H., Lindsey, J. R., Davis, J. K., Davidson, M. K., Brown, M. B., and Mayo, J. G. (1981a). Detection of natural *Mycoplasma pulmonis* infection in rats and mice by an enzyme-linked immunosorbent assay (ELISA). *Lab. Anim. Sci.* **31**, 676–682.

Cassell, G. H., Wilborn, W. H., Silvers, S. H., and Minion, R. C. (1981b). Adherence and colonization of *Mycoplasma pulmonis* to genital epithelium and spermatozoa in rats. *Isr. J. Med. Sci.* **17**, 593–598.

Caster, W. O., Poncelet, J., Simon, A. B., and Armstrong, W. D. (1956). Tissue weights of the rat. I. Normal values determined by dissection and chemical methods. *Proc. Soc. Exp. Biol. Med.* **91**, 122–126.

Cheville, N. (1976). Immunopathology. *In* "Cell Pathology," pp. 197–259. Iowa State Univ. Press, Ames.

Cohen, B. J., DeBruin, R. W., and Kort, W. J. (1970). Heritable hydronephrosis in a mutant strain of brown Norway rats. *Lab. Anim. Care* **20**, 489–493.

Coleman, G. L., Barthold, S. W., Osbaldiston, G. W., Foster, S. J., and Jonas, A. M. (1977). Pathological changes during aging in barrier-reared Fischer 344 male rats. *J. Gerontol.* **32**, 258–278.

Coleman, G. L., Jacoby, R. O., Bhatt, P. N., Smith, A. L., and Jonas, A. M. (1982). Naturally occurring lethal parvovirus infection of juvenile and young-adult rats. *Vet. Pathol.* **20**, 49–56.

Farris, E. J. (1963). Breeding of the rat. *In* "The Rat in Laboratory Investigation" (E. J. Farris and J. Q. Griffith, eds.), pp. 1–18. Hafner, New York.

Fleischman, R. W., McCracken, D., and Forbes, W. (1977). Adynamic ileus in the rat induced by chloral hydrate. *Lab. Anim. Sci.* **27,** 238–243.

Flynn, R. J. (1973). "Parasites of Laboratory Animals." Iowa State Univ. Press, Ames.

Foster, H. L. (1980). Gnotobiology. *In* "The Laboratory Rat" (H. J. Baker, J. R. Lindsey, and S. H. Weisbroth, eds.), Vol. 2, pp. 43–57. Academic Press, New York.

Fox, J. G., Niemi, S. M., Murphy, J. C., and Quimby, F. W. (1977). Ulcerative dermatitis in the rat. *Lab. Anim. Sci.* **27,** 671–678.

Fox, J. G., Beaucage, C. M., and Murphy, J. C. (1979). *Corynebacterium kutscheri* pneumonia in rats on long term tobacco inhalation studies. *30th Ann. Mtng. Am. Ass. Lab. Anim. Sci.* Abs. 104.

Fox, M. H. (1969). Evidence suggesting a multigenic origin of membranous septal defect in rats. *Circ. Res.* **24,** 629–637.

Fujikura, T. (1970). Kidney malformations in fetuses of AXC line 9935 rats. *Teratology* **3,** 245–249.

Gajdusek, D. C., Goldgaber, D., Millard, E., and Ono, S. (1982). "Bibliography of Hemorrhagic Fever with Renal Syndrome (Muroid Virus Nephropathies)." National Institutes of Health, Bethesda, Maryland.

Ganaway, J. R. (1980). Effect of heat and selected chemical disinfectants upon infectivity of spores of *Bacillus piliformis* (Tyzzers disease). *Lab. Anim. Sci.* **30,** 192–196.

Garlinghouse, L. E., and Van Hoosier, G. L. (1978). Studies on adjuvant-induced arthritis, tumor transplantability and serologic response to bovine serum albumin in Sendai virus-infected rats. *Am. J. Vet. Res.* **39,** 297–300.

Gralla, E. J. (1981). Scientific Considerations in Monitoring and Evaluating Toxicological Research." Hemisphere Publ. Corp., Washington, D.C.

Hansen, C. T., Potkay, S., Watson, W. T., and Whitney, R. A. (1981). "NIH Rodents 1980 Catalogue," NIH Publ. 81–606. Dept. of Health and Human Services, Public Health Service, Bethesda, Maryland.

Hebel, R., and Stromberg, M. W. (1972). "Anatomy of the Laboratory Rat," Williams & Wilkins, Baltimore, Maryland.

Hsu, C. K. (1979). Parasitic diseases. *In* "The Laboratory Rat" (H. J. Baker, J. R. Lindsey, and S. H. Weisbroth, eds.), Vol. 1, pp. 307–331. Academic Press, New York.

International Agency for Research on Cancer (1973). "Pathology of Tumors in Laboratory Animals. Tumors of the Rat" (V. S. Turusov, ed.), Int. Agency Res. Cancer, Vol. I, Part 1. Lyon.

International Agency for Research on Cancer (1976). "Pathology of Tumors in Laboratory Animals. Tumors of the Rat" (V. S. Turusov, ed.), Int. Agency Res. Cancer, Vol. I, Part 2. Lyon.

Jacoby, R. O., Bhatt, P. N., and Jonas, A. M. (1979). Viral diseases. *In* "The Laboratory Rat" (H. J. Baker, J. R. Lindsey, and S. H. Weisbroth, eds.), Vol. 1, pp. 272–306. Academic Press, New York.

Jonas, A. M., Percy, D., and Craft, J. (1970). Tyzzer's disease in the rat. *Arch. Pathol.* **90,** 516–528.

Kohn, D. F., and Kirk, B. E. (1969). Pathogenicity of *Mycoplasma pulmonis* in laboratory rats. *Lab. Anim. Care* **19,** 321–330.

Kohn, D. F., Chinookoswong, N., and Chou, S. M. (1981). A new model of congenital hydrocephalus in the rat. *Acta Neuropathol.* **54,** 211–218.

Kraft, L. M., D'Amelio, E. D. and D'Amelio, F. (1982). Morphological evidence for natural poxvirus infection in rats. *Lab. Anim. Sci.* **32,** 648–654.

Lai, Y-L., Jacoby, R. O., Jensen, J. T., and Yao, P. C. (1980). Animal model: Hereditary retinal degeneration in Wag/Rij rats. *Am. J. Pathol.* **98,** 281–284.

Lane-Petter, W. (1972). The laboratory rat. *In* "The UFAW Handbook on the Care and Management of Laboratory Animals," 4th ed. pp. 204–211. Williams & Wilkins, Baltimore, Maryland.

LaVail, M. M., Sidman, R. L., and O'Neal, D. (1972). Photoreceptor pigment epithelial cell relationships in rats with inherited retinal degeneration. *J. Cell Biol.* **53,** 185–209.

Leathem, J. H. (1979). Aging and reproduction in rats. *In* "Development of the Rodent as a Model System of Aging" (D. C. Gibson, R. C. Adelman, and C. Finch, eds.), NIH Publ. 79–161, pp. 143–156. U.S. Dept. of Health, Education and Welfare, Public Health Service, Bethesda, Maryland.

Lentsch, R. H., Wagner, J. E., and Owens, D. R. (1979). Comparison of techniques for primary isolation of respiratory *Mycoplasma pulmonis* from rats. *Infect. Immun.* **28,** 590–593.

Lindsey, J. R. (1979). Historical foundations. *In* "The Laboratory Rat" (H. J. Baker, J. R. Lindsey, and S. H. Weisbroth, eds.), Vol. 1, pp. 1–36. Academic Press, New York.

Lindsey, J. R., Baker, H. J., Overcash, R. G., Cassell, G. H., and Hunt, C. E. (1971). Murine chronic respiratory disease. Significance as a research complication and experimental production with *Mycoplasma pulmonis*. *Am. J. Pathol.* **64,** 675–716.

MacKenzie, W. F., and Garner, F. M. (1973). Comparison of neoplasms in six sources of rats. *JNCI, J. Natl. Cancer Inst.* **50,** 1243–1257.

MacKenzie, W. F., Magill, L. S., and Huise, M. (1981). A filamentous bacterium associated with respiratory disease in wild rats. *Vet. Pathol.* **18,** 836–839.

Maenza, R. M., Powell, D. W., Plotkin, G. R., Formal, S. B., Jervis, H. R., and Sprinz, H. (1970). Salmonella enterocolitis in the rat. *J. Infect. Dis.* **121,** 475–485.

Mitruka, B. M., and Rawnsley, H. M. (1977). "Clinical Biochemical and Hematological Reference Values in Normal Experimental Animals." Masson, New York.

National Academy of Sciences (NAS) (1979). "Animals for Research, A Directory of Sources," 10th ed. Office of Publications, National Academy of Sciences, Washington, DC.

National Institutes of Health (NIH) (1981). "NIH Rodents, 1980 Catalogue," NIH Publ. No. 81-606. U.S. Department of Health and Human Services, Bethesda, Maryland.

National Research Council (NRC) (1978). "Nutrient Requirements of Laboratory Animals," 3rd rev. ed., Vol. 10. NRC, Washington, D.C.

Nguyen, H. T., and Woodward, J. C. (1980). Intranephronic calculosis in rats. *Am. J. Pathol.* **100,** 39–56.

Oldstone, J. N. (1967). Helminths, ectoparasites and protozoa in rats and mice. *In* "Pathology of Laboratory Rats and Mice" (E. Cothchin and F. J. C. Roe, eds.), pp. 641–679. Blackwell, Oxford.

Ringler, D. H., and Dabich, L. (1979). Hematology and clinical biochemistry. *In* "The Laboratory Rat" (H. J. Baker, J. R. Lindsey, and S. H. Weisbroth, eds.), Vol. 1, pp. 105–121. Academic Press, New York.

Robinson, R. (1965). "Genetics of the Norway Rat." Pergamon, Oxford.

Robinson, R. (1979). Taxonomy and genetics. *In* "The Laboratory Rat" (H. J. Baker, J. R. Lindsey, and S. H. Weisbroth, eds.), Vol. 1, pp. 37–54. Academic Press, New York.

Rogers, A. E. (1979). Nutrition. *In* "The Laboratory Rat" (H. J. Baker, J. R. Lindsey, and S. H. Weisbroth, eds.), Vol. 1, pp. 123–152. Academic Press, New York.

Ross, M. H., and Bras, G. (1971). Lasting influence of early caloric restriction on prevalence of neoplasms in the rat. *JNCI, J. Natl. Cancer Inst.* **47,** 1095–1113.

Sher, S. P. (1982). Tumors in control hamsters, rats, and mice: Literature tabulation. *CRC Crit. Rev. Toxicol.* **19,** 49–79.

Shih, T. Y., and Scholnick, E. M. (1980). Molecular biology of mammalian sarcoma viruses. *In* "Viral Oncology" (G. Klein, ed.), pp. 135–160. Raven Press, New York.

Smith, A. L. (1984). State of the art detection methods for rodent viruses. *In* "Complications of Viral and Mycoplasmal Infections in Rodents to Toxicology Research and Testing'' (T. E. Hamm, D. A. Hart, and A. Mc-Carthy, eds.). Hemisphere Publ. Corp., Washington, D.C. (in press).

Smith, E. M., and Calhoun, M. L. (1972). ''The Microscopic Anatomy of the White Rat.'' Iowa State Univ. Press, Ames.

Taguchi, F., Yamada, A., and Fujiwara, F. (1979). Asymptomatic infection of mouse hepatitis in the rat. *Arch. Virol.* **59,** 275–279.

Vogtsberger, L. M., Stromberg, P. C., and Rice, J. M. (1982). Histological and serological response of $B_6C_3F_1$ mice and F344 rats to experimental pneumonia virus of mice infection. *Lab. Anim. Sci.* **32,** 419 (abstr.).

Walberg, J. A., Stark, D. M., Desch, C., and McBride, D. F. (1981). Demodicidosis in laboratory rats (*Rattus norvegicus*). *Lab. Anim. Sci.* **31,** 60–62.

Ward, J. M., and Young, D. M. (1976). Latent adenoviral infection of rats: Intranuclear inclusions induced by treatment with a cancer chem-otherapeutic agent. *J. Am. Vet. Med. Assoc.* **199,** 952–953.

Ward, J. R., and Cole, B. C. (1970). Mycoplasma infections of laboratory animals. *In* ''The Role of Mycoplasmas and L Forms of Bacteria in Disease'' (J. T. Sharp, ed.), pp. 212–329. Thomas, Springfield, Illinois.

Weisbroth, S. H. (1979). Bacterial and mycotic diseases. *In* ''The Laboratory Rat'' (H. J. Baker, J. R. Lindsey, and S. H. Weisbroth, eds.), Vol. 1, pp. 193–241. Academic Press, New York.

Whittlestone, P., Lemcke, R. M., and Olds, R. J. (1972). Respiratory disease in a colony of rats. II. Isolation of *Mycoplasma pulmonis* from the natural disease, and the experimental disease induced with a cloned culture of this organism. *J. Hyg.* **70,** 387–409.

Zeman, W., and Innes, J. R. M. (1963). ''Craigie's Neuroanatomy of the Rat.'' Academic Press, New York.

Chapter 5

Biology and Diseases of Hamsters

G. L. Van Hoosier, Jr. and Warren C. Ladiges

Approximately 1,000,000 hamsters are used in research annually in the United States. Of these, 90% are Syrian (golden), *Mesocricetus auratus,* and the remainder are primarily the Chinese (striped back), *Cricetus griseus;* the Armenian (gray), *Cricetulus migratorius;* and the European, *Cricetus cricetus* (Fig. 1). The family Cricetidae is a member of the order Ro-

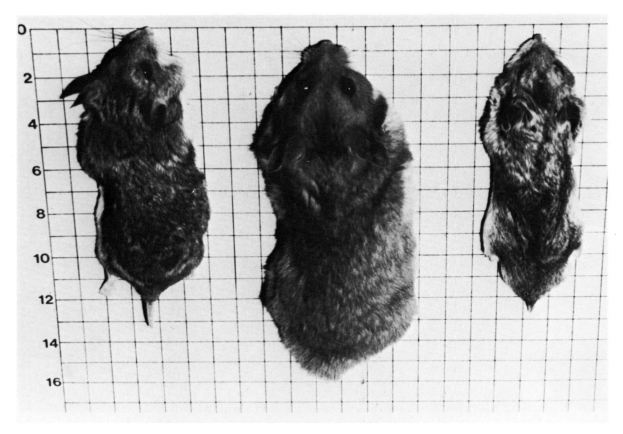

Fig. 1. Appearance and comparative size of, from left to right, Armenian hamster, (age 3 months, weight 45 gm), Syrian hamster (age 11 weeks, weight 103 gm), and Chinese hamster (age 3 months; weight 35 gm).

dentia. Animals in this family are characterized by large cheek pouches that are used to transport and store food, by thick bodies, and by short tails (Fig. 2). They have incisors that grow continuously and cuspidate molars that do not continue to grow (I 1/1, C 0/0, PM 0/0, M 3/3 × 2 = 16).

SYRIAN HAMSTER

I. INTRODUCTION

The publication by Hoffman *et al.* (1968) is a good, comprehensive source of information and reference to the literature prior to 1967 on the biology and experimental uses of the hamster, and is highly recommended to those who have an in-depth interest in the species. Also noteworthy in this regard are the three volumes entitled ''Biology Data Book'' (Altman and Dittmer, 1972, 1973, 1974). The audiotape, 2 × 2 colored

slides, and syllabus prepared by Clark (1975) are recommended as an introduction to the hamster for a variety of learning and instructional settings.

A. Description

The Syrian, or golden, hamster (*Mesocricetus auratus*) is native to the arid, temperate regions of Southeast Europe and Asia Minor. In their natural environment, hamsters live in deep tunnels that ensure a cooler temperature and higher humidity than the general desert environment. They are nocturnal animals. The adult Syrian hamster is larger than a mouse, usually growing to 6 to 8 in. in length (14 to 19 cm) and weighing from 110 to 140 gm. It is virtually tailless, and has smooth, short hair. Normal coloration is reddish-gold, with a grayish-white ventral portion; however, color can range from albino to dark brown, and the dorsal portion may have a dark stripe. The ears are pointed, with darker coloration, and the eyes are small, dark, and bright.

Coarse hair over darkly pigmented skin can be readily ob-

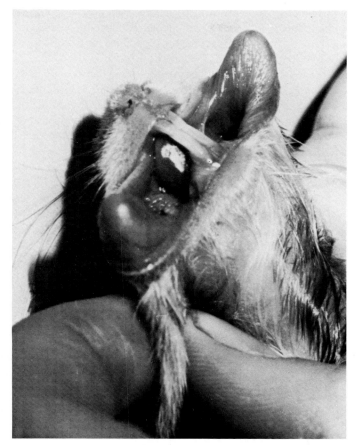

Fig. 2. The cheek pouches of a Syrian hamster, mechanically everted for illustrative purposes.

served in males in the costovertebral area. These are sebaceous glands that produce secretions in response to androgen production. When excited, the hair over these glands will become wet, and the animal will scratch and rub itself as though the glandular area were irritated. There is evidence that the glandular secretions are used for territorial marking. The female also has sebaceous glands on her back, but they are not as easily identified and the secretions are associated with the estrous cycle.

Male hamsters can be identified visually by the prominent glands described above and by large testicles that make the posterior of the body pointed and protuberant. The normal gross anatomy has been described by Magalhaes (1968).

B. Use in Research

Practically all Syrian hamsters now in use as laboratory animals originated from one litter captured in Syria in 1930. Only three members of the litter were retained in captivity, and it is the progeny of these three animals that were first imported to the United States in 1938. By 1971, the hamster had become the third most commonly used laboratory animal in the United States, exceeded only by mice and rats.

The four major reasons given for selection of the Syrian hamster for research are (1) availability and ease of reproduction, (2) relative freedom from spontaneous diseases coupled with susceptibility to many introduced pathogenic agents, (3) anatomical and physiological features with unique potential for study, and (4) rapid development with short life cycle.

1. Availability

The Syrian hamster reproduces readily, with females of some strains able to produce more than 6 litters of 4–12 pups during their breeding life of approximately 1 year. This short breeding cycle has encouraged development of inbred strains that are particularly useful for studying the role of genetic factors in response to some chemical carcinogens and for studying animal models of inherited diseases. Homburger (1972a) describes several significant strains: the BIO (R) 14.6, which shows a dystrophy-like myopathy and cardiomyopathy leading to congestive heart failure in 100% of the animals; BIO (R) 4.24, which is characterized by obesity with multiple endocrine anomalies and a high incidence of benign adenomas of the adrenal cortex; a sex-linked mutation BIO (R) 12.14 with progressive hindleg paralysis; and strains demonstrating cystic prostatic hypertrophy and susceptibility to agents toxic for humans.

2. Freedom from Spontaneous Disease/Susceptibility to Induced Agents

Hamsters have relatively few spontaneous diseases as compared to other laboratory rodents; however, they are susceptible to many experimentally induced diseases and infections (Frenkel, 1979). The increased use of the Syrian hamster during the past 25 years is, in large part, attributable to their susceptibility to tumor induction by viruses of other species, e.g., polyoma virus of mice, SV40 of monkeys, and adenoviruses of humans. Hamsters are susceptible to most deep fungal infections. Histoplasmosis has been studied extensively in hamsters as they are sensitive to small inocula and are useful for diagnostic purposes. Most of the fungi grow in spleen, lymph nodes, and liver. Conchoid Schaumann bodies are produced as during mycobacterial and leishmanial infections. A model for study of generalized dissemination of histoplasmosis after intrapulmonary inocula of hamsters has been developed (Frenkel, 1972). Other infections that can be introduced include tuberculosis, leprosy, atypical mycobacterial infections, leptospirosis, mycoplasmosis, and viral, protozoal, or helminthic infections.

One strain of hamster (the LHC/LAK) has been shown to be susceptible to slow virus diseases such as scrapie and transmissible mink encephalopathy (TME). The human form, Creutzfeldt–Jakob disease, can also be transmitted to the hamster, making it possible to compare these viral diseases in a single host (Marsh and Hanson, 1978).

3. Special Anatomical and Physiological Features

a. Cheek Pouch

The cheek pouches, evaginations of the lateral buccal wall, are devoid of glands, and foreign tissue transplanted to them does not engender the normally expected immunological rejection. This "immunological privilege" is associated with the absence of an intact lymphatic drainage pathway. Billingham (1978) reports that tissue from the cheek pouch can be transplanted to the trunk, and it will maintain its "privileged" characteristics unless lymphatic connections are established bypassing the auto- or isograft.

b. Immunological System

Initial results indicating a failure to reject skin allografts in ways similar to other comparable laboratory rodents, and the relatively enhanced susceptibility to virus infections, especially tumor viruses, and other observations have stimulated a variety of studies aimed at determining whether the hamster has an altered immune system compared to other species. While a detailed analysis and interpretation of these numerous reports (Streilein, 1978; Streilein et al., 1981) is beyond the purview of this chapter, it is apparent that hamsters are atypical in some facets of their immunological system, e.g., the low or absent response to certain antigens, such as alloantigens. While the basic mechanisms responsible for these novel differences are unclear, there is an abiding suspicion that they may be of fundamental biological significance and a major factor accounting for some of the hamster's unique qualities as an experimental animal.

In association with the short gestation period, the ontogeny of the thymic system and associated cellular immunity is delayed as compared to other rodents. Other noteworthy facts are that only 4 of the 5 immunoglobulin classes have been described in the hamster, i.e., IgM, IgG, IgA, and IgE, while IgD remains to be defined, and at least two strains of inbred hamsters are deficient in the sixth component of complement.

c. Hibernation

Hibernation varies among strains and between individual animals; however, exposure to cold stimulates the hamster to gather food, and it will often hibernate at a temperature of approximately 5°C ($\pm 2°$). This feature, which is not common to mice, rats, and guinea pigs, enables hamsters to be used for a variety of unique experimental objectives in experimental physiology or pathology (Lyman, 1979).

Since cold exposure and hibernation in the hamster are associated with desaturation of white adipose tissue, this animal is useful for studies of factors controlling the saturation of fat. Hibernation in the hamster has also been linked to modification of disease entities, such as M. tuberculosis and Treponema pallidum (Lyman and Fawcett, 1954). A hamster does not fatten prior to hibernation, and it will starve unless it wakens periodically to eat. There is evidence that inability to gather a store of food delays hibernation. Hibernating animals remain sensitive to external stimuli and usually are aroused if handled.

d. Radioresistance

The Syrian and Chinese hamster strains are among the most radioresistant mammals ever studied, with respect to lethality and survival time after irradiation (Eddy and Casarett, 1972).

e. Dentition

Syrian hamsters develop dental caries under defined conditions of diet and oral flora, making them useful for the study of etiological factors. Studies show that the caries rate in hamsters is influenced not only by the amount of carbohydrate in the diet but also by the form of carbohydrate. The absence or presence of vitamins in the diet also is suggested as a contributing factor (Shklar, 1972). In addition, hamsters may be useful for testing the anticaries or caries-inhibiting effects of agents such as fluorine and iodoacetic acid.

Jordon and van Houte (1972) state that as result of the work of Keyes and others using hamsters, there is evidence that dental caries may be infectious and transmissible among rodents.

f. Scent Glands

The scent glands of the Syrian hamster are dermal structures in the costovertebral area. They are characterized by large sebaceous glands that are resistant to locally applied carcinogens but susceptible to malignant transformations associated with simultaneous systemic introduction of estrogens and androgens (Homburger, 1972b).

g. Respiratory Tract

The conductive airways of the Syrian hamster contain a limited number of glandular structures, primarily in the upper region of the trachea. This makes M. auratus a potential model for studies of chronic bronchitis (Hayes et al., 1977). The pulmonary vascular bed is similar to human beings in many ways, and hamsters develop pulmonary lesions that resemble centrilobular emphysema in man (Kleinerman, 1972). Spontaneous bronchiogenic and pulmonary cancers are rare, making *M. auratus* a good animal in which to study chemical carcinogenesis in the respiratory tract (Homburger, 1968). Similar to other rodents, the tract appears less sensitive to topical carcinogen applications than skin. The hamster is resistant to pulmonary infection and able to decompose nicotine, making it a good subject for study of long-term inhalation effects.

h. Gastrointestinal System

The digestive process of a Syrian hamster is different from other rodents such as the rat. The esophagus enters between a forestomach and a glandular stomach compartment. The nonglandular forestomach is characteristic of ruminants, containing microorganisms and a high pH suggestive of a fermentation process. Warner and Ehle (1976) suggest that fermentation and deamination may be the initial stage of digestion.

Although mice, rats, and guinea pigs have not been especially useful for studies of naturally occurring neoplasms of the gastrointestinal tract, nonexperimentally induced gastric and intestinal tumors have been reported by Fortner (1957) as common in the hamster. The experimental production of papillomas in the forestomach and intestines, adenomatous polyps in the colon, and adenocarcinomas in the forestomach and intestines suggests that the hamster may be useful for studies of carcinogenesis in the gastrointestinal tract (Homburger, 1968).

Syrian hamsters respond predictably to intragastric administration of purified cholera enterotoxin, presenting with intraluminal accumulation of fluid in the small bowel, cecum, and proximal colon (Lepot and Banwell, 1976). Therefore, it has value for the study of pharmacological agents, such as indomethacin, polymyxin B sulfate, glucose electrolyte solutions, and colchicine, that might inhibit intestinal fluid secretions.

i. Pancreas/Gallbladder/Biliary Tract

The pancreas of the Syrian hamster is similar in function to the mouse and rat, with the major pancreatic ducts joining the common bile duct shortly before it enters the duodenum. This relationship between the ducts distinguishes mice, rats, and hamsters from other mammals, including man.

One of our nation's most urgent health problems is pancreatic cancer, which has a poor prognosis in most cases. The Syrian hamster is the model for pancreatic carcinogenesis, since it is the only animal model in which pancreatic tumors can be induced that are similar to those in humans, both morphologically and clinically (Pour, 1979). The pancreatic carcinogens used effectively in the hamster have not equally affected other laboratory animals.

Induction of exocrine pancreatic tumors has been possible in

Syrian hamsters since 1974, and tumor latency can be as short as 12–15 weeks. The neoplasms that result resemble those found in man (Mohr, 1979). Bile tract carcinomas have been noted following injection of bile from patients with cancer of the bile duct, and a high incidence of adenocarcinoma is produced by implanation of methylcholanthrine pellets into the gallbladder.

j. Kidneys and Urinary Bladder

Next to the dog, the golden or Syrian hamster may be the most reliable model for studying the effect of chemical carcinogens on the urinary bladder. The administration of estrogen to male hamsters causes renal tumors, which may be the best animal model of human renal cancer.

k. Endocrine System

Hamsters are reported to be the first model in which the equivalent of Addisonian adrenal necrosis could be studied. The pituitary gland is of interest because of estrogen-induced adenomas of the intermediate lobe.

4. Short Life Cycle and Rapid Development

The short life cycle of the hamster makes it an excellent animal for study of development and the effect of teratogenic agents. The eighth day of pregnancy is the optimal time for teratogenic studies, when hourly development of the fetal pups can be observed (Ferm, 1967).

Diabetes can be induced in the Syrian hamster in advance of mating, and insulin therapy can be instituted without interrupting the teratogenic effects of the diabetogenic agent, streptozotocin. Connor et al. (1981) showed that conception was not influenced, but the concomitant decrease in birth size and fetal wastage demonstrated effects on the developing young. The hamster thus makes an excellent model for studying embryonic development in diabetic females, to further understanding of the teratogenic effects of the disease.

II. BIOLOGY

A. Development and Physiology

A newborn *M. auratus* pup weighs 2–3 gm. It is hairless, with eyes and ears closed. It does have teeth at birth. On approximately the fifth day, ears will open; at about 15 days, the eyes will open. By age of weaning at 21 days, it will weigh 35–40 gm. By maturity at 6 to 8 weeks, males will weigh

85–110 gm and females 95–120 gm. As the animals age, there is some increase in weight.

The reproductive life span is from 6 to 8 weeks to 15 months of age, and the total life span averages 2 years, with a 3-year maximum expectation. Physiological data, such as heart rate and respiration, can be found in Table I and blood chemistry values in Table II.

B. Genetics

Golden hamsters have a diploid chromosome number of 44. Yoon and Peterson (1979) report on 30 different mutations that have been developed since the introduction of the Syrian hamster as a laboratory animal in the 1930's. Eighteen of the mutations involve coat and eye color, with the earliest being the brown, cream, piebald, and white hamsters; six mutations involve the neuromuscular system, and six are identifiable by

Table I

Normative Data—Syrian (Golden) Hamster[a]

Adult weight	
Male	85–140 gm
Female	95–120 gm
Life span	
Average	2 years
Maximum expected	3 years
Chromosome number (diploid)	44
Water consumption	30 ml/day
Food consumption	10–15 gm/day (adult)
Body temperature	36.2°–37.5°C rectal
Puberty	
Male	6–8 weeks (90 gm)
Female	8–12 weeks (90–100 gm)
Gestation	15–18 days
Litter size	4–12 pups
Birth weight	2–3 gm
Eyes open	15 days
Weaning	21 days (35–40 gm)
Heart rate	280–412
Respiratory frequency	74 (33–127)
Leukocyte counts	
Total	7.62×10^3/mm
Neutrophils	
Segmented polymorphonuclear	21.9%
Nonsegmented polymorphonuclear	8.0%
Lymphocytes	73.5%
Monocytes	2.5%
Eosinophils	1.1%
Basophils	1.1%
Erythrocyte sedimentation rate	1.64 mm/hr
Platelets	670.0×10^3/mm (indirect method)
Red blood cells	7.50×10^6/mm
Hemoglobin g	16.8%

[a]From Aeromedical Review (1975).

quantity or texture of hair. Breeders have also developed inbred strains of hamsters, some of which are of great interest to researchers because of genetically transmitted diseases or conditions and unique susceptibility to teratogenic and carcinogenic agents (Homburger, 1972a).

C. Nutrition

Commercial rat feed is generally used as the basic diet for hamsters, sometimes in combination with rabbit food, to provide a balance of 16–24% protein, 60–65% carbohydrate, and 5–7% fat. Male and female hamsters consume approximately the same amount of food—between 5.5 and 7 gm per day during growth and development.

Although hamsters seem to grow and produce normally with this type of diet, continuing research is demonstrating the need to consider the Syrian hamster in particular as having different nutritional requirements from rats and mice.

The hamster differs from other rodents in that it has a small forestomach in which the first stage of digestion is a fermentation process that affects the utilization of nutrients. For example, when diets of 10% protein, 10% protein supplemented with essential amino acids, and 20% protein were compared for rats and for hamsters, it was found that rats utilized the supplements but hamsters had no better gain with the supplemented 10% diet than with nonsupplemented 10% protein (Banta et al., 1975). For hamsters, unlike other rodents, soybean meal was shown to offer better nutritional efficiency than fishmeal. Simple carbohydrates are not tolerated as well as more complex ones (Ershoff, 1956). Cornstarch provides a good source of energy, and 30–40% in the diet is associated with good growth, reproduction, maintenance, and longevity. Newberne and McConnell (1979) report that the mineral requirements of the Syrian hamster are comparable to the rat's except for zinc, copper, and potassium, for which the hamster requirement is higher. Warner and Ehle (1976) point out that hamsters require a source of many of the B vitamins and also need a source of nonnutritive bulk.

Newberne and Fox (1980) point out that for animals used in research it is imperative that the diet is adequate to ensure that the biological responses obtained are, in fact, related to the experimental procedure. In a recent study by Birt and Conrad (1981) comparing the results of feeding hamsters with five different natural ingredient diets, they found that one commercially available diet showed a significant increase in death rate among males beginning at about 43 weeks. They also showed dietary effects on the number of stillborn and offspring weaned. The present state of nutritional data is reflected in Tables III–V.

Although it is recommended that laboratory animals be fed in a manner that minimizes contamination with excreta, Syrian

Table II

Syrian Hamster Blood Chemistry Values[a]

Test	Unit	Male				Female			
		Mean	Range		N	Mean	Range		N
			SO	Observed			SD	Observed	
Glucose	mg/dl	120.9	33.7	73–173	31	135.4	37.8	37–198	32
BUN	mg/dl	19.4	4.16	12–26	30	18.4	2.3	15–23	32
Total bilirubin	mg/dl	0.4	0.17	0.1–0.8	31	0.3	0.16	0.1–0.9	31
Albumin	gm/dl	4.3	0.22	4.0–4.9	31	4.1	0.27	3.5–4.5	31
Total protein	gm/dl	6.3	0.32	5.8–7.0	31	5.9	0.34	5.2–6.7	32
Calcium	mg/dl	11.1	0.75	9.8–13.2	30	11.0	0.60	9.8–12.4	29
Phosphorus	mg/dl	8.1	1.09	6.2–9.9	31	6.1	1.57	3.0–9.4	32
Uric acid	mg/dl	2.7	0.83	1.4–4.1	31	4.8	0.78	2.7–6.3	31
Creatinine	mg/dl	0.56	0.08	0.4–0.7	30	0.59	0.15	0.4–1.0	32
Cholesterol	mg/dl	94.0	22.6	55–145	31	135.6	20.4	89–181	32
Triglycerides	mg/dl	126.3	42.7	72–227	30	129.8	27.1	79–186	32
Alkaline phosphatase	IU/liter	120.1	16.7	99–151	18	142.7	22.2	86–187	24
SGOT (aspartate)	IU/liter	61.2	39.1	28–122	18	53.3	22.7	33–92	15
SGPT (alanine)	IU/liter	44.7	25.9	22–128	19	50.3	18.3	28–106	21
LDH	IU/liter	222.2	69.7	148–412	18	225.7	79.3	140–372	15
CPK	IU/liter	453	173.8	263–793	18	498	183.6	292–1031	15
a-HBD	IU/liter	337	107.8	224–664	19	320	111.2	193–593	14

[a]From Maxwell *et al.* (1984).

Table III

Nutritional Values[a]

Dietary components	Arrington *et al.* (1966)	Rogers *et al.* (1974)
Casein	18.0	24.0
Sucrose	28.0	21.9
Cornstarch	35.5	40.0
Cellulose fiber	5.0	5.0
Vegetable oil	6.0	3.0
Mineral mix	5.0[b]	5.0[d]
NaCl	2.0[c]	11.0[e]
Protein content	16.1	21.0

[a]From Newberne and McConnell (1979).

[b]USP XIV mixture.

[c]Vitamin diet fortification, Nutritional Biochemicals, Cleveland, Ohio.

[d]Salt mixture gm/kg diet: NaCl, 5.254; potassium citrate, 11.84; potassium phosphate, 3.867; calcium phosphate, 17.777; magnesium carbonate, 2.044; ferric citrate, 0.800; cupric sulfate, 0.027; manganese sulfate, 0.027; aluminum potassium sulfate, 0.0044; potassium iodide, 0.0022; cobalt chloride, 0.0044; zinc carbonate, 0.0176; sodium fluoride, 0.000044.

[e]Vitamin mixture gm/kg diet: cornstarch, 7.94; choline-Cl, 2.0; thiamin-HCl, 0.025; riboflavin, 0.015; niacin, 0.100; calcium pantothenate, 0.040; pyridoxine HCl, 0.006; biotin, 0.0006; folic acid, 0.004; menadione, 0.004; vitamin B_{12} (0.1% trituration with mannitol), 0.050; inositol, 0.20; *p*-aminobenzoic acid, 0.006; DL-tropherd (1.1001 U/g), 0.600; vitamin D_2, 2,484,000 IU.

Table IV

Mineral Content of Satisfactory Hamster Diets and Rat NRC Requirements

Mineral	NRC rat requirement	Banta *et al.* (1975)[a]	Arrington *et al.* (1966)[b]	Rogers *et al.* (1974)[b]
Calcium (%)	0.50	0.54	0.59	0.41
Phosphorus (%)	0.40	0.58	0.30	0.39
Magnesium (%)	0.04	0.13	0.09	0.06
Potassium (%)	0.18	0.79	0.82	0.61
Sodium (%)	0.05	0.19	0.15	0.21
Iron (mg/kg)	35.00	180.00	140.00	154.00
Manganese (mg/kg)	50.00	15.90	3.65	9.00
Copper (mg/kg)	5.00	12.60	1.60	7.00
Zinc (mg/kg)	12.00	9.40	—	9.20
Iodine (mg/kg)	0.15	0.02	1.60	1.70
Cobalt (mg/kg)	—	0.02	—	1.10
Fluoride (mg/kg)	1.00	—	—	0.02

[a]Natural product diet.

[b]Semipurified diet.

hamsters are an exception. If food hoppers are used for hamsters, the feed pellets must be able to fall through the slots to the floor of the cage (Harkness *et al.*, 1977). In a study that began with observations of failing health, decreased conception, and increased cannibalism, the problems were traced to a change in feeders. The feeders that contributed to these problems had 5/16 inch wide slots that prevented the food from dropping to the cage floor. Because hamsters have a broad

Table V

Vitamin Content of Diets Satisfactory for Golden Hamsters

Vitamin	Natural product (mg/kg diet) Banta et al. (1975)	Semipurified diets (mg/kg diet) Arrington et al. (1966)	Rogers et al. (1974)
A	15.3	90.0	2.0
C	—	900.0	—
D	—	5.0	62.1
E	110.0	100.0	600.0
K	5.2	45.0	4.0
Choline	2000.0	150.0	2000.0
PABA	100.0	100.0	6.0
Inositol	100.0	100.0	200.0
B_{12} (μg/kg)	32.0	28.0	50.0
Niacin	92.0	90.0	100.0
Pantothenate	54.0	60.0	40.0
Riboflavin	12.0	20.0	15.0
Thiamin	14.0	20.0	25.0
Pyridoxine	10.1	20.0	6.0
Folic acid	3.1	1.8	4.0
Biotin	0.9	0.4	0.6

muzzle, the animals were forced to chew the food from 2 sides of the metal strips of the feeder. The result was broken teeth and starvation.

Placement of the food directly on the floor of the cage, in addition to or in lieu of the use of a feeder, is preferred for the young, who will begin to eat solid dry food at about 7–10 days of age if they can reach it. Like many other rodents, hamsters are naturally coprophagic.

Fluid requirement is approximately 30 ml/day per animal, generally provided by a stainless steel sipper tube. Glass tubes are especially contraindicated for hamsters, as they are able to bite through glass. The location of the sipper tube must be sufficiently low for the smallest animal caged, as even nursing hamsters need fluids in addition to mother's milk to prevent gastrointestinal disturbances.

D. Pharmacology

Hamsters are apparently more sensitive to the metabolic effects of corticosteroids than some other laboratory animals and are less responsive to histamine. Hamsters are very resistant to morphine; it generally has no sedative or hypnotic effects.

E. Mating and Reproduction

A male hamster is sexually mature when it reaches approximately 90 gm weight. Estrus occurs in the female by 3 months

of age when the animal reaches a weight of 90–100 gm. Reproductivity in both sexes decreases at approximately 14 months of age, but senescent females can often be successfully bred with younger males, although there is a notable increase in defective ova and a decrease in number of young produced. Reproductive activity varies seasonally according to strain.

The female has a 4-day estrous cycle that can be determined by an evaluation of the vaginal discharge. The end of the estrous cycle is marked by the appearance of a copious postovulatory discharge that fills the vagina and may extrude through the vaginal orifice. The discharge is creamy white, opaque, and very viscous, with a distinct odor. If touched, the discharge may be drawn out into a thread 5 to 6 in. in length (Fig. 3). (Clark, 1975). The female can be successfully mated

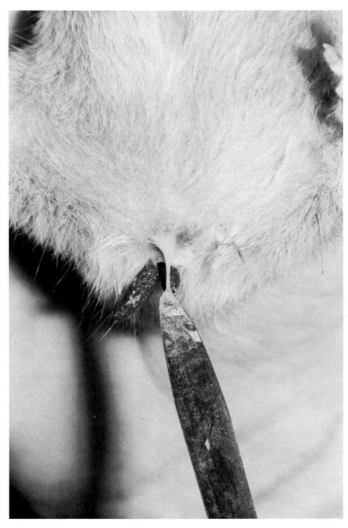

Fig. 3. Postovulatory discharge of a Syrian hamster. (Courtesy of D. Smith, Department of Obstetrics and Gynecology, University of Washington, Seattle, Washington.)

in the evening of the third day after this postovulatory discharge.

A system reported by Ferm (1967) is trial placement. All animals are caged individually for at least 1 week, allowing males to establish cage dominance and the females to cycle normally. Approximately 2 hr after the beginning of the dark cycle, a female is introduced into a male's cage. If she is ready for mating, she will quickly assume a lordotic position with hindlegs spread and tail erect, and will hold it quietly as long as the male is interested. If mating does not occur within 5 min, or if the female is aggressive, she is removed and another is tried. If copulation occurs, the pair can be left together until the following light cycle. With a normal dark cycle, ovulation and fertilization generally occur during the early morning hours, and this (the day of separation) is considered day 1 of gestation. Usually the female is placed in a solid-bottom cage and left undisturbed when pregnant. Colony-raised females can be returned to the colony until the fourteenth day if they do not fight. The female should be moved to a separate cage for at least 2 days before parturition, or she is likely to neglect or destroy her litter.

Another breeding system is to cage one male and one or two females jointly for 7–14 days with the separation of the female to a separate cage for parturition. The use of males older than females or daily checking of animals for wounds can be used to reduce losses from fighting by aggressive females.

Female hamsters may show pseudopregnancy, usually as a result of an infertile mating. The animal can be examined for a postovulatory discharge on days 5 and 9 after mating. If the discharge is present, she is having normal estrous cycles and is not pregnant. A hamster that is pregnant will show a distinct gain in weight after 10 days and will show abdominal distention.

Just prior to parturition, the female becomes restless and alternates between eating, grooming, and nest building. An increase in respiratory rate is also a sign that the young can be expected within the next several hours. The most common time for parturition is on the sixteenth day, although a range of 15–18 days is normal.

Litters range from 4 to 12 pups, with 6 to 8 being the most common. It is possible to sex the pups at birth by comparing the distance from the external urethral orifice to the anus (greater in males), but it is preferable to leave the litter undisturbed until weaning, which begins at 7–10 days. During this time, only fresh water is provided for the mother. If it is necessary to disturb the litter, the probability of cannibalism can be reduced by providing the mother with fresh foods with which she will stuff her cheek pouches.

Pups are left with the mother until they are at least 19 days of age. Normal weaning time is 21–28 days, and the estrous cycle will not resume for the mother until a few days after her young are weaned. Young from different litters can usually be housed together successfully until 40 to 50 days of age, when it usually becomes necessary to separate the females. Males from the same litter can usually be kept together longest.

Copulation activity may begin as early as 4 weeks of age, but it is unusual for pregnancy to occur before 8 weeks of age.

F. Summary of Management and Husbandry

1. Caging and Environment

Hamsters can be maintained in colonies; however, mature animals are usually caged separately because of their tendency to fight. Females to be mated must be given some degree of isolation from adult males and pregnant or lactating females.

A hamster weighing 60 gm or less requires about 10 in.^{2} of space. An animal over 60 gm should have 11–20 in.^{2}, and a female with a litter should have approximately 150 in.^{2}

Caging used for other laboratory rodents is acceptable for hamsters provided it is escape-proof. Hamsters are capable of chewing through thick wood and aluminum. Doors and corners must be close-fitting, and latches must be secure. Plastic shoebox cages with locking lids are recommended. It is essential to have a solid bottom for nesting females and for their young.

Recommended bedding materials include processed hardwood chips, sawdust, shavings, corn cobs, and beet pulp. Pregnant animals will use soft paper for nest building. Urine output is slight, and hamsters tend to consistently use one corner of the cage for elimination. Replacement of bedding materials can be routinely done once or twice weekly and can be left for as long as 10 days to 2 weeks when it is desirable to leave a litter undisturbed.

Cages housing adult hamsters are maintained in an environment of approximately 65°–70°F with 40–60% humidity. Breeding rooms are kept slightly warmer, with a range of 71°–75°F recommended. Hamsters are fairly adaptable to cooler temperatures, and adult animals may actually prefer only a few degrees warmer than their hibernating environment. They are considerably less adaptable to excessively warm temperatures.

A daily light period of 12–14 hr is recommended. The longer 14-hr period is required for breeding colonies. If natural daylight can be eliminated from the room, it is possible for the animals to adapt to an artificial light–dark cycle, which may be more convenient for laboratory management.

2. Handling and Restraint

Hamsters are nocturnal animals, and they tend to be quite inactive during the light cycle in the laboratory.

Magalhaes (1970) reports that the most gentle of the Syrian hamsters is the cream-colored strain, with the mottled, or spotted, variety hardest to manage. Males are more docile and easier to handle than females. Frequent handling seems to reduce wildness, but a startled or awakened hamster is likely to bite.

To move animals, a small can or cup can be placed in the cage. The animal will usually enter the container, and the container with animal can be quickly moved to another cage. The easiest method of hand restraint is to grasp the hamster around the head and shoulders, approaching the animal carefully from the rear. Avoid surprising the animal. Another method is to approach the animal in much the same way, but grasp only the skin. With the loose skin bunched securely in the hand, the skin is taut over the thorax and abdomen. As the animal is lifted, the hand holding it is turned so that the body is supported. An alternate to this method is to approach from the animal's head, so that the thumb and forefinger are gripping the base of the tail and, as before, the loose skin is secured between the fingers and the palmar surface before lifting. Still another method is to approach from the head and enclose the entire body with one hand. The thumb is placed at the base of the rear leg, with the first and second fingers on the opposite side at the base of the tail. The third and fourth fingers restrain the head and forelegs.

III. DISEASES

A. Infectious

1. Bacterial

a. Enteritis

(Proliferative ileitis, regional enteritis, enzootic intestinal adenocarcinoma, transmissible ileal hyperplasia, wet tail)

i. Etiology and prevalence. Enteritis is the most common spontaneous disease of hamsters (Renshaw *et al.*, 1975). The specific cause(s) is unknown. *Escherichia coli* has been associated with the disease (Frisk and Wagner, 1977) and has been reported to experimentally produce diarrhea but not the proliferative lesions (Amend *et al.*, 1976; Frisk *et al.*, 1981). *Campylobacter fetus* subspecies *jejuni* has been isolated from normal hamsters and from hamsters with experimentally induced proliferative ileitis (Fox *et al.*, 1981; La Regina and Lonigro, 1982). However, cultures of the organism failed to produce the disease. Also noteworthy in this regard is the observation of an intracellular antigen demonstrated by electron microscopy (Frisk and Wagner, 1977), and by immunofluorescence (Jacoby, 1978), morphologically compatible with gram-negative rods like *C. fetus*. The possibility that two or more

organisms are acting synergistically is suggested by the failure to reproduce the natural disease by culture of *E. coli* or *C. fetus*. Other bacteria, viruses, parasites, and nutritional deficiencies have also been etiologically incriminated, but there are no recent reports to support these possibilities.

ii. Clinical signs. The signs usually observed are diarrhea and death. The watery diarrhea results in a characteristic moist, matted fur on the tail, perineum, and ventral abdomen (Frisk and Wagner, 1977).

iii. Epizootiology and transmission. The disease can be transmitted experimentally by oral inoculation of tissue homogenate (Jacoby *et al.*, 1975; Amend *et al.*, 1976). The epizootic nature of the disease suggests that it is readily spread by direct contacts and fomites. Animals 3–8 weeks of age are most frequently affected.

iv. Necropsy findings. The gross lesions include a thickening and congestion of the ileum, enlargement of the mesenteric lymph nodes, peritonitis, and adhesions (Fig. 4). Histopathologic changes are characterized by hyperplasia of columnar epithelial cells of the terminal ileum, proliferation of glan-

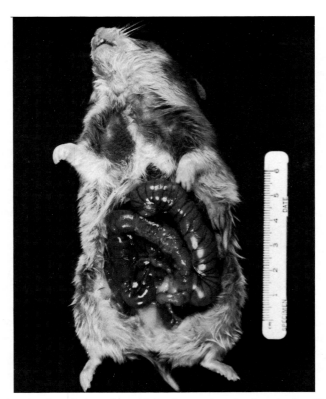

Fig. 4. Gross lesions of enteritis in a hamster inoculated previously with a homogenate of hamster ileum. (From Amend *et al.*, 1976; reprinted with permission from *Laboratory Animal Science*.)

5. BIOLOGY AND DISEASES OF HAMSTERS

133

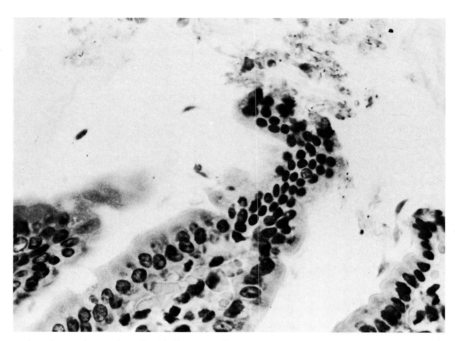

Fig. 5. Villar tip demonstrating abnormal extrusion of epithelial cells. (From Amend *et al.*, 1976; reprinted with permission from *Laboratory Animal Science.*)

dular epithelium, and lymphadenitis with lymphoid hyperplasia, edema, and leukocytic infiltration of sinusoids (Figs. 5 and 6) (Amend *et al.*, 1976; Frisk and Wagner, 1977).

v. Pathogenesis. According to Jacoby (1978), the lesions observed in the ileum develop in two phases following the experimental transmission of disease. The initial phase is characterized by hyperplasia, which begins as a focal lengthening of villi. Approximately 3 weeks following transmission, an inflammatory phase begins associated with focal or segmented necrosis of crypt epithelium. The evolution of the lesions is closely associated with a particulate antigen, presumably bacterial, and detected by immunofluorescence, in the cytoplasm of mucosal epithelial cells. Serum antibodies have been detected that are specific for the intracytoplasmic antigen and which may be of diagnostic value.

vi. Differential diagnosis. Other infectious diseases which should be considered for hamsters with diarrhea are Tyzzer's disease (Section III,A,1,d), cecal mucosal hyperplasia (Barthold *et al.*, 1978), antibiotic-associated diarrhea (Section III,C,3), and salmonellosis. Microbiologic and pathologic findings should distinguish between the various possibilities. In addition, an immunofluorescent test on affected tissue has been described with a high seroconversion rate (Jacoby, 1978).

vii. Prevention, control, and treatment. Before obtaining hamsters, the history of potential suppliers with regard to enteritis should be reviewed. Animals should be purchased from a colony with a minimal disease history and they should not be mixed with animals from other sources. Hamsters with diarrhea should be separated and isolated from other animals. Chemotherapy or chemoprophylaxis is not generally recommended because of the inconsistent and variable results reported. Neomycin was reported by Sheffield and Beveridge (1962) to be effective in the naturally occurring disease, while other have found it to be ineffective in the experimental disease (Jacoby and Johnson, 1981; La Regina *et al.*, 1980). Tetracycline (La Regina *et al.*, 1980) and metronidazole (Jacoby and Johnson, 1981) were found to be moderately effective in prophylaxis of the experimental disease.

viii. Research complications. Enteritis in hamsters can be a major problem with the experimental use of hamsters because of its prevalence, variable morbidity (20–60%), and high mortality (approximately 90%). Because of the report that asymptomatic hamsters may harbor *Campylobacter fetus* subspecies *jejuni*, the potential for transmission to humans should be considered.

b. Pneumonia

i. Etiology and prevalence. In a survey of diseases of the hamster, pneumonia and neoplasia were equally common and were exceeded in frequency only by diarrhea* (Renshaw *et*

*Diarrhea, 10 of 54 respondents; neoplasia, 6 of 54 respondents; pneumonia, 6 of 54 respondents.

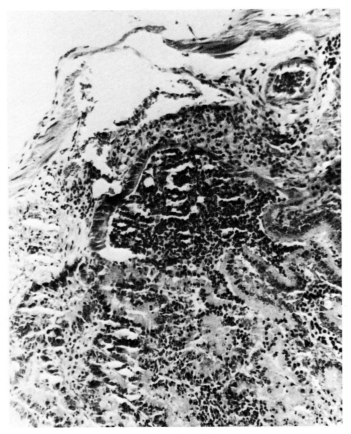

Fig. 6. Ileum of affected hamster demonstrating proliferation of glandular epithelium and penetration of outer muscular layer. (From Amend *et al.*, 1976; reprinted with permission from *Laboratory Animal Science*.)

al., 1975). The respective roles of bacteria, mycoplasma, and viruses or combinations thereof in the etiology of pneumonia are unclear. Possible bacterial etiologies include *Pasteurella pneumotropica, Streptococcus* sp., and *Streptococcus pneumoniae*. While it is clear Sendai virus can be a cause of respiratory disease (see Section III,A,2,C), the role of *Mycoplasma* sp. has not been defined as clearly in hamsters as in rats and mice.

ii. Clinical signs. Overt manifestations of disease may include depression, anorexia, nasal and ocular discharge, and respiratory distress.

iii. Pathogenesis. Various causes of stress, e.g., significant variations from recommended environmental temperatures, may be contributing and predisposing factors in the disease.

iv. Differential diagnosis. A judicious assessment of clinical signs, lesions, and the results of microbiology laboratory reports is essential to definitively diagnose the disease.

v. Prevention, control, and treatment. Stressful situations should be avoided and affected animals should be isolated. If treatment is necessary, the use of antibiotics to which the etiologic organism is sensitive may be appropriate. However, a number of antibiotics e.g., lincomycin, are associated with a fatal enterocolitis (see Section III,C,3).

c. Lymphadenitis

This condition was reported by 2 of 14 respondents to a questionnaire about hamster diseases (Renshaw *et al.*, 1975) and was previously observed, but not reported, by the author in a large colony of hamsters used in tumor virus research (Trentin *et al.*, 1968). One respondent referred to above indicated that *Staphylococcus aureus* was the primary cause, although other agents have been etiologically associated with lymphadenitis in guinea pigs, i.e., β-hemolytic *Streptococcus* Lancefield group C and *Streptobacillus moniliformis*. Until specific reports appear in the literature characterizing the disease in hamsters, it is suggested that a clinician or investigator confronted with the problem consider the disease in hamsters as analogous to the condition in guinea pigs (Ganaway, 1976).

d. Tyzzer's Disease

This condition is most frequently observed in the mouse, but has been reported in several other species including the hamster (Ganaway *et al.*, 1971). The disease is caused by *Bacillus piliformis*, although the organism's exact classification remains uncertain. The only signs reported by Nakayama *et al.* (1975) were diarrhea and death; the lesions observed were enterocolitis, lymphadenitis, and multifocal necrotizing hepatitis. The diagnosis depends on the demonstration of the characteristic organism in affected tissue following special staining with Giemsa or silver techniques. Until additional information is reported about Tyzzer's disease in hamsters, anyone confronted with the problem should consult the description of the condition in mice (Ganaway, 1982).

e. Salmonellosis

The naturally occurring disease is rare in hamsters, as there is apparently only one report in the literature on the subject (Innes *et al.*, 1956). The two outbreaks described were caused by *Salmonella enteritidis* with typical cultural characteristics. According to the report, "Little is known about the clinical aspects of the natural disease other than that it flared up with startling rapidity and that within a few days most of the animals in each batch were dead" (Innes *et al.*, 1956). At necropsy, there was multifocal necrosis of the liver, but no enteritis. Histologically, the disease was characterized by septic thrombi involving the veins and venules, an unusual feature of the infection in hamsters.

Preventive procedures should include the isolation of hamsters from other rodents and quality control procedures to preclude the introduction of contaminated food or bedding.

2. Viral

a. Prevalence

Three different groups have reported on the presence in Syrian hamsters of antibodies to 10 or 11 murine viruses. The results are summarized in Table VI. The most extensive survey was that reported by Parker *et al.* (1967), which involved 850 animals from 14 different colonies. The results by Van Hoosier *et al.* (1970) are for an inbred strain of hamsters designated as LSH; in addition, another 100 animals in a random-bred colony were negative for 13 murine viruses except for 2% which were positive for reovirus type 3. They also reported that type C and "Bernhard" virus particles were present in 2 of 18 primary tumors examined by electron microscopy. It is noteworthy that the infections in the LSH hamster were eliminated by cesarean derivation and foster nursing procedures, except for the Type C and "Bernhard" agents.

While none of the reports summarized in Table VI mentioned any clinical disease associated with the infections, Carthew *et al.* (1978), reported one case of SV5 infection associated with encephalitis and one case of pneumonia virus of mice (PVM) associated with a clinical problem.

The information currently available does not permit one to distinguish between several different options regarding the significance of PVM. Since there are no reports of the isolation of the agent from hamsters, one cannot rigorously exclude the possibility that the serological reactivity detected is non-specific or due to inhibitors instead of antibody. As there is only one anecdotal report of PVM causing natural disease in the mouse, and one similar report in the hamster, it appears that the infection in hamsters is usually subclinical. However, its potential presence may be a complicating factor to the experimental use of the hamster, especially in studies of the respiratory tract.

b. Lymphocytic Choriomeningitis (LCM)

i. Etiology. The infection is caused by a RNA virus of the arena-virus group. Mice are most commonly associated with the disease, although the virus has been isolated from a variety of other species.

ii. Clinical signs. The type of disease observed, if any, may vary with a number of factors, e.g., the age of the hamster at the time of infection, the strain of animal and virus, the route and dose of infection, and the immunologic competence of the host. Approximately half of the animals infected congenitally or as newborns develop a chronic progressive fatal disease characterized by inactivity and weight loss (wasting disease). Other animals infected *in vitro* or as newborns and young adults develop subclinical infections. An impairment of reproductive performance has been reported for chronically infected females (Parker *et al.*, 1976).

iii. Transmission. The implantation of virus-containing tumors has been the principal method of transmission in experimental hamsters. Other methods of spread include direct contact, fomites and aerosols, attributable to the excretion of high titers of virus in the urine. Congenital infections have also been reported.

iv. Necropsy findings. The histopathology of animals developing disease subsequent to congenital or neonatal infection consisted of chronic glomerulonephropathy and widespread vasculitis.

v. Pathogenesis. The experimental infection of young adult hamsters results in a viremia that decreases in titer over a period of 3 months. The amount of virus in the urine exceeds that in the blood and, in some animals, viruria persists longer than viremia. Complement-fixing antibodies appear by the tenth day postinfection, reach peak levels at the sixtieth day, and subsequently decline slowly. Although some animals infected neonatally remain healthy and follow a pattern similar to young adults, others develop disease with a persistent viremia, lower levels of complement fixing, and neutralizing antibodies. The presence of viral antigen and gamma globulin in the glomeruli of affected hamsters suggests an immune complex mechanism for the glomerulonephropathy, analogous to that reported for LCM disease in mice (Parker *et al.*, 1976).

Table VI

Results of Viral Serologic Studies Reported for Hamsters

	Reporting group		
Virus	Parker *et al.* (1967)	Van Hoosier *et al.* (1970)	Reed *et al.* (1974)
Pneumonia virus of mice (PVM)	86%[a]	29% (7/24)[b]	3% (3/100)
Sendai virus	50%	4% (1/27)	1% (1/100)
Simian virus 5 (SV5)	43%	24% (6/25)	ND[f]
Reovirus type 3	29%	0% (0/28)	0% (0/100)
Mouse polio (GDVII)	7%	0% (0/28)	17% (17/100)
Polyoma virus	0%	0% (0/23)	5% (5/100)
Other murine viruses	0%[c]	0% (0/24–29)[c,d]	0% (0/100)[c,e]

[a]Percent positive.
[b]Number positive/number tested.
[c]Mouse hepatitis virus, Kilhams rat virus, mouse adenovirus, K virus.
[d]Lymphocytic choriomeningitis.
[e]Toolans H-1 virus, lymphocytic choriomeningitis.
[f]ND, no data.

vi. Differential diagnosis. Potential causes of wasting disease include graft versus host disease and other procedures resulting in suppression of normal immune responses. The renal lesions should be differentiated from glomerular amyloidosis.

vii. Prevention, control, and treatment. A quality assurance program that includes the regular testing of hamster colonies for antibodies and transplantable tumors for virus, and the elimination of infected animals or tumors, is the principal means of prevention. Since wild house mice can be reservoirs of infection, direct or indirect contact with experimental animal colonies should be excluded.

viii. Research complications. Superinfection of the spontaneous hamster tumor designated Fortner-fibrosarcoma No. 2 resulted in complement fixing antibodies reactive with tumor extracts that mimic specific tumor antigens observed with virus-induced tumors. Especially noteworthy is the fact that three human epidemics with 236 cases occurred in the United States during 1973–74 (Gregg, 1975). Two of the episodes were associated with experimental hamsters bearing transplanted tumors, while the third episode was associated with pet hamsters (see Chapter 22).

c. Sendai Virus

i. Etiology. Parainfluenza 1 (Sendai) is an RNA agent of the paramyxovirus group and is closely related to the human virus hemadsorption type 2 (HA-2). Although mice are believed to be the natural host and are the most common laboratory animal affected, rats, hamsters, and guinea pigs are also susceptible to natural infections. Initial reports of the condition in hamsters were from Japan (Fukumi *et al.*, 1954; Matsumoto *et al.*, 1954).

ii. Clinical signs. Although there is a paucity of relevant reports, the infection is apparently asymptomatic except for occasional deaths in suckling hamsters. However, a variation in susceptibility to clinical disease analogous to that reported in mice (Parker and Richter, 1982) would not be surprising.

iii. Epizootiology and transmission. An enzootic form of the infection has been reported at a research facility in association with the periodic, but continuous, introduction of susceptible hamsters from a commercial vendor (Profeta *et al.*, 1969). Whether the epizootic form, as described in mice (Parker and Richter, 1982), also occurs in hamsters is a matter of conjecture.

iv. Necropsy findings. Consolidation of the lungs has been reported by Profeta *et al.* (1969) during the second serial passage of lung homogenates. In the absence of reports to the contrary, one would anticipate that the gross and microscopic lesions in hamsters would be similar to those reported in mice (Parker and Richter, 1982).

v. Pathogenesis. Specific studies in hamsters have not been reported. In the enzootic form of the disease in mice (Parker and Richter, 1982), virus can be recovered from the lungs for approximately 2 weeks, with 50% of the animals yielding virus between 3 and 6 weeks of age. Complement-fixing and hemagglutinating inhibition antibodies develop in essentially 100% of the animals by 9 weeks of age and persists for 1 year or longer.

vi. Differential diagnosis. The following other causes of pneumonia should be excluded for differential diagnostic purposes: *Streptococcus pneumoniae*, *Pasteurella pneumotropica*, *Streptococcus* sp., and pneumonia virus of mice (Renshaw *et al.*, 1975).

vii. Prevention, control, and treatment. Experimental hamsters should be housed in rooms separate from mice, rats, and guinea pigs as it is possible, if not probable, that other species were the source of infections observed in hamsters. Hamsters from different sources should not be housed in the same room unless all sources are known to be free of the infection. In addition, analogous procedures described for mice should also be applicable to hamsters (Parker and Richter, 1982).

viii. Research complications. The report by Profeta *et al.* (1969) provides a good example of research complications; during an attempt to repeat the work by others on the growth of rubella virus in hamsters, all the experimental animals died within 1 week of the second lung passage. It was subsequently determined that the cause of the mortality was Sendai virus infection.

It has been reported by Garlinghouse and Van Hoosier (1978) that Sendai virus in rats can result in a suppression of the immune response.

d. Type C Virus

i. Etiology. To the extent that this agent has been characterized, the particles observed by Stenback *et al.* (1966) are similar to the family Retrovirdiae, Type C oncovirus group. This viral group includes the murine sarcoma and leukemia virus, the feline sarcoma and leukemia virus, and others, e.g., the guinea pig and bovine type C oncoviruses. The agent was initially observed in human adenovirus-induced hamster tumors.

ii. Clinical signs. Stenback *et al.* (1968) did not demonstrate any oncogenic activity following the inoculation of virus into newborn and weanling hamsters. However, Graffi *et al.* (1968) transmitted lymphomas containing type C particles by cell-free extracts. Of interest in this regard is the observation of an epidemic of lymphomas in hamsters, and the etiological association with an atypical filterable agent (see Section III,B, 3).

iii. Epizootiology, transmission, and pathogenesis. Although specific information is not available for the hamster type C virus, other viruses of this group have copies of genes integrated into normal cells. This viral genetic material may not ordinarily be expressed but can be activated by physical or chemical agents and superinfection by other oncogenic viruses.

iv. Research complications. The presence of hamster agents that resemble known leukemia agents complicates research that involves testing oncogenic potential of other animal or human agents.

3. Parasitic

a. Protozoan

Fecal smears of Syrian hamsters are literally a ''gold mine'' for a protozoologist, as one can observe a large number and variety of organisms. However, their etiologic role in enteric disease remains a matter for speculation as they have been found in similar kinds and numbers in both healthy and diseased animals. The reported prevalence (Wantland, 1955) and usual locations are listed in the tabulation below. In addition, Wagner *et al.* (1974) reported the presence of *Hexamita* sp. as an incidental finding.

Organism	Prevalence	Location
Trichomonas sp.	99%	Cecum, colon
Endamoeba muris	12%	Cecum
Giardia sp.	9%	Small intestine
Chilomastix bettencourti	1%	Cecum

b. Nematodes

The mouse pinworm, *Syphacia obvelata,* has been observed in hamsters (Taffs, 1976). Although the reported prevalence is less than 1% (Wantland, 1955), infection rates can be high in individual colonies. Uninfected hamsters can become infected with *Syphacia muris,* the rat oxyurid, as a consequence of direct contact with infected rats (Ross *et al.,* 1980). The pinworms are predominantly located in the cecum. Eradication

has been reported by two treatment courses with piperazine citrate (10 mg/ml of drinking water) for 7 days separated by 5 days without treatment (Unay and Davis, 1980). For further information on the subject, the description of the condition in mice should be consulted (Foster *et al.,* 1982) (see Chapter 3).

c. Cestodes (Flynn, 1973)

i. Etiology (prevalence, host range). *Hymenolepis nana,* the dwarf tapeworm, is the most important internal parasite found in hamsters, and infection may be common in animals from commercial colonies. They are small to medium in size (25 to 40 mm in length) and usually found in the small intestines. The host range includes mice, rats, nonhuman primates, and man.

ii. Clinical signs. The consequences of infection are usually benign, although the effects depend on the number of parasites and degree of intestinal occlusion, as impactions have been reported.

iii. Epizootiology and transmission. *Hymenolepis nana* is the only known tapeworm with either a direct or an indirect cycle with flour beetles or fleas as the intermediate host. The direct cycle is 14–16 days, while the indirect life cycle is variable. In addition, autoinfection can occur.

iv. Diagnosis. A diagnosis can be made either by the demonstration of eggs in the feces or by the mature worm in the intestines at postmortem. *Hymenolepis nana* can be distinguished from *H. diminuta* by the presence of hooks on the scolex.

v. Prevention, control, and treatment. Preventive and control measures include isolation and quarantine of newly acquired animals, effective insect and wild rodent control, and regular sanitation of cages and ancillary equipment. Yomesan (niclosamide) has been reported as safe and effective for treatment (Ronald and Wagner, 1975).

vi. Research complications. Although any infection associated with morbidity and mortality may interfere with research, the primary significance of *H. nana* is its transmissibility to man. Accordingly, personnel working with animals should be made aware of the potential and receive instruction in hygienic procedures.

d. Mites (Flynn, 1973)

i. Etiology and prevalence. Acariasis in hamsters is predominately associated with two species of the genus *Demodex*

(*D. criceti* and *D. aurati*), although infection with ear mites (*Notoedres* sp.), the tropical rat mite (*Ornithonyssus bacoti*), and a nasal mite (*Spleorodens clethrionomys*) have also been reported. Although infection rates with *Demodex* sp. is high, clinical signs of skin disease are uncommon.

ii. Clinical signs. Alopecia, predominantly of the rump and back, with dry, scaly skin has been reported by Estes *et al.* (1971) in association with *D. aurati* and *D. criceti*. Notoedric mange in the female hamster usually affects the ears only, but in males, lesions may also be observed on the nose, genitalia, tail, and feet.

iii. Pathology. The cases of demodectic mange reported by Estes *et al.* (1971) were characterized by dilated hair follicles that contained debris and mites, loss of the hair shaft, an increase in thickness of the corneus, and little evidence of inflammation.

iv. Pathogenesis. Demodectic mange has been observed in hamsters on a lymphosarcoma transmission study and a chemical carcinogenesis study, although lesions were apparently more related to increasing age than experimental procedures (Estes *et al.*, 1971). Males may be more susceptible to infection and disease than females. Van Hoosier has observed clinical signs in hamsters thymectomized as newborns.

v. Diagnosis. A diagnosis can be established by the demonstration of mites in skin scrapings, although the presence of *Demodex* sp. in association with lesions does not necessarily establish a cause and effect relationship.

vi. Research complications. The presence of a clinical sign is a potential cause of "background noise" in a variety of experimental protocols.

B. Neoplastic

1. Introduction

Spontaneously occurring neoplasia was reported by 11% of the respondents in a survey of hamster diseases and was exceeded only by diarrhea ("wet tail") in the frequency of reporting (Renshaw *et al.*, 1975). The overall incidence of spontaneous malignant neoplasms in Syrian hamsters has been estimated by Van Hoosier and Trentin (1979) at 3.7%, based on 435 tumors in 11,792 animals. However, individual reports in the literature vary considerably. Factors affecting the difference in numbers of tumors reported include the age and sex distribution of the animals observed and the extent to which the animals are examined grossly and microscopically. A distinc-

tion between benign and malignant neoplasms is also useful. For the classification of tumors, the system published by Stewart *et al.* (1959) is employed in Table VII.

2. Benign Neoplasms

The two most commonly reported benign neoplasms of the Syrian hamster are polyps of the intestinal tract and adenomas of the adrenal cortex. Table VIII lists the types of tumors for which ten or more cases have been reported (Van Hoosier and Trentin, 1979).

3. Malignant Neoplasms

Lymphosarcomas are the most frequently reported malignant tumor of the Syrian hamster. Noteworthy is the observation of a 10 : 1 male to female ratio in a total of 30 reported melanomas. Table IX lists the types of malignant tumors for which ten or more cases have been reported (Van Hoosier and Trentin, 1979).

Of special interest in this regard are the outbreaks of horizontally transmitted malignant lymphomas in a hamster colony with an incidence of 50 to 90% in young inbred and random-bred animals (Ambrose and Coggin, 1975). A variety of tests have been negative for Retroviridae agents, e.g., type C virus. Recently, an agent associated with the disease has been demonstrated that has characteristics compatible with a mammalian viroid, i.e., a nonencapsidated, DNase-sensitive low molecular weight, disease-causing, self-replicating, naturally infectious, nucleic acid. Also, enteritis, pyelonephritis, warts, poor breeding efficiency, and intussusception in exposed hamsters

Table VII

Classification of Tumors of Animals According to Site of Origin and Histology

  I. Tumors of epithelial tissue
    A. Tumors presumably of glandular origin
    B. Tumors of postulated but unconfirmed glandular origin
    C. Tumors of nonglandular epithelial origin
  II. Lymphomas and leukemias
    A. Lymphosarcoma or lymphocytic leukemia, or both, with neoplastic involvement of organs and tissues, or leukemic blood, or both, in most instances
    B. Myelocytic leukemia (including chloroleukemia)
    C. Reticulum cell sarcoma
    D. Plasma cell tumor
  III. Tumors of connective tissues
    A. Tumors presumably of connective tissue origin and of specific cell type, e.g., fibrosarcoma
    B. Tumors presumably of connective tissue origin and of nonspecific cell type, e.g., "mixed cell, round cell, or sarcoma"
  IV. Tumors of melanin-forming tissue
  V. Tumors of neural tissue
  VI. Tumors composed of mixed tissues
  VII. Tumors not classified elsewhere

Table VIII

Benign Tumors

Group	Tissue	Tumor type	Site	Relative frequency[a]
I	Epithelial	Polyps	Intestine	+ + + +
I	Epithelial	Adenoma	Adrenal	+ + + +
			Thyroid	+ + +
			Parathyroid	+ +
I	Epithelial	Papilloma	Stomach	+ + +
III	Connective	Hemangioma	Spleen	+ + +
			Liver	+ +
III	Connective	Cholangioma	Bile duct	+ +
III	Connective	Thecoma	Ovary	+ +
IV	Melanin-forming	Cellular blue nevi	Skin	+ +

[a] + + + +, most common; + + +, common; + +, occasional.

were observed (Coggin *et al.*, 1981). These reports are noteworthy because of the epizootic threat to experimental colonies and because the condition may prove to be a valuable model for understanding host–viroid relationships.

C. Miscellaneous

1. Amyloidosis and the Associated Nephrotic Syndrome (Gleiser *et al.*, 1971; Murphy and Fox, 1980)

Amyloidosis is a principal cause of death in hamsters on long-term experiments. The recent report of AA and AP pro-

Table IX

Malignant Neoplasms of the Hamster

Group and site	Type	Relative frequency[a]
Group IA		
Intestines	Adenocarcinoma	+ + +
Adrenal	Carcinoma 26	+ + +
Thyroid	Spindle cell carcinoma 6	+ +
Liver and intrahepatic bile duct	Adenocarcinoma 1	+ +
Uterus	Carcinoma 2	+ +
Kidney	Carcinoma 1	+ +
Group IIA	Lymphosarcoma (lymph nodes)	+ + + +
	Lymphosarcoma (small intestine)	+ + +
	Lymphosarcoma (liver)	+ +
	Lymphosarcoma (kidney)	+ +
	Lymphosarcoma (spleen)	+ +
Group IIC	Reticulum cell sarcoma (lymph nodes)	+ + + +
Group IID	Plasma cell tumor (extramedullary)	+ +

[a] + + + +, most common; + + +, common; + +, occasional.

teins in isolated amyloid fibrils from aged hamsters suggest the condition in hamsters can be classified as the secondary type (Brandwein *et al.*, 1981), although no apparent antecedent disease has been recognized to date. Table X illustrates the increasing prevalence with age in one colony studied that had an overall frequency of 88% in hamsters over 18 months of age. Edema and ascites are sometimes observed in association with the disease. Serum albumin is decreased and the total globulin component is increased in hamsters over 1 year of age. Proteinuria and hypercholesterolemia have been reported. The clinical signs and laboratory findings are consistent with the nephrotic syndrome as described in humans. Conner *et al.* (1983) have reported an apparently analogous condition in CD-1 mice. Histologically, characteristic amyloid deposits are present initially in the glomeruli of the kidney and subsequently in a variety of tissues, especially the spleen, liver, and adrenals. Since amyloidosis in mice is strain associated, genetic factors should be considered in the etiology and pathogenesis of the disease in hamsters.

2. Polycystic Disease (Gleiser *et al.*, 1970)

Cysts have been observed in 76% of hamsters over 1 year of age. The liver is a common site, but they may also be observed at other locations (Table XI). The lesions appear to be due to developmental defects of normal ductal structures, e.g., the bile duct. The condition is not known to be associated with clinical signs, and the typical finding at necropsy is thin-walled hepatic cysts of varying size (0.5–3.0 cm) containing an amber fluid.

3. Antibiotic-Associated Enterocolitis (Small, 1968)

Morbidity and mortality have been observed in hamsters and guinea pigs following the administration of antibiotics generally considered selective for gram-positive organisms. Ampicillin, vancomycin, erythromycin, cephalosporins, and oral gentamicin induced the disease, while the administration of tetracycline and metronidazole were not associated with the disease (Bartlett *et al.*, 1978). In studies reported by Small (1968), anorexia and diarrhea were observed prior to death,

Table X

Amyloidosis in Hamsters[a]

Age (months)	Amyloid	
	Glomerular	Generalized
1–12	0/14	0/14
13–18	5/12 (41.6%)	0/12
19–27	15/17 (88.2%)	15/17 (88.2%)

[a] From Gleiser *et al.* (1971); reprinted with permission from *Laboratory Animal Science*.

Table XI

Frequency of Cysts at Various Sites[a]

Organ	Number	Percent
Liver	17/40	42.5
Epididymis	8/17	47.0
Seminal vesicle	4/17	23.5
Pancreas	5/40	12.5
Renal pelvis	2/40	5.0
Adrenal	1/40	2.5
Esophagus	1/40	2.5
Uterus	1/23	4.35
Ovary	1/23	4.35

[a]From Gleiser *et al.* (1970); reprinted with permission from *Laboratory Animal Care*.

which occurred 4 to 19 days subsequent to the administration of lincomycin hydrochloride. Gross pathology consisted of ileal and cecal distention with hyperemia. In studies of treatment with lincomycin hydrochloride (Lincocin), gram-positive organisms were found in the cecum of only 12% of animals in the treated group as compared with 57% of the control group. Bartlett *et al.* (1978) found that *Clostridium difficile* was consistently present in high concentrations and that cell-free supernatants experimentally reproduced the disease; they concluded that toxin-producing clostridia were responsible for enterocolitis with a variety of antibiotics.

CHINESE HAMSTER

I. INTRODUCTION

The Chinese hamster (*Cricetus griseus*), also known as the striped-back hamster, was first used as a laboratory animal in 1919 (Yerganian, 1958). Since that time, it has been used in several different aspects of biomedical research. It has been shown to be susceptible to a number of infectious disease agents, such as streptococcus, mycobacteria, diphtheria, rabies, influenza, and equine encephalitis (Yerganian, 1958). It was originally thought to have 14 chromosomes, but later studies found the number to be 22. Because of the low incidence of spontaneous and endogenous viral infections, Chinese hamster tissue culture cells have become popular experimental tools for mutagenic and carcinogenic studies. These animals are also used as models for radiobiological research and have been shown to be more radioresistant than the Syrian hamster and many other common laboratory rodents (Corbascio *et al.*, 1962). In 1959 spontaneous hereditary diabetes mellitus with similarities to the human disease was described,

thus providing an animal model which has subsequently been studied extensively (Meier and Yerganian, 1959). The Chinese hamster has also been shown to be susceptible to experimental induction of stomach and esophageal cancer (Baker *et al.*, 1974).

Chinese hamsters can be reared under laboratory conditions and can be purchased commercially. Since their size is comparable to the mouse, similar type caging is adequate. These animals do very well under standard laboratory animal housing conditions. However, animals that are diabetic are very susceptible to stress; special precautions must therefore be taken in their routine care and feeding.

II. BIOLOGY

The Chinese hamster, like the Syrian hamster, has a cheek pouch that can be utilized as an immunologically privileged site (Yerganian, 1958). Another unique biological feature of this animal is the chromosome number of 22, which provides an *in vivo* cell system for cytogenetic studies. The 10 large pairs of autosomes and 2 sex chromosomes can be readily identified, especially since banding patterns were described in bone marrow cells (Lavappa *et al.*, 1973). The rather constant diploidy provides a stable cell system for assessment of agents with known or suspected mutagenic and carcinogenic properties.

Adult animals weigh between 39.3 and 45.7 gm, and are approximately 9 cm long. Newborns weigh 1.5 to 2.5 gm. The average normal life span under laboratory conditions is 2.5 to 3 years. Adult males have exceptionally large testicles; also, the spleen and brain in both sexes are relatively larger than those of the Syrian hamster (Festing, 1972). The normal hemogram is shown in Table XII.

There appears to be no unique dietary needs for the Chinese hamster. They do very well on standard rodent chow, but wheat germ may be used as a supplement for breeders. The average daily water intake was shown to be 11.4 ml per 100 gm body weight for males and 12.9 ml per 100 gm body weight for females (Thompson, 1971).

Early attempts to breed Chinese hamsters under laboratory conditions were unsuccessful. However, in 1958, a reversed illumination schedule was used to successfully establish a production colony (Yerganian, 1958). Normal reproductive data are shown in Table XIII. Sexual maturity is indicated by opening of the vagina, with a mucus-like creamy material frequently secreted at the beginning of estrus. The estrus cycle consists of 4 phases with 16 distinct behavioral characteristics correlating with vaginal and orifice changes (Yerganian, 1958). The use of tester males as well as routine examinations of the vulva, has been used as a means of determining estrus

Table XII

Normal Hemogram of the Chinese Hamster[a]

Hemogram	Absolute value	Percent
WBC	5,500	
Seg	105,600 ± 12,100	19.2 ± 2.2
Bands	935 ± 385	0.17 ± 0.07
Lymphocytes	413,600 ±	75.2 ±
Monocytes	11,500 ± 1760	2.1 ± 0.32
Eosinophils	9350 ± 3850	1.7 ± 0.7
Basophils	825 ± 220	0.15 ± 0.04
RBC	7,120,000 ± 1029	(4.4 – 9.1)
Hgb	12.4 gm/dl	
MCV	59.4 ± 5.8	
MCHC	29.49%	
PCV		
Bleeding time[b]	55 sec	

[a]Data from Moore (1966).
[b]From the cheek pouch (Yerganian et al., 1955).

and optimal breeding time. A copulation plug can be observed immediately following ejaculation, but disintegrates within 3 to 4 hr (Moore, 1965).

Several programs for managing breeding colonies have been described. Since females can become very aggressive immediately following mating and even kill the male, hand mating was originally used. However, when larger, sexually aggressive males are used, along with continual selection of docile females, animals can be used in monogamous pairs (Porter and Lacey, 1969). It has also been shown that Chinese hamsters can be mated in groups. When 3 to 5 female littermates were placed with 3 males from a different litter, less than 3% of the males were lost from injury or death (Cisar et al., 1972).

Temperatures above 82°F may increase the number of runts in litters (Yerganian, 1958). Infertility in young females on

Table XIII

Reproductive Data for the Chinese Hamster

Weaned	21–25 days[a]
Sexually mature	8–12 weeks[a]
Estrous cycle	Polyestrus[b]
Duration of estrous cycle	4 days[a]
Estrus	6–8 hr[a]
Ovulation time	Just before estrus[b]
Copulation	2–4 hr after start of dark period[c]
Implantation	5–6 days[b]
Gestation	20.5 days[a]
Average litter size	4.5–5.2[b]
No. mammae	8[b]
Postpartum estrus	4 days[a]

[a]Yerganian (1958).
[b]Festing (1972).
[c]Moore (1965).

open-bottom cages may be due to an excess growth of hair around the vulva preventing penile penetration during copulation attempts. Progesterone following mating has been reported to increase conception rates (Avery, 1968).

Pregnancy is indicated by a closed vagina with dry, pale, and scaly surrounding tissue at day 4 following mating. Dystocia may occur as a result of fetal wedging in the proximal portion of the vagina during parturition attempts. The fetuses can be saved by surgical removal.

Newborn animals have front incisors. Body hair appears at 3–4 days of age, with complete coverage in 7 days. Eyes and ears are open within 10 to 14 days, and testicles descend in males at about 30 days of age (Moore, 1965). Young animals may leap blindly during first attempts at handling. As animals approach sexual maturity, aggressive females tend to fight. Dominance is established, and separation of the litter may be necessary to prevent trauma and possible deaths.

III. DISEASES

A. Infectious

The Chinese hamster appears to be experimentally susceptible to a number of infectious disease agents. However, very little has been reported concerning spontaneous infections.

Tyzzer's disease occurred in a group of 12 animals (Zook et al., 1977). Signs and lesions were typical for the disease in Syrian hamsters with 9 animals dying. Stress may have been a factor, since these were newly arrived animals that had been without drinking water for a number of hours.

One group has reported on the presence of antibodies to murine viruses (Schiff et al., 1973). Animals 3 months of age were positive for PVM, Kilham rat virus, Reo 3, GD VII, and Toolan's H-1. Animals from the same group, which were 2 or 4 months old, were positive only for PVM and GD VII. All age groups were consistently infected with Trichomonas sp., but no other endo- or ectoparasites were found. In a separate study, demodectic mange was detected in 2 of 157 animals (Benjamin and Brooks, 1977).

B. Metabolic/Genetic

1. Diabetes Mellitus

Spontaneous diabetes mellitus was first recognized in 1957 during the course of inbreeding, and later described in 1959 (Meier and Yerganian, 1959). The disease is similar in a number of aspects to insulin-dependent diabetes of man (Yerganian, 1965).

a. Etiology

The disease is associated with a degranulation of the B cells of the pancreatic islets of Langerhans, resulting in a primary defect in the biosynthesis of insulin (Gerritsen et al., 1970).

b. Clinical Signs

Animals can show signs as early as 18 days, but the disease may occur at any age. There is polydypsia and polyuria with up to 50 to 70 ml of urine passed in 24 hr. The abdomen is usually soiled with urine, and an odoriferous smell may be apparent. At the onset, there is an initial weight gain, but animals soon become sluggish and may die due to dehydration. Occasionally, there is blindness. Nonspecific conjunctivitis and alopecia may be seen. Animals are very susceptible to mild stress of any kind, and sudden death may be triggered by such procedures as cage transfer and changes in environmental temperature. Diabetic females are many times infertile, while animals that do become pregnant show an increased number of abortions and fetal deaths at delivery. It is not unusual for an entire litter to die.

c. Epizootiology

The disease appears to be transmitted as a recessive factor. When glucosuria was used to characterize diabetes, it was shown that four recessive genes were involved (Butler and Gerritsen, 1970). If any pair was homozygous, glucosuria could result. Apparently the duration, severity, and constancy of glucosuria is controlled by modifier genes. It has also been shown that 100% of the offspring become diabetic if the parents are ketotic (Gerritsen et al., 1970). However, it was difficult to predict which offspring would become diabetic when parents were not ketotic. The end result has been the development of sublines by brother–sister inbreeding, which consistently produce diabetic litters (Chang, 1977).

d. Necropsy Findings

Macroscopic lesions are mainly confined to the kidneys. In diabetic animals they are slightly enlarged, spongy, and friable. The pelvis may or may not be dilated. When hydronephrosis is seen, the retained urine is clear and odiferous. The urinary bladder is usually distended with urine. The liver may be moderately enlarged in some animals with a yellow to gray color.

Microscopically, the islets of Langerhans in the pancreas are decreased in number (Meier and Yerganian, 1959). There is a decrease in number of B cells, and those remaining stain lightly basophilic with cytoplasmic granulation and vacuolization. There is PAS-positive material within the cytoplasm that accu-mulates around pyknotic nuclei. Ultrastructural findings in the pancreas have been characterized (Boquist, 1969). In the kidneys, convoluted tubules contain much protein precipitate, with a progressive decrease in staining of lining cells (Lawe, 1962); most glomeruli are hypocellular with marked sclerosis. Intercapillary homogeneous material can be observed that is PAS positive. Bowman's capsule is slightly to moderately thickened, and adhesions between the glomerulus and capsule may be seen. Periodic acid–Schiff-positive material is also found in the basement membrane, which may be wrinkled and slightly thickened. The liver shows an intact lobular arrangement with extensive vacuolization of cells. Perinuclear halos are seen. Intracytoplasmic material is PAS positive but negative for fat. Periodic acid–Schiff-positive material is occasionally found in pericardial adipose tissue.

e. Pathogenesis

The basic defect is a degranulation of B cells, which results in a decreased amount of insulin production with a reciprocal increase in glucagon. In highly inbred glucosuric strains with diabetes, there is a greatly reduced level of pancreatic insulin and a significantly elevated level of glucagon in both pancreas and stomach. A decrease in LDH isozymes appears to be associated with severity of the diabetic condition (Chang et al., 1977). A contributing factor to the renal pathology observed in diabetes may be subnormal levels of specific glycosidases in the kidneys with a resulting change in turnover of tissue glycoproteins (Chang, 1980).

f. Differential Diagnosis

Since diabetes mellitus is a spontaneous disease, it should be ruled out any time a colony illness occurs. If the animals are being used as a model to study diabetes, the experimental protocol will dictate the diagnostic monitoring procedures. Certainly Tyzzer's disease must be considered both in the initial differential phases and also as a secondary complication, since diabetic animals are very susceptible to stress.

g. Treatment and Control

Treatment with hypoglycemic drugs may be indicated in breeding females in an attempt to maintain inbred lines (Meier and Yerganian, 1961).

h. Research Complications

The disease could potentially occur in animals being used in research protocols unrelated to diabetes. Since cellular metabolism is affected, cytogenetic studies could produce adverse, unreliable data.

C. Traumatic

Female littermates can become very aggressive as animals reach maturity. Severe bite wounds, especially about the tail and head area, can be inflicted, and death is not an uncommon occurrence. Litters should, therefore, be separated before fighting becomes a problem. Females following mating can become quite aggressive to the male, so some means of removing the male must be taken into account.

D. Neoplastic

In general, Chinese hamsters have a low incidence of spontaneous tumors, with the liver and reproductive organs mainly involved. The rarity of spontaneous and induced leukemias may be an indication of the absence of innate tumor viruses.

Uterine adenocarcinomas were detected in 30 of 120 females (Ward and Moore, 1969). The growths were firm and whitish with implantation frequently seen on the visceral and parietal peritoneum. Three of the 20 had metastatic foci in the lungs. Another report showed 11 of 77 affected with similar characteristics except that no pulmonary metastasis was seen (Benjamin and Brooks, 1977). Vaginal bleeding was the sign most often seen initially. The incidence of ovarian tumors was significantly increased with radiation exposure, but was rarely reported in control animals (Kohn and Gultman, 1964).

Hepatomas were found in 66 of 253 animals (Ward and Moore, 1969). These were benign and most often occurred as multiple nodules. Nodular hyperplasia, a nonneoplastic lesion, was seen in 111 of 157 animals in another survey (Benjamin and Brooks, 1977).

Pancreatic adenocarcinomas were reported in 3 females 3 years of age that were partially inbred for the development of spontaneous diabetes mellitus (Poel and Yerganian, 1961). Later reports in nondiabetic animals show the tumor to be very rare.

E. Miscellaneous

1. Cerebral Hemorrhage

This lesion occurred in 20% of 253 hamsters given [131]I (Ward and Moore, 1969). Deaths occurred at 1 to 2 years of age. Grossly the hemorrhage was most evident between the cerebral hemispheres with blood often in the lateral ventricles. Microscopically, the hemorrhage was shown to be due to inflammation and necrosis of the anterior cerebral artery. A homogeneous PAS-positive material could be seen within the media of the diseased artery. The vessel wall was greatly thickened in chronic cases.

2. Periodontitis

This condition was found in a strain of Chinese hamster with hereditary diabetes mellitus (Cohen *et al.*, 1961). The lesion is characterized by absorption of alveolar bone, inflammation, and pocket formation due to splitting of the epithelial attachment. The disease corresponds closely to humans with diabetes mellitus in whom periodontitis is seen.

3. Nephrosclerosis

In a study of 157 animals, 46 had evidence of nephrosclerosis (Benjamin and Brooks, 1977). The pathology was reportedly different from the intercapillary glomerulosclerosis associated with diabetes mellitus. Grossly, pitting and a decrease in size were seen when kidneys were severely affected. Microscopically, tubular degeneration, mild interstitial fibrosis, and focal atrophy of the cortex were seen early. In more advanced conditions there was hyaline sclerosis of glomeruli, more severe interstitial fibrosis, and tubular degeneration.

4. Spondylosis

The incidence and appearance were increased in hamsters with spontaneous diabetes mellitus compared to nondiabetic control animals (Silberberg and Gerritsen, 1976).

5. Pulmonary Granulomas

Pulmonary granulomas were observed in 54 of 157 animals (Benjamin and Brooks, 1977). Grossly the lesions appeared as subpleural, yellowish-gray foci, 1 to 3 mm in diameter with variable involvement of the lung parenchyma. Microscopically, lesions consisted of alveolar collections of lipid-filled macrophages, mixed inflammatory cells, septal fibrosis, and occasional cholesterol clefts. The cause is not known. Affected animals were housed in both suspended wire cages and plastic shoebox cages with different types of bedding.

EUROPEAN HAMSTER

I. INTRODUCTION

The European hamster (*Cricetus cricetus*) developed some importance as a laboratory model when several wild-caught animals from a West German industrial area were found to have had bronchogenic squamous cell carcinoma. It has since been found to be susceptible to *N*-diethylnitrosamine with the development of respiratory tumors (Mohr *et al.*, 1973). It was concluded that the European hamster is a more suitable model than the Syrian hamster for highly concentrated and prolonged smoke inhalation studies (Reznik *et al.*, 1975b).

II. BIOLOGY

These animals are nocturnal, and they hibernate during the winter months in the wild. They are the size of a guinea pig, and have white faces and feet and bodies that are reddish-brown dorsally and black ventrally with white patches laterally. Newly weaned animals (25 days postpartum) have an average body weight of 75 gm, with 6-month-old females and males approaching 300 and 400 gm, respectively (Mohr *et al.*, 1973). A normal hemogram is presented in Table XIV.

Water consumption is 5 ml/100 gm body weight, and average food consumption is 2.9 gm/100 gm body weight in summer (August) and 1.8 gm/100 gm body weight in winter (November) (Silverman and Chavannes, 1977). These animals are mainly seed eaters.

Reproductive values are shown in Table XV. Estrus is determined by vaginal smears and by test mating, using a steel mesh divider to keep the pair separated. When no aggressiveness is observed, hamsters may be mated. Pseudopregnancy is seen after mating if conception does not occur. Females will gain 20 to 30 gm in 10 to 12 days, make nest preparations and become very aggressive; but, if not pregnant, begin losing weight over

Table XIV

Normal Hemogram of the European Hamster

Hemogram	Absolute value	Percent
Leukocytes (10^3/ml)		
Total leukocytes	7.4 ± 2.6[a]	
	8.3 ± 2.2[b]	
Neutrophils	1.71 ± 0.065[a]	23.2 ± 2.5
	2.87 ± 3.74[b]	34.6 ± 17
Lymphocytes	5.47 ± 0.059[a]	74 ± 2.3
	5.02 ± 0.387[b]	60 ± 17.6
Monocytes	0.192 ± 0.015[a]	2.6 ± 0.59
	0.083 ± 0.022[b]	1.00 ± 1.00
Eosinophils	0.005 ± 0.002[a]	0.07 ± 0.10
	0.093 ± 0.018[b]	1.13 ± 0.83
Basophils	0.001 ± 0.002[a]	0.02 ± 0.07
	0[b]	
Thrombocytes (10^3/ml)	210 ± 32[b]	
RBC (10^6/ml)	7.64 ± 0.42[a]	
	7.45 ± 0.49[b]	
PCV (%)		49.2 ± 1.6[a]
Hemoglobin (gm/dl)	18.0 0.7 g/dl[a]	

[a]Silverman and Chavannes (1977).
[b]Emminger *et al.* (1975).

Table XV

Reproductive Data for the European Hamster[a]

Sexual maturity	Females, 80–90 days
	Males, 60 days
Estrous cycle	4–6 days
Gestation	18–21 days (captured)
	15–17 days (laboratory born)
Litter size	7–9
Weaning	25 days
	28 days

[a]Data from Mohr *et al.* (1973).

the next 4 days. Estrus will occur 6 days later. Sexual activity is not observed in winter, and females and males are very aggressive toward each other. The vagina is closed in females, and in males the scrotum is decreased in size and the testes are intraabdominal.

Animals are handled similarly to guinea pigs, but handling does not appear to tame them. They are easily frightened and will attack and give a painful bite. Those in captivity are less aggressive toward man.

There is a slight reduction in activity in the winter months. Hibernation affects thrombocytes and leukocyte values, but there was no significant difference in nonhibernating animals during winter or summer (Reznik *et al.*, 1975a).

III. DISEASES

Silverman and Chavannes (1977) tested 8 males and found them to be free of endoparasites, ectoparasites, and blood parasites.

ARMENIAN HAMSTER

The Armenian hamster (*Cricetulus migratorius*), also known as the gray hamster, was first introduced as a laboratory animal in 1963. It was selected for domestication because of its susceptibility to mutagenic and carcinogenic agents and also for studying meiosis because of the semisynchronous meiotic progression beginning at 15 days of age (Yerganian and Lavappa, 1971). It is similar to the Syrian hamster in that it is highly susceptible to oncogenic viruses and has high tolerance to both homologous and heterologous transplantable tumors. The cytological features are comparable to those of the Chinese hamster.

Care and management procedures are similar to that for the Chinese hamster. Body size and weight are also similar. The diploid chromosome number is 22, with the X and Y the same size. Captured animals are aggressive, but if reared in the laboratory, they can be managed to breed successfully. The gestation period is reported to be 18 to 19 days, with an average litter size of 3 to 4 (Hachaturova and Ezbanyan, 1973).

Little has been reported concerning spontaneous infectious diseases in the Armenian hamster. Malformed vertebrae have been reported in 3 of 5 newborn littermates (Akbarzadeh and Arbabi, 1979). A sarcoma in the hip area was found in one animal from a laboratory breeding colony (Hachaturova and Ezbanyan, 1973).

ACKNOWLEDGMENT

The contributions of Ms. Lynn Dahm and Ms. Alice Ruff in the preparation of this chapter are sincerely appreciated.

REFERENCES

Aeromedical Review (1975). The hamster. *In* "Selected Topics in Laboratory Animal Medicine," Vol. 24, pp. 27–35. USAF School of Aerospace Medicine, Brooks Air Force Base, Texas.

Akbarzadeh, J., and Arbabi, E. (1979). Malformed vertebrae in *Cricetus migratorious* (grey hamster): A case report. *Lab. Anim.* **13**, 299–300.

Altman, P. L., and Dittmer, D. S., eds. (1972). "Biology Data Book," 2nd ed., Vol. I. Fed. Am. Soc. Exp. Biol., Bethesda, Maryland.

Altman, P. L., and Dittmer, D. S., eds. (1973). "Biology Data Book," 2nd ed., Vol. II. Fed. Am. Soc. Exp. Biol., Bethesda, Maryland.

Altman, P. L., and Dittmer, D. S., eds. (1974). "Biology Data Book," 2nd ed. Vol. III. Fed. Am. Soc. Exp. Biol., Bethesda, Maryland.

Ambrose, K. R., and Coggin, J. H., Jr. (1975). An epizootic in hamsters of lymphomas of undetermined origin and mode of transmission. *JNCI, J. Natl. Cancer Inst.* **54**(4), 877–879.

Amend, N., Loeffler, D., and Ward, B., and Van Hoosier, G. L., Jr. (1976). Transmission of enteritis in the Syrian hamster. *Lab. Anim. Sci.* **26**(4), 566–572.

Arrington, L. R., Platt, J. K., and Shirley, R. L. (1966). Protein requirements of growing hamsters. *Lab. Anim. Care* **16**, 492–496.

Avery, T. L. (1968). Observations on the propagation of Chinese hamsters. *Lab. Anim. Care* **18**(2), 151–159.

Baker, J. R., Mason, M. M., Gerganian, G., Weisburger, E. K., and Weisburger, J. H. (1974). Induction of tumors of the stomach and esophagus in inbred Chinese hamsters by oral diethylnitrosamine. *Proc. Soc. Exp. Biol. Med.* **146**, 291–293.

Banta, C. A., Warner, R. G., and Robertson, J. B. (1975). Protein nutrition of the golden hamster. *J. Nutr.* **105**(1), 38–45.

Barthold, S. W., Jacoby, R. O., and Pucak, G. J. (1978). An outbreak of cecal mucosal hyperplasia in hamsters. *Lab. Anim. Sci.* **28**(6), 723–727.

Bartlett, J. G., Chang, T., Moon, N., and Onderdonk, A. B. (1978). Antibiotic-induced lethal enterocolitis in hamsters: Studies with eleven agents and evidence to support the pathogenic role of toxin-producing clostridia. *Am. J. Vet. Res.* **39**(9), 1525–1530.

Benjamin, S. A., and Brooks, A. L. (1977). Spontaneous lesions in Chinese hamsters. *Vet. Pathol.* **14**, 449–462.

Billingham, R. E. (1978). Concerning the laboratory career of *Mesocricetus auratus* with special reference to transplantation. *Fed. Proc., Fed. Am. Soc. Exp. Biol.* **37**(7), 2024–2027.

Birt, D. F., and Conrad, R. D. (1981). Weight gain, reproduction and survival of Syrian hamsters fed five natural ingredient diets. *Lab. Anim. Sci.* **31**(2), 149–155.

Boquist, L. (1969). Pancreatic islet morphology in diabetic Chinese hamsters. *Acta Pathol. Microbiol. Scand.* **75**, 399–414.

Brandwein, S. R., Skinner, M., and Cohen, A. S. (1981). Isolation and characterization of spontaneously occurring amyloid fibrils in aged Syrian (golden) hamsters. *Fed. Proc., Fed. Am. Soc. Exp. Biol.* **40**(3), Part I, Abstr. No. 3182, p. 789.

Butler, L., and Gerritsen, G. C. (1970). A comparison of the modes of inheritance of diabetes in the Chinese hamster and the KK mouse. *Diabetologia* **6**, 163–167.

Carthew, P., Sparrow, S., and Verstraete, A. P. (1978). Incidence of natural virus infections of laboratory animals 1976–1977. *Lab. Anim.* **12**(4), 245–246.

Chang, A. Y. (1981). Biochemical abnormalities in the Chinese hamster (*Cricetulus griseus*) with spontaneous diabetes. *Int. J. Biochem.* **13**, 41–43.

Chang, A. Y., Noble, R. E., and Wyse, B. M. (1977). Comparison of highly inbred diabetic and nondiabetic lines in the Upjohn colony of Chinese hamsters. *Diabetes* **26**(11), 1063–1071.

Cisar, C. F., Gumperz, E. P., Nicholson, F. S., and Moore, W., Jr. (1972). A practical method for production breeding of Chinese hamsters (*Cricetulus griseus*). *Lab. Anim. Sci.* **22**(5), 725–727.

Clark, J. D. (1975). The hamster: Introduction and husbandry. *In* "Laboratory Animal Medicine and Science Series" (G. L. Van Hoosier, Jr., coord.). Dist.: HSLRC, University of Washington, Seattle.

Coggin, J. H., Jr., Oakes, J. E., Huebner, R. J., and Gilden, R. (1981). Unusual filterable oncogenic agent isolated from horizontally transmitted Syrian hamster lymphomas. *Nature (London)* **290**, 336–338.

Cohen, M. M., Shklar, G., and Yerganian, G. (1961). Periodontal pathology in a strain of Chinese hamster, *Cricetulus griseus*, with hereditary diabetes mellitus. *Am. J. Med.* **31**(6), 864–867.

Conner, M. W., Conner, B. H., Fox, J. G., and Rogers, A. E. (1983). Spontaneous amyloidosis in outbred CD-1 mice. *Surv. Synth. Pathol. Res.* **1**, 67–78.

Connor, J. S., Lavine, R. L., and Rose, L. I. (1981). *Mesocricetus auratus:* A new animal model for studying diabetes mellitus in pregnancy. *Lab. Anim. Sci.* **31**(1), 35–38.

Corbascio, A. N., Kohn, H. I., and Yerganian, G. (1962). Acute X-ray lethality in the Chinese hamster after whole-body and partial-body irradiation. *Radiat. Res.* **17**, 521–530.

Eddy, H. A., and Casarett, G. W. (1972). Pathology of radiation syndromes in the hamster. *Prog. Exp. Tumor Res.* **16**, 98–119.

Emminger, A., Reznik, G., Reznik-Schuller, H., and Mohr, U. (1975). Differences in blood values depending on age in laboratory-bred European hamsters (*Cricetus cricetus L.*) *Lab. Anim.* **9**, 33–42.

Ershoff, B. H. (1956). Beneficial effects of alfalfa, aureomycin, and cornstarch on the growth and survival of hamsters fed highly purified rations. *J. Nutr.* **59**, 579–585.

Estes, P. C., Richter, C. B., and Franklin, J. A. (1971). Demodectic mange in the golden hamster. *Lab. Anim. Sci.* **21**(6), Part I, 825–828.

Ferm, V. H. (1967). The use of the golden hamster in experimental teratology. *Lab. Anim. Care* **17**(5), 452–462.

Festing, M. (1972). Hamsters. *In* "The UFAW Handbook on the Care and Management of Laboratory Animals" (University Federation for Animal Welfare, ed.), 4th ed., pp. 242–256. Churchill-Livingstone, Edinburgh and London.

Flynn, R. J. (1973). "Parasites of Laboratory Animals." Iowa State Univ. Press, Ames.

Fortner, J. G. (1957). Spontaneous tumors, including gastrointestinal neoplasms and malignant melanomas, in the Syrian hamster. *Cancer* **10**, 1153–1156.

Foster, H. L., Small, J. D., and Fox, G., eds. (1982). "The Mouse in Biomedical Research," Vol. 2. Academic Press, New York.

Fox, J. G., Zanotti, S., and Jordan, H. V. (1981). The hamster as a reservoir of *Campylobacter fetus* subspecies *jejuni*. J. Infect. Dis. **143**(6), 856.

Frenkel, J. K. (1972). Infection and immunity in hamsters. *Prog. Exp. Tumor Res.* **16**, 339.

Frenkel, J. K. (1979). Immunity and infection: Hamster. *In* "Inbred and Genetically Defined Strains of Laboratory Animals," Vol. II, pp. 473–480. Fed. Am. Soc. Exp. Biol., Bethesda, Maryland.

Frisk, C. S., and Wagner, J. E. (1977). Hamster enteritis: A review. *Lab. Anim.* **11**, 79–85.

Frisk, C. S., Wagner, J. E., and Owens, D. R. (1981). Hamster (*Mesocricetus auratus*) enteritis caused by epithelial cell-invasive Escherichia coli. *Infect. Immun.* **31**, 1232–1238.

Fukumi, H., Nishikawa, F., and Kitayama, T. (1954). A pneumotropic virus for mice causing hemagglutination. *Jpn. J. Med. Sci. Biol.* **7**, 345–363.

Ganaway, J. R. (1976). Bacterial, mycoplasma, and rickettsial diseases. *In* "The Biology of the Guinea Pig" (J. E. Wagner and P. J. Manning, eds.), pp. 121–135. Academic Press, New York.

Ganaway, J. R. (1982). Bacterial, mycoplasmal, and mycotic diseases of the digestive system. *In* "The Mouse in Biomedical Research" (H. Foster, J. Small, and J. Fox, eds.), Vol. 2, pp. 1–20. Academic Press, New York.

Ganaway, J. R., Allen, A. M., and Moore, T. D. (1971). Tyzzer's disease. *Am. J. Pathol.* **64**(3), 717–732.

Garlinghouse, L. E., Jr., and Van Hoosier, G. L., Jr. (1978). Studies on adjuvant-induced arthritis, tumor transplantability and serologic response to bovine serum albumin in Sendai virus infected rats. *Am. J. Vet. Res.* **39**(2), 297–300.

Gerritsen, G. C., Needham, L. B., Schmidt, F. L., and Dulin, W. E. (1970). Studies on the prediction and development of diabetes in offspring of diabetic Chinese hamsters. *Diabetologia* **6**, 158–162.

Gleiser, C. A., Van Hoosier, G. L., Jr., and Sheldon, W. G. (1970). A polycystic disease of hamsters in a closed colony. *Lab. Anim. Care* **20**(5), 923–929.

Gleiser, C. A., Van Hoosier, G. L., Jr., Sheldon, W. G., and Read, W. K. (1971). Amyloidosis and renal paramyloid in a closed hamster colony. *Lab. Anim. Sci.* **21**(2), 197–202.

Graffi, A., Schramm, T., Bender, E., Graffi, I., Horn, K.-H., and Bierwolf, D. (1968). Cell-free transmissible leukoses in Syrian hamsters, probably of viral aetiology. *Br. J. Cancer* **22**, 577–581.

Gregg, M. B. (1975). Recent outbreaks of lymphocytic choriomeningitis in the United States of America. *Bull. W.H.O.* **52**, 549–553.

Hachaturova, T. S., and Ezbanyan, B. A. (1973). Inbred breeding lines in grey Armenian hamsters. *Vopr. Onkol.* **19**(11), 86–87.

Harkness, J. E., Wagner, J. E., Kusewitt, D. F., and Frisk, C. S. (1977). Weight loss and impaired reproduction in the hamster attributable to an unsuitable feeding apparatus. *Lab. Anim. Sci.* **27**(1), 117–118.

Hayes, J. A., Christensen, T. G., and Snider, G. L. (1977). The hamster as a model of chronic bronchitis and emphysema in man. *Lab. Anim. Sci.* **27**(5), 762–770.

Hoffman, R. A., Robinson, P. F., and Magalhaes, H., eds. (1968). "The Golden Hamster: Its Biology and Use in Medical Research." Iowa State Univ. Press, Ames.

Homburger, F. (1968). The Syrian golden hamster in chemical carcinogenesis research. *Prog. Exp. Tumor Res.* **10**, 164–237.

Homburger, F. (1972a). Disease models in Syrian hamsters. *Prog. Exp. Tumor Res.* **16**, 69–86.

Homburger, F. (1972b). Chemical carcinogenesis in Syrian hamsters. *Prog. Exp. Tumor Res.* **16**, 152–175.

Innes, J. R. M., Wilson, C., and Ross, M. A. (1956). Epizootic *Salmonella enteritidis* infection causing septic pulmonary phlebothrombosis in hamsters. *J. Infect. Dis.* **98**, 133–141.

Jacoby, R. O. (1978). Transmissible ileal hyperplasia of hamsters. I. Histogenesis and immunocytochemistry. *Am. J. Pathol.* **91**, 433–450.

Jacoby, R. O., and Johnson, E. A. (1981). Transmissible ileal hyperplasia. *In* "Hamster Immune Responses in Infectious and Oncologic Diseases" (J. Streilein, D. Hart, J. Stein-Streilein, W. Duncan, and R. Billingham, eds.), pp. 267–289. Plenum, New York.

Jacoby, R. O., Osbaldiston, G. W., and Jonas, A. M. (1975). Experimental transmission of atypical ileal hyperplasia of hamsters. *Lab. Anim. Sci.* **25**(4), 465–473.

Jordon, H. V., and van Houte, J. (1972). The hamster as an experimental model for odontopathic infections. *Prog. Exp. Tumor Res.* **16**, 539.

Kleinerman, J. (1972). Some aspects of pulmonary pathology in the Syrian hamster. *Prog. Exp. Tumor Res.* **16**, 287–299.

Kohn, H. I., and Gultman, P. H. (1964). Life span, tumor incidence, and intercapillary glomerulosclerosis in the Chinese hamster (*Cricetulus griseus*) after whole-body and partial-body exposure to X-rays. *Radiat. Res.* **21**, 622–643.

La Regina, M., and Lonigro, J. (1982). Isolation of *Campylobacter fetus* subspecies *jejuni* from hamsters with proliferative ileitis. *Lab. Anim. Sci.* **32**(6), 660–662.

La Regina, M., Fales, W. H., and Wagner, J. E. (1980). Effects of antibiotic treatment on the occurrence of experimentally induced proliferative ileitis of hamsters. *Lab. Anim. Sci.* **30**(1), 38–41.

Lavappa, K. S., Fu, M. M., Singh, M., Beyer, R. D., and Epstein, S. S. (1973). Banding patterns of chromosomes in bone marrow cells of the Chinese hamster as revealed by acetic-saline-Giemsa, urea, and trypsin techniques. *Lab. Anim. Sci.* **23**(4), 546–550.

Lawe, J. E. (1962). Renal changes in hamsters with hereditary diabetes mellitus. *Arch. Pathol.* **73**, 166–174.

Lepot, A., and Banwell, J. G. (1976). The Syrian hamster a reproducible model for studying changes in intestinal fluid secretion in response to enterotoxin challenge. *Infect. Immun.* **14**(5), 1167–1171.

Lyman, C. P. (1979). Usefulness of the hamster in the study of hibernation. *In* "Inbred and Genetically Defined Strains of Laboratory Animals," vol. II, p. 431. Fed. Am. Soc. Exp. Biol. Bethesda, Maryland.

Lyman, C. P., and Fawcett, D. W. (1954). Effect of hibernation on growth of sarcoma in hamsters. *Cancer Res.* **14**, 25–28.

Magalhaes, H. (1968). Gross anatomy. *In* "The Golden Hamster; Its Biology and Use in Medical Research" (R. A. Hoffman, P. F. Robinson, and H. Magalhaes, eds.), pp. 91–109. Iowa State Univ. Press, Ames.

Magalhaes, H. (1970). Hamsters. *In* "Reproduction and Breeding Techniques for Laboratory Animals" (E. S. E. Hafez, ed.), pp. 258–272. Lea & Febiger, Philadelphia, Pennsylvania.

Marsh, R. F., and Hanson, R. P. (1978). The Syrian hamster as a model for the study of slow virus diseases caused by unconventional agents. *Fed. Proc., Fed. Am. Soc. Exp. Biol.* **37**(7), 2076–2078.

Matsumoto, T., Nagata, I., Kariya, Y., and Ohaski, K. (1954). Studies on a strain of pneumotropic virus of hamster. *Nagoya J. Med. Sci.* **17**(2), 93–97.

Maxwell, K. O., Murphy, J. C., and Fox, J. G. (1984). Serum chemistry reference values in two strains of Syrian hamsters (in press).

Meier, H., and Yerganian, G. A. (1959). Spontaneous hereditary diabetes mellitus in Chinese hamster (*Cricetulus griseus*). I. Pathological findings. *Proc. Soc. Exp. Biol. Med.* **100**, 810–815.

Meier, H., and Yerganian, G. A. (1961). Spontaneous diabetes mellitus in the Chinese hamster (*Cricetulus griseus*). III. Maintenance of a diabetic hamster colony with the aid of hypoglycemic therapy. *Diabetes* **10**(1), 19–21.

Mohr, U. (1979). The Syrian golden hamster as a model in cancer research. *Prog. Exp. Tumor Res.* **24**, 245.

Mohr, U., Schuller, H., Reznik, G., Althoff, J., and Page, N. (1973). Breeding of European hamsters. *Lab. Anim. Sci.* **23**(6), 799–802.

Moore, W., Jr. (1965). Observations on the breeding and care of the Chinese hamster, *Cricetulus griseus*. *Lab. Anim. Care* **15**(1), 94–101.

Moore, W., Jr. (1966). Hemogram of the Chinese hamster. *Am. J. Vet. Res.* **27**(117), 608–610.

Murphy, J. C., and Fox. J. G. (1980). Nephrotic syndrome in Syrian hamsters. *Am. Assoc. Lab. Anim. Sci., 31st Annu. Sess.* Abstract.

Nakayama, M., Saegusa, J., Itoh, K., Kiuchi, Y., Tamura, T., Ueda, K., and Fujiwara, K. (1975). Transmissible enterocolitis in hamsters caused by Tyzzer's organism. *Jpn. J. Exp. Med.* **45**(1), 33–41.

Newberne, P. M., and Fox. J. G. (1980). Nutritional adequacy and quality control of rodent diets. *Lab. Anim. Sci.* **30**(2), Part 2, 352–365.

Newberne, P. M., and McConnell, R. G. (1979). Nutrition of the Syrian golden hamster. *Prog. Exp. Tumor Res.* **24**, 127–138.

Parker, J. C., and Richter, C. B. (1982). Viral diseases of the respiratory system. *In* "The Biology of the Laboratory Mouse" (H. L. Foster, J. D. Small, and J. G. Fox, eds.), Vol. 2, pp. 107–155. Academic Press, New York.

Parker, J. C., Hercules, J. I., and von Kaenel, E. (1967). The prevalence of some indigenous viruses of rat and hamster breeder colonies. *Bacteriol. Proc.* Abstract, p. 163.

Parker, J. C., Igel, H. J., Reynolds, R. K., Lewis, A. M., Jr., and Rowe, W. P. (1976). Lymphocytic choriomeningitis virus infection in fetal, newborn, and young adult Syrian hamsters. *Infect. Immun.* **13**(3), 967–981.

Poel, W. E., and Yerganian, G. (1961). Adenocarcinoma of the pancreas in diabetes-prone Chinese hamsters. *Am. J. Med.* **31**(6), 861–863.

Porter, G., and Lacey, A. (1969). Breeding the Chinese hamster (*Cricetulus griseus*) in monogamous pairs. *Lab. Anim.* **3**, 65–68.

Pour, P. (1979). The unique role of the hamster in specific carcinogenesis studies. *Prog. Exp. Tumor Res.* **24**, 391.

Profeta, M. L., Lief, F. S., and Plotkin, S. A. (1969). Enzootic Sendai infection in laboratory hamsters. *Am. J. Epidemiol.* **89**(3), 316–324.

Reed, J. M., Schiff, L. J., Shefner, A. M., and Henry, M. C. (1974). Antibody levels to murine viruses in Syrian hamsters. *Lab. Anim. Sci.* **24**(1), 33–38.

Renshaw, H. W., Van Hoosier, G. L., Jr., and Amend, N. K. (1975). A survey of naturally occurring diseases of the Syrian hamster. *Lab. Anim.* **9**, 179–191.

Reznik, G., Reznik-Schuller, H., Emminger, A., and Mohr, U. (1975a). Comparative studies of blood from hibernating and nonhibernating European hamsters (*Cricetus cricetus* L.) *Lab. Anim. Sci.* **25**(2), 210–215.

Reznik, G., Reznik-Schuller, H., Schostek, H., Deppe, K., and Mohr, U. (1975b). Comparative studies concerning the suitability of European hamsters and Syrian golden hamsters for investigations on smoke exposure. *Drug. Res.* **25**(6), 923–926.

Rogers, A. E., Anderson, G. H., Lenhart, G. M., Wolf, G., and Newberne, P. M. (1974). Semisynthetic diet for long-term maintenance of hamsters to study effects of dietary vitamin A. *Lab. Anim. Sci.* **24**, 495.

Ronald, N. C., and Wagner, J. E. (1975). Treatment of *Hymenolepis nana* in hamsters with yomesan (niclosamide). *Lab. Anim. Sci.* **25**(2), 219–220.

Ross, C. R., Wagner, J. E., Wightman, S. R., and Dill, S. E. (1980). Experimental transmission of *Syphacia muris* among rats, mice, hamsters, and gerbils. *Lab. Anim. Sci.* **30**(1), 35–37.

Schiff, L. J., Shefner, A. M., Barbera, P. W., and Poiley, S. M. (1973). Microbial flora and viral contact status of Chinese hamsters (*Cricetulus griseus*). *Lab. Anim. Sci.* **23**(6), 899–902.

Sheffield, F. W., and Beveridge, E. (1962). Prophylaxis of "wet tail" in hamsters. *Nature (London)* **196**, 294–295.

Shklar, G. (1972). Experimental oral pathology in the Syrian hamster. *Prog. Exp. Tumor Res.* **16**, 518.

Silberberg, R., and Gerritsen, G. (1976). Aging changes in intervertebral discs and spondylosis in Chinese hamsters. *Diabetes* **25**(6), 477–483.

Silverman, J., and Chavannes, J. (1977). Biological values of the European hamster (*Cricetus cricetus*). *Lab. Anim. Sci.* **27**(5), 641–645.

Small, J. D. (1968). Fatal enterocolitis in hamsters given lincomycin hydrochloride. *Lab. Anim. Care* **18**(4), 411–420.

Stenback, W. A., Van Hoosier, G. L., Jr., and Trentin, J. J. (1966). Virus particles in hamster tumors as revealed by electron microscopy. *Proc. Soc. Exp. Biol. Med.* **122**, 1219–1223.

Stenback, W. A., Van Hoosier, G. L., Jr., and Trentin, J. J. (1968). Biophysical, biological, and cytochemical features of a virus associated with transplantable hamster tumors. *J. Virol.* **2**(10), 1115–1121.

Stewart, H. L., Snell, K. C., Dunham, L. J., and Schylen, S. M. (1959). "Transplantable and Transmissible Tumors of Animals," p. 378. Armed Forces Institute of Pathology, Washington, D.C.

Streilein, J. W. (1978). Hamster immune responses: Experimental models linking immunogenetics, oncogenesis and viral immunity. *Fed. Proc., Fed. Am. Soc. Exp. Biol.* **37**(7), 2023, 2108.

Streilein, J. W., Hart, D. A., Stein-Streilein, J., Duncan, W. R., and Billingham, R. E., eds. (1981). "Hamster Immune Responses in Infectious and Oncologic Diseases. Adv. Exp. Med. Biol. 134. Plenum, New York.

Taffs, L. F. (1976). Pinworm infections in laboratory rodents: A review. *Lab. Anim.* **10**, 1–13.

Thompson, R. (1971). The water consumption and drinking habits of a few species and strains of laboratory animals. *J. Inst. Anim. Technicians* **22**(1), 29–36.

Trentin, J. J., Van Hoosier, G. L., Jr., and Sanper, L. (1968). The oncogenicity of human adenoviruses in hamsters. *Proc. Soc. Exp. Biol. Med.* **127**, 683–689.

Unay, E. S., and Davis, B. J. (1980). Treatment of *Syphacia obvelata* in the Syrian hamster (*Mesocricetus auratus*) with piperazine citrate. *Am. J. Vet. Res.* **41**(11), 1899–1900.

Van Hoosier, G. L., Jr., and Trentin, J. J. (1979). Naturally occurring tumors of the Syrian hamster. *Prog. Exp. Tumor. Res.* **23**, 1–12.

Van Hoosier, G. L., Jr., Stenback, W. A., Parker, J. C., Burke, J. G., and Trentin, J. J. (1970). The effects of cesarean derivation and foster nursing procedures on enzootic viruses of the LSH strain of inbred hamsters. *Lab. Anim. Care* **20**(2), Part 1, 232–237.

Wagner, J. E., Doyle, R. E., Ronald, N. C., Garrison, R. G., and Schmitz, J. A. (1974). Hexamitiasis in laboratory mice, hamsters, and rats. *Lab. Anim. Sci.* **24**(2), 349–354.

Wantland, W. W. (1955). Parasitic fauna of the golden hamster. *J Dent. Res.* **34**, 631–649.

Ward, B. C., and Moore, W., Jr. (1969). Spontaneous lesions in a colony of Chinese hamsters. *Lab. Anim. Care* **19**(4), 516–521.

Warner, R. G., and Ehle, F. R. (1976). Nutritional idiosyncrasies of the golden hamster (*Mesocricetus auratus*). *Lab. Anim. Sci.* **26**(4), 670–673.

Yerganian, G. (1958). The striped-back or Chinese hamster, *Cricetulus griseus*. *JNCI, J. Natl. Cancer Inst.* **20**(4), 705–727.

Yerganian, G. (1965). Spontaneous diabetes mellitus in the Chinese hamster, *Cricetulus griseus*. *Int. Cong. Ser.—Excerpta Med.* **84**, 612–627.

Yerganian, G., and Lavappa, K. S. (1971). Procedures for culturing diploid cells and preparation of meiotic chromosomes from dwarf species of hamsters. *Chem. Mutagens* **2**, 387–410.

Yerganian, G., Klein, E., and Roy, A. (1955). Physiological studies on Chinese hamster, *Cricetulus griseus*. II. A method for study of bleeding time in hamsters. *Fed. Proc., Fed. Am. Soc. Exp. Biol.* **14**, 424.

Yoon, C. H., and Peterson, J. S. (1979). Recent advances in hamster genetics. *Prog. Exp. Tumor Res.* **24**, 157–161.

Zook, B. C., Albert, E. N., and Rhorer, R. G. (1977). Tyzzer's disease in the Chinese hamster, *Cricetulus griseus*. *Lab. Anim. Sci.* **27**(6), 1033–1035.

Chapter 6

Biology and Diseases of Guinea Pigs

Patrick J. Manning, Joseph E. Wagner, and John E. Harkness

I. INTRODUCTION

A. Taxonomy and General Comments

Guinea pigs are hystricomorph (porcupine-like) rodents closely related to chinchillas and porcupines. The family Caviidae, to which guinea pigs belong, consists of more or less tailless South American rodents that have one pair of mammae, four digits on the forefeet and three on the hindfeet. The domestic guinea pig (*Cavia porcellus*), from which laboratory stocks and strains are derived, has long been identified as a laboratory animal, and indeed the name ''guinea pig'' is often used synonymously with the phrase experimental subject. There are many practical reasons for the popularity of guinea pigs as laboratory animals and as pets. They are readily available, inexpensive to purchase and maintain, tractable, quiet, and have little odor. As laboratory animals guinea pigs have been used extensively in studies of immunology, genetics, otology, infectious diseases, nutrition, and gnotobiology (deWeck and Festing, 1979).

B. Stocks and Strains of Guinea Pigs; Sources

Laboratory animals are identified according to either specific genetic characteristics or an operational definition prevails in which the stock or stain is identified as descendants from a particular commercial, institutional, or other colony. The five most common types of guinea pigs used in research are the short hair (including English and American varieties), Dunkan-Hartley, Hartley, strain 2, and strain 13. The short hair (Fig. 1), Dunkan-Hartley, and Hartley strains are random or outbred stocks. Although short hair varieties have several hair coat colors and Dunkan-Hartley and Hartley guinea pigs are albinos, these strains are genetically heterogeneous. Strains 2 and 13 are inbred, and both strains have tricolor (black, red, and white) hair coats. Two breeds occasionally used in research but more often kept by fancy breeders are the Abyssinian and the Peruvian. Abyssinians (Fig. 2) appear ill kept because their hair grows in a rosette pattern, whereas the Peruvian strain (Fig. 3) has long silky hair. A variety of other stocks and strains of guinea pigs have been described (Shevach *et al.*, 1979; Festing, 1979). A periodically revised listing of sources of several strains of guinea pigs is published by the National Research Council (1979).

C. Clinical Examination of the Guinea Pig

Guinea pigs are among the easiest of the common laboratory animals to examine clinically. They should be observed initially by simply watching them in a walled container. Normal guinea pigs have smooth hair with a low luster and bright alert eyes. When startled, guinea pigs become either immobile or they scatter suddenly and explosively. They can be safely handled by cupping the hand under the sternum and quickly lifting them from their enclosure taking care not to squeeze them if they bolt because we have seen diaphragmatic hernias caused by overzealous handling techniques. Guinea pigs rarely bite, and they are easily palpated and auscultated. All external body orifices are readily examined except the mouth which has a small opening. A thorough examination of the oral cavity requires sedation (ketamine 50 mg/kg im) and the aid of a well-lighted otoscopic speculum or similar device. They have two two inguinal mammary glands. A detailed description of the behavior of guinea pigs is given by Harper (1976), and Breazile and Brown (1976) have complied a brief review of the anatomy of guinea pigs. A comprehensive treatise of their gross anatomy is given by Cooper and Schiller (1975). Several values of clinical interest are provided in Tables I and II.

II. BIOLOGY

A. Unique Physiologic Characteristics

Several aspects of the anatomy, physiology, and metabolism of the guinea pig are unique among domesticated rodents and are reviewed in detail by Wagner and Foster (1976), Festing (1976a), Navia and Hunt (1976), and McCormick and Nuttall (1976).

Fig. 1.   Appearance of a guinea pig typical of the short hair, English and American varieties. The swelling in the ventral neck is an enlarged lymph node infected with *Streptococcus zooepidemicus* the causal organism of most cases of caseous lymphadenitis (see page 160).

Fig. 2. In the Abysinnian guinea pig portions of the hair coat grow in rosette patterns.

Fig. 3. A guinea pig of the Peruvian strain exhibiting the long silky hair characteristic of this strain.

1. Circulatory and Lymphoreticular Systems

The erythrocytic indices (erythrocytes, hemoglobin, and packed cell volume) of the guinea pig are relatively low compared to other laboratory rodents. Lymphocytes are the dominant leukocyte in the guinea pig, as in other rodents, and the morphology of the peripheral blood and marrow is similar to that of man. A unique leukocyte in the guinea pig is the Foa-Kurloff or Kurloff cell (Fig. 4), which is a specialized mononuclear leukocyte containing a large mucopolysaccharide, intracytoplasmic inclusion body. Though usually found within vascular channels and in the thymus gland, their highest density shifts from the lungs and spleen (red pulp) to the thymus and placenta under estrogenic stimulation and pregnancy (Izard *et al.*, 1976). Though the origin and function of these cells are not known, they may function as killer cells in the general circulation or as protectors of fetal antigen in the placenta (Marshall *et al.*, 1971; Eremin *et al.*, 1980; Revell and Soretire, 1980).

The guinea pig, like ferrets and primates, is relatively resistant to steroids, and the numbers of thymic and peripheral lymphocytes are not markedly reduced by corticosteroid injection (Hodgson and Funder, 1978). The guinea pig is an established model for the study of genetic control of the histocompatibility-linked immune response (Chiba *et al.*, 1978). Though the guinea pig's thymus is located in the ventral cervical region and is easily removed surgically, accessory thymic islets exist in contiguous fascia. The thymus apparently has no afferent lymphatic vessels (Ernström and Larsson, 1967). Guinea pig serum, especially from older sows, is a common source of hemolytic complement, but serum concentrations and suitability of the complement for serologic procedures vary.

2. Immediate and Delayed Hypersensitivity

The guinea pig is an established model for antigen-induced, immediate respiratory anaphylaxis (Patterson and Kelly, 1974). Exposure of a sensitized animal to the antigen, usually given intravenously or by aerosol, results in an acute reaction characterized by cyanosis, collapse, and death owing to constriction of bronchial and bronchiolar smooth muscle with consequent asphyxia. Apparently the antigen interacts with IgG_1 and possibly IgE antibodies to release mast cell histamine, which then causes a usually fatal bronchoconstriction. Hartley and strain 13 animals are more susceptible to anaphylaxis than strain 2 guinea pigs (Stone *et al.*, 1964).

Delayed hypersensitivity reactions, usually in reference to intradermal tuberculin injections, occur within 24 to 48 hr in the guinea pig and 48 to 96 hr in the human. Optimal hypersensitivity reactions occur in 2 to 3 month old, 350 to 400 gm animals. Unlike adults in which the cellular response consists of neutrophils and macrophages, neonatal guinea pigs develop a cutaneous basophilic response in delayed sensitivity reactions (Haynes and Askenase, 1977).

3. Ear

The large, accessible guinea pig ear is used for several types of auditory studies, despite the sporadic occurrence of otitis media in this species (McCormick and Nuttall, 1976). Advantages of using the guinea pig ear include the large bullae, ease of entry to the middle and inner ears, and protrusion of the cochlea and blood vessels into the cavity of the middle ear, which allows examination of the microcirculation of the inner ear.

B. Normal Life Cycle and Physiologic Values

Table I lists approximations of life cycle and physiologic data of the guinea pig. These values vary greatly with age, strain, sex, environment, and method of measurement, and the ranges for any given population may be overly broad or even noninclusive. For more detailed information regarding the source of the data and the diversity of subjects measured, the references should be consulted.

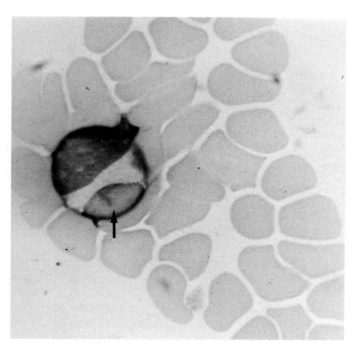

Fig. 4. A Foa-Kurloff cell in a peripheral blood smear of a guinea pig. The intracytoplasmic inclusion body is large and conspicuous (arrow).

Table I

Approximate Physiologic Values for Guinea Pigs[a–c]

General data	
Body weight: adult male	900–1000 gm
Body weight: adult female	700–900 gm
Birth weight	60–115 gm
Body surface area[d,e]	400 gm: 565 cm²
	800 gm: 720 cm²
Rectal temperature[f]	37.2–39.5°C
Diploid number[g]	64
Life span: usual	3–4 years
Life span: extreme	6–7 years
50% survival	60 months
Food consumption	6 gm/100 gm body weight/day
Water consumption	10 ml/100 gm body weight/day
Gastrointestinal transit time[h]	13–30 hr
Critical temperature[f]	30°C
Thermal neutrality range[f]	2–31°C
Cardiovascular and respiratory systems[i–l]	
Respiratory rate	42–104/minute
Tidal volume	2.3–5.3 ml/kg body weight
Oxygen use	0.76–0.83 ml/gm body weight/hr
Plasma CO_2	18–26 mM/liter
CO_2 pressure	21–59 mm Hg
Plasma pH	7.17–7.53
Heart rate	230–380/min
Blood volume	69–75 ml/kg body weight
Cardiac output[m]	240–300 ml/min/kg body weight
Blood pressure	80–94/55–58 mm Hg
Blood cells[m,n]	
Erythrocytes	5.4 × 10⁶/mm³ ± 12%[o]
Hematocrit	43 ± 12%
Hemoglobin	13.4 gm/dl ± 12%
MCV	81 μm³
MCH	25 pg
MCHC	30%
Leukocytes	9.9 × 10³/mm³ ± 30%
Neutrophils	28–34%
Lymphocytes	39–72%
Kurloff cells	3–4%
Eosinophils	1–5%
Monocytes	3–12%
Basophils	0–3%
Platelets	250–850 × 10³/mm³

(continued)

Table I (*Continued*)

Clinical Chemistry[c,n]	
Serum protein	4.6–6.2 gm/dl
Albumin	2.1–3.9 gm/dl
Globulin	1.7–2.6 gm/dl
Serum glucose	60–125 mg/dl
Blood urea nitrogen	9.0–31.5 mg/dl
Creatinine	0.6–2.2 mg/dl
Total bilirubin	0.3–0.9 mg/dl
Serum lipids	95–240 mg/dl
Phospholipids	25–75 mg/dl
Triglycerides	0–145 mg/dl
Cholesterol	20–43 mg/dl
Serum calcium	5.3 mEq/dl
Serum phosphate	5.3 mEq/dl
Magnesium	2.4 mg/dl
Sodium	146–152 mEq/liter
Chloride	98–115 mEq/liter
Potassium	6.8–8.9 mEq/liter

[a]Festing (1976b).
[b]Charles River Breeding Laboratories (1982).
[c]Altman and Dittmer (1974).
[d]Hong *et al.* (1977).
[e]Klaassen and Doull (1980).
[f]Short and Woodnott (1969).
[g]Robinson (1971).
[h]Jilge (1980).
[i]Schalm *et al.* (1975).
[j]Sisk (1976).
[k]Payne *et al.* (1976).
[l]Schermer (1967).
[m]Quillec *et al.* (1977).
[n]Laird (1974).
[o]Coefficient of variation.

C. Diets, Nutrition, and Feeding Behavior

1. Diets and Nutrition

Diets and nutrition are discussed in Section III,B. Comprehensive reviews of guinea pig nutrition have been written by Mannering (1949), Reid and Bieri (1972), and Navia and Hunt (1976). A tabular summary of estimated nutritional requirements of guinea pigs and signs associated with several deficiency states are given in Tables II and IV, respectively.

2. Feeding Behavior

Guinea pigs are strict herbivores. In their native South American habitats they eat mainly grasses, alfalfa roots, seeds, and fruit. When retained as a food supply, they will eat table scraps and fermented corn mash (Escobar and Escobar, 1976). Under natural lighting conditions, guinea pigs feed primarily at dawn and dusk. In laboratory colonies, under artificial lighting conditions, the animals feed both day and night with rest periods between meals.

Guinea pigs discriminate between feed and nonfeed within a few days of birth, and with age they become progressively more fastidious eaters (Ediger, 1976). Dietary preparations for nutritional studies must therefore be introduced within a few days of birth, before the young identify commercial, pelleted chow as the exclusive feed. Pelleting of powdered diets in gelatin, moistening of feed, or transitory mixing of different diets may increase acceptance of unfamiliar feeds (Navia and Hunt, 1976). Guinea pigs usually reject overly bitter, salty, sweet (0.5 *M* glucose), or chemically pure diets. They also

Table II

Estimated Nutrient Requirements for Guinea Pigs[a]

Nutrient	Nutrient/kg diet	Nutrient/kg body weight/day
Dry matter	900 gm	40 gm
Water	100 gm	100 ml total
Fiber	10–18%	4–7.2 gm
Nitrogen-free extract	45–48%	18–19.2 gm
Protein	20–30%	8–12 gm
L-Arginine[b]	1.6%	
L-Tryptophan	0.2%	
L-Sulfur amino acids[c]	0.7%	
Fat[d]		
Methyl linoleate	0.4%	
Calcium	1.0%	400 mg
Phosphate	0.6%	240 mg
Magnesium[e]	0.3%	120 mg
Sodium	0.4%	160 mg
Potassium[f]	0.5%	200 mg
Manganese	40 mg	1.6 mg
Copper	6 mg	0.25 mg
Iron	2.5 mg	
Zinc[c]	20 mg	
Vitamin A acetate	8.0 mg	0.4 mg
Vitamin D[g]	(0.04 mg)	
Vitamin E	100 mg	2.3 mg
Vitamin K[h]	2 μg	0.09 μg
Vitamin C	200 mg	10 mg
Biotin	Not required	
Choline	1500 mg	60 mg
Folic acid	6 mg	0.25 mg
Niacin	15 mg	0.6 mg
Pantothenic acid	15 mg	0.6 mg
Riboflavin	16 mg	0.64 mg
Thiamin	2 mg	0.08 mg
Pyridoxine	3 mg	0.12 mg
Vitamin B_{12}	Or cobalt required	

[a]Approximations based on information in Navia and Hunt (1976) and Reid and Bieri (1972).
[b]With 30% casein diet.
[c]With 30% soybean diet.
[d]Corn oil.
[e]Higher with elevated phosphate.
[f]Higher if cation deficient.
[g]Needed if Ca:P inappropriate.
[h]Intestinal synthesis occurs; includes other B vitamins.

adjust poorly to limited feeding or watering. Guinea pigs tend to spill food and water and often foul the sipper tube and drinking water by injecting a slurry of saliva and ingesta. If crocks or bowls are used for feeding, the guinea pigs may rest, defecate, and urinate in the feed. Therefore, special feeders and waterers have been developed for guinea pigs (Sommers and Betts, 1978; Veall and Wood, 1978).

Guinea pigs are coprophagous (King, 1956; Navia and Hunt, 1976), although the quantity of feces ingested and the contri-bution to total dietary needs are unknown. Fecal pellets are usually eaten directly from the anus, but obese or pregnant animals may ingest the pellet from the floor (Hintz, 1969; Harper, 1976). Young animals eat maternal feces and thereby inoculate their intestine with autochthonous flora. Guinea pigs in which coprophagy is prevented will lose weight, digest less fiber, and excrete higher mineral levels in the feces.

3. Alimentary Tract Anatomy

The anatomy of the guinea pig has been reviewed by Cooper and Schiller (1975) and Breazile and Brown (1976). The teeth are open rooted and erupt continuously (hypsodontic). The four incisors are chisel-like, whereas the premolars and molars have flat, grinding surfaces. Guinea pigs are monogastric and have an intestinal tract typical of herbivores, including a capacious cecum and colon. Lactobacilli are the dominant intestinal microflora, but some streptococci, yeast, and soil bacteria are usually present (Smith and Crabb, 1961). The stomach is undivided and lined with glandular epithelium. The small intestine lies coiled to the right in the abdominal cavity. The cecum, which contains up to 65% of the total gastrointestinal contents, lies in the central and left portions of the abdominal cavity. The cecum is a thin walled, gray-green sac divided into pouches by the action of bands of smooth muscle (taenia coli).

The total pharnyx to anus length, excluding the cecal diversion, is approximately 2.3 m. The cecum is 15 to 20 cm long, with the cecal–colic length only slightly less than the length of the small intestine. The gastric emptying time is approximately 2 hr. The total gastrointestinal transit time, apart from the effects of coprophagy, spans 8 to 30 hr, although most of the meal has been passed by 20 hr postingestion (Jilge, 1980).

D. Behavior

Reviews of guinea pig behavior include that of Harper (1976). Sexual and feeding behaviors are discussed in other sections of this chapter. The laboratory guinea pig housed individually or in groups is quiet but alert, except during short sleep periods or in elevated ambient temperatures. The apparent social or group stability derives from polygymous hierarchies branching from a dominant (alpha) male through subordinate males and, less clearly, through sows. These hierarchies are maintained within territorial limits delineated by anal and supracaudal gland secretions and by urine, vocalization, agonistic displays, and occasional physical combat. Social interactions consist primarily of following and grouping, with the young following adults and males following estrous females. Guinea pigs move, rest, and often eat in groups and then lie in contact under, over, or alongside their cagemates.

Weanling animals often pull hair from dams. Adults, usually boars, frequently barber the hair or nibble the ears of the young, particularly when crowded or stressed. These chewing activities may result in alopecia, skin wounds, and dermatitis on one or more animals within the group. Most agonistic encounters are resolved without combat, although adult animals may fight and severely wound one another. Such encounters are most likely to occur in unstable male hierarchies in competition for territory, food, water, and females. Combat occurs more commonly among newly assembled adults and among retired breeders. Guinea pigs rarely bite human handlers.

A behavior of guinea pigs of both interest and concern to research investigators is the scatter or freeze reactions that occur when the animals are startled by noise, shock, novel environment, or sudden movement (Miller and Murray, 1966). Although the reaction to a specific stimulus is not always predictable, sudden movement generally elicits scattering whereas a noise, shock, or environmental change leads to immobility (Sansone and Bovet, 1970). The scatter response occurs in all directions and can involve stampeding, jumping from a cage, trampling young, or rapid circling. Tonic immobility, on the other hand, may last from several seconds to 20 min. These scatter and immobility responses probably have survival value in the wild, but may affect the interpretation of certain experimental results, especially those involving electrical shock or pain avoidance. Another behavioral trait known as the Preyer or pinna reflex is often used by researchers who use guinea pigs in otologic studies. This reflex involves a cocking of the pinnae of the ears in response to a sharp sound (McCormick and Nuttall, 1976). An impaired or absent Preyer reflex is indicative of hearing dysfunction.

Guinea pig learning occurs progressively over several trials rather than within a single interval, when performance levels may actually decrease. This may be related to some yet poorly known memory consolidation mechanism that requires distributed rather than massed practice (Sansone and Bovet, 1970; Harper, 1976).

E. Biology of Reproduction

Comprehensive descriptions of the reproductive anatomy and physiology of the guinea pig are found in Phoenix (1970), Barnes (1971), Cooper and Schiller (1975), Breazile and Brown (1976), and Sisk (1976). Aspects of guinea pig reproduction of research interest are the nonseasonal, continuous polyestrus pattern, easily recognized estrus with spontaneous ovulation, relatively long estrous cycle and gestation period, precocious young, secretory corpus luteum, ability of ovariectomized sows to maintain pregnancy, hemochorial placenta, and several distinguishing anatomic features. Data are summarized in Table III.

Table III

Reproductive Values for Guinea Pigs[a]

First ovulation	4–5 weeks
First ejaculation	8–10 weeks
Breeding onset: male	600–700 gm (3–4 months)
Breeding onset: female	350–450 gm (2–3 months)
Cycle length	15–17 days
Implantation	6–7 days postovulation
Gestation period	59–72 days
Postpartum estrus	80% fertile
Litter size	2–5
Litter interval	96 days
Weaning age	180 gm (14–28 days)
Breeding life	18–20 months (4–5 litters)
Young production	0.7–1.3/sow/month
Preweaning mortality	5–15%
Milk composition[b]	3.9% fat, 8.1% protein, 3.0% lactose
Milk yield (maximum)[c]	45–65 ml/kg body weight/day

[a]Phoenix (1970), Ediger (1976), Sisk (1976), Peplow et al. (1974), Laird (1974), and Festing (1976b).
[b]Nelson et al. (1951).
[c]Davis et al. (1979).

1. Breeding Systems and Husbandry

Both monogamous and polygamous breeding systems are used. The sow is either left with the boar or removed to a separate cage in late gestation (Ediger, 1976; Festing, 1976b). Continuous cohabitation allows sows to breed during the highly fertile postpartum estrus and results in an average of 5 litters per sow per year. The discontinuous method produces only 3.5 litters per sow per year but allows easier identification of breeding lines, protection of the young from bullying and milk stripping, and better weanling survival. Polygamous systems usually contain 4 to 6 mature sows per boar, although as many as 20 sows per boar have been used.

Husbandry requirements for guinea pig production include an appropriate record keeping system, a controlled light cycle (12 to 16 hr of light), draft-free ventilation, an ambient temperature of 18° to 22°C, and about 50% relative humidity. Cages should be clean and dry, the animals should be handled gently and offered a diet adequate for the demands of pregnancy, lactation, and growth. Cage type and size depends on the breeding method selected. An adult animal should be provided with a floor space from 652 to 940 cm² and a cage height from 17.8 to 25 cm. A recommended floor area for a sow with litter is 1485 cm² (Festing, 1976b; Institute of Laboratory Animal Resources, National Research Council, 1978).

Types of cages include floor pens, bins or drawers, and wire or solid bottom cages. All cage types should have lids or sides high enough to retain the animals, especially mature boars, and exclude intruders. Although both wire and solid bottom cages have been used successfully in guinea pig production colonies, wire floors are more often associated with weight and hair loss among the young, decreased production, cooler ambient temperatures, and fractured limbs (Ediger, 1976).

Clean, nonabrasive, wood shavings up to 5 cm deep are commonly used as bedding in guinea pig breeding cages. Fine shavings, chips, and sawdust cling to moist perineal areas, and the resulting wad may interfere with mating (Plank and Irwin, 1966). By 2 to 3 days of age guinea pigs must be provided with access to solid food and water. The sipper tube must reach the smallest animal in the cage; air should easily enter the tube and bottle to displace the liquid.

2. Sexing

Body size is a poor determinant of sex. There are no coat color or pattern differences between the sexes. Both males and females have a single pair of inguinal nipples, but the sow's nipples are more elongate and overlie glands.

The mature boar has prominent, lateral scrotal pouches containing the testes. Testes are not descended at birth, but they can be felt through the abdominal wall. Digital pressure against the anterior aspect of the prepuce will cause extrusion of the penis. Caudal to the prepuce is the perineal sac opening (area of marking glands) and the anus.

Like other hystricomorph rodents, the female guinea pig possesses an imperforate vaginal closure membrane, which is evident as a translucent membrane across the U-shaped vaginal opening. This membrane opens during proestrus, at parturition, and in some sows during midpregnancy.

3. Male Reproduction

Males begin mounting and thrusting around 30 days of age, and intromissions occur around 45 days. Consistently fertile ejaculations begin when the boar is 90 to 120 days of age and weighs 550 to 600 gm. Breeding usually begins at this time, but maturity is delayed in low weight or undernourished animals (Slob et al., 1979).

The testes remain in the inguinal pouches and exhibit no seasonal weight variations. Inguinal canals remain open for life. The ejaculum contains sperm and secretions of the accessory sex glands. Components of these glands, including enzymes, coagulate in the vagina, forming the copulatory or vaginal plug. This plug, which may retain the ejaculum in the female tract, becomes coated with sloughed vaginal epithelial cells and remains in place for several hours (Stockard and Papanicolaou, 1919).

4. Female Reproduction—Estrous Cycle

Though ovarian follicles are present in sows as young as 2 weeks of age, sexual activity in most females begins at about

68 days of age (range 30–130 days) (Peddie, 1980). The vaginal closure membrane is first perforate around 58 days, and ovulation begins around 60 days (Young et al., 1939). Sows are bred when they weigh 350 to 450 gm or more (2 to 3 months). They are usually used as breeders to around 20 months of age, although sows are often fertile until 4 to 5 years of age. Body weight is not a useful indicator of sexual maturity in guinea pigs because of weight variations among individuals and strains at a given age (Mills and Reed, 1971). The guinea pig's estrous cycle lasts approximately 16 days (range 13–21 days) with minor or inapparent seasonal variation (Phoenix, 1970). Matings that do not result in pregnancy do not alter cycle length, which suggests that pseudopregnancies, if they occur, do not delay estrus. A proestrus of 1 to 1.5 days is characterized by vaginal swelling, increased sexual activity, membrane rupture, and a vaginal smear containing nucleated and cornified epithelial cells (Stockard and Papanicolaou, 1917; Young et al., 1935). The vaginal closure membrane remains open approximately 2–3 days, which includes the time of ovulation (Phoenix, 1970; Harned and Casida, 1972). Estrus lasts 8 to 11 hr and is indicated by the copulatory reflex (assumption of lordosis), a perforate vaginal membrane, vaginal vascular congestion, and cornified cells in the vaginal smear. Estrus occurs more often at night and is of shorter duration in younger animals.

Ovarian follicles mature at days 10 and 11 of the cycle and again at days 14 and 15 (Bland, 1980). Usually 3 or 4 follicles mature for each ovulation. Follicle-stimulating hormone (FSH) and estrogen may be luteotropic, whereas luteolysis appears to result from estrogen-stimulated release of uterine prostaglandin $F_2\alpha$ plus the estrogen inhibition of pituitary FSH (Bland and Donovan, 1970; Marley, 1972). Plasma progesterone concentration has a 5 to 12 hr preovulatory spike, then drops, and then rises again to day 11. The corpora lutea then regress, and progesterone is undetectable in the plasma by day 15 (Feder et al., 1968; Reed and Hounslow, 1971). Follicles mature during the active secretory phase of the corpora lutea and reach maturity in a biphasic pattern (Bland, 1980). Plasma estrogen in the guinea pig is mostly estrone with some 17β-estradiol (Challis et al., 1971).

There is no conclusive evidence of synchronization or entraining of estrous cycles among group-housed guinea pigs (Donovan and Kopriva, 1965; Harned and Casida, 1972). A fertile postpartum estrus lasting approximately 3.5 hr occurs within 12 to 15 hr of parturition in most sows (Rowlands, 1949).

5. Mating

During the sow's estrus, the boar approaches, sniffs, circles, nibbles, licks, and mounts. The female assumes the lordosis posture, with rear quarters elevated (Harper, 1968; Phoenix,

1970). The boar makes one or two intromissions (or as many as 13) and then ejaculates, which terminates coitus with that sow. Coital completion is indicated by grooming, scooting, and perineal marking by the boar, sperm in the vagina, and a copulatory plug (Phoenix, 1970; Ediger, 1976). Sperm capacitation requires 8 to 10 hr. Fertilization occurs within the first few hours of the ovas' 30-hr sojourn in the fallopian ampulla (Hunter *et al.*, 1969). The ova remain fertilizable for approximately 30 hr. Seventy-five to 85% of matings, including those on the postpartum estrus, are fertile.

6. Artificial Insemination

Artificial insemination has been used successfully in guinea pigs. Electroejaculation produces 0.4 to 0.8 ml of semen, which can be placed by bulbed pipette into the vagina or by needle into the peritoneal cavity (Rowlands, 1957; Freund, 1969). Artificial insemination has been successful up to 16 hr postestrus, but injection of sperm too early or too late relative to ovulation can result in no pregnancy, decreased litter size, abortion, or stillbirths. An additional complication is that in some electroejaculated boars, the ejaculum coagulates in the urethra.

7. Detection of Pregnancy

Pregnancy is detected by gentle palpation of firm, oval bodies in the uterine horns (Matthews and Jackson, 1977). At day 15 of pregnancy these swellings are around 5 mm in diameter; at 25 days they are 7 to 15 mm; at 35 days they are 25 mm, and beyond 35 days parts of bodies can be felt. In late pregnancy abdominal distention is apparent, and during the last week the pubic symphysis separates.

8. Gestation

The fertilized ova at the 8- to 12-cell stage enter the uterus on the third day and implant as blastocysts on days 6 or 7 (Deanesly, 1968). Fetal trophoblasts actively penetrate the zona pellucida and establish the placental attachment. Embryonic implantation depends on intact corpora lutea for the first 2 or 3 weeks. After 20 days, the secretory capacity of the corpora lutea declines. The placenta begins actively secreting progesterone around 15 days and assumes the maintenance role for the latter half of pregnancy. This pattern is similar to human gestation (Phoenix, 1970; Csapo *et al.*, 1981). The average total fetal mass (3 fetuses) increases from 92 gm at 45 days to 260 gm near term. During the last 2 weeks of pregnancy fetal weight increases by around 7% daily (Draper, 1920).

Placentation is hemomonochorial, lacunal, and discoidal. A single layer of trophoblastic syncytium arranged in vascularized, fingerlike projections, is bathed in maternal blood (Kaufmann and Davidoff, 1977). Protein molecules, including antibodies, apparently pass through the yolk sac membranes into the fetus but not through the trophoblastic layer. Kurloff cells, described previously, occur in large numbers in the trophoblastic region.

9. Parturition

Primary indicators of advanced pregnancy are the dramatic distension of the abdomen and the separation of the pubic symphysis. Relaxin causes the fibrocartilagenous symphysis to disintegrate between 30 days and term; and during the last week of pregnancy, the separation increases to approximately 3 cm (Zarrow, 1947). No nests are built, and the onset of labor is abrupt. The endocrinologic stimulus for parturition in the guinea pig is unknown, but probably involves progesterone, relaxin, and prostaglandin $F_2\alpha$ (Challis *et al.*, 1971; Csapo *et al.*, 1981). Delivery itself occurs with equal frequency during day or night and requires up to 30 min with 3 to 7 min between births (Boling *et al.*, 1939). The sow squats while delivering, then cleans the pups and eats the placentas. Sows may nibble dead fetuses, but consumption of dead fetuses or neonates is unknown (Wright, 1960).

Dystocia is relatively common and often fatal in obese sows, sows bred for the first time after 7 months of age, and in sows with large fetuses (Hisaw *et al.*, 1944). Indications of dystocia include the clinical history, a bloody or green-brown vaginal discharge, and maternal depression. Because the primary cause of dystocia in guinea pigs is usually the inability to pass a fetus through a confining birth canal, cesarean section is indicated. Young guinea pigs can survive only a few minutes in the isolated uterus or dead mother, and cesarean sections should be done rapidly (Wagner and Foster, 1976). Partially extruded young can be gently pushed or pulled through the tract, and the fetal membranes should be rapidly removed from the faces (Phoenix, 1970).

10. Lactation

Lactation peaks 5 to 8 days postpartum and ceases by days 18 to 30 postpartum. A sow may produce up to 70 ml milk daily, but the quantity varies with fetal mass, strain, parity, and maternal body weight (Mepham and Beck, 1973). Approximately 5 to 15 ml milk can be obtained per milking with a milking device (McKenzie and Anderson, 1979). The quantity of milk is not significantly affected by postpartum demand, as might occur with fostering (Davis *et al.*, 1979). During lactation a sow may lose 50 to 150 gm body weight. Preweanling guinea pigs will nurse other lactating sows in the cage, and they may strip other sows of milk intended for smaller young. Nelson *et al.* (1951) reported that sows not nursed for 24 hr may not return to nursing.

Under conditions of natural rearing, newborn guinea pigs not nursed have a 50% mortality, and those that do survive are small and weak. Survivability increases with the length of nursing time; the first 3 or 4 days are the most critical. Guinea pigs should nurse for at least 15 days for normal growth, reproduction, and disease resistance. Substitute diets of cow's milk, wetted food pellets, or chopped vegetables usually fail to maintain normal growth.

11. Neonatal Development

Guinea pigs are born mobile, fully haired, with teeth, and with eyes and ears open. These precocious characteristics, commensurate with the prolonged gestation, may compensate for the minimal maternal care, passive nursing, and trampling, chewing, and milk stripping often inflicted by larger cagemates. The neonatal maturity makes cesarean derivation, isolation, and artificial rearing of the neonate relatively easy (Wagner and Foster, 1976).

Newborn guinea pigs remain closer to the sow, but may not nurse for the first 12 to 24 hr. During the first 12 to 24 hr the sow may begin to lick the genital area of the offspring, which stimulates defecation and urination. Within the first few days, the young nibble solid food, exhibit withdrawal behaviors, and begin vocal communication (Harper, 1976).

The body weight of newborn guinea pigs is related to parental genetic characteristics, maternal nutrition, litter interval, litter size, and length of gestation. Neonatal weight varies inversely with litter size and directly with the length of gestation. For example, average body weights in litters of 2 have been reported at 112 gm at birth and 292 gm at 28 days, whereas litters of 5 weighed 87 gm per animal at birth and 275 gm at 28 days. Birth weights vary from 45 to 115 gm; animals weighing less than 50 gm usually die.

12. Weaning

Two commonly used criteria for weaning are a 180 gm limit (15 to 28 days) or a 21 day limit (165 to 240 gm). Young males weaned at 15 to 20 days and then isolated may not develop normal adult reproductive behaviors; therefore, weaned males intended for breeding should be weaned late or group housed for an additional 2 to 4 weeks (Harper, 1968; Young, 1969). Guinea pigs should gain 2.5 to 3.5 gm daily to 60 days of age.

III. DISEASES

A. Infectious Diseases

The trend toward production of research guinea pigs by hysterectomy derivation and barrier housing and the continual im-

provement of commercially available, complete diets have markedly reduced the spectrum of infectious disease seen in guinea pigs. Nonetheless similar disease patterns occur year after year in conventionally housed guinea pigs. Guinea pigs are susceptible to many bacterial diseases. The severity of an epizootic is heavily influenced by husbandry practices, concurrent infections, and diet. Following the epizootic, several recovered guinea pigs can be expected to become asymptomatic carriers of the pathogen. The guinea pig's high level of susceptibility to antibiotic-induced fatal enterotoxic cecitis greatly increases the risk associated with most kinds of antibiotic therapy in this species. Many antibiotics selectively inhibit or destroy complex intestinal flora and should be used only when risks and benefits are considered.

Several routine preventive measures can reduce or eliminate the adverse effects of an outbreak of infectious disease in guinea pigs. Whenever possible guinea pigs should be purchased from reputable suppliers that monitor the herd at regular intervals by serologic, microbiologic, and necropsy techniques. The user institution should also frequently surveil their conventionally housed guinea pigs at regular and frequent intervals for selected diseases. Further it is recommended that the animal care and use programs of institutions adhere closely to established principles (Institute of Laboratory Animal Resources, National Research Council, 1978). The design of a disease monitoring program will be influenced by the laboratory and personnel resources of the facility, the origins of the guinea pigs, husbandry programs, and the intended use of the guinea pigs. Newly arrived guinea pigs should be housed in a separate room without other species of animals and observed daily for 2 weeks prior to being housed with resident guinea pigs. Quarantined animals that have signs of infectious disease should be evaluated clinically and microbiologically, and selected guinea pigs should be necropsied. The results of this evaluation will provide essential data regarding the need to alter procurement practices, quarantine and isolation procedures, personnel health programs (in the case of zoonoses), husbandry practices, and treatment regimens. Similarly animals in the holding colony should be observed frequently, and those with clinical signs of infectious disease should be isolated and appropriate procedures taken to establish an etiologic diagnosis.

1. Bacterial/Mycoplasmal/Rickettsial

a. Bordetella bronchiseptica

i. Etiology. *Bordetella bronchiseptica* is a small, motile, non-spore-forming, aerobic, oxidase positive, gram-negative rod that produces 1–2 mm, yellowish-brown nonhemolytic colonies at 48 hr on blood agar. It hydrolyzes urea within 4 hr, utilizes citrate, and grows on potassium tellurite medium.

ii. Clinical signs. *Bordetella bronchiseptica* causes epizootic pneumonia or sepsis with high mortality. Guinea pigs are quite susceptible, and the following descriptions are typical findings in enzootically infected colonies. Young guinea pigs have the highest morbidity and mortality and stresses often trigger outbreaks. In the absence of specific stresses, sporadic deaths occur throughout the year, but particularly in winter. The periodicity of outbreaks is probably also associated with immunity waning to below effective levels. Inappetence, nasal and ocular discharge, and dyspnea are the most common clinical signs. During epizootics pregnant animals frequently die, abort, or have stillbirths. Strain 2 animals are more susceptible than strain 13, which are more susceptible than Hartley guinea pigs.

iii. Epizootiology and transmission. Outbreaks occur 5 to 7 days after exposure of susceptible animals to infected guinea pigs. In enzootically infected colonies, up to 20% of guinea pigs are carriers, and infection is spread from animal to animal by direct contact and through contaminated fomites and respiratory aerosols (Nakagawa *et al.*, 1971). *Bordetella bronchiseptica* is commonly isolated from a variety of research animal species, especially rabbits, cats, and dogs, which may serve as source of infection to guinea pigs.

iv. Necropsy findings. Tracheitis, lung consolidation, and exudate of the external nares and in the middle ears are the most common necropsy findings. Consolidated areas of the lung often appear patchy, dark red or gray, and firm. Histologically, there is a marked purulent bronchopneumonia and pulmonary hyperemia.

v. Differential diagnosis. Infection with *Bordetella bronchiseptica* should always be strongly suspected in cases of epizootic pneumonia of guinea pigs. The agent is readily isolated on blood agar from the infected respiratory tract or middle ears and from the uterus, in which case it may be associated with septicemic metritis. Radiographic examination of the tympanic bullae can be used to screen animals for suppurative otitis media, which is frequently caused by *Bordetella bronchiseptica* (Wagner *et al.*, 1976).

vi. Prevention. As with whooping cough in man, bacterins administered intramuscularily are highly effective in controlling and eliminating infections (Ganaway *et al.*, 1965). Autogenous bacterins are prepared in Freund's incomplete adjuvant. A temperature-sensitive mutant strain of *Bordetella bronchiseptica* that cannot grow at or above 34°C shows promise as a live attenuated vaccine (Shimizu, 1978). This strain grows in the nasal turbinates but not in the lungs.

vii. Control. *Bordetella*-free colonies should be maintained as a closed colony, i.e., entry is restricted to proved-*Bordetella*-free guinea pigs. Enzootically infected colonies can be screened for infected animals by nasal culture on Mac-Conkey's agar. Culture positive animals are culled from the colony. Since *Bordetella bronchiseptica* has a cosmopolitan host range, guinea pigs should not be housed in rooms with other animal species.

viii. Treatment. Almost all isolates are sensitive *in vitro* to several readily available antibiotics, including chloramphenicol, tetracycline, gentamicin, and neomycin (Owens *et al.*, 1975). Antibiotics should be used with caution as advised in the section on antibiotic toxicity. Treatment of clinically affected animals may be less than satisfactory. Ganaway *et al.* (1965) reported that sulfamethazine (4.0 ml of a 12.5% stock solution per 500 ml of drinking water) was effective in controlling death losses, but, shortly after treatment ceased, losses increased gradually until treatment was again necessary. Other antibiotics that may be tried are chloramphenicol 30 mg/kg for 5 days and cephaloridine (200 mg/kg for 5 days). The diet should be reviewed to be sure that the daily intake of ascorbic acid is adequate.

b. Streptococcus zooepidemicus

i. Etiology. *Streptococcus zooepidemicus* of Lancefield's group C is a hemolytic, gram-positive coccus that causes cervical lymphadenitis in guinea pigs.

ii. Clinical signs. "Lumps" (Boxmeyer, 1907) or cervical lymphadenitis, the most common form of this disease, is manifested clinically by the presence of one to several enlarged lymph nodes beneath the mandible and lateral to the cervical trachea (Fig. 1). Affected animals are usually in good condition and show no other signs of illness. Occasionally lymph nodes in other parts of the body are involved. Septicemic and respiratory forms of infection with high mortality occur occasionally as does otitis media (Kohn, 1974; Figs. 5 and 6).

iii. Epizootiology and transmission. The disease is more common in females than males and strain 2 animals are more susceptible than strain 13 guinea pigs (Fraunfelter *et al.*, 1971). The exact means of spread is unknown, but abrasion of the oral mucosa provides a likely route for invasion of the organism (Mayora *et al.*, 1978). The agent is latently carried in the conjunctiva and nasal cavity.

iv. Necropsy findings. In the chronic form of disease, necropsy examination will reveal ventral cervical lymph nodes filled with pus. The acute epizootic form of disease is associated with hydrothorax, sepsis, and pneumonia with widespread necrosis and hemorrhage.

v. Pathogenesis. The lymph node undergoes various stages of inflammation and is eventually destroyed and replaced by an abscess.

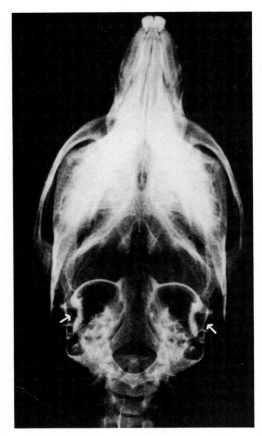

Fig. 5. Radiograph of guinea pig skull showing appearance of normal tympanic bullae (arrows). Compare with Fig. 6.

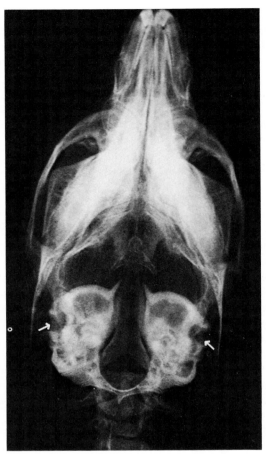

Fig. 6. Radiograph of guinea pig skull. Animal has bilateral chronic otitis media with thickening of the osseous bullae and radiopacity within the middle ear (arrows).

vi. Differential diagnosis. A presumptive diagnosis can be made on the basis of cervical abscesses; however, *Streptobacillus moniliformis* may cause similar lesions (Aldred *et al.*, 1974). Culture of material from the abscess on blood agar reveals β-hemolytic colonies of gram-positive cocci.

vii. Prevention and control. Use of barrier techniques and closed colonies should be considered as an effective preventive measure. Control is effected by culling animals with "lumps" before abscesses can rupture. Colonies with widespread pneumonia and septicemia should be eliminated. Bacterins have failed to prevent disease, but oral scratch inoculations with a laboratory strain of group C *Streptococcus* have produced immunity (Mayora *et al.*, 1978).

viii. Treatment. Cephaloridine, 12.5 mg daily by intramuscular injections for 2 weeks, resulted in recovery (Diaz and Soave, 1973). Spontaneous rupture or surgical incision and drainage of the abscesses is usually followed by recovery.

c. Streptococcus pneumoniae

i. Etiology. *Streptococcus pneumoniae* (Diplococcus or Pneumococcus) is a gram-positive, lancet-shaped coccus that occurs in pairs or short chains. Capsular polysaccharide types IV and XIX are the most common types isolated from guinea pigs and also occur occasionally in man. *Streptococcus pneumoniae* differs from the normal throat streptococci (viridans group) in its bile solubility (lysed by sodium deoxycholate) and its Optochin susceptibility.

ii. Clinical signs. Pneumonia is the most common form of epizootic disease caused by *Streptococcus pneumoniae* in guinea pigs. Acutely affected guinea pigs may have a ruffled hair coat and be relatively inactive. Abortions and stillbirths may occur. High mortality may be associated with epizootics.

iii. Epizootiology and transmission. Clinically normal animals may carry *Streptococcus pneumoniae* in upper respiratory passages. Man, monkeys, rats, and other species may harbor pneumonococci that are potentially virulent for guinea pigs. The ease with which pneumococci are transmitted between these species is often not appreciated.

iv. Necropsy findings.   In acute pneumonic forms of the disease in the guinea pigs the lungs may be edematous or consolidated. Other fibrinopurulent or pyogenic processes associated with enzootic pneumococcal infections in guinea pigs include pericarditis, otitis media, suppurative meningitis, pleuritis, peritonitis, and endometritis (Figs. 5–8).

v. Pathogenesis.   Predisposing factors include stresses, such as those associated with pregnancy, shipment, dietary deficiencies, temperature changes, poor husbandry such as moist bedding, other infections, and experimental procedures.

vi. Differential diagnosis.   In addition to observing the various clinical signs and necropsy findings, diagnosis is facilitated by culture of lesions on blood agar; Gram stain of tissue imprints are particularily helpful. Microscopic examination of direct smear preparations and stained tissue sections of lesions usually reveal typical diplococcal forms (Fig. 7). Some strains of the agent are carboxyphilic, requiring CO_2 for successful *in vitro* isolation.

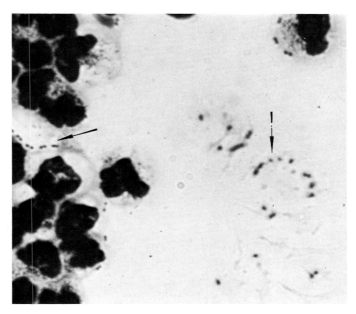

Fig. 7.   Smear from thoracic exudate from a guinea pig with pneumonia and thoracic serositis caused by *Streptococcus pneumoniae*. The diplococci are seen (arrows) within leukocytes and extracellularly.

vii. Prevention, control, and treatment.   Prevention can be accomplished by maintaining a "closed" colony. Several attempts to utilize vaccines in the 1930's were not promising;

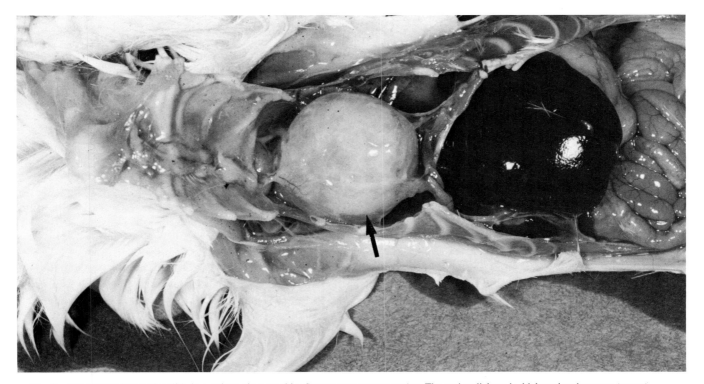

Fig. 8.   Fibrinopurulent pericarditis in a guinea pig caused by *Streptococcus pneumoniae*. The pericardial sac is thickened and opaque (arrow).

however, additional tests based on technology recently applied to preparation of human pneumococcal vaccines holds promise and appear worthy of study in situations where pneumococcal infections are a problem. Sulfadiazine or oral tetracyclines (Wagner and Owens, 1970) may be used to control epizootics, but they will not eliminate carriers.

d. *Salmonella* spp.

i. Etiology. Salmonellae are gram-negative, non-spore-forming, faculative anaerobic bacilli that grow readily on several bacteriologic culture media. The two serotypes most frequently isolated from guinea pigs are *Salmonella typhimurium* and *Salmonella enteritidis* (Habermann and Williams, 1958). Other serotypes are infrequently reported in salmonella infections of guinea pigs.

ii. Clinical signs. Sporadic outbreaks of severe clinical disease characterized by high morbidity and mortality have been observed. Clinical signs, if present, may include rough hair coat, anorexia, weight loss, weakness, conjunctivitis, and abortions. Diarrhea is an uncommon sign. In a colony, the first noticeable sign may be an increased mortality rate. In epizootics, mortality rates from 6 to 90% have been reported (Olfert *et al.*, 1976). In enzootic infections there are chronic carriers with no overt signs of disease.

iii. Epizootiology and transmission. Transmission occurs by ingestion of feces or fecal-contaminated materials. Salmonella may exist latently in the intestinal tract of surviving guinea pigs and be shed intermittently into the environment. Feral rodents or contaminated food or water are other potential sources of salmonella infection. Personnel with contact with other species of mammals, such as cattle, horses, and fowl, may be another source of infection. Guinea pigs of all ages are susceptible to infection, with weanlings and females at the time of parturition being most severely affected. Other predisposing factors include nutritional status, old age, presence of other diseases, salmonella serotype, environmental and experimental stress, and season. Epizootics are reported to occur most frequently in the winter months (Ganaway, 1976).

iv. Necropsy findings. In acute cases of salmonellosis lesions may or may not be found. Lesions seen most frequently are enlargement, focal necrosis, and congestion of the spleen, lymphatic tissues, liver, and intestine. Subacute and chronic infections are characterized by splenomegaly, prominent yellow necrotic foci on the viscera, and enlarged mesenteric lymph nodes and Peyer's patches. Yellow nodules may be found in the lung, pleura, peritoneum, spleen, liver, or in the wall of the uterus (Ganaway, 1976).

v. Differential diagnosis. Necropsy results may be suggestive of salmonellosis, but diagnosis can be made only by isolation of the organism. The feces, cecum, or spleen are the specimens of choice for bacterial isolation procedures. Salmonellae do not hydrolize urea and usually do not utilize lactose. Selective culture media, such as selenite F or tetrathionate broth, are used to enhance growth of salmonella from fecal samples, followed by culture on brilliant green, MacConkey's, or Hektoen enteric agars. In addition, salmonella isolates are further differentiated through serologic testing.

vi. Prevention. The best means of prevention can be accomplished through proper husbandry techniques; fecal culture of all new arrivals; periodic cultures of existing animals, personnel, and equipment; and exclusion of contaminated food and water. Colonies should be housed in a sanitary, closed environment free of feral rodents and birds. A complete pelleted diet should be fed, excluding exogenous food sources such as green leafy vegetables, which are easily contaminated by feral rodent feces. Early detection of salmonella infection may prevent a serious epizootic.

vii. Control. Once infection becomes established in a colony, eradication is extremely difficult. If early detection can be accomplished, elimination of all guinea pigs in close proximity and complete sanitation of the area may prevent a widespread, costly infection (Olfert *et al.*, 1976). Once increased mortality is observed, the infection is often established, and killing, disinfection, and restocking with salmonella-free guinea pigs is the most practical alternative.

viii. Treatment. Treatment of salmonellosis, either through vaccination or antibiotic therapy cannot be generally recommended, though either means may prevent large colony losses. Treatment will not eliminate carrier animals. When evidence of chronic disease is present in a colony, there is no known effective alternative to destroying the colony, thoroughly disinfecting the area, and restocking with salmonella-free guinea pigs.

ix. Research complications. Enzootic salmonella infections may be transformed into an active disease state by the stress of experimental procedures. The mechanisms by which salmonella infections may complicate research results are poorly elucidated, but there is little doubt that a widespread infection could seriously disrupt research as well as present a health threat to personnel (Fish *et al.*, 1968). (See Chapter 22 on zoonoses.)

e. *Yersinia pseudotuberculosis*

Infection with *Yersinia pseudotuberculosis,* a ubiquitous bacterium in nature, was one of the earliest bacterial diseases recognized in guinea pigs. This zoonotic disease appears to be uncommon at present, possibly attesting to the adequacy of preventive measures. The agent does not grow optimally at 37°C. Special media (Paterson and Cook, 1963) and incubation at 20–30°C enhance chances of recovery *in vitro*. Because of yersinia's unusual growth requirements, some infections may be unidentified.

Several types of the disease are recognized in guinea pigs: an acute septicemic form; chronic infections with focal necrotic nodules in many organs, emaciation, and death; a nonfatal infection with lesions of lymph nodes of the head and neck and possibly a latent form in apparently healthy guinea pigs.

Culling of affected animals with palpably enlarged mesenteric lymph nodes and oral pretreatment with an avirulent strain of *Yersinia pseudotuberculosis* have been advocated as control measures (Thal, 1962; Paterson, 1963).

f. Bacillus piliformis

Bacillus piliformis infection (Tyzzer's disease) has been reported in guinea pigs (Zwicker *et al.*, 1978). In all cases the animals were about 4 weeks old and recently arrived from a commercial supplier. Clinical signs included unthriftiness, diarrhea, and sudden death. Necropsy revealed necrosis and inflammation of the ileum, cecum, and colon and multiple necrotic foci in the liver. Tyzzer's disease affects several animal species, including most laboratory animals, which may be sources of infection for guinea pigs. The diagnosis is confirmed by finding intracellular organisms in the intestinal epithelium or liver cells. The agent is an obligate intracellular organism that has not been cultivated *in vitro*.

g. Campylobacter spp.

Marked segmental epithelial hyperplasia of the proximal duodenum with numerous bacilli in the cytoplasm of absorptive epithelial cells was reported in an adult guinea pig that had received cortisone acetate over a long period (Elwell *et al.*, 1981). Several cage mates died acutely and had similar intracytoplasmic bacilli without epithelial cell hyperplasia. Except for its location in the duodenum, the lesions, including intracytoplasmic bacilli, resembled proliferative ileitis of hamsters (Wagner *et al.*, 1973), which has been postulated to be associated with *Campylobacter spp*. Indeed, the pregnant guinea pig is highly susceptible to experimental campylobacteriosis and is used as a model for testing efficacy of *C. fetus* vaccines (Bryner *et al.*, 1979).

h. Mycoplasma spp.

Mycoplasma caviae and acholeplasmas have been isolated from the nose and vagina or uterus of guinea pigs in the United

States and England, and *Mycoplasma pulmonis* has been recovered from guinea pigs in Germany (Stalheim and Matthews, 1975). The isolates were relatively nonpathogenic.

i. Miscellaneous Bacterial Infections

Infections of guinea pigs with the following agents are infrequently encountered and poorly documented in current literature: *Pasteurella, Pseudomonas, Staphylococcus, Klebsiella, Corynebacterium, Streptobacillus, Leptospira, Mycobacterium, Brucella, Escherichia, Listeria,* and mycoplasmal- and rickettsial-like agents. *Klebsiella pneumonia* may be associated with septicemia, pneumonia, peritonitis, and pleuritis (Dennig and Eidmann, 1960). Similar lesions have been associated with *Pasteurella multocida* infections in guinea pigs (Wright, 1936). *Pseudomonas aeruginosa* has been associated with pulmonary botryomycosis (Boström *et al.*, 1969). *Staphylococcus aureus* has been associated with osteoarthritis, chronic pododermatitis (Taylor *et al.*, 1971), exfoliative skin disease, and cheilitis. *Streptobacillus moniliformis* may cause cervical abscesses (Aldred *et al.*, 1974). A variety of bacteria including *Escherichia coli* have been associated with cystitis in parous guinea pigs (Wood, 1981). An hemolytic *Streptococcus* has been associated with mastitis in guinea pigs (Gupta *et al.*, 1970). as have *E. coli, Klebsiella pneumoniae,* and *Streptococcus zooepidemicus* (Kinkler *et al.*, 1976). Guinea pigs recently germfree are very susceptible to infections by *Clostridium perfringens* and possibly other clostridia (Madden *et al.*, 1970). The role of clostridia is discussed further in this chapter in Section III,B,6.

2. Chlymadial/Viral

a. Chlamydia psittaci

Chlamydia psittaci causes inclusion conjunctivitis in guinea pigs (Murray, 1964; Van Hoosier and Robinette, 1976) that is similar to acute trachoma in man. Cyclic outbreaks of severe conjunctivitis are common among guinea pigs in enzootically infected colonies. The disease tends to be self-limiting, and lesions heal spontaneously in 3 to 4 weeks. It is a common disease in many conventional colonies. Staining of conjunctival scrapings with Macchiavello's stain reveals chlamydial intracytoplasmic inclusions. The mode of transmission is not known, although it has been experimentally transmitted by sexual contact.

b. Cytomegalovirus

The cytomegaloviruses (CMV) are species-specific herpesviruses. The guinea pig CMV is frequently used as a model for human CMV infections. The disease is assumed to be widespread in guinea pigs and is usually subclinical and detected as an incidental histologic observation (Connor and Johnson, 1976). Characteristic eosinophilic intranuclear inclusions are seen in ductal epithelial cells of submaxillary salivary glands. Intranuclear inclusions may be seen in a wide variety of other tissues, including the brain. Throughout pregnancy CMV is transmitted to the fetus; thus cesarean rederivation is ineffective in eliminating the disease.

c. Lymphocytic Choriomeningitis (LCM) Virus

LCM occurs as a natural viral disease in guinea pigs as a neurologic syndrome characterized by meningitis and hindlimb paralysis. An RNA arenavirus, LCM virus infects several species of animals and man. The prevalence of infection in guinea pigs is unknown, but several agents isolated from guinea pigs but not identified definitively may have been LCM virus (Van Hoosier and Robinette, 1976). Among laboratory animals the disease is best characterized in the mouse and is discussed more fully in Chapter 3. A provisional diagnosis of LCM may be made on the basis of clinical signs and significant microscopic lesions, including lymphocytic infiltrates in the meninges, choroid plexus, and ependyma. The liver is the recommended organ in which to demonstrate the virus by immunofluorescence or as a source of virus for animal inoculation. LCM virus–free mice injected intracranially with tissue suspensions containing the virus develop clonic convulsions within 7 to 10 days. Persons with LCM usually suffer a mild flu-like illness, rarely contract encephalitis or inflammation of other organs, including heart, testes, lungs, or salivary glands. Fatal lymphocytic meniogencephalomyelitis is very rare in man (Hotchin, 1971).

Most infections of LCM are believed to be contracted by aerosol, but biting insects (mosquitoes, ticks, flies, lice) nematodes, and food, bedding, and fomites contaminated with urine of infected animals have been implicated in transmission of the disease. Wild mice are considered to be the natural reservoir of LCM. Thus prevention and control are aimed at strict control of wild rodents; use of noncontaminated food, bedding, and water; insect and parasite control; surveillance of colony animals for the disease by frequent clinical observation; and necropsy. The contamination of transplantable tumors and other cell lines with LCM should also be considered as a possible source of the virus.

3. Protozoa

A comprehensive review of protozoan parasites of guinea pigs has been written by Vetterling (1976). Only the more commonly encountered infections are discussed here. Also, some infections that are more prevalent in other species of laboratory animals are discussed in more detail in other chapters in this text.

a. Eimeria caviae

Eimeria caviae is the only eimerian species reported in the guinea pig. Diarrhea and death may be the only signs of infection. Grossly the colon appears thickened, and petechial hemorrhages and whitish plaques containing parasites may be seen on the colonic mucosa. Stresses of shipment, immaturity and vitamin C deficiency may increase the severity of coccidiosis in guinea pigs. Treatment with sulfamethazine in the water may be helpful in controlling the disease, as will improved hygiene (Vetterling, 1976).

b. Encephalitozoon spp.

There are several reports of central nervous system and kidney infections caused by *Encephalitozoon* spp. in guinea pigs. The infections resemble those seen in other species and include microgranulomas in the brain and interstitial mononuclear nephritis. The significance of these infections lies in the difficulty in interpreting histologic findings when infected animals are used in research. Lesions of encephalitozoonosis must be differentiated from those of toxoplasmosis which is a zoonotic disease (Moffatt and Schiefer, 1973).

c. Toxoplasma gondii

Though serologic surveys indicate that the incidence of *Toxoplasma gondii* infection is much higher in guinea pigs than in rats (Henry and Beverley, 1977), histologic evidence of toxoplasmosis is rarely seen in conventional guinea pigs. Lesions of toxoplasmosis in guinea pigs include focal or diffuse myocarditis, focal encephalitis with cysts, mild pneumonia without cysts, and relapsing focal histiocytosis in lymph nodes and spleen. Vertical transmission occurs.

d. Cryptosporidium spp.

Cryptospordia are small coccidian parasites recognized in many different animal species. The parasite develops in the striated border of intestinal epithelial cells, and parasites are frequently cited as incidental findings upon histologic examination. They are not associated with clinical signs (Jervis *et al.*, 1966). In other cases ther may be chronic enteritis with changes in small intestinal microvilli, cytoplasmic organelles, and lowered mucosal enzyme activity associated with poor weight gains and an unthrifty appearance. Diarrhea and death may be seen in more severe cases. The infection is transmitted from animal to animal housed in the same cage (Vetterling *et al.*, 1971). Treatment and control methods have not been well studied. Coccidiostats and supportive therapy should have value.

e. Balantidia spp.

Though myriads of *Balantidia* spp. may reside in the ceca of conventional guinea pigs, their association with disease has not been well established. Trophozoites can cause epithelial damage, pass into the lamina propria and submucosa of ceca, and in some cases, make their way to mesenteric lymph nodes.

f. Klossiella cobayae

Klossiella cobayae (caviae) is the kidney coccidium of the guinea pig. Infection is usually discovered as an incidental finding during histologic examination of the kidneys. Schizonts are seen in epithelial cells of the renal tubules in the region of the loops of Henle. Kidney lesions are slight, and no clinical signs are associated with infections. Diagnosis is accomplished through microscopic examinations of infected tissues and demonstration of sporocysts in the urine. There are few recent reports of this parasite, although at one time it was ubiquitous among guinea pig stocks worldwide. Control is accomplished through reducing urine contamination of guinea pig facilities.

4. Nematodes

Paraspidodera uncinata

Paraspidodera uncinata, the cecal worm, is of little clinical consequence. Adult worms, 11 to 28 mm long, are found on the mucosa in the lumen of the cecum and colon. The oval, thick shelled, ascarid-like eggs are found in the feces or in fecal flotation preparations (Fig. 9). Guinea pigs are infected by ingesting infective eggs from which the larva hatch and develop to maturity in 51–66 days. The life cycle is direct, and there is no migration beyond the gut mucosa. Frequent bedding

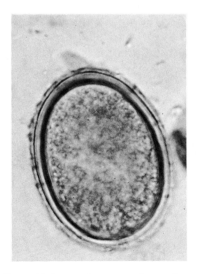

Fig. 9. Paraspidodera uncinata ovum from guinea pig feces.

changes and good sanitation are generally sufficient to eliminate infection. Though no studies are reported, piperazine would appear to be the drug of choice for treatment.

5. Trematodes

Fasciola spp.

Fascioliasis may occur spontaneously in guinea pigs fed forage containing metacercaria of *Fasciola* spp. In some areas of the world it is common for commercial guinea pig raisers to feed forage to guinea pigs. In fluke-infested areas, snails, containing metacercaria, can be chopped into the forage and consumed by the guinea pigs, which may later find their way to research and testing laboratories.

6. Arthropod Parasites

Several different arthropod parasites have been reported to infest guinea pigs (Ronald and Wagner, 1976). Only these ectoparasites believed to be encountered commonly or of clinical significance are discussed in this section. The diagnosis of ectoparasitism is usually not difficult, although considerable persistance may be necessary. Examination of the skin and hair with a magnifying lens and microscopic examination of deep skin scrapings are usually sufficient. Several insecticides used to treat ectoparasiticism in other animals should be effective for control of ectoparasites in guinea pigs. The effect of repeated applications of insecticides on the results of research studies in which the animals are being used must also be considered. Treatments often must be repeated because of reinfestation and lack of susceptibility of ova and larval forms of the parasite to the insecticide.

a. Chirodiscoides caviae

Chirodiscoides caviae is a small, nonburrowing, relatively harmless fur mite commonly found in the pelage of guinea pigs (Wagner *et al.*, 1972; Fig. 10). Infections are easily overlooked because of the small size of the mites and lack of skin or hair lesions. Parasites are frequently joined in pairs. This is a noncopulatory carrying of juvenile males by more mature females, presumably to assure the presence of a mate when both reach sexual maturity. A hand lens aids in viewing these very small mites deep in the fur. Dichlorvos vapor strips or 0.5% carbamyl powder have been used to treat infected guinea pigs.

b. Trixacarus caviae

Trixacarus caviae is a highly pathogenic sarcoptid mite (Fain *et al.*, 1972) that produces mange characterized by hyper-

Fig. 10. Chirodiscoides caviae infestation of a guinea pig. The mites clinging to the hair are about 0.4 to 0.5 mm in length.

keratosis, alopecia, and debilitation. It appears to be widespread in some conventional colonies of guinea pigs. These small mites are difficult to demonstrate without dissolving affected skin in 10% potassium hydroxide. Infection may be associated with intense pruritis and additional skin lesions associated with scratching and secondary bacterial infections. Typical lesions occur on the neck, shoulders, lower abdomen, and inner thighs (Figs. 11–13). Humans may become infected from contact with affected guinea pigs (Dorrestein and Van Bronswijk, 1979). Suggested treatments include a thorough shampoo bath after which the guinea pig is put into a dichlorvos vapor box 1 day a week for 6 weeks. Other treatments include washing weekly for 4 to 6 weeks with a trichlorphon or lindane solution, lime–sulfur baths, and 3 weekly applications of 1% lindane by immersion. A number of other cases of trixacarus mange have since been reported in the United States (Kummel *et al.*, 1980). Untreated animals may lose condition and die.

7. Lice

Gliricola porcelli

Gliricola porcelli and *Gynopus ovalis* are a large, elongated, debris feeding lice commonly found in conventional guinea

pigs stocks (Fig. 14). While the mouth parts of these lice may be used to abrade the skin to obtain cutaneous fluids, most infections (pediculosis) are not accompanied by clinical signs. Severe infestations may be accompanied by alopecia and excessive scratching. The lice and their eggs attached to hair shafts, are readily seen with the unaided eye. Several insecticides as sprays, dips or powders are effective in treating pediculosis. Among effective treatments are dusts containing 0.5% lindane or 4 to 5% malathion. Sprays and dips containing 0.1 to 0.25% malathion or 0.03 to 0.06% Diazinon are also effective.

8. Fungi

a. Trichophyton mentagrophytes

Trichophyton mentagrophytes is almost invariably the cause of epizootic ringworm in guinea pigs. The disease is characterized by a patchy alopecia that starts on the nose and other parts of the head. Lesions may later develop over posterior portions of the back (Fig. 15). Griseofulvin administered orally (15 mg/kg per day) for 14 to 28 days is highly effective in treating dermatophytoses. Infections may spread by contact and fomites to other laboratory animal species and man (Sprouse, 1976).

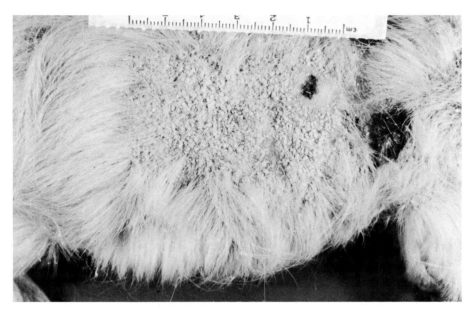

Fig. 11.   Exfoliative dermatitis and alopecia in a guinea pig infested with *Trixacarus caviae*. Dark areas are abrasions due to scratching.

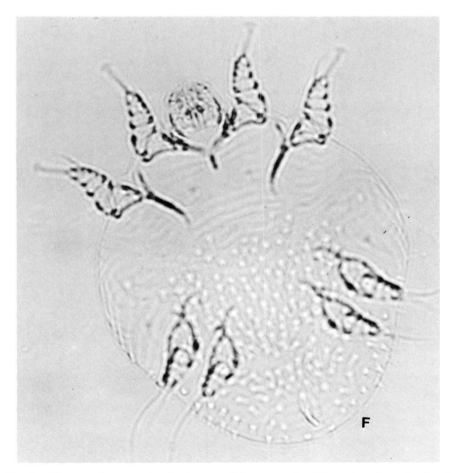

Fig. 12.   The sarcoptic mange mite of guinea pigs, *Trixacarus caviae*. This specimen is an adult female.

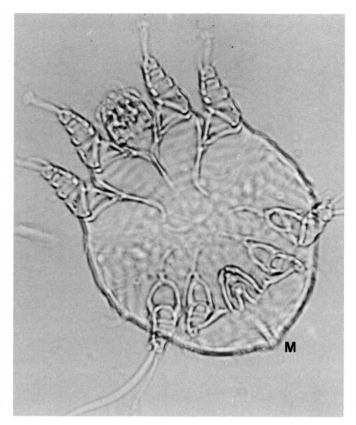

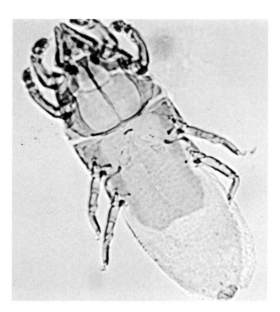

Fig. 14.   *Gliricolla porcelli* shown here is a common cause of pediculosis in guinea pigs.

Fig. 13.   Adult male *Trixacarus caviae*.

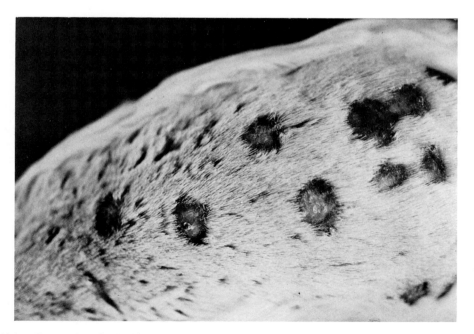

Fig. 15.   Discrete multiple solitary and confluent skin lesions on the dorsum of a guinea pig with generalized dermatomycosis caused by *Trichophyton mentagrophytes*.

b. *Torulopsis pintolopesii*

Torulopsis pintolopesii is generally considered a normal inhabitant of the conventional guinea pig stomach. However, a change in diet or social group may precipate a mycotic enteritis (Kunstyr *et al.*, 1980).

B. Nutritional and Metabolic Diseases

The availability of nutritionally complete diets for laboratory animals has greatly reduced the prevalence of clinically apparent diseases of nutritional origin. Moreover, the diagnosis of a primary nutritional deficiency is seldom confirmed by laboratory tests and rests largely on eliminating other probable causes and a favorable response to nutritional therapy. Marginal nutritional deficiencies (particularly scorbutism) are likely quite common, particularly among guinea pigs fed inappropriate diets or among animals with prolonged partial anorexia due to stressful stimuli, such as shipment, intercurrent diseases, experimental manipulation, or a change in diet.

Guinea pigs are usually fed a pelleted natural diet readily available from several reputable commercial suppliers. These diets are nutritionally complete if properly stored and fed within 90 days of milling. Guinea pigs should not be fed diets containing antibiotics, particularly antibiotics which selectively inhibit gram-positive flora, because there are numerous reports describing untoward "toxicities" resulting from such practices (Rehg and Pakes, 1981). This subject is discussed further in Section III,B,6 antibiotic toxicity. Although used commonly in nutrition research, guinea pigs are notorious for their stubborn refusal to eat experimental diets. The nutritionally limiting ingredients in several types of experimental diets are discussed by Navia and Hunt (1976), and a detailed review of the nutritional and vitamin requirements of guinea pigs is given by Reid and Bieri (1972) and Mannering (1949). A summary of estimated nutritional requirements of guinea pigs is presented in Table II. With the exception of scurvy there are rare reports of naturally occurring vitamin deficiencies in guinea pigs. A tabular summary of the clinical signs of experimentally induced vitamin deficiencies and the recommended theraputic dosages of vitamins is presented in Table IV.

1. Vitamins

a. Vitamin C (Scurvy, Scorbutism)

It is widely known that guinea pigs require dietary vitamin C. Despite the ready availability of adequate diets, scurvy, at least in subclinical form, probably occurs much more often than is generally appreciated in this species (Clarke *et al.*, 1980). Several factors are operative in perpetuating scorbutism in guinea pigs. Although commercial diets for guinea pigs are

formulated to contain ample vitamin C ($\sim$ 800 mg/kg diet) the vitamin is heat-labile, water-soluble (easily leached by moisture), and susceptible to oxidation that is catalyzed by copper (e.g., water pipes). To avoid feeding a nutritionally inadequate diet, feeds should be used within 90 days of milling and should be stored in a cool dry area. If vitamin C is provided in the drinking water, the solution should be prepared fresh daily in distilled or deionized water at a concentration of 200–400 mg/liter. Occasionally guinea pigs are fed supplemental vegetables as a source of vitamin C. Citrus fruits, cabbage, peppers, and kale are rich in ascorbic acid (0.4–0.7 mg/gm), whereas lettuce, which is more commonly fed, is a poor source of the vitamin. Furthermore, guinea pigs cannot store vitamin C to any significant extent and require a virtually continuous intake of the vitamin in especially generous quantities of 15–25 mg/day during growth, pregnancy, lactation, and when stressed.

i. Etiology. Guinea pigs are susceptible to scorbutism or hypovitaminosis C because they lack or are deficient in the microsomal enzyme L-gluonolactone oxidase (Sato and Uderfriend, 1978). This evolutionary abberation, shared with primates and a few species of bats and birds prevents the conversion of glucose to ascorbic acid. The major consequence of a lack of ascorbate is impaired synthesis of collagen owing to defective hydroxylase reactions in the formation of hydroxylysine and hydroxyproline. There is evidence to indicate that there is some variation in the susceptibility of guinea pigs to experimental scorbutism, and some animals apparently require little or no exogenous vitamin C (Ginter, 1976; Williams and Deason, 1967).

ii. Clinical signs. Early clinical signs of scurvy are nonspecific and include rough hair coat, anorexia, dehydration, diarrhea, delayed healing of skin wounds, and possibly intercurrent infection. Hemorrhages particularly about the gingiva and joints (manifested as swelling or bruises and lameness, Fig. 16) are strongly suggestive of the disease. Radiographically demonstrable lesions include enlarged costochondral junctions and abnormal epiphyseal growth centers particularly separation of the proximal epiphyseal cartilage of long bones and pathologic fractures. The specific clinical signs are somewhat influenced by the age, sex, and metabolic status of the animal, with the classic lesions of deficiency being most likely present in animals less than 6 weeks of age, that have been deficient for several weeks. If a breeder colony consisting of adults, juveniles, and nursing animals is involved there may be a variety of clinical signs, including unexplained deaths and stillbirths and often signs no more notable than inactivity and lassitude. A careful analysis of the diet, clinical signs and necropsy of selected animals are essential. Typical necropsy findings include enlarged often separated epiphysis most apparent

Table IV

Experimentally Induced Hypovitaminoses of Guinea Pigs[a]

Vitamin	Major clinical signs, gross and microscopic lesions	Comments and suggested replacement dosage[b]
Thiamin (B$_1$)	Anorexia followed by tremors, ataxia, opisthotonus	Unstable in diets containing oxidizing agents; e.g., K$_2$HPO$_4$; 0.6 mg per os or im daily as necessary
Riboflavin (B$_2$)	Poor growth, rough hair coat, palor of extremities, corneal vascularization	Quantitative requirements not determined; 1 mg per os or im daily as needed
Niacin (nicotinic acid)	Poor growth, palor of extremities, drooling, anemia	No ocular, anal, or skin lesions noted; niacin is produced from tryptophan; 6 mg per os or im daily as needed
Pyridoxine (B$_6$)	No notable clinical signs; poor growth	0.6–1.0 mg per os or im daily as needed
Folic acid (pteroglutamic acid)	Lethargy, weight loss, anemia, and leukopenia; After 5 weeks profuse salivation with terminal convulsions	1 mg per os or im daily
Pantothenic acid	Anorexia, weight loss, rough hair coat, GI and/or adrenal hemorrhage	2.5 mg per os or im daily as needed
Choline	Poor growth, anemia, myasthenia; occasional fatty liver in adults but not young	Turnover of choline is slow because of lack of or low levels of hepatic choline oxidase; choline chloride 150 mg daily
Vitamin C	Weakness, anorexia, anemia, defective collagen synthesis, maintenance and repair, and impaired clotting result in disturbed growth centers in long bones and ribs, and widespread hemorrhages primarily within superficial fascia, gingiva, skeletal muscles and about joints	25–50 mg daily per os or im
Vitamin A (deficiency)	Poor growth, weight loss, desiccation of edge of pinna,	1.5 mg vitamin A acetate daily per os or im

Table IV (*Continued*)

Vitamin	Major clinical signs, gross and microscopic lesions	Comments and suggested replacement dosage[b]
	edema, xerophthalmic keratitis; extensive squamous metaplasia of epithelium of trachea, urinary bladder, and uterus	
Vitamin A (excess)	Degeneration of epiphyseal cartilage; 200,000 USP units/kg (120 mg/kg) in female 14 to 20 days pregnant is teratogenic, producing mainly agnathia, synotia and microstomia	
Vitamin D	Broadened cartilage plates in epiphysis of long bones; enamel hypoplasia of incisors, weight loss; may be unessential if calcium/phosphorus ratio is normal	6 mg (240 IU) per os or im daily
Vitamin E	Myasthenia, paralysis; fetal malformation and resorption skeletal muscle degeneration; testicular atrophy and degeneration	15 mg per os or im daily
Vitamin K	Unknown	None

[a]See Table II for requirements.
[b]Most suggested dosages are based upon providing vitamins in a amount approximately fivefold in excess of that contained in the diet formulated by NIH (Navia and Hunt, 1976) assuming a daily intake of 40 gm/kg for a 750 gm animal.

in the ribs and long bones, loose teeth, subperiosteal hemorrhages, deformed bones resulting from pathologic fractures, and friable, easily removed diaphyseal periosteum. Microscopically pertinent lesions in bones include thin, irregular epiphyseal cartilage with fewer than normal numbers of chondrocytes in irregular columns. Lack of osteoblastic activity results in partial to complete lack of osteoid deposition on the calcified, cartilaginous matrix. Proliferation of fibroblasts in the metaphysis and diaphysis occurs in response to mechanical stress and fractures. Disorganization and atrophy of odontoblasts, irregular dentinogenesis, hemorrhages at sites of trau-

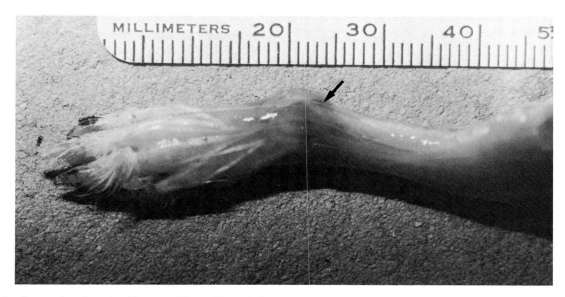

Fig. 16. Carpus of a guinea pig with scurvy. The swelling and discoloration are the result of hemorrhage within the periarticular tissue.

ma, and enlarged adrenals are other common lesions. Should tissue analysis of vitamin C be desirable, those organs that ordinarily contain large amounts of the vitamin and are rapidly depleted during the deficiency state are the spleen, adrenal cortex, and pituitary gland (Fraser *et al.*, 1978).

Treatment of scorbutus should be instituted promptly. Since the animals are often anorectic, supportive therapy in the form of fluids and forced feeding along with administration of 25 mg of vitamin C daily until clinical recovery is achieved is often necessary.

b. Complex Vitamins

Clinical signs of experimentally induced deficiencies of various vitamins within the B complex group and suggested replacement dosages are given in Table IV.

c. Vitamins A, D, E, and K

Primary deficiencies of these vitamins are probably rare in guinea pigs fed natural ingredient diets. Experimental deficiencies and toxicities have been described, and these studies form the basis for the following descriptions as well as much of the data presented in Table IV.

i. Vitamin A. Guinea pigs are herbivorous, and thus a deficiency of vitamin A is very unlikely. Table IV includes the major clinical signs of hypovitaminoisis A and a brief summary of the teratogenic effects of excessive vitamin A.

ii. Vitamin D. Rickets can be induced by deficiencies or imbalances of calcium and phosphorus with or without suffi-

cient vitamin D. The guinea pig can tolerate vitamin D deficiency if the dietary ratio of calcium to phosphorus is normal (Navia and Hunt, 1976). The most useful diagnostic findings are hypocalcemia or hypophosphotemia and radiographs that reveal rachitic metaphyses. The major histologic lesion is lack of deposition of phosphate salts in the cartilagenous matrix between hypertrophic chondrocytes (seen as a lack of purple-blue staining in hematoxylin and eosin stained tissue sections) along with an increase thickness of the zone of mature (large) cartilage cells. Osteoblastic activity and hence osteoid production is normal (unlike the defective osteogenesis of scurvy), unless there is inanition, but the osteoid is not mineralized. The possible effects of excess vitamin D are discussed under metastic calcification.

iii. Vitamins E and K. These deficiencies are not known to occur naturally in guinea pigs. Consult Table IV for further information.

2. Minerals

Calcium, Phosphorus, and Magnesium; Metastatic Calcification

Although guinea pigs have absolute requirements for several minerals, the manifold interactions of magnesium, potassium, phosphate, calcium, and hydrogen ion have been the focus of several studies in guinea pigs (Morris and O'Dell, 1963; Navia and Hunt, 1976). In addition to their nutritional importance, these minerals play a significant role in acid–base balance in guinea pigs. The guinea pig nephron lacks or is deficient in glutaminase and thus glutamine is not used as a source of ammonia to remove hydrogen ions as ammonium ions. The guinea pig uses alternative mechanisms to maintain proper plasma pH, such as removal of hydrogen phosphate ions with divalent cations that reduce an acid load by removing HPO_4 and H_2PO_4 (O'Dell et al., 1956). Guinea pigs can tolerate a wide range of dietary Ca:P if Mg is present in generous amounts, i.e., 0.3%. Generally, recommended amounts of Ca and P are 0.9 and 0.4%, respectively, in which case 0.08% Mg is adequate. An excess of Ca or P will increase the Mg requirement. Naturally occurring rickets is apparently unknown in guinea pigs, but insufficient amounts of dietary Ca and vitamin D have been used to experimentally induce rickets with typical rachitic lesions in the epiphysis of ribs and long bones. Dietary P and Mg, however, appear to be of critical importance in guinea pigs because these minerals influence acid–base balance as well as the occurrence of soft tissue mineralization also known as metastatic calcification.

Metastatic calcification as it occurs in guinea pigs is usually clinically inapparent but could conceivably be accompanied by a variety of clinical signs, most likely muscle stiffness and renal malfunction. The major lesions are seen postmortem in animals over 1 year of age and consist of moderate to severe widespread mineralization of soft tissues especially the kidneys. Initially the renal collecting tubules and interstices are involved, and later phosphate salts are deposited in the convoluted tubules and Bowman's capsule. The trachea, lungs, heart, colon, liver, and stomach are also often involved. Experimentally, a virtually identical disease can be induced by feeding diets with excessive P or deficient in Mg. Although excessive dietary vitamin D is known to play a role in analogous diseases in other animals, its role in this disease of guinea pigs is unknown. Furthermore, the complex interplay among the dietary cations Ca, Mg, Na, and K in the maintenance of acid–base balance is particularly important because, in guinea pigs, cations rather than ammonia are utilized to neutralized H^+ that are then excreted in the urine. Much remains to be learned about this complex and prevalent disease of guinea pigs.

3. Pregnancy Toxemia

Pregnancy toxemia (pregnancy ketosis) is an often fatal syndrome that occurs mostly in adult female guinea pigs during the last 2 weeks of gestation or within a few days of parturition. Several predisposing risk factors have been identified, the most important being obesity ($\geq$ 800 gm), fasting, first or second pregnancy, change in diet, nonspecific stress (handling, shipment, etc.), and heredity (Ganaway and Allen, 1971). Several features of the naturally occurring syndrome can be induced experimentally simply by fasting (or by inducing anorexia by changing the diet) of either obese pregnant sows (primipara) or obese virgin sows. Obesity and fasting appear to

be critical predisposing factors, whereas pregnancy, although increasing the animals' susceptibility, appears not to be an essential feature of this syndrome. Recent evidence indicates that some instances of pregnancy toxemia in guinea pigs may be analogous to preeclampsia in pregnant women and be triggered by uteroplacental ischemia secondary to aortic compression caused by expansion of the gravid uterus (Seidl et al., 1979). It appears that fasting ketosis and the syndrome resembling preeclampsia share many clinical and metabolic features in common, but significant pathogenic differences remain to be elucidated.

The onset of clinical signs is abrupt in that affected animals become sluggish and cease eating and drinking and within 48 hr become prostrate and dyspneic. Death is usually imminent 2 to 5 days after clinical signs appear, and clonic muscle contractions often occur terminally. The most consistent laboratory abnormalities in fasting ketosis of pregnancy are hypoglycemia ($\sim$60 mg/dl), hyperlipidemia, ketonemia, proteinuria, and aciduria (urine pH is normally alkaline). Hematologic and clinical chemistry abnormalities in the preeclamptic variety of this syndrome reflect a different pathogenetic mechanism in that preeclamptic sows are mildly anemic, thrombocytopenic, hyperkalemic, hyponatremic, and hypochloremic and have elevated liver enzymes. Necropsy findings are nonspecific and include generous fat depots, an enlarged fatty liver, enlarged adrenal glands, and an empty stomach. Additional lesions described in preeclamptic females are infarction, hemorrhage, and necrosis of the placentas; renal tubular necrosis; and disseminated intravascular coagulation. Less often the lungs may be congested and the adrenals hemorrhagic.

Systematic studies on the effectiveness of various types of therapy have not been done, but supportive care including fluid therapy is consistently ineffective. The disease can often be prevented or controlled by avoiding obesity through controlled food intake that maintains body weight at less than 500 gm. Pregnant sows should be maintained in a stable environment and fed a nutritionally complete and palatable diet. Supplemental fresh green vegetables, such as cabbage or kale, can be offered daily. Food consumption should be checked daily, and any sign of inappetence should be thoroughly investigated.

4. "Slobbers" (Ptyalism)

"Slobbers" is a descriptive term that probably includes several conditions characterized by wetness around the mouth, chin, and ventral neck. Anorexia, rough hair coat, weight loss, and slobbering are common clinical signs in affected animals. Herbivores salivate profusely, and drooling can occur whenever there is impaired prehension, mastication, or dysphagia. Dental abnormalities, particularly malocclusion, should be considered as the likely cause of slobbering, but other etiologic factors of nutritional origin have been proposed, including folate deficiency, chronic fluorosis (excess dietary fluoride), subacute scurvy, and hereditary factors (Navia and Hunt, 1976).

The oral cavity of guinea pigs with ptyalism should be carefully evaluated. The teeth of guinea pigs grow continuously. Maloccluded incisors deviate rostrally or laterally and sometimes curl into the mouth. Malocclusion of the premolars and molars occurs more often and is less easily diagnosed clinically because of the small mouth and redundant cheek skin. The roots of the maxillary teeth normally incline laterally, while those of the mandibular teeth incline medially. Malocclusion of these teeth results in buccal or lingual deviation, respectively, with concomitant injury to the oral mucosa or impaired mobility of the tongue. Overgrowth of the incisors may also be secondary to malocclusion of the premolars or molars. The incisors and premolars can be clipped to temporarily restore or improve dental occlusion. The diet should be evaluated and only fresh, nutritionally complete, pelleted feed should be fed. Supplemental vegetables especially cabbage or kale may improve the appetite.

5. Diabetes Mellitus

Diabetes mellitus is rarely reported in guinea pigs. A type of diabetes reported to be contagious was originally observed in an Abyssinian breed and was subsequently contracted within 3 months by the majority of Dunkan-Hartley strain guinea pigs housed in the same room (Lang and Munger, 1976). The disease has no sex predilection and occurs mostly in animals over 5 months of age. Affected animals are not clinically ill, although they have considerable glycosuria (urine glucose >100 mg/dl). The concentration of blood glucose is often similar for diabetic and nondiabetic guinea pigs; however, diabetic animals have impaired oral glucose tolerance. Ketoacidosis is not a feature of diabetes in this species in that ketonuria is seen in less than 10% of affected animals and ketonemia has not been observed. Spontaneous remissions apparently occur as a result of regeneration (hyperplasia) of pancreatic islets. Exogenous insulin is not required for survival even when there is marked hyperglycemia and glycosuria. Other clinical or laboratory abnormalities include development of cataracts and mildly increased concentrations of serum triglycerides. Lesions are most severe in animals with marked increases in blood and urine sugar for a period of 6 or more months. Pancreatic abnormalities include degranulated β cells (reduced staining by aldehyde fuchsin), periodic acid–Schiff (PAS)-positive cytoplasmic inclusions, fibrosis, and scarring of the vascular stroma. Thickening of the basal laminae of glomerular and skeletal muscle capillaries has also been described. Attempts to identify a causal agent by inoculation of suspect fluids or tissues into embryonated eggs or by coactivation with established guinea pig cell lines has been unsuccessful. Although

the clinical disease is mild, several aspects of the pathologic alterations resemble those of juvenile diabetes mellitus in children. The reported contagious nature of the disease indicates that it has the potential to become widely distributed. More importantly, however, is the need for further study of this form of diabetes to establish the definitive cause of the disease that would be a significant contribution to experimental biology and might provide further insight into our understanding of some aspects of juvenile diabetes mellitus.

6. Antibiotic Toxicity

Several antibiotics, notably penicillin and clindamycin, can cause a fatal enterotoxic cecitis in guinea pigs (Bartlett, 1979; Lowe *et al.*, 1980; Rehg and Pakes, 1981). Typically, guinea pigs given 10,000 units/kg of penicillin by any of several parenteral routes will become anorectic, dehydrated, hypothermic, and die usually within 4 days. The major consistant lesion at necropsy is a dilated hemorrhagic cecum. The pathophysiology of the toxicity appears to be very similar among several animal species including clindamycin-associated cecitis of hamsters, rabbits and guinea pigs and pseudomembranous colitis in man. Studies in guinea pigs and hamsters have demonstrated that death is caused by a toxin produced by *Clostridium difficile*, although *Clostridium histolyticum* toxin has also been implicated. The causal role of bacterial toxins is strengthened by the observation that germfree guinea pigs are not susceptible to penicillin toxicity (Newton *et al.*, 1964). Regardless of the specific toxin(s) involved, the presumed mechanism of toxemia is that growth of the organism and toxin production are promoted as a result of selective suppression of competing flora by antibiotic therapy. Several antibiotics, especially penicillin, erthryomycin, clindamycin, and perhaps other antibiotics are capable of inducing this toxicity in guinea pigs and should be avoided in this species. Oral administration of 60 mg of chloramphenicol daily was not toxic to guinea pigs in one study (Eyssen *et al.*, 1957). Additional studies are needed to further clarify the pathogenesis of this toxicity and to identify antibiotics that can be safely and effectively used.

C. Neoplastic Diseases

If virus-associated leukemias are excluded, the prevalence of naturally occurring neoplasms is very low in guinea pigs less than 3 years of age. Estimates of tumor prevalence range from 0.4 to 1.4% in guinea pigs less than 3 years of age, and 0.5 to 30% in "aged" animals (Manning, 1976; Ediger and Kovatch, 1976). The prevalence of certain tumors in closed colonies or within specific strains suggests that genetic factors play an important role in the development of neoplasia in guinea pigs. The relative rarity of cancer in this species has been, in part, attributed to a serum factor known as a tumor inhibitory principle that appears to be asparaginase, an enzyme with antitumor activity (Wriston and Yellin, 1973). Asparaginase has also been found in the serum of animal species closely related to guinea pigs and in the serum of some New World primates.

Of further interest is the observation that splenic preparations rich in cells possessing Fao-Kurloff bodies inhibit the *in vitro* growth of carcinomatous cells of human origin (Bimes *et al.*, 1981). A comprehensive review of naturally occurring neoplasms of guinea pigs has been published by Manning (1976).

1. Integumentary System

a. Skin

A significant percentage of tumors affect the skin, subcutis, and mammary glands of guinea pigs. Trichofolliculomas are the most common skin tumor. These neoplasms are considered one of several types of basal cell epithelioma and are benign, solid, or cystic encapsulated masses located within the hypodermis. They are derived from basal cells, which differentiate mainly as abortive hair follicles with scattered sebaceous glands. Presumably these tumors in guinea pigs as in other species seldom metastasize and are easily removed surgically.

b. Mammary Glands

Most mammary tumors in guinea pigs are benign with fibroadenomas being the predominant type (Fig. 17). Malignant mammary tumors comprise about 30% of the reported neoplasms and are mainly adenocarcinomas, some of which have been reported in males (Manning, 1976; Andrews, 1976). These tumors are usually of ductal origin and are large, vascular, and locally invasive. Metastasis is unusual but can occur to local lymph nodes, lungs, or abdominal viscera. Given the relatively high frequency of histologic malignancy in breast tumors in guinea pigs, it is advisable to evaluate breast lesions microscopically to provide a definitive diagnosis as well as to exclude nonneoplastic processes, including mastitis. If surgical excision is attempted, a broad area of normal appearing contiguous tissue should be excised along with some local lymph nodes for microscopic evaluation.

2. Reproductive System

a. Ovarian Tumors

The reproductive tract of the sow has a relatively high prevalence of tumors. The most common are ovarian teratomas, which are believed to arise from parthenogenic development of

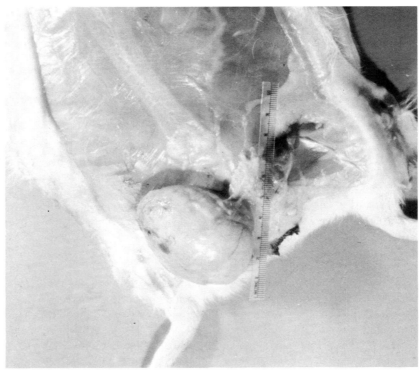

Fig. 17. Mammary tumor (fibroadenoma) in a female guinea pig. (Courtesy of Dr. Ray Ediger.)

oocytes. Teratomas often consist of a mixture of structures derived from ectoderm, mesoderm, or endoderm including neuroepithelial tubules, exocrine and endocrine glandular tissue, smooth or skeletal muscle, hair follicles, teeth, and other elements. Ovarian teratomas may attain considerable size (up to 10 cm) and could presumably be palpated (Fig. 18). These tumors are nearly always unilateral and are seldom malignant, though portions of the tumor may implant on the peritoneal surface of abdominal viscera. The principal clinical complication is intraperitoneal hemorrhage of the tumor, which may be fatal. Diagnosis is based on clinical findings. Radiographs may reveal the tumor if it is large or contains bone or tooth structures. Surgical excision is the preferred treatment.

b. Uterine Tumors

Although uncommon, uterine tumors constitute the second most likely site for tumors of the reproductive tract. These tumors tend to be localized, even though some undergo malignant transformation. The cell types of origin are usually mesenchyme or smooth muscle (e.g., fibroma, fibrosarcoma, leiomyoma, leiomyosarcoma) whereas epithelial tumors are rare.

3. Respiratory System

Pulmonary tumors are the most commonly reported non-hematopoietic tumors of guinea pigs. However, many of these lesions share characteristics of adenomatous hyperplasia as seen in some viral pneumonias of sheep. Experimental evidence also indicates that hyperplastic or adenomatous changes occur in the lungs of guinea pigs in response to a variety of stimuli, including exposure to foreign bodies and diptheroid bacilli. The natural history of these lesions is unknown, and further data are essential before they can be accurately classified.

4. Hemolymphopoietic System

Lymphosarcoma is the most commonly reported malignancy of guinea pigs, and the incidence rate is high in some inbred strains, especially strain 2. Both leukemic and aleukemic forms occur, but the former predominates, and thus a clinical diagnosis can be established by blood cell counts and cytology. Several aspects of this disease have been reported as part of a workshop on the role of an oncogenic RNA retrovirus in some guinea pig leukemias (Rhim and Green, 1977; Nadel, 1977). Type C viral particles are often seen in electron micrographs of lymphoblasts from leukemic guinea pigs, although the role of these viruses is yet unsettled. The disease can be transmitted by inoculation of leukemic blood cell suspensions, plasma pel-

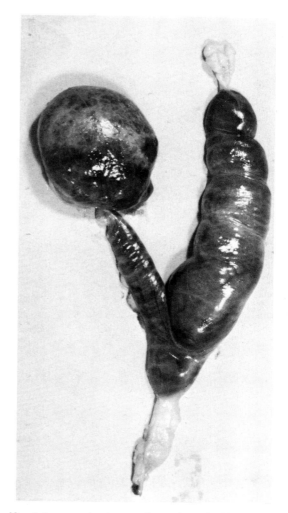

Fig. 18. A large ovarian teratoma in a guinea pig. The enlarged uterine horn contained three fetuses. (Courtesy of Dr. Ray Ediger.)

lets or tissue brei (spleen, lymph node) as well as by cell suspensions per os, but not by cell-free filtrates. Only strain 2 guinea pigs or F_1 hybrids of strain 2X Hartley crosses are uniformly susceptible to this disease. Notably, all tumor cells of this line (L_2C) possess the female karyotype irrespective of the sex of the host harboring the tumors, which indicates that the tumor is transplantable rather than transmissible (Whang-Peng, 1977). The clinical disease, whether spontaneous or experimentally induced, is indistinguishable. Affected animals appear unkept and depressed. Enlargement of peripheral or visceral lymph nodes may be present with or without hepatosplenomegaly (Fig. 19). The diagnosis is usually established by leukocyte enumeration and examination of smears of peripheral blood or bone marrow. Occasionally, lymph node or liver biopsy may be necessary. White blood cell counts usually range from 25,000 to 600,000/mm³. Depending on the extent

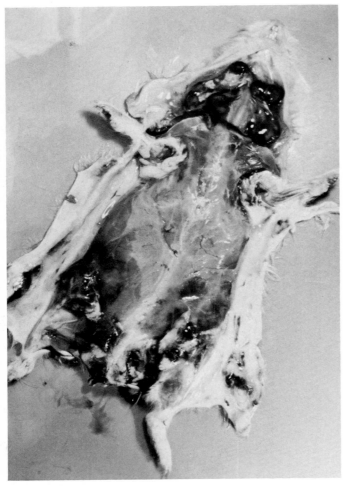

Fig. 19. A strain 2 guinea pig with transplantable L₂C leukeumia. Markedly enlarged, neoplastic lymph nodes are evident along either side of the neck and in the inguinal area. (Courtesy of Dr. Ray Ediger.)

and severity of the disease other changes may include anemia, thrombocytopenia, or impaired function of other organs, such as liver (icterus) or kidney (proteinuria, uremia), owing to infiltrates of tumor cells. Necropsy will often reveal enlarged lymph nodes, liver, and spleen; and microscopically, proliferating lymphoblasts may be seen in the aforementioned organs as well as within the bone marrow and intestinal lymphoid tissues. The leukemic cell is 18 to 40 nm in diameter and has a large nucleus with a sieve-like chromatin structure and scant cytoplasm with few to many vacuoles. Studies have shown this leukemia to represent a rare B cell neoplasm because the cells have surface projections (unlike T cells), surface IgM, and receptors for C_3. The disease has a short course of 2 to 5 weeks (within 10 to 18 days following experimental inoculation), which along with other features such as leukemia, widespread organ infiltration, and favorable response to

chemotherapy offers promise as a model for acute lymphoblastic leukemia of human beings.

5. Cardiovascular System

Mesenchymomas of the right atrium appear to be a tumor peculiar to adult females of the Dunkan-Hartley strain (Ediger and Kovatch, 1976). Although mainly a curiosity, it is possible that these tumors would be detected by routine ECG should they cause cardiac dysrhythmias and perhaps impaired cardiac function.

6. Endocrine System

Endocrine tumors although rare in guinea pigs occur principally in the adrenal cortex, thyroid, and pancreatic islets. These tumors are morphologically benign and appear to be hormonally nonfunctional.

7. Other Tumors

Tumors of other systems, including the alimentary, musculoskeletal and nervous system, are of unusual occurrence even when considered collectively.

D. Other Diseases and Clinical Problems

Several miscellaneous diseases are discussed by Wagner (1976). Only selected conditions are discussed here.

1. Integument

a. Biting, Alopecia, and Hair Pulling (Barbering)

Partial alopecia is common in sows in late gestation particularly among intensively bred animals. Although the cause is unknown, the condition improves following parturition. Transitory partial alopecia occurs in weanling guinea pigs with the emergence of large coarse guard hairs and thinning of the finer undercoat hairs. Hair biting or barbering is a vice commonly observed in guinea pigs and can occur under a variety of circumstances. Pups often bite and ingest hair from the dam during and shortly after lactation (Harper, 1976), and this behavior can persist into adulthood. Subordinate animals may be barbered by dominant animals, and this vice as well as ear biting are often exaggerated under conditions of crowding and stress, such as forming new cohort groups. Self-barbering also occurs. Barbering is easily recognized as patches of unevenly chopped hair; seldom is the underlying skin inflamed. The lesions are usually over the dorsal lumbar area and flanks and less often about the head and thorax. Ear biting, scratching, and other forms of more aggressive activity are more likely to occur among adults, especially mature or elderly males. Secondary infection of scratches and bites occurs rarely. Usually the problems are self-limited when predisposing factors are identified and eliminated.

b. Pododermatitis and Hyperkeratosis of Footpads

Both of these conditions are usually the result of a combination of faulty design of cage floors, poor husbandry, and obesity. Pododermatitis can also result from experimental injection of footpads especially if the injected material contains adjuvants. The lesion (Fig. 20) usually involves only superficial tissues often with distorted nail growth, but on occasion

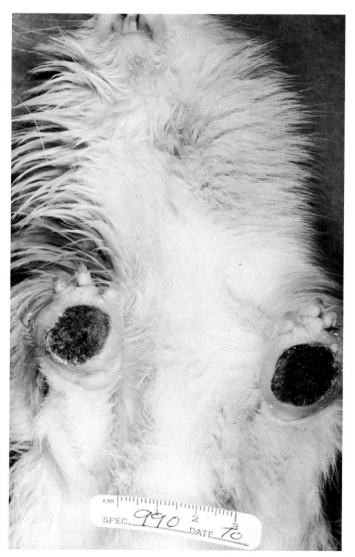

Fig. 20. Chronic severe pododermatitis of the forefeet of a guinea pig. *Staphylococcus aureus* was cultured from the lesions.

tendonitis or periostitis is seen. Weight control, improved husbandry practices, and housing affected animals on solid floors with 2 to 3 in. of soft bedding usually suffice and the lesions heal spontaneously. Local treatment may require clipping the nails, cleansing, application of a topical antibiotic, and perhaps a loose wrapping for a few days.

2. Alimentary System

a. Malocclusion

This is discussed in this chapter in Section III,B,4.

b. Cecitis

Typhlocolitis is a common necropsy lesion in guinea pigs that have died with no antecedent clinical signs. Although antibiotic toxicity (discussed in Section III,B,6) can cause hemorrhagic cecitis in this species, there is often no history that would incriminate an antibiotic (Wagner, 1976). Other predisposing factors include corticosteroid injection, fasting, and late gestation. The cecal lesions and clinical history of sudden unexpected deaths suggest that a number of factors other than antibiotics can induce alterations in enteric microflora that are often fatal to the host. A further understanding of these predisposing factors and the pathogenesis of those diseases awaits the outcome of well-planned research and clinical studies.

ACKNOWLEDGMENT

This work was supported in part by NIH grants RR001234, RR00471, and RR07004.

REFERENCES

Aldred, P., Hill, A. C., and Young, C. (1974). The isolation of *Streptobacillus moniliformis* from cervical abscesses of guinea pigs. *Lab. Anim.* **8,** 275–277.

Altman, P. L., and Dittmer, D. S., eds. (1974). "Biology Data Book, 2nd ed., Vol. III. Fed. Am. Soc. Exp. Biol., Bethesda, Maryland.

Andrews, E. J. (1976). Mammary neoplasia in the guinea pig. *Cornell Vet.* **66,** 82–96.

Barnes, R. D. (1971). "Special Anatomy of Laboratory Animals." University of California, Davis.

Bartlett, J. G. (1979). Antibiotic associated pseudonembranous colitis. *Rev. Infect. Dis.* **1,** 530–539.

Bimes, C., DeGraeve, P., Amiel, S., *et al.* (1981). Demonstration of a principle ensuring destruction of human cancer cells in culture. *C. R. Hebd. Seances Acad. Sci.* **292**(3), 293–298.

Bland, K. P. (1980). Biphasic follicular growth in the guinea pig oestrous cycle. *J. Reprod. Fertil.* **60,** 73–76.

Bland, K. P., and Donovan, B. T. (1970). Oestrogen and progesterone and the function of the corpora lutea in the guinea pig. *J. Endocrinol.* **47,** 225–230.

Boling, J. L., Blandau, R. J., Wilson, J. G., and Young, W. C. (1939). Post-parturitional heat response of newborn and adult guinea pigs. Data on parturition. *Proc. Soc. Exp. Biol. Med.* **42,** 128–132.

Boström, R. E., Huckins, J. G., Kroe, D. J., Lawson, N. S., Martin, J. E., Ferrell, J. E., and Whitney, R. A., Jr., (1969). Atypical fatal pulmonary botryomycosis in two guinea pigs due to *Pseudomonas aeruginosa. J. Am. Vet. Med. Assoc.* **7,** 1195–1199.

Boxmeyer, G. H. (1907). Epizootic lymphadenitis: A new disease of guinea pigs. *J. Infect. Dis.* **4,** 657–664.

Breazile, J. E., and Brown, E. M. (1976). Anatomy. *In* ''The Biology of the Guinea Pig'' (J. E. Wagner, and P. J. Manning, eds.), pp. 53–62. Academic Press, New York.

Bryner, J. H., Foley, J. W., and Thompson, K. (1979). Comparative efficacy of ten commercial *Campylobacter fetus* vaccines in the pregnant guinea pig: Challenge with *Campylobacter fetus* serotype A. *Am. J. Vet. Res.* **40,** 433–435.

Challis, J. R. G., Heap, R. B., and Illingworth, D. V. (1971). Concentrations of oestrogen and progesterone in the plasma of non-pregnant, pregnant and lactating guinea pigs. *J. Endocrinol.* **51,** 333–345.

Charles River Breeding Laboratories (1982). Technical Bulletin 1. Charles River Breeding Laboratories, Wilmington, Massachusetts.

Chiba, J., Otokawa, M., Nakagawa, M., and Egashira, Y. (1978). Serological studies on the major histocompatibility complex of new inbred strains of the guinea pig. *Microbiol. Immunol.* **22,** 545–555.

Clarke, G. L., Allen, A. M., Small, J. D., and Lock, A. (1980). Subclinical scurvy in the guinea pig. *Vet. Pathol.* **17,** 40–44.

Connor, W. S., and Johnson, K. P. (1976). Cytomegalovirus infection in weanling guinea pigs. *J. Infect. Dis.* **5,** 442–449.

Cooper, G., and Schiller, A. L. (1975). ''Anatomy of the Guinea Pig.'' Harvard Univ. Press, Cambridge, Massachusetts.

Csapo, A. I., Eskosa, J., and Tarro, S. (1981). Gestational changes in the progesterone and prostaglandin F levels of the guinea pig. *Prostaglandins* **21,** 53–64.

Davis, S. R., Mapham, T. B., and Lock, K. J. (1979). Relative importance of pre-partum and post-partum factors in the control of milk yield in the guinea pig. *J Dairy Res.* **46,** 613–621.

Deanesly, R. (1968). The effects of progesterone, testosterone and ergocornine on non-pregnant and pregnant guinea pigs. *J. Reprod. Fertil.* **16,** 271–281.

Dennig, H. K., and Eidmann, E. (1960). Klebsielleninfektionen bein meerschweinchen. *Berl. Tieraerztl. Wochenschr.* **73,** 273–274.

deWeck, A. L., and Festing, M. F. W. (1979). Investigations for which the guinea pig is well suited. *In* ''Inbred and Genetically Defined Strains of Laboratory Animals,'' (P. L. Altman and D. D. Katz, eds.), Part 2, pp. 507–509. Fed. Am. Soc. Exp. Biol., Bethesda, Maryland.

Diaz, J., and Soave, O. A. (1973). Cephaloridine treatment of cervical lymphadenitis in guinea pigs. *Lab. Anim. Dig.* **8,** 60–62.

Donovan, B. T., and Kopriva, P. C. (1965). Effect of removal or stimulation of the olfactory bulbs on the estrous cycle of the guinea pig. *Endocrinology* **77,** 213–217.

Dorrestein, G. M., and Van Bronswijk, J. E. M. H. (1979). *Trixacarus caviae* Fain, Howell and Hyatt, 1972 (Acari: Sarcoptidae) as a cause of mange in guinea pigs and papular urticaria in man. *Vet. Parasitol.* **5,** 389–398.

Draper, R. L. (1920). The prenatal growth of the guinea pig. *Anat. Rec.* **18,** 369–392.

Ediger, R. D. (1976). Care and management. *In* ''The Biology of the Guinea Pig'' (J. E. Wagner and P. J. Manning, eds.), pp. 5–12. Academic Press, New York.

Ediger, R. D., and Kovatch, R. M. (1976). Spontaneous tumors in the Dunkan-Hartley guinea pig. *JNCI, J. Natl. Cancer Inst.* **56,** 293–294.

Elwell, M. R., Chapman, A. L., and Frenkel, J. K. (1981). Duodenal hyperplasia in a guinea pig. *Vet. Pathol.* **18,** 136–139.

Eremin, O., Wilson, A. B., Coombs, R. R. A., Ashby, J., and Plumb, D. (1980). Antibody-dependent cellular cytotoxicity in the guinea pig: The role of the Kurloff cell. *Cell. Immunol.* **55,** 312–327.

Ernström, V., and Larsson, B. (1967). Export and import of lymphocytes in the thymus during steroid-induced involution and regeneration. *Acta Pathol. Microbiol. Scand.* **70,** 371–384.

Escobar, G., and Escobar, G. (1976). [Some ethnographic observations on the raising and uses of the cuye (*Cavia porcellus* L.) in the region of Cuzco, Peru.] *Antropol. Andina* **1–2,** 34–49.

Eyssen, H., DeSomer, P., and Van Dijck, P. (1957). Further studies on antibiotic toxicity in guinea pigs. *Antibiot. Chemother. (Washington, D.C.)* **7,** 55–64.

Fain, A., Hovell, G. J. R., and Hyatt, K. H. (1972). A new sarcoptid mite producing mange in albino guinea pigs. *Acta Zool. Pathol. Antverp.* **55,** 56–73.

Feder, H. H., Resko, Ja., and Goy, R. W. (1968). Progesterone concentrations in the arterial plasma of guinea pigs during the oestrous cycle. *J. Endocrinol.* **40,** 505–513.

Festing, M. F. W. (1976a). Genetics. *In* ''The Biology of the Guinea Pig.'' (J. E. Wagner and P. J. Manning, eds.), pp. 99–120. Academic Press, New York.

Festing, M. F. W. (1976b). The guinea pig. *In* ''The UFAW Handbook on the Care and Management of Laboratory Animals'' (University Federation for Animal Welfare, ed.), 5th ed., pp. 229–247. Churchill-Livingstone, Edinburgh and London.

Festing, M. F. W. (1979). Mutants: Guinea pig and non-inbred colonies of special interest: Guinea pig. *In* ''Inbred and Genetically Defined Strains of Laboratory Animals'' (P. L. Altman and D. D. Katz, eds.), Part 2, pp. 507–509. Fed. Am. Soc. Exp. Biol., Bethesda, Maryland.

Fish, N. A., Fletch, A. L., and Butler, W. E. (1968). Family outbreaks of salmonellosis due to contact with guinea pigs. *Can. Med. Assoc. J.* **99,** 418–420.

Fraser, R. C., Pavlovic, S., Kurahara, C. G., Murato, A., Peterson, N. S., Taylor, K. B., and Feigen, G. A. (1978). The effect of variations in vitamin C intake on the cellular immune response of guinea pigs. *Am. J. Clin. Nutr.* **33,** 839–847.

Fraunfelter, F. C., Schmidt, R. E., Beattie, R. J., and Garner, F. M. (1971). Lancefield type C streptococcal infections in strain 2 guinea pigs. *Lab. Anim.* **5,** 1–13.

Freund, M. (1969). Interrelationships among the characteristics of guinea pig semen collected by electro-ejaculation. *J. Reprod. Fertil.* **19,** 393–403.

Ganaway, J. R. (1976). Bacterial mycoplasma and rickettsial disease. *In* ''The Biology of the Guinea Pig'' (J. E. Wagner and P. J. Manning, eds.), pp. 121–135. Academic Press, New York.

Ganaway, J. R., and Allen, A. M. (1971). Obesity predisposes to pregnancy toxemia (ketosis) in guinea pigs. *Lab. Anim. Sci.* **21,** 40–44.

Ganaway, J. R., Allen, A. M., and McPherson, C. W. (1965). Prevention of acute *Bordetella bronchiseptica* pneumonia in a guinea pig colony. *Lab. Anim. Care* **15,** 156–162.

Ginter, E. (1976). Ascorbic acid synthesis in certain guinea pigs. *Int. J. Vitam. Nutr. Res.* **46(2),** 173–179.

Gupta, B. N., Langham, R. F., and Conner, G. H. (1970). Mastitis in guinea pigs. *Am. J. Vet. Res.* **31,** 1703–1707.

Habermann, R. T., and Williams, F. P., Jr. (1958). Salmonellosis in laboratory animals. *JNCI, J. Natl. Cancer Inst.* **20,** 933–948.

Harned, M. A., and Casida, L. E. (1972). Failure to obtain group synchrony of estrus in the guinea pig. *J. Mammal.* **53,** 223–225.

Harper, L. V. (1968). The effects of isolation from birth on the social behavior of guinea pigs in adulthood. *Anim. Behav.* **16,** 58–64.

Harper, L. V. (1976). Behavior. *In* ''The Biology of the Guinea Pig'' (J. E.

Wagner and P. J. Manning, eds.), pp. 31–51. Academic Press, New York.

Haynes, J. D., and Askenase, P. W. (1977). Cutaneous basophil responses in neonatal guinea pigs: Active immunization, hapten specific transfer with small amounts of serum, and preferential elicitation with phytohemagglutinin skin testing. *J. Immunol.* **118,** 1063–1069.

Henry, L., and Beverley, J. K. A. (1977). Toxoplasmosis in rats and guinea pigs. *J. Comp. Pathol.* **87,** 97–102.

Hintz, H. F. (1969). Effect of coprophagy on digestion and mineral excretion in the guinea pig. *J. Nutr.* **99,** 375–378.

Hisaw, F. L., Zarrow, M. X., Money, W. L., Talmage, R. V. N., and Abramowitz, A. A. (1944). Importance of the female reproductive tract in the formation of relaxin. *Endocrinology* **34,** 122–143.

Hodgson, R. J., and Funder, J. W. (1978). Glucocorticoid receptors in the guinea pig. *Am. J. Physiol.* **235,** R115–R120.

Hong, C. C., Ediger, R. D., Raetz, R., and Djurickovic, S. (1977). Measurement of guinea pig body surface area. *Lab. Anim. Sci.* **27,** 474–476.

Hotchin, J. (1971). The contamination of animals with lymphocytic choriomeningitis. *Am. J. Pathol.* **64,** 747–749.

Hunter, R. H. F., Hunt, D. M., and Chang, M. C. (1969). Temporal and cytological aspects of fertilization and early development in the guinea pig, *Cavia porcellus. Anat. Rec.* **165,** 411–430.

Institute of Laboratory Animal Resources, National Research Council (1978). "Guide for the Care and Use of Laboratory Animals." Department of Health and Human Services, Washington, D.C.

Izard, J., Barellier, M. T., and Quillec, M. (1976). The Kurloff cell. Its differentiation in the blood and lymphatic system. *Cell Tissue Res.* **173,** 237–259.

Jervis, H. R., Merrill, T. G., and Sprinz, H. (1966). Coccidiosis in the guinea pig small intestine due to a *Cryptosporidium. Am. J. Vet. Res.* **27,** 408–414.

Jilge, B. (1980). The gastrointestinal transit time in the guinea pig. *Z. Versuchstierkd.* **22,** 204–210.

Kaufmann, P., and Davidoff, M. (1977). "The Guinea Pig Placenta." Springer-Verlag, Berlin and New York.

King, J. A. (1956). Social relations of the domestic guinea pigs after living under semi-natural conditions. *Ecology* **37,** 221–228.

Kinkler, R. J., Jr., Wagner, J. E., Doyle, R. E., and Owens, D. R. (1976). Bacterial mastitis in guinea pigs. *Lab. Anim. Sci.* **26,** 214–217.

Klassen, C. D., and Doull, T. (1980). Evaluation of safety: Toxicologic evaluation. *In* "Casarett and Doull's Toxicology" (J. Doull, C. D. Klaasen, and M. O. Amdur, eds.), 2nd ed., pp. 11–27. Macmillian, New York.

Kohn, D. F. (1974). Bacterial otitis media in the guinea pig. *Lab. Anim. Sci.* **24,** 823–825.

Kummel, B. A., Estes, S. A., and Arlian, L. G. (1980). *Trixacarus caviae* infestation of guinea pigs. *J. Am. Vet. Med. Assoc.* **177,** 903–908.

Kunstyr, I., Niculescu, E., Naumann, S., and Lippert, E. (1980). *Torulopsis pintolopesii* an opportunistic pathogen in guinea pigs? *Lab. Anim.* **14,** 43–45.

Laird, C. W. (1974). Clinical pathology: Blood chemistry. *In* "Handbook of Laboratory Animal Science (E. C. Melby, Jr., and N. H. Altman, eds.), Vol. II, pp. 345–436. CRC Press, Cleveland, Ohio.

Lang, C. M., and Munger, B. L. (1976). Diabetes mellitus in the guinea pig. *Diabetes* **25,** 434–443.

Lowe, B. R., Fox, J. G., and Bartlett, J. G. (1980). *Clostridium difficile*-associated cecitis in guinea pigs exposed to penicillin. *Am. J. Vet. Res.* **41,** 1277–1279.

McCormick, J. G., and Nuttall, A. L. (1976). Auditory research. *In* "The Biology of the Guinea Pig" (J. E. Wagner and P. J. Manning, eds.), pp. 281–303. Academic Press, New YOrk.

McKenzie, W. N., Jr., and Anderson, R. R. (1979). A modified device for collecting milk from guinea pigs. *J. Dairy Sci.* **62,** 1469–1470.

Madden, D. L., Horton, R. E., and McCullough, N. B. (1970). Spontaneous infection in ex-germfree guinea pigs due to *clostridium perfringens. Lab. Anim. Care* **20,** 454–455.

Mannering, G. J. (1949). Vitamin requirements of the guinea pig. *Vitam. Horm.* (N.Y.) **7,** 201–211.

Manning, P. J. (1976). Neoplastic diseases. *In* "The Biology of the Laboratory Guinea Pig" (J. E. Wagner and P. J. Manning, eds.), pp. 211–225. Academic Press, New York.

Marley, P. B. (1972). An attempt to inhibit the uterine luteolysin in the guinea pig. *J. Physiol. (London)* **222,** 169P–170P.

Marshall, A. H. E., Swttenham, K. V., Vernon-Roberts, B., and Revell, P. A. (1971). Studies on the function of the Kurloff cell. *Int. Arch. Allergy Appl. Immunol.* **40,** 137–152.

Matthews, P. J., and Jackson, J. (1977). Pregnancy diagnosis in the guinea pig. *Lab. Anim. Sci.* **27,** 248–250.

Mayora, J., Soave, O., and Doak, R. (1978). Prevention of cervical lymphadenitis in guinea pigs by vaccination. *Lab. Anim. Sci.* **28,** 686–690.

Mepham, T. B., and Beck, N. F. (1973). Variation in the yield and composition of milk throughout lactation in the guinea pig (*Cavia porcellus*). *Comp. Biochem. Physiol. A* **45A,** 273–281.

Miller, D., and Murray, F. S. (1966). Guinea pigs immobility response to sound: Threshold and habituation. *J. Comp. Physiol. Psychol.* **61,** 227–233.

Mills, P. G., and Reed, M. (1971). The onset of first oestrus in the guinea pig and the effects of gonadotrophius and oestradiol in the immature animal. *J. Endocrinol.* **50,** 329–337.

Moffatt, R. E., and Schiefer, B. (1973). Microsporidiosis (encephalitozoonosis) in the guinea pig. *Lab. Anim. Sci.* **23,** 282–283.

Morris, E. R., and O'Dell, B. L. (1963). Relationship of excess calcium and phosphorus to magnesium requirement and toxicity in guinea pigs. *J. Nutr.* **81,** 175–181.

Murray, E. S. (1964). Guinea pig inclusion conjunctivitis virus. *J. Infect. Dis.* **114,** 1–12.

Nadel, E. M. (1977). History and further observations (1954–1976) of the L$_2$C leukemia in the guinea pig. *Fed. Proc., Fed. Am. Soc. Exp. Biol.* **36,** 2249–2252.

Nakagawa, M., Muto, T., Yoda, H., Nakano, T., and Imaizumi, K. (1971). Experimental *Bordetella bronchiseptica* infection in guinea pigs. *Jpn. J. Vet. Sci.* **33,** 53–60.

National Research Council (1979). "Animals for Research," 10th ed. N.R.C., Washington, D.C.

Navia, J. M., and Hunt, C. E. (1976). Nutrition, nutritional diseases, and nutrition research applications. *In* "The Biology of the Guinea Pia" (J. E. Wagner and P. J. Manning, eds.), pp. 235–267. Academic Press, New York.

Nelson, W. L., Kaye, A., Moore, M., Williams, H. H., and Harrington, B. L. (1951). Milking techniques and composition of guinea pig milk. *J. Nutr.* **44,** 585–594.

Newton, W. L., Steinman, H. G., and Brandriss, M. W. (1964). Absence of lethal effect of penicillin in germfree guinea pigs. *J. Bacteriol.* **88,** 537–538.

O'Dell, B. L., Vandepopuliere, J. M., Morris, E. R., and Hogan, A. G. (1956). Effect of a high phosphorus diet on acid-base balance in guinea pigs. *Proc. Soc. Exp. Biol. Med.* **91,** 220–223.

Olfert, E. D., Ward, G. E., and Stevenson, D. (1976). *Salmonella typhimurium* infection in guinea pigs: Observations on monitoring and control. *Lab. Anim. Sci.* **26,** 78–80.

Owens, D. R., Wagner, J. E., and Addison, J. B. (1975). Antibiograms of pathogenic bacteria isolated from laboratory animals. *J. Am. Vet. Med. Assoc.* **167,** 609–609.

Paterson, J. S., and Cook, R. (1963). A method for the recovery of *Pasteu-*

rella pseudotuberculosis from feces. *J. Pathol. Bacteriol.* **85**, 241–242.

Patterson, R., and Kelly, J. F. (1974). Animal models of the asthmatic state. *Annu. Rev. Med.* **25**, 53–68.

Payne, B. J., Lewis, H. B., Murchison, T. E., and Hart, E. A. (1976). Hematology of laboratory animals. *In* "Handbook of Laboratory Animal Science" (E. C. Melby, Jr. and N. H. Altman, eds.), Vol. III, pp. 383–461. CRC Press, Cleveland, Ohio.

Peddie, M. J. (1980). Follicular development in the immature guinea pig. *J. Endocrinol.* **84**, 323–331.

Peplow, A. M., Peplow, P. V., and Hafez, E. S. E. (1974). Parameters of reproduction. *In* "Handbook of Laboratory Animal Science" (E. C. Melby, Jr. and N. H. Altman, eds.), Vol. I, pp. 105–116. CRC Press, Cleveland, Ohio.

Phoenix, C. H. (1970). Guinea pigs. *In* "Reproduction and Breeding Techniques for Laboratory Animals" (E. S. E. Hafez, ed.), pp. 244–257. Lea & Febiger, Philadelphia, Pennsylvania.

Plank, J. S., and Irwin, R. (1966). Infertility of guinea pigs on sawdust bedding. *Lab. Anim. Care* **16**, 9–11.

Quillec, M., Debout, C., and Izard, J. (1977). Red cell and white cell counts in adult female guinea pigs. *Pathol. Biol.* **25**, 443–446.

Reed, M., and Hounslow, W. F. (1971). Induction of ovulation in the guinea pig. *J. Endocrinol.* **49**, 203–211.

Rehg, J., and Pakes, S. P. (1981). *Clostridium difficile* antitoxin neutralization of cecal toxin(s) from guinea pigs with penicillin-associated colitis. *Lab. Anim. Sci.* **31**, 156–160.

Reid, M. E., and Bieri, J. G. (1972). Nutrient requirements of the guinea pig. *In* "Nutrient Requirements of Laboratory Animals" (Subcommittee on Laboratory Animal Nutrition), pp. 9–19. Natl. Acad. Sci., Washington, D.C.

Revell, P. A., and Soretire, E. A. (1980). Kurloff cell levels in the peripheral blood of normal guinea pigs treated with cytotoxic drugs. *Virchows Arch. B* **34**, 77–83.

Rhim, J. S., and Green, I. (1977). Guinea pig L₂C leukemia: Immunological, virological and clinical aspects. *Fed. Proc., Fed. Am. Soc. Exp. Biol.* **36**, 2247–2332.

Robinson, R. (1971). Guinea pig chromosomes. *Guinea Pig Newsl.* **4**, 15–18.

Ronald, N. C., and Wagner, J. E. (1976). The arthropod parasites of the genus *Cavia*. *In* "The Biology of the Guinea Pig" (J. E. Wagner and P. J. Manning, eds.), pp. 201–209. Academic Press, New York.

Rowlands, I. W. (1949). Postpartum breeding in the guinea pig. *J. Hyg.* **47**, 281–287.

Rowlands, I. W. (1957). Insemination of the guinea pig by intraperitoneal injection. *J. Endocrinol.* **16**, 98–106.

Sansone, M., and Bovet, D. (1970). Avoidance learning by guinea pigs. *Q. J. Exp. Psychol.* **22**, 458–461.

Sato, P., and Udenfriend, S. (1978). Scurvy-prone animals, including man, monkey, and guinea pig do not express the gene for gulonolactone oxidase. *Arch. Biochem. Biophy.* **187**, 158–162.

Schalm, O. W., Jain, N. C., and Carroll, E. J. (1975). "Veterinary Hematology." Lea & Febiger, Philadelphia, Pennsylvania.

Schermer, S. (1967). "The Blood Morphology of Laboratory Animals." Davis, Philadelphia, Pennsylvania.

Seidl, D. C., Hughes, H. C., Bertolet, R., and Lang, C. M. (1979). True pregnancy toxemia (preeclampsia) in the guinea pig (*Cavia porcellus*). *Lab. Anim. Sci.* **29**, 472–478.

Shevach, E. M., Festing, M. F., and deWeck, A. L. (1979). Inbred and partially inbred strains: Guinea pig. *In* "Inbred and Genetically Defined Strains of Laboratory Animals" (P. L. Altman and D. D. Katz, eds.), Part 2, pp. 507–509. Fed. Am. Soc. Exp. Biol., Bethesda, Maryland.

Shimizu, T. (1978). Prophylaxis of *Bordetella bronchiseptica* infection in guinea pigs by intranasal vaccination with live strain ts-S34. *Infect. Immun.* **22**, 318–321.

Short, D. J., and Woodnott, D. P., eds. (1969). "The Institute of Animal Technicians Manual of Laboratory Animal Practice and Techniques." Thomas, Springfield, Illinois.

Sisk, O. B. (1976). Physiology. *In* "The Biology of the Guinea Pig" (J. E. Wagner and P. J. Manning, eds.), pp. 63–98. Academic Press, New York.

Slob, A. K., Vreeburg, J. T. M., and van der Werff Ten Bosch, J. J. (1979). Body growth, puberty and undernutrition in the male guinea pig. *Br. J. Nutr.* **41**, 231–237.

Smith, H. W., and Crabb, W. E. (1961). The fecal bacterial flora of animals and man: Its development in the young. *J. Pathol. Bacteriol.* **82**, 53–61.

Sommers, T., and Betts, T. E. (1978). A new guinea pig watering system. *Lab. Anim. Care* **12**, 163–164.

Sprouse, R. F. (1976). Mycoses. *In* "The Biology of the Guinea Pig" (J. E. Wagner and P. J. Manning, eds.), pp. 153–161. Academic Press, New York.

Stalheim, O. H. V., and Matthews, P. J. (1975). Mycoplasmosis in specific-pathogen-free and conventional guinea pigs. *Lab. Anim. Sci.* **25**, 70–71.

Stockard, C. R., and Papanicolaou, G. N. (1917). The existence of a typical oestrous cycle in the guinea pig—with a study of its histological and physiological changes. *Am. J. Anat.* **22**, 225–283.

Stockard, C. R., and Papanicolaou, G. N. (1919). The vaginal closure membrane, copulation, and the vaginal plug in the guinea pig, with further considerations of the oestrous rhythm. *Biol. Bull. (Woods Hole, Mass)* **37**, 222–245.

Stone, S. H., Lacompoulos, P., Lacopoulos-Briot, M., Neven, T., and Halpern, B. N. (1964). Histamine differences in amount available for release in lungs of guinea pig susceptible and resistant to acute anaphylaxis. *Science* **146**, 1061–1062.

Taylor, J. L., Wagner, J. E., Owens, D. R., and Stuhlman, R. A. (1971). Chronic pododermatitis in guinea pigs. *Lab. Anim. Sci.* **21**, 944–945.

Thal, E. (1962). Oral immunization of guinea pigs with avirulent *Pasteurella pseudotuberculosis*. *Nature (London)* **194**, 490–491.

Van Hoosier, G. L., and Robinette, L. R. (1976). Viral and chlamydial diseases. *In* "The Biology of the Guinea Pig" (J. E. Wagner and P. J. Manning, eds.), pp. 137–152. Academic Press, New York.

Veall, D. J., and Wood, M. (1978). Modification of a guinea pig hopper to reduce food wastage. *Lab. Anim. Care* **12**, 229–230.

Vetterling, J. M. (1976). Protozoan parasites. *In* "The Biology of the Guinea Pig" (J. E. Wagner, and P. J. Manning, eds.), pp. 163–196. Academic Press, New York.

Vetterling, J. M., Jervis, H. R., Merrill, T. G., and Spring, H. (1971). *Cryptosporidium wrairi* sp. from the guinea pig *Cavia porcellus* with an emendation of the genus. *J. Protozool.* **18**, 243–247.

Wagner, J. E. (1976). Miscellaneous disease conditions of guinea pigs. *In* "The Biology of the Guinea Pig" (J. E. Wagner and P. J. Manning, eds.), pp. 228–234. Academic Press, New York.

Wagner, J. E., and Foster, H. L. (1976). Germfree and specific pathogen-free. *In* "The Biology of the Guinea Pig" (J. E. Wagner and P. J. Manning, eds.), pp. 21–30. Academic Press, New York.

Wagner, J. E., and Owens, D. R. (1970). Type XIX *Streptococcus pneumonia* (*Diplococcus pneumonia*) infections in guinea pigs. *Proc. 21st Annu. Meet. Am. Assoc. Lab. Anim. Sci.* abstract No. 71.

Wagner, J. E., Al-Rabiai, S., and Rings, R. W. (1972). *Chirodiscoides caviae* infestation in guinea pigs. *Lab. Anim. Sci.* **22**, 750.

Wagner, J. E., Owens, D. R., and Troutt, H. F. (1973). Proliferative ileitis of hamsters: Electron microscopy of bacteria in cells. *Am. J. Vet. Res.* **34**, 249–252.

Wagner, J. E., Owens, D. R., Kusewitt, D. E., and Corley, E. A. (1976). Otitis media of guinea pigs. *Lab. Anim. Sci.* **26**, 902–907.

Whang-Peng, J. (1977). Cytogenetic studies in L₂C leukemia. *Fed. Proc., Fed. Am. Soc. Exp. Biol.* **36**, 2255–2259.

Fig. 1. (*Continued*)

entire group as white-footed mice. Others restrict the use of the term white-footed mouse to a single species, *P. leucopus*. It also appears that some use the terms white-footed mouse and deer mouse interchangeably, but others identify *P. maniculatus* as the true deer mouse. Several species have been maintained in the laboratory including *P. leucopus*, *P. maniculatus*, *P. floridanus*, *P. bairdi*, and *P. polionotus*.

Peromyscus are found over most of North America from Alaska and Labrador southward, and one species extends into South America reaching the extreme north of Colombia. Most species have a limited habitat and consequently are found only in a relatively small part of the total *Peromyscus* range. However, *P. maniculatus* has an enormously broad range, extending from just south of the Arctic circle to the highlands of central Mexico. *Peromyscus leucopus* is found in most of the United States except for the far west and deep south.

Table I

Taxonomic Relationship of Some Species, Order Rodentia

Suborder	Family	Number of species
Sciuromorpha	Sciuridae	366
	Squirrels, woodchuck	
Myomorpha	Muridae	1183
	Norway rat, house mouse,	
	multimammate rat	
	Cricetidae	
	Golden hamster, gerbil,	
	white-footed mouse, vole,	
	cotton rat, white-tailed rat	
Hystricomorpha	Caviidae	180
	Guinea pig	
	Chinchillidae	
	Chinchilla	

Peromyscus vary in color from light brown or gray to dark brown. The ventral surface and feet are white, hence the name white-footed mouse. They characteristically have large protruding eyes and large ears, and some species have small internal cheek pouches. The mature weight varies depending upon the species, ranging from 15 to 50 gm.

Voles (*Microtus*) include almost 50 species and are found in many parts of the world. The common vole (*M. arvalis*) is one of the most common European mammals. The tundra vole (*M. oeconomus*) occurs in some areas of Europe and in Alaska and northern Canada. *Microtus pennsylvanicus* (eastern meadow vole or mouse, field mouse), which is found through much of the upper United States and in Canada and Alaska, has been used in research. It is a small rodent, adults weighing about 40 gm, with pelage of varying shades of brown.

There are two or three species of multimammate rats (*Mastomys*). *Mastomys natalensis* is the commonly used species. There is disagreement regarding taxonomy of the African members of the family Muridae. Some consider *Mastomys* to be a genus; others classify it as a subgenus of *Praomys*. Therefore, both names are used. Commonly, the scientific name of the multimammate rat is written as *Praomys (Mastomys) natalensis*. They are one of the most widely distributed and abundant rodents in Africa. In appearance, multimammate rats are intermediate between rats and mice with adult body weight ranges from 30 to 100 gm. Consequently, they also have been termed multimammate mice. Multimammate rats are characterized by their unusually high number of mammary glands, usually 8–12 pairs but sometimes up to 18 pairs. The fur is soft, and the dorsum of the body is light gray to dark brown; the underside is light gray.

There are two species of chinchilla, the short-tailed (*Chinchilla brevicaudata*) and the long-tailed (*C. laniger*). Some taxonomists only recognize one species and consider the variation in tail length to be due to climatic differences. Nearly all chinchillas that have been domesticated in North America have been *C. laniger*. Chinchillas are best known for their remarkably soft fur. They were once abundant in the Andes Mountains of Argentina, Bolivia, Chile, and Peru. When chinchilla fur became popular in the late 1800s and early 1900s, thousands were killed for their pelts; and, consequently, they almost became extinct in their original habitat. The first chinchillas to leave South America were eleven brought to California by an American engineer in 1923. He may have been responsible for saving the species from total extinction. They were acclimated, a suitable diet formulated, and finally after 3 years, they began to reproduce in captivity. Most commercially bred chinchillas are descended from these 11 animals.

Chinchillas resemble small rabbits, with characteristics including a large head, wide muzzle, squirrel-like tail, long whiskers, and large eyes. The color is bluish-gray, and the weight of adults is 0.5–1.0 kg. Chinchillas should not be confused with chinchilla rats, chinchilla mice, or the chinchilla breed of rabbits.

There are 230 or more species of squirrels, including flying, tree, and ground squirrels. In some instances it is not easy to determine if a species is predominately arboreal or terrestrial. Some tree squirrels spend considerable time on the ground, and some ground squirrels often climb trees. Ground squirrels usually have shorter and less bushy tails than tree squirrels.

Ground squirrels are sometimes called gophers. Of the 32 species, most live in North America, from Mexico to Alaska; however a few are native to Africa, and there are seven species ranging from eastern Europe across nothern and central Asia. Ground squirrels usually are yellowish-gray, some with spots or stripes on the back; the ears are small and the legs are short. Ground squirrels have not been commonly used as research animals, but several species including the California ground squirrel (*Spermophilus beecheyi*) have been maintained in the laboratory. Adult California ground squirrels weigh 0.5–1.0 kg; the back is brown with white or buff spots.

Arboreal squirrels (*Sciurus*) have slender bodies and bushy tails (which are usually as long as the bodies). The muzzle is short and pointed; the forehead is wide. The 55 species of *Sciurus* are distributed all over the world, with the exceptions of Australia, Madagascar, and southern South America. One of the best known and most widespread is the gray squirrel, *S. carolinensis*, which is common in the eastern United States. It is a rather large and powerful squirrel, weighing 350–700 gm.

There are two or three species of cotton rats (*Sigmodon*) found in North and Central America and the tropical areas of South America. *Sigmodon hispidus*, native to the southern United States, is the species commonly used in research. The dorsum is grayish-brown to blackish-brown and the underside is grayish, the ears are short, body length is 12.5–20 cm, and body weight is 70–200 gm.

White-tailed rats or South African hamsters (*Mystromys albicaudatus*) are biologically similar to hamsters but do not have cheek pouches. They are found in South and Central Africa. They are medium-sized (mature weight, 75–185 gm), have relatively short tails, and the fur is grayish-brown and silky smooth. The tail, feet, and ventral surface are creamy white to gray, ears are erect, and the eyes are dark and prominent. *Mystromys* are alert, inquisitive, active, and quick.

The woodchuck, ground hog, whistle-pig, or chuck (*Marmota monax*) is a large, digging marmot found throughout the eastern and midwestern United States and parts of Canada. Its body is thick and 45–70 cm long; the tail is approximately 15 cm long. Its weight varies from 2 to 3 kg when it emerges from hibernation in the spring to 4.5 to 7 kg in the fall. The head is flattened with small ears; the forefeet have long claws used for digging, and the color of the fur is reddish or brownish.

B. Use in Research

Of the rodents discussed in this chapter, the most widely used is the gerbil. Its hardiness, ease of maintenance, good reproductivity, and relative freedom of disease in captivity are some features that have contributed to its use in research. Spontaneous conditions reported in gerbils that may be useful in experimental studies include epileptiform seizures, hyperadrenocorticism, diabetes, obesity, and periodontal disease and dental caries.

Certain gerbils are susceptible to mild to severe spontaneous epileptiform seizures when appropriately stimulated. The seizures frequently are precipitated by environmental changes, including handling, rapid change of temperature or lighting, exposure to testing devices, or confinement in small areas. Sometimes the stimulating factor is not obvious. This is characterized by animals lying passively with limbs extended and body twitching, clonic–tonic seizures, vibrissae and pinnae twitching, and rarely death. Recovery is rapid and apparently complete, and affected gerbils resume normal activity within minutes. However, seizures may recur later in the same animal when appropriately stimulated. Seizure susceptibility may begin as early as 2 months of age and reach an incidence of 40–80% in a colony of gerbils. The trait is apparently inherited and selectively bred; separate seizure-sensitive and seizure-resistant lines of gerbils have been produced. This condition has been proposed as a model of human idiopathic epilepsy.

Gerbils have also proved useful in experimental pathology, pathophysiology, and parasitology studies such as cholesterol absorption and metabolism, cerebral infarction, lead toxicity, inositol lipodystrophy, and filariasis. Gerbils readily develop high serum and hepatic cholesterol levels even on low fat diets, but gross atherosclerotic changes are not seen in the aorta. Gerbils can be infected with *Brugia pahangi, B. malayi, Litomosoides carinii,* or *Dipetalonema viteae* thereby providing an experimental host–parasite system.

Some attributes of white-tailed rats that contribute to their usefulness as laboratory animals include a longer life span than the more common laboratory rodents, absence of special problems associated with care and feeding, and relative freedom from disease. Although they have not been widely used as research animals, some study areas in which they have been used include diabetes mellitus, poliomyelitis virus attentuation, periodontal disease, dental caries, radiobiology, and as experimental host for *Schistosoma* spp., *Mycobacterium* spp., and arthropod-borne viruses.

Chinchillas have characteristics that make them suitable experimental animals. Among these are small size, ease of handling, and long life span (12–20 years). Because of large, easily accessible bulla, absence of presbycusis in long-term studies, and relative freedom of middle ear infections, chin-

chillas are used in hearing research. Also, chinchillas have been reported to be a good experimental model for the study of Chagas' disease. They are very susceptible to infection; large numbers of blood trypanosomes and intracellular forms occur; and the disease is consistently fatal.

Weanling meadow voles have been used in nutrition studies. They are reported to be good bioassay animals for protein content of feeds and for digestibility of forages.

Peromyscus have been used in behavioral experiments including studies of food hoarding, maternal care, feeding patterns, and light preferences. They have been used for genetic and evolutionary studies. Convulsive mutants of several species of *Peromyscus* representing genetic entities of neurological interest with similarities to idiopathic epilepsy in man have been described. The clonic and tonic phases of seizure as well as the recovery and postconvulsive behavior are reported to be almost identical in *Peromyscus* and in man. Associated environmental, biochemical, and behavioral parameters related to seizures can be assessed in these models.

Cotton rats have been used to study behavior, dental caries, calcinosis, and experimental infection with filarids, Venezuelan equine encephalitis virus, poliomyelitis virus, and murine typhus.

Multimammate rats are very susceptible to *Yersinia pestis* and have been used in routine plague diagnostic work and experimental work such as vaccine trials and virulence studies. They have been used as experimental hosts for parasitic infections, such as *Brugia* spp., *L. carinii, Toxoplasma* sp., and *Schistosoma* spp. The multimammate rat is the only known nonhuman natural host of Lassa virus (see Chapter 22). In captivity, this species develops spontaneous adenocarcinoma of the glandular stomach with unusual frequency. Since carcinoma of the glandular stomach is also a common cancer of persons but is rare in other animals, multimammate rats have been used to study this condition. They have also been used to study other gastric ailments: carcinoid tumors and gastric ulcer disease induced by mucosal anaphylaxis in ovalbumin-sensitized animals. The presence of a well-developed prostate in females makes the *Mastomys* a valuable model for the study of hormonal effects on the prostate.

For some studies, the larger size and longer life span of woodchucks, compared with other rodents, are useful features. In captivity, woodchucks develop many of the nutritional and medical problems that affect man, including obesity and vascular and neoplastic diseases. Therefore, woodchucks have been proposed as research animals for several different areas of study such as body weight and energy balance, control of appetite and food consumption, control of endocrine function, and cardiovascular, cerebrovascular, and neoplastic diseases (Young and Sims, 1979). Since woodchucks are hibernating animals, endocrine and metabolic function, appetite, physical activity, and body weight change annually. These annual fluctuations in body weight occur whether or not they hibernate and coincide with cyclic food consumption. During the winter and early spring, weight is lost; but in the late spring and summer, they may double their weight by accumulating body fat. There are a number of similarities in the effects of these weight changes in woodchucks with those seen in man.

There is an unusually high incidence of apparently diet-related arteriosclerosis, aortic rupture, and cerebrovascular and cardiovascular diseases in captive woodchucks. They have also been suggested as a possible naturally occurring model for the study of hepatitis B virus (HBV) and primary hepatocellular carcinoma. Persistent infection with a virus similar to HBV is associated with a naturally occurring primary carcinoma of the liver of the woodchuck. Serum of about 15% of infected woodchucks have particles containing a DNA polymerase and DNA genome that are similar to those of HBV. The virus in woodchucks has been designated woodchuck hepatitis virus (Snyder *et al.*, 1982).

Nonsporadic diabetes has been described in at least ten species of rodents, including gerbils, white-tailed rats, fat sand rats (*Psammonys obesus*), spiny mice (*Acomys cahirinus*), and Chinese hamsters. In most of these rodents, diabetes is accompanied by obesity. In some rodents, hyperglycemia occurs as a response to a high caloric laboratory diet but does not occur when the animal is placed on a diet similar to that present in its natural habitat. The spontaneous diabetes mellitus in white-tailed rats is characterized by hyperglycemia, polyuria, glycosuria, ketonuria, and severe degenerative changes in the pancreatic islets of Langerhans. Obesity is not associated with hyperglycemia in this species, nor has a relationship to laboratory dietary influence been detected. Hyperglycemia is well established in the animals by the age of 4 months and appears to shorten the life expectancy of this normally long-lived rodent. A strong predilection for the disease in males suggests that in this species the disorder may be sex linked.

C. Sources

Gerbils are commercially available from several vendors. Most are outbred stocks, but an inbred strain (MON/Tum) is available. A line of black gerbils has been developed and is commercially available.

Chinchillas are raised commercially for their fur. Research animals can sometimes be obtained from breeders by purchasing animals with fur defects that make then useless for their pelts or as breeding stock.

In general, other rodents discussed in this chapter are not commercially available. Institutional breeding colonies of voles, white-footed mice, multimammate rats, cotton rats, and white-tailed rats have been developed by users of these animals. Squirrels and woodchucks do not reproduce well in captivity, and most must be captured. However, they may be purchased from dealers who sell feral and exotic animals. Voles, white-footed mice, and cotton rats may also be trapped.

D. Management and Husbandry

1. Care

In many respects, the management and husbandry of gerbils, white-footed mice, meadow voles, multimammate rats, cotton rats, and white-tailed rats are similar to common laboratory rodents. Care of chinchillas, squirrels, and woodchucks is somewhat unique. Gerbils, white-footed mice, voles, multimammate rats, cotton rats, and white-tailed rats adapt well to life in captivity and usually present no unusual husbandry problems. As a result, little definitive published information regarding management and husbandry is available for these species. In general, environmental, caging, and bedding requirements acceptable for common rats, mice, and hamsters should be satisfactory for other murine and cricetine rodents. Only known exceptions will be discussed.

Gerbils seem to prefer solid bottom cages and contact bedding rather than cages with wire mesh bottoms. This observation is confirmed by the overall improved well-being of gerbils maintained in solid bottom cages. In addition, gerbils need cages that provide a minimum of 6 in. between the top level of bedding and the cage top. This is necessary because of the gerbils tendency to sit erect on their rear feet and legs. If gerbils are caged together before puberty, there is usually no fighting, and groups can be assembled and maintained uneventfully. However, almost invariably, vicious fighting occurs when adults that have been isolated previously as groups or individuals are caged together. If bedded on pine sawdust or shavings, gerbil fur may become greasy and matted.

Most multimammate rats keep their bodies well groomed and their cages remarkably clean. They will attempt to dispose of wastes by pushing it through any available hole in the cage.

Ground squirrels need a nest box in which they may sleep or use for a retreat when frightened. Groups can be caged together with a minimum of fighting.

Chinchillas seem to do well in a variety of environments and can tolerate cold but are very sensitive to heat stress. Adult chinchillas, especially females, are very aggressive and may have to be caged individually, particularly if the enclosure is small. However, they are commonly maintained in pairs. One male may be caged with several females provided they are in a large enclosure and several nest boxes are available. Chinchillas can be housed satisfactorily in standard rabbit cages. They are rather unique in that they are fanatical in their grooming habits. This behavioral feature necessitates a silver sand mixture in which the chinchilla dusts itself or "bathes" daily. Consequently, chinchillas are usually provided with a box containing a mixture of silver sand and Fuller's earth for a short period every day.

Adult woodchucks can be housed in standard metal cages designed for cats, dogs, or rabbits if several precautions are taken. Doors must be secured, and food containers, water bowls, and cage floors must be attached in such a way that the woodchucks cannot move them. Raised flooring of metal wire or slats is preferable. Use of bedding material is not recommended, since the burrowing instinct of woodchucks will prompt them to quickly scratch the bedding from the cage. Woodchucks have been successfully maintained in outside enclosures such as chain-link fenced dog kennels with concrete floors. When kept outside, nest boxes and hay for nesting material must be provided (Young and Sims, 1979).

Adult male woodchucks must be housed individually, but adult females usually can be housed in large groups without danger of fighting. However, groups of more than two are sometimes unsatisfactory because of food monopolization by dominant animals. Newly captured weanling pups may be housed either in groups or in standard rodent cages. Young males and females can be kept together through their first winter, although they must be separated immediately after the end of hibernation. For studies with woodchucks in which normal physiological responses are required, it is necessary to simulate seasonal changes in day length. Therefore, the photoperiod of lighting must be changed at weekly intervals to correspond with length of natural daylight.

2. Physical Restraint

Some rodents have inherent physiological processes that are defensive adaptations useful in escaping predators. These processes must be considered in physical restraint. Examples of animals with these characteristics include gerbils, white-tailed rats, chinchillas, and spiny rats. When supported by the shaft of the tail, the skin of the tail of gerbils and white-tailed rats may slip off. With the spiny rat, the tail breaks off easily when attempting to restrain or immobilize by this appendage. In the phenomenon called "fur-slip" in chinchillas, animals release some of their fur out of the fur follicles when attacked or frightened during handling. With fur-slip, considerable fur tufts and fur fibers are lost. This condition occurs during times of stress in response to the effect of adrenaline on the arrector pili muscles. It takes about 5 months for the lost fur to be replaced and grow to the normal length.

Aspects of handling for some of these animals are like those of comparable laboratory animals. Some species such as ger-

bils, chinchillas, and white-tailed rats are docile and rather easily handled. Others such as voles, squirrels, and woodchucks are very active, quick, excitable, agile, and rarely become tame. Some are persistent in their attempts to escape and therefore require more care and skill in handling and more secure caging with locking mechanisms. For some animals, such as *Peromyscus*, it is desirable to handle the animals, including cage transfers, on a counter top or table with all sides enclosed to a height of about 30 cm. Escaped animals can be easily retrieved in such an enclosure.

Gerbils are very tame and rarely bite unless mishandled. When excited they will jump, dart about, and resist being caught. The best way to handle a gerbil is to pick it up by the base of the tail and then support the body with the other hand. Gerbils dislike being placed on their back. White-tailed rats should be handled similar to a laboratory rat, i.e., grasp the animal around the thorax. Since the tail is fragile and covered with thin skin, they should not be handled by the tail. Multimammate rats tend to be aggressive to persons, so it may be advisable to wear cotton gloves when handling them. They have a tendency to bite but will become tame with frequent handling so that they can be lifted without using a glove. However, they will quickly revert back to an aggressive behavior when not handled regularly. Chinchillas should be lifted by the base of the tail and the body supported by the other hand. Woodchucks can be physically restrained by placing pressure with a gloved hand on the back and neck forcing the head to the cage floor. At the same time, the tail is grasped with the other hand. The woodchuck can then be lifted and carried by the tail. After being handled several times, a woodchuck may become accustomed to handling and can be carried with one hand grasping the tail and the gloved hand supporting the thorax against the handler's body.

The technique of using dim red light to facilitate handling, cage transfer, and laboratory adaptation has been useful for reducing fear and disturbance in some nocturnal, hard to handle species. Nocturnal animals have retinas composed predominantly of rods that are relatively insensitive to the red end of the color spectrum; thus the red light is not perceived by them. Things are clearly visible to personnel, since human retinas are comprised of both rods and cones.

3. Chemical Restraint and Anesthesia

For chemical immobilization and anesthesia of these rodents, general guidelines used for common laboratory rodents should be followed (see Chapter 18). Agents such as ketamine hydrochloride, sodium pentobarbital, and methoxyflurane are commonly used. According to the literature, specific drugs and dosages that have been successfully used are shown in Table

Table II

Drugs and Dosages Used for Chemical Immobilization and Anesthesia of Some Rodents

Animal	Drug	Route	Dose/kg body weight
Gerbil	Ketamine	im	44 mg
	Sodium pentobarbital	ip	50–60 mg
Chinchilla	Ketamine	im	20–40 mg
	Sodium pentobarbital	ip	30–40 mg
Ground Squirrel	Sodium pentobarbital	ip	40 mg
White-tailed rat	Sodium pentobarbital	ip	40 mg
Woodchuck	Ketamine	im	20 mg
	Droperidol and fentanyl	im	0.3–0.4 ml
Obese	Sodium pentobarbital	ip	35 mg
Thin	Sodium pentobarbital	ip	50 mg

II. Sodium pentobarbital is reported to be comparatively short acting in chinchillas.

II. BIOLOGY

A. Anatomy and Physiology

A characteristic feature of rodents is the four prominent yellow or orange incisor teeth. These teeth are essential for the gnawing habits of these animals. In fact the name "rodent" is derived from the Latin verb *rodere* meaning to gnaw. The incisor teeth continue to grow during the entire lifetime of the animal, which sometimes contributes to a clinical problem when malocclusion or inadequate friction permits the teeth to overgrow. Normally, as incisors wear, the hard enamel front edge on their surface forms a sharp chisel-like edge.

Rodents have no second incisors, canines, or front premolar teeth. After the incisor teeth there is a distinct gap, the diastema, between the incisors and the cheek teeth. The dental formula is reduced, never exceeding (I 1/1 C 0/0 PM 2/1 M 3/3) = 22. In some rodents the molars are rooted and cease growing in mature animals, e.g., gerbils, white-footed mice, squirrels, cotton rats, white-tailed rats, and woodchucks. In other rodents, predominantly hystricomorphs such as chinchillas and guinea pigs, the molars grow throughout the life of the animal.

In murine and cricetine rodents, the stomach is divided into two morphologic zones. The cardiac portion is lined with stratified squamous epithelium, and the pyloric portion is wholly or partly glandular. In some Old World rodents (e.g., *Mesocricetus, Mystromys*) this differentiation is so definitive that the stomach is divided into two distinct compartments. How-

ever, in most New World rodents, the demarcation is less distinct. In *P. leucopus* the separation between cardiac and pyloric portions is less obvious but is grossly visible. The glandular area, or fundus, is a small thick portion of the pyloric wall along the greater curvature of the stomach. Stratified squamous epithelium is infiltrated through much of the glandular mucosa. It has been suggested that cornification of a large portion of the lining of the stomach of these rodents protects the stomach from abrasion by rough food.

Gerbils have several unique anatomical and physiological features. There is a complex gland in the midventral line of the abdomen composed of enlarged sebaceous glands that are associated with hair follicles. In females the gland remains small, but in males it enlarges at puberty, becomes orange in color, and produces an oily, musky scented secretion. This gland is under the influence of sex hormones. The gerbil's adrenal weight per unit of body weight is four times that of the laboratory rat. Gerbils have a great capacity for temperature regulation, a wide range of thermal neutrality, and a high degree of tolerance for heat. They have a relatively short erythrocyte life span as indicated by the continued presence of large numbers of reticulocytes and basophilic erythrocytes in the peripheral blood of nongrowing adult gerbils. Apparently the half-life of erythrocytes is approximately 10 days.

White-tailed rats have several distinguishing anatomic characteristics. Perhaps most striking of these is the stomach, which is subdivided into two grossly and histologically distinct compartments. The forestomach is separated from the glandular stomach by an annular muscular plica. The mucosa of the forestomach is studded with tall keratin papillae, 1.0–1.5 mm high. There are two pairs of inguinal mammae to which the young attach soon after birth and tenaciously remain so until about the third week of life. There is a large sebaceous gland that lies beneath the skin posterior to the mammae and extends to the area of the vulva. This is comparable to the ventral gland described in some other cricetine rodents. Males have an os penis.

The multimammate rat resembles the Norway rat by the absence of a gallbladder. Females of this species have a well-developed prostate gland. Another unique feature is the large number of mammae (8–12 pairs) that are continuously distributed from the pectoral to the inguinal region.

The reproductive system of the female chinchilla is unique in that the vagina has a separate external opening between the rectum and bladder openings. The vagina is closed except during the few days of estrus and at parturition.

Woodchucks have three white nipple-like scent glands located just inside the anus. When the animal is excited, the glands are everted and emit a distinct musky odor. The secretion is not sprayed as in the skunk. No visible liquid is secreted, but the odor can be easily detected. The weight of a woodchuck can vary considerably between 2 and 7 kg. This weight change is seasonal and associated with active and dormant periods.

Woodchucks and some ground squirrels hibernate. White-tailed rats may hibernate during dry periods in their native habitat. Animals that normally hibernate or estivate in their native habitat may not do so when maintained in an environment with constant temperature, light periods, and food supply. However, with these conditions, they may have dormant periods lasting for several days.

In the laboratory, woodchucks will become torpid when deprived of food, placed in darkness, and kept at low temperatures (6°–8°C). However, most woodchucks will hibernate in the presence of light and food. With proper environmental conditions, torpidity can be induced in almost any month of the year. Minor disturbances will not prevent but will delay hibernation and will not arouse hibernating woodchucks. In the laboratory, initiation of torpor requires several weeks. For a variety of reasons including time of year, lack of body fat, and lack of feed, a small percentage of woodchucks die rather than become torpid. During hibernation, periods of torpor are short at first (4–5 days) but become longer (10–12 days). The intermittent arousal periods last about 24 hr, during which time the woodchuck warms up, urinates, and is alert. In nature hibernation lasts 3–4 months. In the laboratory similar times will occur, but some will hibernate for up to 8 months. The stimulus for final arousal is unknown.

Several commonly used antibiotics, although relatively innocuous in many species, produce toxicity in some rodents. Perhaps the best known example involves penicillin-associated mortality in the guinea pig (See Chapters 5 and 6). High mortality in white-tailed rats was thought to be associated with topical application of an ointment containing polymyxin B sulfate, bacitracin, and neomycin sulfate for 2 weeks. It was postulated that antibiotics were ingested when animals groomed themselves after treatment. Death occurred within 14 days of the beginning of treatment. Lesions were similar to those reported in guinea pigs given toxic doses of antibiotics. Acute toxicity to dihydrostreptomycin has been reported in gerbils. Injection of 50 mg of the antibiotic resulted in 80–100% mortality in 55–65 gm gerbils within 1–7 min postinjection. However, this dose is approximately ten times than recommended for clinical use in most species. Chinchillas are said to be sensitive to treatment with some antibiotics, and streptomycin is reported to be toxic in woodchucks. Based on this knowledge in a few rodent species, caution should be exercised with antibiotic treatment.

B. Normal Values

Compared to common laboratory rodents, relatively little normative data has been recorded regarding the lesser used rodent species. Since gerbils have been more widely used, nor-

mative data for this species is more complete. A partial listing of known physiological and reproductive data is given in Tables III and IV.

C. Nutrition

Most of the rodents discussed in this chapter can be maintained satisfactorily on standard commercially available laboratory rodent diets. Some authors recommend supplements of grain and wheat germ. Chinchillas can be fed guinea pig feed or a commercially available chinchilla feed. In addition, they eat relatively large quantities of hay, and some suggest an occasional supplement of fruit or vegetables. The meadow vole is herbivorous and thrives on pelleted alfalfa and salt cubes. A 30 gm vole will consume 7–10 gm of alfalfa pellets per day.

High-fat diets are not recommended for long-term feeding of gerbils. They have a unique fat metabolism and develop high blood serum cholesterol levels even on a diet with a 4% fat level. On a high-fat diet they tend to store fat and become obese. In females fat may accumulate around the ovaries and interfere with productivity.

The exact nutritional requirements of woodchucks are unknown. Some zoos have fed woodchucks fresh vegetables, apples, grains, and peanut and soybean meal. Reportedly, commercial rodent diet is not optimum for long-term maintenance of laboratory held woodchucks. They have been successfully maintained on a commercially available rabbit feed that was prepared in blocks (Young and Sims, 1979). During the summer when woodchucks are most active, they may consume 0.2–0.3 kg of feed at a single feeding and also drink large quantities of water.

Coprophagy has been demonstrated as a normal and important nutritional or behavioral trait in all rodents that have been specifically studied. This practice probably serves two functions: maintenance of a normal gut flora by reinoculation of the upper end of the gastrointestinal tract and capturing of nutrients (especially B vitamins and vitamin K) usually synthesized by microorganisms in the cecum.

Most rodents including chinchillas can be watered by means of a bottle and sipper tube. Woodchucks are watered in bowls securely mounted to the cages. Some rodents have water-conserving renal mechanisms and consequently have small requirements for exogenous water. Adult gerbils and white-tailed rats drink about 5 ml of water per day. Chinchillas also drink very little water.

D. Reproduction

The reproductive characteristics of rodents differ considerably between groups and sometimes between closely related species. For some species, being maintained in captivity alters

Table III

Selected Normative Data[a]

	Chinchilla	Gray squirrel	Cotton rat	White-tailed rat	Woodchuck	Gerbil	White-footed mouse (*P. leucopus*)	Deer mouse (*P. maniculatus*)	Meadow vole	Multimammate rat
Adult weight[b]										
Male	400–500	400–700	70–200	85	3–7 kg	60	22	19–21	40–60	40–45
Female	400–600	400–700	70–200	130	3–7 kg	50–55	22	19–21	40–60	40–45
Life span										
Usual	10Y	14–15Y	23 M	2.4Y	5–6Y	3Y	2–3Y		33 W	2–3 Y
Maximum	20Y	20Y	3 Y	6.2Y	15Y	5Y	38 M		124 W	38 M
Water consumption	—	—	—	5 ml	—	4–7 ml/100 gm	—	—	—	—
Food consumption	—	50 gm	—	—	—	5–8 g/100 gm	—	—	7–10 gm	6 gm
Chromosome No.	64	40	—	32	—	44	—	48	46	36
Body temperature (°C)	36.1–37.8	—	—	—	—	38.2	—	—	—	—
Heart rate (beats/min)	100	390[c]	—	—	180–264[c]	360	—	534	600[c]	—
Puberty										
Male	8–18 M	9–11 M	30–50 D	4–7 M	1–2 Y	9–12 W	42–48 D	35–37 D	42 D	55–75 D
Female	8–18 M	9–11 M	30–50 D	4–5 M	1–2 Y	9–12 W	42–48 D	35–37 D	25 D	55–75 D
Breeding season	Nov.–May	Jan.–Aug.	YR	YR	Mar.–Apr.	YR	YR	YR	YR	YR
Estrous cycle										
Usual	40 D	Seasonal	9 D		Seasonal	4–6 D	4–5 D	4–5 D	5 D	—
Range	16–69 D	—	4–20 D	4–9 D		—	—	—	—	6–8 D
Postpartum estrus	Yes	No	Yes	Yes	No	Yes	Yes	Yes	Yes	Yes
Litters/year	2	2	S	S	1	S	S	S	S	S
Gestation										
Usual	111 D	44 D	27 D	38 D	32 D	24–26 D	22–24 D	23 D	21 D	23 D
Range	105–118 D	—	—	36–39 D	31–42 D	—	—	22–26 D	19–22 D	—
Litter size	D									
Mean	2	4	5	3	4.5	4.5	3.4	4.8	4	8
Range	1–6	1–6	2–10	1–5	1–8	1–12	2–7	1–11	1–11	5–20
Birth weight[b]	35	12.0	7.0	5.0–7.8	45	2.5–3.0	1.5–2.4	1.3–2.2	2.0	2.0–3.0
Eyes open	At birth	4–5 W	1 D	16–20 D	4 W	16–20 D	12–15 D	12–16 D	8 D	13–17 D
Wean	6–8 W	7–10 W	21 D	25 D	35 D	21 D	21–28 D	21–28 D	14 D	19–21 D

[a]Abbreviations used: Y, year; M, month; W, week; D, day; YR, year round; S, several.
[b]Grams unless otherwise noted.
[c]Anesthetized.

Table IV

Selected Hematological Data[a]

	Gerbil	Chinchilla	Multimammate rat	Meadow vole
Leukocyte counts				
Total (× 10³/μl)	11 (4.3–21.6)	7.8 (4.0–25.0)	7.5 (2.8–13.0)	4.3 (2.4–6.2)
Neutrophils (%)	19 (3–41)	43 (9–78)	20 (8–48)	13 (5–23)
Lymphocytes (%)	78 (32–97)	54 (19–98)	75 (48–93)	78 (64–92)
Monocytes (%)	3 (0–9)	1.0 (0–6)	0 (0–1)	5 (0–10)
Eosinophils (%)	1 (0–4)	0.7 (0–9)	1.5 (0–9)	3 (0–7)
Basophils (%)	0.6 (0–2)	0.6 (0–11)	0 (0–1)	0.2 (0–1)
Platelets (× 10³/μl)	638	274 (45–740)	329 (208–754)	—
Packed cell volume (%)	48 (41–52)	38 (27–54)	40	47 (42–52)
Red blood cells (× 10⁶/mm³)	8.5 (7.0–10.0)	6.9 (5.2–10.3)	7.5	13.3 (10.7–16.0)
Hemoglobin (gm/dl)	15 (12.1–16.9)	11.7 (8.0–15.4)	13.0	14.9 (13.5–16.3)

[a]Values in parentheses are normal range.

their biology and instincts, and some may even fail to reproduce. Most murine and cricetine rodents will produce several litters per year. Chinchillas and gray squirrels produce only two litters per year, and woodchucks have only one litter annually. Several of the rodents discussed experience estrus shortly after parturition. This feature along with a high conception rate at the postpartum estrus accounts for the prolific productive characteristics of some species. In general, myomorph and sciuromorph rodents are born naked and helpless and must be cared for in a nest prepared by the parents. In contrast, hystricomorph rodents are born precocious and require much less care by the mother. Sexual maturity may range from as early as 3 weeks in some voles to 1 to 2 years in woodchucks.

The sexing of most rodents, including woodchucks, is essentially the same. In adult males the scrotum may be sufficiently prominent so they can be differentiated from females. A common method that can be used on all animals, regardless of age, is the comparison of the size of the genital papilla and the length of the perineum. The length of the perineum in the male is usually about twice that of the female, and the genital papilla is usually larger in males.

Gerbils can be produced successfully but are not as prolific as most laboratory rodents. The most successful mating scheme involves monogamous pairs that are never separated. Some gerbils are incompatible, and fighting occurs when attempts are made to pair them. Consequently, careful observation is required when introducing new animals; incompatible animals must be separated to avoid serious injury or death. Breeding pairs can be set up more successfully at or before puberty. Also, less aggression occurs when animals to be paired are placed in a clean, neutral (nonterritorially marked by one partner) cage. Once a compatible pair of gerbils is established, they usually will live in harmony indefinitely and should not be separated. Males assist in caring for the young. Mature animals that must be separated may not accept one another when placed together later. Frequently, an older female that has lost her mate will not accept another male. Polygamous mating groups are less successful and must be set up before animals reach puberty.

Examination of vaginal smears is not as reliable an indicator of the stage of the estrous cycle in gerbils as it is in some other rodents. Female gerbils are particularly prone to develop cystic ovaries, which cause decreased litter sizes and earlier failure to reproduce. The incidence and severity of the condition increase with age.

In comparison with rats and mice, the postnatal mortality in gerbils is high. In one study, approximately 20% of young failed to survive to weaning. The principal causes of death were lack of maternal care, including lactation failure, crushing, and suffocation of young. However, maternal cannibalism appears to be low.

White-footed mice reproduce successfully in captivity. Feral white-footed mice have a peak of reproductive activity during the spring, summer, and early autumn when the days are long. In a laboratory environment much better production occurs when the photoperiod is 15–16 hr of light and 8–9 hr of darkness. Reproductive performance tends to decline after three to five litters or in mice more than 18 months old.

Some voles are reputed to be among the most prolific of mammals. An example is the common vole, *M. arvalis*, a common European mammal. Females usually become sexually mature in 2 to 3 weeks, and some mate and conceive at 13–14 days of age. They have a rapid growth rate, attaining 30 gm by 40 days of age. An average of 4–7 young are born per litter, and a female may have a new litter every 3 weeks. Female *M. pennsylvanicus* are induced ovulators and therefore can be bred at any time. Ovulation occurs 12–18 hr after copulation.

In general, sciuromorphs do not produce well in captivity. Most species of Sciuridae are polyestrous during the spring and summer and anestrous during the winter.

With cotton rats, permanent monogamous matings are usually the best system. Newborn cotton rats are precocious, are well haired, and have their eyes open by 18–36 hr after birth. Young can usually begin eating solid food by 1 week of age and may be weaned as early as 10 days. However, the young are stronger and heavier when weaning is delayed until 21 days. Despite their excitable nature, female cotton rats appear to be exceptionally good mothers. They rarely kill or abandon the young, even when the female or the babies are handled shortly after parturition, and will readily accept foster young.

Multimammate rats have a high reproductive rate. Contributing factors include an early onset of breeding and short intervals between litters. In captivity, litters of 6–8 seem to be average. When disturbed, females may cannibalize their litters.

The best mating system for white-tailed rats has been monogamous pairs with the animals remaining together indefinitely. Breeding pairs of white-tailed rats should be set up when young since animals paired after maturity may fight. The average litter size is 3, and the young remain attached to the nipples until 2 or 3 weeks of age. Like many other species, the adults have a tendency to cannibalize their young when excited or stressed.

Female chinchillas are polyestrous between November and May; ovulation is spontaneous, and estrous cycles vary between 30 and 50 days in length. Females are larger than males, dominant, and usually aggressive toward males and other females. Several breeding system have been used. Monogamous pairing has been practiced by some chinchilla breeders; however, this method requires care and housing for additional males, and pairs are frequently incompatible. A variety of polygamous systems have been used. Since the females are aggressive, safety provisions for the male must be provided. When one male and several females are housed in a single large cage, refuge boxes and nest boxes for females should be provided. A variation of this method uses several smaller inter-connected cages. Another method utilizes specially designed cages with holes in the back which open into a runway. Females are restricted to their individual cage by a collar larger than the cage opening. The male is housed in the runway and is free to enter or exit each female's cage.

Like many other hystricomorphs, newborn chinchillas are fully furred, with eyes open and can soon walk. The kits begin eating solid food within a week after birth, but are usually not weaned until 6–8 weeks of age. Pregnant female chinchillas do not make any form of nest. The young are born directly onto the cage floor. Therefore, solid floors minimize the risks of accidental injury to the young.

Woodchucks can be bred and raised in captivity with moderate success. They breed only for a short period in the spring. Male woodchucks should be placed in the cages of females in the fall or shortly before the females arouse from hibernation in the spring. Males should remain with the females until the females have been out of hibernation for 2–3 weeks. To be successful in breeding woodchucks in captivity, environmental conditions in the laboratory should simulate those occurring in nature. The hibernation period and emergence should be synchronized so that males and females will emerge from hibernation at the appropriate times. Most litters are born in March. Males should be removed before females give birth. Females born or raised in the laboratory seldom raise litters successfully, whereas captured pregnant females will usually raise their litters under laboratory conditions. Female woodchucks should be housed in large cages and provided with nest boxes prior to parturition and disturbed as little as possible for the first 2 weeks after the pups are born. After 2 weeks, the females or the pups may be handled successfully.

E. Behavior

Gerbils are docile and usually show very little aggressive behavior, but under certain conditions they will fight. The mixing of groups of adults usually results in vicious fighting with injuries and deaths and should be avoided. Large groups can be assembled safely at weaning and satisfactorily maintained up to and beyond sexual maturity.

Within a family group, meadow voles are compatible but are aggressive when placed with unfamiliar animals. In general, they are not as docile as common laboratory rodents; however, most will adapt to laboratory confinement and tolerate handling.

In the laboratory, the behavior of the multimammate rat is similar to its feral nature, being nervous, quick to escape, and conscious of distractions perceived as danger. Consequently, they are somewhat difficult to handle and may bite without provocation.

Groups of woodchucks can be maintained together, usually with minimal aggression, in large enclosures. When caged in this manner, several nest boxes should be available. Adult male woodchucks must be caged individually. Deer mice and multimammate rats are communal and can be caged in groups. In general chinchillas, gray squirrels, some ground squirrels, and white-tailed rats are solitary animals and should be caged individually.

Gerbils are diurnal with brief periods of intense activity alternating with brief periods of rest or deep sleep. Gray squirrels, ground squirrels, and woodchucks are also diurnal. Deer mice, voles, multimammate rats, chinchillas, and white-tailed rats are nocturnal. However, some captive species tend to become more diurnal.

III. DISEASES

For the rodents discussed in this chapter, with the exception of gerbils, published information regarding biology and care in the laboratory is rather limited. Although several of these rodent species have been used for a number of years as research animals, even less is known regarding incidence of spontaneous disease. Several species are reported to be relatively disease-free. It has been postulated that this situation exists because of the recent advent of use of these animals in the laboratory. With the passage of time, more diseases may be encountered as the opportunity of exposure to disease agents continues. This supposition seems to be supported by the experience with gerbils. As the use of gerbils continues and increases, more diseases and parasites are reported. Nevertheless, when compared with more common laboratory species, these rodents present fewer disease problems.

A. Diseases of More Than One Species

1. Leptospirosis

Leptospirae are primarily saprophytic aquatic organisms that are found in streams, lakes, and sewage. Feral rodents are one of the principal reservoirs of leptospirosis. Woodchucks, cotton rats, deer mice, and meadow voles have been identified as hosts for leptospires. Little is known about clinical disease in these animals since infections are usually latent.

2. Dermatomycosis

Rodents are common hosts for dermatophytes. Frequently, animals are infected but have no lesions. *Trichophyton mentagrophytes* has been recovered from cotton rats, squirrels, and chinchillas. *Microsporum gypseum* has been isolated from cotton rats and gray squirrels. Skin lesions, commonly seen around the nose and on the face, include alopecia and scaly or encrusted patches. Specific diagnosis can be made by culture and identification of the fungus. Dermatomycosis can be successfully treated with orally administered griseofulvin. Doses of 15–60 mg/day have been used in chinchillas. Treatment should continue until the lesions are healed.

3. Adiaspiromycosis

Fungi of the genus *Emmonsia* cause adiaspiromycosis in meadow voles, ground squirrels, and deer mice (*P. maniculatus*). The disease is seen in feral animals or wild-caught captive-held animals and is usually a benign, self-limiting lung infection. In the natural disease, grayish nodules can be seen in the lungs during postmortem examination. Microscopically, the early lesions appear as incipient granulomas. At this stage of the infection, the small adiaspores are completely surrounded by inflammatory tissue composing these granulomas. During further growth of the adiaspore within the alveoli, the surrounding inflammatory tissue becomes more compressed or compact.

4. Toxoplasmosis

The protozoan parasite, *Toxoplasma gondii*, is found rather commonly in many domestic and wild animals including rodents. Animals harboring *Toxoplasma* organisms are frequently asymptomatic. However, some animals have acute, subacute, and chronic stages of the disease, and the clinical signs are nonspecific. The neurologic signs that can accompany acute infections are indistinguishable from those of clinical rabies. Affected animals may behave in an abnormal manner and be aggressive, attempting to bite persons. Gross lesions vary according to the severity of the disease and the organs involved, but may include necrosis, hemorrhage, edema, and ascites. Microscopically, cysts are common in the brain (Fig. 2). Toxoplasmosis has been reported in numerous rodent species, including meadow voles, gray squirrels, chinchillas, cotton rats, *Peromyscus* spp., woodchucks, and ground squirrels.

5. Eperythrozoonosis

Eperythrozoonosis is a noncontagious blood protozoal infection of rodents. It is generally asymptomatic and characterized by the occurrence of coccus, ring- and rod-shaped forms adhering to the red blood cells and free in the plasma. *Eperythrozoon dispar* has been reported from meadow voles and *E. varians* in *P. maniculatus*.

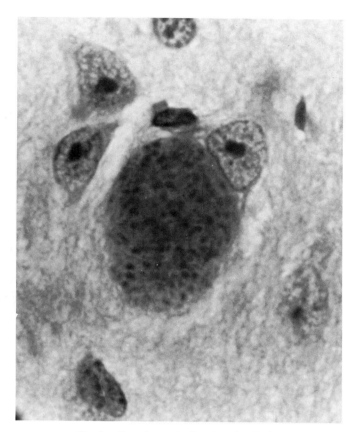

Fig. 2. Toxoplasmosis. Cyst in rodent brain.

6. Nematodes

Cerebral nematodiasis has been reported on numerous occasions in woodchucks and occasionally in ground squirrels. The most common causes of this condition are the migrating larvae of the skunk ascarid, *Baylisascaris columnaris*, and the racoon ascarid, *B. procyonis*. Dog and cat ascarids may occasionally be involved in the etiology. Rodents ingest the infective eggs of these incidental parasites. The larvae then migrate through the tissues. Only a few larvae are needed to cause clinical disease as they migrate through the brain, and the resulting neurological disorder may mimic or be confused with other conditions such as rabies and toxoplasmosis. Affected animals have a history of abnormal behavior; are ataxic, incoordinated, anorexic, aggressive; and have posterior paresis.

Gross lesions often observed at necropsy include white spots, 0.5–1.0 mm in diameter, on the serosa and in the wall of the stomach, small intestine, large intestine, lungs, liver, kidney, heart, and brain. Histologically, parasitic granulomas are observed in affected organs (Fig. 3). Brain sections have eosinophilic perivascular cuffing and focal areas of malacia (Fig. 4).

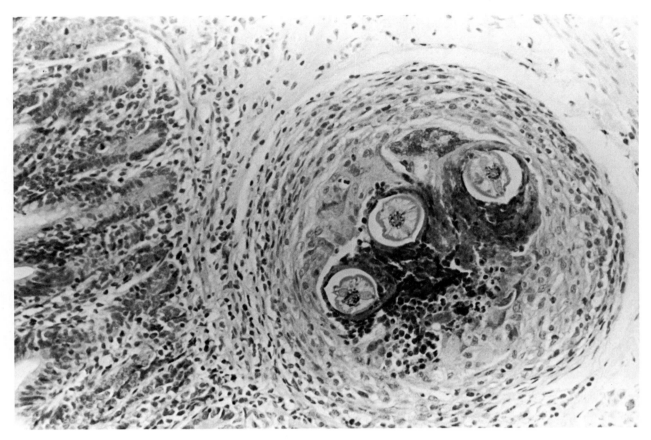

Fig. 3. Cerebral nematodiasis. Circumscribed granuloma containing cross sections of an ascarid larva in the small intestine of woodchuck.

7. Cestodes

Intermediate stages of *Taenia crassiceps* occur in subcutaneous tissue of meadow voles, woodchucks, deer mice, and gray squirrels. Embryonated tapeworm eggs from the definitive host are consumed by these rodent intermediate hosts. Cysticerci develop in subcutaneous locations, most frequently in or adjacent to the neck or axillary region. Swelling may be observed in these areas or in the pleural or abdominal cavities. This condition would be most likely to occur in wild-trapped animals.

Larval stages of *Echinococcus multilocularis* have been found in the livers of *P. maniculatus* and *M. pennsylvanicus.* Individual cysts may range up to 8 mm in diameter. With infections of long duration, the abdomen may become greatly distended and movement is impaired.

8. Malocclusion

In all species of rodents, the incisor teeth and, in some species, the cheek teeth continue to grow during the entire lifetime of the animal. This feature sometimes contributes to a clinical problem when malocclusion or inadequate friction permits the teeth to overgrow. It is characterized by an overgrowth and angularity of incisor or molar teeth. There may be several different causes of malocclusion, including genetic origin, nutritional imbalance, lack of dietary roughage, broken or missing opposing teeth, or deviation of the jaw.

Malocclusion has been reported in several rodent species but can probably occur in all members of the order. Since prehension becomes difficult, affected animals lose weight, eat less, waste feed, become depressed, and emaciated. Overgrown and misaligned teeth may lacerate the tongue or buccal cavity. A constant sign in the latter stages of the condition is excessive drooling of saliva. This causes wet, matted fur around the mouth, chin, chest, neck, and forelegs. Eventually, secondary bacterial infection may occur, and there is hair loss. Unless the condition is corrected, death usually occurs because of starvation or secondary complications. Findings at postmortem examination of animals that die from this condition are emaciation, dehydration, absence of body fat, and overgrown teeth.

To correct the situation, the overgrown teeth must be clipped back to normal length using an instrument such as nail clippers, bone forceps, or side cutting pliers. For examination and

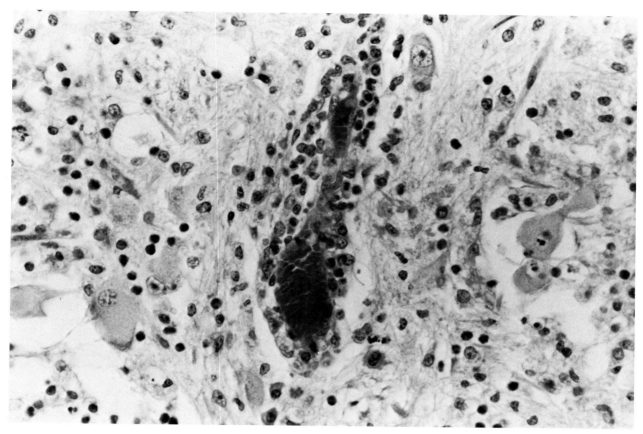

Fig. 4. Cerebral nematodiasis. Perivascular cuffing, malacia, and eosinophilic cellular infiltration in brain of woodchuck.

clipping, the use of a mouth speculum will protect the animal and the handler.

B. Gerbils

1. Tyzzer's Disease

The most commonly reported disease causing significant morbidity and mortality in gerbil colonies is Tyzzer's disease. It is always associated with *Bacillus piliformis*, and, therefore, this bacteria is thought to be the cause. The disease is characterized by sudden onset and sudden deaths. Animals with clinical signs may or may not have diarrhea; are lethargic, listless, and anorexic; huddle in cage corners; and have rough hair coats. Morbidity and mortality may be high in young animals; but adults, if affected at all, have less morbidity and mortality. Gross lesions are observed primarily in the liver. It is enlarged, yellow, and friable; and numerous white to translucent, round, circumscribed areas are scattered over the external and cut surfaces. Microscopically, bile ducts and parenchyma are destroyed by large foci of necrosis filled with nuclear debris,

necrotic hepatocytes, and neutrophils (Fig. 5). Gram-stained tissue sections have pleomorphic bacteria at the periphery of the lesions.

2. Salmonellosis

An outbreak of salmonellosis in a gerbil colony has been reported. The causative organism was *Salmonella enteritidis*. Clinical signs of illness included moderate to severe diarrhea, rough coat, weight loss, listlessness, dehydration, and high mortality. Some deaths were sudden without premonitory signs of illness.

At necropsy, gross lesions included fecal staining around the anus, gastrointestinal tract distension with gas and fluid ingesta, congested liver, and fibrinosuppurative exudation on the peritoneum, with abscesses. Microscopic changes in the liver included multiple foci of inflammation that varied from small accumulations of inflammatory cells to much larger granulomas with caseous necrotic centers. In the granulomas, the central necrotic debris was often partially calcified, and the necrotic tissue was surrounded by epitheloid cells, lympho-

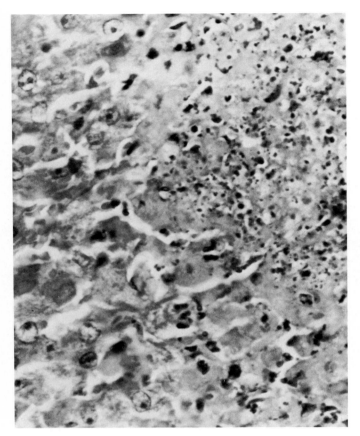

Fig. 5. Focal area of necrosis in liver of rodent with Tyzzer's disease.

Fig. 6. Dermatitis around nares of gerbil.

Lesions begin at the external nares (Fig. 6) and spread to the remainder of the face, the legs and feet, and ventral surface of the body. Chronic dermatitis may persist for weeks. Some affected gerbils may stop eating, lose weight, and die.

This condition responds poorly to treatment. Antibiotics such as chloramphenicol (0.083 gm/100 ml) or tetracycline hydrochloride (0.3 gm/100 ml) in the drinking water and topically applied ointment containing antibiotics have been suggested. Housing on sand allowing animals to sandbathe off excess Harderian gland secretion reportedly leads to partial recovery.

4. Helminths

Several nematodes have been reported from gerbils, but no clinical manifestations of disease have been associated with them. *Dentostomella translucida,* an oxyurid, occurs in the anterior one-third of the small intestine. The prepatent period is between 25 and 29 days. Examination of feces by egg flotation is not a reliable method of diagnosis for this parasite. *Syphacia obvelata,* the mouse pinworm, is readily transmissible between mice and young gerbils. The prepatent period can be as short as 11 days. Older gerbils tend to rid themselves of this parasite. *Syphacia muris,* the rat pinworm, also can be transmitted easily to gerbils.

The tapeworm, *Hymenolepis nana,* occurs in the small intestine. Clinical signs in heavy infections include dehydration and diarrhea. At necropsy, the small intestine may be distended with mucoid material and contain tapeworms. Microscopically, larval forms (cysticercoids) causing extensive inflammation, and degeneration of the surrounding tissue are found in the intestinal mucosa.

5. Ectoparasites

Demodex mites have been identified in skin scrapings from a gerbil with alopecia, scaliness, hyperemia, and multiple focal ulcerations on the tail and rear legs.

cytes, a few neutrophils, and a marginal rim of fibrosis. There was little inflammatory response in the intestine, but crypt abscesses were occasionally observed.

Treatment of animals with salmonellosis is usually unrewarding, and survivors frequently become carriers. Therefore, treatment is not recommended.

3. Dermatitis

One of the most common clinical problems seen in gerbils is dermatitis, usually on the face. Suggested causes include *Staphylococcus aureus, Demodex* mites, and type of caging or bedding. A recent report indicates that either the hypersecretion of the Harderian gland or a failure of gerbils to spread the material by grooming activities may be associated with the dermatitis. Surgical removal of the gland in affected animals stopped secretion and complete recovery occurred (Thiessen and Pendergrass, 1982). Young mature gerbils may be affected more commonly. Lesions progress from an initial erythema to localized alopecia, and finally to an extensive moist dermatitis.

6. Amyloidosis

In gerbils, amyloidosis in an uncommon, spontaneously occurring disease that has been seen in association with an experimental infection with a filarid, *Litomosoides carinii*. Clinical signs included weight loss, lethargy, dehydration, inappetence, and death within 1–2 weeks. Amyloid deposits were most marked in spleen, but also occurred in kidney, liver, heart, pancreas, and intestines.

7. Cystic Ovaries

Gerbils are particularly prone to develop cystic ovaries. The lesions are larger and more prevalent in older females. Animals with cystic ovaries have reduced litter sizes and reproduction ceases prematurely.

8. Neoplasia

The incidence of spontaneous neoplasia in gerbils over 2 years of age is rather high. In decreasing order of frequency, the organs most commonly affected and common types of neoplasia are ovary (granulosa cell tumors, theca cell tumor, leiomyoma), skin [squamous cell carcinoma, melanoma (Fig. 7), sebaceous gland pad adenoma, and carcinoma], kidney (adenoma and hemangioma), adrenal gland (cortical adenoma or carcinoma), cecum (adenocarcinoma), liver (hepatoma, bile duct adenoma, cholangio-carcinoma), uterus (carcinoma, hemangiopericytoma, leiomyoma), pancreas (islet cell adenoma), and testis (teratoma, seminoma).

C. Chinchillas

1. Gastrointestinal Disturbances

The most common clinical condition occurring in chinchillas is enteritis or gastroenteritis. This is not a specific disease but probably represents illness caused by a number of agents all

Fig. 7. Malignant melanoma on skin of gerbil.

causing common clinical signs. Bacterial, viral, or protozoal pathogens have been incriminated; but frequently, the cause is not determined. Specific agents that have been suspected of causing the problem include *Pseudomonas aeruginosa, Salmonella typhimurium, Giardia* sp., and a number of species of coccidia.

Clinical signs of gastroenteritis include sudden death without premonitory signs and high mortality. Chinchillas that have signs of illness may or may not have diarrhea, anorexia, partial paralysis, squeaking noise during expiration, and evidence of pain on palpation of the abdomen. Gross lesions seen at necropsy are gastritis, enteritis, cecitis, and colitis. The stomach may have dark necrotic areas. The small intestine is distended with gas, inflamed, and contains hemorrhagic exudate. The cecum and colon are hemorrhagic and abscessed. Blood tinged transudates in the abdominal cavity may occur. Intussusceptions are common in acute cases. *Giardia* may be observed in low numbers in the gastrointestinal tract contents or stool of clinically healthy chinchillas. High numbers of trophozoites in the small intestine and many cysts in the large intestine are sometimes seen in cases of enteritis. Often, it is difficult to determine if this protozoa is a primary or secondary invader in these cases. It appears that chinchillas have a tendency to develop intestinal impactions, intessusceptions, and rectal prolapse. It has been suggested that these occur with increased frequency in association with gastroenteritis. Others have speculated that lack of roughage in the diet may predispose them to impaction.

Treatment of gastroenteritis is symptomatic unless the cause can be determined so that specific measures can be used.

2. Pneumonia

Lobar or bronchopneumonia is a frequently observed condition in chinchillas but may be secondary to some other condition. Pneumonia may be caused by a number of pathogenic bacteria including staphylococci, streptococci, *Bordetella bronchiseptica*, or *Escherichia coli*. Bacterial pneumonia may be an acute or chronic condition. There is discharge from the eyes and nose, loss of appetite, and rough hair coat. Death is common. Treatment should include parenteral administration of antibiotics and supportive measures.

3. Listeriosis

Apparently, chinchillas are quite susceptible to infection by *Listeria monocytogenes*. The disease is characterized clinically by sudden death or by inappetence, malaise, depression, and sometimes signs of CNS derangement and abortion in pregnant females. The liver is the most commonly affected organ. Lesions may include petechial hemorrhages, slight inflammatory response, and small grayish-white areas of focal coagulation or caseation necrosis. Similar lesions may also occur in other abdominal organs. Slight to moderate congestion of the small intestine with some catarrhal exudate may be observed. Patchy areas of emphysema sometimes occur in the lungs. Specific diagnosis is confirmed by culture and identification of the causative organism. This disease resembles pseudotuberculosis.

Penicillin or broad-spectrum antibiotics are the treatment of choice. However, success rate associated with therapy is not remarkable.

4. Pseudotuberculosis

Pasteurella pseudotuberculosis can cause an acute or chronic fatal contagious disease of chinchillas. The acute form is a fatal septicemia. Chronic disease is manifested by anorexia, depression, progressive loss of weight, and intermittent diarrhea. It may be possible to palpate enlarged mesenteric lymph nodes. Lesions in the chronic form are similar to those described for listeriosis. At necropsy, yellow or white foci, 1–3 mm in diameter, are usually seen in the liver and occasionally in other organs. Microscopically, there is diffuse lymphocytic infiltration and degeneration of liver cells with focal areas of severe degeneration. The spleen may be congested and contain foci of degeneration. Epithelial cells in the kidney may be degenerated, and lungs may be congested. Treatment is usually ineffectual.

5. Enterotoxemia

Disease caused by *Clostridium perfringens,* type D, has been reported in chinchillas (Moore and Greenlee, 1975). Animals may be found dead without signs of illness being observed. Some animals have diarrhea and appear to be experiencing abdominal pain. The highest incidence occurs in 2- to 4-month-old animals. Enterotoxemia has been prevented by immunization with *C. perfringens,* type D, toxoid. It is recommended that two injections be given intraperitoneally 10–14 days apart, another in 6 months, and annually therafter.

6. *Pseudomonas* Infections

Chinchillas appear to be very susceptible to *P. aeruginosa* infections. Reported clinical syndromes include conjunctivitis, otitis, pneumonia, enteritis, metritis, and septicemia (Dall, 1963; Keagy and Keagy, 1951; Larrivee and Elvehjem, 1954; Newberne, 1953). Diagnosis is based upon isolation and identification of the causative agent.

7. Parasites

Adult tapeworms identified as *Hymenolepis* sp. have been reported in chinchillas. With heavy infections, affected chinchillas may be in poor condition and emaciated. Deaths have also been attributed to this tapeworm. Treatment with niclosamide at a dose of 200 mg/kg has been recommended.

Coenurosis caused by the intermediate stage of the tapeworm *Multiceps serialis* occurs in chinchillas. The cysts are observed as subcutaneous swellings. Depending on the location, they may cause mechanical interference with motor function. Pain and other signs are not associated with the swelling. The definitive hosts for this parasite are various members of the family Canidae. Chinchillas ingest the tapeworm eggs in contaminated feed. The eggs embryonate in the digestive tract, and hexacanth larvae are disseminated by the blood throughout the body where they localize and form coenuri. Lesions of the intermediate stage are characteristic, including a firm mass of encysted fluid with little or no inflammation. Autemortem diagnosis is based on palpation of the cysts. The coenurus has a thin membranous wall which upon being opened contains numerous invaginated scolices attached to the wall. The parasitic cysts can be surgically removed. Other lesions likely to be confused with a coenurus are tumors or abscesses.

Coccidial oocysts may be observed in feces of chinchillas. However, direct association with this parasite and disease is frequently difficult to ascertain.

8. Vices

Fur chewing, although not a disease, is a common vice of chinchillas. Animals bite and eat their own fur or that of cage mates. Animals may confine the depilation to a few spots or any part of the body accessible to the mouth. As a result of this vice, fur balls are commonly found in the stomach of animals at necropsy. The specific cause of the condition is not known; consequently, there are no specific recommendations for treatment or prevention.

D. White-Tailed Rats

Little information is available regarding specific diseases of white-tailed rats. Lesions and/or causes of death observed in one colony include pneumonitis, ulcerative enteritis, septicemia, perforated gastric ulcer, cataracts, keratitis, scleritis, hepatitis, nephritis, and otitis externa and media.

1. Ringtail

Ringtail is recognized as a spontaneous disease of neonatal laboratory rats. It is characterized by annular constrictions that may or may not progress to edema, necrosis, and spontaneous amputation of the tail. It is thought to be associated with the type of caging and low relative humidity.

Spontaneous ringtail also has been reported in white-tailed rats. The condition developed only in suckling young less than 1 week of age. Lesions were observed only on the tail and ranged from slight reddening with minimal annulation and swelling to severe discoloration, annulation, and swelling. In severely affected animals, necrosis and subsequent autoamputation of part or all of the tail occurred.

During the episode it was determined that relative humidity in the room ranged from 12–38%. No cases were seen when relative humidity was maintained at 50%.

2. Neoplasia

Spontaneous neoplasms observed in white-tailed rats include perianal squamous cell carcinoma, adnexal tumor of skin, osteosarcoma of the scapula, leiomyosarcoma of the uterus, adenocarcinoma of the liver, hepatoma, and adenoma of the pituitary gland.

E. Multimammate Rats

1. Joint Disease

Among laboratory rodents, the multimammate rat is one of the most susceptible species to spontaneous osteoarthritis (Snell and Stewart, 1975). After 2 years of age, most multimammate rats are reluctant to use their rear legs, and some are paraplegic. This results from severe degenerative joint disease of the diarthroses and intervertebral discs. Most peripheral joints except the hips, shoulders, and sacroiliacs are involved. The elbows and knees are most severely affected. In the diarthrodial joints, there is extensive erosion of articular cartilage and sclerosis of epiphyseal bone. Aseptic necrosis of the secondary center of ossification of the vertebrae is a common finding. Protrusion of the degenerated disc tissue into the vertebral canal occurs at multiple sites along the vertebral column (Fig. 8), particularly in males.

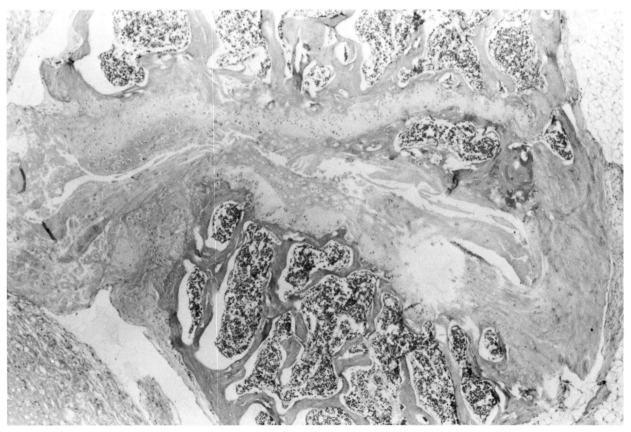

Fig. 8. Degenerative joint disease of vertebral column with degeneration of intervertebral disc in multimammate rat.

2. Kidney Disease

Many multimammate rats develop glomerulonephritis (Snell and Stewart, 1975). It has been suggested that this renal disease is immune mediated. In the membranous form of the disease, 10% or more of the glomeruli may be affected, but the tubules, interstitial tissue, and extraglomerular vascular system are normal. The capillary tufts have hyaline thickening of the glomerular loops and basement membrane of Bowman's capsule. In the proliferative form, in addition to the above changes, there is increased cellularity of glomeruli. A third and chronic form usually occurs in animals over 18 months of age. There may be no gross lesions, or the kidney surface may be granular and pale. Microscopically, all portions of the nephron and interstitial tissue are involved, but the extraglomerular blood vessels are not involved. The capillaries in the glomerular tufts become thickened and the lumen reduced; many contain inflammatory cells; and some glomeruli are destroyed. Tubules first hypertrophy, then atrophy; epithelial cells become flattened and desquamated; and casts are formed. The intersti-

tial tissue becomes thickened, hyperemic, and infiltrated with lymphocytes, plasma cells, and granulocytes (Fig. 9).

3. Neoplasia

Thymic hyperplasia or lymphoepithelial thymoma is common in animals over 2 years old. Other lesions that appear to be associated with this condition are myositis, atrophy of skeletal muscle, and myocarditis. Lymphosarcomas, parathyroid adenomas, prostatic tumors, reticulum cell sarcomas, adenocarcinomas of the glandular stomach, and gastric carcinoids are common. Other neoplasms reported in multimammate rats are hepatomas, granulosa cell tumors of the ovaries, and adenomas of the adrenal gland cortex, pituitary gland, and pancreatic islets.

4. Parasites

Psorergatic mange has been reported in multimammate rats. The skin of affected animals is covered with numerous raised

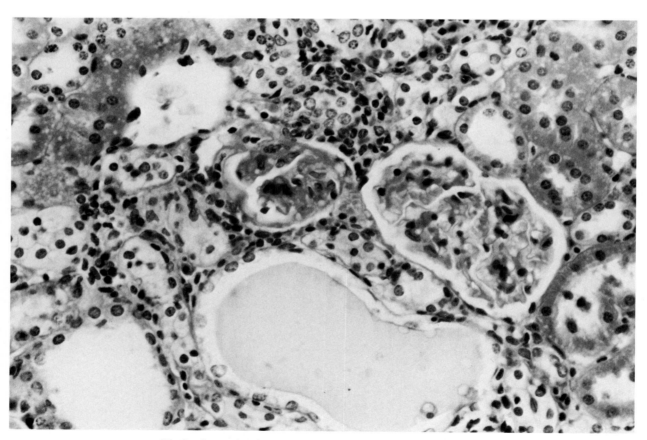

Fig. 9. Severe chronic glomerulonephritis in kidney of multimammate rat.

white crusty nodules. Histologically, these nodules are small cornified pouches filled with mites.

F. Meadow Voles

1. Nephritis

Interstitial nephritis occurs in meadow voles. Kidneys of affected animals have markedly increased amounts of interstitial connective tissue, and hyaline casts are present in many of the renal tubules.

2. Pneumonia

Pulmonary lesions are frequently observed at postmortem examination. Pathologic findings vary from moderate pulmonary edema and congestion to severe pneumonia.

3. Parasites

A mite, *Echinolaelaps* sp., and a follicle-inhabiting mite, *Psorergates simplex,* have been identified from meadow voles.

Three protozoan blood parasites, *Babesia microti, Haemobartonella microti, Trypanosoma microti,* have been reported from feral populations of meadow voles. The parasites apparently have no significant effect on the health of infected animals.

A capillarid helminth occurs in the small intestine.

G. Ground Squirrels

Few diseases have been reported in ground squirrels.

1. Diabetes Mellitus

Spontaneous diabetes mellitus was diagnosed in a laboratory-maintained colony of 13 lined-ground squirrels, (*Citellus tridecemlineatus*). The clinical signs were polydipsia, polyuria, glycosuria, ketonuria, polyphasia, and weight loss. Serum glucose values were increased, and serum insulin values were reduced in diabetic ground squirrels. In affected animals, the proportion of islet tissue to total pancreatic area and the number and size of the islets of Langerhans were reduced.

2. Arthropods

Lung mites, *Pneumocoptes banksi,* occur in ground squirrels, *Citellus beecheyi.* At postmortem examination no gross lesions are observed, but mites are clearly visible by micro-

scopic examination. Histologic lesions of lung tissue include infiltration with granulocytes, reduced functional area, connective tissue proliferation, and thickened alveolar walls.

H. Gray Squirrels

Little is known regarding diseases and parasites of gray squirrels in captivity. Surveys of feral gray squirrels have shown that these animals are relatively free of diseases and parasites that are common in feral populations of other rodents. Nevertheless, a number of parasites have been recorded.

1. Nematodes

Numerous nematodes have been reported from the gray squirrel, but some of them have low prevalence rates. *Heligmodendrium hassalli* and *Strongyloides rubustus* are common in the small intestine. Large numbers of *H. hassalli* in a host may cause mucosal hyperemia and hemorrhagic enteritis and frequently cause hemorrhagic enteritis in the duodenum. Other common nematodes in gray squirrels include *Citellinema bifurcatum, Bohmiella wilsoni, Enterobius scirui,* and *Capillaria americana.*

2. Cestodes

Tapeworms are not very prevalent. *Raillietina bakeri* is the most frequently reported cestode. *Hymenolepis diminuta* and *Catenotaenia dendritica* occur infrequently.

3. Ectoparasites

Sarcoptic mange has been reported in gray squirrels.

4. Protozoa

Coccidian oocysts similar morphologically to *Eimeria ascotensis, E. lancasterensis, E. moelleri,* and *E. neosciuri* are observed commonly in the small intestine. The pathogenicity of coccidia in gray squirrels apparently varies according to the species. Heavy infections could cause clinical signs; however, coccidiosis is not thought to be a major problem in squirrels.

Hepatozoon griseisciuri is also a common protozoan infection in gray squirrels. Circulating gametocytes are found in the blood, and schizogonic stages are observed in the lungs. Pathologic changes associated with infection include thickening of alveolar walls, eosinophil infiltration, pulmonary congestion, and atelectasis. It has been suggested that this parasite may cause death of gray squirrels.

I. Cotton Rats

1. Viral

Encephalomyocarditis virus has been recovered from spontaneous infections in feral cotton rats. No disease has been associated with this infection.

2. Parasites

Listrophorid mites have been reported on cotton rats. Clinical signs include loss of hair, dermatitis, and pruritis.

J. White-Footed Mice

Parasites

A number of different species of tapeworms, including *Hymenolepis*, have been reported in *P. maniculatus*. Nematodes recorded from *P. leucopus* and *P. maniculatus* include parasites in the following genera: *Aspicularis, Syphacia, Capillaria, Nippostrongylus, Nematospiroides,* and *Trichuris*. Other helminths recovered from *P. maniculatus* include *Mastophorus numidica* (from stomach), *Ricturlaria coloradensis* (from small intestine), *Trichuris perognatha* (from cecum), and *Acanthocephala clarki*. Mites, fleas, and lice are common and frequently abundant on feral *Peromyscus*; therefore, laboratory-held *Peromyscus* should be considered susceptible to mite infestation. Ectoparasites can be treated effectively using 5% methyl carbamate insecticide dust.

K. Woodchucks

1. Liver Disease

A common cause of death in captive colonies of woodchucks is primary hepatocellular carcinoma (Fig. 10) and chronic hepatitis (Snyder *et al.*, 1982). Tumors are most often seen in animals more than 3 years old, but a case in a 1-month-old woodchuck has been reported. Most animals with hepatomas also have inflammatory and regenerative changes in the nontumorous liver tissue suggesting preexisting chronic active hepatitis (Fig. 11). Some animals without hepatocellular tumors have lesions of chronic active hepatitis. A viral etiology for this hepatitis has been suggested. Particles with similarities to those associated with human hepatitis B have been found in serum of woodchucks with hepatitis. The virus from woodchucks has been characterized and designated woodchuck hepatitis virus.

2. Vascular Disease

A high incidence of arteriosclerosis, aortic rupture, and cardiovascular and cerebrovascular disease has been reported in woodchucks (Young and Sims, 1979).

3. Helminths

A filarid, *Ackertia marmotae*, occurs in the woodchuck. Adults are located primarily in the lymphatics on the ventral surface of the liver or associated with the gallbladder, cystic duct, or extra hepatic bile ducts. Few microfilaria are found in

Fig. 10. Two large primary nodules typical of hepatocellular carcinomas in woodchuck.

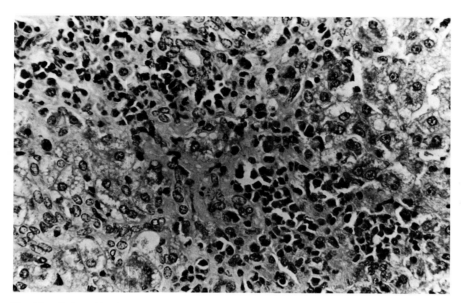

Fig. 11. Chronic hepatitis with abundant mononuclear inflammatory cells and fibroplasia in liver of woodchuck.

the blood, but large numbers are in the reticulum layer of the dermis, especially the skin of the ears. The intermediate host is *Ixodes cookei.*

Several other nematodes occur in woodchucks. The round worm, *Obeliscoides cuniculi,* may be found in the stomach. The oxyurid, *Citellina triradiata,* occurs in the cecum. *Baylisascaris laevis, Capillaria tamiasstriati, Citellinema bifurcatum,* and *Strongyloides* sp. have been reported in the small intestine.

4. Ectoparasites

Woodchucks infested with the mites, *Atricholaelaps glasgowi* and *Haemolaelaps glasgowi,* may scratch themselves, lose weight, lose hair, and have flaky, scaly skin. Fleas, lice, and ticks are also found on woodchucks.

5. Protozoa

Protozoan parasites usually do not cause disease in woodchucks; but several have been identified from this host, namely, *Eimeria, Chilomastix, Trichomonas, Hexamita, Entamoeba, Retortamonas,* and *Hexamastix.*

ACKNOWLEDGMENTS

Figure 1 (b) deer mouse, courtesy of Drs. Wallace D. Dawson and W. Morgan Newton, School of Medicine, University of South Carolina, Columbia, South Carolina. Figure 1 (c) meadow vole, courtesy of Dr. John E. Harkness, The Pennsylvania State University, University Park, Pennsylvania. Figure 1 (d) multimammate rat courtesy of Dr. Stephen Potkay, Division of Research Services, NIH, Bethesda, Maryland. Figure 1 (e) chinchilla courtesy of Dr. William H. Pryor, Jr., and Center for Medical Communication, School of Medicine, East Carolina University, Greenville, North Carolina. Figures 10 and 11 are courtesy of Dr. Robert Snyder, Penrose Research Laboratory, The Zoological Society of Philadelphia, Philadelphia, Pennsylvania.

I wish to thank Dr. Willie L. Chapman, Jr., Ms. Janet Calpin, and Ms. Brenda Sorrow for invaluable assistance in the preparation of this manuscript.

REFERENCES

Clark, J. D., Loew, F. M., and Olfert, E. D. (1978). Rodents. *In* "Zoo and Wild Animal Medicine" (M. E. Fowler, ed.), pp. 457–478. Saunders, Philadelphia, Pennsylvania.

Coetzee, C. G. (1975). The biology, behavior, and ecology of *Mastomys natalensis* in southern Africa. *Bull. W. H. O.* **52,** 637–644.

Cousens, P. J. (1963). The chinchilla in veterinary practice. *J. Small Anim. Prac.* **4,** 199–205.

Dall, J. (1963). Diseases of the chinchilla. *J. Small Anim. Prac.* **4,** 207–212.

Davis, D. H. S. (1963). Wild rodents as laboratory animals and their contribution to medical research in South Africa. *S. Afr. J. Med. Sci.* **28,** 53–69.

Dieterich, R. A., and Preston, D. J. (1977). The meadow vole (*Microtus pennsylvanicus*) as a laboratory animal. *Lab. Anim. Sci.* **27,** 494–499.

Hall, A., III, Persing, R. L., White, D. C., and Ricketts, R. T., Jr. (1967). *Mystromys albicaudatus* (the African white-tailed rat) as a laboratory species. *Lab. Anim. Sci.* **17,** 180–188.

Hamilton, W. J., Jr. (1941). Reproduction of the field mouse *Microtus pennsylvanicus* (Ord.). *Mem.—N.Y., Agric. Exp. Stn. (Ithaca) Mem* **237,** 1–23.

Isaacson, M. (1975). The ecology of *Praomys (Mastomys) natalensis* in southern Africa. *Bull. W.H.O.* **52,** 629–635.

Keagy, H. F., and Keagy, E. H. (1951). Epizootic gastro-enteritis in chinchillas. *J. Am. Vet. Med. Assoc.* **118,** 35–37.

King, J. A. (1968). "Biology of *Peromyscus* (Rodentia)." Am. Soc. Mammal., Shippensburg State Coll., Shippensburg, Pennsylvania.

Larrivee, G. P., and Elvehjem, C. A. (1954). Disease problems in chinchillas. *J. Am. Vet. Med. Assoc.* **124,** 447–455.

Lee, C., and Horvath, D. J. (1969). Management of the meadow vole (*Microtus pennsylvanicus*). *Lab. Anim. Care* **19,** 88–91.

Marsh, R. E., and Howard, W. E. (1971). Care of ground squirrels (*Spermophilus* spp.) in captivity. *Lab. Anim. Sci.* **21,** 367–371.

Marston, J. H., and Chang, M. C. (1965). The breeding, management, and reproductive physiology of the Mongolian gerbil (*Meriones unguiculatus*). *Lab. Anim. Care* **15,** 34–48.

Meyer, B. J., and Meyer, R. K. (1944). Growth and reproduction of the cotton rat, *Sigmodon hispidus hispidus*, under laboratory conditions. *J. Mammal.* **25,** 107–129.

Moore, R. W., and Greenlee, H. H. (1975). Enterotoxaemia in chinchillas. *Lab. Anim.* **9,** 153–154.

Morrison, P., Dieterich, R., and Preston, D. (1976). Breeding and reproduction of fifteen wild rodents maintained as laboratory colonies. *Lab. Anim. Sci.* **26,** 237–243.

Newberne, P. M. (1953). An outbreak of bacterial gastro-enteritis in the South American chinchilla. *North Am. Vet.* **34,** 187–188.

Rich, S. T. (1968). The Mongolian gerbil (*Meriones unguiculatus*) in research. *Lab. Anim. Sci.* **18,** 235–243.

Snell, K. C., and Stewart, H. L. (1975). Spontaneous diseases in a closed colony of *Praomys* (*Mastomys*) *natalensis. Bull. W.H.O.* **52,** 645–650.

Snyder, R. L., Tyler, G., and Summers, J. (1982). Chronic hepatitis and hepatocellular carcinoma associated with woodchuck hepatitis virus. *Am. J. Pathol.* **107,** 422–425.

Soga, J., and Sato, H., eds. (1977). "*Praomys* (*Mastomys*) *natalensis:* The Significance of Their Tumors and Diseases for Cancer Research." Daiichi Printing Co., Meike, Wago-Cho, Niigata, Japan.

Solleveld, H. A. (1981). "*Praomys* (*Mastomys*) *natalensis* in Aging Research." Inst. Exp. Gerontol., TNO, Rijswijk, The Netherlands.

Thiessen, D. D., and Pendergrass, M. (1982). Harderian gland involvement in facial lesions in the Mongolian gerbil. *J. Am. Vet. Med. Assoc.* **181,** 1375–1377.

UFAW Staff with the assistance of W. Lane-Peter, A. N. Warden, B. F. Hill, J. S. Patterson, and H. G. Vevers, eds. (1976). "The UFAW Handbook on the Care and Management of Laboratory Animals," 5th ed. Livingstone, Edinburg.

Vincent, A. L., Porter, D. D., and Ash, L. R. (1975). Spontaneous lesions and parasites of the Mongolian gerbil, *Meriones unguiculatus. Lab. Anim. Sci.* **25,** 711–722.

Vincent, A. L., Rodrick, G. E., and Soderman, W. A., Jr. (1979). The pathology of the Mongolian gerbil (*Meriones unguiculatus*): A review. *Lab. Anim. Sci.* **29,** 645–651.

Weir, B. J. (1967). The care and management of laboratory hystricomorph rodents. *Lab. Anim.* **1,** 95–104.

Williams, C. S. F. (1980). Wild rats in research. *In* "The Laboratory Rat" (H. J. Baker, J. R. Lindsay, and S. H. Weisbroth, eds.), Vol. 2, pp. 245–256. Academic Press, New York.

Young, R. A., and Sims, E. A. H. (1979). The woodchuck, *Marmota monax*, as a laboratory animal. *Lab. Anim. Sci.* **29,** 770–780.

Chapter 8

Biology and Diseases of Rabbits

Alan L. Kraus, Steven H. Weisbroth, Ronald E. Flatt, and Nathan Brewer

I. INTRODUCTION

While previously classified in the order Rodentia, the rabbit differs from rodents in that it has three pairs of incisor teeth (including those two teeth located directly behind the large upper pair) rather than four pair, and has thus been classified in its own order—the Lagomorpha. The order contains two major families: The Ochotonidae (pika) and Leporidae (rabbits and hares). In the family Leporidae, the major genera are the *Lepus* (hares), *Sylvilagus* (cottontail rabbits), and *Oryctolagus* (true rabbits).

The European rabbit, *Oryctolagus cuniculus*, consists of many varieties and over fifty breeds that are used for meat, show, or laboratory animal research and testing. While species other than the true rabbit (*Oryctolagus*) are sometimes used in research, this chapter will concern itself principally with this genus.

According to a survey conducted by the Institute of Laboratory Animal Resources (1980), out of 19.9 million laboratory animals used by nonprofit commercial and federal biomedical research organizations, approximately 440,000 or 2.2% were rabbits. Of those used, 23,000 were bred in the laboratory and 417,000 were acquired from commercial sources. While the

New Zealand white breed of rabbit is by far the most commonly used stock of rabbit, other breeds, such as the Dutch belted, Flemish giant, and Polish, are also used but in relatively small numbers. A number of inbred and otherwise genetically defined strains of rabbits are also available to researchers (Altman and Dittmer-Katz, 1979; Fox, 1975). Sources of inbred rabbits include the Jackson Laboratory, The University of Illinois College of Medicine's Center for Genetics, and the University of Utrecht in the Netherlands. Detailed information on the following characteristics of inbred rabbits is provided by Altman and Dittmer-Katz (1979): gene map, gene names, karyology, coat and eye color mutants, anatomic and physiologic mutants, neurological and behavioral mutants, skeleton and organ structure variations, enzyme and other proteins in serum and tissues, blood group systems, immunoglobulin allotypes, and allogroups of the Ig heavy chain chromosomal region. In addition to inbred strains of rabbits, much information also exists for rabbits with inherited diseases and other biological variations (Lindsey and Fox, 1974).

Rabbits are used in a wide variety of experimental situations, including the following major categories. In addition to experimental uses, the rabbit makes an excellent species for the teaching of principles of mammalian physiology and surgery (Kaplan and Timmons, 1979).

Arteriosclerosis: Rabbits were the first laboratory animal to be used as models for "atherosclerosis," and although they are used in some studies today, the similarities between human atherosclerosis and the disease that affects them cause many to consider them inappropriate models. On the other hand, the medial mineralization of the aorta and the aortic arch that occur naturally resembles Mönkeberg's medial sclerosis in man. The disease occurring following administration of exogeneous cholesterol and fat is more of a generalized lipid storage disorder than atherosclerosis.

Glaucoma: The inherited glaucoma occurring in New Zealand white rabbits has been studied extensively; however, it is linked to a semilethal trait that causes small litters, poor viability, and low fertility that limit its usefulness. There are apparent discrepancies in effect between drugs used to treat glaucoma in man and in rabbits.

C6 Deficiency: Zimmerman and colleagues (1971) have reported on a mutant "strain" of rabbits deficient in the sixth component of complement that is inherited as an autosomal recessive.

Cardiomyopathy: Weber has described a cardiomyopathy occurring in rabbits subjected to severe crowding, four to a cage for 2 weeks followed by 1 week alone in a cage. Thirty-five of 44 rabbits died while on this regimen. Histologically there was myocardial edema and coagulative necrosis (Weber and VanderWatt, 1973).

Hypertension: Thirteen inbred strains of rabbits were surveyed and several were shown to have systolic blood pressure of 30 to 40 mm Hg above "normotensive" rabbits (Fox *et al.*, 1969).

Von Willebrand's disease: Benson reported on a mutant line of Flemish giant, and chinchilla rabbits which expressed characteristics of von Willebrand's disease. The disease is autosomally inherited with variable penetrance (Benson and Dodds, 1977).

Transplantable tumors of the rabbit: Weisbroth has reviewed the principal transplantable neoplasms used as model systems in the rabbit (Weisbroth, 1974). One of the most useful and well known of the transplantable rabbit tumors is the Brown–Pearce carcinoma that originally arose from the scrotal chancre of a rabbit experimentally infected with *Treponema pallidum*. It commonly is implanted and grown in the anterior chamber of the eye.

Teratologic investigation: While at the present time, there is no completely predictable relationship between teratogenesis and chemical structure or pharmacologic activity and no scientific rationale for the selection of any one species for use in teratologic investigation, much work has been performed using the rabbit (Hartman, 1974). Most workers therefore use several species to screen for teratogenic potential.

Immunology: Rabbits are used extensively in immunological studies, particularly those involving the hormonal response to antigenic stimuli. Methods of immunization of rabbits along with lists of commonly used antigen in various techniques for antibody production have been published (Cohen, 1966). Much is known about the blood group and transplantation antigens of the rabbit. This material has been reviewed (Cohen and Tissot, 1974).

Aging research: Rodents and rabbits are important in research on the biology of aging. Two recent publications have reviewed the biology and use of rabbits in aging research (Committee of Animal Models for Research on Aging, 1981; Fox, 1980).

Routine toxicology testing: Rabbits have been used as the animal model of choice in the testing of both prescription and over-the-counter drugs and other products for potential eye irritation. The Draize test, an *in vivo* test in which test materials are placed in the conjunctival sac and physical observations made, has been in use since 1944 (Draize *et al.*, 1944). In addition, rabbits are used extensively to test for skin irritation and other forms of skin toxicity. The Draize eye irritation test has come under attack by animal welfare activists including members of the biomedical community. Several major laboratories are attempting to develop a reliable and otherwise suitable *in vitro* alternative to this test.

In addition, rabbits are used extensively in pyrogen testing. The pyrogen test is designed to limit to an acceptable level the risk of febrile reaction in the patient to the administration, by injection, of the product being tested. The test involves measuring the rise in body temperature over a 3-hr period following intravenous injection of a test solution. A test is considered positive if any of the three rabbits used have temperatures that rise greater than 0.6°C or if the aggregate rise of the three rabbits temperatures exceed 1.4°C (U.S. Pharmacopeial Convention, Inc., 1980).

II. BIOLOGY

A. Comparative Anatomy and Physiology

1. The Gastrointestinal Tract

The mouth of the rabbit is relatively small. The upper lip is divided and the right and left halves are continuous with the respective nostril. The dental formula of the rabbit is

$$I_1^2 C_0^0 P_2^3 M_3^{2-3} = 26-28.$$

There is a small pair of incisors behind the primary incisors in the upper jaw. The teeth of rabbits, like those of rodents, grow continuously throughout life and are kept at their normal length by constant wear on one another. The molars are rootless and present deep enamel folds. The chewing motion is both side to side and front to back. Rabbits have difficulty eating pulverized food, consequently if substances are mixed into ground feed it is best that the feed be again pelleted (Szabo *et al.*, 1974).

There are three pairs of salivary glands: the submaxillary, parotid, and infraorbital. The normal flow rate of saliva from the parotid and submaxillary glands is much lower than in the dog, pig, sheep, or human (Chauncey *et al.*, 1963).

The rabbit is unique among animals reported in that it has three muscle layers in the esophagus. The muscles are striated throughout and extend down to and include the cardiac portion of the stomach (Goetsch, 1910; Alvarez, 1940).

The stomach in a healthy rabbit is never empty. The rabbit, like the rat and horse, cannot vomit.

Rabbits produce relatively large amounts of bile, about seven times as much as a dog on a weight basis. The primary bile salt is cheondeoxycholic acid, which is secreted and reduced in the intestine by bacteria to deoxycholic acid. The pancreas is diffuse and enters the duodenum approximately 30 to 40 cm distal to the biliary duct entrance.

The rabbit intestine, like the human and guinea pig, is largely impermeable to macromolecular compounds (Brambell, 1958). Thus unlike the carnivores, pigs and ruminants, it is difficult for immunoglobulins to be absorbed in nursing rabbits, although there is some evidence that some antibodies can be absorbed in the first few hours of life (Kulangara and Schechtman, 1962). The rabbit therefore gets most of its passive immunity from the doe before birth.

The rabbit has a total dependence on milk up to day 10. A small amount of solid food (5%) is ingested by day 15. Solid food and cecotrophy is started on day 20 (Alus and Edwards, 1977).

The combined length of the small and large intestine is approximately eleven times that of the body. The cecum has ten times the capacity of the stomach. At the terminal portion of the small intestine, a bulblike expansion of lymphatic tissue, the sacculus rotundus, is found. While some consider the sacculus rotundus to be the analogue of the bursa of Fabricius in birds, convincing evidence is lacking. The cecum ends in a narrow blind tube, the appendix. The colon is attached to the duodenum near its terminus and, because of its comparable size, may inadvertently be mistaken for the duodenum.

The colon is divided into a proximal and distal colon, separated by the fusus coli. The fusus coli is probably most important in the production of the two kinds of feces in the rabbit (Bjornhag, 1972). Once or twice daily, the elimination of hard pellets is interrupted by the elimination of soft pellets, called cecotrophs. These cecotrophs are ingested by the rabbit directly from the anus, a process of copraphagy called cecotrophy. The soft feces have a high water, electrolyte and nitrogen content (Ferrando *et al.*, 1970) and are relatively rich in niacin, riboflavin, pathothenate and cyanocobalamin. About one-third of feces are cecotrophs. There is a circadian rhythm to cecotrophy (Hornicke, 1977) with the soft feces or cecotrophy practiced at night by domestic rabbits and by wild rabbits during the day when they are in their burrows.

2. The Respiratory System

The nostrils of rabbits are well equipped with touch cells and they have a well-developed sense of smell. Tactile vibrissae are long, number about 20–25, and are based on the upper lip. Nostril twitching may be absent when the rabbit is totally relaxed, but usually the nose moves up and down at 20–120 times per minute. The up movement occurs with inspiration and probably effectively diverts air over the turbinate bones where the olfactory cells are most concentrated.

Rabbits at rest respire mostly through the activity of the diaphragm, i.e., abdominal breathing. Therefore, holding the animal and alternating its position from a head up to a head down position, stimulating abdominal breathing, is a most effective means of artificial respiration. Manipulating the chest wall is relatively ineffective.

Tracheal intubation of the rabbit is difficult due to the small size of the mouth and larynx, in addition to the propensity of the rabbit to develop laryngospasm. Reflex laryngospasm is common, especially during inhalation of gaseous anesthetics, or during attempts to intubate the trachea.

The rabbit, along with the rat, mouse, and hamster, has relatively few goblet cells in the tracheobronchial tree when compared to man, cat, dog, and guinea pig. The ciliary beat in the rabbit and rat are roughly twice the rate (20 beats/sec) of the cat or human.

There are six lobes of the lung: four right and two left lobes. Flow volumes of air to the left lung is higher than the right due to the lower resistance of the proximal airways per unit volume (Yokoyama, 1979). While in man and dog there is an increase in residual volume (RV) that occurs with age at the expense of vital capacity (VC), in rodents and rabbits lung volume increases with age and RV : VC ratio does not change. Selected respiratory measurements and characteristics are presented in Table I.

3. The Cardiovascular System

The right atrioventricular valve is not tricuspid but rather it consists of one large and one smaller flap or cusp. There are wide variations in the structure of conductive tissue in the heart of different species. The SA and AV nodes in the rabbit are slender and elongated, and the AV node is separated from the annulus fibrosus by a layer of fat (Truex and Smythe, 1965). There is little or no connective tissue in the conductive tissue that for the most part lies in proximity to the endocardium. This simplicity of conductive tissue and the pacemaker structures has made it possible for the rabbit to be the first mammal in which the exact site of the pacemaker has been demonstrated by the use of intracellular microelectrodes (West, 1955).

Table I

Some Respiratory Measurements and Characteristics of Rabbits

Characteristic	Measurement	Reference
Lung weight (2.4 kg rabbit)	9.1 g	Crosfill and Widdicombe (1961)
Mean alveolar diameter (2.4 kg rabbit)	93.97 μm	Crosfill and Widdicombe (1961)
Respiratory rate		
Adult rabbits	30–60/min	Mauderly *et al.* (1979)
2.4 kg rabbits	39/min	Crosfill and Widdicombe (1961)
Tidal volume		
2.4 kg rabbits	15.8 ml	Crosfill and Widdicombe (1961)
1 kg rabbit	10.8 ± 2.3 ml	Caldwell and Fry (1969)
3 kg rabbit	23.9 ± 5.5 ml	Caldwell and Fry (1969)
Minute volume		
Adult rabbits	0.6 liter/min	Guyton (1947)
2.4 kg rabbit	0.62 liter/min	Crosfill and Widdicombe (1961)
Minute volume/kg body weight	0.25 liter/min/kg	Crosfill and Widdicombe (1961)
Lung compliance		
Absolute	6 ml/cm H_2O	Crosfill and Widdicombe (1961)
Per gram lung	0.65 ml/cm H_2O	Crosfill and Widdicombe (1961)
Per millileter lung	0.28 ml/cm H_2O	Crosfill and Widdicombe (1961)
Lung resistance		
Absolute	25 cm H_2O/liter/sec	Crosfill and Widdicombe (1961)
Per gram lung	222 cm H_2O/liter/sec	Crosfill and Widdicombe (1961)
Per milliliter lung	522 cm H_2O/liter/sec	Crosfill and Widdicombe (1961)
Functional residual capacity (2.4 kg. rabbit)	11.3 ml	Crosfill and Widdicombe (1961)
	34.6 ± 10.6 ml	Caldwell and Fry (1969)
	46.9 ± 14.1 ml	Davidson *et al.* (1966)
Total lung capacity	155 ± 25.2 ml	Davidson *et al.* (1966)
	111 ± 14.7 ml	Caldwell and Fry (1969)
Vital capacity		
1 kg rabbits	43 ± 4.9 ml	Caldwell and Fry (1969)
3 kg rabbit	91.7 ± 13.3 ml	Caldwell and Fry (1969)
3.85 ± 0.38 kg rabbits	124 ± 14 ml	Davidson *et al.* (1966)

Additional unique anatomical features of the cardiovascular system of the rabbit had been utilized to advantage. The aortic nerve subserves no known chemoreceptors (Kardon *et al.*, 1974; Stinnet and Sepp, 1979) and responds to baroreceptors only. Because the aortic nerve, which becomes the depressor nerve, runs alongside but separate from the vagosympathetic trunk, it lends itself readily to implantation of electrodes (Karemaker *et al.*, 1980).

The aorta of the rabbit has rhythmic contractions, neurogenic in origin, with a precise phasing pattern related to the pulse wave (Mangel, 1981). This may be true in other animals, but it has only been demonstrated in the rabbit.

The responses of the cerebrovascular flow and resistance to cerebral sympathetic nerves is much more marked in rabbits and monkeys than it is in dogs or cats (Heistad *et al.*, 1978). In rabbits, this response is transient, which suggests a vasomotor escape from sympathetic stimulation (Sercombe *et al.*, 1978).

4. The Urinary System

While most mammals have multipapillate kidneys, the rabbit is unipapillate. The single papilla and calyx, therefore, readily lend themselves to cannulization techniques.

Rabbits urinate infrequently, but the quantity is relatively large (30–35 ml/day). The amount varies with the quantity and type of food ingested and with the availability of drinking water. Does usually urinate more copiously than bucks. Under *ad libitum* feeding, rabbit urine has a pH of about 8.2 (Williams, 1976), but when fasted the pH drops to 6–7 (Kojima and Tanaka, 1974).

The urine of young rabbits is free from precipitate, but it is common for young healthy rabbits to have albuminuria. When rabbits are old enough to eat green feed and cereal grains, the urine takes on a yellowish to brownish tint and becomes cloudy, principally due to the presence of ammonium magnesium phosphate and calcium carbonate monohydrate precipitates (Flatt and Carpenter, 1971). Normal rabbits have little or no epithelial cells, casts, or bacteria in their urine, but both red and white blood cells are occasionally observed.

The maximum urine concentrating ability of the rabbit is approximately 1.9 osmoles/liter (Haupt, 1963). The urine is a major route for both calcium and magnesium excretion. In most mammals, calcium is excreted principally in the bile.

The rabbit and the hamster differ from most mammals in that they are resistant to the phosphoturic effects of parathormone (PTH). In most mammals PTH causes increased phosphate excretion by inhibition of phosphate readsorption (Dennis *et al.*, 1977; Steele, 1976; Knox *et al.*, 1977).

Blood vessels perfusing the medulla (the juxtaglomerular circulation) remain open during many conditions under which vasoconstriction of the cortical tissues takes place, for exam-

Table II

Values for Certain Urinary Excretion Products of Male New Zealand White Rabbits[a]

	No. of animals	Mean ± SEM (range)
Sodium (mEq/kg/day)	19	1.41 ± 0.13 (0.31–2.69)
Potassium (mEq/kg/day)	19	8.67 ± 0.69 (4.46–15.69)
Chloride (mEq/kg/day)	19	3.5 ± 0.4 (0.8–8.2)
Creatinine (mEq/kg/day)	19	44.3 ± 3.7 (22.4–85.0)
Calcium (mEq/kg/day)	19	96.0 ± 11.9.4 (12.1–193.2)
Phosphate (mEq/kg/day)	19	14.0 ± 2.3 (5.0–32.8)

[a] From Kozma *et al.* (1974).

ple, splanchnic stimulation. Thus the medullary tissue is perfused while the cortex is ischemic (Trueta *et al.*, 1947), which results in anuria.

In the rabbit, the clearance of creatinine (Ccr) is identical with the clearance of inulin (Cin) and Ccr may therefore be used to accurately measure the glomerular filtration rate. This is not true for primates, rat or guinea pig, among others.

The rabbit is the only known mammal from which tubules of the kidney can be conveniently dissected with the basement membrane intact. Since the technique was first developed (Burg *et al.*, 1966) every section of the nephron except the glomerulus has been studied. Functional as well as anatomic differences between nephrons and between the three portions of the proximal tubule and the three or more parts of the distal tubules are becoming increasingly clear. New approaches, especially electron probe analysis are adding to our basic knowledge. For more details of growing body information on the kidney of the rabbit, the reader is referred to Lamiere *et al.* (1977), Valtin (1977), Grantham *et al.* (1978), Arruda and Krutzman (1978), and the series edited by Andreoli (1981). Table II presents values for certain excretory products of male NZw rabbits (Kozma *et al.*, 1974).

5. Metabolism

Early studies showed the rabbit to have an extraordinary low metabolic rate (MR) per unit of surface area (SA). When the large ear SA is eliminated however, the results are more comparable to other endotherms. Data on the rabbit MR is presented in Table III.

The neonatal rabbit is an ectotherm until day 7 (Gelineo, 1964). Neonatal rabbits maintain normoglycemia, even when suckling, until their glucose reserves are exhausted, which takes about 6 hr postpartum (Shelley, 1961). In the fasting neonatal rabbit, hypoglycemia develops concomitantly with a marked ketosis (Callikan and Girard, 1979).

The phosphorylation of glucose by glucokinase is relatively

Table III

Metabolic Rate of Rabbit

Weight (kg)	kcal/day
1.52	83
2.46	119
3.57	164
4.33	191
5.33	233

high in the rabbit, pig, dog, and rodents when compared to man and cat. The rate of formation of fatty acids from glucose and adipose tissue is relatively high in rabbits and rodents when compared to humans.

B. Normative Physiological Values

Any discussion of normative physiological values should be prefaced with the proviso that such compendia represent average determinations or observations which, depending on circumstances, may be influenced by a wide range of factors. It is therefore necessary to review briefly both intrinsic and extrinsic factors that bear on normative values. Extrinsic factors relate to variation in values due to the way different laboratories collect and process samples; differences in methodology, equipment, and reagents; and operator variation. Intrinsic factors relate to the differences imposed by species of biological variation. Such factors, as age, sex, breed or other heritable

condition, health and nutritional plane, and even time of collection (circadian rhythm), are examples of circumstances intrinsic to the test group that may have important effects on determined values. For a more thorough review of these factors, as they apply to rabbits, the reader is referred to standard reference works (Mitruka and Rawnsley, 1977; Payne *et al.*, 1976; Kozma *et al.*, 1974). Inasmuch as variations due to breed or strain are not as pronounced as those due to health, age, and sex, the average values reported here for young adult New Zealand white (NZW) random bred rabbits should be considered generally applicable to the other common rabbit breeds.

1. Hematologic Values

Hematologic values for young adult NZW rabbits are summarized in Table IV. It will be seen that male rabbits tend to have higher hematocrit values, thus derivative determinations, e.g., erythrocyte count and hemoglobin will be correspondingly slightly higher, as well. The erythrocyte diameters of neonatal rabbits exceed 9 mm; however, the adult range of 6.5–7.5 mm is attained by 2–3 months of age. Stained films of rabbit blood commonly show polychromasia and reticulocyte counts of 2–4% due to the normally short life span, and high turnover rate of rabbit erythrocytes. Similarly, crenated forms are commonly observed in stained films. The thrombocytes (platelets) occur singly or in groups as oval or roundish forms that are 1–3 mm in diameter and which stain an intense violet color centrally due to azurophilic granules.

Rabbit neutrophils have a predominant granule type that stains red, hence the term "pseudoeosinophil" (also hetero-

Table IV

Hematologic Values of Young Adult New Zealand White Rabbits[a]

Component	Unit	Male			Female		
		Mean	S.D.	Range	Mean	S.D.	Range
Erythrocytes (RBC)	$10^6/mm^3$	6.7	0.62	5.46–7.94	6.31	0.60	5.11–6.51
Hemoglobin (Hb)	gm/dl	13.9	1.75	10.4–17.4	12.8	1.50	9.80–15.8
Mean corpuscular volume	μm^3	62.5	2.00	58.5–66.5	63.1	1.92	57.8–65.4
Mean corpuscular hemoglobin	pg	20.7	1.00	18.7–22.7	20.3	1.60	17.10–23.5
Mean corpuscular hemoglobin concentration	%	33.5	1.85	30.0–37.0	32.3	1.74	28.7–35.7
Hematocrit (PCV)	%	41.5	4.25	33.0–50.0	39.8	4.40	31.0–48.6
Sedimentation rate	mm/hr	2.0	0.50	1.00–3.00	1.75	0.40	0.92–2.55
Platelets	$10^3/mm^3$	480.00	88.00	304–656	450	90.0	270–630
Leukocytes (WBC)	$10^3/mm^3$	9.00	1.75	5.50–12.5	7.90	1.35	5.20–10.6
Neutrophils	%	46.0	4.00	38.0–54.0	43.4	3.50	36.4–50.4
Eosinophils	%	2.00	0.75	0.50–3.50	2.00	0.60	0.80–3.20
Basophils	%	5.00	1.25	2.50–7.50	4.30	0.95	2.40–6.20
Lymphocytes	%	39.00	5.50	28.0–50.0	41.80	5.15	31.5–52.1
Monocytes	%	8.00	2.00	4.00–12.0	9.00	2.20	6.60–13.4

[a]Adapted from Mitruka and Rawnsley (1977).

Table V

Serum Biochemical Values of Young Adult New Zealand White Rabbits[a]

Component	Unit	Male		Female		
		Mean	S.D.	Mean	S.D.	Range
Bilirubin	mg/dl	0.32	0.04	0.30	0.04	0.00–0.74
Cholesterol	mg/dl	26.7	12.9	24.5	11.2	10.0–80.0
Creatinine	mg/dl	1.59	0.34	1.67	0.38	0.50–2.65
Glucose	mg/dl	135.00	12.0	128.00	14.0	78.0–155
Urea nitrogen	mg/dl	19.2	4.93	17.6	4.36	13.1–29.5
Uric acid	mg/dl	2.65	0.88	2.62	0.87	1.00–4.30
Sodium	mEq/liter	146.00	1.15	141.00	1.40	138–155
Potassium	mEq/liter	5.75	0.20	6.40	0.16	3.70–6.80
Chloride	mEq/liter	101.00	1.45	105.00	1.22	92.00–112
Bicarbonate	mEq/liter	24.2	3.15	22.8	3.20	16.2–31.8
Phosphorus	mg/dl	4.82	1.05	5.06	0.93	2.30–6.90
Calcium	mg/dl	10.0	1.11	9.50	1.10	5.60–12.1
Magnesium	mg/dl	2.52	0.24	3.20	0.22	2.00–5.40
Total protein	g/dl	6.90	0.36	6.70	0.41	6.00–8.30
Albumin	g/dl	3.39	0.29	3.04	0.26	2.42–4.05
α_1-Globulin	g/dl	0.60	0.12	0.37	0.08	0.10–0.90
α_2-Globulin	g/dl	0.43	0.09	0.21	0.05	0.15–0.75
β-Globulin	g/dl	1.01	0.20	1.46	0.24	0.50–2.10
γ-Globulin	g/dl	1.46	0.21	1.69	0.22	1.00–2.15
Albumin/globulin		0.97	0.16	0.83	0.15	0.68–1.15

[a]Adapted from Mitruka and Rawnsley (1977).

phil or amphophil). The red granules of neutrophils are noticeably smaller than the corresponding red granules of eosinophils. One may occasionally encounter films demonstrating the Pelger (or Pelger–Huet) anomaly that is reflected as a maturation defect of the neutrophil, in which the nucleus fails to undergo normal segmentation. Eosinophils are distinguishable from neutrophils on the basis of their larger size, 10–15 mm in diameter (compared to 7–10 mm in diameter for neutrophils), and larger granules that fill the cytoplasm obscuring the nucleus. Both large and small lymphocytes are seen in stained films of rabbit blood; the former about the size of the neutrophil, the latter similar in size to the erythrocyte. The monocyte is the largest leukocyte found in rabbit blood films. Normally, granules do not occur in the cytoplasm of monocytes, but they may be seen in toxic conditions.

2. Blood and Serum Chemistry Values

The variation imposed by both extrinsic and intrinsic factors as discussed above for hematologic values, is also true for determined chemical components of rabbit blood, serum, and plasma. Similarly, inasmuch as variation due to extrinsic factors and to age, sex, and health exert a greater effect than those due to breed or strain effects, the results reported for the NZw rabbits should be considered as generally applicable for the other stocks and strains of *Oryctolagus*. For more extensive

tabulations, as well as for fuller treatments of the effect of stock (breed) differences, the reader should consult standard reference works (Mitruka and Rawnsley, 1977; Laird, 1974; Kozma *et al.*, 1974). Values for common serum biochemical, electrolyte and enzyme components are summarized in Tables V and VI.

3. Organ Weights

Organ weights comprise an important aspect of normative anatomic structure assessment because many pathologic and toxic states are reflected by changes in organ size. Organ weights are presented in Table VII.

4. Respiratory, Circulatory, and Metabolic Measurements

Table VIII presents a group of commonly measured aspects of normative physiology. A fuller treatment with more extensive tabulations may be found in the standard reference work on the rabbit (Kozma *et al.*, 1974).

C. Nutrition

Rabbits are herbivorous, monogastric animals with a relatively large cecum. Rabbits regularly eat the soft, moist, fecal

Table VI

Serum Enzyme Activities of Young Adult New Zealand White Rabbits[a]

Component	Unit	Male		Female		Range
		Mean	S.D.	Mean	S.D.	
Amylase	Somogyi units/dl	132.00	16.0	127.00	12.00	90.0–170.00
Alkaline phosphatase	I.U./liter	10.4	2.28	9.96	3.10	4.10–16.2
Acid phosphatase	I.U./liter	1.56	0.53	1.40	0.38	0.30–2.70
Gutamic pyruvic transaminase	I.U./liter	65.7	6.54	62.5	5.85	48.5–78.9
Glutamic oxaloacetic transaminase	I.U./liter	72.3	1.20	68.10	10.5	42.5–98.0
Creatine phospho-kinase	I.U./liter	1.35	0.56	1.30	0.45	0.20–2.54
Lactic dehydrogenase	I.U./liter	84.4	20.5	78.5	22.0	33.5–129.00

[a]Adapted from Mitruka and Rawnsley (1977).

pellets that they produce at night. This practice is called coprophagy or pseudorumination. Conservation of B vitamins, improved utilization of dietary protein, and conservation of water are benefits derived from this practice.

The nutritional requirements of rabbits are not as well understood as some other laboratory animals and domestic livestock. However, reviews of the present knowledge have been published (Hunt and Harrington, 1974; Subcommittee on Rabbit Nutrition, 1977).

Energy requirements have been estimated to be 2500 kcal of digestable energy per kilogram of diet for growth, gestation, and lactation; 2100 kcal/kg are required for maintenance. Approximately 15–17% crude protein is recommended for growth, gestation, and lactation, while 12% is used in a maintenance diet. A dietary fat content of 2% has been found to be adequate,

however, performance was improved with 5% or 10% fat (Subcommittee on Rabbit Nutrition, 1977). Known or estimated nutritional requirements of rabbits for energy, crude protein, fat, fiber, minerals, vitamins, and amino acids are summarized in Table IX.

Most often, commercially prepared pelleted diets are provided for rabbits; however, it occasionally becomes necessary to prepare experimental diets. Experimental diets may be pelleted or nonpelleted. If nonpelleted diets are used, an adjust-

Table VII

Organ Weights of Young Adult New Zealand White Rabbits[a,b]

Component	Male (N = 23)		Female (N = 21)	
	Mean	S.D.	Mean	S.D.
Body weight (kg)	2.775	0.198	2.541	0.235
Brain	0.364	0.035	0.374	0.045
Liver	2.870	0.417	3.275	0.593
Spleen	0.042	0.0040	0.037	0.0070
Adrenal	0.0098	0.0032	0.0095	0.0023
Kidneys	0.521	0.059	0.510	0.055
Ovaries			0.0072	0.0021
Testes (without epididymus)	0.109	0.029		
Thyroid	0.0055	0.0019	0.0063	0.0021
Thymus	0.145	0.0044	0.156	0.053
Heart	0.203	0.021	0.200	0.0044

[a]Adapted from Kozma *et al.* (1974).
[b]All organ weights expressed in grams/100 gm body weight.

Table VIII

Respiratory, Circulatory, and Metabolic Measurements of Young Adult New Zealand White Rabbits[a]

Measurement	Mean	Range
Compliance ratio (chest/lung)	—[b]	0.95–2.43
Respiratory rate	51.0	32.0–60.0
Tidal volume (ml)	21.0	19.3–24.6
Minute volume (liter/min)	1.07	0.37–1.14
Whole blood volume (ml/kg body weight)	—	55.6–57.3
Plasma volume (ml/kg body weight)	38.8	27.8–51.4
Erythrocyte volume (ml/kg body weight)	—[b]	16.8–17.5
Adult blood pressure (mm Hg)		
Systolic	110.0	90.0–130.0
Diastolic	80.0	60.0–91.0
Whole blood pH	7.35	7.21–7.57
Heart rate (beats/min)	—[b]	306.0–333.0
Temperature	39.5°C	38.6–40.1°C

[a]Adapted from Kozma *et al.* (1974).
[b]Not given.

Table IX

Nutrient Requirements of Rabbits Fed *ad Libitium* (Percentage or Amount per Kilogram of Diet)[a,b]

Nutrients	Growth	Maintenance	Gestation	Lactation
Energy and protein				
Digestible energy (kcal)	2500.00	2100.00	2500.00	2500.00
Total digestible nutrients (%)	65.00	55.00	58.00	70.00
Crude fiber (%)	10–12[c]	14[c]	10–12[c]	10–12[c]
Fat (%)	2[c]	2[c]	2[c]	2[c]
Crude protein (%)	16.00	12.00	15.00	17.00
Inorganic nutrients				
Calcium (%)	0.4	—[d]	0.45[c]	0.75[c]
Phosphorus (%)	0.22	—[d]	0.37[c]	0.5
Magnesium (mg)	300–400	300–400	300–400	300–400
Potassium (%)	0.6	0.6	0.6	0.6
Sodium (%)	0.2[c,e]	0.2[c,e]	0.2[c,e]	0.2[c,e]
Chlorine (%)	0.3[c,e]	0.3[c,e]	0.3[c,e]	0.3[c,e]
Copper (mg)	3	3	3	3
Iodine (mg)	0.2[c]	0.2[c]	0.2[c]	0.2[c]
Iron	—[f]	—[f]	—[f]	—[f]
Manganese (mg)	8.5[f]	2.5[f]	2.5[f]	2.5[f]
Zinc	—[d]	—[d]	—[d]	—[d]
Vitamins				
Vitamin A (IU)	580	—[d]	>1160	
Vitamin A as carotene (mg)	0.83[c,d]	—[g]	0.83[c,d]	—[g]
Vitamin D	—[h]	—[h]	—[h]	—[h]
Vitamin E (mg)	40[i]	—[f]	40[h]	40[h]
Vitamin K (mg)	—[j]	—[j]	0.2[c]	—[j]
Niacin (mg)	180	—[k]	—[k]	—[k]
Pyridoxine (mg)	39	—[k]	—[k]	—[k]
Choline (gm)	1.2[c]	—[k]	—[k]	—[k]
Amino acids (%)				
Lysine	0.65	—[h]	—[h]	—[h]
Methionine + cystine	0.6	—[h]	—[h]	—[h]
Arginine	0.6	—[h]	—[h]	—[h]
Histidine	0.3[c]	—[h]	—[h]	—[h]
Leucine	1.1[c]	—[h]	—[h]	—[h]
Isoleucine	0.6[c]	—[h]	—[h]	—[h]
Phenylalanine + tyrosine	1.1[c]	—[h]	—[h]	—[h]
Threonine	0.6[c]	—[h]	—[h]	—[h]
Tryptophan	0.2[c]	—[h]	—[h]	—[h]
Valine	0.7[c]	—[h]	—[h]	—[h]
Glycine	—[h]	—[h]	—[h]	—[h]

[a]From Subcommittee on Rabbit Nutrition (1977). Used with permission.

[b]Nutrients not listed indicate dietary need unknown or not demonstrated.

[c]May not be minimum but known to be adequate.

[d]Quantitative requirement not determined, but dietary need demonstrated.

[e]May be met with 5% NaCl.

[f]Converted from amount per rabbit per day using an air-dry feed intake of 60 gm per day for a 1-kg rabbit.

[g]Quantitative requirement not determined.

[h]Probably required, amount unknown.

[i]Estimated.

[j]Intestinal synthesis probably adequate.

[k]Dietary need unknown.

Table X

Examples of Adequate Diets for Commercial Production[a]

Kind of animal	Ingredients	Percentage of total diet[b]
Growth, 0.5–4 kg	Alfalfa hay	50.00
	Corn, grain	23.50
	Barley, grain	11.00
	Wheat bran	5.0
	Soybean meal	10.00
	Salt	0.5
Maintenance, does and bucks, average 4.5 kg	Clover hay	70.00
	Oats, grain	29.50
	Salt	0.5
Pregnant does, average 4.5 kg	Alfalfa hay	50.00
	Oats, grain	45.50
	Soybean meal	4.0
	Salt	0.5
Lactating does, average 4.5 kg	Alfalfa hay	40.00
	Wheat, grain	25.00
	Sorghum, grain	22.5
	Soybean meal	12.0
	Salt	0.5

[a]From Subcommittee on Rabbit Nutrition (1977). Use with permission.
[b]Composition given on an as-fed basis.

ment period usually is needed before the rabbits will readily eat the diet in meal form. Intake may be low and spillage high during the adjustment period, and some rabbits may refuse to eat the nonpelleted diet (Hunt and Harrington, 1974). Examples of adequate diets for rabbits in commercial production or being held for biomedical use are shown in Table X.

D. Biology of Reproduction

1. Puberty and Sexual Maturity

Puberty and sexual maturity occurs earlier in smaller breeds than larger breeds. The smaller Polish or Dutch rabbits are usually first bred at about 4 months of age; the New Zealand white at 5–7 months; and the large Flemish or checkered giant at from 9 to 12 months.

Since does generally reach sexual maturity earlier than bucks and will copulate prior to being able to ovulate, it is generally advised that breeding be delayed until rabbits are almost fully grown.

The effective breeding life of does may be as long as 5 to 6 years. However, litter sizes generally decrease after several years. Many bucks may be used effectively for up to 6 years.

2. Reproductive Behavior

Rabbits, like the cat, ferret, and some other mustelids, are "reflex" ovulators, i.e., they do not ovulate spontaneously but rather ovulate 10 to 13 hr postcoitus, or after orgasm induced by other female rabbits, or after injection of luteinizing hormone or HCG. Twenty to 25% of does fail to ovulate even after copulation, however.

While rabbits do not have a regular estrous cycle, definite seasonal variations exist in reproductive performance both in wild and domestic rabbits. Wild rabbits are said to have a definite period of anestrus. Domestic rabbits may also have a period of anestrus that varies among colonies and individual rabbits (Hafez, 1970). In addition, domestic does may have a short (1–2 day) period in which they are not receptive every 4 to 17 days.

Information gained by vaginal cytology is not considered to be useful for determination of estrus or receptivity. Full sexual receptivity of does generally occurs when the vulva is congested and purplish in color and when the doe rubs her chin on the hutch or cage and becomes restless.

A doe to be bred should always be placed in the buck's home enclosure. This should be done for between 15 and 20 min during which time one should observe the pair to determine receptivity. If fighting occurs or no breeding takes place one might try another buck or pair the rabbits again the next day. If the doe is receptive to the buck, she will lie in the mating position and raise her hindquarters to permit copulation.

Patterns of exploration, smelling, chin rubbing, mounting, pelvic thrusting, intromission, and orgasm with ejaculation are relatively constant among individual rabbits.

3. Pregnancy and Gestation

The gestation period of the rabbit is from 30 to 33 days in 98% of pregnancies. Seasonal variations in length of gestation have been noted, with conception rates having a close inverse relationship with maximum ambiant temperature and not length of lighting. Larger litters in general are carried a shorter period of time. Litter size in prolonged gestation are usually small and may contain only one or two exceptionally large or stillborn young.

Experienced individuals can palpate developing fetuses *in uiero* as early as 10 to 12 days; however, 14 to 16 day fetuses are readily detected. Does beyond 2 to 3 weeks of gestation will generally refuse a buck. Radiologic confirmation of pregnancy can be performed after the eleventh day of gestation. Does begin hair pulling and nest building activities during the last 3 to 4 days of gestation.

4. Pseudopregnancy

Pseudopregnancy is common in does and may follow sterile matings, injection of LH, by stimulation by other nearby bucks, or mounting by other does. It can also occur when a doe mounts one of her own litter. It results when ovulation is followed by a persistent corpus luteum and generally lasts 15 to 17 days. Toward the end of pseudopregnancy does will perform ritualized hair pulling and nest building activities but will not keep the nest clean. The corpus luteum or corpora lutea secrete progesterone during pseudopregnancy and cause the uterus and mammae to enlarge.

5. Parturition (Kindling)

Does generally kindle in the early morning hours after building a nest and a period of 2 to 3 days of decreased food consumption. Delivery in uncomplicated cases usually takes less than 30 min, although young have been reportedly born anywhere from a few hours to several days apart. Both anterior and breech presentations are normal in the rabbit. Fetuses retained beyond 35 days generally die and if not expelled prevent future pregnancy.

Litter sizes usually range from 4 to 10, with the average size being 7 to 8. Litter size depends upon breed or strain of the rabbit, parity, state of nutrition, and environmental factors. Polish rabbits usually have less than four young; Dutch or Flemish four to five; and New Zealand white eight to ten young per litter.

Cannibalism of young by does is unusual but does occur. The doe normally severs the umbilicus and ingests the placenta. Cannibalism may be due to environmental or hereditary factors or as a result of nervousness of a disturbed doe.

6. Lactation

Mammary glands, usually four pairs, develop rapidly during the last week of pregnancy. Does normally nurse their young only a short period of time each day, usually in the early morning or night regardless of how many young there are or how many times the young attempt to suckle. Normally milk yield ranges from 160 to 220 gm/day. Maximum output occurs at 2 weeks following kindling, when it levels off and then declines during the fourth week. Does lactate for a 6–8 week period. Rabbit milk contains approximately 12.5% protein, 13% fat, 2% lactose, and 2.5% minerals.

7. Postpartum Breeding

While most does are not rebred until after weaning their young at 6–8 weeks, they may be bred immediately after kindling. This period of postpartum receptivity varies, lasting longer in does nursing small (2 to 3 young) litters, but declining rapidly in those nursing large litters. Success at postpartum breeding conception vary; however, one can produce a large number of young in a relatively short period of time by foster nursing young and rebreeding the doe immediately. In this way, one can obtain up to eleven litters from one doe in a year.

Conventional breeding, nursing, and weaning schedules allow only for 4 litters each year. Early weaning, i.e., at 3–4 weeks, can be accomplished by creep feeding, thus allowing up to six or seven litters per doe per year (Fox and Guthrie, 1968).

III. DISEASES

A. Bacterial

1. Pasteurellosis

Pasteurellosis is caused by *Pasteurella multocida*, and it is one of the most common diseases in conventional rabbit colonies. This disease results in economic losses through deaths, failure to gain weight, culling of sick animals, and in some cases failure of affected rabbits to reproduce. Rabbits are more susceptible to infection with *P. multocida* than most other laboratory animals, and several clinical forms of the disease occur. Among these are rhinitis ("snuffles"), pneumonia, otitis media and interna, conjunctivitis, abscesses, vulvovaginitis, pyometra, balanoposthitis and orchitis, as well as a generalized septicemia (Fig. 1).

Infected dams are thought to be responsible for infecting their young shortly after birth (Hagen, 1958). The organism is spread by both direct contact and aerosol. The nasopharynx is thought to be the site of initial colonization and infection, but venereal infection of the genital tract or skin wounds also may occur. Infection of the paranasal sinuses probably results in a life-long infection in most cases. The infection may be clinically silent for varying periods of time with intermittent bouts of mucopurulent nasal discharge. The discharge may be wiped from the nose with the medial aspect of the forepaws, resulting in a wetting and matting down of the fur.

After infection of the nasal cavity, *P. multocida* may spread by several routes including: (a) the eustachian tube to the middle ear and subsequently to the inner ear, meninges, and brain; (b) the nasolacrimal duct to the conjunctival sac; (c) the trachea to the lungs; (d) lymphatics to lymph nodes and the bloodstream; and (e) the blood to organs and tissues throughout the body (Flatt, 1974).

Presumptive diagnosis of pasteurellosis is based on the clinical signs, such as mucopurulent nasal discharge, torticollis, subcutaneous abscesses, and conjunctivitis. Definitive diagnosis requires isolation of the organism. Pasteurella-associated conjunctivitis must be differentiated from entropion seen in rabbits (Fox *et al.*, 1979).

Nearly all of the *Pasteurella* isolates are sensitive to penicillin. Treatment depends in part on the clinical manifestations observed. In rabbits with rhinitis, for example, individual rabbits may be treated with 60,000 IU/kg body weight of procaine penicillin for 10 days. Entire colonies may be treated by

Fig. 1. Saggital section through head of rabbit with snuffles. Mucopurulent exudate is adhering to nasal turbinates. From Flatt (1974).

adding broad spectrum antibiotics to the water (300 mg/liter tetracycline) or by the use of feed additive (sulfaquinozaline at 225 gm/ton of feed or furazolidone at 50 gm/ton of feed). Recurrence of clinical signs frequently accompanies the cessation of treatment.

Prevention of pasteurellosis is best accomplished by procuring rabbits from *Pasteurella*-free colonies. *Pasteurella*-free rabbits must be isolated to protect them from exposure to *P. multocida*. It has been shown the *Pasteurella*-free rabbits can be successfully maintained in a conventional animal facility in which *Pasturella*-infected rabbits were maintained (Scharf *et al.*, 1981). Entrance was restricted into the rooms of the *Pasteurella*-free rabbits, but these rooms were scattered throughout the conventional facility.

Some initial work has been done using a live streptomycin-dependent mutant of *P. multocida* (Chengappa *et al.*, 1980). Vaccinated rabbits were resistant to challenge. Further testing is necessary to determine the value of this vaccine.

2. Tyzzer's Disease

A profuse diarrhea, rapid dehydration, and high mortality characterize Tyzzer's disease in rabbits (Allen *et al.*, 1965).

Most often weanling rabbits are affected, but older animals also may be affected. Death usually occurs 12–48 hr after the onset of diarrhea. Postmortem examination reveals extensive necrosis of the mucosa of the distal ileum, the cecum, and the proximal colon. There may be subserosal hemorrhages in the cecum as well as focal areas of necrosis in the liver and heart. On histologic examination, the etiologic agent, *Bacillus piliformis* may be demonstrated in the cytoplasm of viable cells near the areas of necrosis in the liver, heart, or cecum. However, special staining of the tissue section with periodic acid–Schiff (PAS) or Warthin–Starry stains are needed to demonstrate the organisms (Figs. 2 and 3).

Rabbits may be infected with *B. piliformis* without showing signs of clinical disease. The organism is passed in the feces, and infection occurs by ingestion of spores. The organisms live within the epithelial cells of the cecum and cause little damage until the animal is stressed. Weaning and excessively hot weather are known to trigger the disease, and there are probably numerous other stresses that can initiate the disease process. In stressed rabbits, the organisms proliferate, cause necrosis in the mucosa of the cecum primarily, and gain entrance into the bloodstream. The organisms then colonize the hepatic parenchymal cells and myocardial cells and produce necrosis.

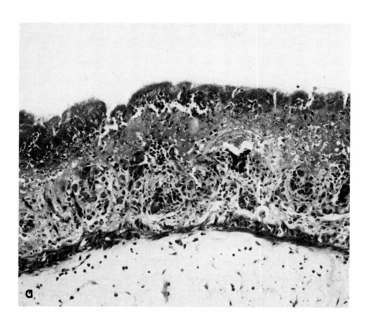

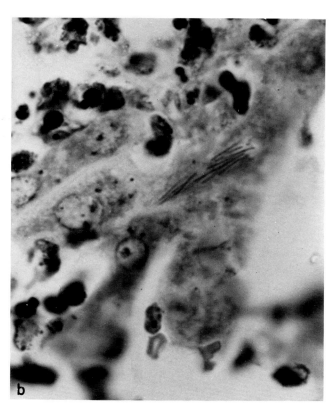

Fig. 2. *(a)* Cecum with extensive mucosal necrosis, bacterial growth on denuded surface, and submucosal edema in a case of Tyzzer's disease. (b) *Bacillus piliformis* growing in hepatic cells adjacent to a focal area of necrosis in Tyzzer's disease. Giemsa stain. From Flatt (1974).

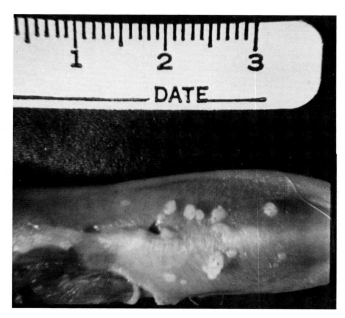

Fig. 3. Oral papillomatosis young, mostly sessile papillomas on ventrum of tongue. From Weisbroth (1974).

Presumptive diagnosis of Tyzzer's disease is usually made on the basis of typical clinical signs (profuse diarrhea) and gross lesions. Histologic demonstration of the organisms in the cytoplasm of viable cells adjacent to areas of hepatic necrosis confirms the diagnosis. *Bacillus piliformis* is an obligate intracellular parasite that has not been cultured on cell-free media, but it has been propagated in embryonating hen's eggs (Ganaway *et al.*, 1971).

Good management is important in preventing Tyzzer's disease. Minimizing the stress of weaning, providing temperature control, preventing overcrowding, and providing good sanitation are helpful practices. The treatment of Tyzzer's disease with antibiotics usually has not proved successful, but in one case the administration of oxytetracycline to an affected colony was associated with the termination of the outbreak (Van Kruiningen *et al.*, 1971).

3. Enterotoxemia

Clostridium perfringens type E iota toxin was demonstrated in the cecal contents of rabbits dying with acute diarrhea (Patton *et al.*, 1978). The postmortem lesions included fecal staining of the hindquarters, dehydration, and petechial and ecchymotic hemorrhages in the subserosa of the cecum. There were few histologic lesions observed other than the subserosal hemorrhage and submucosal edema in the cecum. More information is needed on this disease, but it appears that it may be an important part of the ill-defined mucoid enteropathy complex.

4. Treponematosis

The disease is characterized by dry, crusty exudate overlying ulcers in the skin and mucous membranes. The vulva and prepuce are the sites most commonly affected, but the lesions may occur near the anus, nose, eyelids, and lips (Cunliffe-Beamer and Fox, 1981a). The disease is caused by a spirochete, *Treponema cuniculi,* that can be transmitted venereally or by contact of the young rabbits with an infected dam (Cunliffe-Beamer and Fox, 1981a). Histologically, acanthosis of the epidermis is seen with erosions or shallow ulcers covered by necrotic cellular debris. Plasma cells and macrophages are present in the underlying dermis.

A diagnosis of treponematosis may be made dark field examination of material scraped from the lesions (Cunliffe-Beamer and Fox, 1981a). The rapid plasma reagen test also is used commonly.

Successful treatment of rabbits may be accomplished using three injections of 42,000 IU/kg body weight of benzathine penicillin G–procaine penicillin G given at 7-day intervals (Cunliffe-Beamer and Fox, 1981c). Eradication of treponematosis from an infected colony may require treating all rabbits in the colony (including neonates).

5. Other Bacterial Diseases

Numerous other bacterial infections may occur in rabbits (Flatt, 1974). *Staphylococcus aureus* causes fatal septicemia or suppurative inflammation. *Escherichia coli* or its endotoxin may be responsible for some of the outbreaks of fatal diarrhea. *Proteus, Pseudomonas, Streptococcus,* and other bacteria occasionally are responsible for disease outbreaks. Salmonellosis was recently diagnosed in specific pathogen-free rabbits stressed by surgery and irradiation (Newcomer *et al.,* 1983).

B. Viral

1. Myxomatosis

This disease occurs rarely in the United States, but can be seen in Western United States. It is endemic in the wild European rabbit (*Oryctolagus cuniculus*) population in Europe, parts of South America, and Australia. The disease also affects rabbits of the genus *Lepus* and *Sylvilagus.* In *Oryctologus,* myxomavirus infection causes deaths in up to 99% of the infected animals. Clinically, subcutaneous swelling occurs around body orifices and in the area of the face. As a result, the eyelids may be nearly closed, and subcutaneous edema in the ears causes them to droop. Histologically, the affected subcutaneous tissue contains undifferentiated mesenchymal cells

and some inflammatory cells lying in a matrix of mucinous material and edema.

The presumptive diagnosis of myxomatosis can be made on the basis of typical gross and microscopic lesions. Definitive diagnosis requires isolation of the virus. Since the disease is transmitted by arthropods as well as by direct contact, prevention usually depends on control of the arthropod vectors, principally mosquitoes and fleas. Some attempts have been made to vaccinate susceptible rabbits, and both the rabbit fibroma virus (Fenner and Ratcliffe, 1965) and a modified-live myxomatosis virus (McKercher and Saito, 1964) have been used as vaccines.

2. Rabbit Pox

There have been only six outbreaks of naturally occurring rabbit pox throughout the world since the disease was first recognized in 1932 (Maré, 1974). The entity occurs as a severe, highly fatal disease that may or may not be accompanied by cutaneous lesions. Rabbit pox is caused by a poxvirus that is closely related antigenically to the vaccinia virus. The virus appears to be transmitted through nasal secretions that may be inhaled or ingested by susceptible rabbits.

Clinically, affected rabbits are febrile, and they usually have a profuse nasal discharge. There may be enlargement of inguinal and popliteal lymph nodes. Skin lesions may occur about 5 days after infection, and they are characterized by a cutaneous rash with subsequent formation of nodules that are covered by crusty exudate (Christensen *et al.,* 1967). Edema of the face, oral cavity, and scrotum or vulva has been observed. Blepharitis, keratitis, and purulent conjunctivitis usually are present. Histologically, the skin lesions usually consist of necrosis surrounded by mononuclear cell infiltration. Edema and hemorrhage are present around these affected areas. Cytoplasmic inclusion bodies are seldom observed. There may be extensive necrosis of the lymphoid tissue throughout the body, and focal areas of necrosis may be present in the liver and other organs.

Diagnosis of rabbit pox depends upon isolation of virus or fluorescent antibody demonstration of viral antigen in affected organs. Since so few outbreaks of this disease have occurred, little is known about control and prevention. Isolation of sick animals has been of little value in preventing the spread of rabbit pox through a colony. Vaccination with vaccinia virus has been recommended to protect susceptible populations of rabbits during an epidemic (Maré, 1974).

3. Rabbit Papilloma

Wartlike growths have been recognized on the skin of cottontail rabbits from the midwestern United States (Shope and Hurst, 1933). This disease also has been recognized in domestic rabbits (Hagen, 1966). The etiologic agent is a DNA virus belonging to the genus *Papillomavirus,* and this agent can be transmitted from infected cottontail rabbits by arthropods such as ticks and mosquitoes.

The diagnosis of rabbit papilloma may be made on the basis of typical clinical and histopathological appearance. Little work has been done concerning the prevention of this disease because it has not been of economic importance. Protection from the arthropod vectors is an obvious control measure that could be taken in endemic areas (see Section III,G).

4. Oral Papillomatosis

There have been occasional reports of oral papillomatosis occurring in domestic rabbits. The disease is similar in appearance to that seen in some other domestic animals. It is caused by a DNA virus belonging to the same genus as the rabbit papilloma virus, *Papillomavirus*, but it is immunologically distinct from the rabbit papilloma virus (Maré, 1974).

This disease is usually discovered as an incidental finding, since it seldom causes clinical disease (Weisbroth and Scher, 1970; Dominqez *et al.*, 1981).

Diagnosis is based on typical gross and microscopic appearance. Since the lesions regress spontaneously, treatment usually is not necessary (see Section III,G).

5. Other Viral Diseases

There are a number of other viral agents that have been isolated from rabbits (Maré, 1974). Some of these agents are known to produce clinical disease (e.g., rabbit fibroma, hare fibromatosis, and rabies) and some apparently do not [e.g., *Herpesvirus cuniculi, Herpesvirus sylvilagus*, rabbit kidney vacuolating (RKV) virus, rabbit synctial virus, and snowshoe hare virus].

Many have suspected that some of the common outbreaks of diarrhea in domestic rabbits are caused by viruses. Thus far, there is little evidence upon which to make a judgement. Rotaviruses have been isolated or demonstrated by electron microscopy in some cases of diarrhea in rabbits (Bryden *et al.*, 1976). Also, antibodies to rotaviruses have been demonstrated in clinically healthy rabbits (Petric *et al.*, 1978). An adenovirus also has been isolated from rabbits with diarrhea (Bonden and Prohaska, 1980). It is clear that there will be new evidence forthcoming in the next few years concerning the role of viral agents in producing diarrhea in rabbits.

C. Mycotic

1. Superficial Mycoses (Ringworm, Favus)

Dermatophytoses are uncommonly encountered in the laboratory rabbit. They are more frequent in pet or backyard rabbitries where the marginal husbandry conditions favoring dermatophytosis are more prevalent, and rare under laboratory conditions or from commercial rabbitries with satisfactory hygienic standards. Dermatophytic infections are most often observed as individual, sporadic afflictions, although epizootics have been reported (Flatt *et al.*, 1974). As in other species, younger animals appear more susceptible than adults.

The dermatophyte most commonly isolated from cases of rabbit ringworm is *Trichophyton mentagrophytes*, both in the United States and abroad. Other isolates have included *Microsporum gypseum, M. Canis, M. audouini*, and *Trichophyton schoenleini;* however the validity of the latter two isolates is doubted (Flatt *et al.*, 1974).

Infection occurs through contact with infective forms of the fungus, e.g., macroconidia or arthrospores in the environment, or by direct contact with infected individuals. Dermatophytosis usually arises on or about the head in rabbits. The lesions are pruritic, and spread to the paws and other areas of the body occurs as a secondary phenomenon. The lesions are irregularly coin-shaped and may have ridge of acute inflammation around the periphery (simulating a worm, hence ringworm). Spread occurs by radial extension, with central healing. Affected skin has a patchy alopecia and a crusty or scurfy appearance. Histologic sections reveal hyperkeratosis and acanthosis with diffuse infiltration of the underlying dermis with both acute and chronic inflammatory cells. Abscessation of skin adnexa by secondarily invading bacteria is common. Mycotic elements may not be seen in hematoxylin and eosin stained preparations; however sections stained with periodic acid–Schiff or Gridley fungus stain will show abundant forms associated with hair shafts and epidermal structures.

Diagnosis of dematophytosis involves differentiation from other clinical entities giving a general appearance of crusty alopecia of the head and ears. Other such entities can include *Pasteurella*-related conjunctivitis with exudate scalding of maxillary skin, sarcoptic and notoedric mange, moist dermatosis ("slobbers"), fur pulling (self-induced or by cagemates), molting, and heritable hairlessness. Diagnosis is assisted by a carefully taken clinical history and physical examination. Dermatophytosis can be differentiated from mite infestation, and clinically diagnosed by skin scrapings mounted in 10% KOH. The scrapings should be taken from the periphery of the lesion and allowed to macerate in the KOH for at least 30–60 min prior to microscopic evaluation for fungal forms (mycelia or arthrospores). Finding fungal forms in the scraping should be followed up by Wood's light illumination of lesions for fluorescent forms, and culture of scrapings on a medium suitable for dermatophytic isolation. Diagnosis cannot be sustained by cultural retrieval alone; invasive fungal forms must be demonstrated in scrapings or histologic sections.

Treatment of dermatophytosis involves consideration of its zoonotic potential; active cases are infectious for humans, as well as other rabbits. Confirmed cases should be promptly isolated. Untreated individuals should be euthanatized, and their cages and immediate environment disinfected. Rabbit epizo-otics have been successfully treated with medicated diets containing 0.375 gm griseofulvin per lb of diet (Hagen, 1969). Individual rabbits may be treated with griseofulvin at a dose level of 25 mg/kg administered orally by gavage. In both cases, treatment should be continued for 14 days.

2. Deep Mycoses

The deep mycoses are extremely rare in rabbits. A recent review listed only one naturally occurring infection of significance of *Oryctolagus* laboratory rabbits; aspergillosis (Flatt *et al.*, 1974). Even aspergillosis is of historical interest, as husbandry standards have advanced in recent decades. Pulmonary aspergillosis caused principally by *Aspergillus fumigatus* is well documented in the older European literature, but has not apparently been reported from the Western Hemisphere.

D. Protozoan

As pointed out by Pakes (1974), while it is true that disease-free laboratory rodents are readily available, there are relatively few sources of rabbits of comparable quality. For this reason, several of the protozoan diseases remain of considerable clinical importance in laboratory rabbits. Preeminent among such conditions are coccidiosis and encephalitozoonosis. A more complete listing of protozoan agents infecting *Oryctolagus cuniculus* is to be found in Table XI.

1. Coccidiosis

Coccidiosis commonly occurs in laboratory rabbits in the United States, although both the incidence and severity of *Eimeria*-related disease have declined in recent years. The decline has been attributed to generally elevated husbandry standards for laboratory rabbits, and also to the effectiveness of chemotherapy (Pakes, 1974); however, increasing sophistication in diagnosis of enteric disease in rabbits has also played a role in reducing coccidiosis. Coccidiosis in rabbits occurs in the hepatic form caused by *E. stiedae*, and the intestinal form caused by several *Eimeria* species.

a. Hepatic coccidiosis. *Eimeria stiedae* parasitizes the bile duct epithelium of *Oryctolagus*, and is also considered infectious for *Sylvilagus* and *Lepus* (Pakes, 1974). The condition is common in the United States and is distributed globally. Hepatic coccidiosis, however, is increasingly less common in laboratory rabbits from modern, well-managed rabbitries. The life cycle is initiated by sporulation of oocysts shed in the feces. Natural transmission occurs by ingestion of sporulated oocysts that undergo excystation in the duodenum. The liberated spo-

Table XI
Major Protozoan Parasites of Laboratory Rabbits[a]

Class	Genus	Species	Disease
Sporozoa	*Eimeria*	*steidae*	Hepatic coccidiosis
		irresidua	Intestinal coccidiosis
		magna	Intestinal coccidiosis
		media	Intestinal coccidiosis
		perforans	Intestinal coccidiosis
		exigua	Intestinal coccidiosis
		intestinalis	Intestinal coccidiosis
		matsubayashii	Intestinal coccidiosis
		nagpurensis	Intestinal coccidiosis
		neoleporis	Intestinal ciccidiosis
		piriformsis	Intestinal coccidiosis
Toxoplasmida	*Toxoplasma*	*gondii*	Toxoplasmosis
	Sarcocystis	*cuniculi*	Sarcocystosis
Microsporida	*Encephalitozoon*	*cuniculi*	Encephalitozoonosis
Flagellates and amoebae	*Giardia*	*duodenalis*	Nonpathogenic
	Chilomastix	*cuniculi*	Nonpathogenic
	Monocercomonas	*cuniculi*	Nonpathogenic
	Retortamonas	*cuniculi*	Nonpathogenic
	Entamoeba	*cuniculi*	Nonpathogenic

[a]Adapted from Pakes (1974).

rozoites penetrate the intestinal mucosa and, by means uncertain at present, reach the epithelial cells of the bile ducts and then invade to undergo schizogony. As with other *Eimeria*, merozoites are produced that invade contiguous epthelial cells and undergo gametogeny to produce micro- and macrogametes and, finally oocysts. The oocysts rupture from the epithelial cells and enter the bile duct lumens to be conveyed out of the body with the feces. The prepatent period is about 15–18 days.

Light to moderate infestations by *E. stiedae* may be clinically silent. Heavy infestations are accompanied by clinical signs related to hepatic dysfunction and blockade of bile ducts. Anorexia, weight loss, diarrhea, and hepatomegaly are common clinical signs. Retention icterus and deaths are seen infrequently. Serum chemistry changes include increase in bilirubin, β- and γ-globulins, β-lipoprotein, and succinic dehydrogenase. Gross lesions are limited to the liver, which may or may not be enlarged and studded with irregular yellow-white foci of various sizes. The latter represent parasitized bile ducts filled with exudates consisting of oocysts, leukocytes, and necrotic debris (Fig. 4). An intense inflammatory infiltrate of initially acute, but later, chronic cell types surround infected bile ducts. The duct epithelium undergoes destruction as the oocysts mature, but later regenerate, and in the healing stages often appear hyperplastically reduplicated. Late stage lesions become fibrotic due to scar formation and may mineralize.

Diagnosis is made clinically by fecal flotation and identification of the characteristic oocysts (Pakes, 1974). During necropsy, wet mounts of gallbladder contents may be helpful in

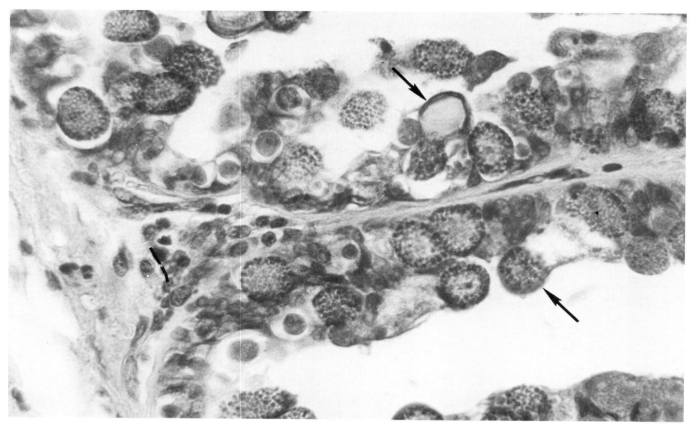

Fig. 4. Hepatic coccidiosis. Various developmental stages of *Eimeria stiedae* in bile duct epithelium. Arrows indicate macrogametes and developing oocytes. A few mononuclear inflammatory cells are evident. Hematoxylin and eosin. ×900. From Pakes (1974).

rapid diagnosis. Histologic sections (stained with hematoxylin and eosin) of parasitized liver may also be used to establish diagnosis by demonstration of typical life stage forms in bile duct epithelium.

The most effective agents for treatment and prevention of both hepatic and intestinal coccidiosis in rabbits are the sulfonamides. Sulfamerazine at the 0.02% level in drinking water and sulfaquinoxaline at the 0.05% level in drinking water, or 0.03% level in the diet for prolonged periods act as coccidiostats in preventing coccidial development. Coccidiostatic agents in the rabbit appear to be effective by retarding multiplication until host immunity develops sufficiently to be effective. Infection immunity resulting from *Eimeria* infestation in rabbits is quite effective and may be life long, inasmuch as coccidiosis is primarily a disease of the young. It is difficult to overstate the importance of good sanitation, cages with suspended floors, bottled or automatic drinking water systems, and other means of preventing, or limiting, contact of rabbits with fecal contamination. Rigid sanitization regimes can break the cycle of transmission and reduce or eliminate the means by which the organism is maintained and spread in the colony.

b. Intestinal Coccidiosis. At least ten *Eimeria* species are recognized as causes of intestinal coccidiosis in rabbits (Table XI). In the United States, four of these species are commonly encountered: *E. irresidua*, *E. magna*, *E. media*, and *E. perforans*. The life cycles of the intestinal coccidia are quite similar and are initiated by ingestion of infective (i.e., sporulated) oocysts. The latter undergo excystation in the small intestine, with liberation of sporozites that enter the intestinal epithelium to undergo multiplication by schizogony. Merozoites, thus produced, undergo gametogeny to produce macro- and microgametes, which in turn, produce oocysts by sporogony. The oocysts rupture the epithelial cells within which they develop and are released into the lumen to pass out of the body with the feces. The prepatent period for the intestinal forms is about 5–8 days, depending on the species.

The clinical signs in intestinal coccidiosis are variable and depend on the infective dose level, age of the host, *Eimeria* species involved, and other factors bearing on susceptibility. Light to moderate infestations are usually clinically silent. Heavy infestations may be accompanied by reduced weight gains and varying degrees of diarrhea. It is important for the

diagnostician to bear in mind that intestinal (and hepatic) coccidiosis often occurs simultaneously or is superimposed upon
underlying mucoid enteropathy. Thus the tendency to ascribe
etiologic significance to the presence to *Eimeria* oocysts when
diagnosing the cause of diarrhea must be carefull evaluated.
Few to a moderate number of oocysts in a flotation sample are
unlikely to reflect a level of infectivity sufficient to cause diarrhea, whereas large numbers may be more likely to be of
causal significance.

Lesions of intestinal coccidiosis are seen in the small and/or
large intestine, depending on *Eimeria* species. Parasitized epithelial cells with various stages of the life cycle are easily recognized in the hematoxylin and eosin stained sections. The
condition may be diagnosed by fecal flotation methods, but
considerable experience is necessary to speciate reliably the
oocysts of the various *Eimeria* species (Fig. 5). The pro-

phylaxis and treatment of intestinal coccidiosis is identical as
outline above for the hepatic form.

2. Encephalitozoonosis

Encephalitozoonosis is caused by infection of rabbits with
Encephalitozoon cuniculi, a microsporidan protozoan. The organism was named *E. cuniculi* in 1922, but taxonomic restruc-
ture resulted in a genus change to *Nosema* (and the corresponding disease to nosematosis) in 1964 (Pakes, 1974). Taxonomic
changes in 1974 reverted the genus name of the organism to
Encephalitozoon, by which it is known at present. Encephalitozoonosis is chronic, usually latent, and has a deserved repu-
tation over the years as a frequent cause of research complication (Shadduck and Pakes, 1971). The condition is common in
the United States and is globally distributed.

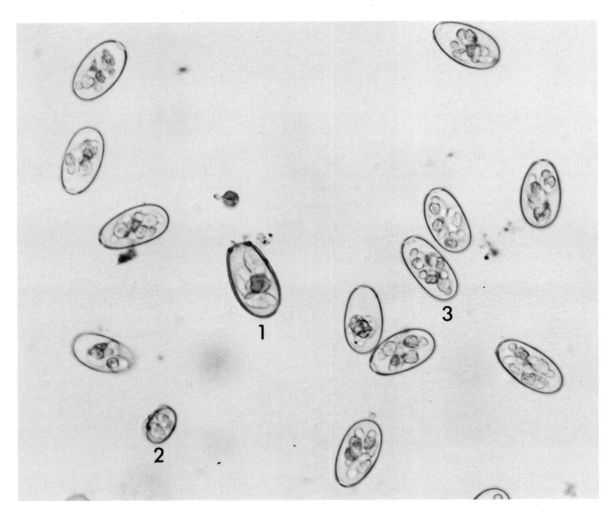

Fig. 5.   Flotation preparation of rabbit feces; a typical mixed *Eimeria* infection. Species include *E. magna* (1), *E. perforans* (2), and *E. stiedae* (3), ×600.
From Pakes (1974).

The life cycle of *E. cuniculi* is not known with certainty; but may be initiated by ingestion of spores originating from the urine of infected animals. Seroconversion to positive and rising titers indicative of infection occurs in at least half of the progeny born to serologically positive does by the eighth to ninth week of life (Lyngset, 1980). These facts suggest that although circumstantial evidence of transplacental (vertical) transmission has been reported, the dominant mode of transmission is horizontally by the oral route from contamination of the young with spore-bearing urine from infected dams during the fourth to sixth week of life. Susceptibility develops at the latter time as a consequence of the decline of colostral antibodies. The matures spores of *E. cuniculi* are 2.5 × 1.5 mm, oval and thick walled. They may be seen as either intracellular or extracellular parasites. The spores stain well with Giemsa, periodic acid–Schiff, gram, and Goodpasture's carbol fuchsin stains, but poorly with hematoxylin and eosin.

As mentioned above, the condition is usually latent, although neurologic signs have been described (Pakes, 1974). Lesions are consistently seen in the brain and kidney. In the latter, gross lesions are 1–2 mm multiple whitish foci or, more frequently, subcortical indentations. Microscopically, a granulomatous nephritis or the later stages of chronic inflammation, fibrosis, and tubular dilatation are noted (Fig. 6). Encephalitozoa may be seen within granulomatous foci or freely in tubular lumens using the stains cited above (Fig. 7). Mature renal lesions, indicated by fibrosis, may no longer have encephalitozoa. Lesions of the brain cannot be discerned grossly, but are seen microscopically as granulomatous encephalitis or encephalomyelitis. The granulomatous foci frequently contain organisms, but the latter may also be seen as aggregates (pseudocysts) in noninflamed areas. Perivascular cuffs of chronic inflammatory cells consistently accompany the presence of brain granulomas. Microscopic lesions occur in all brain areas, with a perivascular and periventricular distribution (Pakes, 1974).

Until fairly recently, the diagnosis of encepahlitozoonosis was limited to retrospective, histologic criteria. Within the last 5–7 years, however, serologic immunodiagnosis has been generally established. The serendipitous discovery of *E. cuniculi* as a contaminant of rabbit-origin choroid plexus cell cultures was the key development, permitting production of antigens to be used for diagnostic serology (Wilson, 1979; Shadduck, 1969). Two immunologic methods, in particular, the India ink immunoreaction (Waller, 1977) and indirect fluorescent anti body (Wosu *et al.*, 1977), have been widely exploited to identify *E. cuniculi*-infected rabbits for a variety of purposes, e.g., monitoring of commercial rabbit sources and screening of breeding colonies.

No methods for chemotherapeutic treatment of encephalitozoonosis are recommended. The means for prevention by immunologic surveillance and establishment of *E. cuniculi*-

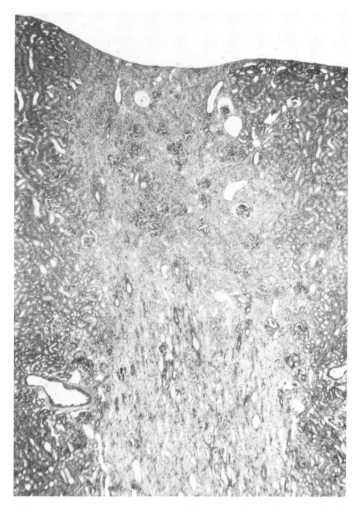

Fig. 6. Encephalitozoonosis. Intense fibrosis of kidney with interstitial infiltration of mononuclear inflammatory cells. Note depression of surface due to scarring. Hematoxylin and eosin. ×47. From Pakes (1974).

free breeding colonies have been described (Cox *et al.*, 1977; Bywater and Kellett, 1978).

E. Helminths

Although rabbits are susceptible to a wide range of naturally occurring and artificially induced infections with helminths, natural infections in laboratory or colony reared rabbits are generally rare and of minor significance if appropriate husbandry practices are followed.

Because rabbits are susceptible to many species of helminths, they are as important as an animal host that can be used in the laboratory to study these agents, which either naturally occur in the wild or are artificially induced in the laboratory.

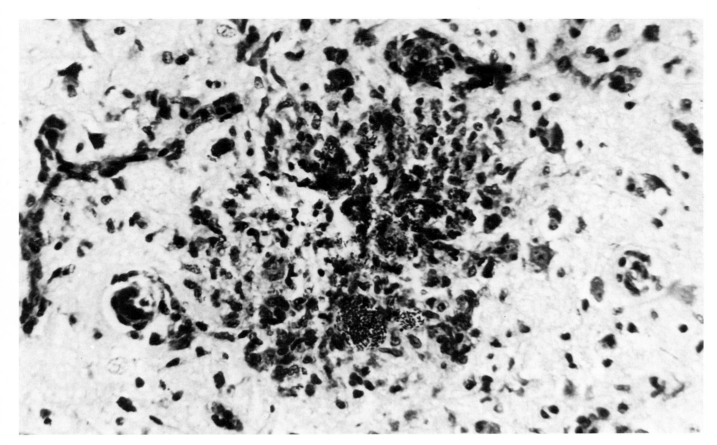

Fig. 7. Encephalitozoonosis. Granuloma with intra- and extracellular *E. cunicula* organisms. Goodpasture stain. ×475. From Pakes (1974).

The helminths afflicting rabbits include representatives of the nematodes, cestodes, and trematodes. Table XII presents a summary of the major helminths of both wild and laboratory or domestic rabbits. No mention is made of those parasites affecting exclusively wild rabbits, however.

Passalurus ambiguus is a typical oxyurid worm with a bulbar enlargement of the posterior portion of the esophagus and with females exhibiting a characteristic finely pointed tail. The ova are approximately 43 × 103 μm and are flattened on one side. The life cycle is direct with eggs deposited in the morula stage. Adults are found in the cecum and large intestines. Rabbits even with heavy infestation with this parasite are usually asymptomatic. Treatment, however, is generally attempted when this parasite is diagnosed. Piperazine adipate has been proved quite effective in eliminating this parasite from rabbits.

Dermatoxys veligera, another oxyurid, is found fairly frequently in wild rabbits and occasionally in domestic animals. They may be distinguished from *Passalurus* in that the males have much shorter posterior spicules and females have a much more posteriorly located vulva. Ova are slightly larger but very similar to *Passalurus*.

For a description of the biology of other helminth parasites of the rabbit, the reader is referred to Wescott (1974).

F.   Arthropods

Only two classes of arthropods—the Arachnida and the Hexapoda—are of significance as parasites of the rabbit.

The Arachnida consists of several important mites and ticks, including *Psoroptes cuniculi*, *Cheyletiella parasitovorax*, and *Haemophysalis leporis-palustris*. The Hexapoda or insects affecting rabbits are many. Several important insects affecting rabbits include *Haemodipsus ventricosus*, a sucking louse, various species of flies especially those in the family *Cuteribridae*, and several species of fleas and mosquitoes whose principal importance to rabbits are as vectors of disease.

Probably the most common ectoparasite of the domestic rabbit is the mite, *Psoroptes cuniculi*. This mite causes a disease that has been described by a variety of terms—ear mange, otoacariasis, ear canker, and psoroptic scabies, among others. This obligate, nonburrowing parasite usually first inhabits the inside of the ears at the bottom of the concha. The lesion then generally extends higher up on the inside of the ear and appears as a dry whitish-gray to tan crusty exudate that consists of desquamated epithelial cells, serum, inflammatory cells, mites, and mite feces. When the exudate is removed, the skin surface is moist, red, and very painful. The rabbit itself may be seen

Table XII

Helminth Parasites of the Laboratory of Domestic Rabbit

	Species	Common name	Life cycle	Prevalence	Experimental	Additional comments
I. Nematodes						
Family: Trichostrongylidae	*Obeliscoides cuniculi* (Graybill, 1923)	Rabbit stomach worm	Direct; adults in stomach	Primarily wild species; reported from domestic	Model for *Trichostrongylus* and *Ostertagia* spp. of ruminants	
	Nematodirus leporis (Chandler, 1924)		Direct; adults in small intestine	Primarily wild; reported from domestic		Several other species of *Nematodirus* are infrequently found in wild rabbits
	Trichostrongylus calcaratus (Ransom, 1911)		Direct; adults in small intestine	Commonly in wild; reported from domestic	Readily transmitted in laboratory	Several other species of *Trichostrongylus* are also found in wild rabbits
Family: Oxyuridae	*Passalurus ambiguus* (Rudolphi, 1819)	Rabbit pin worm	Direct; adults in caecum and large intestine	Common in both wild and domestic rabbits		Treat with piperazine adipate (0.5 gm/kg/day for 2 days for adults or 0.75 gm/kg/day for 2 days for young)
	Dermatoxys veligeria (Rudolphi, 1819)		Probably direct; adults in caeum	Common in wild; occasional in domestic		Can be confused with *Passalurus*
Family: Trichuridae	*Trichuris leporis* (Froelich, 1789)	Rabbit whip worm	Probably direct; adults in caecum and large intestine	High in wild; uncommon in domestic		Other trichurids reported in wild rabbits
II. Cestodes	*Cittotaenia variabilis* (Stiles, 1895)		Indirect; rabbit is the definitive host; oribatid mites are the intermediate host; found in small intestine	Primarily in wild; rare in domestic rabbits		Other species of *Cittotaenia* reported from wild rabbits
	Taenia pisiformis (Bloch, 1780)		Indirect; rabbit is the intermediate host; cysticerci found in liver and attached to mesentery; dog is the definitive host	Common in wild, occasionally found in domestic	Commonly used	
	Taenia serialis (Gervais, 1847)		Indirect; rabbit is the intermediate host; coenurus found in connective tissue of muscle, dog is the definitive host	Occasionally found in wild; rare in domestic		
III. Trematodes	No species of trematode has been reported as naturally occurring in the laboratory or domestic rabbit					

shaking its head or scratching its head or ear with its rear feet, leading to self-mutilation and secondary bacterial infections.

Diagnosis of psoroptic otoacariasis can readily be made from the clinical signs and positive identification of the mite itself in the detritis or on the raw oozing surface if the debris is removed. While the mites are large and can be seen with the unaided eye, an otoscope is useful. Specimens may be readily identified when placed on a microscope slide and examined under the low power objective of a microscope (Fig. 8).

Since under favorable environmental conditions, a complete life cycle may take as little as three weeks, the numbers of mites increase tremendously over time. Treatment with commercial preparations containing an acaricide or an acaricide and antibiotic-antiinflammatory agent may be used. In mild cases or as a prophylactic measure, mineral oil alone may be used to treat the ears.

While chorioptic mites do not presumably affect the rabbit sarcoptic (*Sarcoptes scabiei*) and notoedric (*Notoedres cati*)

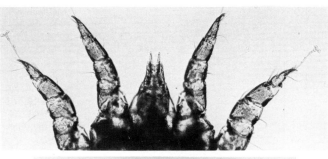

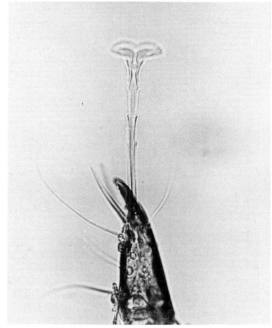

Fig. 8. *Psoroptes cuniculi.* Mouth parts and anterior two pair of legs (top). Jointed or segmented pedicel characteristic of the genus (bottom). From Kraus (1974).

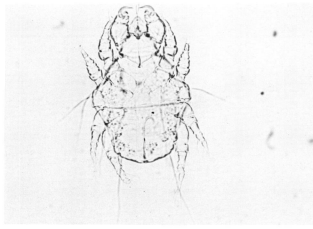

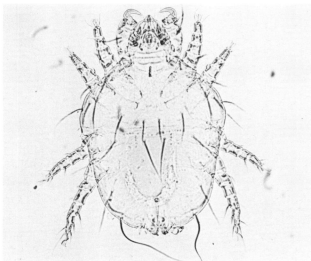

Fig. 9. *Cheyletiella parasitovorax* fur mite. Male (top). Female (bottom). From Kraus (1974).

mites are reported to effect the laboratory rabbit, but much less frequently than *Psoroptes cuniculi.*

The cheyletid fur mite (*Cheyletiella parasitovorax*) affects the rabbit, as well as man. Many infected rabbits do not have overt signs of disease, however. In heavy infestations, lesions are generally located principally on the dorsal trunk and interscapular region, although lesions are sometimes found elsewhere, for example, the ventral abdomen. This parasite is also an obligate, nonburrowing mite and lives in the keratin layer of the epidermis (Fig. 9). The affected area may be partially alopecic with a fine grayish-white slightly oily scale. The lesions are not pruritic as are the lesions caused by psoroptic, sarcoptic, and notoedric mites. Mites may be readily collected and identified from the lesions. Silica gel acaricides are effective in treated affected rabbits.

For tabular information on the arthropods affecting the rabbit, see Table XIII. Most of the other arthropod parasites are

	Species	Common name	Location on host	Prevalence	Comments
I. Class: Arachnida **A. Mites**	*Psoroptes cuniculi* (Delafond, 1859)	Rabbit ear mite	Large nonburrowing obligate parasite of inner surface of ear; also can be found outside ear	Common in domestic and laboratory; not reported in wild rabbits	Spend entire life cycle in host; host specific life cycle completed in as little as 3 weeks. Treat with mineral oil and/or commercial acaricide with/without antibiotics
	Notoedres cati, Sarcoptes scabiei	Sarcoptic mange mite	Burrowing obligate parasites of nose, lips, face external genitalia and outer surface of pinna	Once common; now infrequent in either domestic or wild rabbits	Spend entire life cycle in host; not host specific; zoonotic; life cycle completed in as little as 3 weeks; recommend culling; may however be treated with many preparations
	Cheyletiella parasitivorax (Megnin, 1878)	"Fur mite"	Small nonburrowing obligate parasites of abdomen, back, and scapular regions of dorsal trunk; may be on other areas as well	*C. parasitivorax* common in domestic; other species rare	Affects the rabbit and man, other cheyletid mites reported from rabbits may not produce disease; a known vector of myxomatosis, may be treated with topical silica-gel acaracides
	Listrophorus gibbus (Pagenstecher, 1862)		Small nonburrowing mites found primarily on hair of back and abdomen; occasionally elsewhere	Infrequent in domestic rabbits; true prevalence unknown; found in wild *Oryctolagus* and *Lepus* also	Considered non-pathogenic
	Linguatula serrata (Froelich, 1789)	Pentastome or "tongue worm"	Indirect life cycle. The one reported case in domestic rabbits had liver and lung lesions	Rare in domestic; uncommonly reported in wild, but wild rabbits considered as natural intermediate host; definitive host is the dog or wild carnivore	Generally asymptomatic; incidental finding; no treatment reported
B. Ticks	*Haemaphysalis leporis-palustris*	Continental rabbit tick	Three host ixodid tick feeds for only 3–5 days each time on rabbits; therefore it is rarely encountered; found primarily on head, ears, and back of neck; sometimes around eyes and back of neck	Common in wild rabbits infrequent in domestic	Larval and nymphal stages are not host specific and can infest birds, domestic animals, dog, cat, and even man; prefers the rabbit as definitive-host; Rocky Mountain spotted fever and Q fever among wild rabbit populations; other ixodid and argasid ticks affect wild rabbits

(continued)

Table XIII (*Continued*)

	Species	Common name	Location on host	Prevalence	Comments
II. Class: Hexapoda A. The Anoplura	*Haemodipsus ventricosus* (Denny, 1842)	Sucking louse of rabbits	Obligate parasites, no specific anatomical predilections reported	Occasionally reported in domestic	Voracious blood suckers that can cause serious disease if heavily infested; treat with organophosphates or pyrethrins
B. The Diptera 1. The Cuterebrids	*Cuterebra horripilum, C. buccata*	Bot flies, warble flies	Larvae burrows into a subcutaneous site; prefer ventral cervical regions (*C. horripilum*) or anywhere on trunk (*C. buccata*)	No data available for domestic; occurs in domestic rabbits housed outdoors; common in wild rabbits	Cuterebriasis, myiasis opthalmomyiasis; can kill if multiple larvae
	Cochliomyia	Screwworm	Invades preexisting wounds; most common on face	Similar to *Cuterebra* species	Affects primarily nestling rabbits
	Callitroga	Blowfly or screwworm	Invades preexisting wounds	Similar to *Cuterebra* species	
	Wohlfahrtia vigil	Blowfly or screwworm	Invades preexisting wounds but larvae can penetrate sparsely haired areas of skin	Similar to *Cuterebra* species	Affects primarily nestling rabbits; adults are larviparous; screen all outdoor hutches
2. Fleas	*Cediopsylla simplex*	Common eastern rabbit flea	Primarily on ears, around the face, top of head, and back of the neck	Common in wild and domestic	Relatively few on any given animal; generally little clinial signs; can transmit RMSF[a], plague and a host of other diseases
	Odontopsyllus multispinosis	Giant eastern rabbit flea	Back and hind end of the rabbit	Wild only, but very common	Similar to *C. simplex*
	Spilopsyllus cuniculi	Common European rabbit flea	Similar to *C. simplex*	Common in Europe; not United States	Similar to *C. simplex*
3. Mosquitoes	Many species			Uncommon in domestic, common in wild	Important vectors of myxomatosis, other diseases in the wild

[a]RMSF, Rocky Mountain spotted fever.

relatively uncommon or unimportant in properly managed domestic or laboratory rabbit colonies. For details on all of these parasites which afflict rabbits see the review by Kraus (1974).

G. Neoplasia

1. Naturally Occurring Neoplasms

In this section, a brief review will be made of factors bearing on age, sex, and incidence of rabbits with "naturally occur-

ring" neoplastic tumors. Naturally occurring defines tumors without known antecedent triggers (carcinogens).

Tumors, of all types, increase in incidence as a function of age. It is important to bear in mind that the life span of normal, healthy *Oryctolagus* laboratory rabbits is 6–7 years. Most rabbits in university or industrial research laboratories will be in the 2–18 month old age group; essentially adolescent or young adult. Following completion of experiments, these animals are usually euthanatized and thus do not ordinarily reach a tumor-prone age group. The same factors are operative in the management of commercial breeding colonies, as well, because the

Table XIV

Age–Sex Factors in Neoplastic Incidence in Laboratory Rabbits[a,b]

Category	N	Neoplasms	Percentage of total N with neoplasm	Percentage of age–sex category with neoplasm	Percentage of age category with neoplasm	
M(−)	217	1	0.16	0.46	(+)	(−)
M(+)	54	5	0.83	9.26		
Total M	271	6	0.99			
F(−)	275	6	1.00	2.16		
F(+)	53	4	0.66	7.54		
Total F	328	10	1.66			
Total M + F	599	16	2.65		8.4	1.4

[a]Adapted from Weisbroth (1974).

[b]N = number of rabbits in series; M, male; F, female; (−), <2 years of age; (+), >2 years.

reproductive efficiency of both sexes declines after the second year of life. This fact alone imposes great constraints on the variety and frequency with which naturally occurring neoplasia are likely to be seen in the practice of laboratory animal medicine, and this is underscored by the frequency distribution of tumors in rabbits < 2 years of age (1.4%) versus those > 2 years of age (8.4) (see Table XIV).

Sex also influences the age and frequency of various tumor types. Uterine adenocarcinoma appears to be the most common tumor type in rabbits (Weisbroth, 1974; Stedham, 1976), thus the incidence, viz male vs. female, is skewed in the direction of females by this single tumor type (see Table XV). Although Table XV may to some degree be biased by the Weisbroth's experience, the same general relationships have been reported for several similar reviews, i.e., that tumors are

Table XV

Incidence of Various Tumor Types in Laboratory Rabbits[a,b]

Tumor type	No.	M(−)	M(+)	F(−)	F(+)
Uterine adenocarcinoma	5			4	1
Lymphosarcoma	4		1	2	1
Embryonal nephroma	2	1		1	
Bile duct adenoma	2		1		1
Osteosarcoma	1		1		
Osteochondroma	1				1
Leiomyoma	1		1		
Basal cell adenoma	1				1
Rectoanal papilloma	1		1		
Total	18	1	5	7	5

[a]Adapted from Weisbroth (1974).

[b]M, male; F, female; (−) <2 years of age; (+) >2 years of age.

more frequent in females than males and that the four most common naturally occurring tumor types—(in ranked order) uterine adenocarcinoma, lymphosarcoma, embryonal nephroma, and bile duct adenoma of the liver—accounted for more than 80% of the tumors in the series.

2. Neoplasms Induced by Oncogenic Viruses

Proliferative and truly neoplastic changes associated with certain poxviruses and papovaviruses in rabbits have played an important role in furthering the understanding of viral oncogenesis as a basic biological process. The natural history, virology, and pathology of these viruses has been reviewed earlier in this chapter. Interestingly, all of them with the exception of oral papilloma virus, namely, myxomatosis, the Shope fibroma viruses, Shope papilloma virus, and the Hinze herpevirus lymphoma, have been discovered and limited in natural distribution to the American continents. The viruses and the lesions induced are of great historical interest because early concepts of viral oncogenesis in mammalian species were formed principally on the basis of experimentation with these agents.

a. Rabbit Oral Papilloma. The rabbit oral papilloma virus (ROPV) is the only virus of the group having the laboratory rabbit (*Oryctolagus*) as the natural reservoir host. The virus causes wartlike growths (papillomas) not only on the ventral aspect of the tongue anterior to the frenulum, but may also be seen on the nearby epithelium of the oral cavity and gingiva (Weisbroth, 1974). The virus is transmitted in oral secretions containing virus-bearing sloughed cells from the warts. These secretions are ingested by susceptible rabbits, especially those prior to weaning, and infection occurs in the abraded epithelium of the tongue (Weisbroth and Scher, 1970). The papillomas grow slowly over a period of 3–9 months, during which time they mature from a domelike shape to peduncu-lated, cauliflowerlike masses about 4–6 mm in size. Basophilic intranuclear inclusion bodies (demonstrable with hematoxylin and eosin or Giesma stains) are found in epithelial cells below the cornified strata. Rejection occurs when the rabbit becomes sufficiently immune and proceeds by chronic inflammation at the base of the papilloma, causing sloughing of the tumor, ulcer formation, and, finally, reepithelization. Oral papillomas of the rabbits are not known to undergo the carcinomatous transformation characteristic of the Shope papilloma or of oral papillomas in the canine.

b. Shope Papilloma and Papilloma-Derived Carcinomas. The Shope papilloma was originally limited to *Sylvilagus floridanus* rabbits of the American Midwest and Mississippi River basin. Richard Shope, who characterized this virus, did so with naturally infected *Sylvilagus* rabbits from Iowa. The condition is not natural to eastern cottontails, but is now known to be enzootic in California cottontails, *S. bachmani*. The disease may occasionally occur naturally in laboratory *Oryctolagus* rabbits.

The natural disease is characterized by the formation of cutaneous, pigmented warts (papillomas) 0.5–1.0 cm wide at the base and which project 0.5–1.0 cm or higher from the skin. Naturally infected cottontails may carry from 1–10 warts with some predilection for the skin of the inner thighs, abdominal ventrum, neck, and shoulders, whereas natural infections of *Oryctolagus* rabbits occur on the ears and eyelids. In cottontails, approximately 36% of naturally acquired papillomas regress within 12 months, although 25% of papillomas undergo transformation to squamous cell carcinoma. Inclusion bodies visible with the light microscope, typical with other wart viruses, e.g., ROPV, have not been observed with Shope papilloma virus infection.

Shope papilloma virus (SPV) is usually recoverable from papillomas in *Sylvilagus* up until the time they undergo regression or malignant transformation. Shope papillomas in *Oryctolagus*, however, contain SPV in a masked or latent form that is noninfectious. Masked SPV is present in the carcinomas that develop from transformed papillomas. That SPV antigen is present in the carcinomas has been demonstrated immunologically. This finding led to the concept of masked or "immature" noninfective, but antigenic SPV. Masking is not the final stage of the papilloma–carcinoma sequence, however. One much-investigated transplantable malignant carcinoma derived through this sequence (the VX2, V-2 or V_2 carcinoma) was shown after the forty-sixth transplant generation to have lost all masked antigens to SPV.

c. Shope Fibroma.

Few host–virus interactions have been characterized within the context of coevolution to the extent that they have between the various species of rabbits, on the one hand, and the poxviruses of the myxoma-fibroma subgroup, on the other (Weisbroth, 1974). The reader is referred to several excellent monographs on this subject (Fenner and Ratcliffe, 1965; Gross, 1970; Fabvre, 1962). Fibromalike proliferative lesions of the skin are found in the wild among the three principal species of American *Sylvilagus* rabbits. Although the histology of such fibromas is quite similar in the *Sylvilagus* hosts, the biological character differs, as may be determined by inoculation into *Oryctolagus* rabbits. Thus, the fibroma of the South American *S. brasiliensis* causes classic myxomatosis when inoculated into *Oryctolagus*. In similar fashion, myxomatosis is carried enzootically in California by *S. bachmani* in fibromas of the skin. Epizootics of myxomatosis in Californian *Oryctolagus* rabbits occur by mosquito-borne transfer from fibroma-bearing *S. bachmani* cottontails.

The enzootic fibroma of the eastern and midwestern cottontail, *S. floridanus* does not cause a rapidly fatal myxomatosis-like disease when inoculated into *Oryctolagus*, but, rather, a fibroma much like that in the *Sylvilagus* host. It has been established that any of a variety of hematophagous arthropods, e.g., fleas and mosquitoes, may transmit disease under natural conditions. It remains uncertain whether the insect vector is merely a "flying pin" or if the virus replicates in the vector. Skin-piercing vectorism is necessary since infected rabbits can be maintained in the laboratory without risk of contagion to nearby cages.

Newborn rabbits of both *Sylvilagus* and *Oryctolagus* are more susceptible than those 2–3 weeks of age or older. Rabbits older than 2–3 weeks of age effectively respond immunologically to Shope fibroma virus (SFV) to limit consequences of infection. The resistance can be modified by any of a range of immunomodulating factors that suppress the immunologic response and which induce generalized metastatic fibromatosis. Resistance can be enhanced by concurrent infection with other viruses. A recent report, apparently the first, has described lesions intermediate in character between myxomatosis and Shope fibromas in naturally infected *Orytolagus* rabbits in Texas (Maré, 1974), underscoring again the spectrum of viruses in this group.

d. Hinze Herpesvirus Lymphoma.

Hinze reported in 1968 the isolation of a new herpesvirus from *S. floridanus* cottontails in Wisconsin. In a series of publications, it was shown that the CHV-1 viral isolate, recovered initially from pooled kidney tissue cultures, would induce cytopathogenic effects in rabbit tissue cultures and a lymphoproliferative disease in *Sylvilagus* rabbits (Hinze, 1971). The virus subsequently called *Herpesvirus sylvilagus*, was shown to be distinct from other herpesviruses, including *Herpesvirus cuniculi* (Herpesvirus III). Cottontail rabbits inoculated with the agent develop a persistent viremia, lymphocytosis, leukemia, lymphadenopathy, splenomegaly, and renal enlargement (lymphosarcoma). *Oryctolagus* laboratory rabbits were refractory to inoculation.

H. Heritable

At the present time, there are approximately 70 known genetic loci in the rabbit (Lindsey and Fox, 1974). About one-third of them are related to pelage types and colors, one-third to blood group and tissue antigen types, and the remainder to anatomic variants and heritable diseases. It is beyond the scope of this chapter to fully summarize the scope of genetics in the rabbit, thus the discussion here will emphasize the more important entities of the latter group, *viz.*, heritable diseases.

The heritable diseases may be conveniently divided into two broad groups; the mutants (those controlled by single genes) and the so-called familial or polygenic conditions. Heritable conditions are summarized in Table XVI; and those more commonly encountered in laboratory animals settings are discussed below.

1. Hydrocephalus

Hydrocephalus is considered here as a mutant, although it has been pointed out (Lindsey and Fox, 1974) that this does not fully suffice to explain many instances when hydrocephalus is encountered clinically. In both reports describing autosomal inheritance of hydrocephalus, the condition always occurred in association with other anatomic variants (Lindsey and Fox, 1974). Similarly, hydrocephalus has been observed sporadically on many occasions without appearing to segregate in a heritable pattern. The anatomic condition may reflect a more profound and underlying defect in vitamin A metabolism. It is well recognized that hydrocephalus of the young may be readily produced in rabbits by inducing hypovitaminosis A in the pregnant doe (Lindsey and Fox, 1974).

2. Buphthalmia (Congenital or Infantile Glaucoma)

This mutant occurs quite commonly in laboratory rabbits from contemporary New Zealand white stocks. It is important to recognize that the buphthalmic eye is both enlarged as well as turgid (hard) due to increased intraocular pressure. Thus, while the condition has received some attention as a model of the analogous condition in man, it differs as a result of the less mature scleral coat (in rabbits) at the time of increasing pressures, which permits enlargement of the globe (Lindsey and Fox, 1974). It is well established that most of the ophthalmic lesions of the rabbit condition are related to abnormalities of production and impedance to circulation of aqueous humor from the anterior chamber.

Buphthalmia segregates as an autosomal recessive with incomplete penetrance. The latter attribute is used to explain variation in age of onset, severity (including phenotypic normals in some *bu/bu* individuals), and the fact that the condition may be uni- or bilateral. Although onset may be as early as 2–3 weeks after birth, more commonly it is seen at 3–4 months of age. The condition initially presents as an increase in anterior chamber size with gradual cloudiness (opacity) and bluish tint in the cornea. Later changes include progression of opacity to a milky white color, prominence of the globe and conjunctiva, and scarring and vascularization of the cornea. Structural changes in the eye include widening of the angle, thickening of Descemet's membrane, increased corneal diameter, vascularization and opacification of the cornea, atrophy of the ciliary process, and excavation of the optic disc. Such individuals should be removed from breeding colonies.

Table XVI

Common Heritable Conditions of the Laboratory Rabbit[a]

I. Conditions controlled by single genes (mutants)
  A. Behavioral mutants
    1. Acrobat (*ak/ak*)
    2. Epilepsy (audiogenic seizures) (*ep/ep*)
  B. Neuromuscular mutants
    1. Ataxia (*as/ax*)
    2. Tremor (*tr/tr*)
    3. Paralytic tremor (*pt/−*, male; *pt/pt*, female)
    4. Hydrocephalus (*hy/hy*)
    5. Lethal muscular contracture (*mc/mc*)
    6. Syringomyelia (*sy/sy*)
  C. Ocular mutants
    1. Buphthalmia (congenital glaucoma) (*bu/bu*)
    2. Cataracts (*cat-1* or *cat-2*)
    3. Cyclopia (*cy/cy*)
    4. Red eye (*re/re*)
    5. French rex keratitis (r_1/r_1)
  D. Oral cavity mutants
    1. Mandibular prognathism (*mp/mp*)
    2. Absence of second incisors (I^2/I^2, I^2/i)
    3. Supernumerary second incisors (*isup/isup*)
  E. Skeletal mutants
    1. Achondroplasia (*ac/ac*)
    2. Chondrodystrophy (*cd/cd*)
    3. Dachs (*Da/Da*)
    4. Dwarf (*Dw/Dw* or *nan/nan*)
    5. Hereditary distal foreleg curvature (*fc/fc*)
    6. Brachydactylia (*br/br*)
    7. Spina bifida (*sb/sb*)
    8. Hypoplasia pelvis (*hyp/hyp*)
    9. Femoral luxation (*lu/lu*)
    10. Osteopetrosis (*os/os*)
  F. Hematologic mutants
    1. Atropinesterase (*As/As; As/as*)
    2. Red cell esterases
    3. Pelger (*Pg/pg*); super-pelger (*Pg/Pg*)
    4. Lymphosarcoma (*ls/ls*)
    5. Hemolytic anemia (*ha/ha*)
  G. Genitourinary mutants
    1. Hypogonadia (*hg/hg*)
    2. Renal agenesis (*na/na*)
    3. Renal cysts (*rc/rc*)
  H. Other mutants
    1. Yellow fat (*y/y*)
    2. Adrenal hyperplasia (*ah/ah*)
II. Familial or polygenic conditions
  A. Abnormal ear carriage
  B. Abnormalities of calvarial shape
  C. Acromegaly
  D. Aortic arteriosclerosis
  E. Hereditary premature senescence
  F. Scoliosis
  G. Splay leg
  H. Waltzing (Circling)
  I. Hypertension

[a] Adapted from Lindsey and Fox (1974).

3. Mandibular Prognathism (Malocclusion, Walrus Teeth, Buck Teeth)

In a manner analogous to hydrocephalus, malocclusion is considered here as a mutant (mandibular prognathism), although it commonly occurs as a sporadic condition in rabbit stocks without appearing to segregate as an autosomal recessive. It is commonly encountered in most commercial and institutional breeding colonies.

The single pair of large, chisel-shaped mandibular incisors is normally opposed, as well as a pair of much smaller secondary incisors. The primary maxillary incisors undergo attrition along an arc formed by biting movements of the lower incisors, while the maxillary secondary incisors wear at right angles to the mandibular incisors. Inasmuch as the enamel of rabbit (and rodent) incisors has a heavier layer anteriorly, and a lighter layer posteriorly, normal attrition wears the posterior aspect faster, thus maintaining a sharp biting edge.

Rabbit teeth, both cheek teeth (molars, premolars) and incisors, grow through life and are regulated by the attrition of occlusion to the normal length. The incisors, for example, grow at the rate of 2.0–2.4 mm/week (4–5 in. per year!). When normal occlusion is impeded from any cause, overgrowth of the incisors (or cheek teeth) results. Hereditary mandibular prognathism (mp/mp) is the name given to a common heritable form of this condition (Fox and Crary, 1971). Heritable malocclusion may make its appearance as early as 2–3 weeks after birth, but more commonly is not observed until the 8th–10th week of life, or later. Once out of occlusion, the mandibular incisors come to lie anterior, and at a flat angle to the uppers; while the later curl around within the oral cavity and, unless relieved, grow dorsally to pierce the gingival or buccal mucosa, and may progress to abscessation.

Impaired closure of the mouth and inability to apprehend and masticate food usually leads to malocclusion of the cheek teeth and progressive starvation (Fig. 10). When encountered, it is common practice to carefully cut the incisors back to approximately the normal length (which does not open the pulp cavity) with bone or wire cutters. Such animals need to be trimmed at 2–3 week intervals. Rabbits with malocclusion should not be used for breeding purposes.

4. Splay Leg

The name "splay leg" is given to a rather common condition observed in nursling rabbits during the first few weeks of life. Rabbits so affected cannot adduct the affected limbs and assume normal postural movements. The condition may be uni- or bilateral and may affect the anterior of posterior limbs or all four. Rabbits with this condition "splay" the affected limbs horizontally and can make only weak, ineffectual movements,

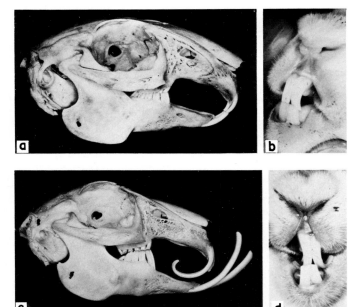

Fig. 10. Skull (a) and mouth (b) of a 10-week-old *mp/mp* rabbit with edge-to-edge occlusion of primary incisors. As such affected animals become a few weeks older, the lower incisors move past the uppers, resulting in a full expression of mandibular prognathism. (c) and (d) Rabbits approximately 3 months of age with extreme mandibular prognathism (*mp/mp*). Impaired closure of the mouth due to overgrown incisors has resulted in reduced wear of jaw teeth. The upper large incisor on the distal side of the skull has broken off. Also, the skull lacks "peg teeth," possibly representing an example of the dominant trait known as "absence of second incisors" (I^2/I^2 or I^2/i^2). Note the normal grooves of the anterior surface of the upper incisors of rabbits in (a) and (d). From Lindsey and Fox (1974).

although they appear mentally normal. Inanition due to inability to reach food and water is a common terminating event.

At the present time, the cause of "splay leg" is not understood. It is not even clear that it is a heritable condition. It may represent the leading clinical manifestation of one of several morphological mutants (Lindsey and Fox, 1974), or perhaps one of several teratologic (congenital) malformations.

I. Management Related

1. Mastitis

Inflammation of the mammary glands (mastitis) occurs most commonly in lactating does, but may occasionally be seen in does in pseudocyesis. It is thought that heavy milk producers are predisposed to mastitis (Flatt *et al.*, 1974), but injury to the glands, inoculation of the teats by nursing pups, and poor sanitary standards are predisposing as well. Affected does have fever (104°–105°F), inappetence but increased thirst, and become depressed. Death may supervene due to the bacterial tox-emia or septicemia. Involved mammary glands are swollen, firm, and hot. The overlying skin is initially pink due to hyper-emia, but progressively becomes bluish due to stasis. Cases are usually sporadic, but may also be quite contagious. Foster nursing of pups withdrawn from an affected doe is not recom-mended because of the risk of spreading the infection to the healthy doe. The microscopic features of mastitis in rabbits is similar to that of acute bacterial mastitis in other species and include congestion, hemorrhage, intense infiltration of mam-mary structures with acute inflammatory cells, and exudation with microcolonies of bacteria. Staphyloccoal species are most commonly encountered as isolates from field cases of mastitis, but streptococci may be causal as well (Flatt *et al.*, 1974). If treatment is attempted, the affected doe should be isolated and given 50,000–100,000 units of penicillin intramuscularly twice daily for 3–5 days.

2. Ulcerative Pododermatitis (Sore Hocks)

Sore hocks is the common name given to trauma-related ul-cerative pododermatitis of rabbits (Flatt *et al.*, 1974). It is a form of pressure-induced, chronic decubitus in the rabbit. Sore hocks is also somewhat a misnomer, since the hock is rarely involved; most commonly the area involved is the plantar sur-face of the metatarsal area, less commonly, volar metacarpal (Fig. 11). The cause of this disease is a pressure necrosis of the affected, weight-bearing metatarsal (or metacarpal) skin of the feet. Predisposing factors include constitutional susceptibility, stamping of the hindfeet, and the metal or wire dimensions and crimp of suspended cage floors. Pododermatitis is most fre-quently seen in mature, heavy rabbits, particularly males.

The lesions consist of roundish, initially ulcerated, but later elevated hyperkeratotic or with scabbed surfaces. The lesions are most commonly dry, fibrotic foci of chronic inflammation, but occasionally become abscessed. Staphylococci are usually isolated from secondarily infected lesions. Affected rabbits often appear otherwise healthy, but some may show signs of painful, awkward movements and, later, inanition.

Treatment of pododermatitis is directed toward relieving the feet of affected animals from the suspended cage floor. This may be accomplished with an impervious resting board, e.g., a sheet of plexiglass, that covers part of the suspended metal floor, or, relocation of the animal to a solid-bottom cage type with contact litter. Palliative treatment with antiseptic or anti-biotic ointments can speed recovery.

3. Traumatic Vertebral Fracture (Broken Back)

Posterior paralysis as a consequence of injury to the spinal cord occurs quite commonly in laboratory rabbits. The onset is sudden and usually is referable to an incident in which the

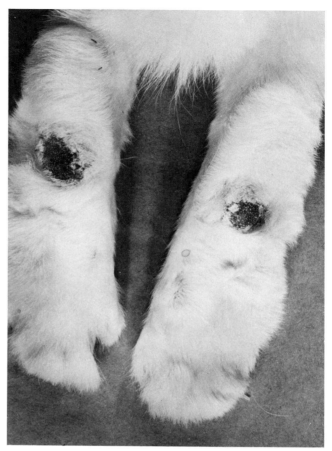

Fig. 11. Plantar surface of the metatarsal area of a rabbit with ulcerative pododermatitis. From Flatt *et al.* (1974).

rabbit was unsupported or struggled during handling. Occa-sionally the incident is not reported (or recognized), and the rabbit is later found with hindquarter paralysis in the cage. The heavy hindquarters twist, when unsupported, about the lum-bosacral junction, which acts as a fulcrum in applying leverage to the vertebral column (Flatt *et al.*, 1974). The L7 vertebral body and its articular processes are the most common sites of fracture. Vertebral fracture is more common than dislocation.

Diagnosis is established on the basis of clinical criteria, but should be confirmed by radiography or necropsy dissection. Clinical criteria include partial or complete motor paralysis with loss of skin sensation. If the cord is transected, paraplegia ensues, along with motor control of the anal sphincter and uri-nary bladder. If recognition of the condition is delayed, over-filling of the bladder, elevated BUN, uremic signs, decubitus ulcers, and staining of the perineum with liquid feces are com-mon findings. Such cases are untreatable, and the rabbit should be euthanized. Frequently, however, the cord is not severed, but merely compressed by posttraumatic edema. If the fractured bone ends remain fixed, the edema may resolve in 1–2 weeks with partial recovery of motor function to a degree compatible with life in a cage. Euthanasia need not be recommended in cases where control over the anal sphincter and urinary bladder remains unimpaired and where partial–complete motor recovery of skeletal muscle occurs within a 2-week period.

4. Moist Dermatitis (Slobbers)

Moist dermatitis is the descriptive name given to a condition with several names, vis. slobbers, ptyalism, and wet dewlap. The condition is a dermatosis of the skin of the chin, intermandibular space, and cervical ventrum resulting from secondary microbial invasion of chronically wet skin. Three general conditions are recognized as playing a role in chronic hydration of this area; (a) drooling (ptyalism) from chronic dental conditions, especially malocclusion; (b) continual wetting of the skin during drinking from water bowls or crocks; and (c) poor husbandry conditions, e.g., cold, damp contact bedding (Flatt *et al.*, 1974). The condition is rarely seen in rabbits with bottled water or automatic drinking valves.

Any of a variety of bacteria may invade the skin that is predisposed by chronic hydration. As the condition progresses, the gross lesions consist of inflammation, alopecia, ulceration, and necrosis of affected skin. Lymphogenous extension may occur to regional lymph nodes.

Unless bacterial invasion has become extensive the prognosis of individual cases is good if the underlying causes are rectified and the animal is treated. Treatment consists of clipping nearby hair and cleansing the affected skin with surgical soap and water. A topical broad spectrum antibiotic ointment should be administered for an 8–10 day course.

5. Hair Balls (Trichobezoars)

Rabbits ingest hair by normal grooming maneuvers (licking), by fur chewing of cage mates, and by pulling fur as a maternal gesture (females). Masses of hair may accumulate in the stomach forming roundish masses called hair balls or trichobezoars. Generally the masses do not form obstructions and are found at necropsy as incidental findings. Occasionally, partial to complete obstruction may result with corresponding clinical signs of inappetence that can progress to inanition and death. The gastric mass may be palpable. While usually individual and sporadic, large numbers of rabbits in clean colonies may be affected.

The reason for exaggerated ingestion of hair is not known, although lack of roughage in the diet and boredom have been suggested. Conservative treatment consists of changing the diet or supplementation with autoclaved hay, as well as treatment with mineral oil by gastric intubation to stimulate evacua-

tion. Daily use of a curry comb to remove loose hair has been recommended as prophylactic. Surgical intervention physically to remove the hair ball by gastric resection has been reported.

6. Mucoid Enteropathy (Rabbit Diarrhea Complex)

The authors of this chapter coined the term "mucoid enteropathy" (ME) in 1974 (Flatt *et al.*, 1974) to describe a clinical complex in young rabbits characterized by a spectrum of enteric conditions varying from constipation with mucous hypersections at one extreme, to profuse, watery diarrhea at the other. Over the years, many synonyms have been used to refer to this complex (bloat, enteritis, mucoid enteritis, mucoid diarrhea, scours, and hypoamylasemia), and they reflect the confusion over these conditions that is only slowly yielding to an etiologic understanding.

As research has progressed in delineation of the components of this complex, it has become evident that a number of distinct infectious entities may act as mediators of, or be superimposed upon, a now more clearly defined condition (ME). Certain factors have acted in combination to blur the distinctions between the component entities. Such factors include the common denominators of dehydration and diarrhea as the chief clinical signs, with a high rate of mortality as a common terminal event. Additionally, they share the commonality of the 7- to 10-week-old age group as the most susceptible for expression of the disease. Also tending to introduce confusion is the tendency of these diseases to occur singly, or in variable combination, which frequently presents to the clinician a symptom complex with a range of expression. Enteric diseases known to occur separately, or in combination with ME to form a complex, include Tyzzer's disease, salmonellosis, intestinal coccidiosis, clostridial enterotoxemia, and colibacillosis. The etiology of the ME is not known at present, but the major pathological features of the disease enable its classification as an enterotoxin-induced secretory diarrhea of multifactorial causation. Etiological understanding of ME has neither yielded to a simple interpretation, e.g., presence of a specific bacterium and/or its toxin, nor has the disease been experimentally reproduced with bacterial isolates. Etiologic investigations have stressed three main avenues of approach: (a) the age-related shift from neonatal to adolescent digestive physiology, (b) the relationship of diet to ME, and (c) the respective role(s) of the microbial forms associated with ME.

A full discussion of such investigations is beyond the scope of this chapter, and the reader is referred to more detailed presentations (Flatt *et al.*, 1974; Whitney, 1976). The suspicion that diet may play a role in the expression of ME was until recently largely inferential, based in part on the observations that ME is not recognized as a disease of wild (or feral) *Oryctolagus* rabbits, and the circumstantial, *de novo* appearance of

ME in association with large-scale, intensive rabbit production facilities feeding scientifically compounded fortified diets. More recently, the relationship of ME-like disease was shown to be influenced directly by the fiber content of the diet; low fiber (0–10%) diets resulted in typical levels of ME that were abolished in higher fiber groups (Cheeke and Patton, 1978).

Two main microbial forms—*Escherichia coli* and *Clostridium* species (*C. perfringens*, *C. difficile*, and perhaps *C. sordelli*)—have been associated with ME (Nekkels *et al.*, 1976; Meshorer, 1976; Patton *et al.*, 1978). Both organisms multiply to a high and often predominating populations in many, but not all, cases of ME. The difficulty of assigning the cause of ME to these organisms is related to the fact that both forms are among the microflora normally resident in rabbit intestines, the fact that Koch's postulates may not be met by inoculating the organisms, and the differences in the disease induced by inoculating large numbers of the organisms or their toxins.

A number of recent publications have documented the pathogenesis of lincomycin/clindamycin-induced enteric disease in rabbits (LaMont *et al.*, 1979; Rehg and Lu, 1981; Thilsted *et al.*, 1981). The mediators of this form of intoxication have been shown to be *E. coli* and *Clostridum* sp., and the disease they induce appears to simulate in most general features the one described here as ME. Antibiotic-associated enteric disease thus appears to involve short-circuiting of the complex factors altering the digestive physiology and microbial populations that culminate in ME; the antibiotic induces the proper microbial milieu and the same final mediators associated with ME produce the disease, viz. *Clostridium* sp.

Measured by any standard, ME ranks as one of the most important causes of mortality in both rabbit production and research facilities. Although the incidence in any given facility tends to wax and wane over the year, mortality statistics of 10–20% are quite common. The disease is globally distributed and probably few, if any, rabbit colonies are entirely free of the condition. There is general agreement that the great majority of cases occur in rabbits 7–10 weeks of age, although cases may be seen as early as 2 weeks of age and as late as 18–20 weeks.

The clinical features of ME indicate a course of 2–4 days in fatal cases, and a more protracted course of 7–14 days in those that survive. The clinical signs include anorexia, polydipsia, and subnormal body temperature (99–102°F). The animal appears depressed, the posture is crouched, the hair coat is roughened, and grinding of the teeth may be heard. Weight loss from diarrhea is rapid, and the rabbit may appear thin, although the abdomen is usually bloated due to gas and a fluid-filled GI tract. The perineum is usually stained with mucous or light yellow, to greenish or brown liquid feces. Paradoxically, the cecum may be obstipated, and hard dry feces excreted simultaneously with copious quantities of gelatinous mucous may be found in the dropping pan. Clinical laboratory findings are mainly referrable to dehydration; an increase is seen in the hematocrit, RBC count, and hemoglobin. The sedimentation rate is increased. The WBC count is also increased with a shift to increased numbers of netrophils. Values for serum lipases, BUN, glucose, proteins, and phosphorus are elevated, while serum electrolytes, e.g., sodium, potassium, chloride, and calcium are decreased.

Gross findings in ME are limited to the GI tract, mesenteric lymph nodes, and gallbladder. The stomach is usually distended with fluids and gas. The duodenum and jejunum are generally filled with watery, bile-stained contents which are recognizable through the thin intestinal wall. The ileum may have pasty contents, and the ileum occasionally is impacted with dry contents and gas. Petechiae are often seen on the serosal aspect of the ileum, cecum, and colon. The colon quite frequently is distended with clear, gelatinous mucus. Distention of the gallbladder is a common finding. Goblet cell hyperplasia is consistently observed in the duodenum, jejunum, and ileum. Apart from these findings, the small intestine is usually histologically normal and without inflammatory changes; although the literature is not consistent on this point. The inconsistencies may be related to the overlay of related conditions in this complex. Goblet cell hyperplasia is usually present in the cecum, sacculus rotundus, and colon but is difficult to assess due to the normal abundance of such cells in these tissues.

Clinical diagnosis of ME is established on the basis of signs of dehydration, diarrhea, bloating of the abdomen, sucussion splash, and associated clinical findings of gas and fluid-filled small intestines, cecal impaction, gelatinous colonic contents, and petechiation of the serosa of the cecum. The diagnosis is confirmed by the finding of goblet cell hyperplasia. The observation of mucosal necrosis, mucosal sloughing, submucosal edema, and/or substantial inflammatory cell infiltration suggests complications with other elements of the complex.

Recommended treatment for ME at present is palliative, aimed at ameliorating the signs and is not specific. Over the years a wide variety of both folk remedies and antibiotics have been used to attempt treatment with generally inconsistent and often disappointing results. The uncertainty surrounding the etiology of this disease precludes treatment or prevention recommendations at the present time.

REFERENCES

Allen, A. M., Ganaway, J. R., Moore, T. D., and Kinard, R. F. (1965). Tyzzer's disease syndrome in laboratory rabbits. *Am. J. Pathol.* **46**, 859–882.

Altman, P. I., and Dittmer-Katz, D., eds. (1979). "Inbred and Genetically Defined Strains of Laboratory Animals," Part 2. Fed. Am. Soc. Exp. Biol., Bethesda, Maryland.

Alus, G., and Edwards, N. A. (1977). Development of the digestive tract of the rabbit from birth to weaning. *Proc. Nutr. Soc.* **36**, 3A.

Alvarez, W. C. (1940). "An Introduction to Gastroenterology" 3rd ed. Hoeber, New York.

Andreoli, T. E. (1981). Renal and electrolyte physiology. *Annu. Rev. Physiol.* **43**, 567–649.

Arruda, J. A. L., and Krutzman, N. A. (1978). Relationship of renal sodium and water transport to hydrogen ion secretions. *Annu. Rev. Physiol.* **40**, 43–66.

Benson, R. E., and Dodds, W. J. (1977). Autosomal factor VIII deficiency in rabbits: Size variation of rabbit for factor VIII. *Thromb. Haemostasis* **38**, 380.

Bjornhag, G. (1972). Separation and delay of contents in the rabbit colon. *Swed. J. Agric. Res.* **2**, 125–136.

Bonden, L., and Prohaska, L. (1980). Isolation of an adenovirus from rabbits with diarrhea. *Acta Vet. Acad. Sci. Hung.* **28**, 247–255.

Brambell, F. W. R. (1958). The passive immunity of the young mammal. *Biol. Rev. Cambridge Philos. Soc.* **33**, 488–531.

Bryden, A. S., Thouless, M. E., and Flewett, T. H. (1976). Rotavirus in rabbits. *Vet. Rec.* **99**, 322.

Burg, M., Grantham, J., Abramow, M., and Orloff, J. (1966). Preparation and study of fragments of single rabbit nephrons. *Am. J. Physiol.* **210**, 1293–1298.

Bywater, J. E. C., and Kellett, B. S. (1978). The eradication of *E. cuniculi* from a SPF rat colony. *Lab. Anim. Sci.* **28**, 402–405.

Caldwell, E. J., and Fry, D. L. (1969). Pulmonary mechanics in the rabbit. *J. Appl. Physiol.* **27**, 280–285.

Callikan, S., and Girard, J. (1979). Perinatal development of glucogenic enzymes in rabbit liver. *Biol. Neonat.* **36**(1–2), 78–84.

Chauncey, H. H., Henriques, B. L., and Tanzer, J. M. (1963). Comparative enzyme activity of saliva from sheep, hog, dog, rat, and human. *Arch. Oral Biol.* **8**, 615–627.

Cheeke, P. R., and Patton, N. M. (1978). Effect of alfalfa and dietary fiber on the growth performance of weanling rabbits. *Lab. Anim. Sci.* **28**, 167–172.

Chengappa, M. M., Myers, R. C., and Carter, G. R. (1980). A streptomycin-dependent live *Pasteruella multocida* vaccine for the prevention of rabbit pasteurellosis. *Lab. Anim. Sci.* **30**, 515–518.

Christensen, L. R., Bond, E., and Matanic, B. (1967). Pockless rabbit pox. *Lab. Anim. Care* **17**, 281–296.

Cohen, C. (1966). Methods in immunization. *In* "Handbook of Biochemistry and Biophysics" (H. C. Damon, ed.). World Publ. Co., Cleveland, Ohio.

Cohen, C., and Tissot, R. B. (1974). Specialized research applications. II. Serological genetics. *In* "The Biology of the Laboratory Rabbit" (S. H. Weisbroth, R. E. Flatt, and A. L. Kraus, eds.), Chapter 7. Academic Press, New York.

Committee of Animal Models for Research on Aging (1981). "*Mammalian Models for Research in Aging*," pp. 170–179. Inst. Lab. Anim. Resour., Nat. Acad. Sci., Washington, D.C.

Cox, J. C., Gallichio, H. A., Pye, D. *et al.* (1977). Application of immunofluorescence to the establishment of *Encephalitozoon cuniculi*-free rabbit colony. *Lab. Anim. Sci.* **27**, 204–209.

Crosfill, M. L., and Widdicombe, J. G. (1961). Physical characteristics of the chest and lungs and the work of breathing in different mammalina species. *J. Physiol. (London)* **158**, 1–14.

Cunliffe-Beamer, T. L., and Fox, R. R. (1981a). Venereal spirochetosis of rabbits: Description and diagnosis. *Lab. Anim. Sci.* **31**, 366–371.

Cunliffe-Beamer, T. L., and Fox, R. R. (1981b). Venereal spirochetosis of rabbits: Epizootiology. *Lab. Anim. Sci.* **31**, 372–378.

Cunliffe-Beamer, T. L., and Fox, R. R. (1981c). Venereal spirochetosis of rabbits: Eradication. *Lab. Anim. Sci.* **31**, 379–381.

Davidson, J. T., Wasserman, K., Lilington, G. A., and Schmidt, R. W. (1966). Effect of aging on respiratory mechanics and gas exchange in rabbits. *J. Appl. Physiol.* **21**, 837–842.

Dennis, V. W., Bello-Reuss, E., and Robinson, R. R. (1977). Response of phosphate transport to parathyroid hormone in segments of rabbit nephron. *Am. J. Physiol.* **233**, F29–F38.

Domingez, J. A., Corella, L., and Auro, A. (1981). Oval papillomatosis in two laboratory rabbits in Mexico. *Lab. Anim. Sci.* **31**, 71–73.

Draize, G. H., Woodward, G., and Calvery, H. O. (1944). Methods for the study of irritation and toxicity of substances applied topically to the skin and mucous membranes. *J. Pharmacol. Exp. Ther.* **82**, 377–390.

Fabvre, H. (1962). The Shope fibroma of rabbits. *In* "Ultrastructure of Tumors Induced by Viruses" (A. J. Dalton and F. Haquenau, eds.), pp. 1–59. Academic Press, New York.

Fenner, F., and Ratcliffe, F. N. (1965). "Myxomatosis." Cambridge Univ. Press, London and New York.

Ferrando, R., Wolter, R., Vilat, J. C., and Megard, J. P. (1970). Teneur en acides amines des deux catégories de fèces du lapin: Caecotrophes et fèces durés. *C. R. Hebd. Seances Acad. Sci.,* 2202–2205.

Flatt, R. E. (1974). Bacterial diseases. *In* "The Biology of the Laboratory Rabbit" (S. H. Weisbroth, R. E. Flatt, and A. K. Kraus, eds.), pp. 193–236. Academic Press, New York.

Flatt, R. E., and Carpenter, A. B. (1971). Identification of crystalline material in urine of rabbits. *Am. J. Vet. Res.* **32**, 655–658.

Flatt, R. E., Weisbroth, S. H., and Kraus, A. L. (1974). Metabolic, traumatic mycotic and miscellaneous diseases in rabbits. *In* "The Biology of the Laboratory Rabbit" (S. H. Weisbroth, R. E. Flatt, and A. K. Kraus, eds.), pp. 435–451. Academic Press, New York.

Fox, J. G., Beaucage, C. B., and Shalev, M. (1979). Congenital entropion in New Zealand white rabbits. *Lab. Anim. Sci.* **24**, 509–511.

Fox, R. R. (1975). "Handbook of Genetically Standardized JAX Rabbits." Jackson Lab., Bar Harbor, Maine.

Fox, R. R. (1980). The rabbit (*Oryctolagus cuniculus*) and research in Aging. *Exp. Aging Res.* **6**(3), 235–248.

Fox, R. R., and Crary, D. D. (1971). Mandibular prognathism in the rabbit: Genetic studies. *J. Hered.* **62**, 23–27.

Fox, R. R., and Guthrie, D. (1968). The value of creep feed for laboratory rabbits. *Lab. Anim. Care* **18**, 34–38.

Fox, R. R., Schlager, G., and Laird, C. (1969). Blood pressure in thirteen strains of rabbits, *J. Hered.* **60**, 312–314.

Ganaway, J. R., Allen, A. M., and Moore, T. D. (1971). Tyzzer's disease of rabbits. Isolation and propagation of *Bacillus piliformis* (Tyzzer) in embryonated eggs. *Infect. Immun.* **3**, 429–437.

Gelineo, S. (1964). Organ systems in adaptation: The temperature regulating system. *In* "Handbook of Physiology" (D. B. Dill, E. F. Adolph, and C. G. Wilber, eds.), Sect. 4, pp. 259–282. Am. Physiol. Soc., Washington, D.C.

Goetsch, E. (1910). The structure of the mammalian oesophagus. *Am. J. Anat.* **10**, 1–40.

Grantham, J. J., Irish, J. M., and Hall, D. A. (1978). Studies of isolated renal tubules *in vitro*. *Annu. Rev. Physiol.* **40**, 249–277.

Gross, L. (1970). "Oncogenic Viruses," 2nd ed. Pergamon, Oxford.

Guyton, A. C. (1947). Measurement of the respiratory volumes of laboratory animals. *Am. J. Physiol.* **150**, 70–77.

Hafez, E. S. E. (1970). Rabbits. *In* "Reproduction and Breeding Techniques for Laboratory Animals" (E. S. E. Hafez, ed.), Chapter 16. Lea & Febiger, Philadelphia, Pennsylvania.

Hagen, K. W., Jr. (1958). Enzootic pasteurellosis in domestic rabbits. I. Pathology and bacteriology. *J. Am. Vet. Med. Assoc.* **133**, 77–80.

Hagen, K. W., Jr. (1966). Spontaneous papillomatosis in domestic rabbits. *Bull. Wildl. Dis.* **2**, 108–110.

Hagen, K. W., Jr. (1969). Ringworm in domestic rabbits: Oral treatment with griseofluvin. *Lab. Anim. Care* **19**, 635–638.

Hartman, H. H. (1974). The fetus in experimental teratology. *In* "The Biology of the Laboratory Rabbit" (S. H. Weisbroth, R. E. Flatt, and A. L. Kraus, eds.), Chapter 5. Academic Press, New York.

Haupt, T. R. (1963). Urea utilization by rabbits fed a low-protein ration. *Am. J. Physiol.* **205,** 1144–1150.

Heistad, D. D., Marcus, M. L., and Gross, P. M. (1978). Effect of sympathetic nerves on cerebral vessels in dog, cat, and monkey. *Am. J. Physiol.* **235,** H544–H552.

Hinze, H. C. (1971). Induction of lymphoid hyperplasia and lymphoma-like disease in rabbits by *Herpevirus sylvilagus. Int. J. Cancer* **8,** 514–522.

Hornicke, H. (1977). Coecotrophy in rabbits—A circadian function. *J. Mammal.* **58,** 240–242.

Hunt, C. E., and Harrington, D. D. (1974). Nutrition and nutritional diseases of the rabbit. *In* "The Biology of the Laboratory Rabbit" (S. H. Weisbroth, R. E. Flatt, and A. L. Kraus, eds.), pp. 403–433. Academic Press, New York.

Kaplan, H. M., and Timmons, E. H. (1979). "The Rabbit: A Model for the Principles of Mammalian Physiology and Surgery." Academic Press, New York.

Kardon, M. B., Peterson, D. F., and Bishop, V. S. (1974). Beat-to-beat regulation of heart rate by afferent stimulation of the aortic nerve. *Am. J. Physiol.* **227,** 598–600.

Karemaker, J. M., Borst, C., and Schreurs, A. W. (1980). Implantable stimulating electrode for baroreceptor afferent nerves in rabbits. *Am. J. Physiol.* **239,** H706–H709.

Knox, F. G., Preiss, J., Kim, J. K., and Dousa, T. P. (1977). Mechanism of resistance to the phosphaturic effect of parathyroid hormone in the hamster. *J. Clin. Invest.* **59,** 675–683.

Kojima, K., and Tanaka, R. (1974). Factors influencing absorption and excretion of drugs. III. Effect of fasting on absorption and excretion of sodium salicylate and aspirin in rabbits. *Chem. Pharm. Bull.* (Tokyo) **22,** 2270–2275.

Kozma, C., Macklin, W., Cummins, L. M., and Mauer, R. (1974). The anatomy, physiology and the biochemistry of the rabbit. *In* "The Biology of the Laboratory Rabbit" (S. H. Weisbroth, R. E. Flatt, and A. L. Kraus, eds.), pp. 50–72. Academic Press, New York.

Kraus, A. L. (1974). Arthropod parasites. *In* "The Biology of the Laboratory Rabbit" (S. H. Weisbroth, R. E. Flatt, and A. L. Kraus, eds.), Chapter 12. Academic Press, New York.

Kulangara, A. C., and Schechtman, A. M. (1962). Passage of heterologous serum proteins from mother into fetal compartments in the rabbit. *Am. J. Physiol.* **203,** 1071–1080.

Kulwich, R., Struglia, L., and Pearson, P. B. (1953). The effect of coprophagy on the excretion of B vitamins by the rabbit. *J. Nutr.* **49,** 639–645.

Laird, C. (1974). Clinical pathology: Blood chemistry. *In* "Handbook of Laboratory Animal Science" (E. C. Melby, Jr., and N. H. Altman, eds.), Vol. II, pp. 347–436. CRC Press, Cleveland, Ohio.

Lamiere, N. H., Lifschitz, M. D., and Stein, J. H. (1977). Heterogeneity of nephron function. *Annu. Rev. Physiol.* **39,** 159–184.

LaMont, J. F., Sonnenblick, E. B., and Rothman, S. (1979). Role of clostridial toxin in the pathogenesis of clindamycin colitis in rabbits. *Gastroenterology* **76,** 356–361.

Lang, C. M., Chairman (1980). "National Survey of Laboratory Animal Facilities and Resources." Committee on Laboratory Animal Facilities and Resources, U.S. Dept. of Health and Human Services, Public Health Source, NIH Publication No. 80-2091.

Lindsey, J. R., and Fox, R. R. (1974). Inherited diseases and variations. *In* "The Biology of the Laboratory Rabbit (S. H. Weisbroth, R. E. Flatt, and A. L. Kraus, eds.), Chapter 15, pp. 377–401. Academic Press, New York.

Lyngset, A. (1980). A survey of serum antibodies to *Encephalitozoon cuniculi* in breeding rabbits and their young. *Lab. Anim. Sci.* **30,** 558–561.

McKercher, D. G., and Saito, J., (1964). An attenuated live-virus vaccine for myxomatosis. *Nature (London)* **202,** 933–934.

Mangel, A., Fahim, M., and van Breemen, C. (1981). Rhythmic contractile activity of the *in vivo* rabbit aorta. *Nature* **289,** 892–894.

Maré, C. J. (1974). Viral disease. *In* "The Biology of the Laboratory Rabbit" (S. H. Weisbroth, R. E. Flatt, and A. L. Kraus, eds.), pp. 237–261. Academic Press, New York.

Mauderly, J. L., Tesarck, J. E., and Sifford, L. J. (1979). Respiratory measurements of unsedated small laboratory mammals using nonbreathing valves. *Lab. Anim. Sci.* **29,** 323–329.

Meshorer, A. (1976). Histological findings in rabbits which died with symptoms of mucoid enteritis. *Lab. Anim.* **10,** 199–202.

Mitruka, B. M., and Rawnsley, H. M. (1977). "Clinical Biochemical Reference Values in Normal Experimental Animals," pp. 82–84 and 133–137. Masson, New York.

Nekkels, R. J., Mullink, M. W. M. A., and Van Vliet, J. C. J. (1976). An outbreak of rabbit enteritis: Pathological and microbiological findings and possible therapeutic regime. *Lab. Anim.* **10,** 195–198.

Newcomer, C. E., Ackerman, J. I., and Fox, J. G. (1983). The laboratory rabbit as reservoirs of *Salmonella mbandaka. J. Infect. Dis.* **147,** 365.

Pakes, S. P. (1974). Protozoal diseases. *In* "The Biology of the Laboratory Rabbit" (S. H. Weisbroth, R. E. Flatt, and A. L. Kraus, eds.), pp. 263–286. Academic Press, New York.

Patton, N. M., Holmes, H. T., Riggs, R. J., and Cheeke, P. R. (1978). Enterotoxemia in rabbits. *Lab. Anim. Sci.* **28,** 536–540.

Payne, B. J., Lewis, H. B., Murchison, T. E. *et al.* (1976). Hematology of laboratory animals. *In* "Handbook of Laboratory Animal Science" (E. C. Melby, Jr., and N. H. Alfman, eds.), Vol. III, pp. 383–461. CRC Press, Cleveland, Ohio.

Petric, M., Middleton, P. J., Grant, C., Tam, J. S., and Hewett, C. M. (1978). Lapine rotavirus: Preliminary studies on epizoology and transmission. *Can. J. Comp. Med.* **42,** 143–147.

Prosser, C. L. (1973). "Comparative Animal Physiology," 3rd ed. Saunders, Philadelphia, Pennsylvania.

Rehg, J. E., and Lu, Y.-S. (1981). *Clostridium difficile* colitis in a rabbit following antibiotic therapy for pasteurellosis. *J. Am. Vet. Med. Assoc.* **179,** 1296–1297.

Rogers, F. T. (1915). Contributions to the physiology of the stomach. XX. The contractions of the rabbit's stomach during hunger. *Am. J. Physiol.* **36,** 183–190.

Scharf, R. A., Monteleone, S. A., and Stark, D. M. (1981). A modified barrier system for maintenance of *Pasteurella*-free rabbits. *Lab. Anim. Sci.* **31,** 513–515.

Sercombe, R., Lacombe, P., Aubino, P., Mamo, H., Pinard, E., Reynier-Rebuffel, A. M., and Seylaz, J. R. (1978). Is there an active mechanism limiting the influence of the sympathetic system on the cerebral vascular bed? Evidence for vasomotor escape from sympathetic stimulation in the rabbit. *Brain Res.* **164,** 81–102.

Shadduck, J. A. (1969). *Nosema cunculi. In vitro* isolation. *Science* **166,** 516–517.

Shadduck, J. A., and Pakes, S. P. (1971). Encephalitozoonosis (nosematosis) and toxoplasmosis. *Am. J. Pathol.* **64,** 657–674.

Shelley, H. J. (1961). Glycogen reserves and their changes at birth. *Br. Med. Bull.* **17,** 137–143.

Shope, R. E., and Hurst, E. W. (1933). Infectious papillomatosis of rabbits. *J. Exp. Med.* **58,** 607–624.

Stedham, M. A. (1976). Naturally occurring rabbit diseases. VI. Rabbit. *In* "Handbook of Laboratory Animal Science" (E. C. Melby, Jr., and N. H. Altman, eds.), Vol. III, pp. 279–305. CRC Press, Cleveland, Ohio.

Steele, T. H. (1976). Renal resistance to parathyroid hormone during phosphorus deprivation. *J. Clin. Invest.* **58,** 1461–1464.

Stinnett, H. O., and Sepe, F. J. (1979). Rabbit cardiovascular responses during PEEP before and after vagotomy. *Proc. Soc. Exp. Biol. Med.* **162,** 485–494.

Subcommittee on Rabbit Nutrition, Committee on Animal Nutrition, Board on Agriculture and Renewable Resources, National Research Council (1977). "Nutrient Requirements of Rabbits," 2nd ed. Nat. Acad. Sci., Washington, D.C.

Szabo, S., Kouraunakis, P., Kovacs, K., Tuchneber, B., and Garg, B. D. (1974). Prevention of organomercurial intoxication by thyroid deficiency in the rat. *Toxicol. Appl. Pharmacol.* **30,** 175–184.

Thilsted, J. P., Newton, W. M., Crandell, R. A. *et al.* (1981). Fatal diarrhea in rabbits resulting from the feeding of antibiotic-contaminated feed. *J. Am. Vet. Med. Assoc.* **179,** 360–362.

Trueta, J., Barclay, A. E., Daniel, P. M., Franklin, K. J., and Prichard, M. M. L. (1947). "Studies on the Renal Circulation." Blackwell, Oxford.

Truex, R. C., and Smythe, M. Q. (1965). Comparative morphology of the cardiac conduction tissue in animals. *Ann. N.Y. Acad. Sci.* **127,** 19–33.

United States Pharmacopeial Convention, Inc. (1980).

United States Pharmacopeia and National Formulary," pp. 902–903. U.S. Pharmacopeial Convention, Inc., Rockville, Maryland.

Valtin, H. (1977). Structural and functional heterogeneity of mammalian nephrons. *Am. J. Physiol.* **233,** F491–F501.

Van Kruiningen, H. J., and Blodgett, S. B. (1971). Tyzzer's disease in a Connecticut rabbitry. *J. Am. Vet. Med. Assoc.* **158,** 1205–1212.

Waller, T. (1977). The inda-ink immunoreaction: A method for the rapid diagnosis of encephalitozoonosis. *Lab. Anim.* **11,** 93–97.

Weber, H. W., and VanderWatt, J. J. (1973). Cardiomyopathy in crowded rabbits. *S. Afr. Med. J.* **47,** 1591–1595.

Weisbroth, S. H. (1974). Neoplastic diseases. *In* "The Biology of the Laboratory Rabbit" (S. H. Weisbroth, R. E. Flatt, and A. L. Kraus, eds.), Chapter 14. Academic Press, New York.

Weisbroth, S. H., and Scher, S. (1970). Spontaneous oral papillomatosis in rabbits. *J. Am. Vet. Med. Assoc.* **157,** 1940–1944.

Wescott, R. B. (1974). Helminth parasites. *In* "The Biology of the Laboratory Rabbit" (S. H. Weisbroth, R. E. Flatt, and A. L. Kraus, eds.), Chapter 13. Academic Press, New York.

West, T. C. (1955). Ultramicroelectric recording from cardiac pacemaker. *J. Pharmacol. Exp. Ther.* **115,** 283–290.

Whitney, J. C. (1976). A review of nonspecific enteritis in the rabbit. *Lab. Anim.* **10,** 209–221.

Williams, C. S. F. (1976). "Practical Guide to Laboratory Animals." C. V. Mosby, St. Louis, Missouri.

Wilson, J. M. (1979). The biology of *Encephalitozoon cuniculi. Med. Biol.* **57,** 84–101.

Wosu, N. J., Shadduck, J. A., Pakes, S. P. *et al.* (1977). Diagnosis of encelpalitozoonosis in experimentally infected rabbits by intradermal and immunfluorescence tests. *Lab. Anim. Sci.* **27,** 210–216.

Yokoyama, E. (1979). Flow-volume curves of excised right and left rabbit lungs. *J. Appl. Physiol.* **46,** 463–468.

Zimmerman, T. S., Arroyave, C. M., and Müller-Eberhard, H. J. (1971). A blood coagulative abnormality in rabbits deficient in the sixth component of complement (C6) and its correction by purified C6. *J. Exp. Med.* **134,** 1591–1607.

Chapter 9

Dogs and Cats as Laboratory Animals

Daniel H. Ringler and Gregory K. Peter

I. INTRODUCTION

Dogs and cats have been used in biomedical research for centuries based on the early belief, then proved fact, that their anatomy, physiology, and response to disease was similar to man's. These studies have made phenomenal contributions to the health and welfare of both man and animals. Most of our knowledge of physiology, pathology, immunology, surgery, biochemistry, nutrition, pharmacology, toxicology, and control of disease is based on animal studies. Many of the discoveries of Harvey, Pasteur, Bernard, and other early leaders in biomedical science resulted from work with dogs and other animals. In past centuries the use of dogs and cats in research was based primarily on size, availability, and simplicity of care. During the early part of this century, extensive use of dogs in physiologic and surgical research was also based primarily on these criteria.

Today, dogs or cats are often the species of choice in animal research because they spontaneously exhibit or can be induced to exhibit aspects of diseases that afflict humans. In effect the animals serve as models of human disease. A comprehensive text deals with the various aspects of spontaneous animal models of human disease (Andrews *et al.,* 1979). The use of the dog as a research model in immunology, hematology, and oncology has recently been summarized (Shifrine and Wilson, 1980). The relevance and appropriateness of selected mammals, including the dog and cat, as models of human aging have been examined (Committee on Animal Models for Research on Aging, 1981). A symposium to examine the past, present, and future contributions of animals to human health and welfare has been held [National Academy of Sciences—National Research Council (NAS—NRC) 1977]. It seems apparent that dogs and cats will continue to play a vital role in biomedical research in the future.

A recent survey of laboratories in the United States indicates that approximately 180,000 dogs and 55,000 cats are acquired for biomedical research and testing each year [Institute of Laboratory Animal Resources (ILAR), 1980]. Of these, only 25% of the dogs and 8% of the cats are bred specifically for research by commercial breeders. The remainder, some 140,000 dogs and 50,000 cats are obtained from pounds and animal shelters. These animals are generally referred to as ''random source'' animals in order to distinguish them from the animals that are bred specifically for research. These random source animals present a particular challenge to laboratory animal veterinarians who often have the responsibility for conditioning these animals for use in research.

II. TYPES AND SOURCES OF DOGS AND CATS

A. Introduction

Dogs and cats for use in research are available from a number of commercial breeders and animal dealers licensed under the Federal Animal Welfare Act. A directory of commercial breeders is published periodically (ILAR, 1979). The directory and additional information on sources of animals and model copies of procurement specifications for dogs and cats are available.* A listing of licensed animal dealers in the United States can be obtained from the United States Department of Agriculture.†

B. Random Source

Dogs and cats that are available for research may be classified into several types according to the source from which they are procured. The types are (1) random source, (2) conditioned for research, and (3) bred for research.

Random source dogs and cats are generally acquired from federally licensed animal dealers, pounds, or individuals. Their history and disease status is generally unknown, and they may be incubating a wide variety of infectious diseases. In the dog, the most common of these include canine distemper, canine parvovirus, parainfluenza SV5, canine adenovirus-type II, and *Bordetella bronchiseptica* infections. Typically, newly received dogs begin to show clinical signs of respiratory disease during the first or second week after arrival in the research facility. The time of onset of these diseases usually is dependent on the time since animals from different sources were first grouped together. If the holding period during which the animals are aggregated prior to receipt in the research facility is as long as 10 days then signs of respiratory infection will tend to occur during the first week after arrival. If the holding period is 4–5 days, then clinical signs are usually seen during the second week after arrival. Dogs that do not exhibit signs of respiratory infection during the first 30 days after arrival rarely have serious respiratory disease later.

Like dogs, cats newly received from pounds and animal dealers may be in the incubative stage of a wide variety of infectious diseases. The most common are feline panleukopenia, rhinotracheitis, calicivirus infection, and pneumonitis (*Chlamydia psittaci*). Infestation with fleas, ear mites, and intestinal parasites also is common. Fox and Beaucage (1979) have shown that newly received cats may carry salmonellosis. Several of these dog and cat pathogens are also hazardous to humans. The zoonotic implications of dog and cat diseases are described in Chapter 22 on zoonoses. Typically, newly received cats show signs of panleukopenia during the first week after arrival, and the respiratory disease complex is manifest during the second or subsequent weeks. Generally, random source cats must be quarantined and conditioned for 30–60 days to ensure that they are healthy, well-nourished, tractable, and adapted to confinement.

---

*Institute of Laboratory Animal Resources, National Research Council, 2101 Constitution Avenue, N.W., Washington, D.C. 20418.

†Senior Staff Veterinarian, Animal Welfare Staff, Veterinary Services, Animal and Plant Health Inspection Service, Room 703, Federal Building, USDA, 6505 Belcrest Road, Hyattsville, Maryland 20782.

C. Conditioned for Research

Random source dogs and cats that have been through a quarantine and conditioning period are available commercially. Alternatively, they can be quarantined and conditioned in the research facility prior to their inclusion in research projects. In either case, the objective of conditioning is to produce stable disease-free animals that will react uniformly in research. The conditioning process is described more fully in Sections IV,A and B. Model procurement specifications for conditioned random source dogs and cats are available from the Institute for Laboratory Animal Resources (see footnote to Section II,A).

D. Bred for Research

Dogs or cats that are bred specifically for research are more uniform and have fewer health problems that random source animals. Other desirable attributes are as follows: pedigrees are generally available, animals may be vaccinated and free of both infectious diseases and parasites, and the animals are accustomed to cage life. Disadvantages of dogs bred for research include cost, shyness, and the small size of the beagle that is most readily available. This size problem has become less severe as hound-type dogs have become increasingly available. Cats that are bred for research may be intractable. Most difficulties in temperment of both dogs and cats can be alleviated by adequate socialization in the supplier's colonies. When procuring dogs or cats that are bred specifically for research, it is important to specify that the animals must be adequately socialized.

III. HOUSING

Construction criteria for research facilities housing dogs or cats are much the same as those for facilities housing other species. These criteria are described in the "Guide for the Care and Use of Laboratory Animals" (ILAR, 1978b) and are treated in some detail in Chapter 17 on design and management of animal facilities. Other useful references include "Dogs—Standards and Guidelines for the Breedings Care and Management of Laboratory Animals" (ILAR, 1973), "Laboratory Animal Management—Cats" (ILAR, 1978d), and "Laboratory Animal Housing" (ILAR, 1978c). In this chapter, only those issues that pertain to primary enclosures (cages or pens) for dogs or cats will be addressed.

The primary enclosure is one of the most important factors in the environment of dogs and cats housed in the research laboratory. It influences the well-being of the animals as well as

their biological responses. Criteria for evaluating a caging system have been identified (ILAR, 1978b) and include animal comfort as a primary consideration. Physical comfort includes such factors as keeping the animal dry, clean, and at an appropriate temperature; providing sufficient space to permit normal postural adjustments; avoiding unnecessary physical restraint; and preventing overcrowding. In addition, the primary enclosure should be designed to facilitate effective sanitation and should be maintained in good repair. Finally, the enclosure should meet the investigator's research requirements.

The need for exercise or additional physical activity especially for dogs has been the subject of controversy in laboratory animal science. Although experimental evidence is scant, it has generally been concluded that confinement in a cage or pen has minimal influence on the amount of exercise that a dog or cat engages in and does not necessarily affect the animal's well-being (ILAR, 1978b). Hite *et al.* (1977) compared beagles housed in standard size cages with those housed in cages three times larger and found no beneficial or adverse effects related to cage size. Newton (1972) studied calcium kinetics and myofibrillar specific enzyme activity and showed no difference between caged and exercised dogs. Since there is a remarkable lack of experimental evidence regarding the effects of cage confinement, the need for supplementary exercise or large enclosures remains a matter of experienced judgement.

In research facilities dogs and cats are usually housed in indoor cages or floor pens. Outdoor housing and indoor–outdoor runs have found less favor in recent years due to a number of factors, including (1) less standardization of environmental conditions; (2) difficulty in preventing access by birds, insects, and wild animals; (3) noise control difficulties in urban areas; (4) inefficient utilization of space; and (5) difficulty in providing suitable outdoor space in association with the usual multistory research facility. However, outdoor or indoor–outdoor housing may be suited to some types of facilities in rural locations.

Within research facilities dogs and cats may be housed individually or in groups. Individual housing is generally preferred during both conditioning and experimental periods because it permits easier observation and handling. Consumption of food and water and production of feces and urine can be monitored more easily.

Generally accepted recommendations for dog and cat cage size are provided in the "Guide for the Care and Use of Laboratory Animals" (ILAR, 1978b) and the legal requirements are specified in the regulations of the Federal Animal Welfare Act [Code of Federal Regulations (CFR), 1982]. The regulations are available from the United States Department of Agriculture (see footnote to Section II,A). Those who design or select commercially available cages should be familiar with applicable portions of these documents.

Table I

Cage Space Recommendations for Dogs and Cats[a]

Animal	Weight (kg)	Type of housing	Floor area/animal	Cage height[b]
Dog[c]	15	Pen or run	0.74 m² (8.0 ft²)	[d]
	15–30	Pen or run	1.11 m² (12.1 ft²)	[d]
	30	Pen or run	2.23 m² (24.0 ft²)	[d]
	15	Cage	0.74 m² (8.0 ft²)	81.3 cm (32 in.)
	15–30	Cage	1.11 m² (12.1 ft²)	91.4 cm (36 in.)
	30	Cage	[c]	[d]
Cat	4	Cage	0.28 m² (3.0 ft²)	61.0 cm (24 in.)
	4	Cage	0.37 m² (4.0 ft²)	61.0 cm (24 in.)

[a]Modified from the "Guide for the Care and Use of Laboratory Animals" [Institute of Laboratory Animal Resources (ILAR) 1978b].

[b]From the resting floor to the cage top.

[c]In order to be in compliance with the regulations of the Animal Welfare Act the required floor area may be computed by the following equation: (length of dog in inches + 6) × (length of dog in inches + 6) ÷ 144 = required square feet of floor space. The length of a dog is the distance from the tip of the nose to the base of the tail.

[d]It is generally accepted that cage or pen height should be at least equal to the height of the dog at the shoulders plus 6 in.

The cage space recommendations from the "Guide for the Care and Use of Laboratory Animals" are provided in Table I. In addition to these recommendations, the Guide emphasizes that cage and pen areas other than those suggested should be considered equally acceptable if they provide equivalent comfort for the animals. However, one must recognize that legal specifications apply to the housing of dogs and cats and are stated in the regulations of the Federal Animal Welfare Act (CFR, 1982). In general, cages that meet the size and construction recommendations of the Guide also meet the regulations of the Animal Welfare Act. However, several additional requirements incorporated into the regulations include (1) limitation to not more than 12 nonconditioned dogs or 12 nonconditioned cats in the same primary enclosure, (2) provision of litter boxes for cats, and (3) solid resting surfaces sufficient to accommodate all cats in the primary enclosure, these resting surfaces must be elevated if the primary enclosure houses two or more cats.

IV. CONDITIONING

A. Conditioning the Random Source Dog

Conditioning is the process whereby a newly arrived random source animal is freed of all parasites and infectious diseases and is acclimated to the laboratory environment. During the conditioning period the animal should achieve a defined health status consistent with the requirements of the particular experimental design.

Conditioning programs for dogs vary widely in length and complexity among research facilities. Most programs consist of physical examination; internal and external parasite control; vaccination for respiratory diseases, infectious canine hepatitis, canine parvovirus, and leptospirosis; laboratory examination for heartworm; and specific treatment of disease present during the conditioning period. In some programs the dogs are held for as little as 14 days prior to release to research projects. In our experience this holding period is insufficient because 5–10% of the dogs begin to exhibit clinical signs of respiratory disease following the 14 day holding period. Most research facilities have been more successful with a conditioning program that lasts at least 30 days. A model conditioning program is provided in Table II.

Table II

Conditioning Program for Random Source Dogs

Arrival procedures
1. Physical examination by veterinarian or animal health technician. Rejection of animals that are immature, aged, emaciated, or ill or those that fail to meet specific criteria (age, sex, disposition, hair length, etc.)
2. Weigh, sex, maintain identification, and initiate individual health record
3. Vaccinate for canine distemper, canine parvovirus, canine adenovirus type I (infectious canine hepatitis), canine parainfluenza and *Bordetella bronchiseptica*
4. Total immersion dip in insecticide solution for ectoparasites
5. House singly in floor pens with wood chip bedding. Alternatively animals can be dried and housed in cages or pens without absorbent bedding
6. Provide food and water

Postarrival procedures
1. Blood sample for heartworm microfilaria examination
2. Identify permanently if research protocol requires
3. Devocalize (optional)
4. Examine feces for intestinal parasites and ova and treat as indicated or, alternatively, administer a broad spectrum anthelmintic at 14-day intervals
5. Monitor food and water consumption daily
6. Observe each animal daily for signs of illness and provide veterinary medical care for ill animals. Alternatively, euthanize ill animals

Prerelease procedures
1. Physical examination
2. Repeat heartworm microfilaria examination if specific studies necessitate.
3. Examine feces for intestinal parasites and ova
4. Hematology and clinical chemistry examinations as required for specific studies

General considerations
Newly received dogs should be isolated as a group. Special precautions should be taken to ensure that personnel, fomites, and the ventilation system do not transmit infectious agents to previously conditioned dogs (see Section IV, C)

Arrival procedures usually include physical examination, dip in insecticide for external parasites, temporary identification, appropriate housing, and provision of food and water. It may be advisable to limit food consumption during the first several days until animals become accustomed to having laboratory diets provided *ad libitum*. At or shortly after arrival, animals should be immunized against canine distemper, canine parvovirus, canine adenovirus, canine parainfluenza, and *Bordetella bronchiseptica*.

Controlled studies have demonstrated that morbidity and mortality is significantly lower in newly received random source dogs that are immunized on arrival. Doyle *et al.* (1979) found that immunization against canine distemper, infectious canine hepatitis, and canine parainfluenza virus reduced the incidence of respiratory disease from 28 to 21% and the mortality from 25 to 14%. Appel (1970) has shown that virulent canine distemper virus spreads to the brain and epithelial tissues on about the ninth day after exposure. In order to be protective, the serum neutralizing antibody titer must be higher than 1 : 100 before the virus reaches the brain and epithelial tissues. Thus, the failure of vaccine to protect newly received dogs from distemper is in all likelihood due to exposure prior to vaccination.

Heartworm infection can interfere with pulmonary arterial circulation (Fig. 1) and result in cor pulmonale. Pulmonary arteries can be completely occluded by intimal proliferation (Fig. 2) and thrombosis induced by embolization of dead parasites. Early in the conditioning period, blood samples should be examined for microfilaria of the canine heartworm *Dirofilaria immitis*. The most commonly used methods are the modified Knott's technique or the filter technique. They are usually repeated during the holding period to help ensure detection of heartworm infection. Difficulties arise in detecting heartworm infestation when dogs harbor adult worms but are amicrofilaremic. These are called occult infections. The prevalence of occult infection has been reported to range between 10 and 60% (Rawlings *et al.*, 1982). Occult infection occurs in prepatent infections, unisexual infections, or where adult worms have been rendered sterile with drugs used in heartworm treatment. In these three types of occult infection no microfilaria are produced. A fourth type, "immune-mediated amicrofilaremia," occurs when microfilaria are produced by the adults but are rapidly cleared from the circulation, presumably due to presence of anti-microfilarial antibody (Wong and Suter, 1979). Currently there are several serologic methods being developed for detecting these antibodies or *Dirofilaria immitis*-related antigens in the serum. For the most part, these tests are still being evaluated, and their sensitivity and specificity has not been documented. Currently, we are utilizing several of these tests to improve our capability to detect heartworms in dogs used in research.

Animals generally are identified permanently with an ear tat-

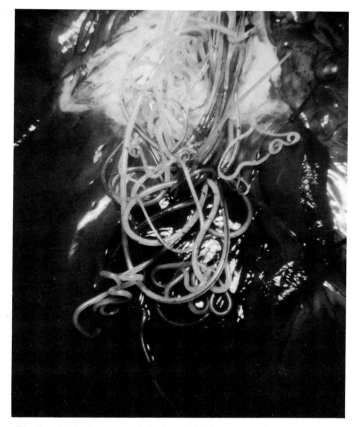

Fig. 1. Canine heartworm infection. Adult heartworms in the right ventricle and pulmonary artery of a newly received random source dog.

too, ear tag, or identification collar. In accord with the regulations of the Federal Animal Welfare Act, the research facility must retain the ability to correlate the permanent identification of the animal with information on the source of each animal.

Diarrhea is an occasional clinical finding in newly received random source dogs and cats, especially those that have been exposed to crowded unsanitary conditions. Many animals improve spontaneously within a few days as they become adjusted to the quarantine facility and laboratory diet. General treatment measures that should be considered include maintenance of fluid, electrolyte, and acid–base homeostasis and dietary restriction. Initially, the GI tract should be rested by withholding food for 24 hr or more. When feeding is resumed, small amounts of bland food should be fed frequently.

If diarrhea persists, a specific etiologic diagnosis should be made to facilitate therapy and to protect other animals in the facility. Many of the bacterial, viral, protozoal, mycotic, and helminth agents that cause disease in newly received dogs and cats are listed in Table III. Differential diagnosis and treatment of diarrhea due to these etiologic agents is beyond the scope of this chapter. For additional information the reader should consult veterinary texts that deal with the subject in more detail

Fig. 2. Canine heartworm infection. Proliferation of intima in pulmonary artery in response to adult heartworms within the lumen. Hematoxylin and eosin. × 25.

(Anderson, 1980; Ettinger, 1983; Kirk, 1980; Kirk and Bistner, 1981).

Several of the enteric pathogens of dogs and cats are also human pathogens. Dogs or cats may serve as a reservoir for salmonella (Morse and Duncan, 1975), yersinia (Wilson *et al.*, 1976), and campylobacter (Fox, 1982). The public health aspects of these diseases are discussed in more detail in Chapter 22 on zoonoses.

Control and elimination of helminth parasites is usually achieved through treatment of all animals in the group with anthelmintics at specified intervals. Alternatively, fecal flotations and sedimentations may be performed and individual animals treated as indicated. A program for group treatment with anthelmintics should be based on an adequate survey of the incidence of intestinal parasites in newly received dogs (Table IV). An effective regimen is dichlorvos on day 3, niclosamide

Table III

Enteric Pathogens of Newly Received Random Source Dogs and Cats

Bacterial
  Bacillus piliformis (Tyzzer's disease)
  Campylobacter jejuni
  Clostridium spp.
  Enteropathogenic *E. coli*
  Salmonella spp.
  Shigella spp.
  Staphylococci
  Yersinia enterocolitica
Viral
  Coronavius
    Canine
    Feline
  Parvovirus
    Canine
    Feline panleukopenia virus
  Rotaviruses
  Astrovirus

Table III (*Continued*)

Protozoal
  Coccidia
  Giardia
  Trichomonads
  Entamoeba histolytica
  Balantidium coli
Mycotic
  Aspergillus spp.
  Candida albicans
  Histoplasma capsulatum
  Phycomycetes
Helminths
  Ascarids
  Cestodes
  Hookworms
  Strongyloides stercoralis
  Whipworms

Table IV

Prevalence of Helminths in Selected Dog Populations

Geographic location	Animal[a] source	Toxocara canis	Toxascaris leonina	Trichuris vulpis	Ancylostoma caninum	Dipylidium caninum	n	Reference
Utah	R	26.0[b]	6.0	—	—	2.0	50	Sawyer *et al.* (1976)
Ohio	R	19.2	10.2	42.2	60.8	—	500	Strietal and Dubey (1976)
New Jersey	R	22.5	5.4	75.6	72.0	28.6	2737	Lillis (1967)
Indiana	R	18.3	8.7	51.9	58.7	16.3	104	Kazakos (1978)
Montreal, Quebec	P	34.0	11.4	1.2	2.5	—	332	Ghadirian *et al.* (1976)
Halifax, Nova Scotia	R	26.0	1.3	1.3	8.0	0.6	474	Malloy and Embil (1978)
New Orleans, Louisiana	P	6.5	—	19.0	46.0	—	325	Vaughn and Jordan (1960)
Indiana, Ohio	R	21.0	4.1	—	—	—	1465	Ehrenford (1957)
Bermuda Islands	P	38	—	3.0	54.1	9.3	366	Williams and Menning (1961)
Iowa	P		3.0	1.2	5.3	—	33594	Lightner *et al.* (1978)
Michigan	R	13	3	13	40	—	74	Peter (1983)

[a] R, random source animals; P, pet animals.
[b] All values are percent.

on day 7, dichlorvos on day 14, and dichlorvos again on day 28.

All dogs should be observed daily by a veterinarian or animal health technician. Any dog exhibiting signs of disease should be subjected to appropriate physical and clinical pathologic diagnostic procedures followed by specific therapy for all conditions except respiratory disease. Because of the high incidence of respiratory disease, up to 70%, and the uniformity in the clinical presentation, it may be advisable to treat all affected animals with broad spectrum antibacterials for several days and perform extensive diagnostic procedures only on dogs that do not respond to treatment. We have found that a combination of sulfadimethoxine (25 mg/kg q24h po, iv, or im) and oxytetracycline (25 mg/kg q8h, po) is an effective antibacterial combination for initial treatment. Other broad spectrum agents may be equally effective.

B. Conditioning the Random Source Cat

Random source cats, those obtained from animal shelters and animal dealers, may be infected with a wide variety of infectious agents. The most life threatening of these is feline panleukopenia virus. In addition, the feline respiratory disease complex produces high morbidity during the first few weeks after receipt. Flea, ear mite, and intestinal parasite infestations also are common. Typically, newly received cats begin to show clinical signs of panleukopenia during the first week after arrival, and the respiratory disease complex is manifest during the second or subsequent weeks. The time of onset of these infectious diseases is in all likelihood dependent on the elapsed time following comingling of animals from different sources.

In an ideal situation, random source cats should be vaccinated for feline panleukopenia, rhinotracheitis, calicivirus, and pneumonitis (*Chlamydia psittaci*) and isolated until immunity developes. Only then should they be assembled into groups from disparate backgrounds. Occasionally arrangements can be made with animal shelter directors or animal dealers to vaccinate cats on arrival in their facilities and then provide some measure of isolation until immunity has developed. In any case, cats should be immunized at the earliest possible time in the hope that immunity will develop prior to exposure to these pathogens.

Conditioning programs are directed toward elimination of infectious agents and parasites and acclimation of the animals to the laboratory environment. Programs for achieving these objectives vary widely in length and sophistication among research facilities. Some programs hold cats for only 7–14 days prior to release to research projects. Most find that this period is insufficient because many cats will exhibit clinical signs of upper respiratory disease following a 14-day holding period. Typically, the respiratory disease complex spreads through a newly received group with a few cats continuing to show upper respiratory signs even after 14 days. Most cat conditioning programs have been more successful if the conditioning period is extended to 30–60 days.

A model conditioning program is illustrated in Table V. Several aspects of the program will be discussed here. The physical examination following arrival should be directed toward selecting the healthiest, most vigorous cats for conditioning. It is usually not cost effective to proceed with diagnosis and treatment of cats that are ill on arrival or those that become ill during the first few days of the conditioning period. These animals should be euthanized. If the newly received cats from different sources have been housed together for several days then one can expect 10–40% mortality due to panleukopenia during the conditioning period. Vigorous fluid, antibiotic, and supportive treatment of individual cats that exhibit signs of

Table V

Conditioning Program for Random Source Cats

Arrival procedures
1. Physical examination by veterinarian or trained technician. Reject animals that are immature, aged, ill, or those that fail to meet specific research criteria (age, sex, disposition, hair length, etc.)
2. Weigh, sex, identify, and initiate individual health records
3. Initiate administration of a broad spectrum antibiotic for 10–14 days
4. Vaccinate for rhinotracheitis, calicivirus, and panleukopenia. (Intranasal vaccine preferred; vaccination for feline pneumonitis, *Chlamydia psittaci*, may also be helpful)
5. Provide food and water

Postarrival procedures
1. Dust or spray each cat with a topical insecticide
2. Examine ears for ear mites and treat as necessary
3. Examine feces for intestinal parasites and ova and treat as indicated or alternatively, administer a broad spectrum anthelmintic at 14-day intervals
4. Monitor food and water consumption daily
5. Observe each animal daily for signs of illness and provide veterinary medical care for ill animals. Alternatively, euthanize ill animals

Prerelease procedures
1. Physical examination
2. Examine feces for intestinal parasites and ova
3. Hematology and clinical chemistry examinations as required for specific studies

General considerations
1. Newly received cats should be isolated as a group. Special precautions should be taken to ensure that personnel, fomites, and the ventilation system do not transmit infectious agents to previously conditioned cats (See Section IV,C)
2. All long-term cats should receive yearly booster vaccinations for panleukopenia, rhinotracheitis, calicivirus. In endemic areas vaccination for pneumonitis (*Chlamydia psittaci*) should be added to this regimen

panleukopenia is often not cost effective, and euthanasia should be considered.

The feline respiratory disease complex affects many newly received cats. The complex consists primarily of feline viral rhinotracheitis virus or feline calicivirus infection. These agents are readily transmitted among cats individually housed in a room, and often a majority of the animals within a group will have respiratory signs. The husbandry staff must observe the precautions outlined in Section IV,C or the disease will spread from room to room. The intensity of treatment depends on the severity of clinical signs and the length of time that the group has been conditioned. Animals that are nearing the end of the conditioning period are more valuable because of the accumulating costs of daily maintenance and conditioning procedures. Often it is not cost effective to institute extensive treatment and nursing care for respiratory disease during the early portion of the conditioning period since the cost of drugs, supplies, laboratory tests, and nursing care can quickly exceed the cost of a replacement animal. Animals that show extensive

clinical signs early in the conditioning period should be euthanized rather than treated. Of course, all ill animals must receive adequate veterinary medical care or be euthanized.

We have found that prophylactic tetracycline (25 mg/kg q12h po or 7 mg/kg q12h im) for 10–14 days beginning on the first day of conditioning is very helpful in reducing the severity of respiratory disease. *Chlamydia psittaci,* the causitive agent of feline pneumonitis, is uniformly sensitive to tetracycline, and secondary bacterial pathogens associated with viral respiratory diseases may also be susceptible.

Treatment of all but the most severely ill cats includes broad spectrum antibiotic therapy, fluid therapy, parenteral vitamins, and ocular care. It is beyond the scope of this chapter to provide detailed recommendations on the treatment of the feline respiratory disease complex. Stein (1980) has described treatment in some detail, and Section VII,B,2 provides additional information on the feline respiratory disease complex.

Ectoparasites, primarily fleas and ear mites, commonly infest newly received cats. Standard methods of treatment can be used during the conditioning period; however, later, when the animals are in experimental use, insecticides must be used with care since exposure, even to low concentrations, can induce or inhibit hepatic microsomal enzyme activity and alter the animal's biologic responses.

Gastrointestinal parasites should be eliminated during the conditioning period. There are two basic methods for eliminating these parasites. The first involves fecal examination of each animal and subsequent therapy at appropriate intervals until a series of negative fecals are obtained. The other involves routine treatment of all newly received cats with a broad spectrum anthelmintic such as dichlorvos at 10–14 day intervals then fecal examinations for parasites and ova at the end of the conditioning period. The choice of methods will be dependent on the prevalence of intestinal parasites in the newly received cat population (Table VI), the cost of laboratory tests, and the costs associated with therapy.

C. Prevention of Respiratory Disease Transmission

The most practical method of avoiding transmission of viral or chlamydial respiratory infections among dogs or cats is to receive, quarantine, and house them in individual cages with only a small number of animals per room. The husbandry staff must use strict rules of hygiene to prevent contamination from animal to animal and room to room. Procedures should include hand washing between handling ill animals, then hand washing and an outer clothing change between rooms. There should be no recirculation of air nor transfer of supplies and equipment between rooms. Cages and equipment should be disinfected between uses. Scott (1980a) tested 35 viricidal disinfectants against three feline viruses. Household bleach (5.6% sodium

Table VI

Prevalence of Helminths in Selected Cat Populations

Geographic location	Animal source[a]	*Toxocara cati*	*Toxascaris leonina*	*Capillaria* sp. and/or *Trichuris* sp.	*Ancylostoma* sp.	*Dipylidium caninum*	*n*	Reference
Utah	R	43[b]	—	—	—	1	100	Sawyer *et al.* (1976)
Ohio	R	25	—	1.3	9.4	—	1000	Christie *et al.* (1976)
New York	P	10	—	—	2	—	100	Dorman and Strand (1958)
New Jersey	R	55.7	—	—	84.6	10.4	1450	Lillis (1967)
Missouri	P		24.4	2.6	6.4	—	1294	Visco *et al.* (1978)
Illinois	P	41	6	3	6	—	34	Guterbock and Levine (1977a)
	L	29	5	2	8	—	124	
	R	35	9	7	14	—	57	
Halifax, Nova Scotia	R	25.1	0.3	0.3	—	—	299	Malloy and Embil (1978)
Bermuda Islands	R	17.9	—	10.3	27.8	12.8	39	Williams and Menning (1961)
Iowa	P		3.3	—	1.7	—	11995	Lightner *et al.* (1978)
Michigan	R	30.0	—	—	41.8	1.8	55	Peter (1983)

[a] R, random source animals; P, pet animals; L, laboratory colony.
[b] All values are percentages.

hypochlorite) diluted 1 : 32 was found to be the most effective and practical broad-spectrum viricidal product. To increase activity, this dilute bleach can be combined with detergents.

Animal husbandry personnel should provide daily care for long-term stable animals early in the day, then proceed to care for conditioning animals next, then newly received animals later in the day. Alternatively, staff members should work only in one colony. Of course, newly received dogs or cats should not be introduced into stable colonies until long-term observation and laboratory tests provide relative assurance that they do not harbor communicable respiratory diseases.

D. Canine Devocalization

In a research animal facility it is important to control environmental variables. One of the variables that is especially difficult to control is the noise of barking dogs. This noise, which can be extremely intense, is undesirable because of its effects on personnel and dogs in the facility as well as on personnel and research animals in adjacent areas. Every effort should be made to reduce sound transmission to other animal housing areas and human occupancy areas such as laboratories and patient-care facilities. Physical and acoustical separation of dog housing areas is usually the best method of reducing noise in adjacent areas. Within the dog housing area, sound control is especially difficult, since the use of acoustical materials in animal housing rooms often presents problems in sanitation and vermin control.

If noise from a dog housing area cannot be controlled through construction and acoustical absorption techniques, it may be necessary to utilize surgical techniques to reduce the intensity of sound produced. Several surgical procedures have been developed to reduce the level of noise produced when dogs bark. These procedures entail sectioning of the vocal cords and removal of the ventricular folds or adjacent mucosa.

The humane aspects of these "devocalization" or "debarking" procedures should be thoughtfully considered. The procedure requires general anesthesia and may cause laryngeal discomfort during the immediate postoperative period. Most dogs eat and drink normally following the immediate postoperative period, and many dogs continue to vocalize with lower intensity and pitch within a few days of the surgery. It is not apparent that there are any long-term detrimental physical or behavioral effects following the procedure.

Removal or sectioning of the vocal cords endoscopically (Anderson, 1955; Young and Sales, 1944) is the least satisfactory and frequently results in only transient devocalization. Transection of the vocal cords with electrocautery (Kraus, 1963) produces more permanent results with less hemorrhage. An extensive surgical method for devocalization that incorporates total removal of the vocal ventricular folds is described by Yoder and Starch (1964). This procedure is the longest lasting and provides the greatest sound reduction, but it requires considerably more surgical time than the other procedures. This can be an important consideration if many dogs must be devocalized.

A procedure developed by Raulston *et al.* (1969) utilizes a combination of transection of the cord with electrocautery and removal of the laryngeal saccular mucosa by means of a burr in a manner used in horses with recurrent laryngeal nerve paralysis. Results are longer lasting and more satisfactory than with cautery alone. It is a very rapid method, allowing the devocalization of 40–50 dogs per hour, excluding anesthetization time.

V. CANINE AND FELINE BREEDING COLONY MANAGEMENT

A. Introduction

In the medical management of breeding colonies, the veterinarian must emphasize as the ultimate goal, the prevention of disease and the production of healthy animals for research. As in producing domestic food animals, a "herd-health" approach is required. The need for treatment of individual cases usually indicates a failure in husbandry or preventive medical programs. The economic costs and the medical benefits of instituting preventive or therapeutic measures must be taken into account when changes in medical care and colony management are contemplated. To aid in diagnosing disease entities and making managerial decisions concerning animal health, access to reliable diagnostic laboratories is essential. The need for diagnosis and treatment of disease should, however, be minimal if sound husbandry and preventive medical programs are implemented, supervised, and properly revised as new problems arise.

B. Preventive Medicine in a Breeding Colony

1. Environment

Although the actual physical plant for a breeding colony will vary depending upon specific needs and circumstances, there are some general principles that have been identified. These principles are discussed in Section III of this chapter and in Chapter 17 on design and management of animal facilities. Special attention must be directed to the adequacy of temperature and humidity control and ventilation in indoor facilities (ILAR, 1973, 1978c,d). Morbidity in the young can often be directly linked to inadequate ventilation or poor temperature and humidity control. There should be at least ten air changes an hour or more depending on temperature, humidity, and population densities. Morbidity can also be reduced by a physical plant that promotes rodent and insect control. Suspended wire cages help eliminate the oral–fecal transmission of intestinal parasites. The underlying theme of these measures is to provide a stable environment that lowers morbidity and mortality from disease by preventing stress and persistent contact with waste matter and potential pathogens.

The facility might include the following separate areas: (1) a housing and mating area for adult males in which each male has an individual territory, (2) a whelping/nursery area to which the female is taken prior to parturition, (3) an area in which weanlings can be raised and adults can be housed, (4) a surgical suite, (5) a diagnostic laboratory, (6) an isolation ward for sick animals and possibly a separate quarantine area for newly arrived animals, and (7) an administrative office area. An additional area for holding animals before shipment is convenient but not necessary.

A proper social environment is also important. The adaptability of a dog or cat to the laboratory setting, as to the home, is dependent upon regular human contact (Fox, 1965). Failure to supply this need can render an animal behaviorally useless as an experimental animal. Animals that exhibit undesirable behavioral traits even after adequate socialization should be culled from the breeding program.

2. Nutrition

Any good quality commercial food meeting the National Academy of Science–National Research Council (NAS–NRC) recommendations for nutrients (1974, 1978) can serve as the maintenance diet for the colony. Adults can be fed one or two times a day or *ad libitum*. Most dogs and cats do well on *ad libitum* feeding. Those that become obese may need to be hand fed. Puppies and kittens can be offered canned food or a moistened gruel of dry food starting at 4 weeks. Weaning usually takes place at 6 weeks, although this may vary depending on litter size. If *ad libitum* feeding is not implemented at this time, more frequent feedings (e.g., three to four times a day) are necessary for the first 3 months followed by decreasing to adult frequency of 1–2 times/day by 8 to 12 months of age.

During gestation nutritional requirements increase. The bitch and queen should be fed approximately 50% more than maintenance levels. During lactation, they should be fed *ad libitum* as they will consume two and one-half to three times as much food due to the energy demand of feeding a litter. They may lose weight, despite this increase in consumption.

Orphaned young or young removed from the mother for any reason should be fostered if possible. If hand-raising is required, a commercial milk replacer such as Esbilac (bitch milk replacer) or KMR (kitten milk replacer), Borden Inc., Norfolk, Virginia, should be fed and the orphan's weight monitored frequently to ensure adequate nutrition (Small, 1980).

Clean water should be made available at all times. If supplied in a bowl or bucket, the container should be rinsed and refilled daily and sanitized weekly. Automatic watering systems obviate this need but require weekly disinfectant flushing and daily inspection to assure proper functioning.

3. Immunization

a. Introduction. A routine vaccination schedule for both adults and young should be implemented. Modified live viral vaccines should be used with caution in pregnant animals since the may cause congenital defects. It is prudent, therefore, to assure that vaccinations with these products are current before mating occurs. This will help ensure adequate colostral antibody to protect the young. All new arrivals in a colony should have current vaccinations. A period of quarantine and conditioning is necessary to prevent the introduction of a diseased or susceptible animal. See Sections IV,A and B for more complete descriptions of quarantining and conditioning programs for newly arrived animals.

b. Canine. In the canine colony, puppies should be immunized against distemper, infectious canine hepatitis (canine adenovirus type 1), leptospirosis, canine parainfluenza virus, and canine parvovirus at 6–8, 10, and 12–14 weeks of age. The exact timing of the first vaccination should coincide with the decline of maternal antibody levels. In most puppies this occurs at approximately 8–9 weeks, but a more precise time can be determined by analyzing serum titers. If the pup has not received colostrum, the vaccine regimen should be begun at 4 weeks of age. The use of measles virus vaccine in pups less than 6 weeks of age for protection against distemper virus is probably not necessary when the dam is revaccinated annually and the pup has received colostrum.

The recent concern over parvovirus infections and its potential disastrous effects in a colony situation warrant its inclusion in a vaccination program. The vaccine is available in combination with the above-mentioned products and can be administered on the same schedule. There is some indication that the canine origin products may provide greater protection than those of feline origin (Carmichael *et al.,* 1981).

Both parenteral and intranasal vaccines are currently available against *Bordetella bronchiseptica.* Although the vaccines may not totally protect dogs against ''kennel cough,'' as there are several etiologic agents implicated, it can decrease the morbidity and severity of this disease complex (Shade and Goodnow, 1979).

In an indoor colony, rabies immunization is not necessary, since the likelihood of exposure is minimal; however, it may be advisable to immunize dogs that are housed outdoors or are being shipped to an outdoor facility. If shipment involves export or interstate transport, rabies immunization may be necessary to comply with shipping regulations.

c. Feline. In the feline colony, kittens should be immunized at 8 and 12 weeks, and adults should be reimmunized annually. Modified live vaccines against feline panleukopenia, feline viral rhinotracheitis (FVR), and feline calicivirus (FCV) are preferred. Intranasal vaccination for FVR and FCV can induce rapid immunity in the face of an outbreak. The inclusion of a vaccine against feline pneumonitis (FPN), a less common chlamydial disease, is warranted if the etiologic agent is present in the colony. Although not affording complete protection, vaccination against FPN can decrease the duration and severity of signs (Mitzel and Strating, 1977) and may eliminate the chronic disease seen in susceptible animals. Vaccination against rabies, as with canines, is not usually necessary unless it is required by shipping regulations.

4. Parasites

Parasite control is an important aspect of preventive medicine in breeding colonies. Parasites constitute a variable in research, and most responsible investigators specify that research animals should be free of both internal and external parasites.

External parasites can be controlled by routine environmental sanitation and regular dipping, spraying, or dusting with appropriate parasiticidal agents. Lindane is commonly used in dogs, but is toxic for cats. Carbaryl or rotenone compounds are safe for cats. Ear mites can be controlled with routine instillation of a 3 : 1 mixture of mineral oil and a commercial rotenone product (Canex, Pittman-Moore, Inc., Washington Crossing, New Jersey). Generalized demodectic mange in dogs possibly reflects a genetic predilection and an associated deficiency in cell-mediated immunity which may necessitate culling the animal (Muller and Kirk, 1976). Astute observation by the colony staff can allow early institution of more effective preventive medical procedures and medical management of these conditions before they become a major problem.

The control of internal parasites also depends on environmental considerations and parasiticidal therapy. Suspended caging, flea, and rodent control and strict sanitation help break the life cycle of these parasites. The various groups of animals in the facility should be routinely surveyed for internal parasites utilizing direct fecal smears and fecal flotations to detect ova or parasites. The frequency should be based on cost-effectiveness. The appropriate therapeutic agent should be administered as dictated by these examinations.

5. Record Keeping

The keeping of thorough, accurate records cannot be overemphasized. The medical and breeding information contained in them is invaluable in formulating and assessing preventive medical measures and breeding practices. A small computer for storing and processing the potentially large amount of data generated can be cost-effective and facilitate information retrieval.

The information on each animal should contain, above all, a record of permanent proper identification such as a tattoo number. Physical characteristics such as size, conformation, congenital abnormalities, and growth rate should be noted. Heritage and the progress of littermates offer valuable information on genetic constitution. For adults, behavioral information on disposition, maternal instinct, mating behavior, and libido are important parameters. Other important breeding and birthing characteristics include date and length of estrus, results of semen examinations, previous gestation lengths, occurrence of dystocia, time of labor, litter size and viability including stillbirths, postparturitional vaginal discharges—character and duration, lactation, and number of pups weaned. Medical histories should include vaccinations, illnesses, diagnoses, therapy, surgical procedures and laboratory data generated by the clinical examination at the time of each illness. Records kept with diligence can be useful in assessing cases of subfertility, disease transmission and prevalence, husbandry deficiencies, and many other problems amenable to improved colony and medical management.

C. Breeding Considerations—Canine

1. Estrus and Mating

The bitch has a monestrous estrous cycle with heats occurring throughout the year, but predominantly in January or February and again in July or August. This is highly variable between breeds and individuals with the time between cycles ranging from 4 to 18 months. For an individual bitch, however, this time interval is usually consistent, and records can help predict the onset of estrus. Estrus is usually detected by regular clinical observation of the vulva for swelling and the serosanguinous discharge typical of proestrus. Vaginal smears to detect estrus, although accurate, are labor intensive and unnecessary in most situations.

Estrus can be induced during the anestrous period with a combination of pregnant mare serum gonadotropin (PMSG) and human chorionic gonadotropin (HCG) or pituitary gonadotropins of sheep origin (Evans, 1980). Induction can be desirable from a management aspect if whelping synchronization is important. Bitches that exhibit prolonged anestrus or incomplete estrus should be culled from the breeding colony.

The prevention of estrus can at times be desirable. Animals to be shipped or those that are group-housed are examples. The blocking of estrus can be accomplished by the use of either a progestational agent such as megestrol acetate (Ovaban, Schering Corp., Kenilworth, New York) (Buke and Reynolds, 1975) or an androgenic agent such as mibilerone (Cheque, Upjohn Co., Kalamazoo, Michigan) (Sokowlowski, 1978; Sokowlowski and Gerg, 1977; Upjohn Co., 1978).

A bitch may be bred on her first heat after 1 year of age and on every cycle thereafter if the duration of anestrus is greater than 8 months. If the intercycle duration is less than this, she should be rested every third cycle (Kirk, 1970).

The male dog is usually brought into service at 1 year but can begin as early as 8 months. He is fertile and able to inseminate females throughout the year.

Mating is accomplished by placing the bitch with the stud every other day during estrus until she refuses to accept him. If natural mating does not occur, artificial insemination can be performed (Seager, 1977). Semen should be collected at regular intervals for semen evaluation. Studs should not service more than one or two bitches per week nor be collected more often than every other day. The usual ratio of males to females is 1 : 10–15 which allows for the relatively long anestral periods.

2. Pregnancy and Parturition

Pregnancy diagnosis is most easily accomplished by abdominal palpation at 21–28 days postmating or radiography after 43 days. A negative finding may indicate failure to conceive, fetal resorption, early abortion, or stud infertility. As always, carefully kept records can assist in delineating the reasons for "misconception."

Gestation length varies between bitches but is consistent for any individual. The average duration is approximately 63 days. The pregnant bitch should be moved to the whelping area 10 to 14 days prior to her expected date of delivery. Adaptation to her surroundings and the avoidance of stress are important to ensure a normal delivery. If she is upset, gestation may be prolonged. A whelping box, preferably disposable and/or nesting paper should be provided.

The first signs of impending parturition are usually behavioral. The bitch seeks seclusion or the company of a keeper to whom she is very attached. She may exhibit nest building behavior and appear apprehensive especially if she is nulliparous. Anorexia and even emesis may be evident. A decrease in body temperature the day before parturition is not as reliable in predicting impending parturition as the palpable relaxation of the pelvic musculature and ligaments. The most reliable method of predicting parturition is to observe these changes in each animal and record them for future reference.

The first stage of parturition, consisting of uterine contractions which increase in frequency and strength, lasts 6–12 hr. This varies by breed and may be longer in nulliparous bitches. The second stage is indicated by obvious tenesmus coincident with the contractions. This increases the intrauterine pressure and causes the rupture of the fetal membranes and expulsion of the pups. The membranes are usually delivered within 15 min of the birth of the pup, although they may be delivered together. Usually, 30–45 min elapse between the birth of subsequent puppies. There may be a green vaginal discharge which is normal as a result of membrane rupture, but if there is an excessive delay before a puppy is delivered, a vaginal exam is indicated. Uncomplicated, the whelping of the entire litter may take only 2 hr in smaller breeds and as long as 12 hr in larger breeds.

If a delay in delivery is evident, determination of the cause dictates the action taken. Malposition may be corrected by digital manipulation *per vaginum*. If the fetus is too large or cannot be rotated into a normal presentation, then a cesarian section must be performed. Smith (1974) describes in detail the surgical procedure. This action should never be postponed until the bitch is exhausted as the surgery is risk enough. If uterine inertia develops, either primary or secondary to obstruction, as many as three oxytocin injections (5–10 U, im or iv) 20 min apart can be administered to stimulate uterine activity. If uterine enertia persists a cesarian section should be performed.

Upon the completion of whelping, abdominal palpation and vaginal examination to ensure that no pups remain should be followed by an oxytocin injection (5–10 U, im or iv) to help control postparturient hemorrhage, contract the uterus, and stimulate milk letdown. The umbilical cord should be dipped in an iodophor disinfectant solution.

Following birth, bitches usually lick their pups vigorously. This stimulates respiration and circulation and removes the membranes and fluids. If pups are delivered by hysterotomy, they should be dried vigorously with a soft terrycloth towel and placed in an incubator until the bitch has recovered. Orphan puppies should receive colostrum if at all possible. They can be maintained on a commercial milk replacer, until they can be fostered to a receptive dam. If they are not fostered, they must be hand raised. This entails the use of an incubator, frequent feedings, and the stimulation of urination and defecation (Olson and Olson, 1971). It is far easier and more efficient to foster whenever possible.

D. Breeding Considerations—Feline

1. Estrus and Mating

The queen is seasonally polyestrous with two to three cycles between mid-winter (January) and early fall (September). Behavioral changes, such as posturing with the pelvis extended and the tail deflected laterally and vocalization, may be evident especially while being handled, but there is no characteristic vulvar swelling or vaginal discharge. Vaginal cytology should be used with care in determining the proper time for mating, because collection of the cytologic sample may actually induce ovulation and pseudopregnancy.

Induction of estrus can be accomplished by housing the anestrous female in close contact with normally cycling females (Colby, 1980) or by administering exogenous gonadotropins (Cline *et al.*, 1980; Colby, 1970; Wildt *et al.*, 1978). Increasing the light to 14–18 hr per 24-hr period may enhance the percentage of cycling queens (Scott and Lloyd-Jacob, 1959).

The queen is always taken to the male's territory, because of the male's strong sense of territoriality. Libido in the male can be easily suppressed by confinement, the presence of another male, unfamiliar surroundings, or an exceptionally aggressive female. Receptivity of the female lasts from 1 to 4 days and induced ovulation occurs 24 hr after mating. Repeated matings are recommended to ensure success. If not mated, the cycle lasts 10 to 14 days followed by another in 2 to 3 weeks.

Semen collected by electroejaculation (Platz and Seager, 1978) can be examined in the laboratory to evaluate fertility of the male and/or it can be used for artificial insemination (Sojka *et al.*, 1970).

2. Pregnancy and Parturition

Pregnancy diagnosis by abdominal palpation is reliable 17–21 days postmating. Implantation sites are firm, marble-sized swellings that can be palpated early in this period. The uterus becomes larger and softer to palpation after day 21. The skeletons of the fetuses can be visualized radiographically after about 43 days.

Gestation requires approximately 63–65 days in the cat. As with the bitch, it is advisable to move the queen to a kindling environment before parturition. It should be quiet, dark, and contain a kindling box.

As parturition draws near, mammary development begins. It is especially noticable during the last 3 days of gestation and milk can be expressed 24 hr before parturition. Other signs of impending birth include mucous vaginal discharge, vulvar enlargement, decreased body temperature, and behavioral changes characterized by seclusion seeking or the strong attachment to a certain handler.

The normal interval between expulsion of kittens can vary from minutes to hours but because of the length of the estrous cycle and the general practice of repeat breedings, kittens may be born in more than one group with days intervening. Thus, delay in expulsion of fetuses does not necessarily mean dystocia. A careful examination of the breeding history and the queen can be helpful in determining whether dystocia exists. If a particular queen has a history of dystocia, a prepartum exam including palpation and radiography may be necessary to determine the need for cesarian section. In the queen, this is usually not necessary as dystocia is not as common as in the dog. Usually digital manipulation is the only assistance necessary. It is not customary to use oxytocin to aid uterine motility and milk letdown in the feline. Oxytocin (0.5–3.0 U, im or iv) may be useful if placental membranes are retained or there is extensive postparturient hemorrhage. A postpartum exam is essential to determine if all kittens have been expelled and the membranes passed. The umbilical cord should be dipped in an iodophor disinfectant solution. Finally, the quality of maternal behavior should be observed and recorded.

E.   Diseases Affecting Reproductive Success—Canine

1.   Introduction

Reproductive success is best measured by the number of healthy pups reaching puberty. Mortality generally occurs before weaning, especially during the first week in which stillbirths, congenital abnormalities, physiological immaturity, and trauma account for a large number of deaths. Thereafter, the major causes of morbidity and mortality are pneumonia and gastrointestinal disease, which may be exacerbated during the stress of weaning.

Neonates are extremely prone to hypoglycemia, hypothermia, and dehydration. Therefore, malnutrition of either the bitch or the neonate or anorexia due to illness can be life-threatening. Much of the treatment of disease in the neonate consists of supportive as well as specific therapy.

Adult animals are culled primarily due to reproductive problems, chronic skin and ear infections, and trauma. The following summary is not meant to be conclusive, but highlights some of the specific health problems seen in breeding colonies.

2.   Diseases of the Young

a. Canine Herpesvirus (Puppy Viremia). Canine herpesvirus infection is an acute fulminant disease that usually affects puppies from 1 to 3 weeks of age. Older dogs are usually unaffected or may exhibit signs of a mild upper respiratory disease. The virus is transmitted *in utero* or as the puppy passes through the birth canal and contacts virus-containing vaginal fluids. Susceptibility to fatal viremia is greatest during the first week of life. The clinical signs include sudden onset, constant crying, cessation of nursing, and death within several hours. The entire litter is commonly affected, and it is usually the primiparous litter. Subsequent litters may not be affected, as self-immunization in the colony usually occurs. Therefore, the occurrence of canine herpes infection in a litter is not a common reason for culling a bitch.

Necropsy reveals pathognomonic renal cortical hemorrhages with necrotic centers. There may be similar hepatic lesions (Fig. 3) and bronchopneumonia. Confirmation of the diagnosis is by histologic observation of basophilic intranuclear inclusions in necrotic areas of parenchymal organs or by viral isolation (Smith *et al.,* 1972).

Successful treatment depends on the early recognition of clinical signs and consists of placing pups in an incubator for 24 hr. The temperature should be 100°F for the first 3 hr and 95°F thereafter. A relative humidity of 60% and frequent oral fluids (5% glucose) help prevent dehydration. Surviving puppies may develop chronic renal disease at an early age (Mosier, 1977).

b. Toxemia and Septicemia. "Toxic milk syndrome" has been associated with uterine subinvolution in the bitch. The

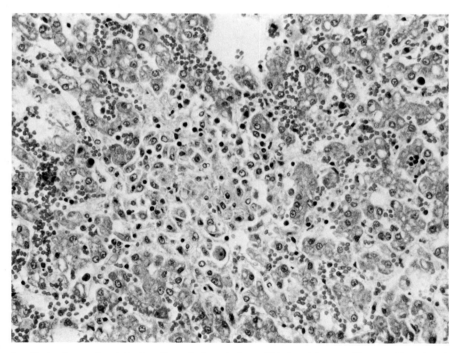

Fig. 3.   Canine herpesvirus infection. Necrotic focus in liver of 11-day-old puppy. Hematoxylin and eosin. × 160.

first indication of a medical problem may be crying, bloated, 4- to 14-day-old puppies with red, edematous rectums. The bitch may exhibit a greenish vaginal discharge indicating metritis. Treatment consists of removing the puppies and placing them in an incubator at 85°–90°F and oral gavage with 5% glucose until the bloat has receded. This is followed by feeding commercially available milk replacer for a 24-hr period. Treatment of the bitch is instituted with an intrauterine antibacterial/estrogen infusion (Utonex, Upjohn Co., Kalamazoo, Michigan) followed by oral ergonovine (0.2 mg t.i.d. for 2–3 days) and systemic antibiotic therapy for 7 to 10 days. The pups can be returned after 24 hr. Routine administration of oxytocin postpartum has been effective in preventing this condition. Good nutrition and regular exercise may also be helpful in ensuring proper uterine involution during subsequent breedings.

Puppy septicemia occurs most frequently during the first 5 weeks of life. During the first 48 hr of life, the syndrome is characterized by crying, bloating, cessation of nursing, with resultant hypothermia, hypoglycemia, and dehydration. Death usually ensues within 12 hr of onset. Older puppies present with distended abdomens, hyperpnea, and crying and die within 18 hr. Signs may appear in only one pup, but many times the entire litter becomes progressively involved. Visceral congestion and intestinal distension with gas are found postmortem. The condition is often associated with metritis and/or mastitis in the bitch. Environmental factors such as high humidity, poor ventilation, and bacterial contamination due to poor sanitation are implicated as predisposing factors. Blood cultures from puppies have revealed *Streptococcus*, *E. coli*, and *Klebsiella* as common agents (Glickman, 1980).

Treatment consists of moving the pups to an incubator, relieving the bloat, administering subcutaneous lactated Ringer's solution, and oral dosing with a broad-spectrum antibiotic. It is best to prevent the problem through improved sanitation and husbandry. Ensuring colostrum consumption may also be helpful. In some colonies where the problem is consistently linked to a particular bacterial agent, autogenous bacterins may be useful. Bitches who have repeated episodes of metritis and/or mastitis should be culled.

c. Gastrointestinal Disease.

In young dogs, morbidity and mortality due to gastrointestinal disease can be significant. Diarrhea and vomiting are commonly seen following weaning and are probably related to dietary changes and stress. These signs usually resolve spontaneously. If not, the diarrhea usually responds to dietary management (addition of cooked rice) and neomycin/anticholinergic therapy.

Persistent cases requiring diagnosis and treatment usually have an infectious etiology that can include parasites, bacteria, and viruses. Although diseases caused by parasites should be minimal in a well-managed colony, initial fecal flotation and direct smears are indicated in diagnosing gastrointestinal disorders. Positive results dictate the appropriate therapy and the need for subsequent fecal examination. The bitch should be similarly examined as she is the probable source of the infestation.

Parvoviral enteritis is potentially a serious threat to young dogs in colonies. Mortality can be very high in 5- to 12-week-old puppies. Lethargy, anorexia, diarrhea, and vomiting with rapid dehydration is the common clinical presentation. Panleukopenia or lymphopenia may be present early in the course of the disease. Coronavirus enteritis presents a similar clinical picture but usually without panleukopenia or lymphopenia. These conditions can be differentiated histologically in the laboratory (see Section VII,B,3 for additional information on canine parvovirus). Parvovirus vaccine is effective and should be included in a routine immunization program for both bitches and puppies.

d. Respiratory Disease.

Prevention of pneumonia requires rigorous environmental control and a complete immunization program. Vaccination against canine distemper, parainfluenza, canine adenovirus I (infectious canine hepatitis) or canine adenovirus II, and *Bordetella bronchiseptica* can prevent much of the respiratory disease in a colony; however, the most critical factor is environmental control.

Tracheobronchitis, or kennel cough, caused by *Bordetella bronchiseptica* and canine parainfluenza in synergy is a common syndrome seen in group-housed dogs. It is usually a mild disease in older dogs requiring little medical attention, but in pups from 4 to 12 weeks of age, the disease can be more severe. Secondary invaders (*Pasteurella, Streptococcus, Klebsiella, Proteus, Hemophilus, E. coli*, mycoplasmas) in addition to the combination of primary etiologic agents (*Bordetella bronchiseptica* and canine parainfluenza virus) can cause life-threatening disease in affected pups. Therapy consists of broad spectrum antibiotic therapy. Severe congestive cases may respond to nebulization with a mucolytic agent and gentamycin. These dogs may suffer permanent lung damage (see Section VII,B,1 for additional information on canine respiratory diseases).

e. Miscellaneous Diseases.

Pustular dermatitis occurs in pups before weaning. It usually affects the entire litter. Predisposing factors, such as high humidity and poor sanitation, allow bacteria, usually staphyococci, to colonize the skin. Lesions occur predominantly on the head and neck. Topical therapy consists of cleansing the skin and administration of an antibiotic ointment.

Puppies are born hypoprothombinemic and can hemorrhage easily. Severing of the umbilical cord too close to the body wall can result in intraperitoneal hemorrhage and death. The prevention of bleeding tendencies in general can be accomplished by the administration of vitamin K_1 to the bitch during the last 30 days of gestation and to the newborn pups.

Trauma in the neonate is usually caused by the inexperienced bitch. If she is traumatic during removal of the fetal membranes, not only hemorrhage but umbilical herniation or evisceration can result. Excitable or inattentive bitches may unknowingly crush their young or provide poor care. They should be culled from the colony.

3. Diseases of the Adult

a. Brucellosis. Disease due to *Brucella canis* is suspected whenever there is a clinical history of abortion, infertility, testicular abnormalities, or lymphadenopathy. The disease, however, can be insidious with very low morbidity.

This systemic disease produces a prolonged bacteremia (up to 2 years) without fever. Transmission can occur at mating and by contact with aborted fetal membranes or vaginal discharges. Early embryonic death may occur and be interpreted wrongly as misconception. Later, abortion may occur in the last one-third of gestation followed by a prolonged vaginal discharge. Stillbirths can occur or live puppies may be weak and die soon after birth. If they survive they are often bacteremic and exhibit lymphadenopathy. Adult males affected with this disease can have epididymitis, orchitis, prostatitis, testicular atrophy, and lymphadenopathy.

The disease also affects humans. The zoonosis is characterized by headaches, weight loss, lymphadenopathy, and flu-like symptoms—chills, weakness, sweating, muscular aches, and malaise (see Chapter 22 for a more comprehensive discussion of this zoonotic disease).

Control of brucellosis in a colony situation requires rigorous serological screening and culling (Moore *et al.*, 1968; Pickerill and Carmichael, 1972). A convenient and rapid slide agglutination test is available (Canine Brucellosis Diagnostic Test, Pittman-Moore, Inc., Washington Crossing, New Jersey), but can give false positive results due to nonspecific agglutination reactions (Brown *et al.*, 1976). If the slide agglutination is positive then tube agglutination and blood culture are indicated to confirm the diagnosis. Suspect dogs should be isolated, and if positive serology is confirmed by the tube agglutination, they should be destroyed. Three consecutive negative monthly tests in all dogs indicate that the colony is free from infection. New arrivals should be quarantined and undergo two or three negative tests at monthly intervals before entering the colony. During an attempt to eradicate the disease, daily cleaning and disinfection of the facility is helpful. Once free of the disease, a colony should be periodically tested with the rapid slide test to ensure that contamination has not occurred.

b. Mastitis. Mastitis usually occurs only when nursing ceases abruptly and the udder becomes engorged. Death of a litter or sudden removal of the pups may precipitate the condition. If mastitis occurs while the pups are nursing, it can be a threat to the puppies' health, especially if the milk is frankly purulent. The puppies should be removed or denied access to the infected glands by the use of bandages. The bitch's milk should be cultured to determine specific antibiotic therapy, and broad spectrum systemic antibiotics should be administered pending laboratory results. Hotpack applications and manual milk removal will hasten recovery and relieve discomfort. Recurrence on subsequent lactations is not uncommon, and repeated mastitis may constitute a reason for culling the bitch.

c. Metritis. Metritis usually occurs in the first week postpartum. Early recognition and prompt therapy shorten the disease course and prevent subsequent decreased breeding efficiency. The toxemic bitch with metritis is usually depressed, anorexic, febrile, and has decreased milk production. An excessive or abnormal vaginal discharge may be present. It is usually fetid, watery, and red. If the placenta is retained, the discharge may be green to black. Nursing pups may be listless, cry, and may have red edematous rectums. They may also exhibit other signs of septicemia/toxemia discussed earlier.

Therapy consists of administration of an ecbolic, either oxytocin or ergonovine; intrauterine infusion of an antibacterial/estrogen solution (Utonex, Upjohn Co., Kalamazoo, Michigan); and systemic antibiotics. Metritis may not necessarily recur after subsequent parturitions. Routine oxytocin injections several hours after whelping may reduce the incidence of metritis in the breeding colony.

F. Diseases Affecting Reproductive Success—Feline

Introduction

Many of the disease conditions affecting reproduction can be lessened by vaccination (panleukopenia, viral rhinotracheitis, calicivirus), and others can be eliminated by testing and culling (feline infectious peritonitis, feline leukemia virus). The prevention of disease in general requires good husbandry practices and prompt recognition and treatment of health problems. The following discussion will deal briefly with disease entities encountered in the feline breeding colony.

a. Panleukopenia. Disease due to feline panleukopenia virus is rare in breeding colonies due to the efficacy of vaccination. Vaccination of queens should occur before mating, since vaccination during pregnancy may have untoward effects on developing fetuses. To be effective, vaccination of the young must occur after maternally acquired antibody has diminished and prior to natural infection. The ideal time may vary from 3 to 16 weeks depending on the level of maternally acquired immunity. It is customary to vaccinate kittens at 8, 12, and 16 weeks and annually thereafter. Affected kittens may die acutely or become progressively depressed, anorectic, and dehydrated. Infected pregnant queens may experience abortions, stillbirths, and neonatal deaths. Kittens may have cerebellar hypoplasia if they are exposed to the virus *in utero*.

b. Respiratory Disease. Vaccination against the common causes of respiratory disease, viral rhinotracheitis, and calicivirus is mandatory. Vaccination against feline pneumonitis is less effective but may be indicated if the disease is enzootic. The major threat to reproductive success is transmission of rhinotracheitis or calicivirus infection to kittens by queens who are carriers. Feline viral rhinotracheitis viremia in the pregnant female may also cause abortions in addition to respiratory signs (Benirschke *et al.*, 1978).

Animals affected with upper respiratory disease should be isolated from the colony while being treated. Carrier queens that repeatedly infect their litters with rhinotracheitis or calicivirus should be culled (see Section VII,B,2 for additional information on feline respiratory diseases).

c. Feline Leukemia Virus. Feline leukemia virus (FeLV) can cause fetal resorption and abortion. Infertility due to persistent anestrus may also occur. Kittens exposed to FeLV may develop thymic atrophy and immunologic incompetence leading to increased susceptibility to other diseases. Methods to rid a colony of animals infected with FeLV are discussed in Section VII,E,1.

d. Feline Infectious Peritonitis. Feline infectious peritonitis (FIP) is an insidious infectious disease caused by a coronavirus which infects cats of all ages. Initial infection may be asymptomatic or manifest as another mild upper respiratory disease. Feline infectious peritonitis is strongly implicated in the kitten mortality complex discussed below. Cats with FIP should be removed from the colony using methods outlined in Section VII,B,4.

e. Kitten Mortality Complex. The kitten mortality complex (KMC) is a relatively newly described disease entity seen in catteries and breeding colonies (Norsworthy, 1979; Scott *et al.*, 1979). The etiology is unknown; however, feline leukemia virus or feline infectious peritonitis virus may play a role. The complex presents as a high frequency of reproductive failures and kitten deaths in successive pregnancies. Reproductive failure is characterized by fetal resorption, abortion in the last half of gestation, and stillbirths. Some queens appear to be repeat breeders, indicating the possibility of early embryonic death.

Commonly kittens are born weak and die soon after birth or are healthy for a few weeks then gradually become anorectic, depressed, lose weight, and die. There are no specific lesions seen other than those attributable to malnutrition. Some kittens

as well as adults may die as the result of acute congestive cardiomyopathy. A histologic diagnosis of FIP has been made in some kittens involved in outbreaks of this complex (Scott *et al.*, 1979). It is usually the granulomatous form. When kittens are affected, adults may also exhibit mild chronic upper respiratory disease. Both adults and kittens exhibit intermittent low grade fevers. Endometritis is commonly seen in breeding colonies afflicted with KMC.

Little is known about this complex at the present time. Prevention may very well depend on the maintenance of an FIP-free colony. At present, treatment is symptomatic.

f. Metritis. Metritis occasionally occurs in the queen following parturition. Causes include retained placental membranes and lack of aseptic technique during assisted delivery. Treatment of the kittens and the queen are similar to that described for canines in Section V,E,3,c.

VI. NUTRITION AND FEEDING

Nutritionally complete commercial diets are available for dogs and cats, and in most research facilities they are fed *ad libitum* unless obesity or research requirements demand controlled dietary intake. Standard diets for dogs should contain the nutrients specified in the National Research Council document "Nutrient Requirement of Dogs" (NAS–NRC, 1974). Nutrient requirements for cats are outlined in the National Research Council document "Nutrient Requirements of Cats" (NAS–NRC, 1978). Laboratory animal veterinarians should have these references available.

Consistent reproducible experimental results require control over as many variables as possible, including the diet. Deficiencies or imbalances of nutrients can influence the manner in which animals respond to a given experimental manipulation. A monograph, "Control of Diets in Laboratory Animal Experimentation," summarizing the most important aspects of dietary control in animal experimentation is available from the National Academy of Sciences (ILAR, 1978a). Some of the most important points include nutritional adequacy, imbalances associated with diet modifications, ingredient variation, contamination, effect of environmental factors, and reporting responsibilities concerning diets.

Most commercially available balanced dog or cat diets are so-called closed-formula diets. These diets must meet labeled specified minimum values for protein and fat and maximum values for ash and fiber, but they do not necessarily provide exact nutrient levels or constancy of ingredient composition from batch to batch. Ingredient composition varies depending on the cost relationships of the various ingredients as the manufacturer attempts to achieve the label requirements at the lowest ingredient cost. If more precise dietary control is required, then open-formula (also called fixed-formula) diets can be purchased from any of a number of commercial manufacturers of special laboratory animal diets. In these diets the ingredients are identified and the percentage of each ingredient is kept constant from batch to batch.

In some experimental designs where nutritional intake must be controlled even more strictly, it may be necessary to provide semipurified diets. These diets are compounded from purified protein or amino acids, lipid, carbohydrate, vitamins, and minerals (NAS–NRC, 1974, 1978). Semipurified diets are also available commercially.

The date of diet manufacture should be known by the user. Preferably the manufacture date should be clearly marked on each bag of diet. In most facilities, diets, if stored under normal room temperature conditions, may be fed up to 6 months after the manufacture date. A system to ensure that diets are properly rotated and only fresh diets are fed should be in operation in each research facility. Of course, shelf-life of food is lengthened if food is stored at refrigerator temperatures; however, refrigeration of standard animal diets is not customary.

Because chemical (heavy metals, pesticides, estrogens, and aflatoxins) and microbiological (*Salmonella* sp. and *E. coli* type 1) contaminants have been found in animal food, some studies may require assays for particular contaminants (Newberne and Fox, 1978). Several commercial companies provide this service. Alternatively, batches of diet can be preanalyzed for contaminants, such as drugs, estrogens, heavy metals, and pesticides. Manufacturers then certify that the batch of diet contains less than certain minimum amounts of these contaminants. This type of preanalyzed diet is often used when animal studies must comply with the Good Laboratory Practices program of the Food and Drug Administration.

The regulations of the Federal Animal Welfare Act provide certain specifications for feeding and watering dogs and cats. They must be fed clean, wholesome palatable food at least once each day from food containers that minimize contamination by excreta. Water must be offered for a 1-hr period twice daily. Food and water containers must be kept clean and must be thoroughly sanitized at least once every 2 weeks. See Chapter 2 for a more complete discussion of the regulations of the Federal Animal Welfare Act.

VII. DISEASES

A. Introduction

Since many dogs and cats in research facilities are newly received from animal shelters, the laboratory animal veterinarian is confronted with the same spectrum of disease that confronts the private veterinary practitioner. The range of disease

encountered extends from infectious diseases and congenital anomalies of the young to disease of the aged and includes metabolic, traumatic, and neoplastic conditions. The diagnosis, medical management, and epidemiology of these diseases is beyond the scope of this chapter and is dealt with in detail in several excellent veterinary texts (Ettinger, 1983; Kirk, 1980; Kirk and Bistner, 1981). While examining newly received dogs and cats the laboratory animal veterinarian should be ever alert for animals with diseases that might serve as animal models of human disease.

In the research facility, the primary veterinary medical focus is on preventive medicine and on dealing with iatrogenic (investigator induced) lesions. Newly received dogs and cats with spontaneous medical difficulties should be identified during the quarantine and conditioning period (see Section IV for a more comprehensive discussion of quarantining and conditioning procedures). The decision to treat or euthanize ill animals is a matter of professional judgment based on prognosis, humane considerations, cost-effectiveness of therapy, and protection of the long-term animals in the facility from communicable diseases.

Once research studies have begun, the decision to isolate ill animals or to treat spontaneous or iatrogenic disease must be made in consultation with the study director. The effect of isolation or therapy on the research study must be weighed against the pain and discomfort that the animal might experience or the threat that the disease might entail for other animals in the study. The decision to treat, withhold treatment, isolate, or euthanize the animal is usually a difficult one that calls for cooperation and trust between the laboratory animal veterinarian and the study director. Resolution of this potential conflict is facilitated if the veterinarian becomes familiar with the study and the study director becomes familiar with the humane and veterinary medical considerations.

In the following sections we will discuss those diseases of the dog and cat that are somewhat unique to the research facility or those that present unusual complications for research. For detailed discussion of the diagnosis, medical management, and epidemiology of disease entities the reader should refer to veterinary texts on the subject (Ettinger, 1983; Kirk, 1980; Kirk and Bistner, 1981).

B. Infectious

1. Respiratory Infections of Random Source Dogs

Severe and often fatal respiratory infection is the most important disease problem encountered in newly received random source dogs (Fig. 4). Studies indicate that up to 70% of newly received dogs may have clinical signs of respiratory disease, and a mortality rate of up to 20% is not uncommon (Bey

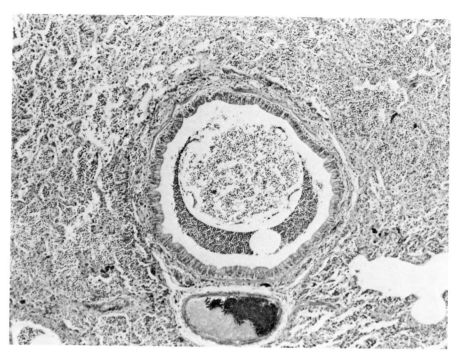

Fig. 4. Canine bronchopneumonia. Typical acute inflammatory exudate, primarily neutrophils, in bronchus and alveoli of a random source dog during quarantine period. Hematoxylin and eosin. × 40.

et al., 1981; Binn *et al.,* 1970b). The affected dogs exhibit mucopurulent nasal discharge, conjuctivitis, cough, dyspnea, depression, and anorexia. Some animals respond favorably to intensive antibiotics and supportive therapy, while others are refractory and the disease progresses to life-threatening bronchopneumonia.

Canine distemper virus has been identified as a major pathogen in this disease complex (Fig. 5). Histopathologic studies have confirmed the diagnosis of canine distemper in most fatal infections (Binn *et al.,* 1979). It has also been shown that dogs seronegative to canine distemper virus or adenovirus are much more likely to succumb to respiratory disease during the conditioning period (Binn *et al.,* 1970b, 1979). In the later study, 74% of the dogs without protective antibody to canine distemper developed respiratory disease, and 33% died. In contrast, only 24% of the dogs with protective antibody became ill and 4% died. Prevention of canine distemper appears to be the key to controlling severe, prolonged, and often fatal respiratory disease. Newly arrived dogs that have antibody to canine distemper also have a prevalence of antibody to infectious canine hepatitis virus (canine adenovirus type 1) twice that of dogs seronegative to canine distemper virus (Binn *et al.,* 1970b). The parallel titers in all likelihood indicate that these dogs have been vaccinated with a bivalent vaccine.

Doyle *et al.* (1979) found that immunization of newly received random source dogs against canine distemper, infectious canine hepatitis, and canine parainfluenza virus on arrival reduced the incidence of respiratory disease from 15 to 2% and the mortality from 16 to 6%. Appel (1970) has shown that to be effective, immunization against canine distemper must occur before exposure to virulent canine distemper virus occurs.

Although canine distemper is the major pathogen of newly received random source dogs, several other agents have been isolated from sick dogs during the conditioning period. Canine parainfluenza virus and canine adenovirus type 2 are the most prevalent. Both of these agents can produce respiratory disease, but it is generally concluded that they are not major pathogens (Binn *et al.,* 1970b). Rosendal (1978) isolated *Mycoplasma cynos* from the lungs of dogs with distemper and demonstrated experimentally that *M. cynos* could cause severe pneumonia.

Streptococcus zooepidemicus can cause acute necrotizing pneumonia in newly received random source dogs (Garnett *et al.,* 1982). The clinical course is often peracute and results in death without clinical signs. Dogs that are less severely affected have cough and moist rales often with purulent nasal discharge and tonsillitis. Necropsy lesions include diffuse hemorrhagic pneumonia and septic thrombi in the kidneys, lymph nodes, spleen, brain, and adrenal glands. Pencillin therapy is often effective if instituted early in the disease course.

Clinical tracheobronchitis (kennel cough) is often seen whenever dogs are group-housed. The clinical syndrome is characterized by a dry hacking cough but in rare instances productive cough, pneumonia, and nasal or ocular discharge may

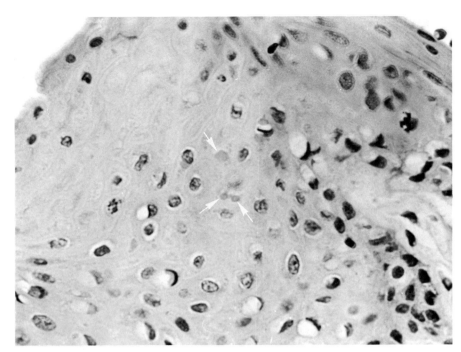

Fig. 5. Canine distemper. Inclusion bodies (arrows) in cytoplasm of epithelium of tongue. Random source dog during the quarantine period. Hematoxylin and eosin. × 400.

occur. The etiology of kennel cough is complex with both a bacterium (*Bordetella bronchiseptica*) and a virus (canine parainfluenza virus) playing a role. The disease is self-limiting unless infection with additional respiratory pathogens, especially canine distemper virus, ensues (Konter *et al.*, 1981). Live avirulent vaccines against both *Bordetella bronchiseptica* (Bey *et al.*, 1981) and canine parainfluenza (Konter *et al.*, 1981) are now commercially available.

Viruses less frequently isolated from the pneumonic canine lung include reovirus type 1 (Massie and Shaw, 1966), canine herpesvirus (Binn *et al.*, 1967), minute virus of canines (Binn *et al.*, 1970a), canine coronavirus (Binn *et al.*, 1975), and reovirus type 2 (Binn *et al.*, 1977). The role of these viruses in respiratory disease of random source dogs is unclear. Canine herpesvirus is rarely isolated, and serological studies indicate little spread of the virus. Minute virus of canines occurs with somewhat greater frequency, but it is rarely isolated from fatal cases, and very few sick dogs have rising titers to the virus. Canine coronavirus is frequently present in the small intestine, but is only rarely isolated from the lungs of sick dogs.

A complete program for immunization of group housed dogs against respiratory pathogens is necessary. In young animals the program should begin at 6–8 weeks or as soon as maternal antibody declines, if that is known. Live avirulent respiratory pathogens that should be included are canine distemper, infectious canine hepatitis (adenovirus type 1) or adenovirus type 2,

canine parainfluenza, and *Bordetella bronchiseptica*. Annual revaccination is recommended.

2. Respiratory Infections of Random Source Cats

Newly received random source cats are beset by a wide variety of respiratory pathogens. The two most important are a herpesvirus, feline viral rhinotracheitis (FVR), and a picornavirus, feline calicivirus (FCV) (Scott, F. W., 1977). Other agents, such as the etiologic agent of feline pneumonitis (*Chlamydia psittaci*) and feline reovirus, may cause primary disease. Palmer (1980) indicates that 40–45% of feline upper respiratory infections are due to FVR, 40–45% are due to FCV, and about 4% are caused by *Chlamydia psittaci*. Numerous bacteria including mycoplasma are secondary invaders. Feline respiratory disease due to one or a combination of these agents is characterized by similar clinical signs, and clinical differentiation of them is not easy. Control of these infections is a difficult problem in research facilities just as it is in catteries and veterinary hospitals.

Sneezing, coughing, fever, and hypersalivation are the first signs observed in FVR, followed by photophobia, chemosis, serous ocular and nasal discharge, and depression. Occasionally, eye involvement is unilateral, but bilateral involvement usually follows in a few hours. As the disease progresses, both ocular and nasal discharges may become purulent and form

Diagnosis can be accomplished by detecting larvae in feces with fecal flotation techniques using $ZnSO_4$ (specific gravity = 1.18) (Ehrenford, 1957), or $NaNO_3$ (specific gravity = 1.2) (Appel, 1970). A large 10- to 15-gm sample of feces should be used with either technique. Bronchial wash may also yield embryonated eggs in heavy infestations.

Most anthelmintics (piperazine, dichlorvos, thiabendazole, levamasole, and dithiazanine iodide) are not effective. Administration of two courses of albendazole (50 mg/kg body weight) for 5 days at 3-week intervals to infected dams is very effective. It prevents vertical transmission and renders subsequent litters free from infection (Georgi *et al.*, 1979). A single course of therapy in experimentally infected dogs killed or sterilized all worms (Georgi *et al.*, 1978). The parasite is not transmitted *in utero*. In infected colonies, transmission can be controlled by use of albendazole, caging animals separately, and reducing fecal contamination of the environment.

D. Iatrogenic

1. Introduction

It is inherent in the nature of experimental work with animals that during experimentation alterations in physiology may be induced. As a consequence, the laboratory animal veterinarian is often presented with clinical cases that result from or are a side effect of experimental procedures. These are called iatrogenic conditions or lesions. Animals in renal failure as a result of experimental hypertension or animals in hypoglycemic shock as a result of induced diabetes are examples of clinical conditions that are a consequence of the experimental alteration. Other clinical problems arise that are not directly attributable to the intended experimental alteration. For example, dogs with experimental gastric fistulas may become alkalotic due to chronic loss of gastric acid or cats with recording electrodes surgically implanted in the central nervous system may develop meningoencephalitis due to postsurgical infection.

If experimental procedures are carefully planned and carried out, many iatrogenic complications can be prevented. The laboratory animal veterinarian can often make a significant contribution by discussing possible iatrogenic complications with investigators prior to initiation of experimental procedures.

2. Medical Management of Chronic Indwelling Central Venous Catheters

Many experimental protocols necessitate continous or repeated access to arteries or veins for blood sampling, drug administration, or pressure measurement. Chronic indwelling central venous catheters that are necessary to achieve these ends can cause many clinical problems. Hysell and Abrams (1967) identified some of the complications seen in laboratory animals with chronic indwelling central venous catheters. Direct physical trauma from the catheter may cause damage of the vascular wall and intravascular thrombus formation at the catheter tip (Fig. 8). If venous thrombi dislodge they become embolic to the lungs and may cause either minor or massive pulmonary infarction. When arterial, especially aortic, thrombi dislodge they may cause infarction of any organ; however, most clinical problems are caused by infarction of the brain, spinal cord, kidney, or gut. Intracardiac catheters may cause valvular insufficiency, vegetative valvular endocarditis, or artrial thrombosis.

Often catheters and resultant thrombi become septic as the catheter provides a direct link between the external environment and the bloodstream of the animal. Bacteria may enter the bloodstream either through the lumen of the catheter or through the tract around the catheter. Maki *et al.* (1976) found that in humans, local inflammation of the skin site was strongly correlated with the incidence of catheter-related septicemia. In animals, we have concluded that the incidence of septicemia due to catheter tract infection is very much dependent on the distance between the skin entry site and the vessel entry site. If the catheter has a relatively long subcutaneous course before entering the vein then there is less likelihood that infection at the skin entry site will result in septicemia. Peters *et al.* (1973) have shown that an intravascular fibrin sleeve develops around

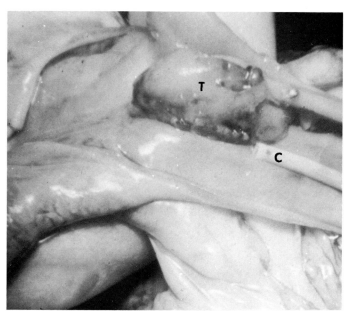

Fig. 8.   Iatrogenic vascular thrombosis. Mural thrombus (T) at the tip of a chronic indwelling vascular catheter (C) in the posterior vena cava.

vascular catheters within 24–48 hr of implantation. This fibrin sleeve can become seeded with bacteria and serve as a locus from which septicemia can result. Several investigators including Maki (1976) have found that the incidence of bacterial colonization of the catheter tip is proportional to the time that the catheter is left in place.

The goal of the laboratory animal veterinarian should be to prevent, not treat, catheter-related sepsis. The education of investigators about the measures that are important in preventing catheter-related septicemia is critical. Sterile catheters must be implanted using sterile technique and extrinsic contamination of the infusion system and catheter tract must be prevented. Maki et al. (1973) and others have found that sources of bacterial contamination include (1) the animal's cutaneous flora, (2) the investigator's cutaneous flora, (3) contaminated infusion fluids, (4) contaminated disinfectants, and (5) autoinfection due to bacteremia seeded from a remote infected locus.

We have found that implantation of sterile catheters utilizing sterile surgical technique is the first important step in preventing septicemia. A healthy infection-free animal is prepared for surgery by shaving, scrubbing, and applying an iodine-containing disinfectant to both surgical site and the site where the catheter enters the skin. A gas sterilized catheter is placed using aseptic surgical technique by personnel that have received training in these techniques. The catheter should be coated with a tridodecylmethyl ammonium chloride (TDMAC)–heparin complex as described by Brenner et al. (1974) to reduce thrombogenesis. The TDMAC–heparin complex coating material is available from Polysciences, Inc., Warrington, Pennsylvania. Following placement, radiography should be utilized to ensure that the location of the catheter tip is correct. Prophylactic antibiotics should be administered one day before and for 3 days after surgery.

Following catheterization, the catheter entry site in the skin should be cleansed, disinfected, and treated with topical antibiotics and bandaged every 2–3 days for the duration of the implantation. Entry site care is especially important if the vessel cannulation site is near the skin entry site.

To prevent bacterial contamination of the catheter lumen the infusion system should be as simple as possible. Fewer connections allow for fewer possible entry sites for bacteria. The system should be opened as infrequently as possible, and all openings should be performed aseptically. During periods when vascular catheters are not being used, they should be flushed daily with heparinized saline solution (10 units heparin/ml), and this solution should remain in the catheter.

Since adypsia, anorexia, and depression are often the first signs of septicemia, the food and water consumption and behavior of animals with vascular catheters should be monitored daily. If adypsia, anorexia, or depression are present, physical examination and blood culture (from a remote site) should be performed promptly. Antibiotic therapy should be instituted immediately and continued for 10 days. Our choice of antibiotic is generally based on previous antibiotic sensitivity tests on organisms cultured from the blood of septicemic animals from that particular laboratory. We find that each laboratory tends to have characteristic organisms that repeatedly cause septicemia.

We have found that removal of the catheter is usually necessary in order to prevent recurrent septicemia once antibiotic therapy is stopped. Our clinical experience with dogs agrees with that of DaRif and Rush (1983) with monkeys. In their study, antibiotic therapy with the catheter left in place resulted in marked clinical improvement; however, blood culture 5 days after cessation of antibiotic therapy demonstrated that the animals were again bacteremic. These findings are in agreement with those of numerous investigators including Maki (1976) who have found that catheter removal is an essential step in successful medical management of septicemia in human hospital patients with vascular catheters.

E. Neoplastic

Feline Leukemia Virus

Feline leukemia virus (FeLV), a retrovirus, widely infects cats. Chronic FeLV infection causes a variety of lymphoproliferative, myeloproliferative and bone disorders, glomerulonephritis, anemia, and neoplasia. Immunosuppression results in increased susceptibility to many other pathogens. These lesions and other as yet unidentified effects may have a significant impact on biological functions in cats used in biomedical research.

Ladiges et al. (1981) have screened newly received random source cats on receipt at a large research institution and have found that 4.9% were viremic. Nine of 20 viremic cats selected for pathologic evaluation had lymphoma or myeloproliferative disease. This provides some indication of the role that FeLV can play in producing lesions even in cats used in short-term studies.

For some long-term studies and in research breeding colonies it may be necessary to identify FeLV-infected cats and eliminate them from the quarantine rooms or colony. According to Hardy et al. (1976) this can be accomplished by identifying viremic cats with the immunofluorescent antibody (IFA) test and removing them from the facility or colony. The IFA test detects FeLV gs antigens in peripheral blood leukocytes and platelets. A positive IFA test means that the cat is viremic. Unfortunately, exposed cats can incubate FeLV for 3 months before viremia is established.

The following procedure as described by Hardy et al. (1976) may be used to free a group or colony from FeLV. All cats housed as a group are tested at 3 month intervals, and all positive cats are eliminated from the group. When all cats in the group have been IFA negative on two consecutive IFA tests after the last infected cat has been removed, the group is considered FeLV free. All new cats introduced into an FeLV free group must first undergo quarantine and two negative IFA tests at 3-month intervals.

Diagnosis and treatment of FeLV-related disorders is beyond the scope of this chapter. Pedersen and Mardewell (1980) have described diagnosis and treatment in some detail, and a text edited by Hardy *et al.* (1980) summarizes recent clinical and experimental information on FeLV.

VIII.   SELECTED NORMATIVE DATA

Parameter	Cat Value	Reference[a]	Dog Value	Reference[a]
Adult weight				
Male	3.5 kg	1	10.5 kg	1
Female	2.5 kg	1	9.9 kg	1
Life span				
Usual	12 years	1	12 years	1
Maximum	28 years	2	20 years	2
Chromosome number	38	1	78	1
Food consumption	24 gm/kg[b]	3	18 gm/kg[c]	4
Body temperature	101.5°F(100.5–102.5°F)	1	102°F(100.2–103.8°F)	1
	38.6°C(38.1–39.2°C)	1	38.9°C(37.9–39.9°C)	1
Puberty				
Male	36 weeks	1	6–12 months	1
Female	5–12 months	1	6–12 months	1
Breeding life	6–8 years	1	6 years	1
Breeding season	January–October	1	All year	1
Ejaculate volume	0.03–0.3 ml	1	2–16 ml	1
Sperm/ejaculation	$15–130 \times 10^6$	1	5×10^8	1
Type of estrous cycle	Polyestrus	1	Monestrus	1
Length of estrous cycle	2–3 weeks[d]	1	16–56 weeks	1
Duration of estrus	3–6 days	1	9–10 days	1
Ovulation mechanism	Induced	1	Spontaneous	1
Ovulation time	25–50 hr postcoitus	1	2–3 days postestrus	1
Ovulation rate	4 ova	1	6–8 ova	1
Pseudopregnancy	Uncommon (6 weeks)	1	Common (60 days)	1
Gestation	58–65 days	1	59–68 days	1
Litter size	3–5	1	1–12	1
Birth weight	110–120 gm	5	250 gm	1
Weaning age	3–8 weeks	1	3–8 weeks	1
Heart rate	110–140 beats/min	1	70–160 beats/min[e]	6
Normal heart axis	—		+40° to +100°	6
Blood pressure[f]				
Systolic	120 mm Hg	2	112 (95–136) mm Hg	2
Diastolic	75 mm Hg	2	56 (43–66) mm Hg	2
Cardiac output	—		1820–2660 ml/min[g]	7
Stroke volume	—		11.3–15.8 ml/min	7
Blood volume				
Plasma	40.7 (34.6–52.0) ml/kg	2	55.2 (43.7–73.0)ml/kg	2
Whole blood	55.5 (47.3–65.7) ml/kg	2	94.1 (76.5–107.3)mg/kg	2
Respiration frequency	26 breaths/min	1	22 breaths/min	2
Tidal volume	34 (20–42) ml	8	251–432 ml[h]	8
Minute volume	0.96 liters/min	8	4.1–6.1 liters/min[h]	8
Whole blood				
pH	7.35 (7.24–7.40)	2	7.36 (7.32–7.42)	2
CO_2	20.4 (17–24) mM/liter	2	21.4 (17–24) mM/liter	2

(continued)

VIII. SELECTED NORMATIVE DATA (*Continued*)

Parameter	Cat Value	Cat Reference[a]	Dog Value	Dog Reference[a]
Plasma				
CO_2 pressure	36 mm Hg	2	38 mm Hg	2
Protein	6.0–7.5 gm/dl[i]	9	6.0–7.5 gm/dl[i]	9
Fibrinogen	150–300 gm/dl	9	150–300 gm/dl	9
Total leukocyte count	5.5–19.5 × 10³/ml	9	6.0–17.0 × 10³/ml	9
Differential leukocyte count				
Neutrophils	(35–75%) 2.5–12.5[j]	9	(60–70%) 3.0–11.4[j]	9
Bands	(0–3%) 0–0.3[j]	9	(0–3%) 0–0.3[j]	9
Lymphocytes	(20–55%) 1.5–7.0[j]	9	(12–30%) 1.0–4.8[j]	9
Monocytes	(1–4%) 0–0.85[j]	9	(3–10%) 0.15–1.35[j]	9
Eosinophils	(2–12%) 0–0.75[j]	9	(2–10%) 0.1–0.75[j]	9
Basophils	rare	9	rare	9
Erythrocyte count	5.0–10.0 × 10⁶/μl	9	5.5–8.5 × 10⁶/μl	9
Hemoglobin	8–15 gm/dl	9	12–18 gm/dl	9
Reticulocyte count	0–1.0%	9	0–1.5%	9
Hematocrit				
PCV (Adult)	30–45%	9	37–55%	9
PCV (Young)	24–34%	9	25–34%	9
MCV	39–55 fl	9	60–77 fl	9
MCHO	30–36 gm/dl	9	19.5–24.5 gm/dl	9
Platelet count	3–7 × 10⁵/μl	9	2–9 × 10⁵/μl	9
Maximum volume of single bleeding	2% of body weight	9	2% of body weight	9
Myeloid/erythroid ratio	0.6–3.9/1.0	9	0.75–2.4/1.0	9
Urine				
pH	5.5–7.5	10	5.5–7.5	10
Normal specific gravity	1.030 (1.018–1.050)	10	1.025 (1.018–1.050)	10
Maximum specific gravity	1.060 +	10	1.080 +	10
Volume	22–30 ml/kg body weight/day	10	25–41 ml/kg body weight/day	10

[a]Key to references: 1, Mather and Rushmer (1979); 2, Altman and Dittmer (1972, 1973, 1974); 3, NAS–NRC (1978); 4, NAS–NRC (1974); 5, Scott (1970); 6, Bolton (1975); 7, Ettinger and Suter (1970); 8, Lumb and Jones (1973); 9, Duncan and Prasse (1977); 10, Osborne *et al.* (1972).

[b]Dry diet, active adult cats.

[c]Dry diet, consumed by a 22.7 kg dog.

[d]If not mated.

[e]Can be as high as 180 beats/min in toy breeds and 220 beats/min in pups.

[f]Aortic.

[g]In 10–31 kg dogs.

[h]Range includes variable sized dogs.

[i]Will be slightly lower in young animals.

[j]Total (× 10³/ml).

REFERENCES

Altman, P. L., and Dittmer, D. S., eds. (1972). "Biology Data Book," 2nd ed., Vol. I. Fed. Am. Soc. Exp. Biol., Bethesda, Maryland.

Altman, P. L., and Dittmer, D. S., eds. (1973). "Biology Data Book," Vol. II. Fed. Am. Soc. Exp. Biol., Bethesda, Maryland.

Altman, P. L., and Dittmer, D. S., eds. (1974). "Biology Data Book," Vol. III. Fed. Am. Soc. Exp. Biol., Bethesda, Maryland.

Anderson, A. C. (1955). Debarking in a kennel: Technique and results. *Vet. Med. (Kansas City, Mo.)* **50**, 409–411.

Anderson, N. V., ed. (1980). "Veterinary Gastroenterology." Lea & Febiger, Philadelphia, Pennsylvania.

Andrews, E. J., Ward, B. C., and Altman, N. H., eds. (1979). "Spontaneous Animal Models of Human Disease," Vols. 1 and 2. Academic Press, New York.

Appel, M. J. S. (1970). Distemper pathogenesis in dogs. *J. Am. Vet. Med. Assoc.* **156**, 1681–1684.

August, J. R., Powers, R. D., Bailey, W. S., and Diamond, D. L. (1980). *Filaroides hirthi* in a dog: Fatal hyperinfection suggestive of autoinfection. *J. Am. Vet. Med. Assoc.* **176**, 331–334.

Barlough, J. E., Jacobson, R. H., Downing, D. R., Marcella, K. L., Lynch, T. J., and Scott, F. W. (1983). Evolution of a computer assisted, kinetics based enzyme-linked immunosorbant assay for detection of coronavirus antibodies in cats. *J. Clin. Microbiol.* **17,** 202–217.

Baxter, M. (1973). Ringworm due to *M. canis* in cats and dogs in New Zealand. *N. Z. Vet. J.* **21,** 33.

Benirschke, K., Garner, F. M., and Jones, T. C., eds. (1978). "Pathology of Laboratory Animals." Springer-Verlag, Berlin and New York.

Bey, R. F., Shade, F. J., Goodnow, R. A., and Johnson, R. C. (1981). Intranasal vaccination of dogs with live avirulent *Bordetella bronchiseptica:* Correlation of serum agglutination titer and the formation of secretory IgA with protection against experimental induced infectious tracheobronchitis. *Am. J. Vet. Res.* **42,** 1131–1132.

Binn, L. N., Eddy, G. A., Lazar, E. D., Helms, J., and Murnane, T. (1967). Viruses recovered from laboratory dogs with respiratory disease. *Proc. Soc. Exp. Biol. Med.* **126,** 140–145.

Binn, L. N., Lazar, E. C., and Eddy, G. A. (1970a). Recovery and characterization of a minute virus of canines. *Infect. Immun.* **1,** 503–508.

Binn, L. N., Lazar, E. C., Helms, J., and Cross, R. E. (1970b). Viral antibody patterns in laboratory dogs with respiratory disease. *Am. J. Vet. Res.* **31,** 697–702.

Binn, L. N., Lazar, E. C., and Keenan, K. P. (1975). Recovery and characterization of a coronavirus from military dogs with diarrhea. *In* "Proceedings of the 78th Annual Meeting of the U. S. Animal Health Association," pp. 359–366. U. S. Anim. Health Assoc., Richmond, Virginia.

Binn, L. N., Marchwicki, R. H., and Keenan, K. P. (1977). Recovery of reovirus type 2 from an immature dog with respiratory tract disease. *Am. J. Vet. Res.* **38,** 927–929.

Binn, L. N., Alford, J. P., Marchivicki, R. H., Keefe, T. J., Beattie, R. J., and Wall, H. G. (1979). Studies of respiratory disease in random source laboratory dogs: Viral infections in unconditioned dogs. *Lab. Anim. Sci.* **29,** 48–52.

Bolton, G. R. (1975). "Handbook of Canine Electrocardiology." Saunders, Philadelphia, Pennsylvania.

Brenner, W. I., Engelman, R. M., Williams, C. D., Boyd, A. D., and Reed, G. E. (1974). Nonthrombogenic aortic and vena caval bypass using heparin-coated tubes. *Am. J. Surg.* **127,** 555–559.

Brown, J., Blue, J. L., Wooley, R. E., Dreesen, D. W., and Carmichael, L. E. (1976). A serologic survey of a population of Georgia dogs for *Brucella canis* and an evaluation of the slide agglutination test. *J. Am. Vet. Med. Assoc.* **169,** 1214–1216.

Buke, T. J., and Reynolds, H. A. (1975). Megesterol acetate for estrus postponement in the bitch. *J. Am. Vet. Med. Assoc.* **167,** 285.

Carmichael, L. E., Joubert, J. C., and Pollock, R. V. H. (1981). A modified-live canine parvovirus strain with novel plaque characteristics. I. Viral attenuation and dog response. *Cornell Vet.* **71,** 408–427.

Christie, E., Dubey, J. P., and Pappas, P. W. (1976). Prevalence of sarcocystis infection and other intestinal parasitisms in cats from a humane shelter in Ohio. *J. Am. Vet. Med. Assoc.* **168,** 421–422.

Cline, E. M., Jennings, L. L., and Soika, N. J. (1980). Breeding laboratory cats during artificially induced estus. *Lab. Anim. Sci.* **30,** 1003–1005.

Code of Federal Regulations (CFR) (1982). Title 9, Animals and Animal Products, Subchapter A–Animal Welfare, Parts 1, 2, and 3.

Colby, E. D. (1970). Induced estus and timed pregnancies in the cat. *Lab. Anim. Care* **20,** 1075.

Colby, E. D. (1980). Suppression/induction of estus in cats. *In* "Current Therapy in Theriogenology" (D. A. Morrow, ed.), pp. 861–865. Saunders, Philadelphia, Pennsylvania.

Committee on Animal Models for Research on Aging (1981). "Mammalian Models for Research on Aging." National Academy Press, Washington, D. C.

Craig, T. M., Brown, T. W., Shefstad, D. K., and Williams, G. D. (1978). Fatal *Filaroides hirthi* infection in a dog. *J. Am. Vet. Med. Assoc.* **172,** 1096–1098.

DaRif, C. A., and Rush, H. G. (1983). Management of septicemia in monkeys with chronic indwelling venous catheters. *Lab. Anim. Sci.* **33,** 90–94.

Dawson, C. O., and Noddle, B. M. (1968). Treatment of *Microsporum canis* ringworm in a cat colony. *J. Small Anim. Pract.* **9,** 613–620.

Dorman, D. W., and Ostrand, J. R. (1958). A summary of *Toxocara canis* and *Toxocara cati* prevalence in the New York City area. *N. Y. State J. Med.* **58,** 2793–2795.

Doyle, R. E., Anthony, R. L., Jr., Jepsen, P. L., Kinkler, R. J., Jr., and Vogler, G. A. (1979). Missouri researchers find that vaccinating random-source dogs with three-part vaccine pays off. *Lab. Anim.* **8,** 39–44.

Duncan, J. R., and Prasse, K. W. (1977). "Veterinary Laboratory Medicine: Clinical Pathology." Iowa State Univ. Press, Ames.

Ehrenford, F. A. (1957). Canine ascariasis as a potential source of visceral larva migrans. *Am. J. Trop. Med. Hyg.* **6,** 166–170.

Ettinger, S. J., ed. (1983). "Textbook of Veterinary Internal Medicine," 2nd ed., Vols. 1 and 2. Saunders, Philadelphia, Pennsylvania.

Ettinger, S. J., and Suter, P. F. (1970). "Canine Cardiology." Saunders, Philadelphia, Pennsylvania.

Evans, L. E. (1980). Induction of estus in the bitch. *In* "Current Therapy in Theriogenology" (D. A. Morrow, ed.), pp. 618–620. Saunders, Philadelphia, Pennsylvania.

Fox, J. G. (1982). Campylobacteriosis—A "new" disease in laboratory animals. *Lab. Anim. Sci.* **32,** 625–637.

Fox, J. G., and Beaucage, C. M. (1979). The incidence of *Salmonella* in random-source cats purchased for use in research. *J. Infect. Dis.* **139,** 362–365.

Fox, M. W. (1965). "Canine Behavior." Thomas, Springfield, Illinois.

Garnett, N. L., Eydelloth, R. S., Svindle, M. M., Vonderfecht, S. L., Strandberg, J. D., and Luzarraga, M. B. (1982). Hemorrhagic streptococcal pneumonia in newly procured research dogs. *J. Am. Vet. Med. Assoc.* **181,** 1371–1374.

Georgi, J. R. (1976). *Filaroides hirthi*: Experimental transmission among beagle dogs through ingestion of first stage larvae. *Science* **194,** 735.

Georgi, J. R., and Theodorides, V. J. (1980). "Parasitology for Veterinarians." Saunders, Philadelphia, Pennsylvania.

Georgi, J. R., Georgi, M. E., and Cleveland, D. J. (1977). Potency and transmission of *Filaroides hirthi* infection. *Parasitology* **75,** 251–257.

Georgi, J. R., Slauson, D. O., and Theodorides, V. J. (1978). Anthelmintic activity of albendazole against *Filaroides hirthi* in dogs. *Am. J. Vet. Res.* **39,** 803–806.

Georgi, J. R., Georgi, M. E., Fahnestock, G. R., and Theodorides, V. J. (1979). Transmission and control of *Filaroides hirthi* lung worm infection in dogs. *Am. J. Vet. Res.* **40,** 829–831.

Ghadirian, E., Viens, P., and Strykowski, H. (1976). Prevalence of *Toxocara* and other helminth ova in dogs and soil in the Montreal metropolitan area. *Can. J. Public Health* **67,** 495–496.

Glickman, L. T. (1980). Preventive medicine in kennel management. *Curr. Vet. Ther.* **7,** 67–76.

Guterbock, W. M., and Levine, N. D. (1977a). Coccidia and intestinal nematodes of east central Illinois cats. *J. Am. Vet. Med. Assoc.* **170,** 720–721.

Guterbock, W. M., and Levine, N. D. (1977b). Coccida and intestinal nematodes of east Central Illinois. *J. Am. Vet. Med. Assoc.* **170** (12), 1411–1413.

Hardy, W. D., Jr., McClelland, A. J., Zuckerman, E. E., Hers, P. W., Essex, M., Cotter, S. M., MacEwen, E. G., and Hayes, A. A. (1976). Prevention of the contagious spread of feline leukemia virus and the development of leukemia in pet cats. *Nature (London)* **263,** 326–328.

Hardy, W. D., Jr., Essex, M., and McClelland, A. J., eds. (1980). "Feline Leukemia Virus." Elsevier/North-Holland, New York.

Hayes, M. A., Russell, R. G., and Babiuk, L. A. (1979). Sudden death in young dogs with myocarditis caused by parvovirus. *J. Am. Vet. Med. Assoc.* **174**, 1197–1203.

Hirth, R. S., and Hottendorf, G. H. (1973). Lesions produced by a new lung worm in beagle dogs. *Vet. Pathol.* **10**, 385–407.

Hite, M., Hansen, N. R., Bohidar, N. R., Conti, P. A., and Mattis, P. A. (1977). Effect of cage size on patterns of activity and health of beagle dogs. *Lab. Anim. Sci.* **27**, 60–64.

Horzinek, M. C., and Osterhaus, A. D. M. E. (1979). Feline infectious peritonitis: A worldwide serosurvey. *Am. J. Vet. Res.* **40**, 1487–1492.

Hysell, D. K., and Abrams, G. D. (1967). Complications in the use of indwelling vascular catheters in laboratory animals. *Lab. Anim. Care* **17**, 273–280.

Institute of Laboratory Animal Resources (ILAR) (1973). "Dogs—Standards and Guidelines for the Breeding, Care, and Management of Laboratory Animals." National Academy Press, Washington, D. C.

Institute of Laboratory Animal Resource (ILAR) (1978a). "Control of Diets in Laboratory Animal Experimentation." National Academy of Sciences, Washington, D. C.

Institute of Laboratory Animal Resources (ILAR) (1978b). "Guide for the Care and Use of Laboratory Animals." DHEW Publ. no. (NIH) 78-23. U. S. Govt. Printing Office, Washington, D. C.

Institute of Laboratory Animal Resources (ILAR) (1978c). "Laboratory Animal Housing." National Academy Press, Washington, D. C.

Institute of Laboratory Animal Resources (ILAR) (1978d). "Laboratory Animal Management—Cats." National Academy of Sciences, Washington, D. C.

Institute of Laboratory Animal Resources (ILAR) (1979). "Animals for Research—A Directory of Sources," 10th ed. National Academy Press, Washington, D. C.

Institute of Laboratory Animal Resources (ILAR) (1980). "National Survey of Laboratory Animal Facilities and Resources." National Academy of Sciences, Washington, D. C.

Kazakos, K. R. (1978). Gastrointestinal helminths in dogs from a humane shelter in Indiana. *J. Am. Vet. Med. Assoc.* **173**, 995–997.

Kirk, R. W. (1970). Dogs. *In* "Reproduction and Breeding Technique for Laboratory Animals" (E. S. E. Hafez, ed.), pp. 224–236. Lea & Febiger, Philadelphia, Pennsylvania.

Kirk, R. W., ed. (1980). "Current Veterinary Therapy VII." Saunders, Philadelphia, Pennsylvania.

Kirk, R. W., and Bistner, S. I. (1981). "Handbook of Veterinary Procedures and Emergency Treatment." Saunders, Philadelphia, Pennsylvania.

Konter, E. J., Wegrzyn, R. J., and Goodnow, R. A. (1981). Canine infectious tracheobronchitis: Effects of an intranasal line canine parainfluenza -*Bordetella bronchiseptica* vaccine on viral shedding and clinical tracheobronchitis (kennel cough). *Am. J. Vet. Res.* **42**, 1694–1698.

Kraus, G. E. (1963). Devocalizing dogs by cautery. *J. Am. Vet. Med. Assoc.* **143**, 979–981.

Ladiges, W. C., DiGiacomo, R. F., Wardrop, K. J., and Hardy, W. D. (1981). Prevalence and sequellae of feline leukemia virus infection in laboratory cats. *J. Am. Vet. Med. Assoc.* **179**, 1206–1207.

Lightner, L., Christensen, B. M., and Beran, G. W. (1978). Epidemiologic findings on canine and feline intestinal nematode infections from records of the Iowa State University Veterinary Clinic. *J. Am. Vet. Med. Assoc.* **172**(5), 564–567.

Lillis, W. G. (1967). Helminth survey of dogs and cats in New Jersey. *J. Parsitol.* **53**(5), 1082–1084.

Lumb, W. V., and Jones, E. W. (1973). "Veterinary Anesthesia." Lea & Febiger, Philadelphia, Pennsylvania.

Maki, D. G. (1976). Sepsis arising from extrinsic contamination of the infusion and measures for control. *In* "Microbiological Hazards of Infusion Therapy" (I. Phillips, P. O. Meeus, and P. F. D'Arcy, eds.), pp. 99–143. Publishing Sciences Group, Littleton, Massachusetts.

Maki, D. G., Goldman, D. A., and Rhame, F. S. (1973). Infection control in intravenous therapy. *Ann. Intern. Med.* **79**, 867–887.

Maki, D. G., Weise, C. E., and Savafin, H. W. (1976). Semi-quantitative method for identifying intravenous catheter-related infection. *Clin. Res.* **24**, 25A.

Malloy, W. F., and Embil, J. A. (1978). Prevalence of *Toxocara* spp. and other parasites in dogs and cats in Halifax, Nova Scotia. *Can. J. Comp. Med.* **42**, 29–31.

Massie, E. L., and Shaw, E. D. (1966). Reovirus type 1 in laboratory dogs. *Am. J. Vet. Res.* **27**, 782–787.

Mather, E. C., and Rushmer, R. A. (1979). Physiological parameters of some species used in reproduction research. *In* "Animal Models for Research on Contraception and Fertility" (N. J. Alexander, ed.), pp. 559–576. Harper & Row, Hagerstown, Maryland.

Merickel, B. S., Hahn, F. F., Hanika-Rebar, C., Muggenburg, B. A., Brownstein, D. G., Rebar, A. H., and DeNicola, D. (1980). Acute parvoviral enteritis in a closed beagle dog colony. *Lab. Anim. Sci.* **30**, 874–878.

Mitzel, J. R., and Strating, A. (1977). Vaccination against feline pneumonitis. *Am. J. Vet. Res.* **38**, 1361–1363.

Moore, J. A., Gupta, B. N., and Conner, G. H. (1968). Eradication of *Brucella canis* infection from a dog colony. *J. Am. Vet. Med. Assoc.* **153**, 523–527.

Morse, E. V., and Duncan, M. A. (1975). Canine salmonellosis: Prevalence, epizootiology, signs, public health significance. *J. Am. Vet. Med. Assoc.* **167**, 817–820.

Mosier, J. E. (1977). Causes and treatment of neonatal deaths. *Curr. Vet. Ther.* **6**, 44–49.

Muller, G. H., and Kirk, R. W. (1976). "Small Animal Dermatology," 2nd ed. Saunders, Philadelphia, Pennsylvania.

National Academy of Sciences—National Research Council (NAS—NRC) (1974). "Nutrient Requirements of Dogs." National Academy Press, Washington, D. C.

National Academy of Sciences—National Research Council (NAS—NRC) (1977). "The Future of Animals, Cells, Models, and Systems in Research, Development, Education, and Testing." National Academy Press, Washington, D. C.

National Academy of Sciences—National Research Council (NAS—NRC) (1978). "Nutrient Requirements of Cats." National Academy Press, Washington, D. C.

Newberne, P. M., and Fox, J. G. (1978). Nutritional adequacy and quantity control of rodent diets. *Lab. Anim. Sci.* **30**, 352–365.

Newton, W. M. (1972). An evaluation of the effects of various degrees of long-term confinement on adult beagle dogs. *Lab. Anim. Sci.* **22**, 860–864.

Norsworthy, G. D. (1979). Kitten mortality complex. *Feline Pract.* **9**, 57–60.

Olson, R. F., and Olson, M. K. (1971). Hand rearing puppies. *Curr. Vet. Ther.* **4**, 53–55.

Osborne, C. A., Low, D. G., and Finco, D. R. (1972). "Canine and Feline Urology." Saunders, Philadelphia, Pennsylvania.

Palmer, G. H. (1980). Feline upper respiratory disease: A review. *VM/SAC, Vet. Med. Small Anim. Clin.* **75**, 1556–1558.

Pedersen, N. C., and Mardewell, B. R. (1980). Feline leukemia virus disease complex. *Curr. Vet. Ther.* **7**, 404–410.

Pedersen, N. C., Boyle, J. F., Floyd, K., Fudge, A., and Barker, J. (1978). An enteric coronavirus infection of cats and its relationship to feline infectious peritonitis. *Am. J. Vet. Res.* **42**, 368–377.

Peter, G. K. (1983). Unit for Laboratory Animal Medicine University of Michigan, Ann Arbor, Michigan (personal communication).

Peters, W. R., Bush, W. H., McIntyre, R. D., and Hill, L. D. (1973). The development of fibrin sheath on indwelling venous catheters. *Surg., Gynecol. Obstet.* **137**, 43.

Pickerill, P. A., and Carmichael, L. E. (1972). Canine brucellosis: Control programs in commercial kennels and effect on reproduction. *J. Am. Vet. Med. Assoc.* **160**, 1607–1615.

Platz, C. C., and Seager, S. W. J. (1978). Semen collection by electroejaculation in the domestic cat. *J. Am. Vet. Med. Assoc.* **173**, 1353–1355.

Pletcher, J. M., Toft, J. D., II, Frey, R. M., and Casey, H. W. (1979). Histopathologic evidence for parvovirus infection in dogs. *J. Am. Vet. Med. Assoc.* **175**, 825–828.

Pollock, R. V. H., and Carmichael, L. E. (1981). Newer knowledge about canine parvovirus. *Proc. Gaines Vet. Symp.* **30**, 36–40.

Raulston, G. L., Swain, S. F., Martin, D. P., and Kerley, W. C. (1969). A method of rapid long-term devocalization of dogs. *Lab. Anim. Care* **19**, 247–249.

Rawlings, C. A., Dawe, D. L., McCall, J. W., Keith, J. C., and Prestwood, A. K. (1982). Four types of occult *Dirofilaria immitis* infection in dogs. *J. Am. Vet. Med. Assoc.* **180**(11), 1323–1326.

Rosendal, S. (1978). Canine mycoplasmas: Pathogenicity of mycoplasmas associated with distemper pneumonia. *J. Infect. Dis.* **138**, 203–210.

Sawyer, T. W., Cowgill, L. M., and Andersen, F. L. (1976). Helminth parasites of cats and dogs from central Utah. *Great Basin Nat.* **36**, 471–474.

Scott, F. W. (1977). "Feline Respiratory Diseases." Feline Information Bulletin, Cornell Feline Research Laboratory, Cornell University, Ithaca, New York.

Scott, F. W. (1980a). Viricidal disinfectants and feline viruses. *Am. J. Vet. Res.* **41**, 410–414.

Scott, F. W. (1980b). Uptake on feline immunization. *Curr. Vet. Ther.* **7**, 1256–1258.

Scott, F. W., Weiss, R. C., Post, J. E., Gilmartin, J. E., and Hoshino, Y. (1979). Kitten mortality complex (Neonatal FIP?). *Feline Pract.* **9**, 44–56.

Scott, P. P. (1970). Cats. *In* "Reproduction and Breeding Techniques for Laboratory Animals" (E. S. E. Hefez, ed.), pp. 192–208. Lea & Febiger, Philadelphia, Pennsylvania.

Scott, P. P., and Lloyd-Jacob, M. A. (1959). Reduction in the anoestrous period of laboratory cats by increased illumination. *Nature (London)* **184**, 2022.

Seager, S. W. J. (1977). Semen collection and artificial inseminations of dogs. *Curr. Vet. Ther.* **6**, 1245–1251.

Shade, F. J., and Goodnow, R. A. (1979). Intranasal immunization of dogs against *Bordetella bronchiseptica*-induced tracheobronchitis (kennel cough) with modified live *Bordetella bronchiseptica* vaccines. *Am. J. Vet. Res.* **40**, 1241–1243.

Shifrine, M., and Wilson, F. D., eds. (1980). "The Canine as a Biomedical Research Model: Immunological, Hematological and Oncological Aspects." Technical Information Center/U.S. Department of Energy, Washington, D. C.

Small, E. (1980). Pediatrics. *Curr. Vet. Ther.* **7**, 77–82.

Smith, A. H., Jones, T. C., and Hunt, R. D. (1972). "Veterinary Pathology," p. 389. Lea & Febiger, Philadelphia, Pennsylvania.

Smith, K. W. (1974). Female genital system. *In* "Canine Surgery," 2nd Archibald ed. (J. Archibald, ed.), pp. 776–779. American Veterinary Publications, Santa Barbara, California.

Sojka, N. J., Jennings, L. L., and Hamner, C. E. (1970). Collection and utilization of cat semen for artificial insemination. *J. Am. Vet. Med. Assoc.* **156**, 1250–1251.

Sokowlowski, J. H. (1978). Evaluation of estrous activity in bitches treated with mibilerone and exposed to adult male dogs. *J. Am. Vet. Med. Assoc.* **173**, 983–984.

Sokowlowski, J. H., and Gerg, S. (1977). Biological evaluation of mibilerone in the female beagle. *Am. J. Vet. Res.* **38**, 1371.

Stein, B. S. (1980). Feline respiratory disease complex. *Curr. Vet. Ther.* **7**, 1279–1284.

Strietal, R. H., and Dubey, J. P. (1976). Prevalence of sarcocystis infection and other intestinal parasitisms in dogs from a humane shelter in Ohio. *J. Am. Vet. Med. Assoc.* **168**(5), 423–424.

Timoney, J. F., Neibert, H. C., and Scott, F. W. (1978). Feline salmonellosis: A nosocomial outbreak and experimental studies. *Cornell Vet.* **68**, 211–219.

Upjohn Co. (1978). "Cheque," Upjohn Vet. Rep. No. 9. Upjohn Veterinary Products, Kalamazoo, Michigan.

Vaughn, J., and Jordan, R. (1960). Intestinal nematodes in well-cared for dogs. *Am. J. Trop. Med. Hyg.* **9**, 29–31.

Visco, R. J., Corwin, R. M., and Selby, L. A. (1978). Effect of age and sex on the prevalence of intestinal parasitism in cats. *J. Am. Vet. Med. Assoc.* **172**, 797–800.

Weiss, R. C., and Scott, F. W. (1980). Feline infections peritonitis. *Curr. Vet. Ther.* **7**, 1288–1292.

Wildt, D. E., Kinney, G. M., and Seager, S. W. J. (1978). Gonadotropin induced reproductive cyclicity in the domestic cat. *Lab. Anim. Sci.* **28**, 301–307.

Williams, R. W., and Menning, E. L. (1961). Intestinal helminths in dogs and cats of the Berumuda Islands and their potential public health significance with a report of a probable case of visceral larva migrams. *J. Parasitol.* **47**, 947–951.

Wilson, H. D., McCormick, J. B., and Freeley, J. C. (1976). *Yersinia enterocolitica* infection in a 4 month old infant associated with infection in household dogs. *J. Pediatr.* **89**, 767–769.

Wong, M. M., and Suter, P. F. (1979). Indirect fluorescent antibody test in occult dirofilariasis. *Am. J. Vet. Res.* **40**, 414–420.

Yoder, J. T., and Starch, C. J. (1964). Devocalization of dogs by laryngofissure and dissection of the thyroarytenoid fold. *J. Am. Vet. Med. Assoc.* **145**, 325–330.

Young, F. W., and Sales, E. K. (1944). The canine debarking operation. *Mich. State Coll. Vet.* **5**, 24–37.

Chapter 10

Ungulates as Laboratory Animals

Dale L. Brooks, Philip C. Tillman, and Steven M. Niemi

I. INTRODUCTION

The term ungulate is used to designate all the large mammals with hooves. The general adaptations of the group include the development of an upright digital gait with most of the animal's weight borne on the central one or two digits, reduction of the medial and lateral digits, elongation of the distal portions of the extremities for running, and, in some cases, modification of the digestive tract for a herbivorous diet. The two orders, Perissodactyla and Artiodactyla, appear to have developed hooves independently and are not closely related to each other.

The use of ungulates as laboratory animals is becoming increasingly popular, in part due to the high cost and decreasing availability of other large laboratory species, such as the dog, cat, and nonhuman primates. All personnel using livestock species must understand husbandry and handling techniques so that they appreciate the potential risks of zoonotic diseases and physical injury.

This chapter is an introduction to the ungulates as laboratory animals. Topics to be addressed include examples of research uses of ungulates, procurement and preventive medicine, husbandry, and facilities. The reader is also referred to the Appendix for information on normative data, nutrition, housing requirements, and infectious diseases. The confines of a single chapter exclude thorough discussion of these topics. The references listed at the end of the chapter are recommended for more information.

II. TAXONOMY

Order: Perrissodactyla (odd-toed ungulates)   This order includes Equidae and other taxonomic families not usually encountered in biomedical research.

Order: Artiodactyla (even-toed ungulates)

     Family: Bovidae (the common ruminants: cattle, sheep, goats, antelopes)

         Species: *Capra hircus* (domestic goats)

                 Ovis aries (domestic sheep)

                 Bos taurus (domestic cattle)

                 Bos indicus (Zebu cattle)

     Family: Suidae (swine)

         Species: *Sus scrofa* (European wild boar and domestic swine)

     Eight other living families are in the order Artiodactyla.

The Bovidae represent a specialized group of mammals in terms of both anatomy and physiology. The presence of horns, a highly compartmentalized stomach, the fact that they are ob-

ligate herbivores and that as adults they derive all their glucose from gluconeogenesis are indications of their uniqueness.

The Suidae are considered somewhat less specialized. The anatomic and functional similarities between pigs and humans derive from the fact that both retain elements of the "basic" mammalian pattern. These elements include brachydont dentition, a simple digestive tract, and an omnivorous habit. Swine are the only large omnivores that are readily available to investigators in many cases.

III. GENERAL COMMENTS AND EXAMPLES OF USE IN RESEARCH

A. Cattle

Agricultural institutions keep cattle for teaching and research of beef and dairy production and husbandry. Adult cattle are seldom used in basic biomedical research, although some genetic anomalies are studied in mature cattle as models of comparable human conditions. These conditions include congenital protoporphyria (Ruth *et al.*, 1977) and GM$_1$ gangliosidosis (Donnelly and Sheahan, 1975). Calves are used in cardiovascular surgery because the size of their bodies and hearts approximates that of humans. In most situations, bovine research subjects are dairy calves, and their management consists of conventional pediatric husbandry (Bath *et al.*, 1978).

B. Sheep

Sheep are frequently maintained for reproductive studies. Their large gravid uterus, long gestation period, and localization of the placenta into discreet caruncles make them convenient animals in which to study the development of the fetus (Kraner and Parshall, 1968). Sheep are used for thoracic surgery research, studies of hemostasis, endotoxic shock, and prostaglandin metabolism, to mention a few examples. A major advantage of the sheep as a research animal is the relative ease with which it adapts to confinement (Hecker, 1983).

C. Goats

Goats are similar in many respects to their close relatives, the sheep, and are used in many similar situations. Because they require no grooming or shearing (except for the Angora) and have large and accessible jugular veins, goats are widely used for production of large quantities of antisera. Congenital hypothyroidism of goats is studied as a model for a similar condition in humans (van Dijk *et al.*, 1979). Other uses of goats include studies of infectious agents, such as caprine arthritis/encephalitis virus (Crawford and Adams, 1981). African pygmy goats are especially suitable for rearing in a research facility (Rogers *et al.*, 1969).

D. Swine

The usefulness of swine in biomedical research arises from the fact that there are many anatomic and physiological similarities between swine and humans. There is close morphological correlation in the eyes, teeth, kidneys, digestive tract, and skin of swine and humans. Swine have often been used to investigate these systems, especially in surgical studies. The pig is a useful model for cardiovascular research, both for surgical manipulation, such as prosthetic heart valves and vascular grafts, and for the study of atherosclerosis, a naturally occurring disease in pigs. Pond and Houpt (1978) have provided an excellent review of the utility of swine as biomedical models.

Newborn swine are an excellent pediatric model and are the animals of choice for comparative study of human infant nutrition. Swine develop gastric ulcers under natural conditions, and the incidence is increased by stress. In addition, swine become obese when fed *ad libitum*, and are used as a model for human obesity.

Familial diseases of swine comparable to human maladies include cutaneous melanoma (Hook *et al.*, 1979) and malignant hyperthermia (Nelson and Anderson, 1976). Another genetic condition of swine used as a model is Von Willebrand's disease (coagulation factor VIII deficiency) (Bowie and Dodds, 1979).

IV. AVAILABILITY, SOURCES, AND ACQUISITION

Laboratory ungulates may be procured from suppliers of laboratory animals or from traditional agricultural sources. One should inspect the animal health records of the prospective supplier and perhaps carcasses of freshly slaughtered animals (e.g., pigs and lambs). It is often helpful to consult the attending veterinarian for additional information regarding preventive medicine practices and animal health status. Having a single source of supply (i.e., a farm) often makes it easier to monitor the health status of the incoming animals and to work with the supplier to control any disease problems.

There are major differences between a farm versus a sales yard as a reliable source of healthy laboratory subjects. An animal often experiences greater stress in a sales yard through handling and shipping, in addition to exposure to a larger number of infectious agents. Regardless of source, stress associated with transportation and potential infection must be minimized prior to arrival at the vivarium. An alternative to purchasing livestock is to establish and maintain a breeding operation. However, this is too costly in most cases for the needs of the institution or investigator.

An investigator's animal requirements should be carefully and clearly ascertained in writing prior to ordering animals. Purchase orders should always define the number of animals needed, their breed, sex, age, an acceptable body weight range, and any other limitations peculiar to the particular study. On arrival at the facility, the supervisor should always make sure the delivery conforms to the specifications of the purchase order. The particulars of acquisition vary somewhat between species and will be discussed separately.

A. Swine

The potential user of laboratory swine may elect to use either conventional domestic swine or laboratory bred miniature swine. The use of swine is limited by their loud and annoying vocalizing, relative difficulty in handling, size of adults, and suitability of secure housing.

For short-term projects, conventional feeder pigs (8–12 weeks old and weighing 12–24 kg) are often satisfactory. These animals can be restrained with moderate effort and can be conveniently housed in most vivaria using slight modifications of housing designed for other large species. Such animals grow rapidly and soon require housing designed specifically for adult swine (Pond and Maner, 1974). If halothane anesthesia is to be used, the investigator should beware of breed-associated sensitivity to halothane and resultant malignant hyperthermia. Pigs can be tested for halothane sensitivity in order to detect susceptible animals (Topel *et al.*, 1975; Pond and Houpt, 1978).

Miniature pigs bred specifically for research are available from several research institutions and commercial vendors (England, 1968). Their adult body weight of 70 kg still presents some difficulties with restraint and housing, but they may be easier to handle than conventional swine weighing two to three times as much (Earl, 1968). Miniature swine are preferred in many cases involving chronic studies on adult pigs, but their characteristics may not always correlate exactly with those of the conventional breeds.

Gnotobiotic (germfree) pigs are used in studies requiring defined microflora, including infectious disease research. These animals require complex housing and husbandry procedures (Meyer *et al.*, 1963; Waxler, 1975). The advent of gnotobiotic swine has enabled establishment of specific pathogen-free

(SPF) herds (Tweihaus and Underdahl, 1975). These animals require less stringent management than gnotobiotic pigs, but should still be maintained as closed herds or procured from a single source.

B. Sheep

The supply of sheep of a particular age, sex, and weight class varies tremendously from season to season, reflecting both the seasonality of the animal's estrous cycle and the agricultural practices of a particular locale. Wethers (castrated males) usually go to market as soon as they reach proper size. Therefore, it is difficult to obtain wethers at times of the year when the past year's lamb crop is shipped to market. Investigators requiring time-dated pregnant animals or a continuous supply of animals of a particular age, sex, and breed must either purchase well in advance of their needs and hold animals until they are used, breed their own sheep, or contact a supplier attuned to the investigator's needs. Otherwise, lambs will only be available from midwinter to late spring, as determined by geographical region.

While a great number of breeds of sheep are available, the majority of commercial animals are cross-bred and can be broadly grouped into the white-faced/wool types, the black-faced/meat types, or range lambs that are usually progeny of white-faced ewes and black-faced rams. Researchers performing vascular surgery or using vascular catheters often prefer the larger, white-faced breeds because of their correspondingly more prominent peripheral blood vessels. There is also considerable variation between breeds in their fecundity, wool length, hardiness to temperature extremes, and other parameters of potential interest (Briggs, 1969; Owen, 1976).

Pregnant or possibly pregnant ewes are to be avoided unless specifically required. Q fever, a zoonotic disease, is disseminated via ovine milk and placental fluids (see Chapter 22 by Fox et al.) More commonly, ewes in late gestation are susceptible to potentially fatal metabolic disorders such as hypocalcemia and pregnancy toxemia (Jensen and Swift, 1982).

C. Goats

Many of the considerations given to sheep also apply to goats. Goats may be more vocal, gregarious, and more apt to escape from their enclosures. Goats are somewhat less seasonal than sheep in their reproductive activity, and management practices can be designed to extend the breeding season. Even so, kids are still difficult to find in the late summer and fall. Nonpregnant does are preferred whenever it is compatible with the experimental design. Intact adult bucks are to be avoided due to their offensive odor, which is most prominent during the

breeding season. Wethers and intact males (either goats or sheep) are at risk in developing urethral calculi. Treatment usually involves extensive therapy or surgery and is laborious (Blood et al., 1979).

Older animals are generally to be avoided because of the cumulative effects of parasites and likelihood of chronic diseases. Very young goats are also to be avoided because they are more likely to develop infectious respiratory and gastrointestinal problems and may require more care and attention than more mature animals.

D. Cattle

Calves may be purchased directly from a local dairy or through a sale yard. The latter source often provides animals that are of poorer quality due to colostrum deprivation or illness. It is an absolute necessity that a research project requiring calves contract with a local dairy to supply calves that are known to have had adequate colostrum intake and are in a good nutritional state before they are acquired for research. Even better is the acquisition of weaned calves or calves that have at least started to consume solids. Calf husbandry and related health considerations are discussed elsewhere (Appleman and Owen, 1974; Blackmer, 1981).

Adult cattle are seldom encountered in animal facilities unless used in agricultural research. Feeder cattle are commonly available at sale yards just after weaning in autumn. Older cattle from sales yards are often culls and were eliminated from herds because of chronic illness. Often an animal that would not be satisfactory in a production situation might be an acceptable blood donor, for example, but any potential research animal should be scrutinized for defects that may limit its usefulness. One should be cognizant of breed differences in size, temperament, production, etc. (Briggs, 1969).

V. EVALUATION OF INCOMING ANIMALS

Disease prevention, through a well-managed immunization and parasite control program at the location where the animals are bred and reared, is the best means to provide investigators with good quality subjects. Animals from livestock dealers and feed lots often have no known history. Depending on the season, the region, and age of the animals, incoming animals may arrive with internal and external parasites, other infectious diseases, or injuries acquired in transit. Even animals that were previously in good health prior to sale may have been substantially stressed (e.g., by transportation, food or water deprivation, cold or heat exposure) and exposed to infectious agents in the process of marketing. The additional stresses of manipulation, weighing, identification, and changes in husbandry imposed by the research facility may further compromise the animal's immune system.

Therefore, it is necessary to establish an effective health screening protocol for an animal when it arrives and periodically thereafter. The exact nature of the screening program will vary with the species, geographic area and associated endemic diseases, the needs of the particular research project, and economic considerations. Therefore, incoming animals should be given a thorough physical examination by a clinical veterinarian who is familiar with the particular species and associated diseases. In addition, samples of blood, feces, and other pertinent biological materials should be obtained for laboratory evaluation. General guidelines are given below. More detailed discussion of this important phase of animal management is presented by Blood *et al.* (1979).

Animals should be examined visually for signs of disease or abnormalities immediately upon arrival. Healthy animals are alert, active, and inquisitive. The skin should be clean and free from scurf, and the haircoat should have a healthy appearance. Pruritus, alopecia, and lichenification may be evident. These conditions can be associated with ectoparasites, dermatophytes, or poor nutrition. The animal's conformation should be bilaterally symmetrical, and there should be good muscle mass and adequate subcutaneous fat. The animal should move about easily, stand squarely on all four feet, and show no signs of lameness. The animal should not show respiratory distress after moderate exercise in the pen and should not cough. The eyes should be clear and bright. Any animal showing obvious abnormalities of conformation or signs of illness such as labored breathing, extension of the neck and panting after minimal exertion, abnormal nasal or ocular discharges, dermal problems, lameness, diarrhea, or emaciation should be rejected immediately. Other, more specific procedures are also indicated for each species and discussed below.

Animals should be permanently identified at the time of the initial physical examination, using ear tags, ear notches, tattoos, freeze brands, or neck chains with identification tags as appropriate for the species (Pond and Maner, 1974; Owen, 1976). It is recommended that a dual identification system be adopted to allow for occasional identification failures. Nonpharmacological methods of restraint may be necessary; these are presented by Conner (1978).

The animal's body weight should be obtained upon arrival. Recording the animal's body weight for several days or weeks after arrival provides an excellent indication of the animal's health status and how well it is adapting to husbandry and surroundings in the facility.

Animals that have been accepted after the initial screening procedures should be housed in isolation for 2 to 4 weeks.

During this conditioning period, animals can become acclimated to their new environment and learn to use the feeding and watering devices, as well as adjust to the husbandry routines. Each animal should be handled daily where appropriate and allowed to adapt to the conditions of the laboratory prior to incorporation in the research project.

Newly arrived ruminants should be provided with good quality hay *ad libitum*. If pelleted feed or concentrates are to be the bulk of their diet, the change from a roughage diet should be made gradually; no concentrates should be offered during the first 48 hr after arrival.

A. Sheep

Nasal exudates in sheep are often caused by *Oestrus ovis* larval infestation of the nasal cavity and sinuses. Such infestation is very common in most parts of the United States during the summer and fall, and is of little significance to the animal. It is often impractical to reject sheep on the basis of a nasal discharge if the animals appear to be in good health otherwise. Sheep with acute pneumonia are usually coughing, febrile, and depressed. Sheep with chronic, low-grade pneumonia are often clinically normal but may produce problems later. Such animals are likely to develop clinical pneumonia after stress (e.g., anesthesia or surgery) and, in any case, have abnormal lungs. Detecting subclinical pneumonia via radiography is usually impractical; accurate diagnosis of pneumonia by auscultation is difficult. Animals with chronic pneumonia may sometimes be identified by exercise intolerance after running the flock around the pen; other signs, such as coughing or labored breathing, may also be evident.

A variety of vaccines are available for sheep. Standard biologics include those for tetanus and enterotoxemia. Consideration should be given to vaccinating incoming animals for contagious ecthyma (orf, soremouth), since the disease is zoonotic, and investigators are likely to handle the sheep's mouth for endotracheal intubation or oral medication.

Sheep should be palpated carefully for superficial abscesses caused by *Corynebacterium psuedotuberculosis,* and affected animals should be rejected. Incoming sheep should also be examined for ectoparasites such as sheep keds and lice, and dipped with an appropriate insecticide/miticide. Products designed for agricultural use (e.g., toxaphene) are often more residual than is necessary or desirable in a vivarium. Less residual insecticides designed for use on dogs and cats (e.g., carbaryl and pyrethrins) are equally effective and preferred for a research facility.

The hooves of incoming sheep should be examined and trimmed if necessary. If foot rot (caused by *Fusobacterium nodosum*) is present, the affected tissue should be trimmed and appropriate topical and parenteral therapy instituted (Jensen and Swift, 1982). If copper sulfate solutions are used, sheep must be closely monitored due to the toxicity of ingested copper compounds for sheep. If damage to the hoof is so extensive that the affected tissue cannot easily be removed or such that the animal will be left with a permanent lameness, the animal should be rejected. Introduction of foot rot organisms via a new sheep may also cause considerable problems for other sheep in the facility, given the durability of the causative bacterium.

B. Goats

Many of the comments made for sheep are also applicable to goats. In general, goats are less likely to harbor chronic pneumonias, are somewhat more hardy, and much cleaner than sheep. Goats are less susceptible to nasal bots and less likely to develop contagious ecthyma. Parasitism and anemia are as great if not more of a problem in goats versus sheep. Foot rot as a specific bacterial infection is less of a problem in goats, although their hooves can become overgrown and require the same care as for sheep. Tetanus and enterotoxemia vaccines should be considered for all goats.

Goats and sheep are easily aged by examining their incisor teeth (see Appendix, Table A.IV) (Habel, 1978). This should be done on all incoming animals and noted on their records. Animals who have lost any incisor teeth are usually too old to provide a good research subject, especially for long-term studies, and should be avoided.

Mandibular, prescapular, prefemoral, and supramammary lymph nodes should be palpated for *Corynebacterium* abscesses, and affected animals rejected. Chronic arthritis in goats can be caused by mycoplasmal or viral infections, and is undesirable in research animals. Animals exhibiting lameness or swollen, painful joints should not be accepted.

C. Swine

As with other species, the most important aspect of a screening program is the health status of the source herd. A thorough physical examination is conducted at the time of arrival. The amount of subcutaneous fat should be appropriate to the age and type of the animal. Common abnormal findings include diarrhea, respiratory infections, and dermatologic disease. Lichenification of the skin is often seen and is indicative of either dietary zinc deficiency or ectoparasites. Pediculosis and mange are particularly common in swine.

Respiratory infections are common; nasal discharge or devia-

tion of the snout (due to atrophic rhinitis) is grounds for rejection. Because a great variety of infectious agents can cause enteric infections in swine, it is highly advisable to separate lots of animals by time of arrival, and animals from different sources. Any pig with diarrhea should be removed from the premise altogether or placed in quarantine as soon as possible. For these reasons alone, procurement of specific pathogen-free (SPF) pigs is recommended, especially if large numbers of swine are used.

Young pigs may be anemic due to dietary iron deficiency, and it is prudent to routinely administer iron dextran intramuscularly to newly arrived young swine. Vaccination regimens should be appropriate for the herd history and geographic locale (Pond and Houpt, 1978; Baker, 1975).

D. Cattle

Young calves should be observed to make certain they can obtain feed from the feeders in the facility. If there is any doubt, the calf should also be fed a milk replacer until it is capable of eating conventional feeds (Appleman and Owen, 1974; Schugel, 1981). Pneumonia and diarrhea are the most common disease entities in young calves. The most important health management factor for a young calf is making certain the animal obtained an adequate supply of colostrum. Colostrum should be force-fed at 50–80 ml/kg within the first 2 hr after birth, and at a total dose of 8–10% of body weight daily for at least the first 48 hr (Blackmer, 1981; Radostits, 1981). In addition, the navel should be disinfected with iodine immediately after birth to avoid other common neonatal conditions, such as navel ill and joint ill. Calves that are febrile or hypothermic, depressed, cough frequently, exhibit severe diarrhea, or are anorectic are poor candidates for research.

Mature cattle should be observed carefully for 10–12 days after arrival for signs of shipping fever complex. Animals that cough or are febrile should be treated immediately with intravenous, broad-spectrum antibiotics (e.g., tetracyclines at 10 mg/kg intravenous). Fecal flotation should be performed for incoming lots of animals. If substantial levels of parasite ova are found (see below), appropriate anthelmintics should be administered. It may be desirable to obtain blood samples for brucellosis testing and to test for tuberculosis via intradermal tuberculin testing, depending on the background of the animal and the prevalence of disease in the region. Brucellosis vaccination of heifer calves may also be necessary, depending on state regulations. Infectious keratoconjunctivitis is common in the summer and fall in animals housed outdoors, and infected animals should be isolated and treated immediately (Blood *et al.*, 1979).

VI. ROUTINE HEALTH MAINTENANCE

Quarterly health maintenance should include recording each animal's condition and body weight, fecal floatation assay and worming, and appropriate vaccinations if needed. Shearing of most sheep breeds and some goat breeds is required at least every spring. Crutching of sheep (shearing the escutcheon and udder) may be indicated even more frequently for optimal cleanliness and health.

Daily monitoring of feed intake is one of the easiest and most cost-effective methods of detecting disease. Anorexia or poor appetite for longer than 1 day, or any weight change greater than 10% from the previous weight (other than weight change due to maturation, pregnancy, or parturition) is sufficient grounds for a physical examination.

Parasite control is particularly important where animals are housed in groups or on pasture. It is seldom possible to maintain ruminants as parasite-free. Ungulates on overcrowded, moist pastures without a grazing rotation scheme are more likely to need frequent worming to maintain parasite loads at subclinical levels. Actual egg counts are more useful in determining levels of parasite load than merely reporting the presence of a particular parasite. Any results greater than the following values are indications for anthelmintic therapy: cattle, 75–150 ova/gm feces (fresh weight); sheep and goats, 300–400 ova/gm feces; swine, 50–100 ascarid ova/gm feces (R. B. Wescott, personal communication). Dosages for anthelmintics and other parasiticides are presented in the Appendix, Table A.XVI.

Good husbandry practices include examination of hoove and feet at least quarterly. Confined animals or those running on soft surfaces will have insufficient hoof wear and will require trimming. Good sanitation, pasture rotation, and maintaining dry surfaces will greatly reduce or eliminate the incidence of hoof problems.

Nutritional diseases can be avoided by providing a balanced diet of good quality feed. Guidelines for nutritional requirements have been established for each species and are updated every several years (National Research Council, 1975, 1976, 1978b, 1979). Daily nutrient requirements are presented in the Appendix, Table A.III. One problem to recognize is copper accumulation and toxicosis in sheep housed indoors or in metabolism cages for long periods (Blood *et al.*, 1979).

Infectious disease is generally not a major problem in a closed system unless the animal turnover rate is high. In designing a vaccine protocol, factors such as the age, physiological and immunological state of the animal, cleanliness of the environment, exposure to recent arrivals of the same or different species, and those diseases endemic to the region where the animals originated as well as where they are housed should all be considered (Todd, 1981). However, vaccines may not provide adequate protection, especially if they are used in response to a disease outbreak as opposed to use in a preventative medicine program. Furthermore, all persons involved in the implementation of a particular preventive medicine program must understand the program's design, purpose, and how to execute it properly. This is accomplished by education, assistance, and monitoring of personnel in a constant effort to maintain the health of the research animal population.

VII. SYNCHRONIZATION OF OVULATION

The purpose of synchronizing ovulation is to provide for more efficient use of artificial insemination or timed pregnancies, as well as preparing animals for embryo transfer. Conventional protocols utilize hormones or hormone analogs that may override endogenous endocrinological patterns (Jochle and Lamond, 1980). Prior to attempting ovulation synchronization, breeding evaluations both for male and female components are necessary in order to maximize success. In addition, the females to be used must be cycling normally.

Two strategies can be employed. One is to suppress ovarian activity with progestins, which will sustain corpora lutea and arrest follicle maturation. After the inhibitory agent is withdrawn at the prescribed time, this allows internally coordinated follicular development and ovulation to occur. This approach is best applied for sheep (Quinlivan, 1980). The second strategy is induction of premature regression of the corpus luteum, using either exogenous or endogenous prostaglandin $F_{2\alpha}$. This promotes the ovary to the proestrous state, with ovulation following at a known interval. This strategy is useful for cattle (Kaltenbach, 1980).

Gonadotropins have been used for estrus induction in swine (Leman and Dzuik, 1975). Progestin regimens can also be applied to swine, but nonpharmacological methods are often just as successful. For immature females (gilts) that are at least 6 months old, estrous ovulation occurs several days after exposure to a boar or its scent or a stressful episode, such as transport. For lactating sows, weaning the litter usually results in estrus several days later.

VIII. FACILITIES

A. General Considerations

Livestock facilities must be designed with the following parameters in mind: ease of containment, handling, and transport of animals; minimization of routes of escape of animals and injury to animals and personnel; ease of maintenance and repairs; allowance for efficient ventilation and avoidance of temperature extremes; and allowance for effective sanitation and waste removal. Several publications are available that address livestock facilities (Jennings, 1974; Sainsbury and Sainsbury, 1979; National Research Council, 1971, 1974b). Recommended minimum space allowances are presented in the Appendix, Table A.VI.

If space is available and the climate is not too severe, most ungulates are more efficiently housed in well-drained outdoor pens, pastures, and barns. Outdoor livestock facilities reduce labor expenses and provide an open fresh air environment where there is space for ample exercise and allowance for individual territorial needs. Animals housed outdoors require the same careful observation and management provided for animals housed indoors. Pastures can be effective means of housing animals that are to be held for long periods of time without significant handling. Pastures require careful management to prevent overgrazing; rotation is necessary to allow the flora to recover (Ensminger and Olentine, 1978; Owen, 1976). Depending on the circumstances, pastures for livestock may not be cost effective when expenses for seeding, fertilizer, weed control, irrigation, fencing, and gastrointestinal parasite control are considered.

High density stocking of either an indoor or outdoor facility leads to more excreta per surface area, as well as competition between animals for space, feed, and water. These results lead to increased incidence of injuries, stress, and disease. In winter, outdoor facilities must provide a dry, draft-free area where animals can keep warm. In hot, arid climates, protection from direct sunlight is very important for confined animals. Ample fresh water and mineral salts must always be available. Since labor is often at least 50% of animal per diem costs, efficiency is a primary consideration for fence placement, building design, equipment utilization, and labor management.

Selection of building materials for fences and buildings will vary with budgets and climates. Although the initial investment may be higher, concrete floors and walls plus metal fences and gates minimize injury to the animals and reduce maintenance costs. Concrete-surfaced pens and corrals facilitate sanitation and, if sloped properly, provide adequate drainage for dry footing for the animals. The slope of flooring surfaces for drainage should be 0.25–0.5 in./foot. Manure in dry lots may be scraped into holding areas or piles or removed. The manure pile is a cost-effective method of cleaning, and the high temperatures generating by the resultant composting activity effectively destroys most potential pathogens. Smaller pens or those with high stocking densities can be cleaned by hosing with water or steam.

Concrete flooring can be finished in a variety of surfaces. The rougher the surface, the more difficult it will be to clean but it also offers more secure footing. Animals are therefore less likely to slip and injure themselves, e.g., fractured femurs

and tuber coxae occur occasionally in excited cattle. The rougher surface may also cause initial, minor lameness problems due to abrasion, especially for animals previously maintained on irrigated pasture.

High-pressure water hoses greatly facilitate cleaning concrete flooring. Sanitation can be enhanced by use of in-line disinfectant and detergent proportionators with the water system. A minimum 6-in diameter sewage drain with trap-type basket is necessary to avoid blockage of the sewage system by hay and bedding materials.

A dry, clean resting area provides comfort for large animals. The area can be made of rubber matting (4 ft × 8 ft × 1 in.), a space with thick straw or wood shavings, or a concrete curbed area containing dirt or sand bedding that the animals may enter or leave easily. The added bedding materials provide extra comfort. Divider panels will provide for individual space, thus minimizing fighting and injuries. An outside lounging area should be shelterd by a 10-ft or higher roof.

Fences constructed of tubular or U-shaped angle iron type metal for pens and corrals are the safest types, and have low maintenance costs. High wooden fences (5 to 6 ft tall) or fences made with galvanized metal or preserved wood posts are less costly to install but are more likely to cause injuries. However, they are satisfactory on pastures where animals have more space and are less likely to challenge the fences. Alternatives include electric fencing and wire mesh fencing. The latter type is commonly used to protect livestock from predators, especially small ruminants from feral dogs.

Wooden rail fences should be avoided. Penned animals that are bored may become avid wood chewers. Most wood preservatives and anti-chewing wood coatings contain pentachlorophenol or creosote compounds that may be toxic if licked or chewed. Fences that are high enough to prevent animals from hanging their heads over them will greatly reduce repairs and injuries, prolonging the lifespan of the fence. Electrically charged wire along the top of the fence will also keep animals off.

Feed bunkers should be made of galvanized metal and should be adjustable so they can be set at the animal's chest to eye level height. The contamination of feed by feces and urine can be greatly reduced if adjustable feeders are hung as high as possible, yet still allowing the animal to eat. For small ungulates, commercial hog self-feeders attached to the fence panels are safe and readily available. These J-type feeders have a flip top cover over the animals feeding trough and another cover over the feed supply keeping the feed dry and clean. Goats and pigs quickly learn how to flip up the lower lid for access to feed. Some animals, especially sheep, may require these lids to be propped open at first. Some animals never learn how to open this lid or how to drink from automatic waterers. Animal caretakers should be aware of such cases so that these animals can be fed and watered by other means.

Free access to clean, fresh, cool water is a necessity at all times. Most domestic animals can survive when inadvertently not fed for a few meals, but lack of water can be fatal after 24–48 hr or less in some climates. "Salt poisoning" is a particularly rapid and fatal syndrome that is seen in newly arrived swine that are not able to gain access to water quickly enough (Smith, 1975).

Some facilities use automatic watering devices of the bowl type with a metal float that must be depressed by the animal's chin and muzzle as it drinks, activating the valve to provide additional water to the bowl. A better device is a small trough with a free open area large enough for the animals to drink and an adjacent, protectively covered area containing a small, toilet bowl-type float that governs the water level. This watering mechanism is easier for animals to use, but unless properly constructed, it is easily broken. Leaking waterers may provide excessive ground moisture around the trough, increasing the incidence of foot rot and providing a breeding place for flies. Drylots and pastures should have concrete pads under and around waterers and feeders to reduce muddy conditions.

Water quality is an important aspect of husbandry that is often overlooked. Groundwater can be a source of essential nutrients, sometimes in inappropriate or harmful concentrations. In addition, there are other substances (e.g., toxic compounds, carcinogens) found in drinking water that may have a substantial impact on physiology and production (National Research Council, 1974a).

Low-roofed animal shelters, especially corrugated metal ones with an interior ceiling height of less than 10 ft, can act as ovens in hot weather. In cold weather, condensation of evaporated water may occur on metal ceilings and thus dampen the animals. Ventilation must be provided by higher roof peaks, roof ridge vents, exhaust fans, eaves vents, or sliding barn doors along the sides of the shelter. Roof insulation helps prevent both overheating in the summer and condensation in the winter. Other publications discuss ventilation and building design in detail (Sainsbury and Sainsbury, 1979; Bates and Anderson, 1979; Anderson and Bates, 1979).

Corrals, pens, shelter, weighing scales, restraint chutes, and stocks should be located on ground higher than the interconnecting paved service and livestock driving lanes to provide adequate drainage. These lanes should, in turn, drain to a collecting dump or ditch or into an ample sewage drain. Local public utilities districts may have regulations restricting the type and quantity of organic materials entering public waste disposal and drainage systems.

Livestock truck and trailer loading chutes and a system of connecting lanes are essential to move livestock, feed, manure trucks, and vehicles between holding areas and the pastures, barns, and dry lots in the facility. Walk-on scales, chutes, and stocks provide safety to livestock and personnel, and are all prime components of a large animal facility.

Feed preparation and storage areas are often needed for ungulate diets that contain a high percentage of hay and bulk feeds. The quantities of foodstuffs may require large hay barns, bulk storage tanks, and special feed choppers and mixers, i.e., much more space than required for the customary feeds used for other laboratory animal species. The summer temperature in improperly dried hay stacks and bulk storage tanks is often very high, so feeds not used quickly may become deficient for heat-labile nutrients. Hay and bulk feeds should be stored where there is adequate ventilation and protection from heat, moisture, and direct sunlight (Ensminger and Olentine, 1978).

B. Indoor Facilities

In some facilities, especially those in urban areas and for research involving infectious or biohazardous materials, specialized containment rooms are needed to control the environment and to prevent the spread of contamination. The large volumes of feed, bedding, and waste present problems of space, labor, and traffic flow.

Indoor housing must provide safety to animals and personnel, be dry, draft-free but adequately ventilated, easily cleaned, and have low maintenance costs. Slatted floors are recommended for swine and sheep. Alternatively, good absorptive bedding helps control ammonia levels from the large volume of urine and feces. In some environments and for some protocols, hosing cannot be used in cleaning because it raises relative humidity to 70–95% for 1–2 hr afterward. This sudden fluctuation in the room's environment predisposes animals to respiratory diseases. Effective air vents and mechanical humidity controls are thus often necessary to maintain low moisture levels, in order to avoid increased incidence of pneumonia and diarrhea. This is especially true for young, colostrum-deprived animals and animals harboring subclinical infections or chronic conditions. Furthermore, ruminants are much more tolerant of cold temperatures due to the amount of heat released by ruminal activity, and thrive at temperatures that are deleterious for other mammals.

Some ungulate diseases are easily transmitted via aerosol to other animals and personnel. Thus, many of the concepts of a controlled rodent room must also be established for the indoor housing of livestock. For example, exposure of personnel via direct or indirect contact in common traffic flow patterns and recirculated air ventilation systems caused Q fever zoonotics in vivaria where pregnant sheep were housed (see Chapter 22 by Fox *et al.*).

Footbaths are often used in livestock facilities as a means of disinfecting the boots of personnel. Although they can be useful if used properly, footbaths should not be relied upon alone to prevent foot traffic contamination. More often than not, these disinfectant foot tubs become contaminated with bedding, manure, and other organic material that can quickly inactivate the disinfectant. If footbaths are to be useful, their contamination with excess organic material must be minimized by having workers wash debris from boots before leaving the animal room. The disinfectant must be monitored and changed once or twice daily in order to maintain its effectiveness.

A better system for sanitation is the use of a separate pair of rubber boots or disposable plastic foot covers for each room. Many experiments are most effectively confined by the use of reusable clothing or disposable clothing only used in the specific animal room. In addition to overalls, other equipment, such as foot covers, head covers, gloves, face masks, and well-fitted respirators may also be required.

Whenever a maximum security (P4) containment area is required for ungulate studies, the usual requirements of smaller animals (e.g., rodents) are further compounded by the increased body size and greater quantities of feed, water, urine, and feces involved. This increased bulk requires larger sewage drains and larger rooms with built-in or portable restraint panels. Sewage systems may require larger waste-liquid holding tanks that can be decontaminated and held for a period of time as needed, prior to pumping or release into the standard sewage system. Solid wastes are usually put into garbage cans with sealable lids that can have a liquid decontaminator added or have the outside sprayed with a disinfectant or covered with two autoclavable bags prior to moving into the dirty side exit room for holding or autoclaving. The use of double bagging protocol helps reduce accidental contamination. Autoclaves must be able to draw a complete vacuum to provide the deep steam penetration necessary for baled hay, straw wood shavings, garbage cans, bagged whole or quartered animals. In some facilities, depending on the type of contaminants, these bagged materials or animals are incinerated instead of autoclaved. Publications are available for further information on biocontainment procedures and facilities (Subcommittee on Arbovirus Laboratory Safety, 1980; U.S. Public Health Service, 1983).

IX. APPENDIX

A. Normative Data for Mature Swine, Cattle, Sheep, and Goats (Tables A.I–A.V)

The parameters in Tables A.I–A.V are expressed as a range of normal values. The values were obtained from data collected by the University of California, Davis, Veterinary Medical, Teaching Hospital, and other sources (Benjamin, 1978; Pond and Houpt, 1978; Blood et al., 1979). B, whole blood; S, serum; P, plasma; HP, heparinized plasma; R, erythrocytes.

Table A.I

General Physiological Parameters

	Units	Pig	Ox	Sheep	Goat
Life span	years	16–18	20–25	10–15	10–15
Rectal temperature	±1°C	39	38.5	39.5	39.0
	±1°F	102.0	101.5	103.0	102.0
Heart rate	beats/min	70–100	60–70	60–120	70–135
Respiration rate	breaths/min	10–20	10–30	10–20	10–20
Fecal output	kg/day	0.5–3	13–35	1–3	0.5–3

Table A.II

Laboratory Values

	Units	Pig	Ox	Sheep	Goat
Blood					
Blood volume	ml/kg	56–69	57–78	58–64	57–90
Plasma volume	ml/kg	36–60	36–41	36–40	42–75
RBC volume	ml/kg	15–18	23–33	22–24	11–18
Packed cell volume	%	32–45	24–48	24–40	24–48
Hemoglobin (B)	gm/dl	10–16	8–14	8–16	8–14
Mean corpuscular volume	fl	50–68	40–60	23–48	15–30
Mean corpuscular hemoglobin	pg	17–23	11–17	9–12	4–7
Mean corpuscular hemoglobin concentration	gm/dl	30–34	30–36	31–38	35–42
Specific gravity					
(B)		1.035–1.055	1.046–1.061	1.046–1.061	1.036–1.050
(P, S)		1.020–1.027	1.025–1.029	1.026–1.029	1.018–1.026
RBC sedimentation rate (Wintrobe) (B)	mm/24 hr	5.5–11	2.2–4	3–8.2	2–2.5

Table A.II (*Continued*)

	Units	Pig	Ox	Sheep	Goat
Blood cells					
Erythrocytes	$10^6/\mu l$	5–7	6–8	8–13	12–14
Leukocytes	$10^3/\mu l$	7–20	5–12	4–12	5–14
Neutrophils	%	30–50	20–40	20–40	20–40
Bands	%	0–10	0–4	0–2	0–2
Lymphocytes	%	40–60	40–70	40–70	50–65
Monocytes	%	2–10	3–10	1–12	1–5
Eosinophils	%	0–10	0–15	0–15	3–8
Basophils	%	0–1	0–1	0–1	0–1
Blood constituents					
Acetylcholine esterase (R)	IU/liter	930	1270–2430	640	270
Acetoacetate	mg/dl	0–1.1			
Acetones	mg/dl	0–10	0–10		
Alanine aminotransferase/SGPT (S,P,HP)	IU/liter	9–17	4–11	10–12	7–24
Alkaline phosphatase (S, HP)	IU/liter	118–395	0–488	68–387	93–387
Arginase (S,HP)	IU/liter		1–30	0–4.5	
Aspartate aminotransferase/SGOT (S,P,HP)	IU/liter	17–45	42–70	68–90	43–132
Bicarbonate ion (B,S,P)	mmole/liter	18–27	17–29	20–25	
Bilirubin (S,P,HP)					
Direct (conjugated)	mg/dl	0–0.3	0.04–0.44	0–0.27	
Indirect (free)	mg/dl	0–0.3	0.03	0–0.12	
Total	mg/dl	0–0.6	0.01–0.47	0.1–0.42	0–0.1
Butyrate (HP)	mg/dl		0–9.0		
Calcium (S,HP)	mg/dl	7.1–11.6	9.7–12.4	11.5–12.8	8.9–11.7
	mEq/liter	3.8–5.8	4.8–6.2	5.7–6.4	4.5–5.8
Carbon dioxide					
Content (B,P)	mmole/liter		21.2–31.2	21–28	25.6–29.6
Partial pressure (B)	mm Hg		35–44	36–46	
Chloride (S,HP)	mmole/liter	94–106	97–111	95–103	99–110
Cholesterol (S,P,HP)					
Total	mg/dl	36–54	80–120	52–76	80–130
Free	mg/dl	6–11	22–52		
Ester	mg/dl	28–48	58–88		
Copper (B)	$\mu g/dl$	133–160	70–170	58–130	
Creatine phosphokinase (S,HP)	IU/liter	2.4–22.5	4.8–12.1	8.1–12.9	0.8–8.9
Creatinine (S,P,HP)	mg/dl	1–2.7	1–2	1.2–1.9	1–1.8
Fibrinogen (P,HP)	mg/dl	100–500	200–500	100–500	100–400
Glucose					
(B)	mg/dl	65–95	35–55	35–60	45–60
(S,P,HP)	mg/dl	85–150	45–75	50–80	50–75
Hydrogen ion (pH) (B)			7.31–7.53	7.32–7.54	
Icterus index (P,HP)	IU	2–5	5–15	2–5	2–5
Iodine, protein-bound (S)	mg/dl	2.6–2.8	2.7–4.1	3.6–4.0	2–5
Iron					
(S)	mg/dl		57–162	166–222	
(P)	mg/dl	0.1–0.3			
Isocitric dehydrogenase (S,HP)	IU–liter		9.4–21.9	0.4–8.0	
Lactic acid (B)	mg/dl		5–20	9–12	
Lactic dehydrogenase (S,HP)	IU/liter	96–160	176–365	60–111	31–99
Isoenzymes					
LDH-1	%	34–62	40–64	46–64	29–51
LDH-2	%	6–9	20–35	0–3	0–5
LDH-3	%	6–12	12–18	16–30	24–40
LDH-4	%	7–16	0–9	4–6	0–6
LDH-5	%	16–35	0–12	10–29	14–36
Lead (B)	mg/dl		4–16	5–25	5–25

(*continued*)

Table A.II (*Continued*)

	Units	Pig	Ox	Sheep	Goat
Magnesium (S,HP)	mg/dl	2.7–3.7	1.8–2.3	2.2–2.8	2.8–3.6
Ornithine carbamyltransferase (S,HP)	IU/liter		4.4–5		
Phosphorus					
(S)	mg/dl	5.3–9.6	5.6–6.5	5.0–7.3	6.5
(HP)	mEq/liter	3.1–5.6	3.2–3.8	4–8	1.7–4.3
Potassium (S,HP)	mmole/liter	4.4–6.7	3.9–5.8	3.9–5.4	3.5–6.7
Protein (S)					
Total	gm/dl	7.9–8.9	6.74–7.46	6–7.9	6.40–7
Albumin	gm/dl	1.8–3.3	3.03–3.55	2.4–3	2.7–3.9
Globulin	gm/dl	5.29–6.43	3.06–3.48	3.5–5.7	2.7–4.1
α_1	gm/dl	0.32–0.44	0.75–0.88	0.3–0.6	0.5–0.7
α_2	gm/dl	1.28–1.54			
β_1	gm/dl	0.13–0.33	0.8–1.12	0.7–1.2	0.7–1.2
β_2	gm/dl	1.26–1.68		0.4–1.4	0.3–0.6
γ_1	gm/dl	2.24–2.46		0.7–2.2	0.9–3
γ_2	gm/dl		1.69–2.25	0.2–1.1	
A/G ratio		0.37–0.51	0.84–0.94	0.42–0.76	0.63–1.26
Protoporphyrin (HP,R)	mg/dl	118	Trace		
Pseudocholinesterase (S)	IU/liter	400–430	70	0–70	110
Sodium (S,HP)	mmole/liter	135–150	132–152	139–152	142–155
Sorbitol dehydrogenase (S,HP)	IU/liter	1–5.8	4.3–15.3	5.8–27.9	14–23.6
Urea (B)	mg/dl		43–64	17–43	21–44
Urea nitrogen (S,P,HP)	mg/dl	10–30	20–30	8–20	10–20
Vitamin A (S)	μg/dl				
Carotenol		10–35	10–30	20–45	
Carotene			25–950	0–20	
Coagulation tests					
Bleeding time	min	1–5	1–5	1–5	1–5
Clotting time					
Capillary	min	1–5	3–15	1–5	1–5
Lee–White	min		4–15	3	
Prothrombin time	sec	9–14	18–28	9–14	9–14
Platelet concentration	$\times 10^4/\mu$l	12–72	10–70	5–55	2–10
Other erythrocyte parameters					
Osmotic fragility					
Initial hemolysis	% Saline	0.70	0.62	0.65	0.74
Complete hemolysis	% Saline	0.45	0.48	0.45	0.60
Erythrocyte survival	days	80–95	135–162	64–118	125
^{51}Cr Half-life in erythrocytes	days	24–32	10.5–13		
^{39}Fe Uptake by erythrocytes	%	72–100	55	74–87	
^{39}Fe Transfer rate (plasma iron)	mg/dl	1.3–4.13	0.27	1.91–2.26	
Urine values					
Urine volume	ml/kg/day	5–30	17–45	10–40	10–40
Specific gravity		1.010–1.030	1.025–1.045	1.015–1.045	1.015–1.045
pH		Acid–alkaline	7.4–8.4	Alkaline	Alkaline
Creatinine	mg/kg/day	20–90	15–20	10	10
Urea	mg/kg/day	430	50–60	210	230
Urea nitrogen	mg/kg/day	201	23–28	98	107
Uric acid	mg/kg/day	1–2	1–4	2–4	2–5
Total nitrogen	mg/kg/day	40–240	40–450	120–350	120–400
Calcium excretion	mg/kg/day		0.1–1.4	2	1
Phosphorus excretion	mg/kg/day		0.2	1	
Magnesium excretion	mg/kg/day	3.7			
Sodium excretion	mEq/liter	11			
	mEq/kg/day	0.2–1.1			
Potassium excretion	mEq/liter		54		
	mEq/kg/day		0.08–0.15		

Table A.II *(Continued)*

	Units	Pig	Ox	Sheep	Goat
Sulfate excretion	mEq/kg/day		3–15		
Chloride excretion	mEq/liter		15		
	mEq/kg/day		0.1–1.1		
Cerebrospinal fluid values					
Calcium	mg/dl		5.1–6.3	5.77	4.6
Chloride	mg/dl		390–435	450–521	460
	mEq/dl		66–74	77–89	78
Glucose	mg/dl	48–57	35–70	52–85	70
Magnesium	mg/dl		2.17	2.86	2.28
pH			7.22–7.26	7.35	7.3–7.4
Phosphorus	mg/dl		1.6–1.9		
Potassium	mg/dl		11.2–13.8		11.6
	mEq/dl		2.9–3.5		2.96
Pressure	mm H_2O	80–145	≤200	60–270	
Protein					
Total	mg/dl	24–29	16–33	29–42	20–45
Albumin	mg/dl	17–24	10–22		
Globulin					
Nonne-Apelt		NEG	NEG	NEG	NEG
Pandy		NEG	NEG	NEG	NEG
Ross-Jones		NEG	NEG	NEG	NEG
Specific gravity			1.005–1.008	1.004–1.008	
Cell concentration	No./μl	1–20	0–10	0–15	1–11
Sodium	mg/dl		650–725	750–868	
	mEq/dl		110–123	128–148	131
Urea	mg/dl		4.6–6.5	5–6.2	5–6.2

Table A.III

Daily Nutrient Requirements of Ungulates[a]

	Units	Pig	Ox	Sheep	Goat
Type/sex		Feeder	Dairy, female, 16 weeks, large breed	Ewe	Dairy, doe
Body weight	kg	20	100	70	70
Activity		Growth	Growth	Maintenance	Maintenance
Daily weight gain	kg	0.5	0.7	0	0
Dry matter intake	kg	1.25	2.8	1.2	1.2
Water intake	kg	2	17	3.6	3.5
Digestible energy	kcal	4370	9260	2900	3010
Crude protein	gm	225	402	107	96
Lysine	gm	9.2	S	S	S
Calcium	gm	7.7	18	3.2	4
Phosphorus	gm	6.5	9	3	2.8
Sulfur	gm	[b]	1.6	1.4–2.6	1.4–2.6
Potassium	gm	2.5	8	5	N/K[d]
Sodium chloride	% of diet	0.25	0.25	0.5	0.5
Magnesium	ppm diet	400	1600	400–800	N/K
Cobalt	ppm diet	[c]	0.1	0.1	N/K
Iron	ppm diet	70	50	30–50	30–50
Copper	ppm diet	4.5	10	5	N/K
Selenium	ppm diet	0.15	0.1	0.1	N/K

(continued)

Table A.III (*Continued*)

	Units	Pig	Ox	Sheep	Goat
Zinc	ppm diet	70	40	35–50	10
Manganese	ppm diet	2.5	40	20–40	90
Iodine	ppm diet	0.14	0.25	0.1–0.8	N/K
Vitamin A	IU	1850	4200	1785	1800
Vitamin D	IU	250	660	388	369
Vitamin E	IU	14	e	e	e
Thiamin	mg	1.4	Sf	S	S
Riboflavin	mg	3.4	S	S	S
Niacin	mg	20	S	S	S
Pantothenic acid	mg	14	S	S	S
Vitamin B_6	mg	1.6	S	S	S
Choline	mg	975	S	S	S
Vitamin B_{12}	mg	16	S	S	S

[a]The exact nutrient requirements of any animal are a function of the animal's age, sex, growth rate, and status of pregnancy or lactation. The information in this table provides requirements for the size and type of each species most often found in research facilities. The values presented for ruminants are calculated on a 100% dry matter basis; for swine, a 90% dry matter basis is used. Investigators formulating rations should consult the official nutrition guidelines established by the National Research Council (1975, 1976, 1978b, 1979).

[b]Elemental sulfur is not required by swine since they utilize sulfur-containing amino acids.

[c]Deficiency rare or absent in natural diets.

[d]N/K, nutrient is required, but exact amount is not known.

[e]Vitamin E may be required for nursing animals, but not for adults in geographic regions where selenium levels are adequate.

[f]S, synthesized by rumen microflora; not required in feed.

Table A.IV

Dental Characteristics

	Deciduous	Permanent
Cow		
Sheep	$2(\mathrm{Di}_3^0 \, \mathrm{Dc}_1^0 \, \mathrm{Dp}_3^3) = 20$	$2(\mathrm{I}_3^0 \, \mathrm{C}_1^0 \, \mathrm{P}_3^3 \, \mathrm{M}_3^3) = 32$
Goat		
Swine	$2(\mathrm{Di}_3^3 \, \mathrm{Dc}_1^1 \, \mathrm{Dp}_3^3) = 28$	$2(\mathrm{I}_3^3 \, \mathrm{C}_1^1 \, \mathrm{P}_4^4 \, \mathrm{M}_3^3) = 44$

Tooth eruption	Cow	Sheep/goat	Pig
Di 1	Before birth	Birth–1 week	2–4 weeks
2	Before birth		6–12 weeks
3	Birth–1 week		Before birth
4	Birth–2 weeks	3–4 weeks	
I 1	1½–2 years		1 year
2	2–2½ years		16–20 months
3	3 years		8–10 months
4	3½–4 years	3–4 years	
Dc			Before birth
C			6–10 months
Dp 2	Birth–3 weeks	Birth–4 weeks	5–6 weeks
3	Birth–3 weeks	Birth–4 weeks	1–4 weeks
4	Birth–3 weeks	Birth–4 weeks	1–4 weeks
P 1			5 months
2	2–2½ years	1½–2 years	12–15 months
3	1½–3 years	1½–2 years	12–15 months
4	2½–3 years	1½–2 years	12–15 months
M 1	5–6 months	3–5 months	4–6 months
2	1–1½ years	9–12 months	8–12 months
3	2–2½ years	1½–2 years	18–20 months

Table A.V

Reproductive Data

Species	Age at puberty	Annual cycle	Estrous cycle length	Duration of estrus	Time of ovulation	Gestation	No. born
Cow	12 months (8–18) (smaller breeds earlier; larger breeds later)	Nonseasonal poly-estrus	21 days (18–24)	18 hr to breed 6 hr postestrus	10–14 hr after *end* of estrus	280 days (279–289)	1 calf
Ewe	9 months (7–12) (two or more ovulations may precede first be-havioral estrus)	Seasonally poly-estrus (Sept.–March) early fall to winter	17 days (15–19)	36 hr breed-ing time	30–32 hr after onset of estrus	147 days (141–151)	1–3 lambs
Doe (Goat)	7 months (3–8) (first autumn following birth; often wait to breed until sec-ond autumn)	Seasonally poly-estrus (Sept.–January) early fall to winter	19 days (18–22)	24–36 hr breeding time	24–27 hr after onset of estrus	147 days (147–156)	1–3 kids
Sow	7 months (4–9) (usually not bred until third es-trus)	Nonseasonal poly-estrus	21 days (18–24)	2 days breed anytime	36–42 hr after onset of estrus	113 days (112–115) nonovulatory postpartum estrus at 2 days	8–12 pig-lets

B. Recommended Minimum Space Allowances for Laboratory Ungulates (Table A.VI)

Table A.VI

Space Allowances[a]

	Cattle							
						Dairy breeds		
	Beef breeds					Heifers and dry cows	Milking cows	
	Calves	Yearlings	Cows	Bulls	Calves			Bulls
Feeder space (in.)								
Concentrates								
Hand-feeding	18–24	24–30	24–30	24–30		12–18		
Self-feeding	6–8	8–12	8–12	12–16				
Roughages (free choice)								
Dry	3–6	6–8	8–12	8–12	24	18–24	24–30	24–30
Silage	8–12	8–12	12–18	12–18				
Roughages (hand-feed)								
Dry or silage	14–16	16–18	18–24	24–30				
Feed bunk specifications								
Height	20–24	24–30	24–30	24–30	30	24–30	30	30
Depth	6–8	8–10	10–12	10–12	10–12	8–12	10–12	10–12
Width—feed one side	24	24–30	24–30	24–30	10	18–24	24–30	24–30
Width—feed two sides	36	36–40	40–48	40–48		36	36	36

(continued)

Table A.VI (*Continued*)

	Beef breeds				Dairy breeds			
	Calves	Yearlings	Cows	Bulls	Calves	Heifers and dry cows	Milking cows	Bulls

Automatic waterer: maximum of 80 head/bowl
Feed lot area (ft²/head)

Paved	50–60	60–70	90–100	100		50–75	100	100–150
Unsurfaced	150–200	150–200	200	150–200		100–150	200	200–300
Sheltered space (ft²/head)	20–25	25–30	40–50	80–100	20–30	20–40	40–50	100–120
Shade (ft²/head) (8–10 ft high)	15–25	25–30	30–40	30–40		20–30	50	50

Outdoor shelter area with solid floor (bedded) 30 ft²/1000 lb
Slatted floor with access underneath to liquid manure tanks (caution: toxic gas poisoning) 20 ft²/1000 lb
Solid floor without bedding or slats 20 ft²/1000 lb
Fan-ventilated indoor barns
   Winter 150 ft³/min/1000 lbs
   Fall, spring 100 ft³/min/1000 lb
   Summer 200 ft³/min/1000 lb

	Weight (lb)	Minimum floor space (ft²/head)
Pens	<150	24
	−400	50
	−1000	100
	−1200	120
	−1400	140
	>1400	151
Stanchions	<700	16
	−900	19
	−1100	21
	−1300	24
	>1300	27

Sheep and goats

	Weight (lb)	Minimum floor space (ft²)
Pens	<60	10
	−110	15
	>110	20

	Feeder lambs	Ewes	Ewes with lambs
Lot (ft²)			
Dirt	20	22	26
Surfaced	16	—	—
Shed space			
Open	6	8	12
Confined	—	14	18
Shade	6	8	12
Self-feeder space (linear measure, no. head per linear foot) concentrates or pellets	3	2	2
Trough space (linear measure, inches per head) hand feeding	12	14	14
Hay feeders (linear measure, inches per head)	—	14	14
Water			
Open trough (no. head/linear foot)	10	8	8
Automatic (no. head/bowl)	20	15	15

Table A.VI (*Continued*)

Swine		
	Weight (lb)	Minimum floor space/pig (ft²)
Pens	<110	6
	−220	12
	>220	30

	To 75 lb	Over 75 lb	Gilts, bred sows	Boars	Sow and litter
Feeder space (in.)					
Self-feeder					
Dry lot	3–4	3	4–6	—	—
Pasture	2–3	3–4	3–4	12	12
Trough	9–10	10–15	18–20	20–24	—
Feed lot area (ft²/head)					
Dry lot					
Paved	5–6	6–8	15–20	20–50	—
Natural	75–100	75–100	100–150	150–200	—
Pasture (No. head/acre)	20	10–20	8–12	5–10	6–8
Sheltered space (ft²/head)					
Outdoor lot	4–8	6–12	—	15–20	—
Total confinement					
Solid floor	8–10	10–16	25	30	48
Slatted floor	3–4	6	—	—	—
Farrowing pen (ft²/sow)					48–64
Farrowing stall (ft²/sow)					30–35
Automatic waterers (no. head/bowl)	25	25	12	3	4

[a]From Department of Animal Resources Service, School of Veterinary Medicine, University of California, Davis; National Research Council, 1974b, 1978a.

C. Infectious Diseases of Ungulates (Tables A.VII–A.XV)

Data in Tables A.VII–A.XV are from Doig *et al.* (1981); Gillespie and Timoney (1981); Jasper (1982); Jensen and Swift (1982); Kahrs (1981); Livingston and Gauer (1982); Mohanty and Dutta (1981); Stalheim (1983); U.S. Department of Agriculture (1978); and pertinent chapters in Dunne and Leman (1975). Asterisked entries indicate diseases that are not encountered in the United States or Canada, but may be found in other developed countries. C, cardiovascular; D, digestive; H, hemolymphatic; I, integument; M, musculoskeletal; Ma, mammary; N, nervous; R, respiratory; U, urogenital.

Table A.VII

Viral Diseases

	Affected host				Systems primarily affected
Disease	Pig	Ox	Sheep	Goat	
Akabane*		x	x	x	U
African swine fever*	x				H

Table A.VII (*Continued*)

	Affected host				Systems primarily affected
Disease	Pig	Ox	Sheep	Goat	
Bluetongue		x	x	x	H,I,D,U
Border disease			x		N,M,I,U
Borna disease*			x		N
Bovine adenovirus infection		x			R,D,U
Bovine coronavirus infection		x			D
Bovine cytomegalovirus infection		x			U
Bovine herpes mammilitis		x			Ma
Bovine parvovirus infection		x			D,U
Bovine rhinovirus infection		x			R
Bovine rotavirus infection		x			D
Bovine viral diarrhea/mucosal disease		x			D,U,R
Caprine arthritis/encephalitis				x	M,N
Caprine herpesvirus infection				x	U,R

(continued)

Table A.VII (Continued)

Disease	Pig	Ox	Sheep	Goat	Systems primarily affected
Contagious ecthyma			x	x	I,D,G
Encephalomyocarditis	x				N,C
Ephemeral fever*		x			R,N,M
Fibropapilloma		x	x	x	I
Foot and mouth disease*	x	x	x	x	I,D
Goatpox*			x	x	I,R,D
Hemagglutinating encephalomyelitis (vomiting-wasting disease)	x				N,D
Hog cholera*	x				H,N,I
Infectious bovine rhinotracheitis		x			R,U,N
Influenza	x				R
Leukemia virus infection		x	x		H,I
Louping ill*	x	x	x		N
Lumpy skin disease*		x			I,R,U
Malignant catarrhal fever		x			N,H,D,R (no virus isolated from North American cases)
Nairobi sheep disease*			x	x	D,H,U,R
Ovine progressive pneumonia (Maedi-visna)			x		R,N
Papular stomatitis		x			D
Parainfluenza-3 virus infection		x	x		R
Peste des petits ruminants*			x	x	R,D,H
Polioencephalomyelitis	x				N
Porcine adenovirus infection	x				R
Porcine cytomegalovirus infection	x				R
Porcine enterovirus infection	x				N,R,D,U,C
Porcine parvovirus infection	x				U
Porcine rotavirus infection	x				D
Poxvirus mammilitis		x			Ma
Pseudorabies	x	x	x		D,N,U,R
Pulmonary adenomatosis (jaagsiekte)			x		R
Rabies	x	x	x	x	N
Reovirus type 1 infection	x	x	x		R (or none)
Respiratory syncytial virus infection		x	x		R
Rift Valley fever*		x	x	x	D,U,R,H
Rinderpest*	x	x	x	x	D,H,R
Scrapie			x	x	N
Sheep pox*			x		I,R,D
SMEDI virus infection	x				U
Swine pox	x				I
Swine vesicular disease*	x				I,D
Teschen/Taflan disease (polioencephalomyelitis)	x				N
Transmissible gastroenteritis	x				D

Table A.VII (Continued)

Disease	Pig	Ox	Sheep	Goat	Systems primarily affected
Ulcerative dermatosis			x		I,U
Vesicular exanthema (no domestic outbreaks since 1956)	x				I,D
Vesicular stomatitis	x	x	x	x	I,D
Wesselsbron disease*			x		U,N

Table A.VIII

Bacterial Diseases

Bacterium/agent	Pig	Ox	Sheep	Goat	Systems primarily affected
Actinobacillus lignieresi	x	x	x	x	D,R,Ma
Actinobacillus seminis			x		U,M
Actinomyces bovis	x	x	x		D,M,U,Ma
Bacillus anthracis	x	x	x	x	H (swine-D,R)
Bacteroides nodosus		x	x	x	I
Bordetella bronchiseptica	x				R
Branhamella (Neisseria) catarrhalis		x			R
Brucella abortus		x	x		U (sheep-U,M, N,R)
Brucella melitensis			x	x	U (sheep-U,M, N,R)
Brucella ovis			x		U
Brucella suis	x				U,M
Campylobacter fetus					
ss. *jejuni*		x			D
ss. *intestinalis*		x	x	x	U
ss. *venerealis*		x			U
C. sputorum ss. *mucosalis*	x				D
Clostridium botulinum	x	x	x		N
Clostridium chauvoei	x	x	x	x	I,M
Clostridium hemolyticum		x	x		H
Clostrididium novyi	x	x	x		C,D,I
Clostridium perfringens					
Type A		x	x		H
Type B*		x	x	x	D
Type C	x	x	x	x	D,N
Type D		x	x	x	H,N
Type E		x	x		D,H
Clostridium septicum	x	x	x		I,M
*Clostridium septicum***			x		D
Clostridium tetani	x	x	x		M,N
Corynebacterium equi	x				H
Corynebacterium ovis			x	x	H,U
Corynebacterium pseudotuberculosis			x	x	H,M
Corynebacterium pyogenes	x	x	x	x	I,M,Ma,R
Corynebacterium renale		x	x		U
*Corynebacterium suis***	x				U
Dermatophilus congolensis		x	x	x	I
Ervsipelas rhusiopathiae	x	x	x		C,I,M (ox, sheep-M)

Table A.VIII (*Continued*)

Bacterium/agent	Pig	Ox	Sheep	Goat	Systems primarily affected
Escherichia coli	X	X	X	X	D,H,Ma
Francisella tularensis			X		D,H,M,U
Fusobacterium nec-rophorum	X	X	X	X	D,I,M
Hemophilus para-hemolyticus	X	X	X		H (R-pig)
Hemophilus parasuis	X				H,R,N
Hemophilus somnus		X			N,R,U
Hemophilus suis	X				H,N,
Klebsiella pneumoniae	X	X			Ma
Leptospira canicola	X	X			D,H,U
Leptospira grippotyphosa	X	X	X		D,H,U
Leptospira hardjo		X	X		D,H,U
Leptospira icterohemor-rhagiae	X	X	X		D,H,U
*Leptospira hyos**	X				D,H,U
Leptospira pomona	X	X	X		D,H,U
Leptospira szwajizak		X			D,H,U
Listeria monocytogenes	X	X	X	X	N,U (N-pig)
Moraxella bovis		X			N (eye)
*Mycoplasma agalactiae**			X	X	M,Ma,N (eye)
Mycoplasma alkalescens		X			Ma
Mycoplasma bovigenitalum		X			Ma,U
Mycoplasma bovirhinis		X			R
Mycoplasma bovis		X			Ma,R
Mycoplasma californicum		X			Ma
Mycoplasma canadense		X			Ma,U
*Mycoplasma capri**				X	R
Mycoplasma dispar		X			R
Mycoplasma hyopneu-moniae	X				R
Mycoplasma hyorrhinis	X				H,R
Mycoplasma hyosynoviae	X				M
*Mycoplasma mucoides**		X			
Mycoplasma mycoides ss. *mycoides*				X	M,R
Mycoplasma ovipneu-moniae			X	X	R
*Mycoplasma strain F38**				X	R
Neisseria ovis			X		N (eye)
Nocardia asteroides		X			Ma
Pasteurella hemolytica	X	X	X	X	R (H,R,Ma-sheep)
Pasteurella multocida	X	X	X	X	H,R
Pseudomonas aeruginosa	X	X			Ma,U
*Pseudomonas pseudomallei**	X	X	X	X	Multi-system
Salmonella cholerasuis	X				D,H
*Salmonella abortusovis**			X		U
Salmonella derby	X				D,H
Salmonella dublin	X	X	X		D,H,U,M
Salmonella typhimurium	X	X	X	X	D,H (D,H,U-sheep)
Salmonella typhisuis	X				D,H
Serratia marcesens		X			Ma
Streptococcus					
Group B: *Streptococcus agalactiae*		X	X	X	Ma

Table A.VIII (*Continued*)

Bacterium/agent	Pig	Ox	Sheep	Goat	Systems primarily affected
Group C: *Streptococcus dysgalactiae*	X	X			Ma (pig-multi-system)
Streptococcus zoo-epidemicus	X	X	X	X	C,Ma,R
Group D: *Streptococcus suis*	X				C,M,N,R
Group E: *Streptococcus uberis*	X				H
	X	X			Ma
Staphylococcus aureus	X	X	X	X	I,Ma
Staphylococcus epidermidis	X	X			I,Ma
Treponema hyodysenteriae	X				D
Ureaplasma urealyticum		X	X		R,U (U-sheep)
Yersinia pseudotuber-culosis			X		U

Table A.IX

Fungal Diseases

Fungus	Pig	Ox	Sheep	Goat	Systems primarily affected
Candida albicans	X	X			D (ox-D,U)
Coccidioides immitis	X	X	X		H
Cryptococcus neoformans		X	X	X	N,R (ox-Ma)
Maduromycosis		X			R
Microsporum canis			X	X	I
Mortierella wolfii		X			U
Mucor spp.		X			U,Ma
Mucormycosis	X	X			D,U
Rhinosporidium seeberi		X		X	R
Sporothrix schenckii		X			H
Trichophyton equinum		X			I
Trichophyton men-tagrophytes	X	X	X	X	I
Trichophyton verrucosum		X	X		I
Trichosporon cutaneum		X			Ma
Aspergillus spp.	X	X	X		U,D,Ma,R

Table A.X

Rickettsial Agents

Agent	Pig	Ox	Sheep	Goat	Systems primarily affected
Anaplasma spp.		X	X	X	H
Chlamydia psittaci	X	X	X	X	U,N,M,D
Colesiota conjunctivae	X	X	X	X	(eye)
*Cowdria rumnantium**		X	X	X	H
Coxiella burnetti		X	X	X	U
Ehrlichia spp.*		X	X		H
Eperythrozoon spp.	X	X	X		H
*Eperythrozoon ovis**			X		H
*Rickettsia phagocytophila**		X	X	X	H,U

Table A.XI

Protozoal Diseases

Agent	Pig	Ox	Sheep	Goat	Systems primarily affected
Babesia spp.	x	x	x	x	H
Balantidium coli	x				D
Cryptosporidium		x	x		D
Eimeria spp.	x	x	x	x	D
Isospora spp.	x				D
Sarcocystis spp.	x	x	x	x	M
Thieleria spp.*		x	x	x	H
Trichomonas fetus		x			U
Toxoplasma gondii	x	x	x	x	R,U,D,N
Trypanosoma spp.*	x	x	x	x	H,C,N

Table A.XII

Nematode Diseases

Agent	Pig	Ox	Sheep	Goat	Systems primarily affected
Ascaris spp.	x				D,R
Ascarops spp.	x				D
Bunostomum spp.		x	x		D,H
Chabertia spp.			x		D,H
Cooperia spp.		x	x		D,H
Dictyocaulus spp.		x	x		R
Elaeophora spp.		x	x		H,I
Globocephalus spp.	x				D
Gongylonema spp.		x	x	x	D
Haemonchus spp.		x	x	x	H
Hyostrongylus rubidus	x				D
Metastrongylus spp.	x				R
Muellerius capillaris			x	x	R
Nematodirus spp.		x	x		D
Neoascaris vitulorum		x			D
Oesophagostomum spp.	x	x	x	x	D
Ostertagia spp.		x	x	x	D
Pelodera (Rhabditis) strongyloides		x			I
Physocephalus sexalatus	x				D
Protostrongylus rufescens			x		R
Setaria spp.		x	x	x	H,N
Stephanofilaria spp.		x			I
Stephanurus dentatus	x				D,U
Strongyloides spp.	x	x	x	x	D,R,I
Trichinella spiralis	x				M
Trichostrongylus spp.		x	x	x	D
Trichuris ovis			x		D
Trichuris suis	x	x			R

Table A.XIII

Cestode Diseases

Agent	Pig	Ox	Sheep	Goat	Systems primarily affected
Echinococcus granulosus	x	x	x	x	R,D,N
Moniezia spp.		x	x	x	D
Multiceps multiceps (Coenurus cerebralis)		x	x	x	N
Taenia spp. (*Cysticerus* spp.)	x	x	x	x	M,D,H
Thysanosoma spp.		x	x	x	D

Table A.XIV

Trematode and Acanthocephalid Diseases

Agent	Pig	Ox	Sheep	Goat	Systems primarily affected
Trematode					
Cotylophoron spp.		x	x	x	D
Dicrocoelium dendriticum	x	x	x	x	D
Fasciola spp.		x	x	x	D
Paramphistomum spp.		x	x	x	D
Schistosoma spp.*	x	x	x	x	H
Fascioloides magna		x	x		D
Acanthocephalid					
Macracanthorhynchus hirudinaceus	x				D

Table A.XV

Arthropod Diseases

Agent	Pig	Ox	Sheep	Goat	Systems primarily affected
Lice					
Damalinia spp.		x	x	x	I
Haematopinus spp.	x	x			I,N
Linognathus spp.		x	x	x	I
Solenoptes capillatus		x			I
Mites					
Chorioptes spp.		x	x	x	I
Demodex spp.	x	x	x	x	I
Psorergates simplex		x			I
Psoroptes spp.		x	x	x	I
Sarcoptes scabei	x	x	x	x	I

Table A.XV (*Continued*)

Agent	Pig	Ox	Sheep	Goat	Systems primarily affected
Insects					
Calliphora spp.		x	x		I
Cochliomyia (*Callitroga*) spp.	x	x	x	x	I,M
Hypoderma spp.		x	x*	x*	I,N
Lucilia spp.		x	x		I
Melophagus ovinus			x		I
Musca domestica		x			I
Oestrus ovis			x	x	R
Phormia spp.			x		I
Sarcophaga haemor-rhoidalis		x	x	x	I
Stomoxys calcitrans	x				I
Fleas					
Echidnophaga gal-linacea	x				I
Pulex irritans	x				I
Ticks					
Amblyomma spp.		x	x	x	I,N
Dermacentor andersoni		x	x	x	I,N
Otobius spp.			x		I,N
Rhipicephalus spp.		x	x		I

D. Drug Dosages (Tables A.XVI and A.XVII)

Although commonly used in livestock, some of the drugs listed in these tables may not be officially approved for use in these species. The use of many of these drugs involves a mandatory withdrawal period for milk or slaughter; refer to the product label for all instructions.

Table A.XVI

Anthelmintics and Other Parasiticides[a]

Agent (generic name)	Indications	Dosage
Amprolium	(C) Coccidia	10 mg/kg
Cambendazole	(C) Intestinal and pulmonary nematodes	40 mg/kg
	(SH, G) Intestinal nematodes and cestodes	25 mg/kg
	(SW) Intestinal nematodes	20 mg/kg
Coumaphos	(C) Intestinal nematodes	2 mg/kg for 6 days
	(BF, SH, G) Ectoparasites	0.125%, spray, dip
	(SW) Ectoparasites	2 qt/100 gal, spray
Crotoxyphos	(C, SH, G, SW) Ectoparasites	0.25% spray

Table A.XVI (*Continued*)

Agent (generic name)	Indications	Dosage
Dichlorvos	(SW) Intestinal nematodes	17 mg/kg, repeat in 4–5 weeks
Dioxathion	(BF, SH, G, SW)	0.15% dip, spray
Famphur 13.2%	(BF) Ectoparasites	1 oz/90 kg
Fenbendazole	(C, SH, G) Cestodes, intestinal and pulmonary nematodes (Trematodes require higher doses at repeated intervals)	10 mg/kg
	(SW) Intestinal and pulmonary nematodes	3 mg/kg for 3 days
Fenthion 20%	(C) Ectoparasites	4 ml/140 kg
Levamisole	(C, SH, G, SW) Intestinal and pulmonary nematodes	8 mg/kg
Lindane	(BF, SH, G) Screwworm wounds	3% liquid, smear
	(BF, SH, G) Ear ticks	3.5% aerosol
	(BF, SH, G) Ectoparasites	0.05% spray, 0.025% dip
Malathion	(BF, SH) Ectoparasites	0.5% spray
Oxfendazole	(C, SH, G) Cestodes, intestinal and pulmonary nematodes	5 mg/kg
	(SW) Intestinal and pulmonary nematodes	3 mg/kg
Oxibendazole	(C, SH, G) Intestinal nematodes	10 mg/kg
Piperazine	(SW) Intestinal nematodes	110 mg/kg
Pyrantel	(SW) Intestinal nematodes	22 mg/kg
Ronnel	(BF, SH, G, SW) Ectoparasites	0.75% spray
Rotenone 5%	(LDC) Ectoparasites	2–4 oz/gal spray
Sulfamethazine	(C, SH, G, SW) Coccida	(refer label)
Toxaphene	(BF, SH, G, SW) Ectoparasites	(refer label)
Thiabendazole	(C) Intestinal nematodes	66 mg/kg
	(SH, G) Intestinal nematodes	44 mg/kg
	(SW) Intestinal nematodes	50 mg/kg
Trichlorfon	(BF) Ectoparasites	0.5 oz/44 kg
Triphenyl phosporothioate	(BF, SH, G) Ectoparasites	0.75% spray
USE Q 335	(C, SH, G, SW) Screwworm wounds	Smear-on

[a] C, cattle; BF, beef cattle; LDC, lactating dairy cattle; SH, sheep; G, goats; SW, swine.

Table A.XVII

Antimicrobial Drugs[a,b]

Drug (generic name)	Swine	Ruminants
Amoxicillin	10 mg/kg BID,PO	None
Ampicillin	4–10 mg/kg BID,IM,IV	Same
Cephaloridine	10 mg/kg BID,IM,SC	Same
Cephapirin	None	Intramammary infusion

(*continued*)

Table A.XVII (*Continued*)

Drug (generic name)	Swine	Ruminants
Chloramphenicol	20–50 mg/kg TID,PO; 10 mg/kg BID,IM,IV; 50–100 mg subconjunctival 1X	Same
Chlortetracycline	6–10 mg/kg IV,IM; 10–20 mg/kg PO	Same
Cloxacillin	None	Intramammary infusion
Dihydrostreptomycin	None	10 mg/kg BID,IM
Erythromycin	None	Intramammary infusion
Erythromycin	2–5 mg/kg IM,SC	Same
Furacin	Topical on lesion	Same
Gentamicin	5 mg/kg IM,IV	Same
Griseofulvin	20 mg/kg PO for 1 wk	
Hetacillin	5–15 mg/kg IM,SC	Same
Hetacillin	None	Intramammary infusion
Kanamycin	6 mg/kg BID, IM; 10–20 mg subconjunctival 1X	Same
Lincomycin-HCl	11 mg/kg IM	None
Neomycin	7–12 mg/kg BID,PO	Same
Nitrofurazone	Topical	Same
Novobiocin	None	Intramammary infusion
Oxytetracycline	6–11 mg/kg IV,IM; 10–20 mg/kg QID,PO	Same
Penicillin		
Pen G, Procaine	40,000 units/kg IM	Same
Pen G and Pen benzathine	40,000 units once	Same
Spectinomycin	10 mg/kg BID, PO	None
Sulfachlorpyridazine	45–75 mg/kg PO	65–95 ng/mg PO
Sulfadimethoxine	55 mg/kg PO (initial dose), 27.5 mg/kg PO	Same
Sulfaethoxypyridazine	200–400 mg/kg PO	Same
Sulfamethazine	200 mg/kg PO (initial dose), 100 mg/kg PO	Same
Tetracycline-HCl	None	11 mg/kg PO
Trimethoprimsulfadiazine	30 mg/kg PO	Same
Trimethoprimsulfamethoxazole	30 mg/kg PO	Same
Tylosin	2–4 mg/kg IM	Same

[a]Data from Howard (1981).

[b]All treatment regimens are daily, unless stated otherwise. BID, 2×/day; TID, 3×/day; QID, 4×/day. Routes of administration: PO, orally; SC, subcutaneous; IM, intramuscular; IV, intravenous.

REFERENCES

Anderson, J. F., and Bates, D. W. (1979). Influence of improved ventilation on health of confined cattle. *J. Am. Vet. Med. Assoc.* **174**, 577–580.

Appelman, R. D., and Owen, F. G. (1974). Breeding, housing, and feeding management. *J. Dairy Sci.* **58**, 447–464.

Baker, D. H. (1975). Swine management. *In* "Diseases of Swine" (H. W. Dunne and A. D. Leman, eds.), 4th ed., pp. 1146–1162. Iowa State Univ. Press, Ames.

Bates, D. W., and Anderson, J. F. (1979). Calculation of ventilation needs for confined cattle. *J. Am. Vet. Med. Assoc.* **174**, 581–589.

Bath, D. L., Dickinson, F. N., Tucker, H. A., and Appleman, R. D. (1978). "Dairy Cattle: Principles, Practices, Problems, Profits," 2nd ed. Lea Febiger, Philadelphia, Pennsylvania.

Benjamin, M. M. (1978). "Outline of Veterinary Clinical Pathology," 3rd ed. Iowa State Univ. Press, Ames.

Blackmer, P. E. (1981). Dairy calf health management. *In* "Current Veterinary Therapy: Food Animal Practice" (J. L. Howard, ed.), pp. 135–140. Saunders, Philadelphia, Pennsylvania.

Blood, D. C., Henderson, J. A., and Radostits, O. M. (1979). "Veterinary Medicine," 5th ed. Lea & Febiger, Philadelphia, Pennsylvania.

Bowie, E. J. W., and Dodds, W. J. (1979). Von Willebrand's disease. *In* "Spontaneous Animal Models of Human Disease" (E. J. Andrews, B. C. Ward, and N. H. Altman, eds.), Vol. 1, pp. 272–275. Academic Press, New York.

Briggs, H. M. (1969). "Modern Breeds of Livestock," 3rd ed. Macmillan, New York.

Conner, G. H. (1978). "Basic Large Animal Clinical Skills." Michigan State Univ. Press, East Lansing.

Crawford, T. B., and Adams, D. S. (1981). Caprine arthritis-encephalitis: Clinical features and presence of antibody in selected goat populations. *J. Am. Vet. Assoc.* **178**, 713–719.

Doig, P. A., Ruhnke, H. L., Waelchli-Suter, Med. R., Palmer, N. C., and Miller, R. B. (1981). The role of *Ureaplasma* infection in bovine reproductive disease. *Compend. Continuing Educ. Pract. Vet.* **3**, S324–S330.

Donnelly, W. J. C., and Sheahan, B. (1975). Bovine GM$_1$ gangliosidosis, cerebrospinal lipidosis of Friesan cattle. *Am. J. Pathol.* **81**, 255–258.

Dunne, H. W., and Leman, A. D., eds. (1975). "Diseases of Swine," 4th ed. Iowa State Univ. Press, Ames.

Earl, F. L. (1968). Housing and handling of miniature swine. *Lab. Anim. Care* **18**, 110–115.

England, D. C. (1968). Genetic basis of and procedures for development of miniature swine. *Lab. Anim. Care* **18**, 99–103.

Ensminger, M. E., and Olentine, C. G., Jr. (1978). "Feeds and Nutrition - Complete." Ensminger Publ. Co., Clovis, California.

Frank, E. R. (1964). "Veterinary Surgery," 7th ed. Burgess, Minneapolis, Minnesota.

Gillespie, J. H., and Timoney, J. F. (1981). "Hagan and Bruner's Infectious Diseases of Domestic Animals," 7th ed. Cornell Univ. Press, Ithaca, New York.

Habel, R. E. (1978). "Applied Veterinary Anatomy," 2nd ed. Published by the author, Ithaca, New York.

Hecker, J. F. (1983). "The Sheep as an Experimental Animal." Academic Press, New York.

Hook, R. R., Jr., Aultman, M. D., Adelstein, E. H., Oxenhandler, R. W., Millikan, L. E., and Middleton, C. C. (1979). Influence of selective breeding on the incidence of melanomas in Sinclair miniature swine. *Int. J. Cancer* **24**, 668–672.

Howard, J. L. (1981). *In* "Current Veterinary Therapy: Food Animal Practice" (J. L. Howard, ed.), pp. 1204–1210. Saunders, Philadelphia, Pennsylvania.

Jasper, D. E. (1982). The role of *Mycoplasma* in bovine mastitis. *J. Am. Vet. Med. Assoc.* **181**, 158–162.

Jennings, L. F. (1974). Housing requirements—large animals. *In* "Handbook of Laboratory Animal Science" (E. C. Melby, Jr. and N. H. Altman, eds.), pp. 87–94. CRC Press, Cleveland, Ohio.

Jensen, R., and Swift, B. L. (1982). "Diseases of Sheep," 2nd ed. Lea & Febiger, Philadelphia, Pennsylvania.

Jochle, W., and Lamond, D. R. (1980). Control of reproductive functions in domestic animals. *Curr. Top. Vet. Med. Anim. Sci.* **7.**

Kahrs, R. F. (1981). "Viral Diseases of Cattle." Iowa State Univ. Press, Ames.

Kaltenbach, C. C. (1980). Control of estrus in cattle. *In* "Current Therapy in Theriogenology" (D. A. Morrow, ed.), pp. 169–174. Saunders, Philadelphia, Pennsylvania.

Kraner, K. L., and Parshall, C. J., Jr. (1968). Experimental procedures and surgical techniques performed on intrauterine fetal animals. *In* "Methods of Animal Experimentation" (W. I. Gay, ed.), Vol. 3, pp. 211–239. Academic Press, New York.

Leman, A. D., and Dzuik, P. J. (1975). Reproductive efficiency and artificial insemination. *In* "Diseases of Swine" (H. W. Dunne and A. D. Leman, eds.), 4th ed. pp. 901–917. Iowa State Univ. Press, Ames.

Livingston, C. W., Jr., and Gauer, B. B. (1982). Effect of veneral transmission of ovine ureaplasma on reproductive efficiency of ewes. *Am. J. Vet. Res.* **43,** 1190–1193.

Meyer, R. C., Bohl, E. H., Henthorne, R. D., Tharp, V. L., and Baldwin, D. E. (1963). The procurement and rearing of gnotobiotic swine. *Lab. Anim. Care, Suppl.* **13,** 655–664.

Mohanty, S. B., and Dutta, S. K. (1981). "Veterinary Virology." Lea & Febiger, Philadelphia, Pennsylvania.

National Research Council (1971). "Swine. Guidelines for the Breeding, Care, and Management of Laboratory Animals." Committee on Standards, Institute of Laboratory Animal Resources, Natl. Acad. Sci., Washington, D.C.

National Research Council (1974a). "Nutrients and Toxic Substances in Water for Livestock and Poultry." Committee on Animal Nutrition, Board on Agriculture and Renewable Resources, Natl. Acad. Sci., Washington, D.C.

National Research Council (1974b). "Ruminants (Cattle, Sheep, and Goats). Guidelines for the Breeding, Care, and Management of Laboratory Animals." Committee on Standards, Institute of Laboratory Animal Resources. Natl. Acad. Sci., Washington, D.C.

National Research Council (1975). "Nutrient Requirements of Sheep," 5th rev. ed. Committee on Animal Nutrition, Board on Agriculture and Renewable Resources, Natl. Acad. Sci., Washington, D.C.

National Research Council (1976). "Nutrient Requirements of Beef Cattle," 5th rev. ed. Committee on Animal Nutrition, Board on Agriculture and Renewable Resources, Natl. Acad. Sci., Washington, D.C.

National Research Council (1978a). "Guide for the Care and Use of Laboratory Animals," rev. ed. Committee on Care and Use of Laboratory Animals, Institute of Laboratory Animal Resources. Natl. Acad. Sci., Washington, D.C.

National Research Council (1978b). "Nutrient Requirements of Dairy Cattle," 5th rev. ed. Committee on Animal Nutrition, Board on Agriculture and Renewable Resources, Natl. Acad. Sci., Washington, D. C.

National Research Council (1979). "Nutrient Requirements of Swine," 8th rev. ed. Committee on Animal Nutrition, Board on Agriculture and Renewable Resources. Natl. Acad. Sci., Washington, D.C.

Nelson, T. E., and Anderson, I. L. (1976). Porcine malignant hyperthermia. *Am. J. Pathol.* **84,** 197–199.

Owen, J. B. (1976). "Sheep Production." Baillière, London.

Pond, W. G., and Houpt, K. A. (1978). "The Biology of the Pig." Cornell Univ. Press, Ithaca, New York.

Pond, W. G., and Maner, J. H. (1974). "Swine Production in Temperate and Tropical Environments." Freeman, San Francisco, California.

Quinlivan, T. D. (1980). Estrous sychronization and control of the estrous cycle. *In* "Current Therapy in Theriogenology" (D. A. Morrow, ed.), pp. 950–954. Saunders, Philadelphia, Pennsylvania.

Radostits, O. M. (1981). Neonatal diarrhea in ruminants (calves, lambs, kids). *In* "Current Veterinary Therapy: Food Animal Practice" (J. L. Howard, ed.), pp. 116–126. Saunders, Philadelphia, Pennsylvania.

Rogers, A. L., Erickson, L. F., Hoversland, A. S., Metcalfe, J., and Clary, P. L. (1969). Management of a colony of African pygmy goats for biomedical research. *Lab. Anim. Care* **19,** 181–185.

Ruth, G. B., Schwartz, S., and Stephenson, B. (1977). Bovine protoporphyria: The first nonhuman model of this hereditary photosensitizing disease. *Science* **198,** 199–201.

Sainsbury, D., and Sainsbury, P. (1979). "Livestock Health and Housing." Baillière, London.

Schugel, L. M. (1981). General pediatric feeding: milk replacers in pediatric feeding. *In* "Current Veterinary Therapy: Food Animal Practice" (J. L. Howard, ed.), pp. 210–213. Saunders, Philadelphia, Pennsylvania.

Smith, D. L. (1975). Sodium salt poisoning. *In* "Diseases of Swine" (H. W. Dunne and A. D. Leman, eds.), 4th ed., pp. 854–860. Iowa State Univ. Press, Ames.

Stalheim, O. H. (1983). Mycoplasmal respiratory diseases of ruminants: A review and update. *J. Am. Vet. Med. Assoc.* **182,** 403–406.

Subcommittee on Arbovirus Laboratory Safety (1980). Laboratory safety for arboviruses and certain other viruses of vertebrates. American Committee on Arthropod-Borne Viruses. *Am. J. Trop. Med. Hyg.* **29,** 1359–1381.

Todd, J. D. (1981). Vaccines and vaccination programs. *In* "Current Veterinary Therapy: Food Animal Practice" (J. L. Howard, ed.), pp. 86–90. Saunders, Philadelphia, Pennsylvania.

Topel, D. G., Christian, L. L., and Ball, R. A. (1975). Porcine stress syndrome. *In* "Diseases of Swine" (H. W. Dunne and A. D. Leman, eds.), 4th ed., pp. 970–977. Iowa State Univ. Press, Ames.

Tweihaus, M. J., and Underdahl, N. R. (1975). Control and elimination of swine diseases through repopulation with specific pathogen-free stock. *In* "Diseases of Swine" (H. W. Dunne and A. D. Leman, eds.), 4th ed., pp. 1163–1179. Iowa State Univ. Press, Ames.

U.S. Department of Agriculture (1978). "Reference Manual, Foreign Animal Disease Courses," 3rd ed. Plum Island Animal Disease Center, Science and Education Administration, U.S.D.A. Press, Beltsville, Maryland.

U.S. Public Health Service (1983). "Biosafety in Microbiological and Biomedical Laboratories." Centers for Disease Control, Atlanta, Georgia (in press).

van Dijk, J. E., de Vijlder, J. J. M., van Voorthuizen, W. F., Belshaw, B. E., and Tegelaers, W. H. H. (1979). Congenital goiter - goats. *In* "Spontaneous Animal Models of Human Disease" (E. J. Andrews, B. C. Ward, and N. H. Altman, eds.), Vol. 1, pp. 111–112. Academic Press, New York.

Waxler, G. L. (1975). Gnotobiotic pigs. *In* "Diseases of Swine" (H. W. Dunne and A. D. Leman, eds.), 4th ed., pp. 1180–1199. Iowa State Univ. Press., Ames.

Chapter 11

Primates

Conrad B. Richter, Noel D. M. Lehner, and Roy V. Henrickson

INTRODUCTION

A. Taxonomic Considerations

Monkeys and prosimians belong to the order Primates (Napier and Napier, 1967), a diverse group of approximately 200 species (Table I). Prosimians are evolutionarily the most primitive of the primates and are classified separately in the suborder Prosimii. They are small- to medium-sized animals with a squirrel or foxlike appearance. Most have a moist, naked, cleft nose much like those of canines. They are highly arboreal, and most are nocturnal and insectivorous. Many are considered endangered, and most are rare or not available in large numbers. Some species, such as tree shrews and galagos (bush babies), are occasionally used in research.

Monkeys and higher primates belong to the suborder Anthropoidea. This suborder is divided into six families, which include two families of New World monkeys, Old World monkeys, lesser apes, greater apes, and man. New World monkeys are found in the Western Hemisphere from southern Mexico throughout the forests of Central America and the northern half of South America. They are alike in many ways and are taxonomically united in a single infraordinal group, the Platyrrhini. A distinguishing characteristic of the New World monkeys is their flat nose, which has large oval nares directed laterally, separated by a broad internarial septum. Other features include a tympanic bulla formed by an expanded middle ear, the absence of cheek pouches and ischial callosities, and the absence of a sigmoid flexure of the colon.

The living New World monkeys are divided by most workers into two families, the Callitrichidae—tamarins and marmosets and the Cebidae—all other New World monkeys. Callitrichids are the smallest and most primitive of the simian primates, although they appear to be evolutionarily quite stable. Hershkovitz (1977) has divided the family Callitrichidae into two major groups: tamarins (which have spatulate incisors and well developed, easily distinguished canines) and marmosets (which have caniniform incisors that blend imperceptibly with the canines, especially on the mandible). A second family, Callimiconidae, having only one species, is frequently included with the callitrichids, but is properly considered separately. It is seldom used as a laboratory animal.

The tamarins include two genera: *Leontopithecus* (the lion tamarins which are seriously endangered and not important as laboratory animals), and *Saguinus* (which has a number of species that are important as laboratory animals).

Marmosets also include two genera: *Cebuella* (which has

only one species and is of relatively limited use as a laboratory animal) and *Callithrix,* possibly the most important callitrichid genus. *Callithrix jacchus jacchus,* the common marmoset, is the most versatile and useful of these species and has proved to be the easiest species to breed and care for in captivity. *Cebuella* is the smallest of the simian primates with adult weights of approximately 150 gm. A complete listing of callitrichid species can be found in Part C.

The Cebidae are a diverse group of animals and have been divided into 11 genera and approximately 30 species (see Part B). Systematics of the Cebidae are still very basic and descriptive, with most species and subspecific designations based on pelage. Some genera, most notably, *Aotus, Saimiri* and *Cebus,* the ones currently most important in medical research, are in need of contemporary investigation and review. The cebids typically weigh from 0.6 to 10 kg.

The Old World monkeys are placed in the infraorder Catarrhini. They are found in Africa and Asia and are divided into two subfamilies, the Colobinae and the Cercopithecinae. The Colobinae have large sacculated stomachs and lack the cheek pouches found in the Cercopithecinae. Their diet consists primarily of leaves, and species of this group are commonly called leaf-eaters. They are difficult to maintain in captivity and are only rarely used in biomedical research.

The subfamily Cercopithecinae includes most of the Old World monkeys found in research laboratories. In addition to having cheek pouches, members of this subfamily have ischial collosities or pads over the ischium that allow these animals to sit for prolonged periods of time.

The species most commonly used for research belong to three genera: *Cercopithecus* (guenons), *Macaca* (macaques), and *Papio* (baboons). Other genera, such as *Cercocebus* (mangabeys) and *Erythrocebus* (patas monkeys), are represented by species used for more specialized projects, such as infectious disease research. A list of the species of Old World monkeys commonly encountered in research laboratories can be found in Part A.

B. Availability

During the past 15 years, the number of nonhuman primates imported into the United States has declined from a high of 126,857 in 1968 to 21,648 in 1980 (Mack, 1982). Factors contributing to this decline include the United States Public Health Service ban on importation of nonhuman primates destined for the pet trade, the inclusion of many nonhuman primate species on the United States Endangered Species Act (Public Law 91-135, 50CFR, Part 17) list of threatened and endangered species, and bans on exportation from countries of origin such as Colombia, India, Brazil, and Peru. These changes have affected both the number of species and the number of animals used in biomedical research.

1. New World Monkeys

The importation of New World monkeys, once great, has declined precipitously since the mid-1960s. In the 5-year period 1968–1972, over 173,000 *Saimiri,* 20,000 *Aotus,* 31,000 *Cebus,* 12,000 *Lagothrix,* and 7000 *Ateles* were imported into the United States (Mack, 1982). Most of these monkeys were imported for resale as pets. Exportation embargoes have dramatically affected cost and availability of these monkeys for research. A cooperative agreement between the Pan American Health Organization (PAHO) and the National Institutes of Health (NIH), has made some New World monkeys available for research from countries that otherwise have an export embargo. Some New World monkeys are still available from commercial sources, principally from Guyana and Bolivia.

Few callitrichids are available for research on an unlimited basis and only small numbers from domestic breeding programs. Presently, 14 callitrichid species are listed on Appendix I of the Convention on International Trade in Endangered Species (CITES) and must be regarded as threatened (Mittermeier *et al.*, 1978). At least 11 species are considered endangered or protected by one or more legislative acts. Fortunately, of those species important to research, only *Saguinus oedipus oedipus* is presently considered endangered.

Four species of callitrichids were available in 1982 through the NIH–PAHO agreement: *Saguinus fuscicollis* ssp., *Saguinus mystax, Saguinus labiatus,* and *Cebuella pygmaea.* Legal acquisition of feral animals by other routes is difficult, and animals acquired by historical trap and export methods are frequently heavily parasitized, injured from fighting, and badly undernourished. High morbidity, and frequently high mortality, can be expected among such groups. Approximately 1700 callitrichids were imported into the United States in 1980; most of these were *S. labiatus.* In addition to the species available through PAHO, *Callithrix jacchus jacchus* and *S. labiatus* are frequently available from commercial sources. The conservation status of callitrichid species was reviewed by Mittermeier *et al.* (1978). Although little change appears to have taken place since then, the greatest threat to most New World species is deforestation of natural habitat, a process that continues unabated.

In 1980, approximately 500 *Aotus,* 250 *Cebus,* and 1800 *Saimiri* were imported for medical research. Very small numbers of other cebids (approximately 200) were imported for exhibit in zoological gardens. Domestic production of cebids has been very limited. Only *Saimiri* have reproduced fairly well under captive conditions, and, even with this species, difficult problems have been encountered. Approximately 500 *Saimiri* were produced domestically in 1981, far short of domestic needs. Less than 50 *Aotus* were produced domestically during that same year. Additional research is needed on the biology of cebids if captive production is to be successful on a reasonable scale.

Several cebid monkeys have been classified as endangered or threatened with extinction according to provisions of the United States Endangered Species Act. Most of these monkeys also have been placed in Appendix I of CITES. The federal law and the international convention have provisions to protect imperiled species by prohibiting capture, interstate commerce, importation, or exportation except under special permit. Cebids currently classified as threatened or endangered are listed in Table II.

2. Old World Monkeys

The ban on exportation of nonhuman primates imposed by India in 1978 eliminated the primary source of the most commonly used laboratory primate, the rhesus monkey. Prior to

Table II

Cebids Currently Classified as Threatened or Endangered

Scientific name	Common name	Source
Alouatta pigra	Black howler monkey	Brazil, Mexico, Guatemala
Alouatta villosa	Howler monkey	Brazil, Mexico, Guatemala
Ateles geoffroyi frontatus	Spider monkey	Costa Rica, Nicaragua
Ateles geoffroyi paramensis	Spider monkey	Costa Rica, Panama
Brachyteles arachnoides	Woolly spider monkey	Brazil
Saimiri oerstedii	Red-backed squirrel	Costa Rica, Panama
Lagothrix flavicauda	Yellow-tailed woolly	Peru
Chiropotes albanansus	White-nosed saki	Brazil
Cacajao (all species)	Uakari	Peru, Brazil, Ecuador, Colombia, Venezuela

1978 approximately 15,000 rhesus monkeys were imported and used annually for research purposes [U.S. Department of Health, Education and Welfare (USDHEW), 1978]. Domestic breeding programs initiated in the mid-1970s provided 3518 rhesus monkeys for research in 1978 and 6049 in 1981. The cost of these domestically produced animals is considerably greater than that paid for the wild caught rhesus monkey. The increased cost and decreased availability of the rhesus monkey has led to a reduction in research use. The wild caught long-tailed macaque,* *Macaca fascicularis,* which is still available from Indonesia, the Phillipines, and Malaysia, appears to be replacing the rhesus monkey as the most commonly used macaque. Imports of this species have increased from 1609 in 1970 to 13,174 in 1980. The number of imports of other species, such as the pigtailed macaque and baboons, while never large, appear to be similar from year to year.

The decline in the number of nonhuman primates introduced into the United States for biomedical research can be attributed to four principal factors: the increased cost of animals, decreased availability, a decrease in federal spending for research projects, and the more effective use of the nonhuman primates that are available. The Interagency Primate Steering Committee, established by the NIH in 1974, has prepared a National Primate Plan (USDHEW, 1978). In addition to expanding domestic production and developing international programs for the production and conservation of nonhuman primates, the committee has recommended research proposal review to assure the most effective use of available animals. The review should address the following questions: Does the research require primates? Is the species selected for the study appropriate? Does the proposal require the minimum number of animals that will produce acceptable results? Will the animals be kept alive un-

---

*Also known as the cynomolgus monkey.

less required by the investigation and, if sacrifice is necessary, will body material be shared with other investigators?

In 1978 the Primate Supply Information Clearinghouse was established as a placement service for surplus primates. This Clearinghouse, located at the Washington Regional Primate Research Center, publishes a weekly list of available primates, which is sent to researchers and institutions. In 1981, 4596 live nonhuman primates were placed in research facilities through this service.

The nonhuman primate will continue to be used in biomedical research, but the patterns of use have changed and will continue to change. Both the number of species used and the number of animals used on specific projects will decline. As wild populations diminish, greater demand will be placed on animals produced domestically. Eventually the laboratory primate will be entirely captive bred and better defined for research use.

REFERENCES

Hershkovitz, P. (1977). "Living New World Monkeys (*Platyrrhini*)," Vol. I. Univ. of Chicago Press, Chicago, Illinois.
Mack, D. (1982). Trends in primate imports into the United States. 1981. *ILAR News* **25**(4), 10–13.
Mittermeier, R. A., Coimbra-Filho, A. F., and van Roosmalen, M. G. M. (1978). Conservation status of wild callitrichids. *In* "Biology and Behaviour of Marmosets" (H. Rothe, H. J. Wolters, and J. P. Hearn, eds.), pp. 17–39. Mecke-Druck, Duderstadt.
Napier, J. R., and Napier, P. H. (1967). "A Handbook of Living Primates." Academic Press, New York.
U.S. Department of Health, Education and Welfare (USDHEW) (1978). "National Primate Plan," Publ. No. (NIH) 80-1520. USDHEW, Washington, D.C.

PART A.   BIOLOGY AND DISEASES OF OLD WORLD PRIMATES

Roy V. Henrickson

I.   RESEARCH CONSIDERATIONS

A.   Research Use

The rhesus monkey has long been the standard laboratory primate. Adaptability to conditions of captivity and an abundant supply from the wild led to the establishment of research colonies of this species in the 1920's. During the 1950's large numbers of rhesus monkeys were imported into the United States for poliomyelitis vaccine production. Also, during this period, considerable interest developed in other species of non-

human primates, and research uses were found for a variety of these species. Colonies were established, and the technology for maintaining such colonies was developed.

Expanding human populations and habitat destruction have led to declining numbers of nonhuman primates in the countries of origin. Concerns over declining numbers have led to restrictions and bans on exportation. Many species of nonhuman primates have assumed threatened or endangered status. These factors have influenced research use by restricting the number of species available for use. Eventually only those species that are captive bred will be available for biomedical research. Currently, only limited numbers of rhesus monkeys are available from the wild; however, domestically reared rhesus monkeys are available from a number of captive colonies. Cynomolgus monkeys and baboons are still available from the wild and some captive colonies do exist. Thus, the pattern of research use has changed rapidly during the past 25 years. A marked expansion in the number of species being used has been followed by a reduction in the number of species being maintained and used. The Old World monkey is used in a wide variety of studies ranging from reproductive physiology to infectious diseases to behavior. The rhesus monkey is still the standard laboratory primate, although the cynomolgus monkey has gained in research use. Some species of Old World monkeys are used because of the occurrence of a spontaneous disease, such as diabetes mellitus in the celebes black ape (*Macaca nigra*), or a susceptibility to a particular disease, such as the susceptibility to leprosy observed in the sooty mangebey (*Cercocebus atys*).

B.   Laboratory Management

Old World primates, especially the macaques and baboons, are adaptable animals that have been maintained in research laboratories under varying environmental conditions for over 50 years. During the past 20 years guidelines have been developed for the care and use of laboratory animals. These guidelines provide recommendations for not only the care and use of laboratory animals but also recommendations relating to the design and construction of animal rooms, supporting facilities, cage sizes, and cage construction. Sound management practices and the maintenance of a proper environment are essential to the health of an animal (see Chapter 17 by Hessler and Moreland).

1.   Housing

Animal rooms housing nonhuman primates should be equipped with a door that opens inward and has a viewing window. This is important when animals as potentially dangerous as male macaques and baboons are being housed. It is good practice to look into the room through the viewing window before entering to make certain that there are no animals that have escaped from their cages.

There are a variety of cages commercially available for individually housing Old World monkeys. These cages are usually constructed of stainless steel or aluminum alloy and features include provision for automatic watering devices and a movable back wall for restraint purposes. The sides and backs of these cages are usually solid to prevent animals from reaching one another. A lockable sliding access door and a detachable feed container are usually included. One of the cage systems commonly encountered uses wall-mounted girders on which cages are hung by bolts over a continuous waste pan. Another common system uses cages mounted in racks that are on wheels. Such systems usually have locking devices to hold the racks in place, but frequently large males will shake the cages so vigorously that the cages will have to be attached to the wall. Animals in cages that are not secured can move their cages across a room and injure themselves while fighting with other animals.

The Guide for the Care and Use of Laboratory Animals (U.S. Department of Health, Education and Welfare 1978) provides space recommendations for animals that are individually caged. These recommendations are listed in Table A.I.

Nonhuman primates are also housed in groups in a variety of cage types including dog runs, indoor–outdoor runs, round cages (corn-cribs), corrals ranging from 0.25 to 5 acres in size, and on islands. A major consideration in group housing Old World monkeys is the trauma that results from aggression. Severe injury and death can result from this aggression. Cage enrichment is important to help reduce trauma. Perches of staggered height, visual barriers, and structures allowing escape of an animal from an aggressor, are examples of cage enrichment.

2. Sanitation

In cages equipped with flush pans, the pans need to be washed and sanitized at least once daily. It is important not to wet the animal during the daily cleaning of these pans. When using disinfectant solutions it is important not to splash the animals as chemical burns of the skin and eyes can occur. If dry bedding is used this should be changed daily. Cages should be changed a minimum of once every 2 weeks and washed in a cage washer attaining temperatures of 82.2°C (180°F). Animals soiling their cages will require more frequent changing. Cages should be cleaned prior to introducing new animals.

3. Vermin Control

Vermin control in primate facilities is extremely important. Primates will eat cockroaches and the feces of wild rodents. Wall-mounted hanging cage systems provide excellent harborage for cockroaches; vermin control systems, therefore, should be instituted and methodically followed. Because of the possible deleterious effects on experimental results, only pesticides of low toxicity should be used in rooms housing animals (see Chapter 23 by Pakes *et al.*).

4. Handling

Animals being moved to new cages or transported to other areas in a colony can be moved in transfer cages. Primates can rapidly be trained to accept this technique. To minimize the spread of disease, each animal room should have a transfer cage for use only in that room. Transfer cages moving throughout a colony should be sanitized after each use.

Animals requiring restraint for management or research purposes should be handled with restraint gloves. These gloves, usually constructed of thick padded leather, provide protection from injury by juvenile and female macaques. Male macaques and baboons should be handled only while under chemical restraint. Ketamine hydrochloride, an injectable dissociative anesthetic, is a safe and extremely effective chemical restraining agent. Injected intramuscularly at 10 mg/kg body weight, adequate restraint is usually attained in 3–5 min.

To minimize the risk of serious injury in group housed animals and to minimize the risk to humans in the event of an

Table A.I

Space Recommendations for Laboratory Animals[a]

Animals[b]	Weight	Type of housing	Floor area/animal		Height	
Group 1	≤1 kg	Cage	0.15 m²	(1.6 ft²)	50.8 cm	(20 in.)
Group 2	≤3 kg	Cage	0.28 m²	(3.0 ft²)	76.2 cm	(30 in.)
Group 3	≤15 kg	Cage	0.40 m²	(4.3 ft²)	76.2 cm	(30 in.)
Group 4	>15 kg	Cage	0.74 m²	(8.0 ft²)	91.4 cm	(36 in.)
Group 5	>25 kg	Cage	2.32 m²	(25.1 ft²)	213.4 cm	(84 in.)

[a]From U.S. Department of Health, Education and Welfare (1978). If primates are housed in groups in pens, only compatible animals should be kept. The minimal height of pens should be 6 ft (1.83 m). Resting perches and appropriate shelter should also be provided. In all cages, the minimal cage height for chimpanzees and brachiating species (orangutans, gibbons, and spider and woolly monkeys) should be such that the animal can swing from the cage ceiling without having its feet touch the floor of the cage when fully extended.

[b]The primates are grouped according to approximate size. Examples of species that may be included in the various groups are the following: Group 1: marmosets, tupaias, and infants of various species; Group 2: cebus and similar species; Group 3: macaques and large African species; Group 4: baboons, and nonbrachiating monkeys larger than 15 kg; Group 5: great apes and brachiating species.

escape, it is standard practice to extract or cut the canine teeth in male Old World monkeys. Canine cutting with aseptic filling of the root canal is the preferred technique (Tomson *et al.,* 1979; Reynolds and Hall, 1979) because extraction leads to malocclusion (see Chapter 19 by Bivin *et al.*).

Animals that are kept in group cages, such as corn cribs and corrals, require special handling techniques. Individual animals can be netted with sturdy hoop nets and injected with immobilizing agents or manually restrained. Some facilities housing nonhuman primates in corrals make use of stationary catch pens and chutes for capture. Restraining cages mounted or attached to capture chutes and equipped with a movable wall for restraint facilitate handling. Large seminatural colonies, such as those kept on islands, are usually captured by trapping animals in corrals where the animals are fed. Following trapping, animals are handled using capture chutes or nets.

Occasionally animals will escape from their cages. Juvenile and female macaques can be hand captured using restraint gloves. Larger animals may require immobilization with equipment such as capture pistols, or blowdarts. The blow-dart system described by Melton (1980) has the advantage of being silent, effective at short range, and relatively atraumatic to the animal. Animals that escape from outdoor cages will frequently remain in the vicinity of the group. These animals can be recaptured using nontraumatic traps baited with fruit.

5. Nutrition

A considerable amount of data has been accumulated on nutrient requirements of the Old World monkey because of its predominant use in research (Table A.II). Recent data on nutritional needs in nonhuman primates in general can be obtained in the most recent NRC publications on the subject as well as publication on primate nutrition, edited by Harris (1970) and Hayes (1979). Table A.III outlines the available data on protein requirements of growing and adult baboons, chimpanzees, rhesus, and cynomolgus (Nicolosi and Hunt, 1979). Investigators generally concur that nonhuman primates adjust readily to 10% total dietary fat provided that essential fatty acids are adequate.

Semipurified diets are available for nonhuman primates that should prove reliable for most nutritional investigations (see Part B). Still, there are deficits in our knowledge of the Old World primates' nutritional needs. Exact mineral requirements in the rhesus monkey, for example, are lacking; nutritional requirements in general for many of the primate species are lacking, and there is a paucity of knowledge about the effects of nutrition on more subtle parameters of health than those commonly used in determining nutrient requirements (National Research Council, 1978).

For most primates, the daily water requirement is 1 milliliter per kilocalorie of gross energy consumed. The water require-

Table A.II

Recommended and Minimum Daily Allowances for Energy and Protein in Primates[a]

| Species and age | Energy (kcal/kg body wt) | Protein[b] | | |
		% Calories (g)	Body weight (kg)	
Human				
Infant	120–60	10 (6.0)[c]	2.4	(2.0)
Adult	30–40	8 (5.0)	0.8	(0.5)
Baboon				
Infant	290	12	5.0	
Adult	53–72	12	2.0	
Chimpanzee				
Infant	120–100	12	5.0	
Adult	50–60	12	2.0	
Rhesus and cynomolgus				
Infant	270–190	12 (6.6)	8.0	(4.0)
Adult	75–120	12 (6.6)	3.0	

[a]From Nicolosi and Hunt (1979).
[b]High-quality protein.
[c]Values in parentheses are minimal requirements; not determined for the baboon and chimpanzee.

Table A.III

Nutrient Requirements for the Growing 3-kg Rhesus Monkey

| Nutrient | Diet | | |
	Per kg body weight[a]	Dry (per kg)	90% Dry matter
Energy: GE (kal)	(70)	?	?
Protein (N × 6.25) (gm)	3	189	170
Linoleic acid (gm)	(0.25)	16	14
Calcium (gm)	0.150	9.6	8.6
Magnesium (gm)	0.040	2.6	2.3
Vitamin A (IU)	400	25,556	23,000
Vitamin D (IU)	25	1,589	1,430
Vitamin E (mg/gm polyunsaturated fatty acids)	0.33–0.83	?	?
Vitamin K (μg)	0.1	6.7	6.0
Ascorbic acid (mg)	25	1,270	1,143
Biotin (mg)	0.01	0.63	0.57
Choline (mg)	Probably required	?	?
Folic acid (mg)	(0.040)	2.6	2.3
Niacin (mg)	(2)	127	114
Pantothenic acid	Required	?	?
Riboflavin (mg)	0.03	1.9	1.7
Thiamin (mg)	0.33	1.9	1.7
Vitamin B_6 (mg)	0.05–0.5	3.2–32	2.9–29
Vitamin B_{12} (mg)	0.070	4.4	4.0

[a]Values in parentheses are tentative estimates of the minimum requirement and contain no margin of safety. From the National Research Council (1978).

ment for rhesus monkeys fed commercial diet is approximately 80 ml/kg/day.

Commercial diets are available, and most Old World monkeys used for biomedical research are maintained on such diets. It is not necessary to supply supplements, such as fruit and vegetables, provided the commercial diets are properly stored and used before the expiration date. Diets should be stored in a cool, dry, vermin-free location and used within 90 days of their milling dates.

Nonhuman primates, like man, require a source of ascorbic acid (vitamin C) in their diet. This vitamin is quite labile, and if diets are improperly stored or stored for prolonged periods of time vitamin C deficiency (scurvy) can occur.

Animals should be fed twice daily to minimize problems such as the possibility of acute gastric dilatation (bloat). For animals housed individually, the amounts fed should be adjusted to meet the animal's requirements. Young growing animals will consume two to three times the amount of food on a per kilogram basis than mature or aged animals require. Animal weight should be monitored on a regular basis to assure that sufficient food is being provided.

Newly imported animals will occasionally require adaptation to commercial diets. Fruit and vegetable supplements are usually provided until animals accept the diet. Older animals with severely worn or missing teeth and animals with periodontal disease may have difficulty masticating the hard, dry biscuits. Such animals can be fed moistened biscuits. It is important that the biscuits be carefully moistened, not soaked in large quantities of fluid that will elute the nutrients.

Water is usually provided by automatic watering devices. These devices should be checked daily to assure patency. When used, water bottles should be cleaned and sanitized each time they are filled. Newly acquired and especially newly imported animals may have to be trained to use an automatic watering device. Pans of water should be provided until such animals become familiar with the devices. Aged animals may have difficulty reaching watering devices mounted high in the cage. Such animals have also been observed to have difficulty in adapting to new watering systems or watering devices in new cage systems. Aged animals require close monitoring when their environment changes. Animals that are ill or debilitated may also have difficulty reaching watering devices. Cages used for hospitalized animals are sometimes modified to place the device lower in the cage thus making it easier for such animals to drink.

6. Preventive Medicine

The single most important management tool used in maintaining a colony of healthy nonhuman primates is a well-defined and closely adhered to preventive medicine program. The essentials of such a program are applicable to any situation in which groups of animals are maintained.

a. Quarantine

Quarantine is the imposition of strict isolation to prevent the spread of disease. When management practices are established or reviewed, it is essential to keep the concept in mind. Outbreaks of disease in primate colonies are frequently traced to breakdowns in established quarantine procedures or poorly designed procedures. In addition to determining if an animal is infected with a disease that may pose a threat to the research colony, the quarantine period also serves as a conditioning period. This is important with newly imported animals that have to be adapted to commercial diets and captivity. The quarantine period is also used to define some of the basic physiological parameters of the individual animal.

Ideally the quarantine area should be located in a building separate from the rest of the colony. Personnel should be restricted to the quarantine area during working hours. There should be no transfer of equipment or supplies between quarantine and the colony unless thoroughly decontaminated. In centralized animal facilities maintaining small numbers of nonhuman primates, such measures are frequently not feasible, and management practices must be carefully monitored to assure isolation between the two groups of animals.

Within the quarantine facility, care must be taken to maintain isolation between groups of animals in different stages of quarantine. It is preferable to have a number of small rooms to house nonhuman primates to assure isolation of animals arriving at different times (shipment groups) and isolation of species. Once a group is established within a room, the entire group should progress through quarantine as an entity. If disease is detected within the group, the quarantine period should be started anew.

On or shortly after arrival, animals should be weighed, tuberculin tested, tatooed or individually identified, and given a physical examination by a veterinarian. Other diagnostic tests, such as a complete blood count, rectal swabs, and stool samples for indentification of enteric pathogens and parasites, and a reference serum sample may be performed at this time. Tuberculin testing at 2-week intervals should continue throughout the quarantine period. The duration of the quarantine period depends upon institutional policy and the source of the animals. The minimum period is 30 days and if possible should be prolonged to 90 days to minimize the risk to the colony. Facilities with large populations at risk usually require a 90-day quarantine.

b. Frequent Observation of the Animals

Animal technicians familiar with individual animals should observe the animals they are responsible for at least once daily. Signs of illness or unusual behavior should be reported to a veterinarian immediately. Nonhuman primates can mask the signs of illness, and frequently, unless subtle signs are recognized, diseases will not be detected until they are well advanced. Animal technicians should be trained to recognize the early signs of diseases commonly observed in the colony and to report their findings no matter how trivial they may seem.

c. Hospitalization of Ill Animals

Animals reported ill should be observed or examined by a veterinarian. Minor health problems are frequently treated in the animal's home cage. Animals more seriously ill should be hospitalized for prompt diagnosis and therapy. Support facilities needed include isolation, intensive care, surgery, and radiology. Laboratory support should include bacteriology, clinical biochemistry, and hematology.

d. Pathology

All animals dying of spontaneous disease should be necropsied. Necropsy reports should be reviewed by the pathologist and the clinical veterinarian to evaluate and monitor the health status and disease problems in the colony. As nonhuman primates share many of the same disease processes as man, attention should be paid to diseases that may serve as models of human disease.

e. Routine Physical Examinations

All animals should be examined on a routine basis. The interval depends on the species and age of the animals, but can vary from semiannual to biennial. Particular attention should be directed towards dental hygiene. Nonhuman primates maintained on commercial diets alone will develop calculi, which predispose to gingivitis and periodontitis. Periodic scaling to remove these calculi will retard the development of dental disease. This is important with the changing trends in primate utilization as animals may spend many years on research projects.

f. Periodic Weighing of Animals

Animals should be weighed on a monthly to semiannual basis to detect individuals with chronic disease. Weighing is done most conveniently at the time of tuberculin testing. Significant weight loss in an animal dictates an examination and evaluation by a veterinarian. Occasionally weight loss may be due to animals not being fed sufficient amounts of food.

g. Tuberculin Testing

The intradermal tuberculin test is the most reliable indicator of tuberculosis in nonhuman primates. Administered correctly and interpreted properly, the test provides an effective means of evaluating new arrivals and monitoring the research colony. New arrivals should be tested upon receipt and at 2 week inter-

vals during the quarantine period. A minimum of three consecutive negative tests should be required prior to entry to the research colony. If possible, five or six consecutive negative tests provide greater assurance that an animal is not infected with *Mycobacterium tuberculosis* or *M. bovis*. The research colony should be tested every 3 to 6 months (see Section III,A).

Thoracic radiographs are frequently used for both the initial evaluation upon receipt of an animal and for the final evaluation just prior to release from quarantine. Radiographic findings of disseminated interstitial pulmonary disease with mediastinal lmyphadenopathy suggests tuberculosis. These animals may be anergic (nonresponsive) to the tuberculin test. Individuals interpreting tuberculin test results should objectively record their observations. Frequently test results are recorded as either positive or negative, thus precluding the retrospective evaluation of changes observed prior to the first recorded positive test. A number of institutions have established standardized tuberculin test reaction grades. A standardized grading system in use at the Oregon Primate Research Center and California Primate Research Center is listed in Table A.IV and Figs. A.2–A.4. Reaction grades 1 and 2 are considered negative, 3 is suspicious, and 4 and 5 are considered positive. Tuberculin tests should be evaluated 48 and 72 hr after the administration of tuberculin.

In colonies with large numbers of animals, there may be frequent movement of animals. Unless sophisticated record systems, such as computerized records are used, animals may be moved often enough to escape routine testing. This is especially true if testing is conducted on a room by room basis. Computer programs to detect individual animals that have not been tested in over a year are a useful management tool. If the animal moves and cage locations are recorded, the tracking of animals that have been in contact with tuberculous animals is facilitated.

Table A.IV

Tuberculin Test Reaction Grades[a] (Intradermal Intrapalpebral Test)

Reaction grade	Description of changes
0	No reaction observed
1	Bruise—extravasation of blood in the eyelid associated with the injection of tuberculin
2	Varying degrees of erythema of the palpebrum without swelling
3	Varying degrees of erythema of the palpebrum with minimum swelling or slight swelling without erythema
4	Obvious swelling of the palpebrum with drooping of the eyelid with varying degrees of erythema
5	Swelling and or necrosis with eyelid closed

[a]Standardized grading system is in use at the California Primate Research Center and the Oregon Primate Research Center.

h. Medical Records

Nonhuman primates are one of the few laboratory animals that are used for multiple simultaneous and multiple sequential research projects. Some animals may spend their entire life at one research institution, and it is not uncommon for facilities to have animals that have been in the colony for 10–20 years. Because of the dwindling numbers of animals available from the wild, captive, self-sustaining breeding colonies have been established. Effective record systems are essential for the efficient management of such colonies. The publication, Laboratory Animal Records (U.S. Department of Health, Education and Welfare, 1979), defines the data elements that are required for an effective record system. The availability of inexpensive computer hardware and software places the use of sophisticated record systems in reach of most facilities maintaining moderate numbers of nonhuman primates.

i. Personnel Health Program

There are a number of bacterial, parasitic, and viral diseases that are transmissible from nonhuman primates to man. It is essential to have a human health surveillance program that includes preemployment physical examinations and periodic tuberculin tests for all individuals who have significant animal contact (Muchmore, 1975). Such individuals include animal technicians and veterinarians. The preemployment physical examination should include a medical history, thoracic radiographs, tuberculin test, baseline laboratory studies such as a complete blood count, immunization against tetanus, and a reference serum sample. Tuberculin tests should be given annually. Personnel having more limited access to the animals should have, as a minimum, an initial reference serum sample and annual tuberculin tests.

Protective clothing should be provided to all animal care personnel and high levels of personal hygiene should be encouraged. Eating or smoking should not be permitted in animal holding areas.

II. BIOLOGY

A. Physiology

There are numerous papers that provide normal values for various physiological parameters. Bourne (1975) is a comprehensive source of information on the anatomy, physiology, management, reproduction, and pathology of the rhesus monkey. Morrow and Terry (1968a,b, 1970a,b, 1972a,b,c) have bibliographies listing publications covering hematologic and clinical chemical values of nonhuman primates. Caminiti (1980a,b) has similar bibliographies on platelets and hepatic enzymes. Hack and Gleiser (1982) provide reference blood values for the baboon. Data on simian cerebrospinal fluid are given in van Bogaert and Innes (1962). Considerable variation of normal values exists that probably reflects differing management schemes, disease status, sampling techniques, and laboratory procedures. Tables A.V–A.VIII provide basic information on the rhesus monkey, the nonhuman primate most widely used in biomedical research.

B. Biology of Reproduction—Breeding Systems

Old World monkeys, such as the rhesus monkey and the baboon, have been bred successfully in the laboratory for a number of years. Research projects frequently require newborns and embryos and fetuses of known gestational age. A considerable body of knowledge has accumulated on the technology of captive breeding (Perkins and O'Donoghue, 1975). The two bibliographies compiled by Caminiti (1979a,b) provide an extensive list of references on the breeding and management of Old World monkeys.

Menarche or the onset of menstruation occurs in the macaque at approximately 2 years of age. All Old World monkeys have menstrual cycles, and many species such as the rhesus monkey

Table A.V

Normal Physiological Values for Rhesus Monkeys

Determination	Value
Body temperature—rectal temperature	37.0°–39.1°C (99.0°–102.5°F)
Heart rate—sedated	120–180 beats/min
Respiration—sedated	32–50/min
Blood volume	55–80 ml/kg body weight
Urine—rate of urine formation	70–80 ml/kg/day

Table A.VI

Rhesus Monkey Hematologic Values[a]

Determination	Mean	SD	Range
Erythrocytes (10^6)	5.02	0.55	3.56–6.95
Leukocytes (10^3)	8.17	3.25	2.5–26.7
Hematocrit (%)	39.60	3.61	26–48
Hemoglobin (gm %)	12.90	1.81	8.8–16.5
Platelets (10^3)	359.3	102.70	109–597
Reticulocytes (%)	0.91	0.81	0.1–5.2
Mean corpuscular volume (MCV) (m^3)	79.48	7.48	58.1–116.9
Mean corpuscular hemoglobin (MCH) (pg)	25.93	2.89	18.5–36.6
Mean corpuscular hemoglobin concentration (MCHC) (%)	32.63	2.11	25.6–40.2

[a]From McClure (1975).

Table A.VII

Rhesus Monkey Differential Leukocyte Values[a]

Determination	Mean	SD	Range
Segmented neutrophils			
Percentage	37.30	9.60	5–88
Absolute (10^3)	2.75	1.97	0.2–14.6
Band neutrophils			
Percentage	0.44	0.90	0–6
Absolute (10^3)	0.03	0.07	0–1.06
Lymphocytes			
Percentage	58.80	15.00	8–92
Absolute (10^3)	4.57	2.31	0.69–14.5
Monocytes			
Percentage	2.15	2.21	0–11
Absolute (10^3)	0.18	0.24	0–1.3
Eosinophils			
Percentage	2.56	2.65	0–14
Absolute (10^3)	0.17	0.24	0–1.4
Basophils			
Percentage	0.20	0.54	0–6
Absolute (10^3)	0.01	0.05	0–0.69

[a]From McClure (1975).

Table A.VIII

Rhesus Monkey Blood Chemistry Values[a]

Determination	Mean	SD	Range
Cholesterol (mg %)	165.50	30.30	108–263
Uric acid (mg %)	0.69	0.42	0.1–1.4
Total protein (gm %)	7.50	0.96	4.9–9.3
Albumin (gm %)	3.60	0.49	2.8–5.2
Globulin (gm %)	4.00	0.96	1.2–5.8
Serum glutamic pyruvic transaminase (SGPT) (S.F. units)	24.60	12.40	0–68
Serum glutamic oxalic transaminase (SGOT) (S.F. units)	43.30	23.30	16–97
Alkaline phosphatase (K.A. units)	57.80	27.20	9.7–89
Bilirubin, total (mg %)	0.40	0.54	0.1–2.0
Bilirubin, direct (mg %)	0.15	0.18	0.05–0.9
Bilirubin, indirect (mg %)	0.26	0.41	0.05–1.6
Blood glucose (mg %)	95.10	25.00	46–178
Amylase (Somogyi units)	128.30	9.30	112–148
Blood urea nitrogen (BUN) (mg %)	17.00	7.70	8–40
Creatinine (mg %)	1.12	0.58	0.1–2.8
Calcium (mg %)	10.05	1.30	6.9–13.0
Phosphorus (mg %)	4.99	0.92	3.1–7.1
Sodium (mEq/liter)	139.10	22.70	102–166
Potassium (mEq/liter)	4.29	0.93	2.3–6.7
Chloride (mgEq/liter)	101.30	8.90	84–126

[a]From McClure (1975).

show external signs of menses. Some species, such as the pigtailed macaque and the baboon, have marked changes in the perineal sex skin that correlate with phases of the menstrual cycle.

The time of first conception in macaques is usually 3–4 years of age but may occur as late as 5–6 years of age. Female baboons attain sexual maturity at $3\frac{1}{2}$–4 years of age. Males generally reach sexual maturity a year later than females.

Mature female macaques and baboons usually have menstrual cycles that are consistent in length from one cycle to the next. Rhesus monkeys have a mean cycle length of 28 days and the baboon has a mean cycle length of 36 days. To provide pregnancies of known gestational length for the rhesus monkey, cycles on individual animals are determined by visual inspection or vaginal swabbing to detect the first day of menses. A simple calculation for determining the optimum day for conception is as follows: The mean duration of the last three cycles is calculated. This value is divided by 2, and 3 days is subtracted from this number to provide the optimal day for conception. For example, if an animal's last three menstrual cycles were 26, 28, and 30 days in duration, the mean cycle length would be 28 days. This value divided by two equals 14. Subtracting 3 from this figure yields 11. Thus, an animal should be mated 11 days after the last onset of menstruation to provide the optimal day of conception. Females are taken to the males' cage for mating.

Pregnancy can be detected by bimanual rectal palpation (Wilson *et al.*, 1970; Mahoney, 1975) hormonal assays (Moore and Henrickson, 1981), or ultrasound (Nyland, 1982). Palpation is an easy and rapid means of detection in macaques and baboons. Pregnancy can be detected as early as 24 days of gestation; however, for the beginner detection at 30–35 days is more reliable. In bimanual palpation, the gloved index finger of one hand is inserted into the rectum, while the other hand is placed on the abdomen grasping the uterus. Uterine size and tone and later in pregnancy fetal head size are aids in determining the stage of gestation.

Gestation in the rhesus monkey and cynomolgus monkey is 161–175 days and 164–186 days in the baboon. Births usually occur at night or in the early morning, and newborn rhesus monkeys weigh 0.40–0.55 kg.

In addition to single caged or timed-mated breeding systems there are also single male or harem systems and multi-male systems. The single male system usually involves groups comprised of 6–12 females and one male. Such groups are housed in a variety of cages, which include indoor rooms, indoor–outdoor runs, and corn cribs. Advantages include flexibility in forming and disbanding groups, the opportunity for close observation, and known parentage. The primary disadvantage is the trauma produced by aggression within these groups. This aggression may be accentuated by the limited space. Multimale groups are housed in corrals or on islands. Group and cage size vary, but the average corral for rhesus monkeys is 0.5–1.0 acre in area and contains 30–100 breeding females and 3–12 adult males. Some island breeding programs have populations as large as 2000 rhesus monkeys. These large cages and islands have the advantage of simulating a more natural free-ranging environment but limit the ability to observe or capture the animals.

C. Infant Rearing

Old World monkeys have complex social organizations. Normal infant development requires exposure to individuals of the same species. Such exposure in captivity is best obtained by rearing infants in large multi–male breeding groups where the natural troop organization is simulated. Occasionally due to experimental design, the death of a mother, or the rejection of an infant, newborns will have to be nursery reared. The technology of nursery rearing is well established, and physically healthy infants can easily be produced (Ruppenthal, 1979). However, the social deprivation that occurs in nursery rearing is more difficult to overcome, and animals with bizarre behavioral patterns may be produced. Such outcomes plus the cost of establishing and running a nursery should be considered.

Newborn infants should be examined, weighed, and identified prior to being placed in the nursery. Permanent identification such as tatooing can be done at a later date unless there is a danger of misidentification. If required, the umbilical cord is ligated close to the body and the remainder cut away. Neonates have poor thermoregulatory ability and must be maintained at temperatures between 30–32°C (85–90°F). This can best be accomplished by keeping infants in human infant incubators or in large plastic rodent bins placed on heating pads. The bins are less expensive and can be thoroughly disinfected at regular intervals. Bath towels serve as a suitable liner and an additional towel, which is rolled or bunched up, serves as a surrogate for the infant to clutch.

Infants are fed a 5% dextrose solution with a small nursing bottle during the first 24 hr after birth. A standard, commercial human milk substitute can be used after the first day. Infants should be fed every 2–3 hr during the average 14–16 hr of daily nursery operation. Food intake and weight should be closely monitored as infants lose approximately 10% of their body weight during the first several days. Birth weight should be reattained by 7 days of age.

A bottle holder can be installed in the incubators for holding the nursing bottle as infants learn to self-feed by the end of the first week. Cereal mixed with formula is provided during the third week and by 1 month of age a pan of milk-soaked biscuits can be given to the infants. By 4–6 weeks of age infants can be housed in individual cages in the nursery. At this time they are placed with other infants of comparable ages for increasing periods of time to provide socialization. By 2 months of age they can be housed continuously in pairs.

In colonies where single-caged females are producing offspring, a weaning program should be established. The age of weaning varies from 1 month to 9 months depending upon the institution. Many facilities are currently weaning infant macaques at 3 months of age. Such infants require close monitoring following weaning as some animals do not eat and develop hypoglycemia and may die (Anderson *et al.*, 1981). Individual caging and close monitoring to include force feeding of infants not eating will minimize losses encountered from the stress of weaning. Once animals are stabilized and have adjusted, they can then be pair housed. After 1–2 weeks, a common management practice is to group animals of comparable ages into "peer groups" of 6–12 animals. Such a scheme provides some of the socialization needed for proper social development.

III. DISEASES

A. Bacterial

Enteric disease is the primary cause of morbidity and mortality in most colonies of Old World monkeys. Frequently the etiology of primate diarrhea is difficult to determine and may be the result of a complex interaction between the host, the environment, and potentially pathogenic microorganisms. It is not uncommon to isolate two or three different organisms with pathogenic potential from rectal swabs and stool samples from an individual animal with diarrhea. The clinician, however, should attempt to determine the primary pathogen and treat with the appropriate antibiotic. Many antibiotics are available in a fruit-flavored pediatric oral suspension. Juvenile macaques readily accept such medication from a syringe thus making treatment easier and less stressful for the animal. The sensitivity patterns of the organisms isolated should be determined and monitered. Antibiotics should be used only in a rational and responsible manner, as intensive use of antibiotics may lead to the emergence of resistant strains of organisms within the colony. The fluid and electrolyte status of the animals should be carefully monitored. Frequently, maintenance of proper fluid and electrolyte balance is all that is required with mild diarrheas. To minimize handling stress in such animals, the provision of a pan of oral electrolyte fluids flavored with a powdered fruit drink should be provided.

Among the most commonly isolated enteric pathogens are organisms of the *Shigella* group (Good *et al.*, 1969; Mulder, 1971; Weill *et al.*, 1971). These organisms, which are probably acquired originally from man, are common in most colonies of macaques. The species most frequently observed are *Shigella flexneri*, *Shigella sonnei*, and *Shigella boydii*. Serological typing distinguishes the serotypes found within the species. Serotyping is important in determining and monitoring the epidemiological characteristics of colony infections. Most colonies have a single serotype responsible for infections. New arrivals, especially wild caught animals, should be sampled to determine if they are infected and if so, by which serotype. *Shigella flexneri* is the most common species encountered in primate colonies.

Clinical signs vary greatly. In its most severe manifestation this organism can produce an acute colitis with the passage of foul smelling liquid stools containing blood, mucus, and necrotic colonic mucosa (dysentery). Animals with such signs are severely ill and require prompt medical attention to correct the life-threatening fluid and electrolyte imbalances. The mucosa of the colon of such animals is usually covered with a fibrino-purulent exudate, and the intestinal wall is edematous and hemorrhagic with focal areas of ulceration (Fig. A.1). Antibiotics such as ampicillin, chloramphenicol, or trimethoprim-sulfa, are also indicated. Epizootics of acute, severe shigellosis can occur especially in animals that are malnourished, under stress, or immature. Frequently, however, not all animals in a group or room will contract the disease, and sometimes the most healthy animals will manifest the most severe symptoms. The more common clinical picture seen with shigellosis is a subacute to chronic diarrhea in which the stool is liquid to semi-solid. Such animals provide a vexing clinical management problem because of the intermittant or episodic nature of the diarrhea. Occasionally animals have firm, mucous-laden feces that are streaked with fresh blood. *Shigella* organisms can be isolated in large numbers from such animals; however the animals do not appear clinically ill. These brief episodes usually resolve spontaneously; however, the animal can serve as a source of infection to other animals.

In colonies where the disease is endemic, young animals become infected during the first 2 years of life and probably become life-long carriers. The organism can frequently be isolated from a small percentage of clinically normal animals. Repeated culturing of the same animals will increase the number of asymptomatic animals diagnosed as shedding *Shigella*.

The recent recognition of *Campylobacter jejuni* as a cause of primate diarrhea (Morton *et al.*, 1983: Bryant *et al.*, 1983) has further confused the search for primary etiological agents. This bacterial pathogen is now recognized as a common cause of acute gastrointestinal disease in children and adults and is now being isolated with increased frequency from colonies of macaques as well as other Old World primates. The organism has probably been present in these colonies for years but is only

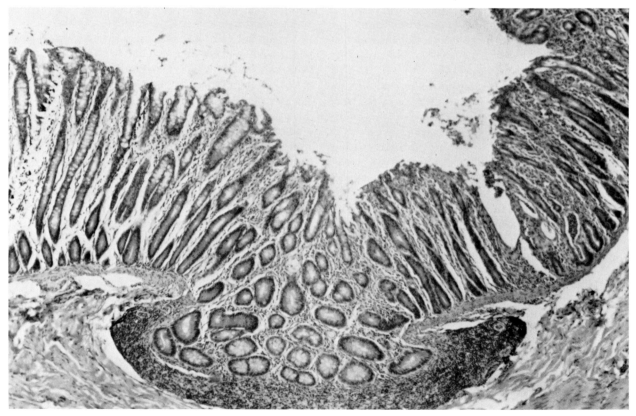

Fig. A.1. Low-power micrograph of the colon of a rhesus monkey. Mucosal herniation commonly seen with shigellosis.

now being recognized because of the introduction of selective culture media and modified incubation procedures. The epidemiological aspects of infection and the role of the organism in the production of diarrhea needs further elucidation. The organism can commonly be isolated both from animals with normal stools and from animals with a watery or hemorrhagic diarrhea (Fox, 1983). A chronic recurrent form of the disease has been recently observed in Patas monkeys; relapses can occur in man also. Animals trapped in a pristine environment have a low prevalence of infection, suggesting that *Campylobacter jejuni* may not be a naturally occuring disease in the wild (Morton *et al.*, 1983). Thus, this organism may be acting similar to *Shigella* in that naturally occurring populations of monkeys living distant from man are not infected. During the course of trapping and exportation, animals become infected, with the establishment of a high prevalence of carrier animals. Such animals serve to maintain the disease in colonies of nonhuman primates. This is similar to what is observed with shigellosis in institutionalized children. Frequently, animals with diarrhea that are positive for *Campylobacter jejuni* will respond to a 7- to 14-day course of erythromycin.

Other bacterial agents producing enteric disease include *Salmonella* spp., *Yersinia pseudotuberculosis,* and *Yersinia enterocolitica.* Salmonellosis is not commonly observed, but may present as a mild to marked gastroenteritis and may become systemic with complications such as endocarditis or meningitis. *Salmonella* infection of the gastrointestinal tract localizes in the ileum in contrast to *Shigella,* which is more commonly found in the colon.

Outbreaks of diarrheal disease, acute deaths, and abortion have been attributed to *Yersinia pseudotuberculosis* (Bronson *et al.*, 1972; Rosenberg *et al.*, 1980) and *Yersinia enterocolitica* (McClure *et al.*, 1971). These organisms infect a wide range of vertebrate hosts, such as reptiles, birds, and various mammals, including man. Wild birds and rodents are considered to be the reservoir hosts. The usual route of infection is through ingestion of contaminated food. In facilities where primates are housed outdoors, these organisms should be considered in the differential diagnosis of diarrhea. Concurrent infections with *Shigella* are common. *Yersinia* sp. is difficult to isolate on routine rectal culture and may be overlooked. Special culture techniques using cold enrichment and culture at 20°C are required for isolation. The clinical signs are quite variable and may include acute deaths, acute to chronic enteritis, and septicemia. At necropsy, ulcerative lesions are found in the small and large intestine. Histologically, lesions are characterized by marked necrosis with the presence of large bacterial colonies. The organisms are usually sensitive to antibiotics, such as chloramphenicol, trimethoprim–sulfa, gentamicin, and kanamycin.

Respiratory disease is the second most common cause of morbidity and mortality in primate colonies. Pneumonia is a serious problem in newly imported animals and may account for a significant portion of deaths. As research facilities rely more heavily on domestically produced animals, the incidence of pneumonia may diminish in importance. Animals with an acute lobar pneumonia may have facial cyanosis, which is especially noticeable on the eyelids and the base of the nose. Such animals may also be observed sitting upright or grasping the cage bars to hold themselves erect to aid in respiration. Rapid, shallow respiration may also be noted. Such animals are usually febrile and will have areas of consolidation radiographically. Prompt antibiotic therapy with a broad-spectrum antibiotic is indicated. The administration of oxygen may be required in severe cases. Outbreaks of respiratory disease are observed in colonies of macaques. These outbreaks are thought to be associated with viral infections, such as measles or other agents. Bacterial pneumonia is a complication of such outbreaks, and organisms isolated include *Klebsiella pneumoniae, Streptococcus pneumoniae, Bordetella bronchiseptica,* and *Pasteurella multocida* (Good and May, 1971). Outbreaks of respiratory disease associated with *Streptococcus pneumoniae,* are frequently accompanied by bacterial meningitis caused by the same organism (Fox and Wikse, 1971). Treatment of all animals at risk with a long-acting penicillin will usually stop the outbreak. An invaluable study on clinical and neurological observations on apes and monkeys provides insight into other simian neurological disorders (van Bogaert and Innes, 1962).

Tetanus has been observed in colonies of macaques maintained outdoors both on island and in corrals (DiGiacomo and Missakian, 1972; Rawlins and Kessler, 1982). Diagnosis is based on clinical signs that include, in the early stages of the disease, depression, a reluctance to interact with other animals, and difficulty in eating. Animals next show a progressive stiffening and adduction of the pectoral limbs. Movement is bipedal with a hopping gait. Trismus, piloerection, and increasing extensor rigidity follows. Spasm of the facial muscles leads to a flattening of the ears, the monkeys' risus sardonicus. Death is due to respiratory paralysis. On the Island of Cayo Santiago, Puerto Rico, 25% of the mortality over a 5 year period was due to this disease, and 85% of the animals afflicted died. The clinical course of disease ranges from 1 to 10 days. Infection is usually associated with trauma or can occur postpartum. Animals that have been treated have recovered; however, it is unknown whether these animals would have recovered spontaneously. Good supportive care is essential until the tetanus toxin has been metabolized. Circulating toxin must be neutralized by administering tetanus antitoxin. Treatment of the infected wound is essential. Antibiotics (penicillin) are also used; however, their value is questionable. Clinical management is difficult, and prevention is the key to treatment. Routine immunization of animals with tetanus toxoid is protective. In some colonies all juvenile monkeys receive two immunizations, 4–12 months apart.

Tuberculosis is a bacterial disease that poses the most insidious threat to colonies of Old World monkeys. This disease, which is caused by *Mycobacterium tuberculosis* or *Mycobacterium bovis,* is usually progressive and fatal (Benson *et al.,* 1955; Keeling *et al.,* 1969; Schmidt, 1969). It is not a naturally occurring disease of nonhuman primates and is usually contracted from humans or in the countries of origin. The source of infection in colonies can usually be traced to a recently imported animal that was infected but eluded detection during quarantine or an infected human who has contact with the colony. There are reports of unusual outbreaks that can be traced to single animals that have been in closed colonies for 1–2 years during which time repeated tuberculin tests were negative. The reasons for such outbreaks usually remain a mystery, but the consequences are evident and severe.

The route of infection is most often by inhalation of the organism. The disease is slowly progressive and may require up to 1 year from infection to death in the rhesus monkey. The clinical signs of disease are usually absent until the disease has progressed. The major problem in controlling an outbreak of tuberculosis is the detection of infected animals. The most reliable diagnostic test is the intradermal tuberculin test. This test detects delayed hypersensitivity to the tuberculin. Delayed hypersensitivity takes a minimum of 3 weeks to develop in a healthy animal infected with an organism of high virulence. Three to 4 weeks following exposure an infected animal can be transmitting organisms to other animals in a room via respiratory droplets. Thus, tuberculosis is easily spread within a colony environment and is difficult to detect and eradicate. Once tuberculosis is discovered in a colony, all animal moves should cease, and animals previously housed in the same room with the infected animals should be located. These animals should be tested biweekly with tuberculin for a minimum of five consecutive negative tests.

Prevention is the key to controlling tuberculosis. An effective quarantine program is the cornerstone of such prevention. A 90-day quarantine with biweekly tuberculin testing is the method of choice when possible. All animals should have thoracic radiographs taken upon arrival in quarantine and just prior to release to detect advanced cases of tuberculosis in which the animal may be anergic. Mammalian old tuberculin is no longer available from the United States Department of Agriculture. Primate facilities are currently using mammalian old tuberculin prepared from human isolates that is available commercially (Tuberculin, Jensen-Salsbery Laboratories, Kansas City, Missouri 64108). The technique most commonly used in primate colonies for tuberculin testing is the intradermal injection of 0.1 ml of undiluted old tuberculin into the upper eyelid. A minimum of 1500 tuberculin units is the recommended dosage for eliciting a positive tuberculin response (Fox *et al.,* 1983). The eyelid is used because of the ease in visualizing the reaction to the tuberculin. A new 27-gauge needle should be used for each injection. Injection sites should be examined at 24, 48, and 72 hr for the presence of hyperemia, edema, and induration. A swelling of the palpebrum with a drooping of the eyelid noted at 48 and 72 hr signifies a positive reaction to the tuberculin test. Such animals should be necropsied immediately and tissues obtained for histopathological examination and culture. Figures A.2–A.4 illustrate negative and positive reactions.

Disease due to infections with *Mycobacterium avium* and other members of the atypical mycobacterial group have been

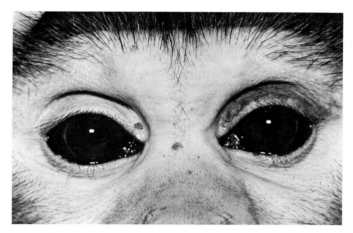

Fig. A.2. Tuberculin test reaction grade 1. Photograph of the face of a rhesus monkey showing a reaction in the left eye to the injection of tuberculin. Reaction 1, a bruise, is considered negative.

Fig. A.3. Tuberculin test reaction grade 4. Photograph of the face of a rhesus monkey showing a reaction in the right eye to the injection of tuberculin. Reaction 4, swelling of the lid and drooping of the lid, is considered a positive reaction to the tuberculin test.

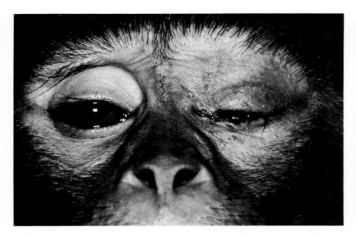

Fig. A.4. Tuberculin test reaction grade 5. Photograph of the face of the rhesus monkey showing a reaction in the left eye to the injection of tuberculin. Reaction 5, swelling and necrosis of the lid, is considered a strong positive reaction to the tuberculin test.

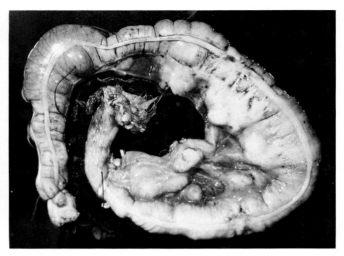

Fig. A.5. Colon of a stump-tailed macaque, *M. arctoides,* observed at necropsy. Note the greatly enlarged mesenteric lymph nodes and the prominent lymphatics.

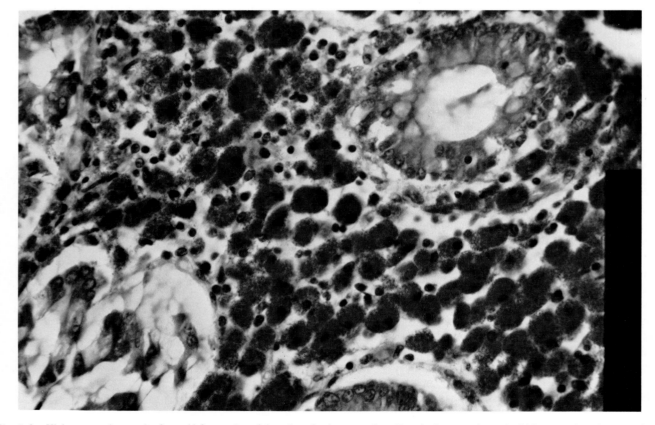

Fig. A.6. High-power micrograph of an acid-fast section of the colon of a rhesus monkey. Note the large numbers of acid-fast organisms in macrophages.

reported in macaques (Sesline *et al.*, 1975; Smith *et al.*, 1973; Sedgwick *et al.*, 1970). These organisms can produce a granulomatous enteritis and mesenteric lymphadenitis (Figs. A.5 and A.6). Such cases are sporadic, and there is no evidence of animal-to-animal transmission. Results of tuberculin testing with avian tuberculin are difficult to interpret, as occasionally animals with disease due to *M. avium* will not react to the tuberculin. Conversely, animals without lesions and that culture negative will occasionally react positively to the tuberculin. Standardized tuberculin test reaction grades are listed in Table A.IV.

B. Parasitic

Parasitic infections are extremely common in wild caught, recently imported nonhuman primates. In stressed or debilitated animals parasitic infections may result in severe clinical disease and death. The more common picture, however, is a balanced host–parasite relationship where the parasite produces little or no damage and the host's reaction to the parasite is minimal. This is observed where parasite and host have evolved together over a long period of time. This compounds the problems for the veterinary clinician, however, for he has to sift through these interactions to try to determine if a parasite is playing a role in the disease he is trying to treat. In cases where the parasite encounters an aberrant host, the picture can be much clearer. In many of these encounters severe disease may occur. This is seen where differing species of nonhuman primates are mixed together or placed into a novel environment.

1. Protozoa

Protozoal parasites are commonly encountered in nonhuman primates. These organisms are extensively covered in the book by Flynn (1973). The hemoflagellate, *Trypanosoma cruzi*, has been reported from macaques; however these infections were probably not naturally occuring infections (Cicmanec *et al.*, 1974). The enteric flagellates occur in large numbers in the intestinal tract of nonhuman primates. Frequently in cases of diarrhea, extremely large numbers of protozoa are observed in stool samples. The species most commonly identified include *Trichomonas* and *Giardia*. Their role in the production of diarrhea is uncertain; however many investigators believe that their presence is a result rather than a cause of the diarrhea. Some clinicians will treat such diarrhea with oral metronidazole at a level of 50 mg/kg given once daily for 5–10 days. Reductions in the numbers of organisms and in the severity of the diarrhea has been observed following treatment. *Entameba histolytica* is another common enteric protozoan. Severe disease with the characteristic flasklike amebic intestinal lesions have been

commonly reported from New World monkeys (see Part B). Reports of disease in Old World monkeys are rare.

Cryptosporidium sp. is a protozoan currently classified in the subphylum Sporozoa. This organism is present in a number of mammalian species, including man where it normally produces an acute self-limited diarrhea. It has been reported as being responsible for diarrhea in patients with diminished immune responses either induced by immunosuppresive therapy or occurring naturally. This organism is assuming greater importance with the appearance of the acquired immunodeficiency syndrome (AIDS) in man. In AIDS patients, *Cryptosporidium* has been associated with profuse watery diarrhea and is unresponsive to treatment (Andreani *et al.*, 1983). Cockrell *et al.* (1974) reported on infections observed in eight young rhesus monkeys. Infections of the duodenum and biliary tree have been noted in rhesus monkeys dying of simian acquired immunodeficiency syndrome (Henrickson *et al.*, 1983).

Nonhuman primates imported from countries where malaria is endemic are frequently found to be infected. Such infections rarely produce clinical disease and may go undetected unless the animal is immunosuppressed or splenectomized. Newly imported cynomolgus monkeys may have prevalence rates as high as 43% (Donovan *et al.*, 1983). *Plasmodium inui*, *P. cynomolgi*, and *P. knowlesi* are some of the species of malarial organisms observed. Diagnosis is based on the presence of malarial organisms in the erythrocytes as observed in thick blood smears.

The presence of enlarged spleens in newly imported animals is another indicator of infection. Clinical disease is manifested by anorexia, fever, weakness, and anemia. Some facilities routinely treat newly imported cynomolgus monkeys with a combination of chloroquine phosphate and primaquine phosphate.

African nonhuman primates are also infected with a malarial organism, *Hepatocystis kochi*. Transmission is by midges. The exoerythrocytic phase is present in the liver, and large cysts are produced that are visible on the surface of the liver.

Pneumocystis carinii is occasionally observed in association with respiratory disease in macaques (Chandler *et al.*, 1976). Animals that are severely debilitated or immunosuppressed may develop a pneumonia. Diagnosis is usually made at necropsy or on histopathologic examination of lung lesions.

2. Helminths

Nematode parasites are common in Old World monkeys, and in most instances there are few if any signs of disease. In young animals or animals that are housed under suboptimum conditions disease may occur. Whipworms, *Trichuris* ssp. are frequently seen in stool samples from baboons. Mebendazole (15 mg/kg given once daily for 2 days) is the drug of choice for treatment of this parasite.

Strongyloides fulleborni can cause serious outbreaks of disease in nonhuman primates. There are three pathways of infection: ingestion of infectious larvae, penetration of the skin by infectious larvae, or autoinfection where larvae developing in the hosts intestine penetrates the bowel and migrates through the body. Autoinfection can cause severe disease and death in young or debilitated animals. Thiabendazole is the drug of choice for eradication of this parasite.

The nodular worm, *Oesophagostomum* ssp. is the commonest nematode of Old World monkeys. Animals are infected by the ingestion of infective larvae, which penetrate the wall of the large intestine producing subserosal nodules. In animals not previously infected, the nodules rupture in 5 to 7 days and the parasite returns to the large intestine where it matures. In previously infected animals, the parasite remains in the nodule for long periods of time. At necropsy or during surgery in heavily infected animals, large numbers of firm, tan nodules 2–4 mm in diameter can be observed in the colon and mesentary. Thiabendazole usually eliminates mature worms found in the intestine, but larval forms will persist in the nodules.

Anatrichosoma cutaneum is found in the nasal mucosa of the rhesus monkey (Ulrich *et al.*, 1981). The life cycle of this parasite is unknown. Normally infection is inapparent and only a mild hyperplasia and parakeratosis of the nasal mucosa occurs. In recent years, however, several laboratories have experienced outbreaks of creeping eruption on the hands, feet, and faces of cynomolgus monkeys.

Diagnosis of nematode infections is through the examination of stool samples and the identification of ova or larvae. A study to determine the efficacy of thiabendazole (100 mg/kg, repeated in 2 weeks) was performed on 43 adult female monkeys. Three pretreatment and two posttreatment stool specimens from each were examined. Posttreatment stool specimens were obtained at least 1 month after the second treatment. Results of the study indicate that while thiabendazole at this dose is effective in reducing infestation, it may not completely eliminate it (Table A.IX) (Valerio *et al.*, 1969). The incidence of heavy infestations has been reduced by the improved husbandry encountered in domestic breeding programs.

3. Cestodes

Cestodes are uncommon in conditioned laboratory primates. Occasionally *Hymenolepis nana,* the dwarf tapeworm, is observed in Old World primates maintained outdoors. The larval form of *Taenia solium* or *Cysticercus cellulosae* may occur in a variety of wild caught nonhuman primates. The larval form of *Echnicoccus granulosus* has been reported from a wide range of wild caught Old World monkeys. The larvae develop into hydatid cysts, which are slowly growing structures found in the abdominal cavity, liver, lungs, and central nervous system.

Table A.IX

Effect of Thiabendazole on Reducing Parasitic Infestation in Adult *M. mulatta*[a]

Parasite species	Pretreatment		Posttreatment	
	No. animals positive	%	No. animals positive	%
Strongyloides sp.	31	74	2	5
Trichostrongylids	29	69	1	2
Oesophagostomum sp.	15	36	4	10
Spiruroids	6	14	0	0
Trichuris trichiura	4	10	1	2
Ternidens sp.	1	2	0	0
Bertiella studeri[b]	1	2	6	14
Anatrichosoma sp.[b]	4	10	10	24

[a]From Valerio *et al.* (1969).

[b]These parasites were not considered in evaluating the effectiveness of thiabendazole. *Bertiella studeri* is a cestode. *Anatrichosoma* is a parasite of the nasal passages; eggs are swallowed and passed through the gastrointestinal tract. There is no conclusive explanation for the increased incidence of these two parasites after treatment.

The clinical signs associated with these cysts are protean depending on location and rate of growth. Surgical removal remains the most effective treatment; however, the results are variable and may actually lead to spread of the infection if a cyst ruptures.

4. Mites

The lung mite, *Pneumonyssus simicola,* is a very common parasite of wild caught rhesus monkeys (Innes *et al.*, 1954; Knezevich and McNulty, 1970). Mild to moderate infestations usually do not produce clinical signs of disease. The mites are found in nodules near the surface of the lung and produce a bronchiolitis (Figs. A.7–A.9). In severe infestations rupture of nodules may occur producing pneumothorax and occasionally hemothorax. Infections have been observed in colony born animals reared with family groups containing imported animals. There is currently no treatment to eliminate this parasite.

C. Fungal

Dermatophytosis, although reported, is not commonly encountered in colonies of nonhuman primates. When observed, the lesions are characterized primarily by scaly patches of alopecia with broken hair shafts. Infections in Old World monkeys are most commonly due to *Microsporum canis* (Baker *et al.*, 1971).

Candida albicans is a common saprophyte of the skin, diges-

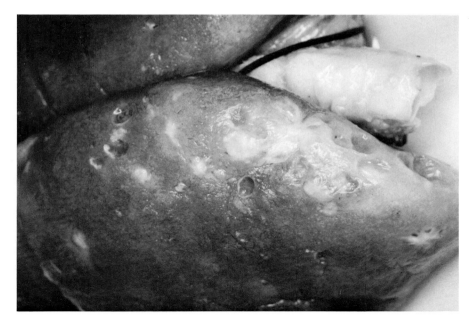

Fig. A.7. Lung mite lesions in the lung of a rhesus monkey observed at necropsy. The lesions are raised 0.5–2.0 cm nodules protruding from the pleural surface.

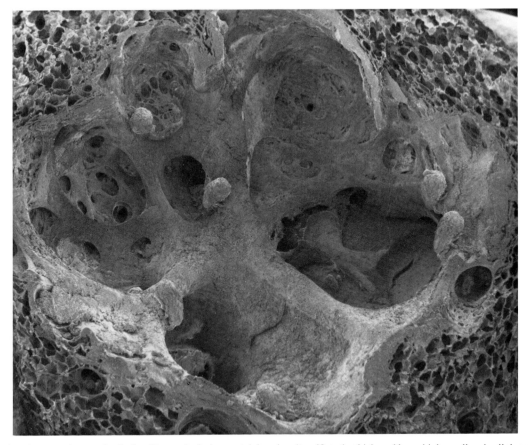

Fig. A.8. Scanning electron micrograph of a focal lung mite lesion containing six mites. Note the thickened bronchiolar wall and cellular accumulation in the lumen. Width of field, 7 mm. (Courtesy of M. S. Brummer, California Primate Research Center.)

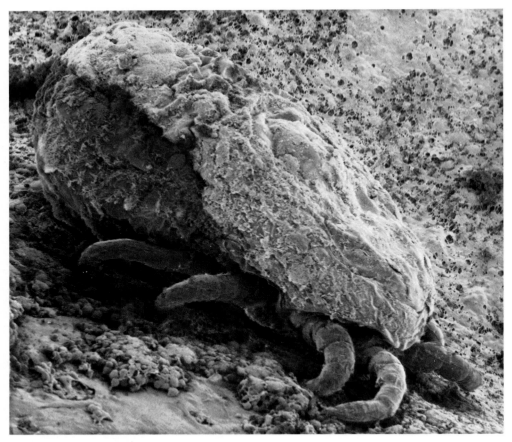

Fig. A.9.   Scanning electron micrograph of a lung mite, *Pneumonyssus simicola,* in the lung of a rhesus monkey. The roughened appearance of the integument is perhaps an artifact of preparation. (Courtesy of M. S. Brummer, California Primate Research Center.)

tive, and reproductive systems. Mild lesions when they occur are usually localized on the dorsum of the tongue and the mucous membrane of the oral cavity. These lesions occur as raised white plaques. In animals that have been on a prolonged course of antibiotic therapy or immunosuppressed the lesions will extend deeper into the underlying tissue and will extend down the esophagus and even involve the digestive tract.

The *Norcardia* sp. are aerobic, gram-positive, partially acid-fast organisms that are important in medical primatology because they can produce a granulomatous lesion that may be confused with tuberculosis. Infection in Old World monkeys is usually due to *Nocardia asteroides,* which is a common soil suprophyte. Isolation of *Nocardia* from lesions is essential for diagnosis; however the histopathologic characteristics of the organism (filamentous, gram-positive, partially acid-fast) should aid in the diagnosis.

Coccidioidomycosis is a systemic mycosis caused by a soil saprophyte, *Coccidioides immitis,* found in the semiarid regions of the southwest. Infection is by inhalation of spores.

Naturally occurring infections of Old World monkeys has occurred, and based on experimental studies the severity of infection depends on the number of arthrospores the animal is exposed to (Breznock *et al.,* 1975).

D.   Viral

In the 1960's and early 1970's, when large numbers of nonhuman primates were being imported for both the pet trade and biomedical research, outbreaks of viral disease were more commonly encountered. During that period, laboratories maintained a greater number of species, and mixing of species within a room frequently occurred. In addition, quarantine and conditioning periods were less controlled than at present. Viral disease causing little or no damage in their natural hosts can cause severe epidemics of diseases in aberrant hosts. The outbreak of Marburg disease, which occurred in Germany and Yugoslavia in 1967, is a good example of this phenomenon.

The outbreak, which was fatal to a large number of African green monkeys and seven humans, was probably contracted from a nonprimate host as no species of nonhuman primate has been found in the wild with antibodies to this virus. The source of the Marburg outbreak and the reservoir of the virus are still unknown (see Chapter 22 by Fox *et al.*).

Measles or rubeola is a highly infectious viral disease of man. It commonly causes epidemics of disease in a wide range of primate species. Large outbreaks with high mortality have been described in New World monkeys. The virus is equally infectious for Old World monkeys, but deaths due to measles is less common. Outbreaks are most commonly seen in newly imported animals. These animals exhibit an exanthematous rash, a mild conjunctivitis, and occasionally gingival ulcerations and diarrhea. Giant cell pneumonia can be a complication (Hall *et al.*, 1971). Measles is also thought to temporarily alter immune function and can interfere with the results of the tuberculin test by causing false negative reactions. Partial protection during an epidemic can be provided by the administration of 1.0 ml of human immune serum globulin (Barsky *et al.*, 1976).

Herpesvirus simiae or B virus is the macaque counterpart of *H. simplex* in man. In the rhesus monkey, infants are probably infected while still protected by maternal immunity. Infection is subclinical but results in a latent infection that persists for life. Occasional animals are observed with labial or lingual ulcerations (Fig. A.10), but these animals may have no other signs of disease. These oral lesions are a rich source of virus and pose a hazard to humans working with the animals, as B virus can produce a highly fatal encephalomyelitis in man. There have been over 20 deaths in humans attributed to this viral disease (see Chapter 22 by Fox *et al.*). Because of the zoonotic potential, animals with oral lesions suggestive of B virus should be destroyed. The rhesus monkey is the most commonly recognized host; however, other macaques may also be infected.

Simian hemorrhagic fever is a highly infectious, fatal, viral disease of macaque monkeys (Palmer *et al.*, 1968; London, 1977). There have been nine outbreaks of this disease in laboratories in the United States and in Europe. In one outbreak, over 500 rhesus and cynomolgus monkeys died during a 4-month period. The patas monkey has been incriminated as a reservoir host. In one study, approximately 50% of a wild population of patas monkeys had antibodies to simian hemorrhagic fever virus. In this species, animals can be viremic for long periods of time without clinical evidence of disease. Patas monkeys were thought to have been the source of an outbreak that occurred in 1972. Baboons and African green monkeys are susceptible to infection, but the disease is usually not fatal. These species may on occasion be the source of the virus. In contrast, species of the genus *Macaca* are susceptible to infection; death due to acute hemorrhagic disease usually ensues.

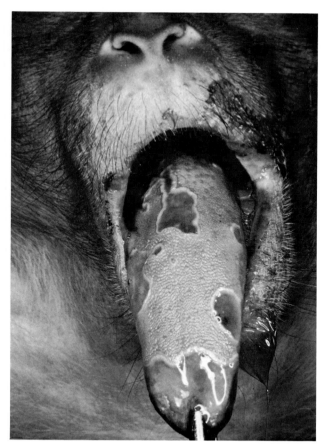

Fig. A.10. Photograph of the face of a bonnet macaque, *M. radiata*. Note the vesicular lesions caused by *Herpesvirus simiae*.

Clinical signs include fever, facial edema, cyanosis, epistaxis, dehydration, and melena. Multiple cutaneous hemorrhages are present. At necropsy, hemorrhages are present throughout the body, and a striking pathognomonic lesion is a hemorrhagic necrosis of the proximal duodenum; this lesion stops abruptly at the pyloris. In colonies maintaining both macaques and patas monkeys, it is recommended that strict isolation of species be maintained. The virus can easily be spread from asymptomatic carrier patas monkeys to macaques by items such as contaminated needles. Once infection occurs in macaques it can be easily spread by direct contact, aerosol, or other indirect means. Epizootics spread rapidly through a colony with devastating results. Detection of Old World monkeys that are asymptomatic carriers of simian hemorrhagic fever is dependent on the inoculation of serum from these animals into susceptible macaques. Peritoneal macrophage cultures from *M. mulatta* provide a promising *in vitro* test for the detection of the virus; however, the test is not always 100% reliable (Gravell *et al.*, 1980).

Yellow fever is a mosquito-borne viral infection occurring in endemic and epidemic forms in tropical American and Africa but not in Asia. It is an important disease of man. In Africa the sylvatic cycle of the yellow fever virus occurs between monkeys and a variety of *Aedes* spp. In African monkeys infection is rarely severe or fatal. Asian monkeys such as the rhesus macaque are susceptible to experimental infection.

Liverpool vervet monkey virus (LVMV), Medical Lake macaque herpesvirus (MLM) and delta herpesvirus are three antigenically related herpesviruses that produce diseases resembling varicella (chicken pox) in monkeys. Liverpool vervet monkey virus produces a highly fatal disease in vervet monkeys that is characterized by vesicular lesions and focal necrosis in parenchymal organs. Delta herpes virus produces a similar disease in patas monkeys. Medical Lake macaque herpesvirus produces a less fatal disease in macaques. The lesions are characterized primarily by a vesicular dermatitis that usually resolves spontaneously.

Hepatitis A is a viral disease of man that is worldwide in distribution. Transmission of this disease is via the fecal–oral route, and there is no persistent viremia as observed in hepatitis B. The chimpanzee is easily infected but rarely shows clinical disease. The patas monkey is reported to be susceptible to experimental infection (see Chapter 22 by Fox *et al.*).

Monkey pox is caused by an orthopox virus that is immunologically similar to smallpox. Outbreaks of disease have been reported from both laboratory and zoo primates and are thought to be due to transmission from recently imported primates. Monkey pox is a zoonosis, and sporadic human cases are reported from the tropical rain forest of west and central Africa (Mutombo *et al.*, 1983). The disease clinically resembles smallpox and is characterized by vesicular lesions. The outbreak of any poxlike disease in a group of primates is to be reported immediately to the Centers for Disease Control in Atlanta, Georgia. Outbreaks of the disease in Old World monkeys, such as the rhesus monkey and cynomolgus monkey, are characterized by a high morbidity and low mortality. The reservoir of this disease in the wild is unknown. Yaba is a viral disease that was reported from an outbreak in a colony of baboons and rhesus monkeys. It is caused by a poxvirus immunologically unrelated to smallpox but capable of infecting and causing lesions in man. In macaques and baboons the lesions are large (4–6 cm), subcutaneous nodules located on the head and limbs. These lesions resolve spontaneously.

Benign epidermal monkey pox (BEMP) or OrTeCa is a pox disease of macaques that was reported in 1967 from three separate primate colonies located in Oregon, Texas, and California. The disease, which is caused by a virus immunologically related to the Yaba virus, is characterized by multiple, raised plaques 0.5–1.5 cm in diameter on the head and limbs. These lesions are not vesicular and involve the epidermis in contrast to Yaba disease, which is located in the subcutis. The lesions resolve spontaneously in 4–6 weeks. The virus is transmissible to man, and lesions similar to those seen in monkeys are produced (see Chapter 22 by Fox *et al.*).

E. Miscellaneous

Neoplastic disease has been considered rare in nonhuman primates. This belief may not be true because of the patterns of primate procurement and use. In the past, only young healthy animals were imported, and these animals were usually killed after use. With the establishment of breeding colonies and recently with the establishment of aged colonies of macaques, the true incidence of neoplasms may become evident. Accurate reporting will also be aided by the establishment of medical record systems. There are a number of good reviews of the incidence of neoplasia in colonies of Old World monkeys (Palotay and McNulty, 1972; McClure, 1973; Seibold and Wolf, 1973).

Trauma, as a result of aggression, is a common clinical problem in group-housed Old World monkeys. Outbreaks of aggression can produce not only severe wounding but also death. Severely wounded animals must be treated vigorously. In rhesus monkeys, animals are frequently presented with severely bitten arms and faces. Intravenous fluid and electrolyte therapy must be rapidly instituted to combat the deleterious effects of muscle crushing that occurs with such injury. Cleaning, debridement, and antibiotic therapy are essential treatment regimens. Caution must be exercised when returning injured animals to the group environment as reinjury may occur.

Acute gastric dilitation or bloat is occasionally observed in colonies of Old World monkeys. The etiology of this condition remains obscure, but theories include improper husbandry practices and an infectious disease basis. In an evaluation of 24 cases of acute gastric dilitation, Bennett *et al.* (1980) were able to isolate *Clostridium perfringens* from the gastric contents of 21 of the 24 animals. Only 2 of 18 normal animals contained this organism. The problem is most commonly seen in animals in restraint chairs and almost never observed in animals in runs or corrals. Once detected, prompt action is required to save the animals. The abdominal distension should be relieved by passing a stomach tube and removing the gas and fluid. Animals recovering from bloat should be placed on a restricted diet and monitored carefully for 7–10 days.

Nonhuman primates ingest a great variety of foreign objects ranging from light bulbs to rocks. Animals born and reared indoors will frequently ingest large quantities of small rocks when placed into outdoor enclosures. Such behavior is less common in wild caught animals. Diarrhea and weight loss are observed in such cases. Abdominal radiographs provide a rapid diagnosis. Kept indoors such animals rapidly rid themselves of such foreign bodies. Trichobezoars present a more difficult diagnostic challenge. Gastric trichobezoars can cause an intermittent anorexia and subsequent weight loss. These foreign bodies are commonly encountered in the rhesus monkey and are the result of ingestion of hair during self-grooming. Radiographs taken after the administration of oral barium usually reveal the presence of trichobezoars. Frequently these objects will have to be removed surgically.

Polychlorinated biphenyls (PCB's) are chlorinated compounds used for a variety of industrial and commercial purposes. One use is as a component of concrete sealers. Old World monkeys are commonly kept in groups in cages having a concrete base. There are reports of toxicity related to housing macaques in cages where PCB's were present in the sealer applied to the concrete (Geistfeld *et al.*, 1982; Altman *et al.*, 1979). Clinical signs reported in affected animals include photophobia, palpebral edema, alopecia, gingivitis, periodontitis, diarrhea, and emaciation. The ducts of the Meibomian glands of the eyelids become distended causing drooping eyelids that produce a characteristic squinty appearance. A common lesion found at necropsy is a mucinous gastric hyperplasia involving both the mucosa and submucosa of the stomach. Concrete runs or cage floors containing PCB's should be either abandoned or resurfaced.

Hypothermia and hyperthermia are frequently encountered in animals kept outdoors in temperate latitudes. Shade must be provided during the summer, especially for animals just recently placed outdoors. Severe sunburn can also occur in such animals. Shelter and supplemental heat is required during the winter in areas where temperatures drop below freezing. Hypothermia and frostbite can readily occur in tropical animals such as the cynomolgus monkey. The rhesus monkey, coming from a country where climatic extremes are common, is more adaptable to living outdoors. When animals are presented with abnormally low or high body temperatures warm or cold water baths should be used to return temperatures to normal. Care must be taken when using heat lamps or hot water bottles because of the possibility of severe skin burns. Frequently animals with hypothermia have underlying disease conditions that predispose them to climatic stress. Such disease conditions, such as dehydration due to diarrhea, must be corrected before returning animals to their cages. In outdoor group cages newborn term infants are frequently found abandoned. They are considered stillbirths and are sent to pathology. These infants should be carefully examined and warmed because frequently they are abandoned at night shortly after birth and become hypothermic.

The list of disease problems not caused by infectious agents is much longer than the few examples provided. Many of these relate to the problems associated with keeping a highly intelligent, curious, and manipulative wild animal in captivity. Caging and management schemes must constantly be evaluated, monitored, and improved to meet the challenges posed by maintaining nonhuman primates in captivity.

REFERENCES

Altman, N. H., New, A. E., McConnell, E. E., and Ferrell, T. L. (1979). A spontaneous outbreak of polychlorinated biphenyl (PCB) toxicity in rhesus monkeys (*Macaca mulatta*): Clinical observations. *Lab. Anim. Sci.* **29**, 661–665.

Anderson, J. H., Rosenberg, D. P., and Henrickson, R. V. (1981). Postweaning hypoglycemia in infant rhesus monkeys. *32nd Annu. Sess., Am. Assoc. Lab. Anim. Sci.* Abstract No. 29.

Andreani, T., Le Charpentier, Y., Bronet, J. C., Lachance, J. R., Modigliani, R., Galian, A., Liance, M., Messing, B., and Vernisse, B. (1983). Acquired immunodeficiency with intestinal cryptosporidiosis: Possible transmission by Haitian whole blood. *Lancet* **1**, 1187–1190.

Baker, H. J., Bradford, L. G., and Montes, L. F. (1971). Dermatophytosis due to *Microsporum canis* in a rhesus monkey. *J. Am. Vet. Med. Assoc.* **159**, 1607–1611.

Barsky, D., Palmer, A. E., London, W. T., and Kerber, W. T. (1976). Use of immune serum globulin (human) to reduce mortality in newly imported rhesus monkeys (*Macaca mulatta*). *J. Med. Primatol.* **5**, 150–159.

Bennett, B. T., Cuasay, L., Welsh, T. J., Beluhan, F. Z., and Schofield, F. (1980). Acute gastric dilatation in monkeys: A microbiologic study of gastric contents, blood and feed. *Lab. Anim. Sci.* **30**, 241–244.

Benson, R. E., Fremming, B. D., and Young, R. J. (1955). A tuberculosis outbreak in a *Macaca mulatta* colony. *Am. Rev. Tuberc.* **72**, 204–209.

Bourne, G. H., ed. (1975). "The Rhesus Monkey," Vols. 1 and 2. Academic Press, New York.

Breznock, A. W., Henrickson, R. V., Silverman, S., and Schwartz, L. W. (1975). Coccidioidomycosis in a rhesus monkey. *J. Am. Vet. Med. Assoc.* **167**, 657–661.

Bronson, R. T., May, B. D., and Reubner, B. H. (1972). An outbreak of infection by *Yersinia pseudotuberculosis* in nonhuman primates. *Am. J. Pathol.* **69**, 289–303.

Bryant, J. L., Stills, H. F., Lentsch, R. H., and Middleton, C. C. (1983). *Campylobacter jejuni* isolated from patas monkeys with diarrhea. *Lab. Anim. Sci.* **33**, 303–305.

Caminiti, B. (1979a). "Colony Breeding of African Monkeys: A Bibliography." Primate Information Center, University of Washington, Seattle.

Caminiti, B. (1979b). "Colony Breeding of Macaques: A Bibliography." Primate Information Center, University of Washington, Seattle.

Caminiti, B. (1980a). "Biology of the Platelets of Nonhuman Primates: A Bibliography 1965–1980." Primate Information Center, University of Washington, Seattle.

Caminiti, B. (1980b). "Hepatic Enzymes of Nonhuman Primates: A Bibliography." Primate Information Center, University of Washington, Seattle.

Chandler, F. W., McClure, H. M., Campbell, W. G., Jr., and Watts, J. C. (1976). Pulmonary pneumocytosis in nonhuman primates. *Arch. Pathol. Lab. Med.* **100**, 163–167.

Cicmanec, J. L., Neva, F. A., McClure, H. M., and Loeb, W. F. (1974). Accidental infection of a laboratory-reared *Macaca mulatta* with *Trypanosoma cruzi*. *Lab. Anim. Sci.* **24**, 783–787.

Cockrell, B. V., Valerio, M. G., and Garner, F. M. (1974). Cryptosporidiosis in the intestines of rhesus monkeys (*Macaca mulatta*). *Lab. Anim. Sci.* **24**, 881–887.

DiGiacomo, R. F., and Missakian, E. A. (1972). Tetanus in a free-ranging colony of *Macaca mulatta:* A clinical and epizootiologic study. *Lab. Anim. Sci.* **22**, 378–383.

Donovan, J. C., Stokes, W. S., Montrey, R. D., and Rozmiarek, H. (1983). Hematologic characterization of naturally occuring malaria (*Plasmodium inui*) in cynomolgus monkeys (*Macaca fascicularis*). *Lab. Anim. Sci.* **33**, 86–89.

Flynn, R. J. (1973). "Parasites of Laboratory Animals." Iowa State Univ. Press, Ames.

Fox, J. G. (1982). Campylobacteriosis: A new disease in laboratory animals. *Lab. Anim. Sci.* **32**, 625–37.

Fox, J. G., and Wikse, S. E. (1971). Bacterial meningoencephalitis in rhesus monkeys. Clinical and pathological features. *Lab. Anim. Sci.* **21** 558–563.

Fox, J. G., Niemi, S. M., and Murphy, J. C. (1983). A comparison of two tuberculins in non-sensitized macaques *J. Med. Primatol.* **11**, 380–388.

Geistfeld, J. G., Bond, M. G., Bullick, B. C., and Varian, M. C. (1982). Mucinous gastric hyperplasia in a colony of rhesus monkeys (*Macaca mulatta*) induced by polychlorinated biphenyl (Aroclor 1254). *Lab. Anim. Sci.,* **32**, 83–86.

Good, R. C., and May, B. D. (1971). Respiratory pathogens in monkeys. *Infect. Immun.* **3**, 87–93.

Good, R. C., May, B. D., and Kawatomari, T. (1969). Enteric pathogens in monkeys. *J. Bacteriol.* **97**, 1048–1055.

Gravell, M., Palmer, A. E., Rodriguez, M., London, W. T., and Hamilton, R. S. (1980). Methods to detect asymptomatic carriers of simian hemorrhagic fever virus. *Lab. Anim. Sci.* **30**, 988–991.

Hack, C. A., and Gleiser, C. A. (1982). Hematologic and serum chemical reference values for adult and juvenile baboons (*Papio* sp.). *Lab. Anim. Sci.* **32**, 502–505.

Hall, W. C., Kovatch, R. M., Herman, P. H., and Fox, J. G. (1971). Pathology of measles in rhesus monkeys. *Vet. Pathol.* **8**, 307–319.

Harris, R. S., ed. (1970). "Feeding and Nutrition of Nonhuman Primates." Academic Press, New York.

Hayes, K. C., ed. (1979). "Primates in Nutritional Research." Academic Press, New York.

Henrickson, R. V., Maul, D. H., Osborn, K. G., Sever, J. L., Madden, D. L., Ellingsworth, L. R., Anderson, J. A., Lowenstine, L. J., and Gardner, M. B. (1983). Epidemic of acquired immunodeficiency in rhesus monkeys. *Lancet* **1**, 388–390.

Innes, J. R. M., Coltran, M. W., Yevich, P. P., and Smith, C. L. (1954). Pulmonary acariasis as an enzootic disease caused by *Pneumonyssus simicola* in imported monkeys. *Am. J. Pathol.* **30**, 813–835.

Keeling, M. E., Froehlich, R. E., and Ediger, R. D. (1969). An epizootic of tuberculosis in a rhesus monkey conditioning colony. *Lab. Anim. Care* **19**, 629–634.

Knezevich, A. L., and McNulty, W. P., Jr. (1970). Pulmonary acariasis (*Pneumonyssus simicola*) in colony-based *Macaca mulatta*. *Lab. Anim. Care* **20**, 693–696.

London, W. T. (1977). Epizootiology, transmission and approach to prevention of fatal simian hemorrhagic fever in rhesus monkeys. *Nature (London)* **268**, 344–345.

McClure, H. M. (1973). Tumors in nonhuman primates: Observations during a six year period in the Yerkes Primate Center Colony. *Am. J. Phys. Anthropol.* **38**, 425–430.

McClure, H. M. (1975). Hematologic, blood chemistry and cerebrospinal fluid data for the rhesus monkey. *In* "The Rhesus Monkey" (G. H. Bourne, ed.), Vol. 2, pp. 409–429. Academic Press, New York.

McClure, H. M., Weaver, R. E., and Kantonann, A. F. (1971). Pseudotuberculosis in nonhuman primates: Infection with organisms of the *Yersinia enterocolitica* group. *Lab. Anim. Sci.* **21**, 376–382.

Mahoney, C. J. (1975). Practical aspects of determining early pregnancy, stage of foetal development, and imminent parturition in the monkey (*Macaca fascicularis*). *In* "Breeding Simians for Developmental Biology" (F. T. Perkins and P. N. O'Donoghue, eds.), pp. 261–274. Laboratory Animals Ltd., London.

Melton, D. A. (1980). Baboon (*Papio ursinus*) capture using a blow-dart. *S. Afr. J. Wildl. Res.* **10**, 67–70.

Moore, A. E., and Henrickson, R. V. (1981). Early pregnancy diagnosis in *Macaca mulatta*. *32nd Annu. Sess., Am. Assoc. Lab. Anim. Sci.* Abstract No. 101.

Morrow, A. C., and Terry, M. W. (1968a). "Enzymes in Blood of Nonhuman Primates Tabulated from the Literature. II. Serum Glutamic Oxalacetic Transaminase and Glutamic Pyruvic Transaminase." Primate Information Center, University of Washington, Seattle.

Morrow, A. C., and Terry, M. W. (1968b). "Enzymes in the Blood of Nonhuman Primates Tabulated from the Literature. III. Oxidoreductases Including Dehydrogenases." Primate Information Center, University of Washington, Seattle.

Morrow, A. C., and Terry, M. W. (1970a). "Hematologic Values for Nonhuman Primates Tabulated from the Literature. I. Erythrocytes, Hemoglobin, Hematocrit and Related Indexes." Primate Information Center, University of Washington, Seattle.

Morrow, A. C., and Terry, M. W. (1970b). "Hematologic Values for Nonhuman Primates Tabulated from the Literature. II. Total Leukocytes and Differential Counts." Primate Information Center, University of Washington, Seattle.

Morrow, A. C., and Terry, M. W. (1972a). Liver function tests in blood of nonhuman primates tabulated from the literature. Primate Information Center, University of Washington, Seattle, Washington.

Morrow, A. C., and Terry, M. W. (1972b). "Urea Nitrogen, Uric Acid and Creatinine in the Blood of Nonhuman Primates: A Tabulation from the Literature." Primate Information Center, University of Washington, Seattle.

Morrow, A. C., and Terry, M. W. (1972c). "Cholesterol and other Plasma Lipids in Nonhuman Primates: A Tabulation from the Literature." Primate Information Center, University of Washington, Seattle.

Morton, W. R., Bronsdon, M., Mickelsen, G., Knitter, G., Rosenkranz, S., Kuller, L., and Sajuthi, D. (1983). Identification of *Campylobacter jejuni* in *Macaca fascicularis* imported from Indonesia. *Lab. Anim. Sci.* **33**, 187–188.

Muchmore, E. (1975). Health programs for people in close contact with laboratory primates. *In* "Biohazards and Zoonotic Problems of Primate Procurement, Quarantine, and Research," Publ. No. (NIH) 76-890, pp. 81–99. U.S. Department of Health, Education and Welfare, Washington, D.C.

Mulder, J. B. (1971). Shigellosis in nonhuman primates. *Lab. Anim. Care* **21**, 734–738.

Mutombo, M. W., Arita, I., and Jazek, Z. (1983). Human monkey pox transmitted by a chimpanzee in a tropical rain-forest area of Zaire. *Lancet* **1**, 735–737.

National Research Council (1978). "Nutrient Requirements of Laboratory Animals." Natl. Acad. Sci., Washington D.C.

Newberne, P. M., and Hayes, K. C. (1979). Semipurified diets for nonhuman primates. *In* "Primates in Nutritional Research" (K. C. Hayes, ed.), pp. 99–119. Academic Press, New York.

Nicolosi, R. J., and Hunt, R. D. (1979). Dietary allowances for nutrients in nonhuman primates. *In* "Primates in Nutritional Research" (K. C. Hayes, ed.), pp. 11–37. Academic Press, New York.

Nyland, T. (1982). School of Veterinary Medicine, University of California, Davis (personal communication).

Palmer, A. E., Allen, A. M., Tauraso, N. M., and Shelokov, A. (1968). Simian hemorrhagic fever. I. Clinical and epizootiologic aspects of an outbreak among quarantined monkeys. *Am. J. Trop. Med.* **17**, 404–412.

Palotay, J. L., and McNulty, N. P. (1972). Neoplasms in nonhuman primates. *Lab. Invest.* **26**, 487–488.

Perkins, F. T., and O'Donoghue, P. N., eds. (1975). "Breeding Simians for Developmental Biology," Lab. Anim. Handb. No. 6. Laboratory Animals Ltd., London.

Rawlins, R. G., and Kessler, M. J. (1982). A five year study of tetanus in the Cayo Santiago rhesus monkey colony: Behavioral description and epizootiology. *Am. J. Primatol.* **3**, 23–39.

Reynolds, J. A., and Hall, A. S. (1979). A rapid procedure for shortening canine teeth of nonhuman primates. *Lab. Anim. Sci.* **29**, 521–524.

Rosenberg, D. P., Lerche, N. W., and Henrickson, R. V. (1980). *Yersinia pseudotuberculosis* infection in a group of *Macaca fasicularis*. *J. Am. Vet.* **177**, 818–819,

Ruppenthal, Gerald C., ed. (1979). "Nursery Care of Nonhuman Primates," Plenum, New York.

Schmidt, L. H. (1969). Control of tuberculosis. *Primates Med.* **3**, 105–112.

Sedgwick, C., Parcher, J., and Durham, R. (1970). Atypical mycobacterial infection in the pigtailed macaque (*Macaca nemestrina*). *J. Am. Vet. Med. Assoc.* **157**, 724–725.

Seibold, H. R., and Wolf, R. H. (1973). Neoplasms and proliferative lesions in 1065 nonhuman primate necropsies. *Lab. Anim. Sci.* **23**, 533–539.

Sesline, D. H., Schwartz, L. W., Osburn, B. I., Thoen, C. O., Terrell, T., Holmberg, C., Anderson, J. H., and Henrickson, R. V. (1975). *Mycobacterium avium* infection in three rhesus monkeys. *J. Am. Vet. Med. Assoc.* **167**, 639–645.

Smith, E. K., Hunt, R. D., Garcia, F. G., Fraser, C. E. O., Merkal, R. S., and Karlson, A. G. (1973). Avian tuberculosis in monkeys. *Am. Rev. Respir. Dis.* **107**, 469–471.

Tomson, F. N., Schulte, J. M., and Bertsch, M. L. (1979). Root canal procedure for disarming nonhuman primates. *Lab. Anim. Sci.* **29**, 382–305.

Tribe, G. W., and Frank, A. (1980). Campylobacter in monkeys. *Vet. Rec.* **106**, 365–366.

Ulrich, P. C., Henrickson, R. V., and Karr, S. L. (1981). An epidemiological survey of wild caught and domestic born rhesus monkeys (*Macaca mulatta*) for *Anatrichosoma* (Nematoda: Trichinellida). *Lab. Anim. Sci.* **31**, 726–727.

U.S. Department of Health, Education and Welfare (1978). "Guide for the Care and Use of Laboratory Animals," Publ. No. (NIH) 78-23. USDHEW, Washington, D.C.

U.S. Department of Health, Education and Welfare (1979). "Laboratory Animal Records," Publ. No. (NIH) 80-2064. USDHEW, Washington D.C.

Valerio, D. A., Miller, R. L., Innes, J. R. M., Courtney, K. D., Palotta, A. J., and Gutt-Macher, R. M. (1969). "Macaca Mulatta: Management of a Laboratory Breeding Colony." Academic Press, New York.

van Bogaert, L., and Innes, J. R. M. (1962). Clinical observations of the nervous system of apes and monkeys. *In* "Comparative Neuropathology" (J. R. M. Innes and L. Z. Saunders, eds.), pp. 139–146. Academic Press, New York.

Weill, J. D., Wark, M. K., and Spertzel, R. O. (1971). Incidence of shigella in conditioned rhesus monkeys (*Macaca mulatta*). *Lab. Anim. Care* **21**, 434–437.

Wilson, J. G., Franklin, R., and Hardman, A. (1970). Breeding and pregnancy in rhesus monkeys used for teratological testing. *Teratology* **3**, 59–71.

PART B.   BIOLOGY AND DISEASES OF CEBIDAE

Noel D. M. Lehner

I.   RESEARCH CONSIDERATIONS

A.   Research Use

Nonhuman primates are desirable animals for use in medical research because of their likeness to humans, and the expectation that results from experimentation with them will have meaning for humans. Cebid monkeys (see Table B.I) have been little used as research subjects. The reasons for this are multiple: cebids are relatively dissimilar to humans compared with the Old World monkeys and apes; cebids are quite rare or, in any case, not available in large numbers; they are difficult to maintain under laboratory conditions; knowledge of the biology of many of the cebids is quite meager.

Among the cebids, only three genera have received much attention as research animals; *Saimiri, Cebus,* and *Aotus. Aotus* have been particularly useful in infectious disease research, being susceptible to human malarial parasites, trachoma ocular infections, and a number of herpesviruses (Barahona *et al.,* 1976).

Cebus monkeys, the hardiest and most intelligent of the Cebidae, have been used mainly in pharmacological and toxicological studies, although their overall use has been small (U.S. Department of Health, Education and Welfare, 1975). *Cebus* monkeys have been used in studies of viral oncogenesis. They are susceptible to genital infection with *Herpesvirus hominis* type 2 and may be useful in the study of the oncogenic potential of this agent (London *et al.,* 1974). *Cebus* monkeys also have been of interest for studies on purine metabolism. They, like several other genera of New World monkeys, have high blood and urine uric acid concentrations (Simkin, 1971). *Cebus* monkeys were among the first nonhuman primates used in studies of atherosclerosis and have subsequently been used in various nutritional studies on this disease (Clarkson *et al.,* 1976).

Saimiri are the New World monkeys that have been used in the greatest numbers. Their extensive use relates more to phylogenetic considerations of being primates and to pragmatic considerations, such as low cost and maintenance in past years, than to specific attributes. They continue to be used widely in pharmacological and toxicological studies and in research on neuroanatomy and neurophysiology. Squirrel monkeys also have been of considerable interest in studies of behavior. They are the New World monkey most extensively used in studies of atherosclerosis and cholesterol metabolism (Clarkson *et al.,* 1976).

Table B.I

Cebidae[a]

Genus	Species	Common name
Alouatta	villosa	Mantled howler monkey
	seniculus	Red howler monkey
	belzebul	Red-handed howler monkey
	fusca	Brown howler monkey
	caraya	Black howler monkey
Aotus	trivirgatus	Douroucouli, night monkey, owl monkey
Ateles	paniscus	Black spider monkey
	belzebuth	Long-haired spider monkey
	fusciceps	Brown-headed spider monkey
	geoffroyi	Black-handed spider monkey
Brachyteles	arachnoides	Woolly spider monkey
Cacajao	calvus	Bald uakari
	melanocephalus	Black-headed uakari
	rubicundus	Red uakari
Callicebus	torquatus	Widow titi
	moloch	Dusky titi
	personatus	Masked titi
Cebus	capucinus	White-throated capuchin
	albifrons	White-fronted capuchin
	nigrivitatus	Weeper capuchin
	apella	Black-capped capuchin
Chiropotes	satanas	Black-bearded saki
	albinasus	White-nosed bearded saki
Lagothrix	lagothricha	Humboldt's woolly monkey
	flavicauda	Hendee's woolly monkey
Pithecia	pithecia	Pale-headed saki
	monachus	Monk saki
Saimiri	sciureus	Common squirrel monkey
	orstedii	Red-backed squirrel monkey

[a]From Napier and Napier (1967).

B. Laboratory Management

General considerations of laboratory management of nonhuman primates may be found in such works as the Institute of Laboratory Animal Resources publication "Laboratory Animal Management: Nonhuman Primates" (ILAR, 1980) and "Guide for the Care and Use of Laboratory Animals" U.S. Department of Health, Education and Welfare (1978). Specific requirements for laboratory maintenance are included in the generic profiles discussed in Section II.

II. GENERIC PROFILES AND BIOLOGICAL CHARACTERISTICS

A. Alouatta (Howler Monkeys; 5 Species, 21 Subspecies)

Howler monkeys are among the largest of the New World monkeys in both size and weight. Adult males may weigh as much as 10 kg and have head–body (crown–rump) lengths of up to 72 cm. Sexual dimorphism is apparent with females at-

taining about 80% of the size and weight of the males. They are widely distributed, having the greatest range of all the New World monkeys, from Mexico to the southern borders of Bolivia, in Paraguay and northern Argentina.

1. Unique Biological Characteristics

One of the most notable features of howler monkeys is specialization of the hyoid apparatus of the larynx. The hyoid bone is expanded into an egg-shaped sacculation forming a resonance chamber. As their common name indicates, howler monkeys can produce loud and long noises that sound like roars or barks. Such vocalizations are emitted in defining their territory. Howler monkeys also have prehensile tails. Their hands have the first two fingers opposable to the other three, imparting an ability for firm grasping and clasping of branches.

2. Normal Values

Hematologic values for howler monkeys are listed in Table B.II. Limited other normative data collected in a rather large field study have been reported. (Malinow, 1968).

3. Nutrition

Howler monkeys are primarily vegetarian and the only New World monkeys known to feed extensively on leaves. They also feed on soft fruits and other plant parts.

4. Reproduction and Behavior

Under natural conditions, howler monkeys live in multiple male–female groups of up to 50 animals. Females become sexually mature at 3–3.5 years, and appear to have regular estrous

Table B.II

Hematologic Values of Alouatta villosa[a]

Determination	Immature females N	Immature females $\bar{X} \pm SD$	Adult females N	Adult females $\bar{X} \pm SD$
Erythrocytes (10^6/mm^3)	33	3.9 ± 0.7	5	3.8 ± 0.7
Leukocytes (10^3/mm^3)	33	11.9 ± 5.5	5	13.2 ± 5.0
Packed cell volume (PCV) (%)	25	36.8 ± 6.7	6	37.0 ± 5.8
Hemoglobin (gm/dl)	25	11.7 ± 1.8	5	11.2 ± 1.7
Neutrophils (%)	7	63.0 ± 18.9	4	60.3 ± 6.4
Lymphocytes (%)	7	34.5 ± 19.3	4	37.3 ± 6.1
Monocytes (%)	7	1.8 ± 1.3	4	2.3 ± 1.0
Eosinophils (%)	7	0.1 ± 0.4	4	0.3 ± 0.5
Basophils (%)	7	0.4 ± 0.5	4	0
MCV (μm^3)	25	97.7 ± 12.7	5	97.6 ± 5.3
MCH (pg)	25	29.7 ± 1.4	5	29.8 ± 0.9
MCHC (%)	21	31.4 ± 5.5	5	30.5 ± 1.7

[a]Adapted from Porter (1971).

cycles of about 16 days and gestation periods of 186 days (Glander, 1980). Such groups are territorial. Territories are defended by howling accompanied by shaking and breaking of branches. Howler monkeys do not participate much in allogrooming and do not spend much time in self-grooming either.

5. Laboratory Management

Howler monkeys do not appear to be very adaptive, and most have not survived very long in captivity. This may in part relate to inadequacies of diet. One laboratory reported successful maintenance after providing their natural foods, leaves and fruits of the trumpet tree, *Cercropia mexicona,* the almond tree, *Terminalia catappa,* and the mango tree, *Mangifera indica.* Prior to providing this natural diet, the monkeys died shortly after arrival in spite of providing a varied diet and medicinal and management efforts (Porter, 1971). They essentially have not been used in research.

B. *Aotus* (Night Monkeys, Owl Monkeys, Douroucoulis; 1 Species, 9 Subspecies)

Aotus are small New World monkeys, weighing 800–1250 gm. They are distributed widely from northern Argentina to western Panama. Considerable differences in morphology are seen among animals from different areas and even among individuals of the same area. Most authors have recognized a single species with numerous subspecies. Pelage, especially the ventral coat color, varies in animals from different regions, ranging from bright orange to buff-white. It remains to be determined if such variation represents different species. They are found in a wide variety of forest habitats from sea level up to 3000 meters.

1. Unique Biologic Characteristics

Aotus are unique in being the only nocturnal simian primate. A dominant feature is their very large eyes. Although early studies suggested that owl monkeys had an all rod retina, it now appears that their retina is complex, having about 50 times more rods than cones. They also appear to have some color vision. These visual characteristics have made owl monkeys intriguing animals for vision research (Jacobs *et al.,* 1979).

Owl monkeys, like some other New World primates, form permanent pair bonds, and males are active participants in rearing of the young.

2. Normal Values

Hematologic values have been reported for owl monkeys (Porter, 1969; Wellde *et al.,* 1971), and combined data from

male and female monkeys of feral origin are listed in Table B.III. Data on serum chemical values are listed in Table B.IV.

Cytogenetic studies have revealed considerable chromosomal polymorphism among *Aotus.* Seven karyotypes, which include differences in chromosomal number as well as morphologic variation (Ma *et al.,* 1976), are listed in Table B.V.

Table B.III

Hematologic Values for *Aotus* Monkeys[a]

Parameter	No. of animals	Mean	SD
Erythrocytes ($10^6/mm^3$)	157	5.2	0.8
Packed cell volume (%)	157	42.0	5.4
Hemoglobin (gm/dl)	60	14.3	1.1
Reticulocytes (%)	78	2.4	1.7
MCV (μm^3)	157	82.4	11.9
MCH (pg)	60	26.9	3.1
MCHC (%)	60	34.1	3.1
Platelets ($10^3/mm^3$)	63	397.1	109.4
Leukocytes ($10^3/mm^3$)	157	12.7	4.7
Differential leukocyte count			
Neutrophils (%)	157	55.4	7.6
Eosinophils (%)	157	9.5	9.2
Mononuclear-lymphoid cells (%)	157	35.5	18.3
Basophils (%)	157	<0.1	—

[a]Adapted from Wellde *et al.* (1971).

Table B.IV

Serum Chemical Values of *Aotus* Monkeys[a]

Parameter	Mean	SD
Na^+ (mEq/liter)	150	
K^+ (mEq/liter)	3.7	
Cl^- (mEq/liter)	102	
CO_2 (mEq/liter)	23	
Ca^{2+} (mg/dl)	9.8	
Total bilirubin (mg/dl)	0.27	
Glucose (mg/dl)	113	40
Uric acid (mg/dl)	0.5	0.4
Urea nitrogen (mg/dl)	14	3
SGOT (Reitman-Frankel units)	153	71
SGPT (Reitman-Frankel units)	47	23
Alkaline phosphatase (King-Armstrong units)	21	6.2
Total protein (gm/dl)	7.0	1.2
Albumin (gm/dl)	2.7	0.7
α_1-Globulin (gm/dl)	0.3	0.1
α_2-Globulin (gm/dl)	1.3	0.3
β-Globulin (gm/dl)	1.0	0.2
α-Globulin (gm/dl)	1.8	0.5

[a]Adapted from Schnell *et al.* (1969) and Wellde *et al.* (1971).

Table B.V

Karyologic Characteristics of *Aotus*[a]

Karyotype	Diploid *N*	Chromosome morphology					Probable origin
		m + sm[b]	st + a[b]	X	Y	Phenotype[c]	
I	54	20	32	sm	m	A	Brazil
II	54	10	42	sm	m	B	Colombia
III	53	11	40	sm	m	B	Colombia
IV	52	12	38	sm	m	B	Colombia
V	46	16	28	sm	a	B	Colmbia
VI female	50	12	36	sm/sm		D	Bolivia
VI male	49	13	35	sm		D	Bolivia
VII	52	16	34	sm		C	Peru

[a]Adapted from Ma *et al.* (1976).

[b]m, metacentric chromosome; sm, submetacentric chromosome; a, acrocentric chromosome; st, subtelocentric chromosome.

[c]Phenotype: A, abdomen and chest hair bright orange, back hair brown; B, abdomen and chest hair yellow, back hair brown; C, abdomen and chest hair bright orange, back hair gray; D, abdomen and chest hair yellow to pale orange, back hair gray.

3. Nutrition

Owl monkeys are thought to be principally frugivorous, but their diet in the wild state also includes foliage, insects, bird eggs, as well as small birds and animals. Their nutritional requirements are not well defined, and they are usually provided a varied diet in captivity. Some zoo monkeys have been adequately maintained for long periods on a high protein meat-based diet consisting principally of ground horse meat and mink chow with added hardboiled eggs, fruits and vegetables, and supplemental vitamins and minerals (Meritt, 1970). In this scheme commercial monkey chow was provided as a supplement.

In research colonies, owl monkeys have been adapted to commercial high-protein monkey chow as the dietary staple. Fresh fruits, such as bananas, oranges, grapes, and apples, along with multiple vitamin formulations are usually provided as supplements. Neonatal mice, eggs, meal worms, milk, and baby cereals also have been used as supplemental foods. Initially, natural foods are fed because they are more readily accepted by the monkeys. Gradually the amount of monkey chow is increased and the amounts of the other foods decreased. *Aotus* are not very good at manipulation with their hands and may drop much of pelleted food such as monkey chow biscuits. If fallen food is out of their reach, such monkeys may starve even though they have been given ample amounts of food.

4. Reproduction

Aotus are monogamous and form strong pair bonds. Attempts to house them in adult groups other than in breeding pairs for long periods invariably result in conflict and injury.

Reproduction in captivity has been very unsuccessful, resulting in only occasional pregnancies and few live births. Recognition of karyotypic variation and pairing of animals of like karyotype has improved the reproductive performance somewhat. Cicmanec and Campbell (1977) reported production of 35 live infants in 3 years from an average daily population of 30 breeding pairs. Much more must be learned about the biology of *Aotus* if breeding results are to be improved in captivity.

Births in captivity usually occur during the light portion of the day with labor lasting about 30 min. The placenta is usually eaten by the female monkey. Newborn infants must crawl to and cling to the mother or be abandoned. Infants are carried exclusively by their mother during the first 3–4 days. After this time the father plays an increasing role in infant care. By 2 weeks of age the father carries the offspring most of the time, sharing this role with the infant's siblings if they are present. The mother takes the infant only during periods of nursing. At about 1 month of age, infants may explore a little on their own. By 4 months of age infants spend most of the time on their own, but may nurse up to 9 months of age. Solid food is first taken from the father's mouth at about 1 month of age, and males may relinquish food from their mouth for their offspring up to 2 years of age (Wright, 1981).

Reproductive cycles of owl monkeys do not appear to be seasonal, and births occur throughout the year. There is some evidence, based on cornification of exfoliated vaginal epithelium, that *Aotus* have estrous cycles of 10–15 days. No evidence of menstruation has been observed (Cicmanec and Campbell, 1977). The gestation period, based on palpation, has been estimated at 120–140 days, and based on detection of chorionic gonadotropin at 126 days (Hall and Hodgen, 1979). Birth weights of live-born owl monkeys have been reported to be 92 ± 21 gm (mean ± SD) (Elliott *et al.*, 1976). Infants

Table B.VI

Food Consumption

Age	Human infant formula (ml)	Solid food
1 day	<10	
3 days	15	
1 week	20	
3 weeks	50	Solid foods offered, fruits and vegetables
4 weeks		Self-feeding
8 weeks	90	Solid food offered 4 times daily
3 months		Solid food fed twice daily
5 months		Stop bottle feeding

weighing less than 80 gm at birth usually do not survive (Meritt and Meritt, 1978).

Owl monkey milk has been found to contain approximately 88% water, 2% protein, 2% fat, 8% lactose, 0.065% calcium, 0.003% magnesium, 0.015% sodium, and 0.066% potassium (Cicmanec and Campbell, 1977). Hand-rearing of infants has been successful using human infant formula. Newborn infants are placed in incubators maintained at 80°–85°F with relative humidity at 30–60%. After 4 weeks, the temperature is reduced to 70°–80°F. This is a temperature range recommended for adults. Infants are initially fed six times daily at about 2.5 hr intervals from 7 AM to 11 PM. Food consumption approximates that shown in Table B.VI. Hand-reared owl monkeys gain about 20 gm weekly and by 20 weeks of age weigh about 450 gm.

5. Behavior

Aotus form strong pair bonds and are normally found in family units. They spend most of their active time feeding, then sleep during the daylight hours. In the wild, sleeping sites consist of holes in hollow trees or dense tangles of vines. Under natural conditions they appear to have permanent sleeping areas and limited territories. Agonistic behavior between paired adults is rare. Dominance by either adult is not obvious. Self-grooming is common, but allogrooming is rare. Owl monkeys have a variety of calls and vocalizations, most of which are low pitched. Olfactory signals appear important in communication. *Aotus* scent mark by rubbing a brown, oily substance from glands located on the ventral surface of the tail base. Marking behavior is exhibited in hostile situations and may be a signal related to sexual function or territorial demarcation.

6. Laboratory Management

Numerous problems have been encountered in the maintenance of owl monkeys. Many investigators have reported high morbidity and mortality in new imports. Provision of a tranquil setting appears to improve their adaptation to captivity. Their natural habitat is the tropical jungle where temperature and humidity are quite stable and relatively high. Adaptation of the animal room environment to simulate the natural environment appears to enhance their well-being and maintenance. The animal room should be kept at 70°–80°F with relative humidity at 30–60%. Diminished illumination is practiced in some colonies by placing red plastic material over light fixtures. This subdued illumination is thought to be less stressful on the monkeys; however, owl monkeys have been maintained on a 12-hr light and 12-hr dark cycle without using red filters. Owl monkeys are naturally reclusive, and attempts should be made to maintain a tranquil environment by restriction of unnecessary noise and activity. Daily routines should be established and followed. Strangers should be excluded. Any handling or restraint is traumatic not only to the animal involved, but to its group and is also upsetting to others in the room. Unnecessary handling should be avoided by use of tunnels for cage transfer. When necessary, capture in nest boxes is less disruptive than handling in the animal room.

Agonistic behaviors have been observed between family groups in natural habitats. These behaviors consist of grunt vocalizations, stifflegged jumping, and rub marking. In these situations the approaching groups usually retreat, and physical contact and fighting are avoided. The frequent rub marking behaviors observed in captive *Aotus* may be an indication of hostility between groups, and visual screening between groups may be helpful in reducing tension.

Owl monkeys may be housed singly, in male–female pairs, or in family groups. Cages should contain one or more perches and be sufficiently high so that the monkeys do not touch the cage floor when sitting on the perch. In general, the more spacious the cages, the better. Cages that measure 18 × 24 × 32 in. have been extensively used. Nest boxes provide secluded sleeping places and if sealable also provide for capture without causing undue disturbance in the colony. Nest boxes commonly used measure about 12 × 7 × 7 in.

C. *Ateles* (Spider Monkeys; 4 Species, 16 Subspecies)

Spider monkeys are long, slender-limbed monkeys with long prehensile tails, squat bodies, and protuberant abdomens. Adults may weigh up to 8 kg but are lankier and much more agile than other New World monkeys of comparable weight. Their thumbs are vestigial or absent, and their hands function as suspending hooks. They move very gracefully in trees, using brachiation-type locomotion. They walk quadripedally or bipedally on the ground. Spider monkeys are widely distributed, ranging throughout most of the lowland tropical forests of Central and South America. Four species and numerous

Table B.VII

Hematologic Values for Immature Female *Ateles geoffroyi*[a]

	N	Mean ± SD
Erythrocytes (10^6/mm^3)	6	4.2 ± 0.6
Packed cell volume (%)	6	40.5 ± 4.9
Hemoglobin (gm/dl)	6	12.4 ± 1.7
MCV (μm^3)	6	97 ± 2.4
MCH (pg)	6	30 ± 0.4
MCHC (%)	6	31 ± 0.8
Leukocytes (10^3/mm^3)	4	13 ± 4.5
Neutrophils (%)	4	39 ± 8
Lymphocytes (%)	4	59 ± 8
Monocytes (%)	4	1 ± 1
Eosinophils (%)	4	1 ± 1

[a]Adapted from Porter (1971).

subspecies are listed by most authors, but there is some feeling that such separation, based mostly on pelage, is artificial.

Spider monkeys are principally frugivorous but also consume appreciable amounts of leafy material. They have been maintained in captivity on varied diets, including high protein commercial monkey chows containing vitamin D$_3$.

Spider monkeys do not appear to be seasonal breeders. Menstrual cycles are about 24–27 days in length. The gestation period in these animals is similar to other large New World monkeys, about 7 months in duration. Infants are carried by their mothers for about 10 months and continue to nurse until over 1 year old.

Spider monkeys under natural conditions are found in groups of varied size, from a few animals to very large troops. These groups contain multiple adult males and females and young of different ages. Adult females usually outnumber adult males.

Spider monkeys are adaptable and have a gentle temperament. They should be provided with spacious pens to allow normal activity. Such cages or pens should allow the animals to hang by their tails or arms and facilitate brachiation-type locomotion. Spider monkeys have been successfully maintained in mixed sex groups in outdoor colonies.

Normal hematologic values for spider monkeys are listed in Table B.VII.

D. *Brachyteles* (Woolly Spider Monkey; 1 Species)

Woolly spider monkeys are essentially similar to the common spider monkeys except for their dense hair coat. They have the same external features including a vestigial or absent thumb. Their diploid chromosome number is 34, the same as *Ateles*. They are quite rare and are found only in forests in southeastern Brazil.

E. *Cacajao* (Uakaris; 3 Species; 2 Subspecies)

Uakaris are medium-sized New World monkeys weighing approximately 4 kg. They are the most peculiar looking of the New World monkeys, having bare faces and bald heads, and otherwise having long silky hair on the remainder of their bodies. They differ from other cebids in having very short tails, which measure only about one-third that of their body length. Their head–body length measures approximately 35–50 cm. Two or three species are listed for the genus depending on whether the red and white uakari are considered separate species. Early studies listed them separately, but they are now considered only subspecifically distinct. Pelage and facial color are important distinguishing features, and the black uakari is considered to be a distinct species. The black uakari has black pelage and a black face, while the others have pink to brilliant scarlet faces. *Cacajao* are distributed in the Amazon basin, Brazil, eastern Colombia, southern Venezuela, and into Peru.

Very little is known about the biology of uakaris. Limited observations indicate that these animals live in fairly sizable groups in the wild, containing up to 30 or more individuals. Tolerance among adults in a seminatural environment suggests that uakaris live in multi-male–multi-female groups. Differing somewhat from other cebids, uakaris exhibit a high rate of allogrooming, principally by adult females that groom their dependent offspring, other infants, juveniles, and adult males.

Cacajao essentially have not been used in medical research.

F. *Callicebus* (Titis; 3 Species, 14 Subspecies)

Callicebus are small- to medium-sized monkeys that weigh approximately 1 kg and have crown–rump lengths of 30–40 cm. The genus contains three species that have different hair color but are otherwise similar. Sexual dimorphism is not readily apparent. Titis are quite widely distributed in South America. They are highly arboreal and are found in the forested areas of Colombia, Venezuela, Ecuador, Peru, Brazil, Bolivia, and Paraguay. They appear to prefer sites along waterways, that are wet or poorly drained.

1. Unique Biologic Characteristics

Unique characteristics of *Callicebus* relate to their behavior. The social group of titis consists of a family unit of an adult male, adult female, and one or two offspring from succeeding pregnancies. Titis appear to be highly territorial, have relatively small home ranges, and defend their territories with vocal activity.

2. Nutrition

Callicebus basically are frugivorous but also consume some leaves, insects, bird eggs, and small birds.

3. Reproduction and Behavior

Callicebus form pair bonds. Male titis participate in the rearing of the young and have been noted to carry the infant as early as 2 days following birth. During early infancy the infants are carried mostly by the male, generally on his back. Infants go to the female only to nurse, and then are returned again to the male. Infants begin to explore on their own at 2–3 months of age. *Callicebus* males appear to care for their offspring even after the juveniles become independent. Relationships between paired adults are usually amiable with very rare agonistic behaviors. Grooming is an important and frequent activity. Grooming occurs particularly between the adult males and females and between males and the youngest animals of the family group. Under natural conditions, births have been recorded for *Callicebus* between the months of November and March. In Louisiana, births in a captive colony occurred in May (Lorenz and Mason, 1971). No information is available on the reproductive cycles in *Callicebus* females.

4. Laboratory Management

Titis appear to be very difficult animals to establish in captivity. Seventy percent of imported animals have been reported to die within the first 3 months of captivity, most dying within the first 3 weeks (Lorenz and Mason, 1971). Although a large number and variety of illnesses have been detected in titis that have succumbed following their importation, a consistent or primary cause for their deaths has not been identified. Death may be stress related. Typically, such monkeys sit on a perch in a hunched-over position. They become increasingly anorectic even though offered a variety of foods that are readily accepted by well-established and stable animals. The depressed animals show little interest in their surroundings and become progressively weaker. Eventually they succumb to pneumonia, gastrointestinal illnesses, or a ketosis-like syndrome. If titis can be maintained during the difficult adjustment period of initial captivity they appear to be quite hardy afterward.

Callicebus are usually housed in adult, mixed sex pairs. Attempts to house them in other kinds of groups results in stress, although not necessarily in overt aggression. These animals also may be housed individually for rather extended periods with little difficulty. Establishment of colony routines and restriction of unnecessary noise, activity, and personnel is important in providing a tranquil environment. Since these animals are highly territorial, consideration might be given to visual screening to reduce stress levels. Titis in captivity may be picky eaters and have been fed commercial monkey chow biscuits along with fresh fruits, nuts, eggs, and dairy products. Daily rations have included hard-boiled eggs, cottage cheese, apples, oranges, grapes, bananas, peanuts, celery leaves, monkey chow, sunflower seeds, meal worms, and even pound cake. This cafeteria style of feeding relates more to attempts to stimulate appetite and food consumption than to nutritional requirements. Among the foods offered, grapes and bananas are usually preferred. Insects also appear to be eagerly eaten. Some zoos have maintained *Callicebus* for long periods on a high protein meat-based diet (Meritt, 1970).

Titis have been housed in wire cages that measure about 3 × 1.5 × 3 ft and in large indoor or outdoor pens. Caging should be sufficiently large to accommodate their normal leaping activity. Perches should be situated with sufficient vertical space to permit normal sitting without having their tail touch the cage floor. Although they do not sleep in nesting holes under natural conditions, titis in captivity will use open-faced nest boxes for sitting and sleeping. Such devices may be used for capture of animals without undue stress. Ambient temperatures should be maintained between 70 and 85°F with 30–60% relative humidity.

Titis have been maintained in outdoor colonies in Louisiana with average daily temperatures in the winter in the mid-50°F range. Night temperatures often dropped below freezing. Under these conditions, infared heat lamps were provided as a source of heat (Lorenz and Mason, 1971).

G. *Cebus* (Capuchins; 4 Species, 33 Subspecies)

Cebus monkeys are the hardiest and most adaptive of all of the Cebidae. They are a diverse group, with most authors including four species and numerous subspecies. They are medium-sized monkeys and weigh up to 4 kg. *Cebus capucinus, C. albifrons,* and *C. nigrivitatus* are slender, long-limbed monkeys, while *C. apella* are much stockier and more heavily muscled. They are widely distributed in Central and South America to northern Argentina. They have opposable thumbs and have greater manual dexterity than the other New World primates. They also have prehensile tails that they use to a greater extent and in more varied ways than other New World monkeys.

1. Unique Biological Characteristics

Capuchins appear to be the most intelligent of the New World primates. They, like humans, have relatively high blood and urine concentrations of uric acid.

Table B.VIII

Hematologic Values and Developmental Changes in *Cebus* Monkeys[a]

	Age					
	0	1 month	2 months	4 months	12 months	Adult female
PCV (%)	55 ± 4[b]	39 ± 6	43 ± 3	46 ± 3	47 ± 3	49 ± 4
Hemoglobin (gm/dl)	19 ± 1	13 ± 1	14 ± 1	15 ± 1	16 ± 1	17 ± 1
Erythrocytes (10^6/mm^3)	5 ± 0.5	4 ± 0.5	5 ± 0.4	6.0 ± 0.4	6.0 ± 0.4	6 ± 0.6
MCV (μm^3)	105 ± 4	97 ± 11	89 ± 7	83 ± 5	79 ± 0.5	79 ± 4
MCH (pg)	37 ± 3	31 ± 3	29 ± 2	27 ± 2	26 ± 2	27 ± 2
MCHC (%)	36 ± 3	32 ± 4	32 ± 2	32 ± 2	33 ± 2	34 ± 2
Leukocytes (10^3/mm^3)	8 ± 2	7 ± 3	8 ± 3	10 ± 4	8 ± 3	8 ± 3
Neutrophils (%)	62	30	26	18	30	
Immature neutrophils (%)	<1	<1	<1	<1	<1	
Lymphocytes (%)	33	60	70	76	62	
Monocytes (%)	4	6	3	4	4	
Eosinophils (%)	<1	4	3	2	3	
Basophils (%)	<1	<1	<1	<1		
Plasma cholesterol (mg/dl)	133 ± 14	164 ± 32	179 ± 37	189 ± 41	165 ± 26	212 ± 42
Total plasma protein (gm/dl)	6.2 ± 0.1	5.4 ± 0.3	5.5 ± 0.6	5.7 ± 0.5	6.7 ± 0.8	8.1 ± 0.6
Albumin (gm/dl)	3.2 ± 0.5	2.9 ± 0.5	3.5 ± 0.3	3.5 ± 0.4	3.7 ± 0.5	4.0 ± 0.5
Globulin (gm/dl)	3.1 ± 0.4	2.5 ± 0.5	2.2 ± 0.4	2.3 ± 0.1	3.1 ± 0.7	3.5 ± 0.5

[a]Adapted from Samonds *et al.* (1974).
[b]Mean ± SD.

Table B.IX

Hematologic Characteristics of Adult Male *Cebus capucinus*[a]

Determination	N	Mean ± SD
Erythrocytes (10^6/mm^3)	14	4.92 ± 0.72
PCV (%)	14	47.0 ± 6.7
Hemoglobin (gm/dl)	14	14.4 ± 1.8
MCV (μm^3)	14	96.7 ± 1.8
MCH (pg)	14	29.7 ± 1.7
MCHC (%)	14	30.9 ± 2.2
Leukocytes (10^3/mm^3)	14	16.0 ± 8.4
Neutrophils (%)	11	55.6 ± 6.6
Lymphocytes (%)	11	40.9 ± 6.7
Monocytes (%)	11	1.8 ± 1.1
Eosinophils (%)	11	1.6 ± 2.2

[a]Adapted from Porter (1971).

2. Normal Values

Developmental changes in hematologic values for *Cebus* monkeys from newborn to adult are listed in Table B.VIII. Values for adult monkeys are listed in Table B.IX.

3. Nutrition

Cebus monkeys are omnivorous. Their diet in their natural state consists mainly of fruits, but other plant parts such as flowers are also eaten. Foods of animal origin include insects, bird eggs, young birds, nesting squirrels, and small lizards. *Cebus* have been utilized in nutritional research including studies of deficiency disease and atherosclerosis. In these studies, *Cebus* monkeys have adapted well to a variety of natural products and purified diets. An example of a nutritionally adequate purified diet is listed in Table B.X. *Cebus* monkeys may be adequately maintained with commercial high protein monkey chow. Fruit supplements are often provided.

4. Reproduction

Under natural conditions *Cebus* monkeys are found in multimale–multi-female groups composed of up to 30 monkeys. These groups have about equal numbers of adults and young monkeys. The number of adult females is generally equal to or possibly greater than the number of adult males. *Cebus* monkeys appear to mate polygamously. Reproduction under natural conditions tends to be seasonal, with most births of C. capucinus in Cental America occurring during the dry season of December to April. In Brazil, C. apella have most births during the dry season of September to December. In captivity births occur throughout the year. Sexual maturity of *Cebus* probably occurs between 3 and 4 years of age. Females have menstrual cycles of 15–21 days (average 17.5). Menstruation is not marked, may not be externally visible, and lasts for 2–4 days. Peak serum estrogenic activity occurs at the sixth day of the cycle (Castellanos and McCombs, 1968).

Table B.X

Purified *Cebus* Monkey Diet[a]

Ingredient	Amount in diet (gm/100 gm)
Casein (vitamin-free)	20
Sucrose	59
Corn oil	15
Salts (Phillips and Hart)	4
Cod liver oil (USP)	2

Vitamin	Amount/100 gm diet
Thiamin HCl	2 mg
Riboflavin	2 mg
Pyridoxine HCl	2 mg
Ascorbic acid	90 mg
Calcium panthothenate	6 mg
Niacin	9 mg
Choline chloride	175 mg
Inositol	10 mg
Folacin	0.2 mg
Biotin	0.003 mg
p-Aminobenzoic acid	10 mg
Menadione	4.5 mg
Vitamin A	1800 units
Vitamin D_3	200 IU
α-Tocopherol	10 mg
Vitamin B_{12}	27 μg

[a]Adapted from Mann (1970).

Breeding in captivity has been successful. *Cebus* monkeys are excitable and do not readily cooperate for vaginal swabbing to chart menstrual cycles. In lieu of timed mating, programs of placing females with males on alternate 14–30 day cycles have been effective. Care must be taken to ensure the compatibility of paired monkeys as fights may take place, resulting in severe injury to the female. Adult males should be maintained in a room restricted for breeding, with the females brought to them. Using these techniques, the majority of adult, acclimated females may become pregnant each year, with at least half of the adult females producing live babies. The gestation period for *Cebus* monkeys is 162 days (Hayes *et al.*, 1972).

Cebus monkeys weigh about 200–240 gm at birth. Hand-rearing has been successful when infants are placed in incubators at 85°F for the first 2–3 weeks and fed human infant formula four times daily. At 2–3 weeks of age the monkeys are able to self-feed from bottles (Ausman *et al.*, 1970). Nursery-reared monkeys ingest about 175 kcal/kg per day during the first week. Caloric intake increases to a peak of 400 kcal/kg per day at 2 months of age and then declines to stable intake of 300 kcal/kg per day for the remainder of the first year. These monkeys grow rapidly, gaining about 100 gm per month for the first 10 months (Samonds and Hegsted, 1973).

Newborn infants cling to their mother's shoulders and back, and the mother provides most of the infant care. Infants may leave their mother for short distances and short periods at about 8 weeks of age. Full independence of movement is attained at about 6 months of age, while nursing may last from 6–12 months. In group situations, other group members, including adult males, females and juveniles, may carry the infants.

5. Behavior

In the wild, social interactions among adult capuchins consist primarily of grooming. Allogrooming is a common behavior. Grooming hierarchies exist with adult females doing most of the grooming, and males, the least. Agonistic behaviors are not frequent and consist mostly of threat gestures and vocalizations. Under natural conditions, *Cebus* commonly associate with other genera, most notably *Saimiri, Alouatta,* and *Ateles.*

6. Laboratory Management

Cebus monkeys are readily maintained in captivity under a variety of environmental conditions. As a minimum, caging should have 3 ft² of floor space and 2.5 ft in height, but larger cages are desirable. Housing may also be provided for unisex or mixed sex groups in large pens. Care should be taken not to overcrowd such groups as fighting and injury may result.

These monkeys have been housed in indoor–outdoor facilities, and under such conditions they frequent the out-of-doors even in freezing weather. Indoor temperatures should be maintained at comfortable levels, 70–80°F.

Cebus monkeys adapt well to natural products and purified diets. They can be maintained on commercial high protein monkey chow. Fruit supplements are often provided. Like other New World monkeys, capuchins require dietary vitamin D_3 as opposed to vitamin D from plant sources.

H. *Chiropotes* (Bearded Sakis; 2 Species, 2 Subspecies)

Chiropotes are medium-sized monkeys weighing 2.7 to 3.2 kg. They measure about 30 to 48 cm in head–body length and have an equally long tail. Bearded sakis are characterized by their distinctive beards, bulbus tufts of hair on either side of their head, long bushy tails, and rather short body fur. Their pelage is principally black. The genus is composed of two species, *Chiropotes albinasus,* which is black with a white nose, and *Chiropotes satanas,* which is all black. *Chiropotes* are probably the least known genus of the New World monkeys. They are thought to be principally frugivorous and/or seed eaters. Captive animals accept meal worms, grasshoppers, and other insects and will even consume young mice and small birds. *Chiropotes* under natural conditions appear to live in multi-male–multi-female groups of 8 to 30 or more animals.

They are very difficult to maintain in captivity and have not survived very long. They essentially have not been used in medical research.

I. *Lagothrix* (Woolly Monkeys; 2 Species, 4 Subspecies)

Woolly monkeys are large, robust New World monkeys that are highly arboreal. They are among the largest of the New World monkeys, weighing up to 10 kg. As indicated by their common name, they have a short, dense, woolly haircoat that varies from dark brown to gray or tan. They are found throughout the Amazon basin and up to 3000 meters above sea level on the eastern slopes of the Andes mountains. They have prehensile tails with hairless tactile pads on the ventral surface near the tip. They are capable of rapid movement through trees with a tendency for brachiation-type locomotion.

1. Normal Values

Not much information has been reported for woolly monkeys. Based on only a few monkeys, hematologic parameters of woolly monkeys are hemoglobin 15 gm/dl, erythrocytes $4.8 \times 10^6/mm^3$, packed cell volume 45%, MCH 89 μm^3, MCHC 32%, leukocytes $15 \times 10^3/mm^3$, neutrophils 45%, bands 1%, lymphocytes 49%, monocytes 1%, and eosinophils 5%. Serum chemical characteristics of woolly monkeys are listed in Table B.XI. Woolly monkeys have relatively high serum uric acid, calcium, and total protein concentrations.

2. Nutrition

Woolly monkeys are thought to be principally frugivorous. In captivity they have been noted to catch and eat insects and small birds. They have been fed a varied diet in captivity including commercial monkey chow, fruits, vegetables, bird eggs, cheese, butter, meat, fish, cow's milk (Williams, 1967).

3. Reproduction

Few observations have been made on reproductive phenomena in woolly monkeys. Breeding activity has been noted throughout the year (Mack and Kafka, 1978). Females have menstrual cycles of about 3 weeks duration with menstruation lasting for 3 days (Castellanos and McCombs, 1968). Mating occurs 10 days after the cessation of menstruation. The gestation period has been estimated at 7.5 months.

4. Behavior

Woolly monkeys under natural conditions have been observed in groups of 15–25 monkeys, presumably containing multiple adult males and females. Such groupings of monkeys have been successful under captive conditions, and fighting rarely occurs if sufficient space is provided.

5. Laboratory Management

Woolly monkeys have been used very little as research subjects. They have been difficult to maintain under laboratory conditions. Newly received woolly monkeys often appear depressed and exhibit a pattern of anorexia and debilitation similar to that described for *Callicebus*. These monkeys should be housed in groups or pairs if possible, as isolation appears to be very stressful to them. Caging should be spacious enough to allow normal activity, hanging by arms or tail and brachiating-type locomotion.

J. *Pithecia* (Sakis; 2 Species, 2 Subspecies)

Pithecia are small- to medium-sized monkeys weighing from 1.4 to 2.25 kg and measure approximately 30 to 48 cm in crown–rump length. Their bodies are rather long and slender, with moderately long arms and legs, and long bushy nonprehensile tails. Their haircoat is thick and coarse. Most authors have divided the genus into two species; however, Hershkovitz (1979) has suggested that four species should be recognized. Hair color varies from black to reddish brown. The face may be white or have whitish stripes extending from the eyes to the corners of the mouth. Sakis are found throughout the Amazon Basin. The innermost two fingers on each hand are opposable to the other three fingers. *Pithecia* remain one of the least known of the South American primates, and little work has been done with them in naturalistic studies or in captivity. What information is available is fragmentary and an-

Table B.XI

Serum Chemical Values of Woolly Monkeys

Determination	Mean ± SD
Uric acid (mg/dl)	3.7 ± 0.7
Sodium (mEq/liter)	152 ± 16
Potassium (mEq/liter)	5.0 ± 0.9
Chloride (mEq/liter)	100 ± 6
Calcium (mg/liter)	12.0 ± 1.7
CO_2 (mEq/liter)	19 ± 6
Total protein (gm/dl)	8.8 ± 1.2
Albumin (gm/dl)	2.6 ± 0.6
Alkaline phosphatase (BLB units)	17 ± 9
SGOT (Reitman-Frankel units)	60 ± 12
Bilirubin (mg/dl)	0.5 ± 0.2
Blood urea nitrogen (mg/dl)	27 ± 18
Glucose (mg/dl)	101 ± 30

ecdotal. They are thought to be basically frugivorous but also eat some seeds, flowers, and leaves. Captive sakis eat insects and meal worms. It is thought that these animals mate monogamously and form lasting pair bonds as has been observed for *Aotus* and *Callicebus*.

Pithecia have been very difficult to maintain in captivity and have seldom survived very long (Murdock, 1978). They essentially have not been used in medical research.

K. *Saimiri* (Squirrel Monkeys; 2 Species, 8 Subspecies)

Saimiri are the most numerous of the New World monkeys. Adult females weigh between 500 and 750 gm and males between 700 and 1100 gm. They are moderately variable in coat color and facial features and exhibit some geographic variation in karyotype (Jones *et al.*, 1973). In the past, as many as five species and 15 subspecies have been described. The current trend in the classification for *Saimiri* is consolidation. Most recent authors list one or two species depending on whether the Central American squirrel monkey is considered a distinct species or a subspecies. Confusion over squirrel monkey taxonomy in the past forced research workers to refer to the animals by their port of export or to identify them by the configuration of circumocular hair patches. These latter characteristics are inconsistent. *Saimiri* are found widely distributed in the Amazon Basin in Brazil, Colombia, Ecuador, Peru, Bolivia, and the Guyanas, extending as far south as Santa Cruz, Bolivia and as far north as Panama and Costa Rica. *Saimiri* appear to be absent from eastern Brazilian coastal forests.

1. Unique Biological Features

Saimiri have been the most used of all the New World monkeys in research. The extensive use of *Saimiri* resulted not so much from unique biological features as from pragmatic considerations such as cost, availability, smallness of size, and ease of maintenance. Unique features of squirrel monkeys include reproductive biology where both males and females have seasonal changes. Females have very short estrous cycles, and males have seasonal enlargement of testes associated with spermatogenesis. Unusual endocrine phenomena include very high blood cortisol concentrations (Ausman and Gallina, 1979; Coe *et al.*, 1978). The natural occurrence of arteriosclerosis and the propensity of squirrel monkeys to develop atherosclerosis when fed dietary cholesterol have also been exploited experimentally.

2. Normal Values

Normal values are listed in Tables B.XII–B.XIX and Figs. B.1–B.4.

Table B.XII

Hematologic Values and Developmental Changes in the Squirrel Monkey[a]

Parameter	0	1 month	2 months	4 months	1 year	2 years
PCV (%)	55 ± 6[b]	34 ± 3	38 ± 3	42 ± 3	43 ± 3	44 ± 3
Hemoglobin (gm/dl)	18 ± 2	11 ± 1	12 ± 1	13 ± 1	14 ± 1	14 ± 1
Erythrocytes (10^6/mm^3)	6 ± 0.7	5 ± 0.5	6 ± 0.5	7 ± 0.6	7 ± 0.6	7 ± 0.5
MCV (μm^3)	89 ± 6	72 ± 4	66 ± 5	63 ± 5	62 ± 3	60 ± 3
MCH (pg)	29 ± 3	23 ± 1	21 ± 2	20 ± 2	20 ± 1	20 ± 1
MCHC (%)	33 ± 2	32 ± 2	32 ± 2	33 ± 2	33 ± 2	33 ± 2
Reticulocytes (%)	4 ± 3	4 ± 2	3 ± 2	1 ± 1	1 ± 0	
Leukocytes (10^3/mm^3)	13	6	8	8	8	
Neutrophils (%)	77	40	40	31	30	
Lymphocytes (%)	13	53	54	62	63	
Monocytes (%)	6	3	4	2	4	
Basophils (%)	1	<1	<1	<1	<1	
Eosinophils (%)	3	4	2	4	2	
Total plasma protein (gm/dl)	5.5 ± 0.7	5.4 ± 0.7	5.6 ± 0.7	6.0 ± 0.6	6.9 ± 0.6	
Plasma albumin (gm/dl)	3.1 ± 0.3	2.9 ± 0.5	3.2 ± 0.5	3.1 ± 0.6	3.4 ± 0.4	

[a]Adapted from Ausman *et al.* (1976).
[b]Mean ± SD.

Table B.XIII

Hematologic Values of Adult Squirrel Monkeys[a]

Parameter	Mean ± SD
Erythrocytes (10^6/mm^3)	7.5 ± 0.8
Hemoglobin (gm/dl)	14.1 ± 1.5
Packed cell volume (%)	42 ± 4.2
MCV (μm^3)	57 ± 4.9
MCHC (%)	34 ± 1.6
Leukocytes (10^3/mm^3)	8.0 ± 2.9
Neutrophils (%)	51 ± 15
Lymphocytes (%)	41 ± 14
Eosinophils (%)	5 ± 7
Monocytes (%)	3 ± 3
Basophils (%)	<1

[a]Adapted from Manning et al. (1969).

Table B.XIV

Serum Chemical Values for Squirrel Monkeys[a]

Parameter	Mean ± SD
Na (mEq/liter)	148 ± 4.9
K (mEq/liter)	4.5 ± 0.9
Cl (mEq/liter)	113 ± 5.4
CO_2 (mEq/liter)	12.4 ± 4.0
Ca (mg/dl)	10.0 ± 0.5
P (mg/dl)	4.9 ± 1.7
Total serum protein (gm/dl)	7.2 ± 0.6
Albumin (%)	56.3 ± 7.0
Globulin (%)	
α_1	5.4 ± 1.8
α_2	12.5 ± 3.7
β	7.0 ± 1.9
γ	18.5 ± 4.3
A/G	1.4 ± 0.4
Serum urea nitrogen (mg/dl)	32 ± 6.4
Serum glucose (mg/dl)	74 ± 17
Serum uric acid (mg/dl)	1 ± 0.3
SGOT (Reitman-Frankel units)	196 ± 53
SGPT (Reitmen-Frankel units)	174 ± 57
Alkaline phosphatase (Bessey-Lowery-Brock units)	22 ± 15
Blood pH	
Nonsedated	7.24 ± 0.06
Sedated	7.36 ± 0.04
Blood pCO$_2$	
Nonsedated	24 ± 3.3
Sedated	31 ± 2.3
Blood pO$_2$	
Nonsedated	85 ± 4.6
Sedated	76 ± 4.3

[a]Adapted from Manning et al. (1969).

Table B.XV

Serum T3 and T4 Concentrations in Squirrel Monkeys[a]

Geographic origin	Sex	T3 (ng/dl)	T4 (μg/dl)	T3:T4
Bolivian	M	57 ± 22	3.0 ± 1.1	23
	F	49 ± 20	2.0 ± 0.9	29
Colombian	M	61 ± 13	3.4 ± 0.8	19
	F	78 ± 35	3.0 ± 1.7	32

[a]Adapted from Kaack et al. (1979).

Table B.XVI

Pulmonary Values of Squirrel Monkeys[a]

Value	Males	Females
Respiratory rate/min	55 ± 1.9[b]	58 ± 1.7
Tidal volume (ml)	8.9 ± 0.37	7.5 ± 0.28
Minute volume (ml)	494 ± 28	434 ± 21
Pulmonary flow resistance (cm H$_2$O/ml/sec)	0.052 ± 0.006	0.086 ± 0.011
Dynamic compliance (ml/cm H$_2$O)	1.78 ± 0.15	1.48 ± 0.12

[a]Adapted from Ulrich et al. (1977).
[b]Mean ± SEM.

Table B.XVII

Organ Weights of Squirrel Monkeys[a]

Organ	Adult male	Adult female
Total body	779 ± 3[b]	662 ± 13
Heart	3.32 ± 0.11	2.81 ± 0.07
Liver	19 ± 0.7	20 ± 0.8
Spleen	1.05 ± 0.11	0.95 ± 0.09
Pancreas	1.19 ± 0.05	1.27 ± 0.04
Brain	24.0 ± 0.3	23.0 ± 0.3
Right gonad	1.58 ± 0.10	0.46 ± 0.09
Left gonad	1.59 ± 0.10	0.45 ± 0.08
Right lung	2.92 ± 0.06	2.59 ± 0.06
Left lung	2.42 ± 0.05	2.16 ± 0.05
Right kidney	1.50 ± 0.06	1.43 ± 0.05
Left kidney	1.50 ± 0.07	1.44 ± 0.05
Right adrenal	0.101 ± 0.003	0.104 ± 0.003
Left adrenal	0.106 ± 0.003	0.110 ± 0.003
Right thyroid	0.061 ± 0.003	0.061 ± 0.004
Left thyroid	0.056 ± 0.003	0.056 ± 0.004

[a]Adapted from Middleton and Rosal (1972).
[b]Weights in grams; Mean ± SEM.

Table B.XVIII

Eruption Sequence of Deciduous Teeth in Squirrel
Monkeys[a]

| Tooth | Age (weeks) | |
	Mean	Range
Maxillary		
Incisor 1	1.3	0–3
Incisor 2	2.2	1–3
Canine	3.4	2–4
Premolar 1	4.9	4–6
Premolar 2	6.1	5–7
Premolar 3	8.9	8–11
Mandibular		
Incisor 1	1.3	1–3
Incisor 2	1.4	1–3
Canine	3.4	3–4
Premolar 1	4.8	4–6
Premolar 2	5.7	5–7
Premolar 3	8.3	7–9

[a]Adapted from Long and Cooper (1968).

Table B.XIX

Eruption Sequence of Permanent Teeth in Squirrel
Monkeys[a]

| Tooth | Age (months) | |
	Mean	Range
Maxillary		
Molar 1	5.5	5–6
Molar 2	8.4	7–9
Incisor 1	9.7	8–12
Incisor 2	12.0	10–14
Premolar 2	13.0	12–15
Premolar 3	13.0	12–15
Premolar 1	14.7	14–15
Molar 3	20.0	19–22
Canine	21.5	21–22
Mandibular		
Molar 1	5.1	5–6
Molar 2	7.0	7
Incisor 1	9.2	8–11
Incisor 2	9.6	8–11
Molar 3	12.3	11–14
Premolar 3	12.8	12–15
Premolar 1	14.3	12–16
Premolar 2	15.0	15
Canine	20.5	19–21

[a]Adapted from Long and Cooper (1968).

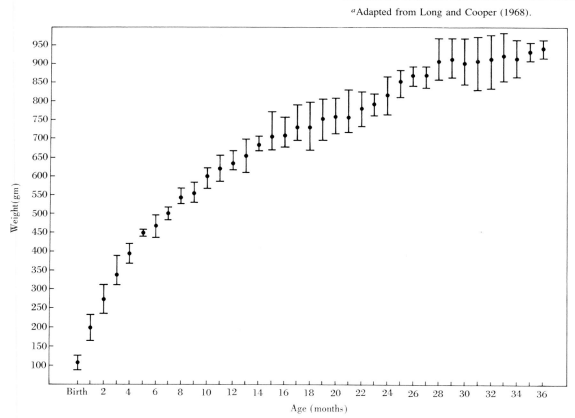

Fig. B.1. Body weights of male squirrel monkeys according to age (range indicated by brackets). From Long and Cooper (1968).

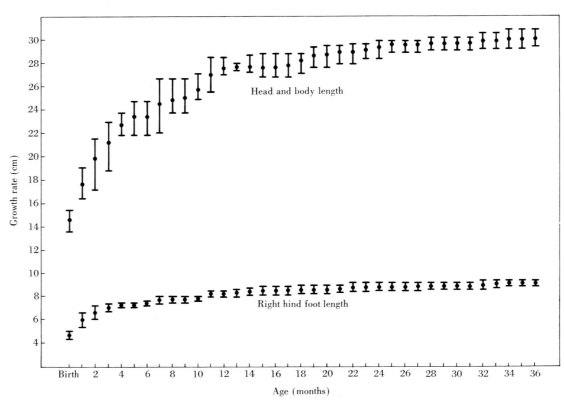

Fig. B.2.   Head–body length and hindfoot length of male squirrel monkeys according to age (range indicated by brackets). From Long and Cooper (1968).

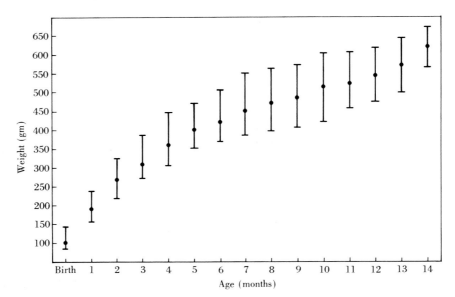

Fig. B.3.   Body weight of female squirrel monkeys according to age (range indicated by brackets). From Long and Cooper (1968).

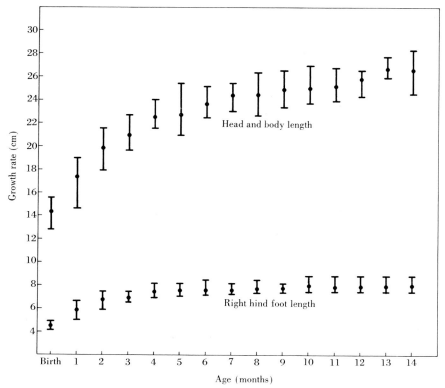

Fig. B.4. Head–body length and hindfoot length of female squirrel monkeys according to age (range indicated by brackets). From Long and Cooper (1968).

3. Nutrition

Under natural conditions *Saimiri* are omnivorous, eating principally fruits and insects. They adapt well to diets composed of natural products or purified ingredients in either solid or liquid form. Acceptable levels of dietary constituents for squirrel monkeys can be found in Table B.XX. An example of a nutritionally adequate semipurified diet for squirrel monkeys is listed in Table B.XXI. Because of their relatively high basal metabolic rate and short gastrointestinal tract, *Saimiri* require a diet of high caloric density.

Infant squirrel monkeys in their rapid growth phase are inefficient in protein utilization, utilizing only about 30 to 35% of the protein ingested. This compares to similarly aged cynomolgus or capuchins that have protein utilization efficiencies of about 65 to 75%. Infant squirrel monkeys require about 13% of their calories as dietary protein, while adult animals appear to be able to maintain themselves on half this amount of dietary protein (Ausman and Gallina, 1979). Squirrel monkeys adapt to various levels of fat in the diet, from 5 to 20% fat by weight.

The requirements of squirrel monkeys for vitamins has been investigated in only a few cases. As in other New World pri-

Table B.XX

Estimate of Nutrient Requirements of Squirrel Monkeys[a]

Nutrient	Estimated need or adequate dietary level
Energy	
Infant	300–500 kcal/kg/day
Adult	100–300 kcal/kg/day
Protein	18–25% diet
Linoleic acid	1% of calories
Vitamins	
A	10,000–15,000 IU/kg diet
D_3	2,000 IU/kg diet
E	50 IU/kg diet
K	10 mg/kg diet
Biotin	0.1 mg/kg diet
Choline	3 g/kg diet
Folacin	0.4 mg/kg diet
Niacin	50 mg/kg diet
Pantothenate	15 mg/kg diet
Riboflavin	5 mg/kg diet
Thiamin	12 mg/kg diet
B_6	2.5 mg/kg diet

(continued)

Table B.XX (*Continued*)

Nutrient	Estimated need or adequate dietary level
B_{12}	0.04 mg/kg diet
C	200 mg/kg diet
Minerals	
Calcium	0.5% diet
Chlorine	0.2–0.5% diet
Magnesium	0.15% diet
Phosphorus	0.47% diet
Potassium	0.8% diet
Sodium	0.2–0.4% diet
Chromium	1.5 mg/kg diet
Copper	15 mg/kg diet
Fluorine	25 mg/kg diet
Iodine	2 mg/kg diet
Iron	180 mg/kg diet
Manganese	40 mg/kg diet
Selenium	0.1 mg/kg diet
Zinc	10 mg/kg diet
Cobalt	0.25 mg/kg diet

[a]The kinds and amount of nutrients listed were estimated from composition of commercial and laboratory formulated diets that appeared to be adequate for growth, maintenance, and reproduction. (National Academy of Sciences, 1978).

Table B.XXI

Semipurified Diet[a]

Ingredient	gm/100 gm
Casein	10
Lactalbumin	10
Dextrin	23
Sucrose	26.45
Corn oil (or lard)	15
Vitamin mix[b]	0.50
Mineral mix (Ausman-Hayes)	4.75
Choline chloride	0.30
Cellulose	10

[a]From Newberne and Hayes (1979).

[b]Vitamin mix (gm/kg): Thiamin HCl, 0.80; riboflavin, 1.60; niacinamide, 8.0; folic acid, 0.8; biotin, 0.04; cyanocobalamin, 0.03; menadione, 0.1; *d*-1α-tocopheryl acetate, 20.0 (500 IU/gm); vitamin A acetate, 5.0 (500,000 IU/gm); vitamin D_3, 1.25 (200,000 IU/gm); ascorbic acid beadlets, 120; inositol, 100; taurine, 50; dextrin, 687.

mates, squirrel monkeys cannot or only very poorly utilize vitamin D from plant sources. These monkeys require a dietary source of vitamin D_3 or irradiation with ultraviolet light. As little as 1 IU/g diet of vitamin D_3 has been found sufficient to prevent rickets in them (Lehner *et al.*, 1967). Squirrel mon-

keys, like presumably all other primates, require exogenous sources of ascorbic acid; 10 mg/kg of body weight per day is sufficient ascorbic acid to correct signs of scurvy (Lehner *et al.*, 1968). Squirrel monkeys have a rather high requirement for dietary folic acid, about 200 μg per day. Deficiencies of this vitamin in pregnant monkeys are associated with megaloblastic anemia, low birth weight infants, and stillbirths (Rasmussen, 1979).

4. Reproduction

Squirrel monkeys under natural conditions and also under stable captive conditions have seasonal reproductive cycles. The mating season has been associated with conditions of low precipitation under natural conditions (DuMond, 1968). The breeding season in Colombia and Peru occurs in July and August with births occurring from mid-December through February. The mating and birth seasons are altered when these monkeys are brought to the Northern Hemisphere, and mating occurs in the spring from March to May with births occurring mostly from August to October.

Female squirrel monkeys do not menstruate but are seasonally polyestrous. The estrous cycle is of short duration, which may facilitate the chance of becoming pregnant during the short breeding season. Most estimates of the estrous cycle length determined by vaginal cytology vary from 7 to 13 days (Lang, 1967; Rosenblum *et al.*, 1967). Plasma estradiol and progesterone have been reported to vary cyclically with 9-day intervals between peak concentrations during the breeding season (Wolf *et al.*, 1977).

Male squirrel monkeys also undergo cyclic changes. Males attain their greatest body weights (fatted male phenomenon) during the breeding season, associated with increased testicular size, spermatogenesis, and elevated plasma testosterone concentration (Baldwin, 1968; DuMond and Hutchinson, 1967; Mendoza *et al.*, 1978).

The gestation period is about 145 to 150 days. Squirrel monkey births invariably occur at night, with labor usually lasting about 1 to 2 hr. Following delivery, the infant immediately clings to the mother's body. Birth weights of squirrel monkey babies are around 100 gm. Newborns that weigh less than 90 gm rarely survive. During the first 2 weeks of life the infant alternates periods of sleeping and nursing. During the next few weeks the infant becomes much more active and attracts the attention of, and may be carried by other females, so-called aunting behavior. After 4 to 5 weeks, the infant may leave its mother's back and explore the environment. Adult males generally do not interact with the infants.

Approximately 20% of colony-born female squirrel monkeys are reproductively active at $2\frac{1}{2}$ years of age, and most are mature by $3\frac{1}{2}$ years of age. A female may be sexually mature at $1\frac{1}{2}$

years, but this is rare. It is estimated that male squirrel monkeys become sexually mature at 3½ years of age.

5. Behavior

Under natural conditions squirrel monkeys associate in large bisexual groups. Adult females form the central social structure of the group. Adult males remain peripheral to the group except during the breeding season, and have little social interaction with the females throughout most of the year. Similar segregation of the sexes has been observed in laboratory situations. Female squirrel monkeys determine social interaction throughout most of the year, while adult males are passive and avoid intersexual interaction. During the mating season, males increase contacts and olfactory investigations of females, and females appear much more receptive to the male presence.

There is no clear-cut dominance hierarchy evident among adult females, but adult males do exhibit a dominance hierarchy during the breeding season. One indicator associated with dominance in males is the penile display, which, along with aggressiveness, is common during the breeding season. Allogrooming does not frequently occur, although autogrooming in the form of scratching is frequent (Baldwin and Baldwin, 1981).

6. Laboratory Management

Once stabilized in a captive environment, squirrel monkeys are quite hardy and can tolerate a wide variety of environmental conditions. Numerous problems may be encountered with the maintenance of newly received animals as they appear to be adversely affected by the stress of capture and shipment. Attempts should be made to reduce stress by providing a warm, quiet, tranquil environment. Further environmental changes should be kept to the minimum. If monkeys have been housed and shipped with cohorts, caging in such groups initially may be less stressful than isolated housing. The diet offered squirrel monkeys that are newly received should reflect the diet to which they have been acclimated. Once established, they accept and can be maintained by commercial foods with or without supplementation with fruits and vegetables.

Squirrel monkeys, once acclimated, tolerate a wide variety of environmental temperatures. They should be provided comfortable quarters with indoor temperatures maintained at 70°–80°F. Relative humidity should be maintained at 50 ± 20%. Squirrel monkeys housed in indoor–outdoor arrangements spend much time outside, even venturing out in freezing weather.

Squirrel monkeys should be provided cages with at least 2–3 ft² of floor area and 2 ft in height to allow them to jump about. Perches within the cages are desirable. Squirrel monkeys may also be housed in large pens either indoors or in indoor–outdoor

type enclosures. Once established, mixed groups are quite stable with little aggressive or agonistic behaviors exhibited. Fighting may occur in group-housed male monkeys during the breeding season, however. New individuals added to stable groups may be the object of much aggression and may be seriously injured.

Breeding paradigms that are most often used for squirrel monkeys are single male–multi-female or multi-female–multi-male groups. One adult male may be housed with up to 12 female monkeys. Monkeys of feral origin reproduce quite well, and over half of breeding females should produce viable progeny each year.

Squirrel monkeys may be successfully hand-reared, and in some colonies it is standard practice to remove squirrel monkeys at birth and to raise them in a nursery. Such infants are placed in incubators maintained at 85°F. Babies are fed diluted human infant formulas as early as 4 to 6 hr after birth. Full-strength formula containing approximately 1 kcal/ml may be fed after the third day. Initially, monkeys are fed seven times daily, usually between 8 AM and 11 PM. At 3 weeks of age, squirrel monkeys consume approximately 500 kcal/kg of body weight per day. The rate of consumption gradually declines thereafter. Squirrel monkeys gain approximately 20 gm/week for the first 8 weeks of life. Infant monkeys may be trained to self-feed from fixed bottles at 4 weeks of age. Hand-reared monkeys are behaviorally abnormal and do not reproduce well.

III. DISEASES

A. Bacterial

1. Respiratory Disease

A number of bacteria have been associated with acute respiratory infections in New World monkeys, including *Bordetella bronchiseptica* (Seibold, *et al.*, 1970), *Pasteurella multocida* (Greenstein *et al.*, 1965; Benjamin and Lang, 1971), *Klebsiella pneumoniae* (Snyder *et al.*, 1970), and *Streptococcus pneumoniae* (Henderson *et al.*, 1970). These organisms may be carried asymptomatically in the nasopharynx. Respiratory disease is often associated with the stress of shipment, other intercurrent infections, or debilitating conditions. Stressful conditions such as shipment may result in multiple animals being affected, suggesting spread of infection. Under stable colony conditions cases of respiratory disease caused by these organisms are usually sporadic.

Respiratory disease may take the form of rhinitis, pharyngitis, and/or pneumonia. Infection of laryngeal air sacs is common in *Aotus* (Giles *et al.*, 1974). Clinical signs vary depending upon the severity of the infection and include serous to

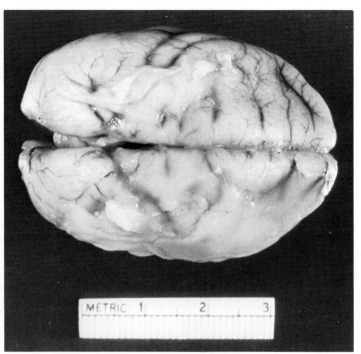

Fig. B.5. Purulent meningitis in squirrel monkey caused by *Pasteurella multocida.*

purulent nasal exudates, coughing and sneezing, dyspnea, lethargy, anorexia, and fever. Peracute septicemia may occur with monkeys found dead without exhibiting prior signs.

The gross lesions induced by infection with these organisms are not specific. Lesions in the lung consist of patchy to diffuse consolidation that may vary from red to deep purple to gray. There may be extension of the pulmonary lesions to include pleuritis and pericarditis with attendant fibrinopurulent exudates in these cavities. In animals that are septicemic, extensive mucosal and serosal petechial hemorrhages may be present. Septicemia may also result in the spread of infection to other sites. *Streptococcus, Klebsiella,* and *Pasteurella* infections may result in suppurative meningitis (Fig. B.5). Diagnosis of these infections is dependent upon isolation and identification of the organisms.

2. Tetanus

Infection with *Clostridium tetani* is rare, but was reported for a group of Bolivian squirrel monkeys that were housed in outdoor pens directly on the ground. Housing directly on the ground probably provided a source of these organisms. Affected monkeys had bite wounds, and clinical signs of tetanus characteristically had an acute onset. Affected monkeys exhibited a slow deliberate and slightly stiffened gait and appeared reluctant to walk. Occasionally such animals would fall over in

a laterally recumbent position. Signs progressed rapidly, usually within 24 hr, to include trismus, extensor rigidity, opisthotonus, and death. Diagnosis of tetanus was based on external injury or wounds and the neurologic signs consistent with tetanus. Prophylaxis consisting of three 0.25 ml intramuscular injections of tetanus toxoid given at monthly intervals appeared effective (Kesler and Brown, 1979).

3. Leptospirosis

Leptospirosis has not been reported as a naturally occurring disease in New World monkeys. Serological surveys also suggest that leptospirosis is not an important or frequent disorder in these animals (Minette, 1966). Squirrel monkeys are, however, susceptible to experimental infection with *Leptospira icterohemorrhagica* and develop acute disease that is sometimes fatal. This experimentally induced infection may be associated with extensive hemolysis, pronounced icterus, and markedly elevated indirect bilirubin concentrations in the serum (Minette and Shaffer, 1968).

4. Tuberculosis

Tuberculosis continues to be one of the most serious infectious diseases of nonhuman primates. The frequency with which tuberculosis has been found in New World monkeys appears to be less than that for the Old World primates, suggesting a lesser susceptibility (Moreland, 1970). Experimental exposure has not been uniformly infective (Bone and Soave, 1970). Regardless of relative susceptibilities, New World primates develop the disease if exposed, and it has been reported in *Saimiri, Aotus, Ateles,* and *Cebus* monkeys. The incidence of tuberculosis in wild populations is probably very low or absent. Infection with this disease occurs in captivity through exposure to infected human or other primates. Tuberculosis in New World monkeys is a progressive disease that appears to be fatal once contracted. Infection may occur by the alimentary or respiratory routes. Infection through the respiratory tract appears most common and results in primary lesions in the lung and bronchial lymph nodes (Fig. B.6). In advanced disease miliary lesions appear as firm white nodules in many organs and tissues, especially in spleen, liver, kidney, and various lymph nodes. In advanced cases the disease may be so generalized that the primary lesion or site of infection cannot be determined. Microscopically, the lesions of tuberculosis consist of granulomas of varying size. Such lesions often have a central focus of caseation necrosis surrounded predominantly by epitheloid cells but including some granulocytes. The lesions in the cebid monkeys are usually nonencapsulated. Tubercle bacilli can be demonstrated in the lesion with acid-fast stains.

The clinical signs of tuberculosis in monkeys are nonspecific

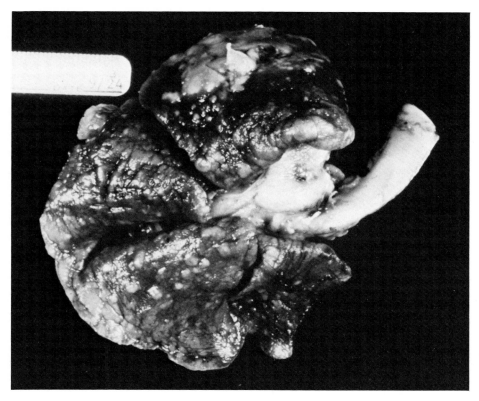

Fig. B.6. Lungs of *Cebus albifrons* with tuberculosis.

and include weight loss, lethargy, coughing, unthrifty appearance, and anorexia. Enlarged lymph nodes may be evident, and hepatic and splenic enlargement may be discernible by palpation in advanced cases. Tuberculosis should be considered in the differential diagnosis of any wasting disorder.

Mycobacterium tuberculosis is the organism that has been reported from tuberculous New World monkeys. These animals most likely are also susceptible to *Mycobacterium bovis.*

Radiography may be useful in identifying pulmonary lesions. Chest radiographs may delineate areas of consolidation in tuberculous lungs, but these lesions cannot be differentiated from other causes by this method. The tuberculin skin test has been the clinical test most widely used to detect tuberculosis in monkeys (ILAR, 1980). The reliability of tuberculin testing in New World monkeys and interpretation of test results are uncertain. Diminished hypersensitivity to the tuberculin test has been reported for tuberculous squirrel monkeys (Moreland, 1970), and unreactivity to the tuberculin skin test has been found in a tuberculous capuchin (Leathers and Hamm, 1976) and a tuberculous owl monkey (Bone and Soave, 1970). Infection of squirrel monkeys with the atypical organism, *Mycobacterium gordoneae,* caused positive tuberculin skin tests; however, the animals had neither gross nor microscopic lesions of tuberculosis (Soave *et al.,* 1981). Similarly, mycobacteria of

the *avium–intracellulare* group have been isolated from lymph nodes of squirrel monkeys that reacted to the tuberculin skin test but had neither gross nor microscopic lesions of tuberculosis.

Diagnosis of tuberculosis is dependent on demonstration of typical lesions containing acid-fast bacilli and isolation and identification of the organism.

5. Gastrointestinal Disease

Enteric diseases caused by bacteria have not been major causes of sickness in New World monkeys. Surveys of cebid monkeys for enteropathic bacteria, including *Salmonella, Shigella,* and *Escherichia,* have revealed low rates of infection, usually in asymptomatic monkeys (Kournay and Porter, 1969; Good *et al.,* 1969).

Yersinia pseudotuberculosis and *Y. enterocolitica* cause gastrointestinal disease in nonhuman primates including New World monkeys (Baggs *et al.,* 1976). These organisms have little host specificity and infect many avian and mammalian species. In the acute disease, affected monkeys are lethargic, anorectic, dehydrated, and have severe diarrhea or dysentery. The disease may also be peracute, with death occurring without prior signs. Affected monkeys invariably have a necrotizing to ulcerative enterocolitis. Lesions consist of numerous irregularly shaped, raised areas in the mucosa surrounded by zones of hyperemia and hemorrhage. Microscopically, the lesions consist of extensive and severe mucosal necrosis with hemorrhage, marked neutrophilic infiltration associated with bacterial colonies.

The disease may also take a more chronic, indolent course with caseous abscess formation of the lymph nodes, liver, spleen, lung, and kidney. Grossly these lesions appear as numerous white to yellow foci that may be minute or up to several centimeters in diameter. Microscopically, such lesions consist of foci of necrosis and acute inflammation, usually containing bacterial colonies.

Most cases of pseudotuberculosis have occurred in animals in indoor–outdoor facilities. Wild rodents and birds may carry these organisms and are a potential source of infection for the monkeys.

6. Streptothricosis

Skin infection with *Dermatophilus congolensis* has been found in owl monkeys (King *et al.*, 1971). Affected monkeys and papillomatous lesions on the chin, eyelid, lips, and extremities including the tail. The lesions were raised and measured 0.5–3 cm in diameter. The lesions were hairless and covered with red-brown crusts that were cracked and fissured. Microscopically, the epidermis of affected skin was greatly thickened due to hyperkeratosis, parakeratosis, and acanthosis. Numerous filamentous, branching, beaded, and septate organisms characteristic of *Dermatophilus* were present in the lesions. Diagnosis may be made by identification of typical organisms from smears or sections or by isolation and identification of the organism on artificial media (see Chapter 22 by Fox *et al.*).

B. Parasitic

1. Protozoa

a. Malaria

Naturally occurring malarial infections in New World monkeys are caused by two parasites, *Plasmodium brazilianum* and *Plasmodium simium*. *Plasmodium brazilianum* is by far the more common of the two, is widely distributed in Central and South America, and has been found in Panama, Venezuela, Colombia, Peru, and Brazil (Garnham, 1966; Young, 1970). *Plasmodium brazilianum* has been found as a natural infection in most of the Cebidae including *Alouatta, Ateles, Brachyteles, Cacajao, Callicebus, Cebus, Chiropotes, Lagothrix,* and *Saimiri*. Malaria has not been found in *Pithecia;* however, they have been studied little. *Aotus* and mem-

bers of the Callithrichidae do not appear to become infected with *Plasmodium brazilianum* in their natural habitats; however, they are susceptible to experimental infection. Humans also may be infected with *Plasmodium brazilianum,* which actually is very similar to, if not the same, as *Plasmodium malariae,* one of the malarial parasites of humans.

Plasmodium simium has only been found as a natural infection in *Alouatta* and *Brachyteles* and has a limited distribution in southeastern Brazil. Experimental infections have been induced in *Aotus, Saimiri, Ateles, Lagothrix,* and *Callithrix*. Attempts to experimentally infect humans have been unsuccessful (Coatney *et al.*, 1971).

All the clinical and pathological manifestations of malarial infections are attributable to the erythrocytic cycle of the parasites. *Plasmodium brazilianum* may cause extensive destruction of erythrocytes. Initial infections are characterized by rising numbers of parasites that peak in 1–3 weeks and may be associated with anorexia, cyclic fever, and anemia (Taliaferro and Taliaferro, 1934). Markedly high numbers of parasites are not usual, and it is rare that as many as 10% of erythrocytes may be affected at one time. Initial infections have sometimes been fatal in *Alouatta, Cebus,* and *Ateles*. Rapid or gradual decline in parasite numbers follow, with periods of low grade parasitemia interspersed with periods of subpatency. Spontaneous relapses of greater or lesser severity may occur. Sharp peaks in body temperature may occur at the time of sporulation provided that parasites are sufficiently numerous. Infections with *Plasmodium brazilianum* may persist for years as low or subpatent infections, the type usually found in laboratory monkeys. Experimental manipulation, such as splenectomy and irradiation with X rays, may activate subpatent infections. Natural infection of *Plasmodium simium* in *Alouatta* and *Brachyteles* have been light parasitemias with little evidence of illness.

The most striking gross pathologic change associated with *Plasmodium brazilianium* infection is splenomegaly (Taliaferro and Cannon, 1936). The spleen may be dark in color and increased in size fourfold. Hyperplasia of lymphoid elements and macrophages occurs in the spleen, liver, and bone marrow. Myeloid hyperplasia of the bone marrow may be associated with erythropoiesis. Malarial pigment accumulates in the macrophages in the red pulp of the spleen, in the Kupffer cells of the liver, and in macrophages in the bone marrow. Malarial pigment must be differentiated from iron, hemosiderin, and formalin pigments. Malarial pigment disappears several months after termination of infection.

Diagnosis of malaria is made by demonstrating the parasites in the erythrocytes. Differential features of the parasites are used to specifically identify them.

Different drugs are needed to treat the erythrocytic and hepatic phases of malarial infections. Chloroquine is effective against the erythrocytic phase of the infection at a dose of 5 mg/kg, which should be continued for 4 days. The hepatic phase of the parasite may be treated with primaquine given at a dose of 0.75 mg/kg for 14 days. These drugs should be given separately because of increased toxicity when given together.

b. Hemoflagellates

New World monkeys may be infected with at least two groups of hemoflagellates, *Trypanosoma cruzi* or *Trypanosoma rangeli* and *rangeli*-like organisms (Marinkelle, 1966; Dunn *et al.*, 1963). Infection of mammalian hosts occurs when infected reduviid bugs feed on them and bug feces containing trypanosomes contaminate bite wounds or abrasions, or the organisms are deposited on and penetrate intact skin, mucous membranes, or conjunctiva. *Trypanosoma rangeli* also infects the salivary glands of the insect vectors and may be innoculated directly by bug bites. *Trypanosoma cruzi* and *T. rangeli* can infect humans, nonhuman primates, and a wide range of other mammals. *Trypanosoma cruzi* and *T. rangeli* are widely found in Central and South America with the distribution of *T. cruzi* extending to southwestern United States. *Trypanosoma rangeli* has been found in many New World monkeys including *Cebus, Saimiri, Ateles, Aotus,* and *Lagothrix. Cebus* monkeys have been the ones most often found infected with *T. cruzi;* however, *Saimiri, Cacajao,* and *Ateles* have also been infected with this organism.

Trypanosoma rangeli is essentially nonpathogenic, being found only in blood with no tissue stages or intracellular multiplication stages being identified in mammalian hosts. *Trypanosoma rangeli* infections abate within a year or so as the host becomes immune. *Trypanosoma cruzi* invades cells, causing necrosis, inflammation, and is most notable as the cause of Chagas' disease in humans. In the acute phase, parasites are disseminated to many organs and tissues with a predilection for striated muscle, neural tissue, and the reticuloendothelial system. Necrosis and inflammation of the heart, brain, lymph nodes, liver, and spleen often occurs. In humans myocardial damage dominates the chronic form of the disease with chronic myocarditis, myocardial fibrosis, cardiac enlargement, conduction defects, especially right bundle branch block, and heart failure. Necrosis of autonomic nervous plexuses may result in marked dilatation of the esophagus, colon, and other tubular organs.

Pathologic changes in New World primates caused by *T. cruzi* mostly are findings of leishmanial forms in cardiac muscle cells and varying degrees of associated myocarditis (Bullock *et al.*, 1967). Inflammatory infiltrates typically consist of lymphocytes and plasma cells. One case of fatal Chagas' myocarditis and heart failure has been reported in a red uakari, *Cacajao rubicundus* (Lasry and Sheridan, 1965). Signs were those of heart failure, cyanosis, dyspnea, hydrothorax, and ascites along with lethargy and anorexia.

Trypanosoma cruzi infections in New World monkeys usually are not associated with clinical disease, however. Squirrel monkeys and capuchins of feral origin have been found to have a high frequency of right bundle branch block, a sequellae commonly found in chronic Chagas' myocarditis in humans (Wolf *et al.*, 1969).

Diagnosis is dependent on identification of trypanosomes in blood or tissues or by serologic methods. Parasites can be demonstrated by the following methods: examination of fresh or stained blood films, animal inoculation with suspect blood and demonstration of trypanosomes in the recipient's blood or tissues, culture of blood on NNN medium, and xenodiagnosis. Complement-fixation is a serologic test used widely to test for infection in people and may be helpful. Histopathological diagnosis is dependent on demonstrating leishmanial forms in the tissues.

Control of the disease is affected by elimination of insect vectors. Currently there is no effective treatment for *T. cruzi* infections.

c. Tissue Cyst-Forming Coccidia

i. Toxoplasma gondii. *Toxoplasma gondii* is a coccidian of cats that is found worldwide (Frankel, 1973). Unlike other coccidia that are host specific, *Toxoplasma* can infect many birds and warm-blooded animals. *Toxoplasma* has an enteric cycle in its definitive host, members of the cat family, but unlike other coccidia, it may parasitize many tissues in definitive as well as intermediate hosts.

Felines become infected by eating raw tissues of animals that contain *Toxoplasma* cysts or by ingestion of oocysts from other cats. Oocysts are excreted in the feces of such cats for 1–3 weeks, beginning 3 days to 3 weeks following ingestion of infective material.

Following ingestion of cysts or oocysts by intermediate hosts, organisms enter the intestinal epithelium, multiply, and are disseminated throughout the body by lymphatics and the systemic circulation. Lymphatic tissue, liver, and lung appear to be primary multiplication sites; however, any tissue may be infected. Proliferative forms give way to thick-walled cysts, as immunity develops. Infections in some animals and humans may be quite prevalent, but usually are asymptomatic. In people, about which most is known, clinical toxoplasmosis, when it occurs, most often is manifested as lymphadenopathy, which may be focal or generalized and which may or may not be associated with fever (Kirck and Remington, 1978). Such infections usually resolve spontaneously within a few months. Rarely in normal persons, more serious illness, such as encephalitis, myocarditis, and pneumonitis, may occur. Contrastingly, in immunologically compromised people, *Toxoplasma* infections may cause fatal encephalitis, myocarditis, and pneumonitis within a few days.

Little is known about the prevalence of *Toxoplasma* infections in nonhuman primates. Information for simian primates consists mainly of case reports of fatal illnesses, all of which have been reported from New World monkeys (deRodaniche, 1954). Attempts at experimental infection in New World monkeys have involved *Aotus* and two marmoset species and resulted in fatal infections (deRodaniche, 1955). Naturally occurring infections with *Toxoplasma* have been found in most of the Cebidae, including *Alouatta, Aotus, Ateles, Cacajao, Callicebus, Cebus, Lagothrix, Pithecia,* and *Saimiri.* Many of these cases were reported from a zoo where raw meat comprised part of the diet (McKissick *et al.,* 1968; Ratcliffe and Worth, 1951). Another case was a pet monkey that also was frequently fed raw meat (Hessler *et al.,* 1971). The association between feeding raw meat to New World monkeys and fatal toxoplasmosis, the fulminant character of the disease in naturally occurring and experimental infections, and the occurrence of the disease in many New World primates attests to the marked susceptibility and lack of resistance of these animals to toxoplasmosis.

In all reported cases of toxoplasmosis in New World monkeys, the animals were found dead with no prior signs or had a short duration of illness of 1 day to 1 week. Signs of illness were nonspecific and included pyrexia, anorexia, lethargy, somnolence, and progressive weakness. Monkeys with pulmonary disease sometimes became dyspneic. Signs of encephalitis included circling, grasping and holding of the head, leaning head against or hitting head on cage, ataxia, and convulsions. Organisms may be present in peripheral blood terminally.

Gross lesions may be variable but often include diffuse reddening of lungs, petechia to ecchymotic pulmonary hemorrhage along with edema and frothy fluid in the bronchi. Dilatation and enlargement of the heart may be present. Lymph node enlargement affecting especially abdominal nodes along with splenomegaly may be evident. Histopathological examination usually reveals lesions and organisms in many organs and tissues. Focal necrosis with or without inflammatory exudate may be found in the lung, liver, lymph nodes, spleen, heart, adrenal, brain and less frequently in other organs. Organisms may be found extracellularly or intracellularly associated with these lesions. Inflammatory exudate may include neutrophils and macrophages.

Clinical diagnosis of toxoplasmosis is difficult. Serological tests, complement fixation, indirect fluorescent antibody, and indirect hemagglutination may be helpful in establishing a diagnosis. The rapid course of the disease in New World monkeys may require therapy before such results are obtained, Similarly, isolation of the organism by mouse inoculation of biopsy specimens may be useful in establishing a diagnosis, but again results may not be available soon enough to help with the decision to initiate therapy for toxoplasmosis. Toxoplas-

mosis should be considered in the differential diagnosis of nonspecific acute illnesses in New World monkeys. A history of feeding raw animal tissues, exposure to cats, lymphadenopathy, or evidence of encephalitis are situations or conditions that may support a decision to initiate antitoxoplasma therapy.

Toxoplasmosis may be controlled by prevention of ingestion of oocysts or cysts. Adequately cooking meat or freezing to $-20°C$ for 2 days will kill cysts in tissues. Cats potentially may play a role in the transmission of toxoplasmosis to New World monkeys and should not be allowed to contaminate the monkeys' environment.

The effectiveness of chemotherapy for toxoplasmosis in monkeys is unknown. In humans, combined treatment with pyrimethamine and a sulfonamide such as sulfadiazine or trisulfapyrimidine is used. Pyrimethamine is a folic acid antagonist and acts synergistically with the sulfa drug. These drugs act on the trophozoites and are suppressive rather than toxoplasmocidal. The dose of pyrimethamine for New World monkeys has not been established. Pediatric human doses may be a good starting point, 1 mg/kg per day in two divided doses, after 2 to 4 days reduced to one-half, and continue for 1 month, along with sulfadiazine treatment, 100 mg/kg per day, two divided doses for 1 month. Pyrimethamine may depress the bone marrow so hematologic examinations should be done to monitor these effects. Folic acid, 1 mg per day, may be given to alleviate bone marrow depression and yet not interfere with the therapeutic action of the drugs.

ii. Sarcocystis sp. *Sarcocystis* are coccidia with a two-host life cycle. Intermediate hosts are herbivores, rodents, primates, and probably many other animals. Definitive hosts include dogs, cats and various other carnivores.

Sarcocystis, although recently demonstrated to cause disease in some animals, appear to be essentially nonpathogenic in New World monkeys (Dubey, 1977). Evidence of infection is an incidental histologic finding of cysts in striated muscle. The cysts are round in cross section, elliptical and up to 100 μm in their long axis. Cysts have a thick wall and contain bradyzoites that measure 3×12 μm, being about twice the size of *Toxoplasma* bradyzoites.

d. Intestinal Flagellates and Ciliates

Listings of intestinal flagellates and ciliates in the Cebidae include occasional reports of *Chilomastix, Embadomanas, Giardia, Trichomonas,* and *Balantidium* (Kuntz and Myers, 1972). These organisms appear to be commensals that are essentially nonpathogenic. Diarrhea has sometimes been attributed to trichomonad infection, but there is no proof that they cause such disease. Trichomonads are found in the fluid contents of the cecum and proximal large bowel and may be found in feces when the host has diarrhea. Fluid stools provide a favorable environment for the trichomonads but are not necessarily caused by them.

Giardia and *Balantidium* are occasionally found in spider monkeys and capuchins. *Balantidium* resides in the large bowel and may cause erosion and ulceration of the mucosa. Diagnosis is made by identification of these large ciliates in the stool by microscopic examination. Treatment regimens for lumenal amoebiasis are effective for balantidiasis.

Giardia are found in the upper small intestine. These organisms may cause diarrhea and steatorrhea. Diagnosis is made by finding trophozoites or cysts in feces or finding trophozoites in intestinal biopsy specimens. Treatment with metronidozale (Flagyl), 30 mg/kg for 5–10 days, is usually effective.

e. Amoebae

Amebiasis in New World primates is caused by infection with *Entamoeba histolytica*. This organism lives in the large bowel where it may cause erosions and ulcers in the mucosa. Occasionally organisms may spread to the liver and induce amoebic abscesses. Two races of *E. histolytica* are recognized and are distinguished by size. The larger pathogenic race has trophozoites that measure 20–30 μm. The smaller, non-pathogenic race measures up to 15 μm. Pathogenic forms also may contain ingested red blood cells. Cysts are formed in the large bowel and may be uninucleate or binucleate. They may mature in the bowel or outside the body, eventually having four nuclei and rodlike chromatoid bodies.

Cysts are readily destroyed by drying but may survive as long as 1 month in water. Infected monkeys pass cysts in feces and are the most important source of infection for other monkeys. Infected humans could potentially be a source of infection, especially if involved with preparation or handling of feed. Flies and cockroaches may serve as mechanical carriers of the parasitic cysts.

Little information is available regarding amebiasis in most New World monkeys; however, in *Ateles* and *Lagothrix,* the disease may be very severe and even fatal (Eichhorn and Gallagher, 1916; Henderson *et al.,* 1970).

Lesions are principally found in the cecum and proximal colon and initially are small, 3–5 mm superficial erosions of the mucosa. Tissue necrosis may extend through the muscularis mucosae and then spread laterally, producing ulcers that are flask-shaped in cross section (Fig. B.7). Ulcers may vary in diameter up to 1 cm or greater. Ulcers have well-defined, raised reddened edges with depressed centers. Lateral extension of the lesions in the submucosa may result in connecting sinuses between adjacent ulcers and irregular trenches of ulceration. Lesions may extend through the muscular coats of the bowel, inducing peritonitis. In fulminant infections,

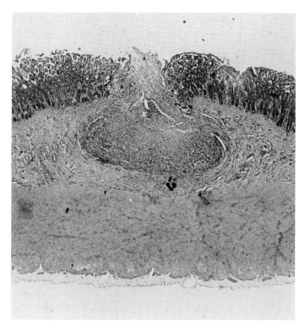

Fig. B.7. Ulcer in colon of woolly monkey caused by *Entamoeba histolytica* Hematoxylin and eosin. ×18.

large areas of the mucosa of the cecum and colon may become necrotic. Such extensive necrotic areas appear gray-white in color, are thickened, and are well demarcated from viable tissue. The affected cecum and colon may be semirigid and appear distended.

Microscopically, the main change is that of necrosis. Amoebae are found at the edges of lesions. The inflammatory reaction is not marked and consists of a few granulocytes, lymphocytes, and scattered macrophages. In tissue sections amoebae appear shrunken and surrounded by a clear space. They stain poorly with hematoxylin and eosin, but stain bright red with periodic acid–Schiff stain. Iron hematoxylin stain is necessary to demonstrate nuclear structure.

With invasive intestinal disease, amoebae may spread to the liver via the portal vein. Amoebae cause focal necrosis of the liver, and trophozoites can be found in the advancing edges of the lesions. As necrotic foci enlarge, contents of the lesions become liquified, forming amoebic abscesses. Necrotic tissue is sharply demarcated from normal tissue by a narrow zone of inflammatory cells, lymphocytes, and granulocytes. Fibrous encapsulation may occur in long-standing lesions.

Infected monkeys may not exhibit signs of illness even though considerable ulceration may be present. Principal signs of disease in monkeys consist of apathy, diarrhea, and gradual weight loss. Affected monkeys are inattentive and dejected. They spend much of the time in a resting attitude, crouched with head between legs and tail curled over the body. For the most part, the monkeys continue to eat and are afebrile. Affected monkeys have diarrhea that may be intermittent. Stools are watery, fetid, and may contain mucus and occult blood, but frank dysentery is not characteristic. The disease usually has a protracted course of several weeks; however, fatalities may occur within a few days in untreated spider and woolly monkeys.

Amebiasis must be considered in the differential diagnosis of diarrheal disease in New World monkeys, especially in spider and woolly monkeys. It must be differentiated from other forms of large bowel disease and other conditions such as dietary change that may cause diarrhea. Diagnosis is dependent on finding and identifying the organism. Amoebae should be looked for in wet mounts of fresh stool specimens. Trophozoites are very sensitive to chilling and only remain active and recognizable for about 30 mins, so only fresh materials should be examined. Fresh specimens should also be fixed, concentrated, and stained to find and identify cysts. These methods for finding amoebae are not very sensitive, so repeated examinations should be made.

Therapy for amebiasis is directed at the trophozoites, and treatment may be necessary for both the lumenal and tissue infections (see Table B.XXII).

f. Encephalitozoon cuniculi

Encephalitozoon cuniculi is a protozoan parasite that causes chronic and usually latent infections in rabbits and rodents. Infections with this organism also have been found in carnivores and primates (Canning, 1977). Infection occurs following the ingestion of spores. The site of initial replication in vertebrate hosts is not known with certainty; however, the parasite has a predilection for cells of the brain and renal tubules.

Reports of infection in primates have been rare, but include several cases in humans and squirrel monkeys (Anver *et al.*, 1972; Brown *et al.*, 1972). Infections in squirrel monkeys have been found in newborns and neonatal monkeys that died at a few months of age. These monkeys were most likely infected

in utero. Affected monkeys may exhibit intermittent petit mal seizures. Infection in adult monkeys is apparently latent.

Lesions in affected squirrel monkeys consist of multiple microscopic foci of granulomatous encephalitis, hepatitis, and nephritis; areas of necrosis surrounded by lymphocytes, plasma cells, and macrophages, or in the case of the brain, microglia. Organisms may be found in the central regions of granulomas or in intact cells without attendant inflammation. Perivascular cuffing with lymphocytes and plasma cells may be present in the brain. The organisms stain poorly with hematoxylin–eosin stain but stain well with Giemsa and Goodpasture stains and are gram-positive. Staining characteristics help to differentiate *Encephalitozoon* from *Toxoplasma*, since *Toxoplasma* stain moderately well with hematoxylin–eosin, are gram-negative, and do not stain well with Goodpasture stains.

Diagnosis may be made by demonstration of lesions and identification of the organisms histologically. The lesions and organisms must be differentiated from toxoplasmosis. Techniques have been reported for antemortem diagnosis of encephalitozoonosis in rabbits that may be applicable to monkeys. Indirect fluorescent antibody tests and intradermal skin tests using *Encephalitozoon* antigens appear to be effective means for identification of latently infected animals.

No chemotherapeutic agent is known that is effective against *Encephalitozoon* in vertebrates. Spread of infection occurs through spores shed intermittently in the urine of infected animals. Control of the spread of the infection requires that latently infected individuals be identified and isolated. Rodents and other feral animals that might be infected and contaminate the environment of captive primates should be controlled.

2. Metazoan Parasites

a. Helminths

i. Nematodes. The majority of helminth parasites of New World monkeys are nematodes. Infestations in newly captured animals may be heavy, producing illness, especially in young animals. Nematode parasites may be a problem in the laboratory setting in circumstances of poor sanitation or in the absence of adequate vector control. In many infestations the worm burdens are light, causing little obvious disease. Nevertheless, they do the host no good, they have the potential for serious illness, may complicate research use, and should be eliminated where possible.

(a) Strongyloides. Strongyloidiasis in New World monkeys is caused by infestations with *Strongyloides cebus* (Little, 1966). These worms infest the duodenum and jejunum, usually producing little pathologic change. In heavy infestations there may be edema and inflammation of the mucosa, affecting absorptive processes and causing diarrhea.

Table B.XXII

Treatment for Amebiasis

Drug	Dose (mg/kg)	Site of action	Duration of treatment (days)
Tetracyclines	25–50	Lumen	5–10
Diiodohydroxyquin (Diodoquin)	30	Lumen	5–10
Metronidazole (Flagyl)	30	Lumen and tissue	5–10
Chloroquin	5	Tissue	14

Parthenogenetic female worms, which are 2–5 mm in length, produce embryonated eggs that are passed in the feces. In 2–3 days, third-stage larvae develop that may infect the host by penetration of the skin or by ingestion. If larvae penetrate the skin, they may migrate to the lungs where they molt to fourth-stage larvae, are coughed up, swallowed, and develop into parthenogenetic females in the small intestine. Ingested larvae may develop directly into adult females in the intestine.

Diagnosis is made by identifying characteristic eggs in the feces or identifying worms in the bowel. Thiabendazole, 50 to 100 mg/kg, is an effective treatment.

(b) Molineus. *Molineus* are trichostrongyles that are common in New World monkeys. They are minute, slender, pale red worms that measure 3–10 mm in length and are found in the duodenum and the pyloric portion of the stomach. *Molineus torulosus* has been found in *Cebus* and *Saimiri*, and *M. elegans* in *Saimiri*. The infestations in *Saimiri* are usually light and appear to cause little clinical disease.

Molineus torulosus may produce severe lesions in *Cebus* monkeys (Brack *et al.*, 1973). Ingested larvae migrate to the submucosa of the bowel, where they mature to adults and deposit eggs. Submucosal granulomas develop that appear as gray, black, brown, or green 5-mm nodules protruding from the serosal surface. Eggs are expelled to the lumen of the intestine through mucosal ulcers. Worms and eggs may be found in submucosal veins associated with fibrinous thrombi. Worms and eggs may also be found in pancreatic ducts, causing inflammation of these structures as well as chronic pancreatitis. Infestation may be diagnosed by finding eggs in the feces or by identifying adult worms in tissues.

(c) Filaroides. *Filaroides* are lung worms commonly found in New World monkeys (Orihel and Seibold, 1972). These worms inhabit bronchioles, alveolar ducts and alveoli (Fig. B.8). They may be associated with peribronchiolar and interstitial mononuclear leukocyte infiltrates but usually do not cause overt clinical disease. Foci of infected lung appear as small elevated nodules on the pleural surface that are dark gray or black. Female worms are ovoviviparous, releasing into the airways larvae that are coughed up and swallowed. Squirrel monkeys born in captivity from infected parents have been free of the disease, presumably because of the absence of a suitable intermediate host.

(d) Trypanoxyuris. *Trypanoxyuris* are pinworms that are common in New World monkeys (Inglis and Dunn, 1964). The life cycle is direct with adults inhabiting the cecum and large

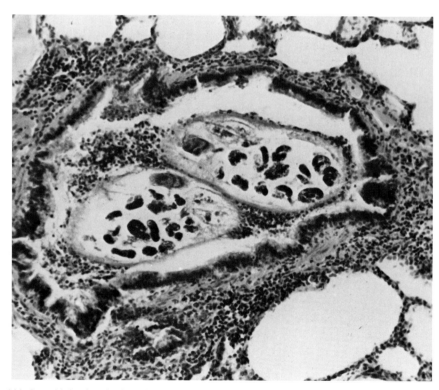

Fig. B.8. *Filaroides* sp. within bronchiole of squirrel monkey associated with peribronchiolar infiltrate of mononuclear leukocytes. Hematoxylin and eosin. ×110.

bowel. Females measure up to 10 mm. Diagnosis is made by finding the typical asymmetrical eggs in the feces or by finding the adult worms in the large bowel. Usually these worms produce little pathogenic change. Thiabendazole, 50–100 mg/kg, is an effective treatment.

(e) Protospirura. *Protospirura muricola* is a spirurid nematode that is principally a parasite of feral rodents but occasionally has been found in various New World monkeys including *Cebus, Aotus,* and *Ateles.* These worms measure up to 7 cm in length and are found in the lumen of the terminal esophagus and in the stomach. Heavy infestations may result in obstruction and tissue invasion in the terminal esophagus with marked inflammation and necrosis. The dusty-tailed or Surinam cockroach, *Leucophaea maderae* may serve as an intermediate host (Foster and Johnson, 1939).

(f) Trichospirura. *Trichospirura leptostoma* are nematodes found in the pancreatic ducts of callitrichids and also in *Callicebus, Saimiri,* and *Aotus* (Fig. B.9). The worms measure up to 2 cm in length and produce typical thick-shelled, spirurid

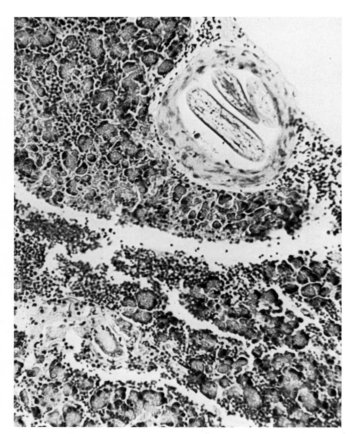

Fig. B.9. *Trichospirura leptosoma* in pancreatic duct of squirrel monkey associated with acute pancreatitis. Hematoxylin and eosin. ×100.

eggs. Most often there is little tissue response to them, but occasionally they may be associated with mild pancreatic acinar atrophy and inflammation. Acute pancreatitis has rarely been reported (Orihel and Seibold, 1972).

(g) Physaloptera. *Physaloptera* are rather large worms, measuring up to 5 cm in length. They have been found in numerous nonhuman primates including several of the New World monkeys, *Cebus, Lagothrix,* and *Ateles. Physaloptera* are found in the stomach where they attach to the gastric mucosa. Light infestations are usual asymptomatic. As with other spiruroids this parasite requires an arthopod intermediate host. Diagnosis may be made by finding typical eggs in the feces or the adult worms in the stomach.

(h) Dipetalonema and Tetrapetalonema. At least eight species of filarid worms have been described in New World monkeys, and all except one belong to the genera *Dipetalonema* or *Tetrapetalonema.* Most of them are found in the subcutaneous tissues or in the serous cavities of the body. *Tetrapetalonema* have been found in the connective tissues over the back, in and around the scapulae. *Dipetalonema* are found in periesophageal connective tissues, in the serous cavities, predominantly in the peritoneal cavity, but also occasionally in the pleural cavities and in the pericardial sac. These worms usually do not induce much of an inflammatory reaction. Heavy infestations in the serous cavities sometimes are associated with a fibrinous peritonitis and many lacelike adhesions.

Larvae are called microfilariae and they circulate in the blood. The differential characters of the filariae in South American monkeys have been described (Chalifoux *et al.,* 1973). These filarids are transmitted by blood-sucking arthropods, probably *Culicoides* flies (Dunn and Lambrecht, 1963). Diethylcarbamazine (50 mg/kg per day for 10 days) has been found to be effective treatment against microfilariae and adult filarid parasites (Eberhard, 1982).

(i) Gongylonema. *Gongylonema* are spirurid parasites that have been found in several New World monkeys including *Saimiri, Callicebus,* and *Cebus* (Orihel and Seibold, 1972). These worms are found in tunnels that they produce in the stratum malpighii of the mucosa of the esophagus. Eggs are also laid in these tunnels, and they eventually are shed at the surface of the mucosa and passed in the feces. An arthropod intermediate host is necessary for the eggs to hatch and develop into infective stages. Little pathologic change in the esophageal mucosa has been associated with these worms.

(ii) Acanthocephalans. *Prosthenorchis* are thorny-headed worms that are probably the most pathogenic of all the helminth parasites of South American monkeys. Two species, *P.*

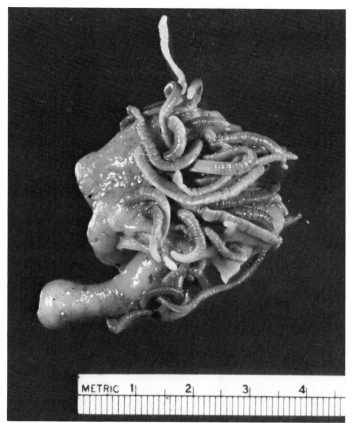

Fig. B.10. *Prosthenorchis elegans* attached to mucosa of terminal ileum and cecum of a squirrel monkey.

elegans and *P. spirula*, have been found. *Prosthenorchis* have a wide host tolerance and in addition to the Callitrichidae, have been found in *Saimiri, Cebus, Callicebus,* and *Ateles.* These worms quickly adapt to new hosts and have been experimentally transmitted to both primate and nonprimate mammals. Originally a South American parasite, these worms also have caused infections in Old World monkeys and apes in zoological gardens.

Adult *Prosthenorchis* inhabit the bowel, attaching to the terminal ileum (Fig. B.10). They have an armed proboscis that penetrates the mucosa and muscularis mucosa to embed in the muscular wall of the intesine where they cause localized inflammation and sclerosis. Occasionally they penetrate the intestine and even invade the peritoneal cavity, causing peritonitis. In heavy infestations these worms may cause obstruction of the bowel lumen. Intussusception of the gut may be associated with the worms and lesions induced by them.

Eggs passed in the feces require intermediate arthropod hosts to become infective larvae. Various cockroaches, including the German roach *Blatella germanica,* and several beetles may serve as the intermediate host.

Diagnosis is made by finding typical thick-shelled eggs in the feces or by finding the adult worm. Infestation may be suspected by palpating nodules transabdominally. Control of these organisms requires interruption of their life cycle by limiting access of intermediate hosts, cockroaches and beetles, to the primates (Schmidt, 1972).

iii. Trematodes. Infestation with trematodes depends upon the habitat, feeding habits, and availability of multiple mollusk intermediate hosts. In general, there has been very little information on trematode infestations in Cebidae, and few have been reported. The prevalence of infection usually has been low and the worm burdens light, with little pathologic change. Three trematodes have been described, two liver flukes and one intestinal fluke. *Phaneropsolus* are intestinal flukes that measure 1×0.5 mm and have been found in *Aotus, Cebus,* and *Saimiri. Controrchis* are 3×1 mm liver flukes found in the bile ducts and gallbladder of howler and spider monkeys. *Athesmia* are also liver flukes found in the bile ducts of *Cebus, Saimiri, Callicebus,* and *Aotus* (Cosgrove, 1966).

iv. Cestodes. Cestodes have indirect life cycles and may infect definitive hosts as adult worms or, in intermediate hosts, infect them as larva forms. Most cestodes that infest New World monkeys are found as adult worms in the intestinal tract associated with little pathologic change (Table B.XXIII). Sparganum, the plerocercoid larvae of *Spirometra* have been found in the connective tissues of *Saimiri.* These larvae migrate through tissues, causing inflammation and edema. Diagnosis of intestinal forms is usually made by finding eggs or segments of the adult worms in the feces or by recovery of adult worms at necropsy (Dunn, 1963).

b. Pentastomes

Pentastomes are the nymphal forms of tongue worms that are arthropod endoparasites of snakes. New World monkeys may become infected with the nymphal stages of the genus *Porocephalus* by ingesting water and food contaminated with excreta of infected snakes. The eggs hatch in the intestine of the monkey, and larvae migrate through the bowel into the tissues. Ultimately these larval forms go through a metamorphosis into a nymphal stage that may be found in any organ or tissue, but are usually found in the lungs, liver, or in the mesentery. Encysted nymphs are about 1 cm in diameter but do not cause a tissue reaction. Snakes become infected when they ingest an infected mammal (Self and Cosgrove, 1972).

c. Ectoparasites

Only a few reports describe ectoparasite infestations in New World monkeys. This may in part be due to the grooming hab-

Table B.XXIII

Cestodes of the Cebidae[a]

Genus	Bertiella	Moniezia	Atriotaenia	Raillietina	Hymenolepis	Spirometra (sparganum)
			Transmitted by			
	Mite	Mite	Arthropod	Arthropod	Rodents	Reptile, amphibians
Alouatta	X	X	X	X		
Ateles		X	X			
Brachyteles		X				
Callicebus	X			X	X	
Cebus	X	X	X			
Saimiri			X		X	X

[a]Adapted from Myers (1972).

its and dextrous nature of monkeys and their ability to remove such parasites from themselves and cohorts. Lice appear to be a frequent and common infestation in spider monkeys and howler monkeys, however. Infestations with a louse, very similar if not identical to the human louse *Pediculus humanus,* have been reported from spider monkeys. Severe anemia has been associated with this infestation (Norman and Wagner, 1973). Biting lice of the genus *Trichodectes* also are common parasites of howler monkeys.

Mites have rarely been reported as being parasitic problems in New World monkeys; however, squirrel monkeys with a generalized scaly dermatitis have been found to be infested with the mites, *Fonsecalges saimirii* (Flatt and Patton, 1969). These mites were numerous in skin scrapings and could be observed with the unaided eye.

C. Rickettsial

Eperythrozoon and Haemobartonella

Eperythrozoon and *Haemobartonella* are rickettsia that are thought to be arthropod-borne, including spread by lice, fleas, and biting flies, although data on vectors are not available in most cases. Transmission can also be effected by injection of blood or blood-containing tissues from infected animals. *Haemobartonella* have been described in *Cebus apella* and *Saimiri sciureus* and *Eperythrozoon* in *Aotus trivirgatus* (Aikawa and Nassenzweig, 1972; Peters *et al.,* 1974). The significance and prevalence of these infections in South American monkeys is not known, however. *Eperythrozoon* and *Haemobartonella* typically produce inapparent infections in normal animals, as patent parasitemia occurs only for a brief time after initial infection. Infection may be exacerbated following splenectomy. Clinical and pathologic changes are the result of erythrocyte destruction and anemia. It is not known if the organisms infecting New World monkeys will produce clinical disease in normal or asplenic hosts.

These organisms appear as red or purple cocci 0.3–1 μm in diameter on or indenting the erythrocyte membrane. *Haemobartonella* also may appear as chains of cocci or be rod-shaped (Gothe and Kreier, 1977). Inoculation of suspect blood into test animals that have been splenectomized 4 weeks previously may identify latent infections. Increasing numbers of parasites should appear in the blood of the test animal with time if the suspect animal was infected. Complement fixation, indirect fluorescent antibody, or indirect hemagglutination inhibition tests that have been used in other animals possibly could be adapted for diagnostic use in New World monkeys.

The effectiveness of treatment for infections with *Haemobartonella* and *Eperythrozoon* in New World monkeys is not known. Antibiotic treatment regimens efficacious in cats for similar infections may be useful: 50 mg/kg of tetracycline or chloramphenicol orally for 10 days.

D. Viral

a. Herpesvirus platyrrhinae

Herpesvirus platyrrhinae causes latent infections in a number of New World monkeys, most notably *Saimiri, Cebus,* and *Ateles* (Holmes *et al.,* 1966). Overt disease has rarely been seen in these reservoir hosts, but when it occurs has consisted

Fig. B.11. Lingual ulcers in a juvenile squirrel monkey consistent with *Herpesvirus platyrrhinae* infection.

of oral, lingual, and labial ulcers in squirrel monkeys (Fig. B.11). Microscopically these lesions contain characteristic intranuclear inclusion bodies. Healing occurs uneventfully in about 1 week. Recurrences of lesions are not known to occur, but the virus is shed periodically. Spread of the infection occurs through direct contact, aerosols, and fomites.

This virus causes pantropic infections in marmosets and owl monkeys, spreading rapidly and causing high morbidity and mortality. Affected monkeys are lethargic, depressed, and anorectic and have oral, labial, and cutaneous ulcers. These monkeys become moribund and die in a few days. Lesions occur in most organs and tissues and consist of hemorrhage and focal necrosis with characteristic intranuclear inclusion bodies (Hunt and Meléndez, 1969). Ulcers are found throughout the gastrointestinal tract.

Prevention of infection in marmosets and owl monkeys in part requires maintaining these monkeys separate from the reservoir hosts. A variant virus vaccine has been developed that appears effective in protecting vaccinated monkeys.

b. Herpesvirus hominis

Herpesvirus hominis causes a disease in owl monkeys that is indistinguishable from that of *Herpesvirus platyrrhinae*. Infection probably occurs from persons shedding the virus. People with active lesions should not be allowed direct or indirect contact with owl monkeys (Meléndez *et al.*, 1969).

c. Herpesvirus saimiri

Squirrel monkeys of feral origin are infected with *Herpesvirus saimiri* (Falk *et al.*, 1972a). As with all known herpesviruses in definitive hosts, the infection is latent but has not been associated with pathological conditions in squirrel monkeys. This virus persists in a latent state in lymphocytes and shares several common properties with Epstein-Barr virus and Marek's disease virus (Falk *et al.*, 1972b). When inoculated into owl monkeys or marmosets, these monkeys develop maglignant lymphoma of the reticulum cell type and die in 3–6 weeks (Hunt and Meléndez, 1974).

d. Measles

Measles (rubeola) is a highly contagious disease of humans, principally children, but does affect a number of nonhuman primates, including New World monkeys. Of the cebid monkeys, *Saimiri* and *Cebus* have been affected. Signs in affected *Cebus albifrons*, *Cebus apella*, and *Saimiri sciureus* include lethargy, dyspnea, serous nasal exudate, and conjunctivitis. Capuchins have a diffuse erythematous rash most notable on the abdomen and inner aspects of the thigh. The disease in capuchins is quite mild, and they recover in about 1 week.

The disease in squirrel monkeys is very severe and sometimes fatal. An extensive maculopapular rash develops over most of the body several days after onset of illness and persists for 5–7 days. Signs of respiratory disease may persist for several weeks. Some affected squirrel monkeys develop acute gastric distension and have tarry stools. The case fatality rate for measles-infected squirrel monkeys is about 5%. Gross lesions in the lung appear as focal red and purple areas. Microscopically the lung lesions include hyperplasia of bronchiolar epithelium with the occurrence of epithelial syncytia and giant cells. Intranuclear eosinophilic inclusion bodies are found in epithelium and giant cells. Gastrointestinal lesions include gastric dilation with serosal and luminal hemorrhages. The small intestine may have extensive submucosal hemorrhage. Microscopic characteristics of the skin lesions include hair follicle necrosis and syncytia formation in the granulosa layer (Abee *et al.*, 1972).

e. Yellow Fever

Yellow fever is a severe disease of humans and nonhuman primates, transmitted by mosquitos, caused by a group B arbovirus. The disease has occurred in cyclic epidemics in Africa and Central and South America. The disease appears to have two phases in its epidemiology, an urban cycle with human to human transmission and a sylvatic or jungle cycle involving principally animals including nonhuman primates. In the Americas, *Aedes* mosquitos are important vectors of the urban form, while *Haemogogus* mosquitos are the vectors for the jungle cycle. Many New World monkeys are susceptible, and epizootics of the disease have occurred. The incubation period in monkeys is 3–8 days. Serological data suggest that many animals recover. The disease may be very severe and is characterized by midzonal necrosis and fatty change of the liver, degeneration of the kidney, and gastrointestinal hemorrhage. Necrotic hepatocytes develop a peculiar intracytoplasmic hyaline change called a Councilman body. The disease may be severe in *Alouatta*, *Saimiri*, *Callicebus*, *Aotus*, *Cacajao*, and *Pithecia* and milder in *Cebus* and *Ateles* (Felsenfeld, 1972).

Table B.XXIV

Tumors of the Cebidae[a]

Organ or system	Alouatta	Aotus	Ateles	Cebus	Lagothrix	Saimiri
Endocrine	20[b]			1		
Skin				1	1	
Musculoskeletal			2	2		
Hematopoietic	1	2			1	1
Respiratory						1
Gastrointesinal				1		2
Liver/gallbladder	1					
Urinary	1	1		2		
Male genital		1				
Female genital			1	1		3
Miscellaneous				1		

[a] Adapted from Squire *et al.* (1978).
[b] Number reported.

E. Neoplastic

Tumors have not been frequently reported in the Cebidae other than lymphoproliferative diseases experimentally induced by infection with some herpesviruses. The frequency of finding tumors in these monkeys undoubtedly has been affected by the relatively small number of animals that have been carefully examined and the small numbers of animals maintained until aged. Recognizing these limitations, the prevalence of tumors in the Cebidae may not be as rare as previously considered, as tumors have been described in many organ systems. Squire *et al.*, (1978) have summarized reports of neoplastic disease in these monkeys. (Table B.XXIV).

REFERENCES

Abee, C. R., Martin, P. D., Lehner, N. D. M., and Falk, L. E., Jr. (1972). Clinical and pathologic characteristics of measles in squirrel monkeys. *23rd Annu. Sess., Am. Assoc. Lab. Anim. Sci.* Abstracts, Publ. 72-3.

Aikawa, M., and Nassenzweig, R. (1972). Fine structure of *Haemobartonella* sp. in the squirrel monkey. *J. Parasitol.* **58**, 628–630.

Anver, M. R., King, M. W., and Hunt, R. D. (1972). Congenital encephalitozoonosis in a squirrel monkey. *Vet. Pathol.* **9**, 475–480.

Ausman, L. M., and Gallina, D. L. (1979). Liquid formulas and protein requirements of nonhuman primates. *In* "Primates in Nutritional Research" (K. C. Hayes, ed.), pp. 39–57. Academic Press, New York.

Ausman, L. M., Hayes, K. C., Lage, A., and Hegsted, D. M. (1970). Nursery care and growth of Old and New World infant monkeys. *Lab. Anim. Care* **20**, 907–913.

Ausman, L. M., Gallina, D. L., Hayes, K. C., and Hegsted, D. M. (1976). Hematologic development of the infant squirrel monkey (*Saimiri sciureus*). *Folia Primatol.* **26**, 292–300.

Baggs, R. B., Hunt, R. D., Garcia, F. G., Hajema, E. M., Blake, B. J., and

Fraser, C. E. O. (1976). Pseudotuberculosis (*Yersinia enterocolitica*) in the owl monkey (*Aotus trivirgatus*). *Lab. Anim. Sci.* **26**, 1079–1083.

Baldwin, J. D. (1968). The social behavior of the adult male squirrel monkey in a seminatural environment. *Folia Primatol.* **9**, 281–314.

Baldwin, J. D., and Baldwin, J. I. (1981). The squirrel monkeys, genus *Saimiri. In* "Ecology and Behavior of Neotropical Primates" (A. F. Coimbra-Filho and R. A. Mittermeier, eds.), Vol. 1, pp. 277–330. Academia Brasileira de Ciências, Rio de Janeiro.

Barahona, H., Meléndez, L. V., Hunt, R. D., and Daniel, M. D. (1976). The owl monkey (*Aotus trivirgatus*) as an animal model for viral diseases and oncologic studies. *Lab. Anim. Sci.* **26**, 1104–1112.

Benjamin, S. A., and Lang, C. M. (1971). Acute pasteurellosis in owl monkeys, *Aotus trivirgatus. Lab. Anim. Sci.* **21**, 258–262.

Bone, J. F., and Soave, O. A. (1970). Experimental tuberculosis in owl monkeys, *Aotus trivirgatus. Lab. Anim. Care* **20**, 946–948.

Brack, M., Myers, B. J., and Kuntz, R. E. (1973). Pathogenic properties of *Molineus torulosus* in capuchin monkeys, *Cebus apella. Lab. Anim. Sci.* **23**, 360–365.

Brown, R. J., Hinkle, D. K., Trevethan, W. P., Kupper, J. L., and McKee, A. E. (1972). Nosematosis in a squirrel monkey (*Saimiri sciureus*). *J. Med. Primatol.* **2**, 114–123.

Bullock, B. C., Wolf, R. H., and Clarkson, T. B. (1967). Myocarditis associated with trypanosomiasis in a cebus monkey (*Cebus albifrons*). *J. Am. Vet. Med. Assoc.* **151**, 920–922.

Canning, E. U. (1977). Microsporidea. *In* "Parasitic Protozoa" (J. P. Kreier, ed.), Vol. 4, pp. 155–196. Academic Press, New York.

Castellanos, H., and McCombs, H. L. (1968). The reproductive cycle of the New World monkey. *Fertil. Steril.* **19**, 215–226.

Chalifoux, L. V., Hunt, R. D., Garcia, F. G., Sehgal, P. K., and Comiskey, J. R. (1973). Filariasis in New World monkeys: Histochemical differentiation of circulating microfilariae. *Lab. Anim. Sci.* **23**, 211–220.

Cicmanec, J. C., and Campbell, A. K. (1977). Breeding the owl monkey (*Aotus trivirgatus*) in a laboratory environment. *Lab. Anim. Sci.* **27**, 512–517.

Clarkson, T. B., Lehner, N. D. M., Bullock, B. C., Lofland, H. B., and Wagner, W. D. (1976). Atherosclerosis in New World monkeys. *Primates Med.* **9**, 90–144.

Coatney, R. G., Collins, W. E., McWilson, W., and Contacos, P. G. (1971). "The Primate Malarias." U.S. Govt. Printing Office, Washington, D.C.

Coe, C. L., Mendoza, S. P., Davidson, J. M., Smith, E. R., Dallman, M. F., and Levine, S. (1978). Hormonal response to stress in the squirrel monkey (*Saimiri sciureus*). *Neuroendocrinology* **26**, 367–377.

Cosgrove, G. E. (1966). The trematodes of laboratory primates. *Lab. Anim. Care* **16**, 23–39.

deRodaniche, E. (1954). Spontaneous toxoplasmosis in the whiteface monkey, *Cebus capucinus*, in Panama. *Am. J. Trop. Med. Hyg.* **3**, 1023–1025.

deRodaniche, E. (1955). Susceptibility of the marmoset *Marikina geoffroyi* and the night monkey, *Aotus zonalis*, to experimental infection with Toxoplasma. *Am. J. Trop. Med. Hyg.* **3**, 1026–1032.

Dubey, J. P. (1977). *Toxoplasma, Hammondia, Besnoitia, Sarcocystis,* and other tissue cyst-forming coccidia of man and animals. *In* "Parasitic Protozoa" (J. P. Kreier, ed.), Vol. 3, pp. 101–239. Academic Press, New York.

DuMond, F. V. (1968) The squirrel monkey in a seminatural environment. *In* "The Squirrel Monkey" (L. A. Rosenblum and R. W. Cooper, eds.), pp 87–145. Academic Press, New York.

DuMond, F. V., and Hutchinson, T. C. (1967). Squirrel monkey reproduction: The fatted male phenomenon and seasonal spermatogenesis. *Science* **158**, 1467–1470.

Dunn, F. L. (1963). Acanthocephalans and cestodes of South American monkeys and marmosets. *J. Parasitol.* **49**, 717–722.

Dunn, F. L., and Lambrecht, F. L. (1963). On some filarial parasites of South American primates with a description of *Tetrapetalonema tamarinae* from Peruvian tamarin marmoset, *Tamarinus nigrocollis. J. Helminthol.* **37**, 261–286.

Dunn, F. L., Lambrecht, F. L., and DuPlessis, R. (1963). Trypanosomes of South American monkeys and marmosets. *Am. J. Trop. Med. Hyg.* **12**, 524–534.

Eberhard, M. L. (1982). Chemotherapy of filariasis in squirrel monkey, *Saimiri sciureus. Lab. Anim. Sci.* **32**, 397–400.

Eichhorn, O., and Gallagher, B. (1916). Spontaneous amebic dysentery in monkeys. *J. Infect. Dis.* **19**, 395–407.

Elliott, M. W., Sehgal, P. K., and Chalifoux, L. V. (1976). Management and breeding of *Aotus trivirgatus. Lab. Anim. Sci.* **26**, 1037–1040.

Falk, L. A., Jr., Wolfe, L., and Deinhardt, F. (1972a). Epidemiology of *Herpesvirus saimiri* infection in squirrel monkeys. *Med. Primatol., Conf. Exp. Med. Surg. Primates, 3rd, 1972* Vol. 3, pp. 151–158.

Falk, L. A., Jr., Wolfe, L. G., and Deinhardt, F. (1972b). Isolation of *Herpesvirus saimiri* from blood of squirrel monkeys, *Saimiri sciureus. JNCI, J. Natl. Cancer Inst.* **48**, 1499–1505.

Felsenfield, A. D. (1972). The arboviruses *In* "Pathology of Simian Primates," (R. N. T-W-Fiennes, ed.), Part II, pp. 523–536. Karger, Basel.

Flatt, R. E., and Patton, N. M. (1969). A mite infestation in squirrel monkeys. *J. Am. Vet. Med. Assoc.* **155**, 1233–1235.

Foster, A. C., and Johnson, C. M. (1939). A preliminary note on the identity, life cycle and pathogenicity of an important nematode parasite of captive monkeys. *Am. J. Trop. Med.* **19**, 265–277.

Frankel, J. K. (1973). Toxoplasmosis: Parasite life cycle, pathology and immunology. *In* "The Coccidia: Eimeria, Toxoplasma, Isospora, and Related Genera" (D. M. Hammond and P. Long, eds.), pp. 343–410. University Park Press, Baltimore, Maryland.

Garnham, P. C. C. (1966). "Malarial Parasites and Other Haemosporidia." Blackwell, Oxford.

Giles, R. C., Jr., Hildebrandt, P. K., and Tate, C. (1974). *Klebsiella* air sacculitis in the owl monkey, *Aotus trivirgatus. Lab. Anim. Sci.* **24**, 610–616.

Glander, K. E. (1980). Reproduction and population growth in free-ranging mantled howling monkeys. *Am. J. Phys. Anthropol.* **53**, 25–36.

Good, R. C., May, B. D., and Kawatomari, T. (1969). Enteric pathogens in monkeys. *J. Bacteriol.* **97**, 1048–1055.

Gothe, R., and Kreier, J. P. (1977). *Aegyptianella, Eperythrozoon,* and *Haemobartonella. In* "Parasitic Protozoa" (J. P. Kreier, ed.), Vol. 4, pp. 251–294. Academic Press, New York.

Greenstein, E. T., Doty, R. W., and Lowy, K. (1965). An outbreak of a fulminating infectious disease in the squirrel monkey, *Saimiri sciureus. Lab. Anim. Care* **15**, 74–80.

Hall, R. D., and Hodgen, G. D. (1979). Pregnancy diagnosis in owl monkeys (*Aotus trivirgatus*): Evaluation of the HAI test for urinary chorionic gonadotropin. *Lab. Anim. Sci.* **29**, 345–348.

Hayes, K. C., Fay, G., Roach, A., and Stare, F. J. (1972). Breeding New World monkeys in a laboratory environment. *In* "Breeding Primates," (W. I. B. Beveridge, ed.), pp. 122–129. Karger, Basel.

Henderson, J. D., Jr., Webster, W. S., Bullock, B. C., Lehner, N. D. M., and Clarkson, T. B. (1970). Naturally occurring lesions seen at necropsy in eight woolly monkeys (*Lagothrix* sp.). *Lab. Anim. Care* **20**, 1087–1097.

Hershkovitz, P. (1979). The species of *Sakis*, Genus *Pithecia* with notes on sexual dichromatism. *Folia Primatol.* **31**, 1–22.

Hessler, J. R., Woodard, J. C., and Tucek, P. C. (1971). Lethal toxoplasmosis in a woolly monkey. *J. Am. Vet. Med. Assoc.* **159,** 1588–1594.

Holmes, A. W., Devine, J. A., Nowakowski, E., and Deinhardt, F. (1966). The epidemiology of a herpesvirus infection of New World monkeys. *J. Immunol.* **90,** 668–671.

Hunt, R. D., and Meléndez, L. V. (1969). Herpesvirus infections of nonhuman primates: A review. *Lab. Anim. Care* **19,** 221–234.

Hunt, R. D., and Meléndez, L. V. (1974). Animal model: Herpesvirus-induced malignant lymphoma. *Am. J. Pathol.* **76,** 415–418.

Inglis, W. G., and Dunn, F. L. (1964). Some oxyurids (Nematoda) from neotropical primates. *Z. Parasitenkd.* **24,** 83–87.

Institute of Laboratory Animal Resources (ILAR) (1980). Laboratory animal management: Nonhuman primates. *ILAR News* **23** (2-3).

Jacobs, G. H., Tootell, R. B. H., and Blakesla, B. (1979). Visual capacities of the owl monkey *Aotus trivirgatus:* Temporal contrast sensitivity. *Folia Primatol.* **32,** 193–199.

Jones, T. C., Thorington, R. W., Hu, M. M., Adams, E., and Cooper, R. W. (1973). Karyotypes of squirrel monkeys (*Saimiri sciureus*) from different geographic regions. *Am. J. Phys. Anthropol.* **38,** 269–277.

Kaack, B., Walker, L., Brizzee, K. R., and Wolf, R. H. (1979). Comparative normal levels of serum triiodothyramine and thyroxine in nonhuman primates. *Lab. Anim. Sci.* **29,** 191–194.

Kesler, M. J., and Brown, R. J. (1979). Clinical description of tetanus in squirrel monkeys, *Saimiri sciureus. Lab. Anim. Sci.* **29,** 240–242.

King, N. W., Fraser, C. E. O., Garcia, F. G., Wolf, L. A., and Williamson, M. E. (1971). Cutaneous streptothricosis (dermatophiliasis) in owl monkeys. *Lab. Anim. Sci.* **21,** 67–74.

Kirck, J. A., and Remington, J. S. (1978). Toxoplasmosis in the adult—An overview. *N. Engl. J. Med.* **298,** 550–553.

Kournay, M., and Porter, J. A., Jr. (1969). A survey for enteropathogenic bacteria in Panamanian primates. *Lab. Anim. Care* **19,** 336–341.

Kuntz, R. E., and Myers, B. J. (1972). Parasites of South American primates. *Int. Zoo Yearb.* **12,** 61–68.

Lang, C. M. (1967). The estrous cycle of the squirrel monkey (*Saimiri sciureus*). *Lab. Anim. Care* **17,** 442–451.

Lasry, J. E., and Sheridan, B. W. (1965). Chagas' myocarditis and heart failure in the red uakari. *Int. Zoo Yearb.* **5,** 182–184.

Leathers, C. W., and Hamm, T. E. (1976). Naturally occurring tuberculosis in a squirrel monkey and a *Cebus* monkey. *J. Am. Vet. Med. Assoc.* **169,** 909–911.

Lehner, N. D. M., Bullock, B. C., Clarkson, T. B., and Lofland, H. B. (1967). Biological activities of vitamins D$_2$ and D$_3$ for growing squirrel monkeys. *Lab. Anim. Care* **17,** 483–493.

Lehner, N. D. M., Bullock, B. C., and Clarkson, T. B. (1968). Ascorbic acid deficiency in the squirrel monkey. *Proc. Soc. Exp. Biol. Med.* **128,** 512–514.

Little, M. D. (1966). Comparative morphology of six species of *Strongyloides* and redefinition of the genus. *J. Parasitol.* **52,** 69–84.

London, W. T., Nahmias, A. J., Naib, Z. M., Fuccillo, D. A., Ellenberg, J. H., and Sever, J. L. (1974). A nonhuman primate model for the study of cervical oncogenic potential of herpes simplex virus type 2. *Cancer Res.* **34,** 1118–1121.

Long, J. O., and Cooper, R. W. (1968). Physical growth and dental eruption in captive-bred squirrel monkeys, *Saimiri sciureus. In* "The Squirrel Monkey" (L. A. Rosenblum and R. W. Cooper, eds.), pp. 193–205. Academic Press, New York.

Lorenz, R., and Mason, W. A. (1971). Establishment of a colony of titi monkeys. *Int. Zoo Yearb.* **11,** 168–175.

Ma, N. S. F., Jones, T. C., Miller, A. C., Morgan, L. M., and Adams, E. A. (1976). Chromosome polymorphism and banding patterns in the owl monkey (*Aotus*). *Lab. Anim. Sci.* **26,** 1022–1036.

Mack, D., and Kafka, H. (1978). Breeding and rearing of woolly monkeys, *Lagothrix lagotricha* at the National Zoological Park, Washington, *Int. Zoo. Yearb.* **18,** 117–122.

McKissick, G. E., Ratcliffe, H. L., and Koestner, A. (1968). Enzootic toxoplasmosis in caged squirrel monkeys *Saimiri sciureus. Pathol. Vet.* **5,** 538–560.

Malinow, M. R. (1968). "Biology of the Howler Monkey." Karger, Basel.

Mann, G. V. (1970). Nutritional requirements of cebus monkeys. *In* "Feeding and Nutrition of Nonhuman Primates." (R. S. Harris, ed.), pp. 143–157. Academic Press, New York.

Manning, P. J., Lehner, N. D. M., Feldner, M. A., and Bullock, B. C. (1969). Selected hematologic, serum chemical, and arterial blood gas characteristics of squirrel monkeys (*Saimiri sciureus*). *Lab. Anim. Sci.* **19,** 831–837.

Marinkelle, C. J. (1966). Observations on human, monkey, and bat trypanosomes and their vectors in Colombia. *Trans. R. Soc. Trop. Med. Hyg.* **60,** 109–116.

Meléndez, L. V., Espana, C., Hunt, R. D., Daniel, M. D., and Garcia, F. G. (1969). Natural herpes simplex infection in the owl monkey *Aotus trivirgatus. Lab. Anim. Care* **19,** 38–45.

Mendoza, S. P., Lowe, E. L., Davidson, J. M., and Levine, S. (1978). Annual cyclicity in the squirrel monkey (*Saimiri sciureus*): The relationship between testosterone, fatting, and sexual behavior. *Horm. Behav.* **11,** 295–303.

Meritt, D. A., Jr. (1970). Edentate diets currently in use at Lincoln Park Zoo, Chicago, *Int. Zoo Yearb.* **10,** 136–138.

Meritt, G. F., and Meritt, D. A., Jr. (1978). Hand-rearing techniques for douroucoulis. *Int. Zoo Yearb.* **18,** 201–204.

Middleton, C. C., and Kosal, J. (1972). Weights and measurements of normal squirrel monkeys. (*Saimiri sciureus*). *Lab. Anim. Sci.* **22,** 583–586.

Minette, H. P. (1966). Leptospirosis in primates other than man. *Am. J. Trop. Med. Hyg.* **15,** 190–198.

Minette, H. P., and Shaffer, M. R. (1968). Experimental leptospirosis in monkeys. *Am. J. Trop. Med. Hyg.* **17,** 202–212.

Moreland, A. F. (1970). Tuberculosis in New World primates. *Lab. Anim. Care* **20,** 262–264.

Murdock, G. K. (1978). Maintenance and breeding of white-faced saki *Pithecia pithecia* at the Denver Zoo. *Int. Zoo Yearb.* **18,** 115–117.

Myers, G. J. (1972). Echinococcosis, Coenurosis, cysticercosis, sparganosis, etc. *In* "Pathology of Simian Primates" (R. N. T-W- Fiennes, ed.), pp. 124–143. Karger, Basel.

Napier, J. R., and Napier, P. H. (1967). "Handbook of Living Primates." Academic Press, New York.

National Academy of Sciences (1978). "Nutrient Requirements of Domestic Animals," No. 14. NAS, Washington, D.C.

Newberne, P. M., and Hayes, K. C. (1979). Semipurified diets for nonhuman primates. *In* "Primates in Nutritional Research" (K. C. Hayes, ed.), pp. 99–119. Academic Press, New York.

Norman, R. C., and Wagner, J. E. (1973). Pediculosis of spider monkeys: A case report with zoonotic implications. *Lab. Anim. Sci.* **23,** 872–875.

Orihel, T. C., and Seibold, H. R. (1972). Nematodes of the bowel and tissues. *In* "Pathology of Simian Primates" (R. N. T-W-Fiennes, ed.), Part 2, pp. 76–103. Karger, Basel.

Peters, W., Molyreux, D. H., and Howells, R. E. (1974). *Eperythrozoon* and *Haemobartonella* in monkeys. *Ann. Trop. Med. Parasitol.* **68,** 47–50.

Porter, J. A., Jr. (1969). Hematology of the night monkey, *Aotus trivirgatus. Lab. Anim. Sci.* **19,** 470–472.

Porter, J. A., Jr. (1971). Hematolgic values of the black spider monkey (*Ateles fusciceps*), red spider monkey (*Ateles geoffroyi*), white-face monkey (*Cebus capucinus*), and black howler monkey (*Alouatta villosa*). *Lab. Anim. Sci.* **21,** 426–433.

Rasmussen, K. M. (1979). Folic acid needs for breeding and growth of nonhuman primates. *In* "Primates in Nutritional Research" (K. C. Hayes, ed.), pp. 73–97. Academic Press, New York.

Ratcliffe, H. L., and Worth, M. D. (1951). Toxoplasmosis of captive wild birds and mammals. *Am. J. Pathol.* **17**, 655–664.

Rosenblum, L. A., Nathan, T., Nelson, J., and Kaufman, I. C. H. (1967). Vaginal cornification cycles in the squirrel monkey (*Saimiri sciureus*). *Folia Primatol.* **6**, 83–91.

Samonds, K. W., and Hegsted, D. M. (1973). Protein requirements of young cebus monkeys (*Cebus albifrons* and *Cebus apella*). *Am. J. Clin. Nutr.* **26**, 30–40.

Samonds, K. W., Ausman, L. M., and Hegsted, D. M. (1974). Hematologic development of the cebus monkey (*Cebus albifrons* and *apella*). *Folia Primatol.* **22**, 72–79.

Schmidt, G. D. (1972). Acanthocephala of captive primates. *In* "Pathology of Simian Primates" (R. N. T-W-Fiennes, ed.), Part 2, pp. 144–156. Karger, Basel.

Schnell, J. V., Siddiqui, W. A., and Geimain, Q. M. (1969). Analyses on the blood of normal monkeys and owl monkeys infected with *Plasmodium falciparum*. *Mil. Med.* **134**, 1068–1073.

Seibold, H. R., Perrin, E. A., Jr., and Garner, A. C. (1970). Pneumonia associated with *Bordetella bronchiseptica* in *Callicebus* sp. *Lab. Anim. Care* **20**, 456–461.

Self, J. T., and Cosgrove, G. E. (1972). Pentastomida. *In* "Pathology of Simian Primates" (R. N. T-W-Fiennes, ed.), Part 2, pp. 194–204. Karger, Basel.

Simkin, P. A. (1971). Uric acid metabolism in cebus monkeys. *Am. J. Physiol.* **221**, 1105–1109.

Snyder, S. B., Lund, J. E., and Bone, J. (1970). A study of *Klebsiella* infections in the owl monkey, *Aotus trivirgatus*. *J. Am. Vet. Med. Assoc.* **157**, 1935–1939.

Soave, O., Jackson, S., and Ghumman, J. S. (1981). Atypical mycobacteria as the probable cause of positive tuberculin reactions in squirrel monkeys, *Saimiri sciureus*. *Lab. Anim. Sci.* **31**, 295–296.

Squire, R. A., Goodman, D. G., Valerie, M. G., Fredrickson, T., Strandberg, J. D., Levitt, M. H., Lingeman, C. H., Harshbarger, J. C., and Dawe, C. J. (1978). Tumors. *In* "Pathology of Laboratory Animals" (K. Benirschke, F. M. Garner, and T. C. Jones, eds.), Chapter 12, pp. 1052–1283. Springer-Verlag, Berlin and New York.

Taliaferro, W. H., and Cannon, P. R. (1936). The cellular reactions during primary infections and superinfections of *Plasmodium brazilianum* in Panamanian monkeys. *J. Infect. Dis.* **59**, 72–125.

Taliaferro, W. H., and Taliaferro, L. G. (1934). Morphology, periodicity and course of infections of *Plasmodium brazilianum* in Panamanian monkeys. *Am. J. Hyg.* **20**, 1–49.

Ulrich, C. E., Monaco, R. A., and Messer, S. M., (1977). Baseline pulmonary physiologic values for the squirrel monkey (*Saimiri sciureus*). *Lab. Anim. Sci.* **27**, 1024–1027.

U.S. Department of Health, Education and Welfare (1975). "Nonhuman Primates—Usage and Availability for Biomedical Programs," Publ. No. (NIH) 76-892. USDHEW, Washington, D.C.

U.S. Department of Health, Education and Welfare (1978). "Guide for the Care and Use of Laboratory Animals," DHEW Publ. No. (NIH) 78-23. USDHEW, Washington, D.C.

Wellde, B. T., Johnson, A. S., Williams, J. S., Langbehn, H. R., and Sadun, E. H. (1971). Hematologic, biochemical, and parasitologic parameters of the night monkey (*Aotus trivirgatus*). *Lab. Anim. Sci.* **21**, 575–580.

Williams, L. (1967). Breeding Humboldt's woolly monkey (*Lagothrix lagotricha*) at Murrayton Woolly Monkey Sanctuary. *Int. Zoo Yearb.* **7**, 86–89.

Wolf, R. C., O'Connor, R. F., and Robinson, J. A. (1977). Cyclic changes in plasma progestins and estrogens in squirrel monkeys. *Biol. Reprod.* **17**, 228–231.

Wolf, R. H., Lehner, N. D. M., Miller, E. C., and Clarkson, T. B. (1969). The electrocardiogram of the squirrel monkey, *Saimiri sciureus*. *J. Appl. Physiol.* **26**, 346–351.

Wright, P. C. (1981). The night monkeys, genus *Aotus*. *In* "Ecology and Behavior of Neotropical Primates" (A. F. Coimbra-Filho and R. A. Mittermeire, eds.), Vol. I, pp. 211–240. Academia Brasileira de Ciencias, Rio de Janeiro.

Young, M. D. (1970). Natural and induced malarias in Western Hemisphere monkeys. *Lab. Anim. Care* **20**, 361–367.

PART C. BIOLOGY AND DISEASES OF CALLITRICHIDAE

Conrad B. Richter

I. RESEARCH CONSIDERATIONS

A complete listing of currently accepted scientific and common nomenclature for the family Callitrichidae can be found in Table C.I. Attention should be paid to species and subspecific names to avoid confusion and misuse in communicating information on this diverse group of animals.

A. Use

Table C.II lists general categories of research for which callitrichids have been used. Additional insight into the usefulness of these species can be found in Gengozian and Deinhardt (1978). It is clear that their use and importance is growing. Nevertheless, much of the potential of these species for biomedical research use remains unfulfilled for a variety of reasons: prior habits of experimentalists; poor record of reproductivity for many species in captivity; physical size; poorly understood metabolic disorders, notably "wasting disease"; poor husbandry techniques; and poorly understood medical problems. Of these, nutrition and husbandry are pivotal problems that must be solved if these species are to fulfill their potential. While physical size is disadvantageous from some standpoints, namely, invasive studies in which extensive manipulation is required, it is markedly advantageous from an economic standpoint. Large numbers of callitrichids can be housed and cared for at a fraction of the cost of larger primates and carnivores. Given better understanding of the problems mentioned, increased use of these species in research can be anticipated.

Table C.I

The Family Callitrichidae[a]

Scientific name	Group	Common name
Family Callitrichidae		
Cebuella pygmaea Spix		Pygmy marmoset
Callithrix jacchus jacchus Linnaeus	Tufted-ear marmosets	Common or Cotton-ear
Callithrix jacchus penicillata E. Geoffroy		Black
Callithrix jacchus geoffroyi Humbolt		Geoffroy's
Callithrix jacchus flaviceps Thomas		Buffy-headed
Callithrix jacchus aurita E. Geoffroy		Buffy
Callithrix humeralifer humeralifer E. Geoffroy	Tassel-ear marmosets	Black and white
Callithrix humeralifer chrysoleuca Wagner		Golden-white
Callithrix humeralifer intermedius ? New subspecies		?
Callithrix argentata melanura E. Geoffroy	Bare-ear marmosets	Black-tailed
Callithrix argentata argentata Linnaeus		Silvery
Callithrix argentata leucippe Thomas		Golden-white
Saguinus nigricollis graellsi Jiminez de la Espada	Black mantle tamarins[b]	Graell's
Saguinus nigricollis nigricollis Spix		Spix's
Saguinus fuscicollis fuscus Lesson	Saddle-back tamarins[b]	Lesson's
Saguinus fuscicollis avilapiresi Hershkovitz		Avila pires's
Saguinus fuscicollis fuscicollis Spix		Spix's
Saguinus fuscicollis nigrifrons I. Geoffroy		Geoffroy's
Saguinus fuscicollis illigeri Pucheran		Illiger's
Saguinus fuscicollis leucogenys Gray		Andean
Saguinus fuscicollis lagonotus Jiminez de la Espada		Red-mantle
Saguinus fuscicollis tripartitus Milne-Edwards		Golden-mantle
Saguinus fuscicollis weddelli Deville		Weddell's
Saguinus fuscicollis cruzlimai Hershkovitz		Cruz Lima's
Saguinus fuscicollis crandalli Hershkovitz		Crandalli
Saguinus fuscicollis acrensis Carvalho		Acre
Saguinus fuscicollis melanoleucus Miranda Ribeiro		White
Saguinus labiatus labiatus E. Geoffroy	Moustached tamarins	Geoffroy's
Saguinus labiatus thomasi Goeldi		Thomas'
Saguinus mystax mystax Spix		Spix's
Saguinus mystax pileatus I. Geoffroy and Deville		Red-cap
Saguinus mystax pluto Lönnberg		White-rump
Saguinus emperator Goeldi		Emperor
Saguinus midas midas Linnaeus	Midas tamarins	Golden-handed tamarin
Saguinus midas niger E. Geoffroy		Black-handed tamarin
Saguinus inustus Schwartz	Mottled-face tamarins	Mottled-face tamarin
Saguinus bicolor bicolor Spix	Bare-face tamarins	Pied
Saguinus bicolor martinsi Thomas		Martin's
Saguinus bicolor ochraceus Hershkovitz		Ochraceous
Saguinus leucopus Gunther		Silvery brown
Saguinus oedipus geoffroyi Pucheran		Red-crested
Saguinus oedipus oedipus Linnaeus		Cotton-top
Leontopithecus rosalia chrysopygus Mikan	Lion tamarins	Golden-rump
Leontopithecus rosalia chrysomelus Kuhl		Gold and black
Leontopithecus rosalia rosalia Linnaeus		Golden
Family Callimiconidae		
Callimico goeldii Thomas		Goeldi's monkey

[a]Adapted from Hershkovitz (1977).
[b]Also referred to as white-lip tamarins.

Table C.II

Principal Species and Common Research Use

Research field	Species						
	C. jacchus	S. oedipus	S. fuscicollis	S. mystax	S. nigricollis	S. labiatus	C. pygmaea
Cancer and viral oncology	+	+	+	+	+		
Behavior and ethology	+	+	+				+
Hematology and immunology	+	+	+		+		
Transplantation			+		+		
Physiology and reproduction	+	+	+				
Radiation research		+	+				
Infectious diseases	+	+	+	+	+	+	
Dental and periodontal disease	+						
Nutrition	+						
Toxicology and teratology	+						
Speech and language	+						+
Metabolism	+						

Of the callitrichid species, *Callithrix jacchus jacchus* (Fig. C.1) has proved to be the most adaptable to captivity, and some notable successes have been achieved with the husbandry of this species. Presently, it is by far the most used, hence has the most extensive baseline and normative data available. This trend seems destined to continue. *Saguinus fuscicollis* ssp. (Fig. C.2), *S. labiatus*, *S. nigricollis*, *S. mystax*, and *S. oedipus oedipus* (Fig. C.3) fall somewhere behind in order of use. Unfortunately, it is not always clear from published work which subspecies has been used, and this is important from the scientific, conservationist, and husbandry standpoints.

Callitrichids have been used extensively in human hepatitis studies (Provost *et al.*, 1977), notably *Saguinus* ssp. These and other speices have been used in studies of slow viruses, spongiform encephalopathy, mycoplasma, chlamydia, cholera, malaria, and viral oncology among others. The ability of Epstein-Barr virus (EBV) to cause malignant lymphoma or regressing lymphoreticular hyperplasia in callitrichids is an important biological link between infectious mononucleosis and neoplastic disease associated with EBV in man (Miller *et al.*, 1977). Levy *et al.* (1969) have described the tumors induced in three callitrichid species with Rous sarcoma virus.

B. Laboratory Management and Husbandry

Early writers on this subject failed to consider many of the unusual behavior and nutrition requirements of these species, hence early misinformation has had far reaching and sometimes unfortunate influences on callitrichid management. In recent years some of the early management practices have been abandoned and a more scientific approach to husbandry has been adopted through the application of targeted behavior and

Fig. C.1. *Callithrix jacchus jacchus*, the common, or cotton-ear marmoset.

Fig. C.2. *Saguinus fuscicollis lagonotus,* the red-mantle saddle-backed tamarin. There are numerous subspecies of *S. fuscicollis.* See Table C.I.

Fig. C.3. *Saguinus oedipus oedipus,* the cotton-top tamarin.

feeding studies. Because of the importance of husbandry to the overall health status and future usefulness of these species, additional space is devoted to this topic. Caminiti's (1980) bibliography on breeding New World primates should be consulted by those interested in the husbandry of callitrichids.

1. Caging

Callitrichids are highly territorial and behave aggressively toward conspecifics of the same sex. Hence it is not possible to cage fully mature adults of the same sex together without risking dominance by one individual and eventually fighting with severe mutilation and even death. Since mixing sexes invariably results in pregnancies, single caging is recommended for adults on experimentation. Cage size is somewhat dependent on species, but in all instances should include the following features: one or more perches, escape proof construction, high location for feed stations, and a nest box, preferably removable with closures for easy capture of inhabitants. Cages should be tall enough so that a completely relaxed individual can posture normally on perches without coming in contact with excrement on the cage floor. Short-tailed *C. j. jacchus* may do well in a 60 × 60 × 60 cm cage, whereas long-tailed and larger *S. o. oedipus* may require a 65 × 80 × 90 cm cage.

Caging for breeders must be considerably larger. Callitrichids are easiest to manage in captivity as nucleus family groups. Such groups consist of the adult male and female and one or more sets of offspring. In prolific species, such as *C. j. jacchus,* these groups may grow to six or more individuals within 1 year's time. Intrafamily competition and play dictates that adequate exercise space and territory be available to permit natural social adjustments. Numerous caging schemes have been attempted: indoor–outdoor caging with a large outdoor room; indoor caging with a common exercise area (used by one family at a time); and double cage units are among the more imaginative and successful types. Table C.III lists some of the more interesting and successful types of breeder caging.

2. Sanitation

Cage sanitation must be rigid and regular. Experimental animal caging should be thoroughly sanitized weekly, and bottom screens or dropping pans should be sanitized more often. With the exception of specific experiments, the argument that repeated washing destroys territorial marking (see Section II,E) and should therefore be limited is not withstanding. Callitrichids can quickly be trained to pass from dirty cages through a wire tunnel to clean cages without handling. Outdoor caging protected from the weather should be hosed daily since feces quickly accumulate.

Table C.III

Housing Styles for Breeding Families of Callitrichids

Type	Dimensions	Reference
Indoor–outdoor	Nest box 1.2 × 2.7 × 2.4 m	Stein *et al.* (1979)
Indoor–outdoor	45.7 × 40.6 × 81.3 cm 3.0 × ~1.0 × 2.3 m	Stein *et al.* (1979)
Indoor–outdoor	133 × 64 × 70 cm island	Mallinson (1978)
Indoor–outdoor	122–183 × 91 × 152 cm 6.7–8.9 m² × 210–260 cm	Mallinson (1975)
Indoor–exercise cage	100 × 50 × 75 cm 150 × 100 × 210 cm	Hearn *et al.* (1975)
Indoor–open wall	91 × 91 × 152 cm × 2	C. B. Richter, unpublished
Indoor (4 cage module)	50 × 50 × 75 cm	W. D. Hiddleston, personal communication
Indoor–outdoor	3.25 × 2.45 × 2.45 m 3.0 × 2.6 × 2.6 m	Epple (1970a)

3. Feeding

Feed should be presented twice daily, and until there is a better understanding of callitrichid nutritional requirements, a varied diet is recommended (see Section II,C). Callitrichids require high energy diets and usually begin feeding shortly after arising (Garber, 1980); hence they should be fed early. A second feeding, in midafternoon, avoids the prolonged periods without food that contribute to the acute episodes of hypoglycemia/hypothermia commonly seen in debilitated animals (see Section III,C).

Most callitrichids are nearly totally arboreal. Although some species such as *S. fuscicollis* ssp. frequent the cage floor as well as the jungle floor more than others, most prefer to feed from higher points within the cage. Small semicircular feed cups suspended high on the inside of the cage work well. Multiple feeding stations, preferably located far apart, are necessary in family groups. S. Davis and C. B. Richter (unpublished) observed that *C. j. jacchus* and *S. o. oedipus* took food significantly more often when food was suspended high outside the cage than when it was placed low inside the cage. *Saguinus f. illigeri* showed no preference. Furthermore, some species are less social feeders (*S. o. oedipus*) than others (*S. f. illigeri*), and casual generalizations about feeding habits should be avoided. Family groups may contain a dominant individual, usually the adult female, who will defend as many feeding stations as possible and claim most of the preferred food (Tardif and Richter, 1981), usually the high energy or sweet foods. As families grow in size and juveniles approach maturity, competition for pre-

ferred food increases and further feeding stations may be necessary.

4. Environment

Because of their natural jungle habitat and small size, callitrichids depend heavily upon vocal communication to maintain contact. Consequently, they have fairly broad vocabularies and are capable of loud and varied sounds. Catching an animal within sight, and to a lesser extent within hearing, of other callitrichids results in raucous behavior and mobbing screams and noises. While this may be acceptable in some experimental groups, it is undesirable in breeding colonies; thus, methods of capture that conceal the act are highly desirable (see Section I,B,5).

Agonistic behavior among conspecifics of the same sex may be so marked (see Sections II,E and III,D) that aggressive individuals, usually males, may spend hours staring at other conspecifics in distant cages and pay little attention to the business of raising their offspring. It is recommended that breeding families of the same species be visually screened from each other. Escapees usually ignore other species and attack conspecifics of the same sex by reaching through the cage wall with damaging results.

To minimize disturbances, breeding callitrichids for production purposes should be accomplished in facilities separate from those used to house experimental animals. All but essential human traffic should be excluded from breeding facilities. Manual capture and handling of infants or adults, which can be observed or heard by other individuals, should be strictly prohibited. Humans with obvious respiratory illness or other evidence of disease that might be communicable to callitrichids (all human infectious diseases should be regarded as such) should be excluded from breeding facilities.

Standard requirements for heating, ventilating, and air conditioning (HVAC) control in animal facilities (see Chapter 17 by Hessler and Moreland) apply for callitrichids. These species are strictly diurnal, and when caged outdoors, they begin to enter the nest box in late afternoon or early evening. In indoor facilities, they enter the nest box shortly after human activity ceases in late afternoon. Although an ambient temperature of 26.6°C is frequently recommended, temperatures of 22.2°–23.8°C are adequate, and in some respects this range is better for indoor daytime housing provided adequate cage space is available for exercise. Tolerance to low temperatures has been observed in outdoor settings by numerous investigators, especially in larger cages where free movement was possible. Humidity levels in indoor facilities should be kept at or above 50% relative; however, levels above 70% may result in inwork environment.

Outdoor facilities should include adequate shade and the possible addition of sprinklers for cooling during very hot days. Sprinklers are also effective means of raising humidity during the heating season in facilities without central humidity control. Closed facilities of all types are rendered more habitable by the addition of ultraviolet sunlamps turned on for 30–60 min/day. This practice has been widely adopted in callitrichid husbandry and has therapeutic as well as behavioral benefits. Ultraviolet light produces native vitamin D₃ by activating 7-dehydrocholesterol, essential to New World primate calcium metabolism. Hence, ultraviolet light is therapeutic in the treatment and prevention of "cage paralysis" (osteodystrophia fibrosa) (Hunt *et al.*, 1967). Ultraviolet light also stimulates grooming activity and "sunning."

5. Nest Box

Callitrichids require a flat surface for sleeping, preferably a nest box where a curled posture can be assumed to conserve body heat. Perches and the cage floor are unacceptable as sleeping places, and flat open shelves are undesirable. The best nest boxes are those that are specifically designed for the purpose. A 17.8 × 17.8 × 30 cm galvanized steel box with a viewing port at one end and an 11.5 × 11.5 cm opening into the cage on one side near the opposite end is excellent for single individuals or pairs. The lateral opening should be closable by a slide, and the nest box removable from the cage so that animals can be driven into the box, the opening closed, and the box removed to the lab or clinic for the withdrawal of the animal. A slightly larger nest box (25 × 25 × 50 cm) is advisable for breeding families (Fig. C.4). The opening to the interior of the cage from the box should be 12.7 × 12.7 cm to permit easier access by adults carrying young. Families as large as six *S. o. oedipus* or eight *C. j. jacchus* have used nest boxes of this size without difficulty. The nest box should be located as high in the cage as possible, and care must be taken to render it escape proof. Nest boxes quickly become unsanitary and should be sanitized frequently.

6. Vermin Control

Standard precautions as described in Chaper 17 apply to callitrichids also; however, certain special precautions are necessary. Feral callitrichids invariably are infected with one or two species of thornyheaded worms (see Section III,B) and the german cockroach, *Blatella germanica*, is an effective intermediate host. Good sanitation, the elimination of hiding and breeding places, and the liberal use of boric acid powder are the most effective control measures. Cockroaches may also serve as potential vectors of other important callitrichid diseases, notably metazoan parasites (Flynn, 1973); however, little evidence for transmission in captivity is available. Where outdoor facilities are contemplated, consideration should be given to potential vectors of the family Dipetalonematidae, filarial

Fig. C.4. Nest box located high inside cage. Entry port for animals (A) and viewing port (B) are shown. Entry port is closed by a sheet metal slide. Box is suspended by top overhang (C), and can be removed from cage.

nematodes. As many as 13 species of microfilaria have been reported in callitrichids (Tankersley *et al.*, 1979), and, in the absence of documented evidence to the contrary, it seems advisable to recommend the selection of breeding stock free of dipetalonematids if potential mosquito vectors are present.

II.   BIOLOGY

A.   Physiology and Anatomy

It is not possible to present a thorough review of this subject; however, significant highlights and important anatomical variations from other species are summarized briefly. Callitrichids are generally accepted as the most primitive of the simian primates and as such fulfill the principles of Mivart's (1873) classical definition. The following notations of important anatomical and physiological characteristics are based on published accounts and observations by the author and others.

Digits: Possess falcula (claws) rather than ungula (nails) with one exception, the hallux. Thumb is not opposable but the hallux is.

Lymphatics: Mesenteric lymphatics empty directly into the posterior vena cava in the region of the renal veins. No cysterna chyli is present.

Dentition: See Section II,B.

Ishial caloses: Absent.

Mammary glands: Axillary rather than anterior.

Large bowel: Rudimentary separation of the ascending, transverse, and descending colon. Nevertheless, a typical pattern of arrangement in this order is found. Only poorly developed hepatic and splenic ligaments are present, and these permit gross displacement of the segments. The large bowel is fully invested by mesocolon to the level of the distal colon or rectal canal. Longitudinal taeniae and sacculi are present but may be overlooked or not visible when bowel distention is present (frequent). There is no sigmoid flexure.

Pancreas: Fairly well-developed head, body, and tail.

Lungs: Five lobes. Right: superior, cardiac, diaphragmatic and postcaval; left: superior and diaphragmatic.

Liver: Four lobes. Right, left, medial, and caudate. Gallbladder is attached to the medial lobe.

Brain: Possesses a calcarine fissure but the cortex is generally smooth and fissures are few (4–8).

Vision: Binocular.

Circumgenital glands: Well-developed sebaceous glands overlying enlarged apocrine glands covering the labia majora and pudendum in the female and the scrotum in the male (see Section II,E).

Sternal glands: Located on the anterior chest but may be difficult to find or absent. May be diffuse or focal in arrangement. Histological type is similar to circumgenital glands.

Chimerism: Normal twinning (see Section II,D) coupled with placental vascular anastomoses (Fig. C.5), and continuous placental hemopoiesis (Jollie *et al.*, 1975) results in stem cell cross-over between developing fetuses; hence all twins and many singletons appear to be permanent chimeras. Hemopoietic and testicular tissues are demonstrably affected, although the functionality of germ cell chimerism has been excluded by Gengozian *et al.* (1980). Free-martinism does not occur in callitrichids in spite of the placental vascular anastomoses, probably because the callitrichid placenta possesses an effective aromatizing enzyme system capable of converting androgens to estrone (Ryan *et al.*, 1981).

Published similarities and dissimilarities in physiological and anatomical characteristics for callitrichids in relation to man and other species are listed in Table C.IV. This table is intended to serve only as a sampling of the literature on these subjects.

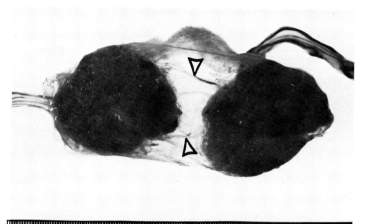

Fig. C.5. Cotton-top placentas. Vascular anastomoses are visible (arrowheads).

Table C.IV

Anatomy and Physiology

Organ/substance	Function/structure/values	Reference
Lymphocytes	Populations	*Cell. Immunol.* **35**, 148 (1978)
Lymphocytes	Glycoproteins, surface	*Int. J. Cancer* **23**, 76 (1979)
Lymphocytes	Antigens	*J. Immunogenet.* **8**, 433 (1981)
Lymphocytes	Antigens, surface T and B	*J. Immunogenet.* **9**, 209 (1982)
Hemoglobin	Structure, primary	*Biochem. Genet.* **14**, 427 (1976)
Hemoglobin	Fetal	*Blood Cells* **5**, 471 (1979)
Erythrocytes	Isohemagglutinins	*Nature (London)* **209**, 722 (1966)
Brain	Purkinje cells	*Neuropathol. Appl. Neurobiol.* **5**, 71 (1979)
Tetrahydrocannabinol	Metabolism	*Psychopharmacologia* **47**, 53 (1976)
Brain	Inferior colliculus	*Brain Behav. Evol.* **13**, 125 (1976)
Brain	Neurons, monoamine	*J. Anat.* **134**, 315 (1982)
Retinal terminations	Geniculate nucleus	*J. Comp. Neurol.* **182**, 517 (1978)
Placenta	Ultrastructure	*Am. J. Obstet. Gynecol.* **122**, 60 (1975)
Placenta	Hemopoiesis	*Anat. Rec.* **85**, 349 (1943)
Ovary	Follicle formation	*Acta Histochem.* **60**, 247 (1977)

(continued)

Table C.IV (*Continued*)

Organ/substance	Function/structure/values	Reference
Urine	Chorionic gonatropin	*Folia Primatol.* **28,** 251 (1977)
Ovary	Enzyme reactivity	*Acta Histochem.* **60,** 247, (1977)
Body	Biochemistry profiles	*Am. Assoc. Zoo Vet. Annu. Proc.* p. 81 (1975)
Tongue	Taste	Rothe *et al.* (1978, p. 83)
Serum	Lipoproteins	*Biochemistry* **18,** 5096 (1979)
Plasma	Reproduction hormones	*J. Reprod. Fertil.* **56,** 23 (1979)
Plasma	Progesterone	*J. Med. Primatol.* **11,** 43 (1982)
Adrenal	DHA-reticular zone	*Endocrinology* **111,** 1797 (1982)
Body	Weights	*Am. J. Phys. Anthropol.* **47,** 211 (1977)
Teeth, peridontium	Development	*Int. Zoo Yearb.* **12,** 51 (1972)
Dental cementum	Growth	*Primates* **23,** 460 (1982)
Diaphragm	Lumbar openings	*Arch. Ital. Anat. Embriol.* **86,** 289 (1981)
Labyrinthine fibers	Origin, efferent	*Arch. Oto-Rhino-Laryngol.* **234,** 139 (1982)
Eye	Morphology	*Folia Primatol.* **29,** 161 (1978)
Eye	Vision	Rothe, *et al.* (1978, p. 89)
Glands, submandibular	Ultrastructure, histo-chemistry	*J. Dent. Res.* **55,** B232 (1976)
Glands, submandibular	Histochemistry	*Acta Anat.* **98,** 361 (1977)
Glands, duodenal	Histology	*Acta Anat.* **93,** 580 (1975)
Glands, scent	Histology	*J. Zool.* **185,** 41 (1978)
Lymphatics	Communications	*Am. J. Anat.* **12,** 447 (1912)

B. Normal Values

1. Body Size and Weight

Body size and weight values are given in Tables C.V and C.VI. Callitrichids exhibit no distinct sexual dimorphism in body size, weight, or appearance, although nonpregnant adult females in captivity may be slightly heavier than males. Care-

Table C.V

Body Length (cm)

Species	Neonate	Adult	Adult tail
C. j. jacchus[a]	7.7	21.2	27.9
Callithrix[b]	8.5	21.9	33.2
S. f. illigeri[a]	8.4	21.6	34.3
Saguinus[b]	10.0	23.5	36.4
S. o. oedipus[a]	8.7	23.9	38.9
Leontopithecus[b]	12.0	26.1	37.0
Callimico[b]	11.0	22.4	30.2

[a]Average of 10 neonates measured in fetal position from the point of the occiput to the rump. Average of three selected healthy adults measured with head erect in normal position from the top of the head to the rump.

[b]From Hershkovitz (1977, p. 445).

Table C.VI

Body Weights (gm)

Species	Neonate	Yearling	Adult	Obese
C. j. jacchus[a]	36.4	300	300–350	450
Callithrix[b]	30.0	—	316	—
S. o. oedipus[a]	40.0	500	450–550	550
Saguinus[b]	43.0	—	450	—
S. fuscicollis[a]	40.2	325	350–450	500
L. rosalia[b]	57.0	—	583	—
C. goeldii[b]	60.0	—	481[c]	—
C. pygmaea[b]	15.0	—	125	—

[a]Average of 10 or more neonates, 6 or more yearlings, and 10 or more adults in full flesh.

[b]From Hershkovitz (1977, p. 445).

[c]Healthy specimens should be significantly heavier.

ful observation and familiarity with the appearance of external genitalia is necessary to distinguish males from females in a cage setting with most species. Well-developed labia majora and a completely scrotal penis make the sexes of *C. j. jacchus* especially difficult to distinguish. Normal organ weights for *S. o. oedipus*, *S. fuscicollis* ssp., *S. nigricollis*, and *C. jacchus* have been published (J. B. Deinhardt *et al.*, 1967; Wadsworth *et al.*, 1981).

2. Dentition

Dental formulae for all callitrichids are the same. Table C.VII shows the pattern of dentition and tooth eruption in *S. fuscicollis* ssp. It should be noted that callitrichids, like other New World primates, have three premolars and two molars, a reversal of the pattern seen in man and Old World primates. Similar information for tooth eruption in other species can be found in Hershkovitz (1977). Winter (1978) has shown a slightly delayed time of eruption for most deciduous teeth in *C.*

Table C.VII

Patterns of Tooth Eruption in *Saguinus fuscicollis*[a]

	Deciduous teeth and age at eruption in weeks					
	I1	I2	C	M1	M2	M3
Maxillary	Birth	Birth	Birth	2	2–5	6–12
Mandibular	Birth	Birth	Birth	Birth-2	2–5	5–8

	Permanent teeth and age at eruption in weeks							
	I1	I2	C	P2	P3	P4	M1	M2
Maxillary	21–23	23–29	39–45	32–39	38–39	29–39	18–23	29–39
Mandibular	21–23	23–29	39–45	32–39	37–39	29–39	16–23	29–31

[a] From Glassman (1982). Reproduced by permission of the author.

jacchus versus *Saguinus* ssp. Dental eruption for *Callimico* lags behind the callitrichids (Lorenz and Heinemann, 1967).

3. Skeletal Growth

Glassman (1982) has done a detailed study of skeletal growth in two species of callitrichids (*S. fuscicollis* and *S. o. oedipus*). Data from his studies on *S. fuscicollis* is reproduced in Fig. C.6 and Table C.VIII. The regression line for the femur (Fig. C.6) may be used to estimate age ±3 months until ~700 days of age.

The potential for increases in stature cease with the last closure of long bone epiphyses. This occurs in the distal femur in the case of *S. fuscicollis,* at slightly more than 2 years of age. Growth may be said to be complete at this time. *Saguinus o. oedipus* shows a similar pattern.

4. Body Temperature

C. B. Richter (unpublished) made several thousand rectal temperature recordings on three species of callitrichids using an electronic thermometer and probe over a 5-year period. Based on these observations normal daytime rectal temperature for callitrichids varies between 38.5° and 40.0°C. This wide range may be accounted for by the excitability of the species, the rigors involved in catching, and nocturnal torpor. Temperatures outside this range suggest that medical attention is necessary. Callitrichids develop a distinct torpor with hypothermia (34.0°C) during sleep (Hetherington, 1978).

5. Karyology

Callitrichid chromosomes generally bear close physical resemblance to those of man, although they are smaller. The following tabulation (Hershkovitz, 1977, p. 432) lists the 2*n*

Table C.VIII

Sequence and Time of Epiphyseal Union in *Saguinus fuscicollis* spp. (months)[a]

Epiphysis	Range of fusion	Initial complete union
Distal humerus	3.1–9.7	3.7
Ischium-pubic ramus	3.6–8.0	3.6
Ischium-pubis to ilium	4.0–9.7	5.2
Ischium to pubis	4.0–14.5	5.2
Medial epicondyle	4.0–14.5	14.5
Coracoid process	5.2–14.5	14.5
Lesser trochanter	5.8–15.8	15.8
Proximal ulna	8.0–15.8	15.8
Distal radius	8.0–22.2	21.0
Greater trochanter	9.7–15.8	15.8
Head of femur	9.7–15.8	15.8
Proximal radius	14.5–15.8	15.8
Distal tibia	15.8–19.6	15.8
Distal fibula	15.8–19.6	15.8
Distal ulna	15.8–22.2	21.0
Proximal tibia	15.8–23.8	23.8
Proximal fibula	15.8–26.9	23.8
Proximal humerus	15.8–26.9	21.0
Distal femur	15.8–26.9	26.9
Ischial epiphysis	19.6–48.6	26.9
Spheno-occipital Syn.	21.0–26.9	21.0
Iliac crest	26.9–75.6	26.9

[a] Adapted from Glassman (1982).

complement for the four callitrichid genera and *Callimico*. Chimerism creates problems for those studying karyology of peripheral leukocytes since it is not possible, except where the chimera is male/female, to establish the identity of the cell source.

Cebuella	44		*Leontopithecus*	46
Callithrix	44 or 46		*Callimico*	48
Saguinus	46			

6. Hematology and Blood Chemistry

Normal hemograms for four species of callitrichids are given in Table C.IX. The data from Richter reported in Table C.IX was taken from healthy parasite-free, manually restrained, fasted adults sampled and examined during a single 1 day exercise. Bush *et al.* (1982) reports significant sexual dimorphism in hematocrit, hemoglobin, and basophil values in *L. rosalia,* but most authorities find no differences between the sexes. Hawkey *et al.* (1982) and Bush *et al.* (1982) report values for subadults at varying age. Both groups of workers found slightly lower values for hemoglobin, hematocrit, red blood cell count, white blood cell count, and neutrophils in the

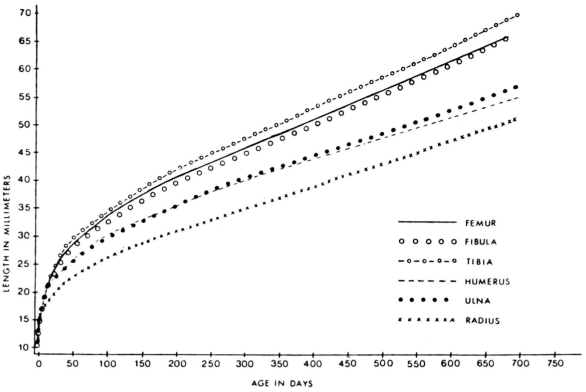

Fig. C.6. Bone growth regression lines for *S. fuscicollis.* From Glassman (1982). Reproduced by permission of the author.

younger animals. Anderson *et al.* (1967) have determined the plasma volume in *S. oedipus* and whitelip tamarins using ^{131}I-labeled albumin. The mean value for eleven animals was 43.77 ml/kg body weight; however, the range of values was quite wide. Total blood volume using the same methods was 82.1 ml/kg. Merritt and Gengozian (1967) determined erythrocyte half-life using ^{51}Cr-labeled red cells in *S. fuscicollis* ssp. and found a mean value of 15.9 days for 14 animals. Small numbers (12) of nucleated red blood cells are commonly seen in callitrichid blood smears.

Blood chemistry values for several species of callitrichids are given in Table C.X. The high value for serum alkaline phosphatase reported by Wadsworth *et al.* (1982) is inconsistent with the observations of others. Original works may be consulted for statistical variations in most cases. Richter's data was taken from 16 hr fasted, healthy adults. McIntosh and Looker (1982) have reported triglyceride levels (1.33 ± 0.06 m*M*/liter) and cholesterol levels (3.57 ± 0.54 m*M*/liter) in *C. jacchus.* Glucose tolerance and insulin response curves in callitrichids are typical of most primates, but normal fasting glucose values are slightly higher. Glucose tolerance data for two normal fasted *C. j. jacchus* are given in Table C.XI. Curves for *S. o. oedipus* and *S. fuscicollis* ssp. are similar in shape and value.

C. Nutrition

Nutrition is of major consequence in the management of callitrichid colonies. Unfortunately, it is poorly understood. Feral species are omnivorous, feeding on insects, fruits, nectars, tree exudates, and whatever small animals they can capture. Marmoset species spend considerable time consuming tree exudates and gums from bark wounds inflicted by themselves or caused by other trauma (Coimbra-Filho and Mittermeier, 1977). Marmosets have marked dental adaptation in the form of procumbent mandibular incisors to accomplish this, but tamarins also use this food source (Garber, 1980). These exudates are rich in sugars and minerals and in some cases, possibly seasonally at least, constitute the major portion of callitrichid diet to the near exclusion of other foods. Richter *et al.* (1981) showed that *C. j. jacchus* preferred mildly sweet (fructose) vitamin-fortified water as a source of energy over other sources, and preferred sweetened food over unsweetened food. King (1978) has examined feeding patterns of marmosets and found higher energy intakes in these species than in apes and lemurs.

High protein requirements have been traditionally associated with callitrichids (King, 1978). However, there is little scien-

Table C.IX

Hemogram

Value	Unit	Mean value						
		a	b	c	d	e		f
Red blood count (RBC)	$10^6/mm^3$	6.59	6.86	6.90	6.95	5.70	10^6	10^{12}/liter
Hemoglobin	gm/dl	15.5	15.1	15.5	17.0	14.9	0.155	mM/liter
Hematocrit	%	45	45	48	52	46	—	—
MCV	m³	69	67	69	74	80	1.0	fl
MCH	pg	23	22	22	25	26	0.0155	fmole
MCHC	%	34	34	32	33	33	0.01	—
Reticulocyte	%	3.5	4.3	3.5	2.6	—	10^6	10^9/liter
Leukocytes	$10^3/mm^3$	12.6	12.8	7.3	11.9	7.1	10^6	10^9/liter
Bands	%	0.4	1.7	—	—	0.1	10^6	10^6/liter
Segmented neutrophils	%	43	28	55	42	62	10^6	10^6/liter
Lymphocytes	%	49	67	43	54	30	10^6	10^6/liter
Monocytes	%	5.0	2.1	0.4	2.8	2.2	10^6	10^6/liter
Eosinophil	%	1.2	0.6	0.5	0.9	4.2	10^6	10^6/liter
Basophil	%	0.1	0.3	1.3	0.7	1.1	10^6	10^6/liter
Platelets	$10^3/mm^3$	—	—	490	—	—	10^6	10^9/liter
Fibrinogen	mg/dl	—	—	300	—	—	0.0293	μM/liter
Prothrombin time	sec	—	—	—	7.1	—	—	—
Partial PT with kaolin	sec	—	—	—	32	—	—	—

[a] C. B. Richter, previously unpublished. Sixteen hr fasted adult *S. oedipus; n,* 10.
[b] C. B. Richter, previously unpublished. Sixteen hr fasted adult *C. jacchus; n,* 10.
[c] Adapted from Hawkey *et al.* (1982). Clinically normal *C. jacchus; n,* 27–30.
[d] Adapted from Wadsworth *et al.* (1982). Young adult *S. labiatus; n,* 33–39.
[e] Adapted from Bush *et al.* (1982). Normal adult *L. rosalia; n,* 160–170.
[f] Conversion factor and SI Units (International System of Units).

Table C.X

Blood Chemistry

	Units	a	b	c	d	e		f
Glucose	mg/dl	134	126	150	228	157	0.0555	mM/liter
Glycohemoglobin	%	2.4	1.6	—	—	—	—	—
Uric Acid	mg/dl	3.0	4.1	—	1.8	2.7	0.0595	mM/liter
Blood urea nitrogen	mg/dl	31	27	—	9.7	9.0	0.714	mM/liter
Urea	mg/dl	—	—	51.8	—	—	0.167	mM/liter
Creatinine	mg/dl	—	—	0.9	1.2	—	76.3	μM/liter
Total bilirubin	mg/dl	0.4	0.5	0.6	—	—	17.1	μM/liter
Direct bilirubin	mg/dl	—	0.2	—	0.1	0.1	17.1	μM/liter
Indirect bilirubin	mg/dl	—	0.3	—	0.1	0.1	17.1	μM/liter
Total protein	gm/dl	6.6	7.0	7.1	—	—	10	gm/liter
Albumin	gm/dl	3.9	3.8	3.9	—	—	10	gm/liter
Globulin	gm/dl	2.7	3.2	3.2	3.6	3.9	10	gm/liter
Alkaline phosphatase	IU/liter	5.4	2.4	606	29.8	11.9	—	—
SGOT	IU/liter	177	160	182	99	113	—	—
SGPT	IU/liter	109	60	32	—	—	—	—
Lactic dehydrogenase (LDC)	IU/liter	883	799	—	—	—	—	—
Gamma glutamyl transpeptidase (GGT)	IU/liter	11	—	—	—	—	—	—
Choline esterase	pHU	—	—	—	0.6	0.3	—	—
Calcium	mg/dl	9.4	9.5	10.2	10.4	10.0	0.50	mM/liter

(*continued*)

Table C.X (*Continued*)

	Units	a	b	c	d	e	f	
Inorganic phosphate	mg/dl	6.0	6.9	—	5.5	4.7	0.323	mM/liter
Sodium	mEq/liter	—	—	158	169	161	1.0	mM/liter
Potassium	mEq/liter	—	—	4.2	5.7	6.0	1.0	mM/liter
Chloride	mEq/liter	—	—	—	114	109	1.0	mM/liter
Magnesium	mEq/liter	—	—	—	2.4	2.1	1.0	mM/liter
CO_2	mEq/liter	—	—	—	13.6	18.7	1.0	mM/liter
Serum iron	μg/dl	118	143	—	—	—	0.179	μM/liter

[a]C. B. Richter, ≥10 fasted *S. oedipus*. Selected healthy adults.
[b]C. B. Richter, ≥10 fasted *C. jacchus*. Selected healthy adults.
[c]Adapted from Wadsworth *et al.* (1982). *S. labiatus;* mean age, 21 months.
[d]Adapted from Holmes *et al.* (1967). White-lip tamarins, adult.
[e]Adapted from Holmes *et al.* (1967). *S. oedipus*, adults.
[f]Conversion factor and SI Units (International System of Units).

Table C.XI

Glucose Tolerance Data for Two Normal Fasted *C. j. jacchus*

Animal	Fasting blood sugar (FBS)	½ hr	1 hr	2 hr
1	95[a] (−)[b]	165 (−)	111 (−)	100 (−)
2	104 (−)	232 (++)	131 (tr)	80 (−)

[a]mg/dl.
[b]Urine glucose tested by dipstick method. −, negative; tr, trace; ++, positive.

tific basis for this, and it is likely that most diets contain considerably more protein than is necessary (Richter *et al.*, 1981). Several attempts have been made to formulate standardized pelleted or biscuited callitrichid diets (NIH-43 marmoset diet; Wirth and Buselmaier, 1982). The major disadvantage of these diets is the manufacturing process forced upon colony managers (baking, pelleting, etc.), while acceptance and results of long term use are uneven or unproved. The reader should consult the works of King (1975, 1978), Wirth and Buselmaier (1982), and Escajadillo *et al.* (1981) for further insight into feeding habits and nutrient intake on callitrichids in captivity.

Most callitrichid husbandry programs have resorted to varied diets in an effort to improve general health status when limited diets fail. Examples of such diets can be found in Hampton (1964), Brand (1981), King (1975), and Levy and Artecona (1964) and are amazingly similar in content. Varied diets suffer from an important defect, namely, the discipline necessary to ensure a consistent feeding practice. As a result, varied diets tend to become unequally weighted in content depending upon availability of ingredient, price, and the whims of animal care personnel unless strict discipline is observed. Changes in diet should only be made cautiously, and the rationale for change

should be based on experimentation. The varied diet fed to three species of callitrichids (*S. o. oedipus, C. j. jacchus, S. fuscicollis* ssp.) over a 2½ year period in a large colony* is presented in Table C.XII. Strict discipline was maintained on time of feeding and amount and substance fed. Varied diets are not difficult to feed, and the process can be made efficient. Nevertheless, varied diets must be looked upon with circumspection. Several important rules of callitrichid nutrition management seem apparent: (1) avoid generalizations with other primate nutrition requirements; (2) diets should contain relatively high, readily available, sources of energy; and (3) behavior (see Section II,E) is an important aspect of callitrichid nutrition. Perhaps the most detailed studies of marmoset nutrition intake are those of King (1975). Further comment on callitrichid diet can be found in Section III,D below.

Captured feral animals are frequently semistarved by the time they reach final destinations and readily accept commercial primate diets initially; however, these diets are excessively bulky and consumption ultimately falls. Few animals thrive on these diets when they are presented as the primary food source over extended periods.

D. Biology of Reproduction

Tardif (1982) has studied the age of onset of progesterone peak cyclicity, e.g., sexual maturity, in *S. o. oedipus* females and concluded that young females paired with unrelated adult males cycle at 417 ± 28.6 days, whereas young females remaining in their natal (family) groups did not cycle until 586.7 ± 53.8 days. This difference is significant, and similar obser-

*Medical and Health Sciences Division, Oak Ridge Associated Universities, Oak Ridge, Tennessee 37830.

Table C.XII

Varied Diet—Oak Ridge Associated Universities

Morning (8:30 AM)	Afternoon (2:30 PM)
Slurry[a,b]	Canned diet[a,d]
3.4 gm ground Monkey Chow[c]	Hard boiled egg[a] (3× weekly)
1.9 ml cold water	Small marshmallows[a] (2× weekly)
2.4 gm bananas	Roasted peanuts in shell (once weekly)
3.1 ml applesauce	Meal worms[a] (once weekly)
1.0 ml Karo syrup	Raisins[a] (once biweekly)
	Shredded coconut (once biweekly)
Feed cups must be removed in 4	Lettuce (once biweekly)
hr	Sweet potatoes, raw (once biweekly)
	Suckling mice (once monthly)

[a]Substances most universally accepted.

[b]Blend daily in large commercial mixer. Slurry was fortified with the following vitamins and minerals on a weekly basis: folic acid (2×); vitamin D_3 (2×); multiple vitamins, $CuSO_4$, and ascorbic acid once weekly.

[c]Monkey Chow, 25% protein. Ralston Purina Co., St. Louis, Missouri.

[d]Zoopreme Marmoset Diet, Hills Division, Riviana Foods, Topeka, Kansas.

vations have been reported in other species (Epple and Katz, 1980; Hearn, 1977). The reason for this delay is debatable, but behavioral influences emanating from the mother are possible causes. Hearn (1977) states that subordinate female *C. jacchus* in peer groups stop cycling until removed from the group. Epple (1970b), among many others, has shown that only the dominant female in peer groups will reproduce, but Tardif's (1982) work clearly demonstrates that young female *S. oedipus oedipus* will commence cycling in the presence of their mothers. Male *C. jacchus* do not reach adult levels of plasma testosterone until approximately 18 months of age (Hearn and Lunn, 1975) even though they are probably capable of impregnating females prior to this age (Epple and Katz, 1980). First pregnancies in callitrichids may occur as early as 1 year of age; however, in view of normal growth and behavior requirements, planned mating before 2 years of age appears to be undesirable. Successful infant rearing by females born in captivity has been inadequate to sustain continuous captive reproduction in some species. This failure is less pronounced when young females are paired with experienced males, but most infant loss in these circumstances appears to be due to failure of young females to accept and nurse their young.

Female callitrichids are biovulatory, and dizygotic twinning is the rule, but singleton births, triplets (Fig. C.7), and even quadruplets are seen (Table C.XIII). Survivability of triplets is poor, and quadruplets fail almost uniformly unless an attempt is made to give all infants a supplemental milk substitute (Ziegler *et al.*, 1981). Alternatively, "collaborative" techniques have been used where infants are rotated off the parents for 24 hr periods of diet supplementation (Hearn and Burden, 1979). These methods are interesting and can be successful, but there are many disadvantages and they are labor intensive.

Estrous cycle varies between 14 and 18 days (Hampton and Hampton, 1977; Hearn and Lunn, 1975; Tardif, 1982) and is characterized by the absence of external signs such as sex skin

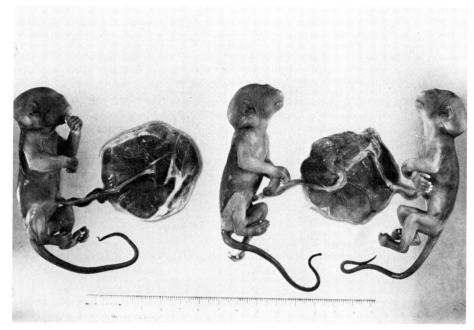

Fig. C.7. Full-term triplet fetuses, *S. o. oedipus*. Breech presentation; mother died.

Table C.XIII

Incidence of Multiple Births

Species	Single	Twins	Triplets	Quadruplets	Total
C. jacchus[a]	19	116	34	0	159
S. fuscicollis spp.[b]	65	238	16	NR[c]	319
Mixed ssp.[d]	35	129	12	2	176
C. jacchus[e]	59	644	329	7	1032

[a]Adapted from Phillips (1976b).
[b]Adapted from Gengozian, Batson, and Smith (1977).
[c]NR, not reported.
[d]Adapted from Hampton and Hampton (1965).
[e]W. D. Hiddleston, personal communication.

Table C.XIV

Pregnant Uterus Size in Callitrichids

Diameter (cm)	Days of gestation—mean values	
	S. fuscicollis spp[a]	C. jacchus jacchus[b]
0.5	29	—
0.6	—	30
0.8	—	46
1.0	65	60
1.2	—	64
1.5	80	79
2.0	89	Not measurable
2.5	109	Not measurable
3.0	117	Not measurable
3.5	131	Not measurable

[a]Adapted from Gengozian et al. (1974) (158 days gestation).
[b]Adapted from Phillips and Grist (1975) (150 days gestation).

changes and menstruation. There appears to be no detectable change in vaginal cytology; however, plasma levels of estrogens, progesterone, and leutinizing hormone vary in a cyclical pattern and serve as the basis for establishing estrous cycle length. Behavioral patterns have been used by Kleiman (1977b) to detect estrus in L. rosalia, but this method requires careful observation and may not be useful in all callitrichids. Lactational anestrus is absent, hence postpartum estrus and short interbirth intervals are common in permanently paired females. Hearn and Lunn (1975) have reported postpartum estrus as early as 3 days (average 7.9 days) in C. jacchus.

Gestation is shortest in the lion tamarins (128 days) (Kleiman, 1977b), but is approximately 145 days for other species. Species variation in interbirth interval is seen, ranging from ~135 days in the lion tamarins (Wilson, 1977) through 178 days in C. j. jacchus (W. D. Hiddleston, personal communication) to 240 days in S. o. oedipus (Hampton and Hampton, 1965). Pregnancy can be diagnosed as early as 2 weeks by plasma or urine measurement of placental chorionic gonadotropin (Hodges et al., 1976); however, callitrichids are ideally suited for digital palpation of the pregnant uterus through the abdominal wall. Several tables have been published for different species (Table C.XIV), and with practice it is possible to detect pregnancy as early as 2 weeks and estimate time of delivery within the same range. Most deliveries occur at night outside the nest box. Neonates receive little help from the mother and must climb to her back and nipple if they are to survive. Rothe (1977) has provided the most elegant observations of callitrichid puerperal behavior.

Callitrichid placentation is monodiscoid hemochorial and hemodynamically labyrinthine (Wynn et al., 1975). As described previously (Section II,A), blood chimerism occurs in callitrichids. Since many singletons are the result of a twin being resorbed, these are commonly chimeric also; however, nonchimeric singletons do occur (Gengozian and Batson, 1975). Benirschke and Miller (1982) have reviewed the physiology

and anatomy of callitrichid placentation. Spontaneous abortion is frequently reported and may be more common than suspected, especially in early pregnancy (Hearn, 1977; C. B. Richter, unpublished observations). Beyond delaying full term pregnancies, the role this may play in overall reproductivity is uncertain; however, caution is indicated in handling females who may be pregnant, and care is necessary during digital pregnancy exams. Abortion with retained placenta constitutes a medical emergency since loss of blood can be severe (see Section III,D).

Reproductive capacity of callitrichids exceeds that of any other simian primate. One female S. f. illigeri produced 37 offspring during a 13½ year period in captivity (Richter and Buyukmihci, 1978), and records of >20 for several species have been commonly reported. There is a tendency for increased multiple births (>2 infants) as captive colonies "mature," but this upward tendency may only be transient. Gengozian and Batson (1975) observed a higher frequency of single births in aging female S. fuscicollis. W. D. Hiddleston (personal communication) has reported dramatic increases in the incidence of triplet and quadruplet deliveries from year to year in C. jacchus. Similar observations have been made by many workers, but no real reason for this phenomenon has been established, and the frequency of such deliveries in feral animals has not been determined. Sex ratios of offspring are 1 : 1, and twins normally show the predicted 1♂♂ : 2♂♀ : 1♀♀ ratio (Gengozian et al., 1977). Some investigators have suggested that seasonality occurs in reproduction in captivity, but the weight of evidence is against this, and pregnancies occur throughout the year. Dawson and Dukelow (1976) observed pregnancies year round in feral S. o. geof-

froyi. Weaning may begin as early as 45 days and is usually complete by 60 days, although some infants may nurse for 1 month or more longer. In addition to supplementation methods for feeding infants from multiple births, hand-rearing of callitrichid infants is sometimes attempted because of parental neglect (Pook, 1976) or for research purposes when it is desirable to eliminate adult contact with young (Wolfe *et al.*, 1972). There have also been attempts to rear callitrichids in the gnotobiotic state for limited time periods (Eichberg *et al.*, 1979). The reader is referred to the original articles on these subjects for details. Hand rearing is fraught with difficulty and accompanied by numerous problems including high failure rate, behavioral aberrations in successfully reared individuals, and very low cost effectiveness. Turton *et al.* (1978) have examined the chemical compositions of marmoset (*C. jacchus*) milk in detail (see Table C.XV). On the basis of this study, human milk substitutes should be enriched with protein, total lipids, and fatty acid content.

Callitrichid ovaries are characterized by the accumulation of luteal (interstitial) bodies in the subcortical stroma throughout life. These bodies can be observed in juvenile females, but in older adult females these bodies may constitute the major substance of the organ. Large luteal bodies are common in other New World primates also, but their role in reproduction is obscure. Ovaries completely devoid of follicles or ova have not been observed in aged females (C. B. Richter, unpublished observation). The embryology of *C. j. jacchus* has been published by Phillips (1976a).

E. Behavior

Much of what is known about callitrichid behavior is the result of studies on captive populations. Small size and natural habitat plus their arboreal nature make field studies of callitrichids difficult. It is quite likely that some behavioral patterns that have been broadly accepted and applied to captive husbandry may well be erroneous and contributory to the difficulties experienced in propagating these species in captivity.

Table C.XV

Chemical Compositions of Marmoset Milk[a]

Component	Value	Element	Value
Protein (gm/dl)	3.6	Na (mg/dl)	21.4
Lactose (gm/dl)	7.5	K (mg/dl)	54.3
Total lipid (gm/dl)	7.5	Ca (mg/dl)	92.2
Total lipid (kcal/dl)	114	P (mg/dl)	22.8
Osmotic pressure	354	Mg (mg/dl)	5.0
(mOsm/kg H_2O)		Cl (mg/dl)	52.2

[a]From Turton *et al.* (1978).

An example would be the almost universal acceptance of strong pair bonding as the normal pattern of breeding male–breeding female relationship (Epple, 1977). Most husbandry programs are predicated on the concept of permanent pair mating and the nuclear family. However, some field studies, notably Neyman's (1977), cast doubt on this modus and suggest that such bondings center on rearing of young and may be transient. Other studies have also demonstrated the transient nature of group or "family" members (Dawson, 1977; Izawa, 1978). Little is known about the rate of reproduction of feral breeding females versus the nearly constant state of pregnancy in captivity. That the latter may be the result of misguided husbandry practices based on the strong pair–bond concept must be considered.

1. Territoriality and Group Size

Feral callitrichids are territorial, occupying a home range area that may have considerable overlapping with the territory of adjacent groups (Izawa, 1978; Neyman, 1977). At least in some circumstances, the home range may be vigorously defended (Dawson, 1977). Size of territory may vary considerably with species: 1 hectare (ha) for *C. j. jacchus* (Stevenson, 1977), 7–10 ha for *S. o. oedipus* (Neyman, 1977), 30–50 ha for *Saguinus nigricollis* (Izawa, 1978). Variation is also seen within species; however, availability of food supply is probably the single most important determinant for group territory size. Group size itself may also vary considerably, and figures from 2 to 40+ individuals are reported (Stevenson, 1977; Dawson, 1977; Neyman, 1977; Izawa, 1978); however, mean group size for most species falls within the 4–7 individuals range. Large groups apparently represent temporary banding of two or more smaller groups and are short-lived (Izawa, 1978). Group size variation is also a reflection of emigration and immigration of transients, usually subadults (Dawson, 1977). The introduction of mature adults into breeding groups in captivity is destabilizing and may result in fierce fighting (Epple, 1970). Group members sleep together, usually in one of a number of familiar nesting trees within the sanctuary area (Neyman, 1977; Coimbra-Filho, 1977). Group activity usually begins 1–1.5 hr after sunrise.

2. Infant Rearing

Infant rearing is among the most complex of the callitrichid behavior patterns and among the ones that fail most often in captivity. There appear to be clear relationships between the state of female parity and the success of infant rearing, at least in captivity. For unexplained reasons, there is a high rate of infant rearing failure among primiparous captive-born females in some species. Even multiparous capture-born females may continue to exhibit this failure. C. B. Richter (unpublished observation) observed a captive-born female *S. o. oedipus* who gave birth to 8 full-term healthy infants (three pregnancies) within a 12-month period and rejected all of them, even though her mate made futile efforts to care for them. Most husbandry practices today are based on the concept that infant care is a learned process, and therefore weanlings remain with adults through at least one subsequent cycle of infant rearing. Virtually all family members, including juveniles, participate in infant carrying, and great variation exists in carrying time for individual family members. Contrary to early reports, males do not singularly perform this duty or even perform it the majority of the time in many instances (Wolters, 1977). C. B. Richter (unpublished observation) had a female cotton-top who lost her mate in late pregnancy but rejected added juvenile helpers and subsequently raised her twins completely by herself. Generally, the adult male and female do most of the carrying (Wolters, 1977). Claims that the adult male or other family members assist the female during parturition are unsubstantiated, but they may act as curious onlookers or nuisances (Rothe, 1977).

Other important aspects of infant rearing include active and passive food sharing with infants by adults and juveniles. Davis and Richter (1980) showed that in *S. o. oedipus* family groups, the male was the giver, both passively and actively. In the latter case a "food" call was emitted. Brown and Mack (1978) have shown the same behavior in *L. rosalia,* but in this case mother, father, and juveniles acted as givers. Izawa (1978) observed similar behavior in feral *S. nigricollis.* Young callitrichids are ambulatory as early as 2 weeks, but do not venture far from adults much before 4 weeks.

3. Intra- and Intergroup Behavior

The work of Stevenson and Poole (1976) should be consulted for a general ethogram of *C. j. jacchus.* Dominance within callitrichid families is usually subtle and may be difficult to identify. Males respond to distress more aggressively than other family members (S. D. Tardif, unpublished observations) and usually dictate the time of entrance into the nest box for sleeping. On the other hand, lactating females may dominate feeding stations and consume most of the preferred food if allowances are not made to accommodate other group members (Tardif and Richter, 1981) (see Section II,C). Females also mark more frequently with the circumgenital glands and experience increased marking behavior with sexual maturation (Tardif, 1982). Both sexes mark by rubbing the genitalia, pubis, or chest on the object being marked. Male *C. j. jacchus* are especially likely to mark in response to disturbances or threats, but female *S. fuscicollis* may mark more than their male partners in the presence of strange conspecifics (Epple, 1977). Social, especially sexual, and territorial messages are thought to be conveyed by marking (Epple, 1973). Both sexes of callitrichids also mark with urine, and marmosets, but not tamarins, "posterior-present" in the face of a perceived threat. In this posture the hair is piloerected, the tail extended upward, and the genitals beared toward the threat. This is accompanied by frequent adjustments in posture and glancing back over the shoulder toward the threat.

The works of Epple (1967), Rothe (1975), and Kleiman (1977b) may be consulted for descriptions of sexual behavior in callitrichid family groups. Sex play begins during the juvenile stage and is manifested by sniffing of genitalia, attempts to mount, and masturbation. Among breeding adults, copulations or attempts at copulation continue through pregnancy in some species (Kleiman, 1977b). Completed attempts to copulate, including intromission and ejaculation, are generally not often witnessed by observers, since many attempts are disrupted by the juvenile family members. Associative behavior between males and females increases during estrus. Only the dominant male and dominant female reproduce (see Section II,D).

Also included in the callitrichid behavior repertoire are tongue displaying, head cocking, lip smacking, and piloerection. Tongue display and lip smacking are frequently associated with copulatory behavior. Allogrooming, play, and aggression are part of normal callitrichid behavior. Allogrooming is observed more among sexually mature individuals, while play is seen among infants and subadults. Aggression can be seen among any family members, but is usually dissociative, e.g., nondestructive (Wolters, 1977), and brief. Destructive agonistic behavior usually results in injury and may require permanent separation of the involved individuals. Males tend to play roles similar to those played by other simian males: defending, stabilizing the position of the adult female, and decision making (Wolters, 1977). It is generally accepted that a dual hierarchial system exists in callitrichid groups: male–male and female–female.

III. DISEASES

General articles on diseases of callitrichids can be found in F. Deinhardt *et al.* (1967), J. B. Deinhardt *et al.* (1967), Nelson *et al.* (1966), Hunt *et al.* (1978), and Cicmanec (1977). Hunt's article should be reviewed by serious workers or clinicians confronted with the problems of callitrichid health care. The article by Nelson *et al.* (1966) is perhaps the earliest comprehensive attempt to examine mortality patterns and causes of death in callitrichids. This article is heavily oriented toward parasitic diseases but lists other diagnoses as well. A shocking 53% of 746 tamarins received in this study over a 3½ year period were dead within 150 days of arrival, 51% of the total within 90 days. Unregulated mass shipments of these species from native habitat is not likely to occur in the future, and considerably more is now known about callitrichid disease.

Clinical examination of callitrichids should be systematic and include the following:

1. Body weight
2. Rectal temperature
3. External examination for evidence of injury, notably the hands, feet, limbs, tail and head
4. Oral cavity, including gums, teeth, tongue, and tonsils. Note the color of the mucous membranes.
5. Urinalysis using dipstick method. Specific gravity determination and microscopic examination of the sediment are advisable.
6. Auscultation of the heart and lungs. Electronic amplification is advisable.
7. Abdominal palpation. Use chemical restraint, such as ketamine, to obtain maximum relaxation if necessary.
8. Fecal examination. Occult blood, parasites, quality, and quantity
9. Eye exam. Include corneal inspection, pupillary response, and fundus examination
10. Complete blood count and clinical chemistry if indicated. Knotts technique for microfilariae is advisable.

The above can be accomplished rapidly and usually without chemical restraint if trained assistance is available. It is also advisable to observe the animal in the cage to determine if locomotion and posturing are normal.

A. Bacterial

With the exception of *Klebsiella pneumoniae,* pathogenic bacteria are relatively uncommon. This organism occurs principally as a cause of lobar penumonia in debilitated or sick individuals and is regarded as an opportunistic pathogen. In an exception, C. B. Richter and W. G. Tankersley (unpublished observation), observed acute septic peritonitis in four otherwise healthy *Callithrix jacchus jacchus* caused by *Klebsiella pneumoniae* serotype 1, the most common *Klebsiella* serotype associated with human disease. This serotype also appears to be the one associated with secondary disease in callitrichids. Animals shedding this organism in their feces can be found in captive populations.

Salmonella ssp. are uncommon in healthy populations but occasionally cause fatal septicemias. J. B. Deinhardt *et al.* (1967) reported on 43 *Salmonella* isolates from a survey population of 750 animals. Fatal cases are characterized by serositis and focal necrosis of the liver and lungs. Antibiotic therapy is not indicated in *Salmonella*-infected individuals unless there is

evidence of sepsis. Therapy should be based on antibiotic sensitivity tests and may be protracted. Chloramphenicol, ampicillin, and trimethoprim–sulfamethoxazole are among the most successful antisalmonella drugs.

Shigellosis is also reported in callitrichids, but the incidence of the organism in fecal surveys exceeds the incidence of disease associated with it. Like *Klebsiella* and *Salmonella,* *Shigella*-associated disease is closely correlated with the general health status of the population in question. Bloody diarrhea, colitis, dehydration, and deaths are attributed to *Shigella sonnei.* Murphy *et al.* (1972) reported 23 *S. nigricollis* and one *S. mystax* infected with *Salmonella* or *Shigella sonnei* D_1 in a total population of 244 animals. Although all of the infected animals died, other complicating diseases makes the role of these organisms in this episode uncertain. Cooper and Needham (1976) reported an acute episode of shigellosis involving *C. jacchus* and *S. nigricollis* in which bloody diarrhea, dehydration, and lethargy were principal signs. Postmortem findings included gas/liquid-filled stomach and small intestine, petechia and erosions of the large bowel, and free blood in the lumen of the large bowel. Neomycin was thought to be effective, but the colony continued to shed the organism long after clinical signs were relieved. As a matter of practice the clinician must consider the possibility of occult klebsiellosis, especially lobar pneumonia, as a complication in sick animals and consider administration of aminoglycoside antibiotics in appropriate cases. Nevertheless, animals shedding *Salmonella* or *Shigella* are apt to experience higher mortality (J. B. Deinhardt *et al.,* 1967).

Tuberculosis is rare in callitrichids, the only reported cases having been associated with an outbreak in Old World species (Moreland, 1970). In a series of over 2500 callitrichid necropsies extending over a 20-year period at a single institution (Oak Ridge Associated Universities), no case of tuberculous mycobacterial disease was observed and only one case of nontuberculous cutaneous mycobacterial infection in a cotton-top tamarin (Richter and Tankersley, 1984). This appeared as a nonhealing purulent sinus track rich in acid-fast bacilli. A single report of erysipelas and a single report of *Corynebacterium equi* lung abscess appear in the literature, but these appear to be curiosities. Hill *et al.* (1978) have reported the bacterial and mycoplasma flora of living and dead *C. jacchus.*

Infraorbital abscess is a common problem in the saddle-back tamarin. Unlike the squirrel monkey, it is usually not possible to associate canine tooth disease as a causative factor; however, when it is a factor, tooth extraction leads to a rapid cure. Careful examination of the canines for splitting, exposed root canal, peridontal disease, and obvious evidence of looseness is necessary, and radiology is advisable. Abscesses should be cultured and probed for evidence of paranasal sinus involvement or osteomyelitis. Where possible, antibiotics selected by sensitivity testing should be injected systemically and intralesionally; however, patience is necessary since most of these lesions heal slowly and may recur. Some abscesses are negative, even when cultured repeatedly, and these can only be treated by draining and irrigation. Care must be taken because the orbit bone is thin and many blood vessels and nerves course beneath the skin of the face.

Osteomyelitis, local or disseminated, is a fairly common problem as a result of puncture wounds or chronic cauditis and is especially common locally in bitten or amputated fingers. These usually respond to systemic antibiotics against gram-negative bacteria but may take extended periods to heal completely. Osteomyelitis of the tail tip is usually secondary to fecolith formation because of inadequate perch-to-floor distance and the inability of long-tailed species to avoid accumulating feces on the tail. Debilitated animals are especially prone to this problem, and the best prevention is a cage with higher perches. The sequence of events involves formation of a fecolith, dermatitis, necrosis of the tail tip, exposure of the terminal coccygeal vertebra, osteomyelitis (as well as tendovaginitis), and a slowly progressive infection. The affected animal should be moved to a larger cage, the tail tip bathed and treated locally with antibacterial or antibiotic ointments. It may be desirable to use systemic antibiotics, especially if the infected member is hot. As healing progresses, the soft tissues will retract exposing half or more of the dry, necrotic distal coccygeal vertebra. This vertebra can be tested periodically by applying tension with forceps and eventually extracted under chemical restraint with minimum trauma. If no residual infection persists, the wound will close completely. Dramatic cures such as partial tail amputation should be avoided. Since it is difficult to protect tail wounds from the animal and from excrement contamination, it is difficult to achieve first intention healing. Tail sutures commonly dehisce or are removed, and second intention healing is slow, requiring constant care.

Systemic osteomyelitis occurs occasionally as a result of fight wounds and is usually accompanied by one or more sinus tracks extending from infected bones or joints. Involved sites should be cultured repeatedly until the associated bacterium is isolated and antibiotic sensitivity determined. Systemic antibiotic therapy should be instituted prior to completion of sensitivity testing in an attempt to reduce patient risk. Since many of these cases involve gram-negative organisms, an aminoglycoside antibiotic coupled with penicillin is recommended until specific sensitivity is determined. Therapy should continue for approximately 1 week after recovery. Bone scans and especially radiography are useful diagnostic tools for osteomyelitis. Surprisingly, most animals recover rather rapidly with antibiotic therapy.

Mycotic infections appear to be uncommon with the exception of *Candida* ssp. Mycotic glossitis, esophagitis, and occasionally gastritis caused by *Candida* are common in debilitated animals. Glossitis is usually manifested as a thickened gray-white coat on the margin or tip of the tongue. The problem is self-correcting when underlying disease is cured. Cryptococcosis has been reported by Takos and Elton (1953) in two juvenile marmosets. C. B. Richter (unpublished observation) has observed a single case of mycetomatous thyroiditis in a cottontop tamarin. J. B. Deinhardt *et al.* (1967) reported a persistent hair shaft infection in *S. oedipus* and white-lip tamarins, but did not speciate the fungus. This episode appears to have been related to other debilitating influences. Skin was not infected.

B. Parasitic

Table C.XVI lists the reported parasites of callitrichids. The most important of these are the acanthocephalan worms *Prosthenorchis elegans* and *P. spirula,* but many others are commonly seen and undoubtedly contribute to the overall health status of colonies and individuals.

Table C.XVI

Parasites of Marmosets and Tamarins

I. Protozoa
  A. Subphylum: Mastigophora (Flagellata)
     Species: *Trypanosoma cruzi*—Visceral organs, CNS, blood
        Trypanosoma minasense—Blood
        Trypanosoma rangeli-like—Blood
        Giardia (lamblia?)—Intestines
        Trichomonas spp.—Intestines, vagina
  B. Subphylum: Sarcodina
     Species: *Entamoeba histolytica*—Cecum, colon
        Entamoeba coli—Cecum, colon
  C. Subphylum: Euspora
     Species: *Toxoplasma gondii*—Visceral organs, CNS
        Sarcocystis spp.—Muscle, skeletal and cardiac
        Isospora arctopitheci—Intestines
        Eimeria spp.—Intestines
  D. Uncertain classification
     Species: *Pneumocystis carinii*—Lung
II. Platyhelminths
  A. Class: Trematoda
     Species: *Athesmia foxi*—Bile duct, gallbladder
        Platynosomum amazonensis—Bile duct, gallbladder
        Platynosomum mamoseti—Bile duct, gallbladder
        Phaneropsolus sp.—Intestines
        Neodiplostomum tamarini—Intestines
  B. Class: Cestoidea
     Species: *Spirometra reptans*—Subcutis   ⎱
        Diphyllobothrium erinacei—Subcutis ⎰ (Sparganum)
        Paratriotaenia pedipomidatus—Intestines
        Atriotaenia megastoma—Intestines
        Bertiella mucronata—Intestines
        Hymenolepis cebidarum—Intestines
        Hymenolepis nana—Intestines
        Raillietina spp.—Intestines

Table C.XVI (*Continued*)

Species: *Echinococcus* (hydatid cyst)—Viscera, CNS
III. Aschelminths
  A. Class: Nematoda
    Species: *Strongyloides* spp.—Intestines
            Angiostrongylus costaricensis—Mesenteric arteries
            Filariopsis sp.—Lungs
            Filaroides barretoi—Lungs
            Molineus vexillarius—Intestines
            Longistriata dubia—Intestines
            Spirura guianensis—Esophagus, stomach
            Trichospirura leptostoma—Pancreatic duct
            Reticularia alphi—Intestines
            Physaloptera dilatata—Stomach
            Dipetalonema gracile—Peritoneal cavity
            Tetrapetalonema spp.—Subcutis, muscle fascia
            Parlitomosa zakii—?
            Trypanoxyuris tamarini—Cecum, colon
            Trypanoxyuris spp.—Cecum, colon
            Subulura jacchi—Intestine, cecum
IV. Acanthocephala
  A. Class: Archiancathocephala
    Species: *Prosthenorchis elegans*—Ileum
            Prosthenorchis spirula—Ileum
V. Arthropoda
  A. Class: Arachnoidea (Arachnida)
    Species: *Fonsecalges saimirii*—Hair
            Listrocarpus cosgrovei—Hair
            Listrocarpus hapalei—Hair
            Rhyncoptes anastosi—Skin (?)
            Dunnalges lambrechti—Skin (?)
            Mortelmansia duboisi—Nasal cavity
  B. Class: Insecta
    Species: *Gliricola pinto*—Skin
            Harrisonia uncinata—Skin
  C. Class: Pentastomida
    Species: *Porocephalus clavatus* (nymph)—Peritoneal and thoracic cavity

1. Protozoa

Trypanosomes, including *T. cruzi*, are common and many species of callitrichids are infected (Marinkelle, 1982), although disease is uncommon. Lushbaugh *et al.* (1969) have reported congenital fatal Chagas' disease in *S. f. lagonotus* fetuses. Visceral (leishmanial) forms were present in the placenta as well as the visceral organs (Fig. C.8). Those working with callitrichids are cautioned that this disease may be present in feral animals.

A single case of toxoplasmosis has been reported by Benirschke and Richart (1960). *Entamoeba histolytica* appears to be fairly common and may produce ulcers in the intestines with submucosal invasion (Fig. C.9). Liver lesions and visceral metastasis have not been reported. Richter *et al.* (1978a) observed widespread colony infection with *Pneumocystis carinii* in a population containing *S. o. oedipus* and *S. fuscicollis* ssp.

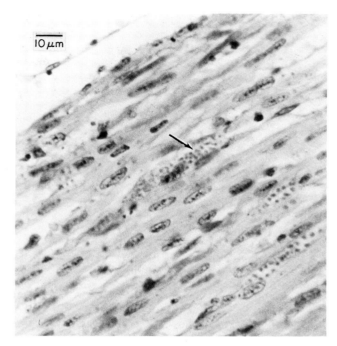

Fig. C.8. Leishmannial *T. cruzi* in heart of *S. fuscicollis lagonotus.* Numerous parasites are visible (arrow). Periodic acid–Schiff stain.

(Fig. C.10). A retrospective examination of 441 necropsies disclosed an infection rate of 11.9%. Only occasionally was the infection judged to have been clinically significant; however, the potential for complicating infection must be considered.

2. Platyhelminths

The role of bile duct trematodes in disease is uncertain, but they are common in feral animals. Local inflammatory response caused by their presence is easily confirmed histologically, and their ability to persist as long as 5 years after capture must be recognized; nevertheless, parenchymal damage associated with their presence is rare or absent. The small intestinal trematodes (Table C.XVI) occasionally invade the lymphatics and may be found in mesenteric lymph nodes (Fig. C.11). Mature cestodes are occasionally seen, but larval forms (spargana) appear to be more common. Spargana are usually found in subcutaneous connective tissue where they cause little inflammation.

3. Aschelminths

Montali and Bush (1980) have reported clinical disease caused by *Riticularia* sp. in *L. rosalia.* The use of mebendazole at 20 mg/kg daily for 3 days was effective in treating infected animals. Filarial worms are troublesome principally

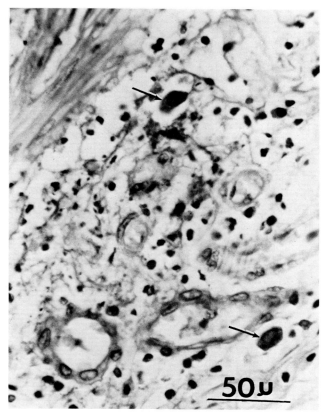

Fig. C.9.   Amoebiasis in *S. fuscicollis*. Amoebae are present in the sub-mucosa of the large bowel (arrows). Inflammatory response is characterized by edema, cellular infiltration, and tissue necrosis. Periodic acid–Schiff stain.

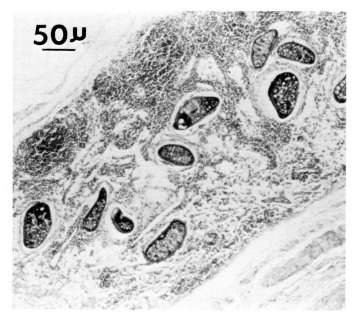

Fig. C.11.   Pancreaticoduodenal lymph node, *S. fuscicollis*. Numerous trematodes have translocated from the small intestine. Both species of trematodes occurring in the small intestine of callitrichids are small in size. Periodic acid–Schiff stain.

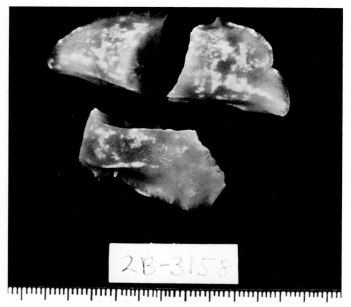

Fig. C.10.   *Pneumocystis carinii* in *S. o. oedipus*. Lesions appear as gray-white plaques beneath the pleura. This animal had concurrent carcinoma of the large bowel.

because microfilaria interfere with experimental procedures such as tissue culture, but their animal health impact is uncertain. Feral animals are commonly infected (Tankersley *et al.*, 1979). Therapy appears to be possible for some species of filarial worms with thiacetarsamide sodium.

4. Acanthocephala

Prosthenorchis spp. are the most common and serious parasite threat to the life and health of callitrichids brought from the jungles. In captivity these worms penetrate the wall of the terminal ileum causing abscesses, peritonitis, and sepsis. It is unknown if similar interactions occur in the jungle. Most infected individuals have 1–4 parasites, 2–2.5 cm long, situated at the ileocecal valve with the head imbedded in the wall of the terminal ileum (Fig. C.12). Treatment is difficult and uncertain because debility is often present and dramatic measures such as surgery may be precluded. There are no reports of the successful use of presently available antihelminthics. The worms may survive 1–3 years in healthy hosts, hence it is important to avoid crises by providing adequate nutrition and antibiotic support when necessary. Richter *et al.* (1976) reported a devastating epidemic of acanthocephaliasis in a marmoset colony mediated by *Blatella germanica*. At least 20 deaths were attributed to this cause, and larval *P. elegans* were readily

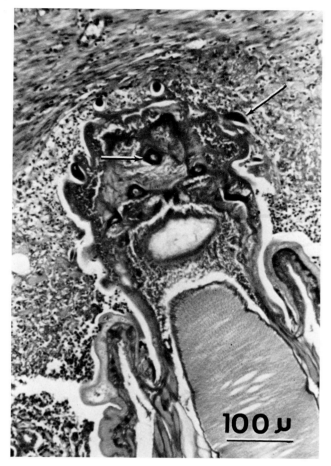

Fig. C.12. *Prosthenorchis elegans* embedded in the wall of the terminal ileum, *S. fuscicollis*. Cellular infiltrate surrounds the head. Several rows of crochets are visible (arrows). Periodic acid–Schiff stain.

found in cockroaches. It is imperative that possible colony-bred episodes caused by completion of the parasite life cycle through ''native'' German cockroaches be eliminated.

5. Arthropoda

The most commonly recognized arthropod is the nymph of *Porocephalus clavatus*. These are found coiled into a ''C'' shape and encapsulated in the peritoneal and thoracic cavities (Fig. C.13). Usually only one or two nymphs are present. They may occasionally be found in parenchymatous organs, notably the lung. Little attention has been paid to other arthropods, and there appears to be no reported instances of disease caused by them.

The lists of parasites published by Kuntz and Myers (1972) and Cosgrove *et al.* (1968) should be consulted for additional information.

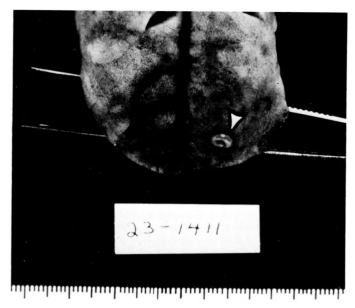

Fig. C.13. *Porocephalus clavatus* nymph adhered to the visceral pleura and diaphragmatic lobe of *S. fuscicollis*.

C. Viral

Viral diseases appear to be common in callitrichids. A survey of *S. nigricollis* and *S. mystax* monkeys taken upon arrival at their laboratory by Murphy *et al.* (1972) disclosed a high incidence of respiratory virus infected animals (Table C.XVII). Thirty percent of 216 animals sampled were harboring one or more viruses. Some infections, such as influenza virus A$_2$, parainfluenza viruses 2 and 3, and reo virus 3 were asymptomatic. Sendai virus infection (parainfluenza virus 1) ranged from asymptomatic to fatal. Combinations of Sendai and *Herpesvirus tamarinus* infection were uniformly fatal. Kalter and Heberling (1972) have shown through serological surveys that many call-

Table C.XVII

Respiratory Viruses Isolated from 216 *Saguinus* sp.[a]

Virus isolated	Number of isolates	Percentage positive
Parainfluenza 1 (Sendai)	29	13.4
Parainfluenza 2	2	0.9
Parainfluenza 3	13	6.0
Herpesvirus tamarinus	11	5.0
Herpesvirus hominis	1	0.4
Influenza A$_2$/Hong Kong/68	9	4.1
Reovirus type 3	14	6.4

[a]From Murphy *et al.* (1972). By permission of the publisher.

itrichids express antibody to a variety of human viruses, but the significance of this remains poorly understood. Hunt's paper (see Hunt *et al.*, 1978) presents a current review of callitrichid virus diseases.

1. Herpesviruses

The most important of these infections is *Herpesvirus tamarinus (platyrrhinae)* (HVT). Natural HVT infection is an often fatal systemic disease with a broad spectrum of organ involvement. The disease is characterized by rhinitis, conjunctivitis, oral and labial vesicles and ulcers, anorexia, lassitude, and ulcerative dermatitis. Numerous accounts of this disease have appeared in the literature since it was first described (Holmes *et al.*, 1964; Melnick *et al.*, 1964). The recent report by Morita *et al.* (1979) is the best clinical–pathological account. *Herpesvirus tamarinus (platyrrhinae)* infected animals sometimes develop hyperesthesia and scratch persistently. Postmortem lesions include ulcerations throughout the gastrointestinal tract from the lips and oral cavity through the esophagus and small and large intestines. Gray-white focal necrosis of the liver and other visceral organs is common. Visceral, neural, and mesenchymal cell intranuclear inclusion bodies are found. Hemorrhage in the adrenal cortex and lymph nodes is common, and multinucleate giant cells may be seen in mucosal cells or adjacent to areas of visceral necrosis. Bronchitis and focal pneumonia are also present. Histological findings include intense monocytic infiltration of the liver, spleen, and lymph nodes. Degeneration of the ganglion cells of the Auerbach and Meisner plexuses may also occur. Encephalitis with perivascular cuffing, focal demyelination, glial cell proliferation and neuronal degeneration are seen, sometimes accompanied by a mild nonsuppurative meningitis (Morita *et al.*, 1979). Degeneration of the muscularis propria of the gastrointestinal tract may also occur. Intranuclear inclusion bodies may appear either basophilic or eosinophilic, and although ultrastructurally different, both contain virus particles. Infected animals may become rapidly emaciated and die. The natural host of HVT appears to be the squirrel monkey, and while natural transmission from *Saimiri* to callitrichids is poorly documented, it must occur if the natural host premise is correct. Squirrel monkeys should be housed separately from callitrichids.

Herpesvirus simplex (HVS), human origin, is capable of causing a disease similar to HVT in callitrichids (Hunt *et al.*, 1978) when inoculated experimentally, but the natural potential for pathogenicity is uncertain. Colony managers would be well advised to exclude humans with active HVS from callitrichid colonies.

Nigida *et al.* (1979) have isolated a cytomegalovirus from *Saguinus fuscicollis* that possessed some immunogenic cross-reactivity with cytomegaloviruses from other species (*Macaca*

and man). Tissue culture transfer of this virus could only be accomplished by whole cell transfer, and serological survey showed antibody present only in wild caught but not colony-born animals. Infection is clinically and pathologically silent as far as is known, but the clinician and pathologist should be alert to the possibility of disease caused by latent virus in debilitated subjects.

At least three herpesviruses have been shown to be oncogenic in one or more species of callitrichids. *Herpesvirus saimiri* (Hunt *et al.*, 1970), *Herpesvirus ateles* (Meléndez *et al.*, 1972), and Epstein-Barr virus (Miller *et al.*, 1977). All produce a rapidly developing fatal, or sometimes regressing, lymphomalike neoplasm. The disease is characterized by lymphocytic leukemia, lymph node and thymus enlargement, splenomegaly, and enlargement of gastrointestinal lymphoid follicles. Essentially all visceral organs are infiltrated, including bone marrow, liver, kidneys, adrenals, heart, lungs, pancreas, etc. There is limited evidence that horizontal spread of *H. ateles* to contacts can occur (Hunt *et al.*, 1972).

2. Poxviruses

Poxvirus infection has been reported in *C. jacchus* by Gough *et al.* (1982). Typical erythematous papules progressing to encrusted plaques, and ulcers were observed (Fig. C.14). Skin lesions characterized by acanthosis, necrosis, and intracytoplasmic eosinophilic inclusion bodies in keratinocytes (Fig. C.15) were seen. Typical poxvirus particles were found in cells examined by electron microscopy. Disease was limited to the skin and ran a clinical course of 4–6 weeks. No deaths

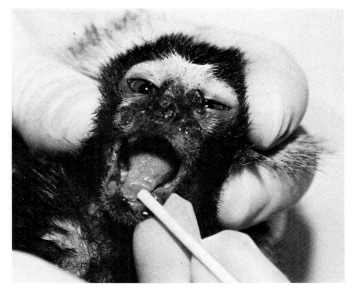

Fig. C.14. Poxvirus encrustations on the face of *C. jacchus*. (From Gough *et al.*, 1982; courtesy of Dr. A. W. Gough, and *Lab. Anim. Sci.*)

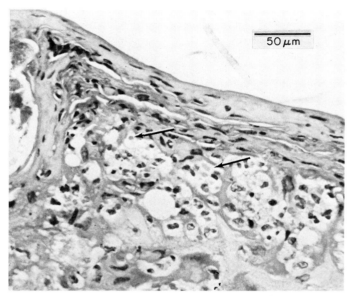

Fig. C.15. Poxvirus dermatitis, *C. jacchus*. Intracytoplasmic eosinophilic inclusion bodies are visible (arrows). Hematoxylin and eosin stain. (From Gough *et al.*, 1982; courtesy of Dr. A. W. Gough and *Lab. Anim. Sci.*)

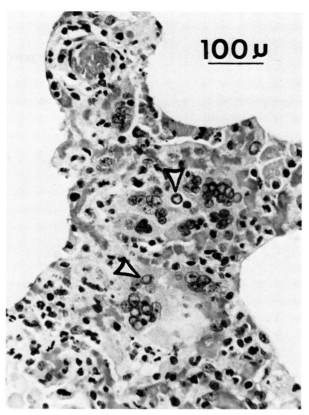

Fig. C.16. Interstitial pneumonitis with giant cell formation and intranuclear inclusion bodies (arrowheads). Hematoxylin and eosin (From Levy and Mirkovic, 1971; courtesy of Dr. Barnet M. Levy and *Lab. Anim. Sci.*)

were attributed to the infection. The virus did not appear to be closely related to vaccinia virus.

3. Myxoviruses

Although only a single well-documented case of measles (rubeola) virus infection has been published (Levy and Mirkovic, 1971), others have observed measleslike disease in callitrichid colonies (Cicmanec, 1977). The devastating episode reported by Levy and Mirkovic caused 326 deaths in a mixed population of *S. oedipus, S. fuscicollis,* and *C. jacchus* over a 7-month period. The duration of the episide was probably prolonged by the continued introduction of new animals into the colony for as long as 5 months after the disease was first observed. Clinical signs included swollen eyelids and lethargy with death ensuing less than a day after first observation of illness. Rhinorrhea, facial edema, and maculopapular exanthema on the lips and skin were also seen. No additional significant observations were made at necropsy. Histopathological findings included interstitial pneumonitis with giant cell formation, multinucleated cells with intranuclear inclusion bodies (Fig. C.16), and focal collections of alveolar giant cells including Warthin–Finkeldey (W–K) types. Bronchopneumonia was irregularly present. Large monocytes with or without intranuclear inclusions were common in the lung parenchyma. Warthin–Finkeldey cells were also found in the spleen, mesenteric lymph nodes, and colon, and inclusion-bearing macrophages were sometimes seen in lymph nodes. Survivors had significant hemagglutination inhibition antibody (HIA) titers to measles (mean, 1 : 62). Nonexposed animals were negative. The findings in Cicmanec's observations (1977) are less clear, but significant titers to rubeola and evidence of positive seroconversion by selected individuals were seen. Callitrichids appear to be exquisitely sensitive to experimental infection with measles virus (Albrecht *et al.*, 1980), and experimental infection produces rapid death and less dramatic but essentially similar lesions to those reported by Levy and Mirkovic (1971). Few animals survived the direct effects of the experimental measles virus infection. Both live attenuated and inactivated vaccines were effective in protecting challenged animals. Inactivated vaccine and human IgG, usually high in rubeola antibodies, are both recommended husbandry practices in callitrichid colonies. The latter can safely be given to infants.

Saguinus mystax has been experimentally infected with parainfluenza virus types 1 (strain Sendai) and 3 (strain HA-1) (Hawthorne *et al.*, 1982). Animals infected with Sendai excreted virus in high titer in throat secretions and transmitted the infection to uninoculated cage mates. Clinical signs in adults included weight loss, sneezing, nasal discharge, coughing, and conjunctivitis. Strain HA-1 virus caused weight loss, sneezing, nasal discharge, and conjunctivitis, all to a lesser degree than parainfluenza virus type 1 (Sendai). Both parainfluenza virus types 1 and 3 are common causes of respiratory disease in children, and the latter should be kept at a distance from large open callitrichid colonies. Clinicians should consider the use of vaccines or antisera in the face of an outbreak. There are no reported cases of natural respiratory paramyxovirus infections in callitrichids; however, C. B. Richter, S. S. Kalter, and D. C. Swartzendruber (unpublished) observed an epidemic virus infection in a mixed population of callitrichids characterized by fever, coryza, hoarseness, sneezing, conjunctivitis, and keratitis. Occasionally Koplik-like spots were seen on the buccal mucosa. Approximately 30 animals, most very young, were lost, and morbidity was high. The epidemic lasted approximately 1 month. Pathology included giant cell-associated bronchopneumonia, necrosis of renal convoluted tubule epithelium, necrosis of hepatocytes and Kupffer cells, and focal necrosis of gastrointestinal epithelium. There was marked necrosis of bronchial epithelium in infected bronchial segments. Inclusion bodies were not seen. Both parainfluenza virus types 1 and 3 were isolated from infected lungs, the former in greater titer. Electron microscopy of infected bronchi showed abundant paramyxovirus particles ranging from 160 to 360 nm containing nucelocapsids measuring approximately 16 nm in diameter. Acute and convalescent sera were inconsistent in demonstrating conversion to parainfluenza viruses 1 and 3 and consistently negative for measles antibodies. No experimental infections were attempted.

Paramyxovirus gastroenterocolitis in *Saguinus oedipus oedipus* has been reported by Frazer *et al.* (1978). Loss of appetite, diarrhea, dehydration, and death were seen. Congestion and hemorrhage of the mucosa were seen at necropsy. Colitis with epithelial sloughing and large epithelial syncytia containing intranuclear inclusions were observed. Gastritis and enteritis were also present . Cholangitis, again with syncytia, was present, and inclusions in these cells were found in the nucleus as well as the cytoplasm. Syncytia were found in pancreatic ducts, hepatic cords, kidney tubules, and even endometrium. Respiratory epithelium was free of lesions. From this and other reports it is clear that paramyxoviruses constitute a threat to callitrichids.

4. Miscellaneous Viruses

Healthy *Callithrix jacchus* may carry a syncytium forming foamy virus in the skin and peripheral leukocytes. The virus possesses an RNA-dependent DNA polymerase (Marczynska *et al.*, 1981) and has not been found in tamarins. There is no known pathology.

An outbreak of hepatitis, presumably viral, has been reported in a zoo collection of callitrichids (Lucke and Bennett, 1982). Twelve animals died in the episode. Principal lesions included modest pleural effusion, swollen liver with leukocyte infiltration, necrosis, anisocytosis, and anisokaryosis of hepatic parenchyma. Moderate fatty change was also present, but inclusion bodies were apparently absent.

Callitrichids have been used extensively in oncornavirus studies employing viruses from other species; however, the only candidate for a native oncornavirus is that described by Seman *et al.* (1975). They observed type-C virus particles budding from placental syncytiotrophoblasts of *S. oedipus* and *S. fuscicollis*. The meaning of this is uncertain.

D.   Miscellaneous

One of the more common causes of morbidity is fighting, almost always among conspecifics. In closely confined experimental animal rooms, escape usually means fighting since patterns of aggressive behavior develop between two individuals in visual contact but in separate cages. Fighting usually produces injury to hands and arms and to a lesser extent feet. Phalanges and whole fingers may be bitten off and severe lacerations are common. An occasional broken arm is also seen. Failure to detect such injuries may lead to abscessation and sepsis with possible osteomyelitis or an occasional death. Fights are much more severe when cage mates engage in combat. Severe facial damage and death are not uncommon. Female–female pairings are almost invariably failures even if caged together from juvenile age. Male–male pairings occasionally work, but only at the price of one subordinate individual who ultimately has to be segregated. Insidious harassment of the subordinate leads to death or severe debility unless properly recognized. Fight wounds can sometimes be sewn successfully, but these require careful attention. Systemic and local antibiotics are remarkably successful in all but the worst lacerations.

The most common single cause of morbidity in most colonies is diarrheal disease, particularly in tamarins. With the exception of the small clusters of cases caused by pathogenic bacteria and viruses, the cause of most of these cases remains uncertain. Since most cases are protracted, are frequently recurrent, and usually associated with debility, it is possible to speculate that nutrition, behavior, and /or mechanical factors are causally associated with the disease. Viral infections may act as initiators, but to date little work has been done in this area. Atony and megacolon are observed clinically, and copious feces are produced by many affected animals. Feces are frequently rich in mucus, and coprophagy is common as reported by numerous observers. Complete recovery is usually slow and best effected by supportive therapy, but many animals never fully recover and eventually die. Histologically this disease in tamarins is typified by crypt atrophy, marked cellular infiltration of the mucosa, residual crypt hyperplasia with reduced or even absent mature mucous cells, crypt branching, crypt abscessation, and cystic crypts, commonly with mucus retention (Fig. C.17). These lesions may be found as early as 1 year of age, and clinical disease may be expressed in a cyclical pattern (Richter *et al.*, 1978b). At necropsy the large bowel is frequently gas filled or alternatively distended with ingesta. There is no known specific cause, but inappropriate dietary regimes are possibly cause associated. In the case of the cotton-top tamarin, chronic colitis may have an inciter relationship to colon cancer; however, this thesis is not supported by the observation of morphologically identical colitis in other *Saguinus* ssp. without carcinoma.

Intussusception is relatively common in callitrichids, and retrospective analysis of clinical and autopsy findings demonstrate that intussusception is frequently associated with rectal prolapse and vice versa. Most intussusceptions originate in the ileum, particularly the terminal ileum, and multiple intussusceptions occur. Rectal prolapse frequently involves the terminal ileum carried through the ascending, transverse, and descending colons. This is possible in callitrichids because of the long complete mesocolon and weak or vestigial hepatic and splenic ligaments. Rectal prolapse independent of intussusception is rare, hence reduction of the prolapse alone usually results in reoccurrence and death. Consequently, rectal prolapse should be treated by laporotomy and cecocolopexy. The cecum is sutured to the peritoneum near the pelvic rim,

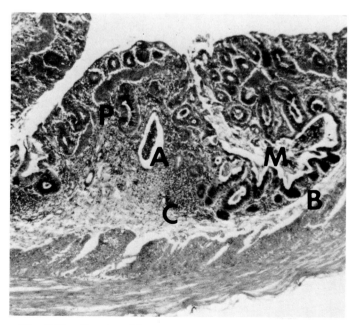

Fig. C.17. Chronic colitis, *S. fuscicollis illigeri.* Crypt atrophy, crypt branching (B), mucus retention (M), crypt abscessation (A), cellular infiltration (C), and crypt penetration of the muscularis mucosa (P) are present. Periodic acid–Schiff stain.

and the hepatic and splenic flexures sutured to the lateral abdominal peritoneum in approximately normal position using chromic catgut. Repeat episodes, one year or more later, may occur and should be repaired in a similar manner. Small intestine intussusception can occasionally be reduced manually by manipulation of the gut under chemical restraint. The small intestine intussusception can sometimes be diagnosed by deep palpation and the acute shocked state of the animal. All intussusceptions constitute an emergency, and the associated shock must also be treated.

The occurrence of acute gastric dilatation has been reported in *Callithrix jacchus*, possibly associated with *Clostridium perfringens* type A (Stein *et al.*, 1981).

Obstetrical problems are uncommon. Richter observed one maternal fatality in 467 deliveries over a $5\frac{1}{2}$ year period, and one other was prevented by caesarean section to remove overlarge, already dead infants. Both of these cases occurred in the cotton-top. W. D. Hiddleston (personal communication) observed 11 dystocias in 1046 deliveries in *Callithrix jacchus*. Placenta praevia has been reported in *S. o. geoffroyi*, and *Callithrix jacchus* (Lunn, 1980). Retained placenta is occasionally observed and is usually associated with toxemia and/or severe blood loss. Retained placentas should be removed as quickly as possible and appropriate treatment instituted. Blood loss may be heavy.

The major cause of infant mortality appears to be parental neglect, although constitutionally weak infants cannot be excluded as a cause. Limited information on infant mortality is available from feral animals (Dawson, 1976). In captivity most infant mortality probably occurs within hours of birth. Mortality after the first week of life is significantly reduced. The majority of full-term "stillborn" callitrichids have inflated lungs when necropsied, which supports the charge of parental (maternal) neglect or indifference.

Congenital disease and birth defects are relatively rare. Hetherington *et al.* (1975) have reported a case of syndactyly in *Saguinus nigricollis.* Bush *et al.* (1980) have described the familial occurrence and surgical repair of diaphragmatic defects in *L. rosalia.* Prevalence in a population of 130 animals was 8.5%, and mode of inheritance appeared to be simple autosomal recessive. Considering the endangered status of this species, this finding only adds to its problems. "Erythroblastosis foetalis" has been reported in *Tamarinus (Saguinus) nigricollis* by Gengozian *et al.* (1966). Icterus was not present, but hepatic hemosiderosis and marked erythroblastosis were observed, and the mother and fetus had positive Coombs and/or saline reactive antibodies to paternal cells. W. D. Hiddleston (personal communication) has observed "imperforate vaginas" in *C. jacchus*, but details and etiology are not clear.

Perhaps the most important and most poorly understood disease syndrome of callitrichids is that characterized by extreme loss of condition with emaciation and death. Dubbed variously "wasting marmoset syndrome" (WMS), "wasting disease" (WD), "wasting marmoset disease" (WMD), and "marmoset wasting disease" (MWD), the condition remains a baffling problem in many colonies. A modest consensus has developed among clinicians and colony managers that the most likely contributing causes are faulty diet and inadequate opportunity for behavioral adjustment (see Sections II,C and II,E). For the purpose of consistency with the most commonly accepted practice, the condition is referred to here as WMS. The common occurrence of this syndrome has led to the establishment of an International Committee on Wasting Marmoset Syndrome.* A computerized data base on WMS, which shows incidence of WMS, husbandry practices, nutrition practices, clinical findings, etc., from 36 callitrichid colonies around the world has been developed.

Among the earliest descriptions of WMS are those of King (1976) and Richter et al. (1978b). The most important clinical sign is a dramatic and persistent loss of weight, principally observed as loss of muscle mass in the legs (thighs), although the phenomenon is generalized. In addition, anemia and hypoproteinemia (hypoalbuminemia) are almost universally found. Cervical or ventral edema may occur as a result of hypoproteinemia. Afflicted animals are weak and express weak vocalizations; they may exhibit alopecia, particularly on the tail; and they are subject to acute episodes of hypoglycemia. Mortality is ultimately high. At necropsy the carcass is emaciated and unkempt, mucous membranes are pale, muscle mass is reduced, there is an almost total absence of body fat, and the liver may be smaller than normal. Total body weight will be well below normal values. Wasting marmoset syndrome should always be considered when weight values for adults fall below normal values in the absence of defined cause. Hence, body weight is a required measurement in the clinical examination of any callitrichid, and a continuous weight record should be maintained on all animals subjected to examination. Current therapy for WMS should be directed toward correcting behavior problems, which may be one of a number of possibilities, and presenting a more palatable diet.

A high incidence of Heinz bodies in erythrocytes from WMS afflicted animals has been reported (W. D. Hiddleston, personal communication; Hawkey et al., 1982). This correlation deserves further attention by clinicians. Anemias of obscure etiology are commonly associated with WMS and other diseases in callitrichids. Many are normochromic/normocytic or normochromic/macrocytic and may respond slowly to testosterone and hematinic therapy.

Wasting marmoset syndrome-afflicted animals are particularly prone to acute episodes of hypoglycemia. These animals are usually found "down" in a profound state of collapse characterized by marked hypothermia (rectal temperatures as low as 30°C) and hypoglycemia (blood glucose levels as low as 40 mg/dl). Therapy should be directed toward restoring body temperature and blood glucose levels and preventing gram-negative shock. In colonies with less well-disciplined care practices, hypoglycemic episodes are commonly seen on Monday mornings. High mortality and recurrences are associated with this disorder.

Vitamin deficiency disease has been reported in callitrichids, the most important of which is vitamin D_3. Hypovitaminosis D_3 is clinically manifested as osteodystrophia fibrosa (Hunt et al., 1967) and was common in almost all early callitrichid colonies. The requirement of New World primates for D_3 versus D_2 is now well established; however, the clinician must constantly be on guard against misadventure through faulty diet preparation, particularly purified diets, where inadvertent substitutions may reproduce the disease. The accepted daily requirement of 500 IU/monkey/week (Hunt et al., 1967) is probably easily met by most diets today, but many colony managers still hedge their bets by supplementing diets and using UV lights (Section I,D).

Experimental folic acid deficiency syndrome has been produced in C. jacchus (Dreizen and Levy, 1969). Angular cheilosis, weight loss, diarrhea, anemia, and mucosal ulceration were seen. There are no reports of natural folic acid deficiency; however, it seems likely that marginal folic acid deficiencies may play a role in general debility syndromes where inappetence is a problem. W. D. Hiddleson (personal communication) has observed scurvy in C. jacchus fed diets stored too long. Affected animals lost weight, became stiff and unwilling to move, sitting principally on the cage floor, and developed petechial hemorrhages in the eyelids. Doses of vitamin C corrected the problem. A report of zinc deficiency can be found in Chadwick et al. (1979); however, experimental support for the report is lacking. Affected S. mystax were generally debilitated, lost hair, and developed scaly, thickened skin. Although no reports appear of "fading infant" (white monkey) syndrome in the callitrichid literature, such infants are occasionally seen in callitrichids. The role of excess zinc as a copper chelator is uncertain, but some evidence exists to support this (Obeck, 1978), and copper therapy may be useful in specific cases. Anemia, steatitis, and muscle necrosis associated with vitamin E deficiency have been reported (Baskin et al., 1983); however, therapeutic vitamin E was not entirely effective. Most nutrition/vitamin deficiency diseases are complex, and this case appears to have been no exception.

In a systematic study, Buyukmihci and Richter (1979) examined the eyes of 526 callitrichids by biomicroscopy and ophthalmoscopy for ocular diseases. Of these, 109 were found to have some form of abnormality, although most were considered to be relatively inconsequential. Trauma appeared to have been the major causative factor.

Brack (1982) has reported two cases of noma in S. o. oedipus. One of these responded to topical application of eth-

*Coordinator, Dr. Martin L. Morin, Veterinary Resources Branch, National Institute of Health, Bethesda, Maryland 20205.

acridine lactate. Noma is poorly understood, and the diagnosis somewhat subjective. Deep seated underlying causes, including undernutrition, must be considered when attempting therapy. Antibiotics may be of questionable value.

Organophosphate poisoning associated with the use of diazinon and chlorpyriphos in a mixed colony of *S. oedipus, C. jacchus,* and *Callimico goeldii* has been reported by Brack and Rothe (1982). Clinical signs included acute bloody mucoid gastroenteritis, trembling, stiff gait, pale mucous membranes, high pitched voices, and mixed miosis or mydriasis reactions. Salivation was not prominent. Two abortions and 19 fatalities were observed in a population of 98 at risk. Petechia in the CNS and hemorrhages in the bone marrow, lungs, and liver were the most common postmortem findings. Atropine and a cholinesterase activator were used in therapy, but with only marginal success.

E. Neoplastic

Relatively few neoplasms have been reported in nonhuman primates, including callitrichids, but this may be an aberration more attributable to inadequate experience with older primates rather than failure of primates to express neoplasia. The cotton-top tamarin is unique among all primates in its susceptibility to mucinous adenocarcinoma of the large bowel. First observed by Lushbaugh *et al.* (1978), it is now apparent that this disease has considerable potential for the study of cancer biology. In well nourished *S. o. oedipus,* these carcinomas are

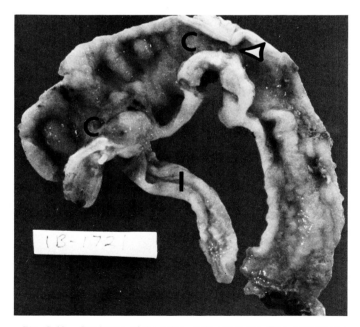

Fig. C.18. Carcinoma of the colon, male cotton-top. Two primary neoplasms are present (C). A stricture is present at one site (arrowhead). Hyperplasia of the muscularis is present at both sites. Terminal ileum (I) is included.

Table C.XVIII

Neoplasms in Callitrichids

Type	Location	Species
Carcinoma	Large bowel	S. o. oedipus
Carcinoma	Ileum	S. fuscicollis
Carcinoma	Adrenal	S. f. illigeri
Carcinoma	Squamous, skin	S. f. lagonotus
Carcinoma	Thyroid	S. nigricollis
Adenoma	Adrenal cortex	S. o. oedipus
Adenoma	Adrenal cortex	?
Adenoma	Pancreatic, islet	S. f. illigeri
Adenoma	Renal	S. mystax
Adenoma	Salivary	S. o. oedipus
Lymphoma (malignant)		S. nigricollis
Lymphoma (malignant)		S. o. oedipus
Lymphoma	Thymic	S. f. illigeri
Hemangiosarcoma	Adrenal (possibly induced)	S. o. oedipus
Fibrochondroma	Larynx	?
Melanoma	Skin, thorax	S. o. oedipus
Myelolipoma	Juxtarenal	S. o. oedipus
Papilloma	Eyelid	?
Cyst	Ovary	S. o. oedipus
Fibrocystic disease	Mammary	?
Cystic	Lungs (neonate)	S. o. oedipus

the principal cause of death and may occur in familial patterns. The cancers are multiple in a single host and readily metastasize to lymphatics, lungs, and occasionally other viscera (Richter *et al.,* 1980; Swartzendruber and Richter, 1980) (Fig. C.18). Clinical diagnosis is usually possible (Richter, 1981), and affected animals may be kept alive for considerable periods, although cachexia is a common aspect of the disease. Primary consideration should be given to this diagnosis when *S. o. oedipus* is presented from an otherwise healthy colony. The earlierst age at which a carcinoma was observed was in a 15-month-old male, but almost all occur in mature animals of either sex. Table C.XVIII of reported and observed neoplasms in callitrichids serves only as an indicator of the breadth of tumor types and a reminder that the clinician must consider neoplastic disease in the differential diagnosis. The table is remarkable for the absence of marmoset species, all listings being from the genus *Saguinus.*

REFERENCES

Albrecht, P., Lorenz, D., Klutch, M. J., Vickers, J. H., and Ennis, F. A. (1980). Fatal measles infection in marmosets pathogenesis and prophylaxis. *Infect. Immun.* **27,** 969–978.
Anderson, E. T., Lewis, J. P., Passovoy, M., and Trobaugh, F. E., Jr. (1967). Marmosets as laboratory animals. II. The hematology of laboratory kept marmosets. *Lab. Anim. Care* **17**(1), 30–40.
Baskin, G. B., Wolf, R. H., Worth, C. L., Soike, K., Gibson, S. V., and

Bieri, J. G. (1983). Anemia, steatitis, and muscle necrosis in marmosets. *Lab. Anim. Sci.* **33**, 74–80.

Benirschke, K., and Miller, C. J. (1982). Anatomical and functional differences in the placenta of primates. *Biol. Reprod.* **26**, 29–53.

Benirschke, K., and Richart, R. (1960). Spontaneous acute toxoplasmosis in a marmoset monkey. *Am. J. Trop. Med. Hyg.* **9**, 269–273.

Brack, M. (1982). Noma in *Saguinus oedipus:* A report of 2 cases. *Lab. Anim.* **16**, 361–363.

Brack, M., and Rothe, H. (1982). Organophosphate poisoning in marmosets. *Lab. Anim.* **16**, 186–188.

Brand, H. M. (1981). Husbandry and breeding of a newly established colony of cotton-topped tamarins (*Saguinus oedipus oedipus*). *Lab. Anim.* **15**, 7–11.

Brown, K., and Mack, D. S. (1978). Food sharing among captive *Leontopithecus rosalia. Folia Primatol.* **29**, 268–290.

Bush, M., Montali, R. J., Kleiman, D. G., Randolph, J., Abramowitz, M. D., and Evans, R. F. (1980). Diagnosis and repair of familial diaphragmatic defects in golden lion tamarins. *J. Am. Vet. Med. Assoc.* **177**, 858–862.

Bush, M., Custer, R. S., Whitly, J. C., and Smith, E. E. (1982). Hematologic values of captive golden lion tamarins (*Leontopithecus rosalia*): Variations with sex, age, and health status. *Lab. Anim. Sci.* **32**, 294–297.

Buyukmihci, N., and Richter, C. B. (1979). Prevalence of ocular disease in a colony of tamarins and marmosets. *Lab. Anim. Sci.* **29**, 800–804.

Caminiti, B. (1980). "Colony Breeding of New World Monkeys: A Bibliography." Primate Information Center, Regional Primate Research Center, University of Washington, Seattle.

Chadwick, D. P., May, J. C., and Lorenz, D. (1979). Spontaneous zinc deficiency in marmosets, *Saguinus mystax. Lab. Anim. Sci.* **29**, 482–485.

Cicmanec, J. L. (1977). Medical problems encountered in a callitrichid colony. *In* "The Biology and Conservation of the Callitrichidae" (D. G. Kleiman, ed.), pp. 331–336. Smithson. Inst. Press, Washington, D.C.

Coimbra-Filho, A. F. (1977). Natural shelters of *Leontopithecus rosalia* and some ecological implications (Callitrichidae, Primates). *In* "The Biology and Conservation of the Callitrichidae" (D. G. Kleiman, ed.), pp. 79–89. Smithson. Inst. Press, Washington, D.C.

Coimbra-Filho, A. F., and Mittermeier, R. A. (1977). Tree-gouging exudate-eating and the short-tusked condition in *Callithrix* and *Cebuella. In* "The Biology and Conservation of the Callitrichidae" (D. G. Kleiman, ed.), pp. 105–115. Smithson. Inst. Press, Washington, D.C.

Cooper, J. E., and Needham, J. R. (1976). An outbreak of shigellosis in laboratory marmosets and tamarins (Family Callitrichidae). *J. Hyg.* **76**, 415–424.

Cosgrove, G. E., Nelson, B., and Gengozian, N. (1968). Helminth parasites of the tamarin, *Saguinus fuscicollis. Lab. Anim. Care* **18**, 654–658.

Davis, S., and Richter, C. B. (1980). Food transfers between family members in *Saguinus oedipus. 3rd Annu. Meet, Am. Soc. Primatol.* Program Abstract.

Dawson, G. A. (1976). Behavioral ecology of the Panamanian tamarin, *Saguinus oedipus* (Callitrichidae, Primates). Ph.D. Dissertation, Michigan State University, East Lansing.

Dawson, G. A. (1977). Composition and stability of social groups of the tamarin, *Saguinus oedipus geoffroyi,* in Panama: Ecology and behavioral implications. *In* "The Biology and Conservation of the Callitrichidae" (D. G. Kleiman, ed.), pp. 23–37. Smithson. Inst. Press, Washington, D.C.

Dawson, G. A., and Dukelow, W. R. (1976). Reproductive characteristics of free-ranging Panamanian tamarins (*Saguinus oedipus geoffroyi*). *J. Med. Primatol.* **5**, 266–275.

Deinhardt, F., Holmes, A. W., Devine, J., and Deii ʳrdt, J. B. (1967). Marmosets as laboratory animals. IV. The microbiᵒgy of laboratory kept marmosets. *Lab. Anim. Care* **17**, 48–70.

Deinhardt, J. B., Devine, J., Passovoy, M., Pohlman, R., and Deinhardt, F. (1967). Marmosets as laboratory animals. I. Care of marmosets in the laboratory, pathology and outline of statistical evaluation of data. *Lab. Anim. Care* **17**, 11–29.

Dreizen, S., and Levy, B. M. (1969). Histopathology of experimentally induced nutritional deficiency cheilosis in the marmoset (*Callithrix jacchus*). *Arch. Oral Biol.* **14**, 577–582.

Eichberg, J. W., Moore, G. T., Kalter, S. S., Rodriguez, A. R., and Berchelmann, M. L. (1979). Rearing of conventional and gnotobiotic nonhuman primates (*Pan troglodytes, Papio cynocephalus, Saguinus nigricollis*). *J. Med. Primatol.* **8**, 69–78.

Epple, G. (1967). Vergleichende untersuchungen über sexual and sozialverhalten de krallenaffen (Hapalidae). *Folia Primatol.* **7**, 37–65.

Epple, G. (1970a). Maintenance, breeding, and development of marmoset monkeys (Callitrichidae) in captivity. *Folia Primatol.* **12**, 56–76.

Epple, G. (1970b). Quantitative studies on scent marking in the marmoset (*Callithrix jacchus*). *Folia Primatol.* **13**, 48–62.

Epple, G. (1973). The role of pheromones in the social communication of marmoset monkeys (Callitrichidae). *J. Reprod. Fertil., Suppl.* **19**, 447–454.

Epple, G. (1977). Notes on the establishment and maintenance of the pair bond in *Saguinus fuscicollis. In* "The Biology and Conservation of the Callitrichidae" (D. G. Kleiman, ed.), pp. 231–237. Smithson. Inst. Press, Washington, D.C.

Epple, G., and Katz, Y. (1980). Social influences on first reproductive success and related behaviors in the saddle-back tamarin (*Saguinus fuscicollis,* Callitrichidae). *Int. J. Primatol.* **1**, 171.

Escajadillo, A., Bronson, R. T., Sehgal, P., and Hayes, K. C. (1981). Nutritional evaluation in cotton-top tamarins (*Saguinus oedipus*). *Lab. Anim. Sci.* **31**, 161–165.

Flynn, R. J. (1973). "Parasites of Laboratory Animals." Iowa State University Press, Ames, Iowa.

Frazer, C. E. O., Chalifoux, L., Sehgal, P., Hunt, R. D., and King, N. W. (1978). A paramyxovirus causing fatal gastroenterocolitis in marmoset monkeys. *Primates Med.* **10**, 261–270.

Garber, P. A. (1980). Locomotor behavior and feeding ecology of the Panamanian tamarin (*Saguinus oedipus geoffroyi,* Callitrichidae, Primates). *Int. J. Primatol.* **1**, 185–201.

Gengozian, N., and Batson, J. S. (1975). Single-born marmosets without hemopoietic chimerism: Naturally occurring and induced. *J. Med. Primatol.* **4**, 252–261.

Gengozian, N., and Deinhardt, F. (1978). "Marmosets in Experimental Medicine." Karger, Basel.

Gengozian, N., Lushbaugh, C. C., Humason, G. L., and Kniseley, R. M. (1966). "Erythroblastosis foetalis" in the primate, *Tamarinis (Saguinus) nigricollis. Nature (London)* **209**, 731–732.

Gengozian, N., Smith, T. A., and Goslee, D. G. (1974). External uterine palpation to identify stages of pregnancy in the marmoset, *Saguinus fuscicollis* ssp. *J. Med. Primatol.* **3**, 236–243.

Gengozian, N., Batson, J. S., and Smith, T. A. (1977). Breeding of tamarins (*Saguinus* ssp.) in the laboratory. *In* 'The Biology and Conservation of the Callitrichidae" (D. G. Kleiman, ed.), pp. 207–213. Smithson. Inst. Press, Washington, D.C.

Gengozian, N., Brewen, J. G., Preston, R. J., and Batson, J. S. (1980). Presumptive evidence for the absence of functional germ cell chimerism in the marmoset. *J. Med. Primatol.* **9**, 9–27.

Glassman, D. M. (1982). Skeletal age changes in *Saguinus fuscicollis* and *Saguinus oedipus* (Callitrichidae, Primates). Ph.D. dissertation, University of Tennessee, Knoxville.

Gough, A. W., Barsoum, N. J., Gracon, S. I., Mitchell, L., and Sturgess, J. M. (1982). Poxvirus infection in a colony of common marmosets (*Callithrix jacchus*). *Lab. Anim. Sci.* **32**, 87–90.

Hampton, J. K., Jr. (1964). Laboratory requirements and observations of *Oedipomidas oedipus*. *Am. J. Physical Anthropol.* **22**, 239–243.

Hamptom, J. K., Jr., and Hampton, S. H. (1965). Marmosets (Hapalidae: Breeding, seasons, twinning, and sex of offspring. *Science* **150**, 915–917.

Hampton, S. H., and Hampton, J. K., Jr. (1977). Detection of reproductive cycles and pregnancy in tamarins (*Saguinus* ssp.). *In* "The Biology and Conservation of the Callitrichidae" (D. G. Kleiman, ed.), pp. 173–179. Smithson. Inst. Press, Washington, D.C.

Hawkey, C. M., Hart, M. G., and Jones, D. M. (1982). Clinical hematology of the common marmoset *Callithrix jacchus*. *Am. J. Primatol.* **3**, 179–199.

Hawthorne, J. D., Lorenz, D., and Albrecht, P. (1982). Infection of marmosets with parainfluenza virus types 1 and 3. *Infect. Immun.* **37**, 1037–1041.

Hearn, J. P. (1977). The endocrinology of reproduction in the common marmoset *Callithrix jacchus*. *In* "The Biology and Conservation of the Callitrichidae" (D. G. Kleiman, ed.), pp. 163–171. Smithson. Inst. Press, Washington, D.C.

Hearn, J. P., and Burden, F. J. (1979). "Collaborative" rearing of marmoset triplets. *Lab. Anim.* **13**, 131–133.

Hearn, J. P., and Lunn, S. F. (1975). The reproductive biology of the marmoset monkey, *Callithrix jacchus*. *In* "Breeding Simians for Developmental Biology" (F. T. Perkins and P. N. O'Donoghue, eds.), Lab. Anim. Handb. No. 6, pp. 191–202. Laboratory Animals Ltd., London.

Hearn, J. P., Lunn, S. F., Burden, F. J., and Pilcher, M. M. (1975). Management of marmosets in biomedical research. *Lab. Anim.* **9**, 125–134.

Hershkovitz, P. (1977). "Living New World Monkeys (*Platyrrhini*)," Vol. I. Univ. of Chicago Press, Chicago, Illinois.

Hetherington, C. M. (1978). Circadian oscillations of body temperature in the marmoset, *Callithrix jacchus*. *Lab. Anim.* **12**, 107–108.

Hetherington, C. M., Cooper, J. E., and Dawson, P. (1975). A case of syndactyly in the white-lipped tamarin *Saguinus nigricollis*. *Folia Primatol.* **24**, 24–28.

Hill, A. C., Turton, J. A., and Bleby, J. (1978). Bacterial and mycoplasma flora of a laboratory colony of the common marmoset (*Callithrix jacchus*). *Vet. Rec.* **103**, 824–827.

Hodgen, G. D., Wolfe, L. G., Ogden, J. D., Adams, R. M., Descalzi, C. C., and Hildebrand, D. F. (1976). Diagnosis of pregnancy in marmosets: Hemagglutinin inhibition test and radioimmunoassay for urinary chorionic gonadotropin. *Lab. Anim. Sci.* **26**, 224–229.

Holmes, A. W., Caldwell, R. G., Dedmon, R. E., and Deinhardt, F. (1964). Isolation and characterization of a new herpesvirus. *J. Immunol.* **92**, 602–610.

Holmes, A. W., Passovoy, M., and Capps, R. B. (1967). Marmosets as laboratory animals. III. Blood chemistry of laboratory-kept marmosets with particular attention to liver function and structure. *Lab. Anim. Care* **17**, 41–47.

Hunt, R. D., Garcia, F. G., and Heysted, D. M. (1967). A comparison of vitamin D_2 and D_3 in New World primates. I. Production and regression of osteodystrophia fibrosa. *Lab. Anim. Care* **17**, 222–234.

Hunt, R. D., Meléndez, L. V., King, N. W., Gilmore, C. E., Daniel, M. D., Williamson, M. E., and Jones, T. C. (1970). Morphology of a disease with features of malignant lymphoma in marmosets and owl monkeys inoculated with *Herpesvirus saimiri*. *JNCI, J. Natl. Cancer Inst.* **44**, 447–465.

Hunt, R. D., Meléndez, L. V., Garcia, F. G., and Trum, B. F. (1972). Pathologic features of *Herpesvirus ateles* lymphoma in cotton-topped mosets (*Saguinus oedipus*). *JNCI, J. Natl. Cancer Inst.* **49**, 1631–1639.

Hunt, R. D., Anderson, M. P., and Chalifoux, L. V. (1978). Spontaneous infectious diseases of marmosets. *Primates Med.* **10**, 238–253.

Izawa, K. (1978). A field study of the ecology and behavior of the black-mantle tamarin (*Saguinus nigricollis*). *Primates* **19**, 241–274.

Jollie, W. P., Haar, J. L., and Craig, S. S. (1975). Fine structural observations on hemopoiesis in the chorioallantoic placenta of the marmoset. *Am. J. Anat.* **144**, 9–38.

Kalter, S. S., and Heberling, R. L. (1972). Serologic evidence of viral infection in South American monkeys. *JNCI, J. Natl. Cancer Inst.* **49**, 251–259.

King, G. (1975). Feeding and nutrition of the callitrichidae at the Jersey Zoological Park. *12th Annu. Rep., Jersey Wildl. Preservation Trust*, pp. 81–90.

King, G. J. (1976). An investigation into "Wasting Marmoset Syndrome" at Jersey Zoo. *13th Annu. Rep. Jersey Wildl. Preservation Trust* pp. 97–107.

King, G. J. (1978). Comparative feeding and nutrition in captive, nonhuman primates. *Br. J. Nutr.* **40**, 55–62.

Kleiman, D. G., ed. (1977a). "The Biology and Conservation of the Callitrichidae." Smithson. Inst. Press, Washington, D.C.

Kleiman, D. G. (1977b). Characteristics of reproduction and socio-sexual interactions in pairs of lion tamarins (*Leontopithecus rosalia*) during the reproductive cycle. *In* "The Biology and Conservation of the Callitrichidae" (D. G. Kleiman, ed.), pp. 181–190. Smithson. Inst. Press, Washington, D.C.

Kuntz, R. E., and Myers, B. J. (1972). Parasites of South American primates. *Int. Zoo Yearb.* **12**, 61–68.

Levy, B. M., and Artecona, J. (1964). The marmoset as an experimental animal in biological research: Care and maintenance. *Lab. Anim. Care* **14**, 20–27.

Levy, B. M., and Mirkovic, R. R. (1971). An epizootic of measles in a marmoset colony. *Lab. Anim. Sci.* **21**, 33–39.

Levy, B. M., Taylor, A. C., Hampton, S., and Thoma, G. W. (1969). Tumors of the marmoset produced by Rous sarcoma virus. *Cancer Res.* **29**, 2237–2248.

Lorenz, R., and Heinemann, H. (1967). Beitrag zur morphologie und Körperlichen jungendentwicklung des springtamarin (*Callimico goeldii*) (Thomas, 1904). *Folia Primatol.* **6**, 1–27.

Lucke, V. M., and Bennett, A. M. (1982). An outbreak of hepatitis in marmosets in a zoological collection. *Lab. Anim.* **16**, 73–77.

Lunn, S. F. (1980). A case of placenta praevia in a common marmoset (*Callithrix jacchus*). *Vet. Res.* **106**, 414.

Lushbaugh, C. C., Humason, G., and Gengozian, N. (1969). Intra-uterine death from congenital Chagas' disease in laboratory-bred marmosets (*Saguinus fuscicollis lagonotus*). *Am. J. Trop. Med. Hyg.* **18**, 662–665.

Lushbaugh, C. C., Humason, G. L., Swartzendruber, D. C., Richter, C. B., and Gengozian, N. (1978). Spontaneous colonic adenocarcinoma in marmosets. *Primates Med.* **10**, 119–134.

Mallinson, J. J. C. (1975). The design of two marmoset complexes at the Jersey Zoological Park. *12th Annu. Rep., Jersey Wildl. Preservation Trust*, pp. 21–26.

Mallinson, J. J. C. (1978). Acclimatization of tropical mammals in the northern hemisphere with special reference to the family Callitrichidae. *In* "Biology and Behavior of Marmosets" (H. Rothe, J.-J. Wolters, and J. P. Hearn, eds.), pp. 49–56. Mecke-Druck, Duderstadt.

Marczynska, B., Jones, C. J., and Wolfe, L. G. (1981). Syncytium-forming virus of common marmosets (*Callithrix jacchus*). *Infect. Immun.* **31**, 1261–1269.

Marinkelle, C. J. (1982). The prevalence of *Trypanosoma* (*Schizotrypanum*) *cruzi* infection in Colombian monkeys and marmosets. *Ann. Trop. Med. Parasitol.* **76**, 121–124.

McIntosh, G. H., and Looker, J. W. (1982). Development of a marmoset colony in Australia. *Lab. Anim. Sci.* **32**, 677–679.

Meléndez, L. V., Hunt, R. D., King, N. W., Barahona, H. H., Daniel, M.

D., Fraser, C. E. O., and Garcia, F. G. (1972). A new lymphoma virus of monkeys: *Herpesvirus ateles*. *Nature (London)* **235**, 182–184.

Melnick, J. L., Midulla, M., Wimberly, I., Barrero-Oro, J. G., and Levy, B. M. (1964). A new member of the herpesvirus group isolated from South American marmosets. *J. Immunol.* **92**, 596–601.

Merritt, C. B., and Gengozian, N. (1967). "Survival of ⁵¹Cr Labeled Red Cells in Marmosets," Med. Div. Res. Rep., pp. 208–211. Oak Ridge Associated Universities, Oak Ridge, Tennessee.

Miller, G., Shope, T., Coope, D., Waters, L., Pagano, J., Bornkamm, G. W., and Henle, W. (1977). Lymphoma in cotton-top marmosets after inoculation with Epstein-Barr virus: Tumor incidence, histologic spectrum, antibody responses, demonstration of viral DNA, and characterization of viruses. *J. Exp. Med.* **145**, 948–967.

Mittermeier, R. A., Coimbra-Filho, A. F., and van Roosmalen, M. G. M. (1978). Conservation status of wild callitrichids. *In* "Biology and Behaviour of Marmosets" (H. Rothe, H.-J. Wolters, and J. P. Hearn, eds.), pp. 17–39. Mecke-Druck, Duderstadt.

Mivart, St. G. (1873). On *Lepilemur* and *Cheirgaleus*, and on the zoological rank of the lemuroidea. *Proc. Zool. Soc. London* pp. 484–510.

Montali, R. J., and Bush, M. (1980). Diagnostic exercise. *Lab. Anim. Sci.* **30**, 33–34.

Moreland, A. F. (1970). Tuberculosis in New World primates. *Lab. Anim. Care* **20**, 262–264.

Morita, M., Iida, T., Tsuchiya, Y., and Aoyama, Y. (1979). Fatal *herpesvirus tamarinus* infection in cotton-topped marmosets (*Saguinus oedipus*). *Exp. Anim.* **28**, 537–550.

Murphy, B. L., Maynard, J. E., Krushak, D. H., and Berquist, K. R. (1972). Microbial flora of imported marmosets: Viruses and enteric bacteria. *Lab. Anim. Sci.* **22**, 339–343.

Nelson, B., Cosgrove, G. E., and Gengozian, N. (1966). Diseases of the imported primate *Tamarinus nigricollis*. *Lab. Anim. Care* **16**, 255–275.

Neyman, P. F. (1977). Aspects of the ecology and social organization of free-ranging cotton-top tamarins (*Saguinus oedipus*) and the conservation status of the species. *In* "The Biology and Conservation of the Callitrichidae" (D. G. Kleiman, ed.), pp. 39–71. Smithson. Inst. Press, Washington, D.C.

Nigida, S. M., Falk, L. A., Wolfe, L. G., and Deinhardt, F. (1979). Isolation of a cytomegalovirus from the salivary glands of white-lipped marmosets. *Lab. Anim. Sci.* **29**, 53–60.

Obeck, D. K. (1978). Galvanized caging as a potential factor in the development of the "fading infant" or "white monkey" syndrome. *Lab. Anim. Sci.* **28**, 698–704.

Phillips, I. R. (1976a). The embryology of the common marmoset (*Callithrix jacchus*). *Adv. Anat., Embryol. Cell Biol.* **52**, 1–47.

Phillips, I. R. (1976b). The reproductive potential of the common cotton-eared marmoset (*Callithrix jacchus*) in captivity. *J. Med. Primatol.* **5**, 49–55.

Phillips, I. R., and Grist, S. M. (1975). The use of trans-abdominal palpation to determine the stage of pregnancy in the marmoset (*Callithrix jacchus*). *J. Reprod. Fertil.* **43**, 103–108.

Pook, A. G. (1976). Some notes on the development of hand-reared infants of four species of marmoset (Callitrichidae). *13th Annu. Rep. Jersey Wildl. Preservation Trust* pp. 38–46.

Provost, P. J., Villarejos, V. M., and Hilleman, M. R. (1977). Tests in rufiventer and other marmosets of susceptibility to human hepatitis A virus. *Primates Med.* **10**, 288–294.

Richter, C. B. (1981). Diagnostic exercise. *Lab. Anim. Sci.* **31**, 137–138.

Richter, C. B., and Buyukmihci, N. (1979). Squamous cell carcinoma of the epidermis in an aged white-lipped tamarin (*Saguinus fuscicollis leucogenys* Gray). *Vet. Pathol.* **16**, 263–265.

Richter, C. B., Humason, G. L., and Tankersley, B. (1976). Acanthocephaliasis in caged marmosets. *27th Annu. Sess., Am. Assoc. Lab. Anim. Sci.* Abstract.

Richter, C. B., Humason, G. L., and Godbold, J. H., Jr. (1978a). Endemic *Pneumocystis carinii* in a marmoset colony. *J. Comp. Pathol.* **88**, 171–180.

Richter, C. B., Tankersley, W. G., and Webb, A. (1978b). Chronic recurrent colitis: A wasting syndrome in marmosets and tamarins. *29th Annu. Sess., Am. Assoc. Lab. Anim. Sci.* Abstract.

Richter, C. B., Lushbaugh, C. C., and Swartzendruber, D. C. (1980). Cancer of the colon in cotton-topped tamarins. *In* "The Comparative Pathology of Zoo Animals" (R. J. Montali and G. Migaki, eds.), pp. 567–571. Smithson. Inst. Press, Washington, D.C.

Richter, C. B., Tankersley, W. G., Davis-Tardif, S., and Carson, R. L. (1981). Preference for high sugar diets by the common marmoset. *Am. J. Primatol.* **1**, 313 (abstr.).

Richter, C. B., Tankersley, W. G., and Gangaware, B. L. (1984). Primary cutaneous infection with a scotochromogenic mycobacterium in a cotton-top tamarin (*Saguinus oedipus oedipus*). *Lab. Anim. Sci.* (accepted for publication).

Rothe, H. (1975). Some aspects of sexuality and reproduction in groups of captive marmosets (*Callithrix jacchus*). *Z. Tierpsychol.* **37**, 255–273.

Rothe, H. (1977). Parturition and related behavior in *Callithrix jacchus* (Ceboidea, Callitrichidae). *In* "The Biology and Conservation of the Callitrichidae" (D. G. Kleiman, ed.), pp. 193–206. Smithson. Inst. Press, Washington, D.C.

Rothe, H., Wolters, H.-J., and Hearn, J. P., eds. (1978). "Biology and Behaviour of Marmosets." Mecke-Druck, Duderstadt.

Ryan, K. J., Benirschke, K., and Smith, O. W. (1981). Conversion of androstenedione-4-¹⁴C to estrone by the marmoset placenta. *Endocrinology* **69**, 613–618.

Seman, G., Levy, B. M., Panigel, M., and Dmochowski, L. (1975). Type-C virus particles in placenta of the cotton-top marmoset (*Saguinus oedipus*). *JNCI, J. Natl. Cancer Inst.* **54**, 251–252.

Stein, F. J., Sis, R. R., and Levy, B. M. (1979). Indoor-outdoor housing systems for a self-sustaining marmoset breeding colony. *Lab. Anim. Sci.* **29**, 805–808.

Stein, F. J., Lewis, D. H., Stott, G. G., and Sis, R. F. (1981). Acute gastric dilatation in a common marmoset (*Callithrix jacchus*). *Lab. Anim. Sci.* **31**, 522–523.

Stevenson, M. F. (1977). The behavior and ecology of the common marmoset (*Callithrix jacchus jacchus*) in its natural environment. *Primate Eye* **9**, 5–6.

Stevenson, M. F., and Poole, T. B. (1976). An ethogram of the common marmoset (*Callithrix jacchus jacchus*): General behavioral repertoire. *Anim. Behav.* **24**, 228–251.

Swartzendruber, D. C., and Richter, C. B. (1980). Mucous and argentaffin cells in colonic adenocarcinomas of tamarins and rats. *Lab. Invest.* **43**, 523–529.

Takos, M. J., and Elton, N. W. (1953). Spontaneous cryptococcosis of marmoset monkeys in Panama. *Arch. Pathol.* **55**, 403–407.

Tankersley, W. G., Richter, C. B., and Batson, J. S. (1979). Therapy of filariasis in tamarins. *Lab. Anim. Sci.* **29**, 107–110.

Tardif, S. D. (1982). Sexual maturation of female *Saguinus oedipus oedipus*. Ph.D. Dissertation, Michigan State University, East Lansing.

Tardif, S. D., and Richter, C. B. (1981). Competition for a desired food in family groups of the common marmoset (*Callithrix jacchus*) and the cotton-top tamarin (*Saguinus oedipus*). *Lab. Anim. Sci.* **31**, 52–55.

Turton, J. A., Ford, D. J., Bleby, J., Hall, B. M., and Whiting, R. (1978). Composition of the milk of the common marmoset (*Callithrix jacchus*) and milk substitutes used in hand-rearing programmes, with special reference to fatty acids. *Folia Primatol.* **29**, 64–79.

Wadsworth, P. F., Budgett, D. A., and Forster, M. L. (1981). Organ weight data in juvenile and adult marmosets (*Callithrix jacchus*). *Lab. Anim.* **15**, 385–388.

Wadsworth, P. F., Hiddleston, W. A., Jones, D. V., Fowler, J. S. L., and Ferguson, R. A. (1982). Haematological, coagulation and blood chemistry data in red-bellied tamarins (*Saguinus labiatus*). *Lab. Anim.* **16**(4), 327–330.

Williams, J. B. (1981). "Behavioral Observations of Feral and Free-ranging New World Monkeys (Cebidae and Callitrichidae): A Bibliography." Primate Information Center, Regional Primate Research Center, University of Washington, Seattle.

Wilson, C. G. (1977). Gestation and reproduction in golden lion tamarins. *In* "The Biology and Conservation of the Callitrichidae" (D. G. Kleiman, ed.), pp. 191–192. Smithson. Inst. Press, Washington, D.C.

Winter, M. (1978). Investigation on the sequence of tooth eruption in hand-reared *Callithrix jacchus*. *In* "Biology and Behavior of Marmosets" (H. Rothe, J.-H. Wolters, and J. P. Hearn, eds.), pp. 109–124. Mecke-Druck, Duderstadt.

Wirth, H., and Buselmaier, W. (1982). Long-term experiments with a newly developed standardized diet for the New World primates *Callithrix jacchus jacchus* and *Callithrix jacchus penicillata* (marmosets). *Lab. Anim.* **16**, 175–181.

Wolfe, L. G., Ogden, J. D., Deinhardt, J. B., Fisher, L., and Deinhardt, F. (1972). Breeding and hand-rearing marmosets for viral oncogenesis studies. *In* "Breeding Primates," (W. I. B. Beveridge, ed.), pp. 145–157. Karger, Basel.

Wolters, H.-J. (1977). Some aspects of role taking behavior in captive family groups of the cotton-top tamarin *Saguinus oedipus oedipus*. *In* "Biology and Behavior of Marmosets" (H. Rothe, ed.), pp. 259–278. Mecke-Druck, Duderstadt.

Wynn, R. M., Richards, S. C., and Harris, J. A. (1975). Electron microscopy of the placenta and related structures of the marmoset. *Am. J. Obstet. Gynecol.* **122**, 60–69.

Ziegler, T. E., Stein, F. J., Sis, R. F., Coleman, M. S., and Green, J. H. (1981). Supplemental feeding of marmoset (*Callithrix jacchus*) triplets. *Lab. Anim. Sci.* **31**, 194–195.

Chapter 12

Biology and Diseases of Ferrets

S. L. Bernard, J. R. Gorham, and L. M. Ryland

I. INTRODUCTION

A. General Comments and Taxonomy

Man probably domesticated ferrets, which are descendants of the European polecat, as early as the fourth century BC to kill snakes and rodents and later to hunt rabbits. Unlike other Mustelidae, i.e., minks, skunks, weasels, otters, and badgers, the ferret is not wild in the true sense but has been raised for centuries in captivity. The common European ferret (*Mustela putorius furo*) should not be confused with (*Mustela nigripes*) the nearly extinct native North American black-footed ferret (Anonymous, 1973).

B. Uses

Frederich and Babish (1983) have surveyed ferret research publications since 1977 and have determined that the ferret is a well established and useful laboratory animal. In excess of 486 research citations were identified. Ferrets are employed for a variety of research purposes, involving bacteriology, virology, reproduction, teratology, pharmacology, physiology, endocrinology, gastroenterology, embryology, and neurology (Hahn and Wester, 1969; Marshall and Marshall, 1973). The ferret is also an ideal animal model for cardiac research (Truex *et al.*, 1974). Ferrets are widely used for toxicology trials because they represent carnivorous species. The ferret has been used for distemper and influenza research for many years.

The ferret is also farmed for its pelt. Fur pieces are marketed as fitch. Farms having 1000 or more ferrets are not uncommon. Also, the ferret is a popular companion animal.

C. Varieties

There are two varieties of ferrets, based on coloration: fitch ferrets are buff with black masks, feet, and tails; albino ferrets are white with pink eyes. The albino phenotype and color mutants (e.g., Siamese, silver, and silver mitt) are recessive to the fitch. Female ferrets are called ''jills'' and males, ''hobs.'' Baby ferrets are ''kits'' (Roberts, 1977).

D. Restraint

The ferret can be restrained by holding above the shoulders, with one hand gently squeezing the forelimbs together with the thumb under the ferret's chin. Most handlers wear a leather glove because many ferrets become apprehensive in strange surroundings.

Table I

Sedatives, Preanesthetics, and Anesthetics Suitable for Use in Ferrets

	Dosage	Route
Sedatives		
Acepromazine	0.2–0.5 mg/kg	im, sc
Xylazine	1.0 mg/kg	im, sc
Preanesthetics		
Atropine	0.05 mg/kg	im, sc
Acepromazine	0.1–0.25 mg/kg	im, sc
Anesthetics		
Ketamine	20–30 mg/kg (ketamine)	im, sc
(9 parts:1 part acepromazine)	0.2–0.35 mg/kg (acepromazine)	
Xylazine	1–4 mg/kg	sc
(followed by ketamine 20–30 mg/kg)		
Halothane		Mask
Methoxyflurane		Mask
Pentobarbital	30 mg/kg	ip

The drugs and dosages commonly used to provide chemical restraint and sedation in ferrets are given in Table I. We routinely use a simultaneous intramuscular injection of xylazine HCl (1 mg/kg) and ketamine HCl (20–30 mg/kg) for 10–20 min periods of chemical restraint. For longer periods of anesthesia, gas anesthesia (methoxyflurane or halothane) delivered by mask can be employed (Adams, 1979).

E. Housing

The cages used for dogs, cats, and rabbits can be used to house ferrets. On commercial farms they are raised in cages constructed with 1 × 1 in., 14 gauge wire netting. Usually a small nest box is provided. Other than at breeding time, male ferrets can be penned together. Female ferrets are often penned together unless they have kits.

II. BIOLOGY

A. Normal Life Cycle

Ferrets have a 42-day gestation period and the litter size averages 8 (range 2–17). Kits are born deaf and blind; the eyes and ears open 3–5 weeks of age. Deciduous teeth begin to erupt at 14 days, at which time the kits begin to eat solid food; their permanent canines erupt at 47 to 52 days of age, and kits are weaned at 6–7 weeks old. They reach their adult weight at about 4 months of age (Shump and Shump, 1978). Males are about twice the size of females (range 600 to 2000 gm). Both sexes undergo photoperiodic weight fluctuations of 30 to 40%. Subcutaneous fat is increased in the fall and lost in the spring (Hammond, 1974). The average life span in laboratory colonies is 5–6 years, but pet ferrets commonly live 9 or 10 years.

B. Physiologic and Laboratory Data

Blood samples may be collected by toenail clipping (< 0.5 ml), venipuncture of the caudal tail vein, jugular venipuncture, or orbital bleeding (Fox *et al.*, 1984). Cardiac puncture is possible by both transthoracic and transdiaphragmatic approaches, but care should be taken.

Hematologic values for ferrets are shown in Table II (Thornton *et al.*, 1979) and are generally similar to those of the cat. The ferret's higher hematocrit (mean 52.3%), more numerous erythrocytes (mean 9.17 × 10⁶/mm³), and higher percentage of reticulocytes (mean 4.6%) are exceptions. Ferret females in estrus tend to have lower platelet and leukocyte counts.

Normal serum chemistry values for ferrets are given in Table III (Thornton *et al.*, 1979). No major differences from those of the cat or dog are apparent. Normal ferrets have a mild to moderate proteinuria. It is postulated that the proteinuria might result from the ferret's relatively high systolic blood pressure and thicker arterial walls within the kidney parenchyma. The naturally dark urine of the male may give false positive colorimetric values for ketonuria.

The normal heart rate for a ferret is about 250 beats/min (Thornton *et al.*, 1979). Electrocardiographic measurements of wave amplitude are similar to those of the normal domestic cat

Table II

Hematologic Values of Normal Ferrets of Both Sexes[a]

Hematology	Mean	Range
Leukocytes		
Lymphocytes (%)	34.5	12–54
Neutrophils (%)	58.3	11–84
Monocytes (%)	4.4	0–9.0
Eosinophils (%)	2.5	0–7.0
Basophils (%)	0.1	0–2.0
Reticulocytes (%)	4.6	1–14
Platelets (10³/mm³)	499	297–910
Erythrocytes (10⁶/mm³)	9.17	6.8–12.2
Total protein (gm/dl)	6.0	5.1–7.4
Hematocrit (%)	52.3	42–61
Hemoglobin (gm/dl)	17.0	15–18
Leukocytes (10³/mm³)	10.1	4.0–19

[a]Modified from Thornton *et al.* (1979).

Table III

Serum Chemistry Values of Normal Ferrets of Both Sexes[a]

Serum chemistry values	Mean	Range
Glucose (whole blood) (mg/dl)	136	94–207
Urea nitrogen (mg/dl)	21.5	10–45
Albumin (mg/dl)	3.25	2.6–3.8
Alkaline phosphatase (IU/liter)	22.5	9–84
Aspartate transaminase (SGOT) (IU/liter)	65	28–120
Total bilirubin (mg/dl)	<1.0	
Cholesterol (mg/dl)	165	64–296
Creatinine (mg/dl)	0.6	0.4–0.9
Sodium (m*M*)	148.5	137–162
Potassium (m*M*)	5.9	4.5–7.7
Chloride (m*M*)	116	106–125
Calcium (mg/dl)	9.15	8.0–11.8
Inorganic phosphorus (mg/dl)	5.9	4.0–9.1

[a]Modified from Thornton *et al.* (1979).

ECG, although the height of the ferret's R wave may approach but should not exceed 2.0 mV. Radiographically, the cardiac silhouette may be interpreted incorrectly as having dilation of the left ventricle. Andrews *et al.* (1979a) found that the arterial pressure in anesthetized ferrets was 140/110 ± 35/31 mmHg. The ferret's normal respiratory rate is 33 to 36 breaths/min, and the mean rectal temperature is about 39°C.

C. Biology of Reproduction

Ferrets become sexually mature during the spring following their birth or at 9–12 months of age (Shump and Shump, 1978). The female is seasonally polyestrous. The usual breeding season is from March to August. However, females may remain in heat for up to 6 months if not bred. The breeding season is light dependent and can be induced by artificial lighting. The beginning of estrus can be recognized by enlargement of the vulva (Fig. 1), which regresses to normal size within 2 to 3 weeks after copulation. Coitus induces ovulation within 30 to 35 hrs, and if fertilization fails to occur, a pesudopregnancy of 42 days will occur. Implantation occurs in 12–13 days postfertilization. The average gestation length is 42 days. Kits are born with their eyes closed and are hairless. Normal neonatal weight of a kit is 7–10 gm. Two litters per year can be obtained if the females are bred early in the season.

Male ferrets' breeding readiness is determined by descent of the testes into the scrotum. Males are sexually active from December until July. At this time spermatogenic activity is present within the seminiferous tubules. During the nonbreeding season, spermatogenesis ceases and gonadal size decreases to a level seen in prepubertal males. During sexual inactivity, the

Fig. 1. An enlarged vulva signifies a female ferret in estrus.

male ferrets plasma testosterone levels are decreased (Neal and Murphy, 1977).

D. Sexing

Male ferrets have an os penis. The preputial opening is on the ventral abdomen. The female ferret's vulva is very close to the anus and swells during estrus (Fig. 1).

E. Unique Anatomical Characteristics

Gross anatomical features of the ferret reveal the absence of a cecum, appendix, seminal vesicles, and prostate gland (Thornton *et al.*, 1979). Ferrets have a vertebral formula of $C_7T_{14}L_6S_3Cd_{14-18}$ with 14 pairs of ribs (Owen, 1973). The ferret's permanent dental formula is $2(I\frac{3}{3} \ C\frac{1}{1} \ P\frac{4}{3} \ M\frac{1}{2})$ (Owen, 1973). Supernumerary incisors are common in adults. Ferrets do not have well-developed sweat glands and are prone to heat exhaustion at a temperature approaching 32°C (90°F). Ferrets have a pair of musk-producing glands lateral to the anus that emit a characteristic odor. Normal histologic features include splenic extramedullary hematopoiesis, lymphoid nodules in the intestinal submucosa, and marked seasonal spermatogenic variations.

F. Behavior

Ferrets usually have genial personalities and get along well with humans, particularly if they are raised from kits with human contact. On the other hand, a female with a litter can be the "triple distilled essence of red-eyed fury." They are inquisitive and playful and when allowed free to satisfy their innate curiosity and inclination to burrow, they require no special exercise equipment. Because they usually urinate and defecate in one place, they can be easily trained to use a litter box.

G. Nutrition

The nutritional requirements of ferrets have not been determined. Ferrets at Washington State University have been maintained adequately for over 25 years on a standard wet feed mink diet (Ensminger and Olentine, 1978). Ferrets can be maintained using commercial dry cat food supplemented with 5% fresh liver. Since we have never recognized any nutritional deficiencies in our colony and there are few reports in the literature, we assume the nutritional requirements of ferrets are similar to those of mink (Ensminger and Olentine, 1978). Because ferrets will eat to their caloric requirement, they can eventually get into a situation of low protein intake on cat food that can cause poor reproduction (Evans, 1982). The ferret has little, if any, capacity to digest fiber. Food passes through the ferret in about 3–4 hr. Fresh water, in either a cup or drinking bottle, should be available at all times.

III. ROUTINE VETERINARY MEDICAL CARE

A. Vaccinations

Ferrets must be vaccinated against canine distemper with modified live virus of chick embryo cell culture origin. The first dose of vaccine should be given at 6–8 weeks of age (4–5 weeks if kits are from unvaccinated dams) and a second vaccination at 9–12 weeks. Thereafter, boosters should be given every 3 years. If the kit is from a vaccinated female and is over 10 weeks of age, only one dose is required. Inactivated distemper vaccine provides only short-term immunity that is slow to develop and may not induce protection. Distemper vaccine prepared from ferret cell cultures should not be used because the vaccine virus may retain its virulence for the ferret.

The ferret is assumed to be highly susceptible to rabies and capable of transmitting the virus. However, vaccination against this disease is not recommended because of the lack of a rabies vaccine licensed for use in wild animals. Vaccine efficacy trials have not been conducted in ferrets, and their response to even killed vaccine is unknown. If a pet ferret must be vaccinated against rabies, only a killed vaccine should be administered. Two cases of ferret rabies have been reported in the United States since 1954, and in one of those, the possibility exists that clinical disease followed vaccination with modified live rabies virus (Anonymous, 1980).

Ferrets are not susceptible to feline panleukopenia, mink virus enteritis, canine hepatitis, feline rhinotracheitis, or feline calicivirus, and do not require vaccination against these diseases. There is no solid evidence that ferrets are susceptible to canine parvovirus. While ferrets are susceptible to human influenza, vaccination is only done in exceptional circumstances. A schedule of vaccinations and prophylactic procedures is outlined in Table IV.

B. Surgical Procedures

Nonbreeder female pet ferrets should be spayed at 6 to 8 months of age. Estrous females have high endogenous estrogen levels that cause a 30–50% prevalence of fatal bone marrow depression (see metabolic diseases) (Bernard *et al.*, 1983). The use of one intramuscular (im) injection of 100 IU

Table IV

Schedule of Vaccinations and Routine Prophylactic Care

Age	Plan
6–8 weeks (4–5 weeks if dam unvaccinated)	First CDV,[a] fecal exam
9–12 weeks	Second CDV, fecal exam
6–8 months	Spay/castrate, fecal exam
	Remove musk glands (optional)
3 years	CDV booster (triennial)

[a]CDV, canine distemper vaccine (nonferret origin).

of human chorionic gonadotropin (HCG) 10 or more days after onset of estrus causes ovulation. Females will cycle out of heat in 20–25 days postinjection and remain anestrous for 40–50 days (Bernard *et al.*, 1983). The use of megestrol acetate to delay or prevent estrus in ferrets greatly increases the risk of pyometra and should not be used.

Nonbreeder male pet ferrets should be castrated at 6 to 8 months of age to reduce their aggressiveness and desire to roam. Castration reduces the musky odor emanating from sebaceous secretions. After descent, the testes are located in the subcutis of the caudoventral abdomen. Fluctuation in size and function is dependent on the breeding season.

Ferrets have paired musk-producing glands lateral to the anus that secrete when the animal is angry, excited, or at estrus. Anal glands may be removed at the time of neutering or spaying using standard procedures for canine anal sac excision or a modified protocol designed specifically for ferret musk gland resection (Creed and Kainer, 1981). Anal gland removal will reduce, but will not eliminate, the ferret's musky odor.

IV. INFECTIOUS DISEASES

A. Bacterial Diseases

1. Botulism

a. Etiology. *Clostridium botulinum* is a gram-positive bacillus that produces a potent toxin. Ferrets are moderately susceptible to toxin types A and B, but they are highly susceptible to type C (Quortrup and Gorham, 1949).

b. Clinical Signs. Signs of botulism include dysphagia, ataxia, and paresis within 12 to 96 hr after eating contaminated food. Death follows due to paralysis of respiratory muscles.

c. Transmission. Botulism occurs after the ingestion of the preformed toxin in contaminated food. Botulism is more likely to occur on farms where ferrets are raised for their pelts than in laboratory or pet situations.

d. Necropsy. Other than the antemortem clinical signs of a flaccid paralysis, there are no apparent gross lesions associated with botulism.

e. Pathogenesis. *Clostridium botulinum* is a strict anaerobe and, in the presence of the proper substrate and environment, will multiply and produce toxin. After the toxin is ingested, it is absorbed into the circulatory and lymphatic system and distributed throughout the body. The toxin acts on nervous tissue by blocking the release of acetylcholine at the neuromuscular junction, producing a flaccid paralysis.

f. Differential Diagnosis. The clinical presentation of flaccid muscle paralysis at about the same time in several ferrets is usually pathognomonic of botulism. Demonstration of the botulism toxin in the food or ingesta is necessary to confirm the diagnosis. Other toxicities may produce paralysis but usually show concurrent central nervous system or systemic signs.

g. Prevention and Control. Annual vaccination with type C toxoid prevents botulism in ferrets. The disease may also be prevented by excluding from the diet ingredients of questionable origin.

h. Treatment. There is no successful treatment for botulism in ferrets. Antitoxin is not available.

2. Proliferative Colitis

a. Etiology. *Campylobacter fetus* subspecies *jejuni* has been isolated from ferrets with proliferative colitis (Fox *et al.*, 1982). *Campylobacter* spp. are gram-negative comma to S-shaped flagellated organisms.

b. Clinical Signs. Clinical signs of proliferative colitis in ferrets include green mucohemorrhagic feces, anorexia, dehydration, and partial rectal prolapse.

c. Transmission and Pathogenesis. Little is known about the transmission or pathogenesis of campylobacterial-associated proliferative colitis in ferrets. It is believed to be similar to intestinal proliferative diseases of swine and hamsters. Transmission is probably by the fecal–oral route. The bacteria may be present in inadequately cooked feed.

d. Necropsy. Macroscopic lesions of thickened descending colon with areas of reddening are found (Fig. 2). Microscopic changes include muscular hypertrophy and mucosal proliferation varying with the degree of inflammation (Fig. 3).

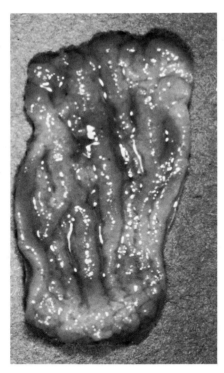

Fig. 2. Colonic mucosa with thick longitudinal folds associated with proliferative colitis in ferrets. (Courtesy of Dr. J. Fox.)

Special stains (Warthin–Starry) demonstrate groups of comma-shaped organisms in the apical portion of mucosal epithelial cells.

e. Differential Diagnosis. Other enteric conditions such as salmonellosis should be ruled out. Endocrine disorders, long-standing bacterial infections (especially pyometra), hepatic problems, or pancreatitis may cause diarrhea. Clinical evaluation and clinical pathology may aid in diagnosis.

f. Treatment, Control, and Prevention. There are no reported successful treatments of proliferative colitis in ferrets. Symptomatic and antimicrobial therapy may be beneficial. Oral erythromycin therapy failed to eliminate the organism from the feces (Fox *et al.*, 1983). Prevention by adequately cooking raw feed, particularly chicken, and decreasing the fecal–oral cycle by placing ferrets in wire floored cages may be beneficial.

The ferret could serve as a reservoir for human campylobacteriosis. A survey revealed that 61% of 168 ferrets were culturally positive from two commercial breeders (Fox *et al.*, 1983).

3. Tuberculosis

a. Etiology. Avian, bovine, and human strains of mycobacteria cause tuberculosis in ferrets (Symmers *et al.*,

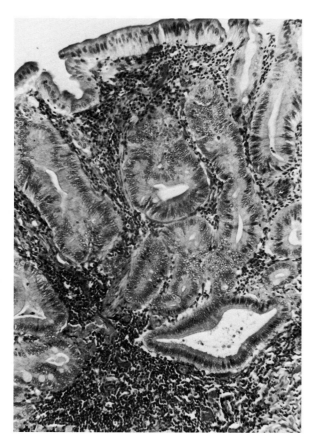

Fig. 3. Hyperplastic colonic mucosa and moderate mononuclear cell infiltrate in the lamina propria. Hematoxylin and eosin × 240. (Courtesy of Dr. J. Fox.)

1953). *Mycobacterium tuberculosis* is a gram-positive acid-fast rod.

b. Clinical Signs. Signs may not appear until the late stages of infection. The ferret may become emaciated and exhibit paralysis of the pelvic adductor muscles. The paralysis may progress to affect all the limbs. The mesenteric lymph nodes may be palpably enlarged.

c. Transmission. The organism can be transmitted by inhalation, ingestion, or wound infection.

d. Necropsy. The adductor muscles appear yellow and soft. Multifocal areas of coagulative necrosis throughout the gastrointestinal tract, hepatosplenomegaly, and mesenteric lymphoadenopathy may be found. Histologically, massive accumulations of histiocytes with acid-fast bacilli in abdominal organs are seen.

e. Differential Diagnosis. Tuberculosis is not common in ferrets. Other diseases that cause posterior paralysis include hemivertebrae, vertebral fractures, intervertebral disc disease, hematomyelia, or myelitis. A thermal response to the subcutaneous tuberculin test aids in diagnosis of tuberculosis.

f. Prevention, Control, and Treatment. Isolate or preferably kill affected ferrets. There is no recommended treatment of tuberculosis in ferrets. Once diagnosed, the animal should be humanely destroyed to control the spread of the disease from ferret to ferret and to human contacts.

4. Abscesses

a. Etiology. *Staphylococcus* spp., *Streptococcus* spp., *Pasteurella* spp., *Corynebacterium* spp., *Actinomyces israeli,* and *Escherichia coli* cause abscesses and localized infection of the uterus (estrous jills), vulva, skin (infected bite wounds during mating), and oral injuries by ingested bones.

b. Clinical Signs. A fluctuant swelling, local draining tract, fever, neutrophilia, vaginal discharge, or decreased feed intake may be noted depending on the location of the lesion.

c. Transmission and Pathogenesis. The primary mode of transmission is by a penetrating wound that leads to a localized infection. In most cases, the infection will be walled off and few systemic signs will be seen. Pyometra may be accompanied by systemic signs (depression, diarrhea, anorexia, etc.).

d. Differential Diagnosis. Biopsy, aspiration, or draining the affected swelling differentiates neoplasia from inflammation.

e. Prevention and Control. Prevention of penetrating wounds by minimizing the animal's exposure to sharp objects in the cage and feed is beneficial. Limiting the time the female is with the male will decrease the bite wounds associated with breeding. Good husbandry practices can reduce the incidence of abscess formation.

f. Treatment. Treatment of localized abscesses by drainage of the area is usually adequate. If this procedure does not eliminate the infection, then a culture and sensitivity should be done and the animal administered appropriate systemic antibiotics. Pyometra should be treated with systemic antibiotics and if the female is in heat, she should be given 100 IU of HCG to cycle out of heat.

5. Mastitis

Mastitis frequently occurs in nursing jills. *Streptococcus* spp., *Staphylococcus* spp., and *Escherichia coli* are the most frequent causes. Examination reveals enlarged, firm, slightly reddened mammary glands. Milk is expressed with difficulty. Gentamycin has been used successfully as a treatment regimen.

Liberson and co-workers (1983) described fulminating mastitis in ferrets caused by hemolytic *E. coli*. In addition to gangrenous mastitis, the condition was characterized by pyrexia, lethargy, acute septicemia, and death. Surgical resection of the involved gland combined with ampicillin (10 mg/kg bid) and gentamycin (5 mg/kg sid) was the most successful treatment.

6. Salmonellosis

Ferrets do not appear to be highly susceptible to salmonellosis (Gorham *et al.*, 1949). *Salmonella typhimurium* has been isolated from ferrets with dysentery, rapid weight loss, and fluctuating body temperature. Macroscopic postmortem findings include gastroenteritis, splenic necrosis, and hepatic necrosis. Diagnosis should be confirmed by isolation of the organism in the feces.

B. Viral Diseases

1. Distemper

a. Etiology. Ferret distemper is caused by canine distemper virus (CDV).

b. Clinical Signs. Clinical signs of distemper appear 7 to 10 days after exposure and include anorexia and mucopurulent ocular and nasal discharge (Fig. 4). A rash appears under the chin, around the anus, and in the inguinal area 10 to 12 days after exposure. The foot pads may swell and become hyperkeratotic. The ferret usually succumbs 12–16 days after exposure with ferret-adapted CDV strains and 21–35 days with canine strains. Ferrets that survive the catarrhal phase may die during a central nervous phase of distemper, signs of which include hyperexcitability, excess salivation, muscular tremor, convulsions, and coma.

c. Transmission and Pathogenesis. An outbreak of distemper can rapidly spread throughout a susceptible ferret colony, since the virus is transmitted by aerosols (Fig. 5). The case–fatality rate approaches 100% in a susceptible population. Canine distemper virus is a pantropic virus infection, infecting and replicating in all epithelial and lymphoid organs.

d. Necropsy. Macroscopic postmortem findings include bilateral nasal and ocular discharge. Often the ferrets die acutely, and few other macroscopic lesions are evident. Histologic lesions include intranuclear or intracytoplasmic inclusion

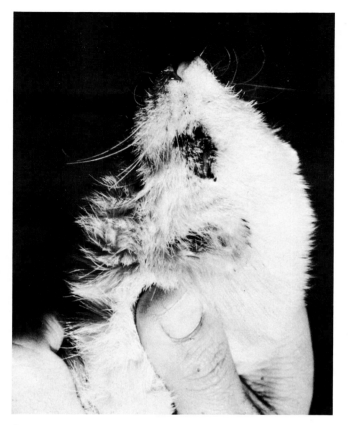

Fig. 4. Ferret infected with canine distemper virus. Note conjunctival exudate.

bodies in a variety of cell types. A productive area to examine for inclusion bodies is the epithelium lining the trachea and urinary bladder.

e. Differential Diagnosis. Because the signs of distemper are so typical, the disease is rarely confused with other condi-

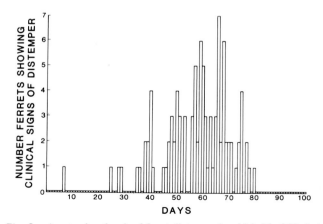

Fig. 5. A natural outbreak of ferret distemper in which 96 of 98 died of distemper.

tions. Clinical diagnosis can be confirmed with fluorescent antibody techniques or histopathology. Serologic procedures are of little value because a rapid diagnosis is necessary to control outbreaks.

f. Prevention, Control, and Treatment. With the high case–fatality rate, treatment of ferret distemper is not warranted. Euthanasia is a practical alternative to treatment. Prophylaxis by vaccination using a modified live CDV vaccine of nonferret cell culture origin is effective. In the face of an outbreak in a susceptible colony, sick animals should be removed from apparently healthy ferrets, and the healthy ferrets should be vaccinated immediately.

2. Influenza

a. Etiology. Ferrets are susceptible to infection with several strains of human influenza virus (orthomyxovirus).

b. Clinical Signs. Within 48 hr after exposure, the ferret becomes listless, febrile, anorectic, and exhibits a serous nasal discharge. Paroxysmal sneezing may develop and may be accompanied by a purulent nasal discharge. The case–mortality rate is low.

c. Transmission. Influenza virus is transmitted from humans to ferrets by droplet infection.

d. Necropsy. Gross postmortem lesions of a mild pneumonia may be recognized.

e. Differential Diagnosis. Initially, the clinical signs are similar to those of distemper. However, recovery usually occurs within 4–5 days after the onset of signs. Other causes of pneumonia, such as bacteria and mycotic infections, can be ruled out on the basis of culture, duration, and response to treatment.

f. Prevention, Control, and Treatment. Respiratory tract congestion may be relieved with antihistamines. Treatment is usually not necessary. Recovery from infection confers immunity for at least 5 weeks against the homologous strain of influenza. Vaccination with attenuated live virus will protect ferrets for a similar duration, but is not routinely done as commercial vaccines are not available or recommended (Potter *et al.*, 1972).

3. Aleutian Disease

a. Etiology. Aleutian disease (AD) is usually subclinical in ferrets but disease signs may occur. Aleutian disease is caused by a parvovirus. There appears to be a biologic dif-

ference between Aleutian disease virus (ADV) isolates of mink or ferret origin (Porter *et al.*, 1982).

b. Transmission and Pathogenesis. Both horizontal (fecal–oral and aerosol) and vertical transmission occur. The virus induces a systemic proliferation of lymphoid cells and a vasculitis (Daoust and Hunter, 1978). The plasma cell proliferation is associated with a hypergammaglobulinemia (Ohshima *et al.*, 1978).

c. Clinical Signs. Signs in affected ferrets range from cachexia to melena. Stressed animals may die acutely. Hypergammaglobulinemia (the serum γ-globulin usually exceeds 20% of the total serum proteins) can be detected in affected animals.

d. Necropsy. Lymphocytic and plasmacytic infiltration of liver, spleen, lungs, and kidneys and follicular hyperplasia of lymph nodes are recorded. Macroscopically these changes are recognized as hepatosplenomegaly (Ohshima *et al.*, 1978).

e. Differential Diagnosis. Aleutian disease may be differentiated from acute episodes of nonspecific enteritis, campylobacteriosis, and gastric ulcers. For definitive diagnosis, serum can be examined by counterimmunoelectrophoresis for ADV antibody.

f. Treatment, Prevention, and Control. There is no treatment. Affected ferrets should be isolated or humanely destroyed to decrease the spread of the virus. Aleutian disease virus probably persists in infected ferrets for years.

4. Rabies

a. Etiology. Ferrets are susceptible to rabies (rhabdovirus). Rabies was diagnosed in one ferret that might have been vaccinated with a modified live rabies vaccine (Anonymous, 1980).

b. Differential Diagnosis. There are other conditions that can cause central nervous system diseases in ferrets, particularly distemper. Usually, the ferret's history includes exposure to a rabid carnivore.

c. Prevention and Control. Affected ferrets should be killed and their brains submitted for laboratory confirmation. Decreasing the risk of exposure to wild carnivores markedly decreases the chance of ferrets contracting the disease.

5. Rotavirus

Over the past 30 years, we have occasionally recorded high mortalities in young ferret kits with diarrhea 2–5 weeks of age.

While we have suspected an infectious agent, we have blamed maternal lactating failure, poor diets, or dirty nest boxes. Other ferret raisers have reported similar losses in this age group. There has been a recent outbreak in Finland in which thousands of ferret kits 2 to 6 weeks of age died. The case–fatality rate approached 100%. The pathologic features were those of a severe enteritis. Veijalainen (1983) has isolated a rotavirus from the dead kits in the outbreak.

6. Other Viral Diseases

Feline leukemia virus (FeLV) does not appear to produce clinical disease in ferrets. We have recognized FeLV positivity by enzyme-linked immunosorbent assay (ELISA) in clinically healthy ferrets. The reason for the ELISA positivity is not known; however, cross-reacting antigens to FeLV or non-specific serum factors are suggested, since we failed to evoke FeLV infection in normal ferrets by inoculating FeLV.

Ferrets are not susceptible to feline panleukopenia or mink virus enteritis (Burger, 1961). Furthermore, there is no experimental evidence that ferrets are susceptible to canine parvovirus. Ferrets can be experimentally infected with pseudorabies (Goto *et al.*, 1968), mink encephalopathy (Eckroade *et al.*, 1973), and infectious bovine rhinotracheitis (Smith, 1978). Natural outbreaks of pseudorabies, mink encephalopathy, or infectious bovine rhinotracheitis have not been recognized.

C. Parasitic and Fungal Diseases

1. Protozoa

Toxoplasmosis and coccidiosis have been reported in ferrets (Lainson, 1957). The clinical significance of these agents is not currently known.

2. Nematodes

Ferrets can be infected with *Toxascaris leonina*. Clinical signs of dull hair coat, pot belly, and wasting are usually seen in young ferrets. Ferrets are also susceptible to *Dirofilaria immitis* (Miller and Merton, 1982). Because of their small heart, 6–10 adult worms are sufficient to produce severe respiratory distress, ascites, and death in an adult ferret. Avermectins suppress the maturation of *Dirofilaria immitis* in ferrets (Blair and Campbell, 1978).

3. Trematodes and Cestodes

To our knowledge, there has been no reported naturally occurring trematode or cestode disease in ranch-raised ferrets.

4. Mites and Fleas

Ferrets can be infested with most of the external parasites of domestic dogs and cats. Infestation with *Sarcoptes scabiei* may take two forms: (1) in cases where skin lesions predominate, focal to generalized alopecia and intense pruritus associated with the mite, which can be found in skin scrapings or (2) the sarcoptic lesions are confined to the toes and feet where the feet are swollen, scabby, and if not treated, clawless. Treatment consists of trimming the claws and removing the scabs after softening them in warm water. The mites may be eliminated by application of sulfur ointment or lime and sulfur dips and washes. Organophosphates and carbamates should be used with caution, as safe levels for ferrets have not been established, but we have treated sarcoptic mange successfully with carbaryl (0.5%) shampoos applied weekly for 3 weeks. Pruritus may be reduced by concomitant administration of corticosteroids. Ferrets may also be infected with fleas (*Ctenocephalides* spp.) and ear mites (*Otodectes coynotis*) (Nie and Pick, 1978), which may be treated with rotenone or pyrethrin products (Ryland and Gorham, 1978).

5. Fungal Disease

Ringworm (*Microsporum canis*) has been reported in young ferrets (Hagen and Gorham, 1972). Cats frequenting stored litter are often the source of infection. The lesions are similar to those described for cats.

V. NUTRITIONAL AND METABOLIC DISEASES

A. Vitamins

It is reasonable to assume that ferrets are subject to some if not most of the vitamin deficiencies seen in mink, including thiamin (Chastek's paralysis), biotin (achromotrichia), vitamin E (anemia), and vitamin D coupled with a calcium–phosphorus imbalance (rickets) (Ensminger and Olentine, 1978).

B. Minerals

Zinc in the form of zinc oxide at levels in excess of 500 ppm is toxic to ferrets (Straube *et al.*, 1980; Straube and Walden, 1981). Ferrets decrease their feed intake and lose weight on high-zinc diets. Greater than 3000 ppm of zinc is lethal to ferrets within 2 weeks. Affected animals become anemic and rapidly lose condition. Postmortem findings include large, pale, soft kidneys and a pale, fatty liver. Histologic characteristics of zinc toxicosis include diffuse nephrosis (primarily cortical), macrocytic hypochromic anemia, and fatty infiltration of the liver. Treatment by decreasing dietary zinc and supportive therapy is indicated.

C. Eclamptogenic Toxemia and Nursing Sickness

Pregnant and postparturient females are predisposed to several nutrition-dependent diseases, such as eclamptogenic toxemia and "nursing sickness syndrome." Neither of these syndromes has been well documented. They are characterized by anorexia, muscular incoordination, weakness, weight loss, and death. Vitamin and mineral supplementation suitable for cats should be instituted during pregnancy and lactation. Ensuring that the dam's dietary needs are met during this critical period (including 0.5% NaCl in the diet) will help reduce preweaning deaths among kits due to lactational failure.

D. Bone Marrow Depression

a. Etiology. Bone marrow depression is due to high endogenous estrogen levels during estrus (Kociba and Caputo, 1981; Bernard *et al.*, 1983).

b. Clinical Signs. Clinical signs of bone marrow depression vary. The ferrets are pancytopenic and therefore are predisposed to bleeding disorders and secondary bacterial infections. Common clinical findings are pale mucous membranes, bilaterally symmetrical alopecia, melena, pyometra, subcutaneous petechiae, anorexia, pneumonia, and depression. One animal presented as a posterior paralysis due to hemorrhage into the spinal cord. Affected animals are anemic (PCV less than 20%), thrombocytopenic (less than 50,000 platelets/μl), and leukopenic (less than 2000 cells/μl).

c. Necropsy. Gross pathologic changes include thin watery blood, pale fatty bone marrow, subcutaneous petechiae to ecchymoses, hemorrhages throughout the gastrointestinal tract, and hydrometra or pyometra. The major histopathologic changes are hypocellular bone marrow and decreased splenic extramedullary hematopoesis.

d. Pathogenesis. Prolonged high levels of endogenous estrogens depress the bone marrow, resulting in pancytopenia. The exact mechanism is not known. Thrombocytopenia predisposes the ferret to bleeding disorders (subcutaneous ecchymoses, gastrointestinal hemorrhages, and hematomyelia), while neutropenia predisposes ferrets to secondary bacterial infections (pyometra, pneumonia, and septicemia).

e. Differential Diagnosis. Diagnosis of bone marrow depression is based on clinical presentation of a female in estrus with a PCV of less than 20%. Since the clinical presentation of these animals can vary from lethargy to pneumonia, the clinician should pay strict attention to the history and estrous cycle.

f. Prevention and Control. Prevention of bone marrow depression is not difficult. All female ferrets not intended for breeding purposes should be ovariohysterectomized prior to their first estrus (at about 8 months of age). After a female is in estrus, she can be treated with one im injection of 50–100 IU of human chorionic gonadotropic hormone (HCG) or by breeding.

g. Treatment. Treatment of bone marrow depression is difficult, as most animals have a PCV of less than 10% and a platelet count under 20,000/mm^3. Two animals are known to have survived this condition using two different treatment regimes. One female was ovariohysterectomized despite a PCV of 6%. The other ferret was treated successfully with ovariohysterectomy and 15 intravenous blood transfusions (10 ml each) over a 5-month period, coupled with anabolic steroids, corticosteroids, force feeding, and oral vitamin supplementation (Ryland, 1982). A bone marrow transplant was also done but is of questionable value. The use of whole blood transfusions (if available) and surgery may be the most logical treatment. If surgery is not elected, the animal can be cycled out of heat with HCG as described previously. During this period, anabolic steroids, vitamins, and transfusions are recommended.

E. Urolithiasis

Urolithiasis is frequently seen in ferrets. The signs are consistent with those observed in cats. Renal and cystic calculi composed of magnesium ammonium phosphate (struvite) have been reported (Nguyen *et al.*, 1979). There is often an accompanying pyelonephritis. The treatment is the same as that for feline urolithiasis and cystitis. Reducing dietary ash and providing adequate water may aid in decreasing the recurrence of uroliths.

F. Posterior Paralysis

Posterior paralysis accompanied with incontinence is sometimes noted (Frederick, 1981; Ryland *et al.*, 1983). This paralysis may be due to hemivertebrae, vertebral fractures, intervertebral disc disease, hematomyelia, or myelitis. Ferrets with no radiographic lesions and ferrets with intervertebral disc problems have responded variably to steroid therapy. This condition can recur, and treatment rarely is successful.

G. Gastric Ulcers

Ulcers of unknown etiology on the antrum and lesser curvatures of the stomach are sometimes seen in ferrets. The diagnosis is based on melena and postmortem findings.

H. Congenital and Heritable Problems

A variety of spontaneous and congenital malformations that may be hereditary have been observed. These include anencephaly, neuroschisis, gastroschisis, cryptorchidism, anury, amelia, corneal dermoids, and cataracts (Willis and Barrow, 1971; Utroska and Austin, 1979).

Cannibalism is more pronounced in certain families and may be an undesirable genetic trait, although normally most dams or kits will consume a dead neonate.

I. Endocrinopathies

Endocrine imbalances similar to those of dogs and cats have been reported in ferrets. Carpenter and Novilla (1977) have satisfactorily treated at least one black-footed ferret for diabetes mellitus with insulin (1–2 units/kg).

A bilaterally symmetrical alopecia of the tail and/or ventral abdomen is common in estrous females and has been observed in males during the normal breeding season (Fig. 1). The cause of the alopecia is unknown.

VI. NEOPLASTIC DISEASES

There are few reports in the literature regarding neoplastic diseases in ferrets. A variety of spontaneous epithelial and mesenchymal neoplasms have been reported in ferrets. Many of these reports are incidental findings or the clinical presentation of the ferrets was not reported. The following is a brief overview of the cases reported.

Two cases of lymphosarcoma have been observed (Andrews et al., 1979b; Chesterman and Pomerance, 1965). The relationship between FeLV and ferret lymphosarcoma is not known. Two other cases of lymphosarcoma have been brought to the authors' attention. One was a lymphoblastic lymphosarcoma (FeLV negative) and the second a mediastinal lymphosarcoma (not tested for FeLV).

Ovarian leiomyomas occur in ferrets. Apparently these tumors do not appear to interfere with ovulation and pregnancy and are incidental findings at necropsy (Cotchin, 1980). Bilateral ovarian thecomas with endometrial hyperplasia have been reported in a ferret as an incidental finding (Chesterman and

Pomerance, 1965). An adrenal adenoma was reported (Chesterman and Pomerance, 1965). Cutaneous squamous cell carcinomas have been reported in ferrets (Chesterman and Pomerance, 1965; Zwicker and Carlton, 1974). Cutaneous collections of mast cells in the skin have been observed (Symmers and Thomson, 1953). Mast cell tumors grow slowly and do not usually metastasize. Two ferret adenocarcinomas have been described. One was pancreatic and was considered an incidental postmortem finding. The second was hepatocellular and was from a cachetic ferret with a chronic draining abdominal fistula (Andrews et al., 1979b). Chowdhury and Shillinger (1982) have diagnosed megakaryocytic myelosis in an adult ferret.

REFERENCES

Adams, C. E. (1979). Anaesthesia of the ranch mink (Mustela vison) and ferret (Mustela putorius furo). Vet. Rec. **105,** 492.

Andrews, P. L. R., Bower, A. J., and Illman, O. (1979a). Some aspects of the physiology and anatomy of the cardiovascular system of the ferret, Mustela putorius furo. Lab. Anim. **13,** 215–220.

Andrews, P. L. R., Illman, O., and Mellersh, A. (1979b). Some observations of anatomical abnormalities and disease states in a population of 350 ferrets (Mustela furo L.). Z. Versuchstierkd. **21,** 346–353.

Anonymous (1973). "Threatened Wildlife of the United States," Publ. No. 114. U. S. Dept. of the Interior, Bureau of Sport Fisheries and Wildlife, U. S. Govt. Printing Office, Washington, D. C.

Anonymous (1980). "Veterinary Public Health Notes." U. S. Dept. of Health and Human Services, Public Health Service, Centers for Disease Control, Atlanta, Georgia.

Bernard, S. L., Leathers, C. W., Brobst, D. F., and Gorham, J. R. (1983). Estrogen-induced bone marrow depression in ferrets. Am. J. Vet. Res. **44,** 657–661.

Blair, L. S., and Campbell, W. C. (1978). Trial of Avermectin B₁a, Mebendazole and Melarsoprol against pre-cardiac Dirofilaria immitis in the ferret (Mustela putorius furo). J. Parasitol. **64,** 1032–1034.

Burger, D. (1961). The relationship of mink virus enteritis to feline panleukopenia virus. M.S. Thesis, Washington State University, Pullman.

Carpenter, J. W., and Novilla, M. N. (1977). Diabetes mellitus in a black-footed ferret. J. Am. Vet. Assoc. **171,** 890–893.

Chesterman, F. C., and Pomerance, A. (1965). Spontaneous neoplasms in ferrets and polecats. J. Pathol. Bacteriol. **89,** 529–533.

Chowdhury, K. A., and Shillinger, R. B. (1982). Spontaneous megakaryocytic myelosis in a four-year-old domestic ferret (Mustela furo). Vet. Pathol. **19,** 561–564.

Cotchin, E. (1980). Smooth-muscle hyperplasia and neoplasia in the ovaries of domestic ferrets (Mustela putorius furo). J. Pathol. **130,** 169–171.

Creed, J. E., and Kainer, R. A. (1981). Surgical extirpation and related anatomy of anal sacs of the ferret. J. Am. Vet. Med. Assoc. **179,** 575–577.

Daoust, P. Y., and Hunter, D. B. (1978). Spontaneous Aleutian disease in ferrets. Can. Vet. J. **19,** 133–135.

Eckroade, R. J., Zurhein, G. M., and Hanson, R. P. (1973). Transmissible mink encephalopathy in carnivores: Clinical, light and electron microscopic studies in raccoons, skunks and ferrets. J. Wildl. Dis. **9,** 229–240.

Ensminger, M. E., and Olentine, C. J. (1978). Mink nutrition. In "Feeds and Nutrition—Complete," pp. 997–1017. Ensminger Publ. Co., Clovis, California.

Evans, R. H. (1982). Ralston Purina Co., St. Louis, Missouri (personal communication).

Fox, J. G., Murphy, J. C., Ackerman, J. I., Prostak, K. S., Gallagher, C. A., and Rambow, V. J. (1982). Proliferative colitis in ferrets. *Am. J. Vet. Res.* **43**, 858–864.

Fox, J. G., Ackerman, J. I., and Newcomer, C. E. (1983). Ferret as a potential reservoir for human campylobacteriosis. *Am. J. Vet. Res.* **44**, 1049–1052.

Fox, J. G., Hewes, K., and Niemsi, S. M. (1984). Retro-orbital technique for blood collection from the ferret (*Mustela putorius furo*). *Lab. Anim. Sci.* (in press).

Frederick, M. A. (1981). Intervertebral disc syndrome in a domestic ferret. *VM/SAC, Vet. Med. Small Anim. Clin.* **76**, 835.

Frederick, K. A., and Babish, J. G. (1983). Compendium of recent literature on the ferret. *Fundam. Appl. Toxicol.* (in press).

Gorham, J. R., Cordy, D. R., and Quortrup, E. R. (1949). Salmonella infections in mink and ferrets. *Am. J. Vet. Res.* **10**, 183–192.

Goto, H., Gorham, J. R., and Hagen, K. W. (1968). Clinical observation of experimental pseudorabies in mink and ferrets. *Jpn. J. Vet. Sci.* **30**, 257–263.

Hagen, K. W., and Gorham, J. R. (1972). Dermatomycoses in fur animals. *VM/SAC, Vet. Med. Small Anim. Clin.* **67**, 43–48.

Hahn, E. W., and Wester, R. C. (1969). "The Biomedical Use of Ferrets in Research." Marshall Research Animals Inc., North Rose, New York.

Hammond, J. (1974). "The Ferret: Some Observations on Photoperiod and Gonadal Activity, and Their Role in Seasonal Pelt and Bodyweight Changes." Heffer, Cambridge, England.

Kociba, G. J., and Caputo, C. A. (1981). Aplastic anemia associated with estrus in pet ferrets. *J. Am. Vet. Med. Assoc.* **178**, 1293–1294.

Lainson, R. (1957). The demonstration of toxoplasma in animals with particular reference to members of the mustelidae. Symposium on toxoplasmosis. III. *Trans. R. Soc. Trop. Med. Hyg.* **51**, 111–117.

Liberson, A. J., Newcomer, C. E., Ackerman, J. I., Murphy, J. C., and Fox, J. G. (1983). Mastitis due to hemolytic *Escherichia coli* in the ferret. *J. Am. Vet. Med. Assoc.* **183**, 1179–1181.

Marshall, K. R., and Marshall, G. W. (1973). "The Biomedical Use of Ferrets in Research." Suppl. 1. Marshall Research Animals Inc., North Rose, New York.

Miller, W. R., and Merton, D. A. (1982). Dirofilariasis in a ferret. *J. Am. Vet. Med. Assoc.* **180**, 1103–1104.

Neal, J., and Murphy, B. D. (1977). Response of immature, mature nonbreeding and mature breeding ferret testis to exogenous LH stimulation. *Biol. Reprod.* **16**, 244–248.

Nguyen, H. T., Moreland, A. F., and Shields, R. P. (1979). Urolithiasis in ferrets (*Mustela putorius*). *Lab. Anim. Sci.* **29**, 243–245.

Nie, I. A., and Pick, C. R. (1978). Infestation of a colony of ferrets with ear mite (*Otodectes coynotis*) and its control. *J. Inst. Anim. Technicians* **29**, 63–65.

Ohshima, K., Shen, D. T., Henson, J. B., and Gorham, J. R. (1978). Comparison of the lesions of aleutian disease of mink and hypergammaglobulinemia in ferrets. *Am. J. Vet. Res.* **39**, 653–657.

Owen, R. (1973). "On the Anatomy of Vertebrates," Vols. II and III. AMS Press, New York.

Porter, H. G., Porter, D. D., and Larsen, A. E. (1982). Aleutian disease in ferrets. *Infect. Immun.* **36**, 379–386.

Potter, C. W., Oxford, J. S., Shore, S. L., McLaren, C., and Harris, C. H. (1972). Immunity to influenza in ferrets. I. Responses to live and killed virus. *Br. J. Exp. Pathol.* **53**, 153–167.

Quortrup, E. R., and Gorham, J. R. (1949). Susceptibility of furbearing animals to the toxins *Clostridium botulinum* Types A, B, C, and E. *Am. J. Vet. Res.* **10**, 268–271.

Roberts, M. F. (1977). "All about Ferrets," TFH Publications Inc., Ltd., Neptune City, New Jersey.

Ryland, L. M. (1982). Remission of estrus-associated anemia following ovariohysterectomy and multiple blood transfusions in a ferret. *J. Am. Vet. Med. Assoc.* **181**, 820–822.

Ryland, L. M., and Gorham, J. R. (1978). The ferret and its diseases. *J. Am. Vet. Med. Assoc.* **173**, 1154–1158.

Ryland, L. M., Bernard, S. L., and Gorham, J. R. (1983). A clinical guide to the pet ferret. *Compend. Continuing Educ. Practicing Vet.* **5**, 25–32.

Shump, A. U., and Shump, K. A. (1978). Growth and development of the European ferret (*Mustela putorius*). *Lab. Anim. Sci.* **28**, 89–91.

Smith, P. C. (1978). Experimental infectious bovine rhinotracheitis virus infections of English ferrets (*Mustela putorius furo* L.). *Am. J. Vet. Res.* **39**, 1369–1372.

Straube, E. F., and Walden, N. B. (1981). Zinc poisoning in ferrets (*Mustela putorius furo*). *Lab. Anim.* **15**, 45–47.

Straube, E. F., Schuster, N. H., and Sinclair, A. J. (1980). Zinc toxicity in the ferret. *J. Comp. Pathol.* **90**, 355–361.

Symmers, W. S. C., and Thomson, A. P. D. (1953). Multiple carcinomata and focal mast-cell accumulations in the skin of a ferret (*Mustela furo* L.) with a note on other tumors in ferrets. *J. Pathol. Bacteriol.* **65**, 481–493.

Symmers, W. S. C., Thomson, A. P. D., and Iland, C. N. (1953). Observations on tuberculosis in the ferret (*Mustela furo* L.). *J. Comp. Pathol.* **63**, 20–30.

Thornton, P. C., Wright, P. A., Sacra, P. J., and Goodier, T. E. W. (1979). The ferret, *Mustela putorius furo*, as a new species in toxicology. *Lab. Anim.* **13**, 119–124.

Truex, R. C., Belej, R., Ginsberg, L. M., and Hartman, R. L. (1974). Anatomy of the ferret heart: An animal model for cardiac research. *Anat. Rec.* **179**, 411–422.

Utroska, B., and Austin, W. L. (1979). Bilateral cataracts in a ferret. *VM/SAC, Vet. Med. Small Anim. Med.* **74**, 1176.

Veijalainen, P. (1983). National Veterinary Institute, Helsinki, Finland (personal communication).

Willis, L. S., and Barrow, M. V. (1971). The ferret (*Mustela putorius furo* L.) as a laboratory animal. *Lab. Anim. Sci.* **21**, 712–716.

Zwicker, G. M., and Carlton, W. W. (1974). Spontaneous squamous cell carcinoma in a ferret. *J. Wildl. Dis.* **10**, 213–216.

Chapter 13

Biology and Diseases of Birds

Everett Bryant

Copyright © 1984 by Academic Press, Inc.
All rights of reproduction in any form reserved.
ISBN 0-12-263620-1

I.   INTRODUCTION

A.   Taxonomy

Most avian species used for research purposes fall within three orders: Galliformes, which includes domestic fowl, quail, pheasants, turkeys, partridges, grouse, guinea fowl and brush turkeys; Anseriformes, which includes ducks, geese, swans, and screamers; and the Columbiformes, which includes pigeons, doves, and sand grouse.

The bird fits into the overall taxonomic scheme as follows: Kingdom, Animal; Phylum, Chordata; Class, Aves; Order, 27 orders with approximately 8600 species. A clear outline of the avian orders listing the common names of birds in each may be found in Steiner and Davis (1981).

B.   The Bird as a Research Animal

The domestic fowl has been the most useful in experimental trials. The fast maturing and highly inbred chicken can be used with great statistical confidence. Diseases can be prevented easily, and isolation rearing is neither difficult nor costly.

C.   Availability of Chickens for Research

Specific pathogen-free (SPF) fertile eggs, day-old chicks, or started pullets are available for use in research. Specific pathogen-free chickens or eggs come from breeding stock negative to diseases caused by mycoplasmas, Newcastle disease, infectious bronchitis, avian encephalomyelitis, infectious bursal disease, quail bronchitis, salmonellosis, pasteurellosis, and infectious coryza.

Somes (1981) lists the sources of specialized lines, strains, mutations, breeds, and varieties of chickens, turkeys, and Japanese quail. Chromosome linkage maps for the three species and modes of inheritance are also given for most genetic traits.

D. General Poultry Husbandry (see Whiteman and Bickford, 1979; Woodard *et al.*, 1973; Schwartz, 1977)

Quality animal welfare and optimal comfort for birds is necessary to insure reliable research data. The ability periodically to observe birds on experimental test through a glass window is ideal; however, if an observation point is not available, the animal technician should check the birds at least twice a day for paleness, dehydration, and other signs indicative of diseases.

Death before 2 weeks of age may be due to chilling, overheating, crowding, and omphalitis. Mortality at 2–6 weeks in research chickens may be due to coccidiosis or one of the enteritides.

Brooding temperature for 1-day-old chicks should be between 90°–95°F and gradually reduced to 75°–80°F by the third week. Chicks will constantly peep if too hot or too cold. Lowered humidity will cause poor feathering, and if the environment is too moist, disease problems will be more pronounced. A relative humidity of 30–40% is ideal. Therefore, frequent monitoring to control temperature and humidity should be practiced. Fresh air, without drafts, helps to minimize moisture accumulation. Stale air retards growth of birds and enhances respiratory diseases.

E. Space Requirements

The density of broilers and layer replacements may vary with the experimental procedure, but Table I will serve as a guideline.

F. Equipment Requirements

There are many kinds of feeders, waterers, and brooders available. The simplest and least expensive may be the best for experimental work. While adjusting to the feeders newly hatched chicks and poults should be fed on papers from day 1 to day 7. Trough feeders work well for birds raised on the floor. It is important to have an adjustable feeder. The level of the feed in the trough should be equal to the level of the back of

Table I

Space Needs for Avian Species

Bird	Age	Floor space/bird (ft²)	Cage space/bird (ft²)
Chicken	1-day	0.30	0.10
	5 weeks	0.60	0.25
	10 weeks	1.25	0.40
	20 weeks	1.50	0.50
Turkey	1-day	1.25	1.25
	10 weeks	2.50	2.50
	25 weeks	4.50	—

a standing chicken. This allows the bird access to the feed and prevents waste of feed.

Tube feeders that hang from the ceiling have the advantage of holding more feed without waste. They are adjustable, and the feed should be at the same level, as described for trough feeding. Too many feeders of either type impede available floor space for the chickens. Two 4-ft trough feeders or two tube feeders are adequate for 100 birds at any age.

Waterers should provide clean, fresh water *ad libitum*. Birds should never be without water. There are many types, but a simple gallon jug with a plastic bottom containing a groove circling the jug is ideal for birds from 1 to 10 days of age. An automatic trough waterer, one 4-ft long per hundred birds, is adequate. Automatic waterers can stop working, run over, and make it difficult to measure water consumption. Therefore, large can-type waterers or open pan-type waterers may be better for some types of trials. They are more easily cleaned and control water intake better. Incidence of candidiasis in a flock of birds is directly related to the level of water sanitation, and proper attention to water quality greatly reduces this disease in research birds.

Commercially available cage equipment will also have feeders and waterers and sometimes a heat source. Daily cleaning of fecal material from cages enhances the level of sanitation. External and internal parasitism is increased in birds reared in dirty cages.

For heat and ventilation, brooder stoves fueled by gas, oil, coal, and electricity are available. For small groups of birds, heat lamps are adequate. Supplementary room heat is necessary for small groups in colder climates. Air conditioning may be necessary during extremely hot weather. An exhaust fan is necessary to ensure removal of stale air. For young chicks, ½ ft³/min will exchange air quite well and for older birds, 1 ft³/min is adequate. Modern animal facilities meeting the criteria for the NIH "Guide for the Care and Use of Laboratory Animals" will provide acceptable heat, ventilation, and air conditioning needs (see Chapter 17).

Nests are needed for laying birds. Adequate nests should be provided to ensure one nest for every four birds. Nests should be constructed of metal for ease of cleaning and should be well ventilated. Lack of sharp corners minimizes trauma, and dry materials, such as shavings, sawdust, or peanut hulls, make excellent nest material.

G. Sanitation

A completely sanitized and disinfected room with sterilized equipment is necessary for research purposes. Birds are coprophagic, and many diseases of the digestive tract are transmitted in this manner. Physical cleaning of room surfaces with high pressure of hot water will remove fecal material, mucus, and other debris from walls and floors. Cages should be sanitized in appropriate cage washers. While the pen is still warm and wet, fumigation with formaldehyde gas is quite effective. The temperature should be 80°F and the humidity 70% or higher to be most efficient. Formaldehyde vapors have a potential health risk to humans, and the reader should refer to the discussion of disinfectants by Hofstad et al. (1978) before using this product.

Once the area and equipment are cleaned, iodines, quaternary ammonium compounds, chlorines, or other agents can be used to disinfect the premises.

H. Isolation

No one should be admitted to isolation areas without a shower, clean clothing, and clean boots. Doors should be locked at all times, and animal technicians should not be assigned to care for any other birds.

II. BIOLOGY

A. Applied Anatomy

Birds have pneumatic bones, with air spaces and channels, which add buoyancy in flight. This system of pneumatic bones provides direct channels from the air sacs through the bones so that air carrying infecting organisms may travel throughout the bird rather rapidly. Certain respiratory infections and parasites use these channels for distribution of the microorganisms.

Small birds lack teeth; the organ of mastication is the gizzard which contains grit or gravel, thereby enabling grinding of hard grains. If the feed is an all-mash or all-pellet diet, no grit is needed in the diet.

Most birds possess a crop, a dilatation in the midesophageal region. Pigeons and pet birds regurgitate and feed their young on crop milk. This process allows transmission of parasites to the young from healthy adult carriers.

Only the left ovary and oviduct are functional in the hen. Frequently a cyst, the remnant of the right oviduct, is found; however, this causes no pathological disturbance.

B. Applied Physiology (see Sturkie, 1976)

Birds can stand extreme variations of environmental temperature for short periods of time. As the relative humidity increases, the temperature tolerance of the chicken decreases. Normal heart rate for the chicken is about 300 beats/min and will increase substantially during excitement and periods of stress. The normal respiration for the chicken is approximately 100/min, and will also increase rapidly as the temperature goes up. Artificial light is important. Fourteen to 16 hr of light per day is needed for optimum egg production. The optimum temperature for egg production is between 50°–70°F. Any variation thereof may cause a decrease in egg production.

C. Normal Values

Growth rate has improved so rapidly that textbooks cannot keep up-to-date. Feed conversions of less than 2.0 are being reported on a 4 lb average weight broiler at 49 days of age. These are under excellent field management with optimum nutrition, genetics, and freedom from disease.

Egg production flocks have also improved their efficiency. The average flock today will peak above 90% by the time they reach 30 weeks of age.

For a complete treatment of avian hematology, the reader is referred to Sturkie (1976). Blood volume in the bird is about 10% of total body weight. This will vary according to age, sex, and species. An atlas on avian hematology is also available (Lucas and Jamroz, 1961).

Data shown in Table II were taken from Olson (1937).

D. Nutrition

Chickens, turkeys, and other birds require the six major nutrients: carbohydrates for energy, fats for energy and essential fatty acids, protein for meat and egg production, minerals for bones and shells, vitamins for chemical catalysts, and water.

Birds should have a complete and balanced ration with very little supplementation. Unnecessary supplementation may create other deficiencies.

Age and functional status of the flock will determine the formula needed. Table III lists approximate types of formulas needed for various ages and classes of birds.

Table II[a]
Chicken Blood Cell Counts and Differentials

Cell type	$10^3/mm^3$
Erythrocytes	3400
Thrombocytes	29
Total leukocytes	23

Differential Leukocyte Counts

Cell type	Percent
Lymphocytes	63.0
Heterophils	24.0
Eosinophils	2.0
Basophils	2.4
Monocytes	9.1

[a]Data from Olson (1937).

High quality protein is necessary for growth and maintenance of birds. All the essential amino acids must be present in the diet as the bird cannot convert them from raw protein. Minerals and vitamins must be provided in balance. Calcium and phosphorus should be in a ratio of 2:1 for young growing chickens.

Antioxidants should always be used to prevent oxidation of fats and fat-soluble vitamins. Feed additives, especially arsenicals and certain drugs, should be fed according to the manufacturer's directions. Consumption of eggs from these birds during this time should be avoided. For a more detailed account of

Table III
Approximate Types of Formulas Needed for Various Ages and Types of Birds[a]

	Protein (%)	Fat (%)	Fiber (%)	Calcium (%)	Metabolizable energy (cal/lb)
Chicken starter (0–5 weeks)	20.0	4.5	2.7	0.9	1360
Chicken grower (5–15 weeks)	15.0	4.0	3.4	0.9	1300
Chicken breeder (20 weeks +)	16.5	4.8	2.7	3.0	1325
Chicken layer (20 weeks +)	18.0	4.4	2.2	3.3	1300
Turkey starter (0–8 week)	27.0	5.3	2.6	1.2	1325
Turkey grower (8–16 weeks)	21.0	5.4	2.5	1.0	1370
Turkey finisher (16 weeks +)	16.0	5.8	2.3	0.9	1440
Turkey breeder (20 weeks +)	17.0	4.9	3.3	2.3	1300

[a]Data from various authors.

nutrition in birds, the reader is referred to the specific texts by Scott *et al.* (1976), by the National Research Council of the National Academy of Sciences (1977), and to the chapter in Hofstad *et al.* (1978).

Specific nutritional deficiencies will be discussed in Section VIII.

E. Behavioral Aspects of Birds

Poultry and other birds are easily startled. Flashing lights, loud noises, and shadows will cause birds to pile up on each other causing the pen mates underneath to suffocate.

Hysteria, caused by sudden fright, will result in birds flying against the wall for no apparent reason. Many experts argue that a sudden change in air pressure will cause hysteria. When the author has encountered this disease, the birds have usually been raised in a filtered-air, positive-pressure facility.

III. DISEASES CAUSED BY BACTERIA (see Hitchner *et al.*, 1975)

A. Colibacillosis and Coligranuloma (Hjarre's Disease)

a. Etiology. At least three serotypes of *Escherichia coli* have been found to be pathogenic for birds. These are 01:K1 (L), 02:K1 (L), and 078:K80 (B).

b. Clinical Signs and Lesions. *Escherichia coli* may be a primary infectious agent; clinical onset is rapid and daily mortality can approach 0.5–1.0%. The usual lesions noted are air sacculitis, fibrinous pericarditis, and fibrinous perihepatitis. Rarely is the trachea involved. Coligranuloma (Hjarre's disease) may be caused by a pathogenic *E. coli* serotype. Avian tuberculosis may be confused with colibacillosis. Granulomas of the intestine are differentiated from tuberculosis by the use of the acid-fast stain. *Escherichia coli*, of course, is not acid fast. Other lesions caused by *E. coli* include arthritis leading to lameness and death, navel infection leading to high mortality, and foot pad infections.

c. Epizootiology. Hens with *E. coli* infection, especially of the oviduct, contaminate the egg shell. Bacteria may then penetrate the egg shell and the chick is hatched with a yolk sac infected with *E. coli*.

d. Diagnosis. *Escherichia coli* should always be considered a potential pathogen in birds. If an air sacculitis syndrome develops, with no isolation of *Pasteurella* sp., *Mycoplasma* sp., or *Hemophilus* sp., it is essential to consider *E. coli* as the potential pathogen. Isolations of *E. coli* from yolk sacs, joints, or bone marrow should always be considered a primary cause of disease.

e. Prevention.   Good management of the chicks must include proper ventilation and the proper care of hatching eggs. This is the most important means of preventing navel ill. Dirty, contaminated eggs should never be used for hatching.

f. Treatment.   As with all bacterial infections in poultry, antimicrobial susceptibility testing should be used to determine the treatment of choice for any generalized *E. coli* infection.

g. Research Implications.   *Escherichia coli* infection will damage experimental flocks, causing uneven growth.

B. Infectious Coryza

a. Etiology.   The cause of coryza is *Hemophilus paragallinarum*. It is a gram-negative, bipolar staining rod and can easily be demonstrated in sinus exudate.

b. Clinical Signs.   Signs include a sudden onset with high morbidity, reduction in feed intake and growth, and depressed egg production. Uncomplicated coryza is a disease of the upper respiratory tract causing ocular and nasal discharge, facial edema, and swollen infraorbital sinuses.

c. Epizootiology and Transmission.   Transmission is accomplished mainly by ill or recovered carrier birds. Inhalation of infectious material and ingestion of contaminated feed and water provide the best means of transmission.

d. Necropsy Findings.   At necropsy, there is conjunctivitis, cheesey exudate in the conjunctival sac, nasal discharge, and a catarrhal inflammation of the nasal passages and sinuses. The infraorbital sinuses may be filled with exudate.

e. Diagnosis.   History, signs, and lesions will suggest coryza. A gram-stained smear of sinus exudate showing gram-negative bipolar rods warrant a strong presumptive diagnosis of coryza. Incubation in a CO_2 atmosphere of a culture of the exudate on blood agar with a *Staphylococcus* nurse colony will confirm the presence of *H. paragallinarum*.

f. Prevention.   Depopulation of all birds before introducing chicks is essential, and a complete sanitizing and disinfecting of the brooder facilities is also necessary. A bacterin is available and can be used in birds 14 weeks of age or older with a booster every 3 weeks. Several injections are necessary to confer protective immunity.

g. Treatment.   Continuous medication appears to suppress clinical infection, but when removed relapses occur. Medications used with success include streptomycin, erythromycin, spectinomycin, tylosin tartrate, and sulfadimethoxine.

h. Research Complications.   A diagnosis of *H. paragallinarum* infection in a research unit would warrant termination of the experiment. Salvage of an experiment with this disease would provide questionable research data at best. The chronicity of the disease and the recovered carrier bird would further jeopardize a research colony.

C. Gangrenous Dermatitis

a. Etiology.   The important etiological agents appear to be *Clostridium septicum* and *Staphylococcus aureus* (Frazier *et al.*, 1964). The host is likely to be immunosuppressed, and probably undergoing a simultaneous outbreak of infectious bursal disease. (See Section IV,B.)

b. Clinical Signs.   Signs in 4- to 16-week old chickens and turkeys include a sudden increase in morbidity and mortality. Wet, gangrenous areas can be seen in the skin of the wings and legs.

c. Epizootiology and Transmission.   Skin wounds of chickens are a common occurrence, and secondary infections are expected. However, gangrenous dermatitis appears to be a specific disease complicated by the presence of heavily contaminated environment containing *Clostridium* spp. and *Staphylococcus* spp. The syndrome was rare until immunosuppression caused by bursal damage from infectious bursal disease virus, mycotoxins, and possibly other factors became more common.

d. Necropsy Findings.   Usual findings include necrosis of the skin and underlying areas of the thigh, breast, wing tips, hip, and back. Hemorrhagic, necrotizing myositis has also been described.

e. Pathogenesis.   It would appear that the toxin of the *Staphylococcus* species causes necrosis and anerobiosis; the *Clostridium* then multiplies and contributes to the gangrene.

f. Differential Diagnosis.   Signs and bacterial cultures from the lesions will give a presumptive diagnosis. Selenium deficiency that causes an exudative diathesis may mimic gangrenous dermatitis.

g. Prevention.   Eliminating all causes of skin trauma and improving the sanitation plus instituting an immunization program for infectious bursal disease will help prevent the disease.

h. Treatment. A susceptibility test for the *Staphylococcus* isolated will suggest the antibiotic of choice. Most broad-spectrum drugs will aid the recovery of affected birds.

i. Research Complications. Widespread gangrenous dermatitis in birds may compromise the study. Because this disease is often a complication of infectious bursal disease immune competence may have been altered; therefore research data collected on affected bird may be unreliable.

D. Necrotic Enteritis

a. Etiology. The cause of necrotic enteritis is thought to be *Clostridium perfringens.* Other inciting factors may be necessary, but have eluded investigators studying this disease. The disease is seen mostly when high energy, broiler-type rations are used.

b. Clinical Signs. Necrotic enteritis is usually seen in broilers; many times only the males, at about 2–8 weeks of age, are affected. Birds show profound depression, and mortality may start at 1% per day and last about 1 week (Helmboldt and Bryant, 1971).

c. Epizootiology and Transmission. It appears that transmission is direct and via the feces.

d. Necropsy Findings. A fibrinonecrotic enteritis is noted, and the entire small intestine becomes dilated with fluid debris and becomes very friable.

e. Pathogenesis. There is evidence that a toxin damages the villus tips, and eventually the integrity of the rest of the intestinal wall is lost.

f. Differential Diagnosis. This disease can be differentiated from ulcerative enteritis (quail disease) by isolation and identification of the *Clostridium perfringens. Eimeria brunetti* may be present as a secondary problem.

g. Prevention. Sanitation, disinfection, and isolation will provide the best protection against necrotic enteritis. Low level feeding of antibiotics at 100 gm per ton (50 mg/lb) of feed from 0 to 10 weeks of age will totally prevent the disease.

h. Treatment. The organism is very easily treated with high levels of tetracyclines, furazolidone, streptomycin, or bacitracin.

i. Research Complications. Diagnosis of disease in a pen warrants instituting a low level antibiotic therapy for succeeding flocks.

E. Ulcerative Enteritis (Quail Disease)

a. Etiology. Clostridium colinum, a gram-positive, spore-forming bacillus, is the cause of ulcerative enteritis. This bacterium is best recovered from fresh liver cultured on tryptose agar or the yolk sac of 5–7 day embryos.

b. Clinical Signs. Affected quail exhibit high morbidity with mortality sometimes approaching 100%. Chickens and turkey poults, on the other hand, may experience low mortalities of only 2–10%. Listlessness, drooping wings, ruffled feathers, and diarrhea are typical signs of this disease.

c. Epizootiology and Transmission. The disease is highly contagious among quail. It spreads mainly through feces from ill or recovered carrier birds.

d. Necropsy Findings. Deep ulcers along the full length of the intestine is the most common finding. Perforation of the intestine may occur, with resulting peritonitis. The cecae have detectable ulcers early in the disease. The liver often has large, diffuse areas of necrosis.

e. Pathogenesis. Intestinal ulcers and necrosis are the direct result of infection with *Clostridium colinum.*

f. Differential Diagnosis. Coccidiosis caused by *Eimeria brunetti* may produce a necrotizing enteritis of the large intestine, which resembles ulcerative enteritis. Scrapings of the intestinal mucosa will usually reveal oocysts of *E. brunetti.* Necrotic enteritis caused by *Clostridium perfringens* can also mimic quail disease. Histologically, necrotic enteritis produces no ulcers, and the intestinal lesion is villus tip loss and cystic dilatation of the deeper glands.

g. Prevention. Birds should be reared in a clean, disinfected and isolated environment. Low level of antibiotics, 50–100 gm per ton of feed, will prevent the disease. Streptomycin, bacitracin, terramycin, aureomycin, and furazolidone are therapeutically effective.

h. Treatment. All of the above antibiotics at 200 gm per ton of feed (100 mg/lb) will curtail and in some cases terminate the mortality.

i. Research Complications. Substantial morbidity and mortality would terminate most experiments. Preventive levels of antibiotics are indicated from 0 to 10 weeks in birds housed in a building where the disease was previously encountered.

F. Botulism (Limberneck: Western Duck Sickness)

a. Etiology. The ingestion of the toxin produced by *Clostridium botulinum* is the cause of botulism. The lethal toxin, which must be produced under anaerobic conditions, is found in decaying feed, dead birds, and maggots. *Clostridium botulinum* organisms per se are not considered pathogenic.

b. Clinical Signs. Drowsiness, weakness, and a progressive paralysis of the legs, wings, and neck are all suggestive of botulism.

c. Epizootiology and Transmission. The ingestion of preformed toxin is necessary to produce the disease. Poultry or wild birds should not be allowed to ingest wet, moldy feed. Accidental water spillage in a research unit could result in conditions necessary for the development of the toxin.

d. Necropsy Findings. The absence of lesions may be the best indication that the birds are dying from botulism. The offending decayed matter containing the toxin may be found in the crop, proventriculus, or intestine.

e. Differential Diagnosis. History, signs, and the presence of decaying feed may help identify the disease. Gizzard or intestinal contents may be washed and some of the fluid inoculated into mice, half of which have been previously protected with *C. botulinum* antiserum, for a definitive diagnosis. Transient paralysis, a mild form of Marek's disease, can produce signs similar to botulism. In pheasants, paralysis caused by eastern encephalomyelitis virus could be confused with botulism.

f. Prevention. If birds are raised under clean conditions, botulism is rare.

g. Treatment. Type A and C antitoxin may be indicated when valuable animals are affected. Laxatives may also be used with some success. It is important to provide fresh, clean water in a botulism outbreak.

h. Research Complications. It would be a rare to encounter this disease under research condition.

G. Erysipelas

a. Etiology. The cause of erysipelas is *Erysipelothrix rhusiopathiae,* a gram-positive, pleomorphic rod. It primarily affects turkeys, though chickens can be infected.

b. Clinical Signs. Usually the disease starts with sudden onset followed by depression, diarrhea, lameness, and a rapidly increasing morbidity and mortality. The snood of the male turkey often becomes swollen.

c. Epizootiology and Transmission. Fecal shedding of *E. rhusiopathiae* occurs in recovered turkeys and continues for about 6 weeks. The oral–fecal route is a significant source of transmission. Cutaneous injuries of the snood and head also may allow entry of the bacterium, and fighting among males is a common method of transmitting this disease.

d. Necropsy Findings. The lesions are those of a septicemia and vasculitis. There is a degeneration and hemorrhage in pericardial fat. The heart muscle, kidney, spleen, and liver may also show hemorrhage. Fibrinopurulent exudate in the joints, vegetative endocarditis, and dark, crusty skin lesions are also common.

e. Differential Diagnosis. History, signs, lesions, and culture of *E. rhusiopathiae* will aid in a diagnosis of erysipelas. It should be differentiated from colibacillosis, fowl cholera, salmonellosis, and possibly a velogenic strain of Newcastle disease.

f. Prevention. Turkey poults should be raised away from older turkeys, sheep, and swine. Poults should be vaccinated with a bacterin at 16–20 weeks of age. Raw fish meal may harbor *E. rhusiopathiae* organisms and should be avoided.

g. Treatment. Penicillin and erysipelas bacterin may be injected simultaneously in an outbreak. Tetracyclines in the feed or water will also help reduce losses.

h. Research Complications. Erysipelas is a septicemic disease; therefore, it will significantly retard growth and weight gain in animals recovering from the disease.

H. Omphalitis (Navel Ill)

a. Etiology. Predisposing causes of omphalitis may include excessive humidity and excessive fecal contamination of the incubator. Fecal contamination of eggs used for hatching purposes will also contribute to omphalitis. As the navel fails to close at hatching, a route of entry for bacteria, such as *E. coli, Pseudomonas,* and *Proteus,* as well as staphylococci is available. Chilling or overheating baby chicks may also precipitate or exacerbate this condition.

b. Clinical Signs. Depression, drooping of the head, and huddling near a heat source are usual signs. A scab over the unhealed navel is often present, and mortality usually is highest on the third and fourth days posthatching, sometimes reaching 1–10% during the first week.

c. Epizootiology and Transmission. The disease is transmitted through infection of the yolk sac prior to hatching. It is not contagious from chick to chick.

d. Necropsy Findings. Infected navels, large unabsorbed yolks, and extensive peritonitis are the common lesions commonly noted.

e. Differential Diagnosis. Isolation and identification of the bacteria from infected yolk will pinpoint omphalitis and its cause.

f. Prevention. Proper hatching egg management, emphasizing care and sanitation of the incubators will reduce the incidence of omphalitis.

g. Treatment. There is no treatment for omphalitis except removing affected chicks. Effective drug levels cannot eradicate the infection in the yolk sac.

h. Research Complications. Affected chicks usually die during the outbreak. Those that do survive, appear to grow well.

I. Fowl Cholera (Pasteurellosis)

a. Etiology. Fowl cholera, an acute septicemic disease of poultry, turkeys, waterfowl, and wild birds, is caused by *Pasteurella multocida,* a gram-negative, bipolar rod (Panigraphy and Glass, 1982). Thirteen serotypes of this bacterium have been identified.

b. Clinical Signs. Fowl cholera usually affects young adults. Birds affected with the acute form of fowl cholera may show no signs other than sudden deaths. Hens may die while laying an egg and will be found dead on the nest. In less acute cases, birds will show depression, anorexia, and cyanosis. A white-green diarrhea is usually present. In the chronic form, localized infections of the joints, wattles, foot pads, sinuses, and middle ear is clinically manifested by lameness and torticollis.

c. Epizootiology and Transmission. Ingestion of decaying carcasses, contaminated feed and water, and feces from recovered carriers are sources of *P. multocida*. Wild birds, and rodents are also potential transmitters. Transovarian transmission does not play a role in transmission.

d. Necropsy Findings. Peracute deaths may preclude the presence of gross lesions. Pinpoint hemorrhages are common

in the coronary fat suggesting an acute septicemia. In birds that die less acutely, pericarditis and perihepatitis are usually found. Cheesy exudate may also be seen around the oviduct. Chronic lesions consist of inflammation of the wattles, joints, conjunctival sac, infraorbital sinus, middle ear, and sometimes the bones of the skull. Fibrinous pneumonia is common in turkeys.

e. Pathogenesis. Fowl cholera is a septicemia with the agent spreading via the bloodstream to many organs.

f. Differential Diagnosis. Isolation and identification of the organism is necessary to differentiate this disease from erysipelas in turkeys and colibacillosis in hens.

g. Prevention. Birds reared in clean, isolated premises usually are free of *Pastuerella* infections. Addition of birds with unknown health status to a clean flock should be avoided. Healthy-appearing birds can carry *P. multocida* and are a common source of introduction of the organism to a premise. Bacterins are protective in hens. The key to successful production of immunity is to use the correct serotype for immunization. Two doses, about 1 month apart, are necessary for adequate protection. Live oral cultures of the *Pasteurella* are available for use in endemic areas; however, use of this product may cause disease in unvaccinated birds.

h. Treatment. Drugs will arrest mortality, but will not cure the disease. Sulfaquinoxaline will work well but will depress egg production. Therapeutic results are obtained with tetracycline at 200 gm per ton (100 mg/lb) of feed for about 2 weeks. However, it is important to note that not all *Pasteurella* isolates are susceptible to the tetracyclines. Improved sanitation, immediate removal of sick and dead birds, and prevention of cannibalism are necessary to control the disease.

i. Research Complications. Research birds affected with *Pasteurella* sp. do not yield reliable research data; termination of the experiment, therefore, is recommended.

J. Fowl Typhoid

a. Etiology. The cause of fowl typhoid is *Salmonella gallinarum;* it also cross-agglutinates with *Salmonella pullorum*. *Salmonella gallinarum* is a gram-negative rod with no spores or capsule.

b. Clinical Signs. Affected chickens have ruffled feathers, pale heads, and shrunken combs. In turkeys, listlessness and a greenish diarrhea are often recognized clinically.

c. Epizootiology and Transmission. Egg transmission via

the yolk or shell surface often perpetuates the disease from dam to offspring.

d. Necropsy Findings. A bile-stained liver, bronzed in appearance and greatly swollen, is the most distinguishing lesion of fowl typhoid. All other lesions are similar to pullorum disease.

e. Differential Diagnosis. This disease is now rare in the United States and must be differentiated from pullorum disease and paratyphoid infections. Isolation and identification of *Salmonella* is necessary for a diagnosis.

f. Prevention. The purchase of pullorum–typhoid-clean chicks from the National Poultry Improvement Program hatcheries is necessary for research.

g. Treatment. No treatment is effective. It is a reportable disease to the state veterinarian, and depopulation will be recommended.

h. Research Complications. *Salmonella*-infected birds should not be used for research purposes.

K. Pullorum Disease

a. Etiology. The cause of pullorum disease is *Salmonella pullorum*. It is a gram-negative bacillus and cross-reacts with *S. gallinarum*.

b. Clinical Signs. Infected adults may show no signs but transmit the agent through their eggs. Young chicks and poults will show white pasting around the vents, and huddling near the heat source. Mortality may reach 90–100%.

c. Epizootiology and Transmission. Egg transmission through the yolk and hatchery transmission via the infected eggs and chicks in the incubators and hatchers are the usual methods of transmitting *S. pullorum*.

d. Necropsy Findings. Myocarditis, pericarditis, and atrophied ovaries are classical lesions of pullorum disease in adults. In young birds, gray nodules may appear on the spleen, peritoneum, lung, liver, heart, and intestine. As with any *Salmonella* infection, cecal cores are commonly seen. An enlarged spleen is present, but this is a nonspecific lesion since it is usually seen with many bacterial infections.

e. Differential Diagnosis. A positive serum agglutination plus isolation, identification, and typing of the agent from the yolk sac, gallbladder, spleen, or cecal tonsil will confirm the diagnosis of *S. pullorum*.

f. Prevention. The purchase of pullorum–typhoid-clean chicks and poults is essential for research.

g. Treatment. No treatment is available. The disease is reportable to the state veterinarian, who when notified will probably insist on depopulation.

h. Research Complications. *Salmonella*-infected birds should not be used for research purposes.

L. Paratyphoid Infection (Salmonellosis)

a. Etiology. Any species of a large group of *Salmonella* sp. may infect birds and mammals and are not host specific. The most common one isolated from birds is *Salmonella typhimurium*. Other common serotypes include *S. enteritis*, *S. oranienberg*, *S. montivideo*, *S. newport*, *S. anatum*, *S. derby*, and *S. bredeny*. Many others exist and may cause mortality.

b. Clinical Signs. Dehydration, pasting of the vents, and huddling near the heat source are signs of a *Salmonella* outbreak. High morbidity and mortality are usually the first signs.

c. Epizootiology and Transmission. Diarrhea in the breeders will cause contamination of the egg shells, and *Salmonella* is often transmitted in this way. Contaminated feed ingredients such as meat scraps may also transmit infection. Transovarian transmission is possible, and human carriers may also be a factor in transmission (See Chapter 22 by Fox *et al.*).

d. Necropsy Findings. Dehydration, enteritis, and focal necrosis of the intestinal mucosa are common lesions. Pigeons usually have joint infections and conjunctivitis. Cheesy cores in the ceca, common in all *Salmonella* infections, will usually be found.

e. Differential Diagnosis. The isolation and identification of the offending bacterium is essential for diagnosis. All *salmonella* infections may have similar signs and lesions. In addition, coccidiosis and blackhead will produce cheesy cores in the cecae.

f. Prevention. Use clean birds free of *Salmonella*. Isolate the new brood and feed only pelleted feed or crumbles to prevent introduction of *Salmonella* via the feed.

g. Treatment. Attempts to treat the infection usually have been discouraging. Most cases do not respond to therapy and carrier birds are produced. Furazolidone in pigeons is partially effective. Sulfa drugs will suppress mortality but will not cure the disease.

h. Research Complications. *Salmonella*-infected birds should not be used for research purposes.

M. Tuberculosis

a. Etiology. Avian tuberculosis is caused by *Mycobacterium avium,* an acid-fast, very resistant bacterium. It is separate and distinct from human and cattle types, but will infect swine and sensitize cattle to the bovine tuberculin test.

b. Clinical Signs. Emaciation in the presence of good feed intake may suggest tuberculosis. Diarrhea and lameness is common. Birds appear very pale, especially the comb and wattles.

c. Epizootiology and Transmission. Transmission is by ingestion of *M. avium.* Association of young chicks with their dams during the growing period in backyard flocks ensures transmission of the bacteria to susceptible birds.

d. Necropsy Findings. Extreme emaciation, nodules along the intestinal tract, and discrete granulomas in liver and spleen are typical of tuberculosis. Granulomas may also be found in the bone marrow and more rarely in the lung.

e. Differential Diagnosis. Coligranuloma can mimic avian tuberculosis. The presence of acid-fast bacilli in liver or spleen impression smears will confirm the presence of *M. avium.*

f. Prevention. An all-in, all-out system of poultry rearing will prevent avian tuberculosis.

g. Treatment. Because of drug resistance and the possible public health significance of this disease, treatment is discouraged. Depopulation is recommended by the state's veterinary authorities.

h. Research Complications. No tuberculosis-affected birds should be used for research purposes.

N. *Mycoplasma gallisepticum* Infection [Chronic Respiratory Disease (CRD) in Chickens: *M. gallisepticum* Infection; PPLO Infection; Infectious Sinusitis of Turkeys] (see Jordan, 1979)

a. Etiology. The primary cause of CRD is *Mycoplasma gallisepticum,* but the following agents will enhance effects of *M. gallisepticum* and cause a complicated respiratory syndrome with high mortality: infectious bronchitis virus, infec- tious laryngotracheitis virus, Newcastle disease virus, *Escherichia coli, Pasteurella multocida, Hemophilus gallinarum,* and others.

b. Clinical Signs. Hens reaching the age of full egg production (35 weeks) will not reach peak egg production referred to as genetic potential for that specific strain. Coughing and sneezing may be noticed, particularly at night. The disease is more pronounced in broilers. Chronic respiratory disease plus secondary bacterial infection, will produce high mortality, high morbidity and high condemnations due to air sacculitis, pericarditis, and perihepatitis. Turkeys may have sinusitis, with exudate in the infraorbital sinuses.

c. Epizootiology and Transmission. Transovarian transmission is the most important method of transmitting *M. gallisepticum.* Lateral spread through aerosols from infected offspring occurs in the hatchers or brooders. Wild birds and fomites can also be a source of agent transmission.

d. Necropsy Findings. Thickened air sacs, mucus in the trachea, sinusitis, and air sacculitis are the common lesions in uncomplicated chronic respiratory disease. In chronic respiratory disease with secondary bacterial infection, the classic lesions include air sacculitis with fibrinous hepatitis and perciarditis.

e. Differential Diagnosis. Many of the respiratory diseases of poultry appear similar and are differentiated with difficulty. A positive *M. gallisepticum* agglutination test in at least ten birds from a previously known *M. gallisepticum* negative flock would strongly suggest a diagnosis of chronic respiratory disease. Histology of the trachea, lungs, and air sacs will show hyperplastic lymphoid follicles if *M. gallisepticum* is present; however, this lesion is not pathognomonic. Isolation of the *Mycoplasma* is the best means of establishing a diagnosis of chronic respiratory disease.

f. Prevention. *Mycoplasma gallisepticum*-free chicks are easy to obtain; therefore use of positive chicks in experimental work is not warranted. Controlled exposure using the live culture of the "Chick F" strain is not recommended except in multi-age, egg-producing flocks. It is not recommended in breeders or experimental trials.

g. Treatment. Improved management and broad-spectrum antibiotics may help control the losses due to secondary bacteria. Tylosin is the most effective drug against *M. gallisepticum.*

h. Research Complications. Positive *Mycoplasma* birds should not be used in research trials.

O. *Mycoplasma meleagridis* Infection

a. Etiology. The cause is *Mycoplasma meleagridis,* an egg-transmitted mycoplasmal agent of turkeys. The usual manifestation is air sacculitis in young poults, but since organisms are shed in semen as well as yolk it can also be considered a veneral disease of this species.

b. Clinical Signs. The signs are usually found only in young growing turkeys. Mild respiratory signs, poor growth, crooked necks, and leg weaknesses are considered signs of this infection.

c. Epizootiology and Transmission. Transovarian transmission is of primary importance. Semen contains the *Mycoplasma* organism, so toms as well as hens serve to transmit *M. meleagridis.* Lateral spread is similar to other mycoplasmas in that stresses and/or other infections seem to enhance the severity of this disease.

d. Necropsy Findings. Air sacculitis is seen in the pipped, unhatched embryos. In young adults, sinusitis, synovitis, and air sacculitis are common.

e. Differential Diagnosis. *Mycoplasma meleagridis* will not affect chickens, but *M. gallisepticum* and *M. synoviae* infections, as well as *M. meleagridis,* will produce similar lesions in turkeys. Negative plate agglutination tests for *M. gallisepticum* and *M. synoviae* helps differentiate the mycoplasmas. However, isolation and identification of *M. meleagridis* is the only way to positively diagnose the disease.

f. Prevention. Commercially reared poults are now available that are claimed to be free of *M. meleagridis.*

g. Treatment. Mortality may be controlled by the use of tylosin. Other antibiotics appear to be of little value in treating an *M. meleagridis* outbreak.

h. Research Complications. Turkeys infected with *M. meleagridis* should not be used for research since birds free of the disease can be purchased.

P. *Mycoplasma synoviae* Infection

a. Etiology. The cause of infectious synovitis is *Mycoplasma synoviae.* It can be isolated in broth media or in 5- to 7-day-old embryonating chicken eggs. Isolates of *M. synoviae* vary in their pathogenicity and also in their susceptibility to drugs.

b. Clinical Signs. Lameness and crouching are the most common signs. Usually morbidity and mortality are low, but retarded growth is a common finding.

c. Epizootiology and Transmission. The transovarian route is the most important means of transmitting the agent from dam to progeny. It is thought that a small number of infected eggs are laid and positive offspring spread infection to pen mates during the growing period.

d. Necropsy Findings. Lesions include yellow exudate in the leg and wing joints but may also be found in the tempormandibular joint, shoulder, and keel bursa. Foot pads are usually swollen and hot. The liver will be greenish due to bile retention. Air sacculitis may also be present.

e. Differential Diagnosis. *Mycoplasma gallisepticum* and *M. meleagridis* must be differentiated from *M. synoviae.* A positive-plate agglutination test for *M. synoviae* will aid in this differentiation.

f. Prevention. *Mycoplasma synoviae*-free flock are available for the supply of clean chicks. *Mycoplasma gallisepticum*- and *M. synoviae*-free stock should always be purchased when available. A low level antibiotic fed continuously will prevent the severe clinical signs and lesions of this disease, but will not eliminate *M. synoviae.*

g. Treatment. It appears that tetracyclines are effective as a treatment with most isolates of *M. synoviae,* but a few isolates are resistant to the tetracyclines and respond to treatment with furazolidone.

h. Research Complications. Only *Mycoplasma*-free chicks should be used for research trials.

Q. Avian Chlamydiosis (Psittacosis; Ornithosis)

a. Etiology. The cause of chlamydiosis is *Chlamydia psittaci.* It is an obligate intracellular organism and can be grown in chick embryos, cell culture, mice and guinea pigs. The host range is wide and includes turkeys, ducks, pigeons, parrots, parakeets, cockatoos, and macaws. Humans are highly susceptible to infection with avian chlamydial agents (See Chapter 22 by Fox *et al.*).

b. Clinical Signs. In mild cases, slight respiratory signs and diarrhea will be seen. In severe turkey cases, depression, weakness, anorexia, and reduced weight are noted. Mortality may reach 25%. In pigeons, conjunctivitis is a feature of the disease, while in the psittacine caged birds, depression, anorexia, diarrhea, rales, and death may be seen.

c. Epizootiology and Transmission. Transmission appears to be by direct contact with infected carriers. Inhalation and ingestion of feces are the two most important means of transmission of *Chlamydia psittaci.*

d. Necropsy Findings. An enlarged congested spleen is almost always found in chlamydiosis in all avian species. Lesions of air sacculitis, pericarditis, and perihepatitis are commonly noted. Unfortunately, these are the same lesions noted in turkeys suffering from mycoplasmosis.

e. Differential Diagnosis. History, clinical signs, and gross lesions aid in the diagnosis. The findings of intracytoplasmic inclusion (LCL bodies) in impression smears of the air sacs or cut surface of the spleen stained by the Macchiavello or Giemsa method will suggest a diagnosis of chlamydiosis. Isolation and identification of the agent should not be attempted by an inexperienced person or without proper hoods and facilities to protect laboratory personnel. This agent, in laboratory culture, is easily transmitted to research technicians.

f. Prevention. Avoid any cross-exposure of research birds to caged pet birds or wild birds. An all-in, all-out system will successfully prevent chlamydiosis. All new arrivals of psittacine birds should be treated for 60 days at 2 gm per gallon of drinking water with tetracycline or with pellets containing an equivalent level of antibiotic.

g. Treatment. Chlamydiosis is a reportable disease and all infected flocks and their treatments should be reported to the proper authorities.

h. Research Complications. No birds infected with *Chlamydia psittaci* should be used for research purposes.

IV. DISEASES CAUSED BY VIRUSES

A. Infectious Bronchitis

a. Etiology. Infectious bronchitis affects only chickens and is caused by a coronavirus. It does not hemagglutinate erythrocytes. Many serotypes exist, but the two most common are Massachusetts and Connecticut.

b. Clinical Signs. In young chicks, a sudden onset of coughing, sneezing, and rales occurs. The birds are weak and crowd toward the source of heat. Feed intake may drop by 50% in broiler-type birds. In adults, severe respiratory signs accompanied by a severe drop in egg production are evident. Egg quality, as evidenced by watery albumen and soft shells with sandpaperlike ends, renders the flock economically useless for egg production.

c. Epizootiology and Transmission. Aerosol transmission over long distances is common. Recovered carriers and contaminated premises may serve as a source of the virus for 30 days or more.

d. Necropsy Findings. Air sacculitis and excessive tracheal mucus are the most common findings. In very young chicks, bronchi can be filled with yellow exudate (Fig. 1).

e. Differential Diagnosis. Isolation of the virus in 9- to 12-day-old chick embryos will confirm the diagnosis. Determination of infectious bronchitis immunity with a serum neutralization test using acute and convalescent serum will also help determine the presence of infectious bronchitis virus.

Fig. 1. Infectious bronchitis. Primary bronchi (arrow) are filled with exudate.

f. Prevention. Isolation of research birds may prevent introduction of infectious bronchitis. However, it is one of the most widespread and contagious diseases of chickens. Vaccines are very effective. They are given at 10 days to 2 weeks of age in the drinking water. Correct serotypes must be selected for the strain of agent present in the area.

g. Treatment. Antibiotics will usually stimulate feed consumption. Increased pen temperature and improved ventilation are helpful in aiding recovery of young birds.

h. Research Complications. Adult laying flocks should be terminated because they will not return to normal egg production.

B. Infectious Bursal Disease (Gumboro Disease)

a. Etiology. Infectious bursal disease is caused by a virus loosely placed in the orbivirus category. It damages primarily B lymphocytes in the bursa of Fabricius, thymus, spleen, and cecal tonsil. Infection with this agent reduces and sometimes totally eliminates humoral immunity.

b. Clinical Signs. The age incidence is usually 3–6 weeks, but it may occur at a later age. Birds are depressed and droopy, with ruffled feathers. Morbidity is high, and mortality also may be high. Vent picking and a bloody diarrhea are sometimes noted.

c. Epizootiology and Transmission. Transmission by direct contact is rapid. The agent is resistant and lives in the environment for long periods of time.

d. Necropsy Findings. The bursa of Fabricius will first enlarge and later atrophy (Fig. 2). Serum-colored edema may be present on the surface of the bursa. The bursa may also contain yellow exudate and hemorrhages. Hemorrhages also may be present in the skeletal muscles, and the kidneys can contain urate crystals.

e. Differential Diagnosis. Signs and lesions provide a provisional diagnosis. A serum neutralization test and histopathology will usually confirm a diagnosis of the infectious bursal disease.

f. Prevention. Drinking water vaccines are available and give adequate protection. A program for highly susceptible, antibody-free, 1-day-old research chicks require vaccination or possibly none at all depending on exposure risk and degree of isolation. Under field conditions, maternal antibodies interfere with vaccines administered early in life. To overcome this, some of the vaccines are quite pathogenic and should not be used in highly susceptible chicks.

g. Research Complications. An outbreak would dictate termination of most experiments, especially when immunologic data is important.

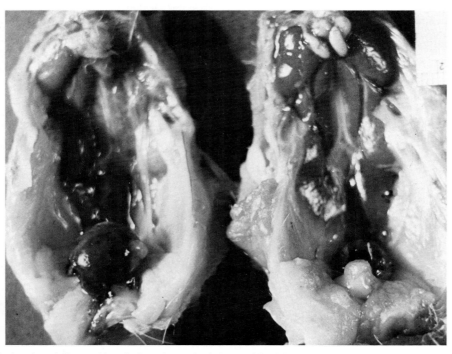

Fig. 2. Infectious bursal disease. Note the large hemorrhagic bursa of Fabricius on the right and the atrophied bursa on the left.

C. Avian Encephalomyelitis (Epidemic Tremor)

a. Etiology. Epidemic tremor is caused by a picornavirus. It will grow in the embryos from a susceptible flock, but will not cause lesions unless it becomes egg "adapted."

b. Clinical Signs. Signs of ataxia, paralysis, and fine tremors of the head and neck will usually be seen between 1 and 3 weeks of age. Turkeys, quail, and pheasants, as well as chickens, may be affected. Growing birds, past 3 weeks of age, show no clinical signs, although recovered birds may sometimes have cataracts.

c. Epizootiology and Transmission. The virus is usually spread through fecal shedding. Adult hens may shed virus in a small percentage of their eggs during infection. Chicks hatch infected with the agent or become infected by fecal contact.

d. Necropsy Findings. There are no gross lesions, but histologically a nonpurulent encephalomyelitis is present. Sections of the brain, proventriculus, and pancreas should be examined microscopically. Central chromatolysis and lymphoid aggregates are seen in the cerebellum and proventriculus, respectively (Butterfield *et al.,* 1969).

e. Differential Diagnosis. Signs, especially head tremors, will suggest avian encephalomyelitis. Histopathology will usually confirm the diagnosis. Direct fluorescent antibody tests, serum neutralization tests, and isolation of the virus are used to diagnose the disease. Avian encephalomyelitis signs are similar to those of Newcastle disease and encephalomalacia. In pheasants, Avian encephalomyelitis must be differentiated from eastern equine encephalomyelitis.

f. Prevention. To protect baby chicks, the dams must be immunized. Both live and inactivated virus vaccines are available.

g. Research Complications. Chicks from immune dams will be satisfactory research birds. However, chicks encountering an outbreak of avian encephalomyelitis are not reliable for research purposes.

D. Eastern Equine Encephalomyelitis (Sleeping Sickness; EEE)

a. Etiology. Eastern equine encephalomyelitis is caused by a togavirus, originally classified as an arbovirus. Wild birds appear to be the natural host of this agent, and serve as a reservoir of infection for man, horses, and swine, in addition to pheasants.

b. Clinical Signs. Pheasants and horses are the primary animals affected, but in epizootics, chickens, wild turkeys, sparrows, domestic turkeys, quail, and doves may be infected. Incoordination, paresis, and progressive paralysis are seen in about 5–20% of a flock. If the flock is debeaked, mortality and morbidity will be markedly reduced. If the flock is not debeaked, feather picking and cannibalism are common.

c. Epizootiology and Transmission. Clinical cases are usually found near swamps and lowland areas, especially during a season with a high mosquito population. Birds are bitten by the mosquito *Culiseta melanura,* whereas the disease in mammals is usually transmitted by *Aedes* and *Mansonia* spp. Horses are a monitor of the impending danger of eastern equine encephalomyelitis to humans, since horses have equal susceptibility but a much greater risk of exposure.

d. Necropsy Findings. No gross lesions are seen; microscopically, there is a nonsuppurative encephalitis characterized by perivascular infiltrates, diffuse gliosis, and vasculitis with vascular infiltrates.

e. Differential Diagnosis. Signs, histopathology, and virus isolation are required for a diagnosis as other viruses may produce encephalitis in domestic fowl. Newcastle disease, avian encephalomyelitis in poults, and Marek's disease in poults and pheasants must also be considered in the differential diagnoses.

f. Prevention. It is important to protect against mosquitoes with screens and sprays and to discourage cannibalism with debeaking and spectacles. Vaccination at 5–6 weeks may reduce mortality, but this method has not been as successful in birds as it has in horses.

g. Research Complications. Research in game birds should be terminated if an outreak of EEE develops.

E. Infectious Laryngotracheitis (ILT)

a. Etiology. Infectious laryngotracheitis (ILT) is caused by a herpesvirus producing type A intranuclear inclusion bodies in epithelium lining the trachea. Similar inclusions are produced in epithelium of the chorioallantoic membrane of experimentally infected fertile eggs. Domestic fowl, pheasants, and pea fowl may be naturally infected (Crawshaw *et al.,* 1982).

b. Clinical Signs. Marked dyspnea with "pump handle" breathing followed by coughing and moist rales are the prominent signs. Shaking of the head and expectoration of blood also are common. Mortality in adults rarely exceeds 10%.

c. Epizootiology and Transmission. Recovered carriers are a common source of infection for susceptible birds. It appears that fomites are more important in ILT transmission than most of the other respiratory diseases.

d. Necropsy Findings. The larynx and trachea are acutely inflamed and sometimes filled with blood clots, or later, caseous cores (Fig. 3). There is usually sinusitis and conjunctivitis. Histopathology of the trachea will reveal intranuclear inclusion bodies in the epithelial tags pulled from the necrotic lining.

e. Differential Diagnosis. Signs, lesions, and virus isolation are the main methods of arriving at a diagnosis. The presence of inclusion bodies in tracheal epithelium are pathognomonic.

f. Prevention. A modified live ILT vaccine is available and is relatively safe. It will spread to susceptible birds in the same pen but usually not to other pens. All birds should be vaccinated at the same time and quarantined for 2–4 weeks (Bryant, 1973).

g. Treatment. For valuable individuals, the tracheal plugs may be removed surgically to prevent dyspnea and death associated with tracheal obstruction.

h. Research Complications. Infectious laryngotracheitis will usually devastate a research trial.

F. Marble Spleen Disease of Pheasants (MSD)

a. Etiology. The cause of marble spleen disease of pheasants is an avian adenovirus (Iltis *et al.*, 1977). It is closely related or identical to the virus of hemorrhagic enteritis of turkeys.

b. Clinical Signs. The disease is peracute with young adults dying after a very short illness. Birds of 4–8 months are more susceptible.

c. Epizootiology and Transmission. Infection by the oral route is most likely.

d. Necropsy Finding. Large, mottled spleens and severe pulmonary edema are the only gross lesions.

e. Differential Diagnosis. Clinical signs and gross lesions suggest the diagnosis. Intranuclear inclusions in the spleen will further support MSD as a diagnosis (Wyand *et al.*, 1972). Agar gel precipitin tests may be used to identify antibodies as an indication of prior infection.

f. Prevention. No commercial vaccine is available, but avirulent strains of hemorrhagic enteritis virus are being utilized in experimental vaccines.

g. Treatment. No treatment is available.

h. Research Complications. This disease is usually found on range flocks. Infected birds would not contribute valid experimental data.

G. Quail Bronchitis (QB)

a. Etiology. The causative agent is an avian adenovirus, the type strain referred to as CELO (chick embryo lethal orphan) virus. It causes very high mortality in young bobwhite quail.

Fig. 3.   Infectious laryngotracheitis in the pheasant. Caseous core occluded the airway.

b. Clinical Signs. In quail under 4 weeks, coughing, sneezing, rales, and huddling are common. Mortality may range from 10 to 100%, but usually ranges between 50 and 60%.

c. Epizootiology and Transmission. Chickens, turkeys, and other birds may be inapparent carriers. Airborne and mechanical transmission are thought to be important means of spread.

d. Necropsy Findings. Tracheal and bronchial mucus is present. Cloudy air sacs, conjunctivitis, and infraorbital sinusitis are quite characteristic lesions of this disease.

e. Differential Diagnosis. Isolation of virus is necessary for a definitive diagnosis. Pulmonary aspergillosis may present similar signs but the microscopic lesion would be a fungal granuloma and easy to differentiate from that produced by quail bronchitis virus.

f. Prevention. Isolation, sanitation, and good husbandry should prevent quail bronchitis.

g. Research Complications. The mortality in quail would cause termination of any research trial.

H. Newcastle Disease (ND; Avian Pneumoencephalitis)

a. Etiology. Newcastle disease is the result of a paramyxovirus infection. There are three major strains as distinguished by their pathogenicity: (a) lentogenic, mild disease, used as vaccine, an example is serotype B-1; (b) mesogenic, moderate disease, domestic Newcastle disease usually seen in the United States; (c) velogenic, very pathogenic, exotic Newcastle disease. Newcastle disease virus hemagglutinates chicken erythrocytes and forms the basis for the hemagglutination inhibition test for anti-NDV antibodies.

b. Clinical Signs. A sudden onset of anorexia and respiratory signs will develop in adults. The eggshells of brown-egg-laying hens will often turn white. Egg production and egg quality will be diminished. In young chicks with no parental immunity, central nervous system signs are seen. Mortality will be high. Torticollis ("star-gazers") is very common. Severe dypsnea, diarrhea, paralysis, and acute death are the signs of velogenic strains. Mortality may reach 100%. Caution must be advised regarding velogenic Newcastle disease since species of pet birds may be inapparent carriers of velogenic strains.

c. Epizootiology and Transmission. Newcastle disease has a wide host range. Many wild birds can transmit Newcastle virus to domestic birds. It affects many wild birds, pet birds, and most game birds. Aerosol transmission is the important means of spread. Contaminated feed, water, and equipment are almost as important. Egg transmission occurs but only for the first few hours during the acute onset of infection in breeders. Velogenic Newcastle disease may be brought into the country with pet birds and fighting cocks. The latter are often smuggled into the country.

d. Necropsy Findings. The mesogenic strains produce lesions characteristic of a tracheitis and a "frothy type" air sacculitis. Velogenic strains, on the other hand, produce massive hemorrhages in all the visceral organs especially the GI tract.

e. Differential Diagnosis. Signs, lesions, and HI tests are usually adequate for diagnosing Newcastle disease. Immunofluorescence tests may be done very quickly to detect the presence of the virus at the time of disease onset. All avian respiratory diseases must be differentiated from Newcastle disease. Histopathology will serve to differentiate the CNS form of Newcastle disease from those produced in avian encephalomyelitis and encephalomalacia.

f. Prevention. Excellent Newcastle disease vaccines are available for immunization of laying hens, breeders, and broilers. One-day-old chicks are given water vaccines and are given boosters several times before sexual maturity.

g. Control. Eradication has been practiced by the United States Department of Agriculture for the velogenic viscerotropic type of Newcastle disease. All Newcastle disease outbreaks should be reported if high mortality is a feature. Pet birds should be kept separate from poultry, turkey, or game bird flocks.

h. Research Complications. A researcher may immunize the flock or raise birds in strict isolation. The latter is ideal but not always successful as Newcastle disease can be transmitted through aerosols. If an outbreak occurs, the trial should be terminated.

I. Pox

a. Etiology. Pox is caused by a large DNA virus. The following commonly encounted strains infect their respective species: fowl pox, turkey pox, pigeon pox, and canary pox. Pox virus is very resistant in the environment and spreads slowly through a flock.

b. Clinical Signs. Two forms exist. Skin pox affects the comb, head, and vent regions; wet pox affects mucous membranes of the larynx and can cause choking. Viremia exists in both forms, and decreased egg production is present for extended periods.

c. Epizootiology and Transmission.   Scabs and skin debris that contaminate the litter feed and water are the usual source of virus. Mosquitoes and cannibalism are secondary means of spread.

d. Necropsy Findings.   Scabs on the comb and other unfeathered areas of the body are suspect pox lesions. Diphtheritic lesions of the pharynx, larynx, and trachea are common in wet pox. The laryngeal opening is usually plugged with a tightly adhered plaque. If cannibalism is a simultaneous problem, the vents may show severe skin necrosis.

e. Differential Diagnosis.   Demonstration of intracytoplasmic inclusion bodies in cutaneous and oral lesions are pathognomonic for pox. Virus inoculation onto the chorioallantoic membrane of embryonating eggs will also produce pocks and cytoplasmic inclusions. Pox must be differentiated from cannibalism, infectious laryngotracheitis and mycoplasmosis. Tracheal plugs are seen in the latter if the strain is highly pathogenic.

f. Prevention.   Vaccination by the wing-web stick method is recommended for pigeons and chickens. Turkeys should be vaccinated by the thigh-stick method. An experimental vaccine is available for canaries. The pigeon pox strain of vaccine is preferred in areas of low incidence, thus preventing spread of the fully virulent fowl pox vaccine to susceptible birds.

g. Research Complications.   It would be unwise to collect data from pox-infected birds.

V. DISEASES CAUSED BY PARASITES

A. Coccidiosis

a. Etiology.   Coccidiosis is caused by a protozoan parasite. There are nine species of coccidia affecting chickens, six affect turkeys, but only three are pathogenic. In geese, three species affect the gastrointestinal tract, and one affects the kidney. The duck and pheasant, each have at least one species that is pathogenic for the gastrointestinal tract. All species in the genus *Eimeria* appear to be host specific with no cross-transfer. Each species has a characteristic size and shape and specific life cycle. Immunity is usually specific for one species only and affords no cross-protection against other species of coccidia.

b. Clinical Signs.   Signs may include pale legs and beaks, ruffled feathers, and listlessness. Birds infected with certain species of coccidia may have bloody fecal droppings. Poor growth and high mortality are the effects of coccidia multiplying in intestinal epithelial cells.

c. Epizootiology and Transmission.   Ingestion of the sporulated oocysts from feces and litter is necessary for transmission. Each gram of feces may contain about 500,000 oocysts. Each sporulated oocyst will produce eight sporozoites, 150 first generation merozoites, 300 second generation merozoites, and will damage about 1,000,000 intestinal host cells.

d. Necropsy Findings.   Table IV illustrates the complexity of diagnosing coccidiosis by gross lesions (see also Fig. 4).

e. Pathogenesis.   Disease is based on the interrelation of age of host, species of coccidia, and dose of organism (degree of exposure). An unsporulated oocyst in the presence of oxygen, 30% moisture, and room temperature will become an infective oocyst in 24–72 hr. The infective oocyst is ingested by a susceptible host, and the life cycle begins. The prepatent period varies by species from 4 to 7 days.

f. Differential Diagnosis.   A differential diagnosis in the chicken and turkey must include various species of coccidia in addition to salmonellosis and histomoniasis. The signs and le-

Table IV

Diagnosing Coccidiosis by Gross Lesion[a]

Bird and species	Area affected	Clinical signs
Chicken		
Eimeria necatrix	Duodenum, jejunum (Fig. 4)	Red and white spots
Eimeria acervulina	Duodenum	White transverse lines
Eimeria mivati	Small intestine, rectum, cecum	Enteritis
Eimeria maxima	Jejunum	Catarrhal enteritis
Eimeria brunetti	Ileum, rectum	Necrotizing enteritis
Eimeria tenella	Cecum	Hemorrhage
Turkey		
Eimeria adenoides	Ileum, rectum, cecum	Catarrhal enteritis
Eimeria meleagrimitis	Small intestine	Catarrhal enteritis
Eimeria gallopavonis	Small intestine	Catarrhal enteritis
Goose		
Eimeria truncata	Kidney	Yellow, pinhead spots
Duck		
Tyzzeria perniciosa	Small intestine	Hemorrhage
Pheasant		
Eimeria phasianus	Cecum	Hemorrhage

[a]From Bryant and Helmboldt (1973).

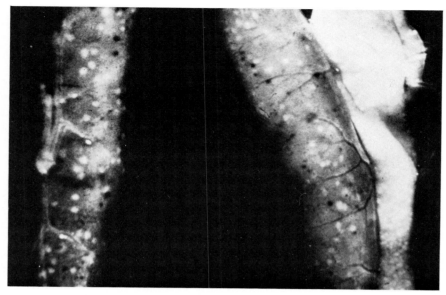

Fig. 4. Coccidiosis (*Eimeria necatrix*). The duodenum and jejunum show discrete black and white serosal foci.

sions of ulcerative and necrotic enteritis are similar to coccidiosis. Direct scrapings of the gut accompanied by histopathology will aid in the diagnosis.

g. Prevention. Prevention of coccidiosis must include reducing the number of infective oocysts. Dry pens and good sanitation practices and/or wire floors will prevent coccidiosis. Immunity is species specific. Vaccines will protect birds but require skillful maintenance of litter at 30% moisture postvaccination to ensure cycling of coccidia for immunization. Coccidiostats are efficacious. Amprolium, coban, and others are available. They are administered continuously at a dose of 0.0125% to allow low level infection but protect against severe disease.

h. Treatment. Amprolium, sulfa drugs, and the nitrofurans have all been used with good success.

i. Research Complications. All birds from 0 to 12 weeks used for research purposes should be fed a low level coccidiostat.

B. Histomoniasis (Blackhead; Infectious Enterohepatitis)

a. Etiology. The cause of histomoniasis is *Histomonas meleagridis,* a protozoan flagellate that inhabits the cecal lumen.

b. Clinical Signs. Turkeys, chickens, and pea fowl with this disease exhibit anorexia, droopiness, listlessness, yellow diarrhea, and cyanosis of the head and face. Morbidity and mortality may be high in young poults.

c. Epizootiology and Transmission. Fresh feces contain the histomonad and may be ingested. More often, the histomonad is contained within the cecal worm (*Heterakis gallinarum*) and the worm is ingested. A third possiblity is *Heterakis* larvae contained within earthworms which in turn may also be ingested.

d. Necropsy Findings. The liver will show discrete, irregular, depressed areas of focal necrosis. The thickened walls of the cecal lumen surround laminated, yellow, caseous cores (Fig. 5).

e. Differential Diagnosis. Hematoxylin and eosin sections of the liver and cecal wall are best for demonstrating the histomonads. Direct smears and scrapings may also reveal the organism.

f. Prevention. The continuous use of antihistomonal drugs is the most effective way to prevent blackhead. Hepzide, histostat and emtryl will all prevent the disease.

g. Treatment. The above drugs serve as effective therapeutic agents at the recommended treatment level.

h. Research Complications. Poults and game birds, especially pea fowl, should be administered anithistomonal drug until 12–15 weeks of age.

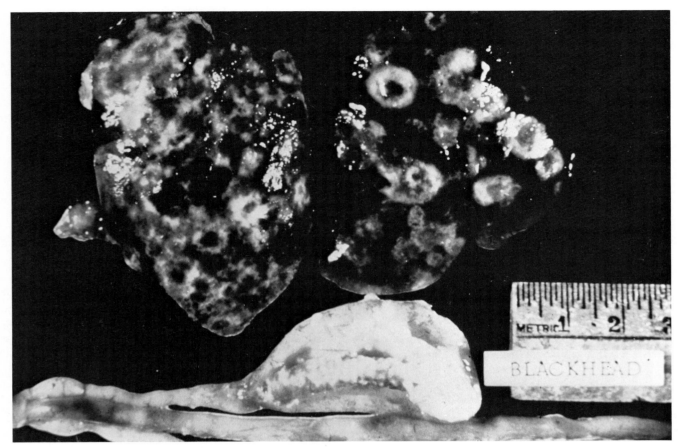

Fig. 5. Histomoniasis (blackhead). Liver shows discrete craterlike necrotic lesions and ceca contains a plug of exudate.

C. Ascaridiasis (Roundworms)

a. Etiology. Ascaridia species are found in most birds. The following species have been reported: *Ascardia galli,* chicken; *Ascaridia numidae,* guinea fowl; *Ascaridia columbae,* pigeon; *Ascaridia dissimilis,* domestic and wild turkeys; *Ascaridia compar,* bobwhite quail.

b. Clinical Signs. Decreased egg production, poor growth and vigor, diarrhea, and visible presence of worms in the feces are often noted.

c. Epizootiology and Transmission. The life cycle is direct. Infective eggs are swallowed and hatch in the proventriculus or anterior portion of the intestinal tract. Larvae live for 9–10 days in the lumen and then enter the mucosal wall and produce damage. On the seventeenth to eighteenth day, young worms again enter the lumen and remain there until maturity, usually 50 days postingestion.

d. Necropsy Findings. Evidence of anemia and retarded growth is seen. The presence of *Ascardia* sp. is the most outstanding feature. A complete intestinal blockage is sometimes present (Fig. 6).

e. Differential Diagnosis. The presence of intestinal roundworms may not be the primary cause of clinical disease, especially when mortality is a common feature. Further examination may reveal capillariasis, coccidiosis, or bacterial infections in combination with ascaridia infestation.

f. Prevention. A complete depopulation and disinfection will usually prevent recurrence of the disease in the next flock. Dirt floors or ranges are impossible to clean and should be rotated annually.

g. Treatment. Piperazine for 24 hr in the drinking water and coumaphos for 18 days in the feed are effective.

h. Research Complications. *Ascardia* sp. usually suggest unsanitary conditions and should therefore be rarely found in properly maintained research facilities.

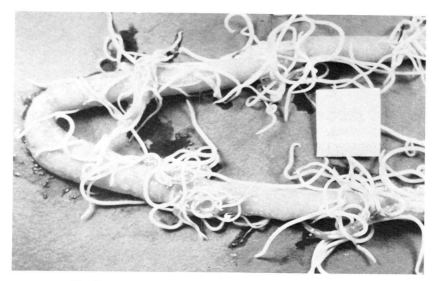

Fig. 6. *Ascaridia galli.* Adult roundworms in the small intestine.

D. Capillariasis (Hairworms)

a. Etiology. *Capillaria obsignata* infests the chicken, turkey, and probably other bird hosts. *Capillaria columbae* is associated with pigeons. In game birds, *C. contorta* is found in the crop and esophagus.

b. Clinical Signs. A sudden drop in egg production and an unthrifty appearance of the flock may often be the result of *Capillaria* infestation.

c. Epizootiology and Transmission. The life cycle is direct. Eight-day embryonated capillaria eggs are ingested by the host, and larvae penetrate the duodenal mucosa. Adult worms are found within 18–20 days postingestion of embryonated eggs (Fig. 7).

d. Necropsy Findings and Diagnosis. Slight enteritis may be present. Nematodes washed from mucosal scrapings will confirm the diagnosis.

e. Prevention. Clean facilities will prevent *Capillaria* infestations.

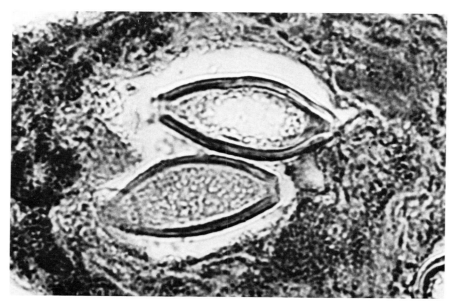

Fig. 7. *Capillaria* eggs. These are located in the crop and are identified by the double operculum.

f. Treatment. Meldane in the feed for 18 days is the usual treatment. Additional vitamin A (12,000 IU/lb feed) may be beneficial in the healing of damaged gut epithelium.

g. Research Complications. *Capillaria* sp. are not usually encountered in a clean research facility.

E. *Syngamus trachea* Infestation (Gapeworms)

a. Etiology. *Syngamus trachea* is a nematode that causes severe respiratory disease in game birds, especially pheasants and quail.

b. Clinical Signs. Tracheal lumen obstruction by adult worms causes dyspnea and death.

c. Epizootiology and Transmission. The life cycle is direct but may also involve earthworms and snails as vectors.

d. Necropsy Findings and Diagnosis. Obstruction of the trachea with worms occluding the air passage usually confirms the diagnosis.

e. Prevention and Treatment. The continuous feeding of thiabendazole has been very effective. More recently, levamizole hydrochloride in the water has been used with success as a treatment. Rotating ranges for pheasants will also help minimize infection.

f. Research Complications. Game birds maintained for research should be fed a preventive level of thiabendazole.

VI. DISEASES CAUSED BY FUNGI/YEASTS (see Chute *et al.*, 1962)

A. Aspergillosis (Brooder Pneumonia)

a. Etiology. *Aspergillus fumigatus* is the causative agent and is easily isolated on Sabouraud's agar.

b. Clinical Signs. Gasping and deep, labored breathing are commonly noted. Ataxia may be seen in a few chicks or poults when the CNS is involved. Blindness may develop because of the infection localizing in the anterior chamber of the eye.

c. Epizootiology and Transmission. Spores of *A. fumigatus* are found in great numbers in old litter. When the spores are inhaled, they produce typical respiratory signs and lesions.

d. Necropsy Findings. Yellow plaques are found in the lungs, air sacs, tracheas, and peritoneal surfaces (Fig. 8). The brain also may contain foci of fungal growth. Histopathology will reveal the presence of hyphae in granulomas stained with periodic acid–Schiff or Gridley stain.

e. Differential Diagnosis. Signs and typical histologic lesions of lung and air sacs will confirm the presence of the *Aspergillus* spp.

f. Prevention. Use clean, dry litter. Wet sawdust may contain numerous spores of *Aspergillus fumigatus,* and its use is contraindicated.

g. Treatment. Commercial poultry and turkey flocks are not usually treated. Individual, valuable birds may be treated with nystatin or amphotericin B.

h. Research Complications. Flocks infected with aspergillosis should be depopulated.

B. Candidiasis (Moniliasis; Crop Mycosis)

a. Etiology. *Candida albicans,* a yeastlike fungus, is the causative agent of crop mycosis. It grows well on Sabouraud's agar at room temperature. It is cautioned, however, that *C. albicans* can be a normal floral inhabitant of the crop and lower digestive tract.

b. Clinical Signs. Depressed egg production and diarrhea are the two most commonly observed signs.

c. Epizootiology and Transmission. Unsanitary waterers and warm humid weather will often combine to produce the necessary environmental conditions for the development of disease. Indiscriminate use of antibiotics may alter normal bacterial flora allowing the yeast to multiply and ellicit disease.

d. Necropsy Findings. The crop and esophagus are covered with a diffuse, white exudate. Often the pharyngeal mucosa is involved. Enteritis with a white, pseudomembrane covering duodenal and jejunal mucosa may sometimes be found.

e. Differential Diagnosis. Culture of the yeast and histopathology will confirm the diagnosis.

f. Prevention. Optimum sanitation of waterers and feeders especially during hot, humid weather is necessary. Disinfectants such as chlorine, iodine, and quaternary ammonium compounds are needed to keep the yeasts and fungi from multiplying in the drinking water.

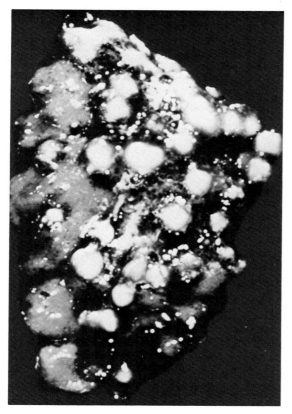

Fig. 8. Aspergillosis. Granulomas on the exterior and cut surfaces of the duck lung are shown.

g. Treatment. Nystatin will effectively handle most outbreaks.

h. Research Complications. Research facilities practicing proper sanitation should not experience this disease in birds.

VII. DISEASES RESULTING IN NEOPLASTIC CHANGES

A. Marek's Disease (MD; Range Paralysis)

a. Etiology. Marek's disease is caused by a herpesvirus. This virus produces inflammatory and neoplastic changes in the CNS, various peripheral nerves, skin, and visceral organs.

b. Clinical Signs. Since Marek's disease may affect any organ system of the body, signs are variable. Paralysis of the wings and legs are common. A gray eye with irregular pupil size is a feature of the ocular form of Marek's disease. Feather follicles of the skin from the leg and breast region are swollen.

Depression and emaciation characterize visceral involvement. Palpable muscle tumors may also be noted.

c. Epizootiology and Transmission. The virus is released in association with dander from feather follicles, which then contaminates the environment.

d. Necropsy Findings. Tumors of various visceral organs, skin, or muscle may be seen. Swollen sciatic and brachial nerves are common. Microscopic lesions feature pleomorphic cellular infiltrates of plasma cells, macrophages, lymphocytes, and lymphoblastic cells. Nerves may have inflammatory or proliferative lesions or a combination of both types.

e. Differential Diagnosis. Marek's disease must be differentiated from lymphoid leukosis (LL). For many years, both these diseases were thought to be caused by a common agent. Marek's disease occurs prior to sexual maturity, but both may occur after sexual maturity. Only Marek's disease affects the nerves and thus causes paralysis. Both produce tumors and enlarged visceral organs. Gray eyes, feather follicle inflammation, and muscle tumors are found only in Marek's disease. Tumors of the bursa of Fabricius are rare in MD, but are an invariable finding in LL.

f. Prevention. Depopulation and thorough disinfection before introducing a new flock is recommended. Baby chicks should be vaccinated at 1 day of age with the nononcogenic turkey herpesvirus if it is not possible to provide strict isolation. Genetic resistance varies among strains of chickens, and is effective in reducing losses in Marek's disease virus-exposed birds.

g. Treatment. No treatment is available.

h. Research Complications. Herpesvirus of turkeys (HVT) Marek's vaccine should be administered to all birds at 1 day of age to prevent the disease.

B. Lymphoid Leukosis (LL; Big Liver Disease)

a. Etiology. Lymphoid leukosis is caused by an oncogenic retrovirus. It is a member of the leukosis–sarcoma group. Subgroups A and B occur in commerical chickens. Various virus strains may cause a variety of neoplasms of mesodermal tissues in addition to typical LL.

b. Clinical Signs. Sudden death of adult chickens may be the only sign. An enlarged abdomen due to a swollen liver is also a common finding. Thickening of the leg bones as a result of osteopetrosis is another possible consequence of infection with some strains of leukosis virus. Other lesions may include erythroid or myeloid leukemias, hemangiomas, or nephroblastomas.

c. *Epizootiology and Transmission.* Egg transmission is the most important means of introducing the virus into a flock. Lateral spread from infected saliva and feces then ensues. Early infection is generally required for tumors to develop.

d. *Necropsy Findings.* An enlarged liver, sometimes significantly displacing other abdominal organs, is a common finding. Tumors in the spleen and kidney are also seen. The bursa of Fabricius will usually reveal a discrete, nodular tumor that constitutes the primary lesion. Other sites are thought to become involved as the result of metastases from the bursa.

e. *Differential Diagnosis.* Lymphoid leukosis does not develop until 4–5 months of age. In contrast, Marek's disease may be seen as early as 3–4 weeks of age. Both diseases can simultaneously occur in the same bird. The presence of a bursal tumor frequently indicates LL, whereas nerve, skin, or muscle involvement indicates MD.

f. *Prevention.* Lymphoid leukosis-free chicks are available. A rigid isolation rearing system should be followed.

g. *Treatment.* No treatment is available.

h. *Research Complications.* If birds are not kept until the age of sexual maturity, little or no LL should be encountered.

VIII. DISEASES CAUSED BY NUTRITIONAL DEFICIENCIES

A. Vitamin A Deficiency

a. *Etiology.* Vitamin A is necessary for growth, normal vision, and integrity of epithelium lining the respiratory, digestive, urinary, and genital tracts. Severity of deficiency signs and age of incidence will vary with the vitamin A levels in the feed and the maternal reserve of vitamin A passed onto the chick. Breeders on adequate vitamin A diets store this vitamin in the egg yolk, which provides the chick with sufficient vitamin A to last for several weeks. Chicks hatched with marginal reserves of vitamin A and placed on a deficient diet will have clinical signs by 7–10 days of age. However, if chicks are hatched from hens receiving adequate vitamin A, they have no signs or lesions until about 6 weeks.

b. *Signs.* Signs of vitamin A deficiency include pale head, muscles of the thigh, and skin. Weakness, lethargy sometimes

associated with ataxia, and lacrimation with conjunctivitis are other signs noted. Decreased sperm counts and reduced sperm motility are reported to be a common finding in vitamin A-deficient roosters. Cartilage and bone development may be depressed. The incidence and severity of blood spots in eggs are increased as dietary vitamin A is reduced. Recovery from intestinal diseases is improved when the level of vitamin A in the feed is increased.

c. *Gross Lesions.* Lesions of vitamin A deficiency consist of small white pustules in the nasal passages, mouth, esophagus, pharynx, and crop. In addition, there may be marked accumulation of urates in the renal tubules.

d. *Histopathology.* Histopathology reveals cytoplasmic atrophy and loss of cilia of the respiratory tract epithelium. There is marked karyorrhexis of nuclei. Chronic vitamin A deficiency results in squamous metaplasia in epithelial lining of the nasal cavities, trachea, bronchi, and submucous glands (Bryant and Helmboldt, 1974).

e. *Prevention.* The normal requirement for vitamin A in chicks from 1 day old to 8 weeks of age is 5000 IU/lb of feed. For growing chickens and turkeys (8–22 weeks of age), 3000 IU/lb of feed is recommended. For laying hens, 4000 IU/lb of feed, and for breeding hens and turkeys, 5000 IU/lb of feed is needed. Five thousand IU/lb feed is also recommended for pheasants, pigeons, and quail.

f. *Treatment.* Treatment levels for vitamin A deficiency is 5000 IU/lb of feed.

B. Vitamin D Deficiency

a. *Etiology.* Vitamin D is needed for metabolism of calcium and phosphorus in the formation of bones, egg shells, beaks, and claws. Vitamin D stimulates the gastrointestinal absorption of calcium. Serum alkaline phosphatase is elevated when vitamin D-deficient diets are fed even in the presence of sufficient calcium and phosphorus.

b. *Signs.* Retarded growth, poor feathering, and soft bones (rickets) are features of vitamin D deficiency in young chicks 2–3 weeks of age. The chicks are unsteady and walk only a short distance before falling, or affected chicks may sit on their hocks and rest for several minutes (Bryant, 1972). Adults will lay a large percentage of thin-shelled and soft-shelled eggs leading eventually to a decrease in total egg production. Signs from a deficiency of vitamin D usually occur about 2 months after vitamin D is withheld. Hatchability is also reduced.

c. Lesions.   Characteristic lesions in young birds include soft pliable bones, beaded ribs, and S-shaped keel bone. The beak and long bones can be bent without breaking. An indented rib cage is evident and causes pressure on the lungs and the heart. In adults, the lesions seen in young birds are present but usually to a lesser degree. Pathological fractures of the ribs and vertebrae may be a feature. The histological features of vitamin D deficiency include an enlarged parathyroid gland with a diffuse hyperplasia. A decrease of normal calcified bone with an excess of osteoid tissue in the long bones is also seen. Poor calcification is best identified at the epiphysis of the tibia or femur.

d. Prevention.   Turkeys of all ages should receive vitamin D in the amount of 600 IU/lb of feed. Hens and young chicks require 500 IU/lb, while growing chickens, 8–22 weeks of age, should have 300 IU/lb of feed. Pheasants, quail, and pigeons will grow well on 600 IU/lb of feed.

e. Treatment.   The feeding of a one-time, very high dose of Vitamin D_3, approximately 15,000 IU/lb of feed, will provide the most effective therapy. This should be given in a single dose, as hypervitaminosis D can result. Dystrophic calcification of the aorta and certain arteries, in addition to kidney tubules, is one complication of hypervitaminosis D. Vitamin D_2 (irradiated ergosterol) is a poor form of vitamin D for birds and can even be toxic at high levels.

C.   Vitamin E Deficiency

a. Etiology.   Vitamin E is necessary for normal egg production, fertility, and hatchability. Multiple problems result if a deficiency of vitamin E occurs in birds.

i. Encephalomalacia ("crazy chick disease").   This condition is a vitamin E deficiency characterized by ataxia, backward retraction of the head, increased incoordination, and death. It is usually seen between 2 and 4 weeks of age. If the breeder flock is deficient in vitamin E, signs may be seen during the first week posthatching. Grossly, the cerebellum appears wet with petechial hemorrhages. Histologically, ischemic necrosis as a result of capillary thrombosis is the main feature. This disease must be differentiated from avian encephalomyelitis (AE) in which the neurons show central chromatolysis. This condition is characterized by cells appearing pale in the center due to loss of the Nissl substance (Helmboldt), 1972).

ii. Exudative diathesis.   This condition is characterized by subcutaneous blood-tinged edema of young birds caused by an abnormal permeability of capillary walls. It appears that tissue peroxides increase causing damage to capillary membranes. Vitamin E and glutathione peroxidase, a selenium-containing enzyme, protect the capillary endothelium against damage by peroxides, thus explaining the dual role of vitamin E and selenium in preventing exudative diathesis.

iii. Nutritional muscular dystrophy.   Dystrophic muscles in chickens, ducklings, and turkeys at approximately 4 weeks of age is caused by vitamin E deficiency, as well as by a deficiency of sulfur-containing amino acids. It is primarily seen in breast muscle but may be found in any of the skeletal muscles. Grossly, there are separated muscle fiber bundles that appear light in color. The microscopic lesions consist of typical Zenker's necrosis, which include hyaline degeneration, proliferation of muscle nuclei and fibroblasts, disruption of muscle fibers, and edema containing heterophiles. Vitamin E deficiency and selenium deficiency in the turkey may show heart and gizzard myopathy.

iv. Enlarged hocks.   Turkeys on a vitamin E-deficient diet have enlarged hocks and bowed legs at 2–3 weeks of age. The same condition is produced by a deficiency of phosphorus, choline, glycine, nicotinic acid, zinc, and biotin.

b. Prevention.   The recommended level of vitamin E for chickens and laying hens is 5 IU/lb of feed; for breeding hens 7.5 IU/lb is recommended. Poults from 0 to 16 weeks require 7 IU/lb, and breeder turkeys need 15 IU/lb of feed.

c. Treatment.   Therapeutic levels of vitamin E and selenium, 10 IU/lb and 0.1 ppm, respectively, will prevent encephalomalacia, exudative diathesis, and gizzard myopathy.

D.   Vitamin K Deficiency

a. Etiology.   Vitamin K is needed to synthesize prothrombin, an important component of the blood clotting mechanism. If vitamin K is deficient, prolonged blood clotting time is noted.

b. Signs and Lesions.   Diets deficient in vitamin K can cause cutaneous hemorrhage of the legs, wings, and breast in addition to hemorrhage into body cavities. Anemia is often seen due to blood volume loss and a hypoplastic nonregenerative bone marrow. Embryos will die during egg incubation from hemorrhages.

c. Prevention.   Vitamin K deficiency is rare and hard to reproduce. The dietary requirement for all birds at any age is 1 mg/lb of feed.

d. *Treatment*. A normal clotting time returns within 5 hr after treatment with menadione sodium bisulfite. Cutaneous hemorrhages and anemia usually take several days to resolve once adequate vitamin K levels are fed.

E. Thiamin Deficiency (Vitamin B₁)

a. *Etiology*. Thiamin is important in carbohydrate metabolism. Signs of thiamin deficiency include polyneuritis, extreme anorexia, and death.

b. *Signs and Lesions*. Vitamin B₁ deficiency can be seen in chicks and adults. Usual signs are weight loss, ruffled feathers, leg weakness, and unsteady gait followed by muscle paralysis. Due to anterior neck muscle paralysis, retraction of the head results in a typical posture referred to as ''star gazing.''

c. *Treatment*. The requirement of thiamin is 1 mg/lb of feed for birds of all ages. Oral therapy with thiamin results in a rapid clinical recovery if the birds are eating; however if anorexia is present, force feeding or injection of thiamin may be necessary.

F. Riboflavin Deficiency (Vitamin B₂)

a. *Etiology*. Riboflavin is an important component of many enzymes associated with oxidation–reduction reactions and cell respiration.

b. *Signs and Lesions*. Signs in chicks may include poor growth, weakness and emaciation, diarrhea, and refusal to walk. Appetite is not affected. Toes will curl medially (Bryant, 1972). In laying hens decreased egg production and hatchability occurs. Increased fat content of the liver is also noted. Embryos are dwarfed and edematous. In turkeys, riboflavin deficiency results in a dermatitis with defective down feathers similar to pantothenic acid deficiency in chickens. Microscopic nerve lesions in chicks show degenerative changes in the myelin nerve sheath.

c. *Prevention*. The normal requirement for riboflavin in all birds is 2 to 2.5 mg/lb of feed.

d. *Treatment*. The administration of adequate riboflavin will usually cure the deficiency rapidly.

G. Pantothenic Acid Deficiency

a. *Etiology*. Pantothenic acid is part of coenzyme A, which is involved in the metabolism of carbohydrates, pro-

teins, and fats. It is required for hatchability and will prevent edema and subcutaneous hemorrhages in the embryos.

b. *Signs and Lesions*. Changes noted in pantothenic acid deficiency include a severe dermatitis, perosis, broken feathers, poor growth, and death. Poor hatchability with the embryos dying during the last 3 days of incubation is typically noted.

c. *Prevention and Treatment*. Six to 7 mg/lb of feed is needed for a normal diet. For therapeutic effect, 10–12 mg/lb of feed administered for a few days either orally or by injection is needed.

H. Biotin Deficiency

Biotin deficiency causes a dermatitis around the beak, eyelids, and feet. Biotin deficiency also can cause perosis. Approximately 0.1 mg/lb of feed will prevent the dermatitis and perosis caused by biotin deficiency.

I. Vitamin B₁₂ (Cobalamin) Deficiency

a. *Etiology*. Vitamin B₁₂ is involved in nucleic acid synthesis, methyl synthesis, and metabolism of carbohydrates and fats.

b. *Signs and Lesions*. Signs include retarded growth, poor feed efficiency, mortality, small egg size, and reduced hatchability. Vitamin B₁₂-deficient embryos may have hemorrhages, edema, perosis, and fatty livers.

c. *Prevention and Treatment*. The inclusion of 0.005 mg/lb of feed will prevent deficiency of Vitamin B₁₂. Therapy includes the addition of 4 mg/ton of feed into the breeding ration.

J. Calcium and Phosphorus Deficiency

a. *Etiology*. Calcium is essential for bone formation, eggshell production, normal blood clotting, normal striated muscle, and maintenance of acid–base balance. Phosphorus is required for carbohydrate and fat metabolism, calcium transport, bone formation, and eggshell production.

b. *Signs and Lesions*. Calcium and/or phosphorus deficiency will cause signs and lesions of rickets, decreased egg production, and increased numbers of soft shelled and shellless eggs (also see Section VII,B on vitamin D deficiency).

c. Prevention and Treatment. The recommended levels are as follows: for nonlaying chickens, 1% calcium and 0.5% available phosphorus; for laying and breeding hens, 3.5% calcium and 0.5% available phosphorus; for nonlaying turkeys, 1.5% calcium and 0.7% phosphorus; and for laying turkeys, 2.0% calcium and 0.6% phosphorus. The above levels used therapeutically will correct any deficiency.

K. Sodium and Chlorine Deficiency

a. Etiology. Sodium is found in the blood and body fluids and is associated with regulation of the hydrogen ion concentration of blood. It is also necessary for normal physiological activity of the heart.

b. Signs. Poor growth, softening of the bones, corneal keratinization, gonadal inactivity, decrease in cardiac output, hypotension, hemoconcentration, and uremia are signs of sodium deficiency. In adults, a sodium deficiency results in decreased egg production and egg size in addition to loss of weight and cannibalism.

c. Prevention and Treatment. Levels of 0.15% sodium and 0.15% chlorine are recommended for optimum growth and egg production. Salt toxicity develops at a level of 4 gm/kg of body weight. Intense thirst, weakness, convulsions, and death are the signs associated with salt toxicity.

L. Manganese Deficiency

a. Etiology. Manganese is essential for growth, egg production, and prevention of perosis.

b. Signs. Results of a manganese-deficient diet include poor quality eggshells, thickened and shortened leg bones, lowered hatchability, and chondrodystrophy in embryos.

c. Prevention and Treatment. The normal requirement of manganese is 50–60 ppm. This level is also used therapeutically to treat manganese-deficient birds.

M. Iodine Deficiency

a. Etiology. Iodine is necessary for the normal function of the thyroid gland. Thyroxine is 65% iodine and regulates body metabolism. With iodine deficiency, the thyroid gland enlarges and is referred to as "goiter."

b. Prevention and Treatment. The recommended level of iodine is 0.35 ppm. Iodized salt provides an adequate source of supplementary iodine.

SELECTED REFERENCES

Bryant, E. S. (1972). Differential diagnoses in avian medicine, I. Diseases of the central nervous system. *Proc. 76th Annu. Meet., U.S. Anim. Health Assoc.* p. 509.

Bryant, E. S. (1973). A program for eradication of infectious laryngotracheitis from New England poultry flocks. *Proc. 77th Annu. Meet., U.S. Anim. Health Assoc.* p. 242.

Bryant, E. S., and Helmboldt, C. F. (1973). Differential diagnosis in avian medicine. II. Diseases of the digestive system. *Proc. 77th Annu. Meet., U.S. Anim. Health Assoc.* p. 543.

Bryant, E. S., and Helmboldt, C. F. (1974). Differential diagnoses in avian medicine. III. Diseases of the respiratory system. *Proc. 17th Annu. Meet., Am. Assoc. Vet. Lab. Diagn.* p. 43.

Bryant, E. S., Anderson, C. R., and van der Heide, L. (1973). An epizootic of eastern equine encephalomyelitis in Connecticut. *Avian Dis.* **17,** 861–867.

Butterfield, W. K., Helmboldt, C. F., and Luginbuhl, R. E. (1969). Studies on avian encephalomyelitis. IV. Early incidence and longevity of histopathologic lesions in chickens. *Avian Dis.* **13,** 53–57.

Chute, H. L., O'Meara, D. C., and Barden, E. S. (1962). "A Bibliography of Avian Mycosis," Misc. Publ. 655. Dept. Anim. Vet. Sci., Orono, Maine.

Crawshaw, G. J., and Boycott, B. R. (1982). Infectious laryngotracheitis in pea fowl and pheasants. *Avian Dis.* **26,** 397–401.

Frazier, M. N., Parizek, W. J., and Garner, E. (1964). Gangrenous dermatitis of chickens. *Avian Dis.* **8,** 269–273.

Helmboldt, C. F., and Bryant, E. S. (1971). The pathology of necrotic enteritis in domestic fowl. *Avian Dis.* **15,** 775–780.

Helmboldt, C. F. (1972). Histopathologic differentiation of diseases of the nervous system of the domestic fowl (*Gallus gallus*). *Avian Dis.* **16,** 229–240.

Hitchner, S. B., Domermuth, C. H., Purchase, H. G., and Williams, J. E. (1975). "Isolation and Identification of Avian Pathogens." Arnold Printing Corp., Ithaca, New York.

Hofstad, M. S., Calnek, B. W., Helmboldt, C. F., Reid, W. M., and Yoder, H. W., Jr. (1978). "Diseases of Poultry." Iowa State Univ. Press, Ames.

Iltis, J. P., Daniels, S. B., and Wyand, D. S. (1977). Demonstration of an avian adenovirus as the causative agent of marble spleen disease. *Am. J. Vet. Res.* **38,** 95–100.

Jordan, F. T. W. (1979). Avian microplasmas. *In* "The Mycoplasmas" (J. G. Tully and R. F. Whitcomb, eds.), Vol. II, pp. 1–48. Academic Press, New York.

Lucas, A. M., and Jamroz, C. (1961). *U.S. Dep. Agric., Agric. Monogr.* **25.**

National Research Council (1977). "Nutrient Requirements of Poultry," 7th ed., no. 1. National Academy of Sciences, Washington, D.C.

Olson, C. (1937). Variations in the cells and hemoglobin content in the blood of the normal domestic chicken. *Cornell Vet.* **27,** 235–263.

Panigraphy, B., and Glass, S. E. (1982). Outbreaks of fowl cholera in quail. *Avian Dis.* **26,** 200–203.

Petrak, M. L. (1982). "Diseases of Cage and Aviary Birds." Lea & Febiger, Philadelphia, Pennsylvania.

Schwartz, L. D. (1977). "Poultry Health Handbook." Pennsylvania State University, Univeristy Park.

Scott, M. L., Nesheim, M. C., and Young, R. J. (1976). "Nutrition of the Chicken." M. L. Scott and Associates, Ithaca, New York.

Somes, R. G. (1981). "International Registry of Poultry Genetic Stocks." Bull. 460. Storrs Agric. Exp. Stat., Storrs, Connecticut.

Steiner, C. V., and Davis, R. B. (1981). "Caged Bird Medicine." Iowa State Univ. Press, Ames.

Sturkie, P. D. (1976). "Avian Physiology." Springer-Verlag, Berlin and New York.

Whiteman, C. E., and Bickford, A. A. (1983). "Avian Disease Manual" Colorado State University, Fort Collins.

Woodard, A. E., Abplanalp, H., Wilson, W. O., and Vohra, P. (1973). "Japanese Quail Husbandry in the Laboratory." Bull. Dept. Avian Sci., Univ. Calif., Davis.

Wyand, D. S., Jakowski, R. M., and Burke, C. N. (1972). Marble spleen disease in ring-necked pheasants: Histology and ultrastructure. *Avian Dis.* **16,** 319–329.

Chapter 14

Biology and Diseases of Amphibians

Miriam R. Anver and Cynthia L. Pond

I. INTRODUCTION

Amphibians have long been used in biological education and research. "Amphibian" means "double life," referring to the ability of these cold-blooded (ectothermic) vertebrates to live both in water and on land during different parts of their lives. They exhibit an aquatic, gill-breathing phase before metamorphosis into land-dwelling lung-breathers. Their habitats as adults can vary from aquatic to arboreal, but most return to water to breed.

II. BIOLOGY

A. Taxonomy

The class Amphibia is divided into three major orders: Gymnophiona (Apoda), Caudata (Urodela), and Salientia (Anura). Members of Gymnophiona are slender wormlike burrowers without limbs or bony girdles. They have greatly reduced eyes that are often covered by skin or bone. Small scales are embedded in their skin and, as in snakes, Gymnophiona have a reduced left lung. They are found primarily in tropical or warm temperate regions and are not widely used in research (Cochran, 1961; ILAR, 1974).

The order Caudata is composed of the urodeles or salamanders, which are found in all temperate and tropical areas but show the greatest speciation in Appalachia. These amphibians have slim bodies and tails and two or four limbs. Most are terrestrial, living in litter, trees, or caves. Some members of this order remain aquatic or return to water as newts. This includes species that fail to metamorphose; their larvae grow to a large size, retain gills, and breed without attaining an adult stage. This process is called neoteny. *Ambystoma* (whose land forms are called axolotls) and *Necturus* (mudpuppy) exhibit this pattern of development (Cochran, 1961; ILAR, 1974).

Members of the largest order, Salientia (or anurans), which includes frogs and toads, range through nearly all temperate and tropical lands. These animals have well-developed hindlimbs and pelvic girdles, which are well suited for swimming, jumping, or crawling. *Xenopus laevis,* the African clawed toad, more properly called a platanna, is highly aquatic, with a streamlined body, muscular legs, and webbed toes for swimming. Its name derives from the presence of small curved black claws on the inner three toes of its hind feet. It is a popular research animal. It feeds by fanning food into its mouth or by using its fingers; members of this family, Pipidae, lack tongues. In dry hot months it may estivate in a mucus-lined cocoon in the caked mud of dry lake beds. When rains return, it is stimulated to come out of estivation and renew feeding and reproduction (Cochran, 1961; ILAR, 1974).

The amphibians most frequently used for research and teaching are anurans of the family Ranidae, the true "frogs." They have bullet-shaped bodies, protruding eyes, large eardrums, and well-developed legs for leaping (Cochran, 1961). They are semiterrestrial, preferring shallow water and swampy habitats. Some species spend little time in the water except to breed. Wide differences in body size, coloration, and patterning are observed among members of this family. The most popular American anuran in research is the leopard frog. *Rana pipiens,* an inhabitant of the northern tier of states, is the leopard frog most frequently named; however eight leopard frog species are now recognized. These species come from different, sometimes overlapping, geographic areas and are not functionally equivalent. *Rana catesbeiana,* the bullfrog, is more aquatic in nature. It is the largest native anuran in North America, measuring a maximum 8 in. from snout to vent. Both *Rana pipiens* and *Rana catesbeiana* have been bred and raised in the laboratory for research use (ILAR, 1974).

The true toads, of the family Bufonidae, have large bodies and short legs. Their slow gait makes them easy prey. To some extent their parotoid skin glands, which emit poisonous secretions when they are frightened or injured, provide protection against predators. *Bufo marinus,* the giant or marine toad, *Bufo*

bufo, the European or common toad, and *Bufo americanus,* the American toad, have been used as research animals. *Bombina orientalis,* the Asiatic fire-bellied toad, has been raised in the laboratory for research use (Cochran, 1961; ILAR, 1974).

B. Use of Amphibians in Research

Amphibians have been used in biological education for well over a century (Culley, 1973). Many experimenters have also chosen amphibians as research animals. Table I lists species commonly used in research and their scientific and common names. A study by Nace (1970) detailed sources of support of research using urodeles and anurans, where the research took place, and the topics of the research. In addition to their classic use in developmental biology, a few of the present research areas involving amphibians include comparative, developmental, and transplantation immunology (Barlow *et al.,* 1981; Cohen *et al.,* 1980; DuPasquier *et al.,* 1979; Horton and Horton, 1975; Katagiri, 1978; Rollins and Cohen, 1980); susceptibility to toxicants (Slooff and Baerselman, 1980); teratogen screen-

Table I

Species of Amphibians Commonly Used in Research[a]

Scientific name	Common name
1. Caudata (urodeles)	
Ambystoma maculatum	Spotted salamander
Ambystoma mexicanum	Mexican axolotl
Ambystoma opacum	Marbled salamander
Ambystoma tigrinum (eastern subspecies)	Tiger salamander
Ambystoma tigrinum (western subspecies)	Axolotl
Necturus maculosus	Mudpuppy
Notophthalmus viridescens	Red-spotted or common newt
Salamandra salamandra	Fire (European) salamander
Taricha granulosa	Rough-skinned newt
Triturus vulgaris	Smooth newt
2. Salientia (anurans)	
Bombina orientalis	Asiatic fire-bellied toad
Bufo americanus	American toad
Bufo bufo	European or common toad
Bufo marinus	Giant or marine toad
Rana catesbeiana	Bullfrog
Rana clamitans	Bronze or green frog
Rana grylio	Pig frog
Rana palustris	Pickerel frog
Rana pipiens	Leopard frog
Rana sylvatica	Wood frog
Rana temporaria	Common or grass frog
Xenopus laevis	South African clawed toad (platanna)

[a]Nace *et al.* (1974); ILAR (1979).

ing (Thant-Shima *et al.*, 1979); limb regeneration (Cochran, 1961; Stock and Bryant, 1981); osmoregulation (Shoemaker and Nagy, 1977); physiology and endocrinology of amphibian metamorphosis (King and Millar, 1981; Platt and LiCause, 1980; Sawin *et al.*, 1978; Wright *et al.*, 1979); embryology; and hormone assays (Hobson, 1965).

C. Sources of Amphibians for Research

Wild-caught and laboratory-raised amphibians are available for research. Wild-caught amphibians are often in extremely poor health. A mortality rate of more than 60% within the first week after capture has been reported, often attributable to conditions of confinement, handling, and sanitation before reaching the laboratory (Gibbs *et al.*, 1971). Inadequacy of care in the laboratory often leads to additional deaths before animals can be used (Nace and Rosen, 1979). As they are genetically undefined, vary in age and source, and are in different physiological states, these animals are extremely variable research subjects. The validity of research results has been questioned when experimental animals are used that are not standardized as to nutrition, health, genetics, or phase of reproductive or seasonal physiological cycles (Emmons, 1973; Nace, 1968; Nace, 1970). The number of wild caught amphibians available for research is rapidly decreasing. Nature can no longer meet the need for *Rana pipiens* used in teaching and research (Nace and Rosen, 1979). Causes of this population decline are multiple. Pollution by insecticides, fertilizers, and chemical or biological wastes has had a marked effect on various frog populations. Wetland destruction and the interference with spawning and seasonal migration caused by drainage, housing, and highways have taken their toll (Gibbs *et al.*, 1971; Hine *et al.*, 1981).

Laboratory breeding and rearing of amphibians were begun in an effort to provide healthy, genetically defined, quality research animals. Programs for large-scale culture and management of several species were initiated at Louisiana State University and the University of Michigan Amphibian Facility. The program at the University of Michigan has since been terminated (Culley, 1976; Nace, 1968; Nace *et al.*, 1971). Efforts to raise germfree *Rana pipiens* have also been made (Timmons *et al.*, 1977). At the present time, listings of amphibian sources are available (ILAR, 1979; Nace and Rosen, 1979).

D. Laboratory Management and Husbandry

Care of amphibians in the laboratory setting requires considerable knowledge and skill. Long-term care is complex, as there is variation in temperature requirement and type of environment, be it aquatic or terrestrial, with the stage of development.

Before amphibians are brought into a facility, areas reserved for caging, isolation, breeding, cage cleaning, storage, office, and insectarium (if needed) should be provided. A description of such facilities and optimum holding temperatures is available in ILAR (1974).

Care upon arrival is often crucial for amphibian survival. Aquatic amphibians should be quickly transferred into clean containers containing water from the shipment container. They can be slowly acclimated to dechlorinated laboratory water by gradual dilution of their shipment water. A temperature change of only 1°C/hr should be made. Terrestrial amphibians may be placed on arrival into large containers filled with 6–10 in. of water to facilitate removal of packing material. Any animals that appear ill or damaged after arrival should be isolated from others and treated (ILAR, 1974).

Several types of long-term caging systems have been designed for amphibians. The following is a brief summary of some of the caging systems described and illustrated in Nace (1968) and ILAR (1974). Choice of caging design rests on the facility needs, stage of development of amphibian, and the water supply system. Once-through continuous water flow systems allow high animal density and lower labor, but large amounts of high-quality constant-flow water with supply and drainage lines free of stoppage must always be available. Flow of water through the cage must be low enough to permit animals to condition their environment, but adequate to keep bacterial counts low. Where water is less plentiful, recirculating water systems can be used. These systems require the use and maintenance of filters and water treatment devices. Water quality must be carefully monitored to prevent accumulation of waste products. Balanced aquarium conditioned-water systems can be used that attempt to recreate the natural environment, but these cause difficulty in the laboratory setting if large numbers of animals are held.

With each system, close attention must be paid to chlorine levels of the water. Chlorine is tolerated by nonhibernating lung-breathing adults and may be beneficial by retarding bacterial growth, but it is lethal to larval stages and aquatic species (Nace, 1968; ILAR, 1974). The reader is referred to ILAR (1974) for a detailed discussion of water quality and monitoring.

During the embryo stage, from fertilization until feeding begins, shallow enamel pans, glass or fiberglass trays, or sided frames can be used. A large surface area or air line should be provided to ensure adequate gas exchange. Artificial medium such as De Boers solution or reconstituted pond water is recommended, which should be changed every third day. After initiation of feeding, larvae may be held in inverted 1 gal plastic bottles with the bottoms removed, equipped with flow-through and flushing devices.

Juveniles nearing metamorphic climax require semiterrestrial caging, as they may drown if terrestrial areas are not provided. Plastic vegetable crispers with solid core neoprene mesh flooring have been used. One end of the cage must be raised to provide a dry resting space at one end and a pool at the other. Continuous flow nonchlorinated water may be used in this system, or the water can be changed three to four times per week. For larger operations, epoxy-coated cattle watering tanks or circular concrete block enclosures can be used. They can be equipped with inclined floors and continuous flow water systems.

After metamorphosis, juveniles' cages can be leveled and chlorinated water used to flood the floor. Deeper water can be provided for species that feed in water, such as *Rana catesbeiana*. Pieces of pottery placed in the bottom of the cage provide a dry resting area. Juveniles can be adapted to adult caging systems as soon as they have adjusted to the terrestrial environment. Figure 1 illustrates a rack of cages for juveniles and adults. In this system a transparent plastic mouse container is inserted into a larger opaque plastic container. The opaque container holds water deep enough for the animal to maintain normal resting positions. The transparent container remains dry and an opening in the bottom of this container allows frogs to move between the aquatic and terrestrial environments; a high dry shelf and pottery pieces are available to the frogs. All floor surfaces are covered by neoprene mesh. Food is placed on the upper compartment floor and insect-proof lids are used. Aquatic amphibians, such as *Xenopus*, can be held in this type of caging system by removing the transparent cage insert.

Terrestrial species, such as *Bufo* and *Bombina*, may be held in shallow containers with well-fitting lids. Shallow water with shelves or pottery pieces should be provided. Axolotls are best maintained in separate gallon aquaria or bowls containing dechlorinated water at 20°–22°C. Terraria can be used for metamorphosed salamanders.

If animals will be held in hibernation, the cold temperatures required can be provided by holding cages in a cold room or using containers with self-contained cooling and water circulating systems. Twenty gallon plastic garbage containers equipped with bottom drains, water filters, and aerators have been utilized for hibernating frogs (ILAR, 1974).

E. Comparative Physiology

Members of the class Amphibia are characterized by smooth, moist, glandular skin without external scales. This skin is shed, often in one piece, under the influence of thyroid hormone. Present in the skin are two types of glands, mucous and granular glands. Mucous glands secrete a transparent substance that serves to lubricate the skin in water and moisten it while on land. Granular glands secrete irritant or toxic sub-

Fig. 1.   A rack of cages for juvenile and adult amphibians. Both a terrestrial resting area and aquatic area are provided, due to the placement of a transparent cage into an opaque cage. Continuous flow water system is used with water supplied at the front of the cage and drained to a collecting trough at the rear.

stances. These secretions are injurious to mucous membranes; the secretion of the toad *Bufo* can cause nausea, weakening of respiration, or muscle paralysis if ingested (Noble, 1954). Small clusters of sense cells in rows on the skin of amphibians compose the lateral line. These cells respond to vibrations of low frequency in aquatic medium and may control the mechanism of equilibrium and posture. The lateral line organ is found in aquatic larvae, adult salamanders, and adults of some anuran species (Cochran, 1961; Marcus, 1981). Ribs in amphibia are reduced or absent. In anurans, the radius and ulna are fused to form a single bone in the forelimb, as are the tibia and fibula in the hindlimb (Noble, 1954).

Several forms of respiration are present in amphibians. Pulmonary and buccopharyngeal respiration are used by most adult amphibians. Cutaneous respiration is used by amphibians during hibernation and submersion. Members of the families Plethodontidae, Salamandra, Sirenidae, Ambystomidae, and Proteidae of the order of Caudata have reduced or absent lungs when adults and must obtain oxygen through their skin in conjunction with mouth lining or gills (Cochran, 1961; Marcus, 1981; Noble, 1954; Shield and Bentley, 1973).

The heart of tadpoles is two-chambered, with one atrium and one ventricle. A septum develops within the atrium during metamorphosis, to form the three-chambered adult heart. A spiral valve in the heart acts to direct the flow of oxygenated and unoxygenated blood to the proper vessels (Marcus, 1981).

As amphibians evolved to live on land, several changes developed in the gastrointestinal tract. Most anurans have tongues attached anteriorly. Many urodeles, however, have mushroom-shaped tongues attached in the center that can be projected several times the length of the head. Aquatic anurans have reduced tongues, and tongues are lacking in the family Pipidae (*Xenopus laevis*). The esophagus and stomach are not sharply demarcated, and cilia are present in both to aid peristalsis. The stomach has secretory, mixing, and storage functions. It can greatly expand to store food for periods of food scarcity. The intestines are a uniform tube, arranged in tightly spiralled double loops. The absorptive area of the intestines can be increased by an increase in length. Its greatest length is observed in frog tadpoles, who require a maximum absorptive area due to their vegetarian diet. Adult amphibians and tadpoles on insectivorous diets have much shorter intestinal tracts (Noble, 1954).

The liver of amphibians elaborates fatty substances, stores glycogen, forms urea, and removes red blood cells from circulation. Liver glycogen is important as an energy source during hibernation and in the mating season. Fingerlike structures for storing fat, the fat bodies, are found attached near the anterior poles of the gonads (Marcus, 1981; Noble, 1954). A large amount of melanin is found in the liver, stored in macrophages (Tanaka *et al.*, 1974).

The renal tubules can concentrate certain substances such as urea, but urine is hypotonic with respect to the blood due to the absence of Henle's loop. Reabsorption of sodium chloride and glucose is accomplished by the tubules. The rate of urinary excretion is dependent on the environmental temperature; during the winter, kidney function is minimal. An essential function of the skin of amphibians is water absorption (Noble, 1954). The amphibian bladder is permeable to water; hypoosmotic urine stored in the bladder can be used to prevent dehydration when the animal is on land (Skadhauge, 1977). The adrenal gland of Salientia is comprised of a strip of yellow tissue adherent to the ventral surface of each kidney; in urodeles, the adrenals lie along the medial edge of the kidneys (Noble, 1954).

The lymphatic system of Salientia is remarkable due to the formation of large sinuses under the skin of the back. These sinuses, the dorsal lymph sacs, may function to prevent rapid drying of the skin. Lymph is pumped into veins by pairs of lymph hearts, simple sacs of endothelium covered by striated muscle and connective tissue (Noble, 1954). These sinuses can be used to advantage for injections.

F. Normal Values

Values for hematologic parameters are available for several amphibian species (Table II). There can, however, be considerable variation in these values, even within a single species (Carmena-Suero *et al.*, 1980; Harris, 1972; Roofe, 1961). Many factors have been identified that can influence hematologic parameters. These include species, age, sex, season, environmental variables (i.e., light–dark cycle, humidity, temperature), altitude and geographic region of origin, habitat (wild or laboratory-housed), level of activity, stress (i.e., shipping, dehydration, crowding), and nutritional and disease status (Gatten and Brooks, 1969; Harris, 1972; Roofe, 1961). These factors should be identified and taken into account when using published values. Unless similar factors affected both the sampled population and the population in the literature, comparison of hematological values would be of little help in the determination of disease status.

An experiment was completed by Harris (1972) to determine variants in several hematological parameters of *Rana pipiens* due to two such factors, sex of the animal and season. Animals were sampled 4 days after capture to eliminate variation caused by shipping and prolonged laboratory housing. Experiments were conducted over five consecutive seasons. Significant seasonal variation was apparent in blood hematocrit, blood hemoglobin concentration, erythrocyte count, and leukocyte count. A significant sex difference in one or more parameters was observed each season. Erythrocyte counts, hematocrit, and hemoglobin were highest in the fall, decreased during the winter, reached a low point by spring, and began to rise in summer. Leukocyte counts were low in fall and winter, highest in spring, and intermediate during the summer. A seasonal influence on hematology was also reported by Robertson (1978); however, he found the hematocrit in *Rana pipiens* to be highest in winter and lowest in summer.

Blood chemistries in several amphibian species are also subject to seasonal variation. Changes in blood glucose due to season and temperature have been reported (Farrar and Frye, 1979; Hermansen and Jorgensen, 1969; Maitrya *et al.*, 1970; C. L. Smith, 1950). Robertson (1978) has reported seasonal variations in plasma sodium and calcium levels.

Data on longevity of various amphibian species has been published by Bowler (1977).

Table II

Normative Data for Selected Amphibians

Species

Parameter	Rana pipiens	Rana pipiens	Rana pipiens	Rana pipiens	Rana pipiens	Rana pipiens	Rana pipiens	Rana pipiens
No. of samples	36	23	30	30	34	23	30	30
Season	Winter	Spring	Summer	Fall	Winter	Spring	Summer	Fall
Sex	M	M	M	M	F	F	F	F
Body weight (gm)	36.4 ± 5.11	33.1 ± 4.27	30.3 ± 5.16	35.7 ± 3.68	37.1 ± 6.28	30.6 ± 6.01	33.9 ± 8.53	41.0 ± 5.30
RBC count (cells/mm^3)	457,000 ± 80,000	317,000 ± 90,000	604,000 ± 110,000	657,000 ± 110,000	406,000 ± 100,000	264,000 ± 90,000	461,000 ± 80,000	591,000 ± 110,000
WBC count (cells/mm^3)	5,975 ± 2,851	16,085 ± 6096	11,565 ± 3,173	6,380 ± 1,902	5,710 ± 2,945	17,900 ± 8,028	10,150 ± 3,824	7,580 ± 3,385
Hematocrit (%)	41.5 ± 5.59	29.0 ± 9.51	44.9 ± 7.47	47.4 ± 6.76	37.1 ± 6.40	24.1 ± 8.43	33.7 ± 4.81	44.4 ± 6.43
Hemoglobin (gm%)	10.08 ± 1.54	6.59 ± 2.80	11.04 ± 1.84	12.72 ± 1.84	9.65 ± 1.52	4.80 ± 2.12	8.44 ± 1.14	12.30 ± 1.70
Mean corpuscular volume (μm^3)	908	916	743	722	915	916	730	752
Mean corpuscular hemoglobin (pg)	221	208	182	184	238	182	183	208
Mean corpuscular hemoglobin concentration (%)	24.3	22.7	24.6	26.8	25.9	19.9	25.1	27.7
Total blood volume (ml/100 gm body weight)	—	—	—	—	—	—	—	—
Plasma volume (ml/100 gm body weight)	—	—	—	—	—	—	—	—
Reference	Harris, 1972	Harris, 1972	Harris, 1972	Harris, 1977	Harris, 1972	Harris, 1972	Harris, 1972	Harris, 1972

Parameter	Rana catesbeiana	Rana catesbeiana	Necturus maculatus	Ambystoma tigrinum	Taricha granulosa	Leptodactylus fallax	Leptodactylus fallax	Hyla septentrionalis
No. of samples	15	—	—	87	18	5	5	17
Season	Spring	—	—	Summer	Hibernation	Summer	Summer	Spring
Sex	M,F	—	—	M,F	M,F	M	F	M,F
Body weight (gm)	285 ± 21	—	—	34.5	7	—	—	31.9 ± 3.5
RBC count (cells/mm^3)	—	450,000	20,000	1,657,000	40,000	600,000 ± 39,000	744,000 ± 39,000	—
WBC count (cells/mm^3)	—	—	—	4,600	—	—	—	—
Hematocrit (%)	40.4 ± 1.2	32.0	21.4	40	—	—	—	22.4 ± 2.0
Hemoglobin (gm%)	9.5 ± 0.2	8.2	4.6	9.38	4.5	10.8 ± 1.8	11.1 ± 1.3	6.2 ± 0.6
Mean corpuscular volume (μm^3)	—	716	10,070	—	—	—	—	—
Mean corpuscular hemoglobin (pg)	—	184	2160	—	—	—	—	—
Mean corpuscular hemoglobin concentration (%)	23.5 ± 2.4	26	22	—	—	—	—	27.7 ± 3.1
Total blood volume (ml/100 gm body weight)	3.35 ± 0.26	—	—	—	—	—	—	7.5 ± 0.3
Plasma volume (ml/100 gm body weight)	2.00 ± 0.24	—	—	—	—	—	—	5.8 ± 0.3
Reference	Carmena-Suero et al., 1980	Wintrobe, 1933	Wintrobe, 1933	Roofe, 1961	Friedmann et al., 1969	Gatten and Brooks, 1969	Gatten and Brooks, 1969	Carmena-Suero et al., 1980

G. Nutrition

Amphibians are opportunistic feeders in nature and will consume any food they can fit into their oversized mouths. After a period of adequate intake, they can survive days to weeks without food. Little is known about the nutrient requirements of amphibians, except that gained from analysis of the ingesta of wild caught amphibians. A comprehensive description of diets of wild anurans and urodeles, from larval stage through adult, has been written by Nace and Buttner (1984).

Due to interest in recent years in laboratory rearing of amphibians, increasing attention has been paid to their nutritional needs in captivity. A few attempts have been made to design laboratory diets according to information from diets in nature. Meeting their nutritional needs, which vary with species and stage of development, is a very diffficult task. This problem has been a great deterrent to successful laboratory culture efforts (Nace and Buttner, 1984).

Guidelines for feeding several laboratory species have been developed (ILAR, 1974). Nace and Buttner (1984) present an excellent review of laboratory diets employed by investigators for anurans and urodeles. The reader is referred to these publications for supplementation of the dietary information presented here.

Young anuran larvae, such as *Rana pipiens,* are herbivorous, eating a diet of softened romaine or escarole lettuce. This can be supplemented with or replaced by a rabbit chow–agar–gelatin preparation (ILAR, 1974). As they mature, the larger tadpoles become omnivorous. A protein supplement, such as raw or boiled liver, should be added to the lettuce diet two to three times per week. Spinach must be avoided, as it causes kidney stones (Berns, 1965). After metamorphosis, *Rana pipiens* prefers a carnivorous diet. They require living food, which is placed in the terrestrial section of the cage. Suitable dietary components include sowbugs, crickets, earthworms, mosquitos (*Culex pipiens*), and greenbottle flies, dusted with a commercial vitamin and protein supplement.

Rana catesbeiana as larvae have been reported to grow well when fed the rabbit chow–gelatin mixture supplemented with boiled leafy lettuce. Postmetamorphic frogs can be fed tadpoles on a short-term basis; fish, especially minnows, are nutritious and accepted on a long-term basis. In the laboratory, a combination of crickets, earthworms, and golden shiners each day can also be used.

Xenopus laevis larvae grow well if fed finely ground food such as dried green pea soup. For maximum growth, the food is mixed with water and the tadpoles are fed the decanted supernatant twice daily. During metamorphosis, the larvae will eat finely diced meat, mosquito larvae, ground beef heart, *Tubifex* (redworms), and powdered milk. Adults will accept nonliving food. such as pieces of beef heart cut to resemble earthworms, which have been soaked in vitamin mixture, or earthworms. A commercial frog ration is also available (Frog Brittle, Nasco, Fort Atkinson, Wisconsin). Feeding the adults more than twice a week promotes rapid growth.

Bufo will eat any of the dietary components of the *Rana pipiens* diet, and *B. marinus* can adapt to a nonliving diet of canned cat or wet dog food. *Bombina orientalis* show little preference for different types of food; crickets and sowbugs are recommended because the appropriate size can be selected for feeding.

Urodele larvae are carnivorous, and can be fed such living items as tadpoles, earthworms, and arthropods. Axolotls, neotenic larvae of *Ambystoma mexicanum,* are best fed brine shrimp or mosquito larvae when newly hatched. Mosquito larvae are readily produced in the laboratory and can provide food source for all ages of larvae (ILAR, 1974). For adequate nutrition, *Daphnia, Tubifex,* earthworms, and sliced beef liver dusted with bone meal should be fed as part of the axolotl diet. Older axolotls prefer earthworms and insects, though thin strips of beef and lamb liver can be fed. Prepared diet mixtures have been also described (Nace and Buttner, 1984). Axolotls should be fed on alternate days or three times a week due to their slow digestion of food. They often will regurgitate if overfed. *Necturus* do well on earthworms and crayfish, with occasional supplementation with raw meat or fish.

H. Reproduction

Distinguishing the sex of members of the class Amphibia can be accomplished by the presence of several sex characteristics (ILAR, 1974). These characteristics are best identified during the breeding season. Of the common laboratory species, *Ambystoma mexicanum* and *A. tigrinum* males have marked enlargement of the lateral margins of the cloaca due to increased size of cloacal glands. Villous papillae are present in the cloaca. Females of the species have plumper bodies due to the presence of eggs and enlarged oviducts. *Xenopus laevis* males are identified by their smaller body size (as compared to gravid females), small cloaca, and dark nuptial pads; females by their enlarged cloacal lips. Mature *Rana* and *Bufo* males also have thickened nuptial pads, which undergo seasonal change under the influence of testosterone, as the testes do. During the summer breeding season, these enlarged thumb pads have a papillate surface with a keratinized layer. At this point of maximal testosterone secretion, spermatogenesis is at its maximum (Saxena and Zal, 1981). In *Rana catesbeiana* and *Rana clamitans,* the size of the tympanic membrane is used as an indicator of sex. In females, the membrane is the same size as the eye, while in males it is nearly twice the size.

Male amphibians have two testes attached by mesorchia to

the ventral kidney. Sperm and urine follow the same course to the cloaca; sperm are then stored in the seminal vesicle until amplexus, or breeding. Females have two ovaries, attached by mesovaria to the kidneys. Eggs break through their follicles, are discharged into the coelomic cavity and are conveyed to the ostia, or opening in the oviduct, by cilia on the parietal peritoneum. The eggs travel single file down the oviduct, gathering a gelatinous covering from glands lining the oviduct. Eggs are retained until oviposition in the ovisac, which is an enlargement of the oviduct as it enters the cloaca (Noble, 1954).

The majority of amphibians are oviparous (egglayers). Fertilization of eggs is external in frogs and toads. During the process of amplexus, the male grasps the female and places sperm directly over the eggs as they are laid. Internal fertilization has been reported in one frog, *Eleutherodactylus coqui* (Townsend *et al.*, 1981). Fertilization and internal development of larvae occur in some members of Caudata and in the family Nectophrynoides of Salientia (Marcus, 1981). During complicated breeding displays in Caudata, females pick up sperm packets (spermatophores) deposited by males and store them on the roof of the cloaca. These stored sperm then fertilize the eggs as they pass through the cloaca; eggs are released several hours later due to the stimulus of breeding (ILAR, 1974).

Natural breeding in the wild can occur immediately after hibernation or some weeks later, following a period of nutritional intake. Many amphibians can be induced to breed in the laboratory by injection of chorionic gonadotropin, pituitary suspension, or pituitaries and progesterone. The publication by ILAR (1974) presents detailed descriptions of hormonal injections and housing procedures used for induction of breeding in several common laboratory species. It also describes detailed techniques for parthenogenesis, gynogenesis, androgenesis, polyploidy, and artificial insemination in amphibians.

Development, or metamorphosis, of amphibian larvae requires the action of thyroid hormone. Retardation of metamorphosis can occur if iodine is not present in the water. In Salientia, this process can be divided into several stages (Marcus, 1981). During the first stage, premetamorphosis, emergence from the egg occurs. The tadpole grows, and hindlegs appear. During the followng stage, prometamorphosis, forelimbs develop, organ systems mature, and growth continues. Several changes occur during climax, the final stage of metamorphosis. There is widening of the mouth and absorption of the tail and gills. The frog begins to breathe air with lungs.

Metamorphosis in Caudata corresponds to the climax phase in Salientia (Etkin, 1964). Larvae resemble adult forms, including limb structure; during maturation in most species, gills are lost. Neotenous members of Caudata, such as axolotls, can be induced to metamorphose by the administration of thyroid extract (Marcus, 1981).

III. DISEASES

A. Infectious

1. Bacterial

Bacterial infections of amphibians are associated with organisms normally present in the animals' environment, i.e., tank surfaces, water, food, or on the amphibian itself (intestinal contents and skin surface). As is true for fish, disease produced by these agents occurs only when there is a disturbance of homeostasis. Such disturbances include changes in water quality, temperature, population density, diet, "stress," trauma, alternation of the amphibian's immune status, metamorphosis, etc. (Amborski and Amborski, 1978). Environmental influences acting upon a bacterium, such as conditions favoring proliferation or exotoxin production, also must be considered. Maintenance of optimal husbandry conditions and correction of deficiencies during a disease outbreak are the surest methods of reducing amphibian morbidity and mortality due to bacterial infectious. Antibiotic therapy can be a useful adjunct, but in itself is rarely sufficient to terminate disease outbreaks, particularly epizootics of gram-negative bacterial septicemia.

a. "Red Leg" Syndrome (Bacterial Septicemia). A number of gram-negative bacteria, principally Enterobacteriacae can cause this condition. Organisms isolated alone or in combination from outbreaks of "red leg" include *Aeromonas hydrophila, Pseudomonas* spp., *Mima* spp., *Proteus (Providentia)* spp., *Flavobacterium,* and *Citrobacter* (Glorioso *et al.,* 1974b). Anurans and urodeles have been infected. Gram-positive organisms have not been implicated in this syndrome, although *Staphylococcus* spp. and *Corynebacterium* spp. have been isolated on occasion in conjunction with gram-negative organisms.

In peracute infections, animals are found dead without premonitory clinical signs. In animals surviving for several days, signs include one or more of the following: anorexia (often the first sign noted), lethargy, subcutaneous edema, coelomic effusion giving a bloated appearance, neurologic disorders, cutaneous ulcers, dull discoloration or pallor of skin, and cutaneous hemorrhages, the latter being the lesion responsible for the name of this syndrome (Fig. 2). In anurans these hemorrhages tend to be on the flexor surface of the thighs and on the foot webs.

Hematologic findings consist of macrocytic anemia, leukopenia, thrombocytopenia, increased clotting time, increased erythrocyte sedimentation rate, and decreased total protein (Marcus, 1981).

"Red leg" may occasionally affect individual animals but more often occurs in epizootic form with high morbidity and

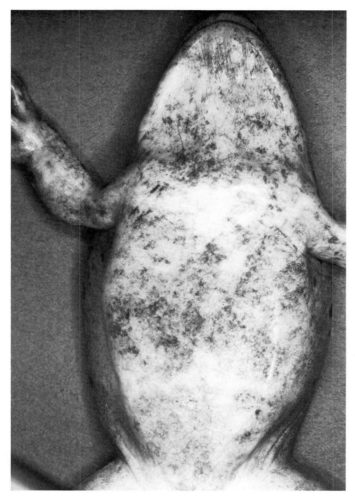

Fig. 2. Petechial and ecchymotic hemorrhages and diffuse congestion on the ventral skin of a *Rana pipiens* with gram-negative bacterial septicemia.

mortality. Outbreaks often occur following shipment of amphibians. Infections have been reported both in larval and adult amphibians; captive and wild populations have been affected (Gibbs *et al.*, 1966; Hird *et al.*, 1981). The postulated routes of infection are percutaneous through traumatized epithelium (Elkan, 1976) and via the intestinal tract (van der Waaj *et al.*, 1974).

Necropsy findings consist of multiple petechial or ecchymotic hemorrhages in the skin, skeletal muscle, and digestive tract; subcutaneous (dorsal lymph sac) edema and coelomic effusions; splenomegaly; and a pale, anemic appearance to the carcass (Boyer *et al.*, 1971; Gibbs *et al.*, 1966).

Histologically, there are multiple foci of coagulative necrosis (principally karyorrhexis) in liver, spleen, other viscera, and skin. These necrotic areas may be accompanied by small numbers of leukocytes associated with clumps of bacteria (Abrams, 1969).

A diagnosis of "red-leg" syndrome is based on typical clinical signs, necropsy and histopathological findings, and culture of the organism(s). For the last, knowledge of bacterial pathogens of aquatic animals and incubation of culture media at 25°C as well as 37°C are essential (Glorioso *et al.*, 1974a). In *Xenopus laevis*, chlamydiosis (Newcomer *et al.*, 1982) can produce high morbidity and mortality with clinical signs and gross lesions identical to *Aeromonas* septicemia (Hubbard, 1981). However, histologic lesions of chlamydiosis are dissimilar, and chlamydiae do not grow on culture media (also see Section III,A,4).

Prevention of "red-leg" is best accomplished by housing animals under optimal husbandry conditions in clean cages with circulating, filtered water and by avoiding overcrowding. Amphibians that have any clinical signs suggestive of bacterial septicemia should be removed from the cage and isolated.

A variety of treatments have been proposed for bacterial septicemia but tend to be ineffective once an epizootic is under way and if husbandry conditions remain uncorrected. Ideally, antibiotic therapy is based on culture and sensitivity results, and a number of antibiotics have been used with some degree of success. In general, antibiotics must be given parenterally or orally, either in the feed or by intubation; adding antibiotics to water is ineffective. The treatment regimens shown in Table III have been proposed (Gibbs, 1963; Riviere *et al.*, 1979; S. W. Smith, 1950).

During epizootics, some investigators also recommend the prophylactic addition of copper sulfate (1 : 5000) or potassium permanganate (1 : 1000) to tanks; for completely aquatic amphibians, raising the salt concentration to 4% has been advocated (Reichenbach-Klinke and Elkan, 1965).

b. Salmonellosis. In some Third World countries, frogs (Sharma *et al.*, 1974) and toads (Singh *et al.*, 1979) may harbor numerous *Salmonella* serotypes in their intestinal tracts. These findings may reflect poor public health practices and sewage contamination of ponds and water channels in the countries of origin. Disease in amphibians associated with

Table III

Antibiotic Regimens Proposed to Treat Bacterial Septicemia

Antibiotic	Administration	Dose
Tetracycline	Gavage	5 mg/30 gm body weight
Chloramphenicol	Oral	5 mg/100 mg body weight initially followed by 3 mg/100 mg twice a day for 5 days
Streptomycin	Parenteral(?)	1000 U for 9 days
Gentamycin[a]	Bath	1 mg/ml

[a] Therapeutic blood levels are attained but clinical efficacy has not been documented.

these isolates has not been documented. Numerous serotypes of *Salmonella* have been isolated in North America from imported aquarium frogs and newts (Trust *et al.*, 1981). Several of the serotypes were those associated with human salmonellosis, and a number were multi-drug-resistant.

Transmission of *Salmonella* to animal handlers via contaminated secretions or tanks could present a public health hazard.

c. Diplobacterium ranarum. This gram-negative bacterium is considered to be a harmless commensal in the amphibian environment, causing disease (septicemia) only in stressed or injured animals. Clinical signs are similar to the "redleg syndrome"—infection with high mortality and mortality (Elkan, 1976).

Grossly, frogs have cutaneous ulceration, enteritis, and anasarca or a hemorrhagic exudate in the lymph sacs. Histologic lesions correspond to the gross findings. Culture is the best method for organism identification, and antibiotic sensitivity aids in determining a course of therapy.

d. Pseudomonas reptivora. This organism is associated with septicemia and mortality in *Necturus maculosus* (mudpuppy). Gross lesions are cutaneous hyperemia and gray foci of necrosis on the gills.

A proposed topical treatment is 0.2 gm of furacin powder sprinkled on the gills daily for 10 days. This can be done in association with chloramphenicol given by gavage at the following dosages: 8 mg in 2 ml of water initially, followed by 4 mg in 1 ml water daily for 6 days, then 4 mg in 1 ml water once per week until there is clinical cure (Marcus, 1981).

e. Mycobacteriosis (Tuberculosis). A number of Runyon's group IV mycobacteria, including *Mycobacterium marinum (balnei)*, *Mycobacterium ranae (fortuitum)*, and *Mycobacterium xenopi*, can be etiologic agents of mycobacteriosis of amphibians and other ectotherms. These "cold water" mycobacteria are ubiquitous in aquatic environments, including pond, sea, and occasionally tap water. They are also common saprophytes on amphibians' skins. The acid-fast organisms grow rapidly on appropriate media when incubated at 23°C (Marcus, 1981).

Mycobacterial infections have been reported principally in anuran amphibians (Abrams, 1969; Elkan, 1976; Marcus, 1981; Shively *et al.*, 1981). Clinical signs may consist of subcutaneous nodules or cutaneous ulcers; animals also may die without dermal changes. Infection occurs principally in animals that are already weakened and debilitated, and, hence, mycobacteriosis is not a colony problem. The portal of entry is postulated to be the skin, facilitated through injury (Elkan, 1976) or iatrogenically, e.g., due to injection technique (Joiner and Abrams, 1967).

Gross lesions are cutaneous ulcers, coalescent verrucous

masses, or yellow-white dermal nodules. Nodules occur in visceral organs, principally the liver, and can become quite extensive. Histologically, the nodules are granulomas composed of macrophages and epithelioid cells. Multinucleated giant cells can be present but are rare. Granulomas are surrounded by a loose fibrous capsule. Central caseation is uncommon in frogs (Abrams, 1969) and is present though not extensive in toads (Shively *et al.*, 1981). In the lung, lobar tuberculous pneumonia occurs rather than granulomas; animals with pulmonary lesions frequently die before organisms spread to other viscera (Elkan, 1976). Calcification does not occur. Acid-fast organisms are present within the granulomas and in pneumonic lungs.

Ulcers and granulomas are also found in mycotic infections; differential diagnosis is based on histologic demonstration of fungal forms and by culture results.

Affected animals, although they do not pose a threat to tankmates, should be culled. Mycobacteriosis can be prevented by using healthy stock and maintaining a clean environment with adequate sterilization or disinfection of tanks.

Cold water mycobacteria have been incriminated as a cause of cutaneous ulcers and nodules in humans ("aquarist's nodules") (Shively *et al.*, 1981). *Mycobacterium xenopi* and *M. marinum* have been isolated from human pulmonary lesions (Marcus, 1981). Thus, an environment contaminated by these organisms may pose a public health hazard as well as a hazard to amphibians.

2. Viral

a. Herpesvirus of Lucké Renal Adenocarcinoma. This agent is associated with the induction and transmission of renal adenocarcinoma in the leopard frog, *Rana pipiens*. The virus has the typical icosahedral morphology and DNA core of herpesvirus on electron microscopy; its size is 95–110 nm and it has 162 capsomers. It has been propagated in frog pronephric cells, and cell fractions from this cell line have transmitted the neoplasm (Tweedell and Wong, 1974) as have virus-containing tumor cells in the algid ("winter tumor") and transitional tumor phases (McKinnell and Ellis, 1972).

Clinical signs are absent until the tumor has reached considerable size and/or metastasized. These latter stages are associated with abdominal bloating, lethargy, and death.

Renal adenocarcinoma can be found in both wild and laboratory-housed *Rana pipiens*. The epizootiology is closely tied to temperature sensitivity of the virus. At temperatures of 22°–25°C and above, viral replication does not occur, and inclusion bodies and viral particles are not present. In this "summer phase" or calid tumor, the herpesvirus genome is incorporated into the tumor cell DNA (Collard *et al.*, 1973), and virus-specific antigen can be detected on the tumor cell surface

(Naegele and Grannoff, 1977). When frogs enter the hibernating state at environmental temperatures of 5°–10°C, the tumor is in the algid phase. There is viral replication, appearance of mature virions, formation of intranuclear inclusion bodies, and massive tumor cytolysis (McKinnell and Ellis, 1972). Viral particles also are present in metastatic tumors (McKinnell and Cunningham, 1982). Virus is transmitted to embryos and larvae in the breeding ponds when cells are lysed in algid phase tumors; vertical transmission also may occur.

Necropsy findings consist of single or multiple white nodules uni- or bilaterally in the kidneys. Their size varies from 1 mm to large masses causing abdominal distension. Metastatic foci can occur in lung, liver, and other viscera. Microscopically, the summer phase tumor is a well-differentiated papillary adenocarcinoma sharply demarcated from adjacent renal parenchyma but not encapsulated. The tumor is composed of tall columnar to pseudostratified cells with hyperchromatic nuclei. Mitotic figures are rare. Cells are located on a basement membrane, and there is a variable connective tissue stroma (Stewart *et al.*, 1959).

Lucké renal adenocarcinoma is not a problem in laboratory-bred as opposed to wild-caught frogs. The condition is of interest to the research community as a model of herpesvirus-induced cancer.

b. Polyhedral Cytoplasmic Viruses (FV1, FV2, FV3, LT, TEV, etc.). These DNA viruses were initially isolated in tissue culture during attempts to grow the Lucké herpesvirus from renal tumors of *Rana pipiens.* Subsequently, they were found in *Rana catesbeiana* and *Diemictylus viridescens* and were transmitted experimentally to *Bufo* sp. All isolates are antigenically related and may actually represent strains of the same virus (Clark *et al.*, 1969; Wolf *et al.*, 1969). They are not associated with tumorigenesis, and infections are generally not clinically apparent in amphibians except for several strains pathogenic to bullfrog tadpoles [tadpole edema virus (TEV)]. The viruses are related to iridoviruses of arthropods and infectious pancreatic necrosis virus of fish. Amphibian polyhedral cytoplasmic viruses are nonpathogenic for fish (Wolf *et al.*, 1969). When inoculated into mice, the virus does not multiply but causes acute hepatotoxicity and death in 24 hr (Kirn *et al.*, 1975).

In bullfrogs, disease is seen in tadpoles 3–60 days old, with mortality decreasing with age, ranging from 55% at 18 days to 3.7% at 49 days. Adults do not develop signs of disease but are thought to transmit the infection to tadpoles. Other than acute death, clinical signs consist of subcutaneous edema.

Pathologic findings are subcutaneous edema; petechial hemorrhages in the stomach, kidney, and skeletal muscles; and foci of coagulative to liquefactive necrosis with edema in the liver, kidney, and gut (Wolf *et al.*, 1969). The condition can be differentiated from gram-negative septicemia on the basis of negative cultural results and the extent of necrosis and edema.

Prevention consists in separating tadpoles from adult frogs and by isolating and culling tadpoles with the disease.

c. Frog Virus-4 (FV4). This herpesvirus was isolated from kidneys of frogs with Lucké renal adenocarcinoma. It is not oncogenic and does not produce disease in *Rana pipiens* (Wong and Tweedell, 1974).

d. Frog Adenovirus-1 (FAV-1). This adenovirus was isolated in turtle cell tissue culture from a granuloma-bearing kidney of a *Rana pipiens.* The virus could not be grown in amphibian or other cell lines and was not associated with any disease syndrome in the host species (Clark *et al.*, 1973).

e. Encephalitis Viruses. Populations of wild frogs have been implicated as reservoir hosts for western equine encaphalitis virus and Japanese B encephalitis (frogs in Taiwan). These animals may be viremic or have serum-neutralizing antibodies but do not have signs or lesions (Burton *et al.*, 1966).

3. Rickettsial

Rickettsial infections are rare in amphibians. *Hemobartonella batrachorum* and *H. ranarum* have been found in frogs (*Leptodactylus* spp.). The infection is subclinical, detected only by examining blood films for pleomorphic coccoid or rod forms that stain well with Giemsa. The organisms are present within or on the surface of erthrocytes (Marcus, 1981).

In a survey of ectotherms in India for evidence of *Coxiella burnetti* infection, no titers or isolates were obtained from *Bufo* sp. or *Rana tigrina* (Yadav and Sethi, 1979).

4. Chlamydial

Chlamydia psittaci infection has been reported in *Xenopus laevis* (African clawed frog) (Newcomer *et al.*, 1982). Clinical signs, which occurred shortly before death in both spontaneous and experimental cases, were lethargy, disequilibrium, patchy cutaneous depigmentation, cutaneous petechiae, and anasarca. Some frogs died peracutely. Morbidity and mortality in the spontaneous outbreak were high, and transmission occurred between animals; the source of *C. psittaci* infection was not identified.

At necropsy, lesions were nonspecific and similar to those of bacterial septicemia: petechial hemorrhages, cutaneous ulceration, subcutaneous edema, coelomic effusion, and hepatosplenomegaly. Histologically, there was multifocal pyogranulomatous inflammation in the liver, kidney, lung, spleen, and heart (Fig. 3). Multiple intracytoplasmic basophilic inclusion bodies were present in hepatic and splenic sinusoidal lining cells (Fig. 4) and in glomerular tufts. Although in-

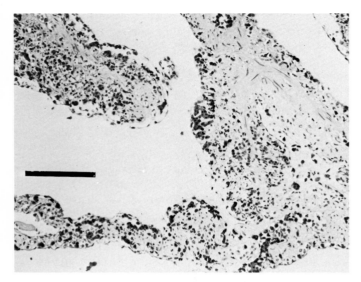

Fig. 3. Lung of *Xenopus laevis* with chlamydiosis. Alveolar septa are thickened by interstitial edema and pyogranulomatous inflammation. (From Newcomer *et al.*, 1982. Reprinted with permission of *Laboratory Animal Science*.)

terstitial pneumonia and fibrinopurulent vegetative endocarditis were common, inclusion bodies were less frequent in these organs. Inclusion bodies were composed of multiple coccoid elements; they were positive to the Giemsa stain and did not react to a tissue Gram stain. Ultrastructurally, inclusions had typical *C. psittaci* morphology of initial, intermediate and elementary bodies.

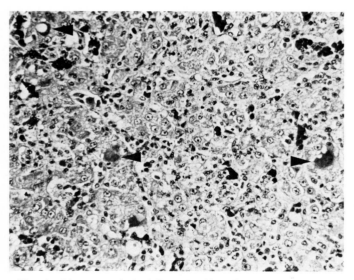

Fig. 4. Liver extensively infiltrated with diffuse purulent and granulomatous inflammation. Intracytoplasmic chlamydial inclusion bodies (arrowheads) stain diffusely and have variable morphology. Hematoxylin and eosin. (From Newcomer *et al.*, 1982. Reprinted with permission of *Laboratory Animal Science*.)

Based on its characteristic histologic lesions, including cytoplasmic inclusion bodies, chlamydiosis can be readily differentiated from bacterial septicemia.

Although the efficacy of treatment is not known, oral tetracycline at doses recommended for "red leg" may prove effective. Prevention of exposure of *Xenopus* to possible sources of chlamydia may be accomplished by feeding liver certified for human consumption or a commercial diet preparation and by prevention of contact of amphibians with feral animals.

5. Parasitic

Amphibians are intermediate and final hosts for many protozoan and metazoan parasites. There are marked differences in incidences of parasitic diseases between laboratory-bred and wild-caught frogs. In the former population, prevention of contact with infected adults and with vectors interrupts the host–parasite cycle; thus, only parasites with direct life cycles are found. In contrast, wild-caught amphibians may be infested with a variety of cestodes, trematodes, nematodes, and protozoa. Clinical signs, mortality, and lesions occur only when the parasitic burden is heavy.

Comprehensive reviews of amphibian parasites are those of Reichenbach-Klinke and Elkan (1965), Kaplan (1973), Elkan (1976), and Marcus (1981).

Table IV lists, by organ system, parasites of clinical significance or interest to laboratory animal medical specialists. Many of these organisms will be detected in blood films, at necropsy, or during microscopic examination of amphibian tissues. In the last instance, the monograph of Chitwood and Lichtenfels (1972) can aid the pathologist in metazoan parasite identification.

a. Protozoan. Most protozoan infestations are subclinical. They are detected either histopathologically or, in the case of blood-borne parasites, on examination of blood smears (Levine and Nye, 1977). Some genera, such as the trypanosomes, are ubiquitous in wild amphibian populations in all major geographic zones (Bardsley and Harmsen, 1973).

b. Nematodes. Nematodes are the most common helminths found in wild-caught amphibians. Many species occur in the gut; others reside in the lungs, blood vessels, and lymphatics. Encysted nematode larvae may also be present in various tissues (Elkan, 1976).

c. Cestodes. Adult cestodes occur in the gut of amphibians; their life cycles have not been as well defined as trematodes. Similar to trematodes, tapeworms may utilize amphibians as intermediate hosts (Reichenbach-Klinke and Elkan, 1965) with larvae encysted in tissue.

Table IV

Amphibian Parasites of Clinical Significance

System/organ	Parasite[a]	Species affected	Comment
Respiratory/lung	(N) *Rhabdias* spp.	Frogs and toads	Direct life cycle—like *Strongyloides*. Can be found in laboratory-bred frogs; microscopic lesion: smooth muscle hypertrophy Prevention: sanitation
Lung	(DT) *Haematoloechus variegatus, H. complexus*	*Rana* spp.	Indirect life cycle—snail, dragonfly are intermediate hosts. Microscopic lesion: smooth muscle hypertrophy
Respiratory/gills	(MT) *Sphyranura osleri, S. oligorchis*	*Necturus maculosus*	Direct life cycle. Gill damage in heavy infestations
Gills	(P) Ciliata *Charchesium polysinum, Oodinium pillularis, Trichodina* spp.	Tadpoles, mudpuppies, salamanders	Direct life cycle. Gill damage in heavy infestations, also parasitize skin. Diagnosis: impression smear. Treatment: copper sulfate (2 mg/ml H_2O); trypoflavin (1:1000 to 1:100 concentration)
Digestive/stomach or intestine	(A) *Acanthocephalus ranae*	Frogs, newts, toads	Armed proboscis buried in bowel wall; gut perforation and coelomitis may occur
Intestine	Numerous nematode species	Anurans and urodeles	No clinical signs unless heavy infestation, then weight loss and/or gut obstruction
Intestine	(C) *Nematotaenia dispar*	Frogs, toads, salamanders	Subclinical unless heavy infestation
Intestine	(C) *Chlamydocephalus namaquensis*	*Xenopus laevis*	Treatment: parenteral bromophenol (Reichenbach-Klinke and Elkan 1965—dose not specified)
Intestine	(C) *Ophiotaenia filaroldes, O. gracillus*	*Ambystoma tigrinum, Rana pipiens*	Indirect life cycle—copepods are intermediate host
Intestine	(DT) *Prosotocus* spp.	Frogs	Indirect life cycle—snails, dragonflies are intermediate hosts
Intestine	(DT) *Cercorchis necturi*	Newts	
Intestine	(P) Coccidia, *Eimeria* spp.	Various amphibians	
Intestine, liver	(P) *Entamoeba ranarum*	Tadpoles and adults	Diagnosis: cysts and trophozoites in feces
Urinary/urinary bladder	(MT) *Polystoma* spp.	Frogs and toads	Direct life cycle. Larvae attach to tadpole gills; migrate at metamorphosis through intestine to urinary bladder
Kidney	(P) Coccidia *Isospora* lieberkuhni	*Rana* spp., *Bufo* spp.	Renal tubule, cytoplasm; subclinical infection
Kidney	(P) Myxosporidia *Leptotheca ohlmacheri*	*Rana pipiens*	Renal tubule lumens
Integumentary	(P) Sporozoa *Dermocystidium, Dermosporidium*	Anurans and urodeles	Dermal cysts with organisms; no treatment. Also infects fish
Musculoskeletal	(P) Microsporidia *Plistophora myotrophica*	*Bufo bufo*	Sporoblasts in skeletal muscle; chronic disease with muscle atrophy and degeneration, weight loss, and high mortality
Hematopoietic	(P) Trypanosoma		Indirect life cycle—leeches are intermediate hosts for aquatic amphibians; blood-sucking insects are intermediate hosts for terrestrial amphibians. Extracellular blood parasites. Usually subclinical infection
	T. inopinatum	Old world anurans	Acute fatal illness may occur
	T. diemictyli	Newts	Fatal illness may occur
	T. pipientis	*Rana pipiens*	Mortality after splenectomy
Hematopoietic	(P) Sporozoa		Intraerythrocytic; indirect life cycle—leeches are intermediate host
	Lankesterella spp.	*Rana* spp.	Sporogony and schizogony in vessel endothelium
	Haemogregarina magna	*Rana pipiens*	Meronts in hepatocytes, gamonts in erythrocytes
	(P) Piroplasma—babesoids	Frogs, toads, salamanders	Intraerythrocytic; no clinical signs
Vascular	(N) *Foleyella* spp.	Frogs and toads	Adult filarids in blood vessels and lymphatics; microfiliarae in blood. Mosquitoes are intermediate host
CNS/brain	(P) Sporozoa *Toxoplasma* spp.	Frogs and toads	Cysts or pseudocysts in brain; organisms are not *T. gondii* (do not infect mammals) (Levine and Nye, 1976)

(continued)

Table IV *(Continued)*

System/organ	Parasite[a]	Species affected	Comment
Other	Digenetic trematodes—cerceria, metacerceriae	Anurans and urodeles	Indirect life cycle—amphibians are intermediate hosts. Encysted larvae in viscera, muscle, skin, etc. Host reaction ranges from minimal to absent to marked inflammation. Melanocyte deposits common in skin. Clinical signs depend on number of larvae, site of encystment, and host immune response
Other	Larval cestodes Terathyridea of *Mesocestoides*	Anurans and urodeles	Same as for trematodes
	Larval cestodes Spargana (pleurocercoids) of pseudophyllidean cestodes (*Spirometra, Diphyllobothrium latum*)	Anurans and urodeles	Same as for trematodes

[a]A, acanthocephalid; N, nematode; MT, monogenetic trematode; DT, digenetic trematode; P, protozoa.

d. Trematodes. The majority of trematodes of amphibia are digenetic, i.e., they have one or several hosts during their larval stages. These helminths may utilize amphibians as intermediate or final hosts. Larvae may be encysted in many tissues and are generally observed during histopathological examination. Adult trematodes are usually found in the respiratory, gastrointestinal, and urinary systems (Marcus, 1981; Reichenbach-Klinke and Elkan, 1965). Monogenetic trematodes live only on one host; these flukes are usually found on the gills of fish, but certain species infest amphibians.

e. Other. Acanthocephalid worms can damage the gut wall in heavy infestations. Leeches, parasitic crustaceans such as *Argulus,* and fly (Diptera) depredation are not likely to be encountered in laboratory amphibians.

6. Fungal

Fungi associated with disease in amphibians are opportunistic pathogens. As is the case of bacterial infections, the organisms are saprophytes normally present in the environment, invading host tissues only when there are disturbances of homeostasis. Such alterations include unsanitary housing conditions, temperature variations (which may affect the amphibian's immune and inflammatory defense mechanisms as well as directly affecting fungal growth rate), malnutrition, and intercurrent diseases.

In general, amphibians housed in an aquatic environment have some of the same fungal diseases (saprolegniasis, ichthyosporidiosis) as do fish, while amphibians that are partially or completely terresterial are infected by soil saprophytes (chromomycosis, phycomycosis).

a. Chromomycosis (Chromoblastomycosis). The etiologic agents are brown-pigmented septate fungi that form conidia and conidiopores in culture. They infect not only amphibians but also a variety of mammalian species including humans. Representative genera are *Fonsecaea* (*F. pedrosoi, F. dermatitidis*), *Phialophora gougerotti, Cladosporium* (*C. carrioni, C. herbarum*), *Scolecosbasidium humicola,* and unidentified species (Cicmanec *et al.,* 1973; Elkan, 1976; Frank, 1976; Rush *et al.,* 1974). Organisms grow on the walls of housing tanks and presumably the portal of entry is the traumatized integument.

Clinical signs in frogs and toads are those of a chronic debilitating disease with anorexia, weight loss, and cutaneous ulcers (Fig. 5). At necropsy, there are multiple tan-white to gray nodules of varying sizes in liver, spleen, kidneys, lung, and other viscera. Histologically, skin ulcers have a base of histiocytes, fibroblasts, and occasional multinucleate giant cells (Langhan's and foreign body). The inflammatory reaction may extend into underlying muscle and bone. Visceral lesions consist of multiple, coalescent granulomas of histiocytes, epithelioid cells, and giant cells surrounded by sheets of fibroblasts and, often, large areas of caseous necrosis. Within the inflammatory reaction are pigmented septate fungal hyphae approximately 5-μm thick and/or tissue phase fungal cells (Fig. 6). The latter are elliptical or spherical pigmented fungal structures (approximately 6–17 μm in diameter) divided by septa into two or four cell stages; they are present singly or in short chains. Tissue phase cells but not hyphae are also found in chromomycotic lesions of mammals.

Gross and microscopic lesions of amphibian chromomycosis are similar to mycobacteriosis but can be readily differentiated by identification of pigmented fungal forms. A rapid method of antemortem diagnosis is demonstration of pigmented fungal

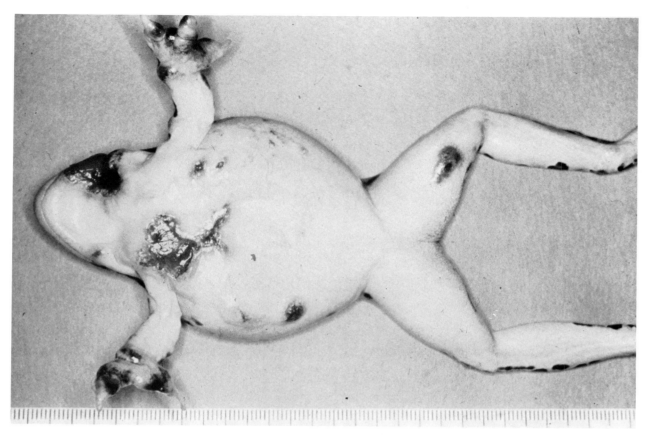

Fig. 5. Cutaneous ulcers and nodules in a *Rana pipiens* with chromomycosis (From Rush *et al.,* 1974. Reprinted with permission of *Laboratory Animal Science.*)

forms in a wet mount preparation collected from a cutaneous ulcer. The fungi also can be isolated on standard laboratory media.

The condition can be prevented by adequate sanitation of tanks. Infected animals should be killed; treatment is not recommended. Because chromomycosis has been reported in humans, generally as a cutaneous infection, handlers should wear gloves to protect against zoonotic transmission.

b. Phycomycosis (Mucormycosis). Phycomycetes in amphibians include the genera *Mucor* spp. and *Basidiobolus ranarum* (Elkan, 1976; Frank, 1976). As well as being present in the environment, the phycomycetes have been isolated from the intestinal tract and feces of toads, frogs, and salamanders. The opportunistic character of phycomycetes and clinical course of the disease resemble chromomycosis.

Mucor infection produces white nodules in liver and spleen. Histologically, there is granulomatous inflammation containing multiple unpigmented fungal spherules. These structures ranged from 10 to 50 μm in diameter. Smaller spherules contain a large vacuole, while larger ones contain up to as many as 12 daughter spherules. Hyphae, which are typically found in

mammalian phycomycosis, do not occur in amphibian tissues but develop along with sporangiophores when the *Mucor* sp. is isolated on culture media (Frank *et al.,* 1974).

Basidiobolus ranarum primarily produces cutaneous ulcers with a base of thick granulation tissue. Fungal hyphae are present in the lesions (Reichenbach-Klinke and Elkan, 1965).

There is no treatment for amphibians with phycomycosis; affected animals should be killed and the environment disinfected. Prevention is best accomplished by using optimal husbandry techniques.

c. Saprolegniasis. Many species of *Saprolegnia* are common inhabitants of fresh water and aquaria. Neotenic urodeles particularly mud puppies (*Necturus maculosus*), newts, aquatic frogs, and tadpoles are the amphibians primarily affected. *Saprolegnia parasitica* and *S. ferax* are two of the most common isolates.

The infection, almost identical to that in fish, is cutaneous with gray-white mycelial mats on the skin surface usually starting on the head. Underlying epithelium may be ulcerated, with necrosis extending into the subcutaneum and muscle. The fungi also grow on dead eggs and may threaten viable eggs.

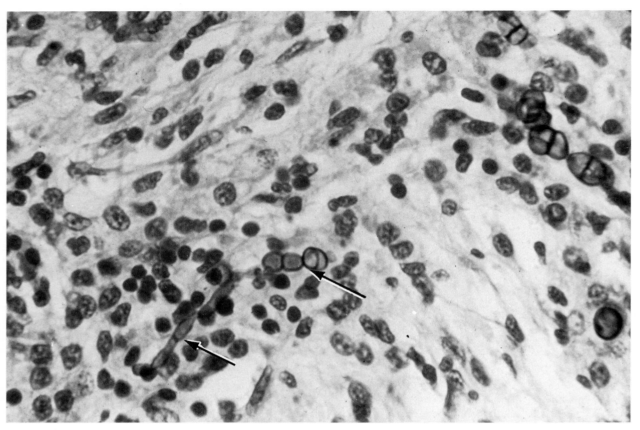

Fig. 6. Histologic lesion of chromomycosis with admixture of histiocytes, fibroblasts, and fungal forms. (Hyphae, short arrow; tissue phase fungal cell, long arrow.)

Treatment of infected amphibians should employ one of the following regimens (Marcus, 1981).

1. Malachite green 1 : 15,000 solution (2 gm malachite green in 3 liters water): Dip amphibians for 15 sec (longer may cause massive epithelial exfoliation) once daily for 2–3 days.

2. Copper sulfate 1 : 2000 solution: Dip for 2 min once per day for 5 days, then once per week until skin lesions regress.

3. 0.001% Chloramine (sodium *p*-toluenesulfonchloramide) bath.

4. 0.002–0.004% Mercurochrome bath.

Less efficacious substances are potassium permanganate, weak tincture of iodine, trypoflavin or methylene blue (Reichenbach-Klinke and Elkan, 1965).

Adequate sanitation procedures should be observed and dead ova removed promptly from tanks.

d. Ichthyosporidiosis. Several species of *Ichthyosporidium* have been isolated from cutaneous lesions of amphibians, particularly urodeles (Elkan, 1976). Description of the disease in amphibians is minimal, but clinical signs and lesions (and treatment) are presumably similar to those in fish.

B. Metabolic/Nutritional and Miscellaneous

a. Management-Related. The problems associated with nutrition of amphibians has been previously discussed in this chapter.

Failure of larvae to undergo metamorphosis, as already mentioned, is often associated with iodine deficiency or goitrogenic substances in the diet. Affected tadpoles are larger than normal; their thyroids have hyperplastic columnar epithelium with minimal colloid. Treatment consists of withdrawal of goitrogens from the diet or, if the etiology is a simple iodine deficiency, supplementation with iodized salt as 0.5% of the total diet (Marcus, 1981).

Bone deformities, seen sporadically in frogs and newts, may be related to vitamin deficiencies or dietary mineral imbalances (Reinchenbach-Klinke and Elkan, 1965).

Rectal prolapse has been reported in many amphibians and

may be associated with diet. Treatment consists of isolating the affected animal and withdrawing food until the prolapse reduces itself, a process taking approximately 1 week (Reichenbach-Klinke and Elkan, 1965).

The time of metamorphosis is a critical point in the life of amphibians. In rearing amphibians from larvae, provisions must be made for the animals to escape from their aquatic environment to a terrestrial millieu. Otherwise, mortality will be considerable (Reichenbach-Klinke and Elkan, 1965).

b. Miscellaneous. Hydrops (anasarca) is a condition found sporadically in axolotls and frogs. Lymph accumulates in the lymph sacs or the coelom. The etiology is unknown, and aspiration or antibiotic therapy is ineffective (Reichenbach-Klinke and Elkan, 1965).

On occasion, laboratory-housed female amphibians fail to ovulate. Eggs are retained within the coelom where they undergo degeneration and resorption. While clinical signs are not obvious, large amounts of melanin are released internally causing marked pigmentation of visceral organs.

c. Genetic. Amphibians can have a variety of congenital anomalies (Reichenbach-Klinke and Elkan, 1965) that are dramatic from a morphologic standpoint but are of minimal research significance. Of more importance to experimental embryologists are mutant genes that can be employed to study the control of organogenesis. Such genes have been recognized in a number of species and most extensively investigated in the axolotl (Malacinski and Brothers, 1974). Also of interest is

recognition of polyploid (as well as diploid) populations of anurans, a situation analogous to that of fishes (Bogart and Tandy, 1976).

Some amphibians, particularly Bufonidae, are potential hermaphrodites. Bidder's organ, a normal anatomic structure located cranial to the gonads, is a rudimentary gonad (testis in females, ovary in males). If the primary gonads are removed or otherwise rendered inactive, Bidder's organ can develop into a fully functional gonad of the opposite sex (Noble, 1954).

C. Neoplastic

The most current reviews of amphibian neoplasms are those of Balls and Clothier (1974), Marcus (1981), and the "Annual Activities Reports" of the Smithsonian Institution's Registry of Tumors in Lower Animals (Harshbarger, 1965–1981). Spontaneous tumors were derived from all germ layers, occurred in all organ systems, and affected both anurans and urodeles. Malignant tumors were reported more frequently than benign tumors (Balls and Clothier, 1974). A diagnosis of malignancy was often made on the basis of local invasion and/or cellular atypia. Malignant tumors also metastasized, although this occurred less frequently.

The majority of reports of amphibian neoplasms have been case reports in a single animal (Squire *et al.,* 1978). Selected series of amphibian tumors are summarized in Table V. Most involved the integument and were histologically similar to their mammalian counterparts (Fig. 7).

Table V

Selected Series of Amphibian Neoplasms

Species	Neoplasm	Comments	Reference
Spontaneous			
Anurans			
Rana pipiens	Renal adenocarcinoma	Discussed under viral diseases	—
Rana pipiens	Skin—squamous cell carcinoma, dermal adenocarcinoma	—	Van Der Steen *et al.,* 1972
Rana temporaria and *Rana ridibunda*	Skin—Mucous gland cystadenoma and cystadenocarcinoma	Malignancy based on atypia	Khudoley and Mizgireuv, 1980
Urodeles			
Triturus vulgaris	Skin (dermis)—fibroma	—	Stolk, 1958
Ambystoma mexicanum	Testicular tumors	Derived from spermatogonia and associated undifferentiated cells	Humphrey, 1969
Triturus cristatus	Skin—squamous cell papilloma, squamous cell carcinoma, mucous gland tumors	Correlation with an endogenous collagenolytic system	Wirl, 1972
Ambystoma tigrinum	Skin—squamous cell papilloma, dermal fibroma and fibrosarcoma, melanoma and malignant melanoma	Neotenic salamanders in sewage lagoon; possible association with jet fuel constituent perylene (run-off contaminant)	Roe and Harshbarger, 1977; Roe, 1977

(continued)

Table V (*Continued*)

Species	Neoplasm	Comments	Reference
Ambystoma mexicanum	Skin—mast cell tumors	Occurred in aged (10–17 years) animals; no visceral involvement or metastases	Delanney *et al.*, 1980; Harshbarger, 1982
Induced			
Rana temporaria	Hepatocellular tumors and "hemocytoblastosis"	Diethylnitrosamine, dimethylnitrosamine in tank water	Khudoley, 1977
Xenopus borealis	Cholangiocarcinoma, hepatocellular carcinoma, renal adenocarcinoma	Dimethylnitrosamine in tank water	Khudoley and Picard, 1980
Rana pipiens	Sarcomas	Methylcholanthrene pellets implanted in forelimbs, greater tumor frequency and earlier appearance in denervated limb	Outzen *et al.*, 1976
Triturus cristatus	Amelanotic melanoma	Methylcholanthrene subcutaneously	Leone and Zanvanella, 1969
Bufo regularis	Ileal adenocarcinoma, hepatoma	Bracken fern by gavage; no tumor descriptions or photographs	El-Mofty *et al.* 1980

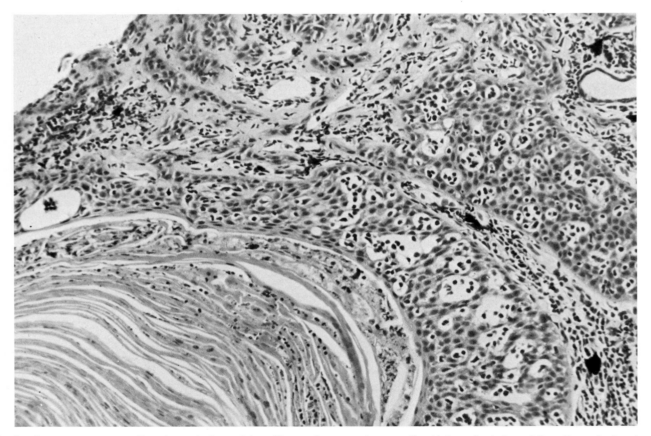

Fig. 7. Cutaneous squamous cell carcinoma in *Rana pipiens*. Cluster of stratum spinosum cells with loss of polarity and fingerlike projections invading dermal connective tissue. Large cyst within tumor containing layered keratin.

"Lymphosarcomas" in *Xenopus laevis* and *Cynops pyrrhogaster* are not listed in Table V. The preponderance of scientific opinion at this time is that these conditions are not neoplasms but rather are a diffuse granulomatous inflammatory response to mycobacteria (Squire *et al.*, 1978).

Amphibians respond to chemical carcinogens in a manner similar to mammals. Some induced tumors are listed in Table V.

REFERENCES

Abrams, G. D. (1969). Diseases in an amphibian colony. *In* "Biology of Amphibian Tumors" (M. Mizell, ed.), pp. 419–428. Springer-Verlag, Berlin and New York.

Amborski, R. L., and Amborski, G. F. (1978). Pathogens and diseases of aquatic animals. *Lab. Anim.* Nov.–Dec. 14–26.

Balls, M., and Clothier, R. H. (1974). Spontaneous tumours in amphibia. A review. *Oncology* **29**, 501–519.

Bardsley, J. E., and Harmsen, R. (1973). The trypanosomes of Anura. *Adv. Parasitol.* **11**, 1–73.

Barlow, E. H., Di Marzo, S. J., and Cohen, N. (1981). Prolonged survival of major histocompatibility complex—Disparate skin allografts transplanted to the metamorphosing frog, *Xenopus laevis*. *Transplantation* **32**, 51–57.

Berns, M. W. (1965). Mortality caused by kidney stones in spinach-fed frogs (*Rana pipiens*). *BioScience* **15**, 297–298.

Bogart, J. P., and Tandy, M. (1976). Polyploid amphibians: Three more diploid-tetraploid cryptic species of frogs. *Science* **193**, 334–335.

Bowler, J. K. (1977). "Longevity of Reptiles and Amphibians in North American Collections" (G. R. Pisani and B. Paschke, eds.), Herpetol. Circ. No. 6, pp. 1–32. Society for the Study of Amphibians and Reptiles, University of Kansas, Lawrence.

Boyer, C. I., Jr., Blackler, K., and Delanney, L. E. (1971). *Aeromonas hydrophila* infection in the Mexican axolotl, *Siredon mexicanum*. *Lab. Anim. Sci.* **21**, 372–375.

Burton, A. N., McLintock, J., and Rempel, J. G. (1966). Western equine encephalitis virus in Saskatchewan garter snakes and leopard frogs. *Science* **154**, 1029–1031.

Carmena-Suero, A., Siret, J. R., Callejas, J., and Arpones-Carmena, D. (1980). Blood volume in male *Hyla septentrionalis* (tree frog) and *Rana catesbeiana* (bull frog). *Comp. Biochem. Physiol. A* **67A**, 187–189.

Chitwood, M., and Lichtenfels, J. R. (1972). Parasitological review: Identification of parasitic metazoa in tissue sections. *Exp. Parasitol.* **32**, 407–519.

Cicmanec, J. L., Ringler, D. H., and Beneke, E. S. (1973). Spontaneous occurrence and experimental transmission of the fungus, *Fonsecaea pedrosoi* in the marine toad, *Bufo marinus*. *Lab. Anim. Sci.* **23**, 43–47.

Clark, H. F., Gray, C., Fabian, F., Zeigel, R., and Zarzon, D. T. (1969). Comparative studies of amphibian cytoplasmic virus strains isolated from the leopard frog, bullfrog, and newt. *In* "Biology of Amphibian Tumors" (M. Mizell, ed.), pp. 310–326. Springer-Verlag, Berlin and New York.

Clark, H. F., Michalski, F., Tweedell, K. S., Yohn D., and Zeigel, R. F. (1973). An adenovirus; FAV-1, isolated from the kidney of a frog (*Rana pipiens*). *Virology* **51**, 392–400.

Cochran, D. M. (1961). "Living Amphibians of the World." Doubleday, Garden City, New York.

Cohen, N., DiMarzo, S. J., and Hailparn-Barlow, E. (1980). Induction of

tolerance to alloantigens of the major histocompatibility complex in the metamorphosing frog, *Xenopus laevis*. *In* "Phylogeny of Immunological Memory" (M. J. Manning, ed.), pp. 225–231. Elsevier/North-Holland Biomedical Press, Amsterdam.

Collard, W., Thornton, H., Mizell, M., and Green, M. (1973). Virus-free adenocarcinoma of the frog (summer phase tumor) transcribes Lucké tumor herpesvirus-specific RNA. *Science* **181**, 448–449.

Culley, D. D. (1973). Use of bullfrogs in biological research. *Am. Zool.* **13**, 85–90.

Culley, D. D. (1976). Culture and management of the laboratory frog. *Lab. Anim.* Sept.–Oct., 30–36.

Delanney, L., Chang, S. C., Harshbarger, J., and Dawe, C. (1980). Mast cell tumors in the caudate amphibian, *Ambystoma mexicanum*. *Adv. Comp. Leuk. Res., Proc. Int. Symp., 9th, 1979* pp. 221–222.

DuPasquier, L., Blomberg, B., and Bernard, C. C. A. (1979). Ontogeny of immunity in amphibians: Changes in antibody repertoires and appearance of adult major histocompatibility antigens in *Xenopus*. *Eur. J. Immunol.* **9**, 900–906.

Elkan, E. (1976). Pathology in amphibia. *In* "Physiology of the Amphibia" (B. Lofts, ed.), Vol. 3, pp. 273–312. Academic Press, New York.

El-Mofty, M. M., Sadek, I. A., and Bayoumi, S. (1980). Improvement in detecting the carcinogenicity of bracken fern using an Egyptian toad. *Oncology* **37**, 424–425.

Emmons, M. B. (1973). Problems of an amphibian supply house. *Am. Zool.* **13**, 91–92.

Etkin, W. (1964). Metamorphosis. *In* "Physiology of the Amphibia" (J. A. Moore, ed.), pp. 427–468. Academic Press, New York.

Farrar, E. S., and Frye, B. E. (1979). Factors affecting normal carbohydrate levels in *Rana pipiens*. *Gen. Comp. Endocrinol.* **39**, 358–371.

Frank, W. (1976). Mycotic infections in amphibians and reptiles. *In* "Wildlife Diseases" (L. A. Page, ed.), pp. 73–88. Plenum, New York.

Frank, W., Roester, U., and Scholer, H. J. (1974). Sphaerule formation by a *Mucor* species in the internal organs of amphibia. *Zentralbl. Bakteriol., Parasitenkd., Infektionskr. Hyg., Abt. 1: Orig., Reihe A* **226**, 405–417.

Friedmann, G. B., Algard, F. T., and McCurdy, H. M. (1969). Determinations of the red blood cell count and haemoglobin content of urodele blood. *Anat. Rec.* **163**, 55–57.

Gatten, R. E., and Brooks, G. R. (1969). Blood physiology of a tropical frog, *Leptodactylus fallax*. *Comp. Biochem. Physiol.* **30**, 1019–1028.

Gibbs, E. L. (1963). An effective treatment for red-leg disease in *Rana pipiens*. *Lab. Anim. Care* **13**, 781–783.

Gibbs, E. L., Gibbs, T. J., and Van Dyck, P. C. (1966). *Rana pipiens*, health and disease. *Lab. Anim. Care* **16**, 142–160.

Gibbs, E. L., Nace, G. W., and Emmons, M. B. (1971). The live frog is almost dead. *BioScience* **21**, 1027–1034.

Glorioso, J. C., Amborski, R. L., Larkin, J. M., Amborski, G. F., and Culley D. C. (1974a). Laboratory identification of bacterial pathogens of aquatic animals. *Am. J. Vet. Res.* **35**, 447–450.

Glorioso, J. C., Amborski, R. L., Larkin, J. M., Amborski, G. F., and Culley, D. C. (1974b). Microbiological studies on septicemic bullfrogs. *Am. J. Vet. Res.* **35**, 1241–1245.

Harris, J. A. (1972). Seasonal variation in some hematological characteristics of *Rana pipiens*. *Comp. Biochem. Physiol. A* **43A**, 975–989.

Harshbarger, J. C. (1965–1981). "Activities Report. Registry of Tumors in Lower Animals." Smithsonian Institution, Washington, D.C. (supplements issued annually).

Harshbarger, J. C. (1982). Epizootiology of leukemia and lymphomas in poikilotherms. *Adv. Comp. Leuk. Res., Proc. Int. Symp., 11th, 1981* pp. 39–46.

Hermansen, B., and Jorgensen, C. B. (1969). Blood glucose in male toads (*Bufo bufo*): Annual variation and hormonal regulation. *Gen. Comp. Endocrinol.* **12**, 313–321.

Hine, R. L., Les, B. L., and Hellmich, B. (1981). Leopard frog populations and mortality in Wisconsin, 1974–76. *Wis. Dep. Nat. Resour., Tech. Bull.* **122.**

Hird, D. W., Diesch, S. L., McKinnell, R. G., Gorham, E., Martin, F. B., Kurtz, W., and Durbrovolny, C. (1981). *Aeromonas hydrophila* in wild caught frogs and tadpoles (*Rana pipiens*) in Minnesota. *Lab. Anim. Sci.* **31,** 166–169.

Hobson, B. M. (1965). Cold blooded vertebrates, including *Xenopus laevis*. *Food Cosmet. Toxicol.* **3,** 209–215.

Horton, J. D., and Horton, T. L. (1975). Development of transplantation immunity and restoration experiments in the thymectomized amphibian. *Am. Zool.* **15,** 73–84.

Hubbard, G. B. (1981). *Aeromonas hydrophila* infection in *Xenopus laevis*. *Lab. Anim. Sci.* **31,** 297–300.

Humphrey, R. R. (1969). Tumors of the testis in the Mexican axolotl (*Ambystoma* or *Siredon mexicanum*). *In* "Biology of Amphibian Tumors" (M. Mizell, ed.), pp. 220–228. Springer-Verlag, Berlin and New York.

ILAR (Institute of Laboratory Animal Resources). (1974). "Amphibians. Guidelines for the Breeding, Care, and Management of Laboratory Animals. A Report of the Subcommittee on Amphibian Standards, Committee on Standards," Natl. Acad. Sci., Washington, D.C.

ILAR (Institute of Laboratory Animal Resources). (1979). Other vertebrates. *In* "Animals for Research. A Directory of Sources," pp. 75–105. Natl. Acad. Sci., Washington, D.C.

Joiner, G. N., and Abrams, G. D. (1967). Experimental tuberculosis in the leopard frog. *J. Am. Vet. Med. Assoc.* **151,** 942–949.

Kaplan, H. M. (1973). Parasites of laboratory reptiles and amphibians. *In* "Parasites of Laboratory Animals" (R. J. Flynn, ed.), pp. 507–644. Iowa State Univ. Press, Ames.

Katagiri, C. (1978). *Xenopus laevis* as a model of the study of immunology. *Dev. Comp. Immunol.* **2,** 5–14.

Khudoley, V. V. (1977). The induction of tumor in *Rana temporaria* with nitrosamines. *Neoplasma* **24,** 249–251.

Khudoley, V. V., and Mizgireuv, I. V. (1980). On spontaneous skin tumors in amphibia. *Neoplasma* **27,** 289–293.

Khudoley, V. V., and Picard, J. J. (1980). Liver and kidney tumors induced by *N*-nitrosodimethylamine in *Xenopus borealis* (Parker). *Br. J. Cancer* **25,** 679–683.

King, J. A., and Millar, R. P. (1981). TRH, GH-RIH, and LH-RH in metamorphosing *Xenopus laevis*. *Gen. Comp. Endocrinol.* **44,** 20–27.

Kirn, A., Birgen, A., Elharrar, M., and Gendrault, J. L. (1975). L'hépatite dégénérative argue de le sowus provoquée par le FV₃ (frog virus 3) un modèle expérimental de toxicité virale. *Bull. Inst. Pasteur, Paris* **73,** 65–94.

Leone, V. G., and Zavanella, T. (1969). Some morphological and biological characteristics of a tumor of the newt, *Triturus cristatus* Laur. *In* "Biology of Amphibian Tumors" (M. Mizell, ed.), pp. 184–194. Springer-Verlag, Berlin and New York.

Levine, N. D., and Nye, R. R. (1976). *Toxoplasma ranae Sp. n.* from the leopard frog *Rana pipiens* Linnaeus. *J. Protozool.* **23,** 488–490.

Levine, N. D., and Nye, R. R. (1977). A survey of blood and other tissue parasites of leopard frogs *Rana pipiens* in the United States. *J. Wildl. Dis.* **13,** 17–23.

McKinnell, R. G., and Cunningham, W. P. (1982). Herpesviruses in metastatic Lucké renal adenocarcinoma. *Differentiation* **22,** 41–46.

McKinnell, R. G., and Ellis, V. L. (1972). Herpesviruses in tumors of postspawning *Rana pipiens*. *Cancer Res.* **32,** 1154–1159.

Maitrya, B. B., Raman, B. N., and Vyas, C. R. (1970). Effect of varying environmental temperature on blood glucose level in Indian frog, *Rana tigrina*. *Indian J. Exp. Biol.* **8,** 339–340.

Malacinski, G. M., and Brothers, A. J. (1974). Mutant genes in the Mexican axolotl. *Science* **184,** 1142–1147.

Marcus, L. C. (1981). "Veterinary Biology and Medicine of Captive Amphibians and Reptiles." Lea & Febiger, Philadelphia, Pennsylvania.

Nace, G. W. (1968). The amphibian facility of the University of Michigan. *BioScience* **18,** 767–775.

Nace, G. W. (1970). The use of amphibians in biomedical research. *In* "Animal Models for Biomedical Research III," pp. 103–124. Natl. Acad. Sci., Washington, D.C.

Nace, G. W., and Buttner, J. K. (1984). Nutritional disorders in amphibians. *In* "Diseases of Amphibians and Reptiles" (G. L. Hoff, F. Frye, and E. Jacobson, eds.). Plenum, New York (in press).

Nace, G. W., and Rosen, J. K. (1979). Sources of amphibians for research. II. *Herpetol. Rev.* **10,** 8–15.

Nace, G. W., Waage, J. K., and Richards, C. M. (1971). Sources of amphibians for research. *BioScience* **21,** 768–773.

Naegele, R. F., and Grannoff, A. (1977). Viruses and renal carcinoma of *Rana pipiens*. XV. The presence of virus-associated membrane antigen(s) on Lucké tumor cells. *Int. J. Cancer* **19,** 414–418.

Newcomer, C. E., Anver, M. R., Simmons, J. L., Wilcke, B. W., Jr., and Nace, G. W. (1982). Spontaneous and experimental infections of *Xenopus laevis* with *Chlamydia psittaci*. *Lab. Anim. Sci.* **32,** 680–686.

Noble, G. K. (1954). "The Biology of the Amphibia." Dover, New York.

Outzen, H. C., Custer, R. P., and Prehn, R. T. (1976). Influence of regenerative capacity and innervation on oncogenesis in the adult frog (*Rana pipiens*). *JNCI, J. Natl. Cancer Inst.* **57,** 79–84.

Platt, J. E., and LiCause, M. J. (1980). Effects of oxytocin in larval *Ambystoma tigrinum:* Acceleration of induced metamorphosis and inhibition of the antimetamorphic action of prolactin. *Gen. Comp. Endocrinol.* **41,** 84–91.

Reichenbach-Klinke, H., and Elkan, E. (1965). "The Principal Diseases of Lower Vertebrates." Academic Press, New York.

Riviere, J. E., Shapiro, D. P., and Coppoc, G. L. (1979). Percutaneous absorption of gentamycin by the leopard frog, *Rana pipiens*. *J. Vet. Pharmacol. Ther.* **2,** 235–239.

Robertson, D. R. (1978). Seasonal changes in plasma and urinary sodium, potassium and calcium in the frog, *Rana pipiens*. *Comp. Biochem. Physiol. A* **60A,** 387–390.

Roe, F. L. (1977). Tissue lesions of tiger salamanders (*Ambystoma tigrinum*): Relationship to sewage effluents. *Ann. N.Y. Acad. Sci.* **298,** 270–279.

Roe F. L., and Harshbarger, J. C. (1977). Neoplastic and possibly related skin lesions in neotenic tiger salamanders from a sewage lagoon. *Science* **196,** 315–317.

Rollins, L. A., and Cohen, N. (1980). On tadpoles, transplantation, and tolerance. *In* "Development and Differentiation of Vertebrate Lymphocytes" (J. D. Horton, ed.), pp. 203–214. Elsevier/North-Holland Biomedical Press, Amsterdam.

Roofe, P. G. (1961). Blood constituents of *Ambystoma tigrinum*. *Anat. Rec.* **140,** 337–340.

Rush, H. G., Anver, M. R., and Beneke, E. S. (1974). Systemic chromomycosis in *Rana pipiens*. *Lab. Anim. Sci.* **24,** 646–655.

Sawin, C. T., Bolaffi, J. L., Callard, I. P., Bacharach, P., and Jackson, I. M. D. (1978). Induced metamorphosis in *Ambystoma mexicanum:* Lack of effect of triiodothyronine on tissue or blood levels of thyrotropin-releasing hormone (TRH). *Gen. Comp. Endocrinol.* **36,** 427–432.

Saxena, P. K., and Zal, S. R. (1981). Seasonal changes in testes and thumb pads of the toad, *Bufo marinus* (Linn.) and their correlation with fluctuations in the environmental factors. *Anat. Anz.* **149,** 337–344.

Sharma V. K., Kaura, T. K., and Singh, I. P. (1974). Frogs as carriers of *Salmonella* and *Edwardsiella*. *Antonie van Leeuwenhoek* **40,** 171–175.

Shield, J. W., and Bentley, P. J. (1973). Respiration of some urodele and anuran amphibia. I. In water, role of the skin and gills. *Comp. Biochem. Physiol. A* **46A,** 17–28.

Shively, J. N., Songer, J. G., Prchal, S., Keasey, M. S., III, and Thoen, C.

O. (1981). *Mycobacterium marinum* infection in Bufonidae. *J. Wildl. Dis.* **17**, 3–8.

Shoemaker, V. A., and Nagy, K. A. (1977). Osmoregulation in amphibians and reptiles. *Annu. Rev. Physiol.* **39**, 449–471.

Singh, S., Sharma, V. D., and Sethi, M. S. (1979). Toads as resevoirs of salmonellae: Prevalence and antibiogram. *Int. J. Zoonoses* **6**, 82–84.

Skadhauge, E. (1977). Excretion in lower vertebrates: Function of gut, cloaca, and bladder in modifying the composition of urine. *Fed. Proc., Fed. Am. Soc. Exp. Biol.* **36**, 2487–2492.

Slooff, W., and Baerselman, R. (1980). Comparison of the usefulness of the Mexican axolotl (*Ambystoma mexicanum*) and the clawed toad (*Xenopus laevis*) in toxicological bioassays. *Bull. Environ. Contam. Toxicol.* **24**, 439–443.

Smith, C. L. (1950). Seasonal changes in blood sugar, fat body, liver glycogen, and gonads in the common frog, *Rana temporaria. J. Exp. Biol.* **26**, 412–429.

Smith, S. W. (1950). Chloromycetin in the treatment of "red-leg." *Science* **112**, 274–275.

Squire, R. A., Goodman, D. G., Valerio, M. G., Frederickson, T. N., Strandberg, J. D., Levitt, M. H., Lingeman, C. H., Harshbarger, J. C., and Dawe, C. J. (1978). Tumors. *In* "Pathology of Laboratory Animals" (K. Benirschke, F. M. Garner, and T. C. Jones, eds.), Vol. 2, Chapter 12, pp. 1255–1257. Springer-Verlag, Berlin and New York.

Stewart, H. L., Snell, K. C., Dunham, L. J., and Schlyen, S. M. (1959). "Transplantable and Transmissible Tumors of Animals," pp. 208–217. Armed Forces Inst. Pathol., Washington, D.C.

Stock, G. B., and Bryant, S. V. (1981). Studies of digit regeneration and their implications for theories of development and evolution of vertebrate limbs. *J. Exp. Zool.* **216**, 423–433.

Stolk, A. (1958). A transplantable fibroma in the skin in the newt, *Triturus taeniatus. Experientia* **14**, 2–3.

Tanaka, Y., Noda, K., Nomaguchi, T., and Yamagishi, H. (1974). Hepatic melanosis and ageing in amphibia. A preliminary observation on two species of anura *Xenopus laevis* and *Rana nigromaculata. Exp. Gerontol.* **9**, 263–268.

Thant-Shima, M., Nakamura, A., Shimazu, H., Murao, S., and Sugiyama, T.

(1979). Possible usefulness of Burmese newt, *Tylototriton Verrucosus Anderson,* in teratogenesis studies. *Kobe J. Med. Sci.* **25**, 193–204.

Timmons, E. H., Olmsted, G. M., and Kaplan, H. M. (1977). The germfree leopard frog (*Rana pipiens*): Preliminary report. *Lab. Anim. Sci.* **27**, 518–521.

Townsend, D. S., Stewart, M. M., Pough, F. H., and Brussard, P. F. (1981). Internal fertilization in an oviparous frog. *Science* **212**, 469–471.

Trust, T. J., Bartlett, K. H., and Lior, H. (1981). Importation of salmonellae with aquarium species. *Can. J. Microbiol.* **27**, 500–504.

Tweedell, K. S., and Wong, W. Y. (1974). Brief communication: Frog kidney tumors induced by herpesvirus culture in pronephric cells. *JNCI, J. Natl. Cancer Inst.* **52**, 621–624.

Van Der Steen, A. B. M., Cohen, B. J., Ringler, D. H., Abrams, G. D., and Richards, C. M. (1972). Cutaneous neoplasms in the leopard frog (*Rana pipiens*). *Lab. Anim. Sci.* **22**, 216–222.

van der Waaj, D., Cohen, B. J., and Nace, G. W. (1974). Colonization patterns of aerobic gram-negative bacteria in the cloaca of *Rana pipiens. Lab. Anim. Sci.* **24**, 307–317.

Wintrobe, M. M. (1933). Variations in the size and hemoglobin content of erythrocytes in the blood of various vertebrates. *Folia Haematol.* **51**, 32–49.

Wirl, G. (1972). Collagenolytic activity and carcinogenesis in the skin of the newt *Triturus cristatus. Arch. Geschwulstforsch.* **40**, 111–115.

Wolf, K., Bullock, G. L., Dunbar, C. E., and Quimby, M. C. (1969). Tadpole edema virus: Pathogenesis and growth studies and additional sites of virus infected bullfrog tadpoles. *In* "Biology of Amphibian Tumors" (M. Mizell, ed.), pp. 327–336. Springer-Verlag, Berlin and New York.

Wong, W. Y., and Tweedell, K. S. (1974). Two viruses from the Lucké tumor isolated in a frog pronephric cell line (37981) *Proc. Soc. Exp. Biol. Med.* **145**, 1201–1206.

Wright, M. L., Majerowski, M. A., Lukas, S. M., and Pike, P. A. (1979). Effect of prolactin on growth, development and epidermal cell proliferation in the hindlimb of the *Rana pipiens* tadpole. *Gen. Comp. Endocrinol.* **39**, 53–62.

Yadav, M. P., and Sethi, M. S. (1979). Poikilotherms as resevoirs of Q-fever (*Coxiella burnetti*) in Uttar Pradesh. *J. Wildl. Dis.* **15**, 15–17.

Chapter 15

Biology and Diseases of Reptiles

Elliott R. Jacobson

I. INTRODUCTION

A. Taxonomic Considerations

The class Reptilia represents a diverse group of animals that is situated at the crossroads of vertebrate evolution. The first reptile evolved from more primitive amphibian ancestors during the upper Paleozoic era, approximately 180–230 million years ago. The development of the cleidoic (amniotic) egg in concert with additional morphologic and physiologic modifications allowed these animals to be truly terrestrial. Today only 4 of the 17 orders present during the Mesozoic remain.

The Chelonia (220+ species) and the Crocodilia (21 species) are the most conservative of the present day reptiles. Whereas the phylogenetic line that gave rise to the chelonians failed to evolve into any other modern group of vertebrates, the line that gave rise to the crocodiles (the same that gave rise to the dinosaurs) progressed to become the class Aves. The remaining orders of reptiles are somewhat closely related and are placed within a single subclass, Lepidosauria. The Rhynchocephalia includes the single monotypic tuatara, *Sphenodon punctatus,* confined to several New Zealand islands. The fourth, and most diversified order of reptiles, is the Squamata including both Lacertilia (lizards: 3000 species) and Serpentes (snakes: 2500 species).

B.   Use in Research

The earliest herpetologists, starting out in the mid-1600s, were interested purely in collecting, naming, and cataloguing new animals. Up to the end of the nineteenth century, the study of reptiles remained one of descriptive cataloguing with much of the work being done in Europe. Starting in the middle part of the 1900s, the United States became a world center for herpetological studies with excellent groups at a variety of universities and museums.

The present-day field of herpetology covers all aspects of reptilian biology, including numerical taxonomy, morphology, histology, physiology, genetics, biochemistry and behavior. As the first group of vertebrates to successfully invade land, reptiles have evolved numerous morphologic and physiologic mechanisms to survive a dehydrating environment. Much of the current basic research is functional in approach. Although relatively few species have been studied in depth, the animal that comes closest to being the laboratory mouse of the reptile world is a lizard, the American anole, *Anolis carolinensis* (Fig. 1). Additional species of reptiles used in research include the American alligator, spectacled caiman, red-eared slider, whiptail lizards, geckos, common iguana, garter snakes, water snakes, kingsnakes, rat snakes, and Burmese python (Figs. 2–12).

C.   Availability and Sources of Research Animals

Investigators working with reptiles as research animals often choose a species for study based upon the needs of the experimental design, accessibility, ease and cost of maintenance, longevity, and reproductive success in captivity. Researchers in the biomedical community should have a knowledge of and a moral commitment to avoid those rare and endangered spe-

*International Union for the Conservation of Nature and Natural Resources, 1110 Morges, Switzerland.

cies that are threatened with extinction. The IUCN Red Data Book lists such endangered species.* Needless to say, the inability to acquire these animals without proper federal permits diminishes their usefulness as research animals.

Research animals are often secured from professional dealers. A listing of animals, including reptiles used in research and sources, can be found in *Animals for Research,* published by the Institute of Laboratory Animal Resources and is available from the National Academy of Sciences. Unfortunately many of the major suppliers of reptiles are not included in this listing. The best sources of information regarding the names of such dealerships and the quality of animal sold are herpetology departments at city zoos and through amateur and professional herpetology clubs and organizations. A listing of these organizations is available (Gilboa, 1972). The quarterly journal, *Herpetological Review,* published by the Society for the Study of Amphibians and Reptiles, often advertises herpetological material for sale.

D.   Laboratory Management and Husbandry

The studies of Cowles and Bogert (1944) demonstrated that behavioral thermoregulation is an important factor in the biology of many reptile species. It is the source of thermal energy that differentiate mammals and birds (endotherms), on one hand, from reptiles (ectotherms), on the other. Whereas birds and mammals (within limits) can control their body temperature within a fairly narrow zone by shifts in metabolic rates, reptiles are dependent upon external sources for regulation of body temperature.

All reptiles have a preferred optimum temperature zone (POTZ) that is fairly characteristic of the species, being regulated by behavioral and physiologic mechanisms. The limits of this zone, particularly with temperate species, may fluctuate with seasons of the year. Many physiologic functions appear to have evolved in unison with this POTZ.

The temperature zone below the POTZ has been termed the critical thermal minimum (CTMin) and is defined as the temperature that causes a cold narcosis and effectually prevents locomotion. The temperature zone above the POTZ is the critical thermal maximum (CTMax) and may be visualized as a value that is the arithmetic mean of the collective thermal points at which locomotory activity becomes disorganized and the animal loses its ability to escape from conditions that will promptly lead to its death (Lowe and Vance, 1955). Table I presents the POTZ, CTMin, and CTMax for a number of reptile species.

Many individuals who maintain reptiles in captivity as research animals, private collections, and zoo collections fail to take into consideration the thermobiologic needs of individual species. Animals maintained at room temperature (22°C) may

Table I

Thermal Profiles of Reptiles

Species	Mean preferred optimum temperature	Voluntary		Critical thermal maximum	Reference[a]
		Minimum	Maximum		
Chelonia					
Clemmys marmorata	24.8[b]	9.0	27.0	—	1
Terrapene ornata	28.0	13.0	35.9	—	1
Chrysemys picta	27.8	8.0	32.0	42.3	1
Sternotherus odoratus	21.2	16.2	28.8	—	1
Gopherus agassizi	30.6	15.0	37.8	39.5	1
Crocodilia					
Alligator mississippiensis	32–35	—	—	38–39	2
Rhynchocephalia					
Sphenodon punctatus	14–18	—	—	—	3
Lacertilia					
Coleonyx variegatus	24.7	15.0	34.0		1
Heloderma suspectum	27.2	24.2	33.7	44.5	1
Amblyrhynchus cristatus	32.9	—	—	—	1
Anolis sagrei	33.1	27.4	36.1	—	1
Basiliscus vittatus	35.0	22.5	38.5	—	1
Crotaphytus collaris	37.2	20.7	43.3	46.5	1
Dipsosaurus dorsalis	40.0	27.0	46.4	47.5	1
Phrynosoma platyrhinos	36.0	26.2	39.5	45.5	1
Sceloporus undulatus	34.8	25.0	38.9	43.7	1
Serpentes					
Boa constrictor	25.1	12.2	34.0	—	1
Coluber constrictor	29.6	15.0	37.4	42.4	1
Lampropeltis getulus	28.1	15.1	31.4	42.0	1
Masticophis flagellum	31.6	24.0	37.0	42.4	1
Pituophis catenifer	26.7	16.4	34.6	40.5	1
Thamnophis sirtalis	25.6	9.0	35.0	38–41	1
Crotalus atrox	27.4	18.0	34.0	39	1
Crotalus cerastes	26.2	17.5	34.5	41.6	1

[a]Key to references: (1) Brattström (1965); (2) Colbert *et al.* (1946); (3) Bogert (1953).
[b]Temperature is in °C.

refuse to eat and, even if the reptile feeds, it may show reduced food conversion and poor digestion. Such animals may regurgitate and slowly decline in body condition. Thus, providing the reptile with minimal thermal needs is essential for long-term maintenance.

The species being maintained will dictate the source of heat provided. Rarely should heat sources be placed within a cage. Incandescent light bulbs (frosted) with a reflector should be directed at basking sites within the cage. Infrared light bulbs provide an excellent source of thermal energy, but bulbs must be placed at a distance from the animal to prevent thermal burns. Heating devices such as heat tape and electric heating pads should be used with caution. Hot spots in these heating devices may result in ventral cutaneous burns. Ideally a cage should be designed to provide a thermal gradient ranging from heated to shaded areas.

Photoperiod has many obvious and esoteric considerations with respect to maintenance and reproduction of captive reptiles. With many reptiles, the lengthening days of spring are correlated with gonadal recrudescence (Crews and Garrick, 1980). Often with this lengthening light to dark cycle there is a concomitant increase in ambient temperature so that several environmental factors may be acting in concert.

The amount of moisture a particular species requires in the form of moisture in the air and available water will depend upon the ecological adaptations of a given species. Some species may fulfill all their water needs in food items, either directly or via metabolism. Still others are active drinkers and may require daily water provided in bowls or droplets on foliage, such as with the dew lapping *Anolis carolinensis*.

For many reptile species a humidity of 50–70% is ideal. If a snake fails to shed during the normal cycle of ecdysis, the skin will become dry and rough in appearance. Often this is attributed to a below optimum humidity level. Such snakes can be soaked in a shallow container of water followed by manual removal of retained skin.

The type of cage and substrate used also will depend upon the ecologic needs of the species. Although natural exhibits are aesthetically attractive, cages incorporating complex habitat design are difficult to clean, and individual animals may be impossible to monitor daily. Ultimately the cage decided upon should incorporate the needs of the investigator, zoo keeper, and private collector as well as the minimal needs of the species being maintained.

Based upon feeding strategies, reptiles can be categorized as either herbivores, omnivores, insectivores, or carnivores. Many disease problems in captive reptiles are related to poor nutrition, either directly or indirectly. Chelonians range from almost strict herbivores to almost strict carnivores, and when provided the proper food will thrive in captivity. Many aquatic species will do well on a diet of trout pellets, diced fish, and good quality vegetables, such as Romaine lettuce and spinach. Tortoises will thrive on mixtures of moistened puppy chow, primate chow, vegetables, alfalfa, and occasional mice. Crocodilians, as adults, are fairly strict carnivores, and most species are easy to maintain on whole animal diets such as rats, rabbits, chickens, and fish.

Of all the reptile groups, lizards show the greatest diversity of feeding strategies. Many species are insectivores and may require a great deal of effort for long-term maintenance. Most of the commercially available insects including mealworms and crickets are deficient in calcium, having inverse Ca : P ratios. Very few lizards in the wild are truly monophagous, and captive animals should be fed a varied insect diet in captivity. Mineral supplementation is often necessary to prevent metabolic bone disease, especially in rapidly growing young animals. The lizards that are carnivores, such as monitors, Gila monsters, and beaded lizards, often do well in captivity and are usually easy to maintain. These animals do well on complete animal diets of mice, rats, and chicks. Omnivorous lizards, such as the common iguana, *Iguana iguana*, will feed upon almost anything including mice, chicks, dog food, and vegetables.

Snakes as a group are carnivores and insectivores, and although several species are highly specialized for feeding upon a specific prey species, many are rather generalized feeders. Snakes that feed upon rodents, including many species of boids, colubrids, viperids, and elapids, are generally easy to maintain in captivity. Most snakes that feed in captivity will accept dead food, and live food should be avoided since rodents are capable of inflicting serious bites.

II. BIOLOGY

A. General Concepts

It is believed by evolutionary biologists that reptiles evolved from amphibians, but there is insufficient evidence to indicate conclusively from which particular group or groups they evolved. The fundamental difference between primitive reptiles and their amphibian ancestor(s) is believed to be in their reproductive strategies: amphibians produce anamniotic eggs, whereas reptiles produce amniotic eggs. The evolution of the amniotic egg was a major advance over more primitive modes of amphibian reproduction for it freed the first reptiles from the constraints of depositing eggs in an aquatic environment. Additional anatomic structures and physiologic mechanisms evolved to allow these animals to survive a dehydrating environment. Several of the major biologic systems of reptiles will be discussed and structure will be related to function.

Fig. 1. *Anolis carolinensis*, the American anole. This lizard is commonly found throughout the southeastern United States. (Courtesy of Mr. David Barker, Dallas Zoological Park.)

Fig. 2. *Alligator mississippiensis*, the American alligator. This crocodilian is found throughout low elevation areas of the southeastern United States.

Fig. 3. *Caiman crocodilus*, the spectacled caiman. This Central/South American crocodilian is commonly sold in the pet trade.

Fig. 4. *Chrysemys scripta elegans*, the red-eared slider. This aquatic turtle is found throughout the Mississippi drainage basin of the United States. (Courtesy of Mr. David Barker, Dallas Zoological Park.)

Fig. 5. *Cnemidophorus sexlineatus*, the six-lined racerunner. This teiid lizard is native to the eastern United States. (Courtesy of Mr. David Barker, Dallas Zoological Park.)

Fig. 6. *Coleonyx brevis*, the Texas banded gecko. (Courtesy of Dr. Joseph LaPointe, New Mexico State University.)

Fig. 7. *Iguana iguana*, the common green iguana. This Central/South American lizard is commonly sold in the pet trade.

Fig. 8. *Thamnophis sirtalis annectens*, the Texas garter snake. (Courtesy of Mr. David Barker, Dallas Zoological Park.)

Fig. 9. *Nerodia rhombifera*, the diamondback water snake. This snake ranges throughout the Mississippi Valley, United States. (Courtesy of Mr. David Barker, Dallas Zoological Park.)

Fig. 10. *Lampropeltis getulus californiae*, the California kingsnake. (Courtesy of Mr. Robert Bader.)

Fig. 11. *Elaphe guttata guttata*, the red rat or corn snake. (Courtesy of Mr. David Barker, Dallas Zoological Park.)

Fig. 12. *Python molurus bivittatus*, the Burmese python. This southeast Asian snake is noted for its ability to regulate egg clutch temperature by skeletal muscle contractions.

1

2

3

4

5

6

7

8

9

10

11

12

B. Integumentary System

The integumentary system plays an important role in the conservation of body fluids by forming a protective barrier between deeper tissues and the dehydrating environment. The stratum corneum is a much thicker portion of the epidermis in reptiles compared to amphibians. The skin is thrown into a series of folds that result in the formation of scales. In snakes and several groups of lizards the palpebrae have become modified evolutionarily into a spectacle that is a highly vascular transparent structure over the cornea. The spectacle protects the cornea against abrasion and trauma and is shed at the time the surrounding body scales are shed.

In contrast to amphibians, which have an abundance of integumentary glands, reptiles almost totally lack such glands. The few integumentary glands that do occur in reptiles are generally scent glands that probably function to ward off predators or function in sex discrimination and species recognition during the breeding season. A complex pigment system consisting of melanophores, erythrophores, xanthophores, and guanophores is found in the dermis and is comparable to that of amphibians.

Different groups of reptiles have evolved different mechanisms to allow for growth of the integumentary system. Snakes appear to be highly specialized in this regard and enter a very discreet cycle of ecdysis that consists of six stages of epidermal development (Maderson, 1965). Clinically, as a snake enters a cycle of ecdysis the skin and spectacle covering the cornea become dull and eventually take on a bluish tinge. Approximately 4–7 days after the height of development of this dullness to the skin, the snake's color rapidly brightens and the spectacle of the eye clears completely, and within 4–7 days shedding occurs.

C. Digestive System

The digestive system contains all structures from the oral cavity to the cloaca. The mouth parts of reptiles are generally more complex than those of amphibians. The oral cavity contains both polystomatic and monostomatic glands producing carbohydrate and carbohydrate-containing secretory products. Specialized monostomatic glands include venom glands of *Heloderma* (Gila monster and beaded lizard) and Duvernoy's and venom glands of poisonous snakes. Reptile venoms are complex mixtures of lethal proteins and enzymes that have been used in biomedical research as a source of pain killing agents for cancer victims, neurotoxins in neuropharmacologic studies, anticoagulants, highly purified RNase and DNases.

The alimentary canal has essentially the same components as in higher vertebrates. The esophagus is a transport system carrying food to the stomach and shows modification from group to group. In marine turtles the esophagus is lined by a series of heavily keratinized papillae that protect the animal from potential lesions resulting from a diet high in spiculated sponges, jellyfish, and silicaceous plants. The esophagus in snakes is rather thin, especially in the cranial portion, and highly distensible, adapted to accommodating large prey items. In the specialized egg-eating snake the anterior esophagus is penetrated by ventral vertebral hypapophyses that perforate and crush the shell of ingested eggs. The esophagus leads into the stomach where digestion commences with mechanical and chemical enzymatic decomposition of food. The largest amount of data accumulated on rates of gastric digestion is in snakes. In the boa contrictor, *Constrictor constrictor,* and the indigo snake, *Drymarchon corais,* the first signs of decalcification of an ingested rat, as demonstrated by radiography, was at 22 hr postfeeding (Blain and Campbell, 1942). Complete digestion under the conditions of the experiment (body temperature was not mentioned) was at 120 hr.

The stomach ultimately empties its contents into the small intestine, which may be highly convoluted in turtles and lizards while relatively straight with minor convolutions in snakes. Digestion is continued and completed in the small intestine. In the duodenum the mixture of stomach, bile, pancreatic enzymes, and intestinal juices is usually alkaline or slightly acidic. From the small intestine digesta passes into the colon. As with the small intestine, the colon of herbivorous reptiles also has a greater volume compared to that of carnivorous species. Feces produced in the colon enter the cloaca, which is a terminal chamber common to both the digestive and urogenital systems.

D. Respiratory System

Reptiles breathe primarily by lungs, but some aquatic turtles have supplemental cloacal, pharyngeal, or cutaneous respiration. The lungs of living reptiles vary in degree of complexity, being generally more complex and efficient than those of amphibians and less complex than those of birds and mammals. While the lungs of most reptiles are sac-like, those of higher lizards show division into a number of chambers. In snakes the lung is elongate, gradually merging into an air sac that terminates in the intestinal mesentery in the vicinity of the gallbladder in terrestrial species and up to the cloaca in aquatic species. In boids and colubrids the respiratory portion of the lung lies between the heart and cranial pole of the liver, while in viperids and elapids it is situated cranial to the heart. In all snakes the right lung is larger than the left, with the left lung being fairly well developed in boids and only vestigal in colubrids.

Whereas the lung volume is quite large in comparison with that of mammals, the surface area is only 1% as large as a comparably sized mammal. This is consistent with basal metabolic rate of reptiles ranging from one-tenth to one-third that of mammals of comparable weight. In snakes, a major portion of this volume is due to the presence of an air sac. This air sac may act as a reservoir for oxygen during periods of apnea. In aquatic forms it may also act as a buoyancy organ.

E. Urinary System

The evolution of reptiles has produced significant differences between their urogenital system and that of amphibians. These differences reflect adaptation for conserving body water in a generally dehydrating environment, the evolutionary trend toward anatomic separation of excretory and reproductive tracts, and development of more precise control over the animal's internal environment. Whereas amphibians have a mesonephric kidney, all higher vertebrates including reptiles have a metanephric kidney.

Although metanephric and mesonephric kidneys consist of the same basic units, the metanephros is more compact than the mesonephros. Compared to those of amphibians, there is a definite reduction in the size of glomeruli in reptiles. This is an adaptation that conserves water by reducing the flow of urine into tubules. Although glomeruli are present in the majority of reptiles, they are missing in some lizards and snakes, so that aglomerular tubules are present. The posterior tubules of some snakes and lizards are sexually dimorphic, with males having an enlarged portion called the sexual segment, which shows seasonal cellular changes with production of a phospholipid secretion that is incorporated into the seminal fluid.

The nitrogenous excretions of reptiles may be in the form of ammonia, urea, or uric acid. The proportions of these products vary with the lifestyle of a particular species. Marine and highly aquatic freshwater turtles and crocodilians excrete up to 25% ammonia as a percent of total urinary nitrogen. Amphibious pond and swamp turtles excrete approximately two to four times as much urea as ammonia or uric acid. Tortoises, the tuatara, lizards, and snakes (especially those found in deserts) excrete primarily uric acid. Reptiles as a group cannot concentrate urine above blood osmolarity, thus the ability to produce and excrete uric acid, which is insoluble in water, was developed as a mechanism for conserving water.

F. Salt Glands

Since reptiles cannot concentrate urine hypersomotic to blood, numerous species have evolved extrarenal sites of salt secretion as a homeostatic mechanism. This is particularly true of marine species (sea turtles, sea snakes, crocodiles) and certain desert species (chuckawalla, desert iguana). These glands have evolved independently among the reptiles at least five times and represent nonhomologous structures between the different groups (Dunson, 1976). In sea turtles and the diamondback terrapin the lacrimal gland has been modified into a salt gland that is associated with the orbit. In lizards (*Amblyrhynchus, Iguana, Dipsosaurus, Sauromalus, Uromastyx, Ctenosaura*) the salt gland is a nasal gland. In sea snakes the location is sublingual, and in the saltwater crocodile (*Crocodilus porosus*) it is in the tongue itself.

G. Circulatory System

Many of the differences between the reptilian and amphibian circulatory system are associated with loss of functional gills and the need for an efficient pulmonary circulation to and from the lungs. Whereas many amphibians meet oxygen needs by cutaneous and pharyngeal routes in addition to pulmonary exchange, in reptiles the main site of oxygen uptake is the lungs. Thus, it is not surprising (as stated earlier) that the reptile lung is larger and more complicated than the amphibian lung. With this increased pulmonary complexity the circulatory system underwent elaborations upon the amphibian design to accommodate this increased oxygen uptake and transport to tissues.

The right and left auricles are completely separate in all reptiles, and, as in higher vertebrates, oxygenated blood returning to the heart from the lungs flows through the pulmonary veins into the left auricle; deoxygenated blood returning from systemic sites flows into the right auricle. There are nearly all degrees of partitioning of the ventricle in living reptiles, varying from the situation (in some lizards) of having practically no interventricular septum to the situation (in crocodilians) of possessing a complete interventricular septum. All reptiles possess two aortae: a right and left.

Based upon the anatomy of the reptile heart (except crocodilians) several patterns of circulation of blood returning to heart from the systemic and pulmonary circulation have been proposed: (1) complete mixing, (2) partial mixing, and (3) a high degree of separation between unoxygenated and oxygenated blood being distributed differentially to the two aortae. Additionally, factors such as temperature, respiratory status, and circuit resistances may alter the pattern within an individual animal. Investigative studies using radiographic techniques have yielded information from complete mixing in some species to a small right to left shunt directed to the left aortic arch in others. Studies measuring oxygen concentration of the atria and great vessels of snakes and lizards suggest that there may be a high degree of separation of blood entering the great vessels (White, 1959).

The major vessels comprising the circulatory system of reptiles also show further modification on the amphibian plan. Reptiles are the first group of vertebrates to have evolved a well-developed coronary artery system. The left aorta is smaller than the right, and both join beyond the heart to form a common dorsal aorta. The esophagus passes through a ring produced by the joining of the aorta, and specialized cardiovascular patterns in snakes must have evolved along with the ability to feed upon large sized prey species.

H. Reproductive System

Reptilian gonads, testes and ovaries, are paired and are situated in the abdominal cavity, either in close proximity to the cranial poles of the kidneys (many lizards) or at a considerable distance cranially (snakes). The gonads arise from the germinal ridge with the testes derived from the medulla and the ovaries from the cortex.

Fertilization is internal in all reptiles and takes place near the cranial end of the oviduct prior to deposition of the egg envelopes by glands lining the oviduct. Glands in the cranial region of crocodilians, turtles, and the tuatara secrete albumen. In those species that produce shelled eggs, specialized shell glands are found in the caudal oviduct, which is often termed the uterus. The uteri open independently into the cloaca. Several species of reptiles have been shown to possess seminal receptacles in the female reproductive tract (Fox, 1956; Conner and Crews, 1980), and it is known that at least some reptiles can store sperm up to several years, producing viable eggs or offspring even though not in direct contact with a male.

The sperm are conveyed by the Wolffian duct to the copulatory organs of crocodilians, turtles, snakes, and lizards and the cloaca of the tuatara. The copulatory organ of crocodilians and turtles is the penis, which lies within the cloaca. The hemipenes of lizards and snakes are paired structures lying inverted in the base of the tail. The tuatara has no specialized copulatory organ.

There are no morphologic features that can be uniformly used for distinguishing the sex of all adult species of reptiles. Even at the familial level, gross external morphologic distinctions between males and females may not be present. In those species where sexual dimorphism exists as adults, this may not necessarily be so for neonates and juveniles.

Adult male tortoises usually have a concave plastron to allow mounting of the female. The plastron of the female is usually flat. In many aquatic emydine turtles the male is smaller than the female, and has elongated forelimb claws which are used for behavioral displays in courtship. Male turtles generally have a larger, broader tail than females with the cloacal vent located further posteriorly. In large turtles the penis can be digitally palpated.

With most crocodilians the adult male has a more massive head than the female, and generally males attain a greater adult size. Following manual or chemical restraint, the penis can be palpated within the cloaca and is fairly easy to extrude.

Almost all species of snakes and many species of lizards are sexually monomorphic, or the differences are so subtle that only the individual specializing in a particular species can accurately determine the sex. In some species of lizards, specialized glandular structures called femoral pores, located along the femoral margins of the legs, are more highly developed in males than females. The copulatory organ of lizards is a paired structure called the hemipenes and is located inverted in the base of the tail. Manual eversion of the hemipenes is possible in small lizards by applying gentle pressure to the tail (care must be taken not to fracture the tail).

Snakes are by far the easiest reptiles to sex using commercially available sexing probes. The probes are made of polished stainless steel and following lubrication can be inserted into the inverted hemipenes. The probes can only be inserted a short distance in females whereas in males it may be passed within the hemipenes to a distance of one-fourth to two-thirds the length of the tail (depending on species). In the family Boidae, many species have remnants of the pelvic apparatus termed spurs. These "claw-like" structures are located in a cutaneous recession near the cloaca at the ventral scale/lateral body scale junction. In many species having spurs, they are larger and more highly developed in the male than in females of the same size.

There is a tremendous amount of information on reproductive cycles of reptiles. Many species show cyclical changes in the development of the reproductive organs, which is timed in each species (or population) to gain maximum benefits from favorable climatic conditions and food sources (Crews and Garrick, 1980). In temperate regions many species breed soon after hibernation ends, with young delivered or hatching later in the season when food resources are plentiful. Many lizards, such as members of the genus *Sceloporus* and *Uta stansburiana*, emerge from hibernation with their testes at maximum size, with testicular atrophy progressing through the summer (Fox, 1977). In other lizards the testes may be small after emergence from hibernation, with maximal size from spring to early summer. In lizards of the genus *Emoia* from New Hebrides, the reproductive cycles are almost continuous, with minimal and maximal periods of May to June and November to December respectively. Among many species of snakes of the genus *Thamnophis* the testes are smallest in December to May and largest in July to October.

Copulatory behavior does not necessarily coincide with height of testicular development. At least in some species of *Thamnophis*, mating takes place in the spring when the testes are small and inactive. Sperm used for reproduction is stored in the epididymis and derived from the previous year's testicular activity during the summer (Cieslak, 1945).

The ovarian cycle of reptiles also shows geographic variation interspecifically and intraspecifically. Follicular enlargement may coincide with testicualr atrophy of males of the same species, sperm being derived from last year's activity. In these species, ovulation coincides with discharge of spermatozoa from the epididymis.

The reptile for which cyclical reproduction patterns and endocrine control mechanism have been best described is the lizard, *Anolis carolinensis*. In this species, as with most animals in general, an array of internal and external factors work in concert to control the patterns of reproductive biology. From late September to late January both males and females are reproductively quiescent. In late January (this varies with the geographic range of this species, which is throughout southeastern United States), the males emerge from hibernation (winter dormancy) and establish breeding territories. About 1 month later, the females emerge, and by May they are laying a single-shelled egg every 10–14 days. The annual reproductive cycle of the female has been divided into three distinct periods (Crews, 1975). Winter ovaries contain both previtellogenic (unyolked) follicles and atretic follicles, with the former being arranged in a stepwise size hierarchy (Jones *et al.*, 1973). Beginning in March, yolk deposition commences in these follicles, which become vitellogenic follicles and are ovulated at a diameter of about 8 mm. Intrafollicular production and secretion of estrogen stimulates follicular hyperemia and subsequent follicular growth. This increased vascularity may lead to the greatest yolk deposition in the most hyperemic follicle (Jones *et al.*, 1975). The "turning on" of vitellogenesis is correlated with rising spring temperatures, which initiate gonadotropin release by the pituitary. Lizards appear to secrete only one gonadotropin, which is similar to mammalian follicle stimulating hormone (FSH) (Licht and Crews, 1975). During the ensuing breeding season in *Anolis carolinensis*, a single follicle matures and is alternatively ovulated between the two ovaries every 10–14 days. In late August, in the last period of the ovarian cycle, vitellogenesis ceases, and the yolking follicles commence a rapid degeneration resulting in the formation of corpora atretica.

There are approximately 27 described reptiles (26 lizards and 1 snake) that have developed parthenogenesis as a successful reproductive strategy (Cole, 1975). These animals provide a unique source of individuals that show little genetic variation between one another and are the only known vertebrates to reproduce normally by this method. The best studied of these reptiles are the teiid whiptail lizards, including *Cnemidophorus uniparens*, *C. velox* and *C. tesselatus*, all from the western United States.

Based upon the type of postfertilization developmental patterns, reptiles can be grouped as those that are egg layers (oviparous) and those that are live-bearers (viviparous) (Packard *et al.*, 1977). Oviparity is believed to represent the an-

cestral mode of reproduction in the class Reptilia. Oviparous reproduction characterizes all living chelonians, crocodilians, the tuatara and most species of lizards and snakes. Nevertheless, many species of lizards and snakes have developed viviparity as a more advanced mode of reproduction. Table II presents the reproductive modes for the major reptile families.

The incubation time and hatching of fertile reptile eggs are temperature dependent. Within limits, incubation time is inversely related to environmental temperature. The preferred incubation temperature for many reptile eggs is 28°–31°C (with the preferred temperature for tuatara eggs being 21°C). In brooding female *Python molurus*, 30.5°C is an ideal temperature. Eggs kept much below this temperature, if they develop at all, seem to have low hatching success and abnormal development or no development at all (Vinegar, 1973).

It has been shown for several species of chelonians, crocodilians, and lizards that sex ratios of neonates hatching from a

Table II

Summary of the Modes of Reproduction Characterizing Families of the Order Squamata[a]

Family	All species oviparous	Both modes of reproduction present	All species viviparous
Lacertilia			
Amphisbaenidae	−	+	−
Gekkonidae	−	+	−
Iguanidae	−	+	−
Agamidae	−	+	−
Chamaeleontidae	−	+	−
Xantusiidae	−	−	+
Teiidae	+	−	−
Lacertidae	−	+	−
Scincidae	−	+	−
Dibamidae	+	−	−
Cordylidae	−	+	−
Gerrhosauridae	+	−	−
Anguidae	−	+	−
Anniellidae	−	−	+
Xenosauridae	−	−	+
Helodermatidae	+	−	−
Varanidae	+	−	−
Serpentes			
Typhlopidae	−	+	−
Leptotyphlopidae	+	−	−
Uropeltidae	−	−	+
Acrochordidae	−	−	+
Boidae	−	+	−
Colubridae	−	+	−
Elapidae	−	+	−
Hydrophiidae	−	+	−
Viperidae	−	+	−
Crotalidae	−	+	−

[a]From Packard *et al.* (1977), with permission of Cambridge University Press.

clutch of eggs are temperature dependent. Although heteromorphic sex chromosomes have evolved independently in reptiles many times, there are many species of the squamata and a few chelonians that lack these chromosomes. Thus, depending upon the clutch temperature, either all females or all males may be produced from a single clutch of eggs in those species lacking heteromorphic sex chromosomes (Vogt and Bull, 1982).

Once oviposition or parturition commences, it is usually complete within a few hours following the delivery of the first young or egg. Reptiles may deposit anywhere from a single egg as with *Anolis carolinensis*, 2 eggs as with many gekonnid lizards, and up to 125 eggs for several species of sea turtles. Dystocia in captive reptiles is not an uncommon event, and for most reptiles this should be considered when signs of straining to pass eggs persists beyond 1 day following the initiation of oviposition. Retained eggs or embryos can be confirmed radiographically.

Oxytocin has been used by several investigators (LaPointe, 1964; Evert and Legler, 1978) for timed collection and for stimulating uterine contractions in animals that apparently are "egg bound." Intraperitoneal administration of 1 to 4 units/100 gm of body weight has been used in turtles, lizards, and snakes. In such cases, if oxytocin is ineffective in inducing parturition or oviposition, surgical intervention must be initiated.

Sexual maturity appears to be more related to size of the individual rather than age. Reptiles appear to grow more rapidly and reach sexual maturity at an earlier age in captivity than in the wild. This is directly attributable to food availability and environmental conditions. For example, green turtles (*Chelonia mydas*) may reach sexual maturity at 9 years of age in captivity (being fed a high protein diet of 42%) compared to 20–40 years in the wild. Captive Burmese pythons (*Python molurus*) may reach 2 m the first year of life, 3 m the second year, and up to 4 m by the third year.

Although numerous reptiles species have successfully bred and reproduced in captivity, there remain many species for which this is not the case. It is only recently that semen collection and artificial insemination of reptiles have been attempted. Since most reptiles are seasonal breeders, semen collection should be attempted during the period of maximal reproductive activity. Manual techniques have been used to evert the hemipenes of snakes and lizards and the penis of turtles and crocodilians. Whereas semen collection by electroejaculation has been attempted with limited success in snakes (Mengden *et al.*, 1980), and crocodilians (Larsen *et al.*, 1982), it has been used successfully in three species of turtles (Platz *et al.*, 1980). In the latter study, attempts at freezing semen from the green turtle, *Chelonia mydas*, using methods reported for the dog and cat were unsuccessful.

A 30% recovery was reported for frozen snake semen using Triladyl Konzentrat as an extender (Mengden *et al.*, 1980). In this study diluted semen was successfully stored at 5°C in an ice bath for periods in excess of 96 hr using modified diluents containing McCoy's media. Using a tuberculin syringe, collected semen was passed into the vaginal fold in the dorsal midline of the cloaca.

In the American alligator, *Alligator mississippiensis*, semen has been collected via aspiration of the penile grove or harvesting it directly from the vas deferens of fresh alligator carcasses (Larsen *et al.*, 1982). In this study semen was successfully maintained in a solution of BES-Tris-yolk; motility was 75% at day 6.

I. Hemopoietic System

The blood of reptiles consists of cellular and acellular components. The whole blood volume shows species variability with 7.3 ml/100 gm body weight for *Alligator mississippiensis* and 9.1 ml/100 gm for the turtle *Pseudemys scripta elegans* (Kaplan, 1974). The cellular components, comprising a packed cell volume of 20–40%, consist of erythrocytes, granulocytes, lymphocytes, monocytes, plasma cells, and thrombocytes. The acellular fraction of the blood, the plasma, comprising 60–80% of blood volume, is a colorless or straw-colored fluid in many species and a yellow deeply pigmented fluid containing carotenoid pigments in others. Suspended within the plasma are a variety of inorganic electrolytes and a variety of organic compounds.

Historically there is confusion in terminology of cell types of reptilian blood and hemopoietic tissues. The inherent problems in naming cells according to staining abilities (basophil, eosinophil, neutrophil) and function (macrophage) have led to discrepancies in naming of cells between different investigators. Although no studies have completely documented the ontogenetic lineage of mature circulating blood cells, Pienaar (1962) in his monograph on "Hematology of Some South African Reptiles" presents the most detailed picture.

There is some variation with regard to hemopoietic centers both specifically (with age) and interspecifically. The centers of blood cell formation include the bone marrow, liver, and spleen. Although erythroid and granulocytic series are predominantly produced in bone marrow, the spleen also maintains a variable granulocytopoietic and erythropoietic activity. The myeloid stem cells are multipotential cells, giving rise to all the cell types found in the bone marrow.

Mature erythrocytes are nucleated and oval. Immature erythrocytes are more rounded with a basophilic cytoplasm and a less chromophilic nucleus than that of the mature cell when stained with Romanowsky stains. Reticulocytes can be demonstrated by supravital staining procedures. Senile red cells are larger than mature cells, with pale staining cytoplasm and pyknotic nuclei. Dimensions for erythrocytes vary both interspecifically and intraspecifically. While the teiid lizard, *Ameiva ameiva,* has a mean least erythrocytic diameter of 7.6 μm, the tuatara, *Sphendon punctatus,* has the largest red blood cells for a reptile, with a mean greatest diameter of 23.3 μm (Saint Girons, 1970).

Erythrocyte counts show a tremendous amount of variation both intraspecifically and interspecifically. Counts have been shown to vary with age, sex, season, altitude, nutritional status, and disease. Techniques for blood collecting and counting of erythrocytes have been described in detail elsewhere (Frye, 1981).

The white blood cell group consists of a variety of cells with unknown homology to high vertebrate cell lines. Much complexity centers around the granulocytes, particularly the eosinophilic and azurophilic granulocytes. Two types of eosinophils have been distinguished based upon the shape of the intracytoplasmic granules, which may be either cylindrical or spheroidal. Eosinophil type I is commonly referred to as the heterophil and is the major white blood cell seen in an acute inflammatory response. While there are reports of both cell types in all the orders of reptiles (Pienaar, 1962), Will (1979) believes this distinction cannot be made for squamates.

The nuclei of eosinophilic granulocytes, although variably shaped, are often round and situated near the periphery of the cytoplasm. Azurophilic granulocytes have relatively small, eccentric nuclei and cytoplasmic granules that stain with pure azure dyes and with pyronin. With Romanowsky stains the cytoplasm is distinctly basophilic. This cell appears to be more closely related to the monocytoid line and may form a lineage with more typical appearing monocytes of reptiles which lack cytoplasmic granules. Reptile basophils show typical staining features of higher vertebrate basophils. Neutrophilic granulocytes are of low frequency in blood and are not homologous to mammalian neutrophils. Optimum staining of these cells is with Leishman mixtures, alone, or in conjunction with Giemsa stain (Pienaar, 1962). The granules are strongly angular and stain deep lavender to magenta.

Thrombocytes show a great deal of morphologic variability, but usually have a high nucleocytoplasmic ratio, with the cytoplasm appearing as a clear band around the nucleus.

Lymphocytes are categorized according to size into large and small forms. The cytoplasm is moderately to weakly basophilic with Romanowsky stains. Plasma cells are somewhat similar in appearance to lymphocytes except that the cytoplasm stains more intensely blue and the nucleus is eccentric.

Reptile plasma is qualitatively similar to that of birds and mammals. Suspended within it are both low molecular weight components including Na^+, K^+, Ca^{2+}, Mg^{2+}, Cl^-, HCO_3^-, P_i, SO_4^{2-}, and organic components including proteins, vitamins, hormones, clotting factors, and enzymes. There is a paucity of information on values for plasma enzymes that are important in evaluating ill reptiles.

Reptiles do not show the precise regulatory mechanisms of plasma constituents as seen in birds and mammals. Blood pH's of 6.5 and 8.1 have been reported (Dessauer, 1970). Many of these substances show tremendous variability with age, reproductive state, season, body temperature, nutritional state, and disease state. Although there are numerous reports documenting values of individual components of plasma in reptiles, there are few complete studies.

III. DISEASES

A. Infectious

1. Viral

a. Herpes-like Virus of Green Sea Turtles

A virus that is morphologically compatible with herpesvirus was found to be the causative agent of epizootics of skin lesions termed grey patch disease in young green turtles (*Chelonia mydas*), between 56 and 90 days posthatching, in mariculture bred and reared turtles at Cayman Turtle Farm, Grand Cayman, British West Indies (Rebell *et al.,* 1975). In nine groups studied, up to 100% of the animals in each group were affected. Secondary bacterial skin disease often followed the initial viral insult to the integument.

Skin lesions first noticed were circular papular lesions that would eventually coalesce into diffuse epidermal gray patches affecting both the soft integument of the flippers, axillary regions, and head and the hard integument of the plastron and carapace. Lesion extracts applied to scarified flippers resulted in the formation of similar lesions within 3–4 weeks.

Light microscopic examination of skin lesions revealed hyperkeratosis, hyperplasia with acanthosis, and the presence of basophilic intranuclear inclusions. Electron microscopic examination of lesions revealed intranuclear enveloped particles 160–180 nm in diameter with an electron-dense core. Based upon location and morphology the particles were consistent with herpesvirus.

The exact route of transmission of gray patch disease is unknown, but studies to date suggest the route is via the egg. Sea turtles hatched from eggs collected immediately from ovipositing females and raised under controlled laboratory conditions developed characteristic skin lesions. Additionally, a water route of transmission could not be discounted.

b. Herpes-like Virus of Pond Turtles

The first report of viral associated hepatic necrosis in reptiles involved two captive Pacific pond turtles, *Clemmys marmorata* (Frye *et al.*, 1977). At necropsy the only gross changes seen were swollen livers. On histopathology, intranuclear hepatic inclusions were associated with areas of acute necrosis. Similar inclusions were found, with less frequency, in renal convoluted tubules and splenic cells. Electron microscopic examination of liver and spleen revealed intranuclear particles with an electron-dense core and measuring 100 nm in diameter. Intracytoplasmic particles were enveloped and measured approximately 140 nm.

In a second report (Cox *et al.*, 1980) an adult male painted turtle, *Chrysemys picta*, at the Metropolitan Toronto Zoo, which died 6 days after being treated for an abscess, was found on necropsy to have a friable liver with an inspisated gallbladder and pulmonary edema. Histopathology revealed randomly scattered microfoci of coagulation necrosis throughout the liver with the presence of large intranuclear inclusions. Similar appearing inclusions were also seen in metaplastic epithelial cells in the lungs. Electron microscopic examination of liver revealed numerous hexagonal capsids measuring 85–115 nm in diameter. The virus particles were presumptively categorized as herpesvirus based upon size, structure, and shape.

Recently a die-off of emydine pond turtles (*Graptemys* spp.) was attributed to a virus that also morphologically resembled herpesvirus (Jacobson *et al.*, 1982a). The examined turtles showed severe hepatic necrosis with intranuclear inclusions present in liver, kidney, and pancreas. Typical herpesvirus particles were identified by electron microscopy in sections of liver. The die-off followed the introduction of western painted turtles (*Chrysemys picta*) into a long-term captive groups of map turtles (*Graptemys spp.*).

It is likely that the herpesviruses identified in the above turtles represent either the same or closely related viruses. The liver lesions are classic for other hepatotropic herpesviruses. Future research on pathogenesis and relationships with other herpesviruses will be dependent upon successful isolation in a tissue culture system.

c. Papilloma-like Virus of Side-Neck Turtles

Recently (Jacobson *et al.*, 1982b), papilloma-like viral crystalline arrays were observed in skin lesions of Bolivian side-neck turtles (*Platemys platycephala*). The turtles were submitted with circular papular skin lesions that in some animals coalesced into patches. All lesions were noted to involve the integument of the head. Light microscopic evaluation of skin lesions revealed hyperkeratosis and hyperplasia with acanthosis; no inclusions were noted. Electron microscopic examination of skin biopsies revealed intranuclear crystalline arrays of hexagonal particles measuring approximately 42 nm. These aggregates morphologically resembled papillomavirus seen commonly in a variety of mammalian wart lesions.

d. Caiman Pox

The only report (Jacobson *et al.*, 1979) of a virus associated disease in a crocodilian is that of a pox-like virus demonstrated in skin lesions of captive caimans, *Caiman sclerops*. Several juvenile captive spectacled caimans were submitted with gray-white circular skin lesions scattered over the body surface and particularly prominent on the palpebrae, tympanic membranes and integument overlying the maxillae and mandible (Fig. 13). While in some animals severe lesions resulted in digital sloughing, other individuals had only focal lesions, often involving only the palpebrae. Light microscopic evaluation of skin lesions showed large eosinophilic intracytoplasmic inclusions within hypertrophied epithelial cells. Numerous inclusions were present in an extremely thickened keratin layer. Electron microscopy demonstrated the inclusions to consist of myriads of viral particles that were morphologically typical of poxvirus. The size of 200 × 100 nm was smaller than previously reported poxviruses of vertebrates and insects.

e. Erythrocyte Viruses of Reptiles

This disease represented the first reptilian viral disease to be documented, although initially this intraerythrocytic agent was considered to be a protozoan and originally named *Pirhemocyton tarentolae* (Chatton and Blanc, 1914). The organism was seen in circulating red blood cells as intracytoplasmic inclusions. The viral nature of the inclusions was elucidated in subsequent studies by Stehbens and Johnston (1966) in which electron microscopic examination of red blood cells of an infected gecko, *Gehyra variegata*, revealed that the inclusions represented virus "assembly pool" or "factor" areas. This was presumptively considered to be an iridovirus based upon structure and morphology.

f. Paramyxovirus-like Virus of Viperid Snakes

The first reported viral associated snake epizootic involved a colony of fer-de-lance, *Bothrops atrox*, housed at a snake farm in Switzerland (Foelsch and Leloup, 1976). Fer-de-lance was the only species affected, although other species were housed in the same room. Snakes initially exhibited a loss of muscle tone characterized by resting prostrate followed by a terminal respiratory disease. Postmortem evaluation revealed fluid filled lungs and body cavity. Lung homogenates injected into false water cobra eggs resulted in egg death 7 days postinocu-

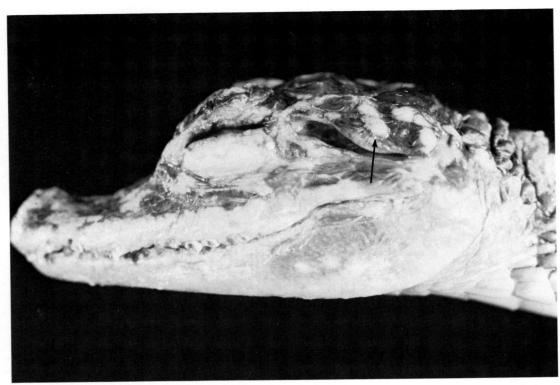

Fig. 13. Juvenile spectacled caiman with gray-white circular lesions on palpebra, margins of upper and lower jaws, and tympanic scale (arrow) typical of pox infection. (From Jacobson *et al.*, 1979.)

lation (Clark *et al.*, 1979). Suspensions of these eggs caused cytopathic effects in gecko embryo and rattlesnake fibroma cell cultures incubated at 30°C. This cytopathic agent recovered from the cell cultures was designated fer-de-lance virus and, based on electron microscopic appearance and biophysical properties, was categorized as a paramyxovirus.

The second reported die-off involved a breeding colony of rock rattlesnakes (*Crotalus lepidus*) that were maintained in a private collection in Tallahassee, Florida (Jacobson *et al.*, 1980a). A new breeder male was introduced into the collection without being quarantined and ultimately was in contact with all other rattlesnakes in the colony. On day 3 following introduction, this newly acquired snake developed head tremors and loss of equilibrium and died 11 days later. Over the next 2 months, 3 females and 3 of the males died after manifesting similar clinical signs.

One of the rattlesnakes showing clinical signs of central nervous system disease was euthanatized and necropsied. The only gross abnormality found was a small amount of exudate in the respiratory tract. Histologic examination of lung tissue demonstrated a marked interstitial infiltrate of inflammatory cells with squamous metaplasia and a mild proliferation of lining pneumocytes (Fig. 14). Brain lesions included multifocal areas of gliosis, minimal perivascular cuffing, and moderate

ballooning of axon sheaths with demyelination and degeneration of axon fibers. While no pathogenic microorganisms were isolated from lung tissue on blood and MacConkey's agar, inoculated viper heart cells showed synctial formation by day 3 after subculturing. This agent was ether sensitive and hemagglutinated chicken erythrocytes at 5°C. Ultrastructural examination of infected viper heart cells on day 5 revealed virus particles budding from cell membranes and extracellular mature particles. These particles were of similar morphology as previously reported fer-de-lance virus and were designated rock rattlesnake virus.

The third reported die-off involved the snake collection at Louisiana Purchase Gardens and Zoo in which 35 viperid snakes (8% of the total viperid collection) died during February and March, 1980 (Jacobson *et al.*, 1981). The genera affected were all members of the family Viperidae including *Crotalus*, *Vipera*, *Bothrops*, *Trimeresurus*, and *Bitis*.

Clinical signs included a sudden gaping of the mouth followed within 1 day by a violent convulsion, with spiraling of the entire body. Several snakes expelled a brownish fluid from the glottis. The only gross lesion seen in the necropsied snakes was a mucoid, or sometimes caseous, exudate within the lung. Often the lung appeared edematous. A broad spectrum of mostly gram-negative microorganisms were isolated from lung

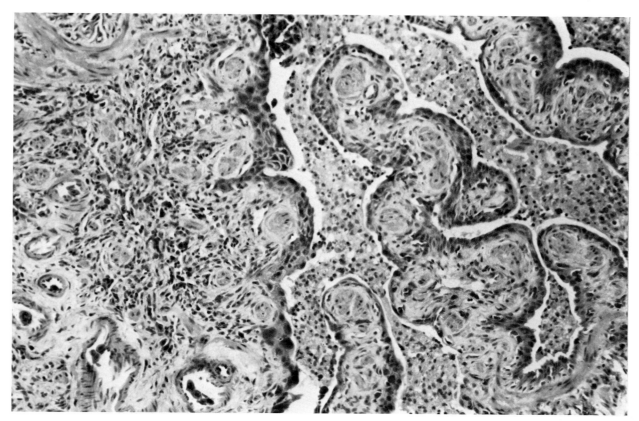

Fig. 14. Photomicrograph of lung tissue from a rock rattlesnake with paramyxo-like viral infection. Pathologic changes include proliferation of the respiratory epithelium and diffuse thickening of the interstitium caused by fibrosis, with infiltration of mixed inflammatory cells. Hematoxylin and eosin stain. × 175. (From Jacobson *et al.,* 1980a.)

tissue, including *Pseudomonas, Proteus,* and *Aeromonas.* No antibiotic treatment was effective in altering the course of disease in ill snakes.

Histologic examination of lung tissue revealed an interstitial inflammatory response with large numbers of degenerating epithelial cells, heterophils, bacterial colonies, and necrotic debris within air passageways. Occasional eosinophilic intracytoplasmic inclusions were seen within lining pneumocytes.

Virus was recovered from lung tissue inoculated into viper heart cells. In Giemsa-stained coverslip preparations of infected monolayers, giant cell formation and intracytoplasmic inclusions were prominent. Ultrastructural examination of infected viper heart cells on day 5 revealed a small number of extracellular mature virus particles and presumptive virions budding from cell membranes. These particles were morphologically similar to fer-de-lance virus and rock rattlesnake virus, and was designated puff adder virus.

In all three reported outbreaks, only viperid snakes were involved. In a survey of snakes at Louisiana Purchase Gardens and Zoo, several nonviperid snakes had titers to the virus, with a reticulated python having the highest titer. This raises the possibility that a nonviperid snake may be a reservoir for this virus without showing clinical disease. Viperid snakes may represent an abnormal host that is exquisitively sensitive to this agent.

g. *Elapid Venom Herpesvirus*

Padgett and Levine (1966) in attempting to elucidate the substructure of murine Rauscher leukemia virus with venoms of the elapid snakes including the Indian cobra, *Naja naja,* and the banded krait, *Bungarus fasciata,* demonstrated the presence of morphologic forms similar to capsids of herpesvirus in preparations of the above venoms incubated with mouse oncornavirus. Monroe *et al.* (1968) in a subsequent electron microscopic study of venoms from Indian cobras and banded kraits, demonstrated the presence of similar appearing viral particles. These particles were from 100 to 125 nm in diameter and possessed a central electron-dense core typical of herpesvirus. Recently, similar appearing particles were associated with venom

gland infections in Siamese cobras, *Naja naja kaouthia* (Simpson et al., 1979). Light microscopic examination of venom glands of 2 of 25 cobras with a history of low-grade venom production revealed degeneration and necrosis of patches of columnar glandular epithelial cells with infiltration of the subepithelium with mixed inflammatory cells. By electron microscopy, naked and enveloped herpesvirus-like particles were seen in necrotic and ruptured cells.

2. Bacterial

a. Pseudomonas

There are relatively few reports of pseudomonad infections in turtles (Dieterich, 1967; Jacobson, 1978), with most reports involving lizards and snakes. In squamates, *Pseudomonas* infections may be manifested as either local integumentary lesions, oral lesions, pneumonia, and septicemias. Although Page (1961) isolated *P. aeruginosa* from several cases of ophidian ulcerative stomatitis, *Aeromonas hydrophila* was considered the most significant pathogen in his study. Whereas *Pseudomonas* spp. have been commonly isolated from the oral cavity of a variety of clinically healthy snakes (Ledbetter and Kutscher, 1969; Parrish et al., 1956), in a recent survey (Draper et al., 1981) of a zoologic collection of snakes, healthy snakes had predominantly gram-positive oral flora with snakes with infectious stomatitis having predominantly gram-negative bacteria including *Pseudomonas aeruginosa*.

Adverse environmental conditions, poor nutritional status, and traumatic damage to tissues include several of the most important factors that may influence the onset and course of disease. Functionally, the immune system of reptiles is temperature dependent, with below optimum environmental conditions leading to depressed immune function. Poor nutritional status may also result in a decline in immune function. The recently imported reptile being kept under suboptimal conditions and traumatizing itself in shipment is ideally suited for invasion by opportunistic pathogens, and indeed it is in these animals that we see some of the most serious infections.

There are no specific clinical signs for *Pseudomonas* infections in reptiles. Squamates with infectious stomatitis may show swelling of the infected tissues with gaping of the labial margins. The mucous membranes may exhibit either multifocal areas of petechiation to ulcerative lesions with a buildup of caseous material. Local lesions involving the oral mucosa may proceed into osteomyelitis, and pneumonia may result from aspiration of necrotic cellular debris. Septicemia may be seen externally as hemorrhagic body scales with coagulative necrosis of the dermis and epidermis.

In treating an active *Pseudomonas* infection in a reptile, different authors have used different approaches. Not only must the infection be treated but also the underlying problem that predisposes the animal to infection must be managed. A severely compromised animal may not be responsive to antibiotics with favorable sensitivity patterns. The use of immunopotentiating agents such as levamisole (10 mg/kg body weight every 5 days for five treatments) may have value in the compromised host.

Lesions involving the oral cavity should be debrided and cleaned daily with 10% hydrogen peroxide. Sulfamylon cream is an ideal topical ointment for treating integumentary lesions. The antibiotics of choice for treating severe cases are the aminoglycosides including kanamycin, gentamicin, amikacin, and tobramycin. Reptiles are treated with gentamicin at 2–4 mg/kg body weight every 72 hr for three to five treatments. Fluid therapy (20 ml/kg) is recommended in those animals receiving potentially nephrotoxic antibiotics such as the aminoglycosides. Gentamicin in combination with carbenicillin (200 mg/kg) therapy is commonly used in severe cases. Response to therapy is often slow, and a minimum of 4 weeks may be necessary to evaluate the animal.

b. Aeromonas

Aeromonas has been associated with ulcerative stomatitis in a variety of lizards and snakes. As with *Pseudomonas*, severe or even benign appearing mouth lesions may allow systemic showering of organisms via the circulatory system, resulting in septicemia. Septicemic reptiles may show multifocal areas of ecchymotic hemorrhages involving the integumentary system. Animals are often anorectic, listless, and steadily decline in condition. Aspiration of necrotic debris from the oral cavity into the respiratory system may result in pneumonia. Reptiles may be presented with gaping of the mouth, labored respiration, and harsh respiratory sounds.

At necropsy, animals dying of *Aeromonas* septicemias may show dramatic gross lesions. In pneumonic infections, the lung vorbronchus and secondary air passageways may be filled with exudate. Pulmonary edema is often a consistent gross finding. Viceral organs and the gastrointestinal tract may show marked congestion with areas of hemorrhage. Histopathology often reveals multifocal to diffuse areas of necrosis with infiltrates of mixed inflammatory cells. In more chronic cases there is a granulomatous inflammatory component.

Reptiles suffering from *Aeromonas* infections need to be treated both locally and systemically. Necrotic tissue in the oral cavity should be debrided followed by irrigation with 10% hydrogen peroxide. The use of broad spectrum antibiotics, including carbenicillin, chloramphenicol, and gentamicin, is often indicated.

c. Salmonella/Arizona

Many reptiles appear to harbor both *Arizona* and *Salmonella* as a normal component of the intestinal flora, without being associated with clinical disease. The major concern is reptiles, particularly turtles, acting as reservoirs for human infections.

Some authors believe that *Salmonella* indigenous to reptiles are more virulent than those derived from birds and mammals (Schroder and Karasek, 1977). Federal regulations have prohibited the interstate sale of baby turtles (particularly the red-eared slider, *Pseudemys scripta elegans*) and have essentially eliminated the interstate shipment of millions of turtles. More than 300,000 human salmonellosis cases have been attributed to contact with these animals and their environmental water (Lamm *et al.*, 1972). Following the ban on interstate shipment of these animals, there has been a substantial decline in reptile-associated salmonellosis in man (D'Aoust and Lior, 1978).

Although a chlorine–gentamicin combination treatment procedure for turtle eggs has been proposed as a method to produce a *Salmonella/Arizona*-free animal, current literature suggests that the ability to culture these organisms may change with time and with stress factors. DuPonte *et al.* (1978) found that after stressing turtles by individual maintenance in 250 ml beakers on moist tissue paper, *Salmonella* organims were isolated from tissues of 7 of 27 stressed turtles previously free of these agents. This surely points to the possibility that the stress of shipment in water-free plastic containers may result in latent *Salmonella/Arizona* shedding.

Both *Arizona* and *Salmonella* appear to be normal components of the gastrointestinal flora of both free-ranging and captive reptiles (Jackson and Jackson, 1971; Kaura and Singh, 1968; McInnes, 1971). Ten isolations of *Salmonella* representing six serotypes were cultured from the intestinal tract of 124 free-ranging Florida lizards (Hoff and White, 1977). Of 317 reptiles cultured, while live or at necropsy, at the National Zoological Park, 117 were positive for 1 or more *Salmonella* or *Arizona* serotype, for an overall infection rate of 37% (Cambre *et al.*, 1980). In the latter study both food and water sources were culture negative for these organisms.

Although in most cases, reptiles infected with *Salmonella* and *Arizona* appear free of overt disease and are clinically normal, still *Salmonella* and *Arizona* have been cultured from lesions in reptiles at necropsy. In the above study at the National Zoological Park, these organisms were cultured from lesions in the gastrointestinal tract, liver, spleen, and vessels of two blood pythons, *Python curtus*. *Arizona* was also cultured from a variety of tissue sites from three of four red-tailed boas (*Contrictor* sp.) to die in a zoologic collection (Boever and Williams, 1975). Numerous yellow necrotic plaquelike lesions of fibrinonecrotic membranes were seen over the mucosa of the small and large intestine. In one animal multiple abscesses protruded from the serosal surface of the gastrointestinal system with additional involvement of the lungs, spleen, and gallbladder. No other pathogens were identified. *Salmonella* and *Arizona* were isolated from hepatic lesions in a hognose snake, *Heterodon* sp., and twin-spotted rattlesnake, *Crotalus pricei*, and from an epi- and myocardial granulomata in a Mexican cantil, *Agkistrodon bilineatus* (Frye, 1981).

Salmonella/Arizona may be transmitted from animal to animal or from animal to man via urine and feces. In turtles, *Salmonella* has been shown to penetrate turtle eggs (Feeley and Treger, 1969) with turtles being infected at hatching (Kaufmann *et al.*, 1972). Transovarian passage of *Salmonella* has also been demonstrated for snakes (Chiodini, 1982). Animals may also become populated with *Salmonella/Arizona* organisms via food such as slaughterhouse offal.

It is the feeling of many investigators that treatment for complete elimination of *Salmonella/Arizona* from reptile collections is not feasible. Therapy should be limited to those animals showing signs of disease. In such cases broad spectrum antibiotics, such as chloramphenicol, Septra, gentamicin, and amikacin are often the antibiotics of choice. Fluid therapy, to counteract the loss of fluids through the feces, should also be instituted.

d. *Citrobacter*

The majority of reports documenting *Citrobacter* infections in reptiles involve chelonians. Septicemic cutaneous ulcerative disease (SCUD) was described for a variety of pond turtles, including the genera *Pseudemys*, *Chrysemys*, and *Emys* (Kaplan, 1957). This disease appears to be management related, since predisposing conditions are necessary to initiate an epizootic and include poor nutrition and maintenance in stagnant filthy water coupled with skin abrasions. Turtles will ultimately become lethargic with reduced muscle tone, limb paralysis, and necrosis of digits with hemorrhage and cutaneous ulcerations. There may be septicemia with necrotic foci in the liver, heart, kidney, and spleen. In a subsequent report (Jackson and Fulton, 1970), *Serratia* was incriminated in the pathogenesis of SCUD by its lipolytic, proteolytic action on tissues, thereby allowing *Citrobacter* to more effectively invade affected tissue. Chloramphenicol was considered the antibiotic of choice.

e. *Serratia*

Serratia has been isolated from numerous reptile species, including both healthy and ill animals. *Serratia* was recovered from rattlesnake venom and swabs of fangs (Ledbetter and Kutscher, 1969), from the oral cavity of captive caimans, and from the oral cavity and cloaca of free-ranging rock iguanas, *Cyclura reylei* (E. R. Jacobson and G. V. Kollias, unpublished findings). Although a nonchromogenic *Serratia marcescens* was isolated from pyogenic arthritis in a teju lizard (*Tupinambis teguixin*), the organism was not isolated from systemic granulomatous lesions that were found throughout the body (Ackerman *et al.*, 1971). *Serratia anolium* (*Serratia marcescens* biotype A$_4$) was isolated from tumorlike lesions in the lizard *Anolis equestris* (Durand-Reynolds and Claussen, 1937). This organism was experimentally demonstrated to cause infection when inoculated subcutaneously in a variety of poikilothermic vertebrates (Claussen and Durand-Reynolds, 1937). *Serratia marcescens* was also cultured from subcutaneous abscesses in a green iguana, *Iguana iguana,* and a spiny tailed iguana, *Ctenosaura acanthura* (Boam *et al.,* 1970).

The pathogenesis of *Serratia* abscesses in reptiles is poorly understood. Since this organism appears to be a part of the normal oral flora of several species, organisms possibly become introduced subcutaneously through a bite wound or possibly other traumatic damage to the integument. Since *Serratia* appears to be an opportunistic pathogen, exogenous and endogenous factors responsible for depressing the immune system may be involved.

Diagnosis of *Serratia* infection is based upon isolation on artificial media and identification by appropriate biochemical tests. The production of pigment is not a characteristic of all isolates, with up to 75% of *Serratia* isolates being nonchromogenic (Ackerman *et al.,* 1971). The nonchromogenic isolates must be differentiated from *Escherichia coli, Enterobacter* spp., and *Aeromonas.*

Treatment of animals with subcutaneous lesions should include the surgical excision of the caseous necrotic debris including the fibrous capsule. The area should be flushed with a dilute organic iodine solution and the dead space reduced with sutures. Following culture and antibiotic sensitivity determinations, appropriate broad spectrum chemotherapy should be administered. Often chloramphenicol, gentamicin, and amikacin are the drugs of choice.

f. *Dermatophilus*

Dermatophilus congolensis is a unique bacteria that forms filamentous structures with transverse segmentation resulting in the formation of packets of coccoid cells that become motile spores. The first report of dermatophilosis in a lizard involved a bearded dragon, *Amphibolurus barbatus barbatus* (Simmons *et al.,* 1972). The animal was submitted with several subcutaneous nodules. An abdominal nodule consisted of a central necrotic core that was demonstrated by histopathologic gram-stain to contain positive branching filamentous structures consistent with *D. congolensis. Dermatophilus congolensis* was isolated in pure culture from the lesion and was transmitted by inoculation to four bearded dragons, one blue tongue lizard (*Tiliqua scincoides scincoides*), and one sheep. Dermatophilosis was subsequently identified in three Australian bearded lizards (Montali *et al.,* 1975) and two marble lizards, *Calotes mystaceus* (Anver *et al.,* 1976).

There are no reports of successful treatment for dermatophilosis in reptiles. In man, intravenous and oral sodium iodide solutions have been used. In one report (Simmons *et al.,* 1972), *D. congolensis* was found sensitive *in vitro* to pen-

icillin, streptomycin, chloramphenicol, oxytetracylcine, and bacitracin, while in a second report (Montali *et al.,* 1975) the organism was found sensitive to a variety of broad-spectrum antibiotics, including chlortetracyline, but was resistant to penicillin, lincomycin, and triple sulfonamides. In the later study, chlortetracyline soaks failed to alter the course of the disease. A rational approach to treatment should be the parenteral administration of a tested antibiotic coupled with local debridement and topical application of an organic iodine ointment.

g. *Mycobacterium*

Cultured reptilian mycobacterial organisms have all been within Runyon groups I and IV (Brownstein, 1978). The most common isolate for reptiles is *M. marinum,* with cases of *M. chelonei* and *M. thamnopheos* also being reported.

Griffith (1939) isolated *Mycobacterium marinum* from 71% (20 of 28 cases) of the mycobacterial cases at London Zoological Park from 1924 to 1933. *Mycobacterium marinum* is a ubiquitous, saprophytic, slow-growing photochromagen in warm aquatic systems and is potentially pathogenic for a wide variety of poikilotherms. *Mycobacterium thamnopheos* and *M. chelonei* are less frequently isolated from reptiles and are rapid growers. Aronson (1929) isolated *M. thamnopheos* from garter snakes at the Philadelphia Zoologic Park. *Mycobacterium chelonei* is an opportunistic pathogen of turtles and is distributed widely in nature, being found in soil and water.

Mycobacteriosis is a sporadic disease in well-managed reptile collections, having an annual incidence of 0.1–0.5% (Brownstein, 1978). Since the major mycobacterial agents infecting reptiles are ubiquitous organisms in nature and the incidence of infection is low, it appears that a commensal relationship probably exists under most situations, with actively infected animals being predisposed to disease.

Natural cases of reptilian mycobacteriosis are generally seen as chronic diseases, although acute episodes have been described. Aside from the cases presented with cutaneous lesions, those animals systemically infected generally show nonspecific signs, including anorexia, listlessness, and chronic weight loss. Dyspnea may be seen in animals having pulmonary involvement. Involvement of the integumentary system is usually seen as either subcutaneous nodules or ulcerative skin lesions. These lesions are easy to biopsy, and acid-fast organisms can be demonstrated in impression smears or in histologic section.

In acute mycobacteriosis, there is extracellular proliferation of acid-fast organisms with associated caseous necrosis and an infiltrate of heterophils; macrophages and fibroplasia are minimal. In chronic mycobacteriosis, there is typical granulomatous inflammation with development of tubercles. Granulomas may range in development from compact nests of macrophages containing intracellular acid-fast bacilli, to more complex granulomas containing caseated centers, and finally to expansion of the caseous region to involve the entire tubercle with large numbers of extracellular bacilli (Brownstein, 1984).

Since mycobacteriosis in reptile collections is not a colony problem, but rather a problem of the individual, elimination of known positive animals is recommended. Correcting predisposing factors such as debris build-up in aquatic systems, thoroughly cleaning water bowls, and ensuring animals are being maintained on a good level of nutrition, should be a part of any prophylactic program. There are no reports of successful chemotherapeutic treatment of reptilian mycobacteriosis. Appropriate toxicity studies of antituberculous drugs have not been performed for reptiles.

3. Fungal

a. Phycomycetes

Members of the class Phycomycetes comprise a large group of branching, rarely septate fungi with broad hyphae. The most commonly reported genera in reptiles include *Mucor, Rhizopus,* and *Basidiobolus.* Most cases are documented in tissue section without concomitant isolation on artificial media.

Most phycomycete infections have been found to involve the integument, respiratory system, and digestive system. In a survey of turtles in a zoologic collection (Hunt, 1957), organisms resembling members of the order Mucorales were associated with plastronal lesions in a variety of turtles. *Mucor* sp. was also identified in ulcerative necrotic skin and shell lesions (Fig. 15) in a series of young Florida soft-shell turtles, *Trionyx ferox* (Jacobson *et al.,* 1980b). *Mucor* has also been isolated from cutaneous lesions (Frank, 1966) in a bearded dragon, *Amphibolurus barbatus* and from pneumonic lesions (Silberman *et al.,* 1977) in three species of crocodilians. *Basidiobolus ranarum* was identified from a granulomatous lesion in the oral cavity of an Aldabra tortoise, *Geochelone gigantea* (Blazek *et al.,* 1968). Zwart (1968) isolated *Rhizopus arrhizus* from skin pustules and pneumonic lesions in captive garter snakes, *Thamnophis sirtalis.*

Integumentary lesions involving phycomycetes are generally necrotizing with minimal to moderate mixed inflammatory cell response. In many cases, the fungi fail to penetrate the dermis. Internally, lesions are generally granulomatous with organisms identified in the caseated center; giant cells are often seen. This granulomatous response probably represents the stage at which the infection is identified, since most reports are based upon postmortem evaluation of chronically debilitated animals.

As is true of other mycotic infections in reptiles, infections with phycomycetes are generally attributed to predisposing factors. Without being repetitive for other fungal agents to be

Fig. 15. Florida softshell turtle with mucormycosis. Note the ulcerative carapace lesions. (From Jacobson *et al.,* 1980b.)

described in this section, these factors will be discussed for reptilian mycoses in general.

Many of the fungal agents associated with disease in captive reptiles are ubiquitous in the animal's environment. With integumentary pathogens, fungi may gain access through breaks in the skin. Reptiles housed in filthy cages, under suboptimal environmental conditions, are primed for infection. As stated earlier, the immune system of reptiles is temperature dependent, and below optimum environmental temperatures may result in an immunocompromised animal. Fungi in the environment may easily invade and become established in such reptiles.

Aquatic species are continually bathed in a sea of pathogens, and reptiles maintained in aquatic environments that are not properly filtered may eventually develop mycotic disease. The first few animals to become infected may act as a nidus for infection of the entire collection.

There are few reports of treatment of reptiles with mycotic disease, since most cases are based upon postmortem identification. Florida soft shell turtles with *Mucor* dermatitis were treated by soaking in a solution of 0.15 mg/liter malachite green. Reptiles with limited focal skin disease can be treated by surgical removal of the lesion. Snakes with mycotic skin disease have been successfully treated with oral ketoconazole (25 mg/kg body weight once per day for 2 weeks). There are no reports of chemotherapeutic treatment of reptiles with systemic mycotic disease. Often by the time the disease is recognized, the lesions are so extensive that chemotherapy is impractical.

Once mycotic disease is identified in a group of reptiles, the predisposing environmental problems should be identified and corrected. In some cases, simply raising environmental temperature may result in a regression in active lesions. The immediate environment should be kept clean and the humidity should be optimum for the species.

Diagnosis of phycomycoses and mycotic disease in general is based upon identification of characteristic hyphae in tissue section and isolation on artificial media. Special stains including periodic acid–Schiff, Grocott's methenamine silver (GMS), and Gridley's fungus stain may be necessary for identification in tissue section. Fluorescent antibody techniques developed for identifying human mycotic agents may have application for reptile mycoses.

b. Aspergillus

As with phycomycetes, *Aspergillus* is ubiquitous in the environment. Most reported infections involve the integumentary and respiratory systems. In a survey of deaths in a zoologic collection of reptiles, Hunt (1957) reported that pulmonary mycosis comprised 3% of the total number of deaths. In this report *Aspergillus* was identified as the major causative agent that resulted in consolidation and gangrene of the lungs. There are several reports (Georg *et al.*, 1962; Andersen and Eriksen, 1968) of respiratory tract aspergillosis in Aldabra tortoises, *Geochelone gigantea elephantina*. *Aspergillus fumigatus* and *A. ustus* were isolated (Jasmin *et al.*, 1968) from lung lesions in 2- to 6-week-old American alligators. *Alligator mississippiensis*.

c. Beauveria

Members of this genus rarely produce disease in higher vertebrates, primarily being an insect pathogen. Insects may serve as a reservoir for infecting susceptible reptiles.

Beauveria infections in reptiles have only been reported from tortoises and alligators. *Beauveria bassiana* was isolated from pulmonary lesions in an Aldabra tortoise and a Galapagos tortoise in a zoologic collection (Georg *et al.*, 1962). In this study, two box turtles, *Terrapene carolina*, were challenged via the intrapulmonary route with spores of this isolate. While a turtle maintained at 60°F died 4 days postinoculation, a turtle maintained at 72°F remained normal. *Beauveria bassiana* was isolated from extensive pulmonary lesions in several alligators in a zoologic park enclosure that experienced a heating system failure during a period of extreme cold (Fromtling *et al.*, 1979). Fungal colonies were found growing within air spaces and on the pleural surfaces. The pulmonary parenchyma contained multifocal granulomas.

d. Penicillium

Most of the documented cases of penicilliosis in reptiles involve chelonians. Since *Penicillium* is commonly isolated as a surface and environmental contaminant, culture must be correlated with histopathologic verification to make an accurate diagnosis. Hamerton (1934) reported on systemic penicilliosis in a tortoise. *Geochelone nigrita*. Multifocal granulomas were found in stomach, pancreas, liver, and lungs, and *Penicillium* was cultured from the lungs and identified in tissue section. *Penicillium lilacinum* has been cultured from granulomatous lung lesions in a yellow-legged tortoise, *G. denticulata* (Bemmel *et al.*, 1960), and a loggerhead turtle, *Caretta caretta*, and a green turtle, *Chelonia mydas* (Keymer, 1974).

e. Geotrichum

In reptiles there is a single report of spontaneous geotrichosis in a tortoise. *Geotrichum candidum* (in addition to *Aspergillus amstelodami*) was isolated from granulomatous lung lesions in a Galapagos tortoise (Georg *et al.*, 1962). Experimentally, generalized infections were induced in snapping turtles, *Chelydra serpentina*, and red-eared sliders, *Chryemys scripta*, given pleuroperitoneal injections of plant and human isolates.

There are several reports of *Geotrichum* dermatitis in snakes. Karstad (1961) cultured *G. candidum* from subcutaneous nodules in a common water snake, *Nerodia sipedon*, and *Geotrichum* was isolated from subcutaneous granulomatous lesions (Fig. 16) in a rosy rat snake, *Elaphe guttata* (Jacobson, 1980). In the latter study, the snake was maintained under below optimum environmental conditions, and simply exposing the snake to an optimum temperature range resulted in regression of lesions. *Geotrichum candidum* was also identified as the causative agent of mycotic dermatitis in captive carpet pythons, *Morelia spilotes variegata* (McKenzie and Green, 1976). The initial lesion commenced at the hinge region between adjacent ventral scales and progressed to involve distal portions of ventral and lateral body scales. Eventually the lesions became necrotic.

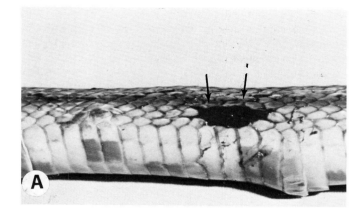

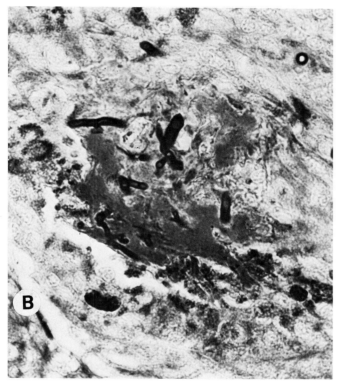

Fig. 16. (A) Rosy rat snake with firm subcutaneous masses bulging the overlying skin; one mass has been biopsied (arrows). (B) Hyphae are seen within the caseated center of the biopsied mass. GMS stain. × 440. (From Jacobson, 1980.)

f. Fusarium

Most cases of fusariosis in higher vertebrates involve the skin and cornea. This also appears the case for reptiles. One of the first reports of mycotic disease in a reptile involved a European green lizard, *Lacerta viridis,* with "tumorous" integumentary lesions (Reichenbach-Klinke and Elkan, 1965). The lesions contained septate conidia that resembled *Fusarium. Fusarium* was also isolated from shell lesions in a radiated

tortoise, *Testudo radiata* (Frank, 1970), and from skin lesions in a caiman, *Caiman crocodilus fuscus* (Kuttin *et al.,* 1978), and a Burmese python, *Python molurus* (Jacobson, 1980). *Fusarium oxysporum* was isolated from an infected spectacle in a rainbow boa, *Epicrates cenchris maurus,* and was treated by complete enucleation (Zwart *et al.,* 1973).

4. Parasitic

a. Entamoeba invadens

This protozoan represents the most significant ameba to cause disease in captive reptiles. Although for a long time crocodilians and turtles were considered inapparent carriers with mortality in snakes and lizards, several cases of chelonian amebiasis including an epizootic involving a colony of red-footed tortoises, have been seen. Reptiles having active amebic infection may exhibit nonspecific signs, including lethargy, anorexia, dehydration, and diarrhea, either mucoid or bloody. In some cases no clinical signs have been noted up to 24 hr prior to death.

Postmortem findings often include a hemorrhagic necrotizing gastroenteritis. In many cases the liver is swollen with diffuse areas of necrosis. Trophozoites are readily identifiable on histologic examination of affected intestine and liver. Periodic acid–Schiff (PAS) is the stain of choice for identifying trophozoites.

In almost all cases of amebiasis there is concomitant infection with opportunistic gram-negative microorganisms, including *Salmonella* and *Arizona.* These organisms appear to act in concert with *E. invadens,* and treatment must include the use of broad spectrum antibiotics, such as chloramphenicol and gentamicin.

Antemortem diagnosis is based upon demonstration of trophozoites (16–19 μm) in the feces. Trophozoites released in feces rapidly dessicate outside the host. The amebic cyst represents the infective stage with oral transmission from animal to animal via contaminated water, food, or the immediate environment. It is often the recently acquired reptile that is shedding cysts that becomes the nidus of infection for a stable, disease-free collection.

Amebiasis can be treated by modifying environmental conditions and/or chemotherapy. With regard to the former, it has been found that under experimental conditions snakes were unable to maintain an active infection when maintained at temperatures above 33°C. The drug of choice for treating infected reptiles is metronidazole at 120–250 mg/kg body weight as a single dose repeated in 2 weeks.

b. Cryptosporidium

Cryptosporidium has been found associated with severe hypertrophic gastritis in zoologic collections of snakes (Brownstein *et al.,* 1977). A consistent clinical sign seen with snakes having severe cryptosporidial gastritis is postprandial reg-urgitation, generally within 3 days following feeding. The stomach is generally thickened and bulges the overlying body wall. Even in mild cases a firm stomach is readily palpated through the body wall.

A diagnosis of cryptosporidiosis can be made by demonstrating oocysts (2.6–6.0 μm) and trophozoites (1.8 × 1.2 μm) in PAS or Wright–Giemsa stained fecal smears or demonstrating organism in smears of biopsy specimens stained by the Wright-Giemsa method.

Histopathology often reveals edema and thickening of the gastric mucosa with mucosal petechiae, brush hemorrhages, and focal necrosis. Early lesions include hyperplasia of mucous neck cells with focal atrophy of granular cells. As the disease progresses there may be a total loss of the granular cells with a distortion of the normal cytoarchitecture resulting from cystic dilatation of glandular elements and a loss of the parallel arrangement of adjacent gastric glands.

The life cycle of reptilian cryptosporidiosis is unknown. A direct life cycle, such as that present in guinea pigs, may be involved, although an indirect cycle using a rodent intermediate host cannot be discounted. As part of a control program, suspect animals should be isolated and cages disinfected. Unfortunately there is no chemotherapeutic agent found to be effective against reptilian cryptosporidiosis.

c. Hemogregarines

Members of this group commonly infect all species of reptiles, except marine turtles, and are transmitted by either arthropod or annelid vectors. The most important general are *Haemogregarina, Hepatozoan, Karyolysus, Schellackia,* and *Lainsonia.* These are all intracellular sporozoan hemoparasites found in both red and white blood cells, which often result in a distorted appearance.

Hemogregarines are commonly reported in both healthy and ill free-ranging and captive reptiles. The pathogenic significance in reptiles is unknown. Usually the identification of a hemogregarine infection in a reptile patient is an incidental finding. Likewise there is no information available on treatment that, in any case, would be an academic pursuit.

d. Trematodes

Reptiles are hosts for numerous species of both monogenetic and digenetic trematodes. Although the vast majority of these parasites appear to be clinically insignificant, there are several documentations of trematode-associated disease problems in reptiles including infection with spirorchids, renifers, and *Styphlodora.*

Adult members of the family Spirorchidae most commonly inhabit the circulatory system of turtles. Although focal mild lesions of endothelial hyperplasia are of limited clinical significance, the inflammatory response to released eggs is another situation. Eggs shed into the circulatory system generally become trapped in small arterioles in almost any organ, where an intense granulomatous inflammatory response may develop.

The renifers comprise a group of trematodes that are commonly found in the oral cavity of a wide variety of snakes. The most commonly encountered genera include *Dasymetra, Zeugorchis, Ochestosoma,* and *Stomatrema.* Adult parasites are generally found within the lung(s) of snakes and may produce focal ulcerative lesions in areas of attachment. Secondary gram-negative bacterial infections have been found associated with renifer infections.

Adult renifers produce eggs that are expelled through the glottis and ultimately pass out in the feces. Eggs are yellow–orange in color, measure approximately 40 × 25 μm and have a pale polar cap. Eggs may be seen in lung washings of infected snakes or demonstrated on direct fecal smear or fecal sediment. Since intermediate hosts include snails and amphibians, infections are self-limiting in captivity. No safe chemotherapeutic is available to successfully treat reptiles with renifer infections. Although oral tetrachlorethylene at a dose of 0.2 ml/kg body weight was found effective in common water snakes, *Natrix spideon* (Nelson, 1950), hepatotoxic effects of this drug limit its usefulness.

Adult flukes in the genus *Styphylodora* have been found within terminal renal tubules and ureters of several species of harmless and poisonous snakes. Probably most infections in captive snakes are subclinical and go undiagnosed or are reported as incidental findings at necropsy. In severe infections, reported pathologic changes included renal tubular dilatation, occlusion of tubules by cellular debris, and chronic interstitial nephritis. Diagnosis is dependent upon demonstration of adult parasites and/or eggs in renal tissue or ureters. Eggs measure approximately 40 × 18 μm, have a bright yellow egg shell, and a faint polar cap. The life cycle of this parasite is unknown and no chemotherapeutic agent is known to be effective for *Styphlodora* infections.

e. Cestodes

Cestodes are known to infect all major groups of reptiles except crocodilians. Reptiles may act as both definitive and intermediate hosts for cestodes. The orders of greatest clinical importance to reptiles are Pseudophyllidea, Mesocestoididea, and Proteocephalidea.

The most significant pseudophyllidean tapeworms include the genera *Bothridium, Bothriocephalus,* and *Spirometra.* Adults of the first two parasites commonly infect the intestine of pythons and, although most infections in captive pythons are subclinical, *Bothridium* sp. infection in green tree pythons was associated with mild chronic enteritis (Toft and Schmidt,

1975). Snakes not uncommonly act as intermediate hosts for *Spirometra*, with plerocercoids (spargana) being found in subcutis or viscera. Snakes may be presented with subcutaneous nodules and larvae may be seen in crushed preparations of surgically excised nodules.

Lizards and snakes act as intermediate hosts for the genus *Mesocestoides*. As with *Spirometra*, animals may be presented with subcutaneous nodules harboring these parasites. In animals with heavy infections, hundreds of encysted larvae may be seen throughout the coelomic mesentery. *Mesocestoides* larvae may also be found in a variety of visceral organs, particuarly the liver and pancreas.

The genus *Ophiotaenia* is the most significant member of the order Proteocephalidea known to infect reptiles. It is the most frequently reported tapeworm found to infect North American snakes. Snakes serve as definitive hosts, with frogs acting as intermediate hosts. In most cases, infections are subclinical. Antemortem diagnosis is dependent upon demonstrating adult parasites and/or eggs in feces. Eggs containing onchospheres measure 50 × 40 μm and may be seen in fecal sediments or direct fecal smears.

There are no drugs found to be effective against migrating larval cestodes. For adult tapeworms within the intestinal tract, the drugs of choice are oral niclosamide at 150 mg/kg body weight and oral bunamide at 50 mg/kg body weight.

f. Nematodes

i. Ascarids. This group of nematodes has been found to infect all major groups of reptiles. A majority of reptilian ascarids have been described from crocodilians and snakes, with *Ophidascaris* being one of the most significant helminth parasites of snakes, and amphibians serving as intermediate hosts. In snakes, ascarids may produce significant lesions as a result of larval migrations. Adult parasites embed into the submucosa of the terminal esophagus and stomach where a tremendous inflammatory response may be elicited. Concomitant secondary bacterial infections may be followed by systemic disease. Most cases of ascaridiasis are subclinical, but in severe infections snakes may exhibit postprandial regurgitation. Diagnosis may be made on demonstration of characteristic thick-shelled eggs measuring 80–100 × 60–80 μm in fecal floatation specimens. Oral mebendazole at 20–25 mg/kg body weight and oral thiabendazole at 50 mg/kg body weight administered as a single dose repeated in 2 weeks are the drugs of choice.

ii. Strongyles. The most significant strongyle found in reptiles is the snake hookworm *Kalicephalus*. This nematode has a direct life cycle with infection resulting from either oral ingestion or percutaneous penetration. *Kalicephalus* is a small nematode most commonly found in the intestinal tract and,

although a majority of infections most likely go undiagnosed, still anorexia, debility, lethargy, and death have been reported in snakes with heavy infections. In such cases, gastrointestinal lesions include hemorrhage, erosions, and ulceration of the musoca. Eggs are thin walled, measure 70–100 × 40–50 μm, and may be larvated by the time the feces is expelled. Larvated *Kalicephalus* eggs, except for being slightly smaller, are indistinguishable from lungworm (*Rhabdias*) eggs. Oral mebendazole and thiabendazole at the same dose described for ascarid infections are the drugs of choice.

iii. Lungworms. The most important lungworm of snakes and lizards belongs to the genus *Rhabdias*. Adult parasites in the lung(s) are parthenogenetic females, and embryonated eggs are expelled via the glottis into the oral cavity, swallowed, and passed in the stool. Eggs measure 60 × 35 μm and hatch into free-living rhabditiform larvae that produce infective filariform larvae that enter a susceptible reptile either via oral or percutaneous routes; larvae ultimately migrate to the respiratory system.

Rhabdias lung infections have been found in snakes with severe clinical respiratory disease. Often there is a concomitant gram-negative bacterial respiratory infection. Snakes may be presented with gaping of the mouth and accumulation of exudate within and around an extended glottis. Exudate from the respiratory system may be expelled into the internal nares and ultimately collect around the external nares. Eggs may be demonstrated in samples obtained from lung washings or directly in exudate accumulating around the nares. At necropsy, adult parasites are often found in the vorbronchus of the lung, which may show fovalveolar thickening and accumulation of exudate in interfovalveolar airways. Intraperitoneal levamisole hydrochloride at 10 mg/kg body weight as a single dose repeated in 2 weeks is the drug of choice. Because of concomitant bacterial infection, therapy often includes the administration of broad spectrum injectable antibiotics.

iv. Filarids. Although filarid infections are not uncommon in reptiles, there is only a single report involving pathology. The filarid *Macdonaldius oschei* was found to cause extensive dermal lesions in several pythons in a German zoo (Frank, 1964). Whereas in natural hosts, colubrid and viperid snakes of Mexico, *Macdonaldius* is normally found in the posterior vena cava and renal veins, in the above pythons it was found in mesenteric arteries. In this site a granulomatous inflammatory response was elicited and aneurysms developed; microfilariae were found to occlude dermal capillaries. Arthropods including mosquitoes and ticks serve as intermediate hosts and are necessary for transmission. Antemortem diagnosis can be made by demonstrating microfilariae in a peripheral blood sample. Although chemotherapy for reptilian filarid infections has not been reported, *M. oschei* can be killed by maintaining snakes at an ambient temperature of 35°–37°C for 24–48 hr. Since this temperature range is near the lethal maximum temperature for many species of snakes, snakes must be monitored closely.

g. Acanthocephala

A number of genera of spiny headed worms have been reported from a variety of reptiles, with aquatic turtles being commonly infected. Invertebrates, particularly crustacea and snails, serve as intermediate hosts, with reptiles acting mostly as either paratenic hosts (many species of snakes) and definitive hosts (turtles). Whereas adult parasites within the intestinal tract produce focal ulcerating lesions, large numbers of encapsulated immature parasites within the coelomic cavity may result in more substantial problems, including mechanical obstruction of the intestine and mechanical damage to the liver. Antemortem diagnosis can be made in animals harboring adult parasites in the intestine by demonstration of characteristic multienveloped eggs containing an acanthor larva with rostellar hooklets. The drugs found to be effective and safe in treating reptilian acanthocephalan infections include 20 mg/kg body weight dithiazine iodide for 10 days, 8 mg/kg intraperitoneal levamisole phosphate, and 5 mg/kg intraperitoneal levamisole hydrochloride repeated in 2–3 weeks (Frye, 1981).

h. Pentastomes

These wormlike parasites, except for *Linguatula* in domestic mammals and *Reighardia* in gulls, are found as adults in reptiles and as juveniles in mammals. The most important genera are *Sebakia* (crocodilians); *Rallietiella* (lizards and snakes); and *Kericephalus*, *Porcephalus*, and *Armillifer* (snakes). Man may serve as an intermediate host for the genera *Rallietiella* and *Armillifer*. Adult pentastomes in reptiles are usually found within the respiratory system, although sexually mature parasites have also been found in the coelomic cavity and subcutis. Juvenile parasites located in the subcutis may elicit an inflammatory response, which is seen clinically as palpably firm nodules. Adult parasites in the lung(s) may result in focal epithelial necrosis with hemorrhage and exudate filling the air passageways. Adults may mechanically obstruct the tracheal system. Secondary bacterial invaders often synergize with the pentastomids. Antemortem diagnosis is dependent upon identifying parasites in subcutaneous sites and/or in the respiratory system (with the aid of fiber optic scopes) or demonstration of characteristic encapsulated eggs measuring 130 μm (*Armillifer*) in feces or lung washings. There are no chemotherapeutics known to be effective against pentastomids. Since intermediate hosts are involved in the life cycle, infections in captivity are self-limiting.

i. Mites and Ticks

Free-ranging reptiles are infested by numerous species of mites and ticks. The most thoroughly studied acarine of reptiles is the snake mite, *Ophionyssus natricis*. While snakes in the wild generally maintain minimal infestations, burdens in captivity may be severe. Mites are commonly found under scales and in the angle formed by the spectacle and the periocular scales. Eggs are deposited in the snake's environment with the life cycle completed in 10–32 days. Disease may result from both anemia due to blood loss and by the mite having the potential to mechanically transmit the gram-negative pathogen, *Aeromonas hydrophila*, from animal to animal. Dichlorovinyl dimethylphosphate (DVDP) pest strips (used for 3 days) will effectively eliminate mite infestations. When pest strips are used, water bowls should be removed from the animal's cage, there should be adequate ventilation, and the reptile should never be allowed to come in direct contact with the strip.

Additional mites that may be encountered in free-ranging and captive reptiles are lung mites (*Entonyssus* and *Ophiopneumicola*) and chiggers (*Eutrombicula*), both of which are generally innocuous.

Ticks commonly found to infest reptiles include the genera *Amblyomma*, *Aponomma*, and *Ornithodoros*. Ticks may be found almost anywhere on the animal's body, but are often particularily abundant under gular scales on snakes and in periocular locations in lizards. As with mites, disease may result directly from blood loss or indirectly from pathogens transmitted by ticks. Most ticks can be either manually removed from reptiles or eliminated with DVDP strips.

B. Metabolic/Nutritional Diseases

1. Nutritional Bone Disease

Included within this broad category of bone disease is rickets, secondary (nutritional) hyperparathyroidism, and nutritional osteodystrophy. These diseases have a nutritional basis, with affected animals being maintained on diets deficient in vitamin D_3 and having either inverse Ca : P ratios or low absolute levels of calcium. Most cases of nutritional bone disease have been seen in young rapidly growing reptiles, such as the green iguana and aquatic turtles, that have been fed all vegetable diets of low nutritional value with inverse Ca : P ratios and in young rapidly growing crocodilians being fed all red meat diets. Such nutritional bone disease has never been reported in reptiles being fed whole animal diets of good nutritional content.

Affected animals are characterized by a progressive difficulty in locomotion, fragile soft bones, teeth that grow out at acute angles (caimans), curvatures of the vertebral column (caimans), and deformities of the shell (turtles). In hatchling green iguanas fed a beef heart diet, osteoporosis developed within 1–2 months, whereas the same lesions required 6–10 months to develop in juvenile iguanas (Anderson and Capen, 1976a). The long bones are characteristically affected in green iguanas with diaphyseal enlargements due to a cuff of proliferating cartilage that either compresses or replaces the overlying cartilage (Fig. 17).

Therapy should include the administration of injectable 10% calcium gluconate (1 ml/kg) in severe cases and modification of the diet to be nutritionally complete. Unfortunately many of these animals have been conditioned to feeding on poor quality food, and it may take a considerable amount of energy and ingenuity to change the animal's food preference. Depending upon the species, the diet should include whole animals such as mice and rats. Softened monkey chow is an ideal food source for many of these species. Pet Cal tablets (Beecham) can be crushed and added to the diet. Where possible, these animals should be exposed to natural sunlight, otherwise fluorescent ultraviolet lamps can be used.

2. Vitamin A Deficiency

Vitamin A is essential for the maintenance of epithelial integrity, and many cases of vitamin A deficiency in reptiles have been seen clinically as epithelial related problems, particularly disease of the adnexal structures of the eye in aquatic turtles. Turtles are often presented with palpebral edema, anasarca, hyperkeratosis of the skin, and overgrowth of the mandibular and maxillary horny mouth parts. Classic histopathological changes seen are squamous metaplasia and

Fig. 17. Chondromatosis of the radius and ulna of a spiny tailed iguana. The primary diet of this lizard was banana. (From Jacobson, 1981a.)

hyperkeratosis of lacrimal and Harderian gland tubules (Elkan and Zwart, 1967).

Therapy should include treatment of the immediate problem and improvement of the deficient diet. In severe cases in which the animal's palpebrae are swollen and closed, injectable vitamin A can be administered intramuscularly at 2000 units/kg body weight every 3 days for 2 weeks. Trout pellets are nutritionally complete and readily accepted by most aquatic chelonians.

3. Vitamin E Deficiency

Vitamin E acts as an antioxidant along with glutathione peroxidase in protecting cell membranes and cytoplasmic constituents against damage by lipid peroxides and other free radicals. Peroxidation of ingested unsaturated fatty acids will occur in the absence of significant levels of vitamin E, resulting in the accumulation of a yellow to brown pigmented metabolite called ceroid, which may accumulate almost anywhere in the animal's body, but most commonly in fat deposits. This condition has been reported in captive crocodilians fed fishes high in unsaturated fatty acids (Frye and Schelling, 1973; Wallach and Hoessle, 1968) and in snakes fed obese laboratory rats (Frye, 1981). At necropsy these ceroid deposits will emit a yellow–orange fluorescence under ultraviolet light, and microscopic lesions consist of acid-fast positive, basophilic amorphous material surrounded by fibroblasts, and macrophages.

Animals having pansteatitis may exhibit nonspecific clinical signs of muscle weakness, inability to locomote, and anorexia. Diagnosis is ultimately dependent upon gross and microscopic findings, although a history of an all fish diet high in unsaturated fatty acids is presumptive evidence for this disease. Therapy should include improvement of the diet to reduce polyunsaturated fatty acid intake and supplementation with exogenous vitamin E at 100 IU per day (Wallach and Hoessle, 1968).

4. Vitamin C Deficiency

Since vitamin C is essential for adequate formation of collagen and capillary integuity, gross vitamin C deficiencies have been associated in mammals with a variety of lesions involving mucous membranes and bone. Since many reptiles may have bacterial production of vitamin C in the colon and cecum, Wallach (1971) believed that stress or disruption of the intestinal flora can produce vitamin C deficiency. Frye (1981) also believed that vitamin C deficiencies occurred in captive reptiles and found that snakes, lizards, and chelonians with infectious stomatitis responded faster when these animals were medicated with vitamin C (25 mg in small reptiles up to several grams in large animals) in addition to antibiotic therapy. Recently, Vosburgh *et al.* (1982) found that plains garter snakes,

Thamnophis radix, and eastern garter snakes, *T. sirtalis sirtalis,* synthesize their requirements for vitamin C in the kidney, with snakes receiving a dietary supply of vitamin C decreasing renal tissue synthesis. In their study, the stresses of confinement and tube feeding did not increase requirements beyond that supplied by tissue synthesis alone. Thus, if other reptiles similarly produce vitamin C in the kidney, a deficiency would be expected to be seen only in cases of renal disease.

5. Iodine Deficiency and Hypothyroidism

Iodine deficiencies in reptiles were first reported by Reichenbach-Klinke and Elkan (1965) and subsequently described in reptiles by other investigators (Zwart and Kok, 1978; Frye, 1981). In such cases, iodine deficiency resulted in goiter. Wallach (1969) and Frye and Dutra (1974) also reported that goitrogenic factors present in hay, lettuce, kale, and spinach, which are commonly fed to tortoises, may result in hypothyroidism. The reported clinical manifestations of hypothyroidism in reptiles included anorexia, lethargy, fibrous goiter, and, often, swelling of subcutaneous tissues (Frye, 1981). On histopathology the thyroids have scant amounts of colloid, and epithelial cells are columnar in appearance. Treatment consists of adding exogenous iodine to the diet (for dosing, see Frye, 1981) and reducing goitrogenic vegetables offered to the animals.

6. Hypervitaminosis D

This condition has been identified in common green iguanas oversupplemented with commercial vitamin D (Wallach, 1966). Pathologic lesions included medial calcification of elastic blood vessels with severe cases being diagnosed radiographically. Metastatic mineralization may occur in a wide variety of visceral structures. On necropsy affected arteries are fairly rigid and when cut have a gritty consistency. Vitamin supplementation, although of benefit in many situations, must be judiciously used, especially with regard to fat-soluble vitamins.

7. Visceral and Articular Gout

Hyperuricemia resulting in visceral gout has been reported as a cause of death in captive reptiles (Appleby and Siller, 1960; Wallach and Hoessle, 1967). Although visceral gout secondary to renal failure has been conclusively demonstrated, the evidence of high protein intake resulting in this condition is less convincing, although often hypothesized as a cause. Coulson and Hernandez (1964) did demonstrate that the intraperitoneal injections of the amino acid, D-serine, to alligators resulted in extreme elevations of blood uric acid and subse-

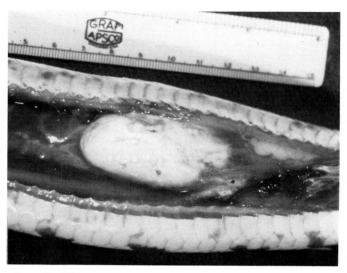

Fig. 18. Visceral gout in a ball python resulting from gentamicin nephrotoxicity. Uric acid precipitate is seen in the pericardial sac. (From Jacobson, 1976.)

quent visceral gout. Water deprivation also has been suggested as a cause of visceral gout, although in most severely dehydrated reptile cases this condition has not been identified.

The aminoglycosidic antibiotic, gentamicin sulfate, has potential nephrotoxic effects in a wide variety of vertebrates, and cases of gentamicin-related visceral gout (Fig. 18) have been described for boid snakes (Jacobson, 1976). Typical lesions of visceral gout followed treatment with gentamicin dosages of 2 mg/lb body weight twice per day for 2 days followed by once per day for 5 days. Subsequently theapeutic dosages for snakes were determined to be at 2.5 mg/kg body weight every 72 hr (Bush, 1980). Classic lesions of visceral gout consist of uric acid accumulations almost anywhere in the animal's body, but particularly prominent in the pericardial sac.

Articular gout has also been reported in reptiles, with most cases seen in chelonians (Frye, 1981). In such cases, the animal's joints appear swollen and are palpably firm. Uric acid crystals can be demonstrated in tissue section and in biopsy specimens of joint lesions.

C. Traumatic

A great variety of traumatic injuries have been identified for captive reptiles, including abrasions, bite wounds, crushing injuries to turtle shells, damaged mouth parts (especially in turtles), and thermal burns. Several of the important traumatic conditions commonly observed in captive reptiles will be discussed.

Since reptiles are ectotherms, ultimately dependent upon environmental infrared radiation as a heat source, high wattage infared and influorescent light bulbs are used as heat sources in captivity. Often the bulb is placed within the cage where the animal may make direct contact, or too "hot" a bulb is used resulting in thermal burns to the animal. Ideally, captive reptiles should be exposed to a gradient of temperature that spans the preferred optimum range for the species. Heat sources should ideally originate outside the cage or at a distance where contact cannot be made. Reptiles receiving thermal burns will develop epidermal vesicles followed by necrosis.

Electric heating pads used to provide heat for captive reptiles often develop hot spots and result in ventral burning of the reptile's integument. Severe necrotizing burn lesions often become invaded with *Pseudomonas* spp. and further add to the problems. Such animals require intensive chronic treatment with broad spectrum injectable antibiotics in addition to organic iodine soaks and topical application of organic iodine ointments.

Trauma in captive reptiles may result from direct aggression between cagemates or indirectly when reptiles come together at feeding time. Two snakes feeding upon the same rodent or fish may result in the larger animal consuming not only the prey species but also the smaller animal. With snakes, often when an animal is fed a prey species a feeding frenzy is elicited whereby anything smelling of the prey species will be attacked and consumed. This includes cagemates. Some individuals may have to be separated from cagemates at feeding time.

For carnivorous reptiles it is recommended that prey items be killed prior to feeding, especially when food items are rodents. Live mice and rats can inflict serious bites upon their reptile predator. Snakes have been seen that have had large areas of skin and underlying musculature removed by prey rodent species left in the cage overnight. Even a constrictor in the act of suffocating its prey species can sustain serious injury. Most reptiles will feed upon dead rodents as voraciously as upon live ones.

D. Neoplastic

Neoplastic diseases of reptiles have been extensively reviewed elsewhere (Jacobson, 1981b), and the purpose of this section will be to briefly discuss some of more common and pertinent neoplastic conditions. Of 159 reports of neoplasms in reptiles, 28 involved chelonians (17 benign and 11 malignant), 4 involved crocodilians (3 benigh and 1 malignant), 30 involved lizards (18 benign and 12 malignant), and 97 involved snakes (29 benign and 68 malignant). Rather than reflecting a biological basis for the neoplasia, the difference in the descriptive incidence of neoplasms within the various groups of reptiles most likely reflect the types of species maintained in reptile collections and the longevity of these animals in captivity. Snakes are the most commonly exhibited species, and thus it is not surprising that the majority of described reptile noeplasms involve this group. It is interesting that crocodilians, which are also popular in zoos and long lived, have few reported neoplasms.

All of the major groups of neoplasms found in mammals have been reported in reptiles except primary noeplasms involving the central nervous sytem. Additionally, reptiles are known to develop tumors of pigment cell origin, including chromatophoromas, xanthophoromas, and iridosarcomas, which are not seen in mammals.

Infectious agents have been associated with several reptilian neoplasms. Papillomas and fibropapillomas commonly encountered in free-ranging and mariculture maintained adult green sea turtles are commonly associated with surface leeches and spirorchid trematode eggs in dermal vessels. The trematode eggs may be elicitng a tremendous connective tissue response ultimately resulting in fibropapilloma development. No viral particles have been found associated with these lesions.

Virus particles have been associated with, and incriminated in, the pathogenesis of neoplasms in several reptile species. Electron microscopic evaluation of papillomas from European green lizards have revealed three distinct virus particles, including herpesvirus, reovirus, and papovavirus (Raynaud and Adrian, 1976). C-Type viral particles have been demonstrated in either tissue cultures derived from neoplastic tissue (Zeigel and Clark, 1969) or within the primary tumor tissue itself (Lunger *et al.*, 1974). The detection of several oncornaviruses in the course of very limited studies in snakes suggests an unusual prevalence in this group of vertebrates.

E. Congenital Abnormalities

A great diversity of congenital defects have been reported or observed in all major groups of reptiles. Except for temperature-related congenital abnormalities in snapping turtles, crocodilians, garter snakes, and Burmese pythons, the etiologic basis for the majority of these conditions remains unknown. Viral, toxicologic, and genetic factors may be involved.

Reports of two-headed reptiles, most commonly lizards and snakes, are scattered throughout the literature. A hatchling Florida red bellied turtle, *Chrysemys nelsoni,* was seen which, along with two heads, had a pair of forelimbs associated with each head; the carapace was partly doubled. Such abnormalities probably result from partial division of a single embryo at the stage of cleavage or gastrulation. Along with this condition there is often a doubling of internal organs including heart, esophagus, stomach, and lung.

REFERENCES

Ackerman, L. J., Kishimoto, R. A., and Emerson, J. S. (1971). Nonpigmented Serratia marcescens arthritis in a teju (*Tupinambis teguixin*). *Am. J. Vet. Res.* **32,** 823–826.

Andersen, S., and Eriksen, E. (1968). Aspergillose bei einer elephantenschildkröte (*Testudo gigantea elephantina*). *Int. Symp. Erkr. Zootiere, 10th, 1968* pp. 65–67.

Anderson, M. P., and Capen, C. C. (1967a). Fine structural changes of bone cells in experimental nutritional osteodystrophy of green iguanas. *Virchows Arch. B* **20,** 169–184.

Anver, M. R., Park, J. S., and Rush, H. G. (1976). Dermatophilosis in the marble lizard (*Calotes mystaceus*). *Lab. Anim. Sci.* **26,** 817–823.

Appleby, E. C., and Siller, W. G. (1960). Some cases of gout in reptiles. *J. Pathol. Bacteriol.* **80,** 427–430.

Aronson, J. D. (1929). Spontaneous tuberculosis in snakes. *J. Infect. Dis.* **44,** 215–223.

Bemmel, A. C. V., Peters, J. C., and Zwart, P. (1960). Report on births and deaths occurring in the Gardens of the Royal Rotterdam Zoo during the year 1958. *Tijdschr. Diergeneeskd.* **85,** 1203–1213.

Blain, A. W., and Campbell, K. N. (1942). A study of the digestive phenomena in snakes with aid of the roentgen ray. *Am. J. Roentgenol.* **48,** 229–239.

Blazek, K., Jaros, Z., Otcenasek, M., and Konrad, J. (1968). Zum vorkommen und zur histopathologie der tiefen organmykosen bei den zootieren. *Int. Symp. Erkr. Zootiere, 10th, 1968* pp. 189–192.

Boam, G. W., Sanger, V. L., Cowan, D. F., and Vaughan, D. P. (1970). Subcutaneous abscesses in iguanid lizards. *J. Am. Vet. Med. Assoc.* **157,** 617–619.

Boever, W. L., and Williams, J. (1975). *Arizona* septicemia in three boa constrictors. *Vet. Med. Small Anim. Clin.* **70,** 1357–1359.

Bogert, C. M. (1953). Body temperatures of the tuatara under natural conditions. *Zoologica (N.Y.)* **38,** 63–64.

Brattstrom, B. H. (1965). Body temperatures of reptiles. *Am. Midl. Nat.* **73,** 376–422.

Brownstein, D. G. *VM/SAC,* (1978). Reptilian mycobacteriosis. *In* "Mycobacterial Infections of Zoo Animals" (R. J. Montali, ed.), pp. 665–668. Smithson. Inst. Press, Washington, D.C.

Brownstein, D. G. (1984). Mycobacterial disease. *In* "Diseases of Amphibians and Reptiles" (G. L. Hoff, F. F. Frye, and E. R. Jacobson, eds.). Plenum, New York (in press).

Brownstein, D. G., Strandberg, J. D., Montali, R. J., Bush, M., and Fortner, J. (1977). Cryptosporidium in snakes with hypertrophic gastritis. *Vet. Pathol.* **14,** 606–617.

Bush, M. (1980). Antibiotic therapy in reptiles. *Curr. Vet. Ther.* **7,** 647–649.

Cambre, R. C., Green, D. E., Smith, E. E., Montali, R. J., and Bush, M. (1980). Salmonellosis and Arizonosis in the reptile collection at the National Zoological Park. *J. Am. Vet. Med. Assoc.* **177,** 800–803.

Chatton, E., and Blanc, G. (1914). Sur un hématozaire nouveau, *Pirhemocyton tarentolae,* du gecko, *Tarentola mauritanica,* et sur les alterations globulaires qu'il détermine. *C. R. Seances Soc. Biol. Seo Fil.* **77,** 496–498.

Chiodini, R. J. (1982). Transovarian passage, visceral distribution, and pathogenicity of *Salmonella* in snakes. *Infect. Immun.* **36,** 710–713.

Cieslak, E. S. (1945). Relations between the reproductive cycle and the pituitary gland in the snake, *Thamnophis radix. Physiol. Zool.* **18,** 299–329.

Clark, H. F., Lief, F. S., Lunger, P. D., Waters, D., LeLoup, P., von Fölsch, D. W., and Wyler, R. W. (1979). Fer-de-lance virus: A probable paramyxovirus isolated from a reptile. *J. Gen. Virol.* **44,** 405–418.

Claussen, H. J., and Durand-Reynolds, F. (1937). Studies on the experimental infection of some reptiles, amphibia, and fish with *Serratia anolium. Am. J. Pathol.* **13,** 441–451.

Colbert, E. H., Cowles, R. B., and Bogert, C. M. (1946). Temperature tolerances in the American alligator and their bearing on the habits, and extinction of the dinosaurs. *Bull. Am. Mus. Nat. Hist.* **86,** 329–373.

Cole, C. J. (1975). Evolution of parthenogenetic species of reptiles. *In* "Intersexuality in the Animal Kingdom" (R. Reenboth, ed.), pp. 340–355. Springer-Verlag, Berlin and New York.

Conner, J., and Crews, D. (1980). Sperm transfer and storage in the lizard, *Anolis carolinensis. J. Morphol.* **163,** 331–348.

Coulson, R. A., and Hernandez, T. (1964). "Biochemistry of the Alligator." Louisiana State Univ. Press, Baton Rouge.

Cowles, R. B., and Bogert, C. M. (1944). A preliminary study of the thermal requirements of desert reptiles. *Bull. Am. Mus. Nat. Hist.* **83,** 265–296.

Cox, W. R., Rapley, W. A., and Barker, I. K. (1980). Herpesvirus-like infection in a painted turtle. *J. Wildl. Dis.* **16,** 445–449.

Crews, D. (1975). Psychobiology of reptilian reproduction. *Science* **189,** 1059–1065.

Crews, D., and Garrick, L. D. (1980). Methods of inducing reproduction in captive reptiles. *In* "Reproductive Biology and Diseases of Captive Reptiles" (J. B. Murphy and J. T. Collins, eds.), pp. 49–70. Society for the Study of Amphibians and Reptiles, Lawrence, Kansas.

D'Aoust, J. Y., and Lior, H. (1978). Pet turtle regulations and abatement of human salmonellosis. *Can. J. Public Health* **69,** 107–108.

Dessauer, H. C. (1970). Blood chemistry of reptiles: Physiological and evolutionary aspects. *In* "Biology of the Reptilia" (C. Gans. and T. S. Parsons, eds.), Vol. 3, pp. 1–72. Academic Press, New York.

Dieterich, R. A. (1967). What's wrong with your turtle? *Int. Turtle Tortoise Soc.* **1,** 20–21.

Draper, C. S., Walker, R. D., and Lawler, H. E. (1981). Patterns of oral bacterial infection in captive snakes. *J. Am. Vet. Med. Assoc.* **179,** 1223–1226.

Dunson, W. A. (1976). Salt glands in reptiles. *In* "Biology of the Reptilia" (C. Gans and W. R. Dawson, eds.), Vol. 5, pp. 413–445. Academic Press, New York.

DuPonte, M. W., Nakamura, R. M., and Chang, F. M. L. (1978). Activation of latent *Salmonella* and *Arizona* organisms by dehydration in red-eared turtles, *Pseudemys scripta elegans. Am. J. Vet. Res.* **39,** 529–530.

Durand-Reynolds, F., and Claussen, H. J. (1937). A contagious tumor like condition in lizard (*Anolis equestris*) as induced by a new bacterial species, *Serratia anolium* (Sp. N.). *J. Bacteriol.* **33,** 369–379.

Elkan, E., and Zwart, P. (1967). The ocular disease of young terrapins caused by vitamin A deficiency. *Pathol. Vet.* **4,** 201–222.

Evert, M. A., and Legler, J. M. (1978). Hormonal induction of oviposition in turtles. *Herpetologica* **34,** 314–318.

Feeley, J. C., and Treger, M. D. (1969). Penetration of turtle eggs by *Salmonella braenderup. Public Health Rep.* **84,** 156–158.

Foelsch, D. W., and Leloup, P. (1976). Fatale edemische infektion in einem serpentarium. *Tieraerztl. Praxis* **4,** 527–536.

Fox, H. (1977). The urogenital system of reptiles. *In* "Biology of the Reptilia" (C. Gans. and T. S. Parsons, eds.), Vol. 6, pp. 1–157. Academic Press, New York.

Fox, W. (1956). Seminal receptacles of snakes. *Anat. Rec.* **124,** 519–539.

Frank, W. (1964). Die pathogenen wirkungen von *Macdonaldius oschei* Chabaud et Frank 1961 (Filaroidea, Onchocercidae) bei verschiedenen arten von schlangen (Reptilian, Ophidia). *Z. Parasitenkd.* **24,** 249–275.

Frank, W. (1966). Multiple hyperkeratose bei einer Bartagame, *Amphibolurus barbatus* (Reptilia, Agamidae), hervorgerufen durch eine Pilzinfektion; zugleich ein Bertrag zur Problematik von Mykosen bei Reptilien. *Salamandra* **2,** 6–12.

Frank, W. (1970). Mykotische erkrankungen der haut und der inneren organe bei amphibien and reptilien. *Int. Symp. Erk. Zootiere, 12th, 1970* pp. 231–235.

Fromtling, R. A., Jensen, J. M., Robinson, B. E., and Bulmer, G. S. (1979). Fatal mycotic pulmonary disease of captive American alligators. *Vet. Pathol.* **16,** 428–431.

Frye, F. L. (1981). "Biomedical and Surgical Aspects of Captive Reptile Husbandry." Vet. Med. Publ. Co., Edwardsville, Kansas.

Frye, F. L., and Dutra, F. R. (1974). Hypothyroidism in turtles and tortoises. *VM/SAC, Vet. Med. Small Anim. Clin.* **69,** 990–993.

Frye, F. L., and Schelling, S. H. (1973). Steatitis in a caiman. *VM/SAC, Vet. Med. Small Anim. Clin.* **68,** 143–145.

Frye, F. L., Oshiro, L. S., Dutra, F. R., and Carney, J. D. (1977). Herpesvirus-like infection in two Pacific pond turtles. *J. Am. Vet. Med. Assoc.* **17,** 882–884.

Georg, L. E., Williamson, W. M., Tilden, F. B., and Getty, R. F. (1962). Mycotic pulmonary disease of captive giant tortoises due to *Beauvaria bassiana* and *Paecilomyces fumoso-roseus. Sabouraudia* **2,** 80–86.

Gilboa, I. (1972). Herpetological associations and societies. *Herpetol. Rev.* **4,** 77.

Griffith, A. S. (1939). Infections of wild animals with tubercle bacilli and other acid-fast bacilli. *Proc. R. Soc. Med.* **32,** 1405–1412.

Hamerton, A. E. (1934). Report on the deaths occurring in the Society's garden during the year 1933. *Proc. Zool. Soc. London* **104,** 389–403.

Hoff, G. L., and White, F. H. (1977). *Salmonella* in reptiles: Isolation from free-ranging lizards (Reptilia, Lacertilia) in Florida. *J. Herpetol.* **11,** 123–129.

Hunt, T. J. (1957). Notes on diseases and mortality in testudines. *Herpetologica* **13,** 19–23.

Jackson, C. G., Jr., and Fulton, M. (1970). A turtle colony epizootic apparently of microbial origin. *J. Wildl. Dis.* **6,** 466–468.

Jackson, C. G., Jr., and Jackson, M. M. (1971). The frequency of *Salmonella* and *Arizona* microorganisms in zoo turtles. *J. Wildl. Dis.* **7,** 130–132.

Jacobson, E. R. (1976). Gentamicin-related visceral gout in two boid snakes. *VM/SAC, Vet. Med. Small Anim. Clin.* **71,** 361–363.

Jacobson, E. R. (1978). Diseases of the respiratory system of reptiles. *VM/SAC, Vet. Med. Small Anim. Clin.* **73,** 1169–1175.

Jacobson, E. R. (1980). Necrotizing mycotic dermatitis in snakes: Clinical and pathologic features. *J. Am. Vet. Med. Assoc.* **177,** 838–841.

Jacobson, E. R. (1981a). Diseases of reptiles. Part I. Noninfectious diseases. *Compend. Contin. Educ. Pract. Vet.* **3,** 122–126.

Jacobson, E. R. (1981b). Neoplastic diseases. *In* "Diseases of the Reptilia" (J. E. Cooper and O. F. Jackson, eds.), Vol. II, pp. 429–468. Academic Press, New York.

Jacobson, E. R., Popp, J., Shields, R. P., and Gaskin, J. M. (1979). Poxlike virus associated with skin lesions in captive caimans. *J. Am. Vet. Med. Assoc.* **175,** 937–940.

Jacobson, E. R., Gaskin, J. M., Simpson, C. F., and Terrell, T. G. (1980a). Paramyxo-like virus infection in a rock rattlesnake. *J. Am. Vet. Med. Assoc.* **177,** 796–799.

Jacobson, E. R., Calderwood, M. R., and Clubb, S. L. (1980b). Mucormycosis in hatchling Florida softshell turtles. *J. Am. Vet. Med. Assoc.* **177,** 835–837.

Jacobson, E. R., Gaskin, J. M., Page, D., Iverson, W. O., and Johnston, J. W. (1981). Paramyxo-like virus associated illness in a zoologic collection of snakes. *J. Am. Vet. Med. Assoc.* **179,** 1227–1230.

Jacobson, E. R., Gaskin, J. M., and Wahlquist, H. (1982a). Herpesvirus-like infection in map turtles. *J. Am. Vet. Med. Assoc.* **181,** 1322–1324.

Jacobson, E. R., Gaskin, J. M., and Clubb, S. (1982b). Papilloma-like virus infection in Bolivian side-neck turtles. *J. Am. Vet. Med. Assoc.* **181,** 1325–1328.

Jasmin, A. M., Carroll, J. M., and Baucom, J. N. (1968). Pulmonary as-

pergillosis of the American alligator (*Alligator mississippiensis*). *Am. J. Vet. Clin. Pathol.* **2,** 93–95.

Jones, R. E., Roth, J. J., Gerrard, A. M., and Kiely, R. G. (1973). Endocrine control of clutch size in reptiles. I. Effects of FSH on ovarian follicular size-gradation in *Leiolopisma laterale* and *Anolis carolinesis. Gen. Comp. Endocrinol.* **20,** 190–198.

Jones, R. E., Tokarz, R. R., and LaGreek, F. T. (1975). Endocrine control of clutch size in reptiles. V. FSH-induced follicular formation and growth in immature ovaries of *Anolis carolinensis. Gen. Comp. Endocrinol.* **26,** 354–367.

Kaplan, H. (1974). Reptiles in laboratory animal science. *In* "Handbook of Laboratory Animal Science" (E. C. Melby and N. H. Altman, eds.), pp. 285–406. CRC Press, Cleveland, Ohio.

Kaplan, H. M. (1957). Septicemic, cutaneous ulcerative disease of turtles. *Proc. Anim. Care Panel* **7,** 273–277.

Karstad, L. (1961). Reptiles as possible reservoir hosts for eastern encephalitis virus. *Trans. North Am. Wildl. Nat. Resour. Conf.* **26,** 186–202.

Kaufmann, A. F., Fox, M. D., Morris, G. K., Wood, B. T., Feeley, J. C., and Frix, M. K. (1972). Turtle associated salmonellosis. III. The effects of environmental salmonellae in commercial turtle breeding ponds. *Am. J. Epidemiol* **95,** 521–528.

Kaura, Y. K., and Singh, I. P. (1968). Prevalence of salmonella in some common wall lizards, birds and rodents. *Indian J. Med. Res.* **56,** 1174–1179.

Keymer, I. F. (1974). The Zoological Society of London, Scientific Report, 1971–1973. Report of the Pathologist, 1971 and 1972. *J. Zool.* **173,** 62, 78.

Kuttin, E. S., Muller, J., May, W., Albrecht, F., and Sigalas, M. (1978). Mykosen bei krokodilen. *Mykosen* **21,** 39–48.

Lamm, S. H., Taylor, A., Gangarosa, E. J., Anderson, H. W., Young, W., Clark, M. H., and Bruce, A. R. (1972). Turtle-associated salmonellosis. I. An estimation of the magnitude of the problem in the United States, 1970, 1971. *Am. J. Epidemiol.* **95,** 511–517.

LaPointe, J. (1964). Induction of oviposition in lizards with the hormone oxytocin. *Copeia* No. 2, pp. 451–452.

Larsen, R. E., DeSena, R. R., Puckett, M. M., and Cardeilhac, P. T. (1982). Collection of semen for artifical insemination. *In* "Proceedings of the Second Annual Alligator Production Conference" (P. Cardeilhac, T. Lane, and R. Larsen, eds.), pp. 42–43. Gainesville, Florida.

Ledbetter, F. O., and Kutscher, A. E. (1969). The aerobic and anaerobic flora of rattlesnake fangs and venom. Therapeutic implications. *Arch. Environ. Health* **19,** 770–778.

Licht, P., and Crews, D. D. (1975). Stimulation of ovarian and oviducal growth and ovulation in female lizards by reptilian (turtle) gonadotropins. *Gen. Comp. Endocrinol.* **25,** 467–471.

Lowe, C. H., and Vance, V. (1955). Acclimation of the critical thermal maximum of the reptile *Urosaurus ornatus. Science* **122,** 73–74.

Lunger, P. D., Hardy, W. D., and Clark, H. F. (1974). C-type particles in a reptilian tumor. *JNCI, J. Natl. Cancer Inst.* **52,** 1231–1235.

McInnes, H. M. (1971). *Salmonella saintpaul* infection of sheep with lizards as possible reservoirs. *N. Z. Vet. J.* **19,** 163–164.

McKenzie, R. A., and Green, P. E. (1976). Mycotic dermatitis in captive carpet snakes. *J. Wildl. Dis.* **12,** 405–408.

Maderson, P. F. A. (1965). Histological changes in the epidermis of snakes during the sloughing cycle. *J. Zool.* **146,** 98–113.

Mengden, G. A., Platz, C. G., Hubbard, R., and Quinn, H. (1980). Semen collection, freezing and artificial insemination in snakes. *In* "Reproductive Biology and Disease of Captive Reptiles" (J. B. Murphy and J. T. Collins, eds), pp. 71–78. Society for the Study of Amphibians and Reptiles, Lawrence, Kansas.

Monroe, J. H., Shibley, G. P., Schidlovsky, G., Nakai, T., Howatson, A. F.,

Wivel, N., and O'Connor, T. E. (1968). Action of snake venom on Rauscher virus., *JNCI, J. Natl. Cancer Inst.* **40**, 135–145.

Montali, R. J., Smith, E. E., Davenport, M., and Bush, M. (1975). Dermatophilosis in Australian bearded lizards. *J. Am. Vet. Med. Assoc.* **167**, 553–555.

Nelson, D. J. (1950). Treatment for helminthiasis in ophidia. *Herpetologica* **6**, 57–59.

Packard, G. C., Tracy, R., and Roth, J. J. (1977). The physiological ecology of reptilian eggs, and the evolution of viviparity within the class Reptilia. *Biol. Rev. Cambridge Philos. Soc.* **52**, 71–105.

Padgett, F., and Levine, A. S. (1966). Fine structure of the Rauscher leukemia virus as revealed by incubation in snake venom. *Virology* **30**, 623–630.

Page, L. A. (1961). Experimental ulcerative stomatitis in king snakes. *Cornell Vet.* **51**, 258–266.

Parrish, H. M., MacLaurin, A. W., and Tuttle, R. L. (1956). North American pit vipers: Bacterial flora of the mouths and venom glands. *Va. Med. Mon.* **83**, 383–385.

Pienaar, U. de V. (1962). "Hematology of some South African Reptiles." Witwatersrand Univ. Press, Johannesburg.

Platz, C. C., Mengden, G., Quinn, H., Wood, F., and Wood, J. C. (1980). Semen collection, evaluation and freezing in the green sea turtle, Galapagos tortoise, and red-eared pond turtle. *In* "Annual Proceedings of the American Association of Zoo Veterinarians, Washington, D.C.," pp. 47–54.

Raynaud, M. M. A., and Adrian, M. (1976). Lésions cutanées à structure papillomateuse associées à des virus chez le lézard (*Lacerta viridis* Laur). *C.R. Hebd. Seances Acad. Sci.* **283**, 845–847.

Rebell, H., Rywlin, A., and Haines, H. (1975). A herpesvirus-type agent associated with skin lesions of green sea turtles in aquaculture. *Am. J. Vet. Res.* **36**, 1221–1224.

Reichenbach-Klinke, H., and Elkan, F. (1965). "Principal Diseases of Lower Vertebrates. Diseases of Reptiles." T. F. H. Publ., Neptune, New Jersey.

Saint Girons, M. C. (1970). Morphology of the circulating blood cells. *In* "Biology of the Reptilia" (C. Gans and T. S. Parsons, eds.), Vol. 3, pp. 73–91. Academic Press, New York.

Schroder, H. D., and Karasek, E. (1977). Toxicity of salmonellae isolated from reptiles. *Int. Symp. Erkr. Zootiere, 19th, 1977* pp. 87–91.

Silberman, M. S., Blue, J., and Mahaffey, E. (1977). Phycomycoses resulting in death of crocodilians in a common pool. *In* "Annual Proceedings of the American Association of Zoo Veterinarians, Honolulu," pp. 100–101. Hill's, Topeka, Kansas.

Simmons, G. C., Sullivan, N. D., and Green, P. E. (1972). Dermatophilosis in a lizard (*Amphibolurus barbatus*). *Aust. Vet. J.* **48**, 465–466.

Simpson, C. F., Jacobson, E. R., and Gaskin, J. M. (1979). Herpesviruslike infection of the venom gland of Siamese cobras. *J. Am. Vet. Med. Assoc.* **175**, 941–943.

Stehbens, W. E., and Johnston, M. R. L. (1966). The viral nature of *Pirhemocyton tarentolae*. *J. Ultrastruct. Res.* **15**, 543–554.

Toft, J. D., II, and Schmidt, R. E. (1975). Pseudophyllidean tapeworms in green tree pythons (*Chondropython viridis*). *J. Zoo Anim. Med.* **6**, No. 2, 25–26.

Vinegar, A. (1973). The effect of temperature on the growth and development of embryos of the Indian python, *Python molurus* (Reptilia : Serpentes : Boidae). *Copeia* No. 1, pp. 171–173.

Vogt, R. C., and Bull, J. J. (1982). Temperature controlled sex determination in turtles: Ecological and behavioral aspects. *Herpetologica* **38**, 156–164.

Vosburgh, K. M., Brady, P. S., and Ullry, D. E. (1982). Ascorbic acid requirements of garter snakes: Plains (*Thamnophis radix*) and eastern (*T. sirtalis sirtalis*). *J. Zoo Anim. Med.* **13**, 38–42.

Wallach, J. D. (1966). Hypervitaminosis D in green iguana. *J. Am. Vet. Med. Assoc.* **149**, 912–914.

Wallach, J. D. (1969). Medical care of reptiles. *J. Am. Vet. Med. Assoc.* **155**, 1017–1034.

Wallach, J. D. (1971). Environmental and nutritional diseases of captive reptiles. *J. Am. Vet. Med. Assoc.* **159**, 1632–1643.

Wallach, J. D., and Hoessle, C. (1967). Visceral gout in captive reptiles. *J. Am. Vet. Med. Assoc.* **151**, 897–899.

Wallach, J. D., and Hoessle, C. (1968). Steatitis in captive crocodilians. *J. Am. Vet. Med. Assoc.* **153**, 845–847.

White, F. N. (1959). Circulation in the reptilian heart (squamata). *Anat. Rec.* **135**, 129–134.

Will, R. (1979). The structural transformation of phagocytic cells, an old phylogenetic principle. Part II. The heterophilic granulocyte. *Vet. Med. Rev.* No. 1, pp. 49–58.

Zeigel, R. F., and Clark, H. F. (1969). Electron microscopic observations on a "C"-type virus in cell cultures derived from a tumor-bearing viper. *JNCI, J. Natl. Cancer Inst.* **43**, 1097–1102.

Zwart, P. (1968). Parasitäre und mykotische lungenaffektionen bei reptilien. *Int. Symp. Erkr. Zootiere, 10th, 1968* pp. 45–48.

Zwart, P., and Kok, A. G. G. (1978). Endemischer kropf bei reptilien in den Niederlanden. *Int. Symp. Erkr. Zootiere, 20th, 1978* pp. 373–377.

Zwart, P., Verver, G. A., DeVries, G. A., Hermanides-Nijhof, F. J., and DeVries, H. J. (1973). Fungal infection of the eyes of the snake *Epicrates cenchria maurus*: Enucleation under halothane narcosis. *J. Small Anim. Pract.* **14**, 773–779.

Chapter 16

Biology and Diseases of Fish

Alexander H. Walsh

I. INTRODUCTION

Although the vast majority of fish species have not been used in scientific experiments, many species are and can be utilized. Conservative estimates place the number of fish species in the world in excess of 20,000. Readily available species, such as the rainbow trout, blue gill, and goldfish, are commonly used. In addition, numerous very specialized fishes, such as the electric eel, the Amazon molly in which all members are female and genetically identical, and hybrids such as the platy-fish–swordtail hybrid which has a very high incidence of melanoma, are also available. Selection of the most suitable species for any experiment is not limited to those utilized by others in similar studies but can be expanded upon greatly by researchers with imagination.

The choice of species often dictates the type and quality of water to be used. Marine species obviously must be maintained in salt water, some species, such as salmonids, require high quality, highly oxygenated, and relatively cool water while the air-breathing betta can be maintained for a month or more in a small volume of water without aeration or cooling. Chlorinated municipal water supplies may be used successfully if that water is allowed to "cure" for at least 24 hr or is filtered through activated charcoal before reaching the fish. Well, spring, and river water have all been used successfully for the maintenance of experimental fish as well as by hatcheries. A saltwater environment may be produced even in areas remote from the sea by dissolving any of several commercially available products in suitable fresh water. For the most critical experiments it is often necessary to utilize reconstituted deionized or even distilled water. The use of reconstituted water allows scientists to closely approximate the conditions of studies done in the past or at distant laboratories.

The type of housing selected is frequently dictated by the test species and in particular the water quality requirements of that species. In addition, it may be necessary to provide flowing water to introduce and maintain a suitable level of test material in the water. On the other hand, serious problems of waste disposal arise when a flowing system is used with carcinogens or slowly degraded toxins. Many tests substances when introduced to water are quickly adsorbed to the glass or other materials within the aquarium. This can be a very important consideration not only for the current study but for possible contamination of future studies.

Some very specialized fishes are available from only a single source, while others, such as many of the salmonids, centrarchids, and ictalurids, may be obtained from numerous hatcheries. Many, especially the saltwater species, must be captured from the wild. Caution must be exercised particularly with those fishes coming from the wild or poorly run hatcheries not to begin a study with diseased fish. Extensive parasite infestations are common in such fish and frequently compromise the final results of otherwise satisfactory studies. Fish, perhaps more than any other laboratory animal, are frequently stressed when maintained in the laboratory. Stress introduced by overcrowding, oxygen depletion, and nutrient enrichment of the water often leads to remarkable enhancement of preexisting parasitic and infectious conditions. Treatment of these conditions during the course of a scientific experiment is undesirable. Therefore, every effort should be made to utilize disease- and parasite-free fish and to provide an environment as nearly optimal as possible. Although this may seem obvious, it has been overlooked or ignored in the design and implementation of many fish studies.

Nutritionally balanced diets are commercially available for the commonly reared species such as the salmonids and ictalurids. It should be remembered that these feeds usually contain antibiotics and may therefore be undesirable for use in certain experiments. Although a large volume of literature exists on the nutritional requirements of fishes, much of it pertains to a few commonly reared species and the precise requirements of the vast majority of fish species can only be extrapolated. Very specialized fishes or fishes from certain environments can develop signs of deficiency when maintained in the laboratory, e.g., the jawfish which develops thyroid hyperplasia on a diet containing adequate iodine for most marine species. Certain

toxins may increase the fishes need for a specific nutrient, as exemplified by the increased requirement for vitamin C in the presence of low levels of toxaphene.

The individual handling of fish may be both difficult for the investigator and detrimental to the fish. Studies requiring the repeated handling of fish either for examination or sampling should be approached with caution. It is generally preferable to start with sufficient numbers to allow for the disposal of the sampled fish following each sampling. Hematologic and serum chemistry parameters will frequently undergo marked changes in fish that have been stressed by handling. In addition, the combination of trauma and increased corticosteriod levels following handling frequently lead to skin infections, particularly those caused by fungi of the order Saprolegniales.

Due to the diversity of fish species, a wide range of husbandry techniques are utilized to maintain fish in the laboratory. Numerous aquaria systems have been devised, varying from water-filled jars with no supplementary features to elaborate systems in which water is aerated, heated or cooled, and filtered by biological, chemical, and mechanical means. Three basic systems are employed: (1) those using water that is not flowing or filtered, so-called static systems; (2) those that continually filter the water and add additional water only to replace that lost by evaporation or spillage; (3) flowing systems in which some fraction or all of the water is replaced during each full cycle.

Static systems are utilized in acute toxicity and other short experiments. Depending upon the concentration of fish (expressed as kg/m^3 or lb/ft^3) and the species used these systems may or may not require aeration. Temperature control is usually accomplished by maintaining room air temperature within the required range. Fish are not usually maintained in static systems for more than 1 week.

Filtration systems may consist of complex self-contained units each supplied with a means of aeration, filtration, and temperature control or may be arranged in banks of aquaria with central water conditioning equipment. In either case, filtration of organic material is usually dependent upon passage of water through a porous medium, such as sand or plastic beads, in which microorganisms convert ammonia to nitrites and nitrates. Particulate matter is removed before the water enters the biological filter(s) by sedimentation and/or passage through mechanical filters, such as fiberglass or screens. Following passage through the biofilter some to all of the water maybe passed through an activated charcoal filter for further removal of nitrogenous wastes and organic substances. The addition of ozone into the aeration system combined with a means of removing the froth generated by this addition is a very effective way to remove high molecular weight compounds from aquaria water. Combined with biological filtration, the judicious use of ozone allows complete recycling without compromising fish health. The advantages of the system include reduced water cost, reduced water heating or cooling cost, increased growth rate due to higher water temperature, and reduced water pollution by effluents from aquaria.

Flowing systems may incorporate all of the features of the filtration systems while replacing some part of the water during each cycle. However, the introduction of new water may actually reduce the effectiveness of the biofilters by introducing new organisms and upsetting the balance achieved when recycling 100% of the water. Although every conceivable dilution rate has probably been attempted in flowing systems, the use of 100% new water is the most desirable when sufficient water of acceptable quality is available.

Although species requirements vary and the ability of certain species to tolerate poor quality water is remarkable, it is essential that fish be provided with water of adequate quality. Well and spring water are generally preferable to surface water since they are usually free of parasites and microbes and maintain more constant chemical and physical characteristics. However, such water should be analyzed for pH; dissolved gases, including oxygen, nitrogen, and carbon dioxide (deep well water may be devoid of O_2); hardness (milligrams per liter of calcium carbonate); and toxic substances such as pesticides and heavy metals. Water from ponds, lakes, and rivers and recycled water needs these same analyses and in addition should be tested for suspended solids, dissolved solids, biological oxygen demand, ammonia, nitrates, and nitrites. Copper plumbing should be avoided. Toxicologic studies demanding an exact duplication of water quality may require demineralization and reconstitution of water. Obviously, such water is expensive and, when required, a filtration system capable of maintaining water quality is very valuable. Duplication of the marine environment is possible with several readily available commercial preparations. These products are mixed with fresh, preferably demineralized water.

The information in this chapter is necessarily brief and often incomplete. Interested readers are urged to consult references such as Ali (1980), Flynn (1973), Ribelin and Migaki (1975), Roberts (1978), and Wolf (1981).

II. BIOLOGY

It should be apparent that the use of fish as research animals is complicated by a number of factors associated with the vast interspecies differences that are present. The reader should keep this in mind and realize that much of the information in this chapter is true for most of the fish species encountered in research, but probably not for all. An attempt has been made to point out major differences between fishes and mammals, but not to detail the many differences known to exist between fishes.

Fish possess a variety of unique anatomical and physiological characteristics that distinguish them from mammals. These differences reflect the fishes' aquatic environment and their more primitive evolutionary position. All fish are poikilothermic and must be capable of adapting to changes in water temperature. Although fish inhabit waters ranging from <0°C to hot geothermal springs, each species is obligated to live within a specific temperature range. For some species, abrupt temperature changes, particularly elevations of >5°C can be lethal. More gradual changes such as those seen with the changing seasons allow fish to adapt and can usually be tolerated.

The organ systems of fish all vary to some extent from those of mammals. The following section depicts some of those variations.

A. Integument

Fish skin does not have a keratin layer as found in the mammals but is instead covered by a cuticle comprised of mucopolysaccharides, mucous, occasional sloughed cells, immunoglobins, and free fatty acids. The epidermis is made up of living Malpighian cells at all levels and variable numbers of mucous cells, granule cells, lymphocytes, macrophages, and in some species club cells or so-called alarm substance cells. The latter, at least in some species, produce a substance (pheromone) that is released when the skin is damaged and causes a panicky "flight" response in nearby fishes of the same species.

Scales, which are calcified plates covered with epidermis, originate in the dermis. Ctenoid scales of the elasmobranchs have spicules extending from the external surface giving these fish a sandpaperlike texture. Cycloid scales of the teleost fishes have a smooth outer surface and like the ctenoid scales are laid down in concentric rings making them useful in determining the age of some fishes. Pigment-containing cells of the dermis may contain endogenous melanin or guanine or exogenous carotenoids. These materials produce the various colors of fish and can be modified by the fish to vary color intensity and quality.

B. Respiratory System

To fully appreciate the problems associated with the extraction of oxygen from an aqueous environment, one must consider the following. One liter of air saturated fresh water at 10°C contains 11.3 ml of oxygen. One liter of air at 10°C and 1 atm of pressure contains 296 ml or 26 times as much oxygen. The combination of low oxygen content and high weight of water when compared to air requires fish to expend much more energy than mammals during respiration. In addition, it must be remembered that water at 30°C contains approximately one-half as much oxygen as water at 0°C. However, for each 10°C rise in water temperature and, therefore, body temperature, oxygen demand increases by a factor of 2.0–2.3. It should also be remembered that the oxygen-carrying capacity of sea water is from 20–23% less than that of fresh water at the same temperature. This is not normally a problem in free-living marine fish, since the ocean is fully oxygenated, but can become a problem in captive fish. Considering these factors it is not surprising that the maximum oxygen uptake of a 2 kg salmon is approximately equivalent to the basal rate of a 2 kg rabbit. Because of this limitation, fish have developed into excellent sprinters with large masses of white muscle fibers in which there is a rapid build up of lactic acid that takes hours to be metabolized. Sustained oxygen uptake of fish is usually in the range of one-tenth to one-hundredth that of birds and mammals.

When confronted with reduced water oxygen content, teleosts increase their respiratory flow rate (volume of water passing over the gills). The elasmobranchs, on the other hand, either maintain or reduce the respiratory flow rate under hypoxic conditions. Both bony and cartilageous fish increase blood pressure and stroke rate under hypoxic conditions, while actually decreasing the heart rate. Bradycardia appears to allow the trabeculated heart muscle more time to absorb oxygen from the venus blood upon which it is dependent. Under conditions of chronic hypoxia, certain species, such as the goldfish, appear to be able to acclimate to lower oxygen levels. Initially these fish survived for only 11 hr in water containing 0.01 mg O_2/100 ml. After prolonged exposure to hypoxic water, survival time increased to 75 hr. As would be expected the acclimated fish have hemoglobin with a higher oxygen affinity.

The respiratory rate of fishes is controlled by oxygen receptors generally found in the first functional gill arch. The monitoring of carbon dioxide is not important in fishes, since it is some 25–30 times more soluble in water than is oxygen. However, carbon dioxide does influence the blood's affinity for oxygen. The Bohr shift, not unique to fish, reduces the affinity of hemoglobin for oxygen in the presence of increased carbon dioxide content or decreased pH. This mechanism allows for a more rapid release of blood-borne oxygen at sites undergoing rapid oxidative processes, while increasing the blood's affinity for oxygen immediately after it enters the gills where carbon dioxide is rapidly dissipated into the water. The Root effect, which is only known to occur in teleost fishes, depends upon differences in the oxygen loading and unloading characteristics of hemoglobin as influenced by changing pH values. At the high pH levels found in the gills, oxygenation is approximately 4 times faster than deoxygenation. At the lower pH of actively metabolizing tissue, a reversal occurs with deoxygenation 400 times as fast as oxygenation.

Respiratory exchange in fishes is primarily a function of the gills. However, some exchange does occur at other sites including the skin, and tissues which have been modified for this purpose. Oxygen absorbed by the skin appears to be of little benefit for tissues other than the skin. Accessory breathing organs e.g., sacs or pouches extending from the oral cavity, modified gills, and lungs (modified swim bladder), are efficient means of respiratory exchange that effectively supply tissues throughout the body with oxygen derived from air as well as water. The majority of fishes with accessory breathing organs are found in tropical water where oxygen tension is low and oxygen demand is very high. Air breathers have not as yet been widely used in research but have in recent years been studied in an attempt to understand the evolutionary development of the lung.

The countercurrent principle, which is highly developed in the fish, is well exemplified in the gill. Water passing through the oral cavity is forced over the gills from anterior to posterior and expelled through the opercular openings. Blood flow around each gill lamella is counter to that of the flow of water. This allows the least oxygenated blood to exchange with the most poorly oxygenated water. As the blood gains more oxygen, it is supplied by more highly oxygenated water until reaching the anterior limits of the lamella. This mechanism allows fish to remove from 40–80% of the oxygen from water under normoxic conditions.

C. Circulatory System

The fish heart has been described as having either two, three, or four chambers. When compared to the mammal, a fish has a two-chambered heart composed of one ventricle and one atrium. Authors describing more than two chambers include either the sinus venosis or the bulbus arteriosis as the third and/or fourth chambers. Blood flow through the simple circulatory system of the fish progresses from the muscular ventricle into the elastic bulbus, which helps maintain arterial pressure, and hence to the ventral aorta. To this point, it is venous blood, and the heart, of necessity, is supplied with the most poorly oxygenated blood reaching any organ. Blood from the ventral aorta enters the several afferent branchial arteries that ascend into the gill arches and, after traversing the capillary network of the gills, return to the efferent branchial arteries and hence to the dorsal aorta. Some of the highly oxygenated blood from the dorsal aorta enters the modified first gill arch (pseudobranch), where it traverses a countercurrent capillary bed. This blood then enters a second countercurrent capillary bed in the choroid "gland" that is closely applied to the posterioventral aspect of the eye. This mechanism supplies highly oxygenated blood to the retina, which in the fish is the most highly oxygen-demanding tissue in the body. The dorsal aorta then carries blood the entire length of the body, supplying all of the viscera and muscle. The return flow is in general not remarkable, exceptions being the presence of a renal portal system and in some species the presence of muscular lymph hearts within the major lymphatic vessels.

D. Urinary System

The urinary system of the most primitive fishes develops from the embryological pronephros. In the more advanced fishes, the urinary system develops from the pronephros and mesonephros. The metanephros of higher vertebrates is unknown in the fishes. In addition to the excretory function, the kidney is the major site of hematopoiesis in the fish. Typically, hematopoietic tissue is found between the nephrons in the posterior or "trunk" kidney and is the most prominent feature of the anterior or "head" kidney. In the cyprinids, the "head" and "trunk" kidneys are separate organs connected only by the anterior cardinal vein. In salmonids, the kidney is a single organ, and the distinction between trunk and head kidney is somewhat arbitrary. In those species that have separate "head" and "trunk" kidneys, nephrons that are present in the "head" kidney of juveniles regress and disappear in the adult.

The excretory function of fish kidney varies depending upon the fish's environment. Freshwater fish must conserve ions and remove large quantities of water, while saltwater fish, some of which have aglomerular kidneys, must conserve water and excrete large quantities of monovalent ions obtained by ingesting large amounts of seawater. Sharks bypass this problem by maintaining osmolar equivalence to seawater with high levels of urea and trimethylamine oxide in their blood. In teleosts, the kidney is not the sole or even primary means of osmoregulation and waste excretion. The majority of nitrogenous wastes are excreted through the gill epithelium, and the chloride cells of the gills, operculum, and buccal cavity are the primary means of maintaining osmoregulation. In a saltwater environment, monovalent ions are excreted by an active transport mechanism within chloride cells. This system appears to be reversed in fresh water where NH_4 is exchanged for environmental Na^+ and HCO_3^- for Cl^-. A marked hyperplasia of chloride cells occurs in euryahaline teleosts when they move from fresh water to a saline environment.

E. Muscular System

Fish muscle is similar in most respects to mammalian muscle. Major differences do occur in the distribution and arrangement of the trunk musculature and the distribution of red and white muscle fibers. The vast majority of fish muscle is associated with the trunk and only minor components with limb function. The trunk musculature is composed of segmentally arranged myomeres made up of muscle fibers that parallel the logitudinal axis of the body. The myomeres are separated by and attached to connective tissue sheets that then attach to the vertebrae. Dorsal and ventral median septa and a horizontal septum extending from the vertebrae to the lateral line divide the myomeres of the fishes tail into four quadrants. The trunk of most fishes is divided in a similar way, with the exception that the ventral septum is replaced by the body cavity. In most fishes this entire muscle mass is comprised of white muscle fibers. Red muscle fibers are confined to a narrow band that parallels and lays beneath the lateral line. This band of red muscle fibers, although more vascular and separated from the deeper white muscle fibers by a connective tissue band, is in fact segmented and part of the myomere system.

F. Reticuloendothelial System

The reticuloendothelial system of fish is similar but not identical to that of homeotherms. Lymph nodes are absent, and functional Kupffer cells are not apparent in the sinusoidal epithelium of the liver. However, fish do possess facilitative phagocytic cells in the endothelial lining of the atrium and the gill lamallae. A distinctive feature of the fish reticulendothelial system is the presence of accumulations of pigmented macrophages within the spleen, kidney, and liver. These accumulations, sometimes referred to as melanomacrophage centers, have also been observed in amphibians and reptiles and have been shown to contain hemosiderin and ceroid as well as melanin. It has been suggested that melanin serves the dual purpose of inhibiting the propagation of free radicals produced during lipid oxidation and a bactericidal action associated with the production of hydrogen peroxide. These mechanisms may be of particular value to poikilotherms since they are nonenzymatic and therefore are not temperature dependent.

Teleost fishes are immunologically competent, having the ability to produce specific immunoglobins, elict cell-mediated immunity, show both delayed and immediate hypersensitivity, and show immunological memory. However, when compared to mammals, the immune response of fish is deficient or at least differs in several areas. The only class of immunoglobin produced by fish is IgM. IgG has been reported in the goldfish, but this has not been substantiated. Fish IgM is usually tetrameric, and sedimentation coefficients of 19 S, 16 S, 7 S, and 6.4 S have been reported. Virus neutralizing, agglutinating, and precipitating antibodies occur in fish, although the last is uncommon. The primary immune response can be detected in spleen and kidney tissue as early as 3 days after challenge; however, circulating antibodies may not be detectable for several weeks. In general, cold-water fishes take longer than warm-water species to produce detectable antibodies, and fish maintained in cold water require longer periods of time to respond than comparable fish in warmer water. Antibody of the IgM class is present in mucous and appears to be produced locally.

The fish thymus appears to be a central lymphoid organ (lymphocytes migrate from but do not return) that is the source of all lymphocytes. Precursers of B lymphocytes are thought to leave the thymus before identifiable lymphocytes appear. After that only T lymphocytes are found in the thymus, while B and T lymphocytes are present in the spleen and B lymphocytes in the kidney. Mitogen responsiveness to concanavalin and phytohemagglutinin is reported in fish as well as lymphokine production. Nonspecific factors, including complement, C-reactive protein, hemolysin, interferon, and lysozyme, have been reported in fish.

G. Digestive System

The digestive system of most fishes are relatively simple but not markedly different from those of the higher vertebrates. Carniverous fishes have the shortest and simplest digestive systems, with the total length of the alimentary tract usually being less than that of the entire fish. The alimentary tract of omniverous fishes is usually greater than that of the entire body, and herbivores such as the carp may have alimentary tracts three or more times the length of the body. Features of the fish digestive tract that are unique and may be of interest to the investigator include the presence of teeth in the pharynx as well as the oral cavity. In most instances, fishes' teeth are ankylosed directly to bone, although the eel has some teeth embedded in moveable connective tissue pads. Taste buds are numerous within the buccal cavity, the pharynx, and often extend into the esophagus. In the catfish, taste buds are also found in the skin, particularly on the barbels. Many fishes have numerous pyloric ceca that secrete digestive enzymes into the intestine just beyond the pyloric sphincter. In fishes without pyloric ceca, digestive enzyme production is entirely from the liver and pancreas, there being no evidence of glands in the intestine. The pancreas of most fishes is a diffusely distributed through the fat surrounding the pyloric ceca, although in the eel it is a more discrete organ. The architecture of the fish liver is less defined than that of the higher vertebrates, there being little evidence of a lobular structure. In many species, pancreatic tissue, the so-called hepatopancreas, is found in close association with the larger central veins.

Toxins are metabolized in fish liver by mechanisms similar to those of mammals. Mixed function oxidase activity requiring NADPH and molecular oxygen is present in the microsomal fraction of fish hepatocytes. In most cases specific activity in mammals is markedly greater than in the fishes, the difference apparently due to enzyme deficiencies and not inhibitors. Seasonal variation in enzyme levels occurs in wild fish but not hatchery fish. An important difference between fish and mammalian mixed function oxidase systems is their temperature optima, which for fish is approximately 12°C below mammalian. Reduction reactions requiring NADPH and a carbon dioxide-sensitive factor similar to cytochrome P-450 have also been demonstrated in fish. These systems are present in several elasmobranchs as well as teleosts and are considered by some to be more primative than the oxidase systems. Like the oxidase systems, activity is greatest at about 26°C, but unlike them reductase activity resides in the "soluble" fraction. Fish are also capable of forming glucuronides and other conjugates. One of the conjugating enzymes, uridinediphosphate transglucuronylase, appears to be similar to its mammalian counterpart except that it is less thermostable and has a lower temperature optimum. Conjugation of 3-trifluoromethyl-4-nitrophenol (TFM), a lamprey larvacide, by rainbow trout but not the more primitive sea lamprey suggests that an evolutionary trend may be present in this system.

The swim bladder or air bladder found in many, but not all, species arises as an outpouching of the digestive tract and in the physostomous species is in communication with the esophagus. The swim bladder adjusts specific gravity reducing the amount of energy required by the fish to stay at a constant depth. In fishes residing at great depths and therefore under trememdous hydrostatic pressure, the swim bladder may be absent or filled with oil. In many species a countercurrent system in the rete mirabile compresses gases to very high pressure within the swim bladder. In some air-breathing species, the swim bladder is modified into a simple lunglike structure.

H. Endocrine System

Fish possess tissue representative of most of the mammalian endocrine glands as well as glands of internal secretion not known in mammals. The pituitary of fish is similar to the mammalian in location, morphology, and products, although the effect of the product may differ, e.g. prolactin controls the water and ion permeability of gill and kidney epithelium by inducing morphological changes and modifying Na, K-ATPase activity. Neurohypophyseal peptides are capable of producing diuresis, but it is not known if they enter the circulation. Thyroid tissue of fish is diffusely scattered along the ventral aorta and branchial arteries. Suprisingly, there is no parathyroid or other hypercalcemic organ in fish. Calcitonin, which is hypocalcemic, is produced by the small discrete ultimobranchial body located at the juncture of the transverse septum with the esophagus. Additional hypocalcemic activity is supplied by the corpuscles of Stannius situated in the kidney. Cortisol and other corticosteroids are produced by the interrenal tissue located in the "head" kidney in close association with the cardial veins. Cortisol antagonizes the action of prolactin and appears to function as a mineralocorticoid, the later being absent in fish. Interrenal tissue is also associated with a poorly characterized reninangiotensin system.

The urophysis, the terminal organ of a neurosecretory system located near the distal end of the spinal cord, produces peptides known as urotensins, four of which have been described.

Changes in environmental salinity alter the production of these hormones, indicating that they are involved in osmoregulation. The pineal organ of teleost fish is intermediate between the photosensory organ of the very primative vertebrates and the neuroendocrine organ of the higher vertebrates. Production of 5-methoxy-*N*-acetyltryptamine (melatonin) by the pineal is suppressed in the presence of light. In addition to influencing the distribution of melanin in the skin, melatonin is believed to be involved in circadian activity, particularly entrainment to the dark–light cycle. Peptides associated with the hypothalamus and neurohypophysis are also released by the pineal organ.

I. Acousticolateralis System

Mechanical stimuli over a wide range of amplitudes and frequencies are detected by the acousticolateralis system. These include linear and angular acceleration, acoustic stimuli, and hydrodynamic stimuli. Anatomically, the system can be divided into two major parts: the ear, which responds primarily to acoustic stimuli and changes in angular acceleration, and the lateral line system, which is sensitive primarily to hydrostatic stimuli but is also sensitive to sound generated in the near field.

The ear comprises two membranous labyrinths, each consisting of three semicircular canals and three otolith organs. The three semicircular canals and the utriculus (pars superior) are innervated by the anterior portion of the eighth cranial nerve and function primarily to detect changes in angular acceleration and gravity. The other otolith organs, the sacculus and lagena (pars inferior), are innervated by the posterior branch of the acoustic nerve and function primarily as sound detectors.

Lateral line organs are of two types: superficial neuromasts that are found on all species and lateral line canal organs that are generally but not universally present. Superficial neuromasts or pit organs are located within the epidermis and are of two types. Small pit organs are widely distributed over the body, although the majority are located on the head. They are shallow porelike structures that contain nonciliated sensory cells. These structures are thought to be electroreceptors. The second type of superficial neuromasts, the so-called large pit organs, are more complex, having cilia extending from the sensory cells and a mucoid canalized cap the cupula. This latter structure, which is usually lost in processing, attaches to a single kinocilium. Mechanical stimuli moving the cupula cause movement of the kinocilia, which in concert with the steriocilia produce nerve impulses. The stereocilia that surround each kinocilium limit the movement of the kinocilium. Through variations in the length and orientation of the cilia variations in frequency, amplitude, and direction of stimuli can be interpreted.

The lateral line canal system extends the length of the trunk with contiguous extensions over the head. The canal is situated in the dermis and is partially encased in bone in the form of accessory ossicles. Neuromasts situated at regular intervals along the canal lie between pores that allow communication between the canal lumen and the external environment. It is thought that water does not actually enter the canal, movement being transmitted through a mucouslike material that fills the canal.

Morphologic similarities between sensory structures of the lateral line organ and those of the ear suggest an evolutionary progression from the simple superficial neuromasts to the acoustically sensitive maculae of the otolith organs. Although most fishes appear to have only limited hearing capabilities, some, such as the clupeids and the ostariophysi, are quite sensitive. The latter utilizes the swim bladder as a resonator and transfers vibrations from that organ through a chain of small bones, the Weberian ossicles, to the sacculus.

It seems apparent that research will continue on the acousticolateralis system of teleosts fishes, since they are the most primitive vertebrates able to detect sound. In addition, it is enticing to contemplate using the readily accessible and relatively simple lateral line organ as a model for the remote and intricate mammalian inner ear.

J. Normal Values

Hematologic and serum chemistry values have been determined for only a few fish species. Extrapolation to other species is uncertain but often unavoidable. Environmental and nutritional variation make comparison of values within a species but between laboratories suspect. Seasonal variation in hematologic and serum chemistry values cited by several authors may not be a direct effect but a reflection of changes in pH, oxygen saturation, mineral content, and perhaps fish size and age. Tenfold increases have been reported in hemoglobin levels of Neva salmon between the ages of 1 and 4 months. Similarly, growth rate and erythrocyte levels increased in Atlantic salmon transfered from hatchery to river. Stress associated with sampling procedures may have a profound effect upon various blood values (Table I). Impoundment, transport, anesthesia (often required for sampling), and death all alter values.

Differences in sampling methods are often necessitated by size differences. Small fish are often sampled by severing the caudal peduncle allowing water on the fish as well as tissue fluids to dilute the sample. Large fish are often sampled by heart puncture, which provides pure venous blood while blood from the peduncle is a mixture of venous and arterial. Some workers have reported thrombocytes as part of the total leuko-

Table I

Changes in Arterial Blood of Rainbow Trout as a Result of Struggling[a,b]

Variable	Control	Time after struggling (min)			
		0	4	30	1320
pH	7.789	7.204	7.325	7.369	7.834
P_{CO_2} (torr)	1.84	3.29	—	3.23	1.93
Hematocrit (%)	25.6	30.0	34.2	37.2	23.5
Σ of change in Plasma ions (mM)	0	+14.5	+26.6	+31.0	−3.7
Lactate (mM)	1.2	4.22	6.52	10.7	1.45
Na$^+$ (mM)	147.6	154.7	162.9	163.1	145.4
HCO$_3^-$ (mM)	4.09	2.26	—	2.65	4.12

[a]From Ali (1980) with permission.
[b]Fish were allowed to struggle 4 min at 15°C.

cyte count, others (as is correct) have not, leading to extreme variations as evidenced in Table III.

These values expressed in Tables II and III and those presented in Tables IV and V for the rainbow trout can be thought of as representative of the species listed as well as providing insight on the variables one may encounter.

It seems apparent that clinical pathology values for fish are of little diagnostic value, at least at the present state of the art, and must be compared to values from precisely matched controls when interpreting experimental data.

Size and longevity vary greatly between species, and in some cases within a species. As might be expected, size and longevity are closely related. The giant sturgeon often lives over 100 years, while *Nothobranchius geuntheri* weighs only a few grams and completes its life cycle in 6 months. These and other small, short-lived species may prove valuable in car-

Table II

Representative Hematological Values for Fish

Species (adults)	Erythrocytes (10^6/mm^3)	Hematocrit (%)	Leukocytes (10^3/mm^3)
Channel catfish[a] (*Ictalurus punctatus*)	2.44	30–50	164
Carp[b] (*Cyprinus carpio*)	1.36	29	10.4
Brook trout[c] (*Salvelinus fontinalis*)	0.87–1.63	55–65	16–39
Pike[d] (*Esox lucius*)	1.89	20–43	79–137

[a]From Grizzle and Rogers (1976).
[b]From Van Vuren and Hattingh (1978).
[c]From Christensen *et al.* (1978).
[c]From Mulcahy (1970).

cinogenicity research because of these characteristics. Variation within a species as well as the potential for genetic manipulation is well demonstrated in the brook trout. These fish were selected for early sexual maturity over many generations, producing a small, short-lived fish that reached maturity in 2 years. The stock from which they were developed were many times larger, lived many years longer, and reached sexual maturity after 5 years. Unfortunately, the small short-lived brook trout have replaced their "superior" progenitors throughout the greatest part of their range.

K. Nutrition

The nutritional requirements of fishes, particularly the salmonids, have been investigated and are with some exceptions well understood. Examination of Tables VI and VII demonstrates the similarity of the dietary requirements of fish to that of mammals, particularly the amino acids, vitamins, and minerals.

Salmonids during active growth must receive 49% of their dietary metabolizable energy from protein, while adults require 37% from this source. Comparable figures for the rat are 13 and 5%, respectively. As would be expected, salmonids have a remarkable capacity to degrade amino acids and convert them into glucose. Salmonids and probably most fish are not able to reduce the rate of amino acid degradation even when protein intake is reduced. This feature of salmonid metabolism precludes the substitution of any substantial quantity of carbohydrate as an energy source for more expensive protein. In addition, salmonids and probably other fishes readily store dietary carbohydrates as liver glycogen but only slowly release it, explaining the extreme glycogen storage observed in many salmonids fed commercial diets. Wild fish consuming a primarily protein diet do not have this problem. Most fishes appear to have similar nutritional requirements as evidenced from the accompanying tables, although some differences exist. Salmonids, ictalurids, and others have an absolute requirement for ascorbic acid, while carp are able to synthesize this nutrient. Similarly, several species, including the yellow perch, have the ability to utilize β-carotene, while the brook trout is incapable of making the conversion to vitamin A.

The vast number of fish species precludes quantitative determination of nutrient requirements for each. Amino acid requirements of Atlantic salmon have been estimated by analysis of amino acid levels in Atlantic salmon eggs. Atlantic salmon fingerlings fed diets containing amino acid levels based upon egg analysis grew more rapidly than similar fish fed a diet based upon the amino acid requirements of chinook salmon. These data suggest that the amino acid requirements of other species may be estimated by this method, reducing the need for extensive feeding trials.

Table III

Representative Differential Blood Cell Counts of Fish[a]

Species	Thrombocytes	Lymphocytes	Neutrophils	Basophils	Eosinophils
Channel catfish[b] (*I. punctata*)	42.0[e]	54.8 (94.5)[f]	3.2 (5.5)	0	0
Herring[c] (*Clupea* sp.)	62.7	31.8 (85.3)	4.5 (12.1)	0.9 (2.4)	0.1 (0.3)
Goldfish[d] (*Carassius auratus*)		92.5	5.1	0.2	2.2

[a]Values expressed as percentage of total leukocyte count, including thrombocytes.
[b]From Grizzle and Rogers (1976).
[c]From Sherburne (1973).
[d]From Watson *et al.* (1963).
[e]Includes 0.3% hemoblasts.
[f]Values in parentheses are corrected to exclude thrombocytes.

Table IV

Variability of Serum Components Among Fish[a,b]

Serum component	N	Mean	CV[c]	Variance of fish	Variance of precision[d]
Albumin (mg/dl)	10	1.1	18.2	0.041	0.0036
Total protein (mg/dl)	10	3.2	16.4	0.29	0.007
Blood urea nitrogen (mg/dl)	10	6.2	17.7	0.0	1.57
Cholesterol (mg/dl)	10	289	4.0	135.1	10.2
Chloride (mEq/liter)	10	127	1.4	2.74	0.53
Glucose (mg/dl)	10	64.9	10.6	39.5	10.5
Potassium (mEq/liter)	10	2.7	34.8	0.96	0.009
Sodium (mEq/liter)	10	136	4.5	33.6	6.17
Cholinesterase (IU/ liter)	10	158	16.0	619.7	61.9
Alkaline phosphatase (IU/liter)	10	172	50.7	8123	38.4
Lactic dehydrogenase (IU/liter)	10	725	22.3	29362	24.5
α-Hydroxybutyrate dehydrogenase (IU/liter)	10	681	17.9	15984	51.4
Glutamic pyruvic transaminase (IU/ liter)	10	26.8	31.2	33.5	11.0
Phosphohexose isomerase (IU/liter)	10	4169	7.0	164353	1621
Inorganic phosphorous (mg/dl)	10	13	10.0	0.5	1.1
Calcium (mg/dl)	10	11.3	4.4	0.30	0.025
Creatinine (mg/dl)	10	0.7	42.9	0.064	0.021
Creatine phosphokinase (IU/liter)	10	2300	37.4	712063	78610

[a]From Warner *et al.* (1979), with permission.
[b]Rainbow trout (*Salmo gairdneri*).
[c]CV (coefficient of variation) = (S.D./mean) × 100.
[d]Serum sample from each fish was sampled three times for each component.

L. Reproduction

The reproductive behavior of fish varies widely between various species. Among a group comprised primarily of egg producers, the presence of a number of viviparous species may represent the widest diversity. Viviparous fishes used in the laboratory include the guppy (*Poecilia reticulata*), the platyfish (*Xiphophorous maculatus*) swordtail (*X. helleri*) hybrid, *Poeciliopsis* sp., and the Amazon molly (*Poecilia for-*

Table V

Maximum Period of Stability of Fish[a] Blood Serum Components under Various Storage Temperatures for 42 Days[b]

Parameter	No. of days serum component remained stable at		
	25°C	4°C	−10°C
Albumin	42	42	42
Total protein	42	42	42
Blood urea nitrogen	42	42	42
Cholesterol	42	42	42
Chloride	21	42	6
Glucose	2	6	2
Potassium	42	42	42
Sodium	6	42	42
Cholinesterase	42	42	42
Alkaline phosphatase	42	42	42
Lactic dehydrogenase	42	42	42
α-Hydroxybutyrate dehydrogenase	21	42	3
Glutamic-pyruvic transaminase	42	42	11
Phosphohexose isomerase	3	6	42
Inorganic phosphorus	42	42	42
Calcium	6	42	42
Creatinine	42	42	42
Creatinine phosphokinase	2	42	11

[a]Rainbow trout (*Salmo gairdneri*).
[b]From Warner *et al.* (1979), with permission.

Table VI

Nutrient Requirements of Fish[a,b]

Nutrients	Salmo gairdneri	Ictalurus punctatus	Cyprinus carpio
Amino acids			
Alanine	N	N	N
Arginine	R	R 4.3	R 4.3
Aspartic acid	N	N	N
Cysteine	N	N	N
Cystine	N	N	N
Glutamic acid	N	N	N
Glycine	N	N	N
Histidine	R	R 1.5	R 2.1
Isoleucine	R	R 2.6	R 2.6
Leucine	R	R 3.5	R 3.9
Lysine	R	R 7.3	R 5.7
Methionine	R 3.0[c]	R 2.9	R 3.1
Phenylalanine	R	R 4.5[d]	R 6.5[d]
Proline	—	N	N
Serine	—	—	N
Threonine	R	R 2.2	R 3.9
Tryptophan	R	R 0.5	R 0.8
Tyrosine	N	N	N
Valine	R	R 3.0	R 3.6
Water-soluble vitamins			
Thiamin	R 1–10 mg	R	R
Riboflavin	R 5–15 mg	R	R 4–10 mg
Pantothenic acid	R 10–20 mg	R	R 40–50 mg
Niacin	— 1–5 mg[e]	R	R 22–28 mg
Vitamin B$_6$ (pyridoxine)	R 5–15 mg	R	R 5 mg
Biotin	R 0.05–0.25 mg	—	R 1 mg
Choline	R 50–100 mg	R	R ≤2000 mg
Folacin	R 1–5 mg	R	—
Vitamin B$_{12}$	—	R	[c]
myo-Inositol	R[f]	N	R
L-Ascorbic acid	R 100 mg	R 25–50 mg	N

[a]Adapted from Ketola (1976, 1977, 1980).
[b]Abbreviations: R, required; N, not required. Amino acid values expressed as percent of protein; vitamins as mg/kg of dry diet.
[c]Methionine plus cystine.
[d]Phenylalanine plus tyrosine.
[e]Requirement has been reported for similar species.
[f]Data are conflicting.

mosa). All of these fish are easily maintained and reared in the laboratory. Mating pairs may be placed together or one to several males placed with a group of females. In either case, gravid females should be separated prior to parturition into aquaria with screening to allow the young a means of escape from the cannibalistic parent. In addition to ease of rearing, the ovoviviparous species offer some unique features to the researcher. The Amazon molly is an all female species in which the sperm of closely related species stimulate the development of the ova but do not contribute to the genetic makeup of the produce. In effect they are clones. Homozygous strains of *Poeciliopsis lucida* and *P. monacha* that have been inbred for up to 50 generations are available. The *Xiphophorus* hybrids offer the oncologist a system for improving our understanding of neoplasia.

The reproductive diversity of the oviparous fish preclude any generalizations pertaining to reproductive biology. Salmonids produce masses of eggs and sperm and following spawning leave the developing embryos to fend for themselves. Centrachids show more parenteral concern by building nests and guarding the eggs, but find their offspring quite tasty soon after hatching. The Cyprinodonts go even further, attaching eggs to stones or plants where the male steadfastly attempts to guard them from the female that so recently produced them.

For fish, reproduction is the most sensitive of the life functions. Stress induced by changes in pH, salinity, temperature, dissolved oxygen, and toxins will prevent reproduction before stopping other life functions, including growth. Therefore, conditions that appear favorable for adult fish may be incapable of supporting reproduction. This is well demonstrated in lakes acidified by "acid rain" and rivers where dam construction has lowered the water temperature. In both cases, re-

Table VII

Nutrient Requirements of Fish[a,b]

Nutrient	Salmo gairdneri	Ictalurus punctatus	Cyprinus carpio
Fat-soluble vitamins			
Vitamin A	R	—	—
Vitamin D	R	—	—
Vitamin E	—	R	R
Vitamin K	R	—	—
β-Carotene	—[c]	—	—
Minerals			
Calcium	—	U	—
Phosphorus	—	R 8 gm (available P)	—
Iron	—[d]	—	—
Iodine	—[d] 0.6–1.1 mg	—	—
Selinium	—[d]	—	—
Magnesium	R	—	—
Fatty acids			
18:2w6 (linoleic acid)	R	U	R
18:3w3 (linolenic acid)	R	U	R

[a]Adapted from Ketola (1976, 1977, 1980).
[b]Abbreviations: R, required; U, utilized. Mineral values expressed as mg or gm/kg of dry diet.
[c]Species listed have not been studied. Data suggest that salmonids cannot utilize β-carotene, but some species are able to convert it to vitamin A.
[d]Species listed have not been studied. Requirement has been shown in other species.

production of certain species ceases while fish of that species continue to live and grow. Unfortunately, fish acclimated to such stress are not better able to reproduce than unacclimated fish. Oogenesis appears to be most sensitive to stress, followed by newly spawned eggs, hardened eggs, larvae, and finally adult fish.

Sex determination is difficult in many species since they have similar markings. In species that are sexually dimorphic, subordinate males may take the female color pattern requiring separate housing of each fish to accurately ascertain sex. A common practice is to perform studies without regard to sex, then determine sex at necropsy.

In general, adult size and time to maturity are directly related. Many small species (<4 cm in length) are sexually mature at 3–4 months of age. In contrast, the salmonids are at least 2 years of age and often weigh over 1 kg when mature. The expense of maintaining such fish is excessive and except for special purposes unnecessary.

Small oviparous fishes suited for laboratory use include: zebrafish (*Brachydanio rerio*) reaches 3 cm in length and matures in 3–4 months producing large numbers of eggs frequently; desert pupfish (*Cyprinodon n. nevadensis*) produces several eggs each day throughout the year; fathead minnow (*Pimephales promelas*) up to 3 cm long, matures at 3–4 months producing eggs daily from April to August; *Rivulus marmoratus*, 2–5 cm long, matures at 4–6 months producing up to 10 eggs daily, they are hermophroditic, reproducing by internal self-fertilization; Medaka (*Oryzias latipes*) small oriental fish, matures at 3 to 4 months, producing 20 to 30 eggs daily all year, very tolerant of temperature variation and salinity, several inbred lines are available.

All of these small egglayers are suitable for routine toxicological and carcinogenicity research and can be raised in the laboratory with a minimum of equipment.

M. Behavior

Fish behavior is influenced by various exogenous stimuli, including light, temperature, sound, odor, pressure, hypoxia, and possibly geomagnetic forces. Circadian rhythms appear to influence the behavior of all fish, even those at great depths in the sea and in caves. Schooling activity of an Arctic coregonid, which schools during the day and disperses at night, continues for at least 10 days in constant darkness. The periodicity of the circadian cycles is not exactly 24 hr. As with other groups, fish are entrained by repeated exposure to light, temperature, tidal, and feeding cycles causing their circadian cycles to coincide with the environmental cycle.

Hypoxia will stimulate frantic swimming activity in many fishes. Often these fish seek the water surface with its oxygen-rich film or where those that are capable can breath air. Similarly, fish having no thermoregulatory capability will seek water within their thermal preferendum. In the laboratory this choice is lost to the fish, requiring that water temperature be maintained within the preferred range.

The sense that appears to exert the greatest influence on fish behavior is odor. Fish have exquisite olfactory sensitivity, the eel being able to detect as little as 2–4 molecules (2×10^3 molecules/ml) of β-phenylethanol within its nasal cavity. Feeding activity can be evoked by introducing extracts of prey or even amino acids, such as tryptophan and arginine. Of particular importance are pheromones; substances secreted by one individual that have the ability to prompt a specific reaction in a second individual of the same species. There is some evidence that pheromones may be involved in the migration of salmon to their native rivers. Ovarian fluid, ova, and even synthetic estrogens have been shown to attract male fish and evoke breeding activity. Schooling, although primarily a sight-dependent response, is influenced by pheromones as demonstrated in the herring that will continue to "school" in complete darkness. Subunits of catfish eel schools will return to their own school but not other catfish eel schools after prolonged separation. Recognition of young is often dependent upon pheromones. Chichlid parents are attracted to water occupied by their offspring, and some species will even eat young of the same species but not their own. Alarm substance, a pheromone produced in "club cells" of the skin of many species, is released following trauma and causes the flight of other fish of the same species from the injured fish. Geomagnetic forces are thought to influence the migration of fish at sea, but this is speculative.

III. DISEASES

All classes of pathogens known to produce disease in other vertebrates have been reported in fish. Most of the viral diseases and many parasites of fish appear to affect only one genus and occasionally only a single species. Table VIII includes examples of disease caused by each of the major classes of pathogens, and by toxins, deficiency, and neoplasia. They were chosen as representative of those occurring in fish and to exemplify the various manifestations seen in fish; for example, of the three cestodes presented one occurs as an adult in one species and a plerocercoid larva in another, the second can exist in both forms in a single fish, and the third is found in fish only in the plerocercoid form. Most of the examples chosen have been reported or might be expected to produce disease in laboratory fishes.

Table VIII

Representative Diseases of Fish

	Species affected	Morphology and culture	Transmission	Clinical signs	Lesions	Diagnosis	Control
				Viral diseases			
Virus							
DNA viruses							
Herpesviruses							
Channel catfish virus disease (CCV) (*Herpesvirus ictaluris*)	Ictalurids		Direct waterborne and probably vertical. Experimental, all routes	Rapid onset, followed by terminal lethargy, high mortality	Exophthalmos, ascites; edema, necrosis of kidney, liver and gut	Growth on ictalurid cells with syncytia and Cowdry type A inclusion bodies. Serum neutralization and fluorescent antibody	Virus free broodstock. During outbreak lower water temperature to 19°C or less to reduce mortality
Herpesvirus disease of Salmonids (*Herpesvirus salmonis*)	Rainbow trout (*Salmo gairdneri*) and probably kokanee (*Oncorhynchus nerka*)		Probably direct	Lethargy, fecal casts, gill pallor	Exophthalmos, ascites, low hematocrit and numerous immature erythrocytes	Growth on salmonid cell lines with syncytium formation and Cowdry type A inclusion bodies. Only salmonid virus to produce syncytia on RTG-2 cells	Avoid exposure. Raise water temperature to 15°C or more to minimize losses
Epitheloma papillosum (fish pox)	Carp (*Cyprinus carpio*) and other cyprinids		Probably by direct contact	Skin lesions appear, slough, and frequently turn black as they heal	Epidermal hyperplasia manifesting as white to yellow plaques	Herpesvirus observed with electron microscopy, virus not isolated but has been carried chronically in tumor cell line	Unknown
Iridoviruses							
Lymphocystis disease (several closely related viruses appear to exist)	Many fresh- and salt-water species but not salmonids. Cross-infection does not usually occur		Contact. Entry facilitated by dermal abrasion	Presence of warty growths. Behavior not changed	Few to innumerable white to yellow nodules on skin, fins and gills (See Fig. 1)	Massive enlargement of fibroblasts (up to 3 mm) with one to many Feulgen-positive intracytoplasmic inclusion bodies and thick hyalin capsule	Usually self-limiting but reinfection may occur. In crowded aquarium treat for secondary bacterial infection. Disease-free stock or disinfect eggs
RNA viruses							
Rhabdoviruses							
Infectious hematopoietic necrosis (IHN)	Rainbow trout (*S. gairdneri*), sockeye salmon (*Oncorhynchus nerka*), and chinook salmon (*O. tschawytscha*)		By contact especially the feeding of salmon cannery waste. Probably shed in semen and eggs. Survivors harbor virus for life	Fry and fingerlings lethargic and/or hyperactive, exophthalmos, dark with abdominal distention and fecal casts. Mortality up to 100% at 12°C or below. Can be severe at 15°C	Gills and viscera pale with hemorrhages in viscera, mesentery and base of fins. Severe necrosis of hematopoietic tissue (see Fig. 2) and of granular cells of stratum compactum (gut)	Presumptive—CPE of virus on RTG-2 cells. Margination of chromatin known only with IHN among salmonid viruses. Positive—serum neutralization or fluorescent antibody (FA)	Virus-free brood stock ideal but difficult. Minimize with iodophore treatment of eggs and water temperature control: 15°C before outbreak is preventive. 18°C after outbreak may control
Spring viremia of carp (SVC) (*Rhabdovirus carpio*)	Carp (*Cyprinus carpio*) other cyprinids and experimentally guppies, northern pike, and grass carp		Experimentally by parenteral routes but not orally. Verticle transmission probable	Loss of coordination and equilibrium, abdominal distention, exophthalmos, and tremor	Vent inflammation, gill pallor, ascites, edema, and hemorrhages in many tissues including the swim-bladder	Presumptive—CPE of virus on many cell lines at 20°–22°C. Positive—serum neutralization	Virus-free water and stock best, if possible. Resistant strains of carp may help. Oral and intraperitoneal vaccines also available. Not reported in North America
Swim-bladder inflammation	Appears to be a slightly different manifestation of disease caused by *R. carpio* characterized by exudation, opacity, and finally collapse of swim-bladder. May be complicated by *Aeromonas* infection						
Viruses of uncertain classification							
Infectious pancreatic necrosis (IPN)	Salmonids except members of the genus *Oncorhynchus*		Direct contact, waterborne and very likely vertically via eggs. Adult carriers have low neu-	Fecal casts, body rotates while swimming. High mortality, sudden onset. Small fish	Liver, spleen, and gills pale, low hematocrit, petechiae of viscera especially area of	Presence of gelatinous material in stomach highly suggestive, as is histopathology.	Select virus-free stock from carriers. Purchase eggs and stock certified as IPN-free

Table VIII (Continued)

	Species affected	Morphology and culture	Transmission	Clinical signs	Lesions	Diagnosis	Control
			tralizing antibody titers. If high titer develops virus is eliminated	become very dark, exophthalmic, and have distended abdomens	pyloric ceca, clear gelatinous material in stomach. Necrosis of pancreas and in some fish, kidney hematopoietic tissue	Confirm CPE on RTG-2 cells or serum neutralization with polyvalent antiserum	
Viruses associated with neoplasia							
Lymphosarcoma of Esocids	Northern pike (*Esox lucius*), muskellunge (*Esox masquinongy*)		Experimentally with cell-free filtrate of tumor	Appearance of tumors some of which regress, most prove fatal	Red ulcerating subcutaneous masses. Microscopically consist of lymphoblasts	Based upon morphology of lesions. Virus not isolated or seen with electron microscopy	None practical. However, regression may occur in water temperature of 21°–30°C
Papilloma of brown bullhead	Brown bullhead (*Ictalurus nebulosus*)	Papillomas on head and lips appear to contain virus particles but virus has not been isolated					
Dermal sarcoma of walleyed pike	Walleyed pike (*Stizostedion vitreum*)	Virus particles resembling a leucovirus seen with electron microscopy, transmission or isolation not reported. A slightly smaller virus has been observed in hyperplastic epithelium of these fish.					

Bacterial diseases

Disease and bacterium	Species affected	Morphology and culture	Transmission	Clinical signs	Lesions	Diagnosis	Control
Furunculosis (*Aeromonas salmonicida*)	Salmonids, particularly Atlantic salmon (*Salmo salar*) and occasionally other freshwater and marine species	Gram negative, nonmotile rod 1.0 x 1.7–2.0 μm. Grow on furunculosis agar or blood agar at 20°–23°C	Contact with diseased fish, contaminated water, and fomites and via infected eggs	Initially, focal subcutaneous swellings that often ulcerate and eventually cavitate. Peracute cases die before lesions develop	In addition to skin lesions; petechiae in muscle, necrosis of kidney and spleen, hemorrhages at many sites. Few if any leukocytes present near lesions due to leukocytolytic exotoxin. Bacterial colonies prominent in tissue sections	Presumptive—history, lesions and presence of gram-negative rods. Confirm by isolation of *A. salmonicida* as sole or predominant organism	Avoid with disease-free brood stock or disease-free or iodophor treated eggs. Immunization with oral or injected antigen elicits circulating antibodies but not satisfactory protection to challenge. Sulfas, nitrofurans, and oxytetracycline effective to control outbreaks but will not eliminate carriers.
Ulcer disease of goldfish (*Aeromonas salmonicida*)	Goldfish (*Carassius auratus*), Roach (*Rutilus rutilus*) and probably others	This organism appears to be a variant of the salmonid form and most of what has been stated above also applies. However, lesions are usually restricted to skin. *A. hydrophila*, often present in lesions, is probably secondary					
Bacterial hemorrhagic septicemia, red sore disease (*Aeromonas hydrophila*) Also known as *A. punctata*, although some consider the latter to be separate subspecies	Many freshwater species including large-mouth bass (*Micropterus salmoides*) in which known as red sore disease	Gram-negative, motile rod 0.7 x 1.0–1.5 μm. Nonmotile strains do occur. Grows well on most media. Rimler–Shotts medium allows isolation and typing in 24 hr	Common water saprophyte but due to great variation of virulence between serotypes introduction of diseased fish or even water from one source to another can cause outbreak. Also possibly via external parasites	Superficial ulcer often ringed in red, erosions and redness in and around mouth, distended abdomen, exophthalmos and in some, cavitated ulcers as seen with *A. salmonicida*. Peracute form may kill before any changes are evident	In addition to superficial changes, abdomen may contain opaque fluid, kidney and liver soft and swollen, and lower intestine swollen red and filled with bloody mucus	Presumptive—signs and lesions combined with presence of typical organisms. Confirm by isolation and positive identification of *A. hydrophila*	Try to avoid rapid increase in water temperature, overcrowding, or introduction of diseased fish or contaminated water. Treat with oxytetracycline, furanace, or chloramphenicol
(*Pseudomonas fluorescens*)	Most freshwater species and on occasion marine species	Gram-negative rod 0.5–0.8 x 2.3–2.8 μm produces fluorescent pigment on iron-deficient media	In general, very similar if not identical to condition produced by *A. hydrophila*				
Vibriosis (*Vibrio anguillarum*)	Very important disease of most marine fishes and occasionally reported in freshwater species	Gram-negative, motile, curved rod 0.5 x 1.0–2.0 μm. Grows on most media fortified with 1–1½% NaCl	Normal inhabitant of intestine of healthy fish. Disease triggered by elevated water temperature and other stress.	Young fish become anorexic, darken, and die. Older fish develop dark areas that may ulcerate, exophthalmus, and	Peritoneal hemorrhage and splenic and renal liquifaction. Survivors develop visceral adhesions, cut-	Presumptive—based on clinical signs and lesions and presence of organisms. Confirmation depends upon	Good management to avoid stress; oral, injectable and hyperosmotic bacterins are in use and appear useful.

(continued)

Table VIII (*Continued*)

	Species affected	Morphology and culture	Transmission	Clinical signs	Lesions	Diagnosis	Control
			Freshwater out-breaks often follow feeding of marine fish offal	dropsy or may show no signs	aneous gran-ulomas, and hemolytic anemia with heavy depos-its of hemosiderin in melanomacrophage centers	positive identifica-tion of organism	Genetic selection of some value with salmon. Oxy-tetracycline, sul-fonamides, and nitrofurans are useful if given ear-ly (before anorexia develops)
Enteric red-mouth (*Yersinia ruckeri*)	Salmonids; rainbow trout (*Salmo gairdneri*) in par-ticular	Gram-negative motile rod 0.7–1.0 x 2.0–3.0 μm. Grows well on simple me-dia. Only common gram-negative fish pathogen which is oxidase negative	Contact with diseased or carrier fish or via contaminated water. Vertical transmission not believed to occur.	Fish become slug-gish, skin darkens, and epithelium of the jaws, palate, and operculum be-come red and may ulcerate	Fat and lower gut hy-peremic; stomach and gut fluid filled. Leukocytes and bacteria promi-nent in most tissues particularly vascular organs. Hematocrit and serum protein lev-els low	Presumptive—based on clinical signs and lesions and presence of typical organisms. Con-firm by isolation and growth of or-ganism and serum agglutination	Identify by serologic testing and remove diseased brood stock. Oral and hy-perosmotic bac-terins available. Combination thera-py with sul-famerazine, oxytetracycline, and chlo-ramphenicol or with tiamulin or tribrissen singly may be useful
Bacterial gill disease (*Flexibacter* sp.) also at-tributed to *Flavobac-terium sp.*	Young salmonids	Gram-negative slender gliding rod. Grows rapidly on cytophaga agar	Organism probably present in water and fish mucus. Outbreaks occur at low water tem-perature and are believed related to factors that induce gill exudate	Respiratory distress with protruding op-ercula	Many gill filaments covered by thick yellow to brown mucoid material. Microscopically, hyperplasia and fu-sion of lamellae may be seen. However, invasion by organisms is not usually evi-dent	Organisms very prominent in wet mounts, but usu-ally washed off in processed sections. This combined with growth on ap-propriate media and typical gross lesions is diag-nostic	Improve environment when possible. Treat with surfac-tants to remove mat of bacteria and mucus
Bacterial kidney disease (*Re-nibacterium salmoninarum*)	Salmonids only. Brook trout (*Sal-velinus fontinalis*) very severely af-fected, rainbow trout (*Salmo gairdneri*) least af-fected	Gram-positive non-motile di-plobacillus 0.4 x 0.8 μm. Very dif-ficult to culture	Experimental evi-dence suggests en-try through skin lesions. Vertical transmission sus-pected	Disease follows slow chronic course not usually manifesting until fish are well grown. Ex-ophthalmos (see Fig. 3), skin dark-ing, hemorrhages at base of pectoral fins. Vesicles pro-gressing to ulcers may develop	Kidney may be pale and nodular or contain miliary white granulomata. Lesions may occur in liver and spleen or even be con-fined to those organs. Extensive cavitation of mus-cle seen in some cases (see Fig. 4). Lesions comprised of numerous mac-rophages engorged with organisms	Presumptive—Pres-ence of gram-posi-tive diplobacilli in typical lesions and particularly within macrophages. Con-firm with immu-nodiffusion or fluorescent anti-body test	Select disease-free brood stock using serum precipitating antibody test. With diseased stock dis-infect ova with iodophors. Avoid feeding fish offal, unless sterilized. Erythromycin has reduced losses in Japanese hatchery
Tuberculosis (*My-cobacterium spp.*) Dif-ferences of opin-ion exist concerning the species affect-fish. *M. mari-num* and *M. for-tuitum* are the most commonly accepted	All species are proba-bly suspectable. Most commonly, infection is seen in tropical saltwater fishes (*M. mari-num*) and occasion-ally freshwater fishes. (*M. for-tuitum* also com-monly infects homeotherms in-cluding man. *M. marinum* has also been reported to infect human beings)	Gram-positive, acid-fast (Ziehl–Neelsen) positive rods 2–6 μm long. Grow on commercial my-cobacteria media. *M. marinum* is slow growing and produces orange pigment. *M. for-tuitum* produces colonies in 3 days at 30°C without pigment	Usually, by ingestion of diseased fish or offal. May be transmitted ver-tically, particularly in ovoviviparous species. Abraided skin allows entry of organism in man	Vary greatly but may include emaciation, ulceration, fin rot, color change, ver-tebral deformities, annorexia, and ex-ophthalmos	Miliary tubercles of liver, kidney and spleen are usual. Other viscera may also be affected. Lesions are usually comprised of a connective tissue wall surrounding numerous mac-rophages, occa-sional giant cells and often a necro-tic central area	Typical lesions con-taining acid-fast organisms, usually in very large num-bers, is diagnostic	Avoid feeding offal from diseased fish. Following outbreak diseased fish must be removed and equipment ster-ilized

Table VIII (*Continued*)

	Species affected	Morphology and culture	Transmission	Clinical signs	Lesions	Diagnosis	Control
Nocardiosis (*Nocardia asteroides*)	Tropical fish, several freshwater species including Rainbow trout (*Salmo gairdneri*) and chinook salmon (*Oncorhynchus tschawytscha*) and cultured marine species	Gram-positive, weakly acid-fast, branching bacillus. Produces aerial mycelium at edges of yellow colonies on nutrient agar	Unknown. Attempts to transmit by oral route have not been successful	Anorexia, emaciation, and swelling of abdomen	Granulomata within oral cavity and protruding from visceral organs and peritoneum. Granulomata are not well circumscribed and contain numerous foci of filamentous hyphae	Based upon the presence of typical lesions and growth characteristics of the organisms	Unknown
Epitheliocystis (chlamydial infection)	Bluegill (*Lepomis macrochirus*), striped bass (*Marone saxatilis*), white perch (*M. americanus*), and others	Obligatory intracellular parasite	Unknown	Usually asymptomatic, although mortalities have been reported	Multiple white cysts up to 0.8 mm in diameter on gill lamellae and skin. Cysts contain large numbers of basophilic organisms that are within distended epithelial cells	Based upon light and electron microscopy	Unknown

Mycotic and algal diseases

Disease and organism	Species affected	Morphology and culture	Transmission	Clinical signs	Lesions	Diagnosis	Control
Branchiomycosis (*Branchiomyces sanguinis* and *B. demigrans*)	Carp (*Cyprinus carpio*), bluegill (*Lepomis macrochirus*), eel (*Anguilla japonica*), an others	Nonseptate, branching hyphae which are confined to blood vessels (*sanguinis*) or penetrate vessel walls (*demigrans*)	Spores are liberated from necrotic gill tissue and are probably ingested. Occurs most commonly in water above 20°C with high organic content	Respiratory distress and frequently death	Gill necrosis with (*demigrans*) or without (*sanguinis*) mycelial growth visible on surface. Thrombosis produced by fungal growth causes necrosis of vessels and gill tissue	Presence of typical lesions and attendant intravascular fungi	Control organic level of water by limiting feeding, removing dead fish, and increasing water flow (if possible) at time of outbreak or high water temperature
Ichthyosporidiosis *Ichthyosporidium* (*Ichthyophonus hoferi*) Originally described as a protozoan and still believed to be by some	Many salt and fresh water species. Herring (*Clupea harengus*) appear to be particularly vulnerable	Exists in several forms. Most common is the large (10–250 μm) PAS-positive spore. Can be grown on Sabouraud dextrose agar with 1% bovine serum	By ingestion of infected fish or copepods	Emaciation, scoliosis (see Fig. 5) and minute raised, black granulomata of skin that may ulcerate	Internally, multiple white granulomata of heart, liver, and most other organs. Microscopically these are often the encysted spores surrounded by a thick connective tissue wall (see Fig. 6). Spores may "germinate" to form macrohyphae	Presence of typical lesions with confirmation by growth of organism	Prevent in cultured fish by sterilization of any fish or fish offal that is fed
Saprolegniasis Usually produced by members of the order Saprolegniales, but other orders may cause similar lesions. Classically attributed to *Saprolegnia parasitica*	Most if not all fresh water and estuarine species. Saprolegniaceaea cannot survive in water with salinity above 2.8%	Nonseptate, branching hyphae that produce motile biflagellate zoospores in terminal sporangia	Organisms are normal water inhabitants that invade traumatized or diseased epidermis. Exogenous and endogenous corticosteroids as well as sex hormones may induce infection	Focal, often circular lesions on skin and gill that have appearance of cotton candy when in water. In fry infected yolk sac extends to and distends coelom	Usually confined to skin and gills but may involve viscera. Superficial lesions contain many hyphae and frequently invade musculature (see Fig. 7)	Often based on gross and microscopic findings. Precise identification of organism requires extensive cultural procedures	Prevent with good husbandry. Treat with external disinfectants, such as malachite green, $CuSO_4$, $KMnO_4$, formalin, or salt. Ova immersed in sea water daily for 2–3 hr have higher survival rate
Blue-green algae Algal infections	"Blooms" of toxic blue-green algae and dinoflagellates have been associated with massive fish kills Algae have been incriminated as the cause of thick-walled granulomata in several species, including the angelfish (not otherwise specified) and bluegill (*Lepomis macrochirus*). Granulomata often contain amorphous pigmented bodies considered to be algae						

Protozoa parasitic to fish and generally located on external surfaces

External protozoan	Species affected	Morphology and culture	Transmission	Clinical signs	Lesions	Diagnosis	Control
Oodinium spp. (Velvet disease)	Numerous aquarium fishes and occasionally cultured fishes	Spherical to piriform, stalked organism with yellow green chromoplasts and	Parasitic stage leaves host, sporulates producing free-swimming flagel-	Heavy infestation produces shimmering yellow brown coating of skin and	Characteristic coating and heavy mucus production over gills and skin	Based upon parasite identification	Treat with methylene blue or copper sulfate

(continued)

Table VIII (*Continued*)

	Species affected	Morphology and culture	Transmission	Clinical signs	Lesions	Diagnosis	Control
		pseudopodia. Vary from 12–150 μm in diameter	lated gymnodinia which attach to suitable host	respiratory distress. In small fish may cause death			
Icthyoboda necatrix (Costiasis)	Most aquarium and hatchery-reared fishes	5–18 μm long flagellate with 2 short and 2 long flagella. Appears piriform and stalked while attached, ovoid when free-swimming	Reproduce by binary fission. Transfer readily from host, to free swimming, to host	Fish produce abundant mucus, hence the name ''blue slime.'' High mortality may occur soon after sac fry placed on feed	Characteristic blue slime with innumerable parasites attached to skin and gills	Based upon microscopic identification of parasite	Treat with formalin, acetic acid, or KMnO₄
Ichthyophthirius multifiliis (''ich'' or white spot disease)	All freshwater species	Largest protozoan parasite of fish. Trophozoites are ciliated, 50 μm to 1 nm in diameter and contain ''horseshoe'' shaped macronucleus. Tomites are oval, ciliated, and 30–45 μm long	Trophozoite leaves fish, settles on bottom or plant, and divides into hundreds to thousands of tomites within cyst. Tomites rupture cyst, swim free until attaching to fish. Cycle takes about 4 days	Initially, fish rub on sides of aquarium or rocks. As trophozoites enlarge they appear as white dots over skin and gills. Severely infested fish are debilitated and often die	Trophozoites are overgrown by hyperplastic epithelium that is ruptured by escaping parasite allowing loss of blood and tissue fluids (see Fig. 8)	Identify parasite by size, characteristic nucleus, and granules in cytoplasm	Interrupt cycle by removing fish for 3 days at 25°C (tomites live only 48 hr at 26°C) quarantine new fish at least 1 week. Treat water containing diseased fish for 10 days at 25°C or 30 days at 10°C with formalin, malachite green, methylene blue, or KMnO₄
Trichodina spp. Other peritrichids are similar in appearance and activity	Most fresh- and salt-water fishes	Bell-shaped organism with basal adhesive disk and aboral ciliary spiral. Appears as ornate disk with dark inner ring when viewed dorsoventrally	Directly from infected fish through water	Inapparent, to white blotches on skin, heavy mucus secretion, scale sloughing, fraying of fins, lethargy, and death. Gill infestation will produce respiratory distress	Hyperplasia of epidermis and/or gill epithelium associated with typical parasite	Based upon parasite identification	Treat with formalin, acetic acid, KMnO₄
Tetrahymena corlissi	Guppy (*Poecilia reticulata*), northern pike fry (*Esox lucius*)	Oval, ciliated protozoan 50–70 μm long that is normally free-living	Under conditions of overcrowding and overfeeding high water organic content leads to massive multiplication of *T. corlissi*	Body wall ruptures and fish eviscerate	Massive invasion of musculature, particularly of ventral ''belly'' wall by parasite	Based upon lesions and presence of masses of *T. corlissi* in water	Reduce feed volume and fish concentration. Treat water with tetracycline

Protozoa parasitic to fish and generally located within tissues

	Species affected	Morphology and culture	Transmission	Clinical signs	Lesions	Diagnosis	Control
Internal protozoan *Plistophora ovariae*	Golden shiner (*Notemigonus crysoleucas*)	Spores 3.5 x 6.5 μm with prominent anterior vacuole	Direct—spores released at ovulation and ingested. Transovarian—lightly infected ova hatch and spores in yolk enter gut	Reproductive failure, particularly in fish above 1 year of age	Markedly enlarged, spore-filled ova with necrosis and in late stages fibrosis of ovary	Based upon identification of parasite in ova	Utilize only yearlings for brood stock. Treatment not effective
Glugea spp. (nosema)	Most freshwater and marine fishes	Spores 3–6 μm long with polar filament. Sporont produces 2 spores	Probably direct by ingestion	Cysts occur in muscle, subcutis, or viscera. Severe infestations may produce deformity and death	Cysts usually walled off. Some cause marked cellular hypertrophy producing xenomas	Based upon lesions and parasite identification	Eliminate infected fish
Eimeria anguillae Other species infect many fishes including carp and goldfish	Eel (*Anguilla* spp.)	Intracellular oocysts spherical, 10 μm in diameter. Microgamonts (6–8 μm in diameter) and macrogametes (12 x 9 μm) attach to cell surface	Oocysts passed in feces and probably transmitted via invertebrate intermediate hosts	Unknown, if any	Usually an incidental finding within the intestine	Based upon host, location in host, and identification of parasite	None
Trypanoplasma (formerly and	Many species of freshwater fishes	Biflagellate, trypanosome-like pro-	Leeches are the intermediate host.	Inapparent to lethargy, anemia, gener-	Increased mucus over skin and gills, as-	Based upon identification of parasite	Eliminate leeches

Table VIII (*Continued*)

	Species affected	Morphology and culture	Transmission	Clinical signs	Lesions	Diagnosis	Control
occasionally still called cryptobia)	including the salmonids and cyprinids	tozoan 10–30 μm long	Transmission occurs while leech is feeding	al debilitation, and death	cites, and organisms in blood, kidney muscle, and ascitic fluid		
Hexamita salmonis	Salmonids, aquarium fishes and others	Binucleate, piriform protozoan with 6 anterior and 2 posterior flagella	Cysts, formed in intestine are carried by feces and ingested by fish	Acute enteritis and death or anorexia, reduced growth rate and gradual debilitation	Catarrhal enteritis (acute form) presence of organisms (both forms)	Based upon identification of organisms	Treat with Enheptin (American Cyanamid Co.) or Carbarsone (Lilly and Co.)
Schizamoeba salmonis	Salmonids	Ameboid organism with multiple nuclei	Probably direct	Inapparent	Presence of multinucleate trophozoites in stomach and intestine	Identification of organism	Unknown and probably unnecessary
Myxosoma cerebralis (whirling disease)	Salmonids, particularly rainbow trout (*Salmo gairdneri*)	10 μm spores are oval with 2 piriform polar capsules	Spores escape decomposing fish and infect tubificid oligochaetes as the actinosporean, Triactinomyxon, which are released and become infectious to trout in 3.5 months	Young fish (< 6 months) swim erratically (whirling), become deformed about head and spine, and die. Older fish become deformed, but experience low mortality	Deformities are caused by necrosis of cartilage. Spores are evident in areas of necrosis and in pockets within bone	Presence of typical mxyosoma spores in the skeleton of salmonids	Maintain fish in concrete raceways. If water source is contaminated raise fry and fingerlings in filtered water which has received ultraviolet irradiation
Henneguya spp.	Channel catfish (*Ictalurus punctatus*) and other freshwater fishes	Spores resemble spermatozoa, have 2 polar capsules and long terminal process that may be single or divided	Probably direct	For most forms presence of cysts is only sign. However, interlamellar form can produce severe gill hyperplasia, respiratory distress, and death, especially in very young	Several forms are reported according to cyst location. These include: (1) branchial—interlamellar, intralamellar; (2) cutaneous—papillomatous, cystic, adipose fin; (3) mandibular; (4) gallbladder	Based upon identification of parasite, location and species of host	Eliminate infected fish and when possible increase water flow or decrease fish population

Parasitic monogenetic trematodes, crustaceans, and miscellaneous parasites

Monogenetic trematodes

	Species affected	Morphology and culture	Transmission	Clinical signs	Lesions	Diagnosis	Control
Gyrodactylus spp.	Virtually all fishes, each having its own species of *Gyrodactylus*	Elongate, viviparous parasite 250–600 x 40–115 μm; larva often visible within adult. Haptor has 2 large anchors and 16 small marginal hooks	Larvae born with anchors and attach immediately, often to same host as parent. Cycle may be as short as 60 hr	Fish rub on side or bottom of aquarium, often become emaciated, may develop blue mucus over skin and gills, become dyspneic, and in severe cases die	Superficial petechia, frayed fins, and presence of organisms on body, fins, and gill	Presence of clinical signs and lesions and identification of parasite	Several chemicals including NaCl, acetic acid, trichlorfon, and formalin are helpful treatments. Rapidly flowing water or frequent water changes are reported to reduce infestation
Dactylogyrus spp.	Common parasite of warm-water fishes, particularly carp (*Cyrinus carpio*) and goldfish (*Carassius auratus*)	Similar to gyrodactylus but without larva, with eyespot and larger, 400–1000 x 85–180 μm	Directly from infested fish. Entire life cycle completed in 1–5 days	Respiratory distress, abduction of opercula, protrusion of gills, and high mortality	Thick mucus over gills, presence of parasite and lamellar epithelial hyperplasia	Presence of clinical signs and lesions and identification of parasite	Same as *Gyrodactylus*

Crustaceans

Branchiuran

	Species affected	Morphology and culture	Transmission	Clinical signs	Lesions	Diagnosis	Control
Argulus spp. (fish louse)	Infects many species worldwide	Green to brown, flattened dorsoventrally with 4 pairs of legs, 2 suction discs and prominent proboscis, 6–25 mm long	Free-swimming larvae attach to host, go through several molts. Fertile adult female deposits eggs on stones or on other objects	Rubbing, parasites and numerous puncture sites visible on skin. Severely infested fish may die	Puncture sites are raised, brown to red papular areas up to 1 cm in diameter	Based upon lesions and identification of parasite	Remove parasites with forceps or treat with trichlorfon or KMnO$_4$

(*continued*)

Table VIII *(Continued)*

	Species affected	Morphology and culture	Transmission	Clinical signs	Lesions	Diagnosis	Control
Copepod *Lernaea cyprinacea* (anchor worm)	Infests many species worldwide	Male and larval female are typical crustaceans. Adult female is pale green to brown, vermiform, with an anchorlike anterior and 2 egg sacs, posterior	Following 3 free-living nauplii (larval) stages and several parasitic larval stages, fertilization occurs and female embeds in host. Cycle requires 20–25 days at 20°–25°C	Parasitic nauplii may cause respiratory distress. However, usual signs are presence of adult female and slow healing, deep ulcers	Parasite penetrates into muscle and on occasion coelom. Embedded portion may be walled off or surrounded by necrotic tissue. Dermal tumors associated with parasite	Based upon parasite identification	Treat with trichlorfon or KMnO₄. Secondary bacterial infection may also require treatment
Mollusks *Margaritifera margaritifera* (Glochidia)	Salmonids	Small (50 μm) bivalve with hooks on inner surface of shell	Nonmotile larvae attach to gill epithelium	Rarely, produces respiratory distress. Mortalities have been reported	Hyperplasia of gill epithelium to form cyst around clam	Based upon identification of glochidia	Avoid using unfiltered pond water during spring and early summer when larval clams may be present
Leeches	Various species of fish are occasionally parasitized by leeches of several genera. Rarely these encounters may be fatal for the fish. Leeches also are vectors for the blood-borne protozoan trypanoplasma and possibly other infections						

	Definitive host	Intermediate host(s)	Organ(s) infested	Clinical signs	Lesions	Diagnosis	Control
Parasitic digenetic trematodes							
Trematode *Diplostomum* spp. (eye fluke)	Gulls and pelicans	1st: Lymnaeid snails 2nd: Numerous fish	Eye (lens and vitreous *D. spathaceum*, retina *D. adamsi* and *D. scudderi*) and brain	Blindness in severe infestations	Initially, may see larvae as white dots. Later, eye becomes opaque. microscopically see typical trematode (see Fig. 9)	Based upon species infected, location and parasite morphology	Eliminate snails and/or definitive host from water source
Uvulifer ambloplitis (blackspot)	Herons and kingfisher	1st: Snails 2nd: Many warm-water fishes	Subcutaneous tissue of body and fins	Black or brown spots up to 4 mm in diameter (see Fig. 10). Some deaths may occur in severe infestations	Black spots contains larval trematode surrounded by thick connective tissue wall. Pigment is within wall not parasite	Based upon gross and microscopic findings and identification of parasite	Same as for *Diplostomum* spp.
Clinostomum marginatum (Yellow grub)	Herons	1st: Snails 2nd: Many warm-water fishes and salmonids	Subcutaneous tissue, muscle, and gill	Yellow to white cysts up to 2.5 cm in diameter. May cause debilatation of smaller fish	Cyst contains metacercaria 1.5–6.5 mm long. Cyst wall is thick and unpigmented	Based upon gross appearance and identification of parasite	Same as for *Diplostomum* spp.
Sanguinicola spp.	Rainbow trout (*Salmo gairdneri*) and several warm-water fishes	Snails	Adults locate in aorta, branchial arteries, renal veins, and choroid gland of eye. Ova lodge in gill capillaries where miracidia emerge and migrate to outside	Stressed fish may evidence respiratory distress. In severe infestations miracidia produce hemorrhage and death	Ova and miracidia visible grossly. Following exodus survivors develop multiple granulomata often with extensive loss of functional gill epithelium	Based upon the presence of ova and miracidia in gill capillaries (see Fig. 11)	Control snails and increase water flow through raceways
Crepidostomum spp.	Many warm- and cold-water fishes including salmonids	1st: Clam 2nd: Mayfly or crayfish	Intestine	Not usually apparent. However, severe infestation may produce emaciation and rarely death	Usually an incidental finding in the intestine	Based on identification of parasite	Interrupt life cycle by eliminating one or both intermediate hosts. May be treated with dibutyltin oxide
Parasitic cestodes, nematodes, and acanthocephalans							
Cestodes *Corallobothrium* spp.	Ictalurids	Copepod and in some cases a small fish. Ova passed in feces ingested by copepod which may in turn be ingested by small fish. Plerocercoid larvae but not adult	Adult resides in intestine of catfish. Plerocercoid larvae develop in coelom, ovary or muscle of small fish	Little known. Do not appear to cause more than minor debility	Presence of adults in intestine, plerocercoid larvae in coelom, muscle, or ovary	Based upon identification of parasite in intestine	Interrupt cycle by eliminating copepods. Treat with dibutyltin

Table VIII (*Continued*)

	Definitive host	Intermediate host(s)	Organ(s) infested	Clinical signs	Lesions	Diagnosis	Control
		parasites develop in small fish					
Proteocephalus ambloplitis	Black bass (*Micropterus* spp.)	Procercoid develops in a copepod, plerocercoid larva develop in ovary and other viscera of many fishes, including bass (*Micropterus* spp.)	Adult found in intestine. In bass plerocercoid larva may migrate from viscera into intestine where they become adults	Emaciation and distortion of body. Frequently produce sterility by migration of plerocercoids through ovary	Massive adhesions of visceral organs. Since plerocercoid larvae do not encyst they elicit marked inflammatory response	Based upon parasite identification	Interrupt cycle by eliminating copepods. No treatment available for plerocercoid larvae
Ligula intestinalis	Fish-eating birds	Procercoid stage in copepod. Plerocercoid larvae develop in coelom of many fishes	Plerocercoid larvae reside in coelom	In severe infestations abdomen becomes distended and may rupture	Presence of plerocercoids and concomitant compression and displacement of organs	Based upon clinical signs and parasite identification	Same as *P. ambloplitis*
Nematodes *Philonema* spp.	Salmonids. Long (up to 300 mm) slender females and smaller (up to 35 mm) males reside in coelomic cavity	Copepod	Following ingestion of parasitized copepod, larvae migrate extensively through viscera	Inapparent to severe emaciation and death	In severe cases massive adhesions of all visceral organs can occur	Presence of parasite in coelom and its identification	Interrupt cycle by control of copepods
Cystidicola spp.	Eels, salmonids, and other fishes. Larvae migrate up pneumatic duct. Some are species specific and will not mature in aberrant host	Amphipods	Swim bladder	Inapparent	Small slender nematode in swim bladder	Identification of parasite	Interrupt cycle by control of intermediate host
Anisakis spp. (anisakiasis in man). Also caused by *Contracaecum* spp. and *Phocanema* spp.	Marine mammals	Herring (*Clupea* spp.) and many other fishes	Larvae found in viscera and musculature	Unreported and probably minimal	Coiled parasites within thick connective tissue capsule. Necrosis and granuloma formation in liver	Identification of parasite	Eating of raw fish can produce eosinophilic granulomata of stomach and intestine in man. Avoid by freezing at $-20°C$ for 24 hr
Acanthocephalans *Pomphorhynchus bulbocolli*	Many predatory fishes	1st: Amphipod 2nd: Small fish	Adult inhabits intestine. Second intermediate stage encysts in liver, spleen or mesentery	Little known. Probably cause some degree of debilitation	Adults penetrate deep into intestinal wall and may perforate and invade other organs. Often produce severe inflammatory response followed by encapsulation. Larvae encyst in viscera	Based upon identification of adults in intestine or larvae in viscera (see Fig. 12)	Unknown
Echinorhynchus spp.	Many fishes	Amphipod	Adult found in intestine of fish following ingestion of infected amphipod	Emaciation and death in heavy infestation	Similar to lesions produced by *P. bulbocolli* adults plus ulceration and obstruction of intestine	Based upon identification of parasite	Unknown

	Species affected	Organ(s) affected	Contributing factors	Clinical signs	Lesions	Diagnosis	Control
			Deficiency diseases				
Nutrient deficient Pantothenic acid	Salmonids and probably others	Gills	Toxins, especially heavy metals, may cause similar conditions	Anorexia, fish congregate at water inlet, and if stressed evidence respiratory distress	Massive hyperplasia of gill lamellar epithelium with fusion of secondary lamellae and anemia	Based upon characteristic gill changes in the absence of other factors, e.g. bacteria, parasites	Supplement diet with milk, dietary yeast or distillers solubles. If diagnosis correct response will be rapid
Vitamin C (ascorbic acid)	Rainbow trout (*Salmo gairdneri*), coho salmon (*On-*	Skin, bone, cartilage, and muscle	Exposure to sublethal levels of toxaphene increase dietary re-	Reduced growth rate, spinal deformities, darkeni of skin	Hemorrhages, soft enlarged vertebrae, fractured spine, de-	Presumptive—based on history signs and lesions. Con-	Remarkable recovery reported for channel catfish fed 60

(*continued*)

Table VIII (*Continued*)

	Species affected	Organ(s) affected	Contributing factors	Clinical signs	Lesions	Diagnosis	Control
	corhynchus kisutch), channel catfish (*Ictalurus punctatus*), eel (*Anguilla japonica*), yellowtail (*Seriola quinqueradiata*), but not carp (*Cyprinus carpio*)		quirement for vitamin C	and erosion of fins, mortality rate increases after 12–15 weeks on deficient diet	formed gill filaments. Wound healing delayed with formation of immature collagen	firm by ascorbic acid analysis of liver and kidney and feed if available	mg vitamin C per kg of feed for 10 days
Vitamin E	Rainbow trout (*S. gairdneri*)	Muscle, fat, bone, and swim bladder	Probably related to high fat content of feed and anti-vitamin E factors associated with rancidity of that fat	Reduced growth rate, irregular swimming, skin darkening, atrophy of posterior myotomes, and high mortality	Muscle atrophy, proliferation of sarcolemmal nuclei, steatitis, and thickening of swim bladder wall with formation of constricting bands	Presence of lesions and conditions such as high cod liver oil content or rancidity of feed	Add synthetic antioxidants and vitamin E to diet
Methionine	Lake trout (*Salvelinus namaycush*), rainbow trout (*S. gairdneri*)	Eye (lens)	Deficiency of riboflavin, zinc, and/or vitamin A may be contributory	Reduced growth rate with grossly visible development of cataracts (bilateral) after 4 months	Increased basophilia and vacuolization at 1 month progressing to epithelial proliferation and liquifaction by 4.5 months	Based upon lesions and establishment of dietary deficiency of methionine	Supplement diet with methionine
Tryptophan	Sockeye salmon (*Oncorhynchus nerka*) rainbow trout (*S. gairdneri*)	Vertebral column, kidney	Low total nitrogen in the presence of apparently adequate essential amino acids may result in deficiency of one or more essential amino acids	Reduced growth rate, scoliosis and/or lordosis	Calcium deposits in kidney and on bones surrounding notochord	In presence of typical signs and lesions confirm by feed analysis	Assure adequate essential amino acids and total protein in diet
Iodine	Salmonids, goldfish (*Carassius auratus*), jawfish (*Opistognathus aurifrons*)	Thyroid	Iodine levels adequate for most fishes may be inadequate for others (jawfish). Environmental pollutants may be contributory	Enlargement of lower jaw and abduction of opercula	Hyperplasia of thyroid tissue that extends the length of ventral aorta and adjacent portions of branchial arteries	Microscopic appearance of thyroid	Supplementary iodine in feed
Linolenic and linoleic acid	Probably all fishes	Heart, skin liver, and fat	This condition appears to be closely related to lipoid liver disease that is associated with fat rancidity and vitamin C and E deficiency	Fin erosion, pigment loss and bulging of anterioventral body wall by enlarged liver	Myxomatous degeneration of fat. Myocardial degeneration and fatty change and ceroid deposition in liver	Suspect from signs and lesions, confirm by feed analysis	Provide essential fatty acids in diet

Toxins known to affect fish

Toxin	Species affected	Organ(s) affected	Contributing factors	Clinical signs	Lesions	Diagnosis	Control
Methyl mercury	Ictalurids, salmonids and probably all others	Liver and kidney	Fish do not have the ability to methylate mercury. Conversion from inorganic to organic Hg can occur in aquatic sediments	Abdominal distention, lethargy, anorexia, and death	Serosanguinous fluid in coelom, focal to diffuse liver necrosis, renal tubular necrosis, and necrosis of hepatopancreas	Depends upon demonstration of high levels of Hg in tissue and feed or environment	Remove source of Hg. Selenium and mercury are mutually protective. Fish chronically exposed to low levels of Hg usually have high levels of tissue mercury and selenium
Acidity: In addition to direct action of acid there may be a secondary toxicity due to the increased avail-	Many species including salmonids and cyprinids	Primary change appears to be acidosis that causes gill and kidney damage and reproductive failure	In carp pond culture, associated with high ammonia levels, growth of algae, and high CO_2 levels. Salmonids and others, related	Carp—respiratory distress and erratic swimming. Salmon—reproductive failure and death of fry. Salmonids disappear from wa-	Carp—gill edema, hyperemia, exudation, necrosis and sloughing. Salmon—failure to ovulate associated with low serum Ca, pre-	Carp—lesions and presence of high ammonia/low pH in water. Salmonids—elimination or decline of population and low	Carp: addition of nontoxic red or yellow dye to ponds; control pH and ammonium levels; treat with chlorinated cal-

Table VIII (*Continued*)

	Species affected	Organ(s) affected	Contributing factors	Clinical signs	Lesions	Diagnosis	Control
ability of aluminum			to acid rain that has been measured as low as 2.32 pH	ters at pH 4.5 or below	cipitation of calcium phosphate in renal tubules	water pH	cium. Salmonids—reduce sulfur, nitrate, and chlorine emmissions from oil and coal burning plants. Lime small lakes and ponds
Petroleum	Many species	Changes have been observed in blood, serum chemistry, skin, and the weight of testes and spleen	Except in experimental situations it is impossible to separate from effects of other pollutants	Inability to detect and retain food, increased susceptibility to parasites, skin necrosis, fin rot, and death	Reduced packed cell volume, alkaline phosphatase, and serum protein. Increased serum ammonia, lower spleen and testis weight	Findings can only be suggestive	Reduce contamination of waterways by oil
Chloramphenicol similar although lesser changes seen with oxytetracycline	Eel (*Anguilla anguilla*)	Hematopoietic system	Possibly other antibiotics administered in close time proximity	Probably not detectable since treated fish will have other signs of disease	Granulocytopenia, thrombocytopenia, monocytosis, and lymphocytosis. Reduction in numbers of erythroblasts and vacuolization of erythrocyte cytoplasm	Presence of these findings following administration of drugs	Reversible following 2 doses of 2 mg/100 gm. Not known if continued treatment will produce irreversible change
Irradiation	Goldfish (*carassius auratus*) (x-ray), rainbow trout (*Salmo gairdneri*) (tritium)	Immune system	Escape of tritium and other radionuclides from nuclear reactors has been suggested	Probably none	Changes reported are increased susceptibility to pathogenic bacteria and parasites and reduction of the immune response	Probably not possible	Necessary to determine acceptable levels and limit escape to that amount
Aflatoxin	Salmonids, particularly rainbow trout (*S. gairdneri*) and probably most fishes	Liver	Toxin is metabolic product of *Aspergillus flavus*. Usual source is oil seed meals, particularly, peanut	Acute—fish darken, become lethargic and die. Chronic—distention of anterior ventral body wall due to liver enlargement	Acute—massive hemorrhagic liver necrosis (see Fig. 13) with moderate bile duct proliferation in fish surviving several days. Chronic—hepatocellular carcinoma (see Fig. 14)	Based upon lesions and identification of toxin in feed	Screen oil meals for toxin before including in diet
Ciguatoxin	Man is usual victim. Over 300 species of fish reported to be toxic. Toxicity usually sporadic and localized	Generalized severe toxicosis	Usually associated with the ingestion of barracuda. Toxin believed to be produced by algae is concentrated first in herbiverous fishes then in predators	Prostration, aches and pains, burning of tongue, vomiting, severe abdominal pain, diarrhea, and occasionally death	Lesions in fish not reported	Clinical findings and history	Avoid consumption of barracuda and other species known to be toxic. In theory young small fish are less apt to be toxic

	Species affected	Organ(s) affected	Etiology	Contributing factors	Macroscopic features	Microscopic features
			Neoplasms			
Neoplasm Melanoma	Platyfish (*Xiphophorus maculatus*) swordtail (*X. hellei*) hybrids	Skin and eye	Genetic spotting pattern of platyfish is not controlled in some hybrids, possibly due to loss of represser genes	Neither platyfish or swordtails develop melanoma. F₁ hybrids with spotting trait may develop only premelanomas. F₁ X swordtail produces frank melanoma	Soft, black raised masses	Some are amelanotic early in development. Later, interlacing spindle cells become intensely black and may invade adjacent tissue
Hepatocellular carcinoma (hepatoma)	Many species but best known in rainbow trout (*Salmo gairdneri*)	Liver with metastasis to kidney and rarely gill (see Fig. 14)	Rarely if ever spontaneously. Most cases associated with carcinogens, aflatoxins in particular	Cyclopropenoid fatty acids found in cotton seed acts as synergist with carcinogens in rainbow trout	Liver, enlarged with rounded edges and numerous raised, often pale nodules. Hemorrhage may be apparent	Neoplastic tissue is usually intensely basophilic and severely anaplastic. Normal liver architec-

(*continued*)

Table VIII (*Continued*)

	Species affected	Organ(s) affected	Etiology	Contributing factors	Macroscopic features	Microscopic features
						ture and fat vacuoles not apparent. Bile duct proliferation often accompanies tumor
Stomatopapilloma (cauliflower disease)	Eel (*Anguilla anguilla*)	Epithelium of head, usually snout. Often prevents feeding	Virus isolation from eels with stomatopapilloma is reported and viruslike particles have been reported in tumors. Initiation of tumors with cell culture fluid, blood, or tumor tissue has been unsuccessful	Incidence is seasonal, being highest in summer and lowest in late winter and spring. Most cases occur in polluted waters	Discoid to large firm fungate papillomas usually attached to mouth. Tumors degenerate in winter	Densely packed basal cells align along thin connective tissue stroma. Cells away from stroma are further apart imparting a spongy appearance
Lymphosarcoma	Northern pike (*Esox lucius*), muskellunge (*E. masquinongy*) and occasionally salmonids	Head and mouth of pike and subcutis of flank area of muskellunge. Metastasis common in both species to spleen, kidney, and liver	Epizootiological evidence suggests virus. Cell-free transmission reported, but virus neither isolated nor seen in tissue	Condition is enzootic in some areas reaching 16% incidence in one Ontario lake	Early lesions appear as raised purple cutaneous masses. These ulcerate and invade muscle and viscera	Neoplastic cells are large, uniform lymphoblasts that infiltrate without stimulating any connective tissue proliferation
Leiomyoma	Yellow perch (*Perca flavescens*)	Testicle	Unknown. Probably spontaneous	None reported	Pale, firm, shiny mass arising from testicle. Largest may have necrotic centers and distend abdomen	Parallel arrays of slender eosinophilic cells with elongate nuclei with rounded ends. Neoplastic cells often entrap islands of spermatazoa

Fig. 1. Verrucose overgrowth. Lymphocystis disease, tail fin, striped bass.

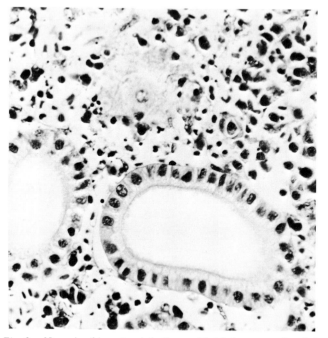

Fig. 2. Necrosis of hematopoietic tissue with tubule preservation. Infectious hematopietic necrosis, Rainbow trout. Hematoxylin and eosin (H & E). × 250.

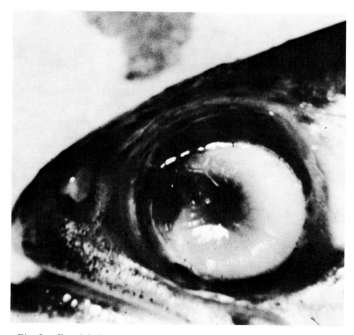

Fig. 3. Exophthalmos (popeye) and panophthalmitis. Bacterial kidney disease, coho salmon.

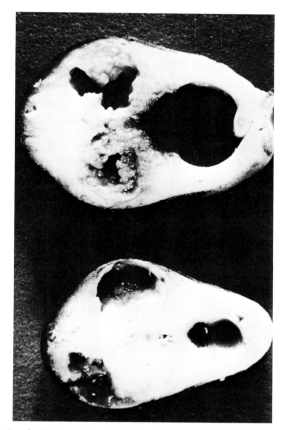

Fig. 4. Cavitations of somatic muscle. Bacterial kidney disease, coho salmon.

Fig. 5. Scoliosis. Ichthyosporidiosis, brown trout.

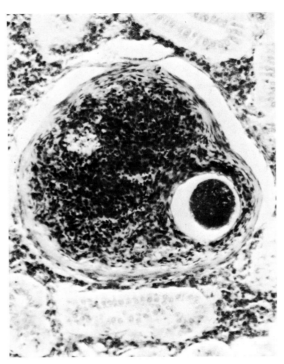

Fig. 7. Nonseptate, branching hyphae. *Saprolegnia* sp. Gridley's stain. × 400.

Fig. 6. Quiescent spore. *Ichthyosporidium hoferi*, trunk kidney, cod. H & E. × 200.

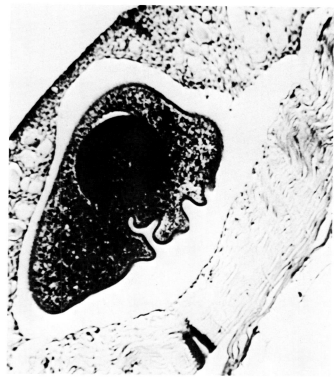

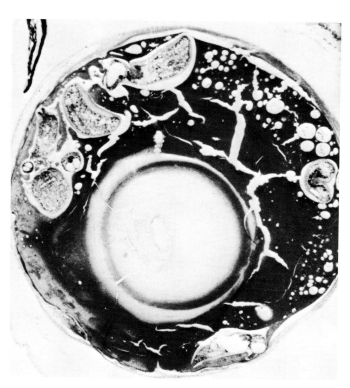

Fig. 8. Trophozoite. *Ichthyophthirius multifiliis*, skin, channel catfish. H & E. × 100.

Fig. 9. Parasitic cataract. *Diplostomum* sp., lens channel catfish. H & E. × 40.

Fig. 10. "Blackspot." Larval digenetic trematode such as *Uvulifer ambloplitis*, bluegill.

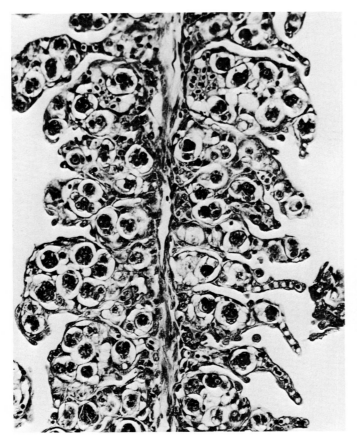

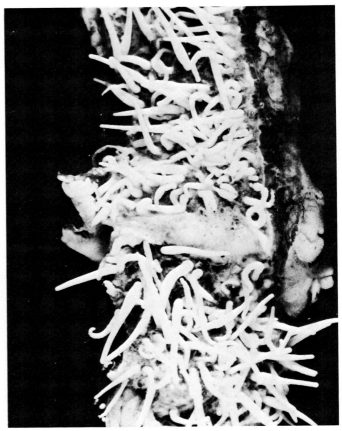

Fig. 11.   Parasitic branchitis. *Sanguinicola* sp., ova and miracidia, gill filament, rainbow trout. H & E. × 40.

Fig. 12.   Acanthocephaliasis. Intestine, striped bass.

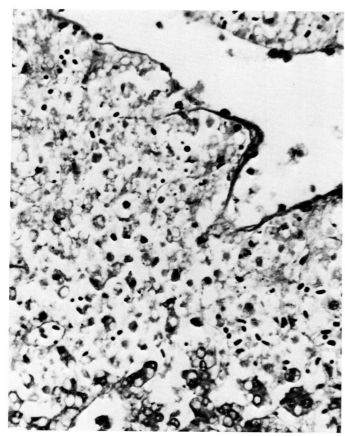

Fig. 13. Centrilobular necrosis. Acute aflatoxicosis, liver, rainbow trout. H & E. × 250.

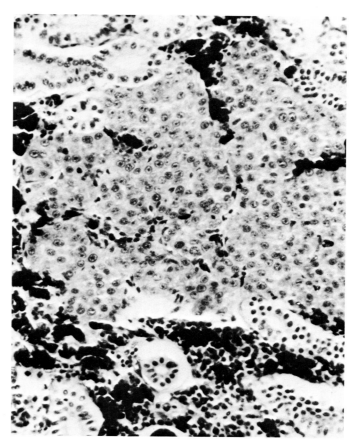

Fig. 14. Metastatic hepatocellular carcinoma. Aflatoxin induced, kidney, rainbow trout. Note intensely pigmented macrophages within kidney tissue. H & E. × 200.

REFERENCES

Ali, M. A., ed. (1980). "Environmental Physiology of Fishes." Plenum, New York.

Christensen, G. M., Fiandt, J. T., and Poeschl, B. A. (1978). Cells, proteins and certain physical-chemical properties of brook trout (*Salvenlinus fontinalis*) blood. *J. Fish Biol.* **12,**51--60.

Flynn, R. J. (1973). "Parasites of Laboratory Animals." Iowa State Univ. Press, Ames.

Grizzle, J. M., and Rogers, W. A. (1976). "Anatomy and Histology of the Channel Catfish." Auburn Printing, Auburn, Alabama.

Ketola, H. G. (1976). Quantitative nutritional requirements of fishes for vitamins and minerals. *Feedstuffs* **48**(7), 42–44.

Ketola, H. G. (1977). Qualitative requirements and utilization of nutrients: Fishes. In "CRC Handbook Series in Nutrition and Food," (M. Rechcigl, Jr., ed.), Sect. D, Vol. 1, pp. 411–418. Chem. Rubber Publ. Co., Cleveland, Ohio.

Ketola, H. G. (1980). Amino acid requirements of fishes: A review. *Proc.—Cornell Nutr. Conf. Feed Manuf., 1980,* pp. 43–47.

Mulcahy, M. F. (1970). Blood values in the pike *Esox lucias* (L.). *J. Fish Biol.* **2,**203–209.

Ribelin, W. E., and Migaki, G., eds. (1975). "The Pathology of Fishes." Univ. of Wisconsin Press, Madison.

Roberts, R. J. (1978). "Fish Pathology." Macmillan, New York.

Sherburne, S. W. (1973). Differential blood cell counts of Atlantic herring *Clupea harengus harengus*. *Fish. Bull.* **71**(4), 1011–1017.

Van Vuren, J. H. J., Hattingh, J. (1978). A seasonal study of the hematology of wild freshwater fish. *J. Fish Biol.* **13,** 305–313.

Warner, M. C., Tomb, A. M., Diehl, S. A. (1979). Variability and stability of selected components in rainbow trout *Salmo gairdneri* serum and the precision of automated analysis in measuring these components. *J. Fish Biol.* **15,** 141–151.

Watson, L. J., Shechmeister, I. L., and Jackson, L. L. (1963). The hematology of goldfish, *Carassius auratus*. *Cytologia* **28,** 118–130.

Wolf, K. (1981). Viral diseases of fish and their relation to public health. In "CRC Handbook Series in Zoonoses: Viral Zoonoses," (G. W. Beran, ed.), pp. 403–437. CRC Press, West Palm Beach, Florida.

Chapter 17

Design and Management of Animal Facilities

J. R. Hessler and A. F. Moreland

I. INTRODUCTION

Reliability of research data is no greater than the least reliable link in the chain of procedures used to derive the data. When animals are a link in the chain, it becomes a formidable challenge to make that link as strong as possible, especially in view of modern, sophisticated technology. Considering that the research animal facility is an extension of the research laboratory, the objective must be to maintain the research animal free of complicating diseases, in a steady state environment, free of chemical and biological contaminants. Although it is not practical, or at this time even possible, to attain this ideal state, the degree to which it can be accomplished is dependent on the design and management of the research animal facility. The purpose of this chapter is to review the salient design features and management procedures of a modern research animal facility.

II. FACILITIES DESIGN AND EQUIPMENT

Planning and designing a research animal facility is a creative process. Each facility must be carefully evaluated according to its programmatic requirements; that is, what type of activity must the facility support. For example, animal production, toxicology testing, and basic biomedical research facili-

LABORATORY ANIMAL MEDICINE

ties all house research animals, but each may have different programmatic requirements. A common requirement is to maintain the animals in a steady state environment free of physical, chemical, and microbiological agents. Efficient management relates to facility design. When reasonable, management considerations should dictate design. It is possible to compensate for poor design with appropriate management procedures, but often at a high operational cost and with built-in inefficiency. While the design of research animal facilities is a creative process resulting in each facility having its unique features, the general requirements are becoming increasingly well defined and have been the subject of several recent papers and symposia [Institute of Laboratory Animal Sources (ILAR), 1976a, 1978a; Lang and Harrell, 1969, 1972; Lang, 1981, 1983; Poiley, 1974; Balk, 1980; Jonas, 1965; Goldstein, 1978; Simmons, 1973].

A. Basic Concepts and Considerations

1. Location

Public health, human comfort, animal husbandry, and environmental control requirements dictate that animal housing be separated from personnel areas. Careful planning must reconcile this necessity with the desirability of locating the animal facilities as near as possible to the research and teaching laboratories. The most efficient animal facility in terms of construction and operational cost is a single story, centralized facility with direct access to ground level transportation. Variations may be acceptable and include a central facility on multiple floors, arranged around or very near dedicated elevators; multiple autonomous units that contain all necessary animal support services; satellite facilities that rely to varying degrees on a primary facility for support services; and a centralized support service with animal housing rooms scattered throughout a building or in multiple buildings. Each individual situation must be evaluated to determine what is best for that particular set of circumstances and requirements (Jonas, 1978). If facilities are properly planned and managed, any arrangement can provide acceptable animal care, although the cost of providing that care varies considerably, depending on the physical arrangement of the facility (ILAR, 1980).

2. Centralized versus Decentralized Units

The advantage of a decentralized facility is that it affords the greatest convenience to the investigator by placing the animals near the research laboratory. Disadvantages include higher construction costs because of the necessity of duplication of space and equipment, the necessity for moving animals and care materials (clean and soiled animal cages, etc.) through

public corridors and on public elevators, and higher animal maintenance costs (ILAR, 1980).

Advantages of centralized facilities include greater flexibility and more efficient utilization of animal housing space (wherein animals may be separated by species, source, microbiological status, and projects and to accommodate fluctuating demands for housing of various species), convenience of having animals near specialized support facilities, and ability to provide specialized animal housing areas, such as barrier, quarantine, and hazardous containment.

3. Traffic Flow Patterns

Flow patterns generally evolve around the cage sanitation facility and the flow of cages to and from it and the animal rooms. Four basic patterns are usually used (Fig. 1). The supe-

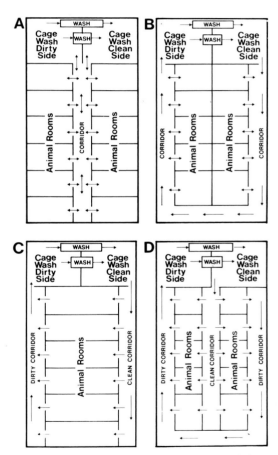

Fig. 1. Four types of traffic flow patterns are illustrated relative to the cage wash area. The arrows indicate the direction of cage traffic. All are drawn within the identical space to the same scale in order to demonstrate the relative "cost" in terms of animal housing space, both square footage and number of spaces. (A) is an example of a "to and fro" traffic pattern, (B) unidirectional, (C) clean–dirty with two corridors, (D) clean–dirty with three corridors.

rior sanitation and infection control offered by the "clean–dirty" corridor systems must be carefully weighed against the higher construction and maintenance costs of the additional corridors.

4. Cubicles

The cubicle concept, which provides maximum flexibility for animal isolation within minimal space, was first described by Dolowy (1961). It has been variously identified as "Illinois Cubicle", "Horsfall Cubicles", "animal modules", and "animal cubicles."

Vertically stacking, three-leaf doors are the most common type of entrance used (Fig. 2A). Another option is two 0.9 m (3 ft) wide standard hinged doors that swing 180° into the aisle, folding flat against the adjacent space (Fig. 3).

The standard air flow within a room with cubicles is from the

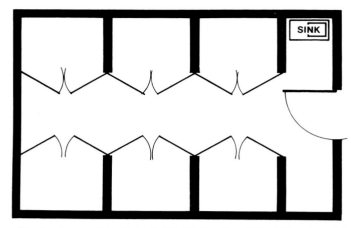

Fig. 3. A room divided into six 1.2 × 1.8 m (4 × 6 ft) cubicles with solid divider walls and conventional hinged doors arranged along a 1.8 m (6 ft) aisle. Space is provided for a sink and food storage.

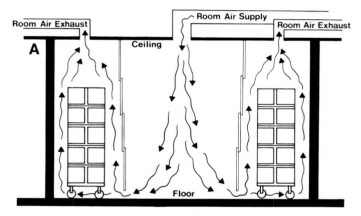

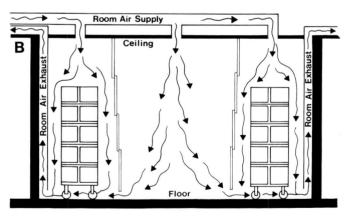

Fig. 2. Two types of air flow patterns for rooms divided into cubicles are illustrated. Also illustrated are stacking type doors which raise vertically. In (A), the air supplied at the ceiling of the aisle between the cubicles flows under the doors of the cubicles, up through the cubicle to be exhausted at the ceiling. In (B), the air is supplied at the ceiling of both the aisle and the cubicles and is exhausted at the floor level in back of each cubicle.

ceiling of the aisle, preferably through a linear diffuser running the length of the aisle, under the cubicle doors and out the ceiling of each cubicle (Fig. 2A). The control of airborne cross-contamination between cubicles may be compromised when the doors to a cubicle are open for servicing; however, extensive experience has not proved this to be of significant concern.

An alternate air flow pattern such as illustrated in Fig. 2B might be considered. With this system, the air pressure in the aisle always remains positive to the cubicle because the air supplied in the ceiling of the aisle is exhausted at floor level in each cubicle.

Because of operational flexibility consideration should be given to including cubicles in new facilities or renovations.

5. Mass Air Displacement "Clean" Rooms (MADC)

The clean room concept has only recently been introduced into animal facilities (Beall *et al.,* 1971; Van der Waaij and Andreas, 1971; McGarrity and Coriell, 1976), having first been used in manufacturing industries requiring a dust-free environment. Mass air displacement "clean" rooms (MADC) rooms are occasionally called laminar flow rooms, but because of the impossibility of creating and maintaining true laminar air flow patterns within the animal room, this term is generally avoided. In MADC rooms, the air is exchanged at the rate of between 200 and 600 times every hour. This is accomplished without creating significant drafts by supplying the room air through the entire area of a perforated ceiling at the rate of between 9.14 m (30 ft) and 27.43 m (90 ft) per minute (Fig. 4A and B). The horizontal distance from the ceiling supply to the floor return should not normally exceed 3.04 m (10 ft). For energy conservation reasons, air exchanged at this rate must be

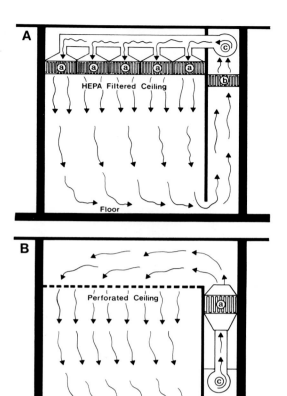

Fig. 4. Two mass air displacement clean (MADC) rooms are illustrated. Each has a similar recirculating air flow pattern. In (A) air supplied at the ceiling through a total ceiling of HEPA filters (a) returns at the floor level to a plenum, being drawn through prefilters (b) by a fan (c) from where it is ducted to the ceiling HEPA filters. (B) differs in that air is supplied to the room from an overhead plenum through a perforated ceiling, after having been passed through remote HEPA filters. The functional difference between the two is that the total HEPA filtered ceiling type allows for greater air flow by virtue of providing greater filter surface.

recirculated within the room. The "clean" room part of the name comes from the fact that the air is passed first through prefilters and then through high efficiency particulate air (HEPA) filters that are defined as capable of filtering out 99.9% of all particles greater than 0.3 μm in diameter. This creates a dust-free, sterile air environment.

Mass air displacement "clean" rooms are classified according to the number of particles greater than 0.3 μm per cubic foot of air. Class 100, class 1000, and class 10,000 are commonly used classes. The amount of air displacement required to achieve a particular class depends on the rate of particle generation in the room. Animal rooms are relatively high particulate generators.

Superimposed over the 200–600 changes per hour recirculating within the same room through HEPA filters is a sufficient

quantity of fresh "makeup" air, possibly up to 30 or more fresh air changes per hour. High efficiency particulate air-filtered MADC rooms are highly desirable in barrier housing facilities and in facilities where animals are fed nonvolatile chemical carcinogens. The HEPA filters must be routinely tested to assure their proper function and appropriately certified if they are to be relied upon for controlling hazardous chemical or infectious agents.

B. Programmatic Areas, Their Function and Interrelationships

1. Support Areas

An animal facility can functionally be divided into support areas and animal housing spaces. Support areas may be appropriately dispersed throughout the animal housing areas. Depending on the size of the animal facility, support functions may require 30–50% of the facility space with the remainder being for animal housing. A small facility may include only those areas essential to support of the basic animal care functions; however, for clarity all of the usual support areas are described.

a. Administrative Support Area. A research animal facility is a service-oriented business. Like any successful business, it must be managed effectively and efficiently and will require the support of a typical business office that processes personnel records, payroll records, health records, income/expense statements, balance sheets, cost accounting, animal records, correspondence, research protocol files, required government records, and purchase orders for equipment, supplies, and animals. Animals should be purchased through the animal facility management office in order to assure the high quality of animals purchased as well as to ascertain housing space availability. Delivery should also be made to the animal facility so that animals receive appropriate care immediately upon arrival. Office space is also required for the office management staff as well as the professional and supervisory staff. A library/conference room is needed for conducting training programs for animal technicians, research technicians, and graduate students. A small reference library stocked with basic veterinary texts and periodicals is a valuable resource to the animal facility staff and investigators. If possible the animal facility office should be located to allow access without passing through animal housing areas.

b. Laboratories. Diagnostic laboratory facilities are an essential management tool (Lang and Harrell, 1972). In a comprehensive animal facility, the full component of laboratory diagnostic services is provided. Smaller animal facilities may provide varying segments of diagnostic support and rely on contracts with other laboratories for the remainder. Diagnostic services are required for the purpose of assuring research animal health before and during experiments, and to provide services to investigators requiring assistance with the interpretation of research data. It is both efficient and convenient for the offices, library/conference room, and laboratories to be arranged in close proximity.

c. Support Areas for Animal Technician Personnel. The personnel support areas should include lavatory, shower, locker room, lounge, and possibly a laundry facility.

To reduce the possibility of transmitting infectious agents between the home and work environment, personnel working with animals should change from street clothes to work clothes, and the work clothes should not be taken home to be laundered. It may be economical and convenient to launder uniforms "in house."

Since smoking, eating, or drinking is not permitted in animal areas, lounge facilities should be provided for personnel breaks and lunch periods.

d. Cage Servicing and Sanitation. The cage servicing area is the center of activity in a research animal facility. All movable cages are transported to and from this area often once per week and occasionally 2–3 times per week, therefore, it is essential that it be conveniently accessible to all animal rooms. Ideally, the cage servicing center is divided into three areas: a "dirty" side where the bedding is removed from the soiled cages, a "clean" side where the sanitized cages may be stored until required, and a cage preparation area where the cages are assembled and the clean bedding is placed in the cages or cage pans. In order to reduce the chances for cross-contamination, the "dirty" and "clean" sides are separated with items passing from one side to the other through a variety of washers (described in Section II,D,1). Sufficient space must be provided on the "dirty" side for soiled cages to stand, for bedding to be dumped from cages and cage pans, and for rinsing the cages and pans to remove debris prior to placing them in the cage washer. The "clean" side should provide sufficient space to store cages equal to at least 20% of the total capacity of the facility. Cage sanitation equipment in large facilities often includes a bedding dispenser that automatically supplies the cages with bedding as they exit the washer in the "clean" side. This reduces labor costs and wastage of bedding. The cage servicing area should also include a bedding storage room adjacent to the cage preparation room. A cage maintenance area is desirable.

e. Housekeeping. Housekeeping facilities are required to facilitate the sanitation of animal rooms, support areas, and corridors; such facilities include a storage room for sanitation equipment and supplies and janitorial closets.

f. Feed Services. Feed services include a feed storage room and a feed preparation room. Most laboratory animal feeds are supplied as dry pellets in 22.5 kg (50 lb) bags. It is recommended that the dry pelleted feed storage room be refrigerated to about 4°C (39°F). This serves to preserve nutritional quality of feeds (Fullerton *et al.,* 1982) and to control vermin. Guidelines recommend that dry laboratory animal feeds be stored at a maximum temperature of 15.5°C (60°F) (ILAR, 1978b). Refrigerated space *must* be provided for perishable feeds such as meat, fresh fruits, and vegetables. Most facilities require a room for the preparation of special diets. Such a room may be equipped with grinders, mixers, blenders, etc. Specially equipped hoods are required if carcinogens or other hazardous chemicals are to be mixed with the feed.

g. Animal Procedure Laboratories. Unnecessary activity within the animal rooms must be discouraged. Multi-user procedure laboratories conveniently spaced throughout the animal housing areas provide space for conducting animal procedures. Such rooms may be equipped with ceiling-mounted examination lights, examination table, animal weighing scales, various types of animal restraint devices, etc. Lockable cabinets and drawers may be provided for the convenience of investigators who share the procedure laboratory. Since animals from different rooms will be brought to the procedure laboratories, appropriate cautions must be taken to avoid the transmission of disease between different species and animals of the same species from different sources.

h. Receiving and Shipping. A strategically located and well-designed receiving and shipping area is needed. This area should be dedicated exclusively to the animal facility. A dock should be located and designed to accommodate a variety of truck sizes and should lead to a receiving and shipping area for temporary holding of animals and/or equipment. A room near the receiving area or dock for short-term holding of animals and shipping containers is recommended.

i. Necropsy and Postmortem Storage. A necropsy laboratory is required both by the veterinary staff and by the investigators. Refrigerated space is required for the storage of animal carcasses. Carcasses that are contaminated with radioisotopes may need to be stored for long periods of time to allow for the decay of the isotope or to provide for the accumulation of quantities sufficient for economic disposal. For this purpose, freezer space should be provided.

j. Incineration Room. Frequently, an incinerator for the disposal of animal carcasses and other wastes is provided within the animal facility or off site.

k. Surgery. The surgical suite should be designed and managed in accordance with accepted standards (ILAR, 1978b). In addition to the operating room, it should include separate areas for dressing rooms, animal preparation, the preparation and supply of sterile materials, and surgeon scrub and gowning.

l. X-Ray, Fluoroscopy, and Whole Body Radiation. A comprehensive animal medicine program and support service will include diagnostic X-ray and fluoroscopy. Whole body radiation may be required for experimental procedures.

2. Animal Housing Areas

Animal housing areas may be programmatically divided into categories. Not all animal facilities will require each of the support areas listed above; likewise, all will not require each of the various types of animal housing areas. The following is a functional description of most of the areas that may be required in a comprehensive animal facility.

a. Conventional. A conventional animal room is generally considered one housing animals that do not require special isolation. With the appropriate caging system, this could include any species although some species, such as dogs, sheep, goats, pigs, and nonhuman primates, are more efficiently housed in rooms especially designed for them. Animal rooms generally should not contain built-in equipment other than a sink. Built-in counter tops and casework are undesirable because they reduce flexibility, impede the maintenance of appropriate sanitation levels and provide harborage for vermin.

b. Quarantine. Every animal entering a research animal facility should be isolated or "quarantined" prior to being entered on a research protocol. The quarantine has a twofold purpose: (1) to allow evaluation of the health status and (2) to allow time for animals to recover from shipping and to acclimate to their new environment. Spurious research data may result from studies on diseased or otherwise metabolically unstable animals (Lang and Vessell, 1976; ILAR, 1978b; Davis, 1978; Landi *et al.*, 1982).

The quarantine space should be near the animal receiving area. Quarantine/isolation can most efficiently be done by providing cubicles such as described in Section II,A,4. Random source animals, such as dogs, cats, sheep, goats, and nonhuman primates, should be quarantined in facilities physically separated from quarantine facilities for rodents and rabbits. The length of quarantine will vary according to the species, stock and strain, source, mode of shipment, observations made during the quarantine, and other factors. Each case requires professional judgment (Weihe, 1965; Flynn *et al.*, 1971; Dymsza *et al.*, 1963; Lowe, 1980).

c. Barrier. Barrier technology was first developed for large-scale commercial production of disease-free rodents (Foster *et al.*, 1963); however, it soon became apparent that research animal facilities required a barrier for the purpose of maintaining the animals free of disease (Brick *et al.*, 1969; Simmons *et al.*, 1967; Christy *et al.*, 1968). Animals that are T cell deficient, immunosuppressed, on lifetime aging studies, and in breeding colonies require a higher level of control of the microbial environment than can be efficiently achieved in a conventional animal room. The purpose of barrier housing is to prevent infectious agents from entering, thereby keeping the animals inside the barrier microbiologically "clean," that is, free from known infectious diseases. This assumes that the animals placed in the barrier are initially "clean." Personnel shower and dress in sterile clothes, including head covers, gloves, and masks, before entering the barrier. Animals enter through ports from isolators or transport containers designed to protect the animals from infectious agents. All other materials are sterilized by chemicals, gas, or steam before entering. Depending on the size of the barrier, a cage washing facility may be included, or cages may be washed outside the barrier and sterilized in a "pass through" autoclave. In addition to the structural barriers, air flow patterns are important. The relative air pressures must be balanced so that small quantities of air move from the cleanest area to the most soiled area. Figure 5 is an example of a typical barrier. An additional level of security can be obtained by designing the animal rooms as rooms with cubicles or as mass air displacement "clean" rooms (Figs. 2–4).

If new animals are to be routinely entered into the barrier it is necessary to provide quarantine space within the barrier. Critical assessment must assure that the animals are "clean" before placement in the barrier. The quarantine space within the barrier provides the opportunity to evaluate the new animals without jeopardizing those already in the barrier. If only small space is needed for barrier housing, plastic isolators of the type used for housing germfree animals may be used in conventional animal rooms (Fig. 6).

A modest barrier, useful for some situations, can be achieved in conventional rooms by covering shoebox type cages with filter material (Kraft *et al.*, 1964; Schneider and Collins, 1966). When augmented with management procedures, such as isolation and cage, bedding, feed, and water sterilization, highly disease susceptible animals, such as thymus-deficient "nude" mice, may be successfully maintained in a conventional room (ILAR), 1976b). Portable laminar flow racks or tents used in a conventional animal room may also provide a barrier when appropriately managed (ILAR, 1976b).

d. Biohazard Containment. If animals are to be exposed to known infectious microorganisms, containment is required [Gerone, 1978; National Institutes of Health (NIH), 1979].

Fig. 5. Typical barrier facility. (Courtesy of Charles River Laboratories, Inc.; from Otis and Foster, 1983.)

Chapter 20 discusses the classification, containment, and management procedures of various microorganisms, based on the degree of risk. Depending upon the degree of risk, containment may be provided for small numbers of animals by housing them in negative pressure isolators with filtered exhaust or reverse flow HEPA filter cage racks or filter tops on shoebox cages (Henkel, 1978). Large numbers of animals require a special area, with multiple animal rooms, designed and managed to contain the hazardous microorganisms within the area. In contrast to the barrier area, all cages and materials are sterilized in a "pass through" autoclave out of the area, and people change clothes before entering and shower before leaving. The physical arrangement and engineering features of the facility should be such as to prevent cross-contamination of the different groups of animals within the area. This may be accomplished by utilizing the "clean–dirty" corridor concept (Fig. 1C). In addition, housing of animals in cubicles (Figs. 2 and 3) within the animal rooms also provides containment. Effective containment relies to a great extent on the relative air pressures and air circulation. The air pressure should be balanced such that the highest pressure is at the least contaminated area (outside of the containment area) to the most contaminated area, which might be considered the soiled corridor of the containment area or the cubicles in which the animals are housed. Careful consideration must be given to the maintenance of mechanical equipment supporting the biohazard containment area. For example, "bag out" type filters, that can be changed without exposing the operator, should be provided, and control panels and mechanical equipment should be located outside the barrier whenever possible. Stainless steel sealed exhaust ducts and decontamination of exhaust air by filtration or incineration may be required for some applications, while dilution of the air from the containment area with air from other areas may be a satisfactory method of contamination control. It is important to recognize that even the most ideal containment facility has limitations (Barkley, 1978).

e. Chemical and Radioisotope Containment. As with infectious containment, the objective is to contain the hazardous agent and to prevent cross-contamination. In addition to management practices, the physical characteristics of the area are important (Newberne and Fox, 1978; Henkel, 1978; NIH, 1979). The cubicle concept is useful for the ability to provide multiple isolated animal housing spaces in a small area, thereby limiting the potential for cross-contamination to a small area. Mass air displacement rooms also may be used for the control of nonvolatile hazardous agents. Relative air pressures

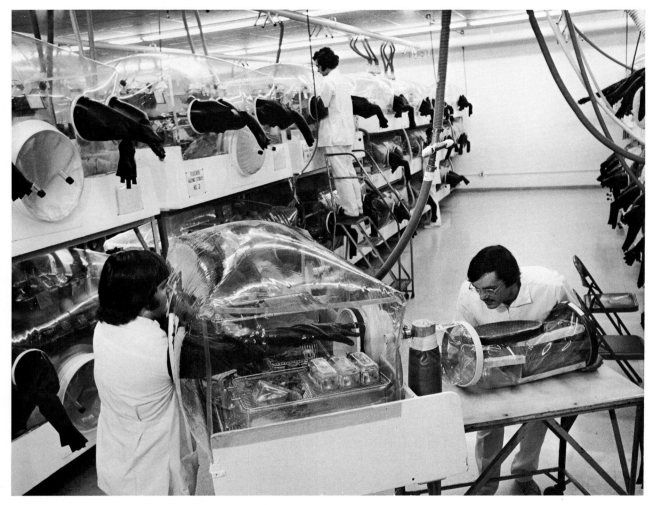

Fig. 6. Isolator facilities such as this have been used by large commercial breeding laboratories since the late 1950's to maintain foundation stock under germfree conditions and to rear gnotobiotic rodents. (Courtesy of Charles River Laboratories, Inc.)

must cause air to flow from outside the containment area to the most contaminated site within the containment, such as the animal housing cubicle.

Decontamination of cages can usually be safely accomplished with the use of conventional mechanical cage washers (Fox and Helfrich-Smith, 1980; Fox *et al.*, 1980). The chemical and radioisotope containment area should be adjacent to the dirty side of the cage sanitation area so that it is not necessary to transport contaminated cages through the corridor. Also recommended is a separate room wherein the contaminated bedding can be removed from the cages or pens inside a laminar air flow cabinet in which the aerosolized contaminant is drawn away from the operator into a HEPA filter (Fig. 7). See Chapter 20 regarding the use of hazardous agents in animal facilities.

f. Nonhuman Primate Housing. Nonhuman primates frequently are housed in conventional animal rooms adjacent to other laboratory animal rooms. However, because of their relatively "dirty" microbiological status as compared with laboratory rodents and their potential for carrying zoonostic diseases (see Chapter 22), the ideal arrangement is to house them in a low risk containment area isolated from other animals and personnel. Care must be taken to assure that the animal room air pressure is maintained negative to the corridor.

g. Canine Housing. Canine housing areas should be isolated from other animal housing areas and from human occupancy areas. This is because of the "dirty" microbiological status of the dog as compared with laboratory rodents and because of the noise level generated by the dog. Special laborato-

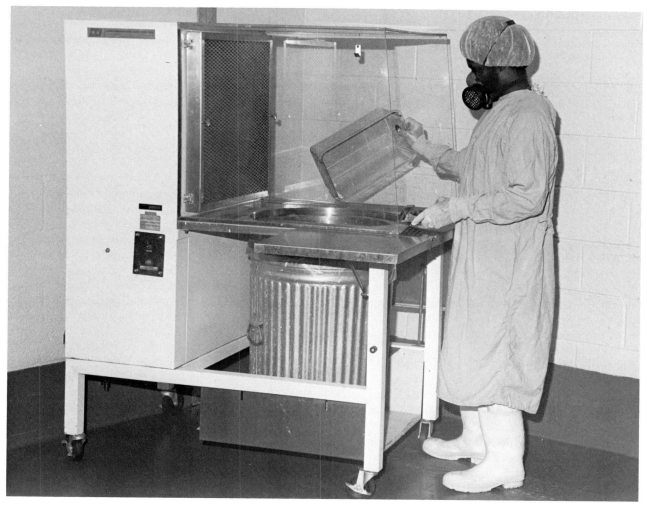

Fig. 7. Portable bedding disposal cabinet that draws aerosols away from the operator into HEPA filters.

ry animal procedure space should also be provided in the canine area. If experimental surgery will involve the use of dogs, it is desirable for the surgery area to be near the canine area. Specially equipped separate rooms should be provided for postoperative housing and care. The canine area should also be near the animal receiving area.

h. Large Animal Housing. Sheep, goats, and pigs are common laboratory animals. While it is possible to house them in conventional animal rooms, with or without floor drains, they can most efficiently be housed in specially planned facilities. A suitable location is adjacent to the canine area.

i. Environmental Control Rooms. Some research requires that animals be housed in rigidly controlled environments at temperature and humidity extremes that cannot be accomplished in conventional animal rooms. Specially designed environmental control rooms may be required for this purpose. Portable environmental cabinets may be satisfactory.

C. Facility Design Specifications

The architectural and engineering specifications must be planned to provide a sanitizable, functional, controlled environment. In this section are some of the more important design features of a research animal facility, most of which are applicable to both new construction and renovation projects (Windman and Zigas, 1978; ILAR, 1978b; Henkel, 1978).

1. Heating, Ventilation, and Air Conditioning (HVAC)

The most important programmatic requirements in an animal facility relate to the HVAC. The objective must be to deliver clean, temperature- and humidity-controlled air to the animal rooms in sufficient quantities and in such a way as to maintain a constant environment relatively free of noxious chemicals and infectious agents (McPherson, 1975).

a. Air Quality. The quality or purity of the air is determined by its source and degree of filtration. Special care must be taken to assure that supply air is not contaminated by exhaust air from the animal facility, incinerator smokestacks, chemical hoods, hospital wards, etc. The degree to which incoming air is filtered varies with the type of facility. For example, the air entering a barrier facility is generally passed through HEPA filters, whereas that supplying a dog kennel may not be filtered at all. Animal facility air handling units should be separate from those that supply human occupancy areas. Ideally, dedicated cooling and heating sources should be provided for the animal facility and, for emergency back-up and maintenance purposes, should be cross-linked into other sources. Air supplied to animal rooms should be 100% fresh air (nonrecycled) in order to avoid cross-contamination by infectious agents and to dispose of irritating and volatile chemicals, primarily ammonia (which is generated from animal urine). The number of fresh air changes required in an animal room has been extensively debated. However, at the time of this writing the "Guide" (ILAR, 1978b) recommends 10–15 fresh air changes per hour. The effectiveness of room ventilation relates not only to the rate of air exchange, but also to the rate of ammonia production and air flow patterns within the room (Briel *et al.,* 1971; Kruckenberg, 1971). Ammonia production depends on factors such as density and species of animals in the room and the sanitation level.

b. Air Flow Patterns. The air flow pattern is a design consideration that must be given careful attention, with the objective being to distribute the air evenly throughout the room without drafts and "dead" air pockets. Although quite expensive this may be effectively accomplished by supplying air through the ceiling (as is done for mass air displacement systems) and exhausting it at floor level. Other methods for supplying, diffusing, and exhausting air to and from a room are numerous and vary in effectiveness (Nevins and Miller, 1972; Woods, 1975). A frequent solution is to supply the air through one or more louvers at the ceiling and exhaust it through one or more louvers at the floor level.

Relative air pressures within the animal facility must be carefully balanced. Air pressures within barriers and hazardous containment areas were previously discussed. In a conventional animal facility with two- or three-corridor arrangements,

air should flow from the clean corridor to the animal rooms to the soiled corridor and then to the dirty side of the cage sanitation facility. With either type of single corridor system, the corridor should be positive to the animal room, although some believe that the room should be positive to the corridor. Convincing arguments can be made on both sides of the issue, with the "right" answer depending on programmatic requirements. In general, support areas should be negative to the corridor, especially the necropsy and cage sanitation areas. Of course, aseptic surgery should be positive to the corridor.

c. Temperature and Humidity Control. The degree of temperature control depends on the programmatic needs of the facility. For a rodent-breeding facility, for example, a desirable temperature may be selected and the system designed for maintaining that temperature around the tolerance of $\pm 1^\circ$C ($\pm 2^\circ$F) in all rooms. It is advisable to provide for individual room control unless it can be guaranteed that animal population densities (and, therefore, heat generation within all rooms controlled by a single thermostat) will remain constant. In a research animal facility where flexibility is required to accommodate a variety of animal species at various population densities, it is essential that the animal room temperatures be individually controlled to within 1°C (2°F) for any temperature in a range of 18°–29°C (65°–85°F).

For most applications, the ability to maintain relative humidity between 30 and 70% is satisfactory (ILAR, 1978b). Individual room humidity control is desirable; however, zone control may be acceptable so long as care is taken to group the zones according to whether a wet or dry sanitation system is to be routinely used. Special attention should be given to humidity control in rooms where flush racks may be used.

d. Energy Conservation. Serious consideration must be given to methods of conserving energy. One way is to reduce the necessary air exchange volumes. This is most effectively accomplished with ventilated racks where the animals' microenvironment, the cage, can be changed. The advantage of such racks is obvious. The disadvantages are reduced flexibility, problems of balancing the ventilation system, inconvenience in working with the racks, difficulty in adequately sanitizing the racks, and the initial high cost of the racks and room hookups. The cubicle arrangement also may reduce ventilation requirements. Considering that the central corridor space in a room with cubicles merely serves as a supply plenum to the cubicles, the total volume of air that needs to be exchanged is only that within the cubicles. Depending on the actual room arrangement, the volume of air within the cubicles may represent only 50% of the total room volume, thus decreasing the ventilation requirements by 50%. Numerous types of energy recovery systems, such as heat pipes, lithium chloride enthropy wheels, "run around" coils, and direct heat transfer loop systems may also be considered for energy conservation (Gorton, 1975, 1978); however incoming air must not be contaminated by outgoing air. Recirculation of animal room air is another possible method of conserving energy, but the air must be carefully cleansed. To use this system the recirculated air is HEPA filtered to remove particulate contaminants, and the gaseous contaminants, primarily ammonia, are removed by passing the air through an oxidant bed such as alumina pellets impregnated with potassium permanganate. Ammonia, being water soluble, also can be scrubbed out of air by saturating it with water and then removing the water (Jeszenka *et al.*, 1981). Because of intensive maintenance requirements and potential for contamination during malfunctions, recirculating type air systems should be carefully studied before a decision is made to use them.

2. Power and Lighting

Emergency power should be available to keep essential services operational in the event the main power fails (ILAR, 1978b). As a minimum, emergency power should be sufficient to maintain ventilation in animal rooms and germfree isolators, and sufficient lighting in corridors and animal rooms to allow for routine care of the animals.

Power outlets to animal rooms, cage wash and other areas where water is used in cleaning, should be waterproof.

Light is a crucial component of the animals' environment that requires careful consideration (Bellhorn, 1980). Lighting in animal rooms should be uniform throughout. Fluorescent lighting is generally the most efficient type with cool white or warm being the most common. Full spectrum tubes more closely mimic natural light, but are more expensive.

The significance of full spectrum lighting for research animals has not been fully documented, but there is evidence that it can alter physiologic responses (Mulder, 1971; Burns *et al.*, 1976; Salterelli and Coppola, 1979). At the time of this writing, "precise lighting requirements for maintenance of good health and physiological stability of animals are not known"; however, the recommended intensity of lighting for animal rooms (ILAR, 1978b) is 75–125 fc (807–11,345 lumens/m²) measured 1.8 m (6 ft) from the floor. There is increasing concern that high levels of lighting, considered adequate for a work environment, are excessive for the animals, causing retinal degeneration in some rodent species (Robinson and Kuwabara, 1976; Robinson *et al.*, 1982). The light–dark cycle of animal rooms should be individually and automatically controlled to assure consistency. The light–dark cycle is known to affect estrous cycles in laboratory rodents (Haus and Halberg, 1970; Haus *et al.*, 1967). Windows in animal rooms are considered undesirable in that they eliminate the ability to maintain constant light levels and light–dark cycles.

3. Interior Surfaces

Interior surfaces must be durable and, above all, sanitizable (ILAR, 1978b; Thibert, 1980). The ideal floor is monolithic, chemical and stain resistant, slip resistant even when wet, yet relatively smooth and easy to sanitize, does not require sealers or waxes to maintain an acceptable appearance, and must be capable of supporting equipment without becoming gouged, cracked, or pitted.

Walls should have curbs or guard rails or be sturdy enough to withstand frequent contact with movable equipment without sustaining serious damage. They should be smooth and capable of withstanding scrubbing, cleaning, and disinfecting agents and impact from high pressure water. Concrete blocks or cement plaster walls coated with epoxy paints have been found to satisfactorily meet these requirements. All penetrations through the walls must be sealed as should the ceiling to wall and wall to floor junctions. Ideally, the monolithic flooring material should form a coved 15 cm (6 in.) wall base.

Ceilings, just as the walls, should be resistant to cleaning and disinfecting agents and capable of withstanding high impact with high pressure water. Cement plaster is most desirable in high moisture areas, such as cage sanitation rooms and animal rooms that receive daily cleaning with water. Dropped ceilings with "laid in" tile are not satisfactory, even in corridors, since such ceilings are not readily sanitized and and because they provide harborage for vermin. Exposed pipes and fixtures are undesirable. Unless there are special programmatic requirements, ceiling heights (for energy conservation reasons) should not be greater than 9 ft. For example, a 360 cm (12 ft) ceiling as compared to a 270 cm (9 ft) ceiling, increases by 33.3% the initial cost for heating and cooling capacity and increases operational cost for the duration of the facility.

4. Doors

For safety reasons doors should open into rooms unless a recessed vestibule is provided. A standard animal room door should not measure less than 107 cm (42 in.) wide and 213 cm (84 in.) high and should seal tight enough to prevent the entrance of vermin. Metal or metal-covered doors are preferred and should be constructed to eliminate vermin access to the interior of the doors. Doors should be protected with armor plates (kick plates) and edge guards. Hardware should be recessed and the doors should be equipped with a self-closing device. A small view panel is desirable. The door frame should be metal and preferably have hospital stops. These are frames on which the stops extend to within 15 cm (6 in.) of the floor.

The doors to the food and bedding storage areas should be of sufficient width to accommodate a freight pallet. Wide [183 cm (6 ft)] automatic doors are recommended in areas of high equipment and materials flow, such as the cage sanitation area, the receiving and shipping area, and any place in corridors where doors are required. Where corridor doors are required for fire safety reasons, they should be equipped with hold-open devices that automatically release when the fire alarm is activated.

5. Plumbing and Drainage

Floor drains in rodent or rabbit rooms are not essential. Adequate room sanitation can be maintained with wet vacuuming and/or mopping. However, for program flexibility, drains are frequently provided. If flush racks are to be used, the drain should be located to accommodate the system, and the floor should be flat and not sloped toward the drain. Drains that will be used intermittently should be equipped with caps and seals to prevent the backflow of sewerage gases. Animal room floor drains should not be less than 10.2 cm (4 in.) in diameter. In heavy use areas, such as the cage sanitation room and dog kennels, rim flush drains at least 15.3 cm (6 in.) in diameter are recommended.

In areas where hoses will be used for routine cleaning and sanitation, hose reels can greatly facilitate sanitation and increase efficiency. Floors in these areas should be pitched a minimum of 0.64 cm/m (0.25 in./yd) and in dog kennels, a pitch of 0.64 cm/0.3 m (0.25 in./ft) is highly desirable.

A sink in every animal room is needed to facilitate room sanitation as well as for hand washing. A small side shelf work area attached to the sink is optional. A hose bibb faucet placed under the sink is useful for filling mop buckets.

If automatic watering is to be used, appropriate cold water supply, pressure reduction stations, and water supply manifolds must be provided near the ceiling of the animal rooms. Otherwise, the undersink faucet may later serve as a cold water supply source to facilitate installation of automatic watering equipment.

6. Vermin Control

A major management challenge is to control vermin without the use of chemicals. Chemicals create an unacceptable variable in the animals' environment. Many insecticides are potent hepatic microsomal enzyme inducers and may alter research data obtained from animals exposed to insecticides (Fouts, 1970). The most effective method of control is to seal vermin out and eliminate hiding and nesting places within. Caulking all cracks, sealing around all utility penetrations, lights, and light switches is essential. Hiding places can be eliminated by building nothing into animal rooms but the essentials, such as a sink, small side shelf, soap and paper towel dispensers. Even these must be installed to eliminate hiding places for vermin by either caulking around them or mounting them away from the wall to facilitate routine cleaning behind them. If rooms require

casework, it should be of an open design that reduces hiding and nesting places for vermin and facilitates cleaning.

7. Noise Control

Excessive sound creates an unacceptable variable for research animals. The stress response that research animals may have to noise has the potential to influence experimental data (Peterson, 1980). Noise production in an animal facility is unavoidable. Therefore, the noise generating areas must be identified and appropriate measures taken to reduce exposure of animals and personnel as much as possible. The primary noise generators are the cage sanitation facility, dogs, and to a lesser extent, nonhuman primates. Effort must be made to contain the noise by isolating the activities away from personnel and the animal housing areas. This may be accomplished by such design features as double entry doors, soundproofed walls, by designing locations of corridors and support areas around the noise generating areas, or by locating the noise generating areas on exterior walls or adjacent to mechanical spaces. Conventional acoustical materials present sanitation and vermin control problems that are best avoided. Cage changing and other activities within animal rooms must be conducted in a manner to reduce the noise generation as much as possible.

8. Communications

Communication considerations include such things as telephone, intercom, hard-copy systems between animal facility offices and satellite or dispersed facilities, and computer connections. Intercom access to every room in the facility is essential. The volume and tone of the sound must be such that it does not stress the animals. Two-way communication capacity in each room is desirable. Two-way communication at least in every corridor of a barrier or biohazard containment area is essential. Hard-copy communication is a valuable management tool for delivering important detailed messages that have a tendency to get lost if delivered by courier or misunderstood over the telephone. Hard-copy communication reduces errors primarily by providing a means of documenting the source of errors.

9. Security

The great value of research animals, along with the fact that animal research facilities are increasingly being maliciously vandalized, demands that adequate attention be given to security. The type and effectiveness of security systems will vary according to the layout of the animal facility and its relationship to other areas. Ideally, the facility should be isolated such that all access doors can be automatically locked at whatever times are appropriate and with access being limited to authorized personnel through any of a number of access identification systems. Locks in individual rooms are also desirable, since frequently it is necessary to assure limited access to animal rooms for reasons related to the research project.

10. Corridors

Consideration must be given to the fact that a great deal of materials and equipment routinely travel through the corridors of a research animal facility. To accommodate this activity, corridors should be at least 2.1 m (7 ft) or, preferably, 2.44 m (8 ft) wide. The walls should be protected with curbs, guardrails, or bumpers and the corners should be appropriately protected with steel corner plates or other durable material.

11. Elevators

If animal facility service elevators are required, they should be dedicated to this purpose. One elevator for soiled cages and one for clean cages and supplies, of sufficient size to hold 3–5 racks of cages, should be provided. Having two elevators also provides for emergencies when one elevator is out of service for repairs or routine maintenance.

D. Equipment

1. Sanitation Equipment

Appropriate equipment is essential in order to maintain the high level of cleanliness and sanitation required. In order to adequately sanitize cages that are hand washed, disinfectants are used which may leave unacceptable chemical residues on the cages. For this reason, mechanical cage washers are considered essential to adequately sanitize cages without use of disinfectants. This is done by exposure of the cages to temperatures in excess of 82.2°C (180°F) for a period of time adequate to destroy vegetative pathogenic organisms (ILAR, 1978b).

Two basic types of cage washers are commonly used. One is a batch type washer in which the cage racks and/or cages are placed in a chamber and subjected to a high-volume, high-pressure spray on all sides through a series of cycles. A final freshwater rinse is needed to decrease the amount of detergent residue on the cages. Depending on program needs, the washer may have the capacity to either hold or to dump the wash water. These machines are available in a variety of sizes. The other commonly used type of washer is a conveyor or "tunnel" type washer, especially useful for sanitizing shoebox cages, cage pans, water bottles, and miscellaneous small items of equipment. In tunnel washers, the items are transported on a continuously moving conveyor belt through various sections

for prerinse, detergent wash, rinse, final freshwater rinse and, possibly drying. In tunnel washers all the water except the prerinse and the final fresh water rinse is usually recycled. Ultrasonic type cleaners are also useful for cleaning cages, especially for removing mineral deposits that accumulate on the cages from urine precipitates. A more common method used in small facilities for removing mineral deposits, is to soak in an acid solution. Animal facilities that do not have a tunnel washer may require a batch type bottle washer. Facilities that utilize a large number of watering bottles may have a separate tunnel type washer dedicated to sanitizing water bottles.

Other types of sanitation equipment commonly used in an animal facility are trash can cleaners, high pressure sprayers, steam generators (primarily for use in outside pens), various types of floor scrubbing equipment, and vacuum cleaners. Conventional portable vacuum cleaners should not be used in animal rooms unless the exhaust is HEPA filtered to avoid the spreading of dust and infectious agents throughout the room. Central vacuum systems that exhaust outside of the animal facility or at least in a separate room, eliminate this problem but present special maintenance problems with keeping the vacuum lines open. An autoclave is also essential for a modern research animal facility. It is required not only for sterilizing surgical instruments and supplies, but also for sterilizing feed, water, bedding, cages, and supplies for the maintenance of immunocompromised animals. An ethylene oxide sterilizer may also be required for the sterilization of items that cannot be autoclaved.

2. Cage Systems

A wide variety of fixed and portable caging systems are used for housing laboratory animals. The most common type of fixed caging is referred to as pens, or "runs," or kennels. The animals are housed either directly on the floor or on raised floors of expanded metal or a similar material. With appropriate modifications for the various species, they may be suitable for housing a variety of animals including dogs, cats, nonhuman primates, sheep, goats, pigs, adult chickens, ducks, and geese. A disadvantage of fixed caging systems is that they are difficult to sanitize.

The majority of laboratory animal caging systems are portable. Cages are either suspended from or resting on racks equipped with casters. This allows for maximum flexibility in use of the animal rooms, facilitates cage cleaning and sanitation, and makes it possible to use the rack and tunnel type washers referred to in the previous section. There are two basic types of cages used with the portable racks. One type has four solid sides and a solid bottom and is generally referred to as the shoebox cage. The other type has perforated bottoms or walk floors, usually wire mesh or metal slats, that allow the urine and feces to fall into a collecting pan or tray. The pan is either flushed regularly with water or filled with a noncontact type bedding material to absorb the urine and fecal moisture. With shoebox cages, the animals are housed directly on the bedding material, which is generally referred to as contact bedding. The shoebox cages may be suspended from a shelf formed of either stainless steel wire mesh or perforated sheet metal which forms the top of the cage (Fig. 8), or be covered with individual tops formed of either stainless steel wire or perforated sheet metal (Fig. 9). The chapters in this volume on the various species detail the type of caging systems most suitable for each species. The "Guide" (ILAR, 1978b) lists the cage size recommendations for each species.

The most common type of material used for constructing cage racks is type 304 stainless steel, although aluminum and galvanized steel are also used. Cages are usually stainless steel or plastic, such as polycarbonate, polypropylene, or polystyrene. Plastic cages, depending on the material, are either transparent or opaque. The advantage of stainless steel shoebox cages is that they are durable and provide seclusion as do opaque plastic cages. The advantage of transparent plastic cages is that they facilitate the observation of the animals. Although undocumented, there is concern that the lack of seclusion in transparent cages may cause distress to the animals. Plastic shoebox cages cost less than stainless steel, although they require more frequent replacement, and, because they are lighter in weight, they are easier to work with. Galvanized metal is undesirable because urine and caustic cleaning agents cause corrosion.

The activity of animals on a contact bedding generates dust and airborne microorganisms. Filter material placed over the shoebox-type cages reduces the spread of airborne disease (Kraft *et al.*, 1964) and has been used extensively with mice and rats (Fig. 9). However, filter tops have a disadvantage in that they alter the microenvironment within the cage, allowing

Fig. 8. Suspended shoebox-type caging system.

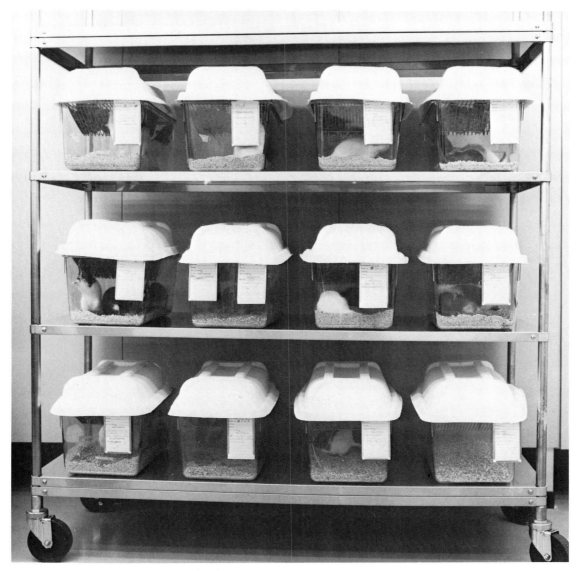

Fig. 9. Shoebox-type caging system with filter tops.

temperature, humidity, and concentration of gases, especially carbon dioxide and ammonia, to gradually increase (Murakami, 1971; Serrano, 1971). The activity of hepatic microsomal enzymes decreases in rodents housed in cages with filter tops (Vesell *et al.*, 1973), and the higher ammonia levels serve as a respiratory irritant leaving the respiratory system more vulnerable to infection (Anderson *et al.*, 1964; Broderson *et al.*, 1976; Gamble and Clough, 1976). Because of this concentration of gases in filter top cages frequent litter changes are necessary. Laminar flow racks (McGarrity and Coriell, 1973), laminar flow tents, and individually ventilated cages are available (see Section II,C,1,d).

Automatic flush racks may be used with suspended wire mesh or slatted bottom cages where the urine and feces is automatically washed away on a timed schedule (Fig. 10). The most common type of flush rack is the "cascade" type. Water supplied at the top of the rack cascades down from shelf to shelf, cleaning each shelf as it flows. After flushing the bottom shelf, the water flows into a floor drain or into a drain trough. For flush racks, the animal room must be equipped with appropriately located drains and the floor must be level. In addition, the HVAC system must be adequate to maintain room humidity at an acceptable level. The advantages of the flush rack are that ammonia does not form, since excreta are removed frequently and the cost of bedding is eliminated.

Germfree, portable isolators are the most effective method of

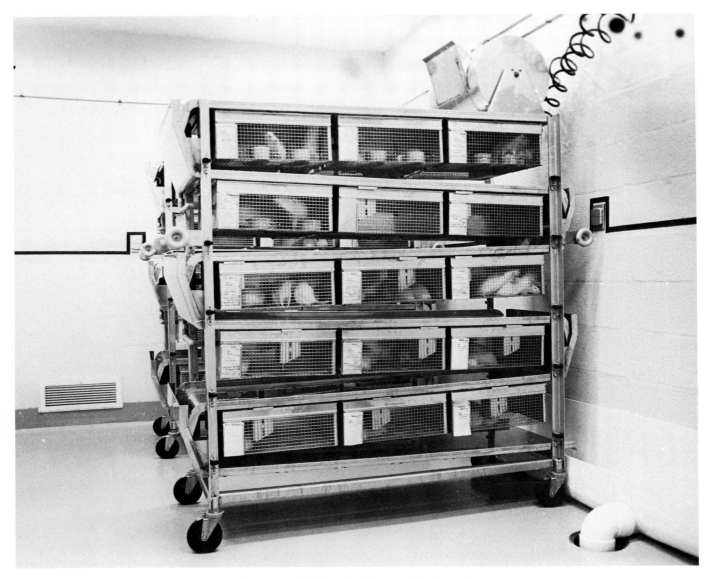

Fig. 10. Automatic flush racks with suspended wire mesh cages.

controlling the microbial environment of animals. These are sealed glove boxes formed of transparent rigid or flexible plastic maintained under positive pressure in which animals are housed in cages (Trexler, 1963). Earlier models were made from stainless steel with glass view ports (Reyniers, 1957). All supplies and equipment as well as the air entering the isolator are sterilized. The operator performs tasks in the isolator from the outside through gloves attached to portholes in the sides of the isolators. In addition to being used for maintaining germfree and defined flora animals, the rigid type isolators can be used as a containment for housing animals exposed to hazardous, infectious, or chemical agents. When used as a con-

tainment, the isolator pressure is maintained negative to the room environment, and the isolator exhaust air is filtered or otherwise sterilized. The maintenance of animals in these types of isolators is labor intensive and, therefore, expensive. With appropriate management, rigid plastic isolators may be effectively and efficiently used for maintaining a pathogen-free (not to be confused with germfree) environment for immunocompromised animals even when opened daily for routine servicing.

Some miscellaneous types of caging systems include metabolism cages designed to collect and separate feces and urine, restraint cages designed to immobilize the animals with minimal physical discomfort, and exposure chambers designed for exposing animals (usually housed in wire cages) to gaseous and particulate airborne chemicals.

E. Facility and Equipment Life Cycles

Both facilities and equipment have a limited life time and will periodically need to be refurbished, updated, or replaced. The best of new equipment and facilities require routine maintenance. Preventative maintenance schedules must be established for all mechanical equipment in an effort to avoid unexpected disruptions of service. Room surfaces will require periodic refinishing in order to maintain them in a sanitizable condition. The floors in certain areas, such as dog pens and the dirty side of cage wash, are particularly difficult to maintain and may require more frequent attention than other areas.

As the degree of sophistication of animal research increases, the animal facility performance specifications evolve to support the research. Eventually, most facilities reach a point where they no longer are able to meet the demands of contemporary research and require major renovations to bring them up to standards.

III. MANAGEMENT

A. Personnel

Qualified personnel are essential in managing a research animal facility. If poorly staffed, the most ideal physical plant cannot be managed successfully. A variety of personnel skills are required, including office and business management, facilities and equipment maintenance, laboratory support, veterinary and veterinary technician, and animal technician skills. Even the smallest facility will require personnel knowledgeable in all of these areas. Often, a single individual may provide service to several areas. Whether the personnel providing these services are on the staff or are supplied from outside the department or even outside the institution will depend on the size and scope of the animal research program, the philosophy of management, and the availability of adequate services outside the department.

Qualifications of office management, facilities maintenance, and laboratory technician personnel are self-evident. Three veterinary specialties are commonly represented on an animal facility staff, i.e., laboratory animal medicine, veterinary pathology, and veterinary surgery. Other specialties, including microbiology, clinical pathology, and anesthesiology, may

also be represented. Laboratory animal medicine is a comprehensive specialty that includes a wide variety of veterinary and management skills. Postdoctoral training and/or experience is essential to becoming a laboratory animal medicine specialist. A comprehensive description of laboratory animal medicine training and institutions where training is provided is available (Cohen *et al.,* 1979). The American College of Laboratory Animal Medicine (ACLAM), a specialty college of the American Veterinary Medical Association, has certified veterinarians as Laboratory Animal Medicine specialists since 1957. Comparative pathologists usually are veterinary pathologists who specialize in pathology of research animals. Their qualifications frequently include certification by the American College of Veterinary Pathology, or ACLAM, or both.

Animal technicians and animal technician supervisors are personnel of great importance in any animal research facility. This group has primary responsibility for control of the animals' environment, which is critical to quality research. Not only are they responsible for the conduct of the feeding, watering, and sanitation programs, they also control, influence, or monitor other environmental factors, such as noise, chemicals (insecticides, deodorants, disinfectants, detergents, and other cleaning agents) temperature, humidity, and lights. They are in position to be the first to detect physical plant or equipment problems, and, in addition, they have more contact with the animals than other individuals, including the investigators. They must be keen observers of animal behavior and disease symptoms and look for signs that indicate an abnormal state that must be communicated to the veterinarian and/or investigator. Senior level animal technician staff, especially, must have knowledge of the biology and behavior of research animals and have the skills and knowledge necessary to administer medication by various routes, administer anesthetic agents, take blood samples, etc. They must also have a working knowledge of the research protocols, especially those protocols that may alter the animals' biology or behavior and/or require individualized care, such as special diets, water, treatments, and light control.

Formal high school training or post–high school training in laboratory animal science is not readily available. Therefore, it is essential that on-the-job training programs be established. The American Association of Laboratory Animal Science (AALAS) has a variety of training materials, including slides and tape presentations, books, manuals, and films. In addition, AALAS conducts a nationwide testing and certification program at three levels: Assistant Laboratory Animal Technician, Laboratory Animal Technician, and Laboratory Animal Technologist. Many animal facility managers effectively utilize this certification program to establish a career ladder for animal technicians whereby promotion and pay increases are based on achieving certification at the various levels. This allows animal technicians to be promoted, based on their technical knowledge and skills, without necessarily requiring performance of supervisory responsibilities. Such a program not only motivates the animal technician to increase knowledge and skills, it also significantly upgrades the quality of animal care. Animal technician supervisory staff frequently come from the ranks of the animal technicians and/or Bachelors and Masters degree programs in animal sicence, biology and zoology. Veterinary animal health technician training is also a valuable educational background for research animal technicians and animal technician supervisors (ILAR, 1978b).

B. Feed, Water, and Bedding

Feed, water, and bedding are important parts of the animals' environment that must be controlled. Of these, feed presents the greatest challenge and is one of the most often neglected variables in the research animals' environment (Newberne and Fox, 1980; Newberne, 1975). Standard laboratory animal diets, like animal production diets, are made up of natural ingredients. They differ in that special effort must be made to use only ingredients that contain no more than minimal acceptable levels of chemical contaminants, such as insecticides, herbicides, PCB's, aflatoxins, heavy metals, antibiotics, and natural biologically active compounds such as estrogen. In addition, the laboratory animal diet must be consistently and carefully balanced. Generally, in order to provide the necessary quality control, commercial laboratory animal diets are prepared in feed mills dedicated to producing only such feeds.

The nutritional requirements of laboratory animals have been published by the National Research Council (NRC, 1978).

There are two basic types of laboratory animal diets, closed formula and open formula. The *closed formula* type is the most common and guarantees a certain level of basic nutritional ingredients according to the nutritional requirements for the intended species (which may be achieved by balancing various available ingredients). Thus, while the level of amino acids, fats, carbohydrates, vitamins, and minerals remain relatively constant, the basic ingredients making up the diet may vary from mix to mix. *Open formula* diets contain a prescribed quantity of each ingredient. This assures consistency of ingredients, but not necessarily a consistent level of basic nutritional value, although a guaranteed analysis is usually given. For example, the nutritional value of feed grains may vary from area to area according to the climate and soil and from season to season. ''*Certified*'' laboratory animal diets are available at a significantly higher cost. These diets are analyzed for various nutritional ingredients and numerous potential contaminants after the diet has been prepared. Results of the analysis is supplied with the feed.

When more rigid control is required than can be supplied with natural product diets, such as for nutritional studies, semi-

purified and chemically defined diets are used. A *semipurified* diet might use casein as a source of protein, sugar or starch for carbohydrates, vegetable oil for fat and some form of cellulose as crude fiber to which chemically pure inorganic salts and vitamins would be added as required. *Chemically defined* diets carry it one step further, using chemically pure compounds, such as amino acids, sugars, triglycerides, essential fatty acids, inorganic salts, and vitamins (Newberne and Fox, 1980). While the chemically defined diet is more expensive than the semipurified diet, both are too expensive for routine use.

The quality of feed also relates to its age and how it is stored between the mill and the time it is fed. Generally, laboratory animal diets are considered acceptable up to 90 days after milling (ILAR, 1978b). After this, there is concern that the nutritional value may have decreased to suboptimal levels. Feed must be stored in a manner that will eliminate contamination with infectious agents, chemicals, insects, and other vermin. Refrigerated storage is highly desirable. Fullerton *et al.* (1982) reported that the vitamin A, thiamin, degree of rancidity, bacteria, and mold content of both semipurified and natural ingredient diets remained at acceptable levels when stored for 168 days at 4°C (39.2°F). Refrigeration not only serves to preserve nutritional quality, it also effectively controls vermin.

Careful consideration must also be given to water quality (Newell, 1980). Municipal water supplies have been considered adequate and continue to be the most frequent source of water for laboratory animals. However, there is increasing concern for the experimental variables that may be introduced through the water, either in the form of minerals or organic chemical contaminants, not generally considered harmful to the public, that are found in municipal water supplies. Water quality may be especially important in long-term, chronic, toxicity studies or life-span studies. Because of this concern, various methods of water purification including reverse osmosis, deionization, and ultrafiltration are being used with increasing frequency. Bacterial contamination of the water must also be controlled. Water bottles and automatic watering systems must be properly sanitized on a regular basis because as animals drink, the drinking devices and the water may become contaminated. Bacterial growth in water bottles may be controlled either by acidification or hyperchlorination of the water (McPherson, 1963; Beck, 1963), although both represent experimental variables that may affect the research data (Fidler, 1977; Hall *et al.*, 1980; Hermann *et al.*, 1982). Bacterial contamination of automatic watering systems may be controlled by recirculating the water through ultraviolet sterilization units or filters or by intermittently flushing the system with fresh water.

Animal bedding must be carefully selected and controlled (Kraft, 1980). The purpose of bedding is to absorb moisture and to provide comfort, insulation, and nesting material. Two basic types of bedding are used: contact and noncontact. *Contact* bedding is a type where the animals are housed directly on the bedding. *Noncontact* bedding is used under cages with perforated floors. Generally, the noncontact type of bedding is more absorbent than the contact. If contact bedding is too hygroscopic, such as diatomaceous earth (frequently used as kitty litter) it may dehydrate newborn animals, especially those born without hair. In addition to being able to absorb moisture without crumbling excessively, bedding must be nonedible, nonabrasive, relatively dust-free, and free of pathogenic organisms, pesticides, other toxic chemicals, droppings from feral animals, and insects and their excreta. Bedding must also be free of natural volatile chemicals that may affect various enzyme systems in the animal.

The most commonly used types of bedding materials include specially processed wood particles and ground corn cobs. Other materials used include sawdust, pelleted peanut hulls, pelleted alfalfa (sometimes mixed with pelleted ground corn cobs or peanut hulls), shredded paper, and specially processed absorbent paper cut into small chips. Generally speaking, aromatic soft woods such as white pine, yellow pine, and red cedar must be avoided as a bedding material since they release volatile hydrocarbons capable of stimulating hepatic microsomal enzymes that, along with other potentially toxic effects, may directly or indirectly influence research data (Ferguson, 1966; Vesell, 1967; Ralls, 1967; Vesell *et al.*, 1973, 1976; Cunliffe-Beamer *et al.*, 1981). Also, soft wood beddings have been shown to increase respiratory infection rates (Seeger *et al.*, 1951), may decrease reproductive productivity (Port and Kallenbach, 1969), and potentially act as carcinogens (Sabine *et al.*, 1973; Schoental, 1973, 1974; Vlahakis, 1977). If wood bedding is used, hardwoods such as birch, beech, or maple are preferable. Aspen also has been used without apparent adverse side effects.

Care must be taken to process, package, ship, and store bedding so as to prevent contamination by chemicals, insects, and animal excreta. The packaged and stored bedding must be dry, not only to retain its absorbent properties, but also to prevent the growth of potentially toxic molds.

C. Sanitation

Maintenance of a high level of sanitation with minimal use of chemicals is a part of stabilizing the animals' environment (Thibert, 1980; Vesell *et al.*, 1973, 1976). The animal room should have surfaces that withstand and facilitate sanitation. The room should have minimal built-in equipment and must be maintained neat and uncluttered in order to facilitate sanitation. The floors should be wet mopped and sanitized daily with a disinfectant solution. Disinfectants, especially phenolics, should not be used on surfaces with which animals have direct contact. Sweeping (which creates dust) should be minimized. When the environment is potentially contaminated with hazardous chemicals, such as animal rooms in which the animals are being fed a carcinogen in the diet, dry sweeping and mopping must not be done because of the hazard of aerosol formation (NIH, 1981). Vacuums, unless a central turbine type or equipped with HEPA filtered exhaust, should not be used in animal rooms. In order to reduce the possibility of cross-contamination, each animal room should be equipped with its own sanitation equipment, such as mops and mop buckets. The mops and other sanitation utensils should be sanitized on a routine basis (Westwood *et al.*, 1971). Optimally, only a freshly laundered mop should be used. Between groups of animals, and at approximately 2-month intervals, animals should be removed from rooms and all surfaces cleaned and sanitized. Ventilation grillwork and light fixtures should also be cleaned at these times. The frequency of cage changing may vary according to the species, density of animals within the cage, and the type of cage system being used. Generally, both contact and noncontact bedding is changed 2–3 times per week and more frequently if indicated by excessive soiling, wetting, or high odor levels. At the time the bedding is changed, the cage or cage pan should also be exchanged for a freshly sanitized one. Since the rate of ammonia production is related to the bacterial count, it makes little sense to place fresh bedding in a cage or pan with a high bacterial count on its surface. Cages with perforated floors and cage racks generally will require cleaning and sanitizing at least one time per week.

Mechanical equipment is highly desirable for sanitizing cages. Cages may be sanitized by hand washing, but this requires soaking them in a disinfectant for an appropriate period of time and then thoroughly rinsing the disinfectant from the cages so that it does not serve as a variable in the experiment. If properly conducted, this is very time consuming and not practical for more than a few cages. Mechanical cage washers in addition to their convenience and labor-saving features sanitize the cages without the use of disinfectants by washing at temperatures above 82.2°C (180°F) (ILAR, 1978b) for an appropriate period of time necessary to destroy vegetative pathogenic organisms. Detergents are usually required to physically clean the cages, but appropriately equipped cage washers provide for adequate rinsing (see Section II,D,1). Water bottles should be sanitized each time they require filling or at the time the cage is sanitized. For some species, such as rabbits and guinea pigs, this may be daily. Automatic watering manifolds and sipper valves attached to the rack are sanitized by the rack washer. For best results, manifolds should be empty. If a rack washer is not available, the manifolds should be flushed with a hyperchlorinated solution. This may be desirable even when they are washed in a rack washer.

Waste materials should be disposed of in a safe and sanitary manner on a regular basis. If it is necessary to store waste, it should be maintained separate from other storage areas, and if the stored waste material is rapidly biodegradable, it should be refrigerated. Soiled bedding represents the largest volume of waste material. To reduce the potential for aerosol spread of infectious organisms, bedding should not be dumped from the cages or cage pans within the animal room. Bedding contaminated with hazardous agents or materials should be dumped into disposal containers within a hood designed to draw the aerosol away from the operator and through appropriate filters (Baldwin et al., 1976; Rake, 1979). Common methods of disposing of bedding include sealing in plastic bags and transport to a landfill, incineration, and flushing into the sanitary sewer system.

Infectious hazardous waste must be sterilized by appropriate means before disposal.

Radioisotope waste disposal requirements vary. Isotopes with short half-lives (less than 65 days) may be stored until the radioactivity reaches a safe level for conventional disposal. When this is not permitted or practical, the radioactive waste materials must be sealed in appropriate containers and buried in designated sites. With an appropriate Nuclear Regulatory Commission license, wastes contaminated with certain isotopes, i.e., ^{14}C and ^3H, may be incinerated in order to reduce the volume of waste material. The radioactive ashes are then collected and appropriately disposed.

Chemical hazardous waste generally is incinerated at temperatures sufficient to destroy the chemical or is buried in properly sealed containers in appropriately designated disposal areas.

All hazardous waste must be disposed of in accordance with municipal, state, and federal regulations pertaining to the specific hazardous material.

D. Environmental Quality Control

Monitoring the research animals' environment is a management tool essential to the successful control of the environment. The best facilities, equipment, and management procedures may fail and, therefore, detection of that failure at the earliest possible time is the goal of the quality-control program. Nearly every aspect of animal facility design and management covered previously in this chapter relates to fulfilling the requirement for controlling and monitoring the animals' environment. This includes the animal room temperature, humidity, air exchanges, relative air pressures, noise levels; and light intensity and cycles; equipment such as cage washers, autoclaves, safety cabinets, HEPA-filtered mass air displacement equipment, and water treatment and dispensing systems; water; feed and bedding; and the effectiveness of sanitation and vermin-control programs. The specific aspects to be monitored, the tolerable limits, and the methods and pro-

cedures for monitoring all the above are complex and may vary considerably for each animal care and use program. A detailed review of the subject of monitoring the research animals' environment has been published (Small, 1982).

IV. CONCLUSION

The maintenance of research animals in a comfortable, stress-free, controlled environment essential to modern biomedical research requires both adequate facilities and proper management. Each must complement the other. The more completely the environmental variables can be controlled, the more reliable and reproducible the research data will be, which in turn reduces the number of animals required to achieve the research goals. Properly designed and managed research animal facilities are essential for high-quality science, sound research economics, and, most importantly, humane care and use of laboratory animals.

REFERENCES

Anderson, D. P., Beard, C. W., and Hanson, R. P. (1964). The adverse effects of ammonia on chickens, including resistance to infection with Newcastle disease virus. Avian Dis. **8,** 369–379.

Baldwin, C. L., Sabel, F. L., and Henke, C. B. (1976). Bedding disposal cabinet for containment of aerosols generated by animal cage cleaning procedures. Appl. Environ. Microbiol. **31,** 322–324.

Balk, M. (1980). Animal-facility design criteria for a toxicologic testing laboratory. Pharm. Technol. **4,** 59–64.

Barkley, W. E. (1978). Abilities and limitations of architectural and engineering features in controlling biohazards in animal facilities. In "Symposium on Laboratory Animal Housing" (Inst. Lab. Anim. Resour.), pp. 158–163. Natl. Acad. Sci.—Natl. Res. Counc., Washington, D.C.

Beall, J. R., Torning, F. E., and Runkle, R. S. (1971). A laminar flow system for animal maintenance. Lab. Anim. Sci. **21,** 206–212.

Beck, R. W. (1963). The control of Pseudomonas aeruginosa in a mouse breeding colony by the use of chlorine in the drinking water. Lab. Anim. Care **13,** Part II, 41–45.

Bellhorn, R. W. (1980). Lighting in the animal environment. Lab. Anim. Sci. **30,** 440–448.

Brick, J. O., Newell, R. F., and Doherty, D. G. (1969). A barrier system for a breeding and experimental rodent colony: Description and operation. Lab. Anim. Care **19,** 92–97.

Briel, J. E., Kruchenberg, S. M., and Besch, E. L. (1971). "Observations of Ammonia Generation in Laboratory Animal Quarters," IER Publ. 72-03. Kansas State University, Manhattan.

Broderson, J. R., Lindsey, J. R., and Crawford, J. E. (1976). The role of environmental ammonia in respiratory mycoplasmosis of rats. Am. J. Pathol. **85,** 115–127.

Burns, E. R., Schering, L. E., Pauly, J. E., and Tsal, T. H. (1976). Effect of altered lighting regimens, time-limited feeding, and presence of Ehrlich ascites carcinoma on the circadian rhythm in DNA synthesis of mouse spleen. Cancer Res. **36,** 1538–1544.

Christie, R. J., Williams, F. P., Whitney, J. R., and Johnson, D. J. (1968). Techniques used in the establishment and maintenance of a barrier mouse breeding colony. *Lab. Anim. Care* **18,** 544–549.

Cohen, B. J., Baker, H. J., Dodds, W. J., Hessler, J., New, A. E., and Grogan, E. W. (1979). Laboratory animal medicine: Guidelines for education and training. *ILAR News* **22,** M1–M26.

Cunliffe-Beamer, T. L., Freeman, L. C., and Myers, D. D. (1981). Barbiturate sleeptime in mice exposed to autoclaved or unautoclaved wood beddings. *Lab. Anim. Sci.* **31,** 672–675.

Davis, D. E. (1978). Social behavior in a laboratory environment. *In* "Symposium on Laboratory Animal Housing" (Inst. Lab. Anim. Resour.), pp. 44–63. Nat. Acad. Sci.—Natl. Res. Counc., Washington, D.C.

Dolowy, W. C. (1961). Medical research laboratory of the University of Illinois. *Proc. Anim. Care Panel* **11,** 267–290.

Dymsza, H. A., Miller, S. A., Maloney, J. F., and Foster, H. L. (1963). Equilibration of the laboratory rat following exposure to shipping stresses. *Lab. Anim. Care* **13,** 60–65.

Ferguson, H. C. (1966). Effect of red cedar chip bedding on hexobarbital and pentobarbital sleep time. *J. Pharm. Sci.* **55,** 1142–1148.

Fiddler, I. J. (1977). Depression of macrophages in mice drinking hyperchlorinated water. *Nature (London)* **270,** 735–736.

Flynn, R. J., Poole, C. M., and Tyler, S. A. (1971). Long distance air transport of aged laboratory mice. *J. Gerontol.* **26,** 201–203.

Foster, H. L., Foster, S. J., and Pfau, E. S. (1963). The large scale production of ceasarean-originated, barrier-sustained mice. *Lab. Anim. Care* **13,** 711–718.

Fouts, J. R. (1970). Some effects of insecticides on hepatic microsomal enzymes in various animal species. *Rev. Can. Biol.* **29,** 377–389.

Fox, J. G., and Helfrich-Smith, M. E. (1980). Chemical contamination of animal feeding systems: Evaluation of two caging systems and standard cage-washing equipment. *Lab. Anim. Sci.* **30,** 967–973.

Fox, J. G., Donahue, P. R., and Essigmann, J. M. (1980). Efficacy of inactivation of representative chemical carcinogens utilizing commercial alkaline and acidic cage wash compounds. *J. Environ. Pathol. Toxicol.* **4,** 97–106.

Fullerton, F. R., Greenman, D. L., and Kendall, D. C. (1982). Effects of storage conditions on nutritional qualities of semipurified (AIN-76) and natural ingredient (NIH-07) diets. *J. Nutr.* **12,** 567–573.

Gamble, M. R., and Clough, G. (1976). Ammonia build-up in animal boxes and its effect on rat tracheal epithelium. *Lab. Anim.* **10,** 93–104.

Gerone, P. J. (1978). Hazards associated with infected laboratory animals. *In* "Symposium on Laboratory Animal Housing" (Inst. Lab. Anim. Resour.), pp. 105–117. Natl. Acad. Sci.—Natl. Res. Counc., Washington, D.C.

Goldstein, S. J. (1978). A theory of architecture: The orchestration of information. *In* "Symposium on Laboratory Animal Housing" (Inst. Lab. Anim. Resour.), pp. 16–24. Natl. Acad. Sci.—Natl. Res. Counc., Washington, D.C.

Gorton, R. L. (1975). System design and energy conservation considerations. *ASHRAE Trans.* **81,** 572–576.

Gorton, R. L. (1978). Energy conservation in water heating and HVAC systems. *In* "Symposium on Laboratory Animal Housing" (Inst. Lab. Anim. Resour.), pp. 179–183. Natl. Acad. Sci.—Natl. Res. Counc., Washington, D.C.

Hall, J. E., White, W. J., and Lang, C. M. (1980). Acidification of drinking water: Its effects on selected biological phenomena in male mice. *Lab. Anim. Sci.* **30,** 643–651.

Haus, E., and Halberg, F. (1970). Circannual rhythm in level and timing of serum corticosterone in standardized inbred mature C mice. *Environ. Res.* **3,** 81–106.

Haus, E., Lakatua, D., and Halberg, F. (1967). The internal timing of several circadian rhythms in the blinded mouse. *Exp. Med. Surg.* **25,** 7–45.

Henkel, C. B. (1978). Design criteria for animal facilities. *In* "Symposium on Laboratory Animal Housing" (Inst. Lab. Anim. Resour.), pp. 142–157. Natl. Acad. Sci.—Natl. Res. Counc., Washington, D.C.

Hermann, L. M., White, W. J., and Lang, C. M. (1982). Prolonged exposure to acid, chlorine, or tetracycline in the drinking water: Effects on delayed-type hypersensitivity, hemagglutination titers, and reticuloendothelial clearance rates in mice. *Lab. Anim. Sci.* **32,** 603–608.

Institute of Laboratory Animal Resources (ILAR) (1976a). Long-term holding of laboratory rodents. *ILAR News* **19**(4), L1–24.

Institute of Laboratory Animal Resources (ILAR) (1976b). "Guide for the Care and Use of the Nude (Thymus-deficient) Mouse in Biomedical Research." Natl. Acad. Sci., Washington, D.C.

Institute of Laboratory Animal Resources (ILAR) (1978a). "Symposium on Laboratory Animal Housing." Natl. Acad. Sci.—Natl. Res. Counc., Washington, D.C.

Institute of Laboratory Animal Resources (ILAR) (1978b). "Guide for the Care and Use of Laboratory Animals," Publ. No. (NIH) 74-23. U.S. Govt. Printing Office, Washington, D.C.

Institute of Laboratory Animal Resources (ILAR) (1980). "National Survey of Laboratory Animal Facilities and Resources," NIH Publ. No. 80-2091. U.S. Govt. Printing Office, Washington, D.C.

Jeszenka, E. V., White, W. J., Lang, C. M., and Hughes, H. C. (1981). Evaluation of a chemical scrubber for the removal of gaseous contaminants from recycled air. *Lab. Anim. Sci.* **31,** 489–493.

Jonas, A. M. (1965). Laboratory animal facilities. *J. Am. Vet. Med. Assoc.* **146,** 600–606.

Jonas, A. M. (1978). Centralized versus dispersed animal care facilities. *In* "Symposium on Laboratory Animal Housing" (Inst. Lab. Anim. Resour.), pp. 11–15. Natl. Acad. Sci.—Natl. Res. Counc., Washington, D.C.

Kraft, L. M. (1980). The manufacturing, shipping and receiving and quality control of rodent bedding materials. *Lab. Anim. Sci.* **30,** 366–374.

Kraft, L. M., Pardy, R. F., Pardy, D. A., and Zwickel, H. (1964). Practical control of diarrheal disease in a commercial mouse colony. *Lab. Anim. Care* **14,** 16–19.

Kruckenberg, S. M. (1971). Control of odoriferous and other gaseous components in laboratory animal quarters. *In* "Proceedings of the Symposium on Environmental Requirements for Laboratory Animals," IER Publ. 71-02, pp. 83–100. Kansas State University, Manhattan.

Landi, M. S., Kreider, J. W., Lang, C. M., and Bullock, L. P. (1982). Effects of shipping on the immune function in mice. *Am. J. Vet. Res.* **43,** 1654–1657.

Lang, C. M. (1981). Special design considerations for animal facilities. *In* "Proceedings of the National Cancer Institute Symposium on Design of Biomedical Research Facilities" (D. G. Fox, ed.), Cancer Res. Saf. Monogr., Vol. IV, NIH Publ. No. 81-2305. pp. 117–127. Natl. Inst. Health, Bethesda, Maryland.

Lang, C. M. (1983). Design and management of research facilities for mice. *In* "The Mouse in Biomedical Research" (H. L. Foster, J. D. Small, and J. G. Fox, eds.), Vol. 8, pp. 37–50. Academic Press, New York.

Lang, C. M., and Harrell, G. T. (1969). An ideal animal resource facility. *Am. Inst. Archit. J.* **52,** 57–61.

Lang, C. M., and Harrell, G. T. (1972). Guidelines for a quality program of laboratory animal medicine in a medical school. *J. Med. Educ.* **47,** 267–271.

Lang, C. M., and Vesell, E. S. (1976). Environmental and genetic factors affecting laboratory animals: Impact on biomedical research. *Fed. Proc., Fed. Am. Soc. Exp. Biol.* **35,** 1123–1124.

Lowe, F. M. (1980). Considerations in receiving and quarantining laboratory rodents. *Lab. Anim. Sci.* **30,** 323–329.

McGarrity, G. J., and Coriell, L. L. (1973). Mass airflow cabinet for control of airborne infection of laboratory rodents. *Appl. Microbiol.* **26,** 167–172.

McGarrity, G. J., and Coriell, L. L. (1976). Maintenance of axenic mice in open cages in mass air flow. *Lab. Anim. Sci.* **26,** 746–750.

McPherson, C. W. (1963). Reduction of *Pseudomonas aeruginosa* and coliform bacteria in mouse drinking water following treatment with hydrochloric acid or chlorine. *Lab. Anim. Care* **13,** 737–744.

McPherson, C. W. (1975). Why be concerned about ventilation requirements of experimental animals? *ASHRAE Trans.* **81,** 539–541.

Mulder, J. B. (1971). Animal behavior and electromagnetic energy waves. *Lab. Anim. Sci.* **21,** 389–393.

Murakami, H. (1971). Differences between internal and external environments of the mouse cage. *Lab. Anim. Sci.* **21,** 680–684.

National Institutes of Health (NIH) (1979). "Laboratory Safety Monograph" (A Supplement to the N.I.H. Guidelines for Recombinant DNA Research). U.S. Dept. of Health, Education and Welfare, Public Health Serv., Natl. Inst. Health, Bethesda, Maryland.

National Institutes of Health (NIH) (1981). "Guidelines for the Use of Chemical Carcinogens," NIH Publ. No. 81-2385. Natl. Inst. Health, Bethesda, Maryland.

National Research Council (NRC) (1978). "Nutritional Requirements of Laboratory Animals," National Academy of Sciences, Washington, D.C.

Nevins, R. G., and Miller, P. L. (1972). Analysis, evaluation, and comparison of room air distribution performance—A summary. *ASHRAE Trans.* **78,** 235–242.

Newberne, P. M. (1975). Influence on pharmacology experiments of chemicals and other factors in diets of laboratory animals. *Fed. Proc., Fed. Am. Soc. Exp. Biol.* **34,** 209–218.

Newberne, P. M., and Fox, J. G. (1978). Chemicals and toxins in the animal facility. *In* "Symposium on Laboratory Animal Housing" (Inst. Lab. Anim. Resour.), pp. 118–138. Natl. Acad. Sci.—Natl. Res. Counc., Washington, D.C.

Newberne, P. M., and Fox, J. G. (1980). Nutritional adequacy and quality control of rodent diets. *Lab. Anim. Sci.* **30,** 352–363.

Newell, G. W. (1980). The quality, treatment and monitoring of water for laboratory rodents. *Lab. Anim. Sci.* **30,** 377–383.

Otis, A. P., and Foster, H. L. (1983). Management and design of breeding facilities. *In* "The Mouse in Biomedical Research" (H. L. Foster, J. D. Small, and J. G. Fox, eds.), Vol. 3, p. 20. Academic Press, New York.

Peterson, E. A. (1980). Noise and laboratory animals. *Lab. Anim. Sci.* **30,** 422–439.

Poiley, S. M. (1974). Housing requirements—General considerations. *In* "Handbook of Laboratory Animal Science" (E. C. Melby, Jr. and N. H. Altman, eds.), Vol. 1, pp. 21–60. CRC Press, Cleveland, Ohio.

Port, C. D., and Kaltenbach, J. P. (1969). The effect of corncob bedding on reproductivity and leucine incorporation in mice. *Lab. Anim. Care* **19,** 46–49.

Rake, B. W. (1979). Microbiological evaluation of a biological safety cabinet modified for bedding disposal. *Lab. Anim. Sci.* **29,** 625–632.

Ralls, K. (1967). Auditory sensitivity in mice: *Peromyscus* and *Mus musculus.* *Anim. Behav.* **15,** 123–128.

Reyniers, J. A. (1957). The control of contamination in colonies of laboratory animals by the use of germfree techniques. *Proc. Anim. Care Panel* **7,** 9–29.

Robinson, W. G., Jr., and Kuwabara, T. (1976). Light-induced alterations of retinal pigment epithelium in black, albino and beige mice. *Ex. Eye Res.* **22,** 549–557.

Robinson, W. G., Jr., Kuwabara, T., and Zwaan, J. (1982). Eye research. *In* "The Mouse in Biomedical Research" (H. L. Foster, J. D. Small, and J. G. Fox, eds.), Vol. 4, pp. 69–95. Academic Press, New York.

Sabine, J. R., Horton, B. J., and Wicks, M. B. (1973). Spontaneous tumors in C3H-A^{vy} and C3H-A^{vyfB} mice: High incidence in the United States and low incidence in Australia. *JNCI, J. Natl. Cancer Inst.* **50,** 1237–1242.

Saltarelli, C. G., and Coppola, C. P. (1979). Influence of visible light on organ weights of mice. *Lab. Anim. Sci.* **29,** 319–322.

Schneider, H. A., and Collins, G. R. (1966). Successful prevention of infantile diarrhea of mice during an epizootic by means of a new filter cage unopened from birth to weaning. *Lab. Anim. Care* **16,** 60–71.

Schoental, R. (1973). Carcinogenicity of wood shavings. *Lab. Anim.* **7,** 47–49.

Schoental, R. (1974). Zearalenone in the diet, podophyllotoxin in the wood shaving bedding, are likely to affect the incidence of "spontaneous" tumors among laboratory animals. *Br. J. Cancer* **30,** 181.

Seeger, K. C., Tomhave, A. E., and Lucas, W. C. (1951). A comparison of litters used for broiler production. *Del., Agric. Exp. Stn., Bull.* **289.**

Serrano, L. J. (1971). Carbon dioxide and ammonia in mouse cages: Effect of cage covers, population and activity. *Lab. Anim. Sci.* **21,** 75–85.

Simmonds, R. C. (1973). Selected topics in laboratory animal medicine—The design of laboratory animal homes. *Aeromed. Rev.* **7-73,** 1–45.

Simmons, M. L., Wynns, L. P., and Choat, E. E. (1967). A facility design for production of pathogen-free, inbred mice. *ASHRAE J.* **9,** 27–31.

Small, J. D. (1982). Environmental and equipment monitoring. *In* "The Mouse in Biomedical Research" (H. L. Foster, J. D. Small, and J. G. Fox, eds.), Vol. 3, pp. 83–100. Academic Press, New York.

Thibert, P. (1980). Control of microbial contamination in the use of laboratory rodents. *Lab. Anim. Sci.* **30,** 339–348.

Trexler, P. C. (1963). An isolator system for control of contamination. *Lab. Anim. Care* **13,** 572–581.

van der Waaij, D., and Andreas, A. H. (1971). Prevention of air-borne contamination and cross-contamination in germfree mice by laminar flow. *J. Hyg.* **69,** 83–89.

Vesell, E. S. (1967). Induction of drug-metabolizing enzymes in liver microsomes of rats and mice by softwood bedding. *Science* **157,** 1057–1058.

Vesell, E. S., Lang, C. M., White, W. J., Passananti, G. T., and Tripp, S. L. (1973). Hepatic drug metabolism in rats: Impairment in a dirty environment. *Science* **179,** 896–897.

Vesell, E. S., Lang, C. M., White, W. J., Passananti, G. T., Hill, R. N., Clemens, T. L., Dai Kee Liu, and Johnson, W. D. (1976). Environmental and genetic factors affecting the response of laboratory animals to drugs. *Fed. Proc., Fed. Am. Soc. Exp. Biol.* **35,** 1125–1132.

Vlahakis, G. (1977). Possible carcinogenic effects of cedar shavings in bedding of C3H-A^{vyfB} mice. *JNCI, J. Natl. Cancer Inst.* **58,** 149–150.

Weihe, W. J. (1965). Temperature and humidity climatograms for rats and mice. *Lab. Anim. Care* **15,** 18–28.

Westwood, J. C. N., Mitchell, M. A., and Legace, S. (1971). Hospital sanitation: The massive bacterial contamination of the wet mop. *Appl. Microbiol.* **21,** 693–697.

Windman, A. L., and Zigas, A. L. (1978). Engineering objectives for laboratory animal housing. *In* "Symposium on Laboratory Animal Housing" (Inst. Lab. Anim. Resour.), pp. 84–88. Natl. Acad. Sci.—Natl. Res. Counc., Washington, D.C.

Woods, J. E. (1975). Influence of room air distribution on animal cage environments. *ASHRAE Trans.* **81,** 559–571.

Chapter 18

Preanesthesia, Anesthesia, Analgesia, and Euthanasia

Donald H. Clifford

I. INTRODUCTION

Knowledge of preanesthesia, anesthesia, analgesia, and euthanasia is essential to those working in laboratory animal medicine. The administration of sedatives, tranquilizers, and analgesic drugs in the postoperative period is complex since (1) it is difficult to judge postoperative pain or discomfort in animals, (2) postoperative ambulation should be encouraged, and (3) many of the sedatives, tranquilizers, and analgesics carry a respiratory and cardiovascular burden that is detrimental to the animal. No single method of euthanasia is acceptable to all. This chapter is limited to the most important techniques and agents in current use with laboratory animals.

II. PREANESTHESIA, ANESTHESIA, AND ANALGESIA

Preanesthesia, anesthesia, and analgesia have parallel uses in veterinary and human medicine. In some instances such as "nonsurvival" procedures, methods and agents may be used in animals that would be unacceptable in animals that would recover from anesthesia or would be used for food. The administration of sodium pentobarbital by intravenous or intraperitoneal routes and the administration of halothane or methoxyflurane by open or closed systems alone or following preanesthetic medication are among the most widely used methods. If preanesthetic drug(s) are required, ketamine and several tranquilizing or narcotic agents are available. The narcotic drugs may be administered to ease postoperative pain and discomfort. Effective anesthesia is both an art and a science, requiring an appreciation of the most suitable anesthetic techniques or agents for a particular animal, procedure, or situation.

The methods and agents used to produce preanesthesia, anesthesia, and analgesia should be designed to minimize pain and discomfort in animals at all times. However, these agents and methods significantly influence the body systems and their effect on physiological functions must be recognized. No study conducted on an anesthetized animal can be considered to be the same as in one that is not anesthetized. Whether variations due to anesthesia are tolerable or the study can be performed without anesthesia must be evaluated in each procedure. This subject is considered separately under complicating factors. Furthermore, the safety to animals and attending personnel should be kept in mind.

Preanesthetic medication, anesthesia, and analgesia are grouped according to the most common laboratory animals. The designations PRE for preanesthesia, ANES for anesthesia, and ANAL for analgesia are intended to denote the principal use(s) of the various agents. Often the use depends on the dose, circumstances, and relationship to other agents. Euthanasia will constitute a separate subject. Anesthetic agents and techniques for rats, guinea pigs, hamsters, and mice are presented separately from those of less commonly used rodents, such as chinchillas, gerbils, and porcupines. This underscores variations that may require different anesthetic methods and agents; however, all rodents are described together under the subject of euthanasia. Unusual or rarely used agents or techniques are not included.

A. Amphibians, Fishes, and Reptiles

1. Physical Method

HYPOTHERMIA[1] *(PRE)*

Hypothermia restricts the activity of fishes, but induces stress and cannot be used effectively for major procedures. Frogs, lizards, and other reptiles can be effectively immobilized by immersion in cracked ice and water for 10–15 min or by placing them in a refrigerator. Normal activity resumes when they are rewarmed to room temperature. Hypothermia may be combined with local anesthesia.

2. Parenteral Methods

a. Intravenous or Intracardiac Route—Pentobarbital[2] Sodium, or Urethane[3] (Ethyl Carbamate) (ANES)

The ventral abdominal vein, which can be exposed by drilling through the plastron of turtles, is utilized to administer pentobarbital (17.5 mg/kg). Intracardiac injection of pentobarbital (16.3 mg/kg) in turtles may be made through the thoracic inlet. Induction time is slow and most turtles are under anesthesia 3 hr later. Preanesthetic administration of chlorpromazine (10 mg/kg) 10 min prior to administration of pentobarbital (10 mg/kg) shortens the induction time of pentobarbital in turtles. Urethane (2.4 gm/kg iv or 1.7 gm/kg ic) is also associated with a long induction and anesthetic period (10 hr) for turtles. The environmental temperature is a very important factor in the time required for recovery. There is a general consensus that the barbiturates and other long-acting parenteral agents should be avoided in reptiles.

b. Intramuscular, Intraperitoneal or Lymph Sac Administration—Amobarbital Sodium,[4] Chloral Hydrate, Hexobarbital Sodium,[5] Paraldehyde, Pentobarbital Sodium, Secobarbital Sodium, Thiopental Sodium,[6] Tricaine Methanesulfonate (MS-222), or Urethane (Ethyl Carbamate) (ANES)

These agents require handling of fish and offer no advantages except cost over agents that are added to water. Paralde-

---

[1]Recommended procedures and agents are capitalized.
[2]The suffix "al" is changed to "one" in the English literature.
[3]Merck and Company, Rahway, New Jersey.
[4]Amytal, Eli Lilly and Company, Indianapolis, Indiana.
[5]Evipal, Winthrop Laboratories, New York, New York.
[6]Pentothal, Abbott Laboratories, North Chicago, Illinois.

hyde (4.2 gm/kg) produces narcosis in frogs in approximately 15 min without excitement when injected into the ventral lymph sac. Injection of 1–2 ml of a 10% solution of chloral hydrate or 0.04–0.12 ml/gm of 5% urethane into the dorsal lymph sac of frogs has been recommended. Frogs are anesthetized by injecting hexobarbital (120 mg/kg) into the dorsal lymph sac. The ventral abdominal vein and ventral lymph sac is utilized to make an injection with a 25 gauge needle. Anesthesia intervenes in 22 min and lasts for approximately 3 hr. Pentobarbital (60 mg/kg) injected via the dorsal lymph sac or intracoelomically has a similar period of induction, with 9 hr for recovery. A high mortality follows barbiturate use in reptiles with the exception of turtles in which pentobarbital produces surgical anesthesia for 2 to 4 hr after an induction period of 65–80 min. Urethane (2.8 gm/kg) lasts over 10 hr in turtles and is considered unsuitable in this species. Thiopental and pentobarbital (15–30 mg/kg) may be injected into the pleuroperitoneal cavity of snakes. Induction is slow and recovery may require 15–3O hr. The long recovery period is not without the risk of cardiorespiratory complications. Tricaine methanesulfonate (15–30 mg/kg) induces anesthesia in 12–40 min in snakes following pleuroperitoneal injection. Anesthesia persists for 60 min with 90 min required for recovery. Recovery from barbiturates and parenteral anesthetic agents requires much longer in amphibians and reptiles than mammals due to their lower metabolic rate.

c. Intramuscular or Dorsal Lymph Sac Route—Ketamine Hydrochloride (PRE, ANES)

Preanesthetic administration of ketamine (100 mg, im) prior to intubation and maintenance with halothane enabled salpingotomy to be performed in a tortoise. Several species of poisonous snakes were tranquilized or anesthetized with ketamine in doses varying from 22 to 132 mg/kg. Doses of 60 to 80 mg/kg results in immobilization for 48 to 96 hr. Most examinations and treatments can be performed with doses from 22 to 66 mg/kg, which are comparable to mammals. However, snakes do not fully recover for 2–3 days. Smaller doses (20–40 mg/kg, im) are satisfactory in most instances and result in chemical restraint for 30–60 min. This smaller dose can be repeated in 30 min if necessary. There is considerable variation in different reptiles.

d. Aqueous Contact—Alcohol, Amobarbital Sodium, or Secobarbital Sodium (ANES)

Frogs can be anesthetized by immersion in 10% ethyl alcohol. Surgical anesthesia is attained in approximately 10 min and lasts for about 20 min. Amobarbital (7–10 mg/liter) or secobarbital (35 mg/liter) and other barbiturates can be added to the water to produce anesthesia in 30–60 min. The recovery period is long.

e. AQUEOUS CONTACT—CARBON DIOXIDE (ANES)

Two hundred parts per million in water is adequate to anesthetize fish. Induction occurs in 1–2 min with recovery in 5–10 min.

f. Aqueous Contact—Chloral Hydrate (PRE)

Chloral hydrate (0.8–0.9 gm/liter) induces anesthesia slowly (8–10 min) and is a poor anesthetic. It is used more effectively as a sedative.

g. Aqueous Contact—Chlorobutanol,[7] Methylpentynol,[8] 2-Phenoxyethanol,[8] Quinidine,[8] 4-Strylpyridine,[8] Tertiary Amyl Alcohol, or Tribromoethanol[9] (ANES)

These and other chemicals have been added to the water to produce anesthesia in frogs and fish. Due to irritation produced, deterioration in water, variation in the response, lack of information relative to carcinogenicity, noxious odor, side effects and lack of published literature, these agents are not widely used.

h. Aqueous Contact—Diethyl Ether (ANES)

Addition of 10–15 ml/liter of water induces anesthesia in fish in 2–3 min with a recovery time of 5–30 min but it causes tissue irritation and is flammable.

i. AQUEOUS CONTACT—TRICAINE METHANESULFONATE (MS-222) (ANES)

MS-222 produces light anesthesia (25–35 mg/liter) or deep anesthesia (50–100 mg/liter). Induction time is 1–3 min with 3–15 minutes for recovery. Tolerance does not develop after repeated exposure. The following factors should be considered prior to the use of tricaine methanesulfonate: (1) Toxic reactions may result in its use in salt water or direct sunlight. (2) The container should be inert. (3) The water containing the chemical should be from the same source and be at the same temperature as the normal habitat. (4) The agent should be well mixed with the water. (5) The fish should be removed and kept moist as soon as swimming subsides and returned if gasping movements or muscle spasms occur. (6) Respiration should be watched closely. (7) Fish should be returned to fresh water as soon as possible and kept moving so that water passes over the gills. (8) The anesthetic water should be discarded if induction time is prolonged.

---

[7]Chloretone, Parke Davis and Company, Detroit, Michigan.
[8]Fischer Scientific Company, Pittsburgh, Pennsylvania.
[9]Avertin, Winthrop Laboratories, New York, New York.

j. Aqueous Contact—Urethane (Ethyl Carbamate) (ANES)

Urethane (5–40 mg/liter) induces anesthesia in fish in 2–3 min and recovery requires 10–15 min.

3. Inhalant Methods

a. Anesthetic Chamber—Ether (ANES)

Frogs may be anesthetized with ether in an anesthetic chamber. Induction time is 5–10 min, and anesthesia may last for hours. Turtles can be administered with a snug-fitted nasal cone. Induction is lengthy (37 min) as is the anesthetic period (10 hr). Induction of turtles in an anesthetic chamber requires 3 hr. Snakes and lizards can be anesthetized safely with ether, although the period of induction is relatively long, 20–60 min. The external temperature is an important factor in the time required for induction in snakes as other animals. Ether can be used with the aid of a jar or chamber followed by a cone to anesthetize amphibians, reptiles, and other exotic animals. Some animals were intubated to make the procedure safer. Ether has gradually been replaced by halothane, since most parenteral drugs act very slowly in poikilothermic animals, and methoxyflurane is associated with a slower induction and recovery than halothane. One should make sure that the animal is receiving an adequate supply of oxygen and that the ether does not come in contact with the skin of the animal.

b. ANESTHETIC CHAMBER, CONE, OR ENDOTRACHEAL ROUTE—HALOTHANE,[10] METHOXYFLURANE,[11] Nitrous Oxide (ANES)

Halothane or methoxyflurane (except in cobras) are used to induce amphibia, snakes, and other reptiles by means of an anesthetic chamber and/or cone. Ketamine is used for preanesthesia with an appropriate reduction in the concentration of exposure of the anesthetic gas(es). Surgical anesthesia in snakes is manifest by lack of resistance when their tongue is withdrawn with forceps. This may be followed by endotracheal intubation. Approximately 5 min is required for induction, while snakes recover in about 10 min from halothane. The concentration of halothane for induction is 3% for most reptiles, while they can be maintained with 1.5% or less. The signs of anesthesia for turtles are loss of (1) neck withdrawal and (2) leg movements. Deaths have been reported in cobras during recovery from methoxyflurane anesthesia. This also has been observed during recovery from ether and may be due to an increased rate of metabolism and hypoxia. Frogs require about 7 hr to recover from methoxyflurane. Halothane and methoxyflurane have replaced ether for inhalant anesthesia in

the laboratory for poikilotherms. Nitrous oxide is not recommended by some anesthetists in reptiles due to delayed release from the various tissue compartments; however, others use halothane, nitrous oxide, and oxygen as in the dog and cat. Snakes require intermittent positive pressure breathing when inhalation anesthesia is used. Special recommendations for anesthetizing reptiles include: (1) ventilation with 100% oxygen before intubation, (2) avoidance of high positive pressure in species that lack a diaphragm; (3) duplication of the normal pattern of breathing for the species, i.e., apneustic breathing for crocodilians; (4) continued administration of oxygen after halothane or methoxyflurane has been discontinued and during the subsequent early phase of recovery; and (5) recovery at normal temperature and humidity for the species. Poisonous snakes require more anesthetic agent(s) than nonpoisonous species.

B. Birds

1. Parenteral Methods

a. Intravenous Route—Methohexital Sodium,[12] Thiopental Sodium (ANES)

Methohexital (5–10 mg/kg, iv) produces general anesthesia for 2 hr in most birds. Recovery requires 20 to 40 min. The anesthetic dose of thiopental is larger (50 mg/kg) in birds than for mammals and surgical anesthesia persists for only about 5 min. A dilution of 0.5–1.0% is recommended. There are indications for the use of these agents, but generally they have been replaced by ketamine and/or halothane.

b. INTRAVENOUS ROUTE—EQUI-THESIN,[13] Glutethimide,[14] Hydroxydione,[15] or PENTOBARBITAL SODIUM (ANES)

The pharmacodynamics of drugs in birds and many aspects of the metabolism in avian species, are poorly understood. Pigeons and larger birds may be anesthetized intravenously, i.e., jugular or alar vein, with Equi-Thesin (1.0–1.5 ml/kg). Each 500 ml contains 21.3 gm chloral hydrate, 4.8 gm pentobarbital, and 10.6 gm of magnesium sulfate in aqueous solution of 35% propylene glycol with 9.5% ethyl alcohol. Pentobarbital alone appears to be reserved for larger birds. After administering half of the dose, the remainder should be injected very slowly and stopped as soon as the comb reflex becomes faint. Sometimes birds enter an excitement phase in which muscular

[10]Fluothane, Ayerst Laboratories, New York, New York.
[11]Metofane, Pitman-Moore Co., Washington Crossing, New Jersey.

[12]Brevane, Elanco Products Company, Indianapolis, Indiana.
[13]Equi-Thesin, Jensen-Salsbery Laboratories, Kansas City, Missouri.
[14]Doriden, Ciba Pharmaceutical Company, Summit, New Jersey.
[15]Viadril, Charles Pfizer and Company, Inc., New York, New York.

tremors occur. The anesthetic dose of pentobarbital (20–30 mg/kg) produces anesthesia for only 10–15 min or comparable to the ultrashort acting thiobarbiturates. Neither Equi-Thesin nor pentobarbital alone are entirely satisfactory. There are large differences in response between the small finches with their rapid metabolic rate and the large domestic birds. Care must be taken to make sure that fluid contents of the crop are not aspirated. The administration of a small supplemental amount of halothane to prolong the period of surgical anesthesia is usually safer than reanesthetizing the bird with pentobarbital. In small birds the technical difficulties and stress of intravenous administration may outweigh the advantages of a more rapid induction and control of the level of anesthesia. Volatile agents appear safer in these birds.

c. Intraperitoneal, Intramuscular or Subcutaneous Route—Equi-Thesin or Pentobarbital Sodium (ANES)

Equi-Thesin (2.0–2.5 ml/kg, im) is relatively safe when administered into the pectoral muscles of small birds, i.e., parakeets and canaries, but produces myositis at the site of injection. Anesthesia occurs in 8–20 min and lasts for 20–90 min. Pentobarbital is not a safe anesthetic agent in budgerigars; however, it can be administered safely in 1-day-old chicks (30 mg/kg). The combination of thiopental and pentobarbital (3:1) produces safe anesthesia in chicks from 1 to 7 days of age. The optimum dose is 0.05 ml (30 mg/ml, ip) for a 1-day-old chick with an additional 0.01 ml for each additional day of age. Induction requires 1–5 min with recovery in 4–6 hr. Death following the intraperitoneal or intramuscular administration of pentobarbital because of individual susceptibility, increased vascularity of the site, or technical errors, such as injecting into the air sac or rupture of one of the venous sinuses in the pectoral muscles, has prompted subcutaneous injection. Basal anesthesia is provided by this technique, which can be supplemented with a reduced dose of volatile agent.

d. Intramuscular Route—Chlorpromazine Hydrochloride,[16] Diethylthiambutene Hydrochloride,[17] Meperidine Hydrochloride, Morphine Sulfate, or Xylazine Hydrochloride[18] (PRE)

Most birds do not appear to experience pain, since caponizing, debeaking, and other minor procedures can be performed in the absence of anesthesia without apparent discomfort.

Smaller pet birds, such as parakeets, seem sensitive to pain around the head, legs, and vent. Chlorpromazine produces only mild sedation in birds while diethylthiambutene in low doses causes vomiting, and in high doses convulsions and death. Xylazine is not tolerated by pigeons but may have uses in other birds.

e. INTRAMUSCULAR ROUTE—KETAMINE HYDROCHLORIDE (PRE, ANES)

Ketamine (33–100 mg/kg, im) has a wide margin of safety and may be used for many procedures that do not involve major surgery. Induction occurs in 3 to 5 min. Anesthesia lasts from 5 to 20 min with 20 to 90 min required for recovery. Ketamine (25 mg/kg, im) may be combined with diazepam (2.5 mg/kg, im). Signs of transient excitement such as flapping the wings may occur during recovery with any anesthetic. Ketamine (15–20 mg/kg, im) was effective as the sole anesthetic for waterfowl, parrots, and birds of prey. In some instances it was necessary to supplement the initial dose with additional amounts (10 mg/kg). Immobilization or anesthesia was apparent in 1–5 min and lasted from 30 min to 6 hr depending upon the dose. A larger dose range (15–60 mg/kg, im) is recommended for domestic poultry. Unilateral wing movements and extensor rigidity are observed. Ketamine may be combined with acepromazine and xylazine, as well as volatile agents such as halothane or methoxyflurane. Birds that are dehydrated or have renal disease may be endangered by ketamine, which is excreted unchanged by the kidneys.

f. Local Anesthesia—Ethyl Chloride, Lidocaine Hydrochloride, Procaine Hydrochloride, Tetracaine Hydrochloride,[19] and Other Local Anesthetics (ANES)

The toxic effects of procaine are due primarily to excessive doses and high concentrations rather than increased susceptibility in birds. Doses of 0.25–1.0 ml in pigeons and 1.0–3.0 ml in chickens and ducks are recommended. Lidocaine, 1 ml of a 2% solution, can be used in pigeons. It is not recommended to use injectable local anesthetics in small birds, i.e., budgerigars, but 0.25% procaine can be used in parakeets. Higher concentrations cause ataxia, seizures and death. Epinephrine should not be added to any local anesthetic for avian use. Ethyl chloride and benzoate-tetracaine[20] sprays are useful for treating prolapse of the rectum and oviduct, abscesses, and warts in small pet birds.

[16]Thorazine, Smith, Kline and French Laboratories, Philadelphia, Pennsylvania.

[17]Thiambutene, Themalon, Parke Davis and Company, Detroit, Michigan.

[18]Rompun, Chemagro Division of Baychem Corporation, Kansas City, Missouri.

[19]Pontocaine, Winthrop Laboratories, New York, New York.

[20]Cetacaine, Haver-Lockhart Laboratories, Kansas City, Missouri.

2. Inhalant Methods

a. *Anesthetic Chamber, Cone or Mask—Ether (ANES)*

Ether can be used satisfactorily for short operations in birds. Spray the anesthetic directly into one nostril by means of a 2 ml syringe equipped with a 26 gauge $\frac{1}{4}$ in. needle from a distance of 1 in. Repeated administration may be required. A small polyethylene tube may be inserted into the esophagus to prevent the aspiration of vomitus. Ether is safe for budgerigars, which are difficult birds to anesthetize. Large parrots are more refractory to ether, and it may be difficult to obtain surgical anesthesia. Irritation of the thoracic and abdominal air sacs may produce secretions that result in subsequent respiratory infection. Although atropine has been advocated, salivation and lachrymal secretion do not appear noticeable. Gaseous and volatile agents are relatively safe when administered as a single exposure. Overdose is avoided by periodically discontinuing the inhalant. The sudden death due to cardiac arrest, which occurs in small birds and wild birds after the administration of ether, is attributed to stress compounded by the sympathomimetic effect of the drug.

b. *ANESTHETIC CHAMBER, CONE, MASK OR ENDOTRACHEAL ROUTE—HALOTHANE WITH OR WITHOUT NITROUS OXIDE, METHOXYFLURANE (ANES)*

The presence of air sacs in addition to the lungs makes rapid control of the level of anesthesia more difficult in birds. Intubation, suction, and forced ventilation are critical when apnea occurs. A semiclosed, nonrebreathing system is recommended rather than open mask or closed circle systems when halothane is used due to the potency of gaseous agents and possibility of accumulation of CO_2. It is not necessary to utilize a humidifier with halothane, and apnea can be treated easily by decreasing the halothane and continuing to administer oxygen. The required concentrations of halothane for induction and maintenance are 3.0 and 0.8%, respectively. Tubing and fittings should be flushed with a low concentration of anesthetic prior to use. A small bird may be fully anesthetized and die in 45 sec. Thus, continuing the concentration needed for induction is hazardous and quickly results in overdosage. Halothane and ether were the only anesthetics that were found to be safe for budgerigars. The birds recovered within 5–20 min after halothane was discontinued. Body temperature should be maintained during and following surgery. After induction in a glass jar, chamber, or plastic bag, anesthesia is maintained with a cone or mask. The depth of anesthesia in the larger birds, such as hens and turkeys, can be evaluated by pinching the skin in the cloacal area, interdigital web, comb, or wattles. Ducks and similar sized birds may be intubated and maintained with a closed or semiclosed system. The preanesthetic administration of O.04–0.10 mg/kg atropine can be used to decrease respiratory secretions, although this had not been found necessary in all birds prior to halothane and oxygen. A microliter syringe and careful attention to the dose is important especially in small birds. The presence of the volatile agent in the air sacs may persist, and continuous administration of oxygen is very helpful in enhancing recovery.

C. Cats and Dogs

1. Oral Method

Oral Route—Acetylsalicylic Acid (ANAL)

The main advantage of the salicylates is that they produce analgesia without respiratory depression. The oral administration of acetylsalicylic acid (25–50 mg/kg/8 hr) is useful in dogs but should be reduced in cats (10–25 mg/kg, S.I.D.) where it may produce gastric irritation, gastric hemorrhage, and other manifestations of acute toxicity.

2. Parenteral Methods

a. *Intravenous Route—α-Chloralose[21] (ANES)*

Chloralose (60–100 mg/kg) is used in physiological studies in which reflexes, such as the baroceptors, must not be abolished. Dogs that are maintained at a constant depth of anesthesia exhibit a progressive increase in heart rate, which stabilizes after $1\frac{1}{2}$ hours, a stable stroke volume for approximately 3 hr, and arterial hypertension. Cats exhibit a decrease in the minute volume and reactivity to carbon dioxide. Thus, chloralose is extremely valuable to the physiologist or pharmacologist but not indicated for routine studies or in clinical applications.

b. *INTRAVENOUS, Intraperitoneal, Intracardiac, Intramuscular, or Intrapleural Route— PENTOBARBITAL SODIUM (ANES) (Fig. 1)*

Pentobarbital (30–35 mg/kg, iv) is inexpensive, does not require an anesthetist after anesthetic induction, is readily available, does not require special equipment, and is not dangerous to attending personnel. The cephalic, saphenous, and femoral veins are convenient routes, while the lingual, jugular, and other veins can be used to administer supplemental amounts of anesthetic. The period of anesthesia (30 to 60 min) can be extended with additional pentobarbital, volatile agents, or other drugs. Preanesthetic agent is unnecessary in manageable dogs that have had food withheld since the previous afternoon. The first half of the calculated dose is given rapidly, while the re-

[21]Fischer Scientific Company, Pittsburgh, Pennsylvania.

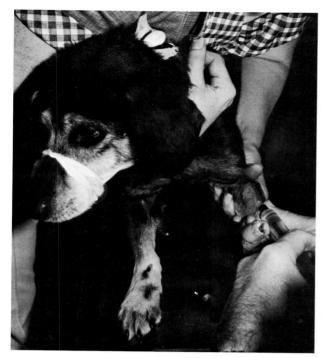

Fig. 1. Intravenous barbiturate anesthesia. Thiopental or thiamylal continues to be used for short surgical procedures or for induction prior to the administration of an inhalant anesthetic. Pentobarbital frequently is preferred for surgical exercises or terminal procedures.

mainder is injected slowly over 2 to 3 min. A respirator should be available for those animals whose respiratory rate decreases below 8 or 10/min. Some breeds, i.e. German Short Haired Pointers, are more prone to excitement during induction and recovery. The difficulty of intravenous administration of pentobarbital in cats and the long recovery period and hypothermia occasionally observed following this drug have led to the widespread use of ketamine as a preanesthetic or anesthetic. The intracardiac, intramuscular, or intrapleural routes of administering pentobarbital are not recommended. The intraperitoneal route is used in small cats, dogs, or vicious animals that are difficult to restrain or have small veins. Since the dangers of halothane, methoxyflurane, and other gaseous agents have become known, pentobarbital, ketamine, and other injectable agents appear to be more popular. A wide range of drugs including chloramphenicol will influence pentobarbital anesthesia in cats and dogs.

c. INTRAVENOUS ROUTE—THIAMYLAL SODIUM[22] OR THIOPENTAL SODIUM (ANES)

Thiopental and thiamylal (15–20 mg/kg) are useful for short surgical procedures, x-rays, and other diagnostic examinations

[22]Surital, Parke Davis and Company, Detroit, Michigan.

in cats and dogs. Generally, animals show evidence of recovery in 10 to 20 min. Induction with a thiobarbiturate is frequently followed by intubation and maintenance with an inhalant anesthetic, such as halothane or methoxyflurane. Recovery from a thiobarbiturate is usually smooth. Slow induction or continued administration of a thiobarbiturate is associated with prolonged recovery.

d. INTRAMUSCULAR, Intravenous, Subcutaneous or Oral Route—ACETYLPROMAZINE MALEATE,[23] Chlorpromazine Hydrochloride, Promazine Hydrochloride,[24] or Triflupromazine Hydrochloride[25] (PRE)

Acetylpromazine (0.5–2.0 mg/kg, iv, im, sc), chlorpromazine (1 mg/kg, iv, 2 mg/kg, im, 3 mg/kg, po), promazine (2–4 mg/kg, iv, 2.5–5.0 mg/kg, im), triflupromazine (1–2 mg/kg, iv, 2–4 mg/kg, im), and other tranquilizers reduce the dose of general anesthetic agents from 20 to 50% and make recovery smoother. The sedative effect of the phenothiazine tranquilizers, i.e., acepromazine alone, is satisfactory for some minor procedures in cats and dogs, but more frequently such drugs are used in conjunction with fentanyl, ketamine, meperidine, xylazine, or some other agent. The tranquilizers decrease many functions of the body, such as the gastric emptying time or gastrointestinal transit time, which may or may not be desirable. Most of the tranquilizers have a long duration of effect and are administered intramuscularly 30 to 60 min before anesthetic induction or restraint for diagnostic procedures.

e. INTRAMUSCULAR or INTRAVENOUS ROUTE— DROPERIDOL AND FENTANYL CITRATE (INNOVAR-VET) (PRE, ANES, ANAL)

Droperidol and fentanyl (1 ml/5–12 kg, im or iv) have been used as a (1) preanesthetic prior to barbiturate or inhalant anesthesia, (2) anesthetic for short or minor procedures, or (3) analgesic in dogs. Premedication with atropine (0.04 mg/kg, im or sc) is advised to decrease salivation and inhibit bradycardia, while acetylpromazine maleate (1.1 mg/kg, im) is useful for its tranquilizing effect and reduction in response to noise. Respiratory depression is minimal in lower doses. Narcotic antagonists, nalorphine (0.2–0.4 mg/kg) and naloxone (0.02–0.04 mg/kg), are effective in reversing respiratory or cardiovascular depression. Droperidol and fentanyl (1 ml/9 kg) produce central nervous system stimulation manifested by restlessness, increased motor activity, and incoordination in cats.

[23]Acepromazine, Ayerst Laboratories, New York, New York.
[24]Sparine, Wyeth Laboratories, Philadelphia, Pennsylvania.
[25]Vetame, E. R. Squibb and Sons, New York, New York.

f. Intramuscular or Intravenous Route—d-Tubocurarine Chloride,[26] Gallamine Triethiodide,[27] or Succinylcholine Chloride (PRE)

d-Tubocurarine (0.4 mg/kg), gallamine (1 mg/kg), or succinylcholine (0.1–0.3 mg/kg) are used infrequently in cats and dogs to supplement anesthesia. Succinylcholine in the dose of 0.1 mg/kg is satisfactory for most cats and provides prolonged relaxation in dogs. It is hazardous to administer any of these drugs without equipment for endotracheal intubation and artificial ventilation. Occasionally succinylcholine is administered as a slow infusion during orthopedic surgery to increase relaxation or during thoracic surgery to facilitate controlled ventilation. Muscular fasciculation is an undesirable side effect of these anesthetic adjuvants.

g. INTRAMUSCULAR, SUBCUTANEOUS, OR INTRAVENOUS ROUTE—KETAMINE HYDROCHLORIDE (PRE, ANES)

Ketamine is useful for minor procedures. Atropine, tranquilizers, such as acetylpromazine (2.0 mg/kg, im or sc), or promazine (2.0 mg/kg, im or sc), and xylazine (2.0 mg/kg, im) can be helpful in minor surgery, but supplemental barbiturates and/or inhalant drugs are recommended for major procedures. Although satisfactory results have been obtained following the administration of ketamine (5–44 mg/kg) in dogs and cats, tonic–clonic convulsions are observed in approximately 5% of the animals about 5 min after administration. Other disadvantages include discomfort at the site of injection, poor relaxation, and salivation. These side effects can be minimized by pretreatment with atropine (0.04 mg/kg, sc or im) and acetylpromazine (0.1 mg/kg, im). The tonic–clonic convulsions are not considered serious, but if they are prolonged they may be stopped by the intravenous injection of pentobarbital (4 mg/kg). Tissue damage at the site of injection is resolved quickly without resultant lameness. Hypertension associated with ketamine can be inhibited by tranquilizers such as chlorpromazine. An advantage of ketamine is its enhancement of cardiovascular function and the fact that it is not a respiratory depressant at the lower dosages. Ease of intramuscular administration, smooth induction, wide margin of safety, minimal side effects, and lack of renal or hepatic toxicity are additional advantages. For these reasons it is commonly used in sick or debilitated animals.

h. INTRAMUSCULAR, INTRAVENOUS, OR SUBCUTANEOUS ROUTE—XYLAZINE HYDROCHLORIDE (PRE, ANES, ANAL)

Xylazine (0.5–4 mg/kg, im, iv, or sc) may be used in place of morphine in dogs. Analgesia (15–30 min) enables minor

procedures to be performed. Atropine is advised to reduce secretions. Amount of pentobarbital required to produce surgical anesthesia is reduced 75% following the intravenous injection of xylazine or 30–40% following the intramuscular administration of this drug. A decrease in the respiratory and cardiac rates and a transient drop of blood pressure are produced by xylazine. Atropine sulfate (0.25, im) followed in 15 min by the simultaneous administration of xylazine (2.0 mg/kg, im) and ketamine (5.5 mg/kg, im) results in satisfactory plane of anesthesia that lasts about 30 min after absorption (10 min). Four of 17 beagles exhibited CNS stimulation resembling mild epileptiform seizures. Caesarean section can be performed in dogs with the aid of xylazine and lidocaine (2%) at site of incision. Xylazine also is used in conjunction with thiamylal, methohexital, and inhalant anesthetics. In conventional doses (2.0–3.0 mg/kg), xylazine produces radiographic changes that are consistent with gastric dilatation and paralytic ileus in dogs. Xylazine also is used as a preanesthetic in cats. Vomiting occurred in 5 of 17 cats following xylazine (0.5 mg/kg, im). Thus, xylazine in cats acts similarly to morphine in dogs and minimizes the chance of vomiting and its complications during general anesthesia. Convulsions are observed with doses of 3.2 mg/kg or more. The administration of xylazine (0.5 mg/kg, im) 20 min prior to ketamine (5.0–10.0 mg/kg, im) eliminates the muscular hypertonicity associated with ketamine, prolongs the duration of anesthesia, and assures a quiet recovery.

i. Subcutaneous, Intramuscular, or Intravenous Route— Atropine Sulfate (PRE)

Atropine (0.03–0.10 mg/kg) is administered with morphine or prior to ketamine, halothane, ether, or droperidol and fentanyl. Sometimes it is administered in combination with a tranquilizer such as acetylpromazine. It reduces secretions, intestinal motility, and vagal effects on the heart and other organs. It is questionable whether atropine should be used unless in conjunction with the above drugs due to its parasympatholytic effects. Cats hydrolyze atropine rapidly.

j. SUBCUTANEOUS or Intramuscular Route— MEPERIDINE HYDROCHLORIDE, MORPHINE SULFATE, Oxymorphine Hydrochloride, or Pentazocine Lactate[28] (PRE, ANAL)

Meperidine (5–10 mg/kg) is used as a preanesthetic in cats as well as for a postoperative analgesia. Meperidine is rapidly metabolized in cats (half-life of 0.7 hr), and larger doses produce excitement and respiratory depression. Morphine (1–2 mg/kg) makes dogs more manageable, produces emesis and defecation, reduces the anesthetic dosage, makes the recovery smoother, and is a potent analgesic. The main disadvantage is

[26]The Upjohn Company, Kalamazoo, Michigan.
[27]Flaxidil, American Cyanamide Company, Princeton, New Jersey.

[28]Talwin, Winthrop Laboratories, New York, New York.

the associated respiratory depression. Oxymorphine (0.2 mg/kg) resembles morphine and is approximately 10 times as potent. Oxymorphine is advocated for cesarean sections. Small doses of morphine (0.1 mg/kg) can be used in cats with minimal risk of excitement. Chlorpromazine appears to enhance the analgesic effect of small doses of morphine in cats. A distinct advantage of morphine is that it may be reversed by a morphine antagonist such as *n*-allylnormorphine, lorfan, or naloxone. Beagles, reportedly more resistant to morphine, require twice (4 mg/kg) the usual dose. Many institutions refuse to use morphine due to the risk of theft.

3. Inhalant Methods

a. Anesthetic Chamber, Cone, Mask, or Endotracheal Route—Diethyl Ether (ANES)

Diethyl ether has been used extensively in cats and dogs. The ether box was indispensible in laboratories and clinics for control of fractious cats. Halothane has replaced ether in this application. Since cats and dogs salivate readily during induction with ether, the preanesthetic administration of atropine (0.03–0.10 mg/kg) and endotrotracheal intubation are advised. An ophthalmic ointment should be applied if the corneas are exposed. Many systems of delivery including open, nonrebreathing, semiclosed, and closed have been used. Cats and dogs may be maintained for hours with minimal mortality even though the tissues of the body contain large amounts of ether and will continue to eliminate it from the body 30 hr or more. Ether stimulates the sympathetic nervous system and is a good analgesic.

b. ANESTHETIC CHAMBER, CONE, MASK, OR ENDOTRACHEAL ROUTE—HALOTHANE OR METHOXYFLURANE WITH OR WITHOUT NITROUS OXIDE (ANES) (Figs. 2 and 3)

Intravenous induction with thiamylal or thiopental (15–25 mg/kg) followed by intubation and the administration of halothane (0.5–1.5%) or methoxyflurane (0.3–1.0%) with or without nitrous oxide is the method of choice for long surgical procedures or for anesthesia in high-risk cats, dogs, and other animals. Cole or small endotracheal tubes (1.5–3.5 mm diameter) should be available for cats. Numerous types of anesthetic machines and techniques have been utilized. Halothane is more rapid in its action, while methoxyflurane is a better analgesic. With controlled ventilation, the mixture of 75% nitrous oxide and 25% oxygen does not alter the circulatory depression observed from that of 100% oxygen following the administration of halothane alone. With spontaneous ventilation, 75% nitrous oxide causes a significant increase in cardiovascular work. Nitrous oxide is less effective in animals than in human patients and must be administered with other agents. Due to the fact that halothane may be vaporized with low flows of

oxygen, it may be effectively combined with nitrous oxide and minimize cardiac depression. Halothane increases the susceptibility of the heart to arrhythmic effects of several adrenergic drugs, such as epinephrine and dopamine, but chlorpromazine, acetylpromazine, and propranolol protect against arrhythmia. The preanesthetic administration of atropine (0.04 mg/kg, sc) is optional. Ketamine may be used as a preanesthetic in fractious cats, while droperidol and fentanyl are preferred to facilitate induction in unruly dogs. Many other tranquilizers and analgesics have been used alone or in combination. The use of a vacuum-powered ventilating system surrounding the inspiratory gas supply in infant animals assists spontaneous respiration while preventing the escape of gases into the environment. Halothane or methoxyflurane are much safer in newborn animals than the barbiturates. Both cats and dogs are subject to malignant hyperthermia following the administration of halothane, but this condition is extremely rare in these species.

D. Ferrets, Foxes, and Mink

1. Parenteral Methods

a. INTRAPERITONEAL, Intramuscular, or Subcutaneous Route—PENTOBARBITAL SODIUM or Thiamylal Sodium (PRE, ANES)

Ferrets may be anesthetized with 35 mg/kg (ip) of pentobarbital. The duration of anesthesia is 30–45 min and recovery

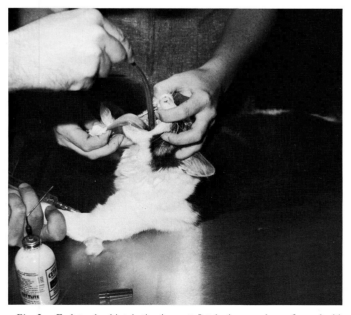

Fig. 2. Endotracheal intubation in a cat. Intubation may be performed with or without a laryngoscope after induction with a barbiturate. Tetracaine hydrochloride (Cetacaine, Haver-Lockhart, Shawnee, Kansas) is sprayed into the larynx to facilitate passage of the endotracheal tube.

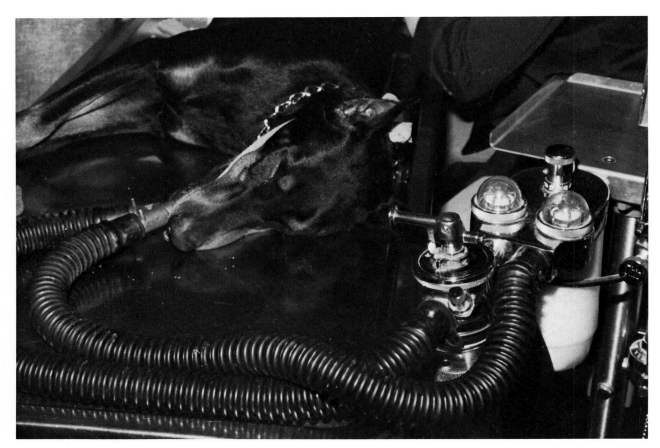

Fig. 3. Gaseous anesthesia. Halothane or methoxyflurane is recommended for long and painful surgical procedures. With some systems, "dead space" in the tubing can be a hazard.

requires 3 to 4 hr. The similar dose of pentobarbital (40 mg/kg, ip) in mink produces light or deep anesthesia, but usually there are no deaths. An intraperitoneal injection of pentobarbital (22–36 mg/kg) in mink while the animal is held by an assistant or pressed against the wire of the cage will permit examination and minor surgery or may be supplemented with volatile agents for more involved surgery.

b. INTRAMUSCULAR OR SUBCUTANEOUS ROUTE— KETAMINE HYDROCHLORIDE (PRE, ANES)

The intramuscular or subcutaneous administration of ketamine hydrochloride (10–40 mg/kg) is valuable as a preanesthetic prior to parenteral or inhalant anesthetics for major surgery or alone to enable minor surgical or diagnostic procedures to be performed.

2. Inhalant Method

ANESTHETIC CHAMBER, CONE, OR MASK—Ether, HALOTHANE, METHOXYFLURANE (ANES)

Most wild or semidomesticated carnivores can be anesthetized safely and quickly if they can be placed in an anesthetic chamber of an appropriate size where they may be viewed during induction with a volatile agent. Halothane can be administered from a cone to ferrets and mink while the animals are being restrained manually. Following induction, the anesthesia may be continued with a cone or mask for short procedures or the animal may be intubated. Atropine (0.04 mg/kg) is recommended prior to halothane or ether, if it can be administered easily and safely. Halothane is accepted much more readily than ether and has replaced ether in most laboratories. Under field conditions, weasels may be anesthetized safely without preanesthetic administration of atropine for short procedures such as weighing, examining, etc. Undescented skunks older than 5–6 weeks can be anesthetized outside with volatile agents in a container with an observation port or piece of glass over the top.

E. Guinea Pigs

1. Physical Method

Hypnosis (PRE)

Hypnosis can be used in guinea pigs as rabbits. Reserpine facilitates the onset of the hypnotic state due to a reduction in the reactivity of the animal to the external stimuli.

2. Parenteral Methods

a. INTRAVENOUS OR INTRAPERITONEAL ROUTE—PENTOBARBITAL SODIUM, or Hexobarbital Sodium (ANES) (Fig. 4)

Pentobarbital (25–40 mg/kg) can be administered intravenously, but the intraperitoneal route is more convenient. Restraining the animal in a "head down position" helps to avoid injection into one of the abdominal organs. It has been reported that 20% of peritoneal injections in guinea pigs are made into abdominal organs rather than in the peritoneal cavity and some animals take 12 hr to recover. The level of anesthesia following intraperitoneal injection of 40 mg/kg of pentobarbital may be unpredictable, and some animals require supplemental morphine (1/16–1/8 grain, sc) while other animals are severely depressed and need an analeptic prior to performing a tracheotomy and attaching them to a positive pressure respirator and administering oxygen. Generally, surgical anesthesia intervenes in 15 min and lasts for 1 hr or more. A 13% mortality has been attributed to pentobarbital anesthesia in guinea pigs. The lethal respiratory arrest dose of pentobarbital in guinea pigs varies from 45 to 70 mg/kg, which is approximately twice the anesthetic dose. Young guinea pigs are less sensitive to the depressant effects of pentobarbital than older animals. If a guinea pig begins to recover from pentobarbital, ether, halothane, or methoxyflurane can be administered with a mask or cone. The potent parasympatholytic effect of the barbiturates is helpful in reducing secretion from the mucous membranes and salivary glands prior to the administration of ether or halothane.

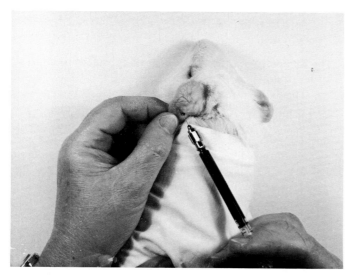

Fig. 4. Intravenous anesthesia in a guinea pig. An aural or penile vein can be used to administer pentobarbital or other parenteral anesthetic agents. (Courtesy of D. H. Going, The University of Texas, Dental Branch, Houston, Texas, 1970.)

b. Intramuscular or Subcutaneous Route—Atropine Sulfate (PRE)

The subcutaneous injection of atropine (0.1–0.2 mg/kg) may be desirable to control salivation and respiratory secretions. It is administered 30 min prior to the administration of inhalant anesthetics such as ether, but reportedly atropine can increase hemorrhage during surgery due to hypertension. Recommended doses of atropine for guinea pigs are larger than for other species. Toxic amounts of atropine dilate the cutaneous vessels in certain vascular beds. It is reported that guinea pigs may suffer laryngeal spasm if exposed to ether in an anesthetic chamber without premedication with atropine.

c. Intramuscular or Subcutaneous Route—Droperidol and Fentanyl Citrate (Innovar-Vet) (PRE, ANES, ANAL)

The administration of a combination of droperidol (20 mg/ml) and fentanyl (0.4 mg/ml) in the amount of 0.66 ml/kg produces adequate anesthesia for a laparotomy, while enucleation of the eye can be performed using a dose of 0.88 ml/kg. Animals are unable to assume sternal recumbancy for 120 min after this dose. The lethal dose is approximately 4.4 ml/kg. Smaller doses (0.08 ml/kg, im) of these combinations are used for minor procedures. In addition to death, lameness and self-mutilation at the site of intramuscular injection in 33% of the animals that received 0.88 ml/kg have led to the conclusion that this dose should not be used in guinea pigs.

d. Intramuscular Route—Ketamine Hydrochloride, Xylazine Hydrochloride (PRE, ANES)

The intramuscular administration of ketamine (44 mg/kg) produces adequate anesthesia for minor surgical procedures lasting 15–25 min. Induction requires 8–10 min, and the animals recover in 30–45 min. It is necessary to administer atropine to inhibit salivation. The range of recommended doses varies from a minimal dose of 22 mg/kg for sedation to 250 mg/kg for anesthesia. Ketamine is less effective in guinea pigs than in monkeys, cats, and certain other animals due to poor analgesia even after large doses (200–250 mg/kg). Ketamine should not be used as the sole anesthetic agent in major surgical procedures that are associated with considerable pain such as orthopedic surgery.

e. INTRAMUSCULAR, SUBCUTANEOUS, or Intraperitoneal Route—MEPERIDINE HYDROCHLORIDE (PRE, ANAL)

Meperidine (2 mg/kg, im) is useful as an analgesic or preanesthetic prior to the administration of pentobarbital (30 mg/kg, im). In this instance, the sleeping time is significantly increased.

3. Inhalant Methods

a. ANESTHETIC CHAMBER—CARBON DIOXIDE (ANES)

Carbon dioxide is used to provide short periods of anesthesia for the purpose of removing blood from the heart, making injections, or performing short diagnostic procedures. The gas is delivered from a cylinder or derived from Dry Ice. General anesthesia is induced in approximately 10 to 20 sec and lasts for about 45 sec. Carbon dioxide anesthesia increases the yield of blood from guinea pigs that are being exsanguinated. In a series of 1460 guinea pigs anesthetized by this method, there was only one death. Another advantage is that the blood drawn from animals under CO_2 anesthesia does not contain a chemical contaminant that may interfere with subsequent laboratory procedures. The immobilization produced by exposure to CO_2 can be supplemented by other intravenous or inhalant anesthetics.

b. ANESTHETIC CHAMBER, CONE OR MASK, OR ENDOTRACHEAL ROUTE—Ether, HALOTHANE, METHOXYFLURANE, NITROUS OXIDE (ANES)

Ether is dangerous in guinea pigs due to respiratory complications. Food should be withheld for 12 hr, and the eyes should be protected with an ophthalmic ointment. The administration of atropine sulfate (0.1–0.2 mg/kg) is recommended prior to exposure to ether, which stimulates a copious secretion of mucus from the mouth and nasopharynx in this species. This mucus is thought to contribute to the subsequent respiratory problems. An 8% mortality has been attributed to ether anesthesia in guinea pigs. The need for atropine is less prior to halothane and unnecessary preceeding methoxyflurane, since this agent does not cause the excessive salivation. A nose cone is suggested for methoxyflurane due to its low volatility. "Squirming" movements are occasionally encountered in animals during stage 3. Initially, one rear leg develops a waving motion that may spread to the entire body. This does not signify a return to consciousness, and a higher percentage of anesthetic is not needed. Increasing the concentration of anesthetic may result in respiratory arrest, which without artificial ventilation is followed quickly by cardiac arrest. It has been noted that premedication with atropine sulfate reduces the incidence of squirming or wriggling movements. Induction is slow, i.e., 4–8 min with methoxyflurane. Induction time may be reduced by administering CO_2 initially, but not without a risk to the animal. Concentrations of 3.0 to 3.5% of methoxyflurane may require 20 to 30 min to anesthetize guinea pigs, but the use of methoxyflurane results in only 2% mortality.

The trachea of the guinea pig is relatively small and inaccessible. Endotracheal intubation can be performed, and

guinea pigs have been maintained under controlled anesthesia with methoxyflurane for 4–6 hr. A variety of open, semiclosed, and closed systems have been utilized with or without an endotracheal tube or tracheostomy. The oxygen requirements of the animal must be considered with all systems as well as the low volatility of methoxyflurane at room temperature. Guinea pigs usually recover from methoxyflurane in 30 to 90 min after the cessation of anesthesia. This will be influenced by the duration of anesthesia and extent of the procedure. Halothane is effective in guinea pigs with an induction concentration of 3 to 5% and maintenance at 0.75 to 1.5%. Hepatic necrosis is produced by repeated administration of halothane to guinea pigs. Nephrotoxicity associated with the use of methoxyflurane in man has not been reported in guinea pigs.

F. Hamsters

1. Parenteral Methods

a. Intramuscular or Intraperitoneal Route—Ketamine Hydrochloride (PRE, ANES)

Ketamine (40–150 mg/kg, im or 100–200 mg/kg, ip) is used as a preanesthetic or anesthetic in hamsters, but the analgesia and muscular relaxation are poor. Concurrent administration of xylazine (10 mg/kg) improves the degree of relaxation and analgesia. The duration of anesthesia varies 30 min to 1 hr depending on the dose.

b. INTRAPERITONEAL, Subcutaneous, or Intravenous Route—PENTOBARBITAL SODIUM, SECOBARBITAL SODIUM, THIOPENTAL SODIUM, OR OTHER BARBITURATES (ANES)

The intraperitoneal route is the most expedient route to administer pentobarbital and other barbiturates, but care must be taken since hamsters are difficult to hold and more aggressive than rats and mice. The intraperitoneal administration of pentobarbital (50–90 mg/kg, 10 mg/ml) has been suggested for surgical anesthesia of 15 to 17 min. If a concentrated solution (60 gm/ml) is used, 90 mg/kg is recommended. The minimum lethal dose (LD_{50}) of pentobarbital is 135 to 150 mg/kg. The intravenous route via the lateral vein of the hindleg is used to administer pentobarbital (30 mg/kg) and thiopental (20 mg/kg) to hamsters. A special holding device facilitates the intravenous injection in the saphenous vein, or exposure to carbon dioxide can be utilized to immobilize the animal. Injection of barbiturates by the subcutaneous route is not recommended. An initial intraperitoneal injection of pentobarbital (30 mg/kg) can be used to facilitate a subsequent intravenous administration, i.e., urethane [50% (w/v)].

c.   Intramuscular or Subcutaneous Route—Morphine Sulfate or Droperidol and Fentanyl (Innovar-Vet) (PRE)

Morphine sulfate in doses as large as (150 mg/kg) by intramuscular, subcutaneous, and intraperitoneal injection is reported to result in analgesia without apparent narcosis or respiratory depression. This appears unique to hamsters. Droperidol and fentanyl are not recommended in hamsters since they result in CNS signs.

2. Inhalant Method

ANESTHETIC CHAMBER, CONE, OR MASK—ETHER, Ethyl Chloride, HALOTHANE METHOXYFLURANE (ANES)

Atropine (0.2–0.5 mg/kg, sc) is recommended prior to ether or halothane anesthesia. The anesthetic techniques that involve halothane and methoxyflurane as described for rats and mice also are useful in hamsters. These include the use of anesthetic chambers, cones, and masks.

G.   Mice

1. Physical Method

Hypothermia (ANES)

Hypothermia has been used prior to thymectomy in newborn, germfree mice. This is accomplished by placing the animals in a test tube sealed in the wall of Plexiglas cooling chamber which is lined with Fiberglas and filled with dry ice. The mouse experiences respiratory arrest, but the technique permits surgery for about 6 min. In 1500 mice that were thymectomized the mortality was approximately 3%, but the mothers killed one-third of the infants that were operated. Newborn animals also may be placed on an ice-cube tray in a freezer until respiration ceases and they lose their pink coloration. This occurs in 6 to 12 min, and the surgical procedure should be performed in 3 min or less, which is usually satisfactory for gonadectomy, thymectomy, nephrectomy, or splenectomy. Only two deaths were directly attributed to the anesthesia in 155 operations.

2. Parenteral Methods

a.   INTRAVENOUS ROUTE—PENTOBARBITAL SODIUM, Thiamylal Sodium, or Thiopental Sodium (ANES)

Pentobarbital can be administered via the tail vein with the aid of a simple holding device made of metal or plastic. After the mouse has been secured, the tail is immersed in warm water or a heat lamp or 100 W bulb is directed toward the animal for 10–15 min to produce hyperemia and vasodilatation. The tail of the mouse is a thermoregulatory organ that is sensitive to humidity and temperature. The ventrolateral veins of the tail become visible as the tail is bent over the index finger and held with the thumb. A 25 to 27 gauge needle attached to a 1 ml tubculin syringe or microsyringe is inserted into the vein. The initial attempt is usually made in the distal portion of the tail so that a subsequent attempt may be made more cranially. A thin flexible polyethylene catheter or tubing (PE-10) can be attached to a 22 gauge needle for perfusion or subsequent injections. No attempt is made to aspirate blood; however, the anesthetist should be careful not to inject air into the vein, since mice are sensitive to emboli of air and 0.1 ml of air may kill the mouse. The absence of retraction of the hindleg in response to pinching the toe pad, absence of corneal reflex, or the tail-pinch reflex are useful indices of surgical anesthesia. The loss of the foot reflex may occur before the loss of the corneal reflex. Metatarsal veins also can be used. The amount of required pentobarbital varies from 35 to 70 mk/kg depending on the sex, strain, etc. The intravenous dose of thiamylal or thiopental is 25–50 mg/kg. There is little advantage to using these drugs, since the duration of anesthesia following pentobarbital is only a few minutes. It is customary to dilute pentobarbital and other parenteral drugs 1 : 10 with physiological saline solution or distilled water prior to injection.

b.   INTRAPERITONEAL ROUTE—PENTOBARBITAL SODIUM (ANES)

This route of administration is used most commonly, although the wide variations between individual mice and strains make it much more hazardous than the intravenous route. The need for extreme caution in extrapolating data from one species to another is illustrated in comparing the sleeping time of rats and mice under pentobarbital anesthesia. Male mice are less susceptible to barbiturates than females, while male Swiss-Webster mice sleep longer than female mice given the same amount of pentobarbital. The amount of pentobarbital required to produce surgical anesthesia in mice (60–90 mg/kg) is much higher than equivalent anesthetic levels in cats and dogs, which is 30–35 mg/kg. Surgical anesthesia occurs 5–10 min after intraperitoneal injection and lasts for 20–30 min. For longer procedures, i.e., 40–50 min, an intraperitoneal dose of 50 mg/kg can be followed by a subcutaneous dose (25 mg/kg). Smaller doses, i.e., 40–60 mg/kg, of pentobarbital are frequently used, but these are used to study the effects of drugs, dehydration, etc., on sleeping time, loss of righting reflex, and other variables. In one study 80 mg/kg produced surgical anesthesia in 30% of the animals with 0% mortality, while 90 mg/kg resulted in anesthesia in 80% with a 20% mortality.

Supplementation of light pentobarbital anesthesia with ether, halothane, or methoxyflurane is helpful in reducing the high mortality of a second injection of pentobarbital. Normal body temperatures should be maintained during recovery. Newborn animals (1–4 days of age) may be anesthetized for approximately 1 hr by administering (5 mg/kg) of pentobarbital intraperitonealy. In order to prevent leakage through the thin abdominal wall the needle is inserted subcutaneously past the ribs, through the diaphragm and into the peritoneal cavity.

c. INTRAMUSCULAR, SUBCUTANEOUS, or Intraperitoneal Route—ACETYLPROMAZINE MALEATE, CHLORPROMAZINE HYDROCHLORIDE, OR PROMAZINE HYDROCHLORIDE (PRE)

Chlorpromazine is administered preanesthetically to extend the anesthetic period and lower the mortality observed following the intraperitoneal injection of a standard dose of pentobarbital. Large doses of chlorpromazine (50 mg/kg, im) followed by the intraperitoneal administration of pentobarbital (50 mg/kg) result in surgical anesthesia for 22 min in 80% of the animals, and 87% are alive 1 week later. Smaller doses of chlorpromazine (5–10 mg/kg) administered subcutaneously are effective and less irritating. Acetylpromazine (0.75 mg/kg im) has been combined with ketamine and xylazine. Promazine also may be administered with ketamine.

d. INTRAMUSCULAR or Intraperitoneal Route— Droperidol and Fentanyl Citrate (Innovar-Vet), MEPERIDINE HYDROCHLORIDE, MORPHINE SULFATE, or Xylazine Hydrochloride (PRE, ANAL)

Premedication with meperidine (20 and 40 mg/kg, ip) and morphine (5 and 10 mg/kg, ip) 15 min prior to pentobarbital (50 mg/kg) in mice prolongs the effects of the barbiturate. Similarly, meperidine (20 mg/kg) and morphine (5 mg/kg) convert a subeffective dose of pentobarbital into an effective one. Chronic administration of meperidine for several weeks does not significantly influence the responses to the barbiturates. Droperidol and fentanyl (0.002 and 0.005 ml/gm of a 10% solution) have been recommended as well as xylazine (2.5 mg/kg) by intramuscular route.

e. Intramuscular, Intraperitoneal or Intravenous Route— Ketamine Hydrochloride (PRE, ANES)

A wide range of doses is recommended for preanesthesia (22–50 mg/kg) and anesthesia (50–400 mg/kg) by the intramuscular, intravenous, and intraperitoneal routes. Atropine (0.04 mg/kg, im) followed by ketamine (44 mg/kg, im) produces catalepsis with surgical anesthesia that enables puncture

of the orbital sinus, exposure of veins, and laparotomy. Recovery occurs in 15 to 30 min. Ketamine (22–44 mg/kg) can be combined with acetylpromazine (0.75 mg/kg) and xylazine (2.5 mg/kg) for intramuscular injection. The combination of ketamine (80 mg/kg, im) and xylazine (16 mg/kg, im) produce excellent sedation and relaxation, but the analgesia is inadequate for major surgery. When these two agents are increased to the lethal range (ketamine, 200 mg/kg, im and xylazine, 16 mg/kg, im), the degree of analgesia continues to be inadequate before death.

3. Inhalant Methods

a. ANESTHETIC CHAMBER—CARBON DIOXIDE WITH OR WITHOUT OXYGEN (ANES)

Exposure of mice to CO_2 is a useful technique to facilitate intracardiac puncture and other procedures that last 1 min or less. Mice lose consciousness within 10 to 15 sec. The use of carbon dioxide may be favored over ether, halothane, or methoxyflurane, since it causes minimal change in blood glucose and other variables.

b. Anesthetic Chamber, Cone or Mask—Chloroform (ANES)

The use of chloroform for anesthesia and/or euthanasia merits special consideration. It has been observed that several strains of mice, i.e., A, C_3H, C_3Hf, DBA, and HR are susceptible to accidental poisoning to chloroform. Exposure for 2 to 3 hr results in lesions in the kidneys of all the males but none of the females. Some animals die soon after exposure, while others die months later. The use or storage of chloroform in an area where mice may directly or indirectly contact it, exhibits gross negligence. Carbon tetrachloride also is extremely toxic to mice, particularly the BALB/c strain. Chloroform, halothane, trichloroethylene, vinyl ether, and ethyl ether are hepatotoxic in mice by esophageal administration. Chloroform was the most toxic, but all agents caused some degree of toxicity 72 hr after exposure.

c. ANESTHETIC CHAMBER, CONE OR MASK—Ether, HALOTHANE, METHOXYFLURANE (ANES) (Fig. 5)

These agents are frequently used to induce anesthesia by means of a chamber or cone. Mice salivate when exposed to ether unless preceded by atropine (0.04 mg/kg, sc or im). For short procedures such as venepuncture ether may be used safely in a ventilated hood without premedication with atropine. Ether causes a loss of the corneal reflex before loss of the ability to retract the foot. Mice have small lungs, which accounts for a rapid induction and recovery. In comparative studies be-

g. *INTRAMUSCULAR or Intravenous Route—*
KETAMINE HYDROCHLORIDE ALONE OR WITH
Acetylpromazine Maleate, CHLORAL HYDRATE,
Diazepam, or XYLAZINE HYDROCHLORIDE (PRE,
ANES, ANAL)

The administration of ketamine hydrochloride in high dosage (44 mg/kg) produces immobilization after 8–10 min, but the degree of analgesia and muscle relaxation is inadequate for major surgical procedures. Ketamine (10–60 mg/kg) was observed to stimulate respiration. Blood pressure is increased by ketamine in rabbits, but this is neutralized by acetylpromazine (0.75–1.0 mg/kg, im) or diazepam, which is administered to potentiate ketamine and prevent convulsions. Xylazine (3, 6, or 9 mg/kg, iv) alone is a suitable analgesia with minimal respiratory depression, but it does not result in adequate immobilization. The results following intramuscular administration of xylazine are more variable. When ketamine (35 mg/kg) and xylazine (5 mg/kg) are administered intramuscularly in a single injection, analgesia and muscle relaxation suitable for orthopedic procedures are obtained. The length of anesthesia varies from 20 to 75 min. Ketamine and xylazine also have been combined with acetylpromazine. Ketamine and chloral hydrate (250 mg/kg, iv) consistently produce anesthesia for 4 hr without marked changes in the respiration or blood pressure. Ketamine (15–20 mg/kg) is compatible with the subsequent administration of halothane or methoxyflurane. Ketamine is administered intravenously with xylazine following premedication with fentanyl and fluanisone (0.1 ml/kg, im).

3. Inhalant Method

ANESTHETIC CHAMBER, MASK, CONE, OR
ENDOTRACHEAL ROUTE—Ether, HALOTHANE,
METHOXYFLURANE, NITROUS OXIDE (ANES)

Ether is satisfactory for short periods of anesthesia when administered with a mask following the injection of atropine and use of an ophthalmic ointment. Most rabbits object to induction by ether with a mask or cone and are induced more smoothly in an anesthetic chamber. Ether causes loss of the corneal reflex before loss of the ability to retract the foot. Ether is administered following barbiturates when the anesthetic period must be extended for a short period. Long exposure is associated with pulmonary edema. Halothane (4–5% for induction and 0.5–1.5% for maintenance) or methoxyflurane (1–3% for induction and 0.3–1.0% for maintenance) following preanesthesia with or without nitrous oxide is preferred. Acetylpromazine maleate (1 mg/kg) is a suitable preanesthetic

in rabbits prior to administering halothane, methoxyflurane, or other volatile agents. Intubation is quite difficult due to the restrictive nature of the oral fissure and small diameter of the glottis and trachea in some animals, but it may be performed in rabbits under barbiturate or inhalant anesthesia. Ketamine or acetylpromazine alone may not produce sufficient relaxation for endotracheal intubation. A modification of the Welch Allyn laryngoscope No. 660 or other small laryngoscope is most useful in rabbits. The cough reflex may be elicited when the endotracheal tube enters the larynx. Plastic endotracheal tubes with an internal bore of 3.0 to 3.5 mm are suitable for rabbits, while nonrebreathing or circle systems have certain advantages over "to and fro" systems in reducing dead space. A simple anesthetic apparatus with soda lime and calcium chloride absorbers has been constructed for use in rabbits. For acute nonsurvival studies a tracheostomy incision can be used to facilitate the administration of volatile agents. A peculiar characteristic of rabbits is their ability to hold their breath when being induced with volatile anesthetic agents. This state, which has been compared to laryngospasm, may be followed by a deep inspiration and result in respiratory or cardiac depression if too much anesthetic is inhaled. This peculiarity usually is not serious but has undoubtedly discouraged some investigators from using volatile agents in rabbits. Rabbits, like guinea pigs, are more susceptible to the hepatotoxic effects of halothane following chronic exposure, than rats and mice.

J. Rats

1. Parenteral Methods

a. *Intravenous or Intraperitoneal Route—α-Chloralose,*
Chloral Hydrate, Chlorobutanol, Paraldehyde,
Tribromoethanol, or Urethane (Ethyl Carbamate)
(ANES)

These agents with the possible exception of α-chloralose (55 mg/kg, ip) possess no advantages over sodium pentobarbital. Intraperitoneal injection of chloral hydrate is associated with peritonitis and subsequent adynamic ileus and should not be used alone in high concentrations for general anesthesia by this route in rats. Rats appear particularly susceptible to developing this condition. Urethane causes less respiratory and cardiac depression than the barbiturates and in doses of 1000–1250 mg/kg is suitable for abdominal surgery. This agent is reserved for long procedures that do not require recovery. The trachea should be cannulated to assist ventilation.

b. INTRAVENOUS ROUTE—Hexobarbital Sodium, PENTOBARBITAL SODIUM, Secobarbital Sodium, Sodium Thiamylal, or Sodium Thiopental (ANES) (Fig. 7)

The tail, saphenous, recurrent tarsal, and metatarsal veins provide convenient routes to administer sodium pentobarbital and other barbiturates. Appropriate devices are available to restrain the animal while dilatation of the vein can be accomplished by immersion in warm water, application of xylazine, acetone, etc. A dose of pentobarbital (30–40 mg/kg) can be individualized and the animal brought to the desired plane of anesthesia. If an indwelling plastic catheter is inserted via the lumen of the needle, supplemental amounts of pentobarbital or other drugs can be administered. As in other small animals, the anesthetic agent should be diluted 1 : 10 with physiological saline or distilled water. Diluted pentobarbital (6.0 mg/ml) can be administered safely at the dose of 50 mg/kg. Factors such as the age, sex, strain, population density, nutrition, the substances present in bedding, and the presence of ammonia will influence the ultimate anesthetic dose. Female rats metabolize some drugs such as pentobarbital more slowly than males. Rats from soiled cages are more refractory or tolerant to barbiturate anesthesia than animals maintained in clean cages. Usually, additional pentobarbital is required in young animals. Small ventilators are available to provide artificial respiration via an endotracheal tube in instances of respiratory arrest or depression. Thiamylal (20 mg/kg, iv) or thiopental (20 mg/kg, iv) is not administered frequently in rats.

c. INTRAPERITONEAL ROUTE-Hexobarbital Sodium, Methohexital Sodium, PENTOBARBITAL SODIUM, or Secobarbital Sodium, etc. (ANES) (Fig. 8)

Pentobarbital, secobarbital, methohexital, hexobarbital, and other barbiturates are administered easily in rats by the intraperitoneal route. Slightly more pentobarbital (35–45 mg/kg) is required intraperitoneally than via the intravenous route, and the response to a fixed dose is variable. Anesthesia following the injection of pentobarbital occur in 5 to 15 min and lasts 30 to 45 min. The additional amount of barbiturate that should be administered, if the initial dose is inadequate or if the animal begins to recover, is subject to judgement, but 20–25% of the original dose is recommended. Intramuscular premedication with longer acting chlorpromazine (1–2 mg/kg) will reduce the dose and extend the period of anesthesia. Other drugs, such as chloramphenicol, also prolong the anesthetic action of pentobarbital in rats. The anesthetic dose of hexobarbital is 100 mg/kg when administered intraperitoneally, while the recommended dose of methohexital is 95 mg/kg by the same route.

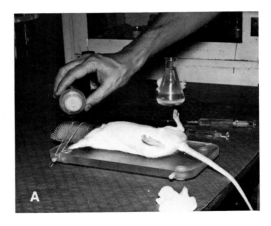

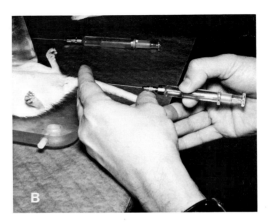

Fig. 7. Inhalant and intravenous anesthesia. Rats may be immobilized with ether (A) prior to performing venepuncture (B) and introducing a polyethylene catheter into the lumen of the vein. The catheter should be ligated in place for long procedures (C). This route can be used to inject supplement amount(s) of pentobarbital or other agent(s). The administration of ether in a hood or in front of an exhaust vent will minimize the chance of an explosion and exposure of attending personnel. (Courtesy of P. Gehring, University of Minnesota, St. Paul, Minnesota, 1964.)

d. Intraperitoneal or Intramuscular Route—Thiamylal Sodium or Thiopental Sodium (ANES)

The intraperitoneal route for the administration of thiamylal (20–25 mg/kg) or thiopental (20–25 mg/kg) in rats is not advised since the tail vein can be used more effectively. For comparative purposes, the median lethal dose (LD_{50}) of thiopental intraperitoneally in rats less than 9 months of age is 125–130 mg/kg, while those of 12 months of age and older is 115–125 mg/kg. The suggested dose of thiopental by the intramuscular route is 40 mg/kg.

e. INTRAMUSCULAR OR SUBCUTANEOUS ROUTE— DROPERIDOL AND FENTANYL CITRATE (Innovar-Vet) (PRE, ANES, ANAL)

Droperidol and fentanyl citrate (0.01–0.05 ml/100 gm) are useful for restraint, anesthesia, or analgesia. If the animal moves or exhibits discomfort, it may be supplemented with ether, halothane, or methoxyflurane. Respiratory depression or arrest is treated by artificial respiration and levallorphan (0.4 ml/100 gm).

f. INTRAMUSCULAR, Subcutaneous, or Intraperitoneal Route—KETAMINE HYDROCHLORIDE (PRE, ANES)

Ketamine (20–40 mg/kg, im) produces a state resembling general anesthesia for 20 to 40 min in rats. About 10 min are required to produce the maximal effect, and rats are fully recovered in approximately 1½ hr. Previous administration of atropine (0.04 mg/kg, sc) is helpful in minimizing salivation. Tranquilizers (acepromazine) or narcotics (meperidine) can be injected simultaneously from the same syringe to lengthen the period of anesthesia and/or reduce pain. The degree of analgesia is variable, and the muscular relaxation usually is poor. The administration of ketamine (60 mg/kg, im) followed by pentobarbital (21 mg/kg, iv) is advocated for anesthesia of 1 hr. Ketamine should be combined with an analgesic for long, painful procedures.

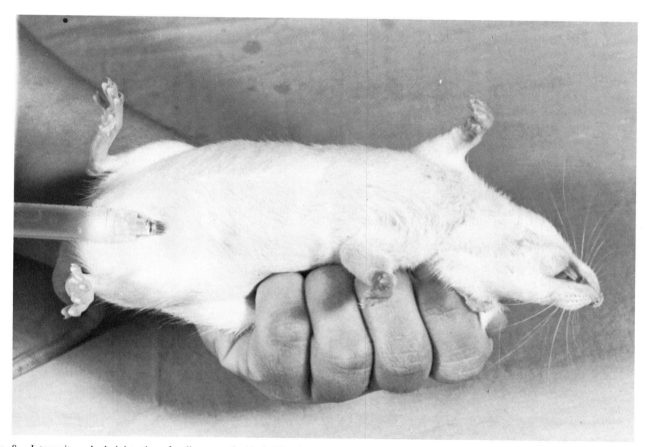

Fig. 8. Intraperitoneal administration of sodium pentobarbital. When it is not contraindicated by the investigational protocol, this method is widely used.

g. *INTRAMUSCULAR OR SUBCUTANEOUS ROUTE—MEPERIDINE HYDROCHLORIDE, MORPHINE SULFATE, OR XYLAZINE HYDROCHLORIDE (PRE, ANAL)*

Morphine (5 mg/kg, sc), meperidine (25–50 mg, im or sc) are used following certain procedures that are associated with pain. Xylazine (2.5 mg/kg) is used in conjunction with acetylpromazine (0.75 mg/kg) and ketamine (22–44 mg/kg).

h. *Subcutaneous, Intramuscular, or Intraperitoneal Route—Atropine Sulfate (PRE)*

Atropine (0.04–0.10 mg/kg, sc) is administered prior to ether anesthesia. It may be administered with ketamine, halothane, droperidol and fentanyl, or pentobarbital. The rat hydrolyzes atropine rapidly.

2. Inhalant Methods

a. *Anesthetic Chamber—Carbon Dioxide with or without Oxygen (ANES)*

Carbon dioxide alone or mixed with oxygen is very effective for short periods of immobilization to enable retro-orbital bleeding, cardiac puncture, anesthetic induction prior to maintenance with halothane, etc.

b. *ANESTHETIC CHAMBER, CONE, MASK, or Endotracheal Intubation—ETHER (ANES)*

The administration of atropine (0.04–0.10 mg/kg) prior to induction in a glass jar and maintenance by means of a small mask or cone is used successfully in rats. The eyes should be protected with an ophthalmic ointment. If the procedure can be performed in front of a hood, spillage of ether into the room is reduced. A 25 ml glass burette with a Teflon stopcock can be used to provide 5–10 drops/min of ether for the maintenance of anesthesia in an open system. Endotracheal intubation or tracheostomy with attachment to a semiclosed or closed system is used. Reasons that this dangerous form of anesthesia is acceptable are the many reports and precedents for using it for the basic procedures, such as the measurement of blood pressure in rats.

c. *ANESTHETIC CHAMBER, CONE, MASK, or Endotracheal Intubation—HALOTHANE OR METHOXYFLURANE (ANES)*

These agents are administered via a cone or mask as well as by a closed system, usually after induction in a jar. The admin-

istration of atropine will reduce salivation, while tranquilizers and/or analgesics will decrease the dose required. Techniques of intubating and maintaining rats on halothane with a closed circuit system are described. Methoxyflurane produces good relaxation and analgesia for long periods of time without any harmful effects in wild rats. No deaths occurred in several hundred rats anesthetized with this agent. A dose-related diabetes insipidus-like syndrome due to the inorganic fluoride occurs following the administration of methoxyflurane to Fischer Strain 344 rats. Resuscitation can be provided by placing one end of a rubber tube over the nose and mouth of the animal and gently breathing into the opposite end.

K. Rodents and Miscellaneous Noncarnivores

1. Parenteral Methods

a. *INTRAVENOUS ROUTE—PENTOBARBITAL SODIUM (ANES)*

The intravenous route is preferred for the administration of pentobarbital to small rodents, such as chinchillas and gerbils when a superficial vein can be located. The dose of 30 mg/kg can be used as a guideline. Most of these animals remain under anesthesia for 30 min or less. Pentobarbital (25 mg/kg, iv) is a suitable anesthetic agent for armadillos. The basal plane of anesthesia my be supplemented with a local infiltration of lidocaine for a cesarean section. Kangaroo rats have lower tolerance to anesthetic agents.

b. *Intravenous Route—Thiamylal Sodium or Thiopental Sodium (ANES)*

Thiopental has been administered intravenously in chinchillas and other small rodents. Preanesthetic medication with ketamine, pentobarbital, or other drugs is frequently necessary to facilitate intravenous injection of a thiobarbiturate. Slow intravenous infusion of thiopental over 1 hr has been used successfully in armadillos.

c. *INTRAPERITONEAL, Intramuscular, or Subcutaneous Route—PENTOBARBITAL SODIUM (ANES)*

These routes are used in chinchillas, gerbils, nutria, and other rodents. Surgical anesthesia is induced in chinchillas by 40 mg/kg (ip) of pentobarbital. Some individuals are not anesthetized by this dose but require the supplemental administration of a volatile agent. The subcutaneous route of administering pentobarbital (0.01 ml of a 60 mg/ml solution per 10 gm of body weight) has been utilized but is not recommended.

d. INTRAMUSCULAR ROUTE—ACEPROMAZINE MALEATE AND KETAMINE HYDROCHLORIDE (ANES)

The intramuscular administration of ketamine (40 mg/kg) and acetylpromazine (0.5 mg/kg) provide a satisfactory level of anesthesia in chinchillas used in auditory surgery. Induction occurred in 4–6 min, and surgery lasted 40–60 min, with recovery in 2–5 hr. There were no deaths attributable to the anesthesia in 80 animals. Intramuscular administration of ketamine (42–83 mg/kg) produced good chemical restraint in agoutis, while xylazine, droperidol-fentanyl, acetylpromazine, chlorpromazine, or promazine were unreliable or ineffective. Ketamine (15–55 mg/kg) alone is not recommended in marmots.

e. INTRAMUSCULAR ROUTE—DROPERIDOL AND FENTANYL (Innovar-Vet) (PRE, ANES)

Droperidol and fentanyl (approximately 0.3 ml/kg, im) are more satisfactory than pentobarbital (10–30 mg/kg, im) or ketamine (approximately 15–55 mg/kg, im) in marmots. They are not recommended for lemmings or ground squirrels but are useful (0.20–0.25 ml/kg, im) to facilitate surgery in armadillos.

f. INTRAMUSCULAR OR SUBCUTANEOUS ROUTE—MEPERIDINE HYDROCHLORIDE (PRE, ANAL)

Meperidine (10–15 mg, im) has been used as a preanesthetic agent in chinchillas prior to local infiltration of hexylcaine[38] along the line of incision. Recovery is rapid without depression of the young.

2. Inhalant Method

ANESTHETIC CHAMBER, CONE, MASK—Ether, HALOTHANE OR METHOXYFLURANE (ANES)

Ether produces considerable salivation in chinchillas if not preceeded by atropine. Halothane following atropine or methoxyflurane appear to be safer drugs in chinchillas, squirrels, and other small animals. Gerbils can be anesthetized with volatile agents in a jar and maintained with a nose cone. Ether is not highly recommended as the sole anesthesic for gerbils due to postanesthetic respiratory problems. Ether is hazardous in armadillos due to their tendency to hold their breath and then breathe rapidly, which leads to respiratory arrest. This is partially prevented by atropine sulfate (1–2 mg). The handling of chinchillas during recovery stimulates them to recover more promptly. Halothane is used successfully in agoutis, nutrias,

opossums, viscachas, and other less common animals. In wallabies, nitrous oxide and halothane provide satisfactory anesthesia following the intraperitoneal administration of pentobarbital (35–40 mg/kg) and intravenous injection of *d*-tubocurarine to facilitate endotracheal intubation. A higher concentration of halothane (2%) is required to maintain opossums under anesthesia.

L. Ruminants (Including Calves, Cattle, Deer, Goats, and Sheep)

1. Parenteral Methods

a. Intravenous Route—α-Chloralose and/or Urethane (ANES)

Chloralose is utilized for nonsurvival in calves, goats, and sheep where baroceptors and other reflexes must remain functional. Some prefer a 1 : 10 mixture of chloralose and urethane by weight. The recommended intravenous dose varies between 45–62 and 450–620 mg/kg, respectively, for these ingredients. Additional injections are required after 1 hr.

b. Intravenous Route—Glyceryl Guaiacolate (Guaifenesen)[39] (PRE, ANES)

This agent (7.7 mg/kg) is used alone or combined with the thiobarbiturates for short procedures or induction prior to the administration of halothane or methoxyflurane. Longer procedures should be performed under halothane or methoxyflurane anesthesia.

c. INTRAVENOUS ROUTE—PENTOBARBITAL SODIUM (ANES)

Pentobarbital has a short duration of action in goats and sheep and, therefore, it must be supplemented for long procedures. The initial anesthetizing dose (25–30 mg/kg), administered during 1–3 min via the jugular vein, is effective for approximately 20 min. The animals recover completely in 40–60 min. Endotracheal intubation with a cuffed tube is recommended to prevent the aspiration of fluid and resuscitate the animal. Fasting for 16–24 hr is important, as well as positioning the animal in a "head down" position. Although pentobarbital is still useful in the surgical laboratory for calves, goats, and sheep where the animal can be intubated and placed on a ventilator, the response is variable and it cannot be recommended where trained personnel and equipment are not available. The use of pentobarbital for intubation prior to long car-

---

[38]Cyclaine, Merck and Company, Rahway, New Jersey.

[39]Gecolate, Summit Hill Laboratories, Summit, New Jersey.

diovascular procedures in calves has been associated with prolonged recovery and is contraindicated. Pentobarbital is useful to enable intubation and maintenance with a gaseous anesthetic or for short procedures. Commercially available solutions of pentobarbital containing propylene glycol should be avoided in goats and sheep, since it may cause hemolysis and hematuria.

d. INTRAVENOUS ROUTE—THIAMYLAL SODIUM OR THIOPENTAL SODIUM (ANES)

An initial dose of 20–25 mg/kg of a 2.0–4.0% solution of either thiobarbiturate (thiamylal[22] or thiopental[6]) via the jugular vein is satisfactory for short procedures or to allow endotracheal intubation in calves, goats, and sheep. Higher concentrations (5–10%) are recommended in large sheep, goats, and cattle to reduce the volume, but these concentrations are more apt to produce a slough if injected perivascularly. Supplemental administration is safe for only a short time. If a long procedure is contemplated, the animal should be intubated and provisions made to administer a volatile agent, such as halothane or methoxyflurane. Recovery from anesthesia from pentobarbital or thiopental in goats and sheep requires about 20 min. Pentobarbital is less irritating if injected perivascularly, less expensive, and more readily available, although thiopental results in less salivation than pentobarbital. In calves being submitted to long cardiovascular procedures, induction with thiopental instead of diethyl ether or halothane requires 24 hr or longer for recovery.

e. Intravenous Route—Xylazine (PRE, ANES, ANAL)

The intravenous administration of xylazine (0.2 mg/kg) produces a state resembling general anesthesia for 45 min in cattle. This route is more rapid and predictable than the intramuscular route. Complications such as cardiac arrhythmia and diarrhea have been observed in cattle.

f. INTRAVENOUS OR INTRAMUSCULAR ROUTE— ATROPINE SULFATE (PRE)

The preanesthetic administration of atropine in ruminants is advocated since these animals salivate excessively. Withholding food for at least 24 hr before induction reduces salivation and tympanites during and after general anesthesia. Water should be withheld for 12 hr. Salivation in adult goats during surgery varies from 100 to 450 ml/hr. Tubes into the rumen will help to avoid bloat. Regurgitation of ruminal contents and salivation appear to be less serious in sheep and goats than in cattle. To be effective in goats a large initial dose (0.7 mg/kg) of atropine must be administered intramuscularly at least 15 min prior to induction, and additional doses (0.2 mg/kg) are indicated every 15 min. Lower doses of atropine are not effective in goats. A total dose of 0.4–0.8 mg of atropine has been

used in calves. In adult cattle the effective dose is greater (13 mg/kg), and it should be administered intravenously. Intubation and aspiration of the forestomachs can be used rather than preanesthetic administration of atropine, which dilates the pupils and deprives the anesthetist of this useful sign. Vagal effects due to handling of viscera may be reduced by the intravenous administration of atropine during surgery.

g. Intramuscular or Intravenous Route—Acetylpromazine Maleate, Chlorpromazine Hydrochloride, Perphenazine, Promazine Hydrochloride, Propiopromazine Hydrochloride, or Triflupromazine Hydrochloride (PRE)

Acetylpromazine (0.6 mg/kg), chlorpromazine (1.0–1.5 mg/kg), perphenazine (0.3 mg/kg), promazine (0.5–2.0 mg/kg), propriopromazine (0.2–1.0 mg/kg), triflupromazine (1.0–2.0 mg/kg), and other ataratic drugs are used in cattle, goats, and sheep that are difficult to handle. Chlorpromazine prevents cattle from rising for 1–2 hr after the administration of a volatile anesthetic agent. The long duration of action of these drugs when used prior to general anesthetic makes their use questionable. These drugs may be useful prior to local anesthesia.

h. INTRAMUSCULAR OR INTRAVENOUS ROUTE— KETAMINE HYDROCHLORIDE (PRE)

The intramuscular route is preferred for the administration of ketamine, since intravenous administration may be accompanied by excitement. Excitement is avoided in pregnant ewes by initially administering 2 mg/kg. This is followed by an infusion. Adult goats and sheep require 22–4 mg/kg, and the major effects last only 15 to 30 min. It is not a good analgesic or muscle relaxant so that it should be combined with other agents for a painful or major procedure. Swallowing, salivation, and pharyngeal and eructive reflexes are not abolished. Supplemental amounts of ketamine with or without xylazine can be given. Diazepam (2 mg/kg, iv) can be given 15 min before ketamine (4 mg/kg, iv), which is followed by an infusion of the drug for minor procedures. This technique may require intubation and the administration of halothane for major surgery.

i. Local Anesthesia—Bupivacaine Hydrochloride,[40] Procaine Hydrochloride, Tetracaine Hydrochloride (ANES)

Many procedures in cattle and other ruminants may be performed under local anesthesia. The preanesthetic administration of tranquilizers is helpful. Sheep are used to study spinal and epidural anesthesia, since the subarachnoid and epidural

---

[40]Abbott Laboratories, North Chicago, Illinois.

spaces are similar to humans, they are similar in size, and easy to handle. Details of these procedures are contained in standard texts on veterinary anesthesia.

2. Inhalant Methods

a. Mask, Cone, or Endotracheal Route—Ether (ANES)

Induction with ether causes salivation and excitement in goats and other ruminants. Following the administration of atropine and induction with thiopental or pentobarbital, ether can be used effectively. It is a very useful agent in cattle, but entails the possibility of postoperative bronchopneumonia. Ether is used also for cardiovascular surgery in calves. Escape of ether and oxygen into the area during surgery on large ruminants produce a risk to attending personnel.

b. MASK, CONE, OR ENDOTRACHEAL ROUTE— HALOTHANE OR METHOXYFLURANE WITH OR WITHOUT NITROUS OXIDE (ANES)

Calves, goats, and sheep may be induced with a 4% concentration of halothane from a mask and maintained on halothane (0.5–2.0%) in oxygen with or without nitrous oxide. The intravenous administration of thiopental or thiamylal is another effective means of inducing ruminants, and prompt intubation increases the margin of safety. Special laryngoscopes facilitate intubation in calves, goats, and sheep. Halothane has been used widely in goats and sheep for fetal or obstetrical procedures, since it is well tolerated by the dam and fetus. Sheep are frequently standing 10–20 min after the volatile agent is discontinued. Methoxyflurane is a suitable agent prior to cardiovascular surgery in calves. The cost of volatile agents is considerable for long procedures in large ruminants, and the use of a closed system should be considered. The problems of hypercapnia, metabolic acidosis, bloat with subsequent respiratory embarrassment and regurgitation, dead space, and special equipment should be carefully evaluated prior to anesthetizing cattle or calves for major cardiovascular surgery. Bloat, respiratory embarrassment, and regurgitation can be minimized in all ruminants by a 24-hr fast and aspiration of the liquid contents of the forestomachs (5 liters in mature goats) after intubation. This fluid may be kept at 38°C and returned at the end of the operation to hasten recovery. Fasting causes a 2.8 to 4.0 kg loss in weight in adult goats.

M. Swine

1. Parenteral Methods

a. INTRAVENOUS ROUTE—α-CHLORALOSE (ANES)

Chloralose (55–86 mg/kg) is used alone or after morphine and/or other preanesthetic agents when the baroceptor and chemoreceptor reflexes must be maintained. A small amount of pentobarbital may be administered to inhibit the paddling associated with chloralose. This anesthetic agent is used primarily in acute studies.

b. INTRAVENOUS ROUTE—PENTOBARBITAL SODIUM (ANES)

The intravenous administration of sodium pentobarbital (20–30 mg/kg) in a 4–6.5% solution is commonly used in swine. The lateral aural or cranial superficial epigastric veins are sites of administration. The preanesthetic administration of ketamine (20 mg/kg, im) facilitates the subsequent administration of pentobarbital at a reduced dose. It is more difficult to assess the level of anesthesia in swine, and deep anesthesia is necessary to abolish all reflexes. Small- to medium-sized swine are apt to recover quickly, (20–30 min) from pentobarbital anesthesia. Continued administration of pentobarbital for long procedures is not without risk to the animal, and this drug is more hazardous in large or older animals. Pentobarbital (8–11 mg/kg) may be used prior to induction, intubation, and maintenance of anesthesia with halothane.

c. INTRAVENOUS ROUTE—THIOPENTAL SODIUM OR THIAMYLAL SODIUM (ANES)

Administration of thiopental or thiamylal sodium (2.5– 5.0%) by means of the ear vein is an excellent means of producing a short period of anesthesia or of facilitating intubation and maintenance with a volatile agent such as halothane or methoxyflurane. The preanesthetic administration of acetylpromazine (0.5 mg/kg, im) and/or ketamine hydrochloride (20 mg/kg, im) will facilitate the subsequent administration of the thiobarbiturate. A dose of approximately 10–15 mg/kg of either thiamylal or thiopental is satisfactory for the initial dose in small- to medium-sized pigs. Swine recover from a single dose of thiopental or thiamylal in approximately 10 to 20 min. Insertion of an indwelling catheter through the ear vein, jugular, or anterior vena cava enables additional anesthetic to be injected. Endotracheal intubation is more difficult in swine, and poor relaxation of the jaws and laryngospasm are encountered. Intubation is facilitated by means of a long No. 4 Magill laryngoscope with a straight blade. Prolonged infusion of a thiobarbiturate results in saturation of the tissues, delayed recovery, and cardiopulmonary depression.

d. INTRAPERITONEAL ROUTE—PENTOBARBITAL (ANES)

The intraperitoneal route with or without preanesthetic medication is a useful route of administration of pentobarbital in small pigs or those in which the ear veins are damaged. The amount of pentobarbital required (25–35 mg/kg) is difficult to assess. Premedication with tranquilizers or supplemental use of volatile agents can be used to extend the usual anesthetic period of 30–45 min. Supplemental amounts of pentobarbital by intravenous injection via the ear or jugular vein may be used following administration of the initial intraperitoneal dose.

e. INTRAMUSCULAR OR INTRAVENOUS ROUTE— ACETYLPROMAZINE HYDROCHLORIDE, CHLORPROMAZINE HYDROCHLORIDE, Diazepam, PROMAZINE HYDROCHLORIDE, or Triflupromazine Hydrochloride (PRE)

Acetylpromazine (O.05–0.1 mg/kg), chlorpromazine (0.5–2.5 mg/kg), or promazine (1.0–2.0 mg/kg) intramuscularly with or without atropine and/or meperidine are advocated for restraint prior to administration of other agents. These agents also may be administered intravenously. The combination of acepromazine 1.5 mg/kg (im), meperidine 4 mg/kg (im), and atropine 0.04 mg/kg (im) is not as rapid or effective as thiopental (iv) prior to halothane anesthesia. Chlorpromazine (2.0–2.5 mg/kg, im), atropine (0.2 mg/kg, sc), and pentobarbital (10–20 mg/kg, iv) are useful in allowing intubation and subsequent maintenance with cyclopropane. The recommended doses of triflupromazine and diazepam are 40 mg/45 kg (iv) and 60 mg/45 kg (im), respectively. Since the phenothiazine tranquilizers are sympathetic alpha blockers and may cause hypotension, their use is contraindicated in shock or prior to long procedures. Acetylpromazine (0.11–0.22 mg/kg, im) appears to be the most predictable member of this phenothiazine group in pigs, but the dose should not exceed 15 mg.

f. INTRAMUSCULAR OR SUBCUTANEOUS ROUTE— ATROPINE SULFATE (PRE)

This agent (0.04–0.40 mg/kg) is helpful in reducing secretions but interferes with parasympathomimetic function and may produce arrhythmias during direct manipulation of the heart. Atropine in doses of 0.04 mg/kg is utilized prior to the administration of ketamine or halothane in swine, but some anesthetists omit its use. Atropine (0.1 mg/kg) is also combined with meperidine (2 mg/kg) and promazine (2 mg/kg) for preanesthetic medication in swine. The small heart of pigs may contribute to anesthetic stress.

g. INTRAMUSCULAR OR INTRAVENOUS ROUTE— DROPERIDOL AND FENTANYL CITRATE (Innovar-Vet) (PRE)

Droperidol and fentanyl in doses of 1 ml/10–15 kg (im) are used for various purposes in swine. These drugs are accompanied by ataxia, CNS stimulation, and squealing. Atropine

(0.04 mg/kg, im), droperidol, and fentanyl (1 ml/13.6 kg, im) and ketamine (11 mg/kg, im) produce satisfactory anesthesia. A small amount of pentobarbital will enhance muscle relaxation and a smooth recovery. Droperidol and fentanyl may be combined with atropine with or without succinylcholine prior to induction with a face mask, intubation, and subsequent maintenance with halothane and oxygen with or without nitrous oxide. In instances in which volatile agents cannot be used, the administration of d-tubocurarine (0.3 mg/kg) followed by an intravenous infusion of droperidol (12.5 mg) and fentanyl (0.5 mg) in 100 ml of saline over a 1 hr period has been useful. Additional administration of droperidol and fentanyl is utilized for longer procedures. Droperidol alone provides some degree of protection against induction of malignant hyperthermia by halothane in swine.

h. Intramuscular or Intravenous Route—d-Tubocurarine Chloride, Gallamine Triethiodide, or Succinylcholine Chloride (PRE)

The doses of d-tubocurarine and gallamine triethiodide, are 0.25 mg/kg (iv) and 2.0 mg/kg (iv), respectively. d-Tubocucarine is much more potent in swine than man. Gallamine is used to aid in endotracheal intubation. Atropine (0.02 mg/kg) and neostigmine (0.05 mg/kg) counteract both d-tubocurarine and gallamine. In one instance, succinylcholine (1 mg/kg im) did not permit consistent relaxation to enable endotracheal intubation, while in another instance a higher dose (2.2 mg/kg) was recommended. Most anesthetists agree that the intravenous administration of succinylcholine (0.8–1.1 mg/kg) through the marginal ear vein causes muscular paralysis and cessation of spontaneous ventilation for 4–7 min. The preanesthetic administration of ketamine (15 mg/kg, im) followed by the intravenous injection of succinylcholine (1 mg/kg) has been found adequate to permit endotracheal intubation.

i. INTRAMUSCULAR ROUTE—KETAMINE HYDROCHLORIDE AND/OR XYLAZINE HYDROCHLORIDE (PRE, ANES)

Ketamine hydrochloride (15–20 mg/kg) with or without xylazine (2 mg/kg) has revolutionized restraint and anesthesia in swine; however, spontaneous, involuntary movements may be exhibited during anesthesia. Xylazine alone is not an effective tranquilizer in swine. Ketamine (20 mg/kg, im) produces anesthesia for 10–20 min, but may be accompanied by excitement during induction and/or recovery unless supplemented with diazepam, pentobarbital, thiopental, halothane, or other agents. Ketamine and xylazine, with or without droperidol and fentanyl can be administered alone or combined with thiopental to enable endotracheal intubation or combined with pentobarbital to extend the period of anesthesia. Supplemental intravenous or volatile agents are administered in reduced amounts. In older or large swine ketamine may cause signs that resemble the porcine stress syndrome.

2. Inhalant Method

MASK, CONE, OR ENDOTRACHEAL ROUTE— HALOTHANE WITH OR WITHOUT NITROUS OXIDE OR METHOXYFLURANE (ANES) (Figs. 9–11)

Halothane in oxygen with nitrous oxide is a very satisfactory anesthetic for swine. Fasting; preanesthetic administration of atropine (0.9 mg/kg) and meperidine (1.0 mg/kg), or ketamine (20 mg/kg, im); intubation following the intravenous use of succinylcholine or thiopental; and maintenance with a semi-closed or closed system have been advocated for thoracic and other major surgery. Spraying a local anesthetic into the pharyngeal area is helpful, since pigs are sensitive to mechanical stimulation and will close the glottis if lightly anesthetized. For short procedures a mask over the snout is adequate. An anesthetic procedure utilizing halothane and nitrous oxide has been developed for surgery laboratories. Acetylpromazine (0.5 mg/kg) and atropine (0.05 mg/kg) are administered intravenously 20 min prior to intubation with an ultrashort-acting barbiturate or halothane via a mask. Anesthesia is maintained by halothane (0.6–0.8%) in nitrous oxide (60%) and oxygen (40%). Nitrous oxide (75%) in oxygen (25%) has been used in a closed system to supplement barbiturate anesthesia for short periods. The use of epinephrine or norepinephrine must be used cautiously in swine as in dogs or man under halothane

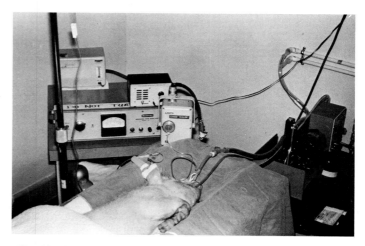

Fig. 10. Maintaining a pig under halothane and nitrous oxide anesthesia. The exhausted gases from the Harvard respirator leave the room by means of fixed tubing which is attached to the wall. The exhaust from surgical laboratories using volatile anesthetic agent(s) should not be recirculated. More halothane and nitrous oxide are used in this open system than a closed system utilizing a CO_2 absorber. (Courtesy of Dr. L. Margolin, Veterans Administration Medical Center, San Diego, California, 1982.)

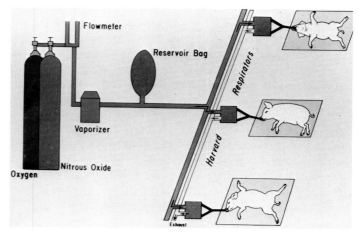

Fig. 11. The use of nitrous oxide, oxygen, and halothane to maintain surgical anesthesia in pigs. Acetylpromazine and atropine are administered prior to the administration of an ultrashort acting barbiturate to permit endotracheal intubation. The pigs are maintained with halothane (0.6–0.8%) in nitrous oxide (60%) and oxygen (40%). This system could also be utilized for other laboratory animals. (Courtesy of Dr. L. Margolin, Veterans Administration Medical Center, San Diego, California, 1981.)

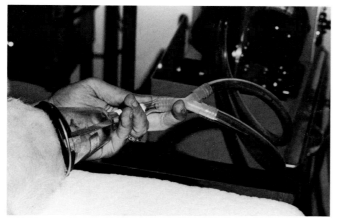

Fig. 9. Anesthetic induction with halothane and nitrous oxide by means of a nose cone in a pig. Preanesthetic medication with an ataractic and atropine is useful. Narcotics or phencyclidine derivatives may also be used with a concomitant reduction in the concentration of halothane. (Courtesy of Dr. L. Margolin, Veterans Administration Medical Center, San Diego, California, 1982.)

anesthesia due to risk of causing ventricular fibrillation and/or cardiac arrest. Swine should be selected from herds free of malignant hyperthermia, even though drugs such as droperidol or dantrolene may provide protection from halothane-induced hyperthermia. Poland China, Landrace, and Pietrain breeds appear most susceptible. Halothane, succinylcholine chloride, and stress can initiate malignant hyperthermia.

III. EUTHANASIA

Euthanasia of laboratory animals is more difficult than the euthanasia of companion and farm animals since it not only must be (1) humane and unassociated with pain, fright or struggling, vocalization or other signs of discomfort; (2) safe for the attending personnel; (3) easy to perform; (4) rapid in its action; (5) appropriate for the age, species, health, and number of the animal(s); (6) esthetically acceptable to attending personnel; (7) reliable and reproducible; (8) irreversible;[41] (9) nonpolluting or without an adverse environmental impact; (10) without a great potential for drug abuse; and (11) economical, but also must not (1) induce chemical changes in the tissues; (2) result in a chemical burden of the tissues; or (3) produce histopathological changes that will interfere with subsequent investigational studies. Usually the investigator will have information as to what method(s) can be used when special studies are required. Concentrated sodium pentobarbital causes congestion of alveolar capillaries and congestive splenomegaly as well as interferring with the microsomal liver enzyme systems, while T-66 was found to produce pulmonary edema and endotracheal necrosis of the lungs and endothelial swelling of the glomerular tuft vessels of the kidneys. Even carbon dioxide, a normal constituent of the body, causes focal alveolar hemorrhage and perivascular edema of the lungs of rats when used to produce euthanasia. Each chemical agent is associated with additional specific effects, such as interference with microsomal liver enzyme assays or the cytochrome system. Physical means, such as decapitation for mice, rats, and guinea pigs and cervical dislocation in mice, are suitable for the examination of abdominal viscera, and similar physical methods would be suitable for killing animals used for food, but they would cause protest if used to euthanatize dogs and cats. The danger of poisoning animals from the meat of animals killed by barbiturates has been reported. When catecholamines are being studied, the handling prior to euthanasia becomes important.

Three documents that are relevant to euthanasia in laboratory animals are (1) The Animal Welfare Act of 1970, (2) "The Guide for the Care and Use of Laboratory Animals," and (3) the "1978 Report of A.V.M.A. Panel on Euthanasia." These publications are subject to periodic review and change. The Animal Welfare Act and supplements, which form the legal basis of laboratory animal care in the United States, are not specific about which methods and agents must be used to euthanatize animals. "The Guide for the Care and Use of Laboratory Animals" contains information on euthanasia. The section on euthanasia is as follows:

> Euthanasia should be performed by trained persons in accordance with institutional policies and applicable laws. The choice of method depends on the species of animal and the project for which the animal was used. The method of euthanasia should not interfere with postmortem examinations or other procedures. Approved procedures for euthanasia should follow guidelines established by the American Veterinary Medical Association Panel on Euthanasia. Animals of most species can be killed quickly and humanely by intravenous or intraperitoneal injections of highly concentrated barbiturate solutions. Mice, rats, and hamsters can be killed by cervical dislocation or by exposure to nitrogen gas or carbon dioxide gas or carbon dioxide in an uncrowded chamber. Ether and chloroform are also effective, but their use is hazardous to personnel: ether is flammable and explosive, and chloroform is toxic and may be carcinogeneic. If animals are killed by ether, special facilities and procedures are required for storage and disposal of carcasses. Storage in non-explosive-proof refrigeration equipment and disposal by incineration can result in serious explosions. Signs indicating the use or presence of these toxic explosive agents should be conspicuously posted.

The Canadian equivalent to the United States Guide contains more detailed information on the methods and agents used to produce euthanasia in laboratory animals.

A. Amphibians, Fishes, and Reptiles

1. Physical Methods[42]

a. CERVICAL CONCUSSION, STUNNING[43]

Most amphibia, fish, and reptiles are euthanatized simply by a sharp blow to the back of the head. Usually this is followed by decapitation or some other physical procedure.

b. CERVICAL DISLOCATION

Small to medium-sized fish are killed by (1) inserting a pencil, the thumb, or a rod into the mouth, (2) holding the fish with the opposite hand, and (3) displacing the head dorsally.

---

[41]The euthanatist must verify that the animal is dead or that cardiac arrest has occurred before discarding the remains. One should remember that some animals that are in respiratory arrest may be minimally ventilated by cardiac action and/or may not have exhausted all their reserves of oxygen and start breathing later. When anesthetic agents are used, twice the anesthetic dose usually produces respiratory arrest, while four times the anesthetic dose produces cardiac arrest in animals that are ventilated artificially. Three times the anesthetic dose usually produces death quickly and uniformly in nonventilated animals. There is considerable variation in anesthetic or lethal doses of drugs, e.g., pentobarbital, in that mice may require twice the amount needed by cats. Thoracotomy following respiratory arrest ensures the irreversibility of the procedure.

[42]Many people consider physical means of euthanatizing animals displeasing or repulsive; however, when the rapidity of the method is considered, they may be more humane.

[43]Recommended procedures and agents are capitalized.

c. DECAPITATION with HEAVY SCISSORS, a KNIFE, or GUILLOTINE

This is an effective method to euthanatize eels. Some people prefer to stun the eels and other animals prior to decapitation.

d. Pithing

Severing the spinal cord behind the head is a rapid and effective method of killing fish, frogs, and other poikilotherms. In some animals, such as the South African clawed frog, *Xenopus laevis,* it is difficult to bend the head forward to expose the atlanto-occipital space, and other methods should be used.

e. HYPOTHERMIA, CONFINEMENT IN A REFRIGERATOR

This technique is used to immobilize animals prior to cervical concussion, dislocation, exsanguination, or other procedures.

2. Parenteral Methods

a. INTRAPERITONEAL OR INTRATHORACIC ROUTES—PENTOBARBITAL SODIUM AND OTHER BARBITURATES

Pentobarbital (60 mg/kg) and other barbiturates can be administered into the abdominal or pleuroperitoneal cavity of most cold-blooded animals. The thoracic cavity of turtles and tortoises is entered through the thoracic inlet or an intraperitoneal injection can be made between the tail and hindlegs.

b. Intrapleuroperitoneal Route—Procaine Hydrochloride or Chlorbutanol

Snakes can be killed by the injection of procaine (0.47 mg/kg) into the pleuroperitoneal cavity. Intraperitoneal administration of a saturated solution of a few milliliters of chlorbutanol kills frogs in a few minutes.

c. Percutaneous Route—Urethane (Ethyl Carbamate)

Fish, frogs, toads, and other aquatic animals are euthanatized by placing them in a 1–2% solution of urethane. The container should be covered and rubber gloves should be worn to keep this carcinogenic substance from being absorbed across the skin of the euthanatist.

d. PERCUTANEOUS ROUTE—TRICAINE METHANESULFATE (MS-222)

This agent may be administered by various routes to produce anesthesia or euthanasia. The addition of 1 to 3 gm/liter of water for aquatic animals restricts their movements prior to decapitation or exsanguination. Longer exposure results in death. This is one of the most expensive methods of killing aquatic animals, but it is very effective and does not appear to pose a hazard to personnel. Many other substances, including formalin and alcohol, may be added to the water of aquatic species to produce euthanasia.

3. Inhalant Methods

a. LETHAL CHAMBER—CARBON DIOXIDE

The animal's death should be verified before it is removed from the chamber if this method is not to be followed by a physical method.

b. Lethal Chamber, Dessicator, or Glass Jar—Ether, Halothane, and Methoxyflurane

This type of euthanasia is relatively rapid and humane for frogs, turtles, snakes, lizards, small caymen, toads, etc., but it results in exposure to personnel. Some agents require a longer exposure to produce death in cold-blooded animals, i.e., tortoises, than is observed in mammals. The anesthetic liquid form should not come in contact with the animal.

B. Birds

1. Physical Methods

a. CERVICAL DISLOCATION, MANUAL (Fig. 12)

Large birds are held in the one hand, while the other is used to overextend the neck. Blood samples may be taken from the heart immediately after death by this method. There is no chemical burden, and it is rapid and inexpensive. It may be objectionable to some personnel. Small birds may be euthanatized by twisting their neck or pressing the neck against the edge of a table. Birds may flap their wings resulting in loss of feathers and dust if the wings are not held. An alternative method is to bend the head caudally. Pressure against the trachea and esophagus until struggling ceases helps to prevent the aspiration of the contents of the crop into the respiratory passages.

Fig. 12. Euthanasia in chicks and small birds. Small chicks can be euthanatized quickly by pressing the neck against an angular object with the thumb.

b. *CERVICAL DISLOCATION with a BURDIZZO EMASCULATOR*

This is similar to the manual method. Cervical dislocation, cervical fractures, and severing the spinal cord are produced by closure of the forceps over the neck. This method is reserved for larger birds such as turkeys or geese where two people may be required. If the forceps are left in place until the bird stops struggling, the contents of the crop will not be aspirated.

c. *DECAPITATION with SCISSORS, a KNIFE, CHOPPER or GUILLOTINE*

Decapitation is a quick and effective method of euthanatizing young and small birds. This technique is useful when tissues are taken from the brain. It may be objectionable to some people.

d. *Cervical Concussion, Stunning*

A piece of wood with metal attached to the edge is used to deliver a sharp blow to the back of the head. This is followed by decapitation, exsanguination, thoracotomy, etc.

e. *Pithing with or without Cervical Exsanguination*

Major cervical vessels may be severed through the mouth without decapitating the bird after the spinal cord is severed.

2. Parenteral Method

INTRAPERITONEAL, *Intravenous, Intracardiac, Intrathoracic, or Intramuscular Routes—BARBITURATES*

Sodium pentobarbital (100 mg/kg) can be injected easily into the abdominal cavity. The administration of pentobarbital by

other routes is more difficult or less effective. Pentobarbital is mixed with chloral hydrate and magnesium sulfate to produce anesthesia and euthanasia in birds.

3. Inhalant Methods

a. *LETHAL CHAMBER—CARBON DIOXIDE*

This is relatively safe and effective means of euthanatizing large numbers of birds or chicks. The cage may be placed in a plastic bag, which is then filled with CO_2. Loss of feathers and the dust from the feathers is minimized by this method.

b. *Lethal Chamber—Ether, Halothane, Methoxyflurane, etc.*

These agents are relatively effective and humane. There may be irritation of mucous membranes and excitement before loss of consciousness. There is increasing reluctance to use any inhalant agent other than carbon dioxide due to exposure of attending personnel.

C. Cats and Dogs[44]

1. Parenteral Methods

a. *INTRAVENOUS, INTRAPERITONEAL, Intrathoracic, or Intracardiac Routes—BARBITURATES*

Sodium pentobarbital at three times the anesthetic dose (90–100 mg/kg, iv) should ensure respiratory and subsequent cardiac arrest in cats and dogs. This drug by the intravenous route is the most recommended means of producing euthanasia in cats and dogs. Premedication with ataractic or sedative drugs by intramuscular or subcutaneous routes facilitates restraint for the intravenous injection. The terminal gasp that may be observed following the intravenous injection of pentobarbital is not thought to be objectionable. Intraperitoneal injection is reserved for fractious animals. If more barbiturate is required, it is less effective and slower by this route but convenient for kittens and small puppies. Intracardiac injection can be used in dogs that are already anesthetized. It may be administered in unanesthetized animals by skilled personnel. The barbiturates are usually available in institutions that utilize cats and dogs, but they are included under the Controlled Substances Act of 1970. Some manufacturers of preparations used for euthanasia, as well as laboratories that prepare their own

---

[44]More criticism is involved in the care and treatment of cats and dogs than of other animals. The institution or attending veterinarian must be assured that the euthanasia of cats, dogs, and other animals is performed properly by either being present or being confident that personnel performing euthanasia are properly trained.

lethal solutions, add blue or red dyes so that these preparations will not be used to produce anesthesia. Other barbiturates such as secobarbital may be used in lieu of pentobarbital.

b. INTRAVENOUS, Intracardiac, Intraperitoneal, or Intrathoracic Route—T-61

This drug (0.3 ml/kg) acts as a narcotic as well as being a respiratory depressant in cats and dogs and is not a controlled or restricted drug. It is associated with pain when administered intraperitoneally, and some animals other than the horse may experience pain following rapid intravenous administration. Intracardiac, intrathoracic, and other routes are acceptable in newborn or anesthetized animals.

c. Intramuscular or Subcutaneous Route—Ketamine Hydrochloride, Narcotics, and/or Tranquilizers

These agents are used for restraint in cats and dogs. They may be combined with the subsequent administration of pentobarbital or physical means such as exsanguination in animals that are difficult to handle. They are not recommended as the sole agent(s) to produce euthanasia.

2. Inhalant Methods

a. LETHAL CHAMBER—CARBON DIOXIDE (CYLINDERS OR DRY ICE) (Fig. 13)

High concentrations (at least 40 and preferably 70%) are required. Some stress may be encountered in cats and dogs depending on the means of delivery of CO_2 and concentrations within the chamber. Filling the chamber before the animal is placed inside will improve the efficiency. Since CO_2 is heavier than air, the opening should be at the top. It appears very suitable for cats, which become unconscious within 90 sec and die in 5 min in concentrations greater than 60%. It may be expensive if an appropriate chamber is not available. Kittens and puppies may be euthanatized by placing them in a plastic bag that is subsequently filled with carbon dioxide. They should remain in the bag for 20 min after respiratory arrest to make certain they do not recommence breathing. This method is favored when an investigator needs to avoid the use of drugs or chemicals. Only one adult cat or dog should be euthanatized in the chamber or enclosure at one time. Dry ice may be used as an alternative to commercial cylinders but contact between the dry ice and the animal should be prevented.

b. LETHAL CHAMBER OR MASK—HALOTHANE OR METHOXYFLURANE

These are satisfactory agents and useful in cats that may be difficult to restrain, small puppies, and kittens. The use of

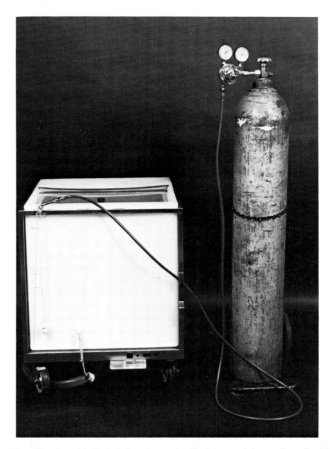

Fig. 13. Portable lethal chamber and cylinder containing carbon dioxide. This equipment is very effective in euthanatizing medium-sized dogs and smaller animals.

these agents is fostered by their wide acceptance as anesthetic agents and consequent availability. Atmospheric pollution and hazards to personnel can be minimized by scavengers and/or special ventilation.

D. Ferrets, Foxes, and Mink

1. Parenteral Method

INTRAPERITONEAL ROUTE—BARBITURATES

The intraperitoneal administration of sodium pentobarbital (100 mg/kg) is a practical means of producing euthanasia in ferrets, foxes, and mink if this drug can be used. The intravenous and intrathoracic routes have been recommended for foxes, but these routes are more difficult or more painful. Sodium pentobarbital also may be given orally (in a capsule) to foxes to produce sedation prior to killing them with other methods.

2. Inhalant Methods

a. LETHAL CHAMBER—CARBON DIOXIDE

The animal's regular cage or a small wire transport cage may be placed in the lethal chamber. Enclosing a cage in a plastic bag and filling it with CO_2 also is used to minimize handling. Only one animal should be euthanatized at a time in the same enclosure.

b. Lethal Chamber—Nitrogen

Flushing a chamber with nitrogen to replace the ambient air has been used successfully to euthanatize mink. The animals are unconscious in less than 1 min and dead in 5 min.

E. Nonhuman Primates (Including Apes, Monkeys, and Prosimians)

1. Parenteral Methods

a. INTRAPERITONEAL, INTRAVENOUS, Intrathoracic, or Intracardiac Route—BARBITURATES

Sodium pentobarbital (100 mg/kg) can usually be injected easily by intraperitoneal or intravenous route while the animal is being held or after it has been squeezed against the front of the cage. Premedication with the intramuscular or subcutaneous administration of ketamine hydrochloride facilitates the procedure. Pentobarbital may be combined with other anesthetic agents or physical means such as exsanguination.

b. INTRAMUSCULAR OR SUBCUTANEOUS ROUTE— KETAMINE HYDROCHLORIDE

The administration of this agent with the aid of a capture gun, pole syringe, blow gun, bow and arrow, or squeeze cage often is an effective first step in euthanatizing large monkeys and apes. This may be followed by the intravenous administration of sodium pentobarbital or use of an inhalant anesthetic agent.

2. Inhalant Method

LETHAL CHAMBER or MASK—CARBON DIOXIDE, HALOTHANE, OR METHOXYFLURANE

These agents are used alone in a chamber or following ketamine and/or general anesthetic agents.

F. Rabbits

1. Physical Methods

a. CERVICAL OR CRANIAL CONCUSSION (STUNNING)

This method involves striking the rabbit behind the head with a wooden or metal object or the heel of the hand as in a karate blow (rabbit punch) and is used when a drug is contraindicated. Some individuals object to this method for esthetic reasons, and this type of stunning requires skill and does not assure death. It is often followed by decapitation, thoracotomy, or exsanguination. It avoids an exogenous drug but may prevent histological studies of the brain. Hunters frequently dispose of small game by this method.

b. CERVICAL DISLOCATION

Rabbits weighing less than 1 kg can be held by the legs at the waist with one hand, while the head is held in the other and the neck is overextended. Agonal movements may cause the rabbit to scratch the person holding the animal. This technique also may be used while the rabbit is held in lateral recumbency on a table or flat surface.

c. Decapitation

This method is reserved for special studies in which drugs are contraindicated. The guillotine is dangerous to personnel. Decapitation of rabbits is objectionable to many people.

d. Exsanguination

The exsanguination of rabbits that have high titers of antibodies is performed after anesthetization. Isolation of the carotid artery or venipuncture via the intracardiac route are means of withdrawing blood.

2. Parenteral Methods

a. Intravenous Route—Air

Air, 5–50 ml/kg, produces rapid death that may be accompanied by convulsions, opisthotonos, pupillary dilatation, and vocalization. It is *not* recommended in unanesthetized animals.

b. INTRAVENOUS, INTRAPERITONEAL, Intrathoracic, and Intracardial Routes— BARBITURATES

The intravenous route is the first choice if the aural veins can be used. The intracardiac route may be painful or slow if the injection is made into the pericardial space. The intracardial route is useful if the animal is already under anesthesia. Sodium pentobarbital (100 mg/kg) is humane, safe, and efficient but consideration should be given to the chemical burden. Barbiturates are controlled substances, but they are cheap and usually available.

3. Inhalant Methods

a. EUTHANASIA CHAMBER—CARBON DIOXIDE

Concentrations of 40–100% are safe and effective for rabbits when used in a well-ventilated room. Since CO_2 is heavier than air the chamber should be opened from the top. This is the best agent used for euthanasia by inhalation of a gas.

b. Euthanasia Chamber or Mask—Ether, Halothane, or Methoxyflurane

The vapors may be irritating to the mucous membranes and causes struggling. The liquid form of the anesthetic should not come in contact with the animal.

G. Rodents (Including Chinchillas, Gerbils, Guinea Pigs, Hamsters, Mice, Porcupines, and Rats)

1. Physical Methods

a. CERVICAL CONCUSSION, STUNNING

Guinea pigs may be stunned with a sharp blow to the back of the neck. This is usually followed by exsanguination, thoracotomy, or some other procedure. This method also is used in mice and other small animals. Rats may be wrapped in a small towel prior to striking them behind the head or against a sink or table. This technique is used primarily in young rats.

b. CERVICAL DISLOCATION, MANUAL (Fig. 14)

This is very satisfactory in mice. It can be used for other animals which weigh less than 250 gm. The thumb and first finger, a pencil, a piece of wood, or metal object is placed against the back of the neck and pressed down against a firm surface such as a table top to produce the dislocation. The effectiveness of dislocation can be verified by separation of cervical tissues. Thoracic vertebrae also may be dislocated. The heart continues to beat after dislocation, which enables terminal sampling. Hamsters and guinea pigs are more difficult to euthanatize by this method due to their short necks, stronger neck muscles, and loose skin over the neck and shoulders. An alternated method of euthanatizing guinea pigs consists of

grasping them over the head from the cranial direction and/or giving the body a quick snap as with a whip.

c. Cervical Dislocation with a Burdizzo Emascultome

This method may be objectionable to personnel. It should be done by trained individuals.

d. DECAPITATION WITH A GUILLOTINE

The head must be completely severed from the body. It is reserved for instances when drugs are contraindicated. When a rat witnesses the death of another rat, it may cause a fivefold increase in plasma corticosterone. Decapitation is unpleasant and dangerous for the operator but commonly used for pharmacologic studies. Equipment should be cleaned before the next animal is decapitated. Blood may be collected in a beaker immediately after decapitation or allowed to run into a sink. This method is more difficult to use in guinea pigs, hamsters, and mice than in rats.

2. Parenteral Method

INTRAVENOUS, INTRAPERITONEAL, Intracardiac, or Intrathoracic Route—BARBITURATES

Sodium pentobarbital is the most commonly used agent. The intraperitoneal route is easier and faster if several animals are

Fig. 14. Cervical dislocation in a mouse. Euthanasia may be produced quickly and efficiently in mice and infant rats by this method. (Courtesy of G. L. Keller, University of Cincinnati, Cincinnati, Ohio, 1981.)

being euthanatized or the animal is difficult to handle such as porcupine. Intravenous injection is more difficult and time-consuming but efficient in the hands of a skilled person for mice, rats, and gerbils. The intrathoracic and intracardial methods may be painful and are recommended only in animals that are already sedated. Respiratory arrest is caused by 150–200 mg/kg by intraperitoneal administration. Animals must be checked carefully to make sure the heart has arrested, especially when pentobarbital is administered intraperitoneally. Some consider this to be the most ideal agent to produce euthanasia in the rat. Pentobarbital may be contraindicated prior to hysterectomy for the purpose of producing specific pathogen free or gnotobiotic animals due to depression of the newborn.

3. Inhalant Methods

a. LETHAL CHAMBER—CARBON DIOXIDE from a CYLINDER or DRY ICE

This is a safe, humane, economical, and rapid method that is considered to be ideal for rodents, since several mice, rats, or other small animals can be euthanatized simultaneously. The cages may be placed in a large chamber or in a plastic bag that is then filled with CO_2. This latter method may utilize more CO_2 or require more time than a chamber, which can be precharged and opened from the top. Newborn rodents are more resistant to CO_2 than older animals. Some institutions save excess rodents for local zoos. It is important that such animals are euthanatized with carbon dioxide rather than barbiturates or other drugs to avoid secondary poisoning of zoological animals and birds. This method should be used in a well-ventilated room. Sublethal exposure to CO_2 may precede exsanguination. The yield of blood has been reported to be increased following this procedure in guinea pigs.

b. Lethal Chamber—Ether, Halothane, or Methoxyflurane

These agents are satisfactory but may be irritating to the mucous membranes (see other species). The chamber must be placed in a well-ventilated area. Ether is explosive and contraindicated if lipid studies are being performed. It stimulates the pituitary adrenal axis. Containers used to euthanatize rodents with ether should be glass rather than plastic. Halothane is the most rapid and more desirable for this reason, with respiratory arrest occurring in 30 sec with minimal struggling. None of these agents in the liquid form should come in contact with the animal. This can be accomplished by separating the animal from the agent by means of a wire screen. Cost and exposure of personnel are disadvantages of these agents.

H. Ruminants (Including Calves, Cattle, Deer, Goats, and Sheep)

1. Physical Methods

a. Electrocution or Electrical Stunning

These methods appear more acceptable with bovine and equine species. The electrodes should be placed so that the current passes through the brain. Techniques in which the current passes through the heart rather than the brain are considered to be inhumane. If the animal is only stunned, exsanguination or some other method must be used. These methods are used in abattoirs where large numbers of animals are slaughtered.

b. SHOOTING WITH A CAPTIVE BOLT

Captive bolts, either penetrating or nonpenetrating, are discharged by gunpowder or compressed air. This is an acceptable method in abattoirs, but may be objectionable to attending personnel in the laboratory setting. The animal must be properly restrained and the muzzle of the gun should be placed at a right angle to the skull and directed toward the center of the brain. The gun should not be fired when the animal is moving its head. The penetrating captive bolt is preferred in sheep due to the bone ridge of the skull and wool on the head that dissipates the force of a nonpenetrating bolt. This procedure is usually followed by exsanguination if the carcass is to be used for food.

c. Exsanguination

Exsanguination is used secondarily to shooting with a captive bolt, cranial concussion, or electrical stunning. It is not recommended as the primary means of producing euthanasia. Federal regulations exempt Kosher exsanguination from previous stunning, electrocution, or shooting.

2. Parenteral Method

INTRAVENOUS ROUTE—BARBITURATES

Sodium pentobarbital (100 mg/kg) injected via the jugular vein is rapid and well-tolerated for calves, sheep, and goats used in surgical research. Cardiorespiratory arrest also can be produced effectively by the intravenous administration of a solution containing chloral hydrate (30 gm), magnesium sulfate (15 gm), and sodium pentobarbital (6.6 gm) in distilled water (1000 ml). This may be combined with exsanguination, thoracotomy, and other means.

3. Inhalant Method

Euthanasia Cone or Mask—Halothane or Methoxyflurane

Small sheep (lambs) or goats (kids) can be euthanatized by placing their head in a cone or mask. This method is not recommended for larger animals, cows, bulls, or animals that are not easy to restrain.

I. Swine

1. Physical Methods

a. Electrical Stunning

It has been demonstrated to be effective when special equipment is available.

b. Shooting with a Captive Bolt

The captive bolt may be discharged by gun powder or compressed air. This method has advantages when the use of drugs is contraindicated. Special enclosures are useful to control the animal. The bolt should be directed toward the brain from a central point slightly above a line between the eyes. Careful placement of the gun is required, since the brain of the pig is small and protected by sinuses. This may be followed by exsanguination.

2. Parenteral Methods

a. INTRAVENOUS, INTRAPERITONEAL, Intrathoracic, or Intracardiac Route—BARBITURATES

Sodium pentobarbital (100 mg/kg) is a very useful anesthetic agent in swine. It is difficult to administer this agent intravenously in small pigs or through veins that have recently been used. The intraperitoneal route can be used or ketamine can be administered prior to the intravenous injection to facilitate the administration of pentobarbital.

b. Intramuscular or Subcutaneous Routes—Ketamine Hydrochloride

This drug is an excellent restraint drug in swine and will facilitate intravenous injection of pentobarbital or euthanasia by physical or other chemical means.

3. Inhalant Method

Euthanasia Chamber or Special Enclosure—Carbon Dioxide

Carbon dioxide in concentrations of 65–90% has been used to immobilize and narcotize swine in abattoirs for several years with varying success. Convulsive seizures may preceed collapse. This method is useful for small pigs in the laboratory.

ACKNOWLEDGMENTS

The author wishes to thank Mrs. Patti Barrett, Mr. Jerry Lubinski, and Ms. Carol Perkins for their assistance in preparing this chapter.

REFERENCES[45]

Anonymous (1970). "Animal Welfare Act of 1970," Public Law 89-544 as amended by 91-579 91st Cong., H.R. 19846, enacted Dec. 29, 1970.

Anonymous (1978). "Guide for the Care and Use of Laboratory Animals," DHEW Publ. No. (NIH) 78-23, Revised. U.S. Dept. of Health, Education and Welfare, Public Health Serv., Natl. Inst. Health, Washington, D.C.

Anonymous (1980). "Guide to the Care and Use of Experimental Animals," Canadian Council on Animal Care, Ottawa, Ontario.

Arnold, J. P., and Usenik, E. A. (1970). Anesthesia. In "Disease of Swine" (H. W. Dunne, ed.), 3rd ed., pp. 981–989. Iowa State Univ. Press, Ames.

Baker, H. J., Lindsey, J. R., and Weisbroth, S. H., eds. (1980). "The Laboratory Rat," Vol. 2. Academic Press, New York.

Bree, M. M., Cohen, B. J., and Abrams, G. D. (1971). Injection lesions following intramuscular administration of chlorpromazine in rabbits. *J. Am. Vet. Med. Assoc.* **159**, 1598–1602.

Bustad, L. K., and McClellan, R. O. (1965). "Swine in Biomedical Research." Frayn Printing, Seattle, Washington.

Calderwood, H. W. (1971). Anesthesia for reptiles. *J. Am. Vet. Med. Assoc.* **159**, 1618–1625.

Callis, J. J., Cohl, E. H., Bustad, L. K., Jennings, L. F., Matthews, P. J., Reisinger, R. C., Teague, H. S., and Watson, D. F. (1971). "Swine Standards and Guidelines for the Breeding, Care and Management of Laboratory Animals," Natl. Acad. Sci., Washington, D.C.

Callis, J. J., Boyd, L. L., Edward, A. G., Flatt, W. B., Jacobson, N. L., Jennings, L. F., Moreland, A. F., Neumann, A. L., Smith, C. K., Terrill, C. E., and Warwick, E. J. (1974). "Ruminants, Cattle, Sheep and Goats," Natl. Acad. Sci., Washington, D.C.

Clifford, D. H., and Soma, L. R. (1964). Anesthesiology. In "Feline Medicine and Surgery. A Text and Reference Work" (E. J. Catcott, ed.), pp. 392–460. Am. Vet. Publ., Inc., Santa Barbara, California.

Cramlet, S. H., and Jones, E. F. (1976). Anesthesiology. In "Selected Topics in Laboratory Animal Medicine," Vol. V, pp. 1–110. U.S.A.F. School of Aerospace Medicine, Brooks Air Force Base, Texas.

Edgson, F. A., and Payne, J. M. (1967). The dangers of poisoning domestic pets with meat from animals subjected to barbiturate euthanasia. *Vet. Rec.* **80**, 364.

Fawell, J. K., Thomson, C., and Cooke, L. (1972). Respiratory artefact produced by carbon dioxide and pentobarbitone sodium euthanasia in rats. *Lab. Animals* **6**, 321–326.

[45]Approximately 2000 books, articles, and reports were used in the preparation of this chapter. Fifty references were selected as a general bibliography. Many excellent older, foreign or specific references were omitted.

Fedde, M. R. (1978). Drugs used for avian anesthesia: A review. *Poult. Sci.* **57,** 1376–1399.

Feldman, D., and Gupta, B. N. (1976). Histopathologic changes in laboratory animals resulting from various methods of euthanasia. *Lab. Anim. Sci.* **26,** 218–221.

Fiennes, R. N. T.-W., Harrison, F. A., Ray, P., and Scott, W. N. (1976). "The UFAW Handbook on the Care and Management of Laboratory Animals," 5th ed. Churchill-Livingstone, Edinburgh and London.

Fleischman, R. W., McCracken, D., and Forbes, W. (1977). Adynamic ileus in the rat induced by chloral hydrate. *Lab. Anim. Sci.* **27,** 238–243.

Frost, W. W. (1977). "Analgesics, Hypnotics, Sedatives and Anesthetics Used in Laboratory Animals," pp. 1–16. Am. Coll. Lab. Anim. Med., Washington State University, Pullman.

Frye, F. L. (1973). Surgery and pathology. *In* "Husbandry, Medicine and Surgery in Captive Reptiles," pp. 112–113. Vet. Med. Publ., Bonner Springs, Kansas.

Gay, W. L., ed. (1965). "Methods of Animal Experimentation," Vol. 1. Academic Press, New York.

Glen, J. B., and Scott, W. N. (1972). CO_2 euthanasia in cats. *Vet. Rec.* **90,** 633.

Graham-Jones, O. (1964). "Small Animal Anaesthesia," Macmillan, New York.

Green, C. J., Knight, J., Precious, S., and Simpkin, S. (1981). Ketamine alone and combined with diazepam or xylazine in laboratory animals: A 10 year experience. *Lab. Animals* **15,** 163–170.

Hall, L. W. (1971). "Veterinary Anesthesia and Analgesia," 7th ed., Bailliere.

Hannah, H. W. (1976). Euthanasia—Some legal aspects. *J. Am. Vet. Med. Assoc.* **168,** 32.

Harkness, J. E., and Wagner, J. E. (1977). "The Biology and Medicine of Rabbits and Rodents," Lea & Febiger, Philadelphia, Pennsylvania.

Heath, R. B. (1980). Chemical restraint and anesthesia. *In* "Bovine Medicine and Surgery" (H. E. Amstutz, ed.), 2nd ed., Vol. 2, pp. 1125–1128. Am. Vet. Publ., Inc., Santa Barbara, California.

Herin, R. A., Hall, P., and Fitch, J. W. (1978). Nitrogen inhalation as a method of euthanasia in dogs. *Am. J. Vet. Res.* **39,** 989–991.

Hughes, H. C. (1981). Anesthesia of laboratory animals. *Lab. Anim.* **10,** 40–56.

Kaplan, H. M., and Timmons, E. H. (1979). "The Rabbit: A Model for the Principles of Mammalian Physiology and Surgery." Academic Press, New York.

Keller, G. L. (1982). Physical euthanasia methods. *Lab. Anim.* **11,** 20–26.

Leash, A. M., Beyer, R. D., and Wilbur, R. G. (1973). Self-mutilation fol-lowing Innovar-Vet injection in the guinea pig. *Lab. Anim. Sci.* **23,** 720–721.

Lumb, W. V., and Jones, E. W. (1973). "Veterinary Anesthesia." Lea & Febiger, Philadelphia, Pennsylvania.

McDonald, L. E., Booth, N. H., Lumb, W. V., Redding, R. W., Sawyer, D. C., Stevenson, L., and Wass, W. M. (1978). Report of the AVMA panel on euthanasia. *J. Am. Vet. Med. Assoc.* **173,** 59–72.

McPherson, C. (1971). Diseases of mice, rats, guinea pigs, and rabbits. *Curr. Vet. Ther.* **4,** 442–450.

Miller, E. V., Ben, M., and Cass, J. S. (1969). Comparative anesthesia in laboratory animals. *Fed. Proc., Fed. Am. Soc. Exp. Biol.* **28,** 1369–1586.

Milligan, J. E., Sablan, J. L., and Short, C. E. (1980). A survey of waste anesthetic gas concentrations in U.S. Air Force veterinary surgeries. *J. Am. Vet. Med. Assoc.* **177,** 1021–1022.

Petrak, M. L. (1982). "Diseases of Cage and Aviary Birds." Lea & Febiger, Philadelphia, Pennsylvania.

Port, C. D., Garvin, P. J., Ganote, C. E., and Sawyer, D. C. (1978). Pathologic changes induced by an euthanasia agent. *Lab. Anim. Sci.* **28,** 448–450.

Riebold, T. W., Goble, D. O., and Geiser, D. R. (1982). "Large Animal Anesthesia. Principles and Techniques." Iowa State Univ. Press, Ames.

Sawyer, D. C. (1965). "Experimental Animal Anesthesiology," U.S.A.F. School of Aerospace Medicine, Brooks Air Force Base, Texas.

Sawyer, D. C. (1982) "The Practice of Small Animal Anesthesia. Major Problems in Veterinary Medicine," Vol. 1. Saunders, Philadelphia, Pennsylvania.

Short, C. E. (1974). "Clinical Veterinary Anesthesia. A Guide for the Practitioner." Mosby, St. Louis, Missouri.

Short, D. J., and Woodnott, D. P. (1963). Humane killing. *In* "ATA Manual of Laboratory Animal Practice and Techniques," pp. 231–243. Thomas, Springfield, Illinois.

Soma, L. R. (1971). "Textbook of Veterinary Anesthesia," Williams & Wilkins, Baltimore, Maryland.

Valerio, D. A., Bullock, B. C., Hall, A. S., Keeling, M. E., Manning, P. J., and Nolan, M. A. (1973). Nonhuman primates. *In* "Standards and Guidelines for the Breeding, Care, and Management of Laboratory Animals," 2nd ed., pp. 1–61. Natl. Acad. Sci., Washington, D.C.

Wagner, J. E., and Manning, P. J., eds. (1976). "The Biology of the Guinea Pig." Academic Press, New York.

Weisbroth, S. H., Flatt, R. E., and Kraus, A. L., eds. (1974). "The Biology of the Laboratory Rabbit," pp. 1–496. Academic Press, New York.

Westhues, M., and Fritsch, R. (1965). "Animal Anaesthesia. General Anaesthesia," Vol. II. Lippincott, Philadelphia, Pennsylvania.

Chapter 19

Techniques of Experimentation

W. Sheldon Bivin and Gerald D. Smith

I. INTRODUCTION

The information presented in the literature covers a wide selection of animal species and a voluminous number of the experimental techniques. The reader should be aware of general reference texts that are available (Gay, 1965-1974; Melby and Altman, 1974–1976; UFAW Staff, 1976). It will be the purpose of this chapter to select and summarize the available information in an attempt to emphasize two major concepts.

1. The biomedical technique employed in animal experimentation is often the critical factor in determining the success or failure of a research protocol.

2. A mastery of selected techniques is extremely useful to the veterinary clinician in performing diagnostic and therapeutic procedures.

The format of this chapter is to discuss or outline by organ system one or more of the following: (1) procedures for administration of drugs and collection of biological specimens, (2) collection of physiological data, (3) surgical procedures, postoperative care, and advantages and disadvantages of alternative ways to perform the same procedure, and (4) references that allow the reader access to detailed descriptions for any described technique. References are cited for complex techniques not described in this text.

Since anesthesia techniques are covered in depth in another chapter we will not discuss anesthesia as it relates to specific surgical procedures. Reference will be made to the type of anesthesia and instrumentation required when it is critical in the performance of the described experimental technique.

II. IDENTIFICATION METHODS

Cage cards can be used to identify groups of rodents or individually housed animals. The information on the card should include the name and location of the responsible investigator, source of the animal, and the strain or stock. In addition to the cage card, individual animals should be identified with a marker. Natural characteristics and coat coloration could be recorded and used to identify individual animals, but this process is cumbersome for large numbers of animals and impossible to use for animals identical in appearance. Table I lists some of the common markers for laboratory animal species.

Animals can be easily marked by the application of dyes or ink to the fur or tail. Felt tip pens work well, but the mark is not permanent and may rub off onto cagemates.

Holes and notches may be placed in the ear with commercially available forceps.* An established code should be fol-

*Michi-Crown, Bay City, Michigan.

Table I

Identification Markers

Animals	Marker
Mouse, rat, hamster	Ear punch, toe clip, dye, tatoo the tail, ear tags
Guinea pig	Ear tags, dye, ear punch, tattoo ear
Rabbit	Tattoo ear, ear tags, dye
Dog	Collar with tag, tattoo, ear tags
Cat	Collar with tags, tattoo
Nonhuman primate	Tattoo, collar with tags

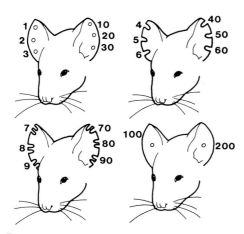

Fig. 1. Ear notch code. (From Harkness and Wagner, 1983; used with permission.)

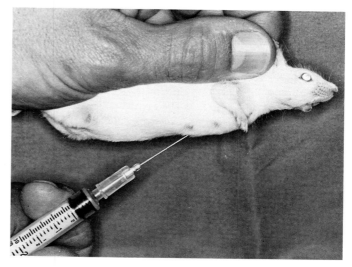

Fig. 2. Cardiac puncture in a mouse. A 22 gauge needle is inserted in the region of the xiphoid cartilage.

lowed (Fig. 1) (Dickie, 1975). This method produces a permanent mark if the ear is not self-mutilated or chewed on by cagemates.

Amputation of toes with scissors according to an established code (Kumar, 1979) will produce a permanent mark. The technique is more traumatic than ear notching, and infection could develop in the wound.

Tattoos are applied with pliers* or electro-vibrators.† The tail or feet are the best sites for tattooing rats and mice (Avery and Spyker, 1977; Schoenborne *et al.*, 1977; Greenham, 1978). The inner surface of the ear can be tattooed in guinea pigs, rabbits, dogs, and cats. Nonhuman primates are usually tattooed on the abdomen or chest. The technique should be performed aseptically, and instruments must be disinfected after each animal.

A variety of types and size ear tags are commercially available. Small aluminum ear studs are used in rats and guinea pigs and larger "ketchum tags" are used in rabbits and dogs. All tags should be placed in the ear according to the manufacturers' directions. It is essential that the tag is not too tight because pressure necrosis and infection will occur. If the tag is too loose it can be easily torn out.

III. BLOOD COLLECTION AND INTRAVENOUS INJECTION

A. Rodents

Cardiac puncture is used for the collection of small serial samples as well as for the collection of a large quantity of blood within a short period of time. However, the technique is difficult, and an unexperienced operator may be frustrated by

*Stone Manufacturing and Supply Co., Kansas City, Missouri.
†Aims Inc., Piscataway, New Jersey.

missing the heart entirely and not collecting a sample, or by causing cardiac tamponade and death when serial samples are needed. From a humane perspective, cardiac puncture should be performed only on anesthetized animals. Cardiac puncture may be accomplished by inserting the needle through the ventral abdominal wall just lateral to the xiphoid process. The needle is inserted at a 10–30° angle above the plane of the abdomen and directed caudocephalicly toward the heart (Ambrus *et al.*, 1951; Falabella, 1967; Kraus, 1980; Simmons and Brick, 1970; Waynforth, 1980) (Fig. 2). Frankenberg (1979) describes insertion of the needle through the thoracic inlet. For animals larger than mice the needle may be inserted through the lateral thoracic wall in the region of maximum palpitation of the heart (Burhoe, 1940; Moreland, 1965) (Fig. 3).

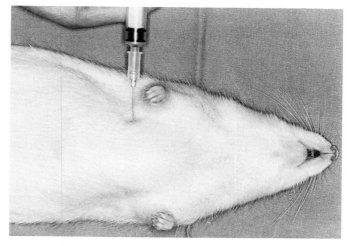

Fig. 3. Cardiac puncture in a rat. A 20 gauge needle is inserted through the right thoracic wall at the point of maximum heart palpitation.

The technique of collecting blood from the orbital sinus is simple to learn, requires minimal equipment, and reliably produces small blood samples. The eye and health of the animal seem to be unaffected when the procedure is properly performed on an anesthetized animal (Cate, 1969; Grice, 1964; Pansky *et al.*, 1961; Simmons and Brick, 1970; Sorg and Buckner, 1964; Stone, 1954). The animal is held on a flat surface and the operator's thumb is used to apply pressure to the external jugular vein immediately caudal to the mandible and thus occlude venous return from the orbital sinus. The forefinger of the same hand is used to pull the dorsal eyelid back and produce slight exophthalmos (Fig. 4). Usually a glass capillary tube or Pasteur pipette is used to penetrate the orbital conjunctiva and rupture the orbital sinus; however, Cate (1969) describes the use of small bore polyethylene tubing cut with a beveled tip. Sorg and Buckner (1964) state that introduction of the tube into the lateral rather than the medial canthus reduces the incidence of epistaxis and eye trauma associated with the technique. However, Timm (1979) describes the orbital venous anatomy of the rat and recommends directing the tube in a caudal and medial direction through the dorsal conjunctiva. This is necessary because the rat has an orbital plexus rather than a venous sinus, and the largest vein of this plexus is located deep within the orbit. Once the sinus or plexus has been ruptured, blood will flow through and around the tube into a collection vessel. Blood flow will cease when the tube is released and pressure is removed from the external jugular vein.

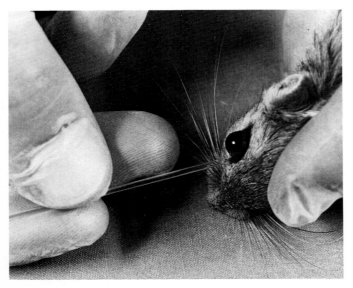

Fig. 4. Collection of blood from the orbital sinus of a gerbil with a capillary tube. Note how traction applied by the forefinger produces exopthalmos.

Collection of blood from the tail is easy to do, minimal equipment is required, serial samples can be collected, hemolysis can be controlled, and it is safe for the animal. Tail bleeding usually involves amputation of the tip of the tail or laceration of the blood vessel within the tail. If a vacuum apparatus is used large samples can be obtained (Levine *et al.*, 1973; Nerenberg and Zedler, 1975; Stuhlman *et al.*, 1972). The primary disadvantages of tail bleeding techniques, which involve laceration or amputation of a portion of the tail, is that blood may not flow freely from the wound, and a clot may form before a sample of adequate volume is obtained. Heparin or citrate solution may be applied directly to the wound to slow clot formation (Ambrus *et al.*, 1951; Lewis *et al.*, 1976). Another technique is to partially lacerate the ventral artery of the tail with a sharp razor blade. This technique prevents constriction of the vessel and improves the yield of blood (Fields and Cunningham, 1976).

Blood samples from and intravenous injection into the tail veins of rodents is another frequently used technique. Proper restraint is essential, and effective restraint chambers are commercially available, or one may be constructed from a plastic syringe (Furner and Mellett, 1975). Transillumination of the tail with a light source will improve visualization of the veins (Kaplan and Wolf, 1972; Keighley, 1966; Mylrea and Abbrecht, 1967) and occlusion of the veins at the base of the tail will also facilitate injection (Barrow, 1968). Compression of the lateral tail vein without compressing the middle coccygeal artery may be accomplished with a pair of forceps designed for wound clips (Bergström, 1971). A 27 gauge or smaller needle is used for venipuncture in the mouse, and a 19 gauge or smaller for the rat. Ambrus *et al.* (1951) and Burhoe (1940) observed that anesthetizing rats and mice with ether seemed to improve blood flow and yielded larger samples. Warming the tail (Ambrus *et al.*, 1951; Fields and Cunningham, 1976; Levine *et al.*, 1973;) or warming the entire animal (Lewis *et al.*, 1976; Stuhlman *et al.*, 1972) seems to increase blood flow. Xylene or xylol applied to the tail will also facilitate blood flow (Ambrus *et al.*, 19751; Burhoe, 1940; Stuhlman *et al.*, 1972). It is essential that the tail be rinsed to remove all chemicals on completion. Other disadvantages of the tail bleeding techniques include stimulation of the sympathetic nervous system due to the pain with resulting vasoconstriction (Carvalho *et al.*, 1975), significant differences between samples from the orbital sinus and tail of mice, sample to sample variation of blood samples in the same animal (Sakaki *et al.*, 1961), and the mixing of venous and arterial blood plus extravascular tissue fluids in samples.

Percutaneous puncture and bleeding from the jugular vein with a needle and syringe are described (Kassel and Leviton, 1953; Phillips *et al.*, 1973). Because the technique is relatively safe it can be used to collect serial samples. Success with the procedure is largely dependent upon proper restraint and positioning of the animal. The neck should be held in hyperextension by fastening a strip of gauze behind the upper incisors and pulling the head back or to the side. A depilator, applied to the ventral neck region, will make identification of landmarks easier. The site for venipuncture is just cephalad to the point where the external jugular vein passes between the pectoral muscle and clavicle. If the needle is inserted through the pectoral muscle it is stabilized better within the vein. Reliability of jugular vein techniques for blood collection can be improved by anesthetizing the animal and surgically exposing the vessel through a skin incision. Once the vessel is dissected, blood can be collected with a needle and syringe, by cannulation, or by severing the vessel and allowing blood to flow directly into a collection vial. Usually one jugular vein is occluded while blood is collected from the opposite vein.

The dorsal metatarsal vein in the rat (Nobunga et al., 1966) is an excellent site for simple intravenous injection with a needle and syringe. By grasping the animal's limb at the stifle joint, the vein is compressed and the leg is immobilized in extension. It is necessary to clip hair from the venipuncture site, and a 27 gauge or smaller needle is used. Other sites well suited for simple intravenous injection in rodents are the sublingual vein (Greene and Wade, 1967; Waynforth and Parkin, 1969) and the penile vein (Grice, 1964; Karlson, 1959). The animals must be anesthetized, and venipuncture performed with a 26 gauge needle.

Decapitation can be used to collect blood from smaller rodents, and the procedure may be accomplished with a commercially available guillotine or autopsy shears (Krause, 1980). Following decapitation, blood flowing from the severed neck is collected in a funnel. The technique is esthetically offensive, dangerous for the operator, and pain may be perceived by the animal for a short period of time following decapitation (Mikeska and Klemm, 1975). Blood collected in this manner will also be contaminated with tracheal and salivary secretions.

Large quantities of blood can be obtained in terminal experiments by severing large vessels and exsanguinating the animal. The inherent disadvantage of such a technique is that only one sample can be collected. Exsanguination techniques should be performed only on anesthetized animals. One to 1.5 ml of whole blood may be collected from mice by incising the brachial vessels (Young and Chambers, 1973). Large amounts of blood can be collected from the abdominal aorta by severing the vessel (Lushbough and Moline, 1961) or by aspirating with a needle and syringe (Grice, 1964).

Blood can be collected from fetal and newborn rodents by severing the jugular and carotid vessels (Smith and McMahon, 1977), by decapitation or amputation of an extremity (Grazer, 1958), or by cardiac puncture through the thoracic inlet (Gupta, 1973).

Blood collection and intravenous injection in the guinea pig is difficult due to the relatively small-sized peripheral veins.

Small amounts of blood can be collected by cutting a toenail close to the nail bed (Vallejo-Freire, 1951). Warming the animal in an incubator (40°C) tends to increase blood flow. The veins of the ear may be punctured with a 25 gauge needle or lacerated with a scalpel blade, and small amounts of blood may be collected with a capillary tube (Bullock, 1983; Enta et al., 1968; Grice, 1964). The auricular vein is also suitable for intravenous injection (Decad and Birnbaum, 1981). A vacuum-assisted bleeding apparatus may be used to collect blood from the lateral or medial metatarsal veins. A small incision is made just distal to the malleolus, and a vacuum of −5 mm Hg is applied (Dolence and Jones, 1975; Lopez and Navia, 1977; Rosenhaft et al., 1971). The lateral metatarsal vein is also suitable for intravenous injection. As noted previously, cardiac puncture by introduction of a needle through the lateral thoracic wall in the region of maximum heart palpitation may also be used to collect blood from anesthetized guinea pigs.

B. Rabbit

Intravenous injection and blood collection techniques commonly utilize the auricular artery or marginal ear veins of the rabbit. Small amounts of blood can be collected from a puncture wound in the vessels produced by a 23 gauge needle. Collection of large amounts of blood and intravenous injection is facilitated if xylene is applied to the tip of the ear (Burke, 1977). Xylene will dilate the vessels, and a 20 gauge needle may then be used, however, the use of xylene is contraindicated if cell counts are required (Grice, 1964). Vasodilatation can also be produced by administration of a droperidol– fentanyl combination (Tillman and Norman, 1983). A vacuum apparatus (Hoppe et al., 1969) or a miniperistaltic pump (Stickrod et al., 1981) may be used to collect even larger quantities (30–50 ml) of blood. Cardiac puncture in the anesthetized rabbit is done with an 18 gauge, 1½-in. needle (Kaplan and Timmons, 1979). The needle is inserted through the lateral chest wall at the cardiac impulse or inserted just caudal to the xiphoid cartilage, holding the needle at a 30° angle above the plane of the abdomen and directing it caudocephalicly (Bivin and Timmons, 1974). Once the needle is within the heart, blood may be collected by aspiration into a syringe by use of a miniperistaltic pump (Stickrod et al., 1981) or by tubing directly into a centrifuge tube (Kaplan and Timmons, 1979). Blood samples may also be collected by penetration of the orbital sinus in anesthetized animals with nonglazed microhematocrit capillary tubes. To facilitate rupture of this sinus, the tubes may be broken to form a sharp edge. The best site for inserting the tube is within the dorsal conjunctival sac midway between the medial and lateral canthi (Lumsden et al., 1974).

C. Ferret and Mink

Small amounts of blood can be collected from the toenail of a ferret. The nail is clipped close to the nail bed and blood is collected with a capillary tube. Larger amounts (3–5 ml) may be safely collected from the caudal artery of the tail (Bleakley, 1980). A 20 or 21 gauge needle is inserted into a groove in the ventral surface of the tail. The artery is superficially located and care must be taken to prevent going through the artery. Blood may also be collected by cardiac puncture, using a syringe and 20 gauge, 1½-in. needle (Baker and Gorham, 1951). The needle is inserted on the midline just caudal to the xiphoid cartilage. When properly done cardiac puncture is safe and yields 10–15 ml of blood. Jugular venipuncture in mink has been described (Fletch and Wabeser, 1970) as well as jugular vein cannulation (Bergman *et al.*, 1972). The jugular vein can be seen if it is occluded by pressing a finger in the area between the sternum and shoulder.

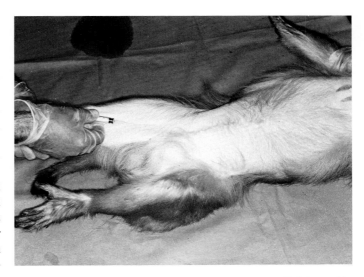

Fig. 5.   Femoral venipuncture in a rhesus monkey. The needle is inserted in the femoral triangle just medial to the femoral pulse.

D. Dog and Cat

Common sites for collection of blood from the dog and cat include the cephalic, recurrent metatarsal, jugular, and femoral veins. Collection of blood from the cephalic vein is usually done while the animal is restrained in sternal recumbency on a table. The vein is stabilized and occluded by grasping the foreleg just behind the elbow. The vein may then be seen or palpated on the dorsal surface of the forelimb. The recurrent metatarsal vein is located on the lateral surface of the hock joint. Collection of blood from the recurrent metatarsal vein is accomplished with the dog restrained on its side. The vein is occluded by grasping the stifle joint and extending the limb. The vein is easily seen as it crosses the lateral surface of the hock joint, but venipuncture can be difficult due to a tendency for the vein to roll away from the needle. The jugular vein is best for collection of blood from dogs with small peripheral veins and when large volumes are required. The animal is restrained with its neck extended and held slightly to one side. Pressure applied at the base of the neck will occlude the vein, and it can then be visualized. Clipping the hair from the neck will aid in visualization of the vein. Usually two people, one to restrain the dog and one to collect the sample, are necessary for jugular bleeding. However, Frisk and Richardsn (1979) describe a technique that requires only one person. Femoral venipuncture is done while the animal is restrained in lateral recumbency with the hindlimb extended. The vein cannot be seen but is located just medial to the femoral pulse.

E. Nonhuman Primates

Intravenous injection and blood collection techniques utilize the cephalic, saphenous, coccygeal, and femoral veins of nonhuman primates (Bowen *et al.*, 1976; Hall, 1966; Whitney *et*

al., 1973). The femoral vein is most commonly used for blood collection, it lies within the femoral triangle just medial to the femoral artery and can be quite superficial in some species of primates (Fig. 5). If the femoral artery is punctured, direct pressure must be applied to the site for a period of about 5 min in order to prevent excess hemorrhage and hematoma formation. This may be especially critical in the owl monkey (*Aotus trivirgatus*) (Loeb *et al.*, 1976). Blood may be collected from the jugular vein of nonhuman primates if the animal is anesthetized. Small amounts of blood may be collected from neonatal primates by lacerating the ear; however, the cellular indices may differ from a sample obtained from the femoral vein (Berchelman *et al.*, 1973).

F. Birds

Common sites for blood collection and intravenous injection in avian species are the brachial veins of the wing, the jugular veins, and the heart. The brachial vein can be easily seen on the medial surface of the wing if the feathers are plucked or separated at the region of the elbow joint. Venipuncture is easily accomplished in the brachial vein if the bird is not too small; however hematoma formation is common following venipuncture (Fredrickson *et al.*, 1958). The right jugular vein, usually largest in birds, is superficially located on the dorsolateral surface of the neck between the dorsal and ventral cervical feather tracts (Law, 1960; Stevens and Ridgeway, 1966). Occlusion of the vein may be accomplished by applying pressure to the base of the neck. Jugular venipuncture is safer than cardiac puncture and hematoma formation rarely occurs. Repeated collections may be done if venipuncture is done first at the base of the neck, and subsequent collections are made from sites nearer to the head. Jugular venipuncture is the blood collection method of choice in Japanese quail (*Coturnix corturnix japonica*). Cardiac puncture may also be used to collect blood from birds, and it has the advantage of yielding large amounts in a short period of time. The needle is inserted 1 in. lateral and 1 in. caudal to the point of the keel in chickens. The syringe and needle are held at a 45° angle above the body and directed toward the opposite shoulder. If properly done, mortality rates are low (Hofstad, 1950). Garren (1959) constructed a vacuum apparatus that maintained constant negative pressure for bleeding from the wing vein or heart of chickens.

IV. VASCULAR CANNULATION

For short-term experiments, the animal may be anesthetized or physically restrained during infusion. However when the catheters are to be used for a long period of time, the technique of cannulation must provide for protection of the catheter and allow freedom of movement for the animal. To accomplish these objectives, many ingenious methods and apparatuses are described in the literature and will be briefly summarized here (Table II).

Following tail vein cannulation in rats and mice, a glass, metal, or plastic tube is placed over the tail and attached to the top of the cage (Born and Moller, 1974; Conner *et al.*, 1980; Plager, 1972; Rhodes and Patterson, 1979; Saarni and Viikari, 1976). This effectively protects the cannulation site and allows freedom of movement for the animal. Conti *et al.* (1979) describes the use of a fiberglass cast applied to the limb of a rhesus monkey to protect a cannula in the saphenous vein.

The most common method of protecting the cannula is by creating a subcutaneous tunnel from the site of vessel cannulation to the dorsum of of the neck. The tunnel may be formed by blunt dissection with scissors or with a modified intramedullary pin (Wingfield *et al.*, 1974). Exiting the catheters from the dorsal surface of the neck minimize the possibility of damage by the animal to the setup. Installation of pouches or boxes over the neck are additional measures taken to protect the infusion setup (Born and Moller, 1974; Wingfield *et al.*, 1974). A pouch may also be formed from skin folds (Hall *et al.*, 1974); Goetz and Hermreck (1972) and Zambraski and DiBona (1976) describe exteriorization devices for chronically implanted catheters that further protect the catheter and facilitate infusion.

Leather or canvas vests that protect implanted catheters are commercially available for a variety of species. Their use in the dog is described by Foss and Barnard (1969), in the rabbit by Knize and Weatherby-White (1974), in the chicken by Hamilton (1978), and in the nonhuman primate by Bryant (1980).

Table II

Blood Vessel Cannulation

Species	Vessel	Reference
Mouse	Tail vein	Conner *et al.*, 1980; Plager, 1972
Rat	Tail vein	Born and Moller, 1974; Rhodes and Patterson, 1979; Saarni and Viikari, 1976
	Jugular vein	Terkel and Urbach, 1974; Waynforth, 1980
	Dorsal aorta	Still and Whitcomb, 1956
	Cranial mesenteric vein	Zammit *et al.*, 1979
Guinea pig	Jugular vein	Christison and Curtin, 1969
	Carotid artery	Shrader and Everson, 1968
Rabbit	Jugular vein	Hall *et al.*, 1974
	Auricular vein	Knize and Weatherly-White, 1974
	Carotid artery	Conn and Lagner (1978)
Dog	Jugular vein	Branham, 1976; Dudrick *et al.*, 1970; Foss and Barnard, 1969; Goetz and Hermreck, 1972; Platts *et al.*, 1972
Nonhuman primate	Jugular vein	Craig *et al.*, 1969
	Coccygeal vein	Stockrod and Pruett, 1979
	Saphenous artery	Munson, 1974
	Saphenous vein	Conti *et al.*, 1979
Swine	Jugular vein	Ford and Maurer, 1978; Wingfield *et al.*, 1974
	Femoral artery and vein	Jackson *et al.*, 1972
	Portal vein	Knipfel *et al.*, 1975
Chicken	Brachial vein	Hamilton, 1978
Pigeon	Carotid artery	Wendt *et al.*, 1982

If the infusion is to be done in a freely moving animal, the tubing, which connects the cannula with the infusion pump, must be shielded with some type of durable flexible sleeve. In addition, swivels and pulleys with counterweights are necessary to prevent the infusion tube from becoming twisted or kinked. Use of this type of apparatus is described by Conn and Langer (1978) and Hamilton (1978).

V. INTRAPERITONEAL INJECTION

Intraperitoneal injection is a common method of administering drugs to rodents. The injection site should be in the lower left quadrant of the abdomen because vital organs are absent from this area. Only the tip of the needle should penetrate the abdominal wall to prevent injection into the intestine (Waynforth, 1980).

VI. SUBCUTANEOUS AND INTRAMUSCULAR INJECTION

The preferred site for subcutaneous injection in most laboratory animals is the back or neck region. The technique is quite simple, a fold of skin is held with one hand and the needle inserted just under skin at the base of the fold.

The thigh muscles are most commonly used for intramuscular injection. When large volumes of irritating substances are to be injected, the quadriceps group rather than the posterior thigh muscles should be used. The sciatic nerve lies posterior to the femur, and any substance injected into a fascial plane of the posterior thigh muscles may be carried directly to the nerve (Leash *et al.*, 1973).

VII. ORAL ADMINISTRATION OF SUBSTANCES

Per os administration of solids and liquids to laboratory animals is an essential technique for a variety of experimental protocols. Gastric intubation ensures that all the material was administered. Flexible catheters may be used for gavage in small rodents, but durable, stainless steel, ball-tipped needles are used more frequently* (Fig. 6). Common complications associated with gastric intubation are damage to the esophagus and administration of substances into the trachea. Careful and gentle passage of the gavage tube will greatly reduce these possibilities. In addition, introduction of the tube from the pharynx into the esophagus is best accomplished when the animal is in the act of swallowing. Usually one can ascertain the tube is not within the trachea by observing the tube's profile as it moves within the esophagus.

Oral speculums such as those constructed from syringe cases or tongue depressors (Fig. 7) may be used to prevent the animal from chewing on the gavage tube when introduced into the oral cavity. Nasogastric intubation prevents tube damage, and it is easily accomplished in nonhuman primates or cats when the animals are anesthetized or severely depressed.

VIII. DIGESTIVE SYSTEM

It is the intent of this section to discuss (1) oral examination techniques, (2) techniques of tooth extraction, (3) pulpectomy, (4) pulpotomy, (5) cannulation of the common bile–pancreatic

*Bio-medical Needles, Popper & Sons, Inc., New Hyde Park, New York 11040.

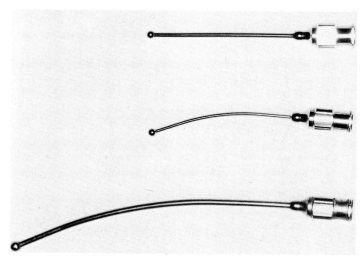

Fig. 6. Stainless steel, ball-tipped needles used for a gavage in rodents.

duct, (6) biopsies of the liver and spleen, (7) intestinal cannulas, and (8) the formation of isolated intestinal loops.

Complex surgical procedures involving gastroscopy, gastric fistulas, pyloric cuff techniques, a denervated Heidenhain pouch, a Pavlov pouch, repeated intestinal biopsies, esophago-gastroscopy, surgical removal of the pylorica antrum, and cecetomy have been described in the literature (Pare *et al.*,

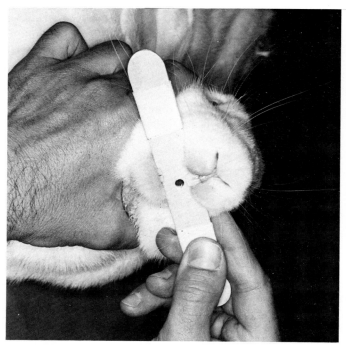

Fig. 7. Use of an oral speculum in a rabbit. This inexpensive, disposable speculum was constructed from two tongue depressors taped together.

1979; Pazin *et al.*, 1978; Cook and Williams, 1978; Bruckner-Kardoss and Worstmann, 1967; Harris and Decker, 1969; Houghton and Jones, 1977). However, because these surgeries require extensive descriptions they will not be discussed in this text.

A. Oral Examination

There are many types of specialized restraining devices for oral examination described in the literature (Evans *et al.*, 1968; Davies and Grice, 1962; Redfern, 1971). Because rodents have a large diastema and a small oral cavity, it is suggested that a tubular device described by Macedo-Sobrinho *et al.* (1978) be used. This technique has proved to be reliable for intraoral examination of mice, rats, hamsters, and guinea pigs, even in the hands of inexperienced technical personnel.

Some species have unique anatomical adaptations that are utilized in research investigations. One example is the hamster cheek pouch. In order to expose this pouch, the hamster is held around the body with a thumb across its cheek. The fifth finger of the opposite hand is placed near the caudal end of the cheek pouch. By pushing cranially, while pulling gently at the corner of the hamster's mouth with the thumb of the free hand, the cheek pouch can be completely everted (Fig. 8). This technique is advantageous because it can be done on an unanesthetized animal, the hamster does not bite its own everted cheek pouch, and traumatizing instruments are not required (Haisley, 1980).

B. Tooth Extraction

Because of their anatomical configuration, the teeth of some animals, such as the primate, are very difficult to remove. The technique of tooth extraction using the dental elevator is as follows: (1) insert the dental elevator between the alveolar bone and the tooth; (2) apply pressure on the elevator and using a rotating wrist action, direct it toward the apex of the root on all sides; (3) do not attempt to pull the tooth until all peridontal ligaments have been severed. Tooth extraction in the dog can be facilitated by using a tooth grasping forceps and a gas-driven oscillating handpiece. The head of this neurological instrument contains a socket designed to oscillate at frequencies up to 417 Hz (Hertz = cycles/sec) and using 120 psi of compressed gas. With this oscillation technique, the extraction time can range from 3 to 115 sec. Little or no hemorrhage results, and there is no trace of attached peridontal ligament (Mumaw and Miller, 1975).

One method of avoiding the risks of trauma, bleeding, and fractures of the mandible is to use a nonsurgical technique for tooth extraction. Serial injections of the peridontal tissues with a 20% solution of potassium hydroxide, which is made into a

Fig. 8. Eversion of the hamster cheek pouch. (Used with permission from Haisley, 1980.)

gel with carbapol 934,* causes a loosening of mandibular and maxillary teeth. Following one to three injections, the tooth is easily removed from the alveolus. Using this same technique for hydrolyzing the membranes, it is possible to loosen malpositioned teeth, reposition them in the alveoli, splint, and allow alveolar reattachment (Patrick *et al.*, 1968).

C. Pulpectomy and Pulpotomy

More recently, the surgical techniques of pulpectomy and pulpotomy have been popularized as alternatives to canine tooth extraction. These procedures have been used in veteri-

*A water-soluble resin, B. F. Goodrich Chemical Co., Cleveland, Ohio.

nary medicine for treating tooth fractures and pulp infections and for disarming dangerous animals. Pulpotomies refer to the opening of the root canal and the partial removal of the pulpal tissue. Pulpotomies have the risk of creating sensitive teeth because of the potential reaction of the remaining pulpal tissue. Pulpectomies, on the other hand, refers to the total removal of the pulpal tissue.

Although variations in techniques and chemicals used are described, the pulpectomy procedure, as outlined by Tomson, is excellent (Tomson *et al.*, 1979). This procedure involves anesthetizing the animal, in this case a nonhuman primate, and placing it on a surgery table in dorsal recumbency. Rubber dams are used on all teeth to prevent oral contamination of the root canal. Each canine tooth is cut off level with adjacent teeth using tapering cross-cut bars or Damascus separating discs* attached to a dental drill. Broaches are used to remove all pulpal tissue from the root canal. The root canal is then irrigated with sterile saline. Paper points are used to dry the canal and check for the presence of blood (Tomson *et al.*, 1979). If bleeding does occur, patience must be used to explore the canal and attempt to remove all pulpal tissue. Other techniques have employed epinephrine flushes or electrocautery to stop the bleeding and dry the root canal. The root canal paste formulas include an RC-2W mixture consisting of 3% $BaSO_4$, 6% Titamicin oxide, 8% bismuth subnitrate, 18% bismuth subcarbonate, 6.5% paraformaldehyde, and 62.5% zinc oxide (Tomson *et al.*, 1979), or a commercial product Hypo-cal[†] which is basically calcium hydroxide in a calibrated disposable syringe with a special applicator needle. After the canal has been filled, excess paste is removed and an amalgam plugger is used to firmly pack the hardening paste into the filled canal. A silver alloy amalgam** filling is used to permanently seal the filled root canal. Excess alloy is removed, and the tooth is sculptured smooth (Tomson *et al.*, 1979) (Fig. 9). Some have advocated a more permanent technique involving the placement of a permanent pulp cap (Reynolds and Hall, 1979). Although this technique is more costly and time consuming, it does give the investigator an opportunity to easily convert troublesome temporary caps to more durable crowns at a later date.

D. Bile Duct Cannulation

Animal models in which bile can be sampled have many applications, especially in pharmacokinetic studies. Some techniques, such as the one described by Spalton and Clifford (1979), involves the oral passage of a duodenal tube into the

*William Dixon, Inc., Carlstadt, New Jersey.

[†]Hellman Dental Manufacturing, Hewlett, New York.

**Caulk Company, Milford, DE or Whitecap capsules type 1–class 1, S.S. White Dental Products International, Philadelphia, Pennsylvania 19102.

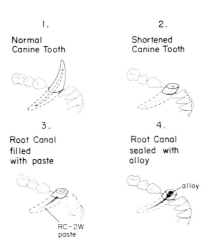

Fig. 9. Diagram illustrates the level at which the canine tooth should be cut in relation to adjacent teeth. Following extraction of the pulpal tissue, the root canal is filled with a root canal paste. (Used with permission from Tomson *et al.*, 1979).

duodenum. After application of a vacuum for several minutes, the animal is given an intravenous injection of pancreozymic enzymes and cholecystokinin which causes contraction of the gallbladder. Aspiration of the bile sample is carried out via the duodenal tube. Attempts to cannulate the common bile duct are described by several authors depending on the species (Boegli and Hall, 1969; Barringer *et al.*, 1982; Soli and Birkeland, 1977; Knapp *et al.*, 1971). Most authors reflect on the difficulty of maintaining an experimental preparation of this type. It is also a well-known fact that some anesthetics interfere with biliary secretion (Berthelot *et al.*, 1970).

To overcome such artifacts, a double recurrent choledo-choledocal biliary fistula is described for rabbits. Through a midline laparotomy, the bile duct is exposed and dissected for a distance of 10 mm from the duodenal wall. Two silastic medical catheters (external diameter 3.16 mm, internal diameter 1.97 mm) are inserted into the bile duct. One is directed toward the liver making sure that its end does not pass the cystic and secondary hepatic ducts. The other catheter is directed toward the duodeneum but is stopped short of the sphincter of Oddi. The catheters are then brought to the body surface through stab wounds in the lateral abdominal wall and connected to taps that allow the bile samples to be taken and solutions infused. The catheters are protected with a specially constructed bandage in which small polyethylene bags are placed for collection of samples (Jimenez *et al.*, 1982).

In smaller species, such as mice and rats, a stereoscopic microscope can be used to identify the common bile–pancreatic duct. Once the junction between the hepatic ducts and cystic duct is individualized, the bile duct is closed immediately below their junction by a nylon ligature. The common bile–pancreatic duct ends in an ampulla that connects it to the duodenal lumen. A small cut is made on the ampulla wall with Belucci scissors, and a polyethylene cannula (internal diameter = 150 μm) is inserted approximately 1 mm into the common bile–pancreatic duct lumen. This cannula is secured by a 7-0 nylon ligature. A calibrated capillary tube is placed in the free end of the cannula to collect the sample (Maillie *et al.*, 1981). Chronic indwelling catheters for continuous collection of bile can be exteriorized through the skin at the base of the skull and protected by a metal coil covering (Balabaud *et al.*, 1981).

E. Liver Biopsy

At times it is necessary to perform frequent liver biopsies. The most simple methods involve a transcutaneous skin puncture, whereas large tissue samples may require abdominal surgery. Voss (1970) describes a simple, rapid, closed biopsy technique for primates utilizing a 16-gauge Klatskin needle and a subcostal approach. Smaller gauge biopsy needles are used in a similar manner on rodent species.

When larger tissue samples are needed, an open liver biopsy procedure may be required. Following anesthesia and surgical preparation of the abdominal area, a midline abdominal incision is made to expose the organ involved. After identification of the appropriate area to be excised, a scalpel or scissors is used to remove the wedge of tissue. The incised area is then packed with a folded piece of absorbable gelatin sponge* (Voss, 1970). Some authors have recommended the use of a TA-90 stapling gun or similar instrument.† The advantage of this technique is that larger samples of biopsy material are provided, and hemostasis of the severed liver lobe is excellent (Nolan and Conti, 1980).

At times, such as nucleotide metabolism studies, liver tissue is required from an unanesthetized animal. Yngner (1975) has designed an instrument that greatly facilitates such a procedure. This instrument is a double blade guillotine which is attached to the neck of a plastic water bottle. As the animal crawls out of the bottle the guillotine blades are closed manually. These blades are adjustable so that decapitation occurs a fraction of a second before the second blade transects the body just caudal to the liver.

F. Intestinal Cannulation

There are many designs and materials used for insertion of rigid or flexible cannulas into alimentary tract fistulas of experimental animals. The choice of cannula depends on several factors, such as the size of the lumen of the organ to be fistu-

*U.S.P. Gelfoam, Upjohn Co., Kalamazoo, Michigan.
†U.S. Surgical Corp., Stamford, Connecticut.

lated, the particle size and viscosity of the material to be sampled or infused, the site of exteriorization of the cannula, and, in the case of reentrant fistulas, the volume and consistency of ingesta expected to flow through the cannulas to avoid producing excessive resistance to flow. Because of these variables, it would appear that the flexible polyvinyl chloride (PVC) cannulas are preferred to rigid cannulas for use in experimental animals (Buttle *et al.*, 1982).

G. Intestinal Loop Isolation

Numerous research protocols for the study of gastrointestinal physiology and pathology require segments of the digestive tract that have no continuity with the fecal stream. These segments are provided by preparation of isolated loops at different levels of the gut. The crucial stage in preparing such a segment is the anastomosis method used on the intestinal wall. For most species, the preferred anastomosis technique is the use of a simple interrupted approximating suture pattern that limits intestinal tissue trauma and minimizes luminal narrowing (Toofanian and Targowski, 1982). Because the guinea pig's small bowel is so friable, this species requires using interrupted 6/0 silk sutures meticulously placed 1 mm apart and 1 mm deep (Bett *et al.*, 1980). Suture patterns in other rodent species would require a similar technique.

IX. Urinary Techniques

The techniques described in this section are limited to (1) methods of urine collection, (2) exteriorization of the ureters, and (3) implantation of the ureters into the intestine. Because meaningful descriptive techniques require more space than allotted for this chapter, the following procedures are referenced only: (1) reconstructive surgery of the ureters, (2) fistula of the urinary bladder, (3) transposing the kidney into the iliac fossa, and (4) denervation and decapsulation of the kidney (Lopukhin, 1976). Techniques are also being refined for kidney transplants in nonhuman primates (Murphy and Grafton, 1973), rabbits (Jacobsen, 1978), and canine (Perper and Najarian, 1964).

Urine is usually removed from the bladder by one of four methods: (a) collection during spontaneous micturation, (b) manual compression of the urinary bladder, (c) catheterization, and (d) cystocentesis. When possible, the first part of the urine stream should not be used for urinalysis or bacterial culture because it may contain debris, bacteria, or exudate flushed from the urethra or genital tract.

Garvey and Alseth (1971) capitalized on the fact that abdominal muscle control is poorly developed in the newborn. Many females stimulate their newborn animals to urinate by stroking the lower abdomen. Although the technique is described for the rabbit, the investigator or clinician can collect urine samples from the newborn of several species by applying gentle strokes and pressure on the lower abdomen. Stroking and pressure are continued until the muscles relax and a drop of urine is observed, then pressure is applied over the bladder until the flow stops. Although individuals vary, up to 5 ml of urine has been collected from newborn rabbits using this method (Garvey and Alseth, 1971). Success in obtaining urine by this method is often greater in females because of reduced urethral resistance.

Other methods of urine collection have required the attachment of a modified pediatric urine bag to fully cover the vulva in swine and thus prevent fecal contamination (Galitzer et al., 1979); or the use of specialized metabolic cages, screens, and baffles to separate the urine and feces (Black and Claxter, 1979; Smith et al., 1981); or the attachment of silicone tubing over the glans penis. In order to accomplish the latter, the prepuce is pushed back on the penis and the tube is positioned over the glans and secured (White, 1971; Rahlmann et al., 1976).

For some species, urethral and/or ureteral catheterization is the method of choice. Diagnostic catheterization is indicated for (1) collecting bladder urine for urinalysis or bacterial culture, (2) studying renal function, (3) instilling contrast media for radiography, (4) evaluating the urethral lumen for strictures and/or obstruction, and (5) surgically repairing the urethra and surrounding structures. These methods are routinely used for dogs and have also been described for calves (Allen, 1974), domestic fowl (Wideman and Braun, 1982), and rats (Cohen and Oliver, 1964).

The technique for catheterization in all species should follow the following format. Regardless of the specific procedure employed, meticulous aseptic and gentle atraumatic technique should be used. Conscious patients should be restrained by an assistant in order to minimize contamination of the catheter as well as trauma to the urethra. Use the smallest diameter catheter that will permit the objective of catheterization. If a stylet is used in the catheter, it should be lubricated before it is inserted into the lumen of the catheter. If it is not lubricated, difficulty may be encountered in removing the stylet after the catheter has been placed in the patient. Regardless of species or sex, the approxiamte distance from the external urethral orifice to the neck of the bladder should be determined and mentally transposed to the catheter. This step will minimize the likelihood of traumatizing the bladder wall due to insertion of an excessive length. Proper lubrication of the catheter with a liberal quantity of sterilized aqueous lubricant will minimize discomfort to the patient and catheter-induced trauma to the urethra. Although usually unnecessary, local anesthesia of the urethra may be provided with a topical anesthetic. Gentle aseptic technique

must be used to advance the catheter into the urethra. If difficulty is encountered in inserting the catheter into the bladder, withdraw the catheter for a short distance and insert it again with a rotating motion. If this does not permit catheterization, a smaller diameter catheter should be used. The tip of the catheter should be positioned so that it is located just beyond the junction of the neck of the bladder within the urethra. Verification of this position may be accomplished by injection of a known quantity of air through the catheter. Inability to remove all of the air indicates improper positioning of the catheter. Proper positioning of the catheter facilitates removal of all of the urine from the bladder. Urine may be aspirated from the bladder with the aid of a syringe and two-way valve. Aspiration must be gentle in order to prevent trauma to the bladder mucosa as a result of sucking it into the eye of the catheter. Catheters that are to remain in the bladder for some time but that are not designed to permit self-retention may be sutured to the skin.

Potential complications associated with catheterization in all species include hematuria and infection. Although the hematuria is usually self-limiting, it may interfere with interpretation of the results of urinalysis. Because of the risk of inducing infection of the bladder as a result of catheterization, this technique should be reserved for diagnostic or therapeutic purposes (Painter et al., 1971; Goodpasture et al., 1982). Procedures that may be used to reduce the incidence of infection following catheterization include: (1) strict adherence to principles of asepsis, (2) administration of oral or parenteral antibiotics, (3) use of catheters impregnated with antibacterial agents, and (4) irrigation of the bladder with antibacterial solutions, such as neomycin or furacin (Osborne et al., 1972).

Urine collection in rodents is generally accomplished by surgical cystotomy (Moreland, 1965) or one of the methods previously described. The small size of the animals is one complicating factor; another is the unique reproductive system of some species. For example, catheterization of the male guinea pig can be accomplished rather easily, however, in almost all cases the male will ejaculate as the catheter is passed and the coagulum will quickly plug the catheter. Some success has been attained with urethral catheterization of the female rat using a No. 4 coude ureteral catheter that has a bend adjacent to the tip of the catheter. This angulation allows the catheter to atraumatically slide by most obstructive areas in the urethra (Cohen and Oliver, 1964).

In order to avoid lower urinary tract infections and facilitate renal function studies, techniques have been developed for catheterizing the ureters. In some species, such as the rabbit, the ureters are very friable and subject to intraluminal bleeding. Reduced urine output during anesthesia combined with bleeding tendencies, will often allow occlusion of the catheter with blood clots. In order to compensate for this problem, Harris and Best (1979) designed a double lumen catheter. Irrigation of blood from the catheter is accomplished by perfusion of heparinized saline through a small inner catheter (Fig. 10).

Other still more invasive techniques have involved relocating or exteriorizing the ureters (Abernathy and Anderson, 1974). The exteriorized ureter can then be easily cannulated, or it can be temporarily obstructed for research purposes. At times, resection of the ureters is required in order to facilitate a surgical procedure. If this is necessary, an elastic catheter with a diameter equal to or a little smaller than the ureter is inserted into both ends of the transected ureter and the ends are then approximated. Interrupted catgut or silk sutures are placed in the periureteral cellular tissue and muscular coat, being careful not to enter the lumen. The catheter is removed through a lateral longitudinal incision in the ureter below the anastomosis. This incision is closed using transverse interrupted sutures, taking care not to narrow the ureteral lumen (Lopukhin, 1976).

Transplantation of the ureter into the intestine is another alternative. The ureter is catheterized and surgically embedded beneath the serous and muscular coats of the lower intestinal wall. The mucosal wall of the intestine is punctured at the most distal point of the ureter. After several days, a ureteral fistula forms and the catheter is removed through the anus (Lopukhin, 1976).

Still another surgical technique involves attaching the ureters to a large Heidenhain's pouch formed from the body and fundus of the stomach. The gastroepiploic vessels are maintained intact, and the ureters are implanted through the wall of this newly formed pouch. Neither hyperchloremia, acidosis,

uremia nor hyperkalemia are encountered. Heidenhain's pouch maintains its secretory function, and the gastric secretions appear to prevent ascending infections (Lopukhin, 1976).

X. RESPIRATORY SYSTEM TECHNIQUES

Bacterial pneumonias can be one of the major causes of death in laboratory animals maintained in captivity (Snyder and Soave, 1970; Ilievski and Fleischman, 1981). When epizootics occur it is necessary to obtain bacteriological specimens from infected animals and establish a differential diagnosis quickly in order to begin prompt and effective therapy. This section will discuss the following techniques as useful tools in the diagnosis and treatment of respiratory diseases: (1) collection of pharyngeal fluids, (2) tracheobronchial washings, (3) endotracheal inoculation and intubation, (4) tracheal pouch formation, (5) tracheostomy, (6) ventriculocordectomy, (7) bronchoscopy, and (8) bronchopulmonary lavage. Other techniques involving the lower respiratory system include lobectomies (Bernstein and Agee, 1964; Markowitz, et al., 1964), bioinstrumentation of the thorax (Harvey and Jones, 1982), and the development of chronic lung–lymph fistulas (Brown et al., 1982). Because these techniques involve extensive written protocols, a description could not be included in this chapter.

A. Collection of Pharyngeal Fluids

A simple, effective and inexpensive method for the collection of pharyngeal cultures without contamination by bacteria is described (Snyder and Soave, 1970). A straight tube formed from a 1 ml tuberculin syringe is used as a speculum for the passage of sterile swabs into the pharynx. The transparent syringe barrel allows for easy visualization of the larynx. Commercial sources for sheathed applicator swabs are also available; however, the described technique can be performed in several species with materials readily available in the laboratory and at a much lower unit cost.

B. Tracheobronchial Washings

Tracheobronchial washing techniques have been described for the anesthetized primate. With the head tilted slightly off the edge of the table, the tongue is grasped with a gauze sponge and traction applied. A fiberoptic light source aids in visualization of the orifice and a sterile straight Kelly forceps is used to grasp the central body of the epiglottis. A pediatric laryngoscope with individual sterile blades will permit easy visualization of the larynx and passing of the tubes. Tra-

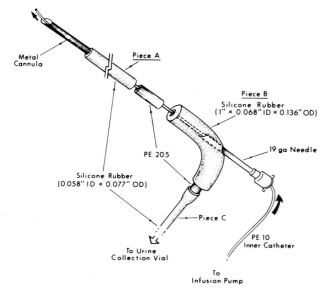

Fig. 10. Catheterization of the ureter using a double lumen catheter. This catheter provides a means to flush with heparinized saline and thus prevent occlusion with blood clots. (Used with permission from Harris, R. H., and Best, 1979.)

cheobronchial washings are obtained using a 3 French, 40.6 cm long feeding tube. To permit sterile placement of the tube into the trachea, the scabbard or outer tube is cut with a blunt taper on one end. This outer tube is lubricated with a sterile lubricant and introduced directly into the trachea on expiration to a position about 5–8 cm posterior to the larynx. The 3 French rubber feeding tube is subsequently passed through the outer tube and introduced to about the level of the carina. A 5 ml syringe containing sterile saline is then attached to the feeding tube and used to infuse saline into the bronchi. Saline is cleared from the tube using 2–3 ml of air. At this point the animal is rotated from side to side several times and light pressure is applied to the chest, while simultaneously aspirating the 5 ml syringe. Generally 1–2 ml of foamy tracheobronchial fluid is obtained (Ilievski and Fleischman, 1981).

A second method is to use the transtracheal aspiration technique that involves passing a needle between the tracheal rings and into the tracheal lumen. A catheter is then directed through the needle and advanced to a bronchus. Sterile saline (10–15 ml for a macaque) is infused and aspirated through the catheter (Stills *et al.*, 1979). This technique, as with the previous technique, circumvents the potential problems of pharyngeal and laryngeal contamination. One advantage of this technique is that it does not require chemical restraint in the acutely ill animal. However the cost of the catheters does make this a more expensive technique.

C.   Endotracheal Inoculation and Intubation

A free airway is useful in reducing mortality during and after surgical procedures. Also some experimental protocols require that chemical substances be administered intratracheally. Because these techniques have already been described for most species, we will only describe techniques applicable to rodents and rabbits.

Acceptable techniques for endotracheal inoculation and intubation in rodents have been described by Nicholson and Kinkead (1982), Yap (1982), and Pena and Cabrera (1980). The equipment required for intratracheal inoculation in small rodents would include two pairs of small curved forceps, 100 μl microsyringe, 38 × 1 mm diameter needle with a blunt tip bent at an angle of 30° 10 mm from the end, and a laryngoscope made from a disposable polypropylene micropipette tip (Yap, 1982). The anesthetized rodent is placed on its back, and a pair of forceps is used to grasp the tongue and hold it to one side. The specially designed laryngoscope (Yap, 1982) is inserted narrow end first into the mouth. The inoculation needle, curved end uppermost, is then inserted through the laryngoscope. On entry, the syringe is held so that the portion of the inoculation needle up to the bend is parallel to the throat. With further insertion, the needle is gradually tilted upward, and,

when resistance is felt, the needle is withdrawn very slightly and tilted up further. On reinsertion a give is felt when the epiglottis is passed (Yap, 1982). Another specialized speculum has been designed by Nicholson and Kinkead (1982) to facilitate intratracheal inoculations. This speculum is inserted into the rat's mouth, keeping the tongue flat beneath the blade and thus providing ease in making inoculations.

Endotracheal intubation in rodents is accomplished by retracting the upper jaw downward with a rubber band passed over the upper incisor teeth, retracting the tongue to one side, and observing the laryngeal region through a surgical microscope under 6× and 10× magnification. After the mucus is cleared with a cotton-tipped applicator to allow visualization of the epiglottis, glottis, and paired arytenoid cartilages, the endotracheal tube is inserted into the trachea. As the end of the tube is advanced past the larynx, water vapor is seen passing from the tube during expiration indicating proper placement (Pena and Cabrera, 1982).

Oral endotracheal intubation in the rabbit has been described by numerous authors; including Davis and Malinin (1974), Bertholet and Hughes (1980), and Alexander and Clark (1980). Although the difficulty of performing the technique varies, the authors feel that the technique by Alexander and Clark (1980) is the simplest and most used. The rabbit is anesthetized with gas anesthesia and placed on the surgery table in an outstretched prone position. Anesthesia is continued until all laryngeal reflexes are abolished. The rabbit's head is tipped and extended into an upright position at right angles to the rest of the body. This position provides a straight passage from the lips to the larynx. A sterile, uncuffed, nylon reinforced, latex endotracheal tube (4.0–4.5 mm internal diameter) lubricated with a water-soluble sterile lubricant is passed into the diastema over the tongue and positioned over, but not touching the larynx. Correct positioning is ascertained by listening to the respiration through the tube and adjusting to obtain maximum ventilation. At inspiration, when the vocal chords are maximally opened, the endotracheal tube is inserted with a straight push to the desired depth (Alexander and Clark, 1980).

D.   Tracheal Pouch Formation

The development of a model for studying tracheal secretions by using a tracheal pouch was first described in the dog (Wardell *et al.*, 1970). The technique involves resecting a segment of cervical trachea, with blood and nerve supply intact, and relocating this closed segment subcutaneously for ease in sampling. Recent studies in the ferret have described a more simplified technique for creating a tracheal pouch. Following anesthesia, an 8 to 10 cm ventral midline skin incision is made three cartilage rings caudal to the larynx to expose the trachea. The trachea is transected at two points between cartilage rings, while exercising care not to disturb the dorsolateral recurrent laryngeal nerves and dorsal vascular supply. Pouches of eight cartilage rings in length can be made without causing discomfort. A stay suture is used to hold the tracheal segment away from the surgical field so that the ferret can easily breathe. Tracheal continuity is reestablished by anastomosis using four No. 000 gut sutures. The cranial end of the tracheal segment is then sutured to the subcutis with four No. 000 silk sutures. The caudal end is closed with three No. 000 chromic gut sutures, thus forming a pouch (Olson, 1974).

E. Tracheostomy

Along this same line, a permanent tracheostomy has been developed for dogs. The procedure consists of dissecting positions of the cartilagenous rings free from the underlying tracheal mucosa, cutting through the mucosa, and suturing the mucosa to the skin. This procedure is performed in approximately 30 min and results in a permanent maintenance-free, mucocutaneous stoma (Dalgard *et al.*, 1979).

F. Ventriculocordectomy

Techniques for long-term devocalization have been extensively employed in laboratory-housed dogs and swine. Almost any damage to the vocal cords, such as biopsy or cautery (Knowles, 1959; Yoder and Starch, 1964) will be effective for a short period (1 month to 6 weeks). The methods for long-term devocalization, which have been described for dogs (Yoder and Starch, 1964; Raulston *et al.*, 1969), are still in use. Sometimes more extensive surgical procedures are required for swine (Mackey *et al.*, 1970).

Although the biopsy technique can be performed with electrocautery or surgical scissors, the most used method is to use a human uterine biopsy punch.* This method involves anesthetizing the animal until all laryngeal reflexes are abolished, grasping the tongue with a 4 × 4 gauze pad to expose the caudal larynx, using a laryngeal speculum and light source to depress the epiglottis, and, finally, inserting the cutting portion of the uterine biopsy punch into the ventricle on either side of the ventrolateral aspect of the larynx. It is important for the surgeon to be familiar with the anatomy of the larynx in order to assure sectioning of the vocal folds and avoid injury to the arytenoid cartilages. Once the punch has been properly placed, the surgeon should attempt to remove as much of the vocal fold on either side as possible. The more vocal fold tissue removed, the longer will be the effect of devocalization.

*Miltex Instrument Co., 300 Park Avenue South, New York, New York 10010.

Perhaps the most accepted technique for long-term devocalization of dogs is the method described by Raulston *et al.* (1969). This procedure involves using a burring instrument designed after the Snell roaring burr. The dog is prepared as in the previous procedure and the burr is inserted into the ventricle. Following placement of the burr, it is rotated clockwise until the mucosa is broken loose from the underlying tissue. The burr is then withdrawn with the mucosa attached and the procedure repeated on the opposite side. To lengthen the effect, an electrocautery knife is used to sever the vocal cord.

G. Bronchoscopy

Flexible fiberoptic bronchoscopy has been well established as a diagnostic and therapeutic tool in human medicine, but has been used infrequently in veterinary medicine. Since the anatomy of the rhesus monkey is quite similar to that of the human, this technique is now utilized as a diagnostic and therapeutic tool in primate medicine. The monkey is anesthetized with ketamine and the vocal cords, larynx, and trachea are desensitized with a supplemental topical anesthesia. Either a pediatric fiberoptic bronchoscope (outer diameter 4.5 mm; length 605 mm) or an adult bronchoscope (outer diameter 5.8 mm; length 605 mm) is used for inspection and photography (Strumpf *et al.*, 1979).

H. Bronchopulmonary Lavage

Bronchopulmonary lavage is used as a therapeutic procedure in man and as a means of recovering cells, surfactant, and inhaled particulates from the lungs of animals (Muggenberg and Mauderly, 1975; Brain and Frank, 1968; Myrvik *et al.*, 1961; Maxwell *et al.*, 1964; Medin *et al.*, 1976). For a more complete description of the mechanical equipment required for this procedure, the reader is referred to the works of Mauderly (1977). Highlights of the technique described by Mauderly are as follows: The subject is deeply anesthetized and placed in a prone position on a lavage platform. A mouth speculum is placed behind the incisor teeth and a high-intensity lamp is used to visualize the epiglottis and trachea. A specially designed tracheal catheter and stylet are moistened and passed into the trachea. The volume of saline to be used for each wash is determined by individual pressure–volume measurements prior to lavage. The subject is hyperventilated with a syringe until either apnea is induced or breathing frequency is significantly reduced. A syringe–manometer system is then connected to the luer fitting, and the lungs are inflated until transthoracic pressure reaches 20 cm H_2O. After multiple determinations, the average syringe volume required to reach that pressure is corrected for manometer displacement and gas compression to calculate the actual volume change of the lung.

Lavage is accomplished by hyperventilating the subject, instilling the calculated volume of warmed normal saline, and immediately withdrawing the saline until a slight resistance is felt on the plunger. Another syringe and the tracheal aspiration tube are used to clear residual fluid from the catheter, and the animal is then ventilated until effective spontaneous breathing is reestablished. A total of four wash sequences are completed, taking care to drain as much fluid as possible from the lungs between washes. After the last wash, the halothane vaporizer is turned off, and the animal is extubated at the first sign of spontaneous movement (Mauderly, 1977).

XI. REPRODUCTIVE SYSTEM

The majority of the laboratory animals have a relatively short gestation and reproduce in large numbers. Others, such as nonhuman primates, have a lengthy gestation and produce limited numbers of offspring (Hafes, 1970). Because there are so many unique animal models available, which possess physiological or anatomical characteristics necessary for a better understanding of reproductive biology in man, studies on the reproductive system are voluminous. Our intent is to elaborate and/or reference the following techniques: (1) laparoscopy as an aid for ovarian biopsy, aspiration of follicle contents, ovarian injections and artificial insemination, (2) testicular biopsy; (3) castration; (4) semen collection; (5) artificial insemination; (6) pregnancy diagnosis; and (7) embryo transfer.

A. Laparoscopy

Techniques involving the ovary are usually related to observation or removal of the organ. In recent years the science of laparoscopy or cystoscopy has achieved a high state of development. The earliest known report of endoscopy occurred in 1806 and described the projection of candlelight through a double lumen urethral cannula. Traditionally, the basic research scientists have made discoveries and developed techniques that were then applied to clinical problems. Laparoscopy evolved in a reverse manner with the extensive development of practical, clinically important techniques relating to ovarian biopsy, cyst removal, pregnancy diagnosis, implantation site quantification, recovery of uterine fluid, and sterilization by ligation of the oviducts in most laboratory animal species (Dukelow *et al.*, 1971; Dukelow and Ariga, 1976; Dukelow, 1978; Wildt *et al.*, 1975; Morcom and Dukelow, 1980).

The basic technique involves anesthetizing the animal and preparing the lower abdominal region for surgery. A Verres cannula is attached to an insufflator and the abdomen is inflated with either 5% carbon dioxide or nitrogen. Sufficient abdominal insufflation is required to prevent collapse of the abdominal wall when the laparoscopic trocar is inserted. Several companies are currently marketing this type of laparoscopy equipment.* A table listing the basic equipment required for endoscopy in the chimpanzee is presented by Graham (1976). It should be recognized that the diameter of the endoscope sleeve will vary with the specie from 3 to 10 mm in most cases. Also there is a marked difference in manuverability and clarity of vision on the part of the investigator depending on the chosen instrument.

Once the abdominal wall has been penetrated, a high intensity light source and the prewarmed endoscope are inserted through the cannula. For ease in observation, the animal is placed in a steep Trendelenberg position. Manipulation and observation of the abdominal organs, in this case the ovary, is a technique that requires repeated examinations in the species selected before a person can produce reliable interpretation. In many cases this has required the insertion of two or more trocars to displace organs or mesentery that are obstructing a clear view with the endoscope. However, perfection of this technique has permitted visualization and color photography of the ovaries, ovarian biopsy, injections into the ovary, aspiration of follicle contents, and artificial insemination (Graham, 1976). Thus, laparoscopy is a technique of great value for the clinician as well as the research investigator. If proper aseptic procedures are used, the technique can be repeated at short intervals with no apparent ill effects. Following laparoscopy, the skin incision is closed with a simple interrupted or mattress suture and a systemic broad-spectrum antibiotic is administered (Dukelow *et al.*, 1971).

B. Testicular Biopsy

Existing methods of sampling testes include castration and biopsy. The most commonly used method of testis biopsy involves incision of the scrotum and tunica albuginea with subsequent removal of a small wedge of testis tissue (McFee and Kenelly, 1964; Simmons, 1952). While this method is preferred over castration, it presents technical difficulties with small testes (Simmons, 1952) and has limited value in repeated sampling from the same individual. The most promising procedure for obtaining multiple samples is described by Martin and Richmond (1972). The instrument for biopsy is constructed from a ¾-in. 16-gauge needle and a 2-in. 25-gauge needle. A small hook is made on the tip of the 25-gauge needle, which is then passed through the lumen of a 16-gauge needle. To effect this procedure, the chosen animal is anesthe-

*Richard Wolf Medical Instrument Corp., Rosemont, Illinois.

tized, the scrotum is surgically scrubbed, and a sterile 16-gauge needle is introduced through the testis parenchyma. The sterile 25-gauge needle probe is then introduced through the lumen of the 16-gauge needle, and the hub is used to rotate the hook in the testis. As the 25-gauge needle is gently withdrawn, the tissue sample is extracted on the hook of the needle. Usually one or two samples of seminiferous tubules provide an adequate sample for paraffin embedding (Martin and Richmond, 1972). The risk of damage by needle biopsy appears to be less than that from surgical excision. Also semen quality and libido do not seem to be impaired by this technique.

C. Castration

Because of illness or perhaps due to the research protocol, it may be necessary to remove the testicle and/or adjacent lymph nodes. This procedure has been well described for most domesticated species. However, selected laboratory animal species have very little description in the literature on castration. Two techniques that may be of value to the surgeon for rabbits and guinea pigs are recorded by Hodesson and Miller (1964) and McGlinn et al. (1976). McGlinn's technique is unique in that it is designed to remove seminiferous tubules and Leydig cells from the tunica albugenia, leaving the epididymis and its nerve and vascular supply intact. This technique is very valuable in studies concerned with reproductive physiology of spermatozoa, the epididymis, and vas deferens of many other mammals.

D. Semen Collection

The task of obtaining a semen sample is not easy. A study of semen is further complicated by a wide variety of ejaculate characteristics between different species. In some species, such as the gorilla, it is very difficult to obtain a usable semen sample. Rectal probe ejaculation (RPE) and electrical stimulation of the penis have been utilized since the 1930's as a means of obtaining semen from domestic agricultural animals. More recently, these techniques have been extended to the common laboratory animals. In some species, such as the bull, dog, and rabbit, it is often more convenient to train individuals to serve an artificial vagina. This method eliminates the need to restrain or tranquilize the animal and also eliminates the risk of urine contamination of the ejaculate (Fussell et al., 1973; Seager and Fletcher, 1972). In these species RPE is more appropriately used for incidental collections from untrained animals. For a more complete description of the instrumentation and techniques required for RPE, the reader is referred to the works of Gould et al. (1978), Van Pelt and Keyser (1970), Lang (1967), and Fussell et al. (1967). Although the majority of the literature supports RPE as the preferred method of electroejaculation, some authors have indicated that direct stimulation of the penis is superior to RPE (Valerio et al., 1969).

Collection of rete testis fluid is reported in monkeys (Waites and Einer-Jensen 1974), rats (Cooper and Waites, 1974), and bulls and rams (Voglmayr et al., 1970). For a detailed description of catheter implantation in ovine, the reader is referred to the works of Ellery and Kinnen (1981). Using this technique, up to 38 ml of rete testis fluid could be collected per testes for periods up to 3 weeks.

E. Artificial Insemination

Artificial insemination has long been recognized as a valuable technique in breeding domesticated mammals and poultry (Jones, 1971). Its application to the breeding of nondomesticated mammals in captivity has been discussed on several occasions, but because of the many difficulties involved, artificial insemination is not used on a routine basis in these species (Jones, 1971; Rowlands, 1974).

Fresh semen, diluted fresh semen, or diluted stored semen can be used for artificial insemination. The osmotic tension, pH, buffer capacity, and electrolyte balance of the diluents must be compatible with the ejaculate. Many types of diluent have been tried including tomato juice, coconut milk, glycerol, egg yolk, lactose, and skim milk solutions. Other diluent extenders have also included DMSO (dimethyl sulfoxide) commercial bovine extender, reconstituted dried skim milk, Locke's solution, and sodium chloride (Seager and Fletcher, 1972).

Prior to insemination, the female is determined to be in a receptive stage for implantation by examination of vaginal smears. The presence of well-defined cornified epithelial cells on the vaginal smear is indicative of ovulation in primates and most other species (Davis et al., 1975; Blakely et al., 1981). Those females that are not naturally receptive can be induced to ovulate by injection of pregnant mare serum gonadotropin. The actual dosages for each species should be carefully selected, as low doses produce unreliable ovulation and high dosages may lead to fragmentation of ova (Wolfe, 1967). Once the female has been determined to be in a receptive state, the insemination pipette is introduced into the cranial portion of the vagina or through the cervical canal. To ensure fertilization of most ova, the number of viable sperm inseminated should equal or exceed 10^6 (Stavy et al., 1978; Wolfe, 1967; Sojka et al., 1970).

In some species, such as swine, it is advantageous to introduce the semen directly into the oviductal lumen using laparoscopy. This procedure may have application for basic studies of sperm and egg physiology and may be adpated to other species (Morcom and Dukelow, 1980).

F. Pregnancy Diagnosis

Pregnancy diagnosis varies from observation of external appearance, to digital palpation, radiographs, and numerous chemical tests. For the most part, rodent species are very receptive to mating and, if copulation does occur, a high percentage of the females will become pregnant. Therefore, it is a common practice to either observe copulation or look for the vaginal plug that forms immediately following copulation. To assure accuracy, vaginal smears can be taken, and if sperm are found, the animal is designated as day 0 of pregnancy (Moler et al., 1979).

A second method, developed primarily for use in primates, is digital palpation. Rectal bimanual palpation of the uterus in the rhesus macaque was described by Hartman (1932), with few improvements added to the technique in subsequent years. The female is restrained by two technicians holding the upper arms and thighs. The palpator inserts a gloved lubricated index finger into the rectum (the "pinky" finger is inserted into the smaller cynomologous macaque, *Macaca fasicularis*). The cervix is encountered upon initial entry, but the finger must be inserted until the anterior free edge of the uterus is palpated (Van Pelt, 1974). The free hand is cupped on the caudoventral abdomen, and the uterus is immobilized between the thumb and first two fingers of the free hand ventrally or dorsally by the rectally positioned finger. The normal nonpregnant uterus can vary in width from 7 to 21 mm or more (Catchpole and van Wagenen, 1975). The first palpable sign of pregnancy is a dorsoventral rounding of the anterior aspect of the body of the uterus (Fig. 11) (Mahoney, 1975a,b). Individual animal records containing approximate uterine sizes on previous palpations and menstrual cycle and breeding records should be reviewed prior to palpating the animal (Moore, 1983).

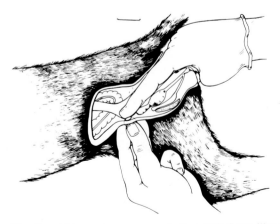

Fig. 11. Rectal bimanual pregnancy palpation. This figure illustrates the positioning of the fingers for digital palpation of the female nonhuman primate reproductive tract. (Used with permission from Van Pelt, 1974.)

Gloves should be changed and relubricated before palpating additional animals. Early embryonic abortions can occur if palpation is performed prior to implantation or if excess manipulation of the uterus occurs soon after implantation. Skilled, cautious palpators can determine pregnancy at 30 days from the onset of the last cycle, with a recheck performed 2 weeks after the initial positive palpation. Females should be palpated prior to assignment to breeding males to determine if they are in fact pregnant, thus freeing the male for use with another cycling female.

Rectal palpation in concert with daily cycle records aids in the evaluation of females as potential breeding stock. A long 7 mm wide "worm" uterus often indicates a poor breeding prognosis. Palpable adhesions of the uterus to abdominal structures may indicate complications in future pregnancies. If spotting of blood is neither observed nor recorded in cycle records, either by technican error or "personal hygiene" practice by the monkey, the female may be erroneously labeled as having irregular cycles. Review of cycle records should reveal the inconsistency, with subsequent cycles occurring at the appropriate cycle interval for that female (Moore, 1983).

Radiographs have also been sued to confirm pregnancy in its latter stages. This technique is most useful following calcification of the skeleton. Due to the variation in gestation periods between species, the effective date at which a diagnostic radiograph can be taken will vary. Therefore the investigator or clinician needs to be knowledgeable of the calcification dates for the species under investigation.

Perhaps the newest techniques for pregnancy determination involves a series of laboratory tests. The nonhuman primate pregnancy test has been used for diagnosis in macaques (Hodgen and Ross, 1974) baboons (Hobson, 1976; Hodgen and Niemann, 1975) marmosets (Hodgen et al., 1976a,b, 1978), chimpanzees, and orangutans (Hodgen et al., 1977). This hemagglutination inhibition test for urinary chorionic gonadotropin uses an antiserum (H-26) that cross-reacts with the chorionic gonadotropin of a variety of primates and provides results within 2 hours. Conventional bioassays and radioimmune assay systems are also useful procedures for detection of chorionic gonadotropin. However, time delays and the cost of instrumentation may make these tests less attractive. Therefore the nonhuman primate pregnancy test, employed in conjunction with uterine palpation, is the method of choice for most nonhuman primates (Hodgen et al., 1978; Lequin et al., 1981; Hall and Hodgen, 1979).

G. Embryo Transfer

Embryo transfer, which involves removing the developing embryo from the reproductive tract of one female and transferring it to another, is used extensively in research dealing with laboratory animals. Experiments involving separation of maternal and fetal genetic effects are possible with this technique. The transfer of preimplantation mouse embryos can be accomplished either surgically by making an incision through the abdominal wall and exposing the uterus or nonsurgically by gaining access to the uterine lumen through the vagina. The later technique appears to be as successful as surgical methods, and it is less painful and requires less time and equipment (Moler *et al.*, 1979).

In some instances, it is advantageous to produce an embryo and hold it for a period of time prior to transfer and implantation. Recent studies in mice have produced a unique research technique. Mouse embryos are now collected and stored for as long as 8 months at $-196°C$. Successful cryoprotectants used for long-term storage of embryos include dimethyl sulfoxide, ethylene glycol, glycerol, and erythritol. Each cryoprotectant has its optimum freezing and thawing rate that must be followed in order to achieve the highest survivability of embryos. Optimal rates appear to be 0.3 to 2°C/min freezing rate and 4 to 25°C/min thawing rate (Leibo *et al.*, 1974; Kasai *et al.*, 1981).

XII. CARDIOVASCULAR TECHNIQUES

Animal models have been utilized to provide a major insights into the techniques of cardiovascular manipulation. These techniques have involved developing surgical methods for organ transplants, vessel transplants and protheses, arteriovenous shunts, blood pressure studies, electromagnetic flow problems, and microangiography (Bishop, 1980). Unfortunately, a detailed description of each of these techniques requires more space than is possible in a text of this magnitude. Therefore, the involved techniques will be referenced, and only the more simplified techniques will be described.

A. Electromagnetic Flow Probe Implantation

In order to measure the effect of diets, drugs, and exercise on abdominal circulation, it is often necessary chronically to implant electromagnetic flow probes on major abdominal vessels. The reader is referred to the works of Mauderly (1970) for specific surgical procedures. The proper surgical placement of these probes requires a thorough understanding of the surgical anatomy involved and much practice. An average procedure is 3 or more hours in length for the skilled surgeon and should not be attempted by a novice.

B. Blood Pressure Techniques

Systolic and diastolic blood pressures are difficult to monitor in animals due to the problems of restraint without the use of anesthetics. Although intraarterial puncture or surgical cut

down procedures can be used, a preferred method is to use a polygraph, pulse transducer, pressure transducer, blood pressure mixer unit, and a pediatric sphygmomanometer cuff.* The procedural sequence consists of (1) setting up and calibration of equipment, (2) immobilization of the animal with chemical restraint, (3) securing the Biocom 6605† adapter over the brachial artery, (4) placement and inflation of the cuff beyond the systolic pressure, and (5) slowly releasing pressure on the cuff while the polygraph tracer records results (Gerbig *et al.*, 1975).

In cardiopulmonary pathophysiology, the measurement of pulmonary arterial (PA) pressure is of considerable importance. This measurement is useful in studying the effects of drugs on pulmonary vasculature and in the diagnosis and treatment of lung and heart diseases. Using a bent-tip 23 gauge lightwall Teflon catheter (PE-50) inserted into the right jugular vein, Carrillo and Aviado (1969) and Hayes and Will (1978) have successfully catheterized the pulmonary artery of the rat. The Teflon catheter is inserted until the "shepherds crook" tip reaches the right atrium and is then manipulated until its bent tip enters the pulmonary artery (Fig. 12).

C. Carotid–Jugular Shunt

The carotid–jugular shunt technique has provided the required research tool for obtaining repeated blood samples from an experimental animal. Following anesthesia, the carotid artery and jugular vein are surgically exposed. The jugular vein is tied off cranially with two strands of No. 2 surgical silk. An incision is made into the vessel and a nontoothed thumb forcep is used to slip the vessel on the plastic connector. The vessel is then secured to the connector and the anterior jugular vein is secured to the shunt. In a similar manner, the carotid artery is attached to the connector and the shunt is secured to both vessels. The shunt is then attached to the skin with a nonabsorable suture passed through an autoclaved clothing button (Belding *et al.*, 1976; Corbitt *et al.*, 1981; Payne *et al.*, 1974).

D. Microangiography

This is a technique in which blood vessels are visualized by the use of roentgen contrast medium and microradiographic techniques. It is a technique that is applicable to a variety of experimental animals, including pigs, dogs, rabbits, and rats, as well as many anatomical regions. The animal is placed under general anesthesia, the artery exposed and cannulated with a corresponding metal cannula, and 1 mg/kg of heparin is given. The level of cannulation and corresponding transection

*Gilson Medical Electronics, Inc., 3000 West Beltline, Middleton, Wisconsin.

†Also obtained from Gilson Medical Electronics, Inc., 3000 West Beltline, Middleton, Wisconsin.

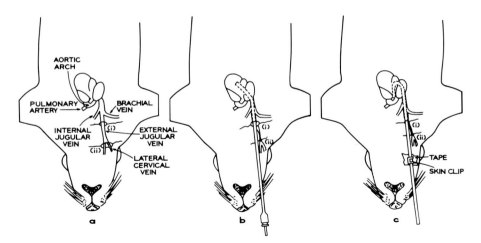

Fig. 12. Pulmonary artery catheterization in the rat. (a) Identification of the vessels involved. (b) Placement of the catheter through the jugular vein into the right ventricle. (c) Catheter is slowly withdrawn until its tip enters the pulmonary artery. (Used with permission from Hayes and Will, 1978.)

of vessels for the efflux of blood and the perfusion of finely ground barium sulfate for each species are well described by Erol *et al.* (1980). Following injection of the contrast medium, rapid film microradiographic techniques are employed.

E. Experimental Heart Surgery

Surgical techniques involving the heart include correction of mitral, pulmonary, and aortic valve insufficiencies; stenosis of the pulmonary trunk; septal defects; infarction; aneurysm; revascularization; electrical stimulation; heart–lung preparations; heart transplants; and implantation of artificial hearts. Because each of these procedures involves a very lengthy and detailed description, the reader is referred to the works of Lopukhin (1976) and Markowitz *et al.* (1964).

XIII. ENDOCRINE SYSTEM TECHNIQUES

Studies involving mechanisms of the endocrine system often require the sacrifice of large numbers of experimental animals in the experimental protocol. Therefore, the rodent species have assumed a major role of importance in these studies. Endocrine techniques discussed in this section include (1) hypophysectomy; (2) pinealectomy; (3) adrenalectomy; (4) thyroidectomy; and (5) parathyroidectomy.

A. Hypophysectomy

Rodents possess an almost flat sella tursica that positions the hypophysis for rapid and fairly easy removal (Sato and Yoneda, 1966). The position of the hypophysis under the midbrain is directly on the midline at a point perpendicular to a line joining openings of the auditory canals (Fig. 13). As the rat grows

older, the hypophysis becomes displaced toward the nose. The transauricular technique of hypophysectomy in the rodent is the most popular and involves anesthetizing the animal, placing the animal in a prone position with the head held toward the surgeon, introducing a modified hypodermic needle sheathed in a steel tube into the left auditory canal, guiding the needle with the right hand along the auditory canal until it reaches the medial wall, and advancing the needle until the outer steel sheath contacts the capsule (a distance of 2.0 to 4.5 mm). Utilizing the correct needle size, the tip of the needle is now located directly beneath the hypophysis. The needle is then connected to a suction device and gentle suction is used to withdraw the hypophysis in 4–7 sec. A 100 ml, two-way flask is interposed for collection of aspirated hypophyses. With proficiency this technique can be completed in 25–35 sec per rodent (Falconi and Rossi, 1964). A modification of this approach is described by Sato and Yoneda (1966). Postoperatively, survival rates are increased if 5% glucose and tetracycline are administered intraperitoneally for 3 days. Also, cortisone can be used for the first 2 days to reduce inflammation and swelling.

Hypophysectomies have been performed in the dog by an extracranial route through the oral cavity (transsphenoidal), by an intracranial route through the anterior or middle cranial fossa, and by the retropharyngeal access. These techniques are described by Lopukhin (1976) and Markowitz *et al.* (1964). Other techniques described include a paraoccular approach for the rabbit (Lopukhin, 1976) and a combination of the transtemporal and parapharyngeal approaches for hypophysectomizing calves (Whipp *et al.*, 1970).

B. Pinealectomy

The rodent pineal body is a median epithalmic structure originating from the dorsal diencephalon and extending dorsally to the superior saggital sinus (Peterborg *et al.*, 1980) (Fig. 13).

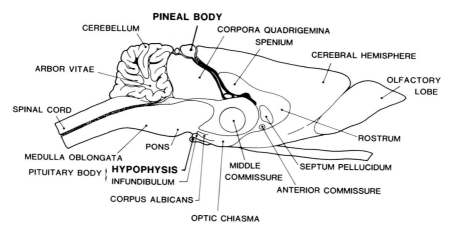

Fig. 13. Anatomical location of the hypophysis and pineal body.

The first technique for pinealectomy was presented in 1910 (Baker *et al.*, 1980). Newer techniques were described by Hoffman and Reiter (1965), Bliss and Bates (1973), and most recently by Kuszak and Rodin (1977) and Jurek (1977). The newer technique involves the following: The anesthesized animal is placed in a stereotaxic instrument; a 1.75 cm longitudinal skin incision is made over the cranium; a dental drill is used to remove a rectangular section of the skull over the saggital and lamboid sutures; the dura is incised; the superior saggital vein is doubly ligated and resected to expose the gland; the pineal gland is freed with a fine pair of curved forceps and, finally; the gland is removed by grasping the base of the gland with the forceps and retracting it. This technique appears to have the advantage of being more direct, produces a minimal amount of hemorrhage and allows for a perfect sham operation since there is no sympathectomy required.

C. Adrenalectomy

Adrenalectomy is the term used to denote the surgical removal of one or both adrenal glands. Removal of both of these glands cause severe physiologic changes that are difficult to correct with replacement therapy. Therefore, in clinical practice it is a rarely used technique. However, from a research standpoint this technique has proved to be very valuable (Lang, 1976).

The technique of adrenalectomy for the dog is well described by Lang (1976) and Lopukhin (1976). Hoar (1966) cites several references indicating the difficulties of performing this procedure in guinea pigs. He also describes a technique for the guinea pig that is shorter, simpler, and less traumatic than most procedures. A brief description of his procedure is as follows: anesthesia, surgical preparation of the thoracoabdominal region, incision of the skin over the penultimate intercostal space; incision of the intercostal muscles (avoid cutting the

peritoneum), incision of peritoneum using a blunt-nosed scissors, saline-soaked gauze is used to pack off the intestines and liver to expose the kidney and adrenal. Mobilization of the adrenal is a four step procedure: (1) free the cranial pole of the kidney; (2) the loosened fascia attached to the adrenal is freed from the diaphragm; (3) the adrenal is now freed by blunt dissection; and (4) the fasical connections between the liver and posterior vena cava are dissected free. When the gland is free enough, the adrenal vein is clamped off, and the gland removed. Gelfoam sponge is used to control hemorrhage over the severed stump of the adrenal vein (Hoar, 1966).

If required, removal of the opposite adrenal is usually accomplished at a later date. The surgical procedure is a duplicate of the one described (Hoar, 1966).

D. Thyroidectomy and Parathyroidectomy

The role of the thyroid in reproduction, body metabolism, myocardial enzyme activity, and tissue catalase activity on pituitary cytology has been investigated in several species (Kromka and Hoar, 1975; Lopukhin, 1976; Lang, 1976). The parathyroid glands are critical for the body to maintain the proper calcium and phosphorus metabolism. Although references to thyroparathyroidectomy procedures are scattered throughout the literature, they are difficult to find and generally are lacking in details necessary for surgical proficiency. The following technique is described for guinea pigs by Kromka and Hoar (1975). The animal is anesthetized and the ventral neck area is surgically prepared. A 1-in. skin incision is made just caudal to the larynx; with the aid of a dissecting microscope, the infrahyoid musculature is separated on a midline exposing the trachea. Retractors are used to visualize the thyroids and parathyroids on either side of the trachea. The anatomy of the thyroid differs from species to species and should be reviewed prior to surgery. The thyroids and/or parathyroids are dissected free, being careful not to damage the recurrent laryngeal nerve as it passes deep and medial to the thyroids. Cautery is helpful in freeing the gland to expose the vascular bed, which lies on the medial aspect of the cranial pole. This vasculature is then ligated and/or cauterized to free the thyroid and parathyroid glands. Surgical thyroidectomy is enough to produce a temporary hypothyroid state, but long-term experimentation requires the additional use of ^{131}I to destroy developing thryoid rests (Kromka and Hoar, 1975). The parathyroids are not always attached to the thyroids in guinea pigs, and so care should be taken to identify these organs before excising (Peterson *et al.*, 1952). Remember the number and the location of the parathyroids vary in each species. This technique is also described for the dog by Lang (1976) and Lopukhin (1976).

XIV. ORTHOPEDIC PROCEDURES FOR LABORATORY ANIMALS

The need for surgical correction of fractures in laboratory animals is usually determined through a thorough physical examination and/or diagnostic radiography. Many of the larger species require anesthesia in order to accomplish these techniques. Because fine details are needed to see fracture lines in bones that may be only a few millimeters in diameter, it is necessary to use special settings and nonscreen x-ray film. For example, the exposure used on a mouse at 30 in. (76.2 cm) could be 55 kV and 15 mA for $\frac{1}{2}$ sec. As the thickness of the body tissue mass increases with other species, new settings would need to be determined.

A. Intramedullary Pinning

The most common fractures in small laboratory animals, such as gerbils, hamsters, and guinea pigs, occur in the tibia, femur, and humerus. If the fracture occurs near the midshaft, it can be repaired by an intramedullary pin. Usually a lateral approach is used for open reduction. A fine Steinman pin, K wire or hypodermic needle is selected for retrograde pinning of the fracture. Postoperative radiographs should be taken to assure proper pin placement. Recovery is usually uncomplicated, and the pin is removed in 4 to 6 weeks (Rickards *et al.*, 1972).

B. External Fixation

When the expense involved exceeds the value of the patient, or perhaps the patient is too weak to undergo major surgery, or if the fracture is comminuted and unsuitable for intramedullary

techniques, external fixation may be the technique of choice. Although tongue depressors, radiographic film and plaster casts have been used, all have their limitations depending on the anatomical conformation of the limb involved. One method, which appears to have universal use in the small species, is the tubular traction splint. The splint is made from any strong plastic cylinder of a suitable diameter, e.g., the barrel of a hypodermic syringe or a syringe case. The tube should be long enough to prevent the extended foot from protruding out the distal end. A piece of tape is attached to the leg and the tape is pulled down the inside of the tubing. Using the tape, traction is applied to the limb to facilitate realignment of the fracture. The tape is then secured to the outside of the tubing. Padding is usually provided around the proximal end and at times inside the tubing. This splint is well tolerated and appears to be a very popular technique for small mammals and birds (Rickards *et al.*, 1972).

C. Modified Kirschner Fixation

A second method of external fixation, which has become popular, is an adaptation of the previously described tubular splint. In this case the plastic tubing is again selected, depending on the size of the limb. Once selected, the tubing is cut lengthwise and a strip is removed. The split plastic tubing is now separated to allow entry of the fractured limb. At this point hypodermic needles or small K wires are placed through the tubing and the bone above and below the fracture to provide a modified Kirschner fixation. The exposed ends of the needle or K wire are cut close to the tubing and usually taped. In some instances when additional strength is required cyanoacrylate glue can be placed at the junction of the pins and the tubing. This technique is adaptable for most small mammals and birds. In the authors opinion, this is the cheapest, quickest, and most effective external fixation technique for those smaller species.

XV. NEUROSURGICAL TECHNIQUES

A. Lumbar Sympathectomy

Lumbar sympathectomy is the operation of choice for Raynaud's disease and has been used for conditions such as arteriosclerosis and thromboangiitis obliterans. Clinically, this operation is not done unless spinal anesthesia causes a definite temporary improvement in blood flow. A second clinical condition, Hirschsprung's disease or megacolon, is also treated with a lumbar ganglionectomy. Those patients in which spinal anesthesia brings relief and evacuation are deemed suitable candidates for this surgical technique. The technique involves anesthesia; surgical preparation of abdomen; a midline incision; use of a self-retaining retractor; packing off the small intestine with body temperature, saline-soaked sterile towels; exposing the lumbar sympathetic chain or in the case of Hirschspung's disease isolating the hypogastric nerve; and finally excising this chain or the hypogastric nerve. Renal sympathectomy is theoretically a sound operation for chronic glomerulonephritis, especially before there is too much destruction of kidney substance. It should be noted that there is regeneration of the excised sympathetic chain in most species. Of further interest is the fact that total extirpation of the sympathetic chains in cats, dogs, and monkeys does not seem to affect the life of the animal (Markowitz et al., 1964).

B. Intracerebral Implantation

A number of experiments have been designed to demonstrate the effects of the direct action of neonatal hormones on a mammal's developing brain (Hayashi and Gorski, 1974). The instruments required for this technique include a converted 5 ml syringe that can be attached to a stereotaxic instrument. The syringe has a central, threaded metal rod that is fitted with a clip to hold the steel wire of the implanting stylet. The stylet (hypodermic needle, 27 gauge), tubing and wire are attached to the syringe and the wire is firmly held by the clip. A hormone implant is fused to the end of the wire and as the wire is pulled through the tubing, the implant is left unattached to the apparatus. This technique of implantation in the gerbil involves anesthetizing the animal; making a scalp incision; placing the head in a stereotaxic apparatus; and, using coordinates, a hole is drilled into the skull. The stylet is now lowered to the proper depth and the implant is detached. The stylet is then removed, and the wound is closed with collodion (Holman, 1980).

Other methods for neurological examination in the gerbil include the implantation of platinum wire recording electrodes and the implantation of a chronic ventricular infusion cannula (Herndon and Ringle, 1969). The collection of ventricular and cisterna cerebrospinal fluid has also been described for cattle (Cox and Littledike, 1978), monkeys (Snead and LaCroix, 1977), horses (Spinelli et al., 1968), rats (Brakkee et al., 1979), and rabbits (Kusumi and Plouffe, 1979). Because there is a paucity of detailed descriptive literature for obtaining rabbit cerebrospinal fluid, we will attempt to describe a relatively simple and safe technique. The rabbit is anesthetized and the dorsal cervical area and occipital area of the skull are shaved. The rabbit is positioned in lateral recumbency and, with the ears firmly secured, the neck is flexed to expose the base of the skull. This area is aseptically prepared and a 22 gauge, 3.81

cm ($1\frac{1}{2}$-in.) needle* is inserted with the free hand approximately 2 mm caudal to the external occipital protuberance (Fig. 14).

The needle is kept parallel to the table and is advanced slowly toward the animals mouth. At times a slight rotation of the head is needed to raise the anterior skull segments to allow this alignment. The needle is advanced through the caudal spinous muscle until a slight decrease in resistance is felt upon entering the fourth ventricle. The stylet is then removed and 1.5 to 2 ml of cerebrospinal fluid can be collected. This procedure can be accomplished in less than 5 min following anesthesia (Kusumi and Plouffe, 1979).

C. Stereotaxic Electrode Implantation

Stereotaxic brain electrode implantation invovles the use of an instrument immobilizing the animal's head and calibrated to identify coordinates in the brain for placement of electrodes in premapped structures. The coordinates are read from horizontal, coronal, and sagittal planes. Stereotaxic atlases are available for most common laboratory animals and can be developed for others (Cain and Dekergommeaux, 1979; Harris and Walker, 1980). The zero point for the coordinate system may be alternately the intersection of the coronal and sagittal skull sutures—"bregma"— or the middle of an interaural line. Atlases have been developed based on these two zero systems. Bregma zero points have the advantage of being grossly visible and introducing less error with different size heads.

Stereotaxic devices come with varying frame sizes for different size animals but interchangeable coordinate bars so that one frame might be used on mouse through dog size heads and another on goats. Parts include a cranial–caudal adjustment for coronal planes, lateral adjustment for sagittal planes, and depth adjustment for horizontal planes, and electrode holder, ear bars, incisor bars, nose clamp, and infraorbital bars for dogs, cats, and monkeys.

Stereotaxic electrode implantation in the rat brain is a surgical procedure. The rat is anesthetized, atropine is given at 1 mg/kg to reduce airway secretions and help keep airways open. The shaved head is steadied in a stereotaxic device by the two ear bars placed in the auditory meatuses and the incisor bar is fitted behind the front incisors with a nose clamp over the snout. Placement must be precise to get the brain aligned symmetrically between these points. Mineral oil is placed in each eye to keep them moist and protect them from alcohol or dental acrylic spills. The scalp is swabbed with alcohol and a midline scalp incision is made from between the eyes to between the ears and to the end of the external occipital crest. The incision is extended to the skull and the periosteum

---

*Stoelting, South Kostner Avenue, Chicago, Illinois 60623.

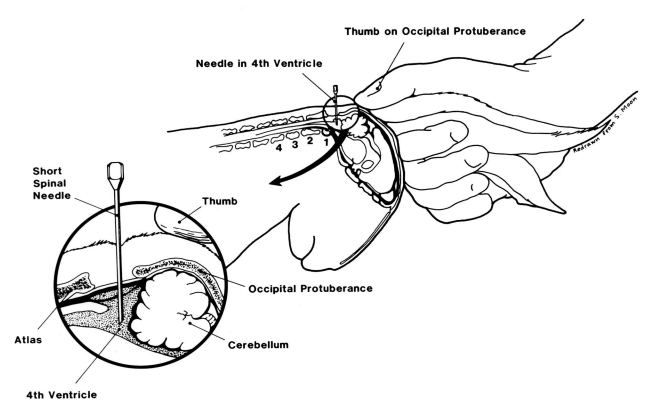

Fig. 14.   Landmarks for cerebrospinal fluid sampling in the rabbit. (Used with permission from Kusumi and Plouffe, 1979.)

scraped back with a bone curette. The periosteum is held back with hemostats on each side of both ends of the incision. Bleeding is controlled with cotton swabs and pressure. An electrode drill is attached to the electrode carrier. The skull is leveled visually or by making horizontal readings at bregma (intersection of sagittal and coronal sutures) and lambda (intersection of sagittal and lambdoidal sutures). Different atlases may specify various horizontal readings above or below zero. If bregma is not used as the zero, this point will have to have been established prior to positioning the ear bars on the rat to get the right interaural depth reading for zero. The coordinates for electrode placement are read off an atlas for the structure being examined as millimeters from zero in each plane. If bregma is zero in the atlas, the coordinates of bregma are read off the stereotaxic device on the rat to an accuracy of 0.1 mm using a vernier scale. The structure site for implantation is identified in the serial sections of an atlas at its largest size, and the coordinates of the atlas added or subtracted from the coordinates attributed to the zero reading in the rat being implanted. Only the bony skull is drilled with the dura left intact. Additional smaller holes are drilled in four corners of the skull within the incision line for placing jeweler's screws to help anchor the dental acrylic cap. The holes should not be placed in the relatively weak suture lines or too close to the electrode

hole. The dura is slit with a sterile needle, and the jeweler's screws are placed 1 mm into the cortex avoiding the dura and brain surface. The electrode is placed using either the surface of the brain or surface of the skull as a reference point for depth. Atlas readings are generally from the brain surface. One millimeter is added for skull thickness if this is used for reference (Cooley and Vanderwolf, 1978). A small piece of Gelfoam dipped in saline and pressed nearly dry is placed around the electrode to keep toxic dental acrylic off the brain (Skinner, 1971). The electrode is fixed in dental cement. Powder and solvent are mixed to syrupy consistency and allowed to harden around the electrode and jeweler's screws. The electrode end(s) are fitted with contacts that attach to a connector base, and they are incorporated in additional dental acrylic layers applied over the skull to create a secure cap. The connector base eases connection of the electrodes to electrical recording instruments. The acrylic cap should be free of sharp edges that could hinder incision healing. The bottom of the connector base should be as close as possible to the skull or the implant cap may be knocked off as the rat bumps its head as in recovery. Skin closure is made with simple interrupted sutures bringing the skin tight against the cement cap. The rat is removed from the stereotaxic device and allowed to recover (Skinner, 1971).

XVI. TUMOR TRANSPLANTATION

Since 1965, stable lines of transplantable tumors have been developed that are well characterized in terms of speed of growth, size, morphology, histology, local invasiveness, tendency to metastasize or regress, and response to chemotherapeutic agents. At least one book cataloging established tumor lines, preferred hosts, and the appropriate models these combinations make is in print (Roberts, 1969).

Serially transplantable tumors are serving several important functions in current research. The most notable service is the use of human tumors transplanted into rodents for the purpose of testing potential anticancer chemotherapeutic agents. Human xenograft tumors are preferred over rodent origin tumors since they retain their human morphology, functional behavior, and chemotherapeutic responsiveness (Shorthouse et al., 1980; Kyriazis et al., 1982). Active research is being conducted using diabetic rats in attempts to prolong survival of transplanted pancreatic tissue (Akimaru et al., 1981; Janney et al., 1982).

A basic understanding of transplantation terminology is necessary when discussing methods of transplantation.

Autograft: A graft of tissue from one site to another on the same individual. Rejection is rare.

Allograft: A graft between genetically dissimilar animals of the same species. This graft may undergo rejection.

Xenograft: A graft between animals of different species. Rejection may be quite violent.

The outcome of any transplantation is dependent upon many factors (Sugiura, 1965). Following the initial transplantation of a tumor, there is a slight tendency toward dedifferentiation. In general, tumors will grow best in the same strain of animal in which they spontaenously appear or can be induced. After many transplantations, the host requirements are less specific. Tumor growth is better in young, vigorous, well-fed rats and mice than in older underfed animals. Sex of the recipient animal will not affect the number of tumor takes, but may influence subsequent growth. Pregnancy has an unpredictable effect on tumor growth. Site of transplantation may also influence many different growth parameters, including percentage of tumor takes, subsequent growth, and response to therapy (Double and Ball, 1975).

The histologic classification of tumors also has a profound effect on transplantability. Malignant melanoma and colon, lung, kidney, and bone neoplasias have been among the more successful donors (Fogh et al., 1980; Shimosato et al., 1976). The difficulty of transplanting human mammary tumors is well documented (Bailey et al., 1980; Outzen and Custer, 1975; Boesen and Cobb, 1974). Endocrine, hormone-dependent, and lymphohematopoietic malignancies are not readily transplanta-

ble (Gazdar et al., 1981). Higher take rates for recurrent and metastatic neoplasms compared to primary tumors has also been observed (Fogh et al., 1980).

One of the most important aspects of assuring a long-lasting transplant take is proper selection and/or preparation of the host to prevent a host versus graft reaction.

Immunosuppression is currently one of the major means of combating the host versus graft reaction. Lance (1976) classifies immunosuppressant procedures or drugs as follows: agents that do not discriminate for lymphocytes include antiinflammatory agents, ionizing radiation, and radiomimetic drugs, and cystostatic agents. Ionizing radiation has been reported to produce a non-imunologic enhancement of tumor growth (Jambaski and Nettersheim, 1977) and is used most frequently in conjunction with other immunosuppressive techniques.

Recent research has been directed toward the development of more types of immunodeficient animals for tumor growth. The asplenic and athymic mutations have been combined to produce the Lasat mouse. Another nude cross has yielded a masked-mouse. Other nude laboratory animals include the nude streaker mouse, nude Rowett rat, nude New Zealand rat, and the nude guinea pigs (Festing, 1980). Preliminary studies utilizing the nude rat demonstrate the same desirable chemotherapeutic responses of transplanted tumors, a broader tumor spectrum, larger host size, more robust animals, and lack of the gnotobiotic environment requirement of nude mice.

The physical techniques involved with tumor transplantation are fairly simple. Careful selection of tumor and host combination is necessary. Healthy host animals must be chosen, and the transplant should be an actively dividing, nonulcerated tumor if possible. High standards of sterile technique are mandatory. The donor is anesthetized or euthanized, and the tumor is removed to a sterile Petri dish. Chilled sterile saline or refrigerated culture medium is used to moisten the tissue and nonhemorrhagic, nonnecrotic areas are aseptically minced into 2–3 mm fragments. If tumor cells are being grown by tissue culture, the cells are dispersed using 0.02% EDTA in phosphate-buffered saline. Cell viability is determined using trypan blue staining. The amount of media necessary to adjust the viable cells to $10^6/0.2$ ml is added (Kyriazis et al., 1982). For optimal results, transplantation should occur quickly following removal from the donor.

Subcutaneous inoculation is the oldest and most commonly used site for transplantation. If tumor fragments are being utilized, a 3–5 millimeter fragment is inoculated subcutaneously, using a trocar. Alternatively 10^6 to 10^7 tumor cells may be injected. A tumor take is usually visible at 7 days. Growth is measured using calipers. The tumor should not be palpated due to the danger of fatal hemorrhage from highly vascular tissues.

Intracranial inoculation is useful, especially since the minimum number of tumor cells necessary for a take is relatively low. The small cell carcinoma invariably grows in the meninges and is very locally invasive compared to subcutaneous inoculation (Gazdar *et al.*, 1981). For intracranial injection, a 26 gauge needle and an insulin syringe are used to inject 0.05 ml of cells. The needle is inserted one-half to two-thirds of the way in at a site just above the midpoint of a line connecting the lateral canthus of the eye and the external auditory meatus. Injected mice seem to go into a state of shock for several seconds and then recover, but a few of the mice will convulse and die within hours. Due to the rich vascular supply of renal tissue, implantation of tumors under the renal capsule may be desirable for enhanced tumor growth. One-millimeter cube fragments are implanted (Bogden *et al.*, 1979).

Induction of cystitis and infusion of tumor cells into the bladder is a technique used for induction of bladder tumors (Edwards *et al.*, 1978). The bladder is infused with Cetavlon, which is left in for 30 min. The bladder is then emptied, irrigated with 0.9% saline, and filled with tumor suspension. This method is very effective for induction of bladder tumors in rabbits and rats. Rats will also show renal tumor growth due to a species characteristic vesicoureteral reflux.

Many obstacles must be overcome in the search for the perfect animal models for human cancer. Differences in growth rate, invasiveness, metastatic tendencies, metabolism, pharmacokinectics, and the use of a stroma provided by the host species must be considered. The potential for xenografts to change biologic properties and chemotherapeutic response complicates all xenografting studies (Steel *et al.*, 1980). These problems can be minimized by the further development of immunodeficient animals, as well as improved immunosuppression techniques.

ACKNOWLEDGMENT

The authors wish to acknowledge the assistance of Dr. Janet Henkel for her library research and contribution to the section on tumor transplantation.

REFERENCES

Abernathy, R. E., and Anderson, C. B. (1974). Exteriorized ureters in the dog. *Lab. Anim. Sci.* **24,** 946–947.

Akimaru, K., Stuhmiller, G. M., and Sugler, H. F. (1981). Allotransplantation of insulinoma into the testis of diabetic rats. *Transplantation* **32,** 227–232.

Alexander, D. J., and Clark, G. C. (1980). A simple method of oral endotracheal intubation in rabbits (*Oryctolagus cuniculus*). *Lab. Anim. Sci.* **30,** 871–873.

Allen, S. A. (1974). A method for bladder catheterization of the male calf. *Lab. Anim. Sci.* **24,** 96–98.

Ambrus, J. L., Ambrus, C. M., Harrison, P. W. E., Leonard, C. A., Moser, C. E., and Cravitz, H. (1951). Comparison of methods for obtaining blood from mice. *Am. J. Pharm.* **123,** 100–104.

Avery, D. L. and Spyker, J. M. (1977). Foot tattoo of neonatal mice. *Lab. Anim. Sci.* **27,** 110–112.

Bailey, M. J., Gazet, J. C., and Peckham, M. J. (1980). Human breast cancer xenografts in immunosuppressed mice. *Br. J. Cancer* **42,** 524–529.

Baker, G. A., and Gorham, J. R. (1951). A technique for bleeding ferrets and mink. *Cornell Vet.* **41,** 235–236.

Baker, H. J., Lindsey, J. R., and Weisbroth, S. H., eds., (1980). "The Laboratory Rat," Vol. 2. Academic Press, New York.

Balabaud, C., Saric, J., Gonzales, P., and Delphy, C. (1981). Bile collection in free moving rats. *Lab. Anim. Sci.* **31,** 273–275.

Barringer, M., Sterchi, J. M., Jackson, D., and Meredith, J. (1982). Chronic biliary sampling via a subcutaneous system in dogs. *Lab. Anim. Sci.* **32,** 283–285.

Barrow, M. V. (1968). Modified intravenous injection technique in rats. *Lab. Anim. Care* **18,** 570–571.

Belding, R. C., Quarles, J. M., Beaman, T. C., and Gerhardt, P. (1976). Exteriorized carotid–jugular-shunt for hemodialysis of the goat. *Lab. Anim. Sci.* **26,** 951–954.

Berchelman, M. L., Kolter, S. S., and Britton, H. A. (1973). Comparison of hematologic values from peripherial blood of the ear and venous blood of infant baboons (*Papio cynocephalus*). *Lab. Anim. Sci.* **23,** 48–52.

Bergman, R. K., Lodmel, D. L., and Hadlow, W. J. (1972). A technique for multiple bleedings or intravenous inoculations of mink at prescribed intervals. *Lab. Anim. Sci.* **22,** 93–95.

Bergström, A. (1971). A simple device for intravenous injection in the mouse. *Lab. Anim. Sci.* **21,** 600–601.

Bernstein, L. I., and Agee, J. (1964). Successful lobectomy in the guinea pig. *Lab. Anim. Care* **14,** 519–523.

Berthelot, P., Erlinger, S., Dhumeaux, D., and Pavaux, A. M. (1970). Mechanism of phenobarbital induced hypercholeresis in the rat. *Am. J. Physiol.* **219,** 809–813.

Berthelot, R. D., and Hughes, C. (1980). Endotracheal intubation: An easy way to establish a patent airway in rabbits. *Lab. Anim. Sci.* **30,** 227–230.

Bett, N. J., Hynd, J. W., and Green, C. J. (1980). Successful anesthesia and small-bowel anastomosis in the guinea pig. *Lab. Anim. Care* **14,** 225–228.

Bishop, S. P. (1980). Cardiovascular research. *In* "The Laboratory Rat" (H. J. Baker, J. R. Lindsey, and S. H. Weisbroth, eds.), Vol. 2, pp. 168–173. Academic Press, New York.

Bivin, W. S., and Timmons, E. H. (1974). Basic biomethodology. *In* "The Biology of the Laboratory Rabbit" (S. H. Weisbroth, R. E. Flatt, and A. L. Kraus, eds.), pp. 73–90. Academic Press, New York.

Black, D., and Claxter, M. (1979). A simple reliable and inexpensive method for the collection of rat urine. *Lab. Anim. Sci.* **29,** 253–254.

Blakely, G. B., Beamer, T. W., and Dukelow, W. R. (1981). Characteristics of the menstrual cycle in nonhuman primates. IV. Timed mating in *Macaca nemistrina*. *Lab. Anim.* **15,** 351–353.

Bleakley, S. P. (1980). Simple technique for bleeding ferrets (*Mustela putorious furo*). *Lab. Anim.* **14,** 59–60.

Bliss, D. K., and Bates, P. L. (1973). A rapid and reliable technique for pinealectomizing rats. *Physiol. Behav.* **11,** 111–112.

Boegli, R. G., and Hall, I. H. (1969). A surgical external biliary fistula for the total collection of bile from rabbits. *Lab. Anim. Care* **19,** 657–658.

Boesen, E. A. M., and Cobb, L. M. (1974). Benign and malignant human breast tumors as an intramuscular xenograft in the hamster. *JNCI, J. Natl. Cancer Inst.* **52,** 281–293.

Bogdon, A. E., Haskell, P. M., Le Page, D. L., Kelton, D. E., Cobb, W. R., and Esber, H. J. (1979). Growth of human tumor xenographs implanted under the renal capsule of normal immunocompetent mice. *Exp. Cell Biol.* **47,** 281–293.

Born, C. T., and Moller, M. L. (1974). A simple procedure for long-term intravenous infusion in the rat. *Lab. Anim. Sci.* **29,** 78.

Bowen, W. H., Coid, C. R., T-W-Fiennes, R. N., and Mahoney, C. J.

(1976). Primates. *In* "The UFAW Handbook on the Care and Management of Laboratory Animals" (UFAW Staff, eds.), 5th ed., pp. 377–427. Churchill-Livingstone, Edinburgh and London.

Brain, J. D., and Frank, N. R. (1968). Recovery of free cells from rat lungs by repeated washings. *J. Appl. Physiol.* **25**, 63–69.

Brakkee, J. H., Wiegant, V. M., and Gispen, W. H. (1979). A simple technique for rapid implantation of a permanent cannula into the rat brain ventricular system. *Lab. Anim. Sci.* **29**, 78–81.

Branham, G. W. (1976). A device for continuous intravenous fluid injections in dogs. *Lab. Anim. Sci.* **26**, 75–77.

Brown, M. J., Erichsen, D. F., Helgerson, R., and Will, J. A. (1982). A modification for preparing the chronic lung–lymph fistula in sheep. *J. Appl. Physiol.* **52**(6), 1664–1666.

Bruckner-Kardoss, E., and Worstmann, B. S. (1967). Cecectomy in germfree rats. *Lab. Anim. Care* **17**, 542–546.

Bryant, J. M. (1980). Vest and tethering system to accommodate catheters and temperature monitors for nonhuman primates. *Lab. Anim. Sci.* **30**, 706–708.

Bullock, L. P. (1983). Repetitive blood sampling from guinea pigs. *Lab. Anim. Sci.* **33**, 70–71.

Burhoe, S. O. (1940). Methods of securing blood from rats. *J. Hered.* **31**, 445–448.

Burke, J. G. (1977). Blood collecting by the ear artery method. *Lab. Anim.* **11**, 49.

Buttle, H. L., Clapham, C., and Oldham, J. D. (1982). A design for flexible intestinal cannulas. *Lab. Anim.* **16**, 307–309.

Cain, D. P., and Dekergommeaux, S. E. (1979). Electrode implantation in small rodents for kindling and long-term brain recording. *Physiol. Behav.* **22**, 799–780.

Carrillo, L., and Aviado, D. M. (1969). Monocrotaline induced pulmonary hypertension and *p*-chlorophenylanaline (PCPA). *Lab. Invest.* **20**, 243–248.

Carvalho, J. S., Shapiro, R., Hopper, P., and Page, L. B. (1975). Methods for serial study of renin–angiotensin system in the unanesthetized rat. *Am. J. Physiol.* **228**, 369–375.

Catchpole, H. R., and van Wagenen, G. (1975). Reproduction in the rhesus monkey, *Macaca mulatta*. *In* "The Rhesus Monkey" (G. H. Bourne, ed.), Vol. 2, pp. 118–140. Academic Press, New York.

Cate, C. C. (1969). A successful method for exsanguinating unanesthetized mice. *Lab. Anim. Care* **19**, 256–258.

Christison, G. E., and Curtin, R. M. (1969). A simple venous catheter for sequential blood sampling from unrestrained pigs. *Lab. Anim. Care* **19**, 259–262.

Cohen, E., and Oliver, M. (1964). Urethral catheterization of the rat. *Lab. Anim. Sci.* **29**, 781–784.

Conn, H., and Langer, R. (1978). Continuous long-term intraarterial infusion in the unrestrained rabbit. *Lab. Anim. Sci.* **28**, 598–602.

Conner, M. K., Dombroske, R., and Cheng, M. (1980). A simple device for continuous intravenous infusion of mice. *Lab. Anim. Sci.* **30**, 212–214.

Conti, P. A., Noland, T. E., and Gehret, J. (1979). Immobilization of a chronic intravenous catheter in the saphenous vein of African green and rhesus monkeys. *Lab. Anim. Sci.* **29**, 234–236.

Cook, R. W., and Williams, J. F. (1978). Surgical removal of the pyloric antrum in weanling rats. *Lab. Anim. Sci.* **28**, 437–439.

Cooley, R. K., and Vanderwolf, C. H. (1978). "Stereotaxic Surgery in the Rat: A Photographic Series," 2nd ed. A. J. Kirby Co., Ontario.

Cooper, T. G., and Waites, G. M. H. (1974). Testosterone in rete testis fluid and blood of rams and rats. *J. Endocrinol.* **62**, 619–629.

Corbitt, R. H., McCormick, K. A. A., and Bowles, C. A. (1981). A modification of the canine carotid–jugular shunt. *Lab. Anim. Sci.* **31**, 516–518.

Cox, P., and Littledike, E. T. (1978). Techniques for sampling ventricular and cisternal cerebrospinal fluid from unanesthetized cattle. *Lab. Anim. Sci.* **28**, 465–469.

Craig, D. J., Trost, J. G., and Talley, W. (1969). A surgical procedure for implantation of a chronic, indwelling jugular catheter in a monkey. *Lab. Anim. Care* **19**, 237–239.

Dalgard, D. W., Marshall, P., Fitzgerald, G. H., and Rendon, F. (1979). Surgical technique for a permanent tracheostomy in beagle dogs. *Lab. Anim. Sci.* **29**, 367–370.

Davies, L., and Grice, H. C. (1962). A device for restricting the movements of rats suitable for a variety of procedures. *Can. J. Comp. Med. Vet. Sci.* **26**, 62.

Davis, L., and Malinin, T. I. (1974). Rabbit intubation and halothane anesthesia. *Lab. Anim. Sci.* **24**, 617–621.

Davis, R. H., Kramer, D. L., Sackman, J. W., and Kyriazis, G. (1975). A simple staining method for vaginal smears using red ink. *Lab. Anim. Sci.* **25**, 319–320.

Decad, G. M., and Birnbaum, L. S. (1981). Noninvasive techniques for intravenous injection of guinea pigs. *Lab. Anim. Sci.* **31**, 85–86.

Dickie, M. M. (1975). Keeping records. *In* "Biology of the Laboratory Mouse" (E. L. Green, ed.), 3rd ed., pp. 23–27. Dover, New York.

Dolence, D., and Jones, H. E. (1975). Pericutaneous phlebotomy and intravenous injection in the guinea pig. *Lab. Anim. Sci.* **25**, 106, 107.

Double, J. A., and Ball, C. R. (1975). Influence of transplant site on the establishment of human tumor xenografts in immunosuppressed mice. *Br. J. Cancer* **54**, 233–234.

Dudrick, S. J., Steiger, E., Wilmore, D. W., and Vars, H. M. (1970). Continuous long-term intravenous infusion in unrestrained animals. *Lab. Anim. Care* **20**, 521–529.

Dukelow, W. R. (1978). Laparoscopic research techniques in mammalian embryology. *In* "Methods in Mammalian Reproduction" (J. C. Daniel, Jr., ed.), pp. 437–460. Academic Press, New York.

Dukelow, W. R., and Ariga, S. (1976). Laparoscopic techniques for biomedical research. *J. Med. Primatol.* **5**, 82–99.

Dukelow, W. R., Jarosz, S. J., Jewett, D. A., and Harrison, R. M. (1971). Laparoscopic examination of the ovaries in goats and primates. *Lab. Anim. Sci.* **21**, 594–597.

Edwards, L., Rosin, D., Leaper, M., Swedan, M., Trott, P., and Vertido, R. (1978). The induction of cystitis and the implantation of tumours in rats and rabbit bladders. *Br. J. Urol.* **50**, 502–504.

Ellery, A., and Kinnen, L. (1981). Operative procedures for the collection of rete testis fluid from conscious sheep. *Lab. Anim.* **15**, 187–188.

Enta, T., Lockey, S. D., Jr., and Reed, C. E. (1968). A rapid safe technique for repeated blood collection from small laboratory animals. The farmer's wife methods. *Proc. Soc. Exp. Biol. Med.* **127**, 136–137.

Erol, O. O., Spira, M., and Levy, B. (1980). Microangiography: A detailed technique of perfusion. *J. Surg. Res.* **29**, 406–413.

Evans, C. S., Smart, J. L., and Stoddart, R. C. (1968). Handling methods for wild house mice and wild rats. *Lab. Anim.* **2**, 29.

Falabella, F. (1967). Bleeding mice: A successful technique of cardiac puncture. *J. Lab. Clin. Med.* **70**, 981–982.

Falconi, G., and Rossi, G. L. (1964). Transauricular hypophysectomy in rats and mice. *Endocrinology* **74**, 301–303.

Festing, M. W. F. (1980). Inherited immunological defects in laboratory animals. *In* "Immunodeficient Animals for Cancer Research" (J. E. Castro, ed.), Oxford Univ. Press, London and New York.

Fields, B. T., Jr., and Cunningham, D. R. (1976). A tail artery technique for collecting one-half milliliter of blood from a mouse. *Lab. Anim. Sci.* **26**, 505–506.

Fletch, S. M., and Wabeser, G. (1970). A technique for safe multiple bleedings, or intravenous injections in mink. *Can. Vet. J.* **11**, 33.

Fogh, J., Thomas, O., Tiso, J., Sharkey, F. E., Fogh, J. M., and Daniels, J.

W. (1980). Twenty-three new human tumor lines established in nude mice. *Exp. Cell Biol.* **48**, 229–239.

Ford, J. J., and Maurer, R. R. (1978). Simple technique for chornic venous catheterization of swine. *Lab. Anim. Sci.* **28**, 615–618.

Foss, M. L., and Barnard, R. J. (1969). A vest to protect chronic implants in dogs. *Lab. Anim. Care* **19**, 113–114.

Frankenberg, L. (1979). Cardiac puncture in the mouse through the anterior thoracic aperture. *Lab. Anim.* **13**, 311–312.

Fredrickson, T. N., Chute, H. L., and O'Meara, D. C. (1958). Simple improved method for obtaining blood from chickens. *J. Am. Vet. Med. Assoc.* **132**, 390–391.

Frisk, C. S., and Richardson, M. R. (1979). Rapid methods for jugular bleeding of dogs requiring one technician. *Lab. Anim. Sci.* **29**, 371–373.

Furner, R. L., and Mellett, L. B. (1975). Mouse restraining chamber for tail-vein injection. *Lab. Anim. Sci.* **25**, 648–649.

Fussell, E. N., Franklin, L. E., and Frantz, R. C. (1973). Collection of chimpanzee semen with an artificial vagina. *Lab. Anim. Sci.* **23**, 252–255.

Fussell, E. N., Roussel, J. D., and Austin, C. R. (1967). Use of the rectal probe method for electrical ejaculation of apes, monkeys and a prosimian. *Lab. Anim. Care* **17**, 528–530.

Galitzer, J., Hayes, R. H., and Oehme, F. W. (1979). A simplified urine collection method for female swine. *Lab. Anim. Sci.* **29**, 404–405.

Garren, H. W. (1959). An improved method for obtaining blood from chickens. *Poult. Sci.* **38**, 916–918.

Garvey, J. S., and Alseth, B. L. (1971). Urine collection from newborn rabbits. *Lab. Anim. Sci.* **21**, 739.

Gay, W. I., ed. (1965–1974). "Methods of Animal Experimentation," Vols. 1–5. Academic Press, New York.

Gazdar, A. F., Carney, D. M., Sims, H. L., and Simmons, A. (1981). Heterotransplantation of small-cell carcinoma of the lung into nude mice: Comparison of intracranial and subcutaneous routes. *Inst. J. Cancer* **36**, 777–783.

Gerbig, C. G., Molello, J. A., and Robinson, V. B. (1975). A simple method for obtaining blood pressure in rhesus monkeys (*Macaca mulatta*). *Lab. Anim. Sci.* **24**, 614–618.

Goetz, K. L., and Hermreck, A. S. (1972). A simple, inexpensive, diaphragm-type skin connector for implanted catheters. *Lab. Anim. Sci.* **22**, 538–540.

Goodpasture, J. C., Cianci, J., and Zanefeld, L. J. D. (1982). Long-term evaluation of the effect of catheter materials on urethral tissue in dogs. *Lab. Anim. Sci.* **32**, 180–182.

Gould, K. G., Warner, H., and Martin, D. E. (1978). Rectal probe ejaculation in primates. *J. Med. Primatol.* **7**, 213–222.

Graham, C. E. (1976). Technique of laparoscopy in the chimpanzee. *J. Med. Primatol.* **5**, 111–123.

Grazer, F. M. (1958). Technique for intravascular injection and bleeding of newborn rats and mice. *Proc. Soc. Exp. Biol. Med.* **99**, 407–409.

Greene, F. E., and Wade, A. E. (1967). A technique to facilitate sublingual vein injection in the rat. *Lab. Anim. Care* **17**, 604–606.

Greenham, L. W. (1978). Tattooing newborn albino mice in life-span experiments. *Lab. Anim. Sci.* **28**, 346.

Grice, H. C. (1964). Methods for obtaining blood and for intravenous injections in laboratory animals. *Lab. Anim. Care* **14**, 483–493.

Gupta, B. N. (1973). Technique for collecting blood from neonatal rats. *Lab. Anim. Sci.* **23**, 559.

Hafez, E. S. E. (1970). *In* "Reproduction and Breeding Techniques for Laboratory Animals" (E. S. E. Hafez, ed.), p. 1–4. Lea & Febiger, Philadelphia, Pennsylvania.

Haisley, A. D. (1980). A technique for cheek pouch examination of Syrian hamsters. *Lab. Anim. Sci.* **30**, 107–109.

Hall, A. S. (1966). Methods and techniques of manipulating laboratory primates. *Lab. Anim. Dig.* **2**, 3–5.

Hall, L. L., DeLopez, O. H., Roberts, A., and Smith, F. A. (1974). A procedure for chronic, intravenous catheterization in the rabbit. *Lab. Anim. Sci.* **24**, 79–83.

Hall, R. D., and Hodgen, G. D. (1979). Pregnancy diagnosis of owl monkeys (*Aotus trivirgatus*): Evaluation of the hemagglutination inhibition test for urinary chorionic gonadotropin. *Lab. Anim. Sci.* **29**, 345–348.

Hamilton, R. M. G. (1978). Intravenous cannulation of hens for long-term infusion. *Lab. Anim. Sci.* **28**, 746–750.

Harkness, J. E., and Wagner, J. E. (1983). "The Biology and Medicine of Rabbits and Rodents," p. 3. Lea & Febiger, Philadelphia, Pennsylvania.

Harris, D. L., and Decker, W. J. (1969). A method for obtaining repeated serial intestinal biopsies from the dog. *Lab. Anim. Care* **19**, 849–852.

Harris, D. V., and Walker, J. M. (1980). A semi-chronic electrode implant for very small animals. *Brain Res. Bull.* **5**, 479–480.

Harris, R. H., Jr., and Best, C. F. (1979). Urethral catheter system for renal function studies in conscious rabbits. *Lab. Anim. Sci.* **29**, 781–784.

Hartman, C. G. (1932). Studies in the reproduction of the monkey *Macacus* (*pithecus*) *rhesus* with special reference to menstruation and pregnancy. *Contrib. Embryol. Carnegie Inst.* **23**, 1–161.

Harvey, R. G., and Jones, E. F. (1982). A technique for bioinstrumentation of the thorax of miniature swine. *Lab. Anim. Sci.* **32**, 94–96.

Hayashi, S., and Gorski, R. A. (1974). Critical exposure time for androgenization by intracranial crystals of testosterone proprionate in neonatal female rats. *Endocrinology* **94**, 1161–1167.

Hayes, B. E., and Will, J. A. (1978). Pulmonary artery catheterization in the rat. *Am. J. Physiol.* **235**, H452–H455.

Herndon, B. L., and Ringle, D. A. (1969). Methods for neurological experimentation in the Mongolian gerbil, *Meriones unguiculatus. Lab. Anim. Care* **19**, 240–243.

Hoar, R. M. (1966). A technique for bilateral adrenalectomy in guinea pigs. *Lab. Anim. Care* **16**, 410–416.

Hobson, B. M. (1976). Evaluation of the subhuman primate tube test for pregnancy of primates. *Lab. Anim.* **10**, 87–89.

Hodesson, S., and Miller, J. N. (1964). Testis and popliteal lymph node removal from the living rabbit using aseptic techniques and other anesthesia. *Lab. Anim. Care* **14**, 494–498.

Hodgen, G. D., and Niemann, W. H. (1975). Application of the subhuman primate pregnancy test kit to pregnancy diagnosis in baboons. *Lab. Anim. Sci.* **25**, 757–759.

Hodgen, G. D., and Ross, G. T. (1974). Pregnancy diagnosis by a hemagglutination inhibition test for urinary macaque chorionic gonadotropin (mCG). *J. Clin. Endocrinol. Metab.* **3**, 927–930.

Hodgen, G. D., Wolfe, L. G., Ogden, J. D., Adams, M. R., Descalzi, C. C., and Hildebrand, D. F. (1976a). Diagnosis of pregnancy in marmosets: Hemagglutination inhibition test and radio immunoassay for urinary chorionic gonadotropin. *Lab. Anim. Sci.* **26**, 224–229.

Hodgen, G. D., Niemann, W. H., Turner, C. K., and Chen, H. C. (1976b). Diagnosis pregnancy in chimpanzees using the nonhuman primate pregnancy test kit. *J. Med. Primatol.* **5**, 247–252.

Hodgen, G. D., Turner, C. K., Smith, E. E., and Bush, R. M. (1977). Pregnancy diagnosis in the orangutan (*Pongo pygmaeus*) using the subhuman primate pregnancy test. *Lab. Anim. Sci.* **27**, 99–101.

Hodgen, G. D., Stolzenberg, S. J., Jones, D. C. L., Hildebrand, D. F., and Turner, C. K. (1978). Pregnancy diagnosis in squirrel monkeys; hemagglutination test, radio immunoassay, and bioassay of chorionic gonadotropin. *J. Med. Primatol.* **7**, 59–64.

Hoffman, R. A., and Reiter, R. J. (1965). Rapid pinealectomy in hamsters and other small rodents. *Anat. Rec.* **153**, 19–21.

Hofstad, M. S. (1950). A method of bleeding chickens from the heart. *J. Am. Vet. Med. Assoc.* **66**, 353–354.

Holman, D. (1980). A method for intracerebrally implanting crystalline hormones into neonatal rodents. *Lab. Anim.* **14**, 263–266.

Hoppe, P. C., Laird, C. W., and Fox, R. R. (1969). A simple technique for bleeding the rabbit ear vein. *Lab. Anim. Care* **19**, 524–525.

Houghton, P. W., and Jones, C. L. (1977). A chronic gastrostomy and test system for evaluation of gastric secretion in rhesus monkeys. *Gastroenterology* **73**, 252–254.

Ilievski, V., and Fleischman, R. W. (1981). A technique for obtaining tracheobronchial washings from rhesus monkeys (*Macacca mulatta*). *Lab. Anim. Sci.* **31**, 524–525.

Jackson, I., Cook, D. B., and Gill, G. (1972). Simultaneous intravenous infusion and arterial sampling in piglets. *Lab. Anim. Sci.* **22**, 552–555.

Jacobsen, L. A. (1978). Renal transplantation in the rabbit: A model for preservation studies. *Lab. Anim.* **12**, 63–70.

Jambaski, R. J., and Nettersheim, P. (1977). Nonimmunological enhancement of tumor transplant-ability in x-irradiated host animals. *Br. J. Cancer* **36**, 723–729.

Janney, C. G., Lacy, P. E., Finke, E. H., and Davis, J. M. (1982). Prolongation of intrasplenic islet xenograft survival. *Am. J. Pathol.* **107**, 1–5.

Jimenez, R., Esteller, A., and Lopez, M. A. (1982). Biliary secretions in conscious rabbits: Surgical technique. *Lab. Anim.* **16**, 182–185.

Jones, R. C. (1971). Uses of artificial insemination. *Nature (London)* **229**, 534–537.

Jurek, F. W. (1977). Antigondal effect of melatonin in pinealectomized and intact male hamsters. *Proc. Soc. Exp. Biol. Med.* **155**, 31–34.

Kaplan, A., and Wolf, I. (1972). A device for restraining and intravenous injection of mice. *Lab. Anim. Sci.* **22**, 223–224.

Kaplan, H. M., and Timmons, E. H. (1979). "The Rabbit: A Model for the Principles of Mammalian Physiology and Surgery." Academic Press, New York.

Karlson, A. G. (1959). Intravenous injections in guinea pigs via veins of the penis. *Lab. Invest.* **8**, 987–989.

Kasai, M., Neva, K., and Tritani, A. (1981). Effects of various dryoprotective agents on the survival of unfrozen and frozen mouse embryos. *J. Reprod. Fertil.* **63**, 175–180.

Kassel, R., and Leviton, S. (1953). A jugular technique for the repeated bleeding of small animals. *Science* **118**, 563–564.

Keighley, G. (1966). A device for intravenous injection of mice and rats. *Lab. Anim. Care* **16**, 185–187.

Knapp, W. C., Leeson, G. A., and Wright, B. J. (1971). An improved technique for the collection of bile in the unanesthetized rat. *Lab. Anim. Sci.* **21**, 403–405.

Knipfel, J. E., Peace, R. W., and Evans, J. A. (1975). Multiple vascular and gastric cannulations of swine for studies of gastrointestinal, liver and peripheral tissue metabolism. *Lab. Anim. Sci.* **25**, 74–78.

Knize, D. M., and Weatherby-White, R. C. A. (1974). Restraint of rabbits during prolonged administration of intravenous fluids. *Lab. Anim. Sci.* **19**, 394–397.

Knowles, A. D. (1959). Debarking–ventriculocordectomy. *Vet. Med. (Kansas City, Mo.)* **54**, 363.

Krause, A. L. (1980). Research methodology. *In* "The Laboratory Rat" (H. J. Baker, J. R. Lindsey, and S. H. Weisbroth, eds.), Vol. 2, pp. 1–42. Academic Press, New York.

Kromka, M. C., and Hoar, R. M. (1975). An improved technique for thyroidectomy in guinea pigs. *Lab. Anim. Sci.* **25**, 82–84.

Kumar, R. K. (1979). Toe clipping procedure for individual identification of rodents. *Lab. Anim. Sci.* **29**, 679–680.

Kusumi, R. K., and Plouffe, J. F. (1979). A safe and simple technique for obtaining cerebrospinal fluid from rabbits. *Lab. Anim. Sci.* **29**, 681–682.

Kuszak, J., and Rodin, M. (1977). A new technique for the pinealectomy of adult rats. *Experientia* **33**, 283–284.

Kyriazis, A. P., Kyriazis, A. A., Scarpelli, D. G., Fogh, J., Rao, S., and Lepers, L. (1982). Human pancreatic adenocarcinoma line Capan-1 in tissue culture and the nude mouse. *Am. J. Pathol.* **106**, 250–260.

Lance, E. M. (1976). Immunosuppression. *In* "Immunology for Surgeons" (J. E. Castro, ed.), pp. 229–257. University Park Press, Baltimore, Maryland.

Lang, C. M. (1967). A technique for the collection of semen from squirrel monkeys (*Saimiri sciureus*) by electroejaculation. *Lab. Anim. Care* **17**, 218–221.

Lang, C. M. (1976). "Animal Physiologic Surgery," pp. 107–120. Springer-Verlag, Berlin and New York.

Law, G. R. J. (1960). Blood samples from jugular vein of turkeys. *Poult. Sci.* **39**, 1450–1452.

Leash, A. M., Beyer, R. D., and Wilber, R. G. (1973). Self-mutilation following Innovar-Vet injections in the guinea pig. *Lab. Anim. Sci.* **23**, 720–721.

Leibo, S. P., Mazur, P., and Jackowski, S. C. (1974). Factors affecting survival of mouse embryos during freezing and thawing. *Exp. Cell Res.* **89**, 79–88.

Lequin, R. M., Elvers, L. H., and Bertens, A. P. M. G. (1981). Early detection of pregnancy in rhesus and sumptailed macaques (*Macaca mulatta* and *Macaca arctoides*). *J. Med. Primatol.* **10**, 189–198.

Levine, G., Lewis, L., and Cember, H. (1973). A vacuum-assisted technic for repetitive blood sampling in the rat. *Lab. Anim. Sci.* **26**, 211–213.

Lewis, V. J., Thacker, W. L., Mitchell, S. H., and Baer, G. M. (1976). A new technique for obtaining blood from mice. *Lab. Anim. Sci.* **26**, 211–213.

Loeb, W. F., Cicmonec, J. L., and Wickum, M. (1976). A coagulopathy of the owl monkey (*Aotus trivirgatus*) associated with high antithrombin III activity. *Lab. Anim. Sci.* **26**, 1084–1086.

Lopez, H., and Navia, J. M. (1977). A technique for repeated collection of blood from the guinea pig. *Lab. Anim. Sci.* **27**, 522–523.

Lopukhin, Y. M. (1976). "Experimental Surgery," pp. 191–236, 271–308, 350–366. M.I.R. Publishers, Moscow.

Lumsden, J. H., Presidenta, P. J. A., and Quinn, P. J. (1974). Modification of the orbital sinus bleeding technique for rabbits. *Lab. Anim. Sci.* **24**, 345–348.

Lushbough, C. H., and Moline, S. W. (1961). Improved terminal bleeding method. *Proc. Anim. Care Panel* **11**, 305–308.

Macedo-Sobrinho, B., Roth, G., and Grellner, T. (1978). Tubular device for introral examination of rodents. *Lab. Anim.* **12**, 137–139.

McFee, A. F., and Kenelly, J. J. (1964). Evaluation of a testicular biopsy technique in the rabbit. *J. Reprod. Fertil.* **8**, 141–144.

McGlinn, S. M., Shepherd, B. A., and Martan, J. (1976). A new castration technique in the guinea pig. *Lab. Anim. Sci.* **26**, 203–205.

Mackey, W. J., Anderson, W. D., and Kubick, W. G. (1970). Ventriculocordectomy technique for use on research swine. *Lab. Anim. Care* **20**, 992–994.

Mahoney, C. J. (1975a). The accuracy of bimanual rectal palpation for determining the time of ovulation and conception in the rhesus monkey (*Macaca mulatta*). *In* "Breeding Simians for Developmental Biology" (F. T. Perkins and P. N. O'Donoghue, eds.), Lab. Anim. Handb. No. 6, pp. 127–138. Laboratory Animals Ltd., London.

Mahoney, C. J. (1975b). Practical aspects of determining early pregnancy, stage of fetal development, and imminent parturition in the monkey (*Macaca fascicularis*). *In* "Breeding Simians for Developmental Biology" (F. T. Perkins and P. N. O'Donoghue, eds.), Lab. Anim. Handb. No. 6, pp. 261–274. Laboratory Animals Ltd., London.

Maillie, A. J., Calvo, E. L., Vaccaro, M. I., Caboteau, L. I., and Pivetta, O. H. (1981). An experimental model to study bile and exocrine pancreatic secretion from mice. *Lab. Anim. Sci.* **31**, 707–709.

Markowitz, J., Archibald, J., and Downie, H. G. (1964). "Experimental Surgery," 5th ed., pp. 326–352, 332–381, 581–598, 630–643. Williams & Williams, Baltimore, Maryland.

Martin, K. H., and Richmond, M. E. (1972). A method for repeated sampling of testis tissue from small mammals. *Lab. Anim. Sci.* **22**, 541–545.

Mauderly, J. L. (1970). Chronic implantation of electromagnetic flow probes on major abdominal vessels in the dog. *Lab. Anim. Care* **20**, 662–669.

Mauderly, J. L. (1977). Bronchopulmonary lavage of small laboratory animals. *Lab. Anim. Sci.* **27**, 255–261.

Maxwell, K. W., Dietz, T., and Marcus, S. (1964). An *in situ* method for harvesting guinea pig alveolar macrophages. *Am. J. Vet. Med. Res.* **37**, 237–238.

Medin, N. I., Osebold, J. W., and Zee, Y. C. (1976). A procedure for pulmonary lavage in mice. *Am. J. Vet. Res.* **37**, 273–238.

Melby, E. C., Jr., and Altman, N. H., eds. (1974–1976). "Handbook of Laboratory Animal Science," 3 vols. CRC Press, Cleveland, Ohio.

Mikeska, J. A., and Klemm, W. R. (1975). EEG evaluation of humaneness of asphyxia and decapitation euthanasia in the laboratory rat. *Lab. Anim. Sci.* **25**, 175–179.

Moler, T. L., Donahue, S. E., and Anderson, G. B. (1979). A simple technique for nonsurgical embryo transfer in mice. *Lab. Anim. Sci.* **29**, 353–356.

Moore, D. M. (1983). Dept. of Comparative Medicine, UTHSC Medical School at Houston, Texas (unpublished paper).

Morcom, C. B., and Dukelow, W. R. (1980). A research technique for the oviductal insemination of pigs using laparoscopy. *Lab. Anim. Sci.* **30**, 1030–1031.

Moreland, A. F. (1965). Collection and withdrawal of body fluids and infusion techniques. *In* "Methods of Animal Experimentation" (W. I. Gay, ed.), Vol. 1, pp. 18–19. Academic Press, New York.

Muggenberg, B. A., and Mauderly, J. L. (1975). Lung lavage using a single lumen endotracheal tube. *J. Appl. Physiol.* **38**, 922–926.

Mumaw, E. D., and Miller, A. S. (1975). The application of a frequency oscillation method for tooth extraction in dogs. *Lab. Anim. Sci.* **25**, 228–231.

Munson, E. S. (1974). Arterial cannulation in awake restrained monkeys. *Lab. Anim. Sci.* **24**, 793–795.

Murphy, G. P., and Grafton, T. (1973). Experimental kidney transplantation. *Lab. Anim. Sci.* **23**, 97–102.

Mylrea, K. C., and Abbrecht, P. H. (1967). An apparatus for tail vein injection in mice. *Lab. Anim. Care* **17**, 602–603.

Myrvik, Q. N., Leake, E. S., and Fariss, B. (1961). Studies on pulmonary alveolar macrophages from the normal rabbit: A technique to produce them in a high state of purity. *J. Immunol.* **86**, 128–132.

Nerenberg, S. T., and Zedler, P. (1975). Sequential blood samples from the tail vein of rats and mice obtained with modified Liebig condenser jackets and vacuum. *J. Lab. Clin. Med.* **85**, 523–526.

Nicholson, J. W., and Kinkead, E. R. (1982). A simple device for intratracheal injections in rats. *Lab. Anim. Sci.* **32**, 509–510.

Nobunaga, T., Nabamura, K., and Imamichi, T. (1966). A method for intravenous injection and collection of blood from rats and mice without restraint and anesthesia. *Lab. Anim. Care* **16**, 40–49.

Nolan, T. E., and Conti, P. A. (1980). Liver wedge biopsy in chimpanzees (*Pan troglodytes*) using an automatic stapling device. *Lab. Anim. Sci.* **30**, 578–580.

Olson, G. A. (1974). A method for surgical construction of a tracheal pouch in the ferret. *Lab. Anim. Dig.* **9**, 47–49.

Osborne, C. A., Low, D. G., and Finco, D. R. (1972). "Canine and Feline Urology," pp. 28–30. Saunders, Philadelphia, Pennsylvania.

Outzen, H. D., and Custer, R. P. (1975). Growth of human normal and neoplastic mammary tissues in cleared mammary fat pad of nude mouse. *J. Natl. Cancer Inst. (U.S.)* **55**, 1461–1463.

Painter, N. W., Borski, A. A., and Trevino, G. S. (1971). Urethral reaction to foreign objects. *J. Urol.* **106**, 227–230.

Pansky, B., Jacobs, M., and House, E. L. (1961). The orbital region as a source of blood samples in the golden hamster. *Anat. Rec.* **39**, 409.

Pare, W. P., Vincent, G. P., and Isom, K. E. (1979). Comparison of pyloric ligation and pyloric cuff techniques for collecting gastric secretion in the rat. *Lab. Anim. Sci.* **29**, 218–220.

Patrick, R. L., Kaplan, C. M., Margetis, P. M., and Pani, K. C. (1968). Nonsurgical extraction of teeth in small primates. *Lab. Anim. Care* **18**, 442–449.

Payne, J. E., Bischel, M. D., and Berne, T. W. (1974). Carotid–jugular arteriovenous fistulas for chronic hemodialysis in dogs. *Lab. Anim. Sci.* **24**, 367–369.

Pazin, G. J., Wu, B. A., and Van Theil, D. H. (1978). Fiberoptic esophagogastroscopy, brushings and biopsy in rabbits. *Lab. Anim. Sci.* **28**, 733–736.

Pena, H., and Cabrera, C. (1980). Improved endotracheal intubation technique in the rat. *Lab. Anim. Sci.* **30**, 712–713.

Perper, J., and Najarian, J. S. (1964). A technique for transplantation of the canine kidney. *Mod. Vet. Pract.* **45**, 31–36.

Peterborg, L. J., Philo, R. C., and Reiter, R. J. (1980). The pineal body and pinealectomy in the cotton rat. (*Sigmodon hispidus*). *Acta Anat.* **107**, 108–113.

Peterson, R. R., Webster, R. C., and Rayner, B. (1952). The thyroid and reproductive performance in the adult male guinea pig. *Endocrinology* **5**, 504–518.

Phillips, W. A., Stafford, W. W., and Stunt, J. (1973). Jugular vein technique for serial blood sampling and intravenous injection in the rat. *Proc. Soc. Exp. Biol. Med.* **143**, 733–735.

Plager, J. E. (1972). Intravenous, long-term infusion in the unrestrained mouse-method. *J. Lab Clin. Med.* **79**, 669–672.

Platts, R. G. S., Wilson, P., and Shaw, K. M. (1972). Design for a chronic right ventricular catheter for dogs. *Lab. Anim. Sci.* **22**, 900–903.

Rahlmann, D. F., Mains, R. C., and Kodama, A. M. (1976). A urine collection device for use with the male pigtail macaque (*Macaca neminstrina*). *Lab. Anim. Sci.* **26**, 829–831.

Raulston, G. L., Swaim, S. F., Martin, D. P., and Kerley, W. C. (1969). A method of rapid long-term devocalization of dogs. *Lab. Anim. Care* **19**, 247–249.

Redfern, R. (1971). Techniques for oral intubation of wild rats. *Lab. Anim.* **5**, 169–172.

Reynolds, J. A., and Hall, A. S. (1979). A rapid procedure for shortening canine teeth of nonhuman primates. *Lab. Anim. Sci.* **29**, 521–524.

Rhodes, M. L., and Patterson, C. E. (1979). Chronic intravenous infusion in the rat: A nonsurgical approach. *Lab. Anim. Sci.* **29**, 82–84.

Richards, D. A., Hinko, P. J., and Morse, E. M. (1972). Orthopaedic procedures for laboratory animals and exotic pets. *J. Am. Vet. Med. Assoc.* **161**, 729–732.

Roberts, D. C. (1969). "Transplanted Tumors of Rats and Mice: An Index of Tumors and Host Strains." Lab. Anim., Ltd., London.

Rosenhaft, M. E., Bing, D. H., and Knudson, K. C. (1971). A vacuum—assisted method for repetitive blood sampling in guinea pigs. *Lab. Anim. Sci.* **21**, 598–599.

Rowlands, I. W. (1974). Artificial insemination of mammals in captivity. *Int. Zoo Yearb.* **14**, 230–233.

Saarni, H., and Viikari, J. (1976). A simple technique for continuous intravenous infusion under neuroleptic tranquillization. *Lab. Anim.* **10**, 69–72.

Sakaki, K., Tanaka, K., and Hirasawa, K. (1961). Hematological comparison of the mouse blood taken from the eye and the tail. *Exp. Anim.* **10**, 14–19.

Sato, M., and Yoneda, S. (1966). An efficient method for transauricular hypophysectomy in rats. *Acta Endocrinol. (Copenhagen)* **51**, 43–48.

Schoenborne, B. M., Schrader, R. E., and Canolty, N. L. (1977). Tattooing newborn mice and rats for identification. *Lab. Anim. Sci.* **27**, 110.

Seager, S. W. J., and Fletcher, W. S. (1972). Collection, storage, and insemination of canine semen. *Lab. Anim. Sci.* **22**, 177–182.

Shimosato, Y., Kameya, T., Nagai, S., Hirohashi, T., Koide, T., Hayashi, H., and Momura, T. (1976). Transplantation of human tumors in nude mice. *J. Natl. Cancer Inst. (U.S.)* **56**, 1251–1255.

Shorthouse, A. J., Smyth, J. F., Steel, G. G., Ellison, M., Mills, J., and Peckham, M. J. (1980). The human tumor xenograft—A valid model in experimental chemotherapy? *Br. J. Surg.* **67**, 715–722.

Shrader, R. E., and Everson, G. J. (1968). Intravenous injection and blood sampling using cannulated guinea pigs. *Lab. Anim. Care* **18**, 214–219.

Simmons, F. A. (1952). Correlation of testicularbiopsy material with semen analysis in male infertility. *Ann. N.Y. Acad. Sci.* **5**, 643–656.

Simmons, M. L., and Brick, J. O. (1970). "The Laboratory Mouse—Selection and Management." Prentice-Hall, Englewood Cliffs, New Jersey.

Skinner, J. E. (1971). "Neuroscience: A Laboratory Animal." Saunders, Philadelphia, Pennsylvania.

Smith, C. J., and McMahon, J. B. (1977). Method for collecting blood from fetal and neonatal rats. *Lab. Anim. Sci.* **27**, 112–113.

Smith, C. R., Felton, J. S., and Taylor, R. T. (1981). Description of a disposable individual mouse collection apparatus. *Lab. Anim. Sci.* **31**, 80–82.

Snead, O. C., III, and LaCroix, J. T. (1977). Lumbar puncture obtaining cerebrospinal fluid in the rhesus monkey (*Macaca mulatta*). *Lab. Anim. Sci.* **27**, 1039–1040.

Snyder, S. B., and Soave, O. A. (1970). A method for the collection of throat specimens for laboratory primates for the rapid identification of respiratory infection. *Lab. Anim. Care* **20**, 518–520.

Sojka, N. J., Jennings, L. L., and Hamner, C. E. (1970). Artificial insemination in the cat (*Felis catus*). *Lab. Anim. Care* **20**, 198–204.

Soli, N. E., and Birkeland, R. (1977). A method for collection of bile in conscious sheep. *Acta Vet. Scand.* **18**, 221–226.

Sorg, D. A., and Buckner, B. (1964). A simple method for obtaining venous blood from small laboratory animals. *Proc. Soc. Exp. Biol. Med.* **115**, 1131–1132.

Spalton, P. N., and Clifford, J. M. (1979). Nasogastric intubation technique for bile sampling in the baboon. *Papio ursinus*. *Lab. Anim. Sci.* **29**, 237–239.

Spinelli, J., Holliday, T., and Homer, J. (1968). Technical notes: Collection of large samples of cerebrospinal fluid from horses. *Lab. Anim. Care* **18**, 565–567.

Stavy, M., Terkel, J., and Marder, U. (1978). Artificial insemination in the European hare (*Lepus europaeus syriacus*). *Lab. Anim. Care* **28**(2), 163–166.

Steel, G. G., Courtenay, V. D., Phelps, T. A., and Peckham, M. J. (1980). The therapeutic response of human tumor xenografts. *In* "Immunodeficient Animals for Cancer Research" (S. Sparrow, ed.), pp. 179–189. Oxford Univ. Press, London and New York.

Stevens, R. W. C., and Ridgeway, G. J. (1966). A technique for bleeding chickens from the jugular vein. *Poult. Sci.* **45**, 204–205.

Stickrod, G., and Pruett, D. K. (1979). Multiple cannulation of the primate superficial lateral coccygeal vein. *Lab. Anim. Sci.* **29**, 398–399.

Stickrod, G., Ebaugh, T., and Garnett, C. (1981). Use of a miniperistaltic pump for collection of blood from rabbits. *Lab. Anim. Sci.* **31**, 87–88.

Still, J. W., and Whitcomb, E. A. (1956). Techniques for long-term intubation of rat aorta. *J. Lab. Clin. Med.* **48**, 152.

Stills, H. F., Jr., Balady, M. A., and Liebenberg, S. P. (1979). A comparison of bacterial flora isolated by transtracheal aspiration and pharyngeal swabs in *Macaca fasicularis*. *Lab. Anim. Sci.* **29**, 229–233.

Stone, S. H. (1954). Method for obtaining venous blood from the orbital sinus of the rat or mouse. *Science* **119**, 100.

Strumpf, I. R., Bacher, J. D., Gadek, J. E., Morin, M. L., and Crystal, R. G. (1979). Flexible fiberoptic bronchoscopy of the rhesus monkey (*Macaca mulatta*). *Lab. Anim. Sci.* **29**, 785–788.

Stuhlman, R. A., Packers, J. T., and Rose, S. D. (1972). Repeated blood sampling of *Mystromys albicaudatus*. *Lab. Anim. Sci.* **22**, 268–270.

Sugiura, K. (1965). Tumor transplantation. *In* "Methods of Animal Experimentation" (W. I. Gay, ed.), Vol. 2, pp. 171–222. Academic Press, New York.

Terkel, L., and Urbach, L. (1974). A chronic intravenous cannulation technique adapted for behavioral studies. *Horm. Behav.* **5**, 141–148.

Tillman, P., and Norman, C. (1983). Droperidol–fentanyl as an aid to blood collection in rabbits. *Lab. Anim. Sci.* **33**, 181–182.

Timm, K. I. (1979). Orbital venous anatomy of the rat. *Lab. Anim. Sci.* **29**, 636–638.

Tomson, F. N., Schulte, J. M., and Bertsch, M. L. (1979). Root canal procedure for disarming nonhuman primates. *Lab. Anim. Sci.* **29**, 382–386.

Toofanian, F., and Targowski, S. (1982). Small intestinal anastomosis and preparation of intestinal loops in the rabbit. *Lab. Anim. Sci.* **32**, 80–82.

UFAW Staff, eds. (1976). "The UFAW Handbook on the Care and Management of Laboratory Animals," 5th ed. Churchill-Livingstone, Edinburgh and Lindon.

Valerio, D. A., Ellis, E. B., Clark, M. L., and Thompson, G. E. (1969). Collection of semen from macaques by electroejaculation. *Lab. Anim. Care* **19**, 250–252.

Vallejo-Freire, A. (1951). A simple technic for repeated collection of blood samples from guinea pigs. *Science* **114**, 524–525.

Van Pelt, L. F. (1974). Clinical assessment of reproductive function in female rhesus monkeys. *Lab. Anim.* **8**, 199–212.

Van Pelt, L. F., and Keyser, P. E. (1970). Observations on semen collection and quality in macaques. *Lab. Anim. Care* **20**, 726–733.

Voglmayr, J. K., Larsen, L. H., and White, I. G. (1970). Metabolism of spermatozoa and composition of fluid collected from the rete testis of living bulls. *J. Reprod. Fertil.* **21**, 449–460.

Voss, W. R. (1970). Primate liver and spleen biopsy procedures. *Lab. Anim. Care* **20**, 995–997.

Waites, G. M. H., and Einer-Jensen, N. (1974). Collection and analysis of rete testis fluid from macaque monkeys. *J. Reprod. Fertil.* **41**, 505–508.

Wardell, J. R., Jr., Chakrin, L. W., and Payne, B. J. (1970). The canine tracheal pouch, a model for use in respiratory mucus research. *Am. Rev. Respir. Dis.* **101**, 741–754.

Waynforth, H. B. (1980). "Experimental and Surgical Technique in the Rat." Academic Press, New York.

Waynforth, H. B., and Parkin, R. (1969). Sublingual vein injection in rodents. *Lab. Anim.* **3**, 35–37.

Wendt, D. J., Normile, H. J., Dawe, E. J., Trompeter, T., and Barraco, R. A. (1982). Chronic intracarotid cannulation of pigeons for administration of behaviorally active peptides. *Lab. Anim.* **16**, 335–338.

Whipp, S. C., Littledike, E. T., and Wangsness, S. (1970). A technique of hypophysectomy of calves. *Lab. Anim. Care* **20**, 533–538.

White, W. A. (1971). A technique for urine collection from anesthetized male rats. *Lab. Anim. Sci.* **21**, 401–402.

Whitney, R. A., Jr., Johnson, D. J., and Cole, W. C. (1973). "Laboratory Primate Handbook." Academic Press, New York.

Wideman, R. F., and Braun, E. J. (1982). Urethral urine collection from anesthetized domestic fowl. *Lab. Anim. Sci.* **32**, 298–301.

Wildt, D. E., Morcom, C. B., and Dukelow, W. R. (1975). Laparoscopic pregnancy diagnosis and uterine fluid recovery in swine. *J. Reprod. Fertil.* **44**, 301–304.

Wingfield, W. E., Tumbleson, M. E., Hicklin, K. W., and Mather, E. C. (1974). An exteriorized cranial vena caval catheter for serial blood sample collection from miniature swine. *Lab. Anim. Sci.* **24**, 359–361.

Wolfe, H. G. (1967). Artificial insemination of the laboratory mouse (*Mus musculus*). *Lab. Anim. Care* **17**, 426–432.

Yap, K. L. (1982). A method for intratracheal innoculation of mice. *Lab. Anim.* **16**, 143–145.

Yngner, T. (1975). A device for sampling liver tissue from unanesthetized mice. *Lab. Anim. Sci.* **25**, 647–648.

Yoder, J. T., and Starch, C. J. (1964). Devocalization of dogs by laryngofissure and dissection of the thyroarytenoid folds. *J. Am. Vet. Med. Assoc.* **145**, 326–327.

Young, L., and Chambers, T. (1973). A mouse bleeding technique yielding consistent volume with minimal hemolysis. *Lab. Anim. Sci.* **23**, 428–430.

Zambraski, E. J., and DiBona, G. F. (1976). A device for the cutaneous exteriorization of chronic intravascular catheters. *Lab. Anim. Sci.* **26**,939–841.

Zammit, M., Toledo-Pereya, L. H., Malcom, S., and Konde, W. N. (1979). Long-term cranial mesenteric vein cannulation in the rat. *Lab. Anim. Sci.* **29**, 364–366.

Chapter 20

Control of Biohazards Associated with the Use of Experimental Animals

W. Emmett Barkley and John H. Richardson

I. INTRODUCTION

The most important initial action in maintaining a safe occupational setting and for reducing the potential for work-associated illness when performing work with experimental animals is to recognize that chemical, physical, and biological hazards are ubiquitously present in the work environment. Biological hazards exist because experimental animals (1) are natural reservoirs for many of the zoonoses and may therefore harbor or be susceptible to an infectious agent capable of causing disease in man; (2) may be used as the host for studies involving pathogenic microorganisms; and (3) are a source of allergens for which susceptible individuals may experience a wide range of responses and reactions. Physical hazards exist because (1) animals can do physical harm to man (e.g., bite, scratch, crush); (2) the work environment can contribute to physical harm (e.g., accidental fall due to wet, slippery floor); and (3) the work practices can cause physical harm (e.g., improper storage of ether leading to formation of dangerously explosive peroxides). Experimentation, by its very nature, often involves the use of chemicals. Chemicals may be found in the animal environment as a cleaning agent or disinfectant, as a means of controlling pests, or as a contaminant in feed or bedding. Chemicals can cause acute and chronic toxic effects as well as physical harm (e.g., fire or explosion). It is only with this recognition that appropriate and adequate safeguards can be selected to control the existing hazards. Recognition of hazards and selection of appropriate safeguards, however, are not sufficient to assure a safe and healthful workplace. This is dependent on the continuing use of these safeguards by all involved in the handling of experimental animals. This chapter will focus primarily on the control of biological hazards. The concepts of biological control are relevant to the control of chemical and physical hazards as well. Specific guidance for physical and chemical safety programs are described by Steere (1980).

II. THE INFECTIOUS PROCESS

Work-associated infections involving experimental animals is the result of a number of factors, all of which are essential to the development of disease. The infectious agent, the experimental animal that serves as the source or reservoir of the agent, and the susceptible host or worker who is involved in handling, using, observing, supporting, or in other ways associated with the experimental animal are three such factors. If the purpose of using experimental animals is to study disease process involving infectious agent, the agent and reservoir will always be present. Where this is not the case, the presence of the agent is dependent on the infection or disease status of the experimental animal. The utilization of conventional or specific pathogen-free animals obtained from reliable sources can reduce the potential that the animal may harbor an infectious agent. Infections in laboratory personnel with salmonellosis, shigellosis, lymphocytic choriomeningitis, Q fever, psittacosis, epidemic typhus, B virus, hepatitis A, and tanapox are examples of documented infections associated with naturally infected laboratory animals. An appropriate quarantine period that allows a veterinarian to observe the animals, to perform diagnostic tests, and to give appropriate treatment or vaccinations can help prevent the introduction of disease into the animal colony (National Institutes of Health, 1978). The susceptibility of the worker is dependent on his or her immune status. This can be a function of the efficacy of artificial immunization, the health of the worker, or prior disease experience.

Three additional factors must also be present for work-associated illness to occur: (1) the agent must be able to escape from the experimental animal; (2) the agent must be transmitted to the worker; and (3) the agent must enter, gain access, or invade the worker. Methods for the control of biological hazards or biosafety practices are primarily utilized to control these factors. An understanding of these factors provides the basis for the selection of appropriate biosafety measures.

A. Mode of Escape

An infectious agent may escape from the experimental animal in either a natural or artificial way. Excretion of the agent in urine, saliva, and feces or release through skin lesions are examples of natural modes of escape. There are numerous mechanisms that allow the artificial escape of infectious agents from experimental animals. Biopsy, drawing a blood sample with a needle and syringe from a viremic animal, or necropsy are obvious mechanisms. Surgical instruments can become contaminated and thereby serve as a means for agent escape. Also, tissues and body fluids that have been removed from the animal may harbor an infectious agent. Vectors present on the experimental animal at the time of procurement or those not prevented from gaining access to the experimental animal by an effective pest control program may facilitate escape of the agent from the animal host.

B. Mode of Transmission

Transmission of an agent to the animal or laboratory worker can occur through numerous routes. The most frequently documented mode of transmission involves contaminated needles and syringes and direct contact with infected animals. These mechanisms account for approximately 40% of all laboratory-acquired infections resulting from documented accidents (Pike, 1979). The formation of aerosols and their easy dissemination is perhaps, however, the most common mode of transmission.

Aerosols are small particles of solids or liquids that are suspended in air and can generally remain airborne for an extended period of time. The smaller the particle, the longer it will remain airborne, the more likely it is to move with air currents, and the more likely it is to be respirable. Aerosolized particles in the <5 μm size range are most likely to be retained in deep pulmonary spaces and therefore establish foci of infection. Infectious aerosols are particles that may contain a single microorganism or clumps of microorganisms that have escaped an experimental animal host or some *in vitro* reservoir (e.g., tissue culture flask). Infectious aerosols may also consist of microorganisms that are attached to inanimate particles, such as dust from animal bedding.

Particles can become aerosolized by any forceful activity. Experimental animals produce aerosols by their activity and movements. Aerosols are produced by most animal care and husbandry practices and many experimental procedures. The worker can also contribute significantly to the process of aerosol production. Vigorous removing of bedding from cages in a typical animal room can increase the concentration of airborne microorganisms by a factor of 10 to 100.

Larger aerosol particles (i.e. <5 μm) often settle out of the air onto surfaces near the source of generation. Smaller particles may travel some distance before they are deposited onto surfaces or are inhaled by someone breathing air containing the aerosolized particle. It is this spreading of contaminants that contributes to secondary contact by personnel who work with animals. Thus the worker himself can become a major transmitter of infectious agents by touching contaminated surfaces and thereby transferring contaminants to himself, other persons, or other surfaces.

C.   Route of Exposure

There are four primary routes of exposure through which an infectious agent can cause disease. These are ingestion, inhalation, contact with mucous membranes, and direct parenteral inoculation. In situations involving work-associated illness, the route of exposure may be similar to that associated with the natural disease process or it may be distinctly different. For example, tularemia is normally acquired through contact of mucous membranes with infectious droplets or by direct percutaneous inoculation with blood or tissue of infected rabbits during skinning and dressing operations. In the laboratory, however, inhalation of aerosols generated while handling cultures has been the principal route of exposure resulting in infection. Wedum (1975) reported that the human ID_{25-50} of tularemia via the respiratory route was of the order of 10 organisms.

The most common mechanism of exposure to infectious agents associated with work involving experimental animals are (1) direct inoculation by needle and syringe, cut or abrasion from contaminated items having sharp edges, and animal bites; (2) inhalation of aerosols generated by accidents, animal care practices, and experimental procedures and manipulations; (3) contact of the mucous membranes of the eyes, nose, or mouth by spills of contaminated materials, contaminated hands, and contaminated surfaces; and (4) ingestion. It is well to emphasize that ingestion is a less likely route of exposure today because of the practices of using pipetting aids and prohibiting the consumption of food and beverages in work areas potentially contaminated with infectious materials.

III.   RISK ASSESSMENT

The laboratory director is responsible for the safe operations under his direction of programs that involve experimental animals. This responsibility requires the knowledge and judgment to assess risks of work-associated illness and to apply appropriate safeguards to reduce the risk of their occurrence. Characteristics of experimental animals, infectious agents that may be involved either directly or indirectly, training and experience of personnel, and activities and procedures required in carrying out the function of the program need to be considered in making decisions regarding risk assessments and the selection of safeguards. Important characteristics that influence the degree of risks, categorized according to the essential factors of the infectious process, are listed in Table I.

Two major concerns in risk assessment are (1) the potential for infection by aerosols and (2) the severity of the disease. These factors for zoonoses that frequently have been accidently acquired by workers in laboratories or experimental animal facilities are summarized in Table II.

IV.   BIOSAFETY PRACTICES AND PROCEDURES

Experimental animal activities involving the use of small mammals may include a variety of species of variable origin (i.e., conventional, specific pathogen-free, or wild caught). Individual animals for any of these categories may have in-

Table I

Risk Assessment Parameters

Infection process factors	Characteristics which influence risk assessments
Infectious agent	Virulence
	Pathogenicity
	Biological stability
Reservoir	Experimental animal species
	Common zoonoses
	Documented source of work-associated illness
Laboratory worker	Severity of disease in man
	Resistance to disease
	Availability of immunoprophylaxis
	Documented work-associated illness
	Training and experience
Mode of escape	Excreted in urine, saliva, or feces of experimental animals
	Agent present in lesion secretions
	Experimental procedures and manipulations
	Nature and function of the operation
Mode of transmission	Aerosols
	Potential for direct contact
	Vectors
	Secondary transfer
	Hands of worker
Route of exposure	Ingestion
	Inhalation
	Direct inoculation
	Cut or abrasion
	Needle puncture
	Bite
	Direct contact with membranes

Table II

Zoonoses of Special Concern to Laboratory Workers

Disease	Animal host	Associated illnesses	Aerosol hazard	Severity of disease in man
Brucellosis	Cattle, sheep, goats, swine	423	+++	++
Q Fever	Cattle, sheep, goats, swine	278	+++	+
Hepatitis	Nonhuman primates	234	+	++
Tularemia	Rabbits	225	+++	++
Tuberculosis	Nonhuman primates, cattle, sheep, goats, swine	176	+++	++
Psittacosis	Parakeets, parrots, pigeons, turkeys, chickens	116	++	+
Murine typhus	Rats, oppossums	68	++	+
Leptospirosis	Rats, dogs, mice, hamsters, guinea pigs, oppossums, skunks, fox, cattle, sheep, goats, swine	67	+	+
Shigellosis	Nonhuman primates	58	+	++
Newcastle disease	Chickens, turkeys	51	+	+
Salmonellosis	Chickens, turkeys, nonhuman primates	48	+	+
Lymphocytic choriomeningitis	Mice, rats, hamsters, guinea pigs	46	++	++
Anthrax	Cattle, sheep, goats, swine	45	++	++
Vesicular stomatitis	Cattle, sheep, goats, swine	40	++	+
Toxoplasmosis	Rats, mice, cats, dogs, cattle, sheep, goats, swine	28	++	++

duced or natural infections with indigenous or exotic agents that represent a threat of infection to personnel or to other animals in the colony.

For those agents associated with zoonotic infections in small mammals commonly used in laboratories, it is convenient to follow the agent risk classification proposed in the CDC-NIH draft publication ''Biosafety in Microbiological and Biomedical Laboratories'' (Centers for Disease Control and National Institutes of Health, 1983). Species-specific agents, such as mouse hepatitis virus, mouse pox virus (ectromelia), pneumonia virus of mice, and minute virus of mice, pose no recognized health hazard to personnel working with infected animals. The risk of infection with these agents to other animals in the colony may, however, be quite high and the consequences disruptive or devestating to experimental studies and colony management. The species-specific agents would meet the general criteria established for Biosafety Level 1 (BSL1) microorganisms (i.e., not associated with disease in healthy human adults).

Induced or natural infections with zoonotic agents, such as salmonellae, *Toxoplasma* spp., and *Leptospira* spp. pose ''moderate'' risks of occupationally associated infections in laboratory or animal care personnel. The majority of indigenous infectious agents that may occur as natural infections or that are induced fall within the ''moderate risk'' (BSL2) risk assessment category. Primary occupational infection hazards are associated with parenteral inoculation (e.g., bites or scratches of infected animals), ingestion of infectious materials, or mucous membrane exposure to infectious droplets.

Infection via exposure to infectious aerosals is not a common means of exposure for moderate risk agents.

Agents that represent a high risk occupational exposure (BSL3) are characterized by having a low infectious dose (ID_{50} < 100 organisms), a high rate of susceptibility in the general population, and transmission via infectious aerosols. Examples of agents with these characteristics are *Mycobacterium tuberculosis*, *Coxiella burnetii*, and lymphocytic choriomeningitis (LCM) virus. These three agents are documented personnel hazards, and each have been introduced into animal colonies by naturally infected but asymptomatic animals. Each has been associated with human infection as a result of exposure to or from working with infected animals.

It is the responsibility of the laboratory or animal facility director to decide who should be allowed to enter laboratory or animal room areas and to ensure that those who are given access have been advised of the potential hazards associated with those work areas. When work is in progress, access should be limited to persons with a need to enter the work areas for program or service purposes. In general, persons who may be highly susceptible to infection or for whom infection might be unusually hazardous should not be allowed to enter the work areas.

A simple barrier to aid in controlling access is the door that separates the laboratory or animal room from the corridor. These access doors should open inward and have self-closing devices so that they are kept closed when infected animals are housed in the room.

The laboratory or animal facility director may determine that the use of a particular infectious agent requires special entry provisions such as prophylactic vaccinations. Where this is necessary, a hazard warning sign (Fig. 1), incorporating the universal biohazard symbol, should be posted on the access doors to the laboratory and animal room areas. The hazard warning sign should identify the infectious agent, list the name and telephone number of a responsible supervisor, and indicate any special requirements for entering the area.

A. Personal Hygiene and Protective Clothing

Hygienic practices followed by the laboratory or animal room worker are among the most important practices for preventing work-associated illness. The most basic of these practices is the thorough washing of hands each time after handling cultures and experimental animals and before leaving the laboratory or animal room. This practice can only be accomplished when the handwashing sink is available in the work area where experimental animals are housed and infectious materials are handled.

Hands are easily contaminated and are effective means for transferring contaminants from one place to another. For this reason, gloves should be worn when feeding, watering, han-

dling, or removing infected animals and when skin contact with infectious materials may be unavoidable. Bare hands should never be placed in the animal cage for any reason. The worker should also develop the habit of not using hands to touch the face, nose, eyes, or mouth in order to avoid exposure to mucous membranes. Exposure to infectious agents through the route of ingestion can also be avoided by prohibiting eating, drinking, smoking, and the storing of food for human use in animal rooms.

Because of the inevitable presence of aerosols, persons entering animals rooms should wear surgical-type masks to reduce exposure to allergens and possible infectious aerosols. This practice is particularly encouraged in animal areas housing nonhuman primates to reduce a potential inhalation exposure to *Mycobacterium tuberculosis*.

Wearing of laboratory coats, gowns, or uniforms will protect personal clothing from contamination that can occur from the settling of aerosol particles and from direct contact with contaminated surfaces and materials. Such clothing can significantly reduce contamination that could occur from an accidental spill of infectious materials. Protective clothing should always be removed before leaving the laboratory or animal facility.

There are important personal hygienic practices that should be followed in response to any accidental spill involving infectious materials. The exposed person should avoid inhaling any potentially infectious aerosols by holding the breath and leaving the spill area immediately. Shut the access door and warn others of the problem so that they may not become exposed. Then go to a wash area and remove the protective clothing with care, folding the contaminated area inward. Discard the clothing into a plastic bag for subsequent decontamination and disposal. Thoroughly wash with soap and copious amounts of water all areas of the body that were potentially exposed as well as the arms, face, and hands.

BIOHAZARD

ADMITTANCE TO AUTHORIZED PERSONNEL ONLY

Hazard identity: _____

Responsible Investigator: _____

In case of emergency call:

Daytime phone _____ Home phone _____

Authorization for entrance must be obtained from
the Responsible Investigator named above.

Fig. 1. Warning sign.

B. Use of Needle and Syringe

The hypodermic needle and syringe is the most dangerous tool used in laboratories and animal facilities that support programs involving infectious microorganisms (Pike, 1976). The documented hazard of accidental autoinoculation has resulted in numerous work-associated illnesses with a variety of infectious agents. Also, removal of the needle from the skin of an animal or diaphragm bottle can cause vibrations that generate aerosols. Hypodermic needles and syringes should only be used for the parenteral injection or aspiration of fluids from the experimental animals and diaphragm bottles. These devices should have needle-locking syringes or they should be of the type where the needle is integral to the syringe. To help avoid a

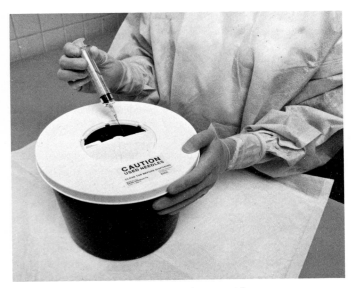

Fig. 2. Puncture-resistant container.

needle puncture when injecting animals with infectious materials, the animal should be restrained or tranquilized. Needles should not be bent, sheared, replaced in the sheath or guards, or removed from the syringe following use. The needle and syringe should be promptly placed in a puncture-resistant container (Fig. 2) and decontaminated, preferably by autoclaving, before discard.

C. Pipetting

Mechanical pipetting devices provide the only safe way to aspirate fluids when using pipettes. It has long been the practice in work areas involving infectious agents and animal materials to prohibit the hazardous practice of pipetting by mouth. Numerous excellent pipetting aids are commercially available to meet every pipetting need.

D. Housekeeping, Decontamination, and Waste Handling

Ideally the animal facility and laboratory areas have been designed and constructed to make it easy to perform routine cleaning and housekeeping operations. These operations are important to prevent accumulation of dust, filth, and contamination. Unclean areas provide harborage for microorganisms and pests, enhance survival of microbial contaminants, retard penetration of disinfectants, contribute to secondary contact, and may cause allergenic sensitization of personnel. Effective housekeeping programs maintain an or-

derly and clean work area that helps in accomplishing program goals, reduces physical hazards, and prevents the accumulation of materials from past experiments or activities.

Floor cleaning should be done in a manner to reduce the generation of aerosols. The use of high pressure water hoses for cleaning cages, pans, and floors should be avoided. Vacuuming with portable filtered exhaust vacuum systems and wet mopping are advisable. Fresh detergent decontamination solution should be applied to the floor from one container. Spent cleaning solution wrung from the mop should be collected in a separate container. A freshly laundered mophead of the cotton string type should be used daily. If floor drains have been installed to remove cleaning solutions, the drain traps should always be filled with water or a suitable disinfectant.

Work surfaces should be decontaminated with a suitable disinfectant after use or following spills of infectious materials. Cages should be decontaminated, preferably by autoclaving, before they are cleaned and washed. All infectious wastes from the animal area should be autoclaved before being disposed of by conventional methods. The autoclave used for this purpose should be located in the animal facility. Infected animal carcasses should be packaged in leakproof containers that are closed before being removed from the animal facility. The contained carcasses should be incinerated. These practices prevent the accumulation of residual contamination and ensure that other persons, not directly associated with the handling of experimental animals and infectious materials, are not exposed to biological hazards.

E. Insect and Rodent Control

It is important that the animal facility maintain an effective insect and rodent control program. Pesticides should be used with caution and only when necessary. The professional animal care staff should be consulted prior to any chemical use to reduce the possibility of adverse interactions that could effect the experimental program. In animal facilities that have operable windows, fly screens would be installed as a physical barrier to insect passage.

F. Aerosol Containment

All procedures should be carefully performed to minimize the generation of aerosols. Biological safety cabinets and other physical containment devices or personal protective devices, such as full face respirators, should be used whenever procedures with a high potential for creating aerosols are conducted (CDC/NIH, 1983). Necropsy of infected animals, dumping of contaminated bedding, harvesting of infected tissues or fluids from animals, and manipulation of high con-

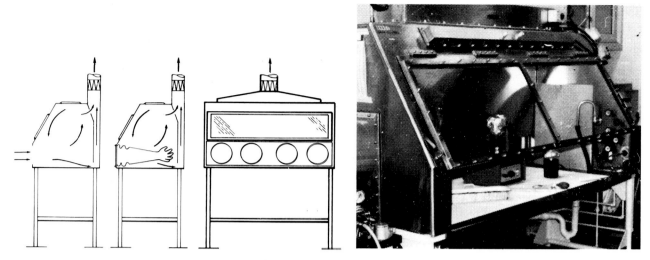

Fig. 3. Class I cabinet.

centration or large volumes of infectious materials are examples of such procedures.

The class I and class II biological safety cabinets are the primary containment barriers for minimizing personnel exposures to infectious aerosols. The class I biological safety cabinet (Fig. 3) is an open-fronted, negative-pressure, ventilated cabinet with a minimum inward face velocity at the work opening of at least 75 ft/min. The exhaust air from the cabinet is filtered by a high-efficiency particulate air (HEPA) filter.

This cabinet may be used in three operational modes: with a full-width open front, with an installed front closure panel not equipped with gloves, and with an installed front closure panel equipped with arm-length rubber gloves.

The class II vertical laminar-flow biological cabinet (Fig. 4) is an open-fronted, ventilated cabinet with an average inward face velocity at the work opening of at least 75 ft/min. This cabinet provides a HEPA-filtered, recirculated mass airflow within the work space. The exhaust air from the cabinet is also

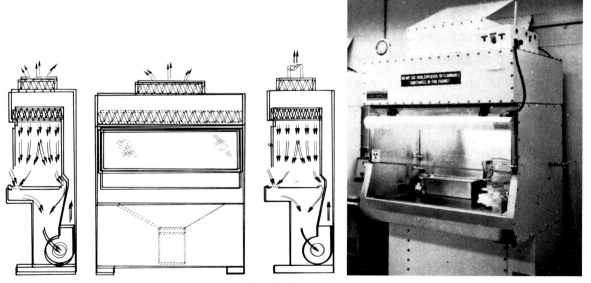

Fig. 4. Class II cabinet.

filtered by HEPA filters. Design, construction, and performance standards for class II cabinets have been developed by and are available from the National Sanitation Foundation, Ann Arbor, Michigan (1983).

Personnel protection provided by class I and class II cabinets is dependent on the inward airflow. Since the face velocities are similar, they generally provide an equivalent level of personnel protection. The use of these cabinets alone, however, is not appropriate for containment of highest-risk infectious agents because aerosols may accidentally escape through the open front.

The use of a class II cabinet in the microbiological laboratory offers the additional capability and advantage of protecting materials contained within it from extraneous airborne contaminants. This capability is provided by the HEPA-filtered, recirculated mass airflow within the work space.

V. SUMMATION

The practices described in this chaper are those recommended by the Centers for Disease Control and the National Institutes of Health for establishing and maintaining a safe work environment when performing activities involving experimental animals that may harbor or be used in experiments with moderate risk pathogens. While the safeguards reflect what is generally accepted as good microbiological practices, additional protective measures would be required for work with agents where the potential for infection by aerosols is real and the disease may have serious consequences. What has been described will be known in the future as "Vertebrate Animal Biosafety Level Criteria for Animal Biosafety Level 2" (Pike, 1979).

REFERENCES

Centers for Disease Control and National Institutes of Health (CDC/NIH) (1983). Draft: "Biosafety in Microbiological and Biomedical Laboratories." CDC/NIH, Atlanta, Georgia.

National Institutes of Health (1978). "Guide for the Care and Use of Laboratory Animals," Publ. No. (NIH) 78-23. NIH, Washington, D.C.

National Sanitation Foundation (1983). "Standard Number 49 Class II (Laminar Flow) Biohazard Cabinetry." NSF, Ann Arbor, Michigan.

Pike, R. M. (1976). Laboratory-associated infections: Summary and analyses of 3,921 cases. Health Lab. Sci. 13, 105–114.

Pike, R. M. (1979). Laboratory-associated infections: Incidence, fatalities, causes, and prevention. Annu. Rev. Microbiol. 33, 41–66.

Steere, N. V. (1980). Physical chemical, and fire safety. In "Laboratory Safety: Theory and Practice" (A. A. Fuscaldo, B. J. Erlick, and B. Hindman, eds.), pp. 3–28. Academic Press, New York.

Wedem, A. G. (1975). History of microbiological safety. Presented at the 18th Biological Safety Conference, Lexington, Kentucky.

Chapter 21

Genetic Monitoring

Chao-Kuang Hsu

I. INTRODUCTION

In performing animal-related investigations, researchers should consider three fundamental factors in order to obtain valid and reproducible scientific data. These factors are the genetic authenticity, health quality, and optimal housing environment of the laboratory animals. Recently, there have been increasing demands for genetically uniform and microbiologically well-defined laboratory animals in biomedical research throughout the world. Animal suppliers and users must make efforts to fulfill these basic and important requirements in order to ensure the validity, accuracy, consistency, and reproducibility of scientific data.

Several hundred inbred strains, congenic strains, and genetically defined stocks of laboratory rodents (mice, rats, guinea pigs, and hamsters) have been developed and used in biomedical research and testing. Each strain or stock possesses unique genetic characteristics that are of some value as animal models in biomedical research. In the selection of animal models for use in research, the genetic characteristics of animals are important factors to be considered and should be carefully evaluated in relation to experimental objectives and design. Also, it is important to many studies to have large numbers of genetically identical individuals.

Most inbred strains of mice, rats and guinea pigs were bred and developed in the 1920s–1930s. They were, subsequently, distributed to and bred in many research institutes and commercial breeding facilities for many generations creating a large number of substrains. During the past 50 years, therefore, variations and changes in the genetic characteristics among sublines of the same strain from different breeding facilities could have occurred. These genetic changes could be due to genetic drift, mutation, residual heterozygosity, and human error, such as differences in breeding methods, poor colony management and husbandry practices, and genetic contamination. These substrains may have significant genetic variation among them and disparity from the common origin. However, these substrains from various sources may still bear the same common strain/stock designation. It is commonly assumed by most investigators that these substrains bearing the same names but obtained from different sources are biologically "identical" and genetically "stable." This "common belief or assumption" is often incorrect and has long been overlooked. Frequently, the substrains are significantly different in their genetic characteristics or biological response. When substrains bearing the same designation from different vendors are used in biomedical research and testing, the data obtained from various laboratories may be different or not reproducible. These differences and nonreproducibility are often and actually due to genetic incompatibility among substrains. Unfortunately, many errors of this kind are mistakenly attributed to the differences in laboratory, technical, environmental, or human factors among research laboratories, and these observations are most likely not published in scientific literature. Therefore, it is imperative to use the proper strain nomenclature and substrain designation in referring to experimental animals and their source in scientific articles.

In contrast with the effects of infections on experimental animals, the manifestations of genetic contaminations or changes in research animals are often not grossly visible or easily detected. Consequently, enormous loss in time, effort, and money and erroneous conclusions resulting from the use of genetically contaminated animals in research are revealed only after painstaking experience. Obviously, the genetic consistency and authenticity of laboratory animals are at least as equally important as their health status in biomedical research. In the past decade, a remarkable achievement has been accomplished in improving the health quality of laboratory animals by eliminating or minimizing major pathogens through gnotobiology and other advanced technology in laboratory animal sciences. To achieve and ensure biomedical research excellence, monitoring both genetic integrity and health quality of research animals should be properly implemented.

The purpose of this chapter is to briefly summarize the significance of the genetic authenticity of research animals in biomedical research and testing, the possible sources of genetic differences in sublines, and the state of the art of the methods used in monitoring the genetic integrity of laboratory animals. Since rodents (mice, rats, guinea pigs, and hamsters) account for the major numbers of animals used in research, this chapter addresses the issues related to inbred strains of mice and rats.

II. SOURCES OF GENETIC INPURITY OF INBRED STRAINS

Changes and variations in the genetic characteristics and biological responses among substrains of rodents produced at different breeding facilities or obtained at different times can be due to a number of factors. Baily (1982) recently reviewed the sources of genetic impurity of inbred mouse strains. These sources include incomplete inbreeding, mutations, inadvertent outcrossing, mislabeling, and epistatic and heterozygote selection during the course of breeding. Generally, the sources of rodent subline divergence or contamination may include one or more of the following elements.

1. Genetic contamination of rodent strains due to improper breeding methods, poor breeding practices (i.e., strain of same coat color bred in the same room or on the same rack), poor colony management, lack of accurate record keeping systems, inadequate animal care personnel training, frequent personnel changes in animal breeding facilities, human error or negligence, and lack of supervision by a mammalian geneticist or other qualified professional.

2. Genetic mutations, substitution, insertion, or deletion of one or more constituents of genetic material. Subsequently, the mutation becomes fixed in the substrain during the course of breeding. Mutation is an ever-recurring event and can be expected to occur in all strains of animals.

3. The derivation and establishment of germfree and specific-pathogen-free (SPF) microflora associated colonies using hysterectomy technique at different commercial breeding or institutional facilities. This procedure requires a foster mother to nurse the hysterectomy-derived litter. There is danger of genetic contamination when the foster mother and nursing litters are the same color coat because the foster mother may give birth to subsequent young as a result of delayed parturition after the foster litter has been placed in the cage (Smith *et al.*, 1982).

4. Incomplete breeding, residual genetic heterozygosity due to a different fixation rate of residual heterozygous loci even after 20 or more brother and sister matings.

5. Failure to conduct a proper genetic monitoring program in breeding colonies for early detection of subline divergence or contamination.

Many resultant genetic changes in animals may not be readily known or be recognized by animal producers or users. Some genetic divergences result in obvious and visible phenotypic changes in characteristics, such as coat and eye color, body size, growth rate, or litter size, whereas many other genetic changes influence less grossly detectable but important biological characteristics, such as life span, physiological function, metabolism, immune responsiveness, susceptibility to diseases, response to drugs, susceptibility to transplantable tumors, naturally occurring tumor incidence rate, aggression, and behavior patterns. The impact and cost due to the failure of timely detection of the genetic variations and differences in research animals of various substrains from different commercial suppliers are enormous and cannot be overemphasized since probably over 60 million inbred rodents are produced and used annually in testing and research.

Major inbred strains of mice and rats were developed five to six decades ago. They were subsequently distributed, bred, and maintained at different research institutions, laboratories, or commercial animal facilities. Presently, there are hundreds of commercial rodent breeders who produce millions of rodents for worldwide use in biomedical research and testing. Commercial breeders among themselves may vary greatly in the practices of colony management, breeding methods, knowledge and experience, genetic monitoring activities, and personnel training related to rodent production. These factors are extremely important to ensure the genetic integrity and uniformity of animals produced in many generations over a long period of time. Like an animal health assurance program, breeders should implement an effective genetic monitoring program to detect any genetic changes that may subsequently be fixed during the breeding and propagation of the particular colonies. Within the same substrain maintained by a particular breeder, genetic consistency also should be guarded longitudinally over time in comparison with the characteristics of the original strain.

III. EVIDENCE OF GENETIC CONTAMINATION

Considering the number of causes that can lead to genetic differences within and among the substrains, the number of rodent breeders, and given time (longitudinally) and space (horizontally), one can expect the occurrence of genetic disparity among substrains produced by many commercial and institutional breeding facilities throughout the world. Many published reports and probably even more unpublished observations have demonstrated enormous economic waste and scientific frustration due to substrain differences and/or disparity between or among sublines of various inbred strains of rodents from a number of animal breeders. These subline differences may be largely due to poor colony management; unintentional and inadvertent contamination; residual heterozygosity, mutations, and heterozygote selection.

A striking genetic difference in ARK mouse substrains from different breeding facilities was documented by Acton et al. (1973). AKR/J and AKR/Cum differ in cell surface antigen (Thy-1.1) detectable by serological tests; in a biochemical marker, malic enzyme-1, which is governed by the gene (*Mod-1*) mapped on chromosome 9; and in the susceptibility to transplanted leukemias. Significant difference in the incidence of spontaneous leukemia and in the electrophoretic patterns of esterase-3 (Es-3) and a major urinary protein (Mup-1) were also noted among AKR/J, AKR/Cum, AKR/FuA, AKR/RuRdA, and AKR/LW substrains. Graff et al. (1975) reported a traumatic experience in the use of AKR substrains from five different commercial vendors in the chemotherapy of syngeneic transplantable leukemia. In their study, the authors found a high percentage of rejection of transplantable leukemia in control groups by most AKR substrains. This observation could be mistakenly interpreted to result from effective treatment by drugs. However, using skin grafting and homologous tumor cell transplantation, the rejection of transplantable leukemia was actually due to the striking substrain difference and significant histoincompatibility between and within the AKR substrains. The authors attributed the subline differences to breeding errors (i.e., outcrossing) in commercial producers' facilities. Obviously, such subline differences were due to genetic contamination rather than mutations. This finding further indicates the implications and serious problems due to histoincompatibility among substrains of rodents from different sources used in biomedical research in various parts of the world.

Genetic differences in histocompatibility (Kahan et al., 1982; Shearer et al., 1972) and isoenzyme (glucosephosphate isomerase) in BALB/c substrains (Green, 1977) were recently reported. Significant differences in response to lipopolysaccharides were observed in C3H substrains (Rosenstreich, 1979). Substrain differences were recently demonstrated in C3H sublines (Heston and Bittner subline) in their susceptibility to murine leprosy (*Mycobacterium lepraemurium*) (Lovik et al., 1982). Morse (1979) stated that the existing differences among the major substrains of the C3H family were due to incomplete inbreeding before distribution. Green and Kaufer (1965) found differences in histocompatibility and radiation sensitivity between CBA/J and CBA/Ca substrains. Additional differences in two genetic loci controlling phosphoglucomutase 1 (*Pgm-1*) and retinal degeneration (*rd*) were also demonstrated between CBA/J and CBA/Ca substrains (Roderick, 1979). The authors questioned if residual heterozygosity would account for the various differences detected in at least two biochemical loci and histocompatibility loci and for differences in radiation sensitivity. The CBA substrain differences were interpreted as the result of inadvertent outcross with C3H with which it had often been housed and with which it is outwardly identical (Roderick, 1979).

Significant differences were reported between CBA/Ki and CBA/St/Ki substrains that have been maintained at the same laboratories after they were transferred from one place to another. These CBA substrains differ remarkably in histocompatibility, tumor transplantations, incidence rate of spontaneous tumors, obesity induction, and retinal degeneration (Liebelt, 1979).

Substrain differences due to genetic contamination, mutation, or residual heterozygosity can greatly influence the pharmacological and behavioral responses to a variety of drugs which can remain unnoticed for a long period of time. Examples include the response of C57BL/6J, C57BL/6BY, C57BL/6A to d-amphetamine (Moisset, 1979), that of BALB/cJ, BALB/cBY to the narcotic effects of ethanol (Moisset, 1979), and the fighting behavior and catecholamine activity of BALB/cJ and BALB/cN (Ciaramello et al., 1974).

Genetic variability was prevalent in a commercial colony of inbred Wistar-Lewis rats as detected by mixed lymphocyte reaction (MLR), hemagglutination assay using specific antisera, and skin grafting (Cramer and Silvers, 1981). This problem is probably the result of genetic contamination by accidental mismatching and breeding of albino animals of different strains, which is a constant problem (Cramer and Silvers, 1981). Heterogeneity in RT 2 antigens on red cells is reported in existing inbred Brown-Norway (BN) rats obtained from various sources (Paul and Carpenter, 1981).

Genetic differences between sublines of inbred strains of rats and mice from various sources are apparently a worldwide serious problem in the scientific community. Significant variations in histocompatibility are reported in the substrains of A, AKR, BALB/c, CBA, C57BL, DBA, and WG (Baily, 1982). Differences in cell membrane alloantigens have been found in the strains of A, AKR, BALB/c, CBA, C3H, C57BL, and C57L. Variant genes affecting immune response have been demonstrated in the strains of C3H and C57BL (Baily, 1982). These facts pointed to the possibility that many animal colonies may have been genetically contaminated from time to time. Many commercial mouse colonies (13%) in the United Kingdom were found to be genetically contaminated, amounting to one genetic contamination per 26 colony-years (Festing, 1974; Festing and Lovell, 1980). A rate of genetic contamination of over 20% of all mouse colonies maintained at many research institutes and universities in Japan was recently reported (Hoffman et al., 1980), indicating a possible erroneous interpretation or even invalidity of scientific results. A confirmed case of genetic contamination was reported among 53 mouse colonies bred in the Netherlands (Green, 1977). Festing (1982) recently listed a number of cases involving the use of genetically contaminated inbred strains of mice and rats that were produced at either institutional or commercial breeding facilities. Most of the problems associated with the nonauthenticity of inbred mouse and rat strains used in research are generally not published in scientific literature. It is commonly noted that inbred mouse and rat strains bred and maintained at universities or research laboratories are often genetically contaminated, and are not subject to routine genetic monitoring

programs. Festing (1982) stated that as many as 4–20% of rodent colonies may well have become genetically contaminated, incurring a considerable financial loss, many unreliable research results, and much wasted effort.

IV.   METHODS OF GENETIC MONITORING

A.   Colony Management

Since most cases of genetic differences between sublines of inbred rat and mouse strains from various rodent breeding facilities appear to be due to genetic contaminations related to human error or negligence, the most effective method in ensuring the genetic authenticity of rodents is the proper management of breeding colonies. In a rodent breeding facility, a proper breeding system, including pedigree foundation colony, pedigree expansion colony, and production colony, should be religiously followed. A proper breeding method should be enforced for each type of colony in order to maintain genuine strain characteristics and integrity. Brother and sister sibling matings should be performed in the foundation and pedigree expansion colonies, whereas random harem breedings can be used in the production colony from which offspring are distributed for research use. The offspring of the production colony should not be used as breeders. Breeding records should be accurately and adequately maintained. Proper and permanent marking is required for all breeders in animal rooms. The breeders of the production colony should be replaced with the offspring of the pedigree expansion colony every three to four generations in order to avoid subline divergence.

The most important element in maintaining the authenticity of research animals is the training and experience of animal technicians for the understanding of breeding systems and methods, genetics of rodent species, record keeping methods, general laboratory animal science, and quality assurance programs. Breedings and maintenance of genetically authentic and microbiologically well-defined strains or stocks of rodents require considerable knowledge, skill, and conscientious effort. Animal technicians should know strain characteristics and be able to recognize and report occurrence of strain variation, mutation, or genetic contamination. Coat color is the commonly used and easily recognized marker as the first line of detecting genetic contamination. Strains of the same or similar phenotype such as coat color, should not be bred or maintained in the same room in order to avoid cross-contamination. Any accidentally escaped animals shall be humanely killed and should never be put back in animal cages in order to avoid erroneous matings. Hybrid (F_1) animals should never be used as breeding stock.

Nonhuman primates are the best animal models of human disease and behavior. The increasing demands for more and better quality nonhuman primate species and the decline of foreign supply have led to the establishment of many primate breeding facilities in many countries, especially the United States.

Extensive genetic variations and polymorphisms are known to exist in wild primate populations. However, in the captive breeding programs in primate centers and zoos, high infant mortality rate, reduced fertility, increased susceptibility to disorders, and changes in behavior have been reported (Smith, 1982). These observations are a result of increased inbreeding depression and reduction or loss of genetic variability in small isolated populations. Inbreeding depression resulting from the mating of closely related animals produces deleterious effects, including reduced viability and fertility. The amount of inbreeding effect and the rate of loss of genetic variability in an isolated breeding colony are functions of the effective breeding size of the population. The effective breeding size is the number of random mating adults that would alone suffice to produce the observed amount of inbreeding or random genetic drift in a given cage.

To monitor or minimize the amount of inbreeding and its deleterious effects in a small isolated primate breeding group, one can improve primate colony management by effective use of genetic markers that include 12 electrophoretically defined protein polymorphisms. These markers are albumin, prealbumin, transferrin, carbonic anhydrase II, phosphohexose isomerase, isocitrate dehydrogenase, 6-phosphogluconate dehydrogenase, NADH diaphorase, NADPH diaphorase, group-specific protein, properdin factor B, and C3 complement component. These genetic markers have been used to identify paternity and reconstruct genealogical relationships in captive breeding groups of rhesus monkeys using paternity exclusion analysis (Smith, 1982). By using this information, effective management can be implemented to maintain production efficiency and vaiability of captive primate breeding colonies.

B. Monitoring Immunogenetic Markers

Immunogenetic markers in this context are referred to as the histocompatibility (*H*) genes that determine cell membrane alloantigens such as erythrocyte antigens, lymphoid tissue antigens, and histocompatibility antigens. The major histocompatibility complex (MHC) in the mouse is *H-2* which consists of a series of genes located on chromosome 17. The major histocompatibility complex in the rat is called *Rt-1*, in the guinea pig *GPLA*. The major histocompatibility complex controls a large array of biological phenomena including allograft rejection, immune responsiveness, susceptibility to diseases, cellular interactions, and cell surface antigens (Bach and Van-

Rood, 1976; Schreffler and David, 1975). The mutation rates are high, 2.25×10^{-4} and 9.33×10^{-4} for *H-2* complex and *H* loci, respectively (Hedrick, 1981). Overall recombination frequency rates ranging from 0.4 to 4% have been reported in MHC loci depending upon strains of mouse (Bach and Van-Rood, 1976; Shiroishi *et al.*, 1982). Subline differences of strains resulting from cross-contamination, recombinations, and mutations can be detected by monitoring genetic markers using the following methods.

1. Skin Grafting

Skin grafting is currently the most commonly used and sensitive genetic monitoring method for detecting differences in histocompatibility genes within or between substrains. The basic procedures and interpretation of the skin grafting method described by Billingham and Medawar (1951) and Billingham and Silvers (1961) are applicable to all animal species. However, a rapid and effective tail skin grafting method is commonly used in mice (Baily and Usama, 1960) and rats (Festing and Grist, 1970). Grafts have to be observed for at least 90–100 days. An acceptance of allografts for longer than 100 days indicates a high degree of uniformity, homozygosity, and inbreeding between exchanged animals. Graft rejections that occur within the first 3 weeks after grafting suggest a significant difference in *H-2* complex. Differences in non-*H-2* regions may result in a chronic graft rejection when using a small graft (Schultz *et al.*, 1976).

The skin grafting method has several merits, including great sensitivity, accessibility, economical, noninvolvement of expensive equipment, easy assessment of the graft survival, relative resistance to transplantation trauma, detection of many *H* genes, and effectiveness in detecting recent genetic contamination and mutations that cause subline divergence. The disadvantages of skin grafting are as follows: (1) the effort is too laborious and time-consuming, requiring observation periods of 100 days in order to detect differences in non-*H-2* histocompatibility genes; (2) requirements of relatively large space for housing animals; (3) some graft losses (10–15%) may be attributable to technical or other nonimmunological factors; and (4) not an ideal method for authenticating a newly established strain (Festing, 1982).

2. Mixed Lymphocyte Reaction (MLR)

Lymphocytes not only participate in the rejection of skin graft or other transplanted tissues, but also have surface antigens coded for by the MHC (*H-2*). When lymphocytes from allogenic animals or genetically different individuals are mixed and cultured together in the cell culture system for an appropriate period, they will begin to enlarge; synthesize DNA, RNA, and proteins; and even divide further. This complex phenomenon is known as the mixed lymphocyte reaction (MLR). Response in MLR is assayed by the incorporation of radioactive labeled thymidine into the responding cells.

Mixed lymphocyte reaction has proved an effective and efficient method for monitoring the authenticity of mice and rats (Cramer and Silvers, 1981; Kahan et al., 1982). A one-way reciprocol mixed lymphocyte culture test is recommended for genetic monitoring. The basic MLR techniques consist of mixing 0.1 ml of 10^5 cells (spleen cells) of responding lymphocytes (from potential recipient) and of irradiated (or mitomycin C-treated) stimulating cells (from potential donor). The mixture in triplicate is placed in a multi-well microplate (U-shape) and incubated at 37°C in a 5% CO_2 incubator for 5 days. At the end of incubation, the culture is pulse labeled with [^{3}H]thymidine and terminated 16–18 hr later by extracting the culture onto a filter paper to collect DNA-precipitable cell lysate. The filter paper is dried and the radioactivity is determined in a liquid scintillation counter. The results are expressed either as counts per minute (cpm) per average culture or as stimulation index, which is the ratio of allogenic combination to the syngeneic or autologous control mixture.

The mixed lymphocyte reaction is known as the in vitro counterpart of the recognition and proliferation phases in the skin graft rejection. The MLR and graft-versus-host (GVH) reactions are controlled by genes located throughout the H-2 complex and MLS locus that is outside and not linked to the H-2 complex. The mixed lymphocyte reaction can be used to determine whether individual mice of substrains from various sources are identical for the MHC or to what degree they differ by measuring the amount of thymidine incorporated by the cell mixture (Bach, 1976). The degree of MLR responsiveness is determined by MHC disparity or phylogenetic distance. Therefore, MLR results are predictive of in vivo GVH reaction. It is conceivable that the genetic differences in MHC among animals detected by skin grafting and GVH will usually also be revealed by MLR. When the average stimulation index is over 2, a significant difference in MHC region between the donor and recipient animals should be suspected (see Table I). Significant MLR stimulations have been observed in various inbred strains of mice presumably identical in their H-2 haplotypes, but differing in single non-H-2 loci, such as H-1, H-3, H-4, or in multiple minor H loci (Klein, 1975). Recently it has been shown that animals differing in other minor loci, such as MLS locus and thy-1 (but not a Tla locus), will have significant MLR stimulation. Therefore, the stimulation of H-2 identical lymphocytes could be the result of the cumulative effect of a difference in multiple minor H loci or of a difference at the MLS or Thy-1 locus.

For the monitoring of the genetic uniformity and isogenecity of various substrains from various sources, the MLR offers many advantages over the skin grafting method that usually complements MLR. Both MLR and skin grafting rejections are determined by the same regions of the major histocompatiblity

Table I

Strength of MLR and GVHR across Various Regions of the H-2 Complex[a]

H-2 region difference in	MLR		GVHR	
	Average ratio of stimulation	Range	Mean spleen index	Range
K	2.0	0.7–4.7	1.4	1.2–1.6
D	1.8	0.8–5.4	1.4	1.2–1.7
IA	6.4	3.7–9.2	1.8	1.8
IB+S	2.7	1.1–4.4	1.5	1.1–1.8
I+S	5.8	2.7–12.8	2.6	2.6
K+I	6.6	3.2–18.3	2.8	2.6–3.1
D+S	2.0	0.7–4.7	1.5	1.3–1.6
K+D	3.4	3.0–3.8	1.8	1.1–2.5
K+S+D	3.3	1.5–8.6	2.3	2.2–2.4
K+I+S+D	7.2	1.2–33.6	2.8	2.4–3.1

[a]Adapted from Klein (1975, p. 461).

complex. Using the in vitro MLR method, one can test and compare the uniformity and isogenecity of as many substrains, colony types, and sources as one wishes. One can monitor animals (by various combinations) between or within substrains, generations, litters, and sexes in the foundation, pedigree expansion, or even in the production colonies by various combinations of lymphocyte cultures. Mixed lymphocyte reaction results can be obtained in a very short period of time (7–9 days). Subline differences resulting from genetic contamination, human error, recombination, or accumulated mutations in MHC can be detected by the consistent and statistically significant MLR stimulation. Therefore, reciprocal MLR performed periodically among a randomly chosen adequate number of mice and rats from animal colonies can safeguard and monitor the genetic quality of research animals.

3. Lymphoid Tissue Transplantation

Lymphoid cells are immunocompetent and have alloantigens on their surface. Histocompatibility between tested substrains or rodents can be monitored by injecting lymphoid tissues or cells of donors into recipient animals. Single cell suspensions are prepared by homogenization of the donor lymphoid organs, such as bone marrow, spleen, lymph nodes, thymus, or fetal liver. Adequate numbers of the viable living cells are then injected intravenously or intraperitoneally into the recipient tested animals. The histoincompatibility is determined by the survival and splenomegaly of the recipient animals (Table I) or is measured radiologically by the survival of the transplanted cells.

4. Tumor Transplantation

The susceptibility of animals to tumor transplants is determined by MHC genes that also code for alloantigens. These

Chapter 22

Selected Zoonoses and Other Health Hazards

James G. Fox, Christian E. Newcomer, and Harry Rozmiarek

I. INTRODUCTION

Derived from the Greek words *zoon,* meaning animals, and *noses,* meaning diseases, *zoonoses* literally refers to diseases transmitted direclty to humans by animals. This chapter reviews selected known or potential zoonotic agents and the disease manifestations produced in humans by exposure to infected laboratory animals. We discuss other health hazards that may be encountered when working with animals, such as bites and allergies. Selected transmission of human infectious agents to animals is also briefly mentioned.

Although the animal, either feral, laboratory, or pet, is not routinely considered as a reservoir for human pathogens, a review of the literature refutes that notion. Furthermore, many of the zoonotic diseases affect many species of laboratory animals. However, with the advent of modern laboratory animal production and management, zoonotic diseases are being curtailed and are nonexistent in many laboratories.

II. VIRAL DISEASES

A. Poxviruses

Many poxvirus infections of laboratory animals are also pathogenic for humans. In humans, these infections are usually manifested by the development of proliferative cutaneous or subcutaneous self-limiting lesions. Clinical signs indicative of systemic disease are occasionally observed in humans, particularly when the poxviruses of nonhuman primates are involved.

1. Nonhuman Primate Poxvirus Infections

Nonhuman primates can harbor several poxviruses that either are known to be zoonotic or are naturally occurring in humans. Smallpox has been transmitted to nonhuman primates experimentally, but there is no evidence of natural infection of nonhuman primates with smallpox (Noble, 1970). Natural transmission of smallpox from monkeys to humans has been advanced on strictly empirical grounds, but has not been corroborated by viral isolation and identification (Cespen, 1975). Other poxviruses may have been responsible for these disease outbreaks occurring concomitantly in monkeys and humans that were ascribed to smallpox. Smallpox is now considered by the World Health Organization to be eradicated on a worldwide basis.

a. Monkeypox

Monkeypox is a zoonotic disease of nonhuman primates that is more closely related to smallpox than vaccinia or cowpox (Rondle and Sayeed, 1972). It produces a serological response and clinical syndrome in humans indistinguishable from smallpox.

i. Reservoir and incidence. At least nine outbreaks of the disease in captive nonhuman primates have occurred, and some have been reviewed in detail (Arita and Henderson, 1968; Soave, 1981). Rhesus monkeys, cynomolgus monkeys, and other Old World monkeys were primarily involved in the outbreaks, but squirrel monkeys also were reported to be susceptible. Serologic studies of nonhuman primates living in areas of human habitation, where cases of human monkeypox had occurred, identified 44 of 206 animals examined to be seropositive for monkeypox (Brenan *et al.,* 1980). However, no human monkeypox infections have been associated with outbreaks of the disease in captive nonhuman primates (Irving, 1974).

ii. Mode of transmission. Disease transmission occurred rapidly in natural outbreaks among susceptible animals producing high morbidity and variable mortality. During both natural outbreaks and experimental inoculation studies, seroconversion without clinical signs was detected in some animals (Irving, 1974). The role that nonhuman primates have in the transmission of monkeypox to humans has not been determined.

iii. Clinical signs. Clinical signs in the nonhuman primate host include fever followed in 4–5 days by cutaneous eruptions, usually on the limbs and, less frequently, on the trunk, face, lips, and buccal cavity.

Humans develop clinical signs of monkeypox very similar to those of the nonhuman primate host. Patients developed fever, malaise, headache, sore throat, and a diffuse maculopustular rash of peripheral distribution. Once humans become infected, person-to-person transmission can occur (Soave, 1981).

iv. Control and prevention. Laboratory-housed nonhuman primates should not be overlooked as potential reservoirs of monkeypox. Sanitation measures, isolation, and vaccination of nonhuman primates with human vaccinia vaccine have been effective in the control of this disease (Irving, 1974).

b. Benign Epidermal Monkeypox

Benign epidermal monkeypox (BEMP) is a poxvirus that has been zoonotic in the laboratory environment on numerous occasions (Irving, 1974; McNulty, 1968).

i. Mode of transmission. Direct transmission of BEMP was suggested by its rapid spread among gang-caged nonhuman primates (Hall and McNulty, 1967). Fomite transmission, presumably by a contaminated tattoo needle, was inferred when lesions were associated with animal identification numbers (Crandel *et al.*, 1969). Immunity to reinfection lasts for at least 6 months (Hall and McNulty, 1967). Prior Yaba virus infection also confers immunity (Kupper *et al.*, 1970).

ii. The disease in nonhuman primates. Benign epidermal monkeypox occurred during 1965–1966 in four research facilities that received animals from a common source (Casey *et al.*, 1967; Valerio, 1971). The disease affected primarily various *Macaca* spp. and *Presbytis entellus*. The African genera, *Cercopithecus* and *Cercocebus*, and South American monkeys are apparently unaffected (Espana, 1971; Hall and McNulty, 1967). In nonhuman primates, circumscribed, oval to circular, elevated red lesions were distributed over the eyelids, face, body, and genitalia (Irving, 1974). The localization of BEMP lesions in the epidermis and adnexal structures differentiates them from Yaba lesions histologically. Similar to Yaba, however, intracytoplasmic eosinophilic inclusions are evident (Kupper *et al.*, 1970). The spontaneous regression of BEMP lesions occurs in 4–6 weeks.

iii. The disease in humans. Benign epidermal monkeypox in humans causes a mild disease similar to that described in monkeys, but lesions regress within 2–3 weeks of onset. Benign epidermal monkeypox is closely related, or possibly identical to, Tanapox virus, an agent isolated from humans in Kenya. This agent is believed to be transmitted from monkeys to humans in Kenya.

c. Yaba Virus

Yaba virus is a closely related agent of a poxvirus subgroup distinct from the vaccinia–variola subgroup. Yaba virus was initially reported in a colony of rhesus monkeys housed outdoors in Yaba, Nigeria, in 1951 (Bearcroft and Jamiesen, 1958); subsequent outbreaks and experimental studies of the disease have also been reported.

i. Reservoir and transmission. Further clarification of the epidemiology of Yaba disease is required. The initial report described a disease in 20 of 35 rhesus monkeys and one baboon. The disease was disseminated quickly among rhesus monkeys, particularly cagemates, but no lesions were recorded in African monkeys housed in the same area (Bearcroft and Jamiesen, 1958). A later natural outbreak and experimental studies clearly extended the host range to include the African monkeys, but did not corroborate the spread of infection from infected monkeys to normal monkeys housed together (Ambrus *et al.*, 1969). The potential role of insect vectors in the spread of this disease has not been determined (Ambrus *et al.*, 1969; Bearcroft and Jamiesen, 1958; Wolfe *et al.*, 1968). Aerosol transmission has been proved experimentally.

ii. The disease in nonhuman primates and humans. The clinical picture and pathogenesis of Yaba virus infection has been very consistent. Infected animals developed subcutaneous benign histiocytomas that reached a maximum size about 6 weeks postinoculation and regressed approximately 3 weeks thereafter.

In rhesus monkeys, tumor regression conferred immunity to reinfection (Niven, 1961). Interruption of natural Yaba tumor regression by surgical resection was speculated to be permissive to reinfection with Yaba virus in a baboon (Bruestle *et al.*, 1981).

Six human terminal cancer volunteers inoculated with Yaba virus developed lesions similar to monkeys; however, Yaba tumors were smaller in humans, and tumor regression was earlier. Accidental inoculation of a laboratory worker with Yaba virus resulted in tumor induction and was resolved completely by surgical resection (Grace, 1963). Transmission to humans directly from monkeys has not been recorded.

2. Contagious Ecthyma (Orf)

Contagious ecthyma, the cause of human orf, is a poxvirus disease of sheep and goats characterized by epithelial proliferation and necrosis in the skin and mucous membranes of the urogenital and gastrointestinal tracts.

i. Reservoir and incidence. Contagious ecthyma occurs worldwide in all breeds of sheep and goats. Although all age groups are susceptible, the disease principally affects young animals, causing high morbidity and predisposing to secondary bacterial infections.

ii. Mode of transmission. Papules evolving to crusted exudative vesiculopustular lesions on the muzzle, eyelids, oral cavity, feet, or external genitalia are laden with virus and serve as a source of infection and environmental contamination. The virus resists dessication and can remain viable for months. Transmission to humans and other animals can occur by direct animal contact or indirectly through contaminated formites.

iii. Disease in humans. Human orf is characterized by firm, large painful nodules that are usually distributed on the hands (Fig. 1). Lesions begin as macules or papules that enlarge in about 2 weeks to firm nodules with weeping red surfaces. The lesions resolve spontaneously with minimal scarring 1–2 months after infection.

B. Hemorrhagic Fevers

The hemorrhagic fevers form a group of acute diseases in which bleeding is the main clinical manifestation. These diseases occur in different parts of the world and may be transmit-

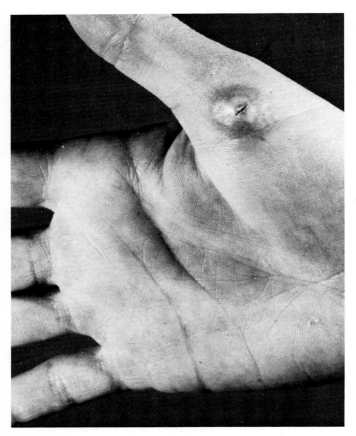

Fig. 1. Human orf. Firm, raised, centrally necrotic lesion on the thumb of an animal technician who handled an infected goat. (Courtesy of Dr. J. Griffith.)

ted to humans by mosquitoes or ticks or by direct contact with excreta of infected rodents. The etiologic agents are RNA viruses (Johnson, 1982).

Transmission to humans is usually seen in rural areas among field workers at harvest time. Detectable levels of antibody to several of the hemorrhagic fever viruses have been found in humans and microtine rodents in southeast Asia, Sweden, and North America. Inapparent infection in humans by various hemorrhagic fever agents is suggested by a lack of correlation between high antibody titers and clinical disease. Natural epidemics of high morbidity and mortality have occurred in small villages in Africa and South America. Experimental animals excrete virus in the urine for prolonged periods and should be handled only with proper precautions. While efforts are being made to produce a vaccine, no acceptable immunizations are available for most of these diseases.

1. Yellow Fever

Yellow fever is an acute mosquito-borne hemorrhagic fever caused by an RNA flavivirus (Hruska, 1979). It is endemic to tropical Central and South America and Africa, where it is a major public health problem. There are two forms of yellow fever: jungle yellow fever (sylvatic or forest) and urban yellow fever. These are distinguished by different arthropod vectors and vertebrate hosts.

Urban yellow fever is transmitted from an infected to a nonimmune person by the bite of the mosquito, *Aedes aegypti.* Urban yellow fever has not occurred in the Americas since 1954 because of the eradication of *A. aegypti* from populated areas; however, the vector remains in areas of the Carribean and northern South America where a real possibility exists for epidemics of urban yellow fever. In Africa large epidemics occur sporadically, with deaths numbering in the thousands. Both humans and monkeys can serve as an intermediate host for urban yellow fever virus in the rural–urban transition areas of tropical Africa.

Jungle or sylvan yellow fever is transmitted in the bite of a forest mosquito and is perpetuated in a monkey– mosquito–monkey cycle. In the Americas, human cases are usually sporadic, with epidemics often corresponding to epizootics within wild monkey populations, or to penetration of forest areas by groups of susceptible individuals, i.e., woodcutting laborers. In Africa the cycle is similar, with *A. africanus* being the principal vector.

i. Clinical signs and diagnosis. African monkeys acquire yellow fever at an early age, pass through a mild disease, and develop immunity. Antibody titers are readily demonstrable. Epizootic disease can occur in New World monkeys and is characterized by fever, vomiting, anorexia, yellow to green urine, albuminuria, and icterus. At necropsy, necrotic, hemorrhagic, and bile-stained organs are observed. The classic lesion is massive hepatic midzonal necrosis with necrotic hepatocytes developing characteristic eosinophilic, intracytoplasmic inclusions, or "Councilman's bodies."

ii. Diagnosis and control. Most nonhuman primates are susceptible to yellow fever, but disease severity varies markedly among species. Imported monkeys must be accompanied by certification that they (1) originate from a yellow fever-free area, (2) have been maintained in a double-screened, mosquito-proof enclosure, or (3) have been immunized against yellow fever (see Chapter 2 by McPherson). Mosquito control and eradication programs, vaccination, and adherence to United States Public Health Service quarantine standards are essential for yellow fever control.

2. Korean Hemorrhagic Fever (Epidemic Hemorrhagic Fever–Osaka)

Korean hemorrhagic fever (KHF) is a febrile disease with renal involvement caused by Hantaan virus (Lee *et al.*, 1978).

Korean hemorrhagic fever was recognized when it appeared among troops serving in the Korean war. Recently, KHF has occurred among researchers and animal technicians working with rats in several research laboratories in Japan (Kawamata *et al.*, 1980; Umenai *et al.*, 1979). Over 100 cases of KHF were detected in 15 Japanese institutions between 1975 and 1981. Hantaan virus is enzootic in wild rodents, *Apodemus* spp. in Korea, and recent evidence suggests that it, or a closely related agent, is present in wild *Rattus norvegicus* (Lee, 1982), *Microtus* spp., and *Clethriomys glareolus* in the United States.

Clinical features of KHF in people include high fever, severe malaise, myalgia, headaches, diarrhea, nausea and vomiting, proteinuria, oliguria or polyuria, and possibly hemorrhagic manifestations. Based on serological evidence, laboratory rats with enzootic KHF are believed to be the source of human KHF in Japan. Clinical signs and lesions have not been reported to occur in laboratory rats, although they develop relatively high antibody titers to the agent.

Monitoring for the KHF agent should be considered when laboratory rats, *Apodemus* spp., other wild rodents, and possibly mice are being moved from Japan and Belgium (or other areas of the world where the agent may exist) to other countries and uninfected areas. Testing should be done in advance of shipment or while animals are held in quarantine. Testing of sera for KHF antibody is done by using immunofluorescent antibody technique (IFA). Although tests are not done on a routine or commercial basis in the United States, it may be possible to obtain IFA tests through select research laboratories or the Center for Disease Control, Atlanta, Georgia.

C. Lymphocytic Choriomeningitis Virus (LCM)

Of the many latent viruses present in the mouse, only LCM naturally infects humans. A review of the literature attests to the ease with which LCM can be transmitted from animals to humans (Lehmann-Grube, 1971).

i. Reservoir and incidence. The natural association of LCM virus and the mouse provides for mutual survival in a symbiotic relationship. Neither the virus nor the host significantly suppresses the other, though each can do so. Lymphocytic choriomeningitis virus exists in the wild mouse population throughout the United States, Europe, Asia, Africa, and probably the world, although it has not been isolated from mice in Australia. Wild mice are the ultimate reservoir of infection for laboratory mice and other susceptible hosts (Maurer, 1964). Mice and, to a lesser extent, hamsters, are the only species in which a long-term, asymptomatic infection is known to exist (Hotchin, 1971; Parker *et al.*, 1976). In an early study, 21.5% of wild mice surveyed in the Washington D.C. area were infected (Armstrong *et al.*, 1940). In a more

recent survey (1967–1970) in the United States, LCM infection was detected in only 2 of 22 production or research colonies (Poiley, 1970). However, this survey was conducted only in retired breeding stock, and the monitoring technique detected only nontolerant infections. Lymphocytic choriomeningitis virus also has been reported in other mouse colonies used for research in the United States (Soave and Van Allen, 1958); LCM existed in colonies in selected institutions in England in 1980, (Skinner and Knight, 1971) and undoubtedly persists in some colonies maintained in the United States. Infection has been eradicated in almost all colonies, however, by cesarean derivation, routine serological monitoring, culling, and prevention of entry of wild mice into laboratory colonies.

In 1965, an epizootic of LCM virus infection of laboratory hamsters in the United States was reported that was responsible for 23 human cases. Since that time, several outbreaks of human cases have been reported that were acquired via contact with infected hamsters. Lymphocytic choriomeningitis virus infections are also frequently found in commercial hamster colonies in Germany, and it is presently estimated that sale of these animals as pets is responsible for approximately 1000 human cases in that country each year (Parker *et al.*, 1976). The first infected commercial hamster colony in the United States was reported in 1974. Control and clinical signs associated with LCM in hamsters are discussed in Chapter 5 by Van Hoosier and Ladiges.

Another source of infection for humans is the presence of LCM virus in experimental mouse tumors. This source was first recognized in a transplantable leukemia of C58 mice, in which inoculation of the tumor produced mild clinical illness. It had been assumed that the illness was due to a leukemia-related toxin; however, LCM was discovered as the etiologic agent (Lindorfer and Syverton, 1953; Taylor and MacDowell, 1949). Subsequently, LCM virus has been found in other commonly used tumor lines (Collins and Parker, 1972). Lymphocytic choriomeningitis virus has also been found as a contaminant of mycoplasma and murine poliovirus.

ii. Mode of transmission. Diagnosis and control of this infection in mouse colonies has been described in Chapter 3 by Jacoby and Fox. Mice that are congenitally infected are normal at birth and appear normal for most of their lifespan, even though they are persistently viremic and viruric. Virtually all cells can be infected with the virus. Most human laboratory infections have been associated with improper handling of infected murine tissues (Baum *et al.*, 1966; Tobin, 1968). Before manipulative procedures begin, all murine tumor lines should be screened for this virus. Humans can also be infected with LCM virus either directly from feces or urine of mice or indirectly by inhaling the dried excreta carried on aerosolized dust originating from the animal cage or room. The wild house mouse, *Mus musculus*, plays an important role in the incidence of human disease from LCM virus (Dalldorf *et al.*, 1946; Mac-Callum, 1949). The original description of human infection with LCM was associated with a reservoir of the virus in the form of persistent latent infections in wild house mice (Armstrong and Lillie, 1934). Although LCM infection can cause death in humans, none of these cases was fatal, nor was there evidence of transmission by human contact. The handling of LCM-infected mice, and being bitten by mice, also appear to be important causes of LCM in humans.

Infection rate in households with infected pet hamsters correlated with the cage type and location of hamsters in the home. Wire cages were associated with the highest infection rate, whereas aquariums or deep boxes with poor ventilation had the lowest infection rate in family members. Cage placement in a common living area was associated with a high infection rate, whereas placement outside the usual areas of human occupancy, i.e., a basement, did not result in household infections (Biggar *et al.*, 1975). This implies that direct physical contact is not important in this setting and supports the concept of aerosol transmission of the virus. In an outbreak of LCM in medical personnel, infection was most common among those who had physical contact with the animals; however, 17 people who denied any physical contact with laboratory animals were also infected (Hinman *et al.*, 1975). The seroconversion rate in these personnel was related to the frequency of entering a room housing LCM-infected hamsters, providing further evidence of air-borne LCM transmission. These studies emphasize the importance of restricting movement of personnel in and out of enclosures having LCM-infected animals. In the latter study, the original source of LCM was probably from contaminated tumor lines.

Control of LCM is related directly to sanitary conditions in homes and laboratories. Careful washing of hands or use of disposable gloves may reduce the chance of infection in humans.

Lymphocytic choriomeningitis can be considered an arthropod-borne virus, having been transmitted experimentally by various blood-sucking insects, including mosquitoes (*Aedes aegypti*), Rocky Mountain wood ticks (*Dermacentor andersoni*), and fleas; all of these vectors could gain entrance into a laboratory animal facility (Hotchin and Benson, 1973). Lymphocytic choriomeningitis virus has also been recovered from cockroaches (Armstrong, 1963).

iii. Clinical signs, susceptibility, and resistance in humans. Although its expression can vary greatly, LCM virus infection appears most frequently as a mild influenza-like syndrome, with or without apparent involvement of the central nervous system (Duncan *et al.*, 1951). The disease cannot be dismissed as a minor flulike illness. Some patients were sufficiently ill to require hospitalization for prolonged periods (Biggar *et al.*, 1975) (Table I). Since LCM has a wide spectrum of clinical

Table I

Symptoms of Persons with Positive Titers for Lymphocytic Choriomeningitis

	Number of cases	
Symptom	49[a]	11[b]
None recognized	3	1
Fever	44	9
Headache	42	7
Myalgia	39	8
Pain on moving eyes	29	7
Nausea	26	9
Vomiting	17	9
Biphasic illness	12	NR[c]
Sore throat	12	NR
Photophobia	12	7
Cough	9	1
Swollen glands	8	NR
Diarrhea	8	1
Rash	6	1
Upper respiratory tract symptoms	6	NR
Orchitis	1	NR

[a]From Biggar *et al.* (1975).
[b]From Maetz *et al.* (1976).
[c]NR, none recognized.

signs, a rise in antibody titers is the most conclusive means of diagnosing the disease, other than by virus isolation.

D. Measles (Giant Cell Pneumonia, Rubeola, Monkey Intranuclear Inclusion Agent)

i. Reservoir and incidence. The reservoir of measles appears to be the human population; a very low incidence occurs in the natural nonhuman primate habitat (Irving, 1974). New World monkeys appear to be somewhat more resistant. Seroconversion in imported macaques often approaches 100% within several weeks of capture, indicating widespread infection early in captivity.

ii. Mode of transmission. Measles is generally believed to be transmitted by contact between nonhuman primates and human populations with endemic measles. Measles is highly contagious and spreads rapidly among animals with no protective immunity, and presumably could spread from monkeys to susceptible humans in the laboratory environment. The virus is shed beginning in the prodromal phase and continuing through the exanthematous phase of the disease. Viral excretion occurs from the mucous membranes of the eye and pharynx and later from the respiratory and urinary tracts.

iii. Clinical signs. The clinical signs of disease in nonhuman primates are strikingly similar. After an incubation period in humans of 9–11 days, the catarrhal prodromal phase begins, evidenced by conjunctivitis. Bluish-white "Koplik spots," which are pathognomonic of measles, develop on the buccal mucosa 2–3 days after onset of disease and a marked leukopenia is concurrently noted. Subsequently, an exanthematous rash begins in the buccal cavity and spreads successively over the cheeks, neck, chest, and body. Interstitial pneumonia in monkeys also develops during the exanthematous phase (Hall *et al.*, 1971). Otitis media and bronchopneumonia are the most common bacterial complications. Rarely, postinfectious encephalitis is observed after the exanthematous phase.

iv. Diagnosis and control. Natural immunity to this disease is generally acquired shortly after capture, rendering preventive measures unnecessary for exposed animals. Vaccination might be considered for naive animals. Diagnosis can be made on the basis of clinical signs, serology, or histopathology.

E. Hepatitis

Over 100 cases of hepatitis in humans have been associated with infected nonhuman primates (Center for Disease Control, 1971).

i. Reservoir and incidence. Hepatitis A virus (HAV, infectious hepatitis) can infect chimpanzees, gorillas, patas monkeys, celebes apes, woolly monkeys, and some tamarins (Deinhardt, 1976). Experimentally, nonhuman primates can be infected with HAV by the oral or parenteral routes and shed the virus in feces (Smetana, 1969). Clinical signs of disease in nonhuman primates frequently are not evident; however, malaise, nausea, vomiting, jaundice, and increased liver enzyme values have been recorded in some chimpanzees and patas monkeys with HAV infection.

Although the cases of HAV infection in humans associated with nonhuman primates represent only a small portion of the national total, the attack rate in persons associated with chimpanzees has been higher than that reported in the general population (Ruddy *et al.*, 1967). Disease in nonhuman primates and zoonotic transmission have primarily involved animals, especially chimpanzees recently introduced into captivity. This suggests that susceptible nonhuman primates acquire the infection after capture and harbor the infection for limited periods of time, with immunity ensuing and the animals no longer a source of virus (Soave, 1981; Irving, 1974).

ii. Clinical signs and prevention in humans. After an incubation period of 15–50 days, HAV causes an illness characterized by abrupt onset of fever, malaise, anorexia, nausea, abdominal discomfort, and jaundice (Fig. 2). The severity of

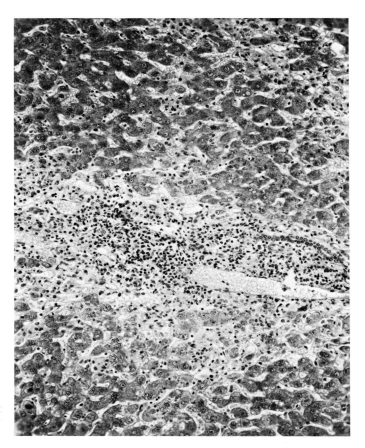

Fig. 2. Human liver with hepatitis A infection. Leukocytic infiltration of portal areas with heptocellular necrosis in the peripheral areas of the lobules. Hematoxylin and eosin. × 250. (Courtesy of Dr. K. Ishak, Armed Forces Institute of Pathology.)

the disease is related to age. The fatality among hospitalized patients is quite low. Anti-HAV IgM and IgG antibody levels can be quantitated to differentiate the acute and convalescent phases of the illness. Development of IgG anti-HAV apparently confers life-long protection against reinfection [Advisory Committee of Immunization Practices (ACIP), 1981].

The control methods observed for other nonhuman primate zoonoses should be effective in the control of HAV infection. Immunoprophylaxis has been recommended by the Public Health Service and has proved advantageous in some institutions (ACIP, 1981; Krusakh, 1970). Immune serum globulin in doses of 0.02 ml/kg body weight administered every 4 months to individuals in close contact with newly imported chimpanzees appears to be an effective prophylactic.

F. *Herpesvirus simiae* (Herpesvirus B)

Of the many reported herpesviruses in primates and other animals species, the one which presents the most serious health hazard to humans is *Herpesvirus simiae*. Originally called herpesvirus B and isolated by Sabin in 1934, it remained the only herpesvirus of zoonotic importance in nonhuman primates for 24 years (Sabin and Wright, 1934). There are now over 35 reported herpesviruses of primates, most of which are not zoonotic, and none are as pathogenic for humans as *Herpesvirus simiae*. For example, herpesvirus T has been incriminated in cases of mild encephalitis in humans (Hunt and Meléndez, 1969).

i. Reservoir and incidence. *Herpesvirus simiae* in its natural host (*Macaca* spp.) presents a mild clinical disease similar to that of *H. simplex* in the human. Lingual and/or labial ulceration occur at the time of primary infection, these lesions usually heal spontaneously in 7–14 days, and may be undetected in the monkey because of the lack of clinical symptoms. Wild-caught rhesus monkeys frequently have circulating antibody titers, and studies have shown that 100% seroconversion in newly trapped wild animals can occur (Hunt and Meléndez, 1969; Shah and Southwick, 1965). The pattern of infection appears analogous with *H. simplex* infection in human populations with prevalence increasing with age.

ii. Mode of transmission. Human disease is usually acquired by the bite of an affected animal or by exposure of the skin to infected saliva or to monkey tissues infected with the virus.

iii. Clinical signs. The disease is seldom seen in humans, except when infected mortality is extremely high (Davidson and Hummeler, 1960). Clinically the patient can have a diphasic illness, very similar in symptomatology to poliomyelitis, with a rapid ascending flaccid paralysis. Most survivors suffer some form of permanent paralysis. One report has described the probable reactivation of a latent case in a 61-year-old male who had not been exposed to the agent for 10 years before development of clinical signs of infection. The trigeminal nerve was most likely the location of the latent virus. *Herpesvirus simiae* has also been isolated from the trigeminal ganglia of a healthy seropositive rhesus monkey.

iv. Control and prevention. All macaques, especially rhesus monkeys, are potential carriers and should be handled accordingly. Protective garments should always be worn when working with these animals and should include face masks and leather gloves when handling the animals to protect against bites and scratches. Ulcers of the tongue and mucous membranes should be treated as suspect lesions until serology and virus isolation methods confirm otherwise. Antiserum is available and should be used in cases of known exposure, but is known not to be fully effective. There is no specific treatment available; a vaccine has been approved for general distribution (Hull,

1971). There is evidence that other simian species have antibodies to this virus, but the agent has been isolated only from macaques. Optimally, procurement of *H. simiae*-negative monkeys for use in biomedical research, is the safest and most effective means of preventing the disease in humans.

G. Marburg Virus Disease (Vervet Monkey Disease, African Hemorrhagic Fever)

Marburg virus disease is an acute, highly fatal disease first described in Germany in 1967 (Martini and Siegart, 1971; Siegart, 1972).

i. Reservoir and incidence. The virus responsible for the disease is a rhabdovirus. The first reported outbreak involved 31 people; all became severely ill and seven died. Many patients had prior contact with blood or tissues from African green monkeys (*Cercopithecus aethiops*) during tissue culture preparation. Six secondary cases occurred in personnel that had contact with these patients. Several other cases occurred in South Africa in 1975 and in Kenya in 1980 (Johnson, 1982).

The natural reservoir host as well as the source of the original infection in all reported cases of Marburg virus remain unknown. The virus has been 100% fatal in experimentally infected African green monkeys, rhesus monkeys, squirrel monkeys, guinea pigs, and hamsters (Hunt *et al.*, 1978).

ii. Mode of transmission. Transmission appears to be from direct contact with infected tissues and close contact with patients. Aerosol transmission is strongly suggested by the development of the disease in uninoculated monkeys housed in the same room as experimentally infected monkeys (Hunt *et al.*, 1978).

iii. Clinical signs, susceptibility, and resistance in humans. The incubation period for Marburg virus disease is 5–7 days. The disease is manifested by the abrupt onset of headache, progressive fever, myalgia, vomiting, and diarrhea. This is followed by a hemorrhagic diathesis. Blood is observed in stool and vomitus. Leukopenia, thrombocytopenia, and proteinuria are also usually observed. The disease often progresses to shock, followed by death in 20–30% of cases. Treatment consists of intensive supportive care and the early administration of plasma containing virus-specific antibody.

iv. Diagnosis and prevention. Diagnosis of Marburg virus disease can be confirmed by isolation of the virus from tissue or blood samples taken during the acute disease. The virus can be recovered from Vero cell tissue culture or guinea pigs inoculated with infected tissues. It can be identified by morphologic characteristics with electron microscopy and by the indirect fluorescent antibody test.

Prevention of Marburg virus disease in the laboratory is probably best accomplished by following strict quarantine procedures on newly imported wild-caught primates. Based on experimental data available, naturally infected monkeys incubating the disease should become ill or die within several weeks. Any suspected primate cases should be handled with extreme caution and placed in strict isolation.

H. Rabies

Rabies is an acute, almost invariably fatal viral disease caused by a member of the rhabdovirus group.

i. Reservoir and incidence. Rabies is worldwide in its distribution. The only exceptions are a few countries that have maintained their rabies-free status because of strict importation and animal control programs and because of the existence of geographical barriers (Gillespie and Timoney, 1981). The primary reservoirs of rabies vary geographically and may involve many wild and domestic species. In lesser developed countries where domestic animal control and vaccination procedures are minimal, dogs and cats persist as prime sources of rabies (Benenson, 1981).

ii. Mode of transmission. Virus-laden saliva introduced into the victim via bite wound, scratch, or abrasion is the most common means of spread of this virus. Rabid dogs shed the virus in their saliva for 1–14 days before developing clinical signs, and viral shedding occurs 3 days before clinical signs in the cat (Gillespie and Timoney, 1981). Skunks, arctic foxes, and bats are suspected of being asymptomatic chronic carriers. Animals exhibiting signs of rabies are usually shedding copious quantities of the virus in their saliva.

Aerosol transmission, although rare, has been documented in laboratory settings and in caves where rabid bats roost (Benenson, 1981).

iii. Clinical signs, susceptibility, and resistance in humans. Humans, as well as all other mammals, are susceptible to this disease. Animal species' susceptibility varies, with the fox being extremely susceptible and the opossum being quite resistant. In humans, natural immunity is unknown.

Clinical rabies is usually divided into five phases: incubation, prodromal, acute neurological, coma, and, rarely, recovery (Hattwick, 1982). The average incubation is from 14 to 60 days, although it may be as short as 10 days, or may take as long as 19 years. Victims are asymptomatic during this period.

During the prodromal phase, nonspecific symptoms noted are fever, headache, malaise, fatigue, and loss of appetite.

Some victims will develop paresthesia at the site of the bite wound. Anxiety, apprehension, irritability, depression, and nervousness also may be noted during this period. The acute neurological phase of the disease is marked by episodes of hyperactivity and bizarre behavior interspersed with periods of normal, cooperative, although anxious, behavior. Gradually, the victim will exhibit signs of increasing paralysis, which eventually lead into the coma phase, respiratory arrest, and death.

Recovery has been reported in only three cases of human rabies. In each of these cases, prophylaxis was begun before the onset of clinical illness.

Intensive medical care was used to prevent or treat most of the complications that are recognized in clinical rabies.

iv. Diagnosis and control. Rabies should be considered as a differential diagnosis in any wild-caught or random-source laboratory animal of unknown vaccination history, exhibiting encephalitic signs.

Diagnosis of rabies can be confirmed by virus isolation from various body fluids and tissues. Diagnosis may also be obtained by using specific fluorescent antibody staining of corneal smears, frozen skin biopsy sections, and mucosal scrapings.

The most important factor in preventing rabies, aside from vigorous first aid for bite wounds, is the control of this disease in domestic animals. Stringent vaccination requirements and enforced animal control measures will greatly decrease the size of the population at risk (Benenson, 1981).

Preexposure immunization should be considered for individuals with high-risk occupations, such as veterinarians, staff of quarantine kennels, personnel working with the virus, and wildlife conservation personnel in areas where the disease is enzootic (Abelseth, 1980). In all cases, when a suspected exposure to the virus has occurred, a physician or local health authority should be consulted for recommended course of action.

III. RICKETTSIAL AND CHLAMYDIAL DISEASES

A. Psittacosis

Chlamydia psittaci is the causative agent of a disease recognized by several names: psittacosis, ornithosis, parrot fever, or chlamydiosis. The chlamydiae are obligately intracellular organisms having a unique development cycle that distinguishes them from other microorganisms.

i. Reservoir and incidence. The genus *Chlamydia* contains only two species, *C. trachomatis* and *C. psittaci*. Humans are the exclusive natural host of *C. trachomatis* with the exception of mice, which have several strains causing pneumonitis; the zoonotic potential of mouse-adapted *C. trachomatis* strains has not been recorded. *Chlamydia psittaci* has a broad host range including mice, guinea pigs, rabbits, cats, lambs, calves, birds, and frogs among the naturally susceptible laboratory animal species (Storz, 1971; Newcomer *et al.*, 1982). The many strains of *C. psittaci* produce a very diverse disease spectrum in animals, including conjunctivitis, pneumonitis, air sacculitis, pericarditis, hepatitis, meningoencephalitis, enteritis, urethritis, arthritis, and endometritis with abortion. Gastrointestinal infection results in enteric shedding of the organism. Upon drying, fecal material produces highly infective aerosols. Latency is a widely recognized feature of chlamydia infection, with the organism causing inapparent infection or fulminant disease in the same host species. In clinically healthy birds, stress can precipitate clinical signs and shedding of the organism (Meyer, 1942).

Although infection with *C. psittaci* has been reported in many laboratory species, the mammalian-derived strains of *C. psittaci* only rarely have been implicated as a source of zoonotic infection (Schachter and Dawson, 1978). Two cases of human conjunctivitis resulting from close association with cats that had chlamydial pneumonitis and conjunctivitis have been observed (Schachter *et al.*, 1969). Other direct evidence for human infection by mammalian-derived *C. psittaci* has been suggested (Enright and Sadler, 1954; Page and Smith, 1974).

Birds constitute the main reservoir of *C. psittaci* infection, and any avian species must be considered a potential source. However, there have been a number of psittacosis outbreaks in humans involving highly virulent *C. psittaci* strains where avian contacts could not be documented and where the source of organism was never identified (Schachter and Dawson, 1978). In addition, approximately 25% of human cases have no history of avian contacts.

ii. The disease in humans. Psittacosis may be either asymptomatic or present clinically after a 1–2 week incubation period. Clinical disease may have an acute or insidious onset and frequently has a respiratory component. Symptoms include fever, chills, myalgia, anorexia, headache, and a nonproductive cough.

Pneumonitis or atypical penumonia may be indicated radiographically by extensive pulmonary involvement. In other cases, respiratory lesions are less prominent, and the disease presents as a toxic or septic condition. Features of this form of the disease include hepatosplenomegaly with hepatitis, meningoencephalitis, and cardiac involvement. Antibiotic regimens of 21 days or longer are generally used in the treatment of the disease as relapses can occur with treatment of shorter duration.

iii. Control. Psittacosis can be controlled in the laboratory environment by introducing birds only from those flocks known to be free of the disease. When wild-caught birds or birds from flocks of unknown disease status must be introduced into the animal facility, chlorotetracycline chemoprophylaxis should be implemented. In addition, animal care personnel and research staff should use protective clothing, particularly masks.

B. Q Fever

Q fever is caused by *Coxiella burnetti*. Laboratory-acquired infections among individuals working with the organism have long been recognized. Recently, however, the importance of Q fever as a serious laboratory hazard has been reemphasized by the contamination of research facilities where asymptomatic, infected ewes were being used in biomedical research.

i. Reservoir and incidence. *Coxiella burnetti* is distributed worldwide among both wild and domestic animals in two intersecting but self-perpetuating cycles of infection (Babudieri, 1959). Wild animals with any of numerous tick hosts comprise one cycle of infection, while the domestic species, especially sheep, goats, and cattle, comprise the other cycle. Epidemiologic studies indicate widespread infection in sheep throughout the United States. Dogs and cats, as well as domestic fowl and chickens, can also be infected. Enzootic infection among the domestic species constitutes the main reservoir of infection for humans.

ii. Mode of transmission. Humans most frequently contract Q fever in areas where infected animals are raised or where rickettsial-laden tissues are handled (Bernard *et al.*, 1982). The organism is shed in the urine, feces, milk, and especially birth products of domestic ungulates, which generally do not have clinical disease. The placenta of an infected ewe can contain up to 10^9 organisms per gram of tissue; large numbers of organisms are also eliminated in amniotic fluid and fetal membranes (Welsh *et al.*, 1958). *Coxiella burnetti* is resistant to desiccation and can persist for months in a variety of animal products and sand or clay, providing for extensive environmental contamination. The principal mode of transmission to humans is by inhalation of infective aerosols, although transmission by ingestion has been recorded (Tiggert *et al.*, 1961). Two recent outbreaks of Q fever in research facilities using sheep as research animals substantiated the importance of the environmental persistence and aerosol transmission of *C. burnetti*. In both instances, many individuals who become seropositive for Q fever did not have direct animal contact but were located along routes followed by sheep transport carts or had other casual exposure to the animals (Bernard *et al.*, 1982). In an outbreak in San Francisco, California five of nine laundry workers whose only exposure to the agent was soiled linens, developed serologic evidence of infection.

iii. The disease in humans. After an incubation period of 2 weeks to 1 month, the disease manifests as an acute febrile systemic illness, or subacute endocarditis, especially in individuals with previous valvular disease (Warren and Hornick, 1979). The cutaneous eruption characteristically seen in the other human rickettsioses is absent. Patients present with fever, chills, profuse sweating, malaise, anorexia, myalgia, and sometimes nausea and vomiting. A severe frontal headache with retroorbital pain is frequently reported. Pharyngitis and conjunctival suffusion may also be found. A mild nonproductive cough coupled sometimes with thoracic pain indicates pulmonary involvement, which is observed radiographically in about one-half the cases.

Many human infections are asymptomatic, and therefore unnoticed. Although frequently characterized as a lung infection, other organ systems also can be involved. Hepatitis, hepatomegaly, or liver function abnormalities may be evident. Encephalitis has been confirmed in several cases. Pericarditis, myocarditis, and endocarditis are seen infrequently, but the latter may be a fatal complication. Most cases of Q fever resolve within 2 weeks, but protracted or relapsing cases can occur, especially in the elderly.

iv. Disease, diagnosis, and control. Recommendations for the control of Q fever in the research facility have recently been formulated (Bernard *et al.*, 1982). Whenever possible male or nonpregnant female sheep should be used in research facilities. Many research investigations, however, require the use of pregnant animals.

The use of Q fever-free sheep to eliminate the zoonotic potential has limited practicality because the identification of Q fever-free flocks requires an intensive surveillance program and frequent serologic testing. Moreover, the serologic status of sheep is not a useful indicator of organism shedding. The experimental vaccination of sheep against Q fever has shown some promise as a means of eliminating organism shedding (Sadecky and Brazina, 1977). However, the feasibility of producing and maintaining a Q fever-free flock by vaccination alone, or in conjunction with other control measures, has not been demonstrated.

Personnel education and control, and physical separation of potentially infected animals from humans are the current methods for controlling Q fever within the research facility. Physical separation of potentially infected animals from all other functions of the research facility is highly desirable. When physical separation is not feasible, the appropriate air handling systems are an essential substitute, and restrictions on animal movement within the facility must be implemented. Personnel not working directly with potentially infected animals or animal products should be protected from inadvertent exposure. This may be accomplished by the conspicuous identification of the sheep-holding area and by the labeling of all material potentially contaminated with the organism. All potentially contaminated material should be sterilized or disinfected.

Additional preventive measures should be applied to individuals working directly with potentially infected animals. The total research plan entailing the use of suspect animals should be defined by the principal investigator and facilities biohazards safety officer or equivalent and communicated to relevant personnel. Personnel working directly with animals should participate in a medical surveillance and health education program.

A formalin-inactivated preparation of *C. burnetti* organisms expressing phase I antigen has been evaluated as a skin test antigen and as a vaccine (Ascher *et al.*, 1983). Although the efficacy of this preparation as a vaccine requires further study, the use of this preparation in the delayed hypersensitivity skin test is efficacious and should be incorporated into medical surveillance programs of high-risk seronegative individuals.

Protective clothing, including surgical masks, disposable gloves, and shoe covers, should be used to contain possible contamination within the biohazard area. Disinfection of surgical and laboratory facilities that have held parturient ewes is warranted. Suitable disinfectants include a 1 : 100 dilution of chlorine bleach containing 5–25% hypochlorine, a 5% solution of hydrogen peroxide, or a 1 : 100 dilution of Lysol. In addition, all individuals using sheep should observe standard microbiologic, laboratory, and animal management practices to limit the exposure of facilities personnel and the community to pathogenic organisms.

C. Other Rickettsial Diseases

With the exception of Q fever and psittacosis, rickettsial infections have not posed a major zoonotic problem in the research animal facility except where they are actively under investigation. Nevertheless, infectious rickettsial species are distributed widely among wild rodents and other wild animals that might be introduced into the laboratory animal facility for other investigative purposes.

1. Rocky Mountain Spotted Fever

Rocky Mountain spotted fever is a serious disease of humans characterized by fever, headache, myalgia, and a generalized maculopapular rash that frequently becomes hemorrhagic. The causative agent, *Rickettsia rickettsii*, is transmitted to humans by various species of ixodid ticks (*Dermacentor andersoni* and *D. variabilis*) and one or more of their host species. Ticks are both biological vectors and true reservoir hosts of the organism by virtue of the phenomenon of transovarial infection of progeny ticks (Burgdorfer and Brinton, 1975); this phenomenon is also evident among the ectoparasitic vectors of other members of the spotted fever group of rickettsiae.

The mammalian host from which the organism has been isolated includes numerous wild rodents and lagomorphs (Burgdorfer, 1979). Single reports of organism isolation have occurred in other species, and many birds and mammals are seropositive, but the role of these species in the propagation of the natural cycle is unknown. Dogs experimentally infected with *R. rickettsii* had an ensuing high and persistent rickettsemia, and thus may be a ready source of infection for ticks (Keenan *et al.*, 1977). Ectoparasite control on newly arriving dogs, cats, and wild caught animals is of paramount importance in preventing this disease from entering the laboratory.

2. Rickettsialpox

Rickettsialpox is caused by *R. akari*, a member of the spotted fever group of *Rickettsia*. In humans rickettsialpox is a mild self-limiting disease characterized by escharlike lesions, fever of 1-week duration, headache, myalgia, lymphadenopathy, and leukopenia, followed by a generalized papulovesicular rash. Domestic mice are the natural hosts of this agent, and rats and moles can also harbor the organism. The disease in humans occurs primarily in rodent-infested urban dwellings where mice, mites, and rickettsia maintain a cycle of infection (Greenberg *et al.*, 1947; Nichols *et al.*, 1953). The natural vector of the disease, *Allodermanyssus sanguineus*, occurs in many parts of the world, but has not been recorded in conventional rodent colonies. *Ornithonyssus bacoti*, a tropical rat mite that occasionally infests laboratory rodents, has been infected experimentally with *R. akari*, but has not been shown to be involved in the natural cycle of rickettsialpox.

Laboratory-acquired infection in humans via the respiratory route has occurred (Sulkin, 1961), but laboratory infection due to mite bites has not been reported. Control and eradication of this disease both in the laboratory and in human dwellings relies upon the elimination of wild mice and the mite vector.

3. Murine typhus

Rickettsia typhi (*mooseri*), the causative agent of murine typhus or endemic typhus, is transmitted to humans by rat fleas (*Xenopsylla cheopis* and *Nasopsyllus fasciatus*). The mite *O. bacoti* can be infected with *R. typhi* and can transmit the infection between guinea pigs (Dove and Shelmire, 1932). Other

mammals and ectoparasites, including cats and cat fleas, have been found infected with *R. typhi*. Although outbreaks of this disease continue to occur in the United States (particularly Texas), natural laboratory rodent infections with *R. typhi* have not been observed. Laboratory-acquired infection among personnel handling experimentally infected mice and performing intranasal inoculations of mice with the agent has been reported (Fox and Brayton, 1982). The clinical signs of murine typhus in humans are similar to those of rickettsialpox.

IV. BACTERIAL DISEASES

A. Systemic Infections

1. Brucellosis

Of the *Brucella* spp., *B. canis* is the most likely zoonotic agent in the laboratory animal facility due to the extensive use of random-source and laboratory-bred dogs in comparison to other large domestic animals. Furthermore, *B. canis* in the general canine population has not been the focus of a large-scale eradication program as have the *Brucella* spp. of large domestic animals. From 1967, when the first human *B. canis* infection was identified, through 1978, 18 cases were reported to the Center for Disease Control, Atlanta, Georgia (Currier *et al.*, 1982); eight (44%) of these cases were attributed to infected dogs.

i. Reservoir and incidence. Canine brucellosis was first recognized when widespread abortions occurred in a breeding colony of beagles. The disease has since been reported in several continents and throughout the United States. The prevalence of *B. canis* varies from 1 to 6%, reflecting the geographic area of the dog population sampled and the diagnostic method used (Flores-Castro and Carmichael, 1980). Dog populations in areas without dog control have the highest prevalence of *B. canis* infection (Flores-Castro and Carmichael, 1980). Random-source dogs, potentially originating from uncontrolled dog populations, may represent the most important source of this organism in the animal facility (Fredrickson and Barton, 1974; Brown *et al.*, 1976). Laboratory-bred dogs may also be a source of *B. canis* infection in the laboratory environment. However, strict implementation of preventive medicine programs including *B. canis* surveillance significantly reduces the prevalence of this pathogen in breeding colonies.

Although *B. canis* is particularly well adapted to dogs and is not readily transmitted to other species, infection has been reported in several wild species of Canidae and raccoons (Carmichael, 1979).

ii. Transmission and clinical signs. Whereas the mode of transmission and portal of entry have been defined in *B. canis* infections resulting from laboratory accidents, they have been identified less frequently in actual zoonotic infections. Laboratory accidents have incriminated the oral and transcutaneous routes of infection. Two individuals became seropositive to *B. canis*, one with concurrent bacteremia, after the oral pipeting of *B. canis*. Another two laboratory technicians preparing *B. canis* antigens developed infection, presumably through skin abrasions on their hands (Morris and Spink, 1969). However, penetration through the intact integument has been demonstrated experimentally for *B. abortus;* this may occur with *B. canis*. Exposure to organism-laden blood and infected tissues constitutes the major mode of zoonotic infection with other *Brucella* spp. The potential for airborne transmission was suggested by the isolation of *B. suis* from the air in an abattoir setting and may occur with *B. canis*.

Zoonotic transmission of *B. canis* was considered a low risk in one study, as evidenced by the lack of seropositivity to *B. canis* among 12 individuals exposed to five infected dogs (Mumford *et al.*, 1975). When human infections have occurred, the source of infectious material has seldom been determined; infected urine was suspected in a case of pet-associated human *B. canis* infection. Canine blood has not been implicated as a source of human infection. However, this potential source warrants consideration as prolonged bacteremia is a prominent feature of canine brucellosis.

Bacteremia has occurred in 10–16 human *B. canis* infections; other systemic involvements included painful generalized lymphadenopathy and splenomegaly. Additional signs included fever, headache, chills, sweating, weakness, malaise, myalgia, nausea, and weight loss. Although *B. canis* produced clinical disease in humans similar to that of other *Brucella* spp., it was generally not as severe. Seroconversion to *B. canis* has been reported in 0.5% of asymptomatic military personnel contacts with infected dogs, indicating that inapparent infection may be a common occurrence (Mumford *et al.*, 1975).

iii. Diagnosis and control. *Brucella canis* infections in dogs should be considered when the history includes abortions, infertility, testicular abnormalities, and poor semen quality. A rapid slide agglutination test that produces presumptive diagnostic information is commercially available. To confirm the results of the slide test, blood cultures and additional serological tests should be performed.

In breeding kennels, dogs with a positive blood culture should be culled. At least three negative consecutive monthly tests of all dogs must be obtained before a kennel can be considered noninfected. Animals introduced into a breeding colony should be maintained in separate quarters until two negative tests, performed at 30-day intervals, are obtained. No immunizing agents are currently available.

Animal technicians and handlers should wear disposable gloves and disinfect their hands before examining each animal. Commercial disinfectants, such as chlorine, organic iodine compounds, and quaternary ammonium compounds, have proved to be rapid bacteriocidal agents for *B. canis*.

2. Leptospirosis

Leptospira microorganisms were discovered in 1914, when isolated from jaundiced patients (Inade *et al.*, 1916), and after further study were named in 1917 (Noguchi, 1918). Leptospirosis is a common zoonotic disease of livestock, pet and stray dogs, and wildlife. Pathogenic serovars have been isolated in 77 countries.

i. Reservoir and incidence. Reservoir hosts of leptospirosis include rats, mice, field moles, hedgehogs, gerbils, squirrels, rabbits, hamsters, other mammals, and reptiles. A particular species of animal will usually act as the primary host of a particular serotype, but most serotypes can be carried by several hosts. *Leptospira* are well adapted to a variety of mammals, particularly wild animals and rodents; clinical manifestations in the chronic form are inconspicuous, with the organism being carried and shed in the urine for long periods of time. Rodents are the only major animal species that can shed leptospires throughout their lifespan without clinical manifestations (Babudieri, 1958; Faine, 1963). Active shedding of leptospires by laboratory animals can go unrecognized until personnel handling the animals become clinically affected. Seven leptospira serovars, including *L. autumnalis, L. australis, L. ballum, L. seirse, L. hebdomadis, L. tarassovi,* and *L. pomona*, are found in wild animals in the United States (Hanson, 1982).

It has been estimated that reactor rates for *Leptospira* antibody titers in livestock in the United States are 15% of the total cattle population and 8% for swine, with higher incidences reported in Central and South America. These estimates indicate nearly 100 million cattle and swine have been infected in the Western Hemisphere alone (Hanson, 1982). These animals serve as a significant source of primary long-term shedding of at least three serovars. Cattle are the natural carriers of *L. hardjo*, whereas swine carry *L. pomona;* each animal can shed the organism for 1 year in urine. Dogs also commonly harbor the organism; stray dogs harbor *L. icterohemorrhagiae* as well as being a natural carrier host of *L. canicola*. Sheep, goats, and horses can also be infected with a variety of serotypes.

Rats and mice are common animal hosts for *L. ballum*, although it has been found in other wildlife, including skunks, rabbits, opossums, and wild cats (Mailloux, 1975). The infection in mice is inapparent and can persist for the animal's lifetime (Torten, 1979). Although earlier reports indicated that several colonies of laboratory mice harbor the organism (Wolf *et al.*, 1949; Yager *et al.*, 1953), no current estimates of the carrier rate among laboratory rodents in the United States are available. In several European laboratories, transmission of leptospires from laboratory rats to laboratory personnel has been reported (Loosli, 1967; Geller, 1979). In a study of leptospiral infections in feral rodents, 2673 rodents of 10 species were collected in Georgia. Of the 933 tested for leptospires (by kidney culture), *L. ballum* was the only serotype cultured. It was isolated from 22% of the house mice and 0.8% of the old field mice, *Peromyscus polionotus* (Brown and Gorman, 1960). Hamsters, young guinea pigs, and gerbils are especially susceptible to leptospirosis, whereas mature rabbits are resistant to infection and are ideally suited for *Leptospira* antibody production.

Since leptospirosis in humans is often difficult to diagnose, the low incidence of reported leptospirosis infection in humans may be misleading. Between 1947 and 1973, only 17 cases of *L. ballum* infection in humans were recorded in the United States (Boak *et al.*, 1960; Center for Disease Control, 1965; Friedmann *et al.*, 1973; Stoenner and Maclean, 1958). Outbreaks in personnel working with laboratory mice in the United States have been documented (Barkin *et al.*, 1974; Boak *et al.*, 1960; Stoenner and Maclean, 1958). In one study, 8 of 58 employees handling the infected laboratory mice (80% of breeding females were excreting *L. ballum* in their urine) contracted leptospirosis.

ii. Mode of transmission. Infection with leptospira most frequently results from handling the infected animals (contaminating the hands with urine) or from aerosol exposure during cage cleaning. Skin abrasions or exposure to mucous membranes may serve as the portal of entry. All secretions and excretions from infected animals should be considered infective. In one instance, it was speculated that a father was infected after his daughter, because of an arugment, used his toothbrush to clean a contaminated pet mouse cage (Friedmann *et al.*, 1973). Also, laboratory or wild mice that are to be used for primary kidney tissue cultures should be ascertained to be free of leptospires (Turner, 1970).

iii. Clinical signs, susceptibility, and resistance in humans. The disease may vary from inapparent to severe infections and death. Infected individuals experience a biphasic disease (Heath and Alexander, 1970). They become suddenly ill with weakness, headache, myalgia, malaise, chills, and fever. Leukocytosis is usually associated with leptospirosis. During the second phase of the disease, a painful orchitis, conjunctival suffusion, and rash are also noted. Unlike the orchitis associated with mumps, leptospirosis caused enlarged testes in only one patient (Friedmann *et al.*, 1973). Two infected personnel in a laboratory mice-associated outbreak required more than 1 month for recovery (Stoenner and Maclean, 1958). Renal, hepatic, pulmonary, gastrointestinal, and conjunctival findings, may be abnormal (Barkin *et al.*, 1974).

iv. Diagnosis. Because of variability in clinical symptoms and lack of pathognomonic pathological findings in humans and animals, serologic diagnosis or actual isolation of leptospires must be undertaken to establish a correct diagnosis (Torten, 1979). As an aid to diagnosis, leptospires can sometimes be observed by examination or direct staining of body fluids or fresh tissue suspensions. A definitive diagnosis in humans or animals is made by culturing the organisms from tissue or fluid samples or by animal inoculation (particularly in 3- to 4-week-old hamsters) and subsequent culture and isolation. Serologic assessment of host infection is accomplished by indirect hemagglutination, agglutination, analysis, complement fixation, microscopic agglutination, and fluorescent antibody techniques (Stoenner, 1954; Torten, 1979). The serologic test most frequently used is the modified microtiter agglutination test. Titers of 1 : 100 or greater are considered of diagnostic importance.

In a survey of trapped wild urban rats, diagnosis of leptospirosis was more accurate by urine or kidney culture, rather than by either indirect fluorescent antibody or macroscopic slide agglutination (Sulzer *et al.*, 1968). Another survey of wild rats confirmed that culture techniques identified more positive rats than did macroscopic slide agglutination (Higa and Fujinaka, 1976).

v. Epidemiology and control. In mouse colonies infected with *L. ballum*, antibodies against *L. ballum* were detected in sera of mice of all ages, but leptospires could be recovered only from mature mice. Progeny of seropositive females had detectable serum antibodies at 51 days of age, but not at 65 days. Progeny of seropositive female mice, which possessed antibody at birth and acquired additional antibody from colostrum, remained free of leptospires if isolated from their mothers at 21 days of age, despite exposure during the nursing period (Stoenner, 1957).

Studies in mice experimentally infected with *L. grippotyphosa* demonstrated that maternal antibodies, whether passed through milk or placental transfer, conferred protection of long duration against the carrier state and shedding of leptospirae. Thus, serologically positive immune mothers do not transmit the disease to their offspring. However, mice born to nonimmune mothers, if infected at 1 day postpartum, become carriers with no trace of antibodies. Thus a population of pregnant carrier mice without antibody could serve as the source for outbreaks among susceptible mouse populations (Birnbaum *et al.*, 1972). Field surveys have supported this data in that the percentage of carrier mice that do not have antibodies is significant. This has led to the diagnostic approach that specifies that both serologic and isolation methods must be used to determine the rate of leptospiral infection in rodents (Galton *et al.*, 1962). Other natural carriers, e.g., dogs with *L. canicola*, may not have detectable serum agglutinins.

Leptospira ballum has been eliminated from a mouse colony by administration of feed containing 1000 gm chlorotetracycline hydrochloride per ton for 10 days (Stoenner *et al.*, 1958). Antibiotic therapy with or without vaccination also has been shown to be effective in prevention of clinical disease in cattle and swine (Stalheim, 1969).

3. Plague

Plague is a recognized zoonotic disease caused by the gram-negative bacteria *Yersinia pestis*. The infection is endemic in scattered populations of wild rodents in the western third of the United States. The incidence of human cases has been steadily increasing in the United States since 1965 (Kaufman *et al.*, 1981; Christie *et al.*, 1980; Sanford, 1982; Poland and Barnes, 1979). Human infections in the United States are sporadic and limited, usually resulting from contact with infected fleas or rodents. However, numerous human cases in recent years have been associated with exposure to infected cats, goats, camels, rabbits, dogs, and coyotes (Rollag *et al.*, 1981). Human infections from nonrodent species have resulted from direct contact with infected tissues, human consumption of infected tissues, scratch or bite injuries, and handling of infected animals. In addition, animals, such as dogs and cats, may serve as passive transporters of infected rodent fleas into the home or laboratory.

i. Clinical signs, susceptibility, and resistance in humans. Bubonic plague in humans is usually characterized by fever and the formation of large tender swollen lymph nodes, or buboes. If untreated, this may progress to severe pneumonic or systemic plague. Inhalation of infective particles may also result in pneumonic plague. Proper antibiotic therapy can significantly reduce the mortality rate, which may exceed 50% in untreated cases. The general population is widely susceptible. A relative immunity is present only after recovery.

ii. Diagnosis and control. A presumptive diagnosis can be made by visualizing bipolar-staining, ovoid, gram-negative rods on the microscopic examination of fluid from buboes, blood, sputum, or spinal fluid; confirmation can be made by culture. Complement fixation, passive hemagglutination, and immunofluorescence staining of specimens can be used for serologic confirmation.

Prevention of plague is generally directed at the control of wild rodents. Feral or random-source animals acquired in known endemic plague areas for use in the laboratory should be carefully quarantined and treated with appropriate insecticides to kill fleas. Routine flea control, as well as rodent control, should be carefully followed in outdoor-housed animals in endemic areas. A vaccine is available for high-risk personnel, such as those doing field work in endemic areas.

B. Respiratory Infections

1. Tuberculosis

i. Reservoir and incidence. Tuberculosis is an important zoonosis associated with laboratory animals. Tuberculosis is caused by acid-fast bacilli of the genus *Mycobacterium*. Natural reservoir hosts for the etiologic agent of this disease correspond to the three most common species of *Mycobacterium*, which are *M. bovis* (cows), *M. avian* (birds), and *M. tuberculosis* (humans). While these are the major reservoir hosts of tuberculosis, many animals, including swine, sheep, goats, monkeys, cats, and dogs, are susceptible and contribute to the spread of disease. Susceptibility of mammalian species is quite varied and relates to host immune responses and to the species of *Mycobacterium* to which exposure has occurred.

The incidence of *Mycobacterium* infection in humans has dramatically decreased in the past few decades (Benenson, 1981). This decrease is a result of the bovine tuberculosis test and slaughter eradication program in the United States and the availability of efficacious therapy. Currently, in humans, the incidence of tuberculosis is sporadic. In nonhuman primates, outbreaks of tuberculosis still occur. This is particularly true in the Old World species of monkeys that are much more susceptible to tuberculosis than New World monkeys and the great apes (Whitney, 1975).

ii. Mode of transmission. *Mycobacterium* bacilli are transmitted from infected animals or infected tissue samples via the aerosol route. Important means of contracting the disease include caring for tuberculous animals or performing autopsies on infected animals. Spread of the disease beyond the natural host range occurs through animal-to-animal and animal-to-human contact, usually by airborne infectious particles. Nonhuman primates in the wild apparently contract the disease only through initial human contact (Whitney, 1975). Transmission of the disease is enhanced in the laboratory by exposure to dusty bedding of infected animals, aerosolization of organisms by the use of high pressure water for sanitation, and coughing of clinically infected animals. Tuberculosis may also be contracted via ingestion of tubercle bacilli.

iii. Clinical signs, susceptibility, and resistance. Clinical signs of tuberculosis in humans are dependent upon the organ system or systems involved. Most familiar are the signs related to pulmonary tuberculosis. While the pulmonary form of disease often remains asymptomatic for months or years, the disease may eventually manifest itself, by cough, sputum production, and hemoptysis. In addition to these local signs, general symptoms of tuberculosis in humans include anorexia, weight loss, lasstidue, fatigue, fever, chills, and cachexia. Tuberculosis may also infect virtually every other organ system. Extrapulmonary forms of the disease include: disseminated (military) tuberculosis, recognized in the very young and very old; hepatic tuberculosis, with a clinical picture that is characterized by elevated liver enzymes; gastrointestinal tuberculosis; tuberculosis meningitis; and tuberculosis bone disease, most commonly affecting the vertebral column.

iv. Diagnosis and control. The positive diagnosis of tuberculosis is often quite difficult. Three widely used tools for a presumptive diagnosis are the intradermal tuberculin test, radiographic analysis, and an acid-fast stained sputum smear. More definitive diagnosis is obtained by culture of the organisms from body fluids or biopsy specimens.

Control of tuberculosis infection, particularly within the biomedical research arena, requires a multifaceted approach. This includes aspects of personnel education, a regular health surveillance program for humans and nonhuman primates, isolation and quarantine of suspect animals, and rapid euthanasia and careful disposal of confirmed positive animals. Certain precautions are necessary when considering vaccination and/or chemoprophylaxis (Muchmore, 1975). Vaccination with *Bacille Calmette-Guèrin* (BCG), a strain of *M. bovis,* has been shown to be an effective means of preventing active tuberculosis and is suggested for use in high-risk groups. However, the BCG vaccination elicits a positive tuberculin test, thereby negating the best diagnostic indicator of early disease. Vaccination in the United States is therefore reserved for demonstrated high-risk individuals and children in locations where 20% or more of school-age children are tuberculin positive. Chemoprophylaxis with an effective antituberculosis agent, such as isoniazid, should be reserved only for special cases of known exposure. A well-conceived tuberculosis control program will include some or all of the above methods tailored to the needs and special circumstances of individual animal resource programs.

C. Enteric Infections

1. Campylobacteriosis

Campylobacter (*vibrio*) have been known as pathogenic and commensal bacteria in domestic animals for many years; however, recently *Campylobacter fetus* ss. *jejuni* has gained recognition as a leading cause of diarrhea in humans.

i. Reservoir and incidence. *Campylobacter fetus* ss. *jejuni* has been recently isolated from laboratory animals, including dogs, cats, hamsters, ferrets, nonhuman primates, and rabbits (Fox, 1982a). In studies performed on farm animals, *C. jejuni* has also been isolated from the intestinal contents of healthy swine, sheep, and cattle and is a common cause of abortion in sheep (Prescott and Munroe, 1982). The bacterium is also observed as a commensal in wild birds, waterfowl, turkeys, and chickens. Like *Salmonella, C. jejuni* can be shed in the stool of asymptomatic carriers.

In most reports citing pet-to-human transmission of *C. jejuni,* diarrheic puppies recently obtained from animal pounds have been the source of infection (Blaser *et al.,* 1982). Other reports implicate pet birds or chickens as a source of infection in humans. In a laboratory animal setting, a technician developed *Campylobacter* enteritis after performing husbandry chores on recently imported nonhuman primates shedding *Campylobacter* in their feces. Incidence studies of pound animals suggest that younger animals are more likely to acquire the infection and hence may more commonly shed the organism. In a somewhat analogous setting, in developing countries where personal hygiene and sanitation are less than optimum, the epidemiology in humans suggests that exposure to *Campylobacter* occurs very early in life.

While *C. jejuni* is widely distributed among animals, it is not known which of these organisms cause diarrhea in humans; different serotypes and biotypes of the organisms have been documented in both animal hosts and humans. Increased awareness of the zoonotic potential and the development of a convenient, reliable serotyping system will allow a more complete understanding of the role that animal hosts play in the epidemiology of *Campylobacter* infection in humans.

ii. Clinical signs. Clinical features of *Campylobacter* enteritis are usually consistent with an acute gastrointestinal illness. Diarrhea, with or without blood and leukocytes, abdominal pain, and constitutional symptoms, especially fever, are routinely noted. The severity of the illness can be variable but in most cases, it is brief and self-limiting. In protracted or severe cases antimicrobial therapy is instituted (Blaser and Reller, 1981).

2. *Salmonella*

Although there are 1600 recognized serotypes, *Salmonella typhimurium* and *S. enteritidus* have been associated most commonly with infections in laboratory animal colonies (Haberman and Williams, 1958). Other serotypes have also been reported in laboratory animals. From 1974 to 1978, the most frequently isolated serotype in the United States was *S. typhimurium.* Other frequently isolated serotypes were *S. newport, S. enteritidis,* and *S. heidelberg.*

i. Reservoir and incidence. *Salmonella* infection in humans and animals, including laboratory animals, occurs worldwide. The organism is an enteric bacteria, inhabiting the intestinal tract of many animals; salmonella is routinely associated with food-borne disease outbreaks, is a contaminant of sewage, and is found in many environmental water sources.

Although the reported incidence of *Salmonella* in laboratory animals has decreased in the last several years because of management practices, environmental contamination with *Salmonella* continues to be a potential source of infection for laboratory animals and, secondarily, for personnel handling these animals. Animal feed containing animal by-products continues to be a source of *Salmonella*, especially if diets consist of raw meal and have not undergone a pelleting process (Stott *et al.*, 1975; Williams *et al.*, 1969). Until animal feeds in the United States and Europe are *Salmonella*-free, laboratory-animal-associated cases of salmonellosis in humans will remain a distinct possibility. The house mouse may also be a reservoir of infection and may play a role in human and animal salmonellosis (Shimi *et al.*, 1979). Endemic salmonellosis in commercially raised guinea pigs has also been a source of infection in personnel working with the animals (Fish *et al.*, 1968).

Both humans and animals are carriers and periodic shedders of *Salmonella*; they may have mild, unrecognized cases or they may be completely asymptomatic. Asymptomatic animals that shed *Salmonella* are particularly important in biomedical research because they are a potential source of infection for other animals, animal technicians, and investigators. The incidence of carrier animals in the colony may vary from a few percent to the majority of animals shedding the organism (Haberman and Williams, 1958); indeed, one investigator suggested that clinically apparent salmonellosis is rare in infected mice (Margard *et al.*, 1963). Dogs, cats, and nonhuman primates may also serve as potential sources of infection in the laboratory. Prevalence data from eight studies conducted in the United States, Europe, and the Sudan indicated that dogs that are culture-positive for *Salmonella* varied from 0.6 to 27.6%. A conservative estimate might set the *Salmonella* prevalence in the United States canine population at 10% (Morse and Duncan, 1975). The incidence of *Salmonella* carriers in newly imported rhesus and cynomolgus monkeys has been found to exceed 20% in some shipments (Fiennes, 1967). A survey of 142 random source cats purchased for a research laboratory found a 10.6% incidence of *Salmonella*-infected animals (Fox and Beaucage, 1979).

Birds and reptiles, especially turtles, are a particularly dangerous source of *Salmonella*; as much as 94% of all reptiles harbor *Salmonella* spp. (Chiodini and Sundberg, 1981). Turtles have received a great deal of zoonotic attention and in 1970 alone may have caused 280,000 human cases of salmonellosis. In the late 1960's, with annual sales of 15 million turtles, zoonotic salmonellosis became a growing problem. In 1972, the United States Food and Drug Administration (FDA) banned importation of turtles and turtle eggs and the interstate shipment of turtles not certified free of *Salmonella* or *Arizona hinshawii* by the state of origin. The effectiveness of this method was variable, and in 1975, the FDA ruled it illegal to sell viable turtle eggs or live turtles with a carapace length less than 10.2 cm, with exceptions made for educational or scientific institutions and marine turtles. Marine turtles have not been shown to be a reservoir of *Salmonella*. There was a 77% decrease in turtle-associated salmonellosis after enactment, indicating the efficacy of the regulation.

The incidence of salmonellosis in humans acquired from lab-oratory animals or vice-versa is unknown. However, those working with laboratory animals, common carriers of *Salmonella*, should be aware that the literature is replete with examples of *Salmonella* infection obtained from pet animals. Also, the widespread use of antibiotics therapeutically and as feed supplements for domestic animals, increases the risk of transmission of R factors from normal enteric flora to enteric pathogens. Direct animal-to-human transmission of *Salmonella* and *Shigella* with R factors has been reported (Fox, 1975).

ii. Clinical signs. The most common clinical sign of salmonellosis in humans is acute gastroenteritis with sudden onset, abdominal pain, diarrhea, nausea, and fever. Diarrhea and anorexia persist for several days. Some cases can lead to febrile septicemia, when organisms invade the bowel wall, even without severe intestinal involvement. In these cases, most clinical signs are attributed to hematogenous spread of the organisms (Robbins, 1974). As with other microbial infections, severity of this disease is related to organism serotype, the number of bacteria ingested, and host susceptibility. Inapparent infections are encountered frequently.

iii. Control and prevention. The control and prevention of salmonellosis in the laboratory should be directed at the rapid detection and treatment of both acute and chronic infections in laboratory animals. Particular attention should be given to animals during the quarantine period. Potential sources of *Salmonella*, such as wild rodents, birds, and feral animals, should be excluded from the laboratory. Animal feed and bedding should be examined as a potential source of infection and pasteurized or autoclaved, if necessary. Finally, consideration should be given to screening animal care personnel for subclinical *Salmonella* infection to prevent the introduction of *Salmonella* into the animal colony from infected workers (Fox and Brayton, 1982).

3. Shigellosis

Shigellosis is a significant zoonotic disease that has frequently been transmitted from nonhuman primates to humans (Mulder, 1971). The disease is caused by gram-negative non-spore-forming bacilli belonging to the genus *Shigella*, with *S. flexneri*, *S. sonnei*, and *S. dysenteriae* being the most common species found in nonhuman priamtes.

i. Reservoir and incidence. Humans are the main reservoir of the disease, which occurs worldwide. Nonhuman primates acquire the disease following capture and subsequent contact with other infected primates or contaminated premises, food, or water. Shigellosis has been reported as the most commonly identified cause of diarrhea in nonhuman primates (Irving, 1974).

ii. Mode of transmission. *Shigella* organisms may be shed from clinically ill as well as asymptomatic humans and nonhuman primates. Transmission to humans occurs by ingestion of fecally contaminated food or water or by direct contact with infected animals. Only minimal contact between humans and nonhuman primates appears necessary for transmission (Fiennes, 1967). Pet monkeys shedding *Shigella* are a particular threat to owners and pet store proprietors who, unless cautious, can contract the disease (Fox, 1975).

iii. Clinical signs and susceptibility. Humans are generally susceptible to shigellosis, with the disease being much more severe in children than in adults. The disease varies from completely asymptomatic to a bacillary dysentery syndrome characterized by blood and mucus in the feces, abdominal cramping, tenesmus, weight loss, and anorexia. More commonly, the disease presents as a clinically mild diarrhea. Fatal shigellosis has been reported in children and adults who have had contact with infected pet or zoo monkeys (Fiennes, 1967; Fox, 1975). Asymptomatic carrier states may persist in recovered individuals. The clinical disease in nonhuman primates is similar to that in humans. However, a high mortality can result from shigellosis in primates.

iv. Diagnosis and control. Shigellosis should be considered in humans or nonhuman primates exhibiting acute diarrhea, especially if blood or mucus is present. A definitive diagnosis requires the isolation of *Shigella* spp. Successful isolation requires the prompt inoculation of fresh rectal swabs onto selective media. Identification can be confirmed by agglutination with polyvalent *Shigella* antisera.

Prevention of shigellosis in the laboratory is directed at the strict quarantine and microbiologic screening for carriers of all newly arrived primates. *Shigella*-infected carriers should be properly treated to ensure elimination of the carrier state. Treatment twice daily for 10 days with a total daily dose of 4.4 mg trimethoprim and 17.6 mg sulfamethoxazole per kilogram of body weight has been shown to be effective in eliminating the *Shigella* carrier state in rhesus monkeys (Pucak *et al.*, 1977).

Nonhuman primates with clinical shigellosis should be isolated, properly treated, and have negative stool cultures before being returned to the animal colony. Every effort should be made to minimize unnecessary human traffic through the primate colony.

4. *Yersinia*

Laboratory animals are susceptible to three *Yersinia* spp. that are potentially zoonotic: *Y. pseudotuberculosis*, *Y. enterocolitica*, and *Y. pestis* (Gillespie and Timoney, 1981). *Yersinia pseudotuberculosis* has been described in numerous animal species and has been encountered as an important spontaneous disease in guinea pigs and nonhuman primates (Ganaway, 1976; Buhles *et al.*, 1981). *Yersinia pseudotuberculosis* and *Y. enterocolitica* produce mesenteric lymphadenitis, appendicitis, and septicemia in humans. Human cases are usually presumed to be related to infected animals; infection can occur through fecally contaminated food or by direct contact with infected animals (Mair, 1969). Direct zoonotic transmission of *Y. pseudotuberculosis* from domestic rabbits has been documented (Splino *et al.*, 1969), but human cases generally have not complemented infections in laboratory animals.

D. Cutaneous Infections

1. Dermatophilosis

Natural disease caused by *Dermatophilus congolensis* has been described in numerous species including horses, cattle, sheep, goats, squirrels, cottontail rabbits, owl monkeys, lizards, and humans (Fox *et al.*, 1973; Anver *et al.*, 1976). Laboratory animals experimentally infected include mice, guinea pigs, and rabbits. Infected animals develop circumscribed areas of alopecia and an exudative dermatitis that progresses to slightly elevated or papillomatous whitish-brown encrustations. Infection in humans is characterized by a pustular desquamative dermatitis and can usually be attributed to direct contact with infected animals. Relapsing dermatophilosis has been reported in an owl monkey after cessation of an apparently appropriate antibiotic regimen (Fox *et al.*, 1973); thus the organism may persist on the pelage of animals after clinical resolution of lesions and may present a continued zoonotic hazard to laboratory personnel.

2. Erysipeloid

Erysipelothrix rhusiopathiae is a pathogen in swine, lambs, calves, poultry, fish and other aquatic species, and both wild and laboratory mice (Gillespie and Timoney, 1981). The organism was discovered by Koch, who called it the "bacillus of mouse septicemia." Septicemia is a feature of the disease in many species. The disease produced by *E. rhusiopathiae* in humans is called erysipeloid and is primarily occupation-related. Erysipeloid usually presents as an inflammatory lesion of the skin with an elevated erythematous edge that spreads circumferentially as discoloration of the central area fades. Septicemia is an infrequent complication in humans, but may develop without preceding cutaneous lesions.

The first report of natural infection in wild animals was of an epizootic among migrating meadow mice and house mice in California. Although the laboratory rodent is susceptible to experimental infection, neither natural disease in laboratory rodents nor zoonotic transmission from them has been reported. Pigs probably represent the most likely source of this organism in the laboratory environment, but other potential sources should not be overlooked.

3. Listeriosis

Listeria monocytogenes has been isolated in association with disease in a wide variety of animals including fish, birds, and mammals (Gray and Killingor, 1966). Ruminants, guinea pigs, rabbits, and chinchillas are the laboratory species most commonly affected. Listeriosis in humans has many manifestations, including cutaneous involvement, coryza, conjunctivitis, metritis with abortion, sepsis, and meningitis. Although primary cutaneous listeriosis appears to be the occupation-related form of the disease most likely to be acquired in the laboratory setting, the potential for more serious listerial infections in the laboratory cannot be discounted.

4. Streptococci

Lancefield serologic group C organisms are the principal streptococci producing disease in laboratory animals, but groups A, B, D, and G also occasionally produce disease in laboratory animals (Williford and Wagner, 1982; Gillespie and Timoney, 1981). Human infections are caused predominantly by the Lancefield group A organism, *Streptococcus pyogenes;* however groups B, C, D, F, and G have also been isolated sporadically from a variety of infections (Facklam, 1980). Infections due to *S. pyogenes,* clinically similar to human impetigo, have been reported in nonhuman primates, but zoonotic transmission of the infection was not mentioned (Boisvert, 1940). *Streptococcus pneumoniae* also causes disease in both animals and humans, but can be isolated in 50–70% of apparently normal humans (Facklam, 1980). Thus, despite this apparent overlap among disease-causing streptococci of animals and humans, zoonotic transmission of disease-causing streptococci appears rare in the laboratory environment.

5. Tularemia (*Francisella tularensis*)

Tularemia in humans is characterized by ulceroglandular or pleuropulmonary disease. Humans usually contract the illness by handling the hides or carcasses of infected animals (Boyce, 1979). Human infections subsequent to a cat bite and a nonhuman primate bite (Nayar *et al.,* 1979) have also been recorded. Ectoparasite vectors are important in natural disease transmission in the sylvatic cycle. Although some rodents, lagomorphs, cats, and dogs are naturally susceptible to infection with *F. tularensis,* the natural infection of these animals in the laboratory and, consequently, the zoonotic transmission of the disease in the laboratory have not been reported.

E. Other

1. Pseudomonas

The zoonotic potential of *Pseudomonas aeruginosa* is difficult to evaluate due to the environmental ubiquity of this normally saprophytic organism. *Pseudomonas aeruginosa* is a potential pathogen of many laboratory animals and is particularly important as an opportunistic pathogen of immunosuppressed animals. Human-to-animal transmission of *P. aeruginosa* has been documented, but conversely, animal-to-human transmission has not been documented. Two outbreaks of *Pseudomonas* infection within a mouse colony were linked to infected animal caretakers (Van der Waaij *et al.,* 1963). The infection was curtailed by transferring the infected personnel who served as reservoirs and by instituting rigid sanitation and hygienic precautions.

2. Staphylococcus

Staphylococcus spp. are ubiquitous in nature and are common inhabitants of the mucocutaneous flora of humans and a variety of animals. *Staphylococcus aureus,* strains of which are differentiated by phage typing, are all potentially pathogenic and frequently cause pyogenic infections in humans and animals. Several studies have indicated that pathogenic *S. aureus,* of human phage types, have emerged as pathogens in specific pathogen-free (SPF), barrier-maintained mouse colonies and in SPF rats and guinea pigs where these organisms were previously absent. The same phage types were isolated from animal caretakers who worked in these areas (Davey, 1962; Blackmore and Francis, 1970; Shults *et al.,* 1973). The lack of an established mucocutaneous flora in these animals may have facilitated the colonization and disease produced by these human phage type staphylococci (Blackmore and Francis, 1970). Other laboratory species have also been shown to harbor pathogenic *S. aureus* of human phage types (Fox *et al.,* 1977; Renquist and Soave, 1969; Rountree *et al.,* 1956). In humans, preexisting nasopharyngeal colonization by normal *S. aureus* strains confers resistance to colonization by virulent strains of *S. aureus* (Eichenwald, 1965). Presumably this phenomenon is also operative in minimizing the zoonotic potential of animal origin *S. aureus* strains.

V. MYCOSES (RINGWORM)

The dermatophytes causing ringworm in animals, including humans, are grouped taxonomically into related fungi that have an affinity for cornified epidermis, hair, horn, nails, and feathers (Rebell and Taplin, 1970). Many—perhaps all—der-

Table II

Dermatophytes in Laboratory and Domestic Animals[a,b]

Species	Cat	Dog	Rodent	Nonhuman primate	Cattle	Sheep	Swine	Avian
Microsporum canis	++++	+++	++	+++	+	++	+	−
Microsporum distorum	+	+	−	+	−	−	−	−
Microsporum audouinii	−	+	−	−	−	−	−	−
Microsporum gallinae	+	+	+	+	−	−	−	++++
Microsporum gypseum	++	+++	+++	++	+	−	+	−
Microsporum nanum	−	−	−	−	−	−	++++	−
Trichophyton simii	−	+	−	+++ (India)	−	−	−	+++ (India)
Trichophyton mentagrophytes	++	+++	++++	+++	++	++	++	−
Trichophyton verrucosum	+	+	−	−	++++	++	+	−
Trichophyton rubrum	−	+	−	+	−	−	−	−
Trichophyton violaceum	+	−	−	−	−	−	−	−
Epidermophyton floccosum	−	+	+	−	−	−	−	−

[a]Adapted from "Dermatophytes" (Rebell and Taplin, 1970).

[b]++++, usual; +++, frequent; ++, occasional; +, reported.

matophytic fungi can digest keratin substrates enzymatically. The species are subcategorized as geophilic, zoophilic, and anthropophilic; geophilic inhabit the soil, zoophilic are primarily parasitic on animals, other than humans; and anthropophilic primarily infect humans. All of the known zoophilic dermatophytes can also produce infection in humans. The dermatophytes have been grouped into three genera, namely, *Microsporum, Trichophyton,* and *Epidermophyton* (Table II).

i. Reservoir and incidence. Dermatophytes are distributed throughout the world, with some species being reported more commonly in certain geographic locations. For example, in a study of small mammals in their natural habitat, *T. mentagrophytes* was isolated from 57 of 1288 animals representing 15 different species (Chmel *et al.,* 1975). In this survey, agricultural workers, exposed to these mammals in granaries and barns, risked contracting *T. mentagrophytes* infection. *Trichophyton mentagrophytes* was isolated from 77% of the 137 agricultural workers infected with ringworm, whereas *T. verrucosum* was isolated from only 23% of the cases. In the same study, of 445 ringworm-infected personnel working with farm animals, 75% were infected with *T. verrucosum* and 28% with *T. mentagrophytes.* Human infection with *T. mentagrophytes* has also followed handling of bags of grain in which mice had been living. Human infection with *M. canis* is common in personnel handling either cats or dogs infected with the dermatophyte. In almost all rodent-associated ringworm infections in humans, *T. mentagrophytes* has been isolated as the etiological agent. Thus, specific exposure to reservoir hosts harboring different dermatophytes determines the type and incidence of infection in humans.

Ringworm infection in laboratory rodents is often asymptomatic and is not recognized until personnel become infected. A prevalence of *T. mentagrophytes* among laboratory mouse stocks as high as 80–90% has been recorded (Davies and Shewell, 1964). During an 8-month period, before infected mice were treated, six of 13 people handling the mice developed ringworm, although less than 1% of the mice showed any clinical signs of the disease. Dipping the animals in an aqueous solution of an acaricide containing a fungicide to remove mites reduced mice carriers from 90 to 21%. Other authors have reported that personnel were infected when handling mice clinically affected with *T. mentagrophytes* infection (Cetin *et al.,* 1965). Dermatophytes are also subclinical in other adult animals, particularly animals from certain kennels, pounds, or catteries (Muller and Kirk, 1976).

ii. Mode of transmission. Transmission of dermatophytes from animals to humans is a well-known and serious public health concern. Transmission from laboratory animals to personnel is often unsuspected, because laboratory animals usually have few visible skin lesions. Transmission occurs via direct or indirect contact with asymptomatic carrier animals or with skin lesions of infected animals, contaminated animal bedding, equipment, or causal fungi present in air, dust, or on surfaces of animal-holding rooms (MacKenzie, 1961). The infection is communicable as long as infected animals are maintained and viable spores persist on contaminated materials. To maintain laboratory animals free of infection, and thus preclude personnel exposure, newly purchased animals should be screened, if practical, for dermatophytes, and the animal facility environment and caging should be sanitized regularly. This preventive practice is not easily achieved with random source dogs and cats, since many of the animals will harbor dermatophytes asymptomatically.

iii. Clinical signs. The disease, dermatomycosis, or ringworm in humans, is often self-limiting, and frequently so mild as to be ignored by the affected person. In general, the dermatophytes cause scaling, erythema, and occasionally vesicles and fissures. Lesions produced in humans by zoophilic and geophilic species are more eczematous or inflammatory and less chronic than those produced by anthropophilic species. The fungi can cause thickening and discoloration of the nails. On the skin of the trunk and extremities, the lesion may consist of one or more circular lesions with a central clearing, forming a ring (Fig. 3) (Mescon and Grots, 1974). Fungal infections in humans are categorized clinically according to location; some examples are tinea capitis, a fungus infection of the scalp and hair; tinea corporis, ringworm of the body; tinea pedis, of the foot; and tinea unguim, of the nails. Human dermatophyte infection acquired from infected laboratory animals usually occurs on the extremities, especially the arm and hand. In rare instances, dermatophytes can produce severe, overwhelming, invasive, or granulomatous disease. For instance, although the zoophilic form of *T. mentagrophytes* is highly inflammatory and often resolves rapidly, the infection may also produce folliculitis, furunculosis, or a widespread tinea corporis.

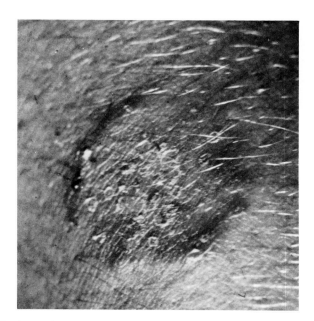

Fig. 3. A circular ringworm lesion on the arm of a man contracted from a rodent infected with *Trichophyton mentagrophytes.* (Courtesy of Dr. W. Kaplan.)

VI. PROTOZOAN DISEASES

A. Systemic Infections

1. Toxoplasmosis

Infection due to *Toxoplasma gondii* is widespread in humans and lower animals (Frenkel, 1973; Levine, 1973). An estimated 500 million humans have been infected with the organism (Kean, 1972).

i. Reservoir and incidence. The life cycle of *T. gondii* consists of definitive and intermediate hosts. *Toxoplasma* infection has radiated throughout the animal kingdom to include hundreds of species of mammals and birds as its intermediate hosts. Mice, rats, hamsters, guinea pigs, and other rodents, rabbits, dogs, sheep, cattle, and nonhuman primates comprise some of the laboratory animals that could serve as intermediate hosts. These laboratory hosts have not proved to be important in zoonotic infection by *T. gondii* in the laboratory environment because the organism is replicating asexually in extraintestinal sites only. Serologic surveys conducted in the United States using the Sabin–Feldman dye test have demonstrated *T. gondii* infection in 30–80% of cats (Ladiges *et al.,* 1982). Presumably, all serologically positive cats have shed *Toxoplasma* oocysts and could reshed organisms by reinfection or by reactivation.

ii. Mode of transmission. Domestic and wild felids develop extraintestinal invasion with *T. gondii* analogous to the nonfelid hosts. In addition, as the definitive hosts in the *T. gondii* life cycle, felines develop intestinal infection with the shedding of oocysts. Thus, the domestic cat predominates as the reservoir for the zoonotic transmission of *T. gondii* in the laboratory environment.

Most postnatally acquired infections in cats are asymptomatic and have a variable prepatent period and pattern of oocyst shedding (Dubey, 1976). The prepatent period can be as brief as 3 days, if the cat has ingested mice or meat containing *T. gondii* cysts, or it can be as long as several weeks if oocysts have been ingested. Shedding of oocysts in the feces occurs for 1–2 weeks, during which cats are considered a public health risk (Ladiges *et al.,* 1982). Oocysts become infectious after sporulation, which occurs in 1–5 days. Oocyst shedding is less likely to occur if the cat was infected by oocysts or tachyzoites than if infection resulted from the ingestion of *Toxoplasma* cysts. Oocyst shedding can be reactivated by induction of hypercortism (Dubey and Frenkel, 1974) or by superinfection with other feline microorganisms such as *Isospora felis* (Chessman, 1972). Oocysts of *T. gondii* have been observed infrequently in the feces of naturally infected cats (Ladiges *et al.*, 1982), and shedding usually precedes the development of antibody titers to *T. gondii*.

iii. The disease in humans. *Toxoplasma* infection in humans is very common, but clinical disease occurs only sporadically and is of low incidence. Congenital infection in humans results in systemic disease, frequently with severe neuropathological changes. Postnatal infection results in disease that is less severe and commonly presents as a generalized lymphadenopathy that may resolve without treatment in a few weeks. Although rare, serious systemic toxoplasmosis can be acquired by older individuals. This is manifested by fever, maculopapular eruption, malaise, myalgia, arthralgia, posterior cervical lymphadenopathy, penumonia, myocarditis, and meningoencephalitis. Ocular toxoplasmosis, usually chorioretinitis, is commonly seen in postnatal infections, but can occur also in infections of older individuals. Clinically severe and progressive illness is most likely to develop in immunocompromised individuals.

iv. Diagnosis and control. Rigorous sanitation should effectively prevent human toxoplasmosis from occurring in the laboratory environment. Since oocysts must sporulate before they are infectious, daily cleaning of litter pans will prevent accumulation of infectious oocysts. Personnel should wear gloves when handling litter pans and wash their hands thoroughly before eating. Pregnant women should completely avoid contact with cat feces.

B. Enteric Infections

1. Amebiasis

Amebiasis is a zoonotic disease caused by the protozoan parasite *Entamoeba histolytica*.

i. Reservoir and incidence. The disease occurs worldwide in humans with a greater prevalence in tropical areas. The parasite is found routinely in the feces of clinically normal monkeys and anthropoid apes, but may cause severe clinical disease in these animals. The reported incidence of *E. histolytica* has ranged from 0 to 31% in rhesus monkeys, 2 to 67% in chimpanzees, and up to 30% in other nonhuman primates (Irving, 1974).

ii. Mode of transmission. *Entamoeba histolytica* exists as either resistant cysts or the more fragile trophozoites. Cysts are the infectious form of the parasite and are usually found in the normal stool of asymptomatic carriers or persons with mild disease (Krogstad, 1982). Cysts may remain viable in moist,

cool conditions for over 12 days and in water for up to 30 days. Epidemics of amebiasis in humans usually result from ingestion of fecally contaminted water containing amebic cysts. Laboratory animal workers handling nonhuman primates are potentially exposed to infections from infected fecal matter transferred to their skin or clothing. The infective cyst forms may be subsequently ingested.

iii. Clinical signs. Most human infections with *E. histolytica* have few or no detectable symptoms (Krogstad, 1982). Clinical signs result when trophozoites invade the large bowel wall causing an amebic colitis. Clinical signs range from a mild watery diarrhea to acute fulminating bloody or mucoid dysentery with fever and chills. The disease may have periods of remission and exacerbation over months to years (Krogstad, 1982). Rarely, extraintestinal amebic abscesses may form in the liver, lung, pericardum, or central nervous system.

iv. Diagnosis and control. The diagnosis of amebiasis requires the microscopic identification of trophozoites or cysts in fresh specimens. The organism must be carefully measured to differentiate it from other nonpathogenic amebae.

Control measures to prevent amebiasis should include strict adherence to sanitation and personal hygiene practices. Water supplies should be protected from fecal contamination, since usual water purification chlorine levels do not destroy the cysts (Benenson, 1981). A chlorine concentration of 10 ppm is necessary to kill amebic cysts (Krogstad, 1982). Cysts may also be killed by heating to 50°C (Irving, 1974). Nonhuman primates should be screened during quarantine to identify carriers of *E. histolytica* and be appropriately treated. Nonhuman primates with acute diarrhea or dysentery should also have stool examined for the presence of *E. histolytica* and be treated as necessary. Metronidazole is the recommended drug for *E. histolytica* infections in both asymptomatic carriers and symptomatic patients (Krogstad, 1982).

2. Balantidiasis

Balantidiasis is a zoonotic disease caused by the large ciliated protozoan *Balantidium coli*.

i. Reservoir and incidence. *Balantidium coli* is distributed worldwide and is common in domestic swine. It may also be found in humans, great apes, and several monkey species. The incidence in nonhuman primate colonies has ranged from 0 to 63%. These infections are usually asymptomatic in most animals, although clinical disease characterized by diarrhea or dysentery may occur (Irving, 1974).

ii. Mode of transmission. Infection results from the ingestion of trophozoites or cysts from the feces of infected animals or humans. Transmission may also occur from ingestion of contaminated water or food.

iii. Clinical signs. Balantidiasis may cause ulcerative colitis characterized by diarrhea or dysentery, tenesmus, nausea, vomiting, and abdominal pain. In severe cases, blood and mucus may be present in the stool. Humans apparently have a high natural resistance and infections are often asymptomatic (Benenson, 1981).

iv. Diagnosis and control. Balantidiasis is diagnosied by the identification of trophozoites or cysts in fresh fecal samples.

Control measures to prevent balantidiasis should be directed at maintaining good sanitation and personal hygiene practices in nonhuman primate and swine colonies. Water supplies should be protected from fecal contamination, especially since usual water chlorination does not destroy cysts (Benenson, 1981). Nonhuman primates exhibiting acute diarrhea should be examined for the presence of *B. coli* organisms in the feces. Positive animals should be isolated and the infection appropriately treated. Tetracyclines, metronidazole, paromomycin, and ampicillin have been used successfully to eliminate *B. coli* infections (Teare and Loomis, 1982).

3. Cryptosporidiosis

Cryptosporidiosis was first described in the mouse (Tyzzer, 1907). The genus *Cryptosporidium* now contains about 11 named species (Levine, 1980), many of which have been incriminated as opportunistic pathogenic parasites.

i. Reservoir and incidence. Cryptosporidiosis has been identified in at least 12 different hosts, including mammals, birds, and reptiles. Although cryptosporidial infection occurs most commonly in the intestinal epithelium, it has been observed also in the stomach of mice and snakes, bursa of Fibricius in chickens, the bile and pancreatic ducts of the rhesus monkey, and the lungs of turkeys (Anderson, 1982). Other laboratory species known to have enteric cryptosporidial infection are pigs, lambs, calves, rabbits, and guinea pigs, but infection in dogs has not been demonstrated. Bovine cryptosporidia from calves can cause infection in newborn pigs, lambs, chicks, mice, rats, and guinea pigs. The relationship between infection with cryptosporidia and clinical disease is equivocal; however, diarrhea has been associated with infection in lambs, calves, pigs, rhesus monkeys, and turkey poults. Severe clinical disease with death occurred in five Arabian foals with combined immunodeficiency complicated by adenovirus and cryptosporidial infection (Snyder *et al.*, 1978).

ii. Transmission. Early studies suggested host specificity for the cryptosporidia, but recent studies have demonstrated infectivity across species lines.

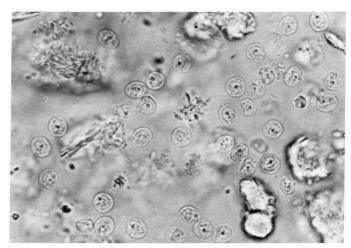

Fig. 4. Cryptosporidial oocysts in an unstained wet mount from calf feces. Note single prominent black dot, which is central or slightly eccentric. Some oocysts are indented. × 1280. (Courtesy of Dr. B. Anderson; reprinted permission of the *Journal of the American Veterinary Medical Association* 178:983, May 1, 1981.)

iii. The disease in humans. Most of the recorded cases of human cryptosporidiosis have occurred in immunodeficient individuals and were regarded as opportunistic endogenous infections (Messel *et al.*, 1976; Lasser *et al.*, 1979; Stemmermann *et al.*, 1980). Disease in these individuals produced low-grade fever, malaise, anorexia, nausea, abdominal cramps, and a protracted watery diarrhea. Repeated intestinal biopsies in a patient have documented endogenous cryptosporidial stages for as long as 1 year; clinical signs also persisted in this patient. These infections have been refractory to antibiotic therapy, and to date no adequate therapeutic regimen is known.

Zoonotic cryptosporidial infection has been suspected in previously healthy individuals. One individual had been performing a survey of *Cryptosporidium* sp. in calves (Reese *et al.*, 1982). In this patient, clinical remission occurred by day 13, and oocysts of cryptosporidium were no longer apparent on fecal flotation (Fig. 4). Experimental transmission studies were conducted in mice and rats to compare this human isolate with calf isolates; indistinguishable infections were produced. This added support to the proposal that cryptosporidiosis should be regarded as a potential zoonosis and that species of *Cryptosporidium* may not be host specific.

4. Giardiasis

Historically, members of the genus *Giardia* have been speciated according to the host of origin. This simplistic taxonomic scheme has recently been challenged, and the taxonomic issue has still to be resolved (Kirkpatrick and Farrell, 1981). Furthermore, studies have clearly refuted the strict host specificity of *Giardia* (Davies and Hilber, 1979; Grant and Woo, 1978).

i. Reservoir and incidence. *Giardia* are found worldwide among all classes of vertebrates and occur among numerous laboratory animals. *Giardia* cysts isolated from a human produced infections when fed to eight different species of test animals, including dogs.

ii. Mode of transmission. Epidemiologic studies and zoonotic transmission have further corroborated the lack of host specificity exhibited by some *Giardia*. Contamination of drinking water with *Giardia* cysts from beavers has been implicated in several human giardiasis outbreaks (Dykes *et al.*, 1980; Keifer *et al.*, 1980), and cysts from this source have infected human volunteers and dogs. Other studies have produced patent infections in dogs using *Giardia* cysts isolated from municipal drinking water known to have infected humans (Dykes *et al.*, 1980; Shaw *et al.*, 1977). Thus, dogs could serve as potential reservoirs of *Giardia* infection in humans. Nonhuman primates also have been implicated. A clinically ill gibbon was presumed to be the source of infection for three zoo attendants and six apes who subsequently developed clinical giardiasis (Armstrong and Hertzog, 1979).

iii. The disease in humans. Giardiasis in humans is characterized by chronic or intermittent diarrhea, steatorrhea, malaise, anorexia, and weight loss. The stool frequently is mucus-laden, light-colored, and soft, but not watery.

iv. Prevention and control. Although many laboratory animals can be infected experimentally with *Giardia* pathogenic for humans, they have not been demonstrated to harbor these organisms naturally. *Giardia* infections of dogs and nonhuman primates possibly represent a greater public health risk and infected animals may warrant treatment. Personnel handling these animals should take appropriate safety measures.

C. Respiratory Infections

Pneumocystis carinii

The etiologic agent of *Pneumocystis* pneumonia (*Pneumocystis carinii*) is a protozoan parasite that had not been propagated *in vitro* until recently (Latorre *et al.*, 1977).

i. Reservoir and incidence. *Pneumocystis carinii* occurs worldwide, and human infections have been recognized in many countries (Hughes, 1981). Laboratory rodents, lagomorphs, nonhuman primates, domestic animals, and a variety of zoo animals have latent infections (Poelma, 1975). *Pneumocystis carinii* (human and animal origin) have identical morphology and tinctoral characteristics and, therefore, currently cannot be differentiated.

ii. Mode of transmission. In post-World War II epidemics of the disease in Europe, the contagious pattern suggested patient-to-patient transmission or acquisition from the institutional habitat (Bommer, 1977), but zoonotic transmission was not implicated. Evidence of the zoonotic potential of *P. carinii* has not been submitted and is unlikely to be forthcoming until our basic understanding of the organism has advanced.

iii. Clinical signs. In both humans and animals clinical disease is seen among infants or individuals with intercurrent debility, poor nutrition, neoplasia, or immunodeficiency. The disease in humans is frequently fatal, and is characterized by a diffuse desquamative alveolitis as the parenchyma fills with an edematous and foamy fluid containing numerous parasites (Hughes, 1981).

VII. HELMINTH INFECTIONS

Many of the helminth parasites common to animals and humans have an indirect life cycle that is interrupted in the laboratory environment, thus precluding cross-infection of animals and humans. Although numerous helminths of laboratory animals should be regarded as zoonoses (Table III) (Soulsby, 1969; Flynn, 1973), the risk of human infection from laboratory-housed animals appears to be negligible. The practices encountered in the properly managed animal facility are not conducive to the transmission of these parasites. Proper quarantine, surveillance, and treatment procedures drastically reduce the endoparasitic burden of laboratory animals. Routine sanitation eliminates most parasitic ova before they have undergone the embryonation necessary for infectivity. Education of personnel on standard hygiene practices further reduces the likelihood of zoonotic infection.

Laboratory-housed nonhuman primates are presumed to be the most likely, albeit infrequent, source of parasitic infection for animal handlers (Irving, 1974; Orihel, 1970). However, a review cited only three cases of zoonotic helminth infections resulting from nonhuman primates. These animals had been kept as pets and not as laboratory subjects (Irving, 1974). Thus, although helminth parasites should be recognized as potentially zoonotic in the laboratory environment, they represent a significantly smaller problem than that posed by possible viral and bacteria zoonoses.

VIII. ARTHROPOD INFESTATIONS

Health hazards to humans due to ectoparasite infestations from arthropods associated with laboratory animals are most

Table III

Zoonotic Helminth Parasites in the Laboratory Environment

Disease	Etiology	Natural host(s)	Aberrant hosts	Comments
Cestodiasis	*Hymenolepis nana*	Rats, mice, hamsters, nonhuman primates	Humans	Intermediate host is not essential to the life cycle of this cestode. Direct infection and internal autoinfection can occur also. Heavy infections result in abdominal distress, enteritis, anal pruritis, anorexia, and headache
Strongyloidiasis	*Strongyloides stercoralis, S. fulleborni*	Nonhuman primates, dogs, cats, humans, Old World nonhuman primates	Humans	Oral and transcutaneous infections can occur in animals and humans. Heavy infections can produce dermatitis, verminous pneumonitis, enteritis. Internal autoinfection can occur
Ternidens infection	*Ternidens deminutus*	Old World primates	Humans	Rare and asymptomatic
Anclystomiasis	*Anclystoma duodenale*	Humans	Nonhuman primates, pigs	Oral and transcutaneous routes of infection occur. Heavy infections produce transient respiratory signs during larval migration followed by anemia due to gastrointestinal blood loss
	Necator americanus	Humans	Nonhuman primates	
Trichostrongylosis	*Trichostrongylus colubriformis. T. axei*	Ruminants, pigs, dogs, rabbits, Old World nonhuman primates	Humans	Heavy infections produce diarrhea
Oesophagostomum	*Oesophagostomum* spp.	Old World primates	Humans	Heavy infections result in anemia. Encapsulated parasitic granulomas are usually an inocuous sequella to infection
Ascariasis	*Ascaris lumbricoides*	Old World primates	Humans	Infection occurs by ingestion of embryonated eggs only. Embryonation, requiring 2 or more weeks, ordinarily would not occur in laboratory. Heavy infections can produce severe respiratory and gastrointestinal tract disease
Enterobiasis	*Enterobias vermicularis*	Humans	Old World primates	Oral and inhalational infection can occur. Disease in humans characterized by perianal pruritis, irritability and disturbed sleep
Trichuriasis	*Trichuris trichiuria*	Humans	Old World primates	Three-week embryonation makes laboratory infection highly unlikely. Heavy infection in humans results in intermittent abdominal pain, bloody stools, diarrhea, and occasionally rectal prolapse
Larval migrans (viscera)	*Toxacara canis*	Dogs and other canids	Humans	Chronic eosinophilic granulomatous lesions distributed throughout various organs. Should not be encountered in laboratory
	Toxacara cati	Cats and other felids	Humans	
	Toxacara leonina	Dogs, cats, wild canids, and felids	Humans	
Larval migrans (cutaneous)	*Anclystoma caninum*	Dogs	Humans	Transcutaneous infection causes a parasitic dermatitis called "creeping eruption"
	Anclystoma braziliense	Dogs, cats	Humans	
	Anclystoma duodenale	Dogs, cats	Humans	
	Uncinaria stenocephala	Dogs, cats	Humans	
	N. americanus	Dogs, cats	Humans	

often mild and limited to manifestations of allergic dermatitis. However, arthropods can serve as vectors to systemic illnesses, such as rickettsialpox and tularemia. Those working with laboratory animals, particularly those species arriving directly from their natural habitat, should be familiar with the arthropods capable of transmitting these diseases.

Mites probably pose the greatest health hazard, not only because they are the most common inhabitant in number and variety of species, but because they also readily transmit agents from almost every major group of pathogens: bacteria, chlamydia, rickettsiae, viruses, protozoa, spirochetes, and helminths (Yunker, 1964) (Table IV). In addition, most of these mites are capable of producing a severe allergic papular dermatitis in humans (Fox and Reed, 1978; Fox, 1982b) (Fig. 5).

Table IV

Ectoparasites[a,b]

Species	Disease in humans	Laboratory host	Agent
Mites			
Obligate skin mites			
Sarcoptes scabiei subspecies	Scabies	Mammals	
Notoedres cati	Mange	Cats, dogs, rabbits	
Nest inhabiting parasites			
Ornithonyssus bacoti	Dermatitis, murine typhus	Rodents and other vertebrates, including birds	WEE,[c] SLE[d] virus *Rickettsia mooseri*
Ornithonyssus bursa	Dermatitis	Birds	WEE, EEE,[e] SLE viruses
Ornithonyssus sylviarum	Dermatitis, encephalitis	Birds	
Dermanyssus gallinae	Dermatitis, encephalitis	Birds	
Allodermanyssus sanguineus	Dermatitis, rickettsialpox	Rodents, particularly *Mus musculus*	*Ricksettsia akari*
Ophionyssus natricis	Dermatitis	Reptiles	
Haemogamasus pontiger	Dermatitis	Rodents, insectivores, straw bedding	
Haemolaelaps casalis	Dermatitis	Birds, mammals, straw, hay	
Eulaelaps stabularis	Dermatitis, tularemia	Small mammals, straw bedding	*Francisella tularensis*
Glycyphagus cadaverum	Dermatitis, psittacosis	Birds	*Chlamydia psittaci*
Acaropsis docta	Dermatitis, psittacosis	Birds	*Chlamydia psittaci*
Trixacarus caviae	Dermatitis	Guinea pigs	
Facultative mites			
Cheyletiella spp.	Dermatitis	Cats, dogs, rabbits, bedding	
Dermatophagoides scheremtewskyi	Dermatitis, urinary infections and pulmonary acariasis	Feathers, animal feed, bird nests	
Eutrombicula spp.	Human pest (chiggers) local pruritis	Chickens, occasional mammals obtained from natural habitat	
Laelaps echidnirus			Potential Argentinian hemorrhagic fever
Ixodids (ticks)			
Rhipicephalus sanguineus	Irritation, RMSF,[f] tularemia, other diseases	Dogs	*Rickettsia rickettsia, F. tularensis*
Dermacentor variabilis	Irritation, RMSF,[f] tularemia tick paralysis, other diseases	Wild rodents, cottontail rabbits, dogs from endemic areas	See above
Dermacentor andersoni	Irritation, Colorado tick fever, Q fever, RMSF,[f] other diseases	Small mammals, uncommon on dogs	See above Ungrouped rhabdoviruses
Dermacentor occidentalis	Irritation, Colorado tick fever, RMSF,[f] tularemia	Small mammals, uncommon on dogs	See above
Amblyomma americanum	Irritation, RMSF,[f] tularemia	Wild rodents, dogs	
Ixodes scapularis	Irritation, possible tularemia	Dogs, wild rodents	
Ixodes spp.			
Ornithodoros spp.	Irritation, relapsing fever	Captive reptiles, wild animals, pigs	*Borrelia recurrentis*
Argas persicus	Irritation, seldom bites humans, but can transmit anthrax, Q fever	Domestic fowl	
Fleas			
Ctenocephalides felis *Ctenocephalides canis* (cat and dog fleas)	Dermatitis, vector of *Hymenolepis diminuta, Dipylidium caninum*	Dogs, cats	
Xenopsylla cheopsis	Dermatitis, plague vector, *H. nana, H. diminuta*	Mouse, rat, wild rodents	
Nasopsyllus fasciatus	Dermatitis, plague vector, *H. nana, H. diminuta* murine typhus	Mouse, rat, wild rodents	
Leptopsylla segnis	*H. diminuta, H. nana,* murine typhus vector	Rat	Harbors salmonella
Echidnophaga gallinacea (stick tight flea)	Potential plague vector	Poultry	
Pulex irritans	Irritation	Domestic animals (esp. pigs) and humans	

[a]Found in laboratory animals which cause allergic dermatitis or from which zoonotic agents have been recovered in nature.
[b]Modified from Yunker (1964).
[c]WEE, western equine encephalitis.
[d]SLE, St. Louis encephalitis.
[e]EEE, eastern equine encephalitis.
[f]RMSF, Rocky Mountain spotted fever.

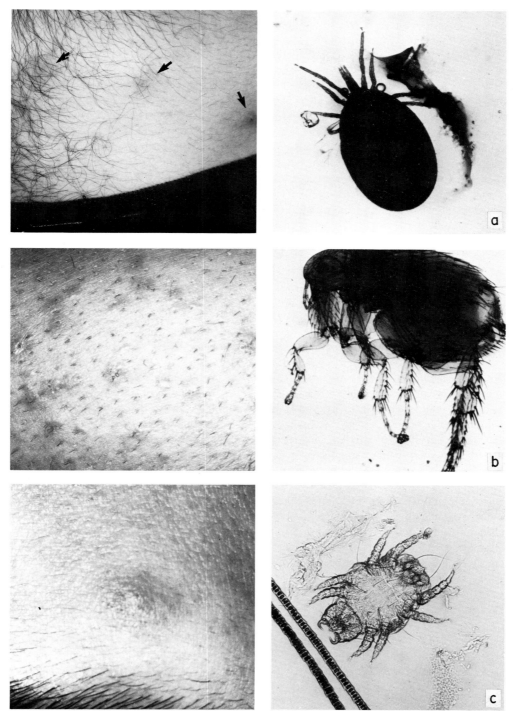

Fig. 5. Maculopapular dermatoses in humans associated with mite and flea bites. (a) Tropical rat mite; (b) flea (courtesy of American College of Laboratory Animal Medicine and Washington State University College of Veterinary Medicine); (c) cheyletiella mite.

Control of mite infestation is basically dependent on their habitats. Some, such as *Sarcoptes* sp. and *Notoedres* sp., are obligate parasites that require treatment of the host. Other mites, such as *Ornithonyssus bacoti,* which live most of the time off the animal, require treatment of the environment with appropriate insecticides.

Ticks, with the exception of those harbored in newly arrived dogs or wild animals brought into the laboratory, are rarely found in the well-managed animal facility. The brown dog tick, *Rhipicephalus sanguineus,* is an exception. It readily infests kennels and vivariums. Ticks, like mites, can transmit a variety of diseases, including Rocky Mountain spotted fever, tick-borne typhus, and others. Control of ticks indoors is aimed primarily at the resting places of the unattached ticks and proper insecticide treatment of newly arrived animals noted for harboring ticks.

Fleas are notorious in their ability to transmit disease to humans, particularly plague and murine typhus. Three rodent fleas, *Xenopsylla cheopis, Nasopsyllus fasciatus,* and *Leptopsylla segnis,* have been found in a high percentage of urban dwellings in certain areas in the United States, and are potential transmitters of disease in the laboratory.

Apparently *X. cheopis* is easily established in animal facilities. At a midwestern United States university, it inhabited rooms housing laboratory mice where, on two separate occasions, fleas bit students (Yunker, 1964). *Leptopsylla segnis,* the mouse and rat flea, bites humans and is a vector for plague and typhus, serious diseases in humans. *Leptopsylla segnis* can also serve as an intermediate host for the rodent tapeworms *Hymenolepis nana* and *H. diminuta,* which can infect humans. The flea's bite can be irritating and cause allergic dermatitis (Fig. 5). The cat flea, *Ctenocephalides felis,* is the most common flea in and around human dwellings in the United States; this flea is capable of experimentally transmitting plague and murine typhus, and therefore the potential exists of transmitting the disease to humans.

Control of fleas consists of treatment of infested areas as well as the primary host; in the case of rodent fleas, the animal facility must be free of feral rodents and their entry to prevent introduction of these arthropods.

IX. BITES AND SCRATCHES

A. Bites

During a 2-year surveillance period (1971–1972), 196,684 animal bite cases were reported from the 15 reporting areas in the United States (Moore *et al.,* 1977). The type of biting animal was reported for 196,117 persons bitten: 84% dogs, 10% cats, 4% rodents, 1% skunks and foxes, and 1% other. By tradition, and public emotion, rabies has been the primary reason for investigating animal bite cases. Rodent bites, especially wild rats, however, present other serious public health hazards, particularly in impoverished areas when feral rodents are plentiful. It is estimated that each year more than 43,000 people in the United States are bitten by rats and mice. About 30,000 bites by other animals, including wild animals, are also estimated (Moore *et al.,* 1977). Other important effects of animal bites that must be considered are pain, anxiety, wound disfigurement, and wound infections caused by bacteria such as *Pasteurella* spp., *Clostridium tetani, Streptobacillus moniliformis,* and *Spirillum minus. Pasteurella multocida* infection in humans due to animal bites, particularly by dogs and cats, occurs frequently. Cat scratches also are important sources of *P. multocida* infection; five other species have been reported to be responsible for *P. multocida* infections in humans: lion, panther, rabbit, rat, and opossum. These data clearly demonstrate that *P. multocida* is part of normal flora of many animals, and this type of zoonosis must be recognized as a common sequel to animal bites. Reported incidences and severity of laboratory animal-associated bites are few, except for published cases of rat bite fever and *Pasteurella* infections (Hubbert and Rosen, 1970). Certainly, bites by certain species of nonhuman primates could pose a threat of herpesvirus B infection. Notification of bite wounds to a physician is warranted. Minimally, the wound should be cleaned thoroughly and an antiseptic applied in the treatment of minor rodent bites. Current tetanus immunization is necessary for personnel working with animals. Bites from animals where exposure to rabies is a possibility must be handled in a prescribed manner by appropriate public health personnel.

B. Rat Bite Fever (Sodoku, Haverhill Fever).

Rat bite fever can be caused by either of two microorganisms: *Streptobacillus moniliformis (Actinomyces muris)* or *Spirillum minus (S. minor, S. morsus muris).*

i. Reservoir and incidence.　These organisms are present in the oral cavity and upper respiratory passages of asymptomatic rodents. Reported incidences of mice as asymptomatic carriers of *Streptobacillus moniliformis* or *Spirillum minus* were not found in an early study. Nearly 50% of the asymptomatic laboratory rats cultured harbored *S. moniliformis* as normal oral flora (Strangeways, 1933). In a more recent study in laboratory Sprague-Dawley rats, *S. moniliformis* was the predominant microorganism isolated from the upper trachea of control animals (Paegle *et al.,* 1976). The lack of reported carrier rates in mice is attributed partly to the usual asymptomatic carrier state and partly to the difficult in isolating the organisms. *Spirillum minus* cannot be cultured *in vitro* and requires inoculation of culture specimens into laboratory animals and subsequent identification of the organism by dark field microscopy. *Streptobacillus moniliformis* grows slowly on artificial media, but only in the presence of 10–20% rabbit or horse serum and when incubated at reduced partial pressures of oxygen (Holmgren and Tunevall, 1970; Rogosa, 1974).

Arkless (1970) and Gilbert *et al.* (1971) described infection caused by the bite of a laboratory mouse and a pet mouse, respectively. The disease is not commonly reported in humans but has been reported in personnel engaged in research involving laboratory rodents, particularly rats (Cole *et al.*, 1969; Gilbert *et al.*, 1971; Holden and MacKay, 1964; Anderson *et al.*, 1983). Historically, however, wild rat bites and subsequent illness have been associated with social conditions of poor sanitation and overcrowding, and almost 50% of all cases have involved children under the age of 12 (Brown and Nunemaker, 1942; Raffin and Freemark, 1979; Richter, 1954; Roughgarden, 1965).

Rat bite fever is not a reportable disease; thus, its incidence, geographic location, or source of infection in humans is difficult to assess. Acute febrile diseases, especially if associated with animal bites, are routinely treated with penicillin or other antibiotics without prior culturing of the bite wound. This therapeutic approach, though successful in aborting cases of potential rat bite fever, does not allow accurate recording of the disease in humans. One would suspect, therefore, because of the high number of rodent bites suffered by humans, that the incidence of rat bite fever is low.

ii. Mode of transmission. The bite of an infected rodent, usually a wild rat but occasionally a laboratory rat or mouse, is the usual source of infection. In some reported cases, infection was attributed to dog, cat, or other animal bites, and rarely, to traumatic injuries unassociated with animal contact (Richter, 1954; Roughgaren, 1965). Outbreaks of febrile illness in humans have been associated with *Streptobacillus*-contaminated milk or food. The disease gained the synonym Haverhill fever after a 1926 outbreak in Haverhill, Massachusetts, attributed to contaminated milk (Place and Sutton, 1934).

iii. Clinical signs, susceptibility, and resistance in humans. Incubation varies from a few hours to 1–3 days in infection with *S. moniliformis* and may range from 1–6 weeks with *S. minus*. Fever is almost always present during the illness, whether it is caused by *S. minus* or *S. moniliformis*. In *S. minus* infection, inflammation often occurs at the site of the bite wound, accompanied by regional lymphadenopathy and the onset of fever. Inflammation and lymphadenopathy are infrequently documented with *S. moniliformis* infection. The fever and local signs may be accompanied by headache, general malaise, myalgia, and chills (Gilbert *et al.*, 1971; Raffin and Freemark, 1979; Anderson *et al.*, 1983). In most cases, a discrete macular rash appears on the extremities, frequently involving the palms and soles; it may become generalized, with pustular or petechial sequelae.

Arthritis has been reported in 50% of the cases of *S. moniliformis* infection and usually affects larger joints. Arthritis occurs less frequently in *S. minus* infection. Prolonged and recurrent joint involvement is noted in untreated patients. Serous to purulent effusion can be recovered from affected joints, with *S. moniliformis* being cultured from the fluid. Complications of the primary infection can result if antibiotic treatment is not instituted early. Pneumonia, hepatitis, pyelonephritis, enteritis, and endocarditis have been reported (McGill *et al.*, 1966). Deaths from *S. moniliformis* endocarditis have occurred, usually in cases with preexisting valvular disease.

C. Cat Scratch Disease

Cat scratch disease (CSD), also known as benign inoculation lymphoreticulosis or nonbacterial regional lymphadenitis, is a nonfatal, usually mild disease whose exact etiology is unknown.

i. Reservoir and incidence. The infectious agent remains unknown despite extensive bacteriological and virological studies. Experimental inoculation of exudate obtained from an affected lymph node, into humans or nonhuman primates, but not lower mammals, results in development of signs and lesions compatible with CSD. Based on controlled studies, in which serological methods have been used (i.e., complement fixation), the suspected cause of the disease falls within the Chlamydiaceae family. The contention that the disease is caused by an atypical *Mycobacterium* has been disproved. Because the etiologic agent is unknown, reported incidence and reservoir data in the reservoir host (or vector), the cat, is not available.

However, in a recent study histopathologic examination of lymph nodes from 39 patients with clinical criteria for CDS revealed pleomorphic gram-negative bacilli in 34 of the 39 nodes. Organisms in lymph node sections exposed to convalescent serum from three patients and to immunoperoxidase stained equally well with all three samples. The authors concluded that the bacilli appear to be the causative agents of CDS (Wear *et al.*, 1983).

ii. Mode of transmission. Of patients with the disease, 90% have a history of exposure to a cat and of these patients, 75% have been either bitten or scratched. Most affected patients are under 20 years of age. In temperate climates, the disease appears seasonally, with 75–80% of the cases being diagnosed between September and February, the highest peak occurring in December.

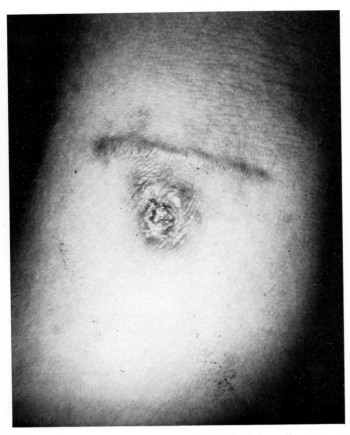

Fig. 6. Cat scratch disease. Ulcerated, circular lesion adjacent to cat scratch. (Courtesy of Dr. J. H. Graham, Armed Forces Institute of Pathology.)

iii. Criteria for diagnosis. If lymphadenitis is present, three of of the four criteria listed below should be fulfilled to diagnose CSD: (1) positive skin-test reaction with an antigen specific for CSD, (2) history of contact with a cat, (3) characteristic histopathologic changes present on involved lymph node biopsy, (4) absence of other disease.

iv. Clinical signs, susceptibility, and resistance in humans. A primary lesion will develop in 50% of the cases about 10 days after a cat bite or scratch; the erythematous pustule will usually persist for 1–2 weeks (Fig. 6). Ten to 14 days after the initial lesion, a regional lymphadenopathy develops in most cases. Lymphadenitis regresses in about 6 weeks, with 30–50% of the nodes becoming suppurative. Of the approximate 65% who develop systemic illness, fever and malaise are the symptoms most often noted. Occasionally observed are generalized lymphadenopathy, thrombocytopenia, encephalitis, osteolytic lesions, and erythema nodosum (Torres *et al.,* 1978). The disease is benign, and most patients recover spontaneously without sequelae within 2 months, although lymphadenopathy can persist up to a year. Controlled trials fail to

support the use of antibiotic therapy for treatment of the disease.

The natural course of CSD with mild or absent fever, few systemic sequelae, and localized lymphadenitis with little or no discomfort probably results in a large number of unrecognized cases (Carithers *et al.,* 1969). Interestingly, the Hager–Rose skin test for CSD is positive in 90% of the patients with the disease and 5% of the unaffected persons, whereas the rate in asymptomatic family members of affected persons is 18% and in veterinary personnel 23% (Warwick, 1967).

X. ALLERGIC SENSITIVITIES

i. Incidence and clinical signs. Allergic skin and respiratory reactions are quite common in personnel working with laboratory animals. Many different mammals have been implicated, including the cat, dog, horse, cow, sheep, goat, pig, rabbit, rat, mouse, hamster, gerbil, and guinea pig (Ohman, 1978). It is probable, however, that a number of other mammalian species will be shown to cause allergic disease in humans. Hypersensitivity reactions to animal allergens are serious occupational health problems. Because of the large number of animals used in biomedical fields, numerous people are constantly exposed to laboratory animal allergens. Hypersensitivity reactions include nasal congestion, rhinorrhea, sneezing, itching of the eyes, angioedema, asthma, and commonly a variety of skin manifestations such as localized uricaria and eczema (atopic dermatitis). Skin wheal and flare reactions were demonstrated in subjects sensitive to rat, mouse, guinea pig, or rabbit when tested with the corresponding animal's pelt extracts (Ohman *et al.,* 1975). Maximal allergenic activity was demonstrated when the skin testing was done using the extract fraction from rat, mouse, and rabbit, which had the electrophoretic mobility of albumin. The fraction of guinea pig extract with maximum allergic activity was of prealbumin mobility.

Levy (1974) studied and quantitated allergic activity of proteins from mice. He found that albumin was the major component of mouse skin extracts and that it was highly allergenic in some patients who were allergic to mice. Furthermore, Siraganian and Sandburg (1979) demonstrated the presence of at least two major allergens in mouse skin, serum, and urine. Further characterization of allergens from urine and animal pelts of inbred laboratory mice identified potent allergens in the mouse urine within the major urinary protein (MUP) complex. This study, which demonstrates cross-reactivity between urinary proteins and antigens in dust from a mouse room, suggests that a possible cause of sensitization in laboratory personnel is the dispersal of urinary protein from litter in mouse cages (Schumacher, 1980). A similar hypothesis may also explain the mode of sensitization from other animal allergens. Allergens have also been found in serums of other animals (horse, cat, dog, rat, guinea pig), saliva (cat, dog, rabbit), urine (rat), and brain tissue (cat) (Ohman, 1978). Also, other antigens in laboratory animal quarters may cause allergic reactions; these include mold spores and proteins in food that might be aerosolized (Patterson, 1964).

ii. Diagnosis. To diagnose occupational allergic disease resulting from working with laboratory animals, one must establish a clinical diagnosis and incriminate the etiologic factors. A careful, detailed history, including patients' complaints as well as clinical symptoms, must be evaluated. The history of the appearance of clinical symptoms concomitant with or following environmental exposure often helps to narrow the number of allergens considered in the differential diagnosis. Nonoccupational exposure to potential allergens must also be considered. The family history of allergy is also important, since atopy predisposes a person to type I allergic reactions.

The physical examination must be thorough and well documented; often a repeat physical examination is helpful if performed when the patient is not having an acute allergic attack. Repeated pulmonary function tests, especially when the lungs are the target organ, are often helpful; radiological examinations are occasionally used. Skin testing with suspected antigens often identifies the hypersensitivity. Skin tests are almost always positive when properly done on a patient who has type I sensitivity to animal dander. Useful laboratory tests include a complete blood count; immunoglobulins and IgE antibody specific to one allergen, as measured by the radio-allergosorbent test (RAST); nasal smears for eosinophilia; and serum precipitants to specific allergens. The *in vitro* RAST, however, is less sensitive and no more specific than the skin test. The direct eosinophil count is another useful laboratory test; it is often elevated in the presence of nasal allergy and is almost always elevated in patients with asthma.

iii. Treatment and prevention After an allergic disorder associated with exposure to laboratory animals has been diagnosed definitely, pharmacological agents are often used to relieve the acute attack. Useful agents include antihistamines, sympathomimetic agents, corticosteroid, and bronchodilators. Some pharmacological agents are somewhat useful on a long-term basis.

Allergen immunotherapy is the systemic administration of etiological antigens in increasing dosages to produce hyposensitivity to animal proteins in the patient. This type of therapy may not be recommended because, in highly sensitive individuals, it can be accompanied by uncomfortable local and systemic reactions. There is also a serious risk of inducing anaphylaxis in the patient (Gupta and Good, 1979). The risk of treatment of patients with animal dander extract, however, is probably no greater than than of treatment with pollen extracts. Newer forms of immunotherapy are still experimental, but some may prove clinically useful. Complete avoidance of the offending antigen is the method of choice for preventing an allergic reaction to certain animal allergens. However, when complete avoidance of the allergen is unfeasible for socioeconomic reasons, such as earning a living, other avenues of treatment and control must be considered (Lutsky and Toshner, 1978). Reduced exposure to the offending allergen is frequently used. Methods include reduction of direct animal contact time, increasing the room ventilation, and using exhaust hoods, filter caps on animal cages, and protective clothing, masks, or respirators when working with laboratory animals.

XI. CONCLUSIONS

With the adoption of modern laboratory animal management, which includes routine disease surveillance, proper sanitary regimens, acceptable personal hygiene, and personnel health monitoring, laboratory animals usually do not present a zoonotic or health hazard. Animal facilities designed to prevent ingress of wild rodents and other vermin help preclude the introduction of animal and human pathogens. Careful attention to design of caging and air-flow dynamics within animal rooms is necessary to minimize exposure to allergens.

ACKNOWLEDGMENTS

The authors wish to acknowledge and thank the following residents at the U.S.A. Medical Research Institute of Infectious Diseases, Frederick, Maryland, for their input and contributions in preparing this chapter: Dr. John C. Donovan, Dr. Jerry P. Jaax, Dr. George A. McNamee, Jr., Dr. William S. Stokes, and Dr. Creighton J. Trahan.

REFERENCES

Abelseth, M. K. (1980). Rabies: Immunization and public health aspects. *Curr. Vet. Ther.* **7**, 1261–1265.

Advisory Committee of Immunization Practices (ACIP). (1981). Immune globulins for the protection against viral hepatitis. *Morbid. Mortal. Week. Rep.* **30**, 423–426.

Ambrus, J. L., Strandstrom, H. V., and Kawinski, W. (1969). "Spontaneous" occurrence of Yaba tumor in a monkey colony. *Experientia* **25**, 64–65.

Anderson, B. C. (1982). Cryptosporidiosis. A review. *J. Am. Vet. Med. Assoc.* **180**, 1455–1457.

Anderson, L. C., Leary, S. L., and Manning, P. J. (1983). Rat-bite fever in animal research laboratory personnel. *Lab. Anim. Sci.* **33**, 292–294.

Anver, M. R., Park, J. S., and Rush, H. G. (1976). Dermatophilosis in the marble lizard (*Calotes mystaceus*). *Lab. Anim. Sci.* **26**, 817–823.

Arita, I., and Henderson, D. A. (1968). Smallpox and monkeypox in nonhuman primates. *Bull. W.H.O.* **39**, 277–283.

Arkless, H. A. (1970). Rat-bite fever at Albert Einstein Medical Center. *Ap. Med. J.* **73**, 49.

Armstrong, C. (1963). Lymphocytic choriomeningitis. *In* "Diseases Transmitted from Animals to Man" (T. G. Hull, ed.), 5th ed., pp. 723–730. Thomas, Springfield, Illinois.

Armstrong, C., and Lillie, R. D. (1934). Experimental lymphocytic choriomeningitis of monkeys and mice produced by a virus encountered in studies of the 1933 St. Louis encephalitis epidemic. *Public Health Rep.* **49**, 1019–1027.

Armstrong, C., Wallace, J. J., and Ross, L. (1940). Lymphocytic choriomeningitis: Gray mice, *Mus musculus*. A reservoir for the infection. *Public Health Rep.* **55**, 1222.

Armstrong, J., and Hertzog, R. E. (1979). Giardiasis in apes and zoo attendants, Kansas City, Missouri. *Vet. Public Health Notes* **1**, 7–8.

Ascher, M. S., Berman, M. A., Ruppanner, R. (1983). Initial clinical and immunological evaluation of a new phase I Q. fever vaccine and skin test in humans. *J. Infect. Dis.* **148**, 214–222.

Babudieri, B. (1958). Animal reservoirs of leptospires. *Ann. N.Y. Acad. Sci.* **70**, 393–413.

Babudieri, B. (1959). Q fever: A zoonosis. *Adv. Vet. Sci.* **5**, 82–182.

Barkin, R. M., Guckian, J. C., and Glosser, J. W. (1974). Infections by *Leptospira ballum*: A laboratory-associated case. *South. Med. J.* **67**, 155–176.

Baum, S. G., Lewis, A. M., Jr., Wallace, P. R., and Huebner, R. J. (1966). Epidemic nonmeningitic lymphocytic-choriomeningitis-virus infection—An outbreak in a population of laboratory personnel. *N. Engl. J. Med.* **17**, 934–936.

Bearcroft, W. G. C., and Jamiesen, W. F. (1958). An outbreak of subcutaneous tumors in rhesus monkeys. *Nature (London)* **182**, 195–196.

Benenson, A. S. (1981). Rabies (hydrophobia, lyssa). *In* "Control of Communicable Diseases in Man" (A. S. Benenson, ed.), 13th ed., pp. 50–52. Am. Public Health Assoc., Washington, D.C.

Bernard, K. W., Parham, G. L., Winkler, W. G., and Helmick, C. G. (1982). Q fever control measures: Recommendations for research facilities using sheep. *Infect. Control* **3**, 6.

Biggar, R. J., Woodall, J. P., Walter, P. D., and Haughie, G. E. (1975). Lymphocytic choriomeningitis outbreak associated with pet hamsters. *J. Am. Vet. Med. Assoc.* **232**, 494–500.

Birnbaum, W., Shenberg, E., and Torten, M. (1972). The influence of maternal antibodies on the epidemiology of leptospiral carrier state in mice. *Am. J. Epidemiol.* **96**, 313–317.

Blackmore, D. K., and Francis, R. A. (1970). The apparent transmission of staphylococci of human origin to laboratory animals. *J. Comp. Pathol.* **80**, 645–651.

Blaser, M. J., and Reller, L. B. (1981). *Campylobacter enteritis. N. Engl. J. Med.* **305**, 1444–1452.

Blaser, M. J., La Force, F. M., Wilson, N. A., and Wang, L. L. (1982). Reservoirs for human campylobacteriosis. *J. Infect. Dis.* **141**, 665–669.

Boak, R. A., Linscott, W. D., and Bodfish, R. E. (1960). A case of *Leptospirosis ballum* in California. *Calif. Med.* **93**, 163–165.

Boisvert, F. L. (1940). Human scarlatinal streptococci in monkey. *J. Bacteriol.* **39**, 727–738.

Bommer, W. (1977). Die interstitielle plasmacellulane *Pneumonie* and *Pneumocystis carinii. Ergeb. Mikrobiol., Immunitaetsforsch. Exp.* **38**, 116.

Boyce, J. M. (1979). *Francisella tularensis. In* "The Principles and Practice of Infectious Disease" (G. L. Mandell, R. G. Douglas, and J. E. Bennett, eds.), pp. 1784–1788. Wiley, New York.

Brenan, J. C., Kalisa-Ruti, J., Steniowoski, M. V., Zanotto, E., Gromyko, A. I., and Arita, I. (1980). Human monkeypox 1970–1979. *Bull. W.H.O.* **58**, 165–182.

Brown, J., Blue, J. C., and Wooley, R. E. (1976). *Brucella canis* infectivity rates in stray and pet dog populations. *Am. J. Public Health* **66**, 889–891.

Brown, R. Z., and Gorman, G. W. (1960). The occurrence of leptospirosis in feral rodents in southwestern Georgia. *Am. J. Public Health* **50**, 682–688.

Brown, T. M., and Nunemaker, J. C. (1942). Rat-bite fever: A review of the American cases with reevaluation of etiology: Report of cases. *Bull. Johns Hopkins Hosp.* **70**, 201–327.

Bruestle, M. E., Golden, J. G., Hall, A., and Bankneider, A. R. (1981). Naturally occurring Yaba tumor in a baboon (*Papio pipi*). *Lab. Anim. Sci.* **31**, 292–294.

Buhles, W. C., Vanderlip, J. E., Russel, S. W., and Alexander, N. H. (1981). *Yersinia pseudotuberculosis* infection: Study of an epizootic in squirrel monkeys. *J. Clin. Microbiol.* **13**, 519–525.

Burgdorfer, W. (1979). The spotted fever-group diseases. *In* "CRC Handbook Series in Zoonoses" (H. J. Steele, ed.), Sect. A, Vol. II. CRC Press, Boca Raton, Florida.

Burgdorfer, W., and Brinton, L. P. (1975). Mechanism of transovarial infection of spotted fever rickettsiae in ticks. *Ann. N.Y. Acad. Sci.* **266**, 61.

Carithers, H. S., Carithers, C. M., and Edwards, R. O., Jr. (1969). Cat-scratch disease. *JAMA, J. Am. Med. Assoc.* **207**, 312–316.

Carmichael, L. E. (1979). Brucellosis (*Brucella canis*). *In* "CRC Handbook Series in Zoonoses" (H. J. Steele, ed.), Sect. A, Vols. I-II, pp. 185–194. CRC Press, Boca Raton, Florida.

Casey, H. W., Woodcruft, J. M., and Butcher, W. I. (1967). Electron microscopy of a benign epidermal pox disease of rhesus monkeys. *Am. J. Pathol.* **49**, 431–447.

Center for Disease Control (CDC) (1965). "Zoonoses Surveillance: Leptospirosis," Rep. No. 7, p. 25. U.S. Public Health Serv., Atlanta, Georgia.

Center for Disease Control (1971). Review of nonhuman primate associated hepatitis. *In* "Primate Zoonoses Surveillance," Rep. No. 6. CDC, Atlanta, Georgia.

Cespen, R. (1975). Relevance of some poxvirus infection in monkeys to smallpox eradication. *Trans. R. Soc. Trop. Med. Hyg.* **69**, 299–302.

Cetin, E. T., Tahsinoglu, M., and Volkan, S. (1965). Epizootic of *Trichophyton mentagrophytes* (*interdigitale*) in white mice. *Pathol. Microbiol.* **28**, 839–846.

Chessman, B. S. (1972). Reactivation of toxoplasm oocyst production in the cat by infection with *Isospora felis. Br. Vet. J.* **128**, 33–36.

Chiodini, R. J., and Sundberg, J. P. (1981). *Am. J. Epidemiol.* **113**, 494.

Chmel, L., Buchvald, L., and Valentova, M. (1975). Spread of *Trichophyton mentagrophytes* Var. Gran. infection to man. *Int. J. Dermatol.* **14**, 269–272.

Christie, A. B., Chen, T. H., and Elberg, S. S. (1980). Plague in camels and goats: Their role in human epidemics. *J. Infect. Dis.* **141**, 724–726.

Cole, J. S., Stroll, R. W., and Bulger, R. J. (1969). Rat-bite fever. *Ann. Intern. Med.* **71**, 979.

Collins, M. J., and Parker, J. C. (1972). Murine virus contaminants of leukemia viruses and transplantable tumors. *J. Natl. Cancer Inst. (U.S.)* **49**, 1139–1143.

Crandel, R. A., Casey, H. W., and Brumlow, W. B. (1969). Studies of a newly recognized poxvirus of monkeys. *J. Infect. Dis.* **119**, 80–88.

Criep, L. H. (1976). "Allergy and Clinical Immunology." Grune & Stratton, New York.

Currier, R. W., Rarthel, W. F., Martin, R. J., and Potter, M. E. (1982). Canine brucellosis. *J. Am. Vet. Med. Assoc.* **180**, 132–133.

Dalldorf, G., Tungeblut, C. W., and Umphlet, M. D. (1946). Multiple cases of choriomeningitis in an apartment harboring infected mice. *JAMA, J. Am. Med. Assoc.* **193**, 12.

Davey, D. G. (1962). The use of pathogen free animals. *Proc. R. Soc. Med.* **55**, 256–262.

Davidson, W. L., and Hummeler, K. (1960). B-virus infection in man. *Ann. N.Y. Acad. Sci.* **85**, 970–979.

Davies, R. B., and Hilber, C. P. (1979). Animal reservoirs and cross-species transmission of *Giardia*. *In* "Waterborne Transmission of Giardiasis" (W. Jakubowski and J. Hoff, eds.), pp. 104–125. U.S. Environ. Prot. Agency, Cincinnati, Ohio.

Davies, R. R., and Shewell, J. (1964). Control of mouse ringworm. *Nature (London)* **202**, 406–407.

Deinhardt, T. F. (1976). Hepatitis in primates. *Adv. Virus Res.* **20**, 113–157.

Dove, W. E., and Shelmire, B. (1932). Some observations on tropical rat mites and endemic typhus. *J. Parasitol.* **18**, 159.

Dubey, J. P. (1976). A review of sarcocystis of domestic animals and of other coccidia of cats and dogs. *J. Am. Vet. Med. Assoc.* **169**, 1061–1078.

Dubey, J. P., and Frenkel, J. K. (1974). Immunity to feline toxoplasmosis: Modification by administration of corticosteroids. *Vet. Pathol.* **11**, 350–379.

Ducan, P. R., Thomas, A. E., and Tobin, J. O. (1951). Lymphocytic choriomeningitis. Review of ten cases. *Lancet* **1**, 956–959.

Dykes, A. C., Juranek, D. D., Lorenz, R. A., Sinclair, S., Jakuboski, W., and Davies, R. (1980). Municipal water-borne giardiasis: An epidemiologic investigation. *Ann. Intern. Med.* **92**, 165–170.

Eichenwald, H. F. (1965). Bacterial interference and staphylococcal colonization in infants and adults. *Ann. N.Y. Acad. Sci.* **128**, 365–380.

Enright, J. B., and Sadler, W. W. (1954). Presence in human sera of complement fixing antibodies to virus of sporadic bovine encephalomyelitis. *Proc. Soc. Exp. Biol. Med.* **85**, 466–468.

Espana, C. (1971). Review of some outbreaks of viral disease in captive nonhuman primates. *Lab. Anim. Sci,* **21**, 1023–1031.

Facklam, R. R. (1980). Streptococci and aerococci. *In* "Manual of Clinical Microbiology" (E. H. Lennette, ed.), 3rd ed., pp. 88–90. Am. Soc. Microbiol., Washington, D.C.

Faine, S. (1963). Antibody in the renal tubules and urine of mice. *Aust. J. Exp. Biol. Med. Sci.* **41**, 811–814.

Fiennes, R., N. T-W (1967). "Zoonoses of Primates," pp. 103–113. Cornell Univ. Press, Ithaca, New York.

Fish, N. A., Fletch, A. L., and Butler, W. E. (1968). Family outbreak of salmonellosis due to contact with guinea pigs. *Can. Med. Assoc. J.* **99**, 418–420.

Flores-Castro, R., and Carmichael, L. E. (1980). Canine brucellosis. *Curr. Vet. Ther.* **7**, 1303–1305.

Flynn, R. J. (1973). "Parasites of Laboratory Animals." Iowa State Univ. Press, Ames.

Fox, J. G. (1975). Transmissible drug resistance in shigella and salmonella isolated from pet monkeys and their owners. *J. Med. Primatol.* **4**, 164–171.

Fox, J. G. (1982a). Campylobacteriosis—A "new" disease in laboratory animals. *Lab. Anim. Sci.* **32**, 625–637.

Fox, J. G. (1982b). Outbreak of tropical rat mite dermatitis in laboratory personnel. *Arch. Dermatol.* **118**, 676–678.

Fox, J. G., and Beaucage, C. M. (1979). The incidence of *Salmonella* in random-source cats purchased for research. *J. Infect. Dis.* **139**, 362–365.

Fox, J. G., and Brayton, J. B. (1982). Zoonoses and other human health hazards. *In* "The Mouse in Biomedical Research" (H. L. Foster, J. D. Small, and J. G. Fox, eds.), Vol. 2, pp. 404–411. Academic Press, New York.

Fox, J. G., and Reed, C. (1978). *Cheyletiella* infestation of cats and their owners. *Arch. Dermatol.* **117**, 1233–1234.

Fox, J. G., Campbell, L. H., Reed, C., Snyder, S. B., and Soave, O. A. (1973). Dermatophilosis (cutaneous streptothricosis) in owl monkeys. *J. Am. Vet. Med. Assoc.* **163**, 642–644.

Fox, J. G., Niemi, S. M. Murphy, J. C., and Quimby, F. W. (1977). Ulcerative dermatitis in the rat. *Lab. Anim. Sci.* **27**, 671–678.

Frederickson, L. E., and Barton, C. E. (1974). A serologic survey for canine brucellosis in a metropolitan area. *J. Am. Vet. Med. Assoc.* **165**, 987–989.

Frenkel, J. K. (1973). Toxoplasmosis: Parasite life cycle, pathology and immunology. *In* "The Coccidia" (D. M. Hammond and P. C. Long, eds.), pp. 343–410. University Park Press, Baltimore, Maryland.

Freidmann, C. T. H., Spiegel, E. L., Aaron, E., and McIntyre, R. (1973). *Leptospirosis ballum* contracted from pet mice. *Calif. Med.* **118**, 51–52.

Galton, M. M., Menges, R. W., Shotts, E. B., Nahmias, A. J., and Heath, C. (1962). "Leptospirosis," Publ. No. 951. U.S. Department of Health, Education, and Welfare, Public Health Service, Center for Disease Control, Atlanta, Georgia.

Ganaway, J. R. (1976). Bacterial mycoplasmal and rickettsial diseases. *In* "The Biology of the Guinea Pig" (J. E. Wagner and P. J. Manning, eds.), pp. 121–135. Academic Press, New York.

Geller, E. H. (1979). Health hazards for man. *In* "The Laboratory Rat" (H. J. Baker, J. R. Lindsey, and S. H. Weisbroth, eds.), Vol. 1, pp. 402–407. Academic Press, New York.

Gilbert, G. L., Caddidy, J. F., and Bennett, N. M. (1971). Rat-bite fever. *Med. J. Aust.* **2**, 1131–1134.

Gillespie, R. H., and Timoney, J. F. (1981). "Hagan and Bruner's Infectious Diseases of Domestic Animals," 7th ed., pp. 170–181, 758–772. Cornell Univ. Press, Ithaca, New York.

Grace, J. T. (1963). Experimental studies of human tumors. *Fed. Proc., Fed. Am., Soc. Exp. Biol.*, **21**, 32–36.

Grant, D. R., and Woo, P. T. K. (1978). Comparative studies of *Giardia* spp. in small mammals in southern Ontario. II. Host specificity and infectivity of stored cysts. *Can. J. Zool.* **56**, 1360–1366.

Gray, M. L., and Killingor, A. H. (1966). *Listeria monocytogenes* and listeric infections. *Bacteriol. Rev.* **30**, 309–382.

Greenberg, M., Pellitteri, O. J., and Jellison, W. L. (1947). Rickettsialpox, a newly recognized rickettsial disease. III. Epidemiology. *Am. J. Public Health* **37**, 860.

Gupta, S., and Good, R. A., eds. (1979). "Cellular, Molecular and Clinical Aspects of Allergic Disorders." Plenum, New York.

Haberman, R. T., and Williams, F. P. (1958). Salmonellosis in laboratory animals. *J. Natl. Cancer Inst. (U.S.)* **20**, 933–941.

Hall, A. S., and McNulty, W. P. (1967). A contagious pox disease of monkeys. *J. Am. Vet. Med. Assoc.* **151**, 833–838.

Hall, W. C., Kovatch, R. M., Herman, P. H., and Fox, J. G. (1971). Pathology of measles in rhesus monkeys. *Vet. Pathol.* **8**, 307–319.

Hanson, L. E. (1982). Leptospirosis in domestic animals: The public health perspective. *J. Am. Vet. Med. Assoc.* **181**, 1505–1509.

Hattwick, M. A. W. (1982). Rabies. *In* "Textbook of Medicine" (J. B. Wyngaarden and L. H. Smith, Jr., eds.), 16th ed., pp. 2097–2100. Saunders, Philadelphia, Pennsylvania.

Heath, C. W., Jr., and Alexander, A. D. (1970). Leptospirosis. *In* "Joseph Brennerman's Practice of Pediatrics" (U. C. Kelley, ed.), Vol. 2, Chapter 26. Harper, New York.

Higa, H. H., and Fujinaka, I. T. (1976). Prevalence of rodent and mongoose leptospirosis on the island of Oahu. *Public Health Rep.* **91**, 171–177.

Hinman, A. R., Fraser, D. W., Douglas, R. D., Bowen, G. S., Kraus, A. L., Winkler, W. G., and Rhodes, W. W. (1975). Outbreak of lymphocytic choriomeningitis virus infections in medical center personnel. *Am. J. Epidemiol.* **101**, 103–110.

Holden, F. A., and MacKay, J. C. (1964). Rate-bite fever—An occupational hazard. *Can. Med. Assoc. J.* **91**, 78.

Holmgren, E. B., and Tunevall, G. (1970). Case report: Rat-bite fever. *Scand. J. Infect. Dis.* **2**, 71.

Hotchin, J. (1971). The contamination of laboratory animals with lymphocytic choriomeningitis virus. *Am. J. Pathol.* **64**, 747–769.

Hotchin, J. E., and Benson, L. M. (1973). Lymphocytic choriomeningitis. *In* "Infectious Diseases of Wild Mammals" (J. W. David, L. N. Karstad, and D. O. Trainer, eds.), pp. 153–165. Iowa State Univ. Press, Ames.

Hruska, J. F. (1979). Yellow fever virus. *In* "Principles and Practice of Infectious Disease" (G. L. Mandell, R. G. Douglas, and J. E. Bennett, eds.), pp. 1253–1257. Wiley, New York.

Hubbert, W. T., and Rosen, M. N. (1970). *Pasteurella multocida* infection due to animal bite. *Am. J. Public Health* **60**, 1103–1117.

Hughes, W. T. (1981). *Pneumocystis carinii. In* "Principles and Practice of Infectious Disease" (G. L. Mandell, R. G. Douglas, and J. E. Bennett, eds.), pp. 2137–2142. Wiley, New York.

Hull, R. N. (1971). B-virus vaccine. *Lab. Anim. Sci.* **21**, 1068.

Hunt, R. D., Carlton, W. W., and King, N. W. (1978). Viral diseases. *In* "Pathology of Laboratory Animals" (K. Benirschke, F. M. Garner, and T. C. Jones, eds.), p. 1313. Springer-Verlag, New York.

Hunt, R. D., and Meléndez, L. V. (1969). Herpes virus infections of nonhuman primates: A review. *Lab. Anim. Care* **19**, 221–234.

Inada, R., Ido, Y., Hoki, R., Kaneko, R., and Ito, H. (1916). The etiology mode of infection and specific therapy of Weil's disease. *J. Exp. Med.* **23**, 377.

Irving, G. W. (1974). "Selected Topics in Laboratory Animal Medicine," Vol. 14, Aeromed. Rev. 2-74. USAF School of Aerospace Medicine, Brooks AFB, Texas.

Johnson, K. M. (1982). Viral hemorrhagic fevers. *In* "Textbook of Medicine" (J. B. Wyngaarden and L. H. Smith, Jr., eds.), 16th ed., pp. 1686–1695. Saunders, Philadelphia, Pennsylvania.

Kaufman, A. F., Mann, J. M., Gardiner, T. M., Heaton, F. M., Poland, J. D., Barnes, A. M., and Maupin, G. O. (1981). Public health implications of plague in domestic cats. *J. Am. Vet. Med. Assoc.* **179**, 875–878.

Kawamata, J., Yamanouchi, T., and Lee, H. W. (1980). Outbreaks of "epidemic hemorrhagic fever" in animal laboratories in Japan. *UCLAS Symp. 7th, 1979* pp. 235–238.

Kean, B. H. (1972). Clinical toxoplasmosis—50 years. *Trans. R. Soc. Trop. Med. Hyg.* **66**, 549–571.

Keenan, K. P., Buhles, W. C., Huxsoll, D. L., Williams, R. G., and Hildebrandt, P. K. (1977). Studies on the pathogenesis of *Rickettsia rickettsii* in the dog: Clinical and clinicopathologic changes of experimental infection. *Am. J. Vet. Res.* **38**, 851–856.

Keifer, A., Lynch, G., and Conwill, D. (1980). Water-borne giardiasis—California, Colorado, Pennsylvania, Oregon. *Morbid. Mortal. Week. Rep.* **29**, 121–123.

Kirkpatrick, C. E., and Farrell, J. P. (1982). Giardiasis. *Comp. Cont. Ed. Prac. Vet.* **4**, 367–377.

Krogstad, D. J. (1982). Amebiasis. *In* "Textbook of Medicine" (J. B. Wyngaarden and L. H. Smith, Jr., eds.), 16th ed., pp. 1736–1739. Saunders, Philadelphia, Pennsylvania.

Krusakh, E. H. (1970). Application of preventive health measures to curtail chimpanzee-associated infectious hepatitis in handlers. *Lab. Anim. Care* **20**, 52–56.

Kupper, J. L., Casey, H. W., and Johnson, D. K. (1970). Experimental Yaba and benign epidermal monkeypox in rhesus monkeys. *Lab. Anim. Care* **20**, 979–988.

Ladiges, W. C., DiGiacormo, R. F., and Yamaguehi, R. A. (1982). Prevalence of *Toxoplasma gondii* antibodies and oocysts in pound-source cats. *J. Am. Vet. Med. Assoc.* **180**, 1334–1335.

Lasser, K. H., Lewin, K. J., and Ryning, F. W. (1979). Cryptosporidial enteritis in a patient with congenital hypogammaglobulinemia. *Hum. Pathol.* **10**, 234–240.

Latorre, C. R., Sulzer, A. J., and Norman, L. G. (1977). Serial propagation of *Pneumocystis carinii* in cell line cultures. *Appl. Environ. Microbiol.* **33**, 1204.

Lee, H. W., Lee, P. W., and Johnson, K. M. (1978). Isolation of the etiologic agent of Korean hemorrhagic fever. *J. Infect. Dis.* **137**, 298–308.

Lee, P. W. (1982). New haemorrhagic fever with renal syndrome-related virus in indigenous wild rodents in the United States. *Lancet* **2**, 1404.

Lehmann-Grube, F. (1971). Lymphocytic choriomeningitis virus. *Virol. Monogr.* **10**, 88–118.

Levine, N. D. (1973). "Protozoan Parasites of Domestic Animals and of Man," 2nd ed. Burgess, Minneapolis, Minnesota.

Levine, N. D. (1980). Some corrections of coccidian (Ampicomplexa: Protozoa) nomenclature. *J. Parasitol.* **66**, 830–834.

Levy, D. A. (1974). Allergic activity of proteins from mice. *Int. Arch. Allergy Appl. Immunol.* **49**, 219–221.

Lindorfer, R. K., and Syverton, J. T. (1953). The characterization of an unidentified virus found in association with line 1 leukemia. *Proc. Am. Assoc. Cancer Res.* **1**, 33–34.

Loosli, R. (1967). Zoonoses in common laboratory animals. *In* "Husbandry of Laboratory Animals" (M. L. Conalty, ed.), Sect. 4, pp. 307–325. Academic Press, New York.

Lutsky, I., and Toshner, D. (1978). A review of allergic respiratory disease in laboratory animal workers. *Lab. Anim. Sci.* **28**, 751–756.

MacCallum, R. O. (1949). The virus of lymphocytic choriomeningitis (LCM) as a cause of benign aseptic meningitis. Laboratory diagnoses of five cases. *Mon. Bull. Minist. Health Public Health Lab. Serv. (G.B.)* **8**, 1771.

McGill, R. C., Martin, A. M., and Edmunds, P. N. (1966). Rat bite fever due to *Streptobacillus moniliformis. Br. Med. J.* **1**, 1213–1214.

MacKenzie, D. W. R. (1961). *Trichophyton mentagrophytes* in mice: Infections of humans and incidence amongst laboratory animals. *Sabouraudia* **1**, 178–182.

McNulty, W. P. (1968). A pox disease of monkeys transmitted to man—Clinical and histologic features *Arch. Dermatol.* **97**, 286–293.

Maetz, H. M., Sellers, C. A., Bailey, W. C., and Hardy, G. E., Jr. (1976). Lymphocytic choriomeningitis from pet hamster exposure: A local public health experience. *Am. J. Public Health* **66**, 1082–1085.

Mailloux, M. (1975). Leptospirosis—zoonoses. *Int. J. Zoonoses* **2**, 45–54.

Mair, N. S. (1969). Pseudotuberculosis in free-living wild animals. *Symp. Zool. Soc. London* **24**, 107–117.

Margard, W. L., Peters, A. C., Dorko, N., Litchfield, J. H., and Davidson, R. S. (1963). Salmonellosis in mice—Diagnostic procedures. *Lab. Anim. Care* **13**, 144–165.

Martini, G. A., and Siegert, R., eds. (1971). "Marburg Virus Disease." Springer-Verlag, Berlin and New York.

Maurer, F. D. (1964). Lymphocytic choriomeningitis. *Lab. Anim. Care* **14**, 414–419.

Mescon, H., and Grots, I. A. (1974). The skin. *In* "Pathologic Basis of Disease" (S. L. Robbins, ed.), pp. 1374–1419. Saunders, Philadelphia, Pennsylvania.

Messel, J. L., Perem, D. R., Meligno, C., and Rubin, C. E. (1976). Overwhelming watery diarrhea associated with *Cryptosporidium* in an immunosuppressed patient. *Gastroenterology* **70**, 1156–1160.

Meyer, K. F. (1942). The ecology of psittacosis and ornithosis. *Medicine (Baltimore)* **21**, 175–206.

Moore, R. M., Jr., Zehmer, B. R., Moultrop, J. I., and Parker, R. L. (1977). Surveillance of animal-bite cases in the United States, 1971–1972. *Arch. Environ. Health* **32**, 267–270.

Morris, R., and Spink, W. W. (1969). Epidemic canine brucellosis due to a new species, *Brucella canis. Lancet* **2**, 1000–1002.

Morse, E. V., and Duncan, M. A. (1975). Canine salmonellosis: Prevalence, epizootiology, signs, and public health significance. *J. Am. Vet. Med. Assoc.* **167**, 817–820.

Muchmore, E. (1975). Health program for people in close contact with laboratory primates. *Cancer Res. Safety Monogr.* **2**, 81–99.

Mulder, J. B. (1971). Shigellosis in nonhuman primates: A review. *Lab. Anim. Sci.* **21**, 734–738.

Muller, G. H., and Kirk, R. W. (1976). "Small Animal Dermatology." Saunders, Philadelphia, Pennsylvania.

Mumford, R. S., Weaver, R. D., Patton, C., Feely, J. C., and Feldman, R. H. (1975). Human disease caused by *Brucella canis:* A clinical and epidemiological study of two cases. *JAMA, J. Am. Med. Assoc.* **231**, 1267–1269.

Nayar, G. P. S., Crawshaw, G. J., and Neufeld, J. L. (1979). Tularemia in a group of nonhuman primates. *J. Am. Vet. Med. Assoc.* **179**, 962–963.

Newcomer, C. E., Anver, M. R., Simmons, J. L., and Nace, G. (1982). Spontaneous and experimental infections of the African clawed frog (*Xenopus laevis*) with *Chlamydia psittaci. Lab. Anim. Sci.* **32**, 6.

Nichols, E., Ridge, M. E., and Russel, G. G. (1953). The relationships of the house mouse and the house mite (*Allodermanyssus sanguineus*) to the spread of rickettsialpox. *Ann. Intern. Med.* **39**, 92–101.

Niven, S. J. F. (1961). Subcutaneous "growths" in monkeys by a pox virus. *J. Pathol. Bacteriol.* **81**, 1–14.

Noble, J., Jr. (1970). A study of New and Old World monkeys to determine the likelihood of a simian reservoir of smallpox. *Bull. W.H.O.* **42**, 509–514.

Noguchi, H. (1918). Morphological characteristics and nomenclature of *Leptospira* (Spirochaeta) icterohemorrhagiae (Inada and Ido). *J. Exp. Med.* **27**, 575–592.

Ohman, J. L. (1978). Allergy in man caused by exposure to mammals. *J. Am. Vet. Med. Assoc.* **172**, 1403–1406.

Ohman, J. L., Lowell, F. C., and Bloch, K. J. (1975). Allergens of mammalian origin. *J. Allergy Clin. Immunol.* **55**, 16–24.

Orihel, T. C. (1970). The helminth parasites of nonhuman primates and man. *Lab. Anim. Care* **20**, 395–401.

Page, L. A., and Smith, P. C. (1974). Placentitis and abortion in cattle inoculated with chlamydiae isolated from aborted human placental tissue. *Proc. Soc. Exp. Biol. Med.* **146**, 269–275.

Parker, J. C., Igel, H. D., Reynolds, R. K., Lewis, A. M., Jr., and Rowe, W. P. (1976). Lymphocytic choriomeningitis virus infection in fetal, newborn, and young adult Syrian hamsters. *Infect. Immun.* **13**, 967–981.

Patterson, R. (1964). The problem of allergy to laboratory animals. *Lab. Anim. Care* **14**, 466–469.

Place, E. H., and Sutton, L. E. (1934). Erythema arthriticum epidemicum (Haverhill fever). *Arch. Intern. Med.* **54**, 659–684.

Poelma, F. G. (1975). *Pneumocystis carinii* infections in zoo animals. *Z. Parasitenkd.* **46**, 61–68.

Poiley, S. M. (1970). A survey of indigenous murine viruses in a variety of production and research animal facilities. *Lab. Anim. Care* **20**, 643–650.

Poland, J. D., and Barnes, A. M. (1979). *In* "CRC Handbook Series in Zoonoses and Mycotic Diseases" (J. H. Steel, ed.), Sect. A; Vol. I, pp. 517–597. CRC Press, Boca Raton, Florida.

Prescott, J. F., and Munroe, D. C. (1982). *Campylobacter jejuni* enteritis in man and domestic animals. *J. Am. Vet. Med. Assoc.* **181**, 1524–1530.

Pucak, G. J., Orecutt, R. P., Judge, R. J., and Rendon, R. (1977). Elimination of the *Shigella* carrier state in rhesus monkeys (*Macacca mulatta*) by trimethoprim–sulfamethoxazole. *J. Med. Primatol.* **6**, 127–132.

Quist, K. D. (1974). Hemorrhagic fevers. *In* "Handbook of Laboratory Animal Science" (E. C. Melby, Jr. and N. H. Altman, eds.), Vol. II, pp. 263–264. CRC Press, Inc., Cleveland, Ohio.

Raffin, B. J., and Freemark, M. (1979). Streptobacillary rat-bite fever; a pediatric problem. *Pediatrics* **64**, 214–217.

Rebell, G., and Taplin, D. (1970). "Dermatophytes: Their Recognition and Identification." Univ. of Miami Press, Miami, Florida.

Reese, N. C., Current, W. L., Ernst, J. V., and Barley, W. S. (1982). Cryptosporidiosis of man and calf: A case report and results of experimental infections in mice and rats. *Am. J. Trop. Med. Hyg.* **31**, 226–229.

Renquist, D., and Soave, O. (1969). Staphylococcal pneumonia in a laboratory rabbit: An epidemiologic follow-up study. *J. Am. Vet. Med. Assoc.* **155**, 1221–1223.

Richter, C. P. (1954). Incidence of rat-bites and rat-bite fever in Baltimore. *JAMA, J. Am. Med. Assoc.* **128**, 324–326.

Robbins, S. L. (1974). Infectious disease. *In* "Pathologic Basis of Disease" (S. L. Robbins, ed.), pp. 396–400. Saunders, Philadelphia, Pennsylvania.

Rogosa, M. (1974). *Streptobacillus moniliformis* and *Spirillum minor. In* "Manual of Clinical Microbiology" (E. H. Lennette, E. H. Spaulding, and J. P. Truant, eds.), 2nd ed., pp. 326–332. Am. Soc. Microbiol., Washington, D.C.

Rollag, O. J., Skeels, M. R., Nims, L. J., Thilsted, J. P., and Mann, J. M. (1981). Feline plague in New Mexico: Report of five cases. *J. Am. Vet. Med. Assoc.* **179**, 1381–1383.

Rondel, C. J. M., and Sayeed, K. A. R. (1972). Studies on monkeypox virus. *Bull. W. H. O.* **46**, 577–583.

Roughgarden, J. W. (1965). Antimicrobial therapy of rat-bite fever. *Arch. Intern. Med.* **116**, 39–54.

Rountree, P. M., Freeman, B. M., and Johnston, K. G. (1956). Nasal carriers of *Staphylococcus aureus* by various domestic and laboratory animals. *J. Pathol. Bacteriol.* **72**, 319–321.

Ruddy, S. J., Mosley, J. W., and Held, J. R. (1967). Chimpanzee-associated hepatitis in 1963. *Am. J. Epidemiol.* **86**, 634–640.

Sabin, A. B., and Wright, A. M. (1934). Acute ascending myelitis following a monkey bite with the isolation of a virus capable of reproducing the disease. *J. Exp. Med.* **59**, 115–136.

Sadecky, E., and Brazina, R. (1977). Vaccination of naturally infected ewes against Q fever. *Acta Virol. (Engl. Ed.)* **21**, 89.

Sanford, J. P. (1982). Plague. *In* "Textbook of Medicine" (J. B. Wyngaarden and L. H. Smith, Jr., eds.), 16th ed., pp. 1521–1523. Saunders, Philadelphia, Pennsylvania.

Schachter, J., and Dawson, C. R. (1978). "Human Chlamydial Infections." PSG Publ. Co., Inc., Littleton, Massachusetts.

Schachter, J., Oster, H. B., and Meyer, K. B. (1969). Human infection with the agent of feline pneumonitis. *Lancet* **1**, 1063–1065.

Schumacher, M. J. (1980). Characterization of allergens from urine and pelts of laboratory mice. *Mol. Immunol.* **17**, 1087–1095.

Shah, K. V., and Southwick, C. H. (1965). Prevalence of antibodies to certain viruses in sera of free-living rhesus and of captive monkeys. *Indian J. Med. Res.* **53**, 488.

Shaw, P. K., Brodsky, R. E., Lyman, D. O., Wood, B. T., Hibler, C. P., Healy, G. R., Macleod, K. I. E., Stahl, W., and Schultz, M. G. (1977). A community-wide outbreak of giardiasis with evidence of transmission by a municipal water supply. *Ann. Intern. Med.* **87**, 426–432.

Shimi, A., Keyhani, M., and Hedayati, K. (1979). Studies on salmonellosis in the house mouse *Mus musculus. Lab. Anim.* **13**, 33–34.

Shults, F. S., Estes, P. C., Franklin, J. A., and Richter, C. N. (1973). Staphylococcal botrymomycosis in a specific-pathogen-free mouse colony. *Lab. Anim. Sci.* **23**, 36–42.

Siegart, R. (1972). Marburg virus. *Virol. Monogr.* **11**, 98.

Siraganian, R. P., and Sandberg, A. L. (1979). Characterization of mouse allergens. *J. Allergy Clin. Immunol.* **63**, 435–442.

Skinner, H. H., and Knight, E. H. (1971). Monitoring mouse stocks for lymphocytic choriomeningitis virus—A human pathogen. *Lab. Anim.* **5**, 73–87.

Smetana, H. F. (1969). Spontaneous and induced viral hepatitis in subhuman primates in relation to human viral hepatitis. *Am. J. Pathol.* **55**, 65a–66a.

Snyder, S. P., England J. J., and McChesney, A. E. (1978). Cryptosporidiosis in immunodeficient Arabian foals. *Vet. Pathol.* **15**, 12–17.

Soave, O. (1981). Viral infections common to human and nonhuman primates. *J. Am. Vet. Med. Assoc.* **179**, 1385–1388.

Soave, O. A., and Van Allen, A. (1958). LCM in a mouse breeding colony. *Proc. Anim. Care Panel* **8**, 135–140.

Soulsby, E. J. L. (1969). "Helminths, Arthropods and Protozoa of Domesticated Animals" (Monnig), 6th ed., Williams & Wilkins, Baltimore, Maryland.

Splino, M., Peychl, L., Kyntera, F., and Kotrlik, J. (1969). Isolierung von *Pasteurella pseudotuberculosis* ans Lerstenlymphknoten. *Zentralbl. Bakteriol. Parasitenkd., Infektionskr. Hyg., Abt. 1: Orig.* **211**, 360–364.

Stalheim, O. V. H. (1969). Chemotherapy of renal leptospirosis in cattle. *Am. J. Vet. Res.* **30**, 1317.

Stemmermann, G. N., Hayashi, T., Glober, G. A., Oishi, N., and Frenkel, R. I. (1980). Cryptosporidiosis report of a fatal case complicated by disseminated toxoplasmosis. *Am. J. Med.* **69**, 637–642.

Stoenner, H. G. (1954). Application of the capillary tube test and a newly developed plate test to the serodiagnosis of bovine leptospirosis. *Am. J. Vet. Res.* **15**, 434–439.

Stoenner, H. G. (1957). The laboratory diagnosis of leptospirosis. *Vet. Med. (Kansas City, Mo.)* **52**, 540–542.

Stoenner, H. G., and Maclean, D. (1958). Leptospirosis (*ballum*) contracted from Swiss albino mice. *Arch. Intern. Med.* **101**, 706–710.

Stoenner, H. G., Grimes, E. F., Thrailkill, F. B., and Davis, E. (1958). Elimination of *Leptospira ballum* from a colony of Swiss albino mice by use of chlortetracycline hydrochloride. *Am. J. Trop. Med. Hyg.* **7**, 423–426.

Storz, J. (1971). "Chlamydia and Chlamydial-Induced Diseases." Thomas, Springfield, Illinois.

Stott, J. A., Hodgson, J. E., and Chaney, J. C. (1975). Incidence of salmonellae in animal feed and the effect of pelleting on content of enterobacteriaceae. *J. Appl. Bacteriol.* **39**, 41–46.

Strangeways, W. I. (1933). Rats as carriers of *Streptobacillus moniliformis. J. Pathol. Bacteriol.* **37**, 45–51.

Sulkin, S. E. (1961). Laboratory acquired infections. *Bacteriol. Rev.* **25**, 203–209.

Sulzer, C. R., Harvey, T. W., and Galton, M. M. (1968). Comparison of diagnostic technics for the detection of leptospirosis in rats. *Health Lab. Sci.* **5**, 171–173.

Taylor, M. J., and MacDowell, C. E. (1949). Mouse leukemia. XIV. Freeing line I from a contaminating virus. *J. Natl. Cancer Inst. (U.S.)* **17**, 233–245.

Teare, J. A., and Loomis, M. R. (1982). Epizootic of balantidiasis in lowland gorillas. *J. Am. Vet. Med. Assoc.* **181**, 1345–1347.

Tiggert, W. D., Benenson, A. S., and Gochenour, W. S. (1961). Air-borne Q fever. *Bacteriol. Rev.* **25**, 285–293.

Tobin, J. O. (1968). Viruses transmissible from laboratory animals to man. *Lab. Anim.* **3**, 19.

Torres, J. R., Sanderes, C. V., Staub, R. L., and Black, F. W. (1978). Cat-scratch disease causing reversible encephalopathy. *JAMA, J. Am. Med. Assoc.* **240**, 1628–1630.

Torten, M. (1979). Leptospirosis. *In* "CRC Handbook Series in Zoonoses" (J. H. Steele, ed.), Sect. A, Vol. I, pp. 363–421. CRC Press, Cleveland, Ohio.

Turner, L. H. (1970). Leptospirosis III. *Trans. R. Soc. Trop. Med. Hyg.* **64**, 623–646.

Tyzzer, E. E. (1907). A sporozoan found in the peptic glands of the common mouse. *Proc. Soc. Exp. Biol. Med.* **5**, 12–13.

Umenai, T., Lee, P. W., Toyoda, T., Yoshinaga, K., Horiuchi, T., Lee, H. W., Saito, T., Hongo, M., Nobunga, T., and Ishida, N. (1979). Korean hemorrhagic fever in staff in an animal laboratory. *Lancet* **1**, 1314–1315.

Valerio, D. A. (1971). Colony management as applied to disease control with mention of some viral disease. *Lab. Anim. Sci.* **21**, 1011–1014.

Van der Waaij, D., Zimmerman, W. M. T., and Van Bekkum, D. W. (1963). An outbreak of *Pseudomonas aeruginosa* infection in a colony previously free of this infection. *Lab. Anim. Care* **13**, 46–51.

Warren, J. W., and Hornick, R. B. (1979). Coxiella burnetti. *In* "Principles and Practices of Infectious Disease" (G. L. Mandell, R. G. Douglas, and J. E. Bennett, eds.), pp. 1516–1520. Wiley, New York.

Warwick, W. J. (1967). The cat-scratch syndrome. Many diseases, or one disease? *Prog. Med. Virol.* **9**, 256–302.

Wear, D. J., Margileth, A. M., Hadfield, T. L., Fisher, G. W., Schlagel, C. J., and King, F. M. (1983). Cat scratch disease: A bacterial infection. *Science* **121**, 1403–1404.

Welsh, H. H., Lennette, E. H., Abinanti, F. R., and Winn, J. F. (1958). Airborne transmission of Q fever: The role of parturition in the generation of infective aerosols. *Ann. N.Y. Acad. Sci.* **70**, 582–590.

Whitney, R. A., Jr. (1975). Important primate diseases (biohazards and zoonoses). *Cancer Res. Safety Monogr.* **2**, 23–52.

Williams, L. P., Vaughn, J. B., Scott, A., and Blanton, V. (1969). A ten-month study of *Salmonella* contamination in animal protein meals. *J. Am. Vet Med. Assoc.* **155**, 167–174.

Williford, C. B., and Wagner, J. E. (1982). Bacterial and mycotic diseases of the integumentary system. *In* "The Mouse in Biomedical Research" (H. L. Foster, J. D. Small, and J. G. Fox, eds.), Vol. 2, pp. 55–72. Academic Press, New York.

Wolf, F. W., Bohlander, H., and Ruys, A. C. (1949). Researches on *Leptospirosis ballum*—The detection of urinary carriers in laboratory mice. *Antonie van Leeuwenhoek* **15**, 1–13.

Wolfe, L. G., Griesemer, G. A., and Farrell, R. L. (1968). Experimental aerosol transmission of the Yaba virus in monkeys. *J. Natl. Cancer Inst. (U.S.)* **41**, 1175–1195.

Yager, R. H., Gochenour, W. S., Jr., Alexander, A. D., and Wetmore, P. W. (1953). Natural occurrence of *Leptospira ballum* in rural house mice and in an opossum. *Proc. Soc. Exp. Biol. Med.* **84**, 589–590.

Yunker, C. E. (1964). Infections of laboratory animals potentially dangerous to man: Ectoparasites and other arthropods, with emphasis on mites. *Lab. Anim. Care* **14**, 455–465.

Chapter 23

Factors That Complicate Animal Research

Steven P. Pakes, Yue-Shoung Lu, and Paul C. Meunier

I. INTRODUCTION

Animal research has reached an unprecedented level of sophistication. The data emanating from animal experiments may be influenced profoundly by a number of environmental and biological factors. The importance of these factors, such as several spontaneous diseases, has been recognized for years. The importance of most factors is subtle, and only recently have we begun to appreciate the influence of these variables on animal research and testing. Therefore, the necessity of presenting as stable an environment for experimental animals as is scientifically and economically feasible is readily apparent.

This chapter will address the salient complicating factors for which we have meaningful information and will focus on principles and concepts that can be applied by investigators to their own unique experimental situations. The factors discussed in this chapter should serve as examples of other variables we will not address and those not yet discovered. We have chosen to categorize these complicating factors as physical, chemical, and microbial. The complex interactions among these and genetic factors should be appreciated. To supplement this discussion, the excellent reviews of Baker *et al.* (1979), Lindsey *et al.* (1978), and Newton (1978) should be consulted.

LABORATORY ANIMAL MEDICINE

II. PHYSICAL FACTORS

A. Environment

Several environmental factors may have profound effects on the biological response of animals to various experimental manipulations. In contrast to the macroenvironment of the room, the caging or housing system constitutes the microenvironment for experimental animals (Clough, 1976; Woods, 1978). The environmental differences between the cage and the animal room must be recognized as these differences may influence experimental results.

1. Cage Design

The microenvironment (cage) is extremely important because it directly influences the biological responses of animals to various experimental procedures. Therefore, one must consider the design of the caging system in the research protocol. Design and construction material influence the amount of air, light, and sound that the animal receives and the amount of heat, humidity, and gaseous waste dissipated into the macroenvironment for elimination via the room ventilation system. Cages should be constructed of materials that cannot be ingested or which do not react with waste products. Publications dealing with the various aspects of cage design and effects on the microenvironment include those of Baker et al. (1979), Clough (1976), Woods (1978), and Serrano (1971).

With regard to the primary enclosure, population density is another important factor to be considered. Overcrowded conditions or the stress of isolation for some species can affect reproduction (Christian and LeMunyan, 1958) and behavior (Davis, 1978) and may seriously compromise the immune system and normal metabolism (Baker et al., 1979).

2. Temperature, Relative Humidity, and Room Ventilation

The temperature and humidity of the animal's microenvironment is dependent on a number of factors, including the temperature and humidity of the macroenvironment, cage design (Serrano, 1971), presence of cage filter top (Besch, 1975; Simmons et al., 1968), cage population (Yamauchi et al., 1965), activity of the animals, and the amount and velocity of air flowing over the cages (Woods, 1978). One strives to attain the thermoneutral temperatures of research animals, which have been compiled by Weihe (1965). Regulatory mechanisms allow homeothermic animals to maintain relatively constant body temperatures over a wide range of ambient temperatures. If ambient temperatures fall below thermoneutral temperatures, animals must adapt to the change or suffer reduced body temperature. In laboratory animals, adaptation mechanisms include peripheral vascular constriction and piloerection. Behaviorally, the animals may huddle and assume postures that reduce body heat loss. If such methods do not suffice, metabolic activity is increased, leading to increased food consumption and associated variations in biological response (Newton, 1978). Weihe (1973) has reviewed the effect of temperature on drug action, emphasizing the importance of temperature regulation for drug toxicity trials. As examples, studies designed to determine optimal group size and environmental conditions for drug testing have shown that the LD_{50} for amphetamine at 27°C was about tenfold less for single housed animals than for groups of 10 mice. Furhman and Furhman (1961) have analyzed patterns of drug toxicity in relationship to temperature and have observed three types of curves: (a) The V- or U-shaped curve with minimum toxicity around thermal neutrality and increasing toxicity at lower and higher temperatures; (b) a linear relationship between increasing toxicity and increasing temperature; and (c) a skewed relationship with constant values of toxicity at or below the thermoneutral zone and increasing toxicity at higher temperatures. The type (a) toxicity curve is characteristic of many centrally acting drugs affecting the thermoregulatory system. In general, acute exposure of small homeothermic animals to heat or cold results in a changed metabolic rate with increased or decreased heat production. Drugs that stimulate heat production, such as sympathomimetic drugs, often show a linear correlation between increasing toxicity and increasing ambient temperatures.

Environmental temperatures also affect other physiologic parameters in laboratory animals. Yagil et al. (1976) observed that lactating rats exposed to 35°C for 8 hr daily produce less milk than rats housed at 22°C. Reproduction in rats decreases markedly at 32°C (Yamauchi et al., 1981). Pucak et al. (1977) reported high mortality and retarded testicular development with infertility in rats accidently exposed to high temperatures (26.6°–37.7°C) for prolonged periods.

High environmental temperatures and humidity may also increase susceptibility to infectious agents, as indicated by Baetjer (1968). She reported that mice housed at 35.6°C at 22% relative humidity (RH) were more susceptible to influenza virus infection than mice housed at 35.6°C at 90% RH. Ringtail is one disease of laboratory animals that has been directly related to temperatures and RH out of the thermoneutral zone. This syndrome has been described in rats, South African hamsters, and mice (Flynn, 1959; Stuhlman and Wagner, 1971; Fox, 1977). Ringtail is characterized by annular constrictions of the tail skin, sometimes resulting in tail sloughing. The syndrome is often seen in young animals. Although the exact cause is unknown, it is thought to be associated with the inability to control heat loss in environments of 40% RH or lower.

3. Light

The intensity, quality, and photoperiod of light are important variables that may influence the biological response. The intensity and quality of artificial light are influenced by the number, power rating, and emission spectra of the room lights (Baker *et al.*, 1979; Clough, 1976).

Light intensity or illumination is measured in foot-candles (fc) as lumens/ft^2 or in lux as lumens/m^2, while quality of light or the wavelength spectrum is expressed in angstrom units (Å). The levels of intensity and quality that are most beneficial or detrimental to laboratory animals are not well known. The recommendation followed by most facilities today is 75–125 fc (807–1345 lux) at cage level [Institute of Laboratory Animal Resources (ILAR), 1978]. Many studies indicate that levels approaching the recommended level of light intensity may cause significant retinal degenerative changes in laboratory rodents (Bellhorn: 1980; LaVail, 1976; O'Steen and Anderson, 1972; O'Steen *et al.*, 1973). The effect of light intensity on retinal structures has been studied extensively in rats (Weihe, 1976). Albino rats exposed to continuous illumination of between 540 and 980 lux, at or below recommended levels, for a period of 65 days, showed complete and irreversible degeneration of the photoreceptors. Studies have shown that the severity of these lesions is more pronounced with increasing age (O'Steen *et al.*, 1974; Weibe and Stotzer, 1974). Similar lesions have been described in mice (Greenman *et al.*, 1982; Robinson and Kuwabara, 1976). Pigmented animals are apparently less susceptible (Reiter, 1973).

The cage position on shelves or in racks is important because light intensity decreases with the square of the distance from the light source. BALB/c mice housed in translucent cages on the top shelves of racks have a higher prevalence of retinal atrophy than mice housed in cages on lower shelves (Greenman *et al.*, 1982; Weihe *et al.*, 1969). It has been shown that the light at the top shelf of a rack may be 80 times more intense than at the bottom shelf of the rack (Weihe, 1976). The cage construction material (opaque, translucent, or transparent) greatly influences the quality and intensity of light to which animals are exposed. It becomes important then to standardize cage designs and to randomize the position of cages on racks and within rooms to reduce the possible complications imposed by light (Clough, 1976; Weihe *et al.*, 1969).

Light is also a powerful stimulant and synchronizer of the reproductive system. This stimulus is thought to be mediated through the hypothalamus, initiating release of neurohormones that control pituitary production of gonadotropic and adrenocorticotropic hormones (Weihe, 1976). Therefore, photoperiod (light–dark cycle, usually stated as L : D in hours) is another important environmental factor that modifies biological response (DeWitt, 1976; Hastings, 1970; Mock *et al.*, 1978). Beyond its influence on reproductive behavior, photoperiodicity has a profound regulatory effect on circadian rhythms (Hastings and Menaker, 1976). A significant number of biochemical, physiological, and behavioral parameters have been shown to have a 24-hr periodicity. Several studies indicate the influence of circadian rhythms, sometimes profound, on blood cells (Berger, 1980, 1981), plasma steroid levels (Alder and Friedman, 1968; Bowman *et al.*, 1970; Ramaley, 1972), and other biochemical data in a variety of species (Cayen *et al.*, 1972; Izquierdo and Gibbs, 1972; Mitropoulos *et al.*, 1972; Zbiesieni, 1980). Photoperiod also greatly influences enzyme activity, drug metabolism, and drug toxicity (Chédid and Nair, 1972; Deimling and Schnell, 1980; Halberg *et al.*, 1973; Haus *et al.*, 1972; Jori *et al.*, 1971; LeBouton and Handler, 1971; Nair and Casper, 1969; Nash and Llanos, 1971; Radzialowski and Bousquet, 1968; Romero, 1976). These observations have significant clinical as well as research implications in animals. For example, Scheving *et al.* (1968) found that phenobarbital sleep times in rats were extended significantly during certain times of the day, depending on the light cycle to which the rats were exposed. Circadian rhythm may also play a role in disease resistance and susceptibility. Studies have shown that resistance of mice to an acute experimental infection by type I pneumococci varied rhythmically in a 24-hr cycle. Mice inoculated during the dark period survived significantly longer than mice inoculated during the light period (Feigin *et al.*, 1969; Shackelford and Feigin, 1973; Wongwiwat *et al.*, 1972).

The recommendations of ILAR (1978) are that animal housing facilities be windowless so a regular diurnal lighting cycle can be controlled through the use of timing devices. A cycle of about 12–14 hr of daily light is usually provided. Daily light of 13–14 hr is thought to be best for rat reproduction (Mulder, 1971). It is important to reiterate once more that intensity, quality, and duration of light should be standardized and controlled for most research and testing regimens.

4. Noise

It has long been recognized that prolonged exposure to high levels of noise can cause auditory lesions in man and lower animals. However, it is only recently that we have become concerned about the possible research complications of very subtle noise levels. Routine noises in animal facilities emanate from the vocalization and activity of animals, from feeding, watering, and cleaning procedures; and from the operation of ventilation systems. Animal technicians and others who work in animal facilities should be aware that noise is best kept to a minimum while performing their activities. It is recommended that noise levels in animal facilities not exceed 85 decibels (db) (Anthony, 1962). The effect of noise on experimental animals is still not well understood, but the review by Fletcher (1976) addresses the salient knowledge in this area. Several studies conducted at decibel levels that exceed the usual background noise in animal facilities show various degrees of destruction of sensory hairs and supporting cells of several animal species. Anthony (1962) states that rats, like man, appear to experience mechanical damage of the auditory system at 160 db, pain at about 140 db, and signs of inner ear damage after prolonged exposures to about 100 db.

The nonauditory effects of noise are of significant concern and may greatly affect biological responses. Some nonauditory responses to noise are genetically determined. Some inbred strains of mice, such as DBA/2, are prone to audiogenic seizures varying in severity and sometimes leading to death (Iturrian, 1971). Audiogenic stress due to pulsed noise exceeding 83 db may also reduce fertility of rodents (Fletcher, 1976; Zakem and Alliston, 1974). Geber et al. (1966) observed eosinophilia, increased serum cholesterol, increased adrenal ascorbic acid level, and increased adrenal weights in rats exposed to 83 db for periods up to 6 min/hr for 3 weeks. Friedman et al. (1967) reported elevation of serum lipids in rats exposed to continuous sound having an intensity of 102 db and intermittent sound of 114 db.

B. Stress Effects on Laboratory Animals, Including Transportation

Stress is a term used for describing emotional and biological responses to novel or threatening stimuli. Experimental and nonexperimental stress-inducing stimuli may have profound physiological consequences (Riley, 1981). Specific biochemical, cellular, and tissue alterations may be associated with activation of the adrenal cortex via the pituitary gland and secretion of adrenocorticotropic hormones (Riley, 1981). Increased serum corticosterone levels may lead to many alterations in experimental data. A direct consequence is injury to elements of the immune system, which may increase susceptibility to latent oncogenic and infectious agents, newly transformed cancer cells, or other potentially pathologic processes normally held in balance by a competent immune system (Riley, 1981; Visintainer et al., 1982).

There are a variety of nonexperimental stressful events that laboratory animals may encounter in animal facilities. Many have already been mentioned. Others include isolation, overcrowding, social rank interaction among animals, and transportation. Because transportation is inevitable in most instances, investigators and students should clearly recognize the possible associated stresses and make appropriate adjustments in protocols to allow stabilization of research animals before beginning experiments. Upon arrival the animals may exhibit weight loss, dehydration, hyperthermia or hypothermia, and/or other signs depending upon the shipping container, water and food supply, and mode and duration of shipment (Baker et al., 1979; Manning and Banks, 1968; Weisbroth et al., 1977). Therefore, it is important to provide an adequate period before experimentation to allow animals to recover physiologically to a stable state. Weisbroth et al. (1977) have defined this recovery period in recently shipped adolescent rats as the time required for recovery of body weight to within one standard deviation of the growth curve of identical unshipped rats maintained by the vendor. This period of equilibration takes 1 to 5 days (Dymsza et al., 1963; Weisbroth et al., 1977).

III. CHEMICAL FACTORS

As previously stated, the biological response is the result of a complex set of interactions involving both genetic and environmental factors. In biomedical research, precise measurements, employing ever more sophisticated equipment, are made of the biological responses to a large number of exogenous agents under carefully "controlled" conditions. In many instances these measurements are made in genetically defined strains of various animal species. Yet the chemical composition of the environment in which these experiments are performed is subject to considerable variation. There is increasing awareness that environmental chemical contamination, and the intentional use of pharmacologic agents (anesthetics, analgesics, tranquilizers, anthelmintics, antibiotics) may measurably influence the biological response of experimental animals (Baker et al., 1979; Cass, 1970; Fouts, 1976; Lindsey et al. 1978; Newberne, 1975; Newberne and McConnell, 1980). These alterations may lead to discrepancies among test results or the loss of experimental animals.

Unintentional exposure of experimental animals to exogenous chemicals (xenobiotics) may occur via the air, water supply, animal feed, and bedding. Caging and equipment may also be contaminated. Exposure is often mediated by complex pathways and uptake may occur by more than one route. Furthermore, many chemicals, such as detergents, disinfectants, and insecticides, are intentionally introduced into the animal environment or the animal is exposed to pharmacological agents as a part of conditioning or the experimental procedure.

Depending upon its nature, a chemical contaminant may exert its toxic effect at various sites in the body. The chemical may have a topical effect at the portal of entry (the skin, respiratory or gastrointestinal tract) or be absorbed across one or more body surfaces into the general circulation. Once absorbed, chemicals may be translocated to storage sites, biotransformation sites, or sites of excretion. The chemical may be inherently toxic, or its metabolism by the host may result in the formation of a toxic product. The parent compound or its metabolite may cause irreversible cell injury resulting in tissue necrosis, or it may interfere with cellular homeostasis, causing the cell to establish new levels of metabolic or functional activity in order to survive. These adaptive measures may have special significance in chronic studies, such as cancer research, the assessment of toxicity, or the study of aging.

The interplay of host and environmental factors determines the ultimate response to exogenous chemical agents. Important host factors include age, state of health, nutritional status, sex, genetic constitution, and immune function. Environmental variables include the concentration of the exogenous agent, its physiochemical properties, potential interactions with other agents in the exposure environment, and the duration, frequency, and route of exposure. A good example of this type of interaction is the accumulation of cadmium, a frequent contaminant of laboratory animal diets, water supply, and ambient air, in the tissues of hypertension-sensitive Dahl rats (Ohanian and Iwai, 1980). This strain of rat accumulates significantly more cadmium in its tissues than hypertension-resistant strains. There are numerous examples of variations in the response to xenobiotic agents associated with differences in genetic factors alone (Nebert and Felton, 1976). When one considers the possible permutations of these host and environmental variables, the need for environmental control is obvious.

Multiple sources and repeated insults may increase the risk of disease from environmental chemicals. In addition to their local irritant effect, xenobiotic agents may produce generalized systemic disease, act as allergens in the induction of an immune response, or alter immune function, thereby enhancing susceptibility to spontaneous disease. They may also induce or inhibit enzyme systems or biotransformation pathways.

A variety of environmental agents and some intentionally administered xenobiotic agents are capable of acting as mutagens, a group of agents which induce changes in DNA. Some of these agents may also act as teratogens by inducing macromolecular changes during embryonic development, resulting in anatomical, functional, or biochemical errors. Dietary contaminants such as aflatoxins (Elis and DiPaolo, 1967) and the salts of certain heavy metals, including lead, mercury, and cadmium, may act in this way, as can organochlorine insecticides and several of the commonly administered anesthetic agents (Gerber et al., 1980; Degraeve, 1981; Ito and Ingalls, 1981).

Intentional exposure of laboratory animals to various levels of environmental contaminants for specified periods has been shown to result in neoplasia. However, laboratory animals may also be unintentionally exposed to carcinogenic agents from a variety of environmental sources. For example, some animal diets are contaminated by nitrosamines (Edwards et al., 1979; Walker et al., 1979) and nitrates (Newberne and Mc-

Connell, 1980), which can be converted to nitrosamines in the gastrointestinal tract. Drinking water is also a potential source of nitrates (Shapiro, 1980). Aflatoxin at levels sometimes found in animal diets have been shown to be carcinogenic in rats, ducks, and trout (Wogan, 1968). Chloroform, which may be carcinogenic under conditions of chronic exposure, is a frequent contaminant of air and water. Animal cages may not be completely decontaminated during wash cycles, allowing for additional environmental exposure to a test agent (Fox et al., 1980; Fox and Helfrich-Smith, 1980). Even if the level of environmental carcinogens does not by itself increase the incidence of neoplasia in a given study, it is possible that environmental carcinogens could act synergistically with other carcinogens (test) or cocarcinogens to increase the incidence of neoplasia or to alter target organs.

The induction of hepatic microsomal enzymes by environmental contaminants may be an example of this type of interaction. Hepatic microsomal enzymes are the primary site of biotransformation, a major way in which rodents eliminate foreign compounds from their body. It is generally accepted that chemical carcinogens induce neoplastic transformation by a multistep process involving initiation and promotion (Farber, 1982). Although some initiators are inherently active, the majority require conversion to an "electrophilic reactant" by microsomal enzymes before they are capable of interacting with DNA. Furthermore, the physiological process of microsomal enzyme induction may itself act as a promoting event in the generation of hepatic neoplasms. The administration of various xenobiotic agents that induce hepatic microsomal enzymes has been associated with an increased incidence of hepatic tumors in mice (Conney, 1967; Tennekes et al., 1981). The consequence of microsomal enzyme induction by environmental agents might be similar.

Although specific xenobiotic agents may induce defined changes in certain biological parameters, the effects of combined exposure to environmental contaminants on the biological response are not easily estimated. This is because the length of exposure and the level of each environmental contaminant may vary widely, the nature of the interaction of those chemicals in the environment and with the host are often unknown, and various environmental chemicals may have opposing effects on a specific host system. For example, dietary factors, aromatic hydrocarbons from a variety of bedding materials, eucalyptol from aerosol sprays, and chlorinated hydrocarbon insecticides have all been shown to induce hepatic microsomal enzymes (Baker et al., 1979; Lindsey et al., 1978; Vessel et al., 1976). However, ammonia, a common contaminant of the laboratory animal environment, is a purported inhibitor of this system (Vessel et al., 1976). There are few data regarding the combined effects of defined levels of environmental contaminants on the biological response. With this in mind, some of the sources of exogenous agents and their potential effects on the biological response will be discussed below.

A.   Ambient Air

The air quality in any animal facility may vary appreciably depending upon a number of factors. In general, these factors fall into two distinct categories: (1) the quality of the air in the macroenvironment, that is, the animal room, and (2) the air quality in the microenvironment or animal cage (Lindsey *et al.*, 1978; Murahami, 1971). Both of these environments contain varying concentrations of particulates, vapors, and fumes that may cause injury to the respiratory system or be absorbed into the general circulation.

Air quality in the macroenvironment may vary depending upon geographic location, proximity to industrial areas, and the intake of air downwind from discharged exhaust wastes. Although the incoming air may contain a variety of chemical contaminants, the quality of the macroenvironment is more directly influenced by the types of chemical agents used in the animal facility. Organic solvents, insecticides, animal room deodorizers, and disinfectants are examples of biologically active agents with impact upon organic processes. Cleanliness of the animal facility is important for the prevention of infectious diseases, but sanitation procedures have implications beyond disease control. For example, room deodorizers are complex mixtures of volatile hydrocarbons and essential oils, which may induce or inhibit hepatic microsomal enzymes (Vessel *et al.*, 1976). Most cleaning agents, surfactants, and organic solvents are volatile compounds with similar effects (Conney and Burns, 1972). Ozone generated by ultraviolet light depresses microsomal enzymes and has pathological effects upon the respiratory epithelium, which could potentially influence the investigation of respiratory diseases.

Pesticides and insecticides have long been recognized as having potent effects upon drug metabolism. Because of their profound effects upon biological systems and their environmental persistence, chlorinated hydrocarbons have essentially been banned from general use, and organophosphates and carbamates are now the most widely used insecticides. Although some of these agents are marketed as plastic strips, they are usually applied as a spray in an oil base or aqueous carrier, which is likely to increase their ambient air concentration and the risk of direct animal contact. Organophosphates are depressants of hepatic microsomal enzymes and may affect the central nervous system, hematopoietic system, and skin. Carbamates are toxic to the central nervous system. Because of these toxic effects, it is preferable to use nonchemical means of vermin control in an animal facility. When absolutely necessary, pesticides and insecticides should be used sparingly, and investigators should be informed of their intended use. Knowl-

edgeable pest control operators should be consulted, and the animal facility surveyed to maximize effectiveness and minimize frequency of chemical use.

Air quality within the animal cage or microenvironment is of equal importance to air quality in the macroenvironment. Although the various components of the gaseous environment in which laboratory animals live are poorly documented, the two most common contaminants derived from the intracage accumulation of animal wastes are carbon dioxide and ammonia (Serrano, 1971). Although little is known about the biological effects of carbon dioxide, consideration must be given to its potential to act as an asphyxiant by displacing oxygen from the ambient air. Ammonia, produced by the action of urease-positive bacteria on urine and feces, is a common biologically active contaminant of the laboratory animal microenvironment. Although an acceptable level of ammonia in the microenvironment is less than 25 ppm, 48% of 368 randomly sampled mouse, hamster, rat, guinea pig, and rabbit cages contained ammonia concentrations above the acceptable level and 21% were above 100 ppm (Gamble and Clough, 1976). The concentration of microenvironmental ammonia may vary with the animal species, number of animals per cage, sex, rate of air flow, temperature, humidity, cage design, type of bedding, and cage cleaning frequency (Serrano, 1971; Lindsey *et al.*, 1978). Ammonia concentrations are lowered by using bacteriostatic cageboard if changed daily and by maintaining adequate intracage air flow rates (14 intrachamber volumes per hour).

The principal adverse effect of environmental ammonia is irritation of the respiratory tract (Schaerdel *et al.*, 1983). Exposure of rats to 25–250 ppm ammonia for 4–6 weeks enhanced the severity of lesions characteristic of murine respiratory mycoplasmosis throughout the respiratory tract and increased the prevalence of lung lesions characteristic of this disease, suggesting that environmental ammonia may be an important pathogenetic factor (Broderson *et al.*, 1976). Environmental ammonia is also a potential coirritant in inhalation toxicology studies, and it may enhance the irritant potency of other environmental contaminants or alter their deposition (Owen, 1969).

B.   Drinking Water

The water consumed by laboratory animals, like the ambient air, is subject to considerable variation depending upon geographic location; area geology; the use of surface or well water; proximity to industrial, agricultural, or major urban centers; and the type of water treatment used. Many of the constituents and contaminants of drinking water occur naturally and enter the water from the rock, soil, and air.

Chemical impurities found in water are generally classified

as suspended solids, and organic and inorganic solutes (Shapiro, 1980). Most, but not all, suspended solids are themselves harmless, although they may act as carriers of biological agents. Of the over 700 organic chemicals identified in drinking water in the United States, over 90% are natural decomposition products of animal and plant origin. However, drinking water also contains low levels of many synthetic organic solutes, such as pesticides and cyclic aromatic and halogenated hydrocarbons. The largest portion of identifiable synthetic organic contaminants are the trihalomethanes. Members of this group of organic compounds are derived from the interaction of a halogen, usually chlorine or bromine, with methane groups from natural organic materials, or they may enter the water supply as contaminants of chlorine per se. These compounds are found in virtually all drinking water that has been disinfected with chlorine. One such compound, chloroform, has biological impact (Vessel et al., 1976) and is found in very high relative concentrations. In addition, high levels of chlorine per se resulting from either residuals in tap water or water acidification may cause variation in the immune response (Hermann et al., 1982).

Inorganic elements may also contaminate the water supply. In typical drinking water, cadmium, barium, selenium, and arsenic are at or exceed the maximum contamination limits established by the Environmental Protection Agency (Shapiro, 1980). Nitrates, frequent contaminants of water systems, are of particular concern because of their potential as procarcinogens.

C. Animal Diets

Many natural and synthetic chemical compounds found in laboratory animal diets may have significant effects upon biochemical and physiological processes without causing overt signs of toxicity. The most widely considered feed contaminants include chlorinated hydrocarbons, organophosphates, polychlorinated biphenyls, heavy metals (such as lead, mercury, cadmium, arsenic, and selenium), aflatoxins, nitrates, nitrosamines, and estrogenic compounds (Newberne and McConnell, 1980; Edwards et al., 1979). Many of these compounds occur naturally in plant materials or remain as residues from agricultural use. Furthermore, raw materials may be contaminated during storage, or food products may be contaminated during formulation. For example, nitrates and amines, which can form nitrosamines in vivo, occur naturally in plant materials, particularly corn (Newberne, 1975). Nitrosamines have been found in many natural product diets in which fish meal was included as a major protein source (Newberne and McConnell, 1980; Edwards et al., 1979). Most staple food products, such as corn, wheat, and other cereals, are subject to contamination during storage with significant amounts of aflatoxin (Wogan, 1968). The inadvertent contamination of laboratory animal feeds with estrogenic compounds during formulation has resulted in serious disturbances in reproductive performance (Hadlow et al., 1955; Wright and Seibold, 1958).

Complete diets formulated from crude natural products, as opposed to diets formulated from semipurified ingredients, are subject to the greatest variation in the type and concentration of contaminants, as well as the greatest variation in nutrient quality (Newberne and McConnell, 1980; Wise, 1982). Variation in dietary constituents may alter the toxicity of chemical contaminants and potentially affect the animal response to specific drugs or chemicals. For example, when corn oil was compared to beef fat as a source of fat in rat diets, it was found that corn oil had a significant enhancing effect on aflatoxin B_1-induced hepatic neoplasms (Newberne et al., 1979).

Excesses or deficiencies in the quantity of available nutrients may also influence experimental results. Variation in the quantity or availability of essential vitamins and minerals may alter drug metabolizing systems, affect membrane integrity, or predispose to the affects of carcinogens (Newberne and McConnell, 1980). Excessive caloric intake may have a serious impact on biomedical research since caloric restriction has been shown to be advantageous to the host in resisting the effects of aging, degenerative and infectious diseases, neoplasia, and the toxicity of chemical agents (Newberne and McConnell, 1980; Ross and Bras, 1973). Excessive caloric intake is perhaps one of the most serious dietary problems affecting the validity of long-term studies.

Some of the most common and important contaminants of laboratory animal diets are the heavy metals (lead, mercury, cadmium, and arsenic). The effects of lead, mercury, and cadmium have been extensively investigated and shown to alter the structure or function of a number of different organ systems. Many of these effects are the result of chronic exposure to subtoxic doses. For example, chronic experimental exposure to low concentrations of dietary cadmium and lead has been associated with morphological and biochemical changes in cardiac tissues, alterations of the cardiac conduction system, accelerated aortic plaque formation, lipid deposition in arterial walls, and hypertension (Kopp et al., 1980; Perry and Erlanger, 1974). Exposure to lead has also resulted in alteration of ATP and total high energy phosphate concentrations in myocardial tissue (Kopp et al., 1980).

It has long been recognized that exposure of rodents to lead and cadmium suppresses their resistance to infectious agents (Hemphill et al., 1971; Cook et al., 1975), and the effects of endotoxin (Cook et al., 1974). Koller (1979) has reviewed the effects of exposure to heavy metals on the immune system. These metals may reduce antibody formation and delayed-type hypersensitivity, effects possibly mediated at the level of the T helper cell. Cadmium may also alter the phagocytic capacity of mouse macrophages and polymorphonuclear neutrophils as well as the microbicidal capacity of alveolar macrophages (Loose et al., 1978). Lead acetate suppresses interferon production in the mouse, and sodium arsenite enhances specific antiviral activity at low concentrations but inhibits interferon at high concentrations (Blakley et al., 1980).

Lead and cadmium may also influence reproductive performance by preventing embryo implantation or delaying fetal growth (Gerber et al., 1980; Degraeve, 1981). Furthermore, both metals are thought to be teratogenic. Lead may induce chromosomal aberrations, while cadmium can modify RNA and DNA metabolism. By altering genetic material these compounds might enhance the mutagenicity or carcinogenicity of other substances.

Maximum recommended levels for dietary contaminants have been published (Environmental Protection Agency, 1979), and the use of a quality controlled diet has been proposed by the Food and Drug Administration as a means of improving reproducibility and the avoidance of artifacts of dietary origin (Food and Drug Administration, 1978). Although the diet is a major route of exposure to environmental contaminants, the combined effects of multiple routes of exposures must also be considered. For example, cadmium, mercury, and lead are also contaminants of the ambient air and water. Insecticides may contaminate the air, water, and feed. Only thorough monitoring of all possible routes of exposure will shed light on the true level of environmental contaminants to which laboratory animals are exposed. Only then can one begin to estimate their possible impact upon experimental results.

D. Drugs

Pharmacological agents are frequently administered to experimental animals as a part of the conditioning program, the treatment of spontaneous disease, or facilitation of the experimental procedure. However, many of these agents may themselves alter physiological processes or affect the pharmacological activity of unrelated chemical agents and thereby alter experimental results. For example, many of the commonly used antibiotics have been shown to induce changes in cardiovascular function of laboratory animals under experimental conditions (Adams, 1976). Some of these agents, such as the aminoglycosides, not only exert a negative inotropic effect on cardiac and arterial muscle, but also alter the positive inotropic effects of other agents such as norepinephrine. Furthermore, neomycin and streptomycin are known to interfere with a number of calcium-dependent membrane phenomena by interfering with calcium conductance (Adams, 1975). This interference not only affects the development of muscle tension, but may also inhibit ganglionic transmission by reducing the release of

acetylcholine (Wright and Collier, 1974). Aminoglycoside antibiotics can also block the cholinergic receptors of skeletal muscle and thereby enhance the effect of anesthetics that act on those receptors at the myoneural junction (Adams, 1976).

Under defined conditions, many antibiotics have been shown to have an adverse effect on a variety of cell functions related to the immune response. Although most of the data are from human studies, similar effects have been noted in several animal species (Hauser and Remington, 1982). For example, tetracyclines have been shown to depress chemotaxis, phagocytosis, lymphocyte transformation, and the delayed hypersensitivity response in humans. Similarly, tetracyclines depress the delayed hypersensitivity response and in vivo chemotaxis of macrophages in mice (Thong and Ferrante, 1980). Numerous other antibiotics have been shown to negatively affect antibody production, microbicidal activity of phagocytes, and oxidative metabolism of neutrophils and to prolong allograft survival time (Hauser and Remington, 1982).

Antibiotics may also alter the metabolic machinery and thereby influence the toxicity or alter the pharmacokinetics of other agents. For example, chloramphenicol inhibits hepatic microsomal enzymes and thereby alters the effects of agents, such as barbiturate anesthetics, that are metabolized by that system (Adams, 1972). In so doing, chloramphenicol may increase or decrease the toxicity of a test agent, depending on whether the compound is inherently toxic or its toxicity is enhanced by metabolic conversion.

Apart from the possible influence of antibiotics on experimental results, the use of these agents must be tempered by knowledge of their toxicity in laboratory animals. Many antibiotics are toxic per se to a variety of laboratory animals, or their use is associated with fatal syndromes such as enterocolitis. For example, penicillin–dihydrostreptomycin is marketed as several different formulations, some of which contain procaine hydrochloride to minimize discomfort at the injection site. Procaine is toxic to guinea pigs, rabbits, and mice (Galloway, 1968), and dihydrostreptomycin causes acute toxicity and death within 2 hr of injection in Mongolian gerbils (Wightman et al., 1980).

The use of penicillin in guinea pigs and hamsters has long been associated with the occurrence of fatal enterocolitis. Most antimicrobial agents are lethal for hamsters and guinea pigs (Bartlett et al., 1978; Lowe et al., 1980), and a similar syndrome has been described in rabbits given clindamycin (LaMont et al., 1979). There has been an association between different toxigenic clostridial organisms and antibiotic-associated enterocolitis in these species, but existing evidence strongly suggests that the toxin-producing agent is Clostridium difficile (Larson, 1980; Lowe et al., 1980). This disease may not simply represent an overgrowth by resistant strains of C. difficile because some investigators have shown that the bacteria producing the toxin were in fact sensitive to the inducing agent (Larson, 1980; Lowe *et al.*, 1980) or they were unable to recover the bacteria or the toxin prior to administering the inducing agent (Lowe *et al.*, 1980). An environmental source of *C. difficile* has been suggested (Larson, 1980). Presumably antibiotic treatment alters the balance of the gastrointestinal flora toward conditions that favor the colonization and proliferation of *Clostridium* spp. Subsequent toxin release results in the induction of enterocolitis.

Anesthetic and tranquilizing agents are another group of compounds that have great potential to influence the results and interpretation of biomedical research data. These compounds are widely used in the implementation of experimental design or as restraining agents. However, the effects of these compounds are not physiologically equivalent, and care must be exercised in the selection of an agent in order to minimize its impact on experimental results.

The influence of various anesthetic agents on cardiovascular function has been reviewed (Parker and Adams, 1978). Some of these agents, for example, pentobarbital and halothane, are thought to have a direct effect on myocardial and vascular smooth muscle by interfering with calcium-dependent processes. Although the majority of these effects have been noted *in vivo*, it has also been suggested that pentobarbital pretreatment may influence myocardial calcium utilization *in vitro* (Rubanyi, 1980). Furthermore, anesthetic agents may influence the results of cardiovascular studies by altering the sensitivity of the myocardium or vascular smooth muscle to a test agent. For example, the halogenated hydrocarbon anesthetics (halothane, etc.) sensitize the heart to the arrhythmogenic activity of the catecholamines, and ketamine may act synergistically to increase their positive inotropic effects. The potential effects of anesthetic agents must also be viewed in light of the use of various preanesthetic agents and anesthetic techniques that may influence results.

Immune function may also be affected by anesthetic agents. Halothane, at concentrations approximately those found in arterial blood, has been shown to decrease the responsiveness of rat lymphocytes to mitogens (Bruce, 1972). Halothane has also been shown to decrease chemotaxis, reduce phagocytosis, and alter the metabolic activity of lymphocytes (Cullen, 1974). It has also been suggested that depressed immune function associated with the use of some anesthetic agents may influence the postoperative progression of tumor metastasis or alter metastatic sites (Shapiro *et al.*, 1981).

Anesthetic agents may be toxic themselves or they may enhance the toxicity of other agents. Reactive intermediates formed during the reductive metabolism of halothane by induced hepatic microsomal enzymes damage hepatocytes of susceptible species under conditions of hypoxia (Harper *et al.*, 1982), and methoxyflurane has been shown to be nephro-

toxic to certain strains of rats (Mazze *et al.*, 1973). Local anesthetics may affect membrane-mediated cellular processes in neural and other cells, and a number of them have been shown to enhance the toxicity of bleomycin for cultured mouse tumor cells (Mizuno and Ishida, 1982). In addition, some barbiturate anesthetics are known inducers of hepatic microsomal enzymes and may therefore affect the metabolism of other agents.

IV. MICROBIAL FACTORS

Microbial agents are often interfering factors in animal research. Pathogenic agents may disrupt studies by causing clinical disease, lesions, and death. Host defenses that normally hold potential pathogens in check may be compromised by a number of factors, many of which have been mentioned, leading to increased susceptibility and subsequent clinical disease and lesions. The detrimental effect on research is obvious. Less obvious and perhaps less appreciated is the more insidious effect of several microbial agents on a number of biological systems. An important consideration is that tumor preparations and other biological materials may be contaminated with infectious agents, and inoculation of such infected tumor lines may contaminate clean colonies. This section will highlight the microbial agents that are most likely to complicate animal research and testing.

A. Viruses

1. Sendai Virus

Sendai virus infection of mice and rats primarily involves the respiratory system. Therefore, infection by this virus may complicate investigations that focus on the respiratory tract. Beyond that, we are beginning to learn that Sendai virus compromises the immune system, which may lead to cascading effects on various studies, i.e., carcinogenesis and host defense mechanisms.

Sendai virus infections in mice and rats may have profound effects on humoral and cell-mediated immune responses. Sendai virus infection depresses the antibody response to sheep red blood cells (SRBC) in rats (Van Hoosier, 1982), reduces mitogenic response of lymphocytes to the T cell mitogens phytohemagglutinin (PHA) and concanavalin A (Con A) in rats and mice (Garlinghouse and Van Hoosier, 1982; Kay, 1978), and reduces the severity of adjuvant-induced arthritis in rats (Garlinghouse and Van Hoosier, 1978). Kay (1978) reported that the long-term effects of Sendai virus infection on the immune system of mice include an increased prevalence of spontaneous autoimmune disease. Sendai virus infection may also influence transplantation immunology studies. Streilein *et al.* (1981) reported that skin isografts from Sendai virus-infected male mice were rejected sooner than those from noninfected donors.

Sendai virus infection in mice may compromise pulmonary carcinogenesis studies using various chemical carcinogens. Nettesheim (1974) reported that Sendai virus infection in BALB/c mice suppresses the induction of pulmonary adenomas by urethane. More recently, Peck *et al.* (1983) showed that Sendai virus infection in strain A mice variably modifies the pulmonary response to chemical carcinogens. Sendai virus infection reduces the number of lung tumors in 10-chloromethyl-9-chloroanthracene-treated mice and increases the number of lung tumors in mice treated with 7,12-dimethylbenz(*a*)anthracene. The mechanism of suppression or enhancement of pulmonary adenoma by Sendai virus is not known. Sendai virus infection may also alter the response to oncogenic viruses and transplantable tumors. Wheelock (1967) observed that following inoculation of Sendai virus-infected DBA/2 mice with Friend leukemia virus (FV), FV did not replicate in the spleen and leukemia did not develop. This is thought to be due to viral interference. Sendai virus infection of transplantable tumors can modify tumor cell surface antigens and alter their tumorigenicity. Matsuya *et al.* (1978) showed that Ehrlich ascites tumor cells treated with Sendai virus were less tumorigenic in mice. Similarly, a L1210 leukemia cell line persistently infected with Sendai virus became less transplantable in normal syngeneic mice (Takeyama *et al.*, 1979).

Pulmonary defense mechanisms against bacterial infections are altered significantly in Sendai virus-infected mice. Antibacterial defenses of the lungs are suppressed, allowing the intrapulmonary multiplication of members of the upper respiratory tract flora, such as *Pasteurella pneumotropica* (Jakab, 1974). The mechanism for the increased susceptibility of the lung to bacterial superinfections is related to defects in the alveolar macrophage bacteriocidal mechanisms instead of the transport mechanisms of the lung. Extensive studies showed that alveolar macrophages obtained from virus-infected lungs are defective in phagocytosis, phagosome–lysosome fusion, intracellular killing, and bacterial degradation (Jakab, 1981).

Sendai virus infection of pregnant rats had profound effects on fetal development and neonatal mortality. Coid and Wardman (1971) reported that infection of pregnant rats at 4–5 days gestation with aerosols of Sendai virus resulted in resorption of embryos. When rats were infected at 11–12 days of gestation (Coid and Wardman, 1972), fetuses and placentas from infected rats weighed less than noninfected controls, and the gestation period was longer in the infected group. In addition, during the first 24 hr after birth, there was increased neonatal mortality in the infected group. Virus was recovered from the lungs and liver of pregnant rats but not from fetuses. Sendai virus has an affinity for fertilized mouse eggs before implantation (Tuffrey *et al.*, 1972) and for early stage embryos. Infected embryos die, resulting in decreased breeding efficiency (Bowen *et al.*, 1978).

2. Mouse Hepatitis Virus (MHV)

Mouse hepatitis virus infections in mice are usually inapparent and are activated by a variety of experimental manipulations (Kraft, 1982). The virus can infect many different organ systems, thus complicating a wide spectrum of animal experiments. Athymic mice are used extensively in research and are very susceptible to MHV infection.

a. Athymic Mice (Nude Mice). Mouse hepatitis virus infection in athymic mice has several effects on the immune system, including T cell differentiation, antibody responses to T-dependent (SRBC) and T-independent (lipopolysaccharide) antigens, macrophage activity, and natural killer (NK) cell activity. Scheid *et al.* (1975) reported that MHV infection in athymic mice can trigger the maturation of prothymocytes to thymocytes. In athymic mice with acute MHV infection, there are enhanced primary and secondary responses to SRBC and an activation of IgM–IgG switching mechanism, which is inactive in noninfected athymic mice. However, the antibody response to a T cell-independent antigen (lipopolysaccharide) is not altered (Tamura *et al.*, 1978). The numbers and phagocytic activity of peritoneal exudate macrophages are increased markedly in infected mice (Tamura *et al.*, 1980). Natural killer cell activity is markedly increased after MHV infection in athymic mice, which has implications in tumor immunology studies (Tamura *et al.*, 1981). Kyriazis *et al.* (1979) reported that human HEP-2 and SW480 tumor cell lines fail to develop into grossly visible tumors when implanted in MHV-infected athymic mice.

b. Euthymic Mice. Mouse hepatitis virus also alters various aspects of the immune system of euthymic mice, including the antibody response to various antigens, macrophage activity, interferon production, and NK cell activity. During acute MHV infection, antibody response to SRBC is either enhanced or depressed, depending on the timing of antigen injection in relation to MHV infection. Persistently infected mice have markedly depressed serum immunoglobulin levels and a depressed humoral response to SRBC and lipopolysaccharide. The persistently depressed response to SRBC is related to the presence of MHV virus, since viral elimination renders the mice immunocompetent (Virelizier *et al.*, 1976; Leray *et al.*, 1982). This suppressed antibody response is believed to be mediated by interferon (Virelizier *et al.*, 1976) and prostaglandin E (Lahmy and Virelizier, 1981). Mouse hepatitis virus infection also affects macrophage numbers as well as their phagocytic and tumoricidal activity (Boorman *et al.*, 1982). Mouse hepatitis virus induces interferon production and activates NK cell activity (Schindler *et al.*, 1982), and there is an inverse relationship between NK cell activity and resistance to MHV infection (Schindler *et al.*, 1982).

Mouse hepatitis virus is also known to modulate some important enzyme systems. Hepatic enzymes, such as isocitric dehydrogenase, glucose-6-phosphate dehydrogenase (Ruebner and Hirano, 1965), and hepatic glutamic-oxaloacetic transaminase (Paradisi *et al.*, 1972), are markedly increased during infection. In contrast, hepatic microsomal levels of cytochrome *P*-450, NADPH oxidase, and aniline hydroxylase (Budillon *et al.*, 1973) and hepatic oxidative enzymes such as succinic dehydrogenase (Ruebner and Hirano, 1965) are markedly decreased in infected mice. Phenobarbital-induced *P*-450 enzymes are reduced in MHV-infected mice compared to noninfected controls (Budillon *et al.*, 1973).

3. Lymphocytic Choriomeningitis Virus (LCMV)

Lymphocytic choriomeningitis virus depresses the humoral and cell-mediated immune responses of mice (Mims and Wainwright, 1968), but stimulates interferon production (Ronco *et al.*, 1981) and activates NK cell and macrophage activity (Welsh, 1978; Blanden and Mims, 1973). Lymphocytic chroimeningitis virus infection also influences the response of animals to oncogenic viruses and increases resistance to development of leukemia induced by Rauscher leukemia virus (Yuon and Barski, 1966) and to polyoma virus-induced tumors (Hotchin, 1962).

4. Lactic Dehydrogenase Virus (LDHV)

Lactic dehydrogenase virus is a common contaminant of transplantable mouse tumors, oncogenic virus preparations, and other biological materials (Riley *et al.*, 1978). The effects of LDHV on host response have been reviewed by Notkins (1971) and Riley *et al.* (1978). The hallmark of LDHV infection in mice is the elevation of lactate dehydrogenase and other serum enzymes. The increased serum enzyme levels are due to impaired clearance of enzymes rather than increased production or cellular leakage. Some enzymes remain persistently elevated for life (Notkins, 1971). Lactic dehydrogenase virus infection enhances the antibody response to several common antigens (Michaelides and Simms, 1980), delays allograft rejection (Riley *et al.*, 1978), depresses graft-versus-host reaction (Notkins, 1971), stimulates interferon production and activates NK cell activity (Koi *et al.*, 1981). Lactic dehydrogenase virus potentiates the induction and progression of Maloney sarcoma virus-induced tumors and suppresses the induction of mammary tumors by the Bittner virus (Riley *et al.*, 1978).

5. Rodent Parvoviruses

The effects of rat parvoviruses on biological research are becoming better understood. The Kilham rat virus can contaminate transplantable tumors and possibly alter their growth pattern. The virus also can inhibit lymphocyte proliferative responses to concanavalin A (Con A), phytohemagglutinin (PHA), and allogeneic lymphoid cells (Campbell *et al.*, 1977) and alters natural cytotoxicity of rat spleen cells to lymphoma target cells (Darrigrand *et al.*, 1984). Prior infection with H-1 parvovirus in hamsters reduces the incidence of fibrosarcoma induced by 7,12-dimethylbenz(*a*)anthracene (Toolan *et al.*, 1982).

B. Mycoplasma and Rickettsiae

1. *Mycoplasma pulmonis* and *Mycoplasma arthritidis*

Murine respiratory mycoplasmosis (MRM) caused by *Mycoplasma pulmonis* is a very common and serious disease of rodents used for biomedical research. The complicating features of MRM have been extensively reviewed (Lindsey *et al.*, 1971; Cassell *et al.*, 1981). *Mycoplasma pulmonis* has effects on various aspects of pulmonary function. There is reduced mucociliary function, ciliostatis, altered tracheobronchial cell proliferation, and increased mucus secretion (Cassell *et al.*, 1981), all of which affect mucociliary clearance. With regard to the immune system, Davis *et al.* (1980) reported that *M. pulmonis*-infected rats had altered lung lymphocyte numbers and subpopulation distribution. *Mycoplasma pulmonis* is mitogenic for rat (Naot *et al.*, 1979) and mouse (Cole *et al.*, 1975) lymphocytes. *Mycoplasma pulmonis* infection may also affect pulmonary carcinogenesis studies. Schreiber *et al.* (1972) reported that MRM can enhance the neoplastic response of the lungs to the chemical carcinogen, *N*-nitrosoheptamethyleneimine. The exact mechanism of the enhanced tumor response in lungs is not known, but it may be due to suppression of immunologic surveillance, increase in the size of carcinogen susceptible cell population, or alteration of metabolism of the carcinogen. Pulmonary chemical carcinogenesis studies in rodents should be interpreted with caution in view of the high prevalence of *M. pulmonis* infection.

It is becoming increasingly apparent that *M. pulmonis* is a major pathogen of the genital tract of rodents. A number of reproductive parameters are altered in infected animals, including spermatozoan motility, *in vitro* egg fertilization, embryo implantation, skeletal development and ossification of the fetus, and vaginal cytology (Cassell *et al.*, 1981).

Mycoplasma arthritidis may alter immune and interferon responses in rats and mice. *Mycoplasma arthritidis* can enhance or depress interferon production (Cole *et al.*, 1975), depress antibody response, and suppress the mitogenic response of lymphocytes to PHA (Kaklamanis and Pavlatos, 1972).

2. *Hemobartonella muris* and *Eperythrozoon coccoides*

These two blood-borne parasites are becoming rare in rodent colonies due to advances in husbandry practices. They may alter susceptibility to transplantable tumors, phagocytosis, interferon production, and susceptibility to other infectious agents (Baker *et al.*, 1971).

C. Bacteria

The magnitude of research complications due to bacterial diseases is significant. For example, Tyzzer's disease (*Bacillus piliformis*) is said to have ruined many cancer research studies in the United Kingdom (Ganaway, 1982). Bacterial diseases interfere with research primarily by causing clinical disease, lesions, and mortality. Subtle changes, while they occur, are not widely recognized, as for viral diseases. Clinical bacterial disease occurs much more frequently when animals are subjected to natural and experimentally induced stresses, such as immunosuppression, surgical manipulation, sudden changes of environment, and poor husbandry (Ganaway, 1982). As an example, *Pseudomonas aeruginosa* is a normal commensal of conventional laboratory rodents and rarely causes clinical disease in immunocompetent animals. However, infected mice and rats die sooner and in larger numbers than uninfected animals when subjected to an array of procedures, including neonatal thymectomy, anti-lymphocyte serum treatment, cortisone treatment, lethal whole body irradiation, thermal trauma, surgical trauma, tumor implantation, and infection with other organisms (Flynn, 1963; Hightower *et al.*, 1966; Weisbroth, 1979). *Streptococcus pneumoniae* infection in rats is generally subclinical, but latent infection can be activated by various experimental manipulations, such as splenectomy, iron deficiency, and pulmonary edema (Weisbroth, 1979). Rats with clinical disease may show altered serum enzymes, hepatic peroxisomes, blood chemistry parameters, and thyroid function (Weisbroth, 1979).

D. Parasites

Parasites may complicate research by (1) decreasing the volume of host blood and body fluids, (2) competing with the host for nutrients, (3) inducing tissue damage, (4) stimulating abnormal tissue growth, (5) altering host physiology, (6) altering immune function, and (7) mechanical interference (Hsu, 1980). Such subtle effects may not be recognized since the morbidity and mortality due to parasitic infections are usually very low. Animals infected with internal or external parasites that feed on red blood cells may be anemic, making them unsuitable for most research endeavors. Some intestinal parasites damage intestinal epithelium, thereby altering absorption and thus compromising the usefulness of infected animals for nutritional studies. Some parasites may stimulate abnormal tissue growth, thereby complicating study of toxic and carcinogenic agents. As examples, *Trichosomoides crassicauda* causes hyperplasia or papillomatosis of urinary bladder epithelium, and *Gongylonema neoplasticum* is associated with esophageal tumors in rats. Rats infected with these parasites are undesirable for carcinogenesis studies. In fact, the established carcinogenicity of saccharin in the rat was questioned due to the complication created by possible *T. crassicauda* infection (Homburger, 1978). Internal and external parasites are known to alter host immune responses. Mice infested with mites may have a significantly reduced cellular infiltration in contact hypersensitivity reactions (Laltoo and Kind, 1979). Rodents infected with the pinworm, *Syphacia obvelata*, may have a depressed antibody response to various antigens (Hsu, 1980). Rats infected with this parasite develop less severe lesions of adjuvant-induced arthritis as compared to noninfected controls (Pearson and Taylor, 1975). *Encephalitozoon cuniculi*, a common parasite of rabbits, may also influence the immune response of laboratory animals. *Encephalitozoon cuniculi*-infected rabbits have a depressed IgG response and an increased IgM response to *Brucella abortus* antigens (Cox, 1977). Niederkorn *et al.* (1983) observed increased NK cell activity in mice experimentally infected with *E. cuniculi*. *Encephalitozoon cuniculi* may also influence tumor transplantation studies. Petri (1966) reported that Yoshida sarcoma tumor cells infected with *E. cuniculi* did not grow in rats as did noninfected tumor cells.

ACKNOWLEDGMENTS

Supported in part by funds from United States Public Health Service Grant RR00890 and the Samuel Roberts Noble Foundation, Ardmore, Oklahoma.

REFERENCES

Adams, H. R. (1972). Prolongation of barbiturate anesthesia by chloramphenicol in laboratory animals. *J. Am. Vet. Med. Assoc.* **157**, 1908–1913.

Adams, H. R. (1975). Cardiovascular depressant effects of the neomycin-streptomycin group of antibiotics. *Am. J. Vet. Res.* **36**, 103–108.

Adams, H. R. (1976). Antibiotic-induced alterations of cardiovascular reactivity. *Fed. Proc., Fed. Am. Soc. Exp. Biol.* **35**, 1148–1150.

Alder, R., and Friedman, S. B. (1968). Plasma cortisone response to environmental stimulation: Effects of duration of stimulation and the 24-hour adrenocortical rhythm. *Neuroendocrinology* **3**, 378–386.

Anthony, A. (1962). Criteria for acoustics in animal housing. *Lab. Anim. Care* **13**, 340–347.

Baetjer, A. M. (1968). Role of environmental temperature and humidity in susceptibility to disease. *Arch. Environ. Health* **16**, 565–570.

Baker, H. J., Cassell, G. H., and Lindsey, J. R. (1971). Research complications due to *Haemobartonella* and *Eperythrozoon* infections in experimental animals. *Am. J. Pathol.* **64**, 625–656.

Baker, H. J., Lindsey, J. R., and Weisbroth, S. H. (1979). Housing to control research variables. *In* "The Laboratory Rat" (H. J. Baker, J. R. Lindsey, and S. H. Weisbroth, eds.), Vol 1, pp. 169–192. Academic Press, New York.

Bartlett, J. G., Chang, T., Moon, N., and Onderdonk, A. B. (1978). Antibiotic-induced lethal enterocolitis in hamsters: Studies with eleven different agents and evidence to support the pathogenic role of toxin-producing clostridia. *Am. J. Vet. Res.* **39**, 1525–1530.

Bellhorn, R. W. (1980). Lighting in the animal environment. *Lab. Anim. Sci.* **30**, 440–450.

Berger, J. (1980). Seasonal influences on circadian rhythms in the blood picture of laboratory mice. I. Leucocytes and erythrocytes. II. Lymphocytes, eosinophils and segmented neutrophils. *Z. Versuchstierkd.* **22**, 122–134.

Berger, J. (1981). Seasonal variations in reticulocyte counts in blood of laboratory mice and rats. Short communication. *Z. Versuchstierkd.* **23**, 8–12.

Besch, E. L. (1975). Animal cage-room dry-bulb and dew-point temperature differentials. *ASHRAE Trans.* **81**, 549–558.

Blakley, B. R., Sisodia, C. S., and Mukkur, T. K. (1980). The effect of methylmercury, tetraethyl lead and sodium arsenite on the humoral immune response in mice. *Toxicol. Appl. Pharmacol.* **52**, 245–254.

Blanden, R. V., and Mims, C. A. (1973). Macrophage activation in mice infected with ectromelia or lymphocytic choriomeningitis virus. *Aust. J. Exp. Biol. Med. Sci.* **51**, 393–398.

Boorman, G. A., Luster, M. I., Dean, J. H., Campbell, J. L., Laner, L. A., Talley, F. A., Wilson, R. E., and Collins, M. J. (1982). Peritoneal macrophage alterations caused by naturally occurring mouse hepatitis virus. *Am. J. Pathol.* **106**, 110–117.

Bowen, R. A., Storz, J., and Leary, J. (1978). Interaction of viral pathogens with preimplantation embryos. *Theriogenology* **9**, 88.

Bowman, R. E., Wold, R. C., and Sackett, G. P. (1970). Circadian rhythms of plasma 17-hydrocorticosteroids in the infant monkey. *Proc. Soc. Exp. Biol. Med.* **133**, 342–344.

Broderson, J. R., Lindsey, J. R., and Crawford, J. E. (1976). The role of environmental ammonia in respiratory mycoplasmosis of rats. *Am. J. Pathol.* **85**, 115–130.

Bruce, D. L. (1972). Halothane inhibition of phytohemagglutinin-induced transformation of lymphocytes. *Anesthesiology* **36**, 201–205.

Budillon, G., Carrella, M., DeMarco, F., and Mazzacca, G. (1973). Effect of phenobarbital on MHV-3 viral hepatitis of the mouse. *Pathol. Microbiol.* **39**, 461–466.

Campbell, D. A., Stall, S. P., Manders, E. K., Bonnard, G. D., Oldham, R. K., Salzman, L. A., and Hereberman, R. B. (1977). Inhibition of *in vitro* lymphoproliferative responses by *in vivo* passaged rat 13762 mammary adenocarcinoma cells. II. Evidence that Kilham rat virus is responsible for the inhibitory effect. *Cell. Immunol.* **33**, 378–391.

Cass, J. S. (1970). Chemical factors in laboratory animal surroundings. *Bio-Science* **20**, 658–662.

Cassell, G. H., Lindsey, J. R., and Davis, J. K. (1981). Respiratory and genital mycoplasmosis of laboratory rodents: Implications for biomedical research. *Isr. J. Med. Sci.* **17**, 548–554.

Cayen, M. N., Givner, M. L., and Kraml, M. (1972). Effect of diurnal rhythm and food withdrawal on serum lipid levels in the rat. *Experientia* **28**, 502–503.

Chédid, A., and Nair, V. (1972). Diurnal rhythm in endoplasmic reticulum of rat liver. Electron microscopic study. *Science* **175**, 176–179.

Christian, J. J., and LeMunyan, C. D. (1958). Adverse effects of crowding on lactation and reproduction of mice and two generations of their progeny. *Endocrinology* **63**, 517–529.

Clough, G. (1976). The immediate environment of the laboratory animal. *In* "Control of the Animal House Environment" (T. McSheehy, ed.), Lab. Anim. Handb. No. 7, pp. 77–94. Lab. Anim. Ltd., London.

Coid, C. R., and Wardman, G. (1971). The effect of parainfluenza type 1 (Sendai) virus infection on early pregnancy in the rat. *J. Reprod. Fertil.* **24**, 39–43.

Coid, C. R., and Wardman, G. (1972). The effect of maternal respiratory disease induced by parainfluenza type 1 (Sendai) virus on foetal development and neonatal mortality in the rat. *Med. Microbiol. Immunol.* **157**, 181–185.

Cole, B. C., Overall, J. C., Lombardi, P. S., and Glasgow, L. A. (1975). Mycoplasma-mediated hyporeactivity to various interferon inducers. *Infect. Immun.* **12**, 1349–1354.

Conney, A. H. (1967). Pharmacological implications of microsomal enzyme induction. *Pharmacol. Rev.* **19**, 317–366.

Conney, A. H., and Burns, J. J. (1972). Metabolic interactions among environmental chemicals and drugs. *Science* **178**, 576–586.

Cook, J. A., Marconi, E. A., and DiLuzio, N. R. (1974). Lead, cadmium and endotoxin interaction: Effects on mortality and hepatic function. *Toxicol. Appl. Pharmacol.* **28**, 292–297.

Cook, J. A., Hoffmann, E. O., and DiLuzio, N. R. (1975). Influence of lead and cadmium on the susceptibility of rats to bacterial challenge. *Proc. Soc. Exp. Biol. Med.* **150**, 741–747.

Cox, J. C. (1977). Altered immune responsiveness associated with *Encephalitozoon cuniculi* infection in rabbits. *Infect. Immun.* **15**, 392–395.

Cullen, B. F. (1974). The effects of halothane and nitrous oxide on phagocytosis and human leukocyte metabolism. *Anesth. Analg. (Cleveland)* **53**, 531–536.

Darrigrand, A. A., Singh, S. B., and Lang, C. M. (1984). The effects of Kilham rat virus on natural cell-mediated cytotoxicity in Brown Norway and Wistar-Furth rats. *Am. J. Vet. Res.* **45**, 200–202.

Davis, D. E. (1978). Social behavior in a laboratory environment. *In* "Symposium on Laboratory Animal Housing" (Inst. Lab. Anim. Resour.), pp. 44–63. Natl. Acad. Sci.—Natl. Resour. Counc., Washington, D.C.

Davis, J. K., Maddox, P. A., Thorp, R. B., and Cassell, G. H. (1980). Immunofluorescent characterization of lymphocytes in lungs of rats infected with *Mycoplasma pulmonis*. *Infect. Immun.* **27**, 255–259.

Degraeve, N. (1981). Carcinogenic, teratogenic, and mutagenic effects of cadmium. *Mutat. Res.* **86**, 115–135.

Deimling, M. J., and Schnell, R. C. (1980). Circadian rhythms in the biological response and disposition of ethanol in the mouse. *J. Pharmacol. Exp. Ther.* **213**, 1–8.

DeWitt, G. H., ed. (1976). "Symposium on Biological Effects and Measurement of Light Sources," DHEW (FDA) Publ. No. 77-8002. U.S. Govt. Printing Office, Washington, D.C.

Dymsza, H. A., Miller, S. A., Maloney, J. R., and Foster, H. L. (1963). Equilibration of the laboratory rat following exposure to shipping stresses. *Lab. Anim. Care* **13**, 60–65.

Edwards, G. S., Fox, J. G., Policastro, P., Goff, U., Wolf, M. H., and Fine, D. H. (1979). Volatile nitrosamine contamination of laboratory animal diets (letter). *Cancer Res.* **39**, 1857–1858.

Elis, J., and DiPaolo, J. A. (1967). Aflatoxin B1. Induction of malformations. *Arch. Pathol.* **83**, 53–57.

Environmental Protection Agency (1979). (Proposed Guidelines.) *Fed. Regist.* **44**, 27353–27354.

Farber, E. (1982). Chemical carcinogens: A biological perspective. *Am. J. Pathol.* **106**, 271–296.

Feigin, R. D., San Joaquin, V. H., Haymond, M. W., and Wyatt, R. G. (1969). Daily periodicity of susceptibility of mice to pneumococcal infection. *Nature (London)* **224**, 379–380.

Fletcher, J. I. (1976). Influence of noise on animals. *In* "Control of the Animal House Environment" (T. McSheehy, ed.), Lab. Anim. Handb. No. 7, pp. 51–62. Lab. Anim. Ltd., London.

Flynn, R. J. (1959). Studies on the etiology of ringtail of rats. *Proc. Anim. Care Panel* **9**, 155–160.

Flynn, R. J. (1963). *Pseudomonas aeruginosa* infection and its effects on biological and medical research. *Lab. Anim. Care* **13**, 1–6.

Food and Drug Administration (1978). Nonclinical laboratory studies. Good laboratory practice regulations. *Fed. Regist.* **43**, 60017.

Fouts, J. R. (1976). Overview of the field: Environmental factors affecting chemical or drug effects in animals. *Fed. Proc., Fed. Am. Soc. Exp. Biol.* **35**, 1162–1165.

Fox, J. G. (1977). Clinical assessment of laboratory rodents on longterm bioassay studies. *J. Environ. Pathol. Toxicol.* **1**, 199–226.

Fox, J. G., Smith, M. E., and Helfrich-Smith, L. (1980). The contamination with chemicals of animal feeding systems: Evaluation of two caging systems and standard cage washing equipment. *Lab. Anim. Sci.* **30**, 967–973.

Fox, J. G., Donahue, P. R., and Essigmann, J. M. (1980). Efficacy of inactivation of representative chemical carcinogens utilizing commercial alkaline and acidic cage washing compounds. *J. Environ. Pathol. Toxicol.* **4**, 97–105.

Friedman, M., Byers, S. O., and Brown, A. E. (1967). Plasma lipid responses of rats and rabbits to an auditory stimulus. *Am. J. Physiol.* **212**, 1174–1178.

Fuhrman, G. J., and Fuhrman, F. A. (1961). Effects of temperature on the action of drugs. *Annu. Rev. Pharmacol.* **1**, 65–78.

Galloway, J. H. (1968). Antibiotic toxicity in white mice. *Lab. Anim. Care* **18**, 421–425.

Gamble, M. R., and Clough, G. (1976). Ammonia build-up in animal boxes and its effect on rat tracheal epithelium. *Lab. Anim.* **10**, 93–104.

Ganaway, J. R. (1982). Bacterial infections. *In* "The Mouse in Biomedical Research" (H. C. Foster, J. D. Small, and J. G. Fox, eds.), Vol. 2, pp. 2–13. Academic Press, New York.

Garlinghouse, L. E., Jr., and Van Hoosier, G. L., Jr. (1978). Studies on adjuvant-induced arthritis, tumor transplantability, and serologic response to bovine serum albumin in Sendai virus infected rats. *Am. J. Vet. Res.* **39**, 297–300.

Garlinghouse, L. E., Jr., and Van Hoosier, G. L., Jr. (1982). The suppression of lymphocytic mitogenesis in Sendai virus infected rats. *Am. Soc. Microbiol. Abstr. Annu. Meet.* p. 258.

Geber, W. F., Anderson, T. A., and Van Dyne, V. (1966). Physiologic responses of the albino rat to chronic noise stress. *Arch. Environ. Health* **12**, 751–754.

Gerber, S. B., Leonard, A., and Jacquet, P. (1980). Toxicity, mutagenicity, and teratogenicity of lead. *Mutat. Res.* **76**, 115–141.

Greenman, D. L., Bryant, P., Kodell, R. L., and Sheldon, W. (1982). Influence of cage shelf level on retinal atrophy in mice. *Lab. Anim. Sci.* **32**, 353–356.

Hadlow, W. J., Grimes, E. F., and Jay, G. E. (1955). Stilbesterol-contaminated feed and reproductive disturbances in mice. *Science* **122**, 643–644.

Halberg, F., Haus, E., Cardoso, S. S., Scheving, L. E., Kuhl, J. F. W., Shiotsuka, R., Rosene, G., Pauly, J. E., Runge, W., Spaulding, J. F., Lee, J. K., and Good, R. A. (1973). Toward a chronotherapy of neoplasia tolerance of treatment depends upon host rhythms. *Experientia* **29**, 909–934.

Harper, M. H., Collins, P., Johnson, B., Egar, E. I., and Biava, C. (1982). Hepatic injury following halothane, enflurane, and isoflurane anesthesia in rats. *Anesthesiology* **56**, 14–17.

Hastings, J. W. (1970). The biology of circadian rhythms from man to microorganisms. *N. Engl. J. Med.* **282**, 435–441.

Hastings, J. W., and Menaker, M. (1976). Physiological and biochemical aspects of circadian rhythms. *Fed. Proc., Fed. Am. Soc. Exp. Biol.* **35**, 2325–2357.

Haus, E., Halberg, F., Scheving, L., Pauly, J. E., Cordoso, S., Kuhl, J. F. W., Sothern, R. B., Shiotsuka, R. N., and Hwang, D. W. (1972). Increased tolerance of leukemic mice to arabinosyl cytosine given on schedule adjusted to circadian system. *Science* **177**, 80–82.

Hauser, W. E., and Remington, J. S. (1982). Effect of antibiotics on the immune response. *Am. J. Med.* **72**, 711–716.

Hemphill, F. E., Kasberle, M. L., and Buck, W. B. (1971). Lead suppression of mouse resistance to *Salmonella typhimurium*. *Science* **172**, 1031–1032.

Hermann, L. M., White, W. J., and Lang. C. M. (1982). Prolonged exposure to acid, chlorine, or tetracycline in drinking water: Effects on delayed-type hypersensitivity, hemagglutination titers, and reticuloendothelial clearance rates in mice. *Lab. Anim. Sci.* **32**, 603–608.

Hightower, D., Uhrig, H. T., and Davis, J. I. (1966). *Pseudomonas aeruginosa* infection in rats used in radiobiology research. *Lab. Anim. Care* **16**, 85–93.

Homburger, F. (1978). Saccharin and cancer. *N. Engl. J. Med.* **297**, 560–561.

Hotchin, J. (1962). The biology of lymphocytic choriomeningitis infection: Virus induced immune disease. *Cold Spring Harbor Symp. Quant. Biol.* **27**, 479–499.

Hsu, C. K. (1980). Parasitic diseases: How to monitor them and their effects on research. *Lab Anim.* **9**, 48–53.

Institute of Laboratory Animal Resources (ILAR) (1978). "Guide for the Care and Use of Laboratory Animals." Natl. Acad. Sci., Washington, D.C.

Ito, T., and Ingalls, T. H. (1981). Sodium pentobarbital-induced mutations in the hamster. *Arch. Environ. Health* **36**, 316–320.

Iturrian, W. B. (1971). Effect of noise in the animal house on experimental seizures and growth of weanling mice. *In* "Defining the Laboratory Animal," pp. 332–352. Natl. Acad. Sci., Washington, D.C.

Izquierdo, J. N., and Gibbs, S. J. (1972). Circadian rhythms of DNA synthesis and mitotic activity in hamster cheek pouch epithelium. *Exp. Cell Res.* **71**, 402–408.

Jakab, G. J. (1974). Effect of sequential inoculations of Sendai virus and *Pasteurella pneumotropica* in mice. *J. Am. Vet. Med. Assoc.* **164**, 723–728.

Jakab, G. J. (1981). Interactions between Sendai virus and bacterial pathogens in the murine lungs: A review. *Lab. Anim. Sci.* **31**, 170–177.

Jori, A., DiSalle, E., and Santini, V. (1971). Daily rhythmic variation and liver drug metabolism in rats. *Biochem. Pharmacol.* **29**, 2965–2969.

Kaklamanis, E., and Pavlatos, M. (1972). The immunosuppressive effect of mycoplasma infection. I. Effect on the humoral and cellular response. *Immunology* **22**, 695–702.

Kay, M. M. B. (1978). Long term subclinical effects of parainfluenza (Sendai) infection on immune cells of aging mice. *Proc. Soc. Exp. Biol. Med.* **158**, 326–331.

Koi, M., Saiato, M., Ebina, T., and Ishida, N. (1981). Lactate dehydrogenase-elevating agent is responsible for interferon and enhancement of natural killer cell activity by inoculation of Erhlich ascites carcinoma cells into mice. *Microbiol. Immunol.* **25**, 565–574.

Koller, L. D. (1979). Effects of environmental contaminants on the immune system. *Adv. Vet. Sci. Comp. Med.* **23**, 267–295.

Kopp, S. J., Glanek, T., Erlanger, M., Perry, E. F., Barany, M., and Perry, H. M. (1980). Altered metabolism and function of rat heart following chronic low level cadmium/lead feeding. *J. Mol. Cell. Cardiol.* **12**, 1407–1425.

Kraft, L. M. (1982). Murine (mouse) hepatitis virus infection. *In* "The Mouse in Biomedical Research" (H. L. Foster, J. D. Small, and J. G. Fox, eds.), Vol. 2, pp. 173–191. Academic Press, New York.

Kyriazis, A. P., DiPersio, J., Michael, J. G., and Pesce, A. J. (1979). Influence of the mouse hepatitis virus (MHV) infection on the growth of human tumors in the athymic mouse. *Int. J. Cancer* **23**, 402–409.

Lahmy, C., and Virelizier, J. L. (1981). Prostaglandins as probable mediators of the suppression of antibody production by mouse hepatitis virus infection. *Ann. Immunol. (Paris)* **132C**, 101–105.

Laltoo, H., and Kind, L. S. (1979). Reduction of contact sensitivity reactions to oxazolone in mite-infested mice. *Infect. Immun.* **26**, 30–35.

LaMont, J. T., Sonnenblink, E. B., and Rothman, S. (1979). Role of clostridial toxin in the pathogenesis of clindamycin colitis in rabbits. *Gastroenterology* **76**, 356–361.

Larson, H. E. (1980). The experimental pathogenesis of antibiotic related colitis. *Scand. J. Infect. Dis., Supp.* **22**, 7–10.

LaVail, M. M. (1976). Survival of some photoceptor cells in albino rats following long-term exposure to continuous light. *Invest. Ophthalmol.* **15**, 64–72.

LeBouton, A. V., and Handler, S. D. (1971). Persistent circadian rhythmicity of protein synthesis in liver of starved rats. *Experientia* **27**, 1031–1032.

Leray, D., Dupuy, C., and Dupuy, J. M. (1982). Immunopathology of mouse hepatitis virus type 3 infection. IV. MHV3-induced immunodepression. *Clin. Immunol. Immunopathol.* **23**, 539–547.

Lindsey, J. R., Baker, H. J., Overcash, R. G., Cassell, G. H., and Hunt, C. E. (1971). Murine chronic respiratory disease: Significance as a research complication and experimental production with *Myocoplasma pulmonis.* *Am. J. Pathol.* **64**, 675–716.

Lindsey, J. R., Conner, M. W., and Baker, H. J. (1978). Physical, chemical and microbial factors affecting biologic response. *In* "Symposium on Laboratory Animal Housing" (Inst. Lab. Anim. Resour.), pp. 37–43. Natl. Acad. Sci.—Natl. Resour. Counc., Washington, D.C.

Loose, L. D., Silkworth, J. B., and Simpson, D. W. (1978). Influence of cadmium on the phagocytic and microbicidal activity of murine peritoneal macrophages, pulmonary alveolar macrophages, and polymorphonuclear neutrophils. *Infect. Immun.* **22**, 378–381.

Lowe, B. R., Fox, J. G., and Bartlett, J. G. (1980). *Clostridium difficile* associated cecitis in guinea pigs exposed to penicillin. *Am. J. Vet. Res.* **41**, 1277–1279.

Manning, P. J., and Banks, K. L. (1968). The effects of transportation on the aged rat. *N.A.S.—N.R.C., Publ.* **1591**, 98–103.

Matsuya, Y., Kusano, T., Endo, S., Takahashi, N., and Yamane, I. (1978). Reduced tumorigenicity by addition *in vitro* of Sendai virus. *Eur. J. Cancer* **14**, 837–850.

Mazze, R. I., Cousins, M. J., and Kosek, J. C. (1973). Strain differences in metabolism and susceptibility to the nephrotoxic effects of methoxyflurane in rats. *J. Pharmacol. Exp. Ther.* **184**, 481–488.

Michaelides, M. C., and Simms, E. S. (1980). Immune responses in mice infected with lactic dehydrogenase virus. Antibody response to a T-dependent and a T-independent antigen during acute and chronic LDV infections. *Cell. Immunol.* **50**, 253–260.

Mims, C. A., and Wainwright, S. (1968). The immunodepressive action of lymphocytic choriomeningitis virus in mice. *J. Immunol.* **101**, 717–724.

Mitropoulos, K. A., Balasubcamaniam, S., Gibbons, G. F., and Reaves, B. E. A. (1972). Diurnal variation in the activity of cholesterol 7-hydroxylase in the livers of fed and fasted rats. *FEBS Lett.* **27**, 203–206.

Mizuno, S., and Ishida, A. (1982). Selective enhancement of bleomycin cytotoxicity by local anesthetics. *Biochem. Biophys. Res. Commun.* **105**, 425–431.

Mock, E. J., Norton, H. W., and Frankel, A. I. (1978). Daily rhythmicity of serum testosterone concentration in the male laboratory rat. *Endocrinology* **103**, 1111–1121.

Mulder, J. B. (1971). Animal behavior and electromagnetic energy waves. *Lab. Anim. Sci.* **21**, 389–393.

Murahami, H. (1971). Differences between internal and external environments in the mouse cage. *Lab. Anim. Sci.* **21**, 680–684.

Nair, V., and Casper, R. (1969). The influence of light on daily rhythm in hepatic drug metabolizing enzymes in rat. *Life Sci.* **8**, 1291–1298.

Naot, Y., Menchav, S., Ben-David, E., and Ginsburg, H. (1979). Mitogenic activity of *Mycoplasma pulmonis.* I. Stimulation of rat B and T lymphocytes. *Immunology* **36**, 399–406.

Nash, R. E., and Llanos, J. M. E. (1971). Twenty-four hour variations in DNA synthesis of a fast-growing and a slow-growing hepatoma: DNA synthesis rhythm in hepatoma. *JNCI, J. Natl. Cancer Inst.* **47**, 1007–1012.

Nebert, D. W., and Felton, J. S. (1976). Importance of genetic factors influencing the metabolism of foreign compounds. *Fed. Proc., Fed. Am. Soc. Exp. Biol.* **35**, 1133–1141.

Nettesheim, P. (1974). Review and introductory remarks: Multifactorial respiratory carcinogenesis. *In* "Experimental Lung Cancer, Carcinogenesis and Bioassays" (K. Eberhard, and J. F. Park, eds.), pp. 157–160. Springer-Verlag, Berlin and New York.

Newberne, P. M. (1975). Influence on pharmacological experiments of chemicals and other factors in diets of laboratory animals. *Fed. Proc., Fed. Am. Soc. Exp. Biol.* **34**, 209–218.

Newberne, P. M., and McConnell, R. G. (1980). Dietary nutrients and contaminants in laboratory animal experimentation. *J. Environ. Pathol. Toxicol.* **4**, 105–122.

Newberne, P. M., Weigert, J., and Kula, N. (1979). Effects of dietary fat on hepatic mixed function oxidases and hepatocellular carcinoma induced by aflatoxin B$_1$ in rats. *Cancer Res.* **39**, 3986–3991.

Newton, W. M. (1978). Environmental impact on laboratory animals. *Adv. Vet. Sci. Comp. Med.* **22**, 1–28.

Niederkorn, J. Y., Brieland, J. K., and Mayhew, E. (1983). Enhanced natural killer cell activity in experimental murine encephalitozoonosis. *Infect. Immun.* **41**, 302–307.

Notkins, A. L. (1971). Enzymatic and immunologic alterations in mice infected with lactic dehydrogenase virus. *Am. J. Pathol.* **64**, 733–746.

Ohanian, E. V., and Iwai, J. (1980). Etiological role of cadmium in hypertension in an animal model. *J. Environ. Pathol. Toxicol.* **4**, 229–241.

O'Steen, W. K., and Anderson, K. V. (1972). Photoreceptor degeneration after exposure of rats to incandescent illumination. *Z. Zellforsch. Mikrosk. Anat.* **127**, 306–313.

O'Steen, W. K., Shear, C. R., and Anderson, K. V. (1973). Retinal damage after prolonged exposure to visible light. A light and electron microscopic study. *Am. J. Anat.* **134**, 5–22.

O'Steen, W. K., Anderson, K. V., and Shear, C. R. (1974). Photoreceptor degeneration in albino rats. Dependency on age. *Invest. Ophthalmol.* **13**, 334–339.

Owen, P. R. (1969). Turbulent flow and particle deposition in the trachea. *In* "Circulation and Respiratory Mass Transport" (G. E. W. Wolstenholme and J. Knight, eds.), pp. 236–255. *Ciba Found. Symp.*

Paradisi, F., Graziano, L., and Maio, G. (1972). Histochemistry of glutamicoxaloacetic transaminase in mouse liver during MHV-3 infection. *Experientia* **28**, 551–552.

Parker, J. L., and Adams, H. R. (1978). The influence of chemical restraining agents on cardiovascular function: A review. *Lab. Anim. Sci.* **28**, 575–583.

Pearson, D. J., and Taylor, G. (1975). The influence of the nematode *Syphacia obvelata* on adjuvant arthritis in rats. *Immunology* **29**, 391–396.

Peck, R. M., Eator, G. T., Peck, E. B., and Litwin, S. (1983). Influence of Sendai virus on carcinogenesis in strain A mice. *Lab. Anim. Sci.* **33**, 154–156.

Perry, H. M., and Erlanger, M. W. (1974). Metal induced hypertension following chronic feeding of low doses of cadmium and mercury. *J. Lab. Clin. Med.* **83**, 541–547.

Petri, M. (1966). The occurrence of *Nosema cuniculi (Encephalitozoon cuniculi)* in the cells of transplantable, malignant ascites tumors and its effect upon tumor and host. *Acta Pathol. Microbiol. Scand.* **66**, 13.

Pucak, G. J., Lee, C. S., and Zaino, A. S. (1977). Effects of prolonged high temperature on testicular development and fertility in the male rat. *Lab. Anim. Sci.* **27**, 76–77.

Radzialowski, F. M., and Bousquet, W. F. (1968). Daily rhythmic variation in hepatic drug metabolism in the rat and mouse. *J. Pharmacol. Exp. Ther.* **163**, 229–238.

Ramaley, J. A. (1972). Changes in daily serum corticosterone values in maturing male and female rats. *Steroids* **20**, 185–197.

Reiter, R. J. (1973). Comparative effects of continual lighting and pinealectomy on the eyes, the Harderian glands and reproduction in pigmented and albino rats. *Comp. Biochem. Physiol.* **44**, 503–509.

Riley, V. (1981). Psychoneuroendocrine influences on immunocompetence and neoplasia. *Science* **212**, 1100–1109.

Riley, V., Spackman, D. H., and Santisteban, G. A. (1978). The LDH virus: An interfering biological contaminant. *Science* **200**, 124–126.

Robinson, W. G., and Kuwabara, T. (1976). Light-induced alterations of retinal pigment epithelium in black, albino and beige mice. *Exp. Res.* **22**, 549–557.

Romero, J. A. (1976). Influence of diurnal cycles on biochemical parameters of drug sensitivity: The pineal gland as a model. *Fed. Proc., Fed. Am. Soc. Exp. Biol.* **35**, 1157–1161.

Ronco, P., Riviere, Y., Thoua, Y., Bandu, M. T., Guillon, J. C., Verroust, P., and Morel-Maroger, L. (1981). Lymphocytic choriomeningitis in the nude mouse. An immunological study. *Immunology* **43**, 763–770.

Ross, M. H., and Bras, G. (1973). Influence of protein under- and overnutrition on spontaneous tumor prevalence in the rat. *J. Nutr.* **103**, 944–963.

Rubanyi, G. (1980). Effect of hemorrhagic shock on the performance, O_2-consumption, and ultrastructure of isolated rat hearts. *Circ. Shock* **7**, 59–70.

Ruebner, B. H., and Hirano, T. (1965). Viral hepatitis in mice. Changes in oxidative enzymes and phosphatases after murine hepatitis virus (MHV-3) infection. *Lab. Invest.* **14**, 157–168.

Schaerdel, A. D., White, W. J., Lang, C. M., Dvorchik, B. H., and Bohner, K. (1983). Localized and systemic effects of environmental ammonia in rats. *Lab. Anim. Sci.* **33**, 40–45.

Scheid, M. P., Goldstein, G., and Boyse, E. A. (1975). Differentiation of T-cells in nude mice. *Science* **190**, 1211–1213.

Scheving, L. E., Vedral, D. F., and Pauly, J. E. (1968). A circadian susceptibility rhythm in rats to pentobarbital sodium. *Anat. Rec.* **160**, 741–749.

Schindler, L., Engler, H., and Kirchner, M. (1982). Activation of natural killer cells and induction of interferon after injection of mouse hepatitis virus type 3 in mice. *Infect. Immun.* **35**, 868–873.

Schreiber, H., Nettesheim, P., Lijinsky, W., Richter, C. B., and Walburg, H. E., Jr. (1972). Induction of lung cancer in germfree specific-pathogen-free and infected rats by *N*-nitrosoheptamethyleneimine: Enhancement by respiratory infection. *JNCI, J. Natl. Cancer Inst.* **4**, 1107–1114.

Serrano, L. J. (1971). Carbon dioxide and ammonia in mouse cages: Effect of cage covers, population and activity. *Lab. Anim. Sci.* **21**, 75–85.

Shackelford, P. G., and Feigin, R. D. (1973). Periodicity of susceptibility to pneumococcal infection: Influence of light and adrenocortical secretions. *Science* **182**, 285–287.

Shapiro, J., Jersky, J., Katzav, S., Feldman, M., and Segal, S. (1981). Anesthetic drugs accelerate the progression of postoperative metastases of mouse tumors. *J. Clin. Invest.* **68**, 678–685.

Shapiro, R. (1980). Chemical contamination of drinking water: What it is and where it comes from. *Lab. Anim.* **9**, 45–51.

Simmons, M. L., Robie, D. M., Jones, J. B., and Serrano, L. J. (1968). Effect of a filter cover on temperature and humidity in a mouse cage. *Lab. Anim.* **2**, 113–120.

Streilein, J. W., Shadduck, J. A., and Pakes, S. P. (1981). Effects of splenectomy and Sendai virus infection on rejection of male skin isografts by pathogen-free C57BL/6 female mice. *Transplantation* **32**, 34–37.

Stuhlman, R. A., and Wagner, J. E. (1971). Ringtail in *Mystromys albicaudatis*: A case report. *Lab. Anim. Sci.* **21**, 585–587.

Takeyama, H., Kawashima, K., Yamada, K., and Ito, Y. (1979). Induction of tumor resistance in mice by L1210 leukemia cells persistently infected with HVJ (Sendai virus). *Gann* **70**, 493–501.

Tamura, T., Machii, K., Ueda, K., and Fujiwara, K. (1978). Modification of immune response in nude mice infected with mouse hepatitis virus. *Microbiol. Immunol.* **22**, 557–564.

Tamura, T., Sakaguchi, A., Kai, C., and Fujiwara, K. (1980). Enhance phagocytic activity of macrophages in mouse hepatitis virus infected nude mice. *Microbiol. Immunol.* **24**, 243–247.

Tamura, T., Sakaguchi, A., Ishida, T., and Fujiwara, K. (1981). Effect of mouse hepatitis virus infection on natural killer cell activity in nude mice. *Microbiol. Immunol.* **25**, 1363–1368.

Tennekes, H. A., Wright, A. S., Dix, K. M., and Koeman, J. H. (1981). Effects of dieldrin, diet, and bedding on enzyme function and tumor incidence in livers of male CF-1 mice. *Cancer Res.* **41**, 3615–3620.

Thong, Y. H., and Ferrante, A. (1980). Effect of tetracycline on immunological responses in mice. *Clin. Exp. Immunol.* **39**, 728–732.

Toolan, H. W., Rhode, S. L., III, and Gierthy, J. F. (1982). Inhibition of 7,12-dimethylbenz(*a*)anthracene-induced tumors in Syrian hamsters by prior infection with H-1 parvovirus. *Cancer Res.* **42**, 2552–2555.

Tuffrey, M., Zisman, B., and Barnes, R. D. (1972). Sendai (parainfluenza 1) infection of mouse eggs. *Br. J. Exp. Pathol.* **53**, 638–640.

Van Hoosier, G. L., Jr. (1982). Modulators of the immune system. *In* "ACLAM Forum: Immunobiology," pp. 110–114. Am. Coll. Lab. Anim. Med.

Vessel, E. S., Lang, C. M., White, W. J., Passanati, G. T., Hill, R. N., Clemens, T. L., Lui, D. K., and Johnson, W. D. (1976). Environmental and genetic factors affecting the response of laboratory animals to drugs. *Fed Proc., Fed. Am. Soc. Exp. Biol.* **35**, 1125–1132.

Virelizier, J. L., Virelizier, A. M., and Allison, A. C. (1976). The role of circulating interferon in the modifications of immune responsiveness by mouse hepatitis virus (MHV-3). *J. Immunol.* **117**, 748–753.

Visintainer, M. A., Volpicelli, J. R., and Seligman, M. E. P. (1982). Tumor rejection in rats after inescapable or escapable shock. *Science* **216**, 437–439.

Walker, E. A., Castegnaro, M., and Griciute, L. (1979). *N*-Nitrosamines in the diet of experimental animals. *Cancer Lett.* **6**, 175–178.

Weibe, I., and Stotzer, H. (1974). Age- and light-dependent changes in the rat eye. *Virchows Arch. A: Pathol. Anat. Histol.* **362**, 145–156.

Weihe, W. H. (1965). Temperature and humidity climatograms for rats and mice. *Lab. Anim. Care* **15**, 18–28.

Weihe, W. H. (1973). The effect of temperature on the action of drugs. *Annu. Rev. Pharmacol.* **13**, 409–425.

Weihe, W. H. (1976). Influence of light on animals. *In* "Control of the Animal House Environment" (T. McSheehy, ed.), Lab. Anim. Handb. No. 7, pp. 63–76. Lab. Anim. Ltd., London.

Weihe, W. H., Schidlow, J., and Strittmatter, J. (1969). The effect of light intensity on the breeding and development of rats and golden hamsters. *Int. J. Biometeorol.* **13**, 69–79.

Weisbroth, S. H. (1979). Bacterial diseases. *In* "The Laboratory Rat" (H. J. Baker, J. R. Lindsey, and S. H. Weisbroth, eds.), Vol. 1, pp. 194–228. Academic Press, New York.

Weisbroth, S. H., Paganelli, R. G., and Salvia, M. (1977). Evaluation of a disposable water system during shipment of laboratory rats and mice. *Lab. Anim. Sci.* **27**, 186–194.

Welsh, R. M. (1978). Cytotoxic cells induced during lymphocytic choriomeningitis virus infection of mice. I. Characterization of natural killer cell induction. *J. Exp. Med.* **148**, 163–181.

Wheelock, E. F. (1967). Inhibitory effects of Sendai virus on Friend leukemia in mice. *JNCI, J. Natl. Cancer Inst.* **38**, 771–778.

Wightman, S. R., Mann, P. C., and Wagner, J. E. (1980). Dihydrostreptomycin toxicity in the mongolian gerbil, *Meriones unquiculatus. Lab. Anim. Sci.* **30**, 71–75.

Wise, A. (1982). Interaction of diet and toxicity—The future role of purified diet in toxicological research. *Arch. Toxicol.* **50**, 287–299.

Wogan, G. N. (1968). Aflatoxin risks and control measures. *Fed. Proc., Fed. Am. Soc. Exp. Biol.* **27**, 932–938.

Wongwiwat, M., Sukapanit, S., Triyanond, C., and Sawyer, W. D. (1972). Circadian rhythm of the resistance of mice to acute pneumococcal infection. *Infect. Immun.* **5**, 442–448.

Woods, J. E. (1978). Interactions between primary (cage) and secondary (room) enclosures. *In* "Laboratory Animal Housing" (Inst. Lab. Anim. Resour.), pp. 65–83. Natl. Acad. Sci.—Natl. Resour. Counc., Washington, D.C.

Wright, J. F., and Seibold, H. R. (1958). Estrogen contamination of pelleted feed for laboratory animals. Effect on guinea pig reproduction. *J. Am. Vet. Med. Assoc.* **132**, 258–261.

Wright, J. M., and Collier, B. (1974). Inhibition by neomycin of acetylcholine

release and ^{45}Ca accumulation in the superior cervical ganglion. *Pharmacologist* **16**, 285.

Yagil, R., Etzion, Z., and Berlyne, G. M. (1976). Changes in rat milk quantity and quality due to variations in litter size and high ambient temperature. *Lab. Anim. Sci.* **26**, 33–37.

Yamauchi, C., Takahashi, H., and Ando, A. (1965). Effects of environmental temperature on physiological events in mice. I. Relationship between environmental temperature and number of caged mice. *Jpn. J. Vet. Sci.* **27**, 471–478.

Yamauchi, C., Fujita, S., Obara, T., and Ueda, T. (1981). Effects of room temperature on reproduction, body and organ weights, food and water intake, and hematology in rats. *Lab. Anim. Sci.* **31**, 251–258.

Yuon, J. K., and Barski, G. (1966). Interference between lymphocytic choriomeningitis and Rauscher leukemia in mice. *JNCI, J. Natl. Cancer Inst.* **37**, 381–388.

Zakem, H. B., and Alliston, C. W. (1974). The effects of noise level and elevated ambient temperatures upon selected reproductive traits in female Swiss-Webster mice. *Lab. Anim. Sci.* **24**, 469–475.

Zbiesieni, B. M. (1980). Diurnal and seasonal changes in ascorbic acid level in adrenal glands of laboratory mice. *Bull. Acad. Pol. Sci., Ser. Sci. Biol.* **27**, 653–656.

Chapter 24

Animal Models in Biomedical Research

George Migaki and Charles C. Capen

I. INTRODUCTION

There is increasing evidence that knowledge gained from the study of an individual pathologic process in different animal species often enhances understanding of the pathogenesis of the disease in humans. It also often suggests valuable new approaches for the investigation of specific diseases in humans.

Animals affected with the same pathologic conditions as humans and having similar or identical clinical signs would be ideal in evaluation of the pathogenesis and therapy of the human counterpart. There is also increasing evidence that the pooling of knowledge of diseases in various animal species may lead to more rapid progress than when only a few examples are studied in isolated institutions. Thus the publication of descriptions of animal models in the scientific literature establishes a means of communication that not only has educational value but contributes to the enhancement of the volume and quality of knowledge in understanding diseases in humans. The objective of this chapter is to introduce the reader to the potential uses of animal models for research on human diseases and to identify additional sources of more specific information in the literature.

Cornelius and Arias (1966) state that, historically, human and veterinary medicine developed as one and that this was documented as early as 1900 BC in the medical literature of ancient Egypt. They also state that both Hippocrates and Aristotle discussed the symptoms and treatment of diseases of animals and humans. They further indicate that although veterinary and human medicine have become separate in modern times, the contributions by veterinarians to human medicine and by physicians to veterinary medicine have increased tremendously the broad knowledge of pathobiology in medicine. Feldman (1963) says that human medicine and animal medicine do complement each other in the task of making the world a healthier place for humans and for animals. Jones (1950) discusses pathologic changes in some spontaneous diseases in animals which may lead to new investigational approaches to the study of similar diseases in humans.

It is difficult to determine who should be credited for recognition of the value of using animals to study human disease, but it is safe to say that modern medical research began in the late 1700s with the successful conquest of a number of human diseases by the use of animals in which diseases comparable to those in humans were discovered and studied comprehensively. Subsequently, biomedical scientists recognized the value of diseases in animals in helping to solve the ''mysteries'' of diseases in humans. In a continuing effort to stimulate biomedical scientists, Leader (1969), during his presentation on animal models of inherited diseases at a meeting of the Federation of American Societies for Experimental Biology (FASEB), reiterated what his colleague, Rene Dubos of The Rockefeller University, had said: ''If we look carefully enough we will eventually find an animal model for every disease.''

II. HISTORY

It is not the intent here to discuss comprehensively the contribution of veterinary medicine to public health but to historically identify and briefly mention selected events relevant to the subject. Some of these events were highlighted by Bustad *et al.* (1976) and Jones (1980) when they discussed the value of animal models. Greep (1970), recognizing the value of animal models, stated that the health of man is directly related to the health of the animals about him and that this relationship had been suspected from ancient times but was not docu-

mented until 1798. That was the year that Edward Jenner, a Gloucestershire physician, published his findings that individuals infected with cowpox were protected against the dreaded disease smallpox. Jenner developed a method of vaccination by using the cowpox virus to infect humans without hazard, and it is said that by 1800 over 100,000 people in several countries had been vaccinated with dramatically beneficial results. Today, Jenner's original technique, with slight modification, is the universal immunization procedure, and smallpox has been virtually eliminated. Although subsequent development of other viral vaccines was relatively slow, such as the rabies vaccine produced by Pasteur and his colleagues in 1884, the immunization method of Jenner represents a landmark in the protection of humans against infectious diseases by use of attenuated virus.

Transmissibility of an infectious agent in neoplasms is an accepted phenomenon today. Ellermann and Bang (1908) were the first to suspect a transmissible agent when they experimentally produced erythromyeloblastosis in healthy chickens. Rous (1911) demonstrated that a cell-free filtrate from a sarcoma in a chicken could reproduce the neoplasm at the site of inoculation in healthy chickens and that filtered extracts from the induced sarcomas would produce neoplastic growths in experimentally injected chickens. These were the initial studies on virus-induced neoplastic diseases in animals. Subsequent investigations have led to the discovery of the mammary tumor virus, mouse leukemia virus, polyoma virus, simian virus 40, and many other oncogenic viruses (Gross, 1970).

The potential research value of *Herpesvirus saimiri,* a latent virus in the squirrel monkey (*Saimiri sciureus*), was recognized by Hunt *et al.* (1970). When this virus was inoculated into owl monkeys (*Aotus trivirgatus*) and several species of marmosets, an acute fatal malignant lymphoma and leukemia developed. This was the first herpesvirus of primate origin that was found to be oncogenic in primates. Although much has been said about viruses as the cause of neoplasms in animals, information gained about other carcinogens, such as chemical (e.g. polycyclic hydrocarbons) and physical (e.g. ionizing radiation) agents, has also contributed much to our understanding of tumorigenesis.

In 1885, D. E. Salmon, a veterinarian, and T. Smith, a physician, isolated an organism that they believed to be the cause of hog cholera (later proved to be caused by a virus) and that was named *Bacillus cholerae suis*. Subsequently, the name *Salmonella* was proposed for this genus in honor of Dr. Salmon, the first Chief of the Bureau of Animal Industry, United States Department of Agriculture. The real significance of Salmon and Smith's work in bacteriology was that their findings of heat-killed cultures of the *Salmonella* organism were the initial steps in the development of a bacterial vaccine. Prior to the establishment of an effective immunization program, hog cholera was responsible for great economic loss.

Absolute proof of the ''germ theory'' was provided in 1876 by a German physician, R. Koch, in his work on the cause of anthrax. He cultured the causative organism, *Bacillus anthracis,* and reproduced anthrax in mice by injecting suspensions of the culture. The same organism was cultured from the infected mice. Thus the specific nature of infectious diseases was understood and the basis formed for ''Koch's postulates.''

In the 1890's, a triumvirate composed of T. Smith, the physician mentioned above, F. L. Kilborne, a veterinarian, and C. Curtice, a veterinarian and entomologist, collaborated on a medical revelation about the transmission of diseases in man and animals. They demonstrated that Texas fever (tick fever) in cattle is transmitted by an arthropod, which serves as an intermediate host (Curtice, 1891; Smith and Kilborne, 1893). Their discovery proved to have long-lasting effects. The problem of infectious diseases, for both humans and animals, could now be resolved, since the mode of transmission had been determined.

Considered to be one of the dramatic events in the field of anticoagulant therapy was the discovery of dicoumarin, which resulted from investigations conducted by F. W. Schofield (1924) and L. M. Roderick (1929), both veterinary pathologists, during the 1920s on cattle fed moldly sweet clover. Affected cattle had large areas of hemorrhage, especially hematomas in the subcutaneous tissues, and they bled easily following minor bruises. The hemorrhages were difficult to control. Sweet clover, belonging to the genus *Melilotus* and widely used to improve soils and as forage, contains coumarin, which, in itself is not toxic to animals. Spoilage of sweet clover, however, causes the coumarin to be converted to dicoumarin, which inhibits the formation of prothrombin, thus resulting in failure of the clotting of blood. To complete this discovery, dicoumarin was isolated and identified by Campbell and Link in 1941.

Perhaps one of the greatest contributions to the fields of metabolism and nutrition was the successful treatment of diabetes mellitus with insulin in the dog. Although it was known as early as 1889 that total pancreatectomy in dogs resulted in severe and fatal diabetes mellitus, Banting and Best (1922) proved that pancreatic extracts were essential for the treatment of diabetic dogs. Their experiments of ligation of the pancreatic ducts resulted in subsequent atrophy of the acinar tissues. The atrophied pancreas was then surgically removed and a saline extract prepared. The pancreatectomized dog became comatose and, following injection of the prepared extract, recovered from the coma. No one can doubt that this is one of the key contributions in the history of medicine, which illustrates again the value of animal models.

K. F. Meyer is another veterinarian who made many valuable contributions to public health. These include his work in experimental typhoid infections, brucellosis, psittacosis, and leptospirosis. A graduate of the University of Zurich in 1908,

Dr. Meyer probably is best remembered for solving the problem of botulism, which caused the deaths of many people in the 1920s and threatened to destroy the canning industry. The cause and means of controlling encephalomyelitis in the horse was another of his accomplishments. He also contributed significantly to the fight against bubonic plague. Feldman (1963) stated, ''Dr. Meyer's unrelenting and successful attacks on a considerable number of serious diseases represent an important basic and substantial increment in the armamentarium that man has slowly forged for his protection against a long list of biologic enemies.''

Direct application of knowledge of a disease in an animal, scrapie in sheep, has led to a better understanding of a human disease. D. C. Gajdusek, a physician, and his colleagues (1965) at the National Institutes of Health, investigating kuru, a disease of the primitive people of New Guinea, considered this fatal degenerative condition of the nervous system to be of congenital origin. W. J. Hadlow (1959), a veterinary pathologist, having had considerable research experience with scrapie in sheep and goats, called attention to the similarities of the histologic lesions in the brains of kuru patients and those of sheep with scrapie (a viral disease with an incubation period of 2 years or longer). Dr. Hadlow suggested experimental inoculation of brain material from kuru patients into animals. Dr. Gajdusek, using chimpanzees, produced a fatal disease in these animals after a long incubation period. With the establishment of a viral etiology for kuru and by stopping the practice of cannibalism, this disease has been virtually eliminated in New Guinea. The study of scrapie in sheep has led to the identification of the cause of kuru and may provide new approaches in solving other poorly understood disorders of the human nervous system such as multiple sclerosis.

A valuable contribution in the field of abnormal development and congenital malformations was the discovery by Binns *et al.* (1959) of the plant *Veratrum californicum* as the cause of cyclopean malformations in sheep. The lambs were alive at birth but died shortly thereafter. Abnormalities were confined to the head and consisted of severe distortion or absence of some bones of the face, cyclopean deformity of the eyes, fusion of the cerebral hemispheres, and hydrocephalus. In searching for the cause, Binns *et al.* (1965) determined that the malformations could be produced in the lamb by feeding of *V. californicum* to the ewe on the fourteenth day of pregnancy. The importance of this ovine model is that it can be used for the distinction between genetic and environmental causes in similar types of congenital malformations in children and can also provide insight into factors that regulate the length of the gestation period. Other examples of congenital malformations and their causes have been described by Kalter (1968).

It is clear from the previously mentioned events in biomedical research that scientists have long recognized the importance of diseases in animals to the understanding of diseases in humans. "Animal models," therefore, should not be considered a recently coined term. A review of the literature (see list of publications at the end of this chapter) indicates that there has been considerable interest in animal models during the past two decades, but more specifically since 1965. One important stimulus was the Workshop Conference of Potential Research in Comparative Pathology at Portsmouth, New Hampshire, October 5–6, 1961, sponsored by the Pathology Training Committee of the National Institutes of Health (NIH). Summaries of this conference were prepared independently by three of the participants, R. E. Stowell, M. D. (1963), Armed Forces Institute of Pathology, Washington, D.C., from the viewpoint of human pathology; T. C. Jones, D.V.M. (1964), Angell Memorial Animal Hospital, Boston, Massachusetts, from the viewpoint of veterinary pathology; and H. L. Ratcliffe, Ph.D. (1962), Philadelphia Zoological Garden, Philadelphia, Pennsylvania, from the viewpoint of comparative pathology. In summary Dr. Stowell said, "Comparative pathology offers one of the broadest horizons and best opportunities of any investigative field in biology and medicine."

The National Institute of General Medical Sciences sponsored a Workshop on Comparative Medicine in 1967 with emphasis on the need for development of comparative medicine in various institutions to further the study of specific animal diseases in a comparative way and to emphasize comparisons to human diseases (Gay, 1967). It was recommended that information on animal disease models be compiled and made available through publications and symposia so that the greatest number of scientists would be informed.

Much of the support for development of animal models comes under the purview of the NIH's Animal Resources Program Branch of the Division of Research Resources. McPherson (1978) stated that the role of NIH in the development of animal models is to identify diseases and research areas where there is a particular need for animal models and to support projects to develop these models.

Leader (1964) presented an excellent treatise on several spontaneous diseases in animals, namely, arteriosclerosis, Aleutian disease of mink, and Chediak–Higashi syndrome, which may have potential value in helping to elucidate similar diseases of man. The main point made in his article is that while comparative pathologists are competent in identifying diseases in animals, skill and imagination of the investigators must be applied to the greatest possible degree to solve the mysteries of pathogenesis by the utilization of the most advanced techniques. Cornelius and Arias (1966), in addressing the problems of biomedical models in veterinary medicine, stated that spontaneous metabolic diseases in animals with similar counterparts in humans are now commonplace. They briefly discussed two new strains of sheep. One was a mutant strain of Southdowns with a defective hepatic uptake of organic anions, such as bilirubin and sulfobromophthalein so-

dium, from the blood (Cornelius and Gronwall, 1968). These sheep have unconjugated hyperbilirubinemia. They are also photosensitive, which results from the retention of phylloerythrin. The disease is similar to Gilbert's syndrome in humans. Recently, the indigo snake was found to have the same defect in bilirubin metabolism as this mutant strain of sheep (Noonan et al., 1979). The other breed discussed by Cornelius et al. (1965) was the Corriedale sheep, which had an abnormality similar to Dubin–Johnson syndrome—a disease characterized by defective transfer of organic anions from the liver to the bile. Animals with inherited defects in hepatic uptake and excretory transport have provided an opportunity for study of ion transport by the liver, the formation and excretion of bile, and the pathogenesis of jaundice.

Another congenital hyperbilirubinemia in animals is common to a mutant strain of the Wistar rat. The Gunn rat has a chronic nonhemolytic unconjugated hyperbilirubinemia in which there is a defect in glucuronyltransferase activity, which prevents the formation of bilirubin glucuronide (Schmid et al., 1958). The rats have jaundice and other characteristics similar to Crigler–Najjar syndrome in humans. Similar to the mutant sheep, the Gunn rat will serve as an excellent model for study of problems relating to bilirubin metabolism, including intestinal glucuronide formation and kernicterus.

In the field of inherited bleeding disorders, Mustard and Packham (1968), in emphasizing the value of animals in the study of human disease processes, described their experiences with dogs with congenital defects in blood coagulation and hemostasis. Because of the close similarities of these bleeding disorders to those in humans, a colony of dogs was established for intensive investigations on the nature of defects in the hemostatic pathway.

III. SYMPOSIA

By now the biomedical community was sufficiently aware of animal models of human disease, and various scientific organizations and institutions began sponsoring symposia. In 1968 for example, the Federation of American Societies for Experimental Biology (FASEB) sponsored a symposium entitled "Choice of Animal Models for the Study of Disease Processes in Man," which included six topics: "Mammalian and Avian Models of Disease in Man" (Jones, 1969); "Models for Cytogenetics and Embryology" (Benirschke, 1969); "Models for Infectious Diseases" (Frenkel, 1969); "Models of Immunologic Diseases and Disorders" (Good et al., 1969); "Models for Obstetrical and Gynecological Disease" (Craig, 1969); and "Utility and Failure of Models in Oncology" (Huseby, 1969). The following year, the FASEB sponsored another symposium, "Living Models for the Study of Disease Processes in Man," which included five topics: "Discovery and Exploitation of Animal Model Diseases" (Leader, 1969); "Invertebrates and Their Diseases as Models for the Study of Disease Processes in Man" (Steinhaus, 1969); "Transformation and Recovery of the Crown Gall Tumor Cell: An Experimental Model" (Braun and Wood, 1969); "Meadow Mushroom Provides a Model for the Study of Mitochondrial DNA" (Vogel, 1969); and "Survey of Some Spontaneous and Experimental Disease Processes of Lower Vertebrates and Invertebrates" (Scarpelli, 1969). The FASEB has since sponsored other symposia on animal models at their annual meetings.

IV. PUBLICATIONS

In response to the biomedical community's need for a centralized source of information on animal models, the Registry of Comparative Pathology began publication of the *Comparative Pathology Bulletin* in the fall of 1969 and of the *Handbook: Animal Models of Human Disease* in 1972. These and other publications of the Registry are listed in Additional Books, Proceedings, and Bulletins.

The Institute of Laboratory Animal Resources (ILAR) of the National Research Council, Washington, D.C., in collaboration with the American College of Laboratory Animal Medicine in 1968, initiated publication of "Animal Models for Biomedical Research." These were publications of proceedings of symposia held in conjunction with the annual meeting of the American Veterinary Medical Association. These and other publications of ILAR are listed in Additional Books, Proceedings, and Bulletins.

Prior to 1972, there were reviews and listings of animal models (Cornelius and Arias, 1966; Leader, 1967; Kitchen, 1968; Doyle *et al.*, 1968; Mulvihill, 1972). Following publication of bibliographic data listing selected references of animal models and their human counterparts by organ systems (Cornelius, 1969; Jones, 1969; Leader and Leader, 1971), it became clear that an abundant amount of documented information existed on animal models. By a careful analysis of these publications, one develops a positive attitude towards Dubos's statement that if we look carefully enough we will eventually find an animal model for every disease (Leader, 1969). Bustad *et al.* (1977) prepared an updating of the previously mentioned reviews on animal models. Approximately 900 new references were included and grouped in tabular form by organ systems. Continuing these efforts to update references on animal models, Hegreberg and Leathers (1982a,b) published a two-volume set, "Animal Models Bibliographies." Volume 1 contains 1289 references of naturally occurring animal models of human disease, and volume 2 contains 2707 references of induced animal models. The references are organized by organ

systems for ease in locating the desired entry. A tremendous amount of bibliographic information is available for scientists interested in both spontaneous and induced animal models of human disease. Another valuable source of information on animal models is a two-volume text edited by Andrews *et al.* (1979).

Symposia, with subsequent publication of the proceedings, are valuable as means of establishing communication and disseminating information to biomedical scientists who are interested in advancing their studies on animal models. The proceedings that have a direct application to animal models are listed in Additional Books, Proceedings and Bulletins. In addition, the list includes descriptions of animal models of human disease that have been published in various scientific journals, textbooks, bulletins, and newsletters.

V. SELECTION OF ANIMALS

The choice or selection of animals to use in the study of human disease is an important consideration, and good judgment must be used. These considerations have been discussed by various investigators (Schmidt-Nielsen, 1961; Jones, 1969; Van Citters, 1973; Leader and Padgett, 1980; Held, 1981). Van Citters states that a model is "a small copy, an imitation or preliminary representation which serves as the plan from which a final larger object is constructed or formulated" and that "a model is intended to be copied or followed." He also states that "most animal models are not, in fact, an accurate representation of the larger object, which in most cases is man." Leader and Padgett (1980) in judging the use of certain animals, raise an important question: "Must the animal model be an exact duplicate of a human condition, or can differences sometimes also be of value?" Their criteria for a good animal model system are that the animal should (a) accurately reproduce the disease or lesion under study, (b) be available to multiple investigators, (c) be exportable, (d) be large enough for multiple biopsy samples, (e) fit into available animal facilities of most laboratories, (f) be capable of being easily handled by most investigators, and (g) survive long enough to be usable and that the disease should (a) if genetic, be in a polytocous species and (b) be available in multiple species. Jones (1969) states that animal diseases of every category are known to occur in so many species that one might wonder why they have been used so seldom as models of human disease. Many reasons may exist for this, not the least of which is the failure of communication between investigators interested in diseases of humans and those interested in diseases of other species.

Other important considerations for biomedical investigators are the selection and availability of the animals. This is sometimes a difficult problem, and suggestions and solutions have been offered by several investigators (Gay, 1967; Jones, 1969; Van Citters, 1973). Gay (1967) believes that companion animals, such as dogs, cats, and horses, which are readily available, could be used in the study of degenerative diseases, age-related diseases, hereditary defects, and behavioral problems; that zoo animals and aquatic animals offer an important opportunity for study of environmental health problems and for gerontologic research; that large colonies of inbred laboratory animals are excellent for study of congenital disorders; and that veterinary medical institutions offer a vast wealth of comparative medical research material. According to Jones (1969), breeding colonies of laboratory animals have yielded many excellent models, the most successful of which are colonies of mice at the Jackson Laboratory, Bar Harbor, Maine (Green, 1968). Additional information on inbred and genetically defined strains of some laboratory animals (mouse, rat, hamster, guinea pig, rabbit, and chicken) has been published by Altman and Katz (1979a,b). Van Citters (1973) indicates that the pig could be used for research on the skin and cardiovascular system and that nonhuman primates, because of their phylogenic qualities, are nearly ideal models for study. Lower vertebrates and invertebrates should not be overlooked, as they are readily available models for a variety of human diseases (Steinhaus, 1969; Scarpelli, 1969; Bulla and Cheng, 1978).

VI. SELECTED ANIMAL MODELS

During the past two decades there have been numerous contributions by investigators on diseases in animals that either mimic or are similar to a human counterpart. Some selected animal models that may be useful in the study of important human diseases are discussed briefly. For more detailed descriptions of specific animal models and comprehensive listings of documented animal models, readers are referred to the publications listed in Additional Books, Proceedings, and Bulletins.

A. Kidney Diseases

Robinson and Dennis (1980) have prepared a listing of spontaneously occurring diseases and functional defects of the kidney in animals that may or may not resemble those seen in humans. The list illustrates a wide spectrum of diseases of interest to those studying nephritic diseases. Some of the diseases that have been well characterized are mentioned briefly here.

An excellent review on the comparative aspects of both primary and secondary glomerulonephritis in animals has been prepared by Slauson and Lewis (1979). Primary glomerulone-

phritis represents a glomerular disease in which the kidney is the only or the predominant organ affected. Glomerulonephritis was induced in sheep by Steblay (1980) by the injection of heterologous glomerular basement membrane (GBM) in complete Freund's adjuvant. Lerner and Dixon (1966) found a high percentage of normal slaughtered sheep to be affected with proliferative glomerulonephritis, but they did not observe anti-GBM antibodies, which suggests that the glomerulonephritis was mediated via the glomerular deposition of antigen–antibody complexes. Similar glomerular lesions were found in cattle and goats (Lerner et al., 1968). A rapidly progressive and fatal glomerulonephritis with immunofluorescent deposits has been observed in young lambs of a flock of Finnish Landrace sheep. The disease has many similarities to mesangiocapillary glomerulonephritis in humans (Angus et al., 1974, 1975).

As for nonhuman primates, membranous glomerulonephritis has been described in galagos by Burkholder and Bergeron (1970). Poskitt et al. (1974) observed a high percentage (41%) of renal biopsy specimens from adult cynomolgus monkeys (Macaca irus) with histologic and immunofluorescent findings typical of glomerulonephritis. A similar glomerulonephritis was observed in owl monkeys (Aotus trivirgatus) by King et al. (1976). Spontaneous cases of immune complex-mediated glomerulonephritis have been observed in dogs (Kurtz et al., 1972; Muller-Peddinghaus and Trautwein, 1977) and in cats (Slauson et al., 1971), thus providing models in companion animal species that may contribute to a better understanding of the human nephritic condition.

Secondary glomerulonephritis represents a glomerular disease in which glomeruli may be injured by a variety of factors and in the course of a number of systemic diseases, including specific viral infections and immunologic and metabolic diseases. In animals, glomerulonephritis is believed to be mediated as an immunologic function of persistent viremia with the formation of antiviral antibodies and subsequent glomerular deposition of circulating immune complexes. Spontaneous glomerulonephritis in animals in association with chronic viral infections has been observed in chronic hog cholera in pigs (Figs. 1 and 2) (Cheville, 1973), lymphocytic choriomeningitis in mice (Buchmeier and Oldstone, 1978), Aleutian disease of mink (Henson and Gorham, 1974), lactic dehydrogenase virus in mice (Porter and Porter, 1971), murine leukemia viruses in mice (Hirsch et al., 1969; Oldstone et al., 1972), and equine infectious anemia in horses (Gorham and Henson, 1976). Further information on other immune complex diseases is provided in an excellent review on this subject by Cochrane and Koffler (1973). The availability of these virus-induced models may assist in establishing a causal relationship between viral infections and the clinical disease in humans.

Autoimmune diseases associated with glomerulonephritis are observed in certain strains of New Zealand mice, especially the

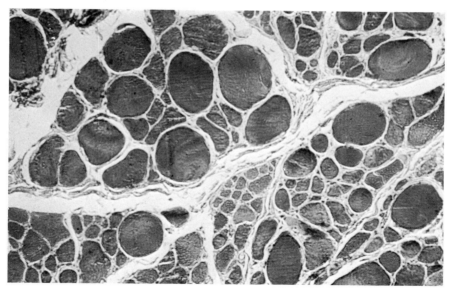

Fig. 3. Muscular dystrophy in a mink. Transverse section of skeletal muscle demonstrating the variation in fiber size and an increase in endomysial and perimysial connective tissue. Gomori's trichrome. ×250. (From Hegreberg, 1976; with permission from the *American Journal of Pathology.*)

joints. It is of interest to note that the vastus intermedius is consistently affected and that other members of the quadriceps group are frequently involved. The histologic lesions of the ovine model have many similarities to muscular dystrophy in humans. The sheep, because of its larger size, may have advantages over other animals in certain surgical and experimental procedures.

6. Dog

Muscular dystrophy in the Labrador retriever begins when the pups are 3 months of age, with clinical signs of muscular disorders in which the animal is unable to hold up its head (Kramer *et al.*, 1976). Later, with involvement of other muscles, the pups display abnormal head and neck posture and a stiff, forced, hopping gait. Histologic and cytochemical examinations of muscle biopsy specimens will reveal relative predominance of type I and a deficiency of type II myofibers. The mode of inheritance is autosomal recessive.

7. Turkey and Chicken

A deep pectoral myopathy characterized by spontaneous necrosis of the suprachoroid muscle has been recognized in breeding age turkeys (Harper *et al.*, 1975) and in broiler-type chickens (Wight *et al.*, 1981). The pathogenesis of this condition is completely different from that of the above-mentioned dystrophic condition in the chicken. It is briefly mentioned here because it represents a different form of muscular dystrophy. No specific human counterpart exists, but this condition closely resembles the gross and histologic lesions seen in ischemic necrosis or infarction of muscles in the anterior tibial compartment of man. These muscles are located in a restricted and almost inelastic compartment, bound by bone and tough fibrous fascia. Similarly, the suprachoroid muscle in the turkey and chicken lies in an osteofascial compartment bound by the carnia, facies muscularis sterni, and a rather strong intermuscular fascia. The susceptibility to ischemic necrosis in turkeys has been increased by genetic selection toward larger breast muscles and by forced wing-flapping. This results in intramuscular vascular dilatation and increased intracompartmental pressure, which can be corrected by fasciotomy to avoid infarction. These avian models are valuable in that they are readily induced and can be used to study muscular necrosis in humans, for which the pathogenesis is undetermined or poorly understood.

F. Hypersensitivity Pneumonitis

This is one of the most important immunologically mediated pulmonary diseases because of the ubiquity of etiologic agents, the severity of the acute onset of symptoms, and its potential for alveolar fibrosis. The disease, also known as extrinsic allergic alveolitis, is caused by the inhalation of small-particle antigenic material of biologic origin, such as fungal spores, dried and fragmented fungal or insect particles, and particulate animal proteins such as those found in avian feces. Many species of fungi are involved, but most belong to the genera *Micropolyspora* or *Thermoactinomyces*. Numerous syndromes have been identified, but the classic representation is farmer's lung, which is a response to the inhalation of dust from moldy hay or other moldy vegetable produce. Olenchock (1977) presents an excellent review of the potential use of animal models for a better understanding of the mechanisms involved in the pathogenesis of hypersensitivity pneumonitis.

Precipitating antibodies or precipitins to *Micropolyspora faeni* and other thermophilic actinomycetes are found in sera of affected cattle (Pirie *et al.*, 1972) and horses (Mansmann *et al.*, 1975; Breeze, 1979), as well as in human serum. The exposure of horses to dust derived from avian feces can produce hypersensitivity pneumonitis in these animals, which suggests an immunologic-related cause rather than cause by an infectious agent. In cattle, the principal clinical signs during the acute stage are coughing, anorexia, and dyspnea, while in the chronic stages, coughing and hyperpnea are the common signs. Histologic lesions consist of diffuse infiltration of the alveolar septa by lymphocytes, plasma cells, and interstitial cells and intraseptal aggregates of lymphocytes without germinal centers. Experimentally, exposure of rats to aerosol extracts of pigeon droppings results in an interstitial pneumonitis with precipitins being detected in the serum (Fink *et al.*, 1970). Exposure of guinea pigs to aerosol extracts of *M. faeni* causes increased respiratory rates and histologic lesions of hypersensitivity pneumonitis (Wilkie *et al.*, 1973). Due to the relative ease with which precipitin can be produced in the rabbit, this animal has been a popular experimental model for hypersensitivity pneumonitis and has been immunized by various antigens, including moldy hay extract (Parish, 1961) and ovalbumin in complete Freund's adjuvant (Richerson *et al.*, 1971). Exposure of monkeys to pigeon serum has caused an increased respiratory rate and histologic lesions of hemorrhage and exudative alveolitis (Hensley *et al.*, 1974). The use of these experimental models would serve to enhance our understanding of the immunologic processes in hypersensitivity pneumonitis.

G. Asthma

In their review of animal models of asthma, Patterson and Kelly (1974) found that the dog is the only animal in which a defined hypersensitivity disease related to aeroallergens occurs. The clinical disease is most commonly ragweed pollenosis, although hypersensitivity to grass, trees, housedust, and cat antigens has been identified. The clinical manifestations include conjunctivitis, rhinitis, and an intensely pruritic dermatitis. The dogs have circulating antibody against ragweed antigen, which is heat-labile, homocytotropic, of long latency, and reaginic and appears to reside in a class of immunoglobulins analogous to human IgE.

H. Emphysema

According to Karlinsky and Snider (1978), emphysema is a condition of the lung characterized by abnormal, permanent enlargement of the air spaces distal to the terminal bronchioles, accompanied by destructive changes of their walls. These investigators describe four types of emphysema in humans: (a) panlobular or panacinar, affecting all alveolar ducts and alveolar sacs in a unit, (b) centrilobular or proximal acinar, affecting mainly respiratory bronchioles with subsequent extension and involvement of the alveoli, (c) paraseptal or distal acinar, involving the subpleural tissue and interlobular septa, and (d) irregular or paracicatricial, involving secondary lobules, usually with scarring. More than one type may be seen in a lung.

Karlinsky and Snider (1978) reviewed the animal models of emphysema, both spontaneous and experimentally induced, and the new information regarding emphysema derived from these models. Although the models may not be exact counterparts to the human disease, they are valuable in the study of the cause and pathogenesis of human emphysema. Following are some of the animal models along with a brief mention of their principal characteristics of emphysema. In studying emphysema, one should be aware of the anatomic differences of the lung between animal species and humans (Slauson and Hahn, 1980).

"Heaves" in horses has lesions similar to the panlobular and centrilobular types of emphysema (Gillespie *et al.*, 1964), and careful pathoanatomical evaluations are necessary for differentiation from other pneumonic conditions. Emphysema in rabbits is usually of the panlobular type and is generally seen after the animal is $2\frac{1}{2}$ years of age. It is generally associated with chronic interstitial fibrosis (Strawbridge, 1960). In rats, the paracicatricial form of emphysema is often associated with chronic murine pneumonia (Paleček and Holusa, 1971). Due to a genetic defect in elastin synthesis, the blotchy mouse develops a spontaneous progressive panlobular emphysema (Fisk and Kuhn, 1976). The importance of this model is that it illustrates a defect in the elastic tissue network of the lung as the primary cause of the emphysema, and this would be of value in studying protease-induced emphysema. These mice have a genetically determined defect that prevents the generation of the lysine-derived aldehyde necessary for cross-linking of collagen and elastin. The intratracheal administration of the proteolytic enzyme papain to rats has produced emphysema (Gross *et al.*, 1965), as has the administration of elastase to hamsters (Hayes *et al.*, 1975). It is suggested that the blotchy mouse and enzyme-induced models offer the best opportunity for understanding of the cellular and molecular events taking place in the lung that give rise to emphysema (Karlinsky and Snider, 1978).

I. Diabetes Mellitus

For a comprehensive discussion on animal models of diabetes mellitus, readers are referred to excellent recent reviews (Like, 1977; Lage et al., 1980; Mordes and Rossini, 1981). Diabetes in humans is a complex metabolic disorder resulting from a variety of factors, including genetics, age, sex, state of nutrition, and physical activity. Diabetes may be divided into two major types: (a) juvenile-onset, ketosis-prone, insulin-dependent, hypoinsulinemia and (b) maturity-onset, nonketotic, non-insulin-dependent, obesity. Although none of the animal models, either spontaneous or experimentally induced, is an exact human counterpart, each has contributed to a better understanding of the cause, pathogenesis, treatment, and prevention of this common and complex human malady. A brief mention is made of some of the animal models and their principal diabetic characteristics.

1. Animal Models for Hypoinsulinemia

The BB rat, a spontaneous mutation of the Wistar rat, undergoes metabolic changes similar to those of humans with the juvenile-onset type of diabetes and could be valuable in the investigation of cause—whether from genetic, viral, or immune factors (Nakhooda et al., 1977). Approximately 30% of the rats develop absolute insulin deficiency between 60 to 120 days of age with clinical signs of hypoinsulinemia, hyperglycemia, and ketosis. Death results if the ketotic animals are not treated with insulin. The administration of tolbutamide or arginine fails to stimulate an insulin response. A significant finding during the early stages of the disease is an intense pancreatic isletitis with cellular infiltrates of lymphocytes and macrophages. The cause of the isletitis remains undetermined; however, both infectious agents (viruses and bacteria) and immunologic injury have been suggested.

The Chinese hamster (Cricetulus griseus), similar to the BB rat, has early-onset diabetes mellitus beginning between 1 and 3 months of age and has low plasma insulin (Gerritsen and Dulin, 1967). Isletitis is not seen, but there is glycogen infiltration, degranulation, and decreased numbers of beta cells. The cause remains undetermined. Although the animals are hyperphagic, they are not obese.

The South African hamster (Mystromys albicaudatus), having histologic lesions in the pancreas similar to the Chinese hamster, was recognized as an animal model of diabetes in 1969 (Stuhlman, 1979). Changes in the beta cells vary directly

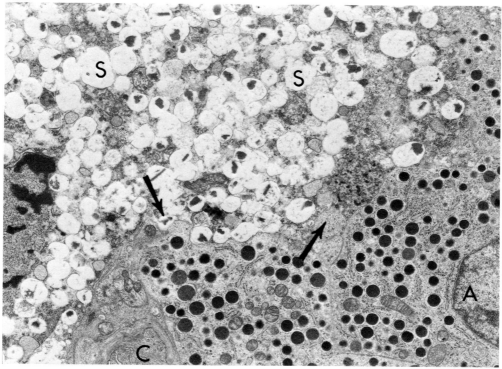

Fig. 4. Experimental diabetes mellitus in a dog following combined alloxan–streptozotocin administration. After 10 hr there was severe degeneration of beta cells, indicated by the clumping of nuclear chromatin and disruption of the plasma membrane (arrows). Secretory granules (S) are swollen and many have lost their internal cores. A heavily granulated alpha cell (A) near a capillary (C) is unaffected. Uranyl acetate and lead citrate. ×6000. (From Black et al., 1980; with permission of the *American Journal of Pathology.*)

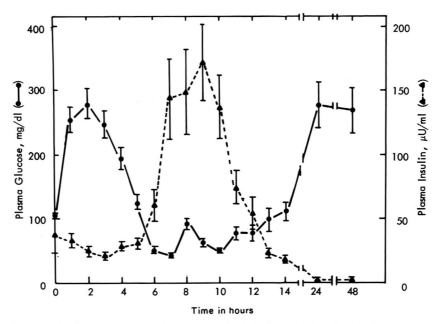

Fig. 5. Triphasic pattern of changes in plasma glucose and insulin for 48 hr following administration of alloxan–streptozotocin to dogs. There was a reciprocal relationship between plasma insulin and glucose levels. By 1 hr, plasma glucose levels had increased from mean values of 56 ± 5 to 260 ± 46 mg/dl and remained elevated for the next 2 hr. Plasma glucose levels declined rapidly after 3 hr, reaching a nadir at 6–7 hr and remained low until 12 hr. By 13 hr, glucose levels were rising to a persistent hyperglycemic level, which was attained at 24 hr. Insulin levels decreased from a baseline of 37 ± 18 to 20 ± 3 U/ml by 3 hr, then increased over the next 6 hr, reaching a maximal level of 173 ± 30 U/ml by 9 hr. This acute rise in plasma insulin was followed by rapid decline. A state of persistent hyposecretion of insulin was reached by 24 hr. (From Black *et al.* 1980; with permission of the *American Journal of Pathology.*)

with the extent and duration of hyperglycemia and include glycogen infiltration, degranulation, and nuclear pyknosis. Other important histologic findings in diabetic *Mystromys* include renal glomerular lesions consisting of mesangial cell proliferation and increased thickness of capillary basement membranes and capillary microangiopathy in skeletal muscle.

Hypoinsulinemia has also been produced experimentally in certain strains of mice by subcutaneous inoculation of the M variant of encephalomyocarditis virus (Craighead, 1976). Overt hyperglycemia develops in 4–7 days. During the acute stage, degranulation and necrosis of the beta cells occurs, and macrophages and lymphocytes infiltrate the islets. Viral antigen can be located in the beta cells by immunofluorescence. Because of the abrupt hyperglycemia and marked hypoinsulinemia and isletitis, this virus-infected mouse model has close resemblance to the juvenile-onset type of diabetes. It also has been suggested that the virus-induced beta cell damage could sensitize the mouse and result in an immune-mediated lesion of the islets.

Experimentally, alloxan and streptozotocin are both cell-specific toxins that destroy beta cells and thus cause hypoinsulinemia (Dulin and Soret, 1977). While these chemicals offer the advantage of rapid and effective necrosis of the beta cells, they have the disadvantage of being highly toxic at the diabetogenic dose levels. However, the combination of alloxan

and streptozotocin each in subdiabetogenic doses reduces the adverse side effects while retaining the high degree of beta cell toxicity (Figs. 4 and 5) (Black *et al.*, 1980). In mice, a single diabetogenic dose of streptozotocin will cause hyperglycemia within 48–72 hr, whereas multiple small doses, none of which are diabetogenic, will produce a pancreatic isletitis with progression to beta cell necrosis and hypoinsulinemia (Like and Rossini, 1976). Alloxan-induced diabetic dogs have reproducible retinal lesions, including microaneurysms, that are similar to those in humans (Engerman, 1976).

A closed colony of New Zealand white rabbits has been described (Roth *et al.*, 1980) as having an 18.5% incidence of insulin-dependent diabetes mellitus. Hypergranulation of the beta cells develops, but no evidence exists of hyperplasia of the beta cells, isletitis, or amyloidosis. A defect in insulin secretion is suspected.

Spontaneous cases in dogs vary considerably in the number and severity of lesions among individuals. It is believed that the most frequent cause of diabetes is chronic pancreatitis in which there is progressive loss of both exocrine and endocrine cells by inflammatory destruction and fibrosis. There is a decreased number of islets of Langerhans and degeneration of the beta cells. Recently Kramer *et al.* (1980) described diabetic pups, 2 to 6 months of age, from a keeshond dog with hypoinsulinemia. The mode of inheritance is autosomal recessive.

Lesions include a reduction in the number of islets and formation of cataracts.

Most of the spontaneous cases of diabetes in domestic cats are associated with amyloid deposits in the islets of Langerhans (Fig. 6) (Johnson and Stevens, 1973). The cause of amyloidosis remains undetermined, and it appears that the islet amyloidosis represents a localized form, as amyloid deposits are not found outside of the pancreas. Untreated cats develop ketosis. The peak incidence in cats occurs at 8–9 years of age; therefore, diabetes in affected cats would correspond to the maturity-onset type.

A similar situation occurs in nonhuman primates in which islet amyloidosis is found in the diabetic animal. The incidence in the black Celebes ape (*Macaca nigra*) is at least 50%, with diabetes increasing with age (Howard, 1972; Howard and Palotay, 1976). The amyloid appears to be islet specific, as it is not found in the extrapancreatic tissue. In addition, aortic atherosclerosis and cataracts are common findings in severely diabetic monkeys.

Spontaneous diabetes in guinea pigs has been observed (Lang *et al.*, 1977) in which the clinical signs are recognized within 6 months of age. Ketonuria is mild and transitory, which may explain why exogenous insulin is not necessary for survival. Histologic lesions include degranulation of the beta cells of the islets and fibrosis of the vascular stroma.

2. Animal Models for Hyperinsulinemia

Animals with hyperinsulinemia are common among laboratory rodents, and they generally are obese. Although the precise metabolic defects are not understood completely, diabetes in most of the animal species listed below is associated with either hyperplastic or hypertrophic changes in the beta cells of the islets of Langerhans. Diabetes has been described in the sand rat (Schmidt-Nielsen *et al.*, 1964), spiny mouse (Gonet *et al.*, 1965), KK mouse, (Iwatsuka *et al.*, 1970), Wellesley hybrid (C3hf×1) mouse (Like and Jones, 1967), obese (C57BL/6J) mouse (Coleman, 1978), diabetes (C57BL/KsJ) mouse (Coleman and Hummel, 1974), New Zealand obese mouse (Herberg and Coleman, 1977), PBB mouse (Walkley *et al.*, 1978), and Djungarian hamster (Herberg *et al.*, 1980).

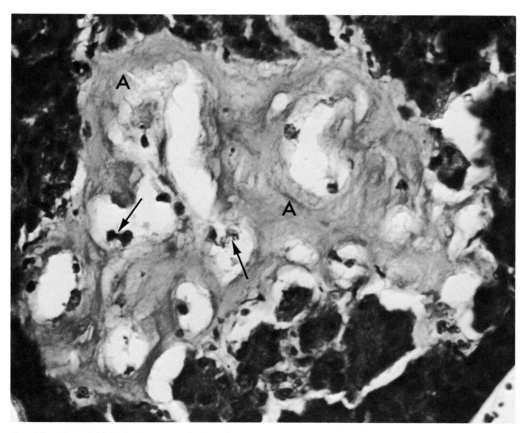

Fig. 6. Selective deposition of amyloid (A) in the pancreatic islet of a cat with diabetes mellitus. Only a few degenerative islet cells (arrows) remain that are surrounded by an extensive fibrillar network of amyloid. ×315. [From "Textbook of Veterinary Internal Medicine," (S. J. Ettinger, ed.), Vol. II. Saunders, Philadelphia, Pennsylvania, 1975.]

Table I

Lysosomal Storage Diseases

Disease	Defective enzyme	Animal	Selected reference
Glycogenoses			
Type II	α-1,4-Glucosidase	Cat	Sandström et al. (1969)
Type II	α-1,4-Glucosidase	Sheep	Manktelow and Hartley (1975)
Type II	α-1,4-Glucosidase	Dog	Mostafa (1970)
Type II	α-1,4-Glucosidase	Cattle	Richards et al. (1977)
Type II	α-1,4-Glucosidase	Quail	Murakami et al. (1980)
Type III	Amylo-1,6-glucosidase	Dog	Ceh et al. (1976)
Type VIII	Phosphorylase b kinase	Mouse	Lyon et al. (1967)
Type VIII	Phosphorylase b kinase	Rat	Clark et al. (1980)
Sphingolipidosis			
GM$_1$ gangliosidiosis	β-Galactosidase	Cattle	Donnelly and Sheahan (1976)
GM$_1$ gangliosidiosis	β-Galactosidase	Cat	Baker and Lindsey (1975)
GM$_1$ gangliosidiosis	β-Galactosidase	Dog	Read et al. (1976)
GM$_2$ gangliosidiosis	Hexosaminidase A and B	Cat	Cork et al. (1977)
GM$_2$ gangliosidiosis	Hexosaminidase A and B	Dog	Karbe (1974)
GM$_2$ gangliosidiosis	Hexosaminidase A and B	Pig	Pierce et al. (1977)
Krabbe's disease	Galactocerebrosidase	Dog[a]	Fletcher and Kurtz (1972)
Krabbe's disease	Galactocerebrosidase	Cat	Johnson (1970)
Krabbe's disease	Galactocerebrosidase	Mouse	Kobayashi et al. (1980)
Gaucher's disease	Glucocerebrosidase	Sheep	Laws and Saal (1968)
Gaucher's disease	β-Glucosidase	Dog	Van De Water et al. (1979)
Gaucher's disease	β-Xylosidase	Mouse	Stephens et al. (1979)
Niemann-Pick Type A	Sphingomyelinase	Cat	Wenger et al. (1980)
Type A	Sphingomyelinase	Dog	Bundza et al. (1979)
Type C	Sphingomyelinase	Mouse	Adachi et al. (1977)
Mucopolysaccharidoses			
MPS I (Hurler, Scheie, Hurler/Scheie)	α-L-Iduronidase	Cat	Haskins et al. (1981)
MPS VI (Maroteaux-Lamy)	Arylsulfatase B	Cat	Jezyk et al. (1977)
Mannosidosis	α-Mannosidase	Cattle	Jolly (1974)
Mannosidosis	α-Mannosidase	Cat	Burditt et al. (1980)
Neuronal ceroid lipofuscinosis	Unknown	Dog	Patel et al. (1974)
Neuronal ceroid lipofuscinosis	Unknown	Cat	Green and Little (1974)
Neuronal ceroid lipofuscinosis	Unknown	Sheep	Jolly et al. (1980)

[a]See Fig. 7.

J. Lysosomal Storage Diseases

Hers (1965) defines an inborn lysosomal storage disease as one in which (a) a single lysosomal enzyme is deficient and (b) abnormal deposits (of substrate) are present within membrane-bound vesicles. This implies that the lysosomes become engorged with undigested material (such as protein, polysaccharides, mucopolysaccharides, and complex lipids) resulting from a deficiency of a specific enzyme or that the degradative enzyme has become sequestered within lysosomes. Since the discovery of glycogenosis by Hers in 1963, more than 50 individual disorders have been identified (Stanbury et al., 1978). The majority of them affects the central nervous system, consequently neurologic signs are helpful clues in the detection of these disorders. Much progress has been made in recent years in the detection of these diseases. This progress is attributed to several factors: (a) electron microscopic examination of peripheral blood cells and skin biopsies, (b) thin-layer chromatography, (c) availability of artificial substrates for lysosomal enzymes, and (d) use of cultured fibroblasts for genetic studies.

Many examples of lysosomal storage diseases have been recognized in animals (Jolly and Blakemore, 1973; Jolly and Hartley, 1977). Table I is a listing of the diseases, defective enzymes, animal species, and selected references (see also Fig. 7). Although these models are valuable to the study of the course of the disease, their greatest potential contribution is in the evaluation of therapy such as enzyme replacement.

K. Erythrocyte Abnormalities

Kitchen (1968), Bannerman (1974), and Smith (1981) have provided excellent reviews of defects in the erythrocytes of

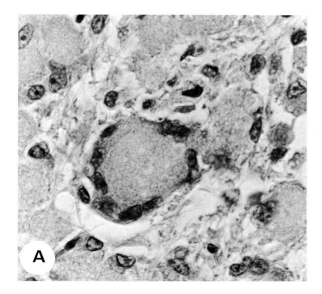

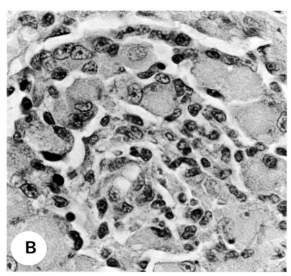

Fig. 7. Globoid cell leukodystrophy (Krabbe's disease) in a dog. (A) A large multinucleated globoid cell is present in region of severely destroyed cerebral white matter. Hematoxylin and eosin. ×500. (B) Perivascular accumulation of globoid macrophages. Hematoxylin and eosin. ×500. (From Fletcher, 1972; with permission of the *American Journal of Pathology.*)

animals that mimic those in humans. Below are brief discussions of some examples of erythrocyte abnormalities.

Pyruvate kinase deficiency was observed (Tasker *et al.*, 1969; Searcy *et al.*, 1979) in young basenji dogs that had congenital nonspherocytic hemolytic anemia. Affected dogs have decreased exercise tolerance, retarded growth, pale mucous membranes, tachycardia, and splenomegaly. The anemia is characterized by shortened survival of the red blood cell, reticulocytosis, and an increased number of nucleated erythrocytes. The erythrocytes from affected basenjis have lowered adenosine triphosphate levels, elevated 2,3-diphosphoglyce-

rate levels, and decreased glucose utilization and lactate production. In the autohemolysis test, hemolysis is not thwarted by glucose supplementation but is inhibited significantly by adenosine triphosphate. These findings are similar to type II autohemolysis of human pyruvate kinase deficiency.

Erythropoietic porphyria is a hereditary defect of porphyrin metabolism and has been observed in cattle (Fourie, 1939). Once considered to be a relatively rare condition, there has been an increase in incidence following artificial insemination programs. Clinically, affected cattle have chronic photosensitivity of the nonpigmented areas of the skin, pink discoloration of teeth and bones, hemolytic anemia, and increased urinary and fecal excretion of porphyrin. Exposure to long-wave ultraviolet light reveals bright red fluorescence. There is a defect in the activity of uroporphyrinogen III cosynthetase, similar to the condition in affected humans, thereby resulting in the overproduction and deposition of uroporphyrin I in tissues.

Protoporphyria also is an inherited defect in porphyrin metabolism and has been described in cattle (Ruth *et al.*, 1977). As in humans, the condition is characterized by high concentrations of protoporphyrin in erythrocytes, plasma, and feces of cattle. Exposure to sunlight results in photosensitivity. The metabolic defect may be a partial deficiency of ferrochelastase, which incorporates iron into protoporphyrin to form heme. Ferrochelastase activity is decreased to about 10% of that of normal cattle.

Methemoglobin reductase deficiency has been observed in various breeds of dogs (Harvey *et al.*, 1974; Letchworth *et al.*, 1977). Clinically, cyanosis of the mucous membranes, lethargy, and general weakness are the principal clinical signs. The activity of the erythrocytic enzyme, nicotinamide adenine dinucleotide (NADH)-methemoglobin reductase, is decreased to 8 to 60% of that in normal dogs. The deficiency of this enzyme inhibits reduction of methemoglobin, allowing it to accumulate. The characteristic chocolate brown blood is suggestive of methemoglobinemia, and electrophoretic studies are necessary to identify the defective enzyme.

Hereditary stomatocytosis has been observed consistently with dwarfism in purebred Alaskan malamute dogs (Fletch and Pinkerton, 1974; Pinkerton *et al.*, 1974). The hemogram is similar to that in humans in that there is mild anemia, erythrocytic macrocytosis, reticulocytosis, increased osmotic fragility, shortened life span of the erythrocytes, and the presence of stomatocytes. In stained blood smears the stomatocytes are red blood cells having a linear, slitlike area instead of a normal circle of pallor.

Hereditary spherocytosis occurs in the deer mouse (Huestis *et al.*, 1956) and results in a mild anemia, reticulocytosis, increased osmotic fragility, and shortened life span of erythrocytes. The erythrocytes have an abnormal spheroidal shape. Splenectomy in the deer mouse prevents hemolysis as it does in humans.

For those interested in studying other hereditary anemias in animals, the work of Bannerman (1974) is recommended. Valuable information may be found in proceedings edited by Kitchen and Boyer (1974) on abnormal hemoglobins in animals.

L. Hemorrhagic Disorders

The causes of spontaneous and excessive bleeding following trauma or surgical procedures are many and varied, but they can be grouped broadly under (a) platelet deficiency or dysfunction and (b) derangements in the coagulation mechanism. The disorders are either inherited or acquired. Dodds (1980) states that during the past 40 years a variety of hemorrhagic diseases has been recognized and studied in humans and animals and that hemorrhagic disorders in animals have been one of the most successfully exploited areas of biomedical research in comparative medicine. Animal models have been recognized for many of the disorders, and they have proved to be reliable and reproducible. In fact, only factor V and factor XIII deficiencies of the coagulation defects remain to be discovered in animals (Dodds, 1979). Excellent reviews and detailed descriptions of animal models of hemorrhagic disorders have been documented (Dodds, 1979, 1981). Only selected examples of animal models of the various hemorrhagic disorders will be briefly mentioned.

Platelet Abnormalities and Coagulation Defects

a. *Thrombasthenia (Glanzmann's disease).* Otterhound dogs have an inherited defect in platelet function similar to thrombasthenia, except for a large proportion of bizarre giant platelets that resemble those seen in Bernard–Soulier syndrome (Dodds, 1967; Raymond and Dodds, 1979).

b. *Thrombopathia.* Conditions in which there are deficiencies or defects in platelet storage granules have been observed in an inbred strain of fawn-hooded rats (Tschopp and Zucker, 1972), in a mutant strain of C57BL/6J mice (Novak et al., 1981), and in cattle with Chediak–Higashi syndrome (Bell et al., 1976).

c. *Immune-Mediated Thrombocytopenia.* This condition, also known as idiopathic thrombocytopenic purpura, has been documented in dogs and horses (Dodds and Wilkins, 1978).

d. *Disseminated Intravascular Coagulation (DIC).* Spontaneous DIC has been recognized as a pathophysiologic state associated with a variety of diseases in many animal species. Readers are referred to a review prepared by Kociba (1976). The major causes of DIC include viral and bacterial systemic infections, endotoxemia, obstetrical complications, heartworm disease, shock, heat stroke, malignancy, and liver disease.

e. *Complement Deficiencies.* C4 deficiency has been documented in guinea pigs (Dodds et al., 1977) and C6 deficiency in rabbits (Rother et al., 1966). Intrinsic coagulation was found to be reduced in both models.

f. *Hereditary Coagulation Defects.* Readers are encouraged to consult the comprehensive review prepared by Dodds (1980).

g. *Factor I Deficiency (Afibrinogenemia, Hypofibrinogenemia, Dysfibrinogenemia).* Afibrinogenemia has been observed in a family of Saanen dairy goats in the Netherlands (Breukink et al., 1972), hypofibrinogenemia in a family of St. Bernard dogs in West Germany (Kammermann et al., 1971), and dysfibrinogenemia in an inbred family of borzoi dogs (Dodds, 1981).

h. *Factor II Deficiency (Hypoprothrombinemia, Dysprothrombinemia).* Acquired hypoprothrombinemia in animals is seen with warfarin poisoning in dogs and moldy sweet clover poisoning in cattle. Inherited dysprothrombinemia has been suspected in dogs (Dodds, 1979).

i. *Factor VII Deficiency.* The defect is common in beagle dogs, especially those from commercial breeding colonies (Mustard et al., 1962; Spurling et al., 1972). The disease is mild with no overt bleeding tendency except that of easy bruising of the affected animals.

j. *Factor VIII Deficiency (Hemophilia A).* This is one of the most common of the severe coagulation factor deficiencies. The defect has been observed in many breeds of dogs (Benson and Dodds, 1976), in standardbred and thoroughbred horses (Graham et al., 1975), and in cats (Cotter et al., 1978). The condition in both humans and animals is inherited as an X-chromosome-linked recessive trait.

k. *Factor IX Deficiency (Hemophilia B, Christmas Disease).* The mode of inheritance is X-chromosome-linked recessive, similar to that for hemophilia A, but the disease is much less common. Hemophilia B has been observed in many breeds of dogs (Mustard et al., 1960; Dodds, 1978). Double hemophilia AB has been observed in dogs (Brinkhous et al., 1973).

l. *Von Willebrand's Disease.* This is the most common inherited bleeding disorder in humans and dogs. The disease has been observed in pigs, in several breeds of dogs (Dodds, 1977), and in a mutant line of Flemish giant-chinchilla rabbits

(Benson and Dodds, 1977). Pigs homozygous for this disease are more resistant than normal pigs to the development of atherosclerosis (Bowie and Fuster, 1980).

m. Factor X Deficiency (Stuart Factor Deficiency). This defect has been found only in a large family of inbred cocker spaniel dogs (Dodds, 1973).

n. Factor XI Deficiency (Plasma Thromboplastin Antecedent Deficiency). This deficiency has been observed in several breeds of dogs (Dodds and Kull, 1971) and in Holstein cattle (Kociba *et al.*, 1969; Gentry *et al.*, 1975).

o. Factor XII Deficiency (Hageman Trait, Hageman Factor Deficiency). This deficiency has been observed in cats (Green and White, 1977; Kier *et al.*, 1980).

M. Autoimmune Disease

Since recognition of the concept of immune reaction against self-antigen, around 1904, a growing number of diseases has been attributed to this phenomenon. Dameshek and Schwartz (1938) substantiated the existence of autoimmune disease by identifying self-generated immune hemolysins in hemolytic anemia patients. Autoantibodies can be found in the serum of normal older individuals. Since innocuous autoantibodies are often formed following tissue damage, a distinction should be made between autoimmunity and autoimmune disease. In other words, the presence of autoantibodies in the serum does not necessarily imply that the individual is suffering from an autoimmune disease. In autoimmunity, there is no pathologic consequence, and the condition is potentially reversible following successful eradication of the causative agent(s). For a disease to be truly autoimmune, the autoimmune reaction

should be the primary cause of the disease. It should be noted that autoimmune diseases are extremely variable in that autoantibodies may be directed against a single organ or tissue, such as in autoimmune thyroiditis, while in other instances, as in systemic lupus erythematosus, a diversity of antibodies results in widespread lesions throughout the body. In the latter, autoantibodies react with nuclear constituents of many different cells. Autoantibodies have also been demonstrated against specific hormone receptors (e.g. anti-acetylcholine) on the surface of cells in myasthenia gravis. For a discussion of autoimmune diseases in animals, readers are referred to a review prepared by Tizard (1982). Table II is a listing of animal models of autoimmune diseases with probable antigen(s), animal species, and selected references (see also Figs. 8 and 9).

N. Primary Immunodeficiency Diseases

Primary immunodeficiency diseases are genetically determined and are seen during early life. Affected individuals are susceptible to recurrent infections. The defect is in the development of the immune system. In this respect, invasion and destruction of lymphoid tissue by neoplasms should also be considered. This could inhibit immunoglobulin synthesis and cell-mediated immune response. Traditionally, immunodeficiency diseases are classified according to the primary component or components involved, such as B or T lymphocytes and mode of inheritance, but the groupings are not always distinct. The inability of the individual to produce antibodies is associated with failure of B lymphocyte differentiation. B Lymphocytes either fail to produce surface immunoglobulins or Ig-coated B lymphocytes are unable to differentiate into antibody-producing plasma cells. In general, disturbances of cell-mediated immunity are associated with altered T lymphocyte function or nonfunctional thymus-dependent lymphocytes, in

Table II

Autoimmune Diseases

Disease	Probable antigen(s)	Animal	Reference
Systemic lupus erythematosus	Multisystem, numerous nuclear and cellular components	Dog	Lewis (1973)
Systemic lupus erythematosus	Multisystem, numerous nuclear and cellular components	(NZB/NZW)F$_1$ mouse	Howie and Helyer (1968)
Rheumatoid arthritis	IgG	Dog	Lewis and Borel (1971)
Autoimmune thyroiditis	Thyroglobulin	Chicken	Wick (1976)
Autoimmune thyroiditis	Thyroglobulin	BUF rat	Hajdu and Rona (1969)
Autoimmune thyroiditis	Thyroglobulin, other colloid components	Dog[a]	Gosselin *et al.* (1981)
Autoimmune hemolytic anemia	Erythrocyte antigens	NZB mouse	Howie and Helyer (1965)
Autoimmune hemolytic anemia	Erythrocyte antigens	Dog	Bull *et al.* (1971)
Pemphigus vulgaris	Squamous epithelium, intercellular	Dog	Hurvitz (1980)
Myasthenia gravis	Acetylcholine receptors, muscle	Dog	Palmer *et al.* (1980)
Idiopathic thrombocytopenic purpura	Platelet surface antigen	Dog, horse	Dodds and Wilkins (1978)

[a]See Figs. 8 and 9.

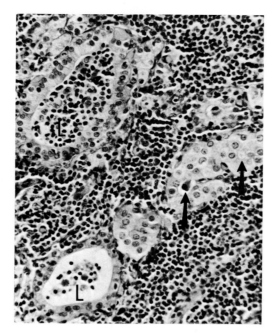

Fig. 8. Lymphocytic thyroiditis in a dog with hypothyroidism characterized by diffuse infiltration of the interstitium and follicular lumen (L) by lymphocytes, plasma cells, and macrophages. The remaining follicles are lined by columnar follicular cells and have narrow lumina (arrows) with little colloid. Periodic acid–Schiff. ×315. (From Gosselin *et al.*, 1981; with permission of *Veterinary Pathology.*)

which case antibody production is normal. A deficiency of both T and B lymphocytes results in impaired humoral and cell-mediated immunity. Excellent reviews on immunodeficiency diseases in animals have been prepared (Gershwin and Merchant, 1981; Perryman and Magnuson, 1982).

1. B Lymphocyte Deficiency

Immunoglobulin is defective, but the cell-mediated immune system is normal with this defect. X-Linked agammaglobulinemia (Bruton's type) has been observed only in male horses that were totally devoid of IgM, IgA, and IgG(T) (Banks *et al.*, 1976). Other inherited conditions in which there are deficiencies of isolated immunoglobulin classes are IgG_2 deficiency in a breed of Red Danish cattle (Nansen, 1972), IgM deficiency in horses (Perryman and McGuire, 1980), and IgA deficiency in chickens (Luster *et al.*, 1977). An inherited dysgammaglobulinemia in chickens having abnormally low serum immunoglobulin levels has some characteristics of common variable immunodeficiency of humans (Benedict *et al.*, 1978).

2. T Lymphocyte Deficiency

Due to a defect in the development of the thymus (from the third and fourth pharyngeal pouches), there is a deficiency in thymus-derived lymphocytes resulting in failure of cell-mediated immunity. Aplastic or hypoplastic thymus and lack of hair

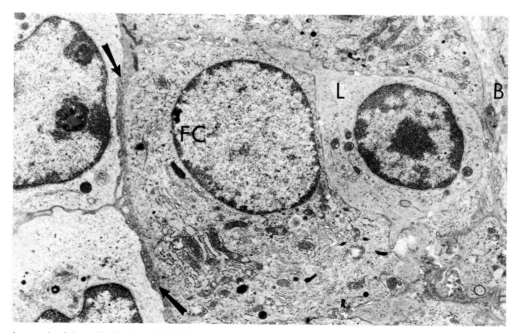

Fig. 9. Electron micrograph of the wall of a thyroid follicle from a dog with lymphocytic thyroiditis and hypothyroidism. A lymphocyte (L) is migrating between follicular cells (FC) in the wall of a thyroid follicle. Microvilli (arrows) on the luminal surface of follicular cells are compressed by numerous lymphocytes in the follicular lumen. B, basement membrane of the thyroid follicle. (From Gosselin *et al.*, 1981; with permission of *Veterinary Pathology.*)

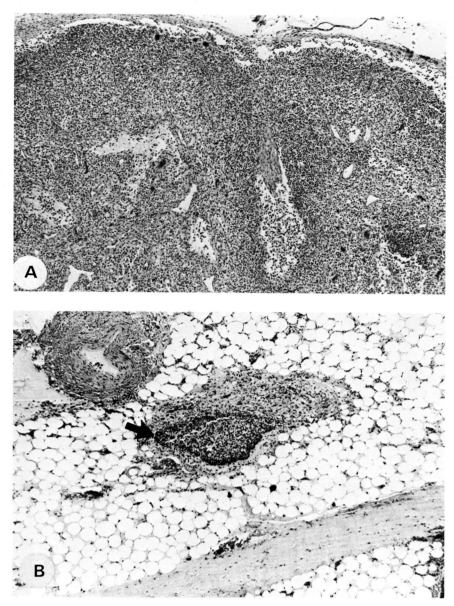

Fig. 10. Combined immunodeficiency (Swiss-type agammaglobulinemia) in a foal. (A) Lymph node lacks organized lymphoid structures including follicles and germinal centers. Hematoxylin and eosin. ×140. (B) Thymus is composed primarily of fat, with a focus of lymphoid tissue (arrow). Hematoxylin and eosin. ×56. (From McGuire, 1975; with permission of the *American Journal of Pathology.*)

are characteristic clinical findings. Animals exhibiting T lymphocyte deficiency are the nude mouse (Fogh and Giovanella, 1978), nude rat (Festing, 1981), guinea pig (O'Donoghue and Reed, 1981), and Black Pied Danish cattle (Brummerstedt *et al.,* 1978).

3. T and B Lymphocyte Deficiency

This is an extremely serious form of immunodeficiency in which there are deficiencies in both T and B lymphocytes. The condition, also known as Swiss-type agammaglobulinemia, has been observed in young Arabian horses (Fig. 10) (McGuire *et al.,* 1976). The deficient foals are highly susceptible to infection produced by an adenovirus or *Pneumocystis carinii.* Hypogammaglobulinemia, lymphopenia, and absence of germinal centers in lymph nodes result from B lymphocyte deficiency. Lymphocyte depletion in arteriolar sheaths of the spleen and paracortical areas of lymph nodes and absence of normal thymic tissue result from T lymphocyte deficiency.

O. Glaucoma

Glaucoma is an ocular disorder due to a persistent increase in intraocular pressure resulting from an obstruction of the flow of aqueous humor within the eyeball. The site of obstruction is most often the pupil, the angle of the anterior chamber, or the outflow pathways that lie between the anterior chamber angle and the episcleral veins. Glaucomas are divided traditionally into congenital, primary, and secondary, with the last type developing as a complication of some other primary disease process. A review on animal models of glaucoma has been prepared by Gelatt (1977). Although there is no exact animal counterpart, spontaneous glaucoma in animals can provide additional information relative to the genesis of the different forms of glaucoma in humans.

Gelatt et al. (1981) have observed beagle dogs with an inherited glaucoma similar initially to primary open-angle glaucoma in man and in later stages to chronic narrow-angle glaucoma. The condition is believed to be an autosomal recessive trait. The canine model may permit essential studies of aqueous humor dynamics, outflow pathways, pharmacologic trials, and of glaucomatous changes in the optic disk and nerve. Rabbits with inherited buphthalmia have been used in glaucoma research during the past 20 years, and the disease has been found to be associated with congenital goniodysgenesis (Kolker et al., 1963).

ACKNOWLEDGMENTS

This work was supported in part by Public Health Service Grant RROO301-17 from the Division of Research Resources, National Institutes of Health, United States Department of Health and Human Services, under the auspices of Universities Associated for Research and Education in Pathology, Inc. The opinions or assertions contained herein are the private views of the authors and are not to be construed as official or as reflecting the views of the Department of the Army or the Department of Defense. The authors thank Sidney R. Jones and Timothy P. O'Neill for their critical reviews of the manuscript and Susan Allen, Charmaine Goetz, and Merryanna Swartz for technical assistance.

REFERENCES

Adachi, M., Volk, B. W., and Schneck, L. (1977). Niemann-Pick disease. In "Handbook: Animal Models of Human Disease" (T. C. Jones, D. B. Hackel, and G. Migaki, eds.), Model No. 117. Registry of Comparative Pathology, Armed Forces Institute of Pathology, Washington, D.C.

Altman, P. L., and Katz, D. D., eds. (1979a). "Inbred and Genetically Defined Strains of Laboratory Animals. Part 1. Mouse and Rat." Fed. Am. Soc. Exp. Biol., Bethesda, Maryland.

Altman, P. L., and Katz, D. D., eds. (1979b). "Inbred and Genetically Defined Strains of Laboratory Animals. Part 2. Hamster, Guinea Pig, Rabbit, and Chhicken." Fed. Am. Soc. Exp. Biol., Bethesda, Maryland.

Andrews, E. J., Ward, B. C., and Altman, N. H., eds. (1979). "Spontaneous Animal Models of Human Disease," Vols. 1 and 2. Academic Press, New York.

Angus, K. W., Gardiner, A. C., Morgan, K. T., and Gray, E. W. (1974). Mesangiocapillary glomerulonephritis in lambs. II. Pathological findings and electron microscopy of renal lesions. J. Comp. Pathol. 84, 319–330.

Angus, K. W., Gardiner, A. C., Mitchell, B., and Aitchison, G. U. (1975). The role of colostrum in glomerulonephritis of Finnish Landrace lambs: Effect of cross-fostering newly born Finnish Landrace and Cheviot lambs. Res. Vet. Sci. 19, 222–224.

Baker, H. J., and Lindsey, J. R. (1975). GM_1 gangliosidosis. In "Handbook: Animal Models of Human Disease" (T. C. Jones, D. B. Hackel, and G. Migaki, eds.), Model No. 52. Registry of Comparative Pathology, Armed Forces Institute of Pathology, Washington, D.C.

Banks, K. L., McGuire, T. C., and Jerrells, T. R. (1976). Absence of B lymphocytes in a horse with primary agammaglobulinemia. Clin. Immunol. Immunopathol. 5, 282–290.

Bannerman, R. M. (1974). Animal models of hereditary hematologic disease. Birth Defects, Orig. Artic. Ser. 10, 278–285.

Banting, F. G., and Best, C. H. (1922). The internal secretion of the pancreas. J. Lab. Clin. Med. 7, 251–266.

Bell, T. G., Meyers, K. M., Prieur, D. J., Fauci, A. S., Wolff, S. M., and Padgett, G. A. (1976). Decreased nucleotide and serotonin storage associated with defective function in Chediak-Higashi syndrome cattle and human platelets. Blood 48, 175–184.

Benditt, E. P. (1977). The origin of atherosclerosis. Sci. Am. 236, 74–85.

Benedict, A. A., Chanh, T. C., Tam, L. Q., Pollard, L. W., Kubo, R. T., and Abplanalp, H. A. (1978). Inherited immunodeficiency in chickens. In "Animal Models of Comparative and Developmental Aspects of Immunity and Disease" (M. E. Gershwin and E. L. Cooper, eds.), pp. 99–109. Pergamon, Oxford.

Benirschke, K. (1969). Models for cytogenetics and embryology. Fed. Proc., Fed. Am. Soc. Exp. Biol. 28, 170–178.

Benson, R. E., and Dodds, W. J. (1976). Immunologic characterization of canine factor VIII. Blood 48, 521–529.

Benson, R. E., and Dodds, W. J. (1977). Autosomal factor VIII deficiency in rabbits: Size variations of rabbit factor VIII. Thromb. Haemostasis 38, 380.

Binns, W., Thacker, E. J., James, L. F., and Huffman, W. T. (1959). A congenital cyclopian-type malformation in lambs. J. Am. Vet. Med. Assoc. 134, 180–183.

Binns, W., Shupe, J. L., Keeler, R. F., and James, L. F. (1965). Chronologic evaluation of teratogenicity in sheep fed Veratrum californicum. J. Am. Vet. Med. Assoc. 147, 839–842.

Bishop, S. P., Kawamura, K., and Detweiler, D. K. (1979a). Systemic hypertension. In "Spontaneous Animal Models of Human Disease" (E. J. Andrews, B. C. Ward, and N. H. Altman, eds.), Vol. 1, pp. 50–54. Academic Press, New York.

Bishop, S. P., Sole, M. J., and Tilley, L. P. (1979b). Cardiomyopathies. In "Spontaneous Animal Models of Human Disease" (E. J. Andrews, B. C. Ward, and N. H. Altman, eds.), Vol. 1, pp. 59–64. Academic Press, New York.

Black, H. E., Rosenblum, I. Y., and Capen, C. C. (1980). Chemically induced (streptozotocin-alloxan) diabetes mellitus in the dog. Am. J. Pathol. 98, 295–309.

Bowie, E. J. W., and Fuster, V. (1980). Resistance to atherosclerosis in pigs with von Willebrand's disease. Acta Med. Scand., Suppl. 642, 121–130.

Braun, A. C., and Wood, H. N. (1969). Transformation and recovery of the crown gall tumor cell: An experimental model. Fed. Proc., Fed. Am. Soc. Exp. Biol. 28, 1815–1819.

Breeze, R. G. (1979). Heaves. Vet. Clin. North Am. 1, 219–230.

Breukink, H. J., Hart, H. C., Arkel, C., Veldon, N. A., and Watering, C. C.

(1972). Congenital afibrinogenemia in goats. *Zentralbl. Veterinaermed., Reihe A* **19,** 661–676.

Brinkhous, K. M., Davis, P. D., Graham, J. B., and Dodds, W. J. (1973). Expression and linkage of genes for X-linked hemophilias A and B in the dog. *Blood* **41,** 577–585.

Brummerstedt, E., Basse, A., Flagstad, T., and Andresen, E. (1978). Acrodermatitis enteropathica. *In* "Handbook: Animal Models of Human Disease" (T. C. Jones, D. B. Hackel, and G. Migaki, eds.), Model No. 137. Registry of Comparative Pathology, Armed Forces Institute of Pathology, Washington, D.C.

Buchmeier, M. J., and Oldstone, M. B. A. (1978). Virus-induced immune complex disease: Identification of specific viral antigens and antibodies deposited in complexes during chronic lymphocytic choriomeningitis virus infection. *J. Immunol.* **120,** 1297–1304.

Bull, R. W., Schirmer, R., and Bowdler, A. J. (1971). Autoimmune hemolytic disease in the dog. *J. Am. Vet. Med. Assoc.* **159,** 880–884.

Bulla, L. S., and Cheng, T. C., eds. (1978). "Invertebrate Models for Biochemical Research." Plenum, New York.

Bundza, A., Lowden, J. A., and Charlton, K. M. (1979). Niemann-Pick disease in a poodle dog. *Vet. Pathol.* **16,** 530–538.

Burditt, L. J., Chotai, K., Hirani, S., Nugent, P. G., Winchester, B. G., and Blakemore, W. F. (1980). Biochemical studies on a case of feline mannosidosis. *Biochem. J.* **189,** 467–473.

Burkholder, P. M., and Bergeron, J. A. (1970). Spontaneous glomerulonephritis in the prosimian primate *Galago. Am. J. Pathol.* **61,** 437–456.

Bustad, L. K., Gorham, J. R., Hegreberg, G. A., and Padgett, G. A. (1976). Comparative medicine: Progress and prospects. *J. Am. Vet. Med. Assoc.* **169,** 90–105.

Bustad, L. K., Hegreberg, G. A., and Padgett, G. A. (1977). "Naturally Occurring Animal Models of Human Disease: A Bibliography." Nat. Acad. Sci., Washington, D.C.

Campbell, H. A., and Link, K. P. (1941). Studies on the hemorrhagic sweet clover disease. IV. The isolation and crystallization of the hemorrhagic agent. *J. Biol. Chem.* **138,** 21–33.

Cardinett, G. H., III, and Holliday, T. A. (1979). Neuromuscular diseases of domestic animals: A summary of muscle biopsies from 159 cases. *Ann. N.Y. Acad. Sci.* **317,** 290–313.

Ceh, L., Hauge, J. G., Svenkerud, R., and Strande, A. (1976). Glycogenosis type III in the dog. *Acta Vet. Scand.* **17,** 210–222.

Cheville, N. F. (1973). Viral glomerulonephritis. *In* "Handbook: Animal Models of Human Disease" (T. C. Jones, D. B. Hackel, and G. Migaki, eds.), Model No. 19. Registry of Comparative Pathology, Armed Forces Institute of Pathology, Washington, D.C.

Clark, D. G., Topping, D. L., Illman, R. J., Trimble, R. P., and Malthus, R. S. (1980). A glycogen storage disease (gsd/gsd) rat: Studies on lipid metabolism, lipogenesis, plasma metabolites, and bile acid secretion. *Metab. Clin. Exp.* **29,** 415–420.

Clarkson, T. B., and Lofland, H. B. (1961). Effects of cholesterol-fat diets on pigeons susceptible and resistant to atherosclerosis. *Circ. Res.* **9,** 106–109.

Clarkson, T. B., Prichard, R. W., Bullock, B. C., St. Clair, R. W., Lehner, N. D. M., Jones, D. C., Wagner, W. D., and Rudel, L. L. (1976). Pathogenesis of atherosclerosis: Some advances from using animal models. *Exp. Mol. Pathol.* **24,** 264–286.

Cochrane, C. G., and Koffler, D. (1973). Immune complex disease in experimental animals and man. *Adv. Immunol.* **16,** 185–264.

Coleman, D. L. (1978). Obese and diabetes: Two mutant genes causing diabetes–obesity syndromes in mice. *Diabetologia* **14,** 141–148.

Coleman, D. L., and Hummel, K. P. (1974). Hyperinsulinemia in pre-weaning diabetes (db) mice. *Diabetologia* **10,** 607–610.

Cork, L. C., Munnell, J. F., Lorenz, M. D., Murphy, J. V., Baker, H. J., and

Rattazzi, M. C. (1977). GM₂ ganglioside lysosomal storage disease in cats with β-hexosaminidase deficiency. *Science* **196,** 1014–1017.

Cornelius, C. E. (1969). Animal models—A neglected medical resource. *N. Engl. J. Med.* **281,** 934–944.

Cornelius, C. E., and Arias, I. M. (1966). Biomedical models in veterinary medicine. *Am. J. Med.* **40,** 165–169.

Cornelius, C. E., and Gronwall, R. R. (1968). Congenital photosensitivity and hyperbilirubinemia in Southdown sheep in the United States. *Am. J. Vet. Res.* **29,** 291–295.

Cornelius, C. E., Arias, I. M., and Osburn, B. I. (1965). Hepatic pigmentation with photosensitivity. A syndrome in Corriedale sheep resembling Dubin–Johnson syndrome in man. *J. Am. Vet. Med. Assoc.* **146,** 709–713.

Cotter, S. M. Brenner, R. M., and Dodds, W. J. (1978). Hemophilia A in three unrelated cats. *J. Am. Vet. Med. Assoc.* **172,** 166–168.

Cowgill, L. D., Goldfarb, S., Goldberg, M., Slatopolsky, E., and Agus, Z. S. (1977). Nature of the renal defect in familial hypophosphatemic rickets. *Clin. Res.* **25,** 505A.

Craig, J. M. (1969). Models for obstetrical and gynecological disease. *Fed. Proc., Fed. Am. Soc. Exp. Biol.* **28,** 206–210.

Craighead, J. E. (1976). Diabetes mellitus. *In* "Handbook: Animal Models of Human Disease" (T. C. Jones, D. B. Hackel, and G. Migaki, eds.), Model No. 70. Registry of Comparative Pathology, Armed Forces Institute of Pathology, Washington, D.C.

Curtice, C. (1891). The biology of the cattle tick. *J. Comp. Med. Vet. Arch.* **12,** 313–319.

Dameshek, W., and Schwartz, S. O. (1938). The presence of hemolysins in acute hemolytic anemia. *N. Engl. J. Med.* **218,** 75–80.

Dodds, W. J. (1967). Familial canine thrombocytopathy. *Thromb. Diath. Haemorrh., Suppl.* **26,** 241–248.

Dodds, W. J. (1973). Canine factor X (Stuart–Prower factor) deficiency. *J. Lab. Clin. Med.* **82,** 560–566.

Dodds, W. J. (1977). Von Willebrand's disease. *In* "Handbook: Animal Models of Human Disease" (T. C. Jones, D. B. Hackel, and G. Migaki, eds.), Model No. 110. Registry of Comparative Pathology, Armed Forces Institute of Pathology, Washington, D.C.

Dodds, W. J. (1978). Inherited bleeding disorders. *Canine Pract.* **5,** 49–58.

Dodds, W. J. (1979). Hemorrhagic disorders. *In* "Spontaneous Animal Models of Human Disease" (E. J. Andrews, B. C. Ward, and N. H. Altman, eds.), Vol. 1, pp. 266–285. Academic Press, New York.

Dodds, W. J. (1980). Hemostasis and coagulation. *In* "Clinical Biochemistry of Domestic Animals" (J. J. Kaneko, ed.), 3rd ed., pp. 671–718. Academic Press, New York.

Dodds, W. J. (1981). Second international registry of animal models of thrombosis and hemorrhagic diseases. *ILAR News* **24,** R3–R50.

Dodds, W. J., and Kull, J. E. (1971). Canine factor XI (plasma thromboplastin antecedent) deficiency. *J. Lab. Clin. Med.* **78,** 746–752.

Dodds, W. J., and Wilkins, R. J. (1978). Immune thrombocytopenia. *In* "Handbook: Animal Models of Human Disease" (T. C. Jones, D. B. Hackel, and G. Migaki, eds.), Model No. 127. Registry of Comparative Pathology, Armed Forces Institute of Pathology, Washington, D.C.

Dodds, W. J., Raymond, S. L., Moynihan, A. C., Pickering, R. J., and Amiraian, K. (1977). Coagulation studies in C4-deficient guinea pigs. *Proc. Soc. Exp. Biol. Med.* **154,** 382–385.

Donnelly, W. J. C., and Sheahan, B. J. (1976). GM₁ gangliosidosis type II. *In* "Handbook: Animal Models of Human Disease" (T. C. Jones, D. B. Hackel, and G. Migaki, eds.), Model No. 87. Registry of Comparative Pathology, Armed Forces Institute of Pathology, Washington.

Doyle, R. E., Garb, S., Davis, L. E., Meyer, D. K., and Clayton, F. W. (1968). Domesticated farm animals in medical research. *Ann. N.Y. Acad. Sci.* **147,** 129–204.

Dulin, W. E., and Soret, M. G. (1977). Chemically and hormonally induced diabetes. *In* "The Diabetic Pancreas" (B. W. Volk and K. F. Wellman, eds.), pp. 425–465. Plenum, New York.

Eicher, E. M., Southard, J. L., Scriver, C. R., and Glorieux, F. H. (1976). Hypophosphatemia: Mouse model for human familial hypophosphatemic (vitamin D-resistant) rickets. *Proc. Natl. Acad. Sci. U.S.A.* **73**, 4667–4671.

Ellermann, V., and Bang, O. (1908). Experimentelle Leukämie bei Hühnern. *Zentralbl. Bakteriol., Parasitenkd., Infektionskr. Hyg., Abt. 1: Orig.* **46**, 595–609.

Engerman, R. L. (1976). Animal models of diabetic retinopathy. *Trans.—Am. Acad. Ophthamol. Otolaryngol., Sect. Ophthalmol.* **81**, 710–715.

Falconer, D. S., Latyszewski, M., and Isaacson, J. H. (1964). Diabetes insipidus associated with oligosyndactyly in the mouse. *Genet. Res.* **5**, 473–488.

Feldman, W. H. (1963). Veterinary medicine and human medicine—Complementary disciplines in public health. *Sci. Proc. Am. Vet. Med. Assoc.* **143**, 200–214.

Festing, M. F. (1981). Athymic nude rats. *In* "Immunologic Defects in Laboratory Animals" (M. E. Gershwin and B. Merchant, eds.), Vol. 1, pp. 267–283. Plenum, New York.

Fink, J. N., Hensley, G. T., and Barboriak, J. J. (1970). An animal model of hypersensitivity pneumonitis. *J. Allergy* **46**, 156–161.

Fisk, D. E., and Kuhn, C. (1976). Emphysema-like changes in the lungs of the blotchy mouse. *Am. Rev. Respir. Dis.* **113**, 787–797.

Fletch, S. M., and Pinkerton, P. H. (1974). Inherited hemolytic anemia. *In* "Handbook: Animal Models of Human Disease" (T. C. Jones, D. B. Hackel, and G. Migaki, eds.), Model No. 41. Registry of Comparative Pathology, Armed Forces Institute of Pathology, Washington, D.C.

Fletcher, T. F. (1972). Globoid cell leukodystrophy. *Am. J. Pathol.* **66**, 375–378.

Fletcher, T. F., and Kurtz, H. J. (1972). Globoid cell leukodystrophy. *In* "Handbook: Animal Models of Human Disease" (T. C. Jones, D. B. Hackel, and G. Migaki, eds.), Model No. 9. Registry of Comparative Pathology, Armed Forces Institute of Pathology, Washington, D.C.

Florentin, R. A., and Nam, S. C. (1968). Dietary-induced atherosclerosis in miniature swine. I. Gross and light microscopy observations: Time of development and morphologic characteristics of lesions. *Exp. Mol. Pathol.* **8**, 263–301.

Fogh, J., and Giovanella, B. C. (1978). "The Nude Mouse in Experimental and Clinical Research." Academic Press, New York.

Fourie, P. J. J. (1939). Bovine congenital porphyrinuria (pink tooth) inherited as a recessive character. *Onderstepoort J. Vet. Sci. Anim. Ind.* **13**, 383–398.

Frenkel, J. K. (1969). Models for infectious diseases. *Fed. Proc., Fed. Am. Soc. Exp. Biol.* **28**, 179–190.

Gajdusek, D. C., Gibbs, C. J., and Alpers, M. (1965). "Slow, Latent, and Temperate Virus Infection," NINDB Monogr. No. 2, PHS Publ. No. 1378. Nat. Inst. Health, Bethesda, Maryland.

Gay, W. I. (1967). Comparative medicine. *Science* **158**, 1220–1237.

Geer, J. C., and Guidry, M. A. (1965). Experimental canine atherosclerosis. *In* "Comparative Atherosclerosis" (J. C. Roberts and R. Straus, eds.), pp. 170–195. Harper & Row, New York.

Gelatt, K. N. (1977). Animal models for glaucoma. *Invest. Ophthalmol. Visual Sci.* **16**, 592–596.

Gelatt, K. N., Glenwood, G. G., Gwin, R. M., Bromberg, N. M., Merideth, R. E., and Samuelson, D. A. (1981). Primary open angle glaucoma. *In* "Handbook: Animal Models of Human Disease" (C. C. Capen, D. B. Hackel, T. C. Jones, and G. Migaki, eds.), Model No. 222. Registry of Comparative Pathology, Armed Forces Institute of Pathology, Washington, D.C.

Gentry, P. A., Crane, S., and Lotz, F. (1975). Factor XI (plasma thromboplastin antecedent) deficiency in cattle. *Can. Vet. J.* **16**, 160–163.

Gerritsen, G. C., and Dulin, W. E. (1967). Characterization of diabetes in the Chinese hamster. *Diabetologia* **3**, 74–84.

Gershwin, M. E., and Merchant, B., eds. (1981). "Immunologic Defects in Laboratory Animals," Vols. I and II. Plenuum, New York.

Gertz, E. W. (1973). Cardiomyopathy. *In* "Handbook: Animal Models of Human Disease" (T. C. Jones, D. B. Hackel, and G. Migaki, eds.), Model No. 28. Registry of Comparative Pathology, Armed Forces Institute of Pathology, Washington, D.C.

Gillespie, J. R., Tyler, W. S., and Eberly, V. E. (1964). Blood pH, O_2, and CO_2 tensions in exercised control and emphysematous horses. *Am. J. Physiol.* **207**, 1067–1072.

Gonet, A. E., Stauffacher, W., Pictet, R., and Renold, A. E. (1965). Obesity and diabetes mellitus with striking congenital hyperplasia of the islets of Langerhans in spiny mice (*Acomys cahirinus*). *Diabetologia* **1**, 162–171.

Good, R. A., Finstad, J., Cain, W. A., Fish, A., Perey, D. Y., and Gatti, R. A. (1969). Models of immunologic diseases and disorders. *Fed. Proc., Fed. Am. Soc. Exp. Biol.* **28**, 191–205.

Gorham, J. R., and Henson, J. B. (1976). Persistent viral infections. *In* "Handbook: Animal Models of Human Disease" (T. C. Jones, D. B. Hackel, and G. Migaki, eds.), Model No. 63. Registry of Comparative Pathology, Armed Forces Institute of Pathology, Washington, D.C.

Gosselin, S. J., Capen, C. C., and Martin, S. L. (1981). Histologic and ultrastructural evaluation of thyroid lesions associated with hypothyroidism in dogs. *Vet. Pathol.* **18**, 299–309.

Gottlieb, H., and Lalich, J. (1954). The occurrence of arteriosclerosis in the aorta of swine. *Am. J. Pathol.* **30**, 851–855.

Graham, J. B., Brinkhous, K. M., and Dodds, W. J. (1975). Canine and equine hemophilia. *In* "Handbook of Hemophilia" (K. M. Brinkhous and H. C. Hemker, eds.), pp. 119–139. Excerpta Medica, Amsterdam.

Green, E. L., ed. (1968). "Handbook on Genetically Standardized JAX Mice." Jackson Laboratory, Bar Harbor, Maine.

Green, P. D., and Little, P. B. (1974). Neuronal ceroid-lipofuscin storage in Siamese cats. *Can. J. Comp. Med.* **38**, 207–212.

Green, R. A., and White, F. (1977). Feline factor XII (Hageman) deficiency. *Am. J. Vet. Res.* **38**, 893–895.

Greep, R. O. (1970). Animal models in biomedical research. *J. Anim. Sci.* **31**, 1235–1246.

Gresham, G. A. (1976). "Primate Atherosclerosis." Karger, Basel.

Gross, L. (1970). "Oncogenic Viruses," 2nd ed. Pergamon, Oxford.

Gross, P., Pfitzer, E. A., Tolker, E., Babyak, M. A., and Kaschak, M. (1965). Experimental emphysema: Its production with papain in normal and silicotic rats. *Arch. Environ. Health* **11**, 50–58.

Hadlow, W. J. (1959). Scrapie and kuru. *Lancet* **2**, 289–290.

Hadlow, W. J. (1973). Myopathies of animals. *In* "The Striated Muscle" (C. M. Pearson and F. K. Mostofi, eds.), pp. 364–409. Williams & Wilkins, Baltimore, Maryland.

Hajdu, A., and Rona, G. (1969). Spontaneous thyroiditis in laboratory rats. *Experientia* **25**, 1325–1327.

Harper, J. A., Bernier, P. E., Helfer, D. H., and Schmitz, J. A. (1975). Degenerative myopathy of the deep pectoral muscle in the turkey. *J. Hered.* **66**, 362–366.

Harris, J. B., and Slater, C. R. (1980). Animal models of muscular dystrophy. *Br. Med. Bull.* **36**, 193–197.

Harvey, J. W., Ling, G. V., and Kaneko, J. J. (1974). Methemoglobin reductase deficiency in a dog. *J. Am. Vet. Med. Assoc.* **164**, 1030–1033.

Haskins, M. E., Jezyk, P. F., Desnick, R. J., and Patterson, D. F. (1981). Mucopolysaccharidosis I. *In* "Handbook: Animal Models of Human Disease" (C. C. Capen, D. B. Hackel, T. C. Jones, and G. Migaki, eds.), Model No. 226. Registry of Comparative Pathology, Armed Forces Institute of Pathology, Washington, D.C.

Hayes, J. A., Korthy, A., and Snider, G. L. (1975). The pathology of elastase-induced panacinar emphysema in hamsters. *J. Pathol.* **117**, 1–14.

Hegreberg, G. A. (1976). Muscular dystrophy. *Am. J. Pathol.* **85**, 233–236.

Hegreberg, G. A., and Leathers, C. (1982a). "Bibliography of Naturally Occurring Animal Models of Human Disease," Vol. I. Student Book Corp., Pullman, Washington.

Hegreberg, G. A., and Leathers, C. (1982b). "Bibliography of Induced Animal Models of Human Disease," Vol. II. Student Book Corp., Pullman, Washington.

Hegreberg, G. A., Norton, S. L., and Gorham, J. R. (1977). Muscular dystrophy. *In* "Handbook: Animal Models of Human Disease" (T. C. Jones, D. B. Hackel, and G. Migaki, eds.), Model No. 119. Registry of Comparative Pathology, Armed Forces Institute of Pathology, Washington, D.C.

Held, J. R. (1981). Animals for medical research and testing. *Bull. Pan. Am. Health Organ.* **15,** 36–48.

Hensley, G. T., Fink, J. N., and Barboriak, J. J. (1974). Hypersensitivity pneumonitis in the monkey. *Arch. Pathol.* **97,** 33–38.

Henson, J. B., and Gorham, J. R. (1974). Persistent viral infections. *In* "Handbook: Animal Models of Human Disease" (T. C. Jones, D. B. Hackel, and G. Migaki, eds.), Model No. 33. Registry of Comparative Pathology, Armed Forces Institute of Pathology, Washington, D.C.

Herberg, L., and Coleman, D. L. (1977). Laboratory animals exhibiting obesity and diabetes syndromes. *Metab., Clin. Exp.* **26,** 59–99.

Herberg, L., Buchanan, K. D., Herbertz, L. M., Kern, H. F., and Kley, H. K. (1980). The Djungarian hamster, a laboratory animal with inappropriate hyperglycemia. *Comp. Biochem. Physiol. A* **65A,** 35–60.

Hers, H. G. (1965). Inborn lysosomal diseases. *Gastroenterology* **48,** 625–633.

Hicks, J. D. (1966). Vascular changes in the kidneys of NZB mice and F NZB × NZW hybrids. *J. Pathol. Bacteriol.* **91,** 479–486.

Hirsch, M. S., Allison, A. C., and Harvey, J. J. (1969). Immune complexes in mice infected neonatally with Moloney leukaemogenic and murine sarcoma viruses. *Nature (London)* **223,** 739–740.

Homburger, F., Baker, J. R., Nixon, C. W., and Wilgram, G. (1962). New hereditary disease of Syrian hamsters. Primary, generalized polymyopathy and cardiac necrosis. *Arch. Intern. Med.* **110,** 660–662.

Howard, C. F., Jr. (1972). Spontaneous diabetes in *Macaca nigra*. *Diabetes* **21,** 1077–1090.

Howard, C. F., Jr., and Palotay, J. L. (1976). Diabetes mellitus. *In* "Handbook: Animal Models of Human Disease" (T. C. Jones, D. B. Hackel, and G. Migaki, eds.), Model No. 81. Registry of Comparative Pathology, Armed Forces Institute of Pathology, Washington, D.C.

Howie, J. B., and Helyer, B. J. (1965). Autoimmune disease in mice. *Ann. N.Y. Acad. Sci.* **124,** 167–177.

Howie, J. B., and Helyer, B. J. (1968). The immunology and pathology of NZB mice. *Adv. Immunol.* **9,** 215–266.

Huestis, R. R., Anderson, R. S., and Motulsky, A. G. (1956). Hereditary spherocytosis in peromyscus. I. Genetic studies. *J. Hered.* **47,** 225.

Hunt, R. D., Meléndez, L. V., King, N. W., Gilmore, C. E., Daniel, M. D., Williamson, M. E., and Jones, T. C. (1970). Morphology of a disease with features of malignant lymphoma in marmosets and owl monkeys inoculated with *Herpesvirus saimiri*. *JNCI, J. Natl. Cancer Inst.* **44,** 447–465.

Hurvitz, A. I. (1980). Pemphigus vulgaris. *In* "Handbook: Animal Models of Human Disease" (C. C. Capen, D. B. Hackel, T. C. Jones, and G. Migaki, eds.), Model No. 203. Registry of Comparative Pathology, Armed Forces Institute of Pathology, Washington, D.C.

Huseby, R. A. (1969). Utility and failure of models in oncology. *Fed. Proc., Fed. Am. Soc. Exp. Biol.* **28,** 211–215.

Iwatsuka, H., Shino, A., and Suzuoki, Z. (1970). General survey of diabetic features of yellow KK mice. *Endocrinol. Jpn.* **17,** 23–35.

Jezyk, P. F., Haskins, M. E., Patterson, D. F., Mellman, W. J., and Greenstein, M. (1977). Mucopolysaccharidosis in a cat with arylsulfatase B deficiency: A model of Maroteaux–Lamy syndrome. *Science* **198,** 834–836.

Johnson, K. H. (1970). Globoid leukodystrophy in the cat. *J. Am. Vet. Med. Assoc.* **157,** 2057–2064.

Johnson, K. H., and Stevens, J. B. (1973). Light and electron microscopic studies of islet amyloid in diabetic cats. *Diabetes* **22,** 81–90.

Jolly, R. D. (1974). Mannosidosis. *In* "Handbook: Animal Models of Human Disease" (T. C. Jones, D. B. Hackel, and G. Migaki, eds.), Model No. 44. Registry of Comparative Pathology, Armed Forces Institute of Pathology, Washington, D.C.

Jolly, R. D., and Blakemore, W. F. (1973). Inherited lysosomal storage diseases: An essay in comparative medicine. *Vet. Rec.* **92,** 391–400.

Jolly, R. D., and Hartley, W. J. (1977). Storage diseases of domestic animals. *Aust. Vet. J.* **53,** 1–8.

Jolly, R. D., Janmaat, A., West, D. M., and Morrison, I. (1980). Ovine ceroid-lipofuscinosis: A model of Batten's disease. *Neuropathol. Appl. Neurobiol.* **6,** 195–209.

Jones, T. C. (1950). Analogous pathologic patterns in different animal species. *Bull. Int. Assoc. Med. Mus.* No. 31, pp. 23–35.

Jones, T. C. (1964). Preparation for research in veterinary and comparative pathology. *J. Am. Vet. Med. Assoc.* **144,** 1105–1112.

Jones, T. C. (1969). Mammalian and avian models of disease in man. *Fed. Proc., Fed. Am. Soc. Exp. Biol.* **28,** 162–169.

Jones, T. C. (1980). The value of animal models. *Am. J. Pathol.* **101,** S3–S9.

Julian, L. M. (1973). Hereditary muscular dystrophy. *In* "Handbook: Animal Models of Human Disease" (T. C. Jones, D. B. Hackel, and G. Migaki, eds.), Model No. 22. Registry of Comparative Pathology, Armed Forces Institute of Pathology, Washington, D.C.

Kalter, H. (1968). "Teratology of the Central Nervous System." Univ. of Chicago Press, Chicago, Illinois.

Kammermann, B., Gmür, J., and Stünzi, H. (1971). Afibrinogenämie beim Hund. *Zentralbl. Veterinaermed., Reihe A* **18,** 192–205.

Karbe, E. (1974). GM_2 gangliosidosis. *In* "Handbook: Animal Models of Human Disease" (T. C. Jones, D. B. Hackel, and G. Migaki, eds.), Model No. 39. Registry of Comparative Pathology, Armed Forces Institute of Pathology, Washington, D.C.

Karlinsky, J. B., and Snider, G. L. (1978). Animal models of emphysema. *Am. Rev. Respir. Dis.* **117,** 1109–1133.

Kier, A. B., Bresnahan, J. F., White, F. J., and Wagner, J. E. (1980). The inheritance pattern of factor XII (Hageman) deficiency in domestic cats. *Can. J. Comp. Med.* **44,** 309–314.

King, N. W., Jr., Baggs, R. B., Hunt, R. D., Van Zwieten, M. J., and MacKey, J. J. (1976). Glomerulonephritis in the owl monkey (*Aotus trivirgatus*): Ultrastructural observations. *Lab. Anim. Sci.* **26,** 1093–1103.

Kitchen, H. (1968). Comparative biology: Animal models of human hematologic disease: A review. *Pediatr. Res.* **2,** 215–229.

Kitchen, H., and Boyer, S. (1974). Hemoglobin: Comparative molecular biology models for the study of disease. *Ann. N.Y. Acad. Sci.* **241,** 1–737.

Kobayashi, T., Yamanaka, T., Jacobs, J. M., Teixeira, F., and Suzuki, K. (1980). The Twitcher mouse: An enzymatically authentic model of human globoid cell leukodystrophy (Krabbe disease). *Brain Res.* **202,** 479–483.

Kociba, G. J. (1976). Spontaneous disseminated intravascular coagulation in animals. *DHEW Publ. (NIH) (U.S.)* **NIH 76-982,** 44–48.

Kociba, G. J., Ratnoff, O. D., Loeb, W. F., Wall, R. L., and Heider, L. E. (1969). Bovine plasma thromboplastin antecedent (factor XI) deficiency. *J. Lab. Clin. Med.* **74,** 37–41.

Koletsky, S. (1977). Hypertension, type IV hyperlipidemia. *In* "Handbook: Animal Models of Human Disease" (T. C. Jones, D. B. Hackel, and G. Migaki, eds.), Model No. 91. Registry of Comparative Pathology, Armed Forces Institute of Pathology, Washington, D.C.

Kolker, A. E., Moses, R. A., Constant, M. A., and Becker, B. (1963). The development of glaucoma in rabbits. *Invest. Ophthalmol.* **2,** 316–321.

Kramer, J. W., Hegreberg, G. A., Bryan, G. M., Meyers, K., and Ott, R. L.

(1976). A muscle disorder of Labrador retrievers characterized by deficiency of type II muscle fibers. *J. Am. Vet. Med. Assoc.* **169**, 817–820.

Kramer, J. W., Nottingham, S., Robinette, J., Lenz, G., Sylvester, S., and Dessouky, M. I. (1980). Inherited, early onset, insulin-requiring diabetes mellitus of keeshond dogs. *Diabetes* **29**, 558–565.

Kurtz, J. M., Russell, S. W., Lee, J. C., Slauson, D. O., and Schechter, R. D. (1972). Naturally occurring canine glomerulonephritis. *Am. J. Pathol.* **67**, 471–482.

Lage, A. L., Mordes, J. P., and Rossini, A. A. (1980). Animal models of diabetes mellitus. *Comp. Pathol. Bull.* **12**, 1, 4–6.

Lambert, D. H., and Dixon, F. J. (1968). Pathogenesis of the nephritis of NZB/W mice. *J. Exp. Med.* **127**, 507–522.

Lang, C. M., Munger, B. L., and Rapp, F. (1977). The guinea pig as an animal model of diabetes mellitus. *Lab. Anim. Sci.* **27**, 789–805.

Laws, L., and Saal, J. R. (1968). Lipidosis of the hepatic reticuloendothelial cells in a sheep. *Aust. Vet. J.* **44**, 416–417.

Leader, R. W. (1964). Lower animals, spontaneous disease, and man. *Arch. Pathol.* **78**, 390–404.

Leader, R. W. (1967). The kinship of animal and human diseases. *Sci. Am.* **216**, 110–116.

Leader, R. W. (1969). Discovery and exploitation of animal model diseases. *Fed. Proc., Fed. Am. Soc. Exp. Biol.* **28**, 1804–1809.

Leader, R. W., and Leader, I., eds. (1971). "Dictionary of Comparative Pathology and Experimental Biology." Saunders, Philadelphia, Pennsylvania.

Leader, R. W., and Padgett, G. A. (1980). The genesis and validation of animal models. *Am. J. Pathol.* **101**, S11–S16.

Lee, K. T., Jarmolych, J., Kim, D. N., Grant, C., Krasney, J. A., Thomas, W. A., and Bruno, A. M. (1971). Production of advanced coronary atherosclerosis, myocardial infarction and "sudden death" in swine. *Exp. Mol. Pathol.* **15**, 170–190.

Lerner, R. A., and Dixon, F. J. (1966). Sponteous glomerulonephritis in sheep. *Lab. Invest.* **15**, 1279–1289.

Lerner, R. A., Dixon, F. J., and Lee, S. (1968). Spontaneous glomerulonephritis in sheep. II. Studies on natural history, occurrence in other species, and pathogenesis. *Am. J. Pathol.* **53**, 501–510.

Letchworth, G. J., Bentinck-Smith, J., Bolton, G. R., Wootton, J. F., and Family, L. (1977). Cyanosis and methemoglobinemia in two dogs due to a NADH-methemoglobin reductase deficiency. *J. Am. Anim. Hosp. Assoc.* **13**, 75–79.

Lewis, R. M. (1973). Systemic lupus erythematosus. *In* "Handbook: Animal Models of Human Disease" (T. C. Jones, D. B. Hackel, and G. Migaki, eds.), Model No. 27. Registry of Comparative Pathology, Armed Forces Institute of Pathology, Washington, D.C.

Lewis, R. M. (1974). Spontaneous autoimmune diseases of domestic animals. *Int. Rev. Exp. Pathol.* **13**, 55–82.

Lewis, R. M., and Borel, Y. (1971). Canine rheumatoid arthritis: A case report. *Arthritis Rheum.* **14**, 67–74.

Like, A. A. (1977). Spontaneous diabetes in animals. *In* "The Diabetic Pancreas" (B. W. Volk and K. F. Wellman, eds.), pp. 381–423. Plenum, New York.

Like, A. A., and Jones, E. E. (1967). Studies on experimental diabetes in the Wellesley hybrid mouse. IV. Morphologic changes in islet tissue. *Diabetologia* **3**, 179–187.

Like, A. A., and Rossini, A. A. (1976). Streptozotocin-induced pancreatic insulitis: New model of diabetes mellitus. *Science* **193**, 415–417.

Like, A. A., Lavine, R. L., Poffenbarger, P. L., and Chick, W. L. (1972). Studies in the diabetic mutant mouse. VI. Evolution of glomerular lesions and associated proteinuria. *Am. J. Pathol.* **66**, 193–224.

Liu, S.-K. (1977). Pathology of feline heart disease. *Vet. Clin. North Am.* **7**, 323–339.

Liu, S.-K., and Tilley, L. P. (1980). Animal models of primary myocardial diseases. *Yale J. Biol. Med.* **53**, 191–211.

Liu, S.-K., Maron, B. J., and Tilley, L. P. (1979). Hypertrophic cardiomyopathy in the dog. *Am. J. Pathol.* **94**, 497–508.

Liu, S.-K., Maron, B. J., and Tilley, L. P. (1981). Feline hypertrophic cardiomyopathy. *Am. J. Pathol.* **102**, 388–395.

Luster, M. I., Bacon, L. D., Rose, N. R., and Leslie, G. A. (1977). Immunogenetic and ontogenetic studies of chickens with selective IgA deficiency and autoimmune thyroiditis. *Cell. Immunol.* **32**, 417–423.

Lyon, J. B., Porter, J., and Robertson, M. (1967). Phosphorylase b kinase inheritance in mice. *Science* **155**, 1550–1551.

McGavin, M. D. (1975). Muscular dystrophy. *In* "Handbook: Animal Models of Human Disease" (T. C. Jones, D. B. Hackel, and G. Migaki, eds.), Model No. 51. Registry of Comparative Pathology, Armed Forces Institute of Pathology, Washington, D.C.

McGuire, T. C. (1975). Combined immunodeficiency (severe), Swiss-type agammaglobulinemia. *Am. J. Pathol.* **80**, 551–554.

McGuire, T. C., Banks, K. L., and Poppie, M. J. (1976). Combined immunodeficiency (severe), Swiss-type agammaglobulinemia. *In* "Handbook: Animal Models of Human Disease" (T. C. Jones, D. B. Hackel, and G. Migaki, eds.), Model No. 83. Registry of Comparative Pathology, Armed Forces Institute of Pathology, Washington, D.C.

McPherson, C. (1978). Role of the National Institutes of Health in seeking animal models. *J. Am. Vet. Med. Assoc.* **173**, 1210–1211.

Malmros, H., and Sternby, N. H. (1968). Induction of atherosclerosis in dogs by a thiouracil-free semisynthetic diet containing cholesterol and hydrogenated coconut oil. *Prog. Biochem. Pharmacol.* **4**, 482–487.

Manktelow, B. W., and Hartley, W. J. (1975). Generalized glycogen storage disease in sheep. *J. Comp. Pathol.* **85**, 139–145.

Mansmann, R. A., Osburn, B. I., Wheat, J. D., and Frick, O. (1975). Chicken hypersensitivity pneumonitis in horses. *J. Am. Vet. Med. Assoc.* **166**, 673–677.

Michelson, A. M., Russell, E. S., and Harman, P. J. (1955). Dystrophia muscularis: A hereditary primary myopathy in the house mouse. *Proc. Natl. Acad. Sci. U.S.A.* **41**, 1079–1084.

Minick, C. R., and Murphy, G. E. (1973). Experimental induction of atheroarteriosclerosis by the synergy of allergic injury to arteries and lipid-rich diet. *Am. J. Pathol.* **73**, 265–300.

Minick, C. R., Fabricant, C. G., Fabricant, J., and Litrenta, M. M. (1979). Atheroarteriosclerosis induced by infection with a herpesvirus. *Am. J. Pathol.* **96**, 673–706.

Mordes, J. P., and Rossini, A. A. (1981). Animal models of diabetes. *Am. J. Med.* **70**, 353–360.

Moss, N. S., and Benditt, E. P. (1970). The ultrastructure of spontaneous and experimentally induced arterial lesions. III. The cholesterol-induced lesions and the effect of a cholesterol and oil diet on the pre-existing spontaneous plaque in the chicken aorta. *Lab. Invest.* **23**, 521–535.

Mostafa, I. E. (1970). A case of glycogenic cardiomegaly in a dog. *Acta Vet. Scand.* **11**, 197–208.

Muller-Peddinghaus, R., and Trautwein, G. (1977). Spontaneous glomerulonephritis in dogs. I. Classification and immunopathology. *Vet. Pathol.* **14**, 1–13.

Mulvihill, J. J. (1972). Congenital and genetic disease in domestic animals. *Science* **176**, 132–137.

Murakami, H., Takagi, A., Nanaka, S., Ishiura, S., and Sugita, H. (1980). Glycogenesis II in Japanese quails. *Jikken Dobutsu* **29**, 475–485.

Mustard, J. F., and Packham, M. A. (1968). The unrealized potential of animal diseases in the study of human diseases. *Can. Med. Assoc. J.* **98**, 887–890.

Mustard, J. F., Rowsell, H. C., Robinson, G. A., Hoeksema, T. C., and Downie, H. G. (1960). Canine hemophilia B (Christmas disease). *Br. J. Haematol.* **6**, 259–266.

Mustard, J. F., Secord, D., Hoeksema, T. D., Downie, H. G., and Rowsell, H. C. (1962). Canine factor-VII deficiency. *Br. J. Haematol.* **8**, 43–47.

Nakhooda, A. F., Like, A. A., Chappel, C. I., Murray, F. T., and Marliss, E.

B. (1977). The spontaneously diabetic Wistar rat. *Diabetes* **26,** 100–112.

Nansen, P. (1972). Selective immunoglobulin deficiency in cattle and susceptibility to infection. *Acta Pathol. Microbiol. Scand., Sect. B* **80B,** 49–54.

Noonan, N. E., Olsen, G. A., and Cornelius, C. E. (1979). A new animal model with hyperbilirubinemia: The indigo snake. *Dig. Dis. Sci.* **24,** 521–524.

Noren, G. R., Staley, N. A., Kaplan, E. L., and Jankus, E. F. (1975). Cardiomyopathy. *In* "Handbook: Animal Models of Human Disease" (T. C. Jones, D. B. Hackel, and G. Migaki, eds.), Model No. 49. Registry of Comparative Pathology, Armed Forces Institute of Pathology, Washington, D.C.

Novak, E. K., Hui, S. W., and Swank, R. T. (1981). The mouse pale ear pigment mutant as a possible animal model for human platelet storage pool deficiency. *Blood* **57,** 38–43.

O'Donoghue, J. L., and Reed, C. (1981). The hairless immune-deficient guinea pig. *In* "Immunologic Defects in Laboratory Animals" (M. E. Gershwin and B. Merchant, eds.), Vol. 1, pp. 285–296. Plenum, New York.

Okamoto, K. (1969). Spontaneous hypertension in rats. *Int. Rev. Exp. Pathol.* **7,** 227–270.

Okamoto, K., Tabei, R., Yamori, Y., and Ooshima, A. (1973). Spontaneously hypertensive rat as a useful model for hypertension research. *Exp. Anim., Suppl.* **22,** 289–298.

Oldstone, M. B. A., Tishon, A., Tonietti, G., and Dixon, F. J. (1972). Immune complex disease associated with spontaneous murine leukemia: Incidence and pathogenesis of glomerulonephritis. *Clin. Immunol. Immunopathol.* **1,** 6–14.

Olenchock, S. A. (1977). Animal models of hypersensitivity pneumonitis: A review. *Ann. Allergy* **38,** 119–126.

Paleček, F., and Holusa, R. (1971). Spontaneous occurrence of lung emphysema in laboratory rats. A quantitative functional and morphological study. *Physiol. Bohemoslov.* **20,** 335–344.

Palmer, A. C., Lennon, V. A., Beadle, C., and Goodyear, J. V. (1980). Autoimmune form of myasthenia gravis in a juvenile Yorkshire terrier × Jack Russell terrier hybrid contracted with congenital (non-autoimmune) myasthenia gravis of the Jack Russell. *J. Small Anim. Pract.* **21,** 359–364.

Parish, W. E. (1961). The response of normal and sensitized experimental animals to products of mouldy hay. *Acta Allergol.* **16,** 78–79.

Patel, V., Koppang, N., Patel, B., and Zeman, W. (1974). *p*-Phenylenediamine-mediated peroxidase deficiency in English setters with neuronal ceroid-lipofuscinosis. *Lab. Invest.* **30,** 366–368.

Patterson, R., and Kelly, J. F. (1974). Animal models of the asthmatic state. *Annu. Rev. Med.* **25,** 53–68.

Perryman, L. E., and McGuire, T. C. (1980). Evaluation for immune system failures in horses and ponies. *J. Am. Vet. Med. Assoc.* **176,** 1374–1377.

Perryman, L. E., and Magnuson, N. S. (1982). Immunodeficiency disease in animals. *In* "Animal Models of Inherited Metabolic Disease" (R. J. Desnick, D. F. Patterson, and D. G. Scarpelli, eds.), pp. 271–305. Alan R. Liss, Inc., New York.

Pierce, K. R., Kosanke, S. D., Bay, W. W., and Bridges, C. H. (1977). GM₂ gangliosidosis. *In* "Handbook: Animal Models of Human Disease" (T. C. Jones, D. B. Hackel, and G. Migaki, eds.), Model No. 104. Registry of Comparative Pathology, Armed Forces Institute of Pathology, Washington, D.C.

Pinkerton, P. H., Fletch, S. M., Brueckner, P. J., and Miller, D. R. (1974). Hereditary stomatocytosis with hemolytic anemia in the dog. *Blood* **44,** 557–567.

Pirie, H. M., Dawson, C. O., Breeze, R. G., Selman, I. E., and Wiseman, A. (1972). Precipitins to *Micropolyspora faeni* in the adult cattle of selected herds in Scotland and northwest England. *Clin. Allergy* **2,** 181–187.

Porter, D. D., and Porter, H. G. (1971). Deposition of immune complexes in the kidneys of mice infected with lactic dehydrogenase virus. *J. Immunol.* **106,** 1264–1266.

Poskitt, T. R., Fortwengler, H. P., Jr., Bobrow, J. C., and Roth, G. J. (1974). Naturally occurring immune-complex glomerulonephritis in monkeys (*Macaca irus*). *Am. J. Pathol.* **76,** 145–164.

Ratcliffe, H. L. (1962). Comparative pathology: An opportunity for biologists? *Science* **137,** 550–552.

Raymond, S. L., and Dodds, W. J. (1979). Platelet membrane glycoproteins in normal dogs and dogs with hemostatic defects. *J. Lab. Clin. Med.* **93,** 607–613.

Read, D. H., Harrington, D. D., Keenan, T. W., and Hinsman, E. J. (1976). Neuronal-visceral GM₁ gangliosidosis in a dog β-galactosidase deficiency. *Science* **194,** 442–445.

Richards, R. B., Edwards, J. R., Cook, R. D., and White, R. R. (1977). Bovine generalized glycogenosis. *Neuropathol. Appl. Neurobiol.* **3,** 45–56.

Richerson, H. B., Cheng, F. H. F., and Bauserman, S. C. (1971). Acute experimental hypersensitivity pneumonitis in rabbits. *Am. Rev. Respir. Dis.* **104,** 568–575.

Robinson, R. R., and Dennis, V. W. (1980). Spontaneously occurring animal models of human kidney diseases and altered renal function. *Adv. Nephrol.* **9,** 315–366.

Roderick, L. M. (1929). The pathology of sweet clover disease in cattle. *J. Am. Vet. Med. Assoc.* **74,** 314–326.

Ross, R., and Glomset, J. A. (1976). The pathogenesis of atherosclerosis. *N. Engl. J. Med.* **295,** 369–377, 420–425.

Roth, S. I., Conaway, H. H., Sanders, L. L., Casali, R. E., and Boyd, A. E. (1980). Spontaneous diabetes mellitus in the New Zealand white rabbit. *Lab. Invest.* **42,** 571–579.

Rother, K., Rother, V., Müller-Eberhard, H. J., and Nilsson, U. R. (1966). Deficiency of the sixth component of complement in rabbits with an inherited complement defect. *J. Exp. Med.* **124,** 773–785.

Rous, P. (1911). Transmission of a malignant new growth by means of a cell-free filtrate. *JAMA, J. Am. Med. Assoc.* **56,** 198.

Ruth, G. R., Schwartz, S., and Stephenson, B. (1977). Bovine protoporphyria: The first nonhuman model of this hereditary photosensitizing disease. *Science* **198,** 199–201.

Sandström, B., Westman, J., and Öckerman, P. A. (1969). Glycogenosis of the central nervous system in the cat. *Acta Neuropathol.* **14,** 194–200.

Scarpelli, D. G. (1969). Survey of some spontaneous and experimental disease processes of lower vertebrates and invertebrates. *Fed. Proc., Fed. Am. Soc. Exp. Biol.* **28,** 1825–1833.

Schmid, R., Axelrod, J., Hammaker, L., and Swarm, R. L. (1958). Congenital jaundice in rats, due to a defect in glucuronide formation. *J. Clin. Invest.* **37,** 1123–1130.

Schmidt-Nielsen, B. (1961). Choice of experimental animals for research. *Fed. Proc., Fed. Am. Soc. Exp. Biol.* **20,** 902–906.

Schmidt-Nielsen, K., Haines, H. B., and Hackel, D. B. (1964). Diabetes mellitus in the sand rat induced by standard laboratory diets. *Science* **143,** 689–690.

Schofield, F. W. (1924). Damaged sweet clover: The cause of a new disease in cattle simulating hemorrhagic septicemia and blackleg. *J. Am. Vet. Med. Assoc.* **64,** 553–574.

Scott, R. F., Daoud, A. S., and Florentin, R. A. (1971). Animal models in atherosclerosis. *In* "The Pathogenesis of Atherosclerosis" (R. W. Wissler and J. C. Geer, eds.), pp. 120–146. Williams & Wilkins, Baltimore, Maryland.

Searcy, G. P., Tasker, J. B., and Miller, D. R. (1979). Pyruvate kinase deficiency. *In* "Handbook: Animal Models of Human Disease" (T. C. Jones, D. B. Hackel, and G. Migaki, Eds.), Model No. 176. Registry of

Comparative Pathology, Armed Forces Institute of Pathology. Washington, D.C.

Shirai, T., Welsh, G. W., and Sims, E. A. H. (1967). Diabetes mellitus in the Chinese hamster. II. The evolution of renal glomerulopathy. *Diabetologia* **3**, 266–286.

Slauson, D. O., and Hahn, F. F. (1980). Criteria for development of animal models of diseases of the respiratory system. *Am. J. Pathol.* **101**, S103–S122.

Slauson, D. O., and Lewis, R. M. (1979). Comparative pathology of glomerulonephritis in animals. *Vet. Pathol.* **16**, 135–164.

Slauson, D. O., Russell, S. W., and Schechter, R. D. (1971). Naturally occurring immune-complex glomerulonephritis in the cat. *J. Pathol.* **103**, 131–133.

Smith, J. E. (1981). Animal models of human erythrocyte metabolic abnormalities. *Clin. Haematol.* **10**, 239–251.

Smith, T., and Kilborne, F. L. (1893). "Investigations into the Nature, Causation, and Prevention of Texas or Southern Cattle Fever," Bull. No. 1. Department of Agriculture, Bureau of Animal Industry, U.S. Govt. Printing Office, Washington, D.C.

Spurling, N. W., Burton, L. K., Peacock, R., and Pilling, T. (1972). Hereditary factor-VII deficiency in the Beagle. *Br. J. Haematol.* **23**, 59–67.

Stanbury, J. B., Wyngaarden, J. B., and Fredrickson, D. S., eds. (1978). "The Metabolic Basis of Inherited Disease," 4th ed. McGraw-Hill, New York.

Steblay, R. W. (1980). Anti-glomerular-basement-membrane glomerulonephritis. *In* "Handbook: Animal Models of Human Disease" (C. C. Capen, D. B. Hackel, T. C. Jones, and G. Migaki, eds.), Model No. 191. Registry of Comparative Pathology, Armed Forces Institute of Pathology, Washington, D.C.

Steinhaus, E. A. (1969). Invertebrates and their diseases as models for the study of disease processes in man. *Fed. Proc., Fed. Am. Soc. Exp. Biol.* **28**, 1810–1814.

Stephens, M. C., Bernatsky, A., Legler, G., and Kanfer, J. N. (1979). The Gaucher mouse: Additional biochemical alterations. *J. Neurochem.* **32**, 969–972.

Stowell, R. E. (1963). Training in comparative pathology. *Lab. Invest.* **12**, 830–845.

Strawbridge, H. T. G. (1960). Chronic pulmonary emphysema. (An experimental study). II. Spontaneous pulmonary emphysema in rabbits. *Am. J. Pathol.* **37**, 309–329.

Strong, J., ed. (1976). "Atherosclerosis in Primates." Karger, Basel.

Stuhlman, R. A. (1979). Diabetes mellitus. *In* "Handbook: Animal Models of Human Disease" (T. C. Jones, D. B. Hackel, and G. Migaki, eds.), Model No. 177. Registry of Comparative Pathology, Armed Forces Institute of Pathology, Washington, D.C.

Tasker, J. B., Severin, G. A., Young, S., and Gillette, E. L. (1969). Familial anemia in the basenji dog. *J. Am. Vet. Med. Assoc.* **154**, 158–165.

Thant, M. (1970). Arteriosclerosis in spontaneously hypertensive rats on high fat diet. *Jpn. Circ.* **34**, 83–107.

Tilley, L. P., Liu, S.-K., Gilbertson, S. R., Wagner, B. M., and Lord, P. F. (1977). Primary myocardial disease in the cat: A model for human cardiomyopathy. *Am. J. Pathol.* **86**, 493–522.

Tizard, I. R. (1982). "Veterinary Immunology." Saunders, Philadelphia, Pennsylvania.

Tschopp, T. B., and Zucker, M. B. (1972). Hereditary defect in platelet function in rats. *Blood* **40**, 217–226.

Valtin, H. (1977). Hypothalamic diabetes insipidus. *In* "Handbook: Animal Models of Human Disease" (T. C. Jones, D. B. Hackel, and G. Migaki, eds.), Model No. 107. Registry of Comparative Pathology, Armed Forces Institute of Pathology, Washington, D.C.

Van Citters, R. (1973). The role of animal research in clinical medicine. *DHEW Publ.* (NIH) (*U.S.*) **NIH 72-333**, 3–8.

Van De Water, N. S., Jolly, R. D., and Farrow, B. R. H. (1979). Canine Gaucher disease—the enzyme defect. *Aust. J. Exp. Biol. Med. Sci.* **57**, 551–554.

Vesselinovitch, D. (1979). Animal models of atherosclerosis, their contributions and pitfalls. *Artery* **5**, 193–206.

Vogel, F. S. (1969). Meadow mushroom provides a model for the study of mitochondrial DNA. *Fed. Proc., Fed. Am. Soc. Exp. Biol.* **28**, 1820–1824.

Wagner, W. D., Clarkson, T. B., Feldner, M. A., and Prichard, R. W. (1973). The development of pigeon strains with selected atherosclerosis characteristics. *Exp. Mol. Pathol.* **19**, 304–319.

Walkley, S. U., Hunt, C. E., Clements, R. S., and Lindsey, J. R. (1978). Description of obesity in the PBB/Ld mouse. *J. Lipid Res.* **19**, 335–341.

Wehner, H., Höhn, D., Faix-Schade, U., Huber, H., and Walzer, P. (1972). Glomerular changes in mice with spontaneous hereditary diabetes. *Lab. Invest.* **27**, 331–340.

Wenger, D. A., Sattler, M., Kudoh, T., Snyder, S. P., and Kingston, R. S. (1980). Niemann-Pick disease: A genetic model in Siamese cats. *Science* **208**, 1471–1473.

Wick, G. (1976). Thyroiditis. *In* "Handbook: Animal Models of Human Disease" (T. C. Jones, D. B. Hackel, and G. Migaki, eds.), Model No. 85. Registry of Comparative Pathology, Armed Forces Institute of Pathology, Washington, D.C.

Wight, P. A. L., Siller, W. G., and Martindale, L. (1981). March gangrene. *In* "Handbook: Animal Models of Human Disease" (C. C. Capen, D. B. Hackel, T. C. Jones, and G. Migaki, eds.), Model No. 224. Registry of Comparative Pathology, Armed Forces Institute of Pathology, Washington, D.C.

Wilkie, B., Pauli, B., and Gygax, M. (1973). Hypersensitivity pneumonitis: Experimental production in guinea pigs with antigens of *Micropolyspora faeni*. *Pathol. Microbiol.* **39**, 393–411.

Wissler, R. W., and Vesselinovitch, D. (1977). Atherosclerosis in nonhuman primates. *Adv. Vet. Sci. Comp. Med.* **21**, 351–420.

Wissler, R. W., and Vesselinovitch, D. (1978). Evaluation of animal models for the study of the pathogenesis of atherosclerosis. *Abh. Rheinisch-Westfael. Akad. Wiss.* **63**, 13–29.

Zook, B. C., Paasch, L. H., Chandra, R. S., and Casey, H. W. (1981). The comparative pathology of primary endocardial fibroelastosis in Burmese cats. *Virchows Arch. A: Pathol. Anat. Histol.* **390**, 211–227.

ADDITIONAL BOOKS, PROCEEDINGS, AND BULLETINS

Alexander, N. J., ed. (1979). "Animal Models for Research on Contraception and Fertility." Harper & Row, Hagerstown, Maryland.

Animal Models for Biomedical Research (1968, 1969, 1970, 1971, 1973, 1976). Vols. I–VI resp. Nat. Acad. Sci., Washington, D.C.

Animal Models of Human Disease. *Am. J. Pathol.* (descriptions published under this heading in each issue of journal since January 1972).

Animal Models of Human Disease. *Yale J. Biol. Med.* (descriptions published under this heading in this journal since 1978).

Cohen, B. J., ed. (1981). "Mammalian Models for Research on Aging." Nat. Acad. Press, Washington, D.C.

Comparative Pathology Bulletin. Registry of Comparative Pathology, Armed Forces Institute of Pathology, Washington, D.C. (quarterly publication since November 1969).

Desnick, R. J., Patterson, D. F., and Scarpelli, D. G., eds. (1982). "Animal Models of Inherited Metabolic Disease." Alan R. Liss, Inc., New York.

Dodds, W. J., ed. (1976). "Animal Models of Thrombosis and Hemorrhagic Diseases." DHEW Publ. No. 76-982. Natl. Inst. Health, Bethesda, Maryland.

Gershwin, M. E., and Cooper, E. L., eds. (1978). "Animal Models of Comparative and Developmental Aspects of Immunity and Disease." Pergamon, Oxford.

Hackel, D. B., ed. (1980). Needs for new animal models of human disease. *Am. J. Pathol., Suppl.*

Harmison, L. T., ed. (1973). "Research Animals in Medicine," DHEW Publ. No. (NIH) 72-333. Natl. Inst. Health, Bethesda, Maryland.

Hommes, F. A., ed. (1979). "Models for the Study of Inborn Errors of Metabolism." Elsevier/North-Holland Biomedical Press, Amsterdam.

ILAR News. National Academy of Sciences, Washington, D.C. (published since 1957).

Jones, T. C., Capen, C. C., Hackel, D. B., and Migaki, G., eds. (1972). "Handbook: Animal Models of Human Disease." Registry of Comparative Pathology, Armed Forces Institute of Pathology, Washington (new fascicles added each year since 1972).

Mitruka, B. M., Rawnsley, H. M., and Vadehra, D. V. (1976). "Animals for Medical Research." Wiley, New York.

Navia, J. M. (1977). "Animal Models in Dental Research." Univ. of Alabama Press, University.

Rose, F. C., and Behan, P. O., eds. (1979). "Animal Models of Neurological Disease." Pitman Medical Ltd., Kent.

Shifrine, M., and Wilson, F. D., eds. (1980). "The Canine as a Biomedical Research Model: Immunological, Hematological, and Oncological Aspects." National Technical Information Service, U.S. Department of Commerce, Springfield, Virginia.

Chapter 25

Developing Research in Laboratory Animal/ Comparative Medicine

Thomas B. Clarkson, David S. Weaver, and Frances M. Lusso

I. INTRODUCTION

Laboratory animal medicine as a specialty of veterinary medicine began as a result of the recognized need by a few medical schools for veterinary medical skills in the care and management of laboratory animals. As these service programs became operational, it became apparent that the individuals involved with the programs had additional skills in comparative biology and medicine that prepared them to make contributions to biomedical research in addition to their contributions to animal care.

It is of interest to observe how the terms laboratory animal medicine and comparative medicine came to be used as the balance between service and academic activities shifted. Today, laboratory animal medicine is usually used to describe the veterinary medical aspects of the care of laboratory animals. Comparative medicine, on the other hand, is now used to describe a group of academic activities, including the characterization of animal models for biomedical research and the use of such models to advance understanding about disease or physiologic processes.

As the fields of laboratory animal medicine and comparative

medicine develop, there is a need for more vigorous efforts directed toward advancing knowledge in those areas. The research progress in both laboratory animal and comparative medicine has been less than one would expect based on the number of fulltime professionals in those fields. The purpose of this chapter is to describe the development and execution of research programs in laboratory animal/comparative medicine in the hope that the description of research programs might increase research productivity. The needs for increased research productivity are equally great in laboratory animal and comparative medicine. Within laboratory animal medicine the needs relate to improved methods of diagnosis, identification of new naturally occurring diseases, further understanding of partially characterized diseases, and more information on the effect of chronic diseases on experimental outcome. The research needs in comparative medicine are the characterization of animal models for certain human diseases, especially for those diseases for which no models exist, improvement in the animal modeling of some of the major disease problems of man, and continuing studies on the relevance of observations made on animal models to the understanding of human disease.

II. THE RESEARCH PROCESS

A. Considerations in Research Design

The design of an experiment has profound consequences on the eventual outcome of the experiment. The most elegant research propositions will remain untested unless consideration is given to the variables to be examined, the hypotheses, the sample sizes, the control of extrinsic and intrinsic sources of experimental error, and the analytic design.

Experimental variables are defined as the items to be measured in a study. Variables are classified with regard to their independence and their scale of measurement. In general, variables used in experiments are classed as either independent or dependent (Sokal and Rohlf, 1981). An independent variable is one whose value is determined independent of the experimental result. For example, the treatment group into which an animal is placed is an independent variable that, in some way, may affect the experimental result (usually the dependent variables). In most experimental designs that concentrate on cause and effect, the causes are considered independent and the effects are considered dependent. In practice, the distinction between independent and dependent may be obscured, occasionally to the extent that what appears to be an independent variable in one segment of a study may be treated as a dependent variable at another time in consideration of another segment of the research problem.

Variables exist in a scale of measurement (Siegel, 1956).

The scale of the variables places restrictions on the types of statistical analysis that can be performed on the experimental results, and scale should be considered in the design of the experiment. It is possible to test an experimental proposition with what would have been the appropriate experiment, but to sacrifice most of the ultimate value of the experiment through the use of an inappropriate scale of measurement, thus making the most appropriate statistical approach difficult or impossible. Variables are classified into three levels of measurement, the nominal, ordinal, and interval scales. In the nominal scale, data are simply classified into named, mutually exclusive categories. No order or other intrinsic relationship between the categories is implied. The ordinal scale classifies data into named categories where an ordered relationship is assumed, at least on some common scale of measurement. Interval scales possess the features of the first two types of scales, with the added property of preserving the absolute differences between measurement intervals. Only on interval scales are the mathematical operations with which most investigators are familiar, interpolation and estimation of intermediate points, for example, possible. Thus, the designation of sex (male or female) is a nominal classification; the classification of experimental animals according to social rank is an ordinal one, while the evaluation of a clinical serum concentration takes place on an interval scale. The determination of measurement scale is important not only because specific types of statistical analysis require the data to be classified on a particular scale, but also because the scale used has an influence (which may be quite powerful but, at the same time, quite elusive) on the type of hypotheses that are perceived as relevant to the experiment. If an experimenter is accustomed to thinking only in terms of interval scales, a more general pattern in the data, which might better fit an ordinal or nominal classification, may well be lost. Contrary to most perceptions of the measurement scales, the level of measurement has no intrinsic value. It is not necessary to force nominal or ordinal variables into an interval scale simply because of a belief that interval scales somehow constitute "better science." The best choice of scale is the one that best aids in answering the research question, not necessarily the one that yields the most elegant statistical analysis.

The design of a study is either experimental or nonexperimental (Spector, 1981). Experimental designs are those in which the conditions of the experiment are manipulated and specified by the investigator. Nonexperimental designs are those in which the investigator has no *a priori* control over at least some of the conditions of the experiment. Most designed research is experimental, in that the investigators at least attempt to control the conditions of the research. By contrast, many epidemiological studies are nonexperimental, inasmuch as the "natural" experiment has already been performed, under existing conditions, by the time the investigation begins. As with scales of measurement, the common view that an experimental design is always somehow "better" than a nonexperimental one is incorrect. Obviously some investigations are simply impractical as experimental designs. Controlled (experimental) studies of the millions of smokers used in epidemiological studies would be unreasonable. Blind dedication to the use of only experimental designs may prevent an investigator from recognizing an important *post facto* nonexperimental result. The experience of realizing an entirely different sort of meaningful result from that which the experiment was designed to examine is not uncommon. Most of those fortuitous discoveries stem from well-controlled experiments, but in terms of the eventual important result, the research was in fact a nonexperimental design.

Experimental designs depend on hypotheses. Hypotheses are sets of propositions about experimental outcomes that can be tested and evaluated. In experimental designs, hypotheses are usually formed *a priori,* that is, at the outset. In fact, the use of any other form of hypothesis in experimental designs is not advisable, since the statistical tests used to test hypotheses in experimental designs depend on *a priori* statements of hypotheses. In nonexperimental designs, it is advisable to approach the data with *a priori* hypotheses in hand, but inspection of the data may well allow hypotheses to develop from the research results. Statistical testing of these *a posteriori* results is often more subtle and may require a good deal more inference than the tests of *a priori* hypotheses. For methodological and perhaps epistemological reasons, then, most investigators have emphasized the use of *a priori* hypothesis in conjunction with experimental research designs.

Experimental research designs can be grouped into several categories, according to the number of experimental groups and number of variables. Designs may involve a combination of one or multiple groups and any combination of one or more variables. The simplest design, using one group measured for a single variable, is also the most restrictive. The most flexible designs, using multiple variables and multiple groups, may prove extremely complex and difficult to interpret. Most experimental research designs fall between these extremes. In general, the simplest design that will address the research question successfully is to be preferred. In practice, this usually means the comparison of several groups for a few (or a few combinations of) variables is an acceptable approach. More complex research designs have become more common and, with the advent of convenient and accessible computer data management and analytic techniques, have become more manageable.

The question of sample size in research design is widely misunderstood. While it may be correct to guess, in the absence of any information bearing on the necessary sample size, at a "reasonable" sample size, in almost all cases a better estimate of the appropriate sample size can be made. An estimate of the required sample size is essential to prevent the use of a sample size that is too small to allow any substantial probability that the experiment will test the hypotheses that have been posed. If there is any basis for an estimate of the required sample size, however, the conventional sample sizes that may be put forth for a particular research design or statistical test may prove unnecessarily large. An estimate of the required sample size, then, is necessary to allow an experiment that is, at the same time, sufficient to test the propositions at hand and as economical in cost and animal use as possible.

A well-constructed research design contains means to control and evaluate the various confounding effects that inevitably occur in experiments. The validity of the measurement instrument must be verified. The investigator should be certain that the measurement is appropriate for the experiment and that the accuracy of the instrument is sufficient to create data that will be useful in testing the hypotheses. Pilot studies are often used to verify the use of a particular instrument and to establish the accuracy of measurement. Measurement bias and random experimental error should be determined, as it is possible for a statistically significant result to occur due to consistent biases or other types of systematic error.

The interaction of variables, which might at first seem to be a severe confounding factor, can, if properly analyzed, often yield important results. Synergisms and other interactive effects may reflect the actual response of the subjects to an experiment, and may be more useful than single "uncontaminated" variables. Statistical techniques abound to examine interactive effects, and those techniques should be considered in the research design.

Most experiments employ one or more control groups and one or more experimental (test) groups. The intention for control groups is, of course, to leave them unaffected by the experimental treatment. This condition should always be tested, rather than assumed, as it is possible for a control group to have been affected in some unforeseen way by the experiment or for both the control and the test groups to have been affected, in different ways, by some external condition. This control/test dichotomy is not always necessary for an adequate research design and may occasionally be impossible. It is important to know as much as possible about all the groups used in an experiment, whether or not control groups are used. A common alternative to the control/test design is the random assignment of experimental subjects to experimental groups. Randomization approaches are often sensitive to unknown and unmanageable biases, however, and the approach should be used with caution. It is important to verify that the groups derived from a randomization procedure do not exhibit any initial patterning that might have an effect on the experimental result.

B. Selection of the Animal Model

Assuming that the investigator about to undertake a research project has the necessary physical facilities to support a proposed project, there are several other considerations that are essential preconditions to the project. It is essential to determine that the proposed animal model is appropriate for the experiment. The choice of an animal model must be based on a reasonable expectation that the desired results are possible in that animal, not on other matters such as simple convenience or the investigators' familiarity with a particular type of animal. The suitability of a particular animal model for a specific type of research should be well established, rather than assumed.

The experiment may demand particular genetic relationships among the experimental animals. If an inbred strain is needed, for example, it is obviously improper to simply purchase an animal sample at random. On the other hand, the unwitting use of inbred strains may well introduce uncontrolled conditions to the experiment. Close cooperation between the research center and the animal supplier is essential to ensure that the appropriate animals are used in an experiment (see Chapter 21).

Careful consideration must be given to the formulation of the experimental diets of the research animals. Many experiments depend on the maintenance of the animals in a state of good general health, while attempting to modify some specific condition in the animals. Because of the interactive effects of some dietary modifications and the possibility that even a diet that seems sufficient may prove harmful to the animals or may otherwise perturb the experiment, the efficacy of the proposed diet should be established, either from previous experience or through pilot studies (see Chapter 23).

Environmental conditions are crucial to the successful conduct of an experiment. Crowding may introduce stress as a significant influence on an experiment. Social animals may respond poorly to individual caging. Animals that require large social distances between social groups in their natural state may respond badly to the typical laboratory condition where groups are housed adjacent to each other. Available natural light and the photoperiod may each have powerful effects on experimental results. Noise may have its effects, particularly if the animals are stressed at important times during the experiment. Investigators should thoroughly understand the conditions under which the experiment will be conducted and should be satisfied that those conditions will not have uncontrolled effects on the experiment. It will also be necessary to remain alert, during the conduct of the experiment, to the possibility that experimental conditions may have unforeseen effects.

C. Statistical Considerations

The best planned, most conscientiously conducted experiment can be severely damaged by an unsatisfactory statistical treatment. Even when due attention has been given questions of measurement scale, experimental design, sample size, and all the other necessary conditions of an experiment, the use of inappropriate statistical techniques or the formation of improper conclusions from the statistical analysis will very much reduce the value of the study. It is, therefore, advisable to retain the services of a qualified applied statistician throughout the planning and conduct of the experiment as well as during the analysis of the results.

Many statistical techniques depend on assumptions about the character of the data. Analysis of variance, for example, depends on the following: (1) the sampling of the individual cases for inclusion in the analysis has been random with respect to the variables to be examined, (2) errors that may occur in the samples are independent of each other, (3) the true variances of the populations from which the experimental samples have been drawn are equivalent, (4) the distribution of the error terms in the experiment is normal, and (5) the interactive effects observed in the experiment are additive (Sokal and Rohlf, 1981). In actual experiments it is unlikely that all of the above assumptions for the analysis of variance will be satisfied, but one goal of the research design should be to attempt to conform to as many of the necessary conditions of an analysis as possible. Some statistical conditions are essential, as are random sampling, independence of error, and additivity of interaction in analysis of variance, while other conditions are quite lenient. A normal distribution of the data is a strong requirement in a number of statistical treatments and should be tested before comparisons of groups are attempted. If normality is not verified, it may be necessary to chose other statistical techniques that are less sensitive to the assumption of normality (Siegel, 1956).

All experiments depend on the accuracy and reliability of the measurements and observations taken. The accuracy and reliability of measurements and observations must be verified early in the study. Periodic checks of the validity of the measures should be conducted during the course of the experiment, as it is not unusual for biases to develop during the experiment that may appear to represent important results, but which are the result of a failure to examine and maintain standards for the data collection. The unconscious rejection of the data that do not "look right," the usually innocent omission of data that "cannot be true," as well as the isolated incidence of intentional modification of experimental data can be minimized by periodic checks on the quality of the data collected. It may be advisable to maintain an independent check on the validity and security of data through the use of a disinterested party to periodically review the data.

While it is not always possible to perform blind studies, the use of such studies is probably the most effective way to minimize bias in experimental studies. In blind studies, the identity of the experimental subjects, and often of the treatments themselves, is concealed, usually through the use of a coding system, from the persons performing the analyses. It may sometimes be desirable to design several levels of coding to prevent biases in data collection while avoiding analytic biases.

D. The Research Protocol

To ensure that a coherent research design has been formulated and to serve as a guide for the conduct of the experiment, a written research protocol is needed. A research protocol is a detailed plan of the design, hypotheses, procedures, proposed analyses, and expectation of the research. A typical protocol begins with a statement of the research problems, including discussions of the background that exists for the research and the relevance of the particular research that is being proposed. The specific hypotheses to be addressed by the research should be presented and the means by which the proposed research will examine the hypotheses should be explicitly stated. A statement of the research design that will be used, including justifications of the type and number of variables used, the sample sizes proposed, the examination and control of experimental error, and the assignment of individuals into various experimental conditions, is essential. The protocol should include detailed discussion of the methods of procedure for the experiment, including the methods and scheduling of data collection, the scheduling and management of the experimental subjects, and the organizations of the data in a form that will allow the data to be retrieved and used in analysis. The most likely methods of laboratory and statistical analysis should be presented in the protocol. In many cases, a budget is included in the protocol. A well-argued, well-constructed protocol is invaluable in the planning and conduct of an experiment and will serve as a powerful aid in the subsequent evaluation and analysis of the experimental results.

For any research involved in the testing of a new drug or drug compound, the Food and Drug Administration has developed a set of standards for the inspection of project management, data collection, data management, and the conduct of the experiment. The Good Laboratory Practice for Nonclinical Laboratory Studies Regulations require periodic reviews of all aspects of a research project by a disinterested party to examine adherence to the study protocol, data collection, data reliability, and data management, as well as the general conduct of the study. As the wisdom of a system of periodic inspection of projects becomes evident, one should expect that many funding agencies will begin to impose similar requirements.

III. TRAINING FOR RESEARCH IN LABORATORY ANIMAL/COMPARATIVE MEDICINE

A. Needs of the D.V.M. Trainee

Professional education in veterinary medicine does not emphasize the clinical problems of laboratory animals nor approaches to research. In order to be productive in research, individuals who have attained the D.V.M. degree must usually obtain additional training. That additional training can be obtained in two ways. It is possible for individuals to plan their own course of research training while engaged in the professional aspects of laboratory animal medicine. In those situations the individual must seek formal courses to supplement professional education and, most important, arrange for tutorial guidance in research by a member of the faculty accomplished in research.

The more common way of obtaining research training is to enroll in a formal program in laboratory animal/comparative medicine at one of several academic institutions. Most of the existing training programs have as their objective the development of professional skills in laboratory animal medicine and the preparation of the individual for a research career in either laboratory animal medicine, comparative medicine, or a combination (Clarkson, 1961, 1965, 1980; Cohen et al., 1979; Lang, 1979). Since many of the research opportunities are to be found in the diagnostic processes of laboratory animal medicine, it seems unwise to dissociate professional training from research training. For that reason, our faculty has viewed training in clinical laboratory animal medicine as essential in preparing individuals for academic careers. The objective is to provide the individual with the training and experience necessary for professional competence in dealing with problems of laboratory animal care and medicine and for the diagnosis of obscure illnesses that might be of comparative interest. The approach used to achieve this objective is to assign the individual for a period of time to a category of training known as the professional core. The professional core consists of courses in diseases of laboratory animals and medical primatology in a series of clinical rotations under the direction of specialists in laboratory animal medicine. Depending upon the interest and resources of the training institution, the clinical rotations may involve nonhuman primate medicine, avian medicine, rodent/ lagomorph medicine, experimental surgery, and participation in the activities of diagnostic laboratories.

The academic and research aspects of post-D.V.M. training are usually included in a component of the program known as the academic core. The objective of the academic core is to provide the individual with formal courses that relate to further capability in comparative medicine research along with appropriate training experiences in research. The courses appropriate to the academic core will vary with the individual's background and the resources of the training institution. Usually such courses involve animal models in biomedical research, some advanced courses in pathologic anatomy and clinical pathology, and courses in experimental design and biostatistics.

Research training is clearly the more important part of the academic core. It is best provided on a tutorial basis by an experienced investigator. Appropriate research training should acquaint the individual with the experimental method of problem solving. Throughout the research training process scholarly curiosity should be encouraged and research skills developed that will equip the individual to advance knowledge and understanding in laboratory animal/comparative medicine.

While there is clearly a need for individuals in laboratory animal/comparative medicine with research training in the sciences basic to medicine and in the clinical specialities, the greatest need is for comparative medicine clinical investigators. The clinical investigator usually utilizes a multidisciplinary approach to further the understanding of a disease process, whether it involves laboratory animals or animal models of a human disease. A comparative medicine clinical investigator would likely use the disciplines of pathology, genetics, behavioral science, endocrinology, clinical pathology, and others.

B. Needs of the M.D. Trainee

Increasingly, individuals whose professional education has been in medicine are interested in becoming investigators in comparative medicine. Like D.V.M. trainees, their medical education has not provided them with knowledge of the diseases or biology of laboratory animals nor the formal education usually required for a research career. The needs of the M.D. trainee are not markedly different from those of the D.V.M. trainee. It is unlikely that such individuals will be responsible for the clinical care of laboratory animals. Therefore emphasis on the professional core of training is of little benefit. It is, however, likely that many of their ideas for research may arise from awareness of naturally occurring clinical problems among laboratory animal populations. For that reason, it is necessary as a part of their training to develop awareness of the clinical approaches used by laboratory animal veterinarians and establish early in their research careers an appreciation for the need for active communication with those responsible for laboratory animal clinical problems.

In order to be productive in comparative medicine research, the M.D. trainee must be exposed either formally or informally to the body of knowledge concerned with animal models of human disease. Usually it is also necessary to provide formal or informal training in such areas as genetics, experimental design, and biostatistics.

C. Needs of the Ph.D. Trainee

The needs of the Ph.D. trainee to prepare for a research career in comparative medicine are similar to the M.D. trainee. It is important early in the individual's research career to establish the usefulness of communication with laboratory animal veterinarians about disease processes that may be useful in comparative medicine research. The Ph.D. trainee usually has a good background in experimental design and biostatistics, but there is need for formal and informal instruction about animal models for biomedical research.

IV. DEVELOPMENT OF RESEARCH RESOURCES

A. Specialized Staff Capabilities

Meaningful research often results from the collaborative efforts of investigators with varied disciplinary skills. These varied skills can become a part of an academic program in laboratory animal/comparative medicine in two major ways. First, since most laboratory animal programs are located within medical schools, research collaborators can be attracted from other departments. Individuals both in the clinical sciences and the sciences basic to medicine have active interests in the use of animal models to gain research objectives. The incorporation of a biostatistician in the research team will encourage appropriate examination and interpretation of the experimental results. Their contributions along with those trained in laboratory animal/comparative medicine can often be brought together for research programs much stronger than could be mounted by either group alone. A second way of developing investigative strengths is by way of thoughtful recruitment to the faculty of laboratory animal/comparative medicine programs. The more productive programs seem to have utilized both of these approaches, since the recruitment to the laboratory animal/comparative medicine faculty of individuals with differing disciplinary skills will usually facilitate the attraction of other scientists within the institution. The kinds of disciplines needed to facilitate research will vary considerably depending upon the interest of the laboratory animal/comparative medicine program. The disciplines usually represented in the more productive programs are often biochemistry, microbiology, physiology, reproductive biology, behavioral sciences, and biostatistics. The distinction between an investigator from one of these disciplines that joins a comparative medicine group as opposed to a department of their own discipline concerns the research approach. The research approach in comparative medicine almost always involves the use of animal models to research a problem rather than basic exploration of phenomena of interest to a particular discipline in the case of those individuals that join a disciplinary department.

To summarize, skilled investigators and consultants are necessary for a strong research program. The laboratory animal/comparative medicine group should define carefully its research goal and the objectives and strategies to be utilized to obtain that goal, and on that basis they should decide the areas of faculty recruitment, with the anticipation that the recruitment of a good faculty will facilitate collaboration with other areas in the institution.

B. Laboratory Capabilities

Suitable laboratory capabilities are also essential for the development of research programs. The diagnostic/investigational laboratory programs supported by the Animal Resources Program of the Division of Research Resources of the National Institutes of Health have been extremely helpful in the establishment of laboratory capability for a large number of programs. Since the research in laboratory animal/comparative medicine usually involves disease, the establishment of diagnostic/investigational laboratories provides the capability for initiating programs in research on disease. The precise kinds of laboratory capability required for developing research depend largely on the interest of the program. It has proved useful in most institutions to utilize the core concept to the fullest possible advantage rather than to encourage duplication of effort among program scientists. Usually, if two or more scientists require the same methodology, the development of a core resource will be useful. In most laboratory animal medicine programs, the core resources exist for such things as necropsy, microbiology, clinical pathology, histology, and data management (see Chapter 26).

C. Data Management Capabilities

The availability and use of effective computerized data management and analysis have become determining factors in the viability of research programs. As such, it is important that any training program that hopes to produce effective and competitive researchers should provide the necessary human and machine resources and training in the use of these resources.

When computerized data management and analysis resources are available and utilized, it is possible to more effectively meet the varied informational needs of an active laboratory animal medicine/comparative medicine facility. More and better information can be collected, organized, and made readily available, and the increasingly important skills needed to use computers in research and animal management can be learned.

Most institutions already have some system of computerized record keeping. The effectiveness of such a system will, in a large part, be determined by the availability of the computer resources to the user community—in this case, the laboratory animal medicine and comparative medicine personnel. Terminals must be conveniently located and easy to use. The computer programs must be responsive to the specialized needs of the medical and research activities and to a user community

with a wide range of computer use skills. They must be easy to use but powerful enough to provide the user with data manipulation, investigation, graphics, and statistical analysis capabilities.

A prime, but often overlooked, factor in the effectiveness of the data management resource is the interaction between the data management staff and the user community. It is important that there be frequent communication between all parties. The data management personnel must understand and be responsive to the activities of a medical and research environment. The medical and research personnel must be willing to adjust some of their thinking and methods to take advantage of the available technology. There must be a commitment and an incentive to learn what tools are available and how to use them.

1. Types of Information Needs

Although there are areas of common interest and need, the data requirements of laboratory animal medicine and comparative medicine are not the same. Because of this, it is extremely important that the information requirements of each group be known and appreciated by the other. In this way, areas of mutual interest can be shared, and information, generated by one group but of interest to the other, will not be lost.

In most instances, data collected in research activities are highly structured and very selective, as specified in a protocol. Often research data will consist of a series of measurements obtained from groups of animals subjected to different conditions over a period of time. Information about the clinical treatment and health of the experimental animals, while not necessarily research data, is important in order to examine any confounding effects on the research.

Animal care and management information needs are more generalized. They include such things as results of diagnostic laboratory tests, administration of drugs, changes of diet and location, breeding performance, and treatment schedules. Data are generated in a less structured way and are used to monitor and describe an animal's condition and treatment for the duration of its stay at a facility. These historical data are a valuable resource for nonexperimental research into diagnostic methodologies or the effect of chronic diseases on experimental outcome. Research generated data may also be pertinent to maintaining the health of the animals in the study or as a potential area of research.

2. Laboratory Animal Medicine Data Needs

Because the nature and uses of the data generated by laboratory animal medicine and comparative medicine are different, it is not surprising that the methods of data collection, storage, and use or analysis also differ.

Data on the health, treatment, and management of laboratory

animals are centered primarily on the individual animal. Data often must be available immediately to be useful in maintaining the health of the animal. To meet this requirement, the data should be stored in an on-line data base with new information being entered into the computer as quickly as possible. In many institutions, a record keeping system is already in place. When that is the case, it is important that the laboratory animal medicine and research staff know about the system and how to use it. If a system does not already exist, careful planning involving laboratory animal medicine, research, and data management personnel should precede any implementation. Current and future needs of both animal care and research must be addressed, and a responsive system must be selected or designed. It is crucial that all parties be involved in the planning and use of the data base. Pertinent information would include the reason for treatment, results of laboratory tests, administration of drugs, type and schedule of treatment, and changes of diet and location. Additional information on breeding activities, parentage, use in experiments, origin, and screening for disease would also be valuable information to both the laboratory animal staff and researchers. The responsibility for recording information would fall to different individuals depending on the source of the data item. The method for retrieving information on the animals must be rapidly accessible by both the laboratory animal medicine personnel and the research staff. A minimum amount of input from the user should be required to obtain data. It should be sufficient to input the identity of the animal and the type of information needed. In some facilities, it would be useful to have a schedule identifying the treatments planned for the day or week. Special reports listing items, such as current location, age, sex, body weight, last disease screening data, experimental assignment, and clinical treatment status, on all animals at the facility should also be available.

Since it is not always possible to predict which items of information will be related, and since the identification of the data items (by individual and date) is less complex than in experimental data, it is likely that all the information can be maintained in one data file. Depending on the number of animals, it might be more manageable to maintain separate files for each species or animal type. All of the data related to an individual would be stored together and updated as needed. A systematic method for ordering the animals in the data base would be used. Data input forms to accommodate the wide variety of laboratory tests and treatment procedures would be needed. Individuals to be responsible for filling out the data input forms would need to be identified and trained. Depending on the size of the facility and the amount of data, this responsibility could rest with the laboratory animal medicine staff or with an individual whose sole activity was preparing treatment and management information for input to the data base.

It should be apparent that, over a period of time, the body of data generated and stored would represent a valuable source for nonexperimental research by both the laboratory animal medicine and comparative medicine staff. This information is also extremely important in interpreting the outcome of any experiment performed on the animals. In order to look at this large body of data in a more generalized way, retrievals need to be based on large groups of animals that would be identified by an unlimited combination of variables. For example, one might want to look at all or some data on animals of a certain species, experiment, housing location, diet, age and/or lineage. On the other hand, the identifying item might be a range of values from a laboratory test or the occurrence of disease. These kinds of requests require a very versatile method of retrieval, but are usually not needed as quickly as information about an individual's health. Therefore, the methods and programs used to respond to this type of request could be different from the individual request programs, providing for complicated data searching without interfering with the more time-critical uses.

Whatever the system used, it is crucial that everyone involved with the data collection and use be fully aware of the system's capabilities and trained to use this resource. The quality of the data in the data base will be directly related to the amount the data base is used.

3. Comparative Medicine Research Data Needs

Experimental research data, whether generated in comparative medicine activities or laboratory animal medicine, require carefully planned and detailed methods of collection, storage, and analysis. Because each experiment seeks to answer different questions, it is necessary to make sure that the data are managed in such a way that the questions can be asked. In storing research data it is rarely sufficient to identify the variables (blood values, body weight, etc.) by just animal and date. It is necessary to know how that variable fits into the experiment in a more detailed way. For this reason, many items may be needed to identify each variable recorded. For example, it might be necessary to identify the experimental group (test, control), the phase of the experiment (baseline, stage 1, stage 2, etc.), the number of weeks or days into each phase, the sex, age, housing condition, diet, and other items that might be pertinent to the research being conducted. Because the identification items will vary with each experiment, the format of the data to be recorded will also vary. The data format must be carefully planned before the experiment begins, as it is very difficult and time consuming, if not impossible, to reconstruct and record old information later in the study. It is important to begin working with the data management staff at this early stage so that they can begin to design a system for gathering, recording, processing, and reporting the data. The individuals who will be responsible for gathering and recording data must be selected and trained in data collection methods and filling out the necessary data input forms. They should be involved, along with the researchers and data management staff, in designing the input forms so as to take into account the logistics of the actual data collection. Because of the differing information needs in each experiment, it is not usually possible or desirable to include data from more than one experiment in one data file. In most cases, it is more effective to set up several related data files to follow or "track" data from one experiment. The data might be grouped in files by logical association and/or the sampling schedule for the experiment. The exact format for a data file should take into account the items needed to identify the data collected, the logistics of data collection and entry (collecting samples, filling out data sheets, keying data), the data analysis methods to be used (statistics, graphics, etc.), the interim and final retrieval requirements of the investigator(s), and the programs or packages to be used for data management. If there are data management software packages available at the facility, selection of the appropriate package and arrangements for its use must be made. Otherwise, additional time may be needed in order to write special programs for the experiment. There are many different systems available to store and manipulate research type data. Some of them have special hardware requirements and, for facilities just establishing computer support centers, it would be advisable to investigate these before making equipment decisions. Both before and throughout the experiment it is imperative that the research and laboratory animal medicine staff and the data management support person meet regularly to discuss the requirements and progress of the project and their contribution to the research endeavor.

V. SUPPORT FOR RESEARCH

One of the most important and sometimes perplexing aspects of developing research programs in laboratory animal/comparative medicine is obtaining financial support for the activities. Usually, universities can provide only modest support for research programs so individual scientists must turn to extramural sources for support of significant programs. A detailed description of support for research is beyond the limitations of this chapter, but we shall attempt to relate a few observations that are important to developing research programs. Thoughtful descriptions of developing research grants have also appeared elsewhere (Eaves, 1972, 1973).

The availability of funds for pilot studies is critical to the planning and defense of grant applications. If modest financial support for research is available from the university, it can probably be best used to finance pilot studies. Given the current high level of competition, it is virtually impossible to obtain grant support without having first done preliminary studies to provide evidence that the ideas are likely workable and the approaches feasible. A key aspect of the merit of a grant application is the way in which the applicant uses preliminary data in both the sections on background information and previous work done to establish the credibility of the projects being proposed.

Another difficult area in developing extramural support is choosing from among several possibilities the research projects to propose that are the most likely to be approved and funded. There are no definitive guidelines to help the investigator in this matter, but there are some generalities that might be useful. Of perhaps the most importance is that the research area be one of genuine deep interest and concern to the applicant. It is only when individual deep interest and concern exist that the application can reflect positive, creative, and excited views and communicate these characteristics to the reviewers. To try to develop a grant application dealing with a subject about which one is only partially interested can be deadly dull, and inevitably this lack of enthusiasm is apparent to reviewers.

Perhaps of next importance in selecting the direction for a grant application is the investigator's appraisal of the level of national peer interest in the questions. Whether it is ultimately good or bad, it is true that in any field the perceived relative importance of certain kinds of questions are continuously either increasing or decreasing in the level of enthusiasm by which they are perceived by one's peers. To be successful in obtaining extramural funding for research, it is essential that the applicant be aware of the current status of changing patterns of thought about various questions related to the proposed research.

Another concern in selecting an area for the preparation of a grant application concerns the significance of the problem being proposed. Over the past several decades, during which financial resources for research were more available than is the case currently, significance could be based on the contribution of the derived data to general knowledge about a particular subject. As financial support has become more limited, however, it is increasingly important that the proposed research relate in a meaningful way to the better understanding of some aspect of either animal or human health. Even among projects that relate clearly to health, those that relate to important health problems are more likely to be reviewed favorably than those that relate to little known or esoteric disorder. Significance should not be overemphasized, however, because the current system of peer review of research applications continues to place major emphasis on scientific merit. Having made these several considerations about the area to be chosen for the grant application, it is good for the applicant to remember that the success or failure will then lie almost entirely with the scientific merit of the application. There are several generalities about the preparation of an application that can help communicate the scientific merit of the work being proposed.

Under general categories of previous work done and background information, the applicant should place in clear context how the work being proposed relates to the existing information about the subject. In doing so the applicant should identify clearly the hypotheses that are to be tested in the experiments being proposed.

The specific aims of a grant application are one of the most important parts of the entire application. The aims should be stated as clearly and concisely as possible, and it is usually helpful if they appear as numbered statements. In enumerating the specific aims it should be clear to the reader how the aims relate to testing the hypotheses that have been generated.

The methods section of a grant application is sometimes frustrating to write because it is not possible to describe in detail each and every method that will be undertaken in the research project. In the preparation of the methods section, it is useful to recall the ways in which that section is used by reviewers. First, although the reviewers recognize that the methods cannot be comprehensive, it is the only opportunity to evaluate whether the applicant has knowledge of current methods and thinks in sufficient detail to carry out a project successfully. In areas in which it is known generally that measurements are difficult, the reviewer also expects to see a reflection in the application of the applicant's realization that measurements are difficult. Method sections are more likely to be deficient in discussions of strains of animals, experimental design, morphometric analyses, and statistical analyses than in most other areas. The applicant should strive to find an intermediate position between being so detailed as to lose the reader and being too sketchy to communicate to the reviewer the depth of knowledge about the problems in the research area being undertaken.

We have indicated earlier the importance of the significance of the research being proposed. Most grant applications contain a section that request anywhere from a half of a page to a page describing the applicant's view of the significance of the research. This section is usually at the end of the application and unfortunately, the applicant is often tempted to deal with it in a hurried and superficial way. It is in fact one of the reviewer's most effective ways of determining whether the applicant does see the real significance of the work to be undertaken. Significance sections should be reasonably specific in their flavor and not global and sweeping.

Discussion of support for research thus far has related to applications for grant support. In developing research programs in laboratory animal/comparative medicine, the availability of contracts to support research should not be overlooked. Contracts are usually of two basic kinds. The more common is the competitive contract in which a federal agency or some other research agency has identified a specific area of research that needs to be undertaken and solicits contract proposals from groups capable of doing the work. By far the less common type of contract is when an individual researcher identifies a specific task that needs to be undertaken and submits to the granting agency an unsolicited proposal to carry out the work. In the case of competitive contract proposals, the investigator must deal with a different group of considerations from those that relate to grant applications. When announcements are made of contracts of interest to a granting agency, it is possible to obtain from that agency a detailed description of what is being requested, and this is commonly called a request for proposals. If the request for a proposal should happen to fit with what the applicant believes is an important and exciting area of research and if the applicant believes the results will hold significance for animals or man or both, then contracts can be as useful as grants in developing research programs. While contracts are considered for approval and funding in many of the same ways as grant applications, there are important differences. First, one of the principal criteria for approval of a contract is the degree of responsiveness of the applicant to the work outlined in request for proposals. Another major difference relates to the budget. Generally, budges for grant applications are reviewed by peers and either approved or modified without significant negotiation. On the other hand, contract budgets are arrived at by a long and often difficult negotiation essentially on an item-by-item basis between the granting agency contracting officer and the institutional representative. Finally, and perhaps of the most importance, while the individual scientist has great latitude under grants to pursue important leads and develop new research directions based on the data as it emerges, one must under a contract do precisely the work outlined in the proposal or negotiate for modification with both a project officer and a contract officer.

Sources of funding can be an important topic. Most institutions maintain a research development office where one can obtain good information about sources of funding. The National Institutes of Health funds the overwhelming majority of biomedical research projects in this country, but one should not overlook industry, the voluntary health agencies, and various foundations in deciding a source for potential funding.

REFERENCES

Clarkson, T. B. (1961). Laboratory animal medicine and the medical schools. *J. Med. Educ.* **36**, 1329–1330.

Clarkson, T. B. (1965). Consideration of composition and educational philosophy of graduate training activities in laboratory animal medicine. *In* "Proceedings of Workshop IV—Graduate Education in Laboratory Animal Medicine," pp. 20–30. Nat. Acad. Sci., Washington, D.C.

Clarkson, T. B. (1980). Evolution and history of training and academic programs in laboratory animal medicine. *Lab. Anim. Sci.* **30**(4), 790–792.

Cohen, B. J., Baker, H. J., Jr., Dodds, W. J., Hessler, J., and New, A. E. (1979). Laboratory animal medicine: Guidelines for education and training. *ILAR News* **22**(2), M1–M26.

Eaves, G. N. (1972). Who reads your project—Grant application to the National Institutes of Health? *Fed. Proc., Fed. Am. Soc. Exp. Biol.* **31**(1), 2–9.

Eaves, G. N. (1973). The project—Grant application of the National Institutes of Health. *Fed. Proc., Fed. Am. Soc. Exp. Biol.* **32**(5), 1541–1550.

Lang, C. M. (1979). Development of a comparative medicine program in medical and graduate education. *J. Am. Vet. Med. Assoc.* **173,** 1221–1223.

Siegel, S. (1956). "Nonparametric Statistics for the Behavioral Sciences." McGraw-Hill, New York.

Sokal, R. R., and Rohlf, F. J. (1981). "Biometry." Freeman, San Francisco, California.

Spector, P. E. (1981). "Research Designs," Sage Univ. Pap. Ser. Quant. Appl. Soc. Sci., 07–023. Sage Publications, Beverly Hills, California.

Chapter 26

Rodent and Lagomorph Health Surveillance—Quality Assurance

J. David Small

I. INTRODUCTION

Quality biomedical research is dependent on the use of disease-free animals. The impact of disease on laboratory animals and the research these animals represent cannot be overlooked. Large sums of money, time, and effort can be wasted by initiating experiments with diseased animals or when animals develop diseases while on experiment. The loss of breeding production in rodent colonies or loss of the entire colony, which may represent valuable genetic stock, is disruptive to research programs. The outbreaks of mousepox beginning in 1979 in several animal facilities in the United States proved to be extremely expensive, not only in dollars but in lost effort and loss of animals (New, 1981). Likewise, in 1973 and 1974, outbreaks of lymphocytic choriomeningitis (LCM) in research institutions affected not only the rodent population but the human population as well (Hinman *et al.*, 1975; Biggar *et al.*, 1977). Epizootics caused by Sendai virus, the rat coronaviruses (RCV/SDA),* and mouse hepatitis virus (MHV) continue to disrupt research though they frequently are overlooked or go unreported. Hsu *et al.* (1980) has summarized

*Abbreviations for the murine viruses are given in Table II.

much of the data pertaining to the effects of murine viruses on biological systems (see Chapter 23).

During the past 25 years, great strides have been made in differentiating and defining the infectious diseases of laboratory animals, identifying the causative agents, and understanding their epizootic nature. From this newer knowledge, coupled with germfree technology developed earlier (Trexler, 1983), has come an effort to establish breeding colonies of rodents and rabbits free of infectious diseases. In concert with the availability of disease-free laboratory animals, programs to assess their health status have been developed. These programs serve the following functions: (1) For the commercial breeder they provide a basis to furnish users of the animals with a statement defining the health status of the animals purchased. (2) For users of laboratory animals, these programs serve to characterize the health status of animals and their tissues and to monitor the health status of animals on experiment or in intramural breeding colonies. These programs evolved along similar lines in many institutions and commercial firms and have been referred to by a number of terms: animal health monitoring, disease surveillance, microbiological monitoring, quality assurance, preventive medicine, and animal health surveillance program. For simplicity throughout the remainder of this chapter the term quality assurance (QA) will be used to refer to health monitoring programs. A QA program is not, in and of itself, a disease prevention program, but rather disease prevention is the result of proper management. As such, a well designed QA program can be a powerful adjunct to managing and maintaining a disease-free animal colony or a colony with a minimal amount of disease. Even in the absence of appropriate physical facilities to maintain disease-free animals, the health status of animal colonies should be characterized and diseases, if present, should be identified.

This chapter discusses QA programs for rodents (primarily mice and rats) and rabbits. Quality assurance programs for other species are described elsewhere in this text and in other species-oriented texts. However, the principles of QA programs are the same for most species, especially when they are used in a laboratory setting. Thus, discussions about rodents and rabbits can be applied to the management of other species. Where possible, readily available references have been selected to encourage further reading. Specifically, it is not the author's intent to present the details of test procedures and describe lesions of each disease one might encounter. That information is available elsewhere in this text and in the species-oriented texts sponsored by the American College of Laboratory Animal Medicine [Weisbroth *et al.*, 1974 (rabbit); Wagner and Manning, 1976 (guinea pig); Baker *et al.*, 1979, 1980 (rat); Foster *et al.*, 1981–1983 (mouse), texts on pathology (Benirschke *et al.*, 1978), and various other publications (e.g., Needham and Cooper, 1984; American Public Health Association (APHA), 1979; American Society for Microbiology

(ASM), 1980a,b)]. The current edition of the "Guide for the Care and Use of Laboratory Animals" [Institute of Laboratory Animal Resources (ILAR), 1978] should be consulted for an extensive bibliography on laboratory animal science.

II. DESIGN OF QUALITY ASSURANCE PROGRAMS

A. Organization and Structure

The animal health QA program is only one facet of the management program of an animal colony, be it a closed breeding colony or a multispecies animal facility serving an integrated university campus or research and development complex. How well a QA program serves an organization is dependent not only on the design and function of the QA program but on how well the management program functions. A critical factor for the success of any animal facility program is that those responsible for managing the animal facility have responsibility for and control of all animals and animal tissues entering the facility. Such control is in the interest of all concerned. Quality assurance begins with defining the microbiological status of the animals or tissues before they enter the research or breeding colony, not after disease becomes apparent. The administrative structure of a surveillance program for introducing animals and tissues into a facility has been described (Small and New, 1981).

Quality assurance programs have as their common theme, the task of characterizing the animal's microbial burden and health status. The degree of characterization and concern with the microbes identified will vary with the species and the nature and purpose of the colony, i.e., germfree, defined flora, or conventional and experimental versus breeding. An animal facility should have access to competent assistance for the diagnosis of both infectious diseases and problems not of a microbial nature. Ideally, this assistance will be an inhouse diagnostic laboratory staffed by one or more trained professionals offering a full array of pathological, microbiological, and parasitological services. Few organizations have, or can justify, total inhouse diagnostic capabilities, and when additional services are required, use can be made of consultants and commercial diagnostic laboratories. More important is that necessary tests be identified and done, results obtained promptly, and the results quickly and correctly interpreted. A theoretical table of organization for a diagnostic laboratory supporting an animal facility is shown in Fig. 1. Environmental and equipment monitoring functions of the diagnostic laboratory are depicted as well. While important to the successful operation of an animal facility, they have been covered elsewhere (Newton,

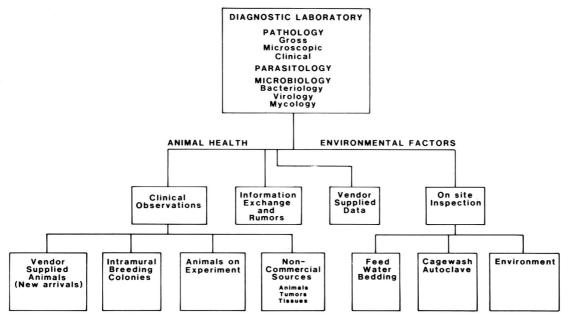

Fig. 1.   Sources of information and specimens for an animal health quality assurance program.

1978; Small, 1983; Chapter 17 by Hessler and Moreland) and will not be discussed further in this chapter. Information developed by the laboratory is passed rapidly to those responsible for clinical care and management. The exact routing of information and the chain of events that occurs prior to action being taken are not as important as are speed and effective communication. Oral communications concerning laboratory results and recommended courses of action should be supported by documented reports. Written documentation of problems is important but should not be stressed to the point that it delays or eliminates communication of problems.

Regularly scheduled staff conferences involving the professional and supervisory members of the animal facilities management team are useful for assuring exchange of animal health information and discussing the consequences of the findings. Written records of these meetings serve as a reference for the staff and assist in keeping issues before management until they are resolved.

The animal technicians in daily contact with the animals are very important members of the animal health team. They should be considered an extension of the veterinarian's role and be trained in the fundamentals of how to observe and report animal health problems. It is frequently the animal technicians who first alert the professional staff to problems. Communications between the professional staff and the technicians must be open, honest, and nonthreatening (Arnold *et al.*, 1978).

Each QA program must be designed for and around the people it is to serve. The needs and interests of the investigators,

capabilities and interests of those responsible for animal care, species present and their origins, and the laboratory and financial resources available to support the program all play a significant role in determining the QA program adopted. Unfortunately, many investigators and administrators remain ignorant of the effects of intercurrent disease on their animals and, furthermore, are unaware that with rodents and rabbits, these disease problems can be avoided by the purchase of disease-free animals and sound management. It is not unreasonable to include portion of the cost of a QA program in the per diem rates for each species. In this way QA and diagnosis of animal health problems become an integral aspect of the animal care program. Infectious diseases and even some noninfectious problems, e.g., incorrect pH or chlorine concentration in drinking water which may initially mimic an infectious disease, can affect the work of more than the investigator with whom the problem is first identified.

B.   Records

Accurate records for all experimental and breeding animals are essential if an animal facility is to function properly. More recently, federal laws have required that certain species be individually identified. The Good Laboratory Practices Regulations (Food and Drug Administration, 1978) require the identification of rodents to at least the cage level throughout a study. With rodents and rabbits, recommended recorded data for each

Table I

Information Suggested to Be Recorded on Cage Cards When Animal Is
Received

Species
Sex
Strain/stock
Date of birth
Date received
Source (vendor and vendor's area designation)
Assigned to (investigator)
Room
Treatment or experiment

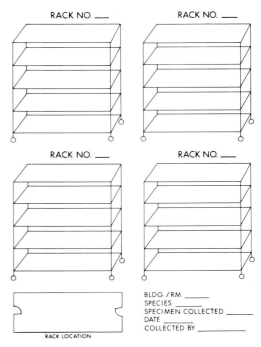

Fig. 2. Form for recording cage and specimen location during epizootiologic investigations.

animal or cage of animals on arrival from the vendor or inhouse breeding colony are shown in Table I.

Morbidity and mortality should be reported daily, and records should be kept of the losses. Clinical records are rarely maintained on individual animals smaller than rabbits, unless they are housed alone. Usually, a cage of rodents is examined as a single unit. This is especially true of mice and rats. Rabbits and larger animals, whether housed singly or in groups, are individually identified. From well-kept records will emerge a pattern of nonexperimentally caused deaths. This information can assist in identifying and correcting animal health problems.

Clinical records should be maintained for rooms and cages of animals just as for individuals. If medication is given, this should be noted by recording the drug, manufacturer, dose, length of treatment, and the reason for medication. Animals on experiment or slated to be used in experiments should not be medicated without the investigator being informed.

When investigating the epizootic pattern of diseases, especially in rodents, it is helpful to keep track of the infected animal's physical relationship with other animals in the room. An excellent method is to use charts depicting empty animal racks (Fig. 2). Cage positions are drawn on the rack and numbered, and positive cages are noted. Such charts can be given to those collecting specimens, and the location of each specimen can be marked on the chart. When test results are available they are added to a master chart, thus indicating the distribution of infected animals within rooms or on individual racks. This technique was used to investigate an epizootic of mousepox (Wallace *et al.*, 1981; Werner *et al.*, 1981), and the author has used it for studies of pinworm infections and outbreaks of mouse hepatitis. Individual room charts should be combined with a facility floor plan depicting affected and nonaffected rooms. The addition of routes of supplies, people, and air flow to the floor plan helps in understanding the epizootiology of the disease and picking up errors in management procedures.

C. Selection of Tests

The selection of meaningful tests to be performed on each species is dependent on a thorough knowledge of the diseases affecting the species, the probability of those diseases being present, and an understanding of the sensitivity and specificity of the tests available. Table II summarizes the diagnostic procedures available commercially or in test kit form for the murine viruses (Parker and Collins, 1984). The accumulation of test results with little idea of how to interpret those results is a waste of effort and money. Animal colonies may be unnecessarily destroyed if results are incorrectly interpreted. Tables III–VII are offered as guidelines for those organisms or diseases that a diagnostic laboratory should be capable of identifying or diagnosing in the respective species, should they be present. The presence of an organism named in these tables should not necessarily be used as the basis for rejecting animals from entry into a colony. These tables are to be used as a guide in establishing screening programs to characterize the animals or tissues received or raised. In addition to examination for microbes or antibody, histologic examination of the major organs should also be considered. The degree of characterization and how it is to be accomplished and the interpretation of test results is a matter of experience and professional judgment.

infection is present in any one animal, it is likely it will be passed to the others. However, with many of the viruses, i.e., Sendai, PVM, and RCV/SDA, the virus is present for only a short time, and passage of infection ceases following the appearance of antibody or shortly thereafter.

A monitoring scheme intended solely for the determination of murine viruses places individually identified sentinel animal in the room. At specified intervals, blood samples are collected and examined for the presence of antibody. By retaining the animal in the room, a questionable titer can be checked by rebleeding the same animal.

In addition to examining sentinel animals, many tests can be done on a regular basis utilizing experimental or breeding animals. Cellophane tape preparations for *Syphacia* spp. can be collected without interfering with most research. Fecal samples can be collected by placing rodents in a clean empty cage for a few minutes or, in many cases, a fresh pellet can be collected when the animal is picked up. Composite fecal samples can also be obtained from soiled cages. Examination of fecal specimens for parasites and culture of fecal pellets for *Salmonella* spp. from a statistically significant number of cages monthly, quarterly, or semiannually should be considered. The presence or absence of *Pseudomonas* spp. should also be noted. In microbiologically defined barrier mouse colonies, consideration should be given to examining for *Citrobacter freundii*. Strain 4280 appears to be the strain of greatest clinical significance (Ganaway, 1982).

All animals in the colony that die or are observed ill should be examined. With small rodents, especially in large colonies, not every animal is observed every day and dead animals will be decomposed and yield little information. However, all deaths should be brought to the attention of the pathologist, who then makes the decision regarding further evaluation. Surviving cagemates should be examined carefully for indications of disease or other problems. The situation in experimental colonies is more complex because many times deaths are a result of the experimental procedure. An understanding of the investigator's work and the expected results is most helpful. One must be cautious, however, not to be lulled into a false sense of security, believing that all illnesses and deaths are due to experimental procedures.

In addition to animal health information derived from sentinel animals and animals submitted to the laboratory, those caring for the animals and the investigators and technicians using the animals can provide valuable information. It is essential for the clinician to establish rapport with these people and to enjoy their confidence.

Investigator's results can sometimes offer the first clue that there is a disease within a colony. For example, mice with active Sendai virus infection have depressed cell-mediated immunity (Jakab, 1981) and those with active MHV infection have stimulated macrophages (Boorman *et al.*, 1982). The clinically inapparent infection of guinea pigs with *Salmonella enteritidis* has resulted in increased activity of peritoneal macrophages (J. D. Small, unpublished observation) (see Pakes *et al.*, Chapter 23).

III. QUARANTINE

A. Introduction

Quarantine procedures for newly arrived rodents and rabbits and their respective tissues will vary depending on the animal's origin. The quarantine of newly arrived animals and their tissues provides an opportunity to examine them for the presence of potential pathogens and, in the case of animals, to characterize their health status by laboratory tests and/or clinical examination. At the same time, the isolation of new arrivals reduces the possibility of transmitting active infections to established colonies of animals. An often overlooked value of the quarantine period is the opportunity it gives animals to recover from the journey to the laboratory and to adjust to their new surroundings (Lang, 1983). Recent evidence indicates that mice have altered immune functions and elevated corticosterone levels for 48 hr following arrival (Landi *et al.*, 1982).

B. Animals

1. Examination

Usually, logistical considerations will not allow for the quarantine of each shipment of rodents and rabbits. However, in some cases, where space permits or as part of the experimental design, animals are quarantined in the room in which they will be used. Most often, commercially reared rodents and rabbits from recently characterized sources with an acceptable health profile are admitted after only a visual examination on arrival (Fox, 1977). This procedure is reasonably satisfactory if confidence exists in the integrity of the vendor and there is an agreement with the supplier immediately to inform customers of problems.

In many cases, only those animals received from noncommercial sources or from sources that have not been recently characterized are quarantined. Quarantine programs may be divided into two major classes: passive and active, based on the testing done during the quarantine period. In a passive quarantine, animals are merely isolated for a period of time, usually 7 to 28 days (more often the former), and observed for signs of illness or mortality. This form of quarantine will not reveal many infectious agents that rodents and rabbits may carry and is considered inadequate. However, a passive quarantine of 4–6 weeks will allow certain of the nonpersistent murine viral diseases, i.e., Sendai virus pneumonia and RCV/SDA, to run their course. During this period no breeding should take place, or if it does, litters should be destroyed. The animals will de-

In contrast, an active quarantine is one during which the animals or tissues are actively examined for the presence of undesirable microorganisms. An active quarantine program developed originally for mice has been described by Small and New (1981). Animals and tissues are presumed to be contaminated and are isolated from other animals in the facility. This usually means housing incoming animals or animals inoculated with tissues in plastic isolators (Trexler, 1983), Illinois cubicles (Lang, 1983), or Horsfall-Bauer units (Horsfall and Bauer, 1940) (Fig. 3). Blood is collected and screened for antibody to the murine viruses appropriate for the species. In addition, animals are thoroughly examined for ecto- and endoparasites, pathogenic bacteria, and mycoplasma as appropriate for the species (Tables III–VII). The specific methods used vary with the diagnostic skills, preferences of the staff, available facilities, and finances. The individual responsible for the quarantine program must understand the limitations of the tests used, the nature of the agents being sought, and the consequences of admitting certain organisms to the animal facility. Parker and Collins (1984) have discussed the assays for the murine viruses. This publication should be consulted for details of the various procedures and the interpretation of the serologic tests. The current procedures used for testing for murine viruses are rapidly changing with the development of ELISA (enzyme linked immunosorbent assay) tests for most of the viruses. Kits are commercially available for performing many of the viral assays and testing for *Mycoplasma pulmonis*. New tests are being developed at a rapid rate. Consultation with experienced

Fig. 3. Horsfall-Bauer type isolation cages. (Photograph courtesy of PLAS ■ LABS®, Lansing, Michigan.)

individuals is suggested when establishing a murine diagnostic laboratory.

2. Cohabitants

When receiving animals from an uncharacterized source, it is helpful if additional animals can be received for QA purposes. These additional animals must come from the same population as those received for experimental or breeding purposes. If additional animals are not available and the experimental animals cannot be bled or otherwise worked with, two to three cohabitant animals of the same species known to be disease-free, can be added to each of the cages of newly introduced animals for 4 weeks. In the case of mice, use of athymic (nude) mice in conjunction with euthymic mice as cohabitants has merit. However, athymic mice should not be depended on to develop antibody. Females should be used to avoid impregnating the animals received, should they be females. Also, in the case of many strains of mice, less fighting will occur among females than among males. Any litters born to cohabitants can be removed and euthanatized or used for test purposes. At the end of the cohabitation period, the cohabitant mice are fully characterized as to their disease status. The basic assumption is that any virus or other infectious agent being shed during this 4-week period will be passed to the cohabitant animals and antibody, clinical disease, or the agent (e.g., parasite, bacterium, or mycoplasma) will be detected. Cohabitant animals will not develop antibody to viruses that the quarantined animals had previously and with which they are no longer infected or no longer shedding. In some cases this is not a problem; however, with this procedure the immunologic status of the cohabitants is determined, not that of the newly introduced animals. A statement of the test results should reflect this important point. During the quarantine period, even though the experimental or breeding animals may not be available for collection of blood, they can be examined for ectoparasites, and cellophane tape preparations for *Syphacia* spp. can be prepared and feces can be examined for parasitic ova and pathogenic bacteria. Three negative cultures of feces at weekly intervals for *Salmonella* spp. are recommended before declaring animals free of *Salmonella* spp. Even if the animals being quarantined are examined directly, the use of cohabitants is a useful procedure as they may be euthanatized and their tissues evaluated.

3. Immunosuppression

Animals under quarantine or cohabitants are sometimes given immunosuppressants, e.g., steroids (cortisone acetate, 200 mg/kg) or cyclophosphamide (200 mg/kg), in order to increase the opportunity to express pathogens. For example, this technique is used to increase the susceptibility of mice to

Tyzzer's disease (*Bacillus pilifomis*) (Ganaway, 1982), and to MHV (Boorman *et al.*, 1982).

C. Tumors and Tissues

1. Antibody Production Test

Most viral infections in rodents are routinely detected by serological procedures and not isolation of the specific virus or viruses. With living animals, blood is collected and assayed for antibody using appropriate serologic tests. In the case of tumors or tissues, an antibody production test is usually done (Rowe *et al.*, 1962; Collins and Parker, 1972). The tumor or tissue is frozen and thawed a few times to break the cells and then inoculated into the appropriate host, usually by the intraperitoneal and/or intranasal route. The recipients should be of the same species as the origin of the tumor or tissue or the host through which it has passed or will be passed, i.e., mouse tissue or mouse tumors should be inoculated into mice as should tumors or tissues of other species that have been passed in mice [usually athymic (nude) mice]. In the case of tumors, a nonsyngeneic strain may be used so the tumor does not kill the recipient prior to completion of the test. The usual term for this test procedure is MAP (mouse antibody production) test. If rats are used, it is an RAP (rat antibody production) test, and if hamsters are used, it is an HAP (hamster antibody production) test, etc. After 28 days the recipients are euthanatized, blood is collected, and titers to the appropriate viruses determined. While this procedure was originally reported for detecting antibody to mouse viruses, antibody to *Mycoplasma* spp. and possibly other organisms may also be determined. Positive results with the ELISA test for *Mycoplasma pulmonis* should be confirmed by culture methods (Lindsey *et al.*, 1982). Likewise, the recipients may be examined further for the presence of other infectious agents. In addition, tumors and tissues, especially tissue cultures, should always be cultured directly for *Mycoplasma* spp. and other undesirable organisms.

The exact details concerning the number of animals inoculated, routes of inoculation, and whether or not dilutions of the inoculum are used will vary among laboratories. Procedures for the MAP test have been recently published (Parker and Collins, 1984).

2. Animal Inoculations

In addition to the MAP test, animals (mice) are inoculated to detect the presence of lactic dehydrogenase (LDH) and lymphocytic choriomeningitis (LCM) viruses in both tumors and animals. Lactic dehydrogenase virus is detected by measuring an increase in the level of LDH enzyme compared to control values following the inoculation of blood or other tissues into mice. The complement fixation test for LCM virus is not particularly sensitive; therefore mice are frequently inoculated with blood. The recent introduction of an ELISA test for LCM will reduce the need for mouse inoculation. Both of these tests are discussed by Parker and Collins (1984). Lymphocytic choriomeningitis virus can cause serious disease in people and appropriate containment facilities and techniques must be used (see Fox *et al.*, Chapter 22).

IV. CONCLUSIONS

Animal health surveillance needs to be an integral part of any biomedical research program. Continued assurance of the quality of the animals used in a research project is fundamental to the integrity of the final research results. At the same time it must be understood that QA programs are evolving as new diseases are identified and the role of old diseases is more clearly defined and understood. The knowledge gained in recent years concerning the epizootiology of Sendai virus pneumonia (Parker and Richter, 1982) and the rat coronaviruses (Jacoby *et al.*, 1979) has been helpful in dealing with these troublesome diseases. Recent information concerning MHV suggests that this virus is not as persistent as previously thought (Barthold and Smith, 1983). Thus, means for dealing with this disease are suggested. Recent advances in test procedures for infectious agents, especially the ELISA test, have raised the precision of diagnoses to new levels. New ELISA tests for many of the organisms affecting rodents and rabbits are being developed at this time. The rapid detection of antibody to *Pasteurella multocida* in rabbits with the ELISA technique is a step forward in identifying carriers or potential carriers (Peter *et al.*, 1983; Matsumoto *et al.*, 1983). The occurrence of *Mycobacterium avium-intracellulare* in a colony of mice (Waggie *et al.*, 1983) and the recent recognition of *Campylobacter fetus* subspecies *jejuni* as a potential pathogen for hamsters and other laboratory animals (Fox, 1982; La Regina and Lonigro, 1982) are of considerable interest. The finding of coronavirus-like particles in rabbits with diarrhea (Lapierre *et al.*, 1980) and the identification of coronavirus-like particles in the serum of rabbits with cardiomyopathy, also known as pleural effusion disease (Small *et al.*, 1979; Osterhaus *et al.*, 1982) lengthens the list of diseases for this species. The recent association of a gliding bacillus with chronic respiratory disease of rats is of interest (Ganaway *et al.*, 1983).

As new knowledge of the diseases of laboratory animals is acquired, QA programs will need to change to meet the challenge. However, the basis for an excellent QA program will remain an in depth understanding of the diseases and their causative organisms coupled with an inquisitive mind.

ACKNOWLEDGMENT

The assistance of Mrs. Jean H. Gordner in preparing this chapter is acknowledged.

REFERENCES

Acha, P. N., and Szyfres, B. (1980). "Zoonoses and Communicable Diseases Common to Man and Animals," Sci. Publ. No. 354. Pan Am. Health Organ., Washington, D.C.

American Public Health Association (APHA) (1979). "Diagnostic Procedures for Viral, Rickettsial and Chlamydial Infections," 5th ed. APHA, Washington, D.C.

American Society for Microbiology (ASM) (1980a). "Manual of Clinical Microbiology," 3rd ed. ASM, Washington, D.C.

American Society for Microbiology (ASM) (1980b). "Manual of Clinical Immunology," 2nd ed. ASM, Washington, D.C.

Anderson, R. M., and May, R. M. (1979). Population biology of infectious diseases. Part I. Nature (London) 280, 361–367.

Anderson, C. A., Murphy, J. C., and Fox, J. G. (1983). Evaluation of murine cytomegalovirus antibody detection by serological techniques. J. Clin. Microbiol. 18, 753–758.

Anonymous (1982). Muroid virus nephorpathies. Lancet 2, 1375–1377.

Arnold, D. L., Fox, J. G., Thibert, P., and Grice, H. (1978). Toxicology studies. I. Support personnel. Food Cosmet. Toxicol. 16, 479–484.

Baker, H. J., Lindsey, J. R., and Weisbroth, S. H., eds. (1979). "The Laboratory Rat," Vol. 1. Academic Press, New York.

Baker, H. J., Lindsey, J. R., and Weisbroth, S. H., eds. (1980). "The Laboratory Rat," Vol. 2. Academic Press, New York.

Bang, F. B. (1978). Genetics of resistance of animals to viruses. 1. Introduction and studies in mice. Adv. Virus Res. 23, 269–348.

Barthold, S. W., and Smith, A. L. (1983). Mouse hepatitis virus S in weanling Swiss mice following intranasal inoculation. Lab. Anim. Sci. 33, 355–360.

Benirschke, K., Garner, F. M., and Jones, T. C., eds. (1978). "Pathology of Laboratory Animals," Vols. 1 and 2. Springer-Verlag, (Berlin and New York.)

Biggar, R. J., Schmidt, T. J., and Woodall, J. P. (1977). Lymphocytic choriomeningitis in laboratory personnel exposed to hamsters inadvertently infected with LCM virus. J. Am. Vet. Med. Assoc. 171, 829–832.

Boorman, G. A., Luster, M. I., Dean, J. H., Campbell, M. L., Lauer, L. A., Talley, F. A., Wilson, R. E., and Collins, M. J. (1982). Peritoneal macrophage alterations caused by naturally occurring mouse hepatitis virus. Am. J. Pathol. 106, 110–117.

Casals, J. (1979). Arenaviruses. In "Diagnostic Procedures for Viral, Rickettsial and Chlamydial Infections" (E. H. Lennette and N. J. Schmidt, eds.), 5th ed., pp. 815–841. Am. Public Health Assoc., Washington, D.C.

Collins, M. J., and Parker, J. C. (1972). Murine virus contaminants of leukemia viruses and transplantable tumors. J. Natl. Cancer Inst. (U.S.) 49, 1137–1143.

Cunliffe-Beamer, T. L., and Les, E. P. (1983). Effectiveness of pressurized individually ventilated (PIV) cages in reducing transmission of pneumonia virus of mice (PVM). Lab. Anim. Sci. 33, 495 (Abstr. No. 58).

Food and Drug Administration (1978). "Good Laboratory Practice Regula-

tions for Nonclinical Laboratory Studies," 21 CFR Part 58; 43 FR 59986-60020. FDA, Washington, D.C.

Foster, H. L., Small, J. D., and Fox, J. G., eds. (1981–1983). "The Mouse in Biomedical Research," Vols. 1–4. Academic Press, New York.

Fox, J. G. (1977). Clinical assessment of laboratory rodents on long term bioassay studies. J. Environ. Pathol. Toxicol. 1, 199–226.

Fox, J. G. (1982). Campylobacteriosis—A "new" disease in laboratory animals. Lab. Anim. Sci. 32, 625–637.

Fujiware, K., Takenaka, S., and Shumiya, S. (1976). Carrier state of antibody and viruses in a mouse breeding colony persistently infected with Sendai and mouse hepatitis viruses. Lab. Anim. Sci. 28, 297–306.

Ganaway, J. R. (1982). Bacterial and mycotic disease of the digestive system. In "The Mouse in Biomedical Research" (H. L. Foster, J. D. Small, and J. G. Fox, eds.), Vol. 2, pp. 1–20. Academic Press, New York.

Ganaway, J. R., Spencer, T. H., Moore, T. D., and Allen, A. M. (1983). Isolation, propagation, and characterization of a newly recognized pathogen, cilia-associated respiratory (CAR) bacillus of the rat: A predicted cause of chronic respiratory disease. Lab. Anim. Sci. 33, 502–503 (Abstr. No. 88).

Gibbs, C. J., Jr., Takenaka, A., Franko, M., Gajdusek, D. C., Griffin, M. D., Childs, J. E., Korch, G. W., and Wartzok, D. (1982). Seroepidemiology of Hantaan virus. Lancet 2, 1407.

Hinman, A. R., Fraser, D. W., Douglas, R. G., Bowen, G. S., Kraus, A. L., Winkler, W. G., and Rhodes, W. W. (1975). Outbreak of lymphocytic choriomeningitis virus infections in medical center personnel. Am. J. Epidemiol. 101, 103–110.

Horsfall, F. L., and Bauer, J. H. (1940). Individual isolation of infected animals in a single room. J. Bacteriol. 40, 569–580.

Hsu, C. K., New, A. E., and Mayo, J. G. (1980). Quality assurance of rodent models. In "7th Symposium of the International Council for Laboratory Animal Science (ICLAS)" (A. Spiegel, S. Erichsen, and H. A. Solleveld, eds.), pp. 17–28. Gustav Fischer Verlag, Stuttgart.

Institute of Laboratory Animal Resources (ILAR) (1976). Long-term holding of laboratory rodents. ILAR News 19, L1–L25.

Institute of Laboratory Animal Resources (ILAR) (1978). "Guide for the Care and Use of Laboratory Animals." Natl. Acad. Sci., Washington, D.C.

Jacoby, R. O., and Barthold, S. W. (1980). Quality assurance for rodents used in toxicological research and testing. In "Scientific Considerations in Monitoring and Evaluating Toxicological Research" (E. J. Gralla, ed.), pp. 27–55. Hemisphere Publ. Corp., Washington, D.C.

Jacoby, R. O., Bhatt, P. N., and Jonas, A. M. (1979). Viral diseases. In "The Laboratory Rat" (H. J. Baker, J. R. Lindsey, and S. H. Weisbroth, eds.), Vol. 2, pp. 272–306. Academic Press, New York.

Jakab, G. J. (1981). Interactions between Sendai virus and bacterial pathogens in the murine lung: A review. Lab. Anim. Sci. 31, 170–177.

Kraft, L. M., Pardy, R. F., Pardy, D. A., and Zwickel, H. (1964). Practical control of diarrheal disease in a commercial mouse colony. Lab. Anim. Care 14, 16–19.

Landi, M., Krieder, J. W., Lang, C. M., and Bullock, L. P. (1982). Effect of shipping on the immune functions of mice. Am. J. Vet. Res. 43, 1654–1657.

Lang, C. M. (1983). Design and management of research facilities for mice. In "The Mouse in Biomedical Research" (H. L. Foster, J. D. Small, and J. G. Fox, eds.), Vol. 3, pp. 37–50. Academic Press, New York.

Lapierre, J., Marsolais, G., Pilon, P., and Descoteaux, J. P. (1980). Preliminary report on the observation of a coronavirus in the intestine of the laboratory rabbit. Can. J. Microbiol. 26, 1204–1208.

La Regina, M., and Lonigro, J. (1982). Isolation of Campylobacter fetus subspecies jejuni from hamsters with proliferative ileitis. Lab. Anim. Sci. 32, 660–662.

Lee, P. W., Amyx, H. L., Gajdusek, D. C., Yanagihara, R. T., Goldgaber,

D., and Gibbs, C. J., Jr. (1982). New haemorrhagic fever with renal syndrome-related virus in indigenous wild rodents in United States. *Lancet* **2**, 1405.

Lehmann-Grube, F. (1982). Lymphocytic choriomeningitis virus. *In* "The Mouse in Biomedical Research" (H. L. Foster, J. D. Small, and J. G. Fox, eds.), Vol. 2, pp. 231–266. Academic Press, New York.

Lindsey, J. R., Cassell, G. H., and Davidson, M. K. (1982). Mycoplasmal and other bacterial diseases of the respiratory system. *In* "The Mouse in Biomedical Research" (H. L. Foster, J. D. Small, and J. G. Fox, eds.), Vol. 2, pp. 21–41. Academic Press, New York.

Loew, F. M. and Fox, J. G. (1983). Animal health surveillance and health delivery systems. *In* "The Mouse in Biomedical Research" (H. L. Foster, J. D. Small, and J. G. Fox, eds.), Vol. 3, pp. 69–82. Academic Press, New York.

McGarrity, G. J., and Coriell, L. L. (1973). Mass airflow cabinet for control of airborne infection of laboratory rodents. *Appl. Microbiol.* **26**, 167–172.

Matsumoto, M., Patton, N. M., Holmes, H. T., and Zehfus, B. (1983). Development of a serological test for detection of pasteurellosis in rabbits. *Lab. Anim. Sci.* **33**, 488 (Abstr. No. 30).

May, R. M., and Anderson, R. M. (1979). Population biology of infectious diseases. Part II. *Nature (London)*, **280**, 455–461.

Needham, J. R., and Cooper, J. E., eds. (1984). "Handbook of Laboratory Animal Health." Academic Press, New York (in press).

New, A. E., ed. (1981). Ectromelia (mousepox) in the United States. *Lab. Anim. Sci.* **31**, 549–635.

Newton, W. M. (1978). Environmental impact on laboratory animals. *Adv. Vet. Sci. Comp. Med.* **22**, 1–28.

Osterhaus, A. D. M. E., Teppema, J. S., and van Steenis, G. (1982). Coronavirus-like particles in laboratory rabbits with different syndromes in The Netherlands. *Lab. Anim. Sci.* **32**, 663–665.

Parker, J. C., and Collins, M. J., Jr. (1984). Microbiological techniques—Viruses. *In* "Handbook of Laboratory Animal Health" (J. R. Needham and J. E. Cooper, eds.). Academic Press, New York (in press).

Parker, J. C., and Reynolds, R. K. (1968). Natural history of Sendai virus infection in mice. *Am. J. Epidemiol.* **88**, 112–125.

Parker, J. C., and Richter, C. B. (1982). Viral diseases of the respiratory system. *In* "The Mouse in Biomedical Research" (H. L. Foster, J. D. Small, and J. G. Fox, eds.), Vol. 2, pp. 110–158. Academic Press, New York.

Parker, J. C., Whiteman, M. D., and Richter, C. B. (1978). Susceptibility of inbred and outbred strains to Sendai virus and prevalence of infection in laboratory rodents. *Infect. Immun.* **19**, 123–130.

Peter, G. K., Ringler, D. H., and Keren, D. F. (1983). An ELISA to detect IgG and IgA to *Pasteurella multocida* in rabbit sera and nasal washes. Abstract No. 33. *Lab. Anim. Sci.* **33**, 489.

Rowe, W. P., Hartley, J. W., and Huebner, R. J. (1962). Polyoma and other indigenous mouse viruses. *In* "Problems of Laboratory Animal Disease" (R. C. J. Harris, ed.), pp. 131–142. Academic Press, New York.

Sedlacek, R. S., Orcutt, R. P., Suit, H. D., and Rose, E. F. (1981). A flexible barrier at cage level for existing colonies: Production and maintenance of a limited stable anaerobic flora in a closed inbred mouse colony. *In* "Recent Advances in Germfree Research" (S. Sasaki, A. Ozawa, and K. Hashimoto, eds.), pp. 65–69. Tokai Univ. Press, Tokyo.

Small, J. D. (1983). Environmental and equipment monitoring. *In* "The Mouse in Biomedical Research" (H. L. Foster, J. D. Small, and J. G. Fox, eds.), Vol. 3, pp. 83–100. Academic Press, New York.

Small, J. D., and New, A. E. (1981). Prevention and control of mousepox. *Lab. Anim. Sci.* **31**, 616–629.

Small, J. D., Aurelian, L., Squire, R. A., Strandberg, J. D., Melby, E. C., Jr., Turner, T. B., and Newman, B. (1979). Rabbit cardiomyopathy associated with a virus antigenically related to human coronavirus strain 229E. *Am. J. Pathol.* **95**, 709–729.

Trexler, P. C. (1983). Gnotobiotics. *In* "The Mouse in Biomedical Research" (H. L. Foster, J. D. Small, and J. G. Fox, eds.), Vol. 3, pp. 1–16. Academic Press, New York.

Waggie, K. S., Wagner, J. E., and Lentsch, R. H. (1983). A naturally occurring outbreak of *Mycobacterium avium-intacellulare* infections in C57BL/6N mice. *Lab. Anim. Sci.* **33**, 249–253.

Wagner, J. E., and Manning, P. J., eds. (1976). "The Biology of the Guinea Pig." Academic Press, New York.

Wallace, G. D., Werner, R. M., Golway, P. L., Hernandez, D. M., Alling, D. W., and George, D. A. (1981). Epizootiology of an outbreak of mousepox at the National Institutes of Health. *Lab. Anim. Sci.* **31**, 609–615.

Weisbroth, S. H., Flatt, R. E., and Kraus, A. L., eds. (1974). "The Biology of the Laboratory Rabbit." Academic Press, New York.

Werner, R. M., Allen, A. M., Small, J. D., and New, A. E. (1981). Clinical manifestations of mousepox in an experimental animal holding room. *Lab. Anim. Sci.* **31**, 590–594.

Index

U